CRC World Dictionary of
PLANT NAMES

Common Names, Scientific Names, Eponyms, Synonyms, and Etymology

Volume III M — Q

CRC World Dictionary of
PLANT NAMES

Common Names, Scientific Names, Eponyms, Synonyms, and Etymology

Volume III M — Q

Umberto Quattrocchi, F.L.S.

CRC Press
Taylor & Francis Group
Boca Raton London New York

CRC Press is an imprint of the
Taylor & Francis Group, an **informa** business

CRC Press
Taylor & Francis Group
6000 Broken Sound Parkway NW, Suite 300
Boca Raton, FL 33487-2742

© 2000 by Taylor & Francis Group, LLC
CRC Press is an imprint of Taylor & Francis Group, an Informa business

First issued in paperback 2019

No claim to original U.S. Government works

ISBN 13: 978-0-367-44751-9 (pbk)
ISBN 13: 978-0-8493-2677-6 (hbk)

Visit the Taylor & Francis Web site at
http://www.taylorandfrancis.com

and the CRC Press Web site at
http://www.crcpress.com

M

Maackia Ruprecht Fabaceae

Origins:
Named for the Estonian naturalist Richard Karlovich (Karlovic) Maack (Maak), 1825-1886, explorer, botanist, plant collector in Siberia. Among his works are *Puteshestvie na Amur ... v 1855 ghodu.* [Journey to Amur ... in 1855] Sanktpeterburgh 1859 and *Puteshestvie po dolinye ryeki Usuri*, etc. [Journey to the valley of the River Usuri, etc.] S.-Peterburgh 1861. See J.H. Barnhart, *Biographical Notes upon Botanists.* 2: 417. 1965; Eduard August von Regel (1815-1892), *Tentamen Florae Ussuriensis*, oder Versuch einer *Flora des Ussuri-Gebietes*. Nach den von Herrn R. Maack gesammelten Pflanzen bearbeitet von E. Regel. St. Petersburg 1861; H.N. Clokie, *Account of the Herbaria of the Department of Botany in the University of Oxford.* 204. Oxford 1964; E.M. Tucker, *Catalogue of the Library of the Arnold Arboretum of Harvard University.* 1917-1933; R. Zander, F. Encke, G. Buchheim and S. Seybold, *Handwörterbuch der Pflanzennamen.* 14. Aufl. 745. Stuttgart 1993; F. Boerner & G. Kunkel, *Taschenwörterbuch der botanischen Pflanzennamen.* 4. Aufl. 129. Berlin & Hamburg 1989; Emil Bretschneider, *History of European Botanical Discoveries in China.* 1981.

Maba Forster & Forster f. Ebenaceae

Origins:
Presumably the Tongan and Fijian native name. See William T. Wawn, *The South Sea Islanders and the Queensland Labour Trade*, a record of voyages and experiences in the Western Pacific, from 1875 to 1891. London 1893; Johann Reinhold Forster and Johann Georg Adam Forster, *Characteres generum plantarum.* 121, t. 61. 1775; C.M. Churchward, *Tongan Dictionary.* Government Printing Press, Nuku'alofa, Tonga 1959.

Mabola Raf. Ebenaceae

Origins:
From a vernacular name, for *Diospyros mabola* Roxb.; see C.S. Rafinesque, *Sylva Telluriana.* 11. 1838.

Macadamia F. Muell. Proteaceae

Origins:
Named in honor of the Australian chemist John Macadam, 1827-1865, lecturer at the University of Melbourne, medical man, M.D. Glasgow University, secretary of the Philosophical Institute of Victoria (later the Royal Society of Victoria) and in 1863 vice-president. His writings include *Chemistry. Tables Introductory to a Course of Testing in Qualitative and Quantitative Analysis.* London, Glasgow 1853, *Report of Chemical Examination and Analyses of Waters*: conducted for the Geelong Water Commission. Melbourne 1857 and *Report on, and Analysis of, the Moffat Mineral Wells.* 1854. See F. von Mueller, *Transactions and Proceedings of the Philosophical Institute of Victoria.* 2: 72. 1857; W. Keddie, *Moffat; Its Walks and Wells*, etc. 1854.

Species/Vernacular Names:
M. heyana (Bailey) Sleumer (after Rev. Nicholas Hey)
Australia: Hey's nut oak

M. integrifolia Maiden & E. Betche
Australia: macadamia nut, smooth-shelled bush nut, Bauple nut (from Mount Bauple, Queensland), bopple nut, Australian bush nut

M. ternifolia F. Muell.
Australia: small-fruited bush nut, Maroochy nut, small-fruited Queensland nut

M. tetraphylla L. Johnson
Australia: rough-shelled bush nut, macadamia nut

M. whelanii (Bailey) Bailey
Australia: Whelan's nut oak

Macairea DC. Melastomataceae

Origins:
Latin *macaerinthe* "another name for rosmarinus," Greek *makaira.*

Macaranga Thouars Euphorbiaceae

Origins:
It is a Madagascan native name for one species in the genus; see Louis-Marie Aubert Aubert du Petit-Thouars

(1758-1831), *Genera nova Madagascariensis*. 26. 1806 (in *Mélanges de Botanique et de Voyages* ... Paris 1811).

Species/Vernacular Names:

M. sp.

Liberia: garfoe, zar-zreh-wehn-ye

Ivory Coast: tofe

Malaya: mahang

M. barteri Müll. Arg.

Nigeria: aragasa, awasa, arasa, ohaha, ohoha, aran, aghasa, agaha; agbaasa, asasa (Yoruba); ohaha (Edo); ohoha (Ishan); aran (Ijaw); owariwa (Igbo); akpap (Efik); akpap (Ibibio)

Yoruba: ararasa

Congo: esomba, mussassa

M. capensis (Baillon) Benth. ex Sim (*Mappa capensis* Baill.; *Macaranga bachmannii* Pax; *Mallotus capensis* (Baill.) Müll. Arg.)

English: wild poplar, macaranga, spiny macaranga, swamp poplar

Southern Africa: wildepopulier; muNyakomo (Shona); umFongamfongo, umFongafonga, umFongofongo, umBhongabhonga, iPhumela, umPhumelele, iPhubane, umPhumela (Zulu); umBengele (Xhosa)

S. Rhodesia: muNyakomo

M. grandifolia (Blanco) Merrill

English: coral tree

M. hurifolia Beille

Nigeria: asa-wewe (Yoruba); ohaha (Edo); owariwa (Igbo)

Yoruba: owariwa, owolewa

M. inamoena Benth.

English: buff macaranga

M. involucrata (Wall.) Müll. Arg. var. *mallotoides* (F. Muell.) Perry

English: brown macaranga

M. kilimandscharica Pax

Southern Africa: mugarahanga (Shona)

M. mappa (L.) Müll. Arg. (*Ricinus mappa* L.; *Croton grandifolius* Blanco; *Macaranga grandifolia* (Blanco) Merr.)

The Philippines: bingabing

M. mellifera Prain

English: mountain macaranga

Southern Africa: muFukusha, muKwirimakoka, muSorgwe, muSozwe (Shona)

S. Rhodesia: muSoswe, muFukusha

M. spinosa Müll. Arg.

Zaire: luibiba, oyala, weenge, yala, yenge

M. subdentata Benth.

English: needlebark

M. tanarius (L.) Müll. Arg. (*Ricinus tanarius* L.)

Australia: blush macaranga, heart leaf

Malaya: hairy mahang, inchong, kundoh, jebat musang, tampu

M. triloba Müll. Arg.

Malaya: landas bukit, mahang serindit, mahang tekukor

Macarthuria Huegel ex Endlicher Molluginaceae

Origins:

After the Australian horticulturist Sir William Macarthur, 1800-1882, agriculturist, botanist, the 5th son of the wool pioneering Captain John Macarthur (see Rum Rebellion), William was knighted in 1856, farmer at Camden Park, from 1849 to 1855 a member of the N.S.W. Legislative Council, commissioner to the Paris Exhibition of 1855, wrote *Letters on the Culture of the Vine, Fermentation, and the Management of Wine in the Cellar*. By Maro (W. Macarthur). Sydney 1844 and *Catalogue of Plants Cultivated at Camden Park, New South Wales*. Sydney 1857, co-author with Charles Moore (1820-1905) of [New South Wales, Australia], *Catalogue des collections de bois indigènes des différents districts de cette colonie*. Paris [1855]. See John H. Barnhart, *Biographical Notes upon Botanists*. 2: 417. 1965; Ethelyn Maria Tucker, *Catalogue of the Library of the Arnold Arboretum of Harvard University*. Cambridge, Massachusetts 1917-1933; S.L. Endlicher et al., *Enumeratio plantarum quas in Novae Hollandiae ... collegit C. de Hügel*. 11. Wien 1837.

Macbridea Elliott ex Nuttall Labiatae

Origins:

After the American physician James Macbride, 1784-1817, botanist. See John H. Barnhart, *Biographical Notes upon Botanists*. 2: 418. 1965; Howard Atwood Kelly and Walter Lincoln Burrage, *Dictionary of American Medical Biography*. Lives of eminent physicians of the United States and Canada, from the earliest times. New York 1928; S. Lenley et al., *Catalog of the Manuscript and Archival Collections and Index to the Correspondence of John Torrey*. Library of the New York Botanical Garden. 463. 1973; William Darlington (1782-1863), *Reliquiae Baldwinianae*. Philadelphia 1843.

Macbridea Raf. Asclepiadaceae

Origins:

For the American physician and botanist James Macbride, 1784-1817; see C.S. Rafinesque, *Am. Monthly Mag. Crit. Rev.* 3: 99. 1818.

Macbrideina Standley Rubiaceae

Origins:

After the American botanist James Francis Macbride, 1892-1976. Among his writings are "Spermatophytes, mostly Peruvian — II." *Field Mus. Nat. Hist., Bot. Ser.* 8(2): 75-130. 1930, "New or renamed spermatophytes mostly Peruvian." *Candollea.* 5(4): 346-402. 1934 and "Andean plants: A new *Astragalus* and new names in *Dalea.*" *Candollea.* 7: 221-223. 1937, joint author and editor of *Flora of Peru.* Chicago 1936 etc. See John H. Barnhart, *Biographical Notes upon Botanists.* 2: 418. 1965; T.W. Bossert, *Biographical Dictionary of Botanists Represented in the Hunt Institute Portrait Collection.* 247. Boston, Massachusetts 1972; Ethelyn Maria Tucker, *Catalogue of the Library of the Arnold Arboretum of Harvard University.* Cambridge, Massachusetts 1917-1933; Joseph Ewan, *Rocky Mountain Naturalists.* 261. The University of Denver Press 1950; Ida Kaplan Langman, *A Selected Guide to the Literature on the Flowering Plants of Mexico.* Philadelphia 1964; R. Zander, F. Encke, G. Buchheim and S. Seybold, *Handwörterbuch der Pflanzennamen.* 14. Aufl. Stuttgart 1993.

Species/Vernacular Names:

M. peruviana Standley

Peru: llausa quiro, yausaquiro, yuasa quiro

Maccoya F. Muell. Boraginaceae

Origins:

After the Irish naturalist Sir Frederick McCoy, 1817-1899 (d. Melbourne), paleontologist, zoologist, geologist, professor of geology and mineralogy at Queen's College at Belfast, 1854 professor of natural sciences at Melbourne University, Director of the National Museum of Victoria, 1880 Fellow of the Royal Society, knighted 1891. Among his numerous publications and scientific papers are *Contributions to British Palaeontology ... from the Tertiary, Cretaceous, Oolitic, etc.* Cambridge 1854, *A Detailed Systematic Description of the British Palaeozoic Fossils in the Geological Museum of the University of Cambridge.* 1854, *Dangerous Snakes of Victoria.* Melbourne [1883] and *Natural History of Victoria. Prodromus of the Zoology of Victoria.* Melbourne 1878-1890. See F. von Mueller, *Fragmenta Phytographiae Australiae.* 1: 127. Melbourne 1859; John H. Barnhart, *Biographical Notes upon Botanists.* 2: 421. 1965.

Macdonaldia Gunn ex Lindley Orchidaceae

Origins:

Named for Mrs. Charlotte Smith (*née* Macdonald, wife of John Grant Smith), d. 1838 (Launceston, Tasmania), collector for Ronald Campbell Gunn (1808-1881), collected orchids and algae for W.H. Harvey; see T.E. Burns and John Rowland Skemp, *Van Diemen's Land Correspondents. Letters from R.C. Gunn [and others] ... to Sir William J. Hooker, 1827-1849.* Queen Victoria Museum. [Launceston] 1961.

Macdougalia A. Heller Asteraceae

Origins:

For the American botanist Daniel Trembly (Trembley) MacDougal, 1865-1958, authority on the vegetation of African and American deserts, professor of botany University of Minnesota, plant collector, 1899-1905 Director Laboratory of the New York Botanical Garden, Desert Laboratory 1905-1928 Tucson, Arizona, wrote *Botanical Features of North American Deserts.* Washington, D.C. 1908. See John H. Barnhart, *Biographical Notes upon Botanists.* 2: 422. 1965; I.C. Hedge and J.M. Lamond, *Index of Collectors in the Edinburgh Herbarium.* Edinburgh 1970; Joseph Ewan, ed., *A Short History of Botany in the United States.* New York and London 1969; T.W. Bossert, *Biographical Dictionary of Botanists Represented in the Hunt Institute Portrait Collection.* 247. 1972; S. Lenley et al., *Catalog of the Manuscript and Archival Collections and Index to the Correspondence of John Torrey.* Library of the New York Botanical Garden. 274-275. 1973; Ida Kaplan Langman, *A Selected Guide to the Literature on the Flowering Plants of Mexico.* Philadelphia 1964; R. Zander, F. Encke, G. Buchheim and S. Seybold, *Handwörterbuch der Pflanzennamen.* 14. Aufl. Stuttgart 1993; E.M. Tucker, *Catalogue of the Library of the Arnold Arboretum of Harvard University.* 1917-1933; Ignatz Urban, *Geschichte des Königlichen Botanischen Museums zu Berlin-Dahlem (1815-1913). Nebst Aufzählung seiner Sammlungen.* Dresden 1916; Ira L. Wiggins, *Flora of Baja California.* Stanford, California 1980; Joseph Ewan, *Rocky Mountain Naturalists.* The University of Denver Press 1950; Irving William Knobloch, compil., "A preliminary verified list of plant collectors in Mexico." *Phytologia Memoirs.* VI. 1983.

Macfadyena A. DC. Bignoniaceae

Origins:

Named for the Scottish botanist Dr. James Macfadyen, 1798 (or 1800)-1850 (in Jamaica), physician, M.D. Glasgow 1821-1822, from 1826 in Jamaica, 1838 Fellow Linnean Society, author of the incomplete *Flora of Jamaica*. London, Edinburgh, Glasgow 1837. See J.H. Barnhart, *Biographical Notes upon Botanists*. 2: 423. 1965; Georg Christian Wittstein, *Etymologisch-botanisches Handwörterbuch*. 361. Ansbach 1852; R. Zander, F. Encke, G. Buchheim and S. Seybold, *Handwörterbuch der Pflanzennamen*. 14. Aufl. 363, 746. Stuttgart 1993; H.N. Clokie, *Account of the Herbaria of the Department of Botany in the University of Oxford*. 204. Oxford 1964; A. Lasègue, *Musée botanique de Benjamin Delessert*. Paris 1845; E.M. Tucker, *Catalogue of the Library of the Arnold Arboretum of Harvard University*. 1917-1933; Mea Allan, *The Hookers of Kew*. London 1967; Ignatz Urban, ed., *Symbolae Antillanae*. 1904; I.C. Hedge and J.M. Lamond, *Index of Collectors in the Edinburgh Herbarium*. Edinburgh 1970.

Species/Vernacular Names:

M. uncata (Andrews) Sprague & Sandwith (*Bignonia uncata* Andrews)

Peru: garra de murciélago, mashuricra, mashuishio, mashudillo, uña de gato

M. unguis-cati (L.) A. Gentry (*Bignonia unguis-cati* L.; *Bignonia tweediana-cati* Lindl.; *Doxantha unguis-cati* (L.) Miers; *Batocydia unguis* (L.) C. Martius ex DC.)

English: cat's claw, cat's claw trumpet, funnel creeper, cat's claw creeper, cat's claw climber

Macgregoria F. Muell. Stackhousiaceae

Origins:

After John McGregor, 1828-1884, Victorian politician and a member of the Parliament, a patron of the sciences; see F. von Mueller, *Fragmenta Phytographiae Australiae*. 8: 160. Melbourne 1874.

Machaeranthera Nees Asteraceae

Origins:

From the Greek *machaira* "a dagger, a short sword" and *anthera* "anther."

Species/Vernacular Names:

M. arida B. Turner & D. Horne

English: Silver Lake daisy

M. canescens (Pursh) A. Gray

English: hoary-aster

M. carnosa (A. Gray) G. Nesom

English: shrubby alkali aster

M. juncea (E. Greene) Shinn.

English: rush-like bristleweed

Machaerina Vahl Cyperaceae

Origins:

From the Greek *machaira* "a dagger, a short sword, a bent dagger," Latin *machaera, ae* for a sword or a weapon.

Species/Vernacular Names:

M. sp.

Hawaii: mau'u, kaluhaluha

M. angustifolia (Gaud.) T. Koyama (*Vincentia angustifolia* Gaud.; *Cladium angustifolium* (Gaud.) Drake; *Cladium vincentia* C.B. Clarke; *Mariscus angustifolius* (Gaud.) Kuntze)

Hawaii: 'uki

M. mariscoides (Gaud.) J. Kern (*Baumea mariscoides* Gaud.; *Baumea meyenii* Kunth; *Cladium meyenii* (Kunth) Drake)

Hawaii: 'uki, 'ahaniu

Machaerium Pers. Fabaceae

Origins:

From the Greek *machaira* "a dagger, large knife," *machairion* "surgeon's knife," referring to the recurved stipular thorns.

Machaerocarpus Small Alismataceae

Origins:

Greek *machaira* and *karpos* "fruit."

Machaerocereus Britton & Rose Cactaceae

Origins:

From the Greek *machaira* "a dagger, a short sword" and the genus *Cereus*, referring to the spines.

Machaerophorus Schltdl. Brassicaceae

Origins:

From the Greek *machaira* "a dagger" and *phoros* "bearing, carrying."

Machairophyllum Schwantes Aizoaceae

Origins:

Latin *machaerophyllon* for a plant, Greek *machaira* "a dagger, a short sword, a bent dagger" and *phyllon* "a leaf," sword-leaf.

Machilus Nees Lauraceae

Origins:

From a Moluccan name, or from *Machilis*, a name of an insect.

Species/Vernacular Names:

M. japonica Sieb. & Zucc. (*Machilus thunbergii* var. *japonica* (Sieb. & Zucc.) Yatabe)

Japan: hoso-ba-tabu

M. thunbergii Sieb. & Zucc.

Japan: tabu, tabu-no-riki, inu-gusa

Okinawa: tabu, tamuru

Mackaya W.H. Harvey Acanthaceae

Origins:

After the Scottish (b. Kirkcaldy, Fife) botanist James Townsend Mackay, 1775-1862 (d. Dublin), gardener, botanical collector, 1804 Dublin, member of the Botanical Society of London, founder and Curator of the Botanic Gardens of Trinity College in Dublin. His works include *A Catalogue of the Plants Found in Ireland*. Dublin 1825 and *Flora hibernica*. Dublin 1836, contributed to the *English Botany* by Sir J.E. Smith. See John H. Barnhart, *Biographical Notes upon Botanists*. 2: 426. Boston 1965; R. Zander, F. Encke, G. Buchheim and S. Seybold, *Handwörterbuch der Pflanzennamen*. 14. Aufl. 746. Stuttgart 1993; Ernest Nelmes and William Cuthbertson, *Curtis's Botanical Magazine Dedications, 1827-1927*. 119-120. [1931]; E.C. Nelson & Eileen May McCracken (1920-1988), *The Brightest Jewel: A History of the National Botanic Gardens, Glasnevin, Dublin*. Kilkenny 1987; T.W. Bossert, *Biographical Dictionary of Botanists Represented in the Hunt Institute Portrait Collection*. 247. 1972; W.J. Hooker, *Companion to the Botanical Magazine*. 1: 158. 1835; H.N. Clokie, *Account of the Herbaria of the Department of Botany in the University of Oxford*. 205. Oxford 1964; S. Lenley et al., *Catalog of the Manuscript and Archival Collections and Index to the Correspondence of John Torrey*. Library of the New York Botanical Garden. 463. 1973; E.M. Tucker, *Catalogue of the Library of the Arnold Arboretum of Harvard University*. 1917-1933; N. Colgan and Reginald William Scully, *Contributions Towards a Cybele Hibernica*. Ed. 2nd. Dublin

1898; Mariella Azzarello Di Misa, a cura di, *Il Fondo Antico della Biblioteca dell'Orto Botanico di Palermo*. 179. Palermo 1988; Ray Desmond, *Dictionary of British & Irish Botanists and Horticulturists*. 453-454. London 1994; H.R. Fletcher and W.H. Brown, *Royal Botanic Garden Edinburgh, 1670-1970*. Edinburgh 1970.

Species/Vernacular Names:

M. bella Harv.

English: river bells, beautiful mackaya

Southern Africa: blouklokkiesbos, mackaya; uZwathi, umZwathi, uPhulule, uMavuthwa, iCaphozi (Zulu); mufhanza (Venda)

Mackinlaya F. Muell. Araliaceae

Origins:

After the South Australian explorer John McKinlay, 1819-1872, 1861-1862 led one of the expeditions in search of the remains of the Burke and Wills Expedition, in 1865-1866 explored Northern Territory. Among his works are *McKinlay's Journal of Exploration in the Interior of Australia*. (Burke Relief Expedition) Melbourne [1862], "Diary of Mr. J. McKinlay, leader of the Burke relief expedition fitted out by the government of South Australia." from the *Journal of the Royal Geographical Society of London* for 1863, *Exploration. McKinlay's Diary of His Journey across the Continent of Australia*. Melbourne 1863 and *J. McKinlay's Northern Territory Explorations, 1866*. [J. McKinlay's Journal and Report of Explorations, 1866.] 1867. See John Davis, *Tracks of McKinlay and Party Across Australia. By John Davis, one of the expedition*. London 1863; Jonathan Wantrup, *Australian Rare Books, 1788-1900*. 234-239, 242-244. Hordern House, Sydney 1987; Duncan Whyte, *Sketch of Explorations by the late John M'Kinlay in the Interior of Australia, 1861-1862*. [A paper read before the Cowal Society, Oct. 28, 1878.] Glasgow 1881; George E. Loyau, *The Gawler Handbook: A record of the rise and progress of that important town; to which are added memoirs of McKinlay the explorer and Dr. Nott*. Adelaide 1880; F. von Mueller, *Fragmenta Phytographiae Australiae*. 4: 119. Melbourne 1864; Tom Bergin, *In the Steps of Burke & Wills*. Sydney 1981.

Macleania Hooker Ericaceae

Origins:

Dedicated to John Maclean, 1832-1854 Peru, plant collector, merchant and patron of botany, sent specimens to W.J. Hooker and W. Herbert; see Ray Desmond, *Dictionary of British & Irish Botanists and Horticulturists*. 457. London 1994; F. Boerner & G. Kunkel, *Taschenwörterbuch der botanischen Pflanzennamen*. 4. Aufl. 130. Berlin & Hamburg 1989.

Macleaya R. Br. Papaveraceae

Origins:

After the Scottish scientist Alexander Macleay (McLeay), 1767-1848 (d. Sydney, N.S.W.), entomologist, established the McLeay Collection, 1794 Fellow of the Linnean Society and in 1809 of the Royal Society, 1798-1825 Secretary of the Linnean Society, arrived at Sydney in January 1826, Colonial Secretary N.S.W., first Speaker of the first representative Assembly of the Colony, father of seventeen children, owner of a garden famous for its rare plants. See A.T. Gage, *A History of the Linnean Society of London*. London 1938; John Blackman, *A Catalogue of an Extensive and Valuable Library ... of Alexander McLeay*, Esq., M.C., who is removing to the country; Which will be sold by auction ... on Tuesday, 1st, Wednesday, 2nd, Thursday, 3rd, and Friday, 4th Days of Aprile. Sydney [1846]; F. Boerner & G. Kunkel, *Taschenwörterbuch der botanischen Pflanzennamen*. 4. Aufl. 130. Berlin & Hamburg 1989; G.C. Wittstein, *Etymologisch-botanisches Handwörterbuch*. 546. 1852; Douglas Pike, ed., *Australian Dictionary of Biography*. 2: 177-180. Melbourne 1967.

Species/Vernacular Names:

M. cordata (Willd.) R. Br. (*Bocconia cordata* Willd.)

English: plume poppy

China: bo luo hui, po lo hui

Maclura Nutt. Moraceae

Origins:

After the Scottish-born (b. Ayr) North American geologist William Maclure, 1763-1840 (d. Mexico), agriculturist, traveler, merchant, 1782 to USA, 1815 met in Paris the naturalist–engraver Charles Alexandre Lesueur (1778-1846), 1817-1839 one of the founders and President of the Academy of Natural Sciences of Philadelphia, 1817-1818 expedition to Spanish Florida and the Sea Islands of Georgia with Lesueur, the naturalist and entomologist Thomas Say (1787-1834), the naturalist Titian Ramsay Peale (1799-1885) and George Ord, 1820-1824 in Spain, supported the ideas on evolution offered by Lamarck. Among his writings are *Observations on the Geology of the United States of America*. Philadelphia 1817 and *Opinions on Various Subjects, Dedicated to the Industrious Producers*. [Three volumes.] New Harmony, Indiana 1831-1838, in New Harmony (Indiana) met and joined the reformer Robert Owen (1771-1858). See G.C. Wittstein, *Etymologisch-botanisches Handwörterbuch*. 546. 1852; National Library of Wales, *A Bibliography of Robert Owen, The Socialist 1771-1858*. Second edition, revised and enlarged. Aberystwyth 1925; Thomas Nuttall, *The Genera of North American Plants*, and catalogue of the species, to the year 1817. 2: 233. Philadelphia 1818; S.G. Morton, *A Memoir of William Maclure, Esq*. Philadelphia 1841; George W. White, in *Dictionary of Scientific Biography* 8: 615-617. 1981; R.W.G. Vail, *The American Sketchbooks of Charles Alexander Lesueur 1816-1837*. Worcester 1938; H.B. Weiss and G.M. Zeigler, *Thomas Say, Early American Naturalist*. Springfield-Baltimore 1931; Clifford Merrill Drury, *Diary of Titian Ramsay Peale*. Los Angeles 1957; George Ord, "A Memoir of Charles Alexandre Lesueur." in *American Journal of Science*. 2nd ser. 8: 189-216. 1849.

Species/Vernacular Names:

M. cochinchinensis (Lour.) Corner (*Vanieria cochinchinensis* Lour.; *Cudrania javanensis* Trécul; *Cudrania cochinchinensis* (Lour.) Kudô)

English: cockspur thorn

Japan: kakatsu-gayu

Okinawa: gamino-tsuma

Vietnam: cay bom, mo qua

Nepal: gai dimmar

M. pomifera (Raf.) C. Schneider

English: bow wood, Osage orange

M. tricuspidata Carrière (*Cudrania triloba* Hance; *Cudrania tricuspidata* (Carr.) Bur. ex Lav.)

English: Chinese silkworm thorn

Maclurea Raf. Moraceae

Origins:

For *Maclura* Nutt.; see Constantine Samuel Rafinesque, *Med. Fl.* 2: 268. 1830.

Macluria Raf. Moraceae

Origins:

For *Maclura* Nutt.; see Constantine Samuel Rafinesque, *Jour. Phys. Chim. Hist. Nat.* 89: 260. 1819.

Maclurodendron T.G. Hartley Rutaceae

Origins:

After William Maclure, 1763-1840. See S.G. Morton, *A Memoir of William Maclure, Esq.* Philadelphia 1841; George W. White, in *Dictionary of Scientific Biography* 8:

615-617. (Editor in Chief Charles Coulston Gillispie.) New York 1981.

Maclurolyra C. Calderón and Söderstrom Gramineae

Origins:

After William Maclure, 1763-1840 and the genus *Olyra* L.

Macnabia Benth. ex Endl. Ericaceae

Origins:

For the Scottish (b. Ayrshire) gardener William McNab, 1780-1848 (d. Edinburgh), Kew 1801-1810, Curator Royal Botanic Gardens Edinburgh 1810-1848. His writings include *Hints on the Planting and General Treatment of Hardy Evergreens, in the Climate of Scotland.* Edinburgh 1831 and *A Treatise on the Cultivation and General Treatment of Cape Heaths*, in a climate where they require protection during the winter months. Edinburgh 1832, he was the father of James McNab (1810-1878) and grandfather of William Ramsay McNab (1844-1889); see M. Hadfield et al., *British Gardeners: A Biographical Dictionary.* London 1980; Ray Desmond, *Dictionary of British & Irish Botanists and Horticulturists.* 459. 1994; H.R. Fletcher and W.H. Brown, *Royal Botanic Garden Edinburgh, 1670-1970.* Edinburgh 1970; E.C. Nelson & Eileen May McCracken, *The Brightest Jewel: A History of the National Botanic Gardens, Glasnevin, Dublin.* Kilkenny 1987; Stafleu and Cowan, *Taxonomic Literature.* 3: 231-232. Utrecht 1981; E.M. Tucker, *Catalogue of the Library of the Arnold Arboretum of Harvard University.* 1917-1933; R. Zander, F. Encke, G. Buchheim and S. Seybold, *Handwörterbuch der Pflanzennamen.* 14. Aufl. Stuttgart 1993.

Macodes (Blume) Lindley Orchidaceae

Origins:

Greek *makros* "large, long," referring to the mid-lobe of the lip.

Macoucoua Aublet Aquifoliaceae

Origins:

A vernacular name for *Ilex guianensis* (Aublet) O. Kuntze (*Macoucoua guianensis* Aublet), this tree is called *macoucou* by the Galibis.

Macowania Oliver Asteraceae

Origins:

After the British (b. Hull, Yorks.) botanist Peter MacOwan, 1830-1909 (Uitenhage, C.P., S. Africa), professor of chemistry, Director of the Botanical Garden at Capetown 1881-1892, 1885 Fellow of the Linnean Society, Government Botanist 1892-1905, Secretary of the South Africa Botanical Exchange Society. Among his writings are "Personalia of botanical collectors at the Cape." *Trans. S. Afr. Philos. Soc.* 4(1): xxx-liii. 1884-1886, "Notulae capenses." *J. Linn. Soc. Bot.* 10: 480-482. 1869, *Catalogue of South African Plants.* Grahamstown 1866, "New Cape plants." *J. Linn. Soc. Bot.* 25: 385-394. 1889, *The Collecting and Preserving of Botanical Specimens.* [Cape Town 1893] and *Report upon the Botanic Gardens Government Herbarium, Cape Town, for the Year 1891.* Capetown 1892, with Harry Bolus wrote "Catalogue of printed books and papers relating to South Africa. Part I, Botany." in *Trans. S. Afr. Philos. Soc.* 2(3): 111-187. 1880-1881, "Novitates capenses: description of new plants from the Cape of Good Hope." *J. Linn. Soc. Bot.* 18: 390-397. 1881 and *The Olive at the Cape.* Wynberg 1897. See J.H. Verduyn den Boer, *Botanists at the Cape.* Cape Town and Stellenbosch 1929; Mia C. Karsten, *The Old Company's Garden at the Cape and Its Superintendents*: involving an historical account of early Cape botany. Cape Town 1951; Gilbert Westacott Reynolds (1895-1967), *The Aloes of South Africa.* 57, 58, 65. Balkema, Rotterdam 1982; John Hutchinson (1844-1972), *A Botanist in Southern Africa.* 643-644. London 1946; Mary Gunn and Leslie E. Codd, *Botanical Exploration of Southern Africa.* 240-242. Cape Town 1981; J.H. Barnhart, *Biographical Notes upon Botanists.* 2: 432. 1965; T.W. Bossert, *Biographical Dictionary of Botanists Represented in the Hunt Institute Portrait Collection.* 248. 1972; H.N. Clokie, *Account of the Herbaria of the Department of Botany in the University of Oxford.* 205. 1964; S. Lenley et al., *Catalog of the Manuscript and Archival Collections and Index to the Correspondence of John Torrey.* Library of the New York Botanical Garden. 278. 1973; E.M. Tucker, *Catalogue of the Library of the Arnold Arboretum of Harvard University.* 1917-1933; A. White and B.L. Sloane, *The Stapelieae.* Pasadena 1937; R. Zander, F. Encke, G. Buchheim and S. Seybold, *Handwörterbuch der Pflanzennamen.* 14. Aufl. Stuttgart 1993; I.C. Hedge and J.M. Lamond, *Index of Collectors in the Edinburgh Herbarium.* Edinburgh 1970.

Macrachaenium Hook.f. Asteraceae

Origins:

From the Greek *makros* "long" and achene, achaena or achaenium.

Macradenia R. Br. Orchidaceae

Origins:
From the Greek *makros* "long" and *aden* "gland," referring to the pollinia, to the long anther appendage.

Macraea Hook.f. Asteraceae

Origins:
From the Greek *makros* "long."

Macrandria Meissn. Rubiaceae

Origins:
From the Greek *makros* "long" and *andros* "man, male."

Macranthera Nutt. ex Benth. Scrophulariaceae

Origins:
From the Greek *makros* "long, large, big" and *anthera* "anther."

Macranthisiphon Bureau ex K. Schumann Bignoniaceae

Origins:
Greek *makros*, *anthos* "flower" and *siphon* "tube."

Macroberlinia (Harms) Hauman Caesalpiniaceae

Origins:
From the Greek *makros* "long, large" plus the genus *Berlinia* Sol. ex Hook.f.

Species/Vernacular Names:
M. bracteosa (Benth.) Hauman
Congo: m'possa
Gabon: gibora, obolo, ebiara
Cameroon: essabem, koum, abem
Zaire: m'bosa, m'posa

Macrobia (Webb & Berthel.) G. Kunkel Crassulaceae

Origins:
From the Greek *makros* "long, large" and *bios* "life."

Macroblepharus Phil. Gramineae

Origins:
From the Greek *makros* "long, large" and *blepharon* "an eyelid," *blepharis* "eyelash."

Macrobriza (Tzvelev) Tzvelev Gramineae

Origins:
From the Greek *makros* "long, large" plus *Briza* L.

Macrocalyx Costantin & J. Poiss. Malvaceae

Origins:
Greek *makros* "long, large" and *kalyx* "calyx."

Macrocalyx Tieghem Loranthaceae

Origins:
From the Greek *makros* "long, large" and *kalyx* "calyx."

Macrocarpaea (Griseb.) Gilg Gentianaceae

Origins:
From the Greek *makros* "long, large" and *karpos* "fruit."

Macrocarpium (Spach) Nakai Cornaceae

Origins:
From the Greek *makros* "long, large" and *karpos* "fruit."

Macrocatalpa (Griseb.) Britton Bignoniaceae

Origins:
From the Greek *makros* "long, large" plus *Catalpa* Scop.

Macrocaulon N.E. Br. Aizoaceae

Origins:
From the Greek *makros* "long, large" and *kaulos* "stem."

Macrocentrum Hook.f. Melastomataceae

Origins:
From the Greek *makros* "long" and *kentron* (*kenteo* "to prick, torture, torment, sting, spur") "a spur, point."

Macrocentrum Philippi Orchidaceae

Origins:
Greek *makros* and *kentron* "a spur, point," large spurs of the lip.

Macrochaeta Steudel Gramineae

Origins:
From the Greek *makros* "long" and *chaite* "bristle, long hair."

Macrochaetium Steudel Cyperaceae

Origins:
Greek *makros* "long" and *chaite* "bristle, long hair."

Macrochilus Knowles & Westcott Orchidaceae

Origins:
From the Greek *makros* "long, large" and *cheilos* "lip," the large lip.

Macrochiton (Blume) M. Roemer Meliaceae

Origins:
Greek *makros* and *chiton* "a tunic, covering"; see Max Joseph Roemer, *Familiarum naturalium regni vegetabilis synopses monographicae.* 1: 84, 104. Vimariae [Weimar] 1846; Arthur D. Chapman, ed., *Australian Plant Name Index.* 1911. Canberra 1991.

Macrochlaena Hand.-Mazz. Umbelliferae

Origins:
From the Greek *makros* "long, large" and *chlaena, chlaenion* "cloak."

Macrochlamys Decne. Gesneriaceae

Origins:
From the Greek *makros* "long, large" and *chlamys* "cloak."

Macrochloa Kunth Gramineae

Origins:
From the Greek *makros* "long" and *chloe, chloa* "a grass."

Macrochordion de Vriese Bromeliaceae

Origins:
From the Greek *makros* "long" and *chordion, chorde* "string," fiber from leaves for rope and thread for sewing leather.

Macrocladus Griff. Palmae

Origins:
From the Greek *makros* "long, large" and *klados* "a branch," leaves distichous.

Macroclinidium Maxim. Asteraceae

Origins:
From the Greek *makros* "long, large" and *kline* "a bed," *klinidion* "a little bed."

Macroclinium Barb. Rodr. Orchidaceae

Origins:
From the Greek *makros* "long, large" and *kline* "a bed," *klinion* "a little bed," the clinandrium.

Macrocnemum P. Browne Rubiaceae

Origins:
From the Greek *makros* "long" and *kneme* "limb, leg," referring to the flower stalks or to the corolla tube.

Species/Vernacular Names:
M. roseum (Ruíz & Pav.) Wedd.
Bolivia: coloradillo canteado

Macrococculus Becc. Menispermaceae

Origins:
From the Greek *makros* "long" and the genus *Cocculus* DC.

Macrocroton Klotzsch Euphorbiaceae

Origins:
From the Greek *makros* "long" plus the genus *Croton* L.

Macrodiervilla Nakai Caprifoliaceae

Origins:
From the Greek *makros* "long" plus *Diervilla* Miller.

Macrodiscus Bureau Bignoniaceae

Origins:
From the Greek *makros* "long" and *diskos* "a disc."

Macroditassa Malme Asclepiadaceae

Origins:
From the Greek *makros* "long" plus the genus *Ditassa* R. Br.

Macroglena (C. Presl) Copel. Hymenophyllaceae

Origins:
From the Greek *makros* "large, long" and *glene* "an eye"; see E.B. Copeland, in *Philippine Journal of Science*. 67: 82. 1938.

Species/Vernacular Names:
M. caudata (Brackenr.) Copel.
English: jungle bristle fern

Macroglossum Copel. Marattiaceae (Angiopteridaceae)

Origins:
From the Greek *makros* "long" and *glossa* "a tongue."

Macrohasseltia L.O. Williams Flacourtiaceae (Tiliaceae)

Origins:
From the Greek *makros* "long" plus the genus *Hasseltia* Kunth.

Macrolenes Naudin ex Miq. Melastomataceae

Origins:
From the Greek *makros* "long" and *olene* "an arm, the elbow, bundle."

Macrolepis A. Rich. Orchidaceae

Origins:
From the Greek *makros* "large, long" and *lepis* "scale," the floral bracts.

Macrolobium Schreber Caesalpiniaceae

Origins:
From the Greek *makros* "large, long" and *lobos* "a lobe."

Species/Vernacular Names:
M. sp.
Nigeria: nya, essabam

Macromeria D. Don Boraginaceae

Origins:
From the Greek *makros* "large" and *meris* "part," referring to the size of the flowers.

Macromyrtus Miq. Myrtaceae

Origins:
From the Greek *makros* "large" plus *Myrtus*.

Macronax Raf. Gramineae

Origins:
See C.S. Rafinesque, *Med. Repos*. II. 5: 535. 1808; E.D. Merrill, *Index rafinesquianus*. 203. 1949.

Macronema Nutt. Asteraceae

Origins:
From the Greek *makros* "large" and *nema* "thread, filament."

Macropanax Miq. Araliaceae

Origins:
From the Greek *makros* "large, long" plus the genus *Panax*.

Macropelma Schumann Asclepiadaceae

Origins:
Greek *makros* "large, long" and *pelma, pelmatos* "sole of the foot, stalk."

Macropeplus Perkins Monimiaceae

Origins:
From the Greek *makros* "large, long" and *peplos* "a robe."

Macropetalum Burchell ex Decne. Asclepiadaceae

Origins:
From the Greek *makros* "large, long" and *petalon* "a petal, leaf," alluding to the length of the corolla lobes.

Macropharynx Rusby Apocynaceae

Origins:
From the Greek *makropharynx* "long-necked."

Macrophloga Becc. Palmae

Origins:
From the Greek *makros* "large, long" and *phlox, phlogos* "a flame."

Macrophthalmia Gasp. Moraceae

Origins:
From the Greek *makros* "large, long" and *ophthalmos* "eye."

Macropidia J.L. Drumm. ex Harvey Haemodoraceae

Origins:
Greek *makros* "large, long" and *pous, podos* "a foot," alluding to a foot of *Macropus*, a genus of kangaroo; see

Hooker's Journal of Botany & Kew Garden Miscellany. 7: 57. London 1855; James A. Baines, *Australian Plant Genera. An Etymological Dictionary of Australian Plant Genera.* 228. Chipping Norton, N.S.W. 1981; F.A. Sharr, *Western Australian Plant Names and Their Meanings.* A glossary. 46. University of Western Australia Press 1996; Helmut Genaust, *Etymologisches Wörterbuch der botanischen Pflanzennamen.* 359. [from the Greek *makros* and *pideeis* "rich in spring"] Basel 1996.

Macropiper Miq. Piperaceae

Origins:
From the Greek *makros* "large, long" and the genus *Piper* L.; see F.A.W. Miquel, *Bulletin des Sciences Physiques et Naturelles en Néerlande.* 1839: 447, 449. Rotterdam 1839.

Species/Vernacular Names:
M. excelsum (Forster f.) Miq.
English: pepper tree
New Zealand: kawa-kawa, kawakawa (maori name)

Macroplacis Blume Melastomataceae

Origins:
From the Greek *makros* "long" and *plakis* "a bench, seat, couch of flowers."

Macroplectrum Pfitzer Orchidaceae

Origins:
Greek *makros* and *plektron* "a spur, cock's spur," lip, sepals and petals form a very long spur.

Macroplethus C. Presl Polypodiaceae

Origins:
Perhaps from the Greek *makros* "large, long" and *plethos* "multitude."

Macropodandra Gilg Buxaceae

Origins:
Greek *makros* "long," *pous, podos* "foot" and *aner, andros* "man, stamen," *macropodus* "stout-stalked."

Macropodanthus L.O. Williams Orchidaceae

Origins:
Greek *makros* "long, large," *pous, podos* "foot" and *anthos* "flower," indicating the column-foot.

Macropodia Benth. Haemodoraceae

Origins:
Orthographic variant of *Macropidia* Drumm. ex Harvey.

Macropodiella Engl. Podostemaceae

Origins:
From the Greek *makros* "long" and *pous, podos* "foot," possibly referring to thallus.

Macropodina R.M. King & H. Robinson Asteraceae

Origins:
From the Greek *makros* "long" and *pous, podos* "foot".

Macropodium R. Br. Brassicaceae

Origins:
From the Greek *makros* "large, long" and *podion* "little foot," referring to the fruits.

Macropsidium Blume Myrtaceae

Origins:
From the Greek *makros* "large, long" and *psidion* "the pomegranate," genus *Psidium* L.

Macropsychanthus Harms ex K. Schumann & Lauterb. Fabaceae

Origins:
From the Greek *makros* "large, long," *psyche* "butterfly, moth" and *anthos* "a flower."

Macropteranthes F. Muell. ex G. Bentham Combretaceae

Origins:
From the Greek *makropteros* "long-winged" and *anthos* "a flower," referring to the winged calyx. See G. Bentham, *Flora Australiensis*. 6: 446. 1873; Arthur D. Chapman, ed., *Australian Plant Name Index*. 1912. Canberra 1991.

Species/Vernacular Names:
M. kekwickii F. Muell. ex Benth. (possibly after Wilhelm Kekwick, associated with the Stuart's Expedition, 1862)
Australia: bullwaddy, bulwaddy
M. leichhardtii F. Muell. ex Benth.
Australia: bonewood

Macroptilium (Benth.) Urban Fabaceae

Origins:
From the Greek *makros* "long" and *ptilon* "feather, wing"; see Ignatz Urban, ed., *Symbolae Antillanae*. 9: 457. Berlin 1928.

Species/Vernacular Names:
M. atropurpureum (DC.) Urban (*Phaseolus atropurpureus* Moçiño & Sessé ex DC.)
English: purple bean
Latin America: siratro, sirato, conchito
M. lathyroides (L.) Urban
English: oneleaf clover, wild bean, cow pea, phasey bean
M. lathyroides (L.) Urban var. *lathyroides* (*Phaseolus lathyroides* L.; *Phaseolus semierectus* L. var. *angustifolius* Benth.)
Japan: tachi-nanban-azuki

Macrorhamnus Baillon Rhamnaceae

Origins:
From the Greek *makros* "long" plus *Rhamnus*.

Macrorungia C.B. Clarke Acanthaceae

Origins:
From the Greek *makros* "large, long" with the genus *Rungia* Nees.

Macrosamanea Britton & Rose Mimosaceae

Origins:
From the Greek *makros* and the genus *Samanea* (DC.) Merr.

Macroscepis Kunth Asclepiadaceae

Origins:
From the Greek *makros* "large, long" and *skepe* "covering," an allusion to the size of the calyx.

Macrosciadium V.N. Tikhom. & Lavrova Umbelliferae

Origins:
From the Greek *makros* "large, long" and *skiadion, skiadeion* "umbel, parasol."

Macroselinum Schur Umbelliferae

Origins:
From the Greek *makros* "large, long" plus *Selinum* L.

Macrosepalum Regel & Schmalh. Crassulaceae

Origins:
From the Greek *makros* "large, long" and Latin *sepalum.*

Macrosiphon Miq. Rubiaceae

Origins:
From the Greek *makros* "large, long, big" and *siphon* "tube," referring to the flowers.

Macrosiphonia Müll. Arg. Apocynaceae

Origins:
Greek *makros* "large, long, big" and *siphon* "tube," referring to the flowers.

Macrosolen (Blume) Reichb. Loranthaceae

Origins:
From the Greek *makros* "large, long, big" and *solen* "a channel, furrow."

Macrosphyra Hook.f. Rubiaceae

Origins:
From the Greek *makros* "large, long, big" and *sphyra* "a hammer," alluding to the style.

Species/Vernacular Names:
M. longistyla (DC.) Hiern
Yoruba: opataba, ikuuku ekun

Macrostachya A. Rich. Gramineae

Origins:
From the Greek *makros* "long" and *stachys* "a spike."

Macrostegia Nees Labiatae (Verbenaceae)

Origins:
From the Greek *makros* "large, long" and *stege* "roof, cover."

Macrostegia Turcz. Thymelaeaceae

Origins:
From the Greek *makros* "large, long" and *stege* "cover"; see Porphir Kiril N.S. Turczaninow (1796-1863), *Bulletin de la Société Impériale des Naturalistes de Moscou.* 25(2): 177. 1852.

Macrostelia Hochr. Malvaceae

Origins:
From the Greek *makros* "large, long" and *stele* "a pillar, column, central part of stem."

Macrostemon Boriss. Scrophulariaceae

Origins:
From the Greek *makros* "large, long" and *stemon* "a thread, stamen."

Macrostigma Hooker Rosaceae

Origins:
From the Greek *makros* "large, long" and *stigma* "stigma."

Macrostigmatella Rauschert Brassicaceae

Origins:
Greek *makros* "large, long" and *stigma* "stigma."

Macrostomium Blume Orchidaceae

Origins:
From the Greek *makros* "large, long" and *stoma* "mouth," referring to the flowers.

Macrostylis Bartling & H.L. Wendland Rutaceae

Origins:
Greek *makros* "large, long" and *stylos* "a pillar, style," referring to the style in anthesis.

Macrostylis Breda Orchidaceae

Origins:
Greek *makros* "large, long" and *stylos* "a pillar, style," referring to the column.

Macrosyringion Rothm. Scrophulariaceae

Origins:
Greek *makros* and *syrinx* "a pipe, tube."

Macrothelypteris (H. Ito) Ching Thelypteridaceae

Origins:
Greek *makros* and the genus *Thelypteris*, 3-pinnatifid fronds; see R.C. Ching, in *Acta Phytotaxonomica Sinica.* 8: 308. 1963

Species/Vernacular Names:
M. torresiana (Gaudich.) Ching (*Polystichum torresianum* Gaudich.; *Dryopteris uliginosa* (Kunze) C. Christensen; *Thelypteris torresiana* (Gaudich.) Alston; *Thelypteris uliginosa* (Kunze) Ching)

English: Mariana maiden fern, Torres's fern

Macrotomia DC. ex Meissner Boraginaceae

Origins:
From the Greek *makros* "large" and *tomos*, *temno* "division, section, to slice," *makrotomos*, referring to the segments of the calyx.

Macrotorus Perkins Monimiaceae

Origins:
From the Greek *makros* "large" and Latin *torus* "a knot, swelling, thickness."

Macrotropis DC. Fabaceae

Origins:
From the Greek *makros* "large" and *tropis*, *tropidos* "a keel."

Macrotyloma (Wight & Arnott) Verdcourt Fabaceae

Origins:
Greek *makros* "large," *tylos* "knob, swelling" and *loma* "border, margin"; see B. Verdcourt, in *Kew Bulletin.* 24(2): 322. 1970.

Species/Vernacular Names:
M. axillare (E. Meyer) Verdc. (*Dolichos axillaris* E. Meyer)

English: perennial horse gram

M. geocarpum (Harms) Marechal & Baudet

English: ground bean, geocarpa groundnut, Kersting groundnut

M. uniflorum (Lam.) Verdc. (*Dolichos uniflorus* Lam.)

English: horse gram

Macrozamia Miq. Zamiaceae

Origins:
Greek *makros* and the genus *Zamia* L.; see James A. Baines, *Australian Plant Genera.* 228-229. Chipping Norton, N.S.W. 1981.

Species/Vernacular Names:
M. communis L. Johnson

English: zamia, common zamia

M. diplomera (F. Muell.) L. Johnson

English: zamia

M. douglasii Bailey

English: zamia

M. dyeri (F. Muell.) C. Gardner

English: zamia

M. fawcettii C. Moore

English: zamia

M. flexuosa C. Moore

English: zamia

M. heteromera C. Moore

English: zamia

M. lucida L. Johnson

English: zamia

M. miquelii (F. Muell.) A. DC. (for the Dutch botanist Friedrich (Frederik) Anton Wilhelm Miquel, 1811-1871, physician, M.D. Gröningen 1833, Director of the Rotterdam Botanical Garden, professor of botany and Director of the Botanical Garden of Amsterdam 1846-1859, from 1862 Director of the Leyden Rijksherbarium. Among his most valuable writings are *Flora van Nederlandsch Indië*. Amsterdam, Utrecht and Leipzig 1855-1859 and *De palmis archipelagi indici* observationes novae. Amstelodami 1868; see Stafleu and Cowan, *Taxonomic Literature*. 3: 508-520. 1981; J.H. Barnhart, *Biographical Notes upon Botanists*. 2: 495. 1965; E.D. Merrill, *Bernice P. Bishop Mus. Bull.* 144: 136-137. 1937; T.W. Bossert, *Biographical Dictionary of Botanists Represented in the Hunt Institute Portrait Collection*. 268. 1972; Ida Kaplan Langman, *A Selected Guide to the Literature on the Flowering Plants of Mexico*. Philadelphia 1964; A. Lasègue, *Musée botanique de Benjamin Delessert*. Paris 1845; E.M. Tucker, *Catalogue of the Library of the Arnold Arboretum of Harvard University*. 1917-1933; F.A. Stafleu, *Wentia*. 16: 1-95. 1966; F.A. Stafleu, in *Dictionary of Scientific Biography* 9: 417. 1981; R. Zander, F. Encke, G. Buchheim and S. Seybold, *Handwörterbuch der Pflanzennamen*. 14. Aufl. Stuttgart 1993; Leonard Huxley, *Life and Letters of Sir J.D. Hooker*. London 1918).

English: zamia

M. moorei F. Muell.

English: springsure zamia

M. pauli-guilielmi F. Muell.

English: zamia

M. platyrachis Bailey

English: zamia

M. riedlei (Gaudich.) C. Gardner (after Anselme Riedle, 1775-1801 (died in Timor), gardener on Baudin's Expedition)

English: zamia

M. secunda C. Moore

English: zamia

M. spiralis (Salisb.) Miq.

English: zamia

M. stenomera L. Johnson

English: zamia

Macrozanonia (Cogn.) Cogn. Cucurbitaceae

Origins:

From the Greek *makros* "large" plus the genus *Zanonia* L.

Macucua Raf. Aquifoliaceae

Origins:

A vernacular name, for *Macucua* J.F. Gmelin = *Macoucoua* Aublet; see Constantine S. Rafinesque, *Sylva Telluriana*. 46. 1838.

Macuillamia Raf. Scrophulariaceae

Origins:

See Constantine S. Rafinesque, *Neogenyton, or Indication of Sixty-Six New Genera of Plants of North America*. 2. 1825 and *Autikon botanikon*. Icones plantarum select. nov. vel rariorum, etc. 44, 69. Philadelphia 1840; E.D. Merrill, *Index rafinesquianus*. 217. 1949.

Macvaughia W.R. Anderson Malpighiaceae

Origins:

For the American botanist Rogers McVaugh, b. 1909, professor of botany, traveler, plant collector and botanical historian.

Macvaughiella R.M. King & H. Robinson Asteraceae

Origins:

Dedicated to the American botanist Rogers McVaugh, b. 1909, traveler, plant collector, lichenologist and professor of botany, *Curator Emeritus* of University of Michigan Herbarium, 1977 Merit Award Winner of the Botanical Society of America. His writings include "Botanical exploration in Nueva Galicia from 1790 to the present time." *Contr. Univ. Mich. Herb.* 9(3): 205-357. 1972, "Galeotti's botanical work in Mexico: The numbering of his collections and a brief itinerary." *Contr. Univ. Mich. Herb.* 11(5): 291-297. 1972, *Flora Novo-Galiciana: A Descriptive Account of the Vascular Plants of Western Mexico.* (general editor William R. Anderson, University of Michigan Herbarium), *Edward Palmer, Plant Explorer of the American West.* University of Oklahoma Press, Norman 1956 and "Compositarum Mexicanarum Pugillus." *Contr. Univ. Mich. Herb.* 9(4): 361-484. 1972, with Stanley A. Cain and Dale J. Hagenah wrote "*Farwelliana*: an account of the life and botanical work of Oliver Atkins Farwell, 1867-1944." *Cranbrook Institute of Science Bulletin* 34. 1953; see John H. Barnhart, *Biographical Notes upon Botanists*. 2: 433. Boston 1965; T.W. Bossert, *Biographical Dictionary of Botanists Represented in the Hunt Institute Portrait Collection*. 261. 1972; S. Lenley et al., *Catalog of the Manuscript and Archival Collections and Index to the Correspondence of*

John Torrey. Library of the New York Botanical Garden. 278. 1973; Ida Kaplan Langman, *A Selected Guide to the Literature on the Flowering Plants of Mexico.* Philadelphia 1964; Irving William Knobloch, compil., "A preliminary verified list of plant collectors in Mexico." *Phytologia Memoirs.* VI. 1983; George Bentham, *Plantae Hartwegianae* (Plantas Hartwegianas imprimis Mexicanas adjectis nonnullis Grahamianis enumerat novasque describit). London 1839-1857. (Reprint, with new introduction by Rogers McVaugh.) 1970.

Madagaster Nesom Asteraceae

Origins:
Aster from Madagascar.

Madarosperma Benth. Asclepiadaceae

Origins:
Greek *madao* "make bald," *madaros* "bald" and *sperma* "seed."

Maddenia Hook.f. & Thomson Rosaceae

Origins:
For the Irish botanist Edward Madden, 1805-1856 (d. Edinburgh), 1830-1849 in the Bengal Army, 1853 President Botanical Society of Edinburgh, plant collector (Egypt), author of *Observations on Himalayan Coniferae.* Calcutta 1850, he was a friend of Michael Pakenham Edgeworth (1812-1881). See John H. Barnhart, *Biographical Notes upon Botanists.* 2: 433. Boston 1965; H.N. Clokie, *Account of the Herbaria of the Department of Botany in the University of Oxford.* 205. Oxford 1964; Ethelyn Maria Tucker, *Catalogue of the Library of the Arnold Arboretum of Harvard University.* Cambridge, Massachusetts 1917-1933; Isaac Henry Burkill, *Chapters on the History of Botany in India.* Delhi 1965; E.C. Nelson & Eileen May McCracken, *The Brightest Jewel: A History of the National Botanic Gardens, Glasnevin, Dublin.* Kilkenny 1987.

Madhuca Buch.-Ham. ex J.F. Gmelin Sapotaceae

Origins:
A Sanskrit name; see Helmut Genaust, *Etymologisches Wörterbuch der botanischen Pflanzennamen.* 360. Basel 1996; M.P. Nayar, *Meaning of Indian Flowering Plant Names.* 214. Dehra Dun 1985.

Species/Vernacular Names:
M. hainanensis Chun & How
English: Hainan madhuca
China: hai nan zi jing mu
M. longifolia (Koenig) Macbr.
English: south Indian mahua, mowra butter tree
India: mahwa, mahua, mohwa, mauwa, maul, mahula, mohwra, mowa, moa, moha, madgi, mahuda, ippa, illupei, elupa, hippe, poonam, ilupa
Nepal: mahuwa
M. pasquieri (Dubard) H.J. Lam (*Madhuca subquincuncialis* Lam & Kerpel)
English: peanut madhuca
China: zi jing mu
Vietnam: sen, sen dua

Madia Molina Asteraceae

Origins:
Madi is the native Chilean name for *Madia sativa* Molina; see Giovanni Ignazio Molina (1737-1829), *Saggio sulla storia naturale del Chili.* 136, 354. Bologna 1782.

Species/Vernacular Names:
M. citriodora E. Greene
English: lemon-scented tarweed
M. doris-nilesiae T.W. Nelson & J.P. Nelson
English: Nile's madia
M. elegans Lindley
English: common madia
M. exigua (Smith) A. Gray
English: threadstem madia
M. glomerata Hook.
English: mountain madia
M. gracilis (Smith) Keck
English: slender tarweed
M. hallii Keck
English: Hall's madia
M. nutans (E. Greene) Keck
English: nodding madia
M. radiata Kellogg
English: showy madia
M. sativa Molina
English: Chile tarweed, coast tarweed
M. stebbinsii T.W. Nelson & J.P. Nelson
English: Stebbin's madia

M. yosemitana A. Gray

English: Yosemite tarweed

Madronella Mill. Labiatae

Origins:

See *Monardella* Benth.

Maelenia Dumortier Orchidaceae

Origins:

For the brothers Van der Maelen, Belgian horticulturists.

Maerua Forsskål Capparidaceae (Capparaceae)

Origins:

An Arabic name *maeru* or *meru*, probably derived from *mehr* "sun."

Species/Vernacular Names:

M. angolensis DC.

English: bead-bean tree, bead maerua

W. Africa: berebere, belebele, bili, bile

South Africa: knoppiesboontjieboom; umEnwayo, umGodithi (Zulu); mugesi, mureri (Shona)

S. Rhodesia: umPoqompoqani

Nigeria: gazare; chichiwa (Yoruba); leggal bali (= sheep tree) (Fula); shegara el zeraf (= giraffe tree) (Arabic)

M. cafra (DC.) Pax (*Maerua triphylla* (Thunb.) Dur. & Schinz; *Niebuhria triphylla* (Thunb.) Wendl.)

English: white-wood, Christmas flower, common bush-cherry

Southern Africa: gewone witbos, witbas, without, withoutboom, wildebashout, lemoentjie, witboshout, witboom, witgatboom; unTswantwsane (Zulu); umPhunzisa (Xhosa)

M. crassifolia Forssk.

Nigeria: gazare, ngizari; jiga (Hausa); sarah (Arabic)

Arabic: maeru, meru, sarh, sarkh, sarha, sarah

M. gilgii Schinz. (*Boscia angustifolia* sensu Harv.; *Maerua angustifolia* (Harv.) Schinz; *Maerua stenophylla* Sprague) (the specific name honors the German taxonomist Ernst Friedrich Gilg, 1867-1933, contributor to H.G.A. Engler and K.A.E. Prantl *Die Natürlichen Pflanzenfamilien* ed. I and ed. 2, to Engler's *Das Pflanzenreich.* (with J. Perkins); see John H. Barnhart, *Biographical Notes upon Botanists.* 2: 48. 1965; T.W. Bossert, *Biographical Dictionary of*

Botanists Represented in the Hunt Institute Portrait Collection. 144. 1972; Stafleu and Cowan, *Taxonomic Literature.* 1: 941-942. Utrecht 1976)

South Africa: Gilg's maerua

M. nervosa (Hochst.) Oliv. (*Niebuhria nervosa* Hochst.) (from the Latin *nervosus, a, um* (*nervus*) "nerved, nervous, fibrous," it possibly refers to the veins prominent beneath the leaves)

English: Natal bush-cherry

Southern Africa: Natalwitbos; iThandana (Zulu)

M. racemulosa (A. DC.) Gilg & Ben. (*Boscia caffra* Sond.; *Capparis racemulosa* A. DC.; *Capparis undulata* Zeyh. ex Eckl. & Zeyh.; *Maerua undulata* (Zeyh. ex Eckl. & Zeyh.) Dur. & Schinz; *Maerua peduncolosa* (Hochst.) Sim; *Niebuhria pedunculosa* Hochst.)

English: forest bush-cherry

Southern Africa: witboshout; umPhunzisa, umPhunziso (Xhosa); umPhunziso, iDungamuzi-elicane (Zulu)

M. rosmarinoides (Sond.) Gilg & Ben. (*Niebuhria rosmarinoides* Sond.)

English: needle-leaved bush-cherry, rosemary maerua, rosemary-like maerua

Southern Africa: naaldlaarwitbos; unTswantwsane (Zulu)

M. schinzii Pax (*Maerua paxii* Schinz; *Maerua arenicola* Gilg) (the specific name after the Swiss botanist and plant collector Hans Schinz, 1858-1941, author of "Durch Südwestafrika." *Verh. Ges. Erdk. Berl.* 14: 322-324. 1887; see John H. Barnhart, *Biographical Notes upon Botanists.* 3: 227. 1965; Mary Gunn and Leslie E. Codd, *Botanical Exploration of Southern Africa.* 311-313. Cape Town 1981; T.W. Bossert, *Biographical Dictionary of Botanists Represented in the Hunt Institute Portrait Collection.* 353. 1972; E.M. Tucker, *Catalogue of the Library of the Arnold Arboretum of Harvard University.* 1917-1933; Elmer Drew Merrill, *Bernice P. Bishop Mus. Bull.* 144: 163. 1937)

Southern Africa: kringboom; omundipu (Herero)

Maesa Forssk. Myrsinaceae

Origins:

From the Arabic *maas*, a common name for the type species of the genus; see Pehr (Peter) Forsskål, *Flora aegyptiaco-arabica.* 66. Copenhagen 1775.

Species/Vernacular Names:

M. chisia Buch.-Ham. ex D. Don

English: greyleaf maesa

China: mi xian du jing shan

M. indica (Roxb.) A. DC.

English: Indian maesa

China: bao chuang ye

India: nagapadhera, kiriti, kirithi, tanipele, jiundali, atki

Indochina: cu den

Malaya: kasi hutan

M. japonica (Thunberg) Moritzi & Zollinger (*Doraena japonica* Thunb.)

English: Japanese maesa

Japan: izu-sen-ry♦

China: du jing shan

M. lanceolata Forssk. (*Maesa angolensis* Gilg; *Maesa lanceolata* Forssk. var. *rufescens* (A. DC.) Taton; *Baeobothrys lanceolata* (Forssk.) Vahl)

English: false assegai

Arabic: máas, arar

Southern Africa: basterassegaai; liGucu, umBohlobohlo, umBhongabhonga (Swazi); uMagupu, uMaququ, umPhongaphonga, uPhongaphonga, iNdende, isiDenda, uPhophopho, inHlavubele, uBhoqobhoqo (Zulu); iNtendekwane, inTentekiwane (Xhosa); muunguri, mutiba-mmela (Venda); muGarapatonora, muDovatova, muDowatowa, mandara, Ndwatwa, muPenenmbi, chiTsamva (Shona)

S. Rhodesia: mhandara, chiTsamva, muDovatova

M. montana A. DC.

English: mountain maesa

China: jin zhu liu

M. ramentacea (Roxburgh) A. DC. (Latin *ramentum, i* "what is grated, scrapings, chips")

English: ramentaceous maesa

Malaya: mengambir, gambir gambir, gambir jantan, gambir-gambir jantan, gegambir, bekaras, telor belangkas, belangkas hutan, kampur, membola, membuloh, pesat, setulang, kecham utan, tulang hutan, patah tulang

China: cheng gan shu

M. rugosa C.B. Clarke

English: wrinkled-leaf maesa

China: zhou ye du jing shan

Maesobotrya Benth. Euphorbiaceae

Origins:
From the genus *Maesa* and *botrys* "cluster, a bunch of grapes," referring to the grape-like fruits.

Species/Vernacular Names:
M. barteri (Baill.) Hutch.

Nigeria: olohun, orowo (Yoruba); oruru (Edo); miri ogu (Igbo); ntum kache (Boki)

Yoruba: olohun, olowun, orowo, odun

Maesopsis Engl. Rhamnaceae

Origins:
From the genus *Maesa* and the Greek *opsis* "resembling, aspect," referring to the leaves.

Species/Vernacular Names:
M. eminii Engl. (for Emin Pacha)

Nigeria: oubiogiekhue; igilogbon (Yoruba); ovbiogiekhue (Edo); awuru (Igbo)

Tropical Africa: musizi

Congo: nabit

Gabon: ken, nkanguelé

Zaire: osongo, ishongo, bosongu

Cameroon: esenge, nkangela, nkala, londo

Ivory Coast: manasati, anschia-sain, sagou-doué

Kenya: musizi

Mafureira Bertol. Meliaceae

Origins:
Mafoureira or *mafurra*, vernacular names for *Trichilia emetica.*

Magnistipula Engl. Chrysobalanaceae

Origins:
From the Latin *magnus, a, um* "great, big" and *stipula, ae* "stipule."

Species/Vernacular Names:
M. butayei De Wild.

Central Africa: bongolu montane, djungu, ehungu lo lowe, penzi

Cameroon: nom asila abim

Ivory Coast: bombi

Magnolia L. Magnoliaceae

Origins:
After the French (b. Montpellier) botanist Pierre Magnol, 1638-1715 (d. Montpellier), physician, an innovator in botanical classification, professor of botany and medicine and Director of the Botanical Garden at Montpellier, in correspondence with all the great botanists of Europe. His writings include *Botanicum monspeliense.* Lugduni [Lyon] 1676, *Novus caracter plantarum*, in duos tractatus divisus. Monspelii 1720, *Prodromus historiae generalis plantarum.*

Monspelii [Montpellier] 1689 and *Hortus regius monspeliensis*. Monspelii 1697. See Paul Jovet & J.C. Mallet, in *Dictionary of Scientific Biography* 9: 17-18. 1981; Frans A. Stafleu (1921-1997), *Linnaeus and the Linnaeans*. The spreading of their ideas in systematic botany, 1735-1789. Utrecht 1971; G.C. Wittstein, *Etymologisch-botanisches Handwörterbuch*. 551. 1852; John H. Barnhart, *Biographical Notes upon Botanists*. 2: 436. 1965; Baron Philippe Picot de Lapeyrouse (1744-1818), *Histoire abrégée des plantes des Pyrénées*. Toulouse 1813; Edmund Berkeley and Dorothy Smith Berkeley, *John Clayton, Pioneer of American Botany*. Chapel Hill 1963; Ethelyn Maria Tucker, *Catalogue of the Library of the Arnold Arboretum of Harvard University*. Cambridge, Massachusetts 1917-1933; T.W. Bossert, *Biographical Dictionary of Botanists Represented in the Hunt Institute Portrait Collection*. 249. 1972; Jonas C. Dryander, *Catalogus bibliothecae historico-naturalis Josephi Banks*. London 1796-1800; H.N. Clokie, *Account of the Herbaria of the Department of Botany in the University of Oxford*. 206. Oxford 1964; A. Lasègue, *Musée botanique de Benjamin Delessert*. Paris 1845; Stafleu and Cowan, *Taxonomic Literature*. 3: 243-245. Utrecht 1981; A.J.C. Grierson & D.G. Long, *Flora of Bhutan*. 1(2): 234. Edinburgh 1984; David G. Frodin & Rafaël Govaerts, *World Checklist and Bibliography of Magnoliaceae*. Royal Botanic Gardens, Kew 1996; David Hunt, ed., *Magnolias and Their Allies*. Proceedings of an International Symposium, Royal Holloway, University of London, Egham, Surrey, U.K., 12-13 April 1996. International Dendrology Society and The Magnolia Society. 1998.

Species/Vernacular Names:

M. acuminata (L.) L.

English: cucumber tree

M. amoena Cheng

English: Tianmu Mountain magnolia

M. campbellii Hook. & Thomson

Nepal: ghoge chanp, lal chanp

M. dealbata Zucc.

Mexico: guie zehe, quije, zehe, yo zaha

M. denudata Desr. (*Magnolia yulan* Desf.; *Magnolia conspicua* Salisb.; *Magnolia heptapeta* (Buc'hoz) Dandy)

English: lily tree, Yulan magnolia, Yulan tree, Yulan, tulip tree

China: xin yi, hsin i, ying chun, hou tao, mu pi (= wood pencil), yu lan

Japan: haku-mokuren

M. fraseri Walter (*Magnolia auriculata* Bartr.)

English: early-leaved umbrella tree

M. globosa Hook. & Thomson

Nepal: kokre chanp

M. grandiflora L. (*Magnolia foetida* Sarg.)

English: large-flowered magnolia, bull bay, southern magnolia, loblolly magnolia, big laurel

Japan: tai-san-boku (Taishan is a mountain in China)

M. hypoleuca Siebold & Zucc. (*Magnolia obovata* Thunb.)

English: Japanese cucumber tree, the big leaved magnolia

Japan: ho-no-ki

China: hou po, mu lan, huang hsin (= yellow heart), mu lien hua

M. kobus DC.

English: the northern Japanese magnolia

Japanese: kobishi (= fist)

M. liliiflora Desr. (*Magnolia discolor* Vent.; *Magnolia gracilis* Salisb.; *Magnolia quinquepeta* (Buc'hoz) Dandy; *Magnolia purpurea* Curtis)

English: woody orchid, mu-lan, the lily flowered magnolia, magnolia tree, red magnolia

China: xin yi

Japan: shi-moku-ren, mokuren

M. macrophylla Michx.

English: large-leaved cucumber tree, great-leaved macrophylla, umbrella tree

M. officinalis Rehder & E. Wilson

English: officinal magnolia, magnolia

China: hou po

M. pterocarpa Roxb.

Nepal: chanp, patpate

M. salicifolia (Siebold & Zucc.) Maxim.

English: anise magnolia, willow-leaf magnolia

M. sieboldii Koch

English: Oyama magnolia

M. x soulangiana Soul.-Bod. (*Magnolia speciosa* Geel)

English: saucer magnolia, Chinese magnolia

M. sprengeri Pamp. (*Magnolia denudata* var. *purpurascens* (Maxim.) Rehd. & Wils.; *Magnolia parviflora* Sieb. & Zucc. non Bl.; *Magnolia oyama* hort.; *Magnolia verecunda* Koidz.)

Japan: sarasa-renge

M. stellata (Sieb. & Zucc.) Maxim. (*Magnolia tomentosa* Thunb.; *Magnolia halleana* Parsons; *Magnolia kobus* var. *stellata* (Sieb. & Zucc.) Blackburn)

English: star magnolia

Japan: shide-kobushi

M. tripetala L. (*Magnolia umbrella* Desr.)

English: umbrella magnolia, umbrella tree, elkwood

M. virginiana L. (*Magnolia glauca* L.)

English: sweet bay, swamp bay, swamp laurel

M. zenii Cheng

English: Zen magnolia

Magodendron Vink Sapotaceae

Origins:

From the Greek *magos* "wizard, magical, enchanter" and *dendron* "tree."

Maguirea A.D. Hawkes Araceae

Origins:

After the American botanist Bassett Maguire, 1904-1991, explorer, plant collector, from 1943 New York Botanical Garden. Among his writings are "Guttiferae." in R.E. Schultes, "Plantae Austro-Americanae VII." *Bot. Mus. Leafl.* 15(2): 55-69. 1951, "Guttiferae." in B. Maguire, J.J. Wurdack and collaborators, "The botany of the Guayana Highland — part IV(2)." *Mem. New York Bot. Gard.* 10(4): 21-32. 1961, "Rapateaceae." in B. Maguire, J.J. Wurdack and collaborators, "The botany of the Guayana Highland — part VI." *Mem. New York Bot. Gard.* 12(3): 69-102. 1965 and "Notes on the Clusiaceae — chiefly of Panama. III." *Phytologia.* 39(2): 65-77. 1978, with Y.-C. Hung wrote "Styracaceae." in B. Maguire, J.J. Wurdack and collaborators, "The botany of the Guayana Highland — part X." *Mem. New York Bot. Gard.* 29: 204-223. 1978, with R.E. Weaver, Jr., wrote "The neotropical genus *Tachia* (Gentianaceae)." *J. Arnold Arbor.* 56(1): 103-125. 1975, with J.A. Steyermark and D.G. Frodin wrote "Araliaceae." in B. Maguire, J.J. Wurdack and collaborators, "The Botany of the Guayana Highland — part XII." *Mem. New York Bot. Gard.* 38: 46-84. 1984; see John H. Barnhart, *Biographical Notes upon Botanists.* 2: 436. 1965; Ida Kaplan Langman, *A Selected Guide to the Literature on the Flowering Plants of Mexico.* Philadelphia 1964; T.W. Bossert, *Biographical Dictionary of Botanists Represented in the Hunt Institute Portrait Collection.* 249. 1972; S. Lenley et al., *Catalog of the Manuscript and Archival Collections and Index to the Correspondence of John Torrey.* Library of the New York Botanical Garden. 279. 1973; J. Ewan, ed., *A Short History of Botany in the United States.* 22. 1969; Laurence J. Dorr, "In memoriam. John J. Wurdack, 1921-1998." in *Plant Science Bulletin.* 44(2): 41. Summer 1998.

Maguireanthus Wurdack Melastomataceae

Origins:

After the American botanist Bassett Maguire, 1904-1991, explorer, plant collector; see John H. Barnhart, *Biographical Notes upon Botanists.* 2: 436. 1965.

Maguireocharis Steyermark Rubiaceae

Origins:

After the American botanist Bassett Maguire, 1904-1991, explorer, plant collector; see John H. Barnhart, *Biographical Notes upon Botanists.* 2: 436. 1965.

Maguireothamnus Steyerm. Rubiaceae

Origins:

After the American botanist Bassett Maguire, 1904-1991, explorer, plant collector; see John H. Barnhart, *Biographical Notes upon Botanists.* 2: 436. 1965.

Magydaris W.D.J. Koch ex DC. Umbelliferae

Origins:

From the Greek *magydaris, magudaris* "inflorescence of the *silphion*" (also its root, sap or seed).

Mahagoni Adans. Meliaceae

Origins:

Obscure origins, probably from the vernacular *mohogoni*, Central America or West Indies; see Helmut Genaust, *Etymologisches Wörterbuch der botanischen Pflanzennamen.* 361. 1996; Ernest Weekley, *An Etymological Dictionary of Modern English.* 2: 879. New York 1967; Manlio Cortelazzo & Paolo Zolli, *Dizionario etimologico della lingua italiana.* 3: 768. [" maogano, magogano, maogani, mogano, magoni, mochogon, makogany, mogogane, magogon"] Bologna 1983; C.T. Onions, *The Oxford Dictionary of English Etymology.* Oxford University Press 1966; N. Tommaseo & B. Bellini, *Dizionario della lingua italiana.* Torino 1865-1879; G.C. Wittstein, *Etymologisch-botanisches Handwörterbuch.* 545. 1852.

Mahernia L. Sterculiaceae

Origins:

More or less an anagram of the genus *Hermannia*, dedicated to the German-born Dutch botanist Paul Hermann, 1646-1695, professor of botany at Leyden 1680-1695, plant collector at the Cape. Among his most numerous writings are *Catalogus Musei Indici.* Lugduni Batavorum [1711], *Florae Lugduno-Batavae Flores.* Lugd[uni] Batav[orum] 1690, *Horti Academici Lugduno-Batavi Catalogus.* Lugduno Batavorum [Leyden] 1687, *Paradisus batavus.* Opus posthumum edidit William Sherard. Lugduni-Batavorum [Leyden]

1698 and *Musaeum Zeylanicum*, sive catalogus plantarum in Zeylana sponte nascentium. Lugduni Batavorum 1717; see John H. Barnhart, *Biographical Notes upon Botanists.* 2: 163. 1965; Peter MacOwan, "Personalia of botanical collectors at the Cape." *Trans. S. Afr. Philos. Soc.* 4(1): xxx-liii. 1884-1886.

Mahonia Nuttall Berberidaceae

Origins:
Named after the Irish-born American horticulturist Bernard M'Mahon (McMahon), c. 1775-1816 (d. Philadelphia), botanist and seedsman, in 1796 went to United States, nurseryman, author of *The American Gardener's Calendar*; adapted to the climate & seasons of the United States, etc., Philadelphia 1806. See J.H. Barnhart, *Biographical Notes upon Botanists.* 2: 430. 1965; J.W. Harshberger, *The Botanists of Philadelphia and Their Work.* 117-119. 1899; Leslie Walter Allen Ahrendt (1903-1969), "*Berberis* and *Mahonia*." *Journal of the Linnean Society, Botany.* 57: 1-410. 1961; Ethelyn Maria Tucker, *Catalogue of the Library of the Arnold Arboretum of Harvard University.* Cambridge, Massachusetts 1917-1933; J. Ewan, ed., *A Short History of Botany in the United States.* 5, 136 [by George H.M. Lawrence, *Horticulture.*]. 1969; Jeannette Elizabeth Graustein, *Thomas Nuttall, Naturalist. Explorations in America, 1808-1841.* 471. 1967; T. Nuttall, *The Genera of North American Plants.* 1: 211. 1818; F. Boerner & G. Kunkel, *Taschenwörterbuch der botanischen Pflanzennamen.* 4. Aufl. 131. Berlin & Hamburg 1989; Georg Christian Wittstein, *Etymologisch-botanisches Handwörterbuch.* 551. Ansbach 1852; Stafleu and Cowan, *Taxonomic Literature.* 3: 229-230. Utrecht 1981; William Darlington, *Reliquiae Baldwinianae.* Philadelphia 1843.

Species/Vernacular Names:
M. aquifolium (Pursh) Nutt. (*Ilex japonica* Thunb.)
English: Oregon grape
M. bealei (Fort.) Carr.
China: shi da gong lao ye
M. fortunei (Lindl.) Fedde
China: shi da gong lao ye
M. japonica (Thunb.) DC. (*Ilex japonica* Thunb.)
Japan: hiiragi-nanten (hiiragi = holly)
China: shi da gong lao ye

Mahurea Aublet Guttiferae

Origins:
The vernacular name.

Maianthemum G.H. Weber ex Wigg. Convallariaceae (Liliaceae)

Origins:
Latin *Maius* "may," Greek *Maios* and *anthemon* "flower."

Species/Vernacular Names:
M. bifolium (L.) F.W. Schmidt
English: May lily
M. dilatatum (Alph. Wood) Nelson & J.F. Macbr.
English: false lily-of-the-valley
M. racemosum (L.) Link
English: false spikenard, American spikenard, wild spikenard

Maidenia Domin Umbelliferae

Origins:
After the botanist Joseph Henry Maiden, 1859-1925; see J.H. Barnhart, *Biographical Notes upon Botanists.* 2: 437. 1965; Karel Domin (1882-1953), *Acta Botanica Bohemica.* 1: 41. 1922.

Maidenia Rendle Hydrocharitaceae

Origins:
After the British-born (Londoner by birth) Australian botanist Joseph Henry Maiden, 1859-1925 (Sydney), studied botany under Prof. R. Bentley and Prof. D. Oliver, 1880 migrated to Australia, member of the Council of the Royal Society and Linnean Society of NSW and President of both Societies, Honorary Secretary of the Australasian Association for the Advancement of Science, Curator and Secretary of the Technological Museum, investigator of the economic botanical resources of Australia, 1896 succeeded the late Mr. Charles Moore as Government Botanist of New South Wales and Director of the Botanic Gardens of Sydney, 1889 Fellow of the Linnean Society (in 1915 was awarded its Gold Medal), from 1896 to 1924 New South Wales Government Botanist, 1916 admitted into the Royal Society of London, 1916 was appointed to the Imperial Service. Among his writings are *The Useful Native Plants of Australia.* (Including Tasmania). London and Sydney 1889, *The Olive and Olive Oil*; being notes on the culture of the tree and extraction of the oil as carried out in South Australia and the Continent of Europe. Sydney 1887, *The Forest Flora of New South Wales.* Sydney [1903-] 1904-1925, *Mount Seaview and the Way Thither.* Sydney 1898 and *Sir Joseph Banks: The "Father of Australia".* Sydney, London 1909, with Ernst Betche (1851-1913) wrote *A Census of New South Wales Plants.* Sydney 1916. See R. Zander, F.

Encke, G. Buchheim and S. Seybold, *Handwörterbuch der Pflanzennamen*. 14. Aufl. Stuttgart 1993; Stafleu and Cowan, *Taxonomic Literature*. 3: 249-255. 1981; J.H. Barnhart, *Biographical Notes upon Botanists*. 2: 437. 1965; I.C. Hedge and J.M. Lamond, *Index of Collectors in the Edinburgh Herbarium*. Edinburgh 1970; Ida Kaplan Langman, *A Selected Guide to the Literature on the Flowering Plants of Mexico*. Philadelphia 1964; E.D. Merrill, *Bernice P. Bishop Mus. Bull*. 144: 129-130. 1937 and *Contr. U.S. Natl. Herb*. 30(1): 202-203. 1947; A.B. Rendle, *The Journal of Botany*. 54: 316, t. 545. 1916; T.W. Bossert, *Biographical Dictionary of Botanists Represented in the Hunt Institute Portrait Collection*. 249. 1972; H.N. Clokie, *Account of the Herbaria of the Department of Botany in the University of Oxford*. 206. Oxford 1964; E.M. Tucker, *Catalogue of the Library of the Arnold Arboretum of Harvard University*. 1917-1933; Frans A. Stafleu, *Linnaeus and the Linnaeans*. The spreading of their ideas in systematic botany, 1735-1789. Utrecht 1971.

Maihuenia (F. Weber) Schumann Cactaceae

Origins:
A local vernacular name, in Chile.

Species/Vernacular Names:
M. poeppigii (Otto) Weber
Chile: maihuén, luanmamell

Maihueniopsis Speg. Cactaceae

Origins:
Resembling the genus *Maihuenia* (F. Weber) Schumann.

Maillardia Frappier & Duchartre Moraceae

Origins:
For the French botanist L. Maillard, engineer, author of *Notes sur l'Île de la Réunion* (Bourbon). Paris 1862; see J.H. Barnhart, *Biographical Notes upon Botanists*. 2: 437. 1965; Stafleu and Cowan, *Taxonomic Literature*. 3: 256. Utrecht 1981.

Maillea Parlatore Gramineae

Origins:
For the French botanist Alphonse Maille, 1813-1865; see J.H. Barnhart, *Biographical Notes upon Botanists*. 2: 437. 1965; H.N. Clokie, *Account of the Herbaria of the Department of*

Botany in the University of Oxford. 206. Oxford 1964; Ethelyn Maria Tucker, *Catalogue of the Library of the Arnold Arboretum of Harvard University*. Cambridge, Massachusetts 1917-1933; Jean-Louis Kralik and J. Billon, *Catalogue des Reliquiae Mailleanae*. Paris 1869; F.N. Hepper and F. Neate, *Plant Collectors in West Africa*. 53. Utrecht 1971.

Maingaya Oliver Hamamelidaceae

Origins:
For the British (b. Yorks.) physician Alexander Carroll Maingay, 1836-1869 (murdered Rangoon), cryptogamist, graduated in Medicine at Edinburgh in 1858, botanist of the East India Company (British Indian Medical Service), plant collector in North China and in the Malaysian region, Superintendent of the jail at Rangoon, author of "Timber trees of Straits Settlements." *Kew Bull*. 112-134. 1890, co-author with William Mudd (1830-1879) of *A Manual of British Lichens*. Darlington 1861. See J.H. Barnhart, *Biographical Notes upon Botanists*. 2: 438. 1965; Henry Nicholas Ridley (1855-1956), *The Flora of the Malay Peninsula*. London 1922-1925; I.H. Vegter, *Index Herbariorum*. Part II (4), *Collectors M*. Regnum Vegetabile vol. 93. 1976; Emil Bretschneider, *History of European Botanical Discoveries in China*. 1981; I.C. Hedge and J.M. Lamond, *Index of Collectors in the Edinburgh Herbarium*. Edinburgh 1970; James Britten (1846-1924) and George E. Simonds Boulger (1853-1922), *A Biographical Index of Deceased British and Irish Botanists*. London 1931; John Dransfield, "*Maingaya malayana* (Hamamelidaceae)." *The Kew Magazine*. 10(1): 81-84. May 1993.

Maireana Moq. Chenopodiaceae

Origins:
Dedicated to the French botanist Antoine Charles Lemaire, 1801-1871, naturalist, specialist in Cactaceae, editor of *L'Horticulteur universel*. Paris 1839-1845, from 1854 to 1869 edited *L'Illustration horticole*. His works include *Manuel de l'amateur de cactus*. Paris [1845], *Cactearum aliquot novarum ac insuetarum in horto monvilliano cultarum accurata descriptio*. Lutetiae Parisiorum [Paris] et Argentorati [Strasbourg] 1838, *Les Cactées. Histoire, ... culture*, etc. Paris [1868], *Iconographie descriptive des cactées*. Paris [1841-1850?] and *Flore des serres et jardins de l'Europe*. Gand 1845-1855, contributed to *Dictionnaire universel d'histoire naturelle*. Paris [1840] 1841-1849; see John H. Barnhart, *Biographical Notes upon Botanists*. 2: 366. 1965; P.G. Wilson, " A taxonomic revision of the genus *Maireana* (Chenopodiaceae)." in *Nuytsia*. 2(1): 2- 83. 1975; Ethelyn Maria Tucker, *Catalogue of the Library of the*

Arnold Arboretum of Harvard University. Cambridge, Massachusetts 1917-1933; Elmer Drew Merrill, *Contr. U.S. Natl. Herb.* 30(1): 187. 1947; Christian Horace Bénédict Alfred Moquin-Tandon (1804-1863), *Chenopodearum monographica enumeratio.* 95. Parisiis 1840; Gordon Douglas Rowley, *A History of Succulent Plants.* Strawberry Press 1997; F.N. Hepper and F. Neate, *Plant Collectors in West Africa.* 53. [this genus dedicated to the French mycologist René Maire, 1878-1949.] Utrecht 1971.

Species/Vernacular Names:

M. aphylla (R. Br.) Paul G. Wilson (*Kochia aphylla* R. Br.; *Salsola aphylla* (R. Br.) Sprengel)

English: cottonbush, leafless bluebush, spiny bluebush

M. astrotricha (L. Johnson) Paul G. Wilson

Australia: low bluebush

M. atkinsiana (W. Fitzg.) Paul G. Wilson

Australia: bronze bluebush

M. brevifolia (R. Br.) Paul G. Wilson (*Kochia brevifolia* R. Br.; *Kochia tamariscina* (Lindley) J. Black; *Suaeda tamariscina* Lindley; *Enchylaena tamariscina* (Lindley) Druce)

Australia: small leaved bluebush, short-leaved bluebush, eastern cottonbush, yarga bush, cottonbush, yanga bush, small leaf bluebush

M. carnosa (Moq.) Paul G. Wilson

Australia: cottony bluebush

M. cheelii (R. Anderson) Paul G. Wilson (after the Australian (b. England, Kent) botanist Edwin Cheel, 1872-1951 (d. Sydney), one of the contributors to *The Flora of the Northern Territory* by Alfred J. Ewart and Olive B. Davies. Melbourne 1917; see Ray Desmond, *Dictionary of British & Irish Botanists and Horticulturists.* 144. London 1994)

Australia: chariot wheels

M. ciliata (F. Muell.) Paul G. Wilson

Australia: hairy fissure-weed, fissure-weed

M. convexa Paul G. Wilson

Australia: mulga bluebush

M. coronata (J. Black) Paul G. Wilson

Australia: crown fissure-weed

M. decalvans (Gand.) Paul G. Wilson

Australia: black cottonbush

M. enchylaenoides (F. Muell.) Paul G. Wilson (*Duriala villosa* (F. Muell.) Ulbr.)

Australia: wingless bluebush, wingless fissure-weed

M. eriantha (F. Muell.) Paul G. Wilson

Australia: woolly bluebush

M. erioclada (Benth.) Paul G. Wilson

Australia: rosy bluebush

M. excavata (J. Black) Paul G. Wilson

Australia: bottle bluebush, bottle fissure-weed

M. georgei (Diels) Paul G. Wilson

Australia: satiny bluebush, golden bluebush, George's bluebush

M. glomerifolia (F. Muell. & Tate) Paul G. Wilson

Australia: ball-leaf bluebush

M. lanosa (Lindley) Paul G. Wilson

Australia: woolly bluebush

M. melanocoma (F. Muell.) Paul G. Wilson

Australia: pussy bluebush

M. microcarpa (Benth.) Paul G. Wilson

Australia: swamp bluebush

M. microphylla (Moq.) Paul G. Wilson

Australia: eastern cottonbush

M. oppositifolia (F. Muell.) Paul G. Wilson

Australia: heathy bluebush

M. pentatropis (Tate) Paul G. Wilson

Australia: erect mallee bluebush

M. planifolia (F. Muell.) Paul G. Wilson

Australia: low bluebush, flat-leaved bluebush

M. platycarpa Paul G. Wilson

Australia: shy bluebush

M. pyramidata (Benth.) Paul G. Wilson

Australia: black bluebush, sago bluebush, shrubby bluebush

M. radiata (Paul G. Wilson) Paul G. Wilson

Australia: grey bluebush

M. sedifolia (F. Muell.) Paul G. Wilson

Australia: pearl bluebush

M. suaedifolia (Paul G. Wilson) Paul G. Wilson

Australia: lax bluebush

M. tomentosa Moq.

Australia: felty bluebush

M. trichoptera (J. Black) Paul G. Wilson

Australia: pink-seeded bluebush, downy bluebush

M. triptera (Benth.) Paul G. Wilson

Australia: three-winged bluebush

M. turbinata Paul G. Wilson

Australia: satiny bluebush

M. villosa (Lindley) Paul G. Wilson

Australia: silky bluebush

Mairetis I.M. Johnston Boraginaceae

Origins:

For the French botanist René Charles Joseph Ernest Maire, 1878-1949, professor of botany in Algeria, physician, mycologist, an authority on N. African flora; see John H. Barnhart, *Biographical Notes upon Botanists*. 2: 438. Boston 1965; R. Zander, F. Encke, G. Buchheim and S. Seybold, *Handwörterbuch der Pflanzennamen*. 14. Aufl. Stuttgart 1993; T.W. Bossert, *Biographical Dictionary of Botanists Represented in the Hunt Institute Portrait Collection*. 249. 1972; S. Lenley et al., *Catalog of the Manuscript and Archival Collections and Index to the Correspondence of John Torrey*. Library of the New York Botanical Garden. 299. 1973; Ethelyn Maria Tucker, *Catalogue of the Library of the Arnold Arboretum of Harvard University*. Cambridge, Massachusetts 1917-1933; I. Urban, *Geschichte des Königlichen Botanischen Museums zu Berlin-Dahlem (1815-1913). Nebst Aufzählung seiner Sammlungen*. Dresden 1916; F.N. Hepper and F. Neate, *Plant Collectors in West Africa*. 53. 1971; A. White and B.L. Sloane, *The Stapelieae*. Pasadena 1937.

Maizilla Schltdl. Gramineae

Origins:

Maize, *Zea mays*.

Majidea J. Kirk ex Oliv. Sapindaceae

Origins:

After Sultan Majid of Zanzibar [now in Tanzania], son of Sayyid Sa'id ibn Sultan. On Sa'id's death in 1856 Majid succeeded to his African dominions, while another son, Thuwayn, succeeded to Oman. Majid died in 1870 and was succeeded by his brother Barghash (1834-1888).

Species/Vernacular Names:

M. fosteri (Sprague) Radlk.

Cameroon: mokombe

Ivory Coast: keremon

Majorana Miller Labiatae

Origins:

Derivation uncertain, Latin *maezuranam*, *amaracum*, Greek *amarakos*; see H. Genaust, *Etymologisches Wörterbuch der botanischen Pflanzennamen*. 362. 1996; E. Weekley, *An Etymological Dictionary of Modern English*. 2: 896. 1967; Manlio Cortelazzo & Paolo Zolli, *Dizionario etimologico della lingua italiana*. 3: 699. 1983; Serapiom, *El libro agregà de Serapiom*. A cura di G. Ineichen. ["maçorana"] Venezia-Roma 1962-1966.

Malabathris Raf. Melastomataceae

Origins:

Greek *malabathron*, used by Plinius and Dioscorides for a leaf of some species of *Cinnamomum*; *malabathrinos* "prepared with *malabathron*"; Latin *malobathron*, *malabathron* for an Indian or Syrian plant from which a costly ointment was prepared, perhaps betel or base cinnamon; see C.S. Rafinesque, *Sylva Telluriana*. 97. 1838.

Malacantha Pierre Sapotaceae

Origins:

Greek *malakos* "soft, soothing" and *anthos* "a flower," some suggest from *akantha* "thorn."

Species/Vernacular Names:

M. sp.

Nigeria: awami

M. alnifolia (Bak.) Pierre

Nigeria: akala, aningueri, awami, baushen kurimi, shoibuzozi; baushen kurmi (Hausa); akala (Yoruba)

Yoruba: oorunmu, afunnikunre, akala, akala odan, osan odan

M. superba Verm.

Zaire: bulanga, yanganga

Malaccotristicha C. Cusset & G. Cusset Podostemaceae

Origins:

From Malacca (Malaya) and *Tristicha* Thouars.

Malachadenia Lindley Orchidaceae

Origins:

Possibly from the Greek *malache* "mallow, tree-mallow" and *aden* "gland," referring to the color of the gland, or from *malakos* "soft, weak," indicating the character of the gland; see R.E. Schultes and Arthur Stanley Pease, *Generic Names of Orchids. Their Origin and Meaning*. 191. 1963; Hubert Mayr, *Orchid Names and Their Meanings*. 157. 1998.

Malache Vogel Malvaceae

Origins:

From the Greek *malache* "mallow, tree-mallow."

Malachra L. Malvaceae

Origins:

A variant of the Greek *malache, maloche* "mallow," Greek *malakos* "soft," Akkadian *lakû* "suckling, young, child," Hebrew *lah* "moist, fresh, green"; see Giovanni Semerano, *Le origini della cultura europea*. Dizionario della lingua Greca. 2(1): 174. Firenze 1994.

Species/Vernacular Names:

M. sp.

Panama: kigia

M. capitata (L.) L.

The Philippines: lapnis, bulubuluhan, paang-baliuis, bulahan, anabo, bakambakes, labog-labog, bulbulin, tambaking

Malacmaea Griseb. Malpighiaceae

Origins:

Greek *malakos* "soft" and *akme* "the top, summit."

Malacocarpus Fischer & C.A. Meyer Zygophyllaceae

Origins:

From the Greek *malakos* "soft, soothing" and *karpos* "a fruit," an allusion to the fleshy fruit.

Malacocarpus Salm-Dyck Cactaceae

Origins:

Greek *malakos* and *karpos* "a fruit."

Malacocera R.H. Anderson Chenopodiaceae

Origins:

Greek *malakos* "soft, soothing" and *keras* "a horn," referring to some differences with the genus *Bassia* All.; see *Proceedings of the Linnean Society of New South Wales*. 51: 382, t. XXV. 1926.

Species/Vernacular Names:

M. tricornis (Benth.) R. Anderson (*Chenolea tricornis* Benth.; *Bassia tricornis* (R. Br.) F. Muell.)

English: goat-head, soft horns, soft-horned saltbush, star saltbush

Malacochaeta Nees Cyperaceae

Origins:

From the Greek *malakos* "soft" and *chaite* "bristle, long hair"; in *Linnaea*. 9: 292. 1834-1835.

Malacomeles (Decne.) Engl. Rosaceae

Origins:

From the Greek *malakos* "soft, soothing" and *melon* "an apple."

Malacothamnus Greene Malvaceae

Origins:

Greek *malakos* "soft, soothing" and *thamnos* "a shrub."

Species/Vernacular Names:

M. abbottii (Eastw.) Kearney

English: Abbott's bush mallow

M. aboriginum (Robinson) E. Greene

English: Indian Valley bush mallow

M. clementinus (Munz & I.M. Johnston) Kearney

English: San Clemente Island bush mallow

M. davidsonii (Robinson) E. Greene

English: Davidson's bush mallow

M. fasciculatus (Torrey & A. Gray) E. Greene

English: chaparral bush mallow

M. jonesii (Munz) Kearney

English: Jones' bush mallow

M. palmeri (S. Watson) E. Greene

English: Santa Lucia bush mallow

Malacothrix DC. Asteraceae

Origins:

From the Greek *malakos* "soft" and *thrix, trichos* "hair."

Species/Vernacular Names:

M. coulteri A. Gray

English: snake's head

M. foliosa A. Gray

English: leafy malacothrix

M. glabrata A. Gray

English: desert dandelion

M. incana (Nutt.) Torrey & A. Gray

English: dundelion

M. indecora E. Greene

English: Santa Crux Island malacothrix

M. squalida E. Greene

English: Island malacothrix (referring to Santa Crux Island)

Malacurus Nevski Gramineae

Origins:
From the Greek *malakos* "soft" and *oura* "tail."

Malagasia L.A.S. Johnson & B.G. Briggs Proteaceae

Origins:
Malagasy "of or pertaining to Madagascar."

Malaisia Blanco Moraceae

Origins:
From the Philippine local plant name, *malais-ís.*

Species/Vernacular Names:

M. scandens (Lour.) Planchon

English: crow ash, burny vine, fire vine

The Philippines: hingi, hingiu, hinguin, malais-ís, sabá (Tagálog); sádak (Ilóko); salimpágot (Tagbanuá); sígid (Panay Bisáya)

Malaloleuca Gand. Myrtaceae

Origins:
Orthographic variant of *Melaleuca* L.

Malanthos Stapf Melastomataceae

Origins:
From the Greek *malasso* "to soften, make soft" and *anthos* "flower."

Malaxis Solander ex Swartz Orchidaceae

Origins:
Greek *malaxis* "a softening, soothing," *malasso, malassein* "to soften, make soft," *malakos* "soft," referring to the soft pleated leaves, to the delicacy and to the succulent nature of the plant; see B.A. Lewis and P. J. Cribb, *Orchids of the Solomon Islands and Bougainville.* Royal Botanic Gardens, Kew 1991.

Species/Vernacular Names:

M. latifolia Smith (*Microstylis latifolia* (Smith) J.J. Smith; *Microstylis kizanensis* Masam.; *Microstylis ishigakensis* Ohwi)

Japan: Kizan-hime-ran

M. paludosa (L.) Swartz

English: bog orchid

M. pectinata (J.J. Smith) P.F. Hunt

Bougainville Island: mengonirung

M. resupinata (Forst.f.) Kuntze

The Solomon Islands: ongi'ongi

Malcolmia R. Br. Brassicaceae

Origins:
For the English nurseryman William Malcolm, d. 1798, seedsman at Kennington and afterwards at Stockwell, and his son (or a nephew? or a descendant?) William, *circa* 1768/1769-1835; William (the Elder) issued *A Catalogue of Hot-House and Green-House Plants, Fruit and Forest Trees, Flowering Shrubs, Herbaceous Plants.* London 1771; see James Edwards, *Tabulae Distantiae ... Companion from London to Brighthelmston, in Sussex.* Dorking 1789-1801; William Curtis (1746-1799), *A Catalogue of the British, Medicinal, Culinary, and Agricultural Plants, Cultivated in the London Botanic Garden.* London 1783; Arthur D. Chapman, ed., *Australian Plant Name Index.* 1925. 1991; Helmut Genaust, *Etymologisches Wörterbuch der botanischen Pflanzennamen.* 363. [dedicated to William Malcolm, 1768-1835] Basel 1996; John Claudius Loudon (1783-1843), *Arboretum et fruticetum britannicum.* London 1838; F. Boerner & G. Kunkel, *Taschenwörterbuch der botanischen Pflanzennamen.* 4. Aufl. 130. Berlin & Hamburg 1989; Ray Desmond, *Dictionary of British & Irish Botanists and Horticulturists.* 464. London 1994; W. T. Aiton, *Hortus Kew.* ed. 2. 4: 121. Dec 1812 ("Malcomia") (nom. et orth. cons.); G.C. Wittstein, *Etymologisch-botanisches Handwörterbuch.* 553. 1852.

Species/Vernacular Names:

M. maritima (L.) R. Br.

English: Virginia stock

Malephora N.E. Br. Aizoaceae

Origins:

Greek *male* "armhole" and *phoros* "bearing, carrying," referring to the seed pocket of the fruits, some suggest from Latin *malus* "apple, apple-tree"; see Christo Albertyn Smith, *Common Names of South African Plants*. Edited by E. Percy Phillips and Estelle Van Hoeppen. Pretoria 1966; Helmut Genaust, *Etymologisches Wörterbuch der botanischen Pflanzennamen*. 363. Basel 1996.

Species/Vernacular Names:

M. crocea (Jacq.) Schwantes var. *crocea*

South Africa: vingerkanna, geelvingerkanna (= geel = yellow), copper vygie, fingerkanna

Malesherbia Ruíz & Pav. Malesherbiaceae

Origins:

For the French (b. Paris) statesman Chrétien-Guillaume de Lamoignon de Malesherbes, 1721-1794 (d. Paris), botanist, a patriot, agriculturist, 1746-1749 studied botany under Bernard de Jussieu, 1774 Secretary of State for the Royal Household, 1794 arrested as a Royalist, he was guillotined in Paris. Among his writings are *Mémoire sur les moyens d'accélérer les progrès de l'économie rurale en France*. Paris 1790 and *Observations de Lamoignon-Malesherbes sur l'histoire naturelle générale et particulière de Buffon et Daubenton*. Paris 1798; see Rhoda Rappaport, in *Dictionary of Scientific Biography* (Editor in Chief Charles Coulston Gillispie.) 9: 53-55. 1981.

Malidra Raf. Myrtaceae

Origins:

Referring to *Eugenia aquea* Burm.f. and *Syzygium aqueum* (Burm.f.) Alston; see Constantine Samuel Rafinesque, *Sylva Telluriana*. 107. 1838; E.D. Merrill, *Index rafinesquianus*. The plant names published by C.S. Rafinesque, etc. 174. Jamaica Plain, Massachusetts, USA 1949.

Malinvaudia Fourn. Asclepiadaceae

Origins:

After the French botanist Louis Jules Ernest Malinvaud, 1836-1913, from 1884 to 1908 *deus ex machina* of the Société Botanique de France; see John H. Barnhart, *Biographical Notes upon Botanists*. 2: 440. 1965; T.W. Bossert, *Biographical Dictionary of Botanists Represented in the Hunt Institute Portrait Collection*. 250. 1972; Ethelyn Maria Tucker, *Catalogue of the Library of the Arnold Arboretum of Harvard University*. Cambridge, Massachusetts 1917-1933; Stafleu and Cowan, *Taxonomic Literature*. 3: 264-265. Utrecht 1981.

Malleastrum (Baillon) J.-F. Leroy Meliaceae

Origins:

Resembling *Mallea* A. Juss.

Malleola J.J. Smith & Schlechter Orchidaceae

Origins:

Latin *malleolus* "little hammer," referring to the shape of the column.

Malleostemon J. Green Myrtaceae

Origins:

Latin *malleus* "a hammer" and Greek *stemon* "a thread, stamen," referring to the form of the geniculate stamens; see J.W. Green, "*Malleostemon*, a new genus of Myrtaceae (subfamily Leptospermoideae, tribe Chamelaucieae) from south-western Australia." in *Nuytsia*. 4(3): 295-315. 1983.

Mallophora Endl. Labiatae (Dicrastylidaceae, Verbenaceae)

Origins:

From the Greek *mallos* "wool" and *phoros* "carrying," the flowerheads are woolly, Latin *mallus* "a lock of wool"; see Stephan F. Ladislaus Endlicher, *Stirpium Australasicarum herbarii Hügeliani decades tres*. 207. Vindobonae 1838.

Mallophyton Wurdack Melastomataceae

Origins:

Greek *mallos* "wool, a lock of wool" and *phyton* "plant."

Mallostoma Karst. Rubiaceae

Origins:

From the Greek *mallos* "wool, a lock of wool" and *stoma* "mouth."

Mallotonia (Griseb.) Britton Boraginaceae

Origins:

Greek *mallos* "wool, a lock of wool," *mallotos*, *malloton* "fleecy, lined with wool."

Species/Vernacular Names:

M. gnaphalodes (L.) Britton

English: iodine bush

Mallotopus Franchet & Sav. Asteraceae

Origins:

From the Greek *mallotos, malloton* "fleecy, lined with wool" and *pous* "foot."

Mallotus Lour. Euphorbiaceae

Origins:

Greek *mallotos* "woolly, fleecy," referring to the villose plants or to the woolly fruits; see J. de Loureiro, *Flora cochinchinensis*. 601, 635. 1790.

Species/Vernacular Names:

M. claoxyloides (F. Muell.) Müll. Arg.

English: green kamala, smell-of-the-bush

M. discolor Benth.

English: yellow kamala

M. japonicus (Thunb.) Müll. Arg. (*Croton japonicus* Thunb.)

Japan: aka-me-gashiwa, tahi

M. mollissimus (Geiseler) Airy Shaw

English: kamala

M. nesophilus Müll. Arg.

English: kamala

M. oblongifolius (Miq.) Müll. Arg.

English: water castor oil

Malaya: jarak utan

M. oppositifolius (Geisel.) Müll. Arg.

Nigeria: kafar mutuwa (Hausa); oju-eja (Yoruba); okpokirinyan (Igbo)

Yoruba: atori igbo, eja, are, iruja

M. paniculatus (Lam.) Müll. Arg. (*Croton paniculatus* Lam.)

English: turn-in-the-wind

Japan: urajiro-akame-gashiwa, Annan-akame-gashiwa

Malaya: balek angin

M. philippensis (Lam.) Müll. Arg. (*Croton philippensis* Lam.)

English: red berry, red kamala, kamala tree

Japan: kusu-no-ha-gashiwa, aka-mamiki, fira-jika

The Philippines: banato, apuyot, darandang, panagisian, panagisien, pangaplasin, panagisen, tafu, sala, pikal, tagusala, tulula, kamala

Malaya: kasirau, minyak madja, rambai kuching

India: kamala, kamala tree

China: lu song qiu mao

Nepal: sano panheli

M. polyadenos F. Muell.

English: climbing kamala

M. repandus (Willd.) Müll. Arg.

English: kamala

China: gang ziang teng

Malaya: chiarek puteh

M. tiliifolius (Blume) Müll. Arg.

English: turn-in-the-wind, linden-leaf

Malaya: balek angin, baru laut besar

Malmea R.E. Fries Annonaceae

Origins:

For the Swedish botanist Gustaf Oskar Andersson Malme (*né* Andersson), 1864-1937, lichenologist, at the Stockholm Riksmuseum, traveled in South America (1901-1903, Argentina, Paraguay). His writings include *Ex herbario Regnelliano*. Stockholm 1898-1901, "Über die Asclepiadaceen-Gattung *Tweedia* Hooker & Arnott." *Ark. Bot.* 2(7): 1-20. 1904, *Die Bauhinien von Matto Grosso*. [Stockholm 1905], "Asclepiadaceae Duseninanae in Paraná collectae." *Ark. Bot.* 21A(3): 1-48. 1927, "Die Compositen der zweiten Regnellschen Reise. III. Puente del Inca und Las Cuevas (Mendoza)." *Ark. Bot.* 24A(8): 58-66. 1932, "Asclepiadaceae austroamericanae praecipue andinae." *Ark. Bot.* 25A(7): 1-26. 1932, "Asclepiadaceae Brasilienses, novae vel minus bene cognitae." *Ark. Bot.* 28A(5): 1-28. 1936, "Über die Gattung *Grisebachiella* Lorentz." *Ark. Bot.* 28B(2): 1-4. 1936, "Beiträge zur Kenntniss der chilenischen Asklepiadazeen." *Ark. Bot.* 28B(6): 1-6. 1936 and "Die in Rio Grande do Sul vorkonmenden Spezies der Gattung *Lathyrus*." *Revista Sudamer. Bot.* 3(1-2): 8-13. 1936, with A.F.M. Glaziou (1828-1906) wrote *Xyridaceae brasilienses*. Stockholm 1898. See J.H. Barnhart, *Biographical Notes upon Botanists*. 2: 440. 1965; I.C. Hedge and J.M. Lamond, *Index of Collectors in the Edinburgh Herbarium*. Edinburgh 1970; T.W. Bossert, *Biographical Dictionary of Botanists Represented in the Hunt Institute Portrait Collection*. 250. 1972; Frederico Carlos Hoehne, M. Kuhlmann and Oswaldo Handro, *O jardim botânico de São Paulo*. 128. 1941; Ida Kaplan Langman, *A Selected Guide to the Literature on the Flowering Plants of Mexico*. Philadelphia 1964.

Species/Vernacular Names:

M. depressa (Baill.) Fries

Spanish: lemoy

Guatemala: yaya

Malmeanthus R.M. King & H. Robinson Asteraceae

Origins:

For the Swedish botanist Gustaf Oskar Andersson Malme (*né* Andersson), 1864-1937; see John H. Barnhart, *Biographical Notes upon Botanists*. 2: 440. 1965.

Malope L. Malvaceae

Origins:

Latin *malope* for mallows (Plinius), resembling the genus *Malache* Vogel or resembling the Greek *malache* "mallow, tree-mallow"; some suggest from the Greek *melon, malon* "an apple, any tree fruit" and *opsis* "resembling, aspect," referring to the fruits.

Malortiea H. Wendland Palmae

Origins:

After Ernst von Malortie, Oberhofmarschall to the King of Hannover during the time of Wendland.

Malosma (Nutt.) Raf. Anacardiaceae

Origins:

Probably from the Greek *melon, malon* "an apple, any tree fruit" and *osme* "smell, odor, perfume," or from the Latin *malus* "bad, evil"; see Constantine Samuel Rafinesque, *Autikon botanikon*. Icones plantarum select. nov. vel rariorum, etc. 83. Philadelphia 1840; E.D. Merrill, *Index rafinesquianus*. 158. 1949.

Species/Vernacular Names:

M. laurina (Nutt.) Abrams

English: laurel sumac

Malouetiella Pichon Apocynaceae

Origins:

The diminutive of the genus *Malouetia* A. DC.

Malperia S. Watson Asteraceae

Origins:

Named for the American (b. Norfolk, England) botanist Edward Palmer, 1831-1911 (Washington, D.C.), ethnologist, to USA 1849, botanical and zoological collector, traveler, naturalist, physician, 1853-1855 La Plata Expedition, botanical explorer, described the ethnobotany and aboriginal medical practice of the American Southwest. His writings include "Food products of the North American Indians." *U.S. Dept. Agr. Rpt.* 1870: 404-428. 1871 and "Plants used by the Indians of the United States." *Amer. Nat.* 12: 593-606, 646-655. 1878. See Rogers McVaugh, *Edward Palmer, Plant Explorer of the American West*. University of Oklahoma Press, Norman 1956; Janice J. Beaty, *Plants on His Back: A Life of Edward Palmer*, adventurous botanist and collector. Pantheon Books, New York 1964; Virgil J. Vogel, *American Indian Medicine*. University of Oklahoma Press 1977; Ella Dales Cantelow & Herbert Clair Cantelow, "Biographical notes on persons in whose honor Alice Eastwood named native plants." *Leaf. West. Bot.* 8(5): 83-101. 1957; Irving William Knobloch, compil., "A preliminary verified list of plant collectors in Mexico." *Phytologia Memoirs*. VI. 1983; John H. Barnhart, *Biographical Notes upon Botanists*. 3: 44. 1965; T.W. Bossert, *Biographical Dictionary of Botanists Represented in the Hunt Institute Portrait Collection*. 298. 1972; S. Lenley et al., *Catalog of the Manuscript and Archival Collections and Index to the Correspondence of John Torrey*. Library of the New York Botanical Garden. 320. 1973; Ida Kaplan Langman, *A Selected Guide to the Literature on the Flowering Plants of Mexico*. Philadelphia 1964; G. Murray, *History of the Collections Contained in the Natural History Departments of the British Museum*. 1: 172. London 1904; Ira L. Wiggins, *Flora of Baja California*. 42. Stanford, California 1980; Ray Desmond, *Dictionary of British & Irish Botanists and Horticulturists*. 532. London 1994; Rogers McVaugh, in *Dictionary of Scientific Biography* 10: 285-286. 1981; http://www.herbaria.harvard.edu/Libraries/archives/PALMER.html.

Species/Vernacular Names:

M. tenuis S. Watson

English: brown turbans

Malpighia L. Malpighiaceae

Origins:

After the Italian scientist Marcello Malpighi, (baptized) 1628-1694 (d. Rome, Italy), anatomist, microscopist, physician and biologist, embryologist, Fellow of the Royal Society, 1646 he entered the University of Bologna, 1653 graduated as doctor of medicine and philosophy at the University of Bologna, 1656 professor of theoretical medicine at the University of

Pisa, teacher of Albertini, Domenico Bottone (Sicilian physician and philosopher, in 1697 a member of the Royal Society) and Valsalva, influenced by the mathematician and one of the most distinguished pupils of Galileo, Giovanni Alfonso Borelli (1608-1679), he also participated in animal dissections in Borelli's home laboratory and entered the Accademia del Cimento (the Accademia was active from 1657 to 1667), 1662 succeeded Castello at Messina, 1666 returned to Bologna to lecture in practical medicine, scientific correspondent with the Royal Society of London, 1691 Malpighi was called to Rome as chief physician to Pope Innocent XII (Antonio Pignatelli, 1615-1700). Malpighi is the author of numerous important treatises and is best known for his *De pulmonibus observationes anatomicae*. 1661, *De lingua*. 1665, *Anatome plantarum*. Londini 1675-1679, *De formatione pulli in ovo*. London 1673 and *De cerebro*. 1665. See Luigi Belloni, in *Dictionary of Scientific Biography* 9: 62-66. 1981; Howard B. Adelmann, *Marcello Malpighi and the Evolution of Embryology*. Ithaca 1966; D. Bertoloni Meli, ed., *Marcello Malpighi, Anatomist and Physician*. Casa Editrice Leo S. Olschki, Firenze 1997; Accademia del Cimento, *Saggi di naturali esperienze fatte nell'Accademia del Cimento* sotto la protezione del serenissimo Principe Leopoldo di Toscana e descritte dal segretario di essa Accademia. Giuseppe Cocchini, Florence 1667; Lorenzo Magalotti (1637-1712), *Lettere scientifiche, ed erudite*. [Quarto, first edition, M. was the secretary to the Accademia del Cimento.] Florence 1721; Antonio Maria Valsalva (1666-1723), *De aure humana tractatus*. Bologna 1704; G.A. Borelli, *Elementa conica Apollonii Paergei et Archimedis opera*. [First edition.] Roma 1679 and *De motu animalium*. Opus posthumum. Rome 1680-1681; Garrison and Morton, *Medical Bibliography*. 469, 762 and 1546. New York 1961; Mariella Azzarello Di Misa, a cura di, *Il Fondo Antico della Biblioteca dell'Orto Botanico di Palermo*. 180. Palermo 1988; Alexander B. Adams, *Eternal Quest. The Story of the Great Naturalists*. New York 1969.

Species/Vernacular Names:
M. glabra L.

Peru: cereza(o), sanango

Tropical America: Barbados cherry, acerola

Mexico: huizaa, yaga nuizaa

Malpighiantha Rojas Acosta Malpighiaceae

Origins:
After the Italian scientist Marcello Malpighi, 1628-1694.

Malpighiodes Nied. Malpighiaceae

Origins:
Resembling *Malpighia*.

Malus Miller Rosaceae

Origins:

Latin *melum, malum* "an apple," *malus* "an apple tree," Greek *melon, malon* "an apple," Akkadian *malum* "fullness, full, to be full, to be filled," *mullu* "to make full, filling," Hebrew *male* "full, abundant"; see Philip Miller, *The Gardeners Dictionary*. Abr. ed. 4. London (Jan.) 1754; A. Stefenelli, *Die Volkssprache im Werk des Petron im Hinblick auf die romanischen Sprachen*. Wien-Stuttgart 1962.

Species/Vernacular Names:
M. floribunda Van Houtte

English: Japanese crab, Japanes flowering crab apple, purple chokeberry, showy crab apple

M. fusca (Raf.) C. Schneider

English: Oregon crab apple

M. halliana Koehne (for the American physician George Rogers Hall, 1820-1899, trader in the China Sea, plant collector)

English: Hall crab apple, flowering crab apple

Japan: hana-kaido

M. ioensis (Wood) Britton

English: prairie crab, prairie crab apple, wild crab apple

M. prunifolia (Willd.) Borkh.

English: Chinese apple

M. pumila Miller

English: apple, paradise apple, eating apple

Japan: ringo

China: ping guo

M. spectabilis (Aiton) Borkh.

English: Asiatic apple, Chinese flowering crab apple

M. sylvestris (L.) Miller

English: apple, wild crab apple, European apple, wild crab, paradise apple, crab apple, wild apple

Malva L. Malvaceae

Origins:

Latin *malva, ae* "mallows," Greek *malache, maloche* "mallows," see also Greek *malakos* "soft," Akkadian *malhu* "plucked branch, *mallahtu* "a plant"; see Carl Linnaeus, *Species Plantarum*. 687. 1753 and *Genera Plantarum*. Ed. 5. 308. 1754; Giovanni Semerano, *Le origini della cultura europea*. Dizionario della lingua Latina e di voci moderne. 2(2): 465-466. Firenze 1994; G.C. Wittstein, *Etymologisch-botanisches Handwörterbuch*. 554. 1852; Ernest Weekley, *An Etymological Dictionary of Modern English*. 2: 885. Dover Publications, New York 1967; Giovanni Semerano,

Le origini della cultura europea. Dizionario della lingua Greca. 2(1): 174. 1994; M. Cortelazzo & P. Zolli, *Dizionario etimologico della lingua italiana.* 3: 707. 1983.

Species/Vernacular Names:

M. moschata L.

English: musk mallow

M. neglecta Wallr.

English: common mallow, cheeses

M. nicaeensis All.

English: bull mallow

M. parviflora L.

English: small mallow, little mallow, bread-and-cheese, small-flowered mallow, marshmallow, small-flowered marshmallow, cheese weed, Egyptian mallow

Arabic: khobbeiza, khobbeiza reziza

Southern Africa: brood-en-botter, kasies, kasiesblaar, kesieblaar, kiesieblaar, kissieblaar, wildepampoenkies; mo-oratsatsi (Sotho); thibapitsa (Shona); unomolwana (Xhosa)

M. sinensis Cavan.

English: Chinese mallow

M. sylvestris L.

English: tall mallow, high mallow, cheeses, high malva, common mallow, marsh mallow

Arabic: khobbiza, khobbeiza

French: grande mauve

Italian: malva

Japan: zeni-aoi

Tibetan: ma-ning nyi-dga'

M. verticillata L.

English: mallow, curled mallow, cluster mallow, curly-leaved mallow, farmer's tobacco, cheeseweed

China: dong kui zi

Tibetan: molcam, tshod-ma, pholcam, nyi-dga', lcampa

Malvastrum A. Gray Malvaceae

Origins:
False mallow, from the Latin *malva, ae* "mallow" and the suffix *-aster* meaning false or an incomplete resemblance; see A. Gray, in *Memoirs of the American Academy of Arts and Sciences*, ser. 2. 4: 21. 1849.

Species/Vernacular Names:

M. americanum (L.) Torrey (*Malva americana* L.; *Malva spicata* L.; *Malva brachystachya* F. Muell.; *Malvastrum spicatum* (L.) A. Gray)

English: malvastrum, spiked malvastrum

M. coromandelianum (L.) Garcke (*Malva coromandeliana* L.; *Malvastrum tricuspidatum* A. Gray; *Sida fauriei* H. Lév.; *Sida oahuensis* H. Lév.)

English: prickly malvastrum, false mallow

Japan: enoki-aoi

Yoruba: asa, olowonransansan, sekuseku, asa orisa, aboris-awaye

Malvaviscus Cav. Malvaceae

Origins:
From the generic names *Malva* and *Hibiscus*; some suggest from the Latin *malva, ae* "mallow" and *viscum, i* or *viscus, i* "bird-lime, glue," possibly referring to the sticky pulp of the seeds or the sticky nature of the whole plant or to the pulp around the seeds; see Antonio José Cavanilles, *Monadelphiae classis dissertationes decem.* 131. Madrid 1787; Georg Christian Wittstein, *Etymologisch-botanisches Handwörterbuch.* 554. Ansbach 1852; Helmut Genaust, *Etymologisches Wörterbuch der botanischen Pflanzennamen.* 364. 1996; F. Boerner, *Taschenwörterbuch der botanischen Pflanzennamen.* 2. Aufl. 136. Berlin & Hamburg 1966; F. Boerner & G. Kunkel, *Taschenwörterbuch der botanischen Pflanzennamen.* 4. Aufl. 131. 1989.

Species/Vernacular Names:

M. sp.

English: firecracker hibiscus

Indonesia: kembang lampu

Bali: bungan pucuk tabia-tabia (tabia = chili pepper)

Mexico: goba ya laza, coba ya laza

M. arboreus Cav.

English: fire dart, wax mallow

M. penduliflorus DC. (*Malvaviscus arboreus* Cav. var. *penduliflorus* (DC.) Schery)

English: Turk's cap

Hawaii: aloalo pahupahu

Malvella Jaub. & Spach Malvaceae

Origins:
Diminutive of the Latin *malva, ae* "mallow."

Species/Vernacular Names:

M. leprosa (Ortega) Krapov. (*Malva leprosa* Ort.; *Malva hederacea* Douglas ex Hook.; *Sida leprosa* (Ort.) Schumann; *Sida leprosa* (Ort.) Schumann var. *hederacea* (Douglas ex Hook.) Schumann; *Sida hederacea* (Douglas ex Hook.) Torrey ex A. Gray)

English: alkali sida, ivy-leaved sida, alkali mallow, white weed

Malveopsis C. Presl Malvaceae

Origins:
Resembling the genus *Malva* L.

Malya Opiz Gramineae

Origins:
After the Bohemian–Austrian botanist Joseph Karl (Carl) Maly, 1797-1866, physician, M.D. Prague 1823. His works include *Enumeratio plantarum phanerogamicarum imperii austriaci universi*. Vindobonae [Wien] 1848 and *Oekonomisch-technische Pflanzenkunde*, etc. Wien 1864. See J.H. Barnhart, *Biographical Notes upon Botanists*. 2: 441. 1965; T.W. Bossert, *Biographical Dictionary of Botanists Represented in the Hunt Institute Portrait Collection*. 251. 1972; Ethelyn Maria Tucker, *Catalogue of the Library of the Arnold Arboretum of Harvard University*. Cambridge, Massachusetts 1917-1933; R. Zander, F. Encke, G. Buchheim and S. Seybold, *Handwörterbuch der Pflanzennamen*. 14. Aufl. Stuttgart 1993; I.C. Hedge and J.M. Lamond, *Index of Collectors in the Edinburgh Herbarium*. Edinburgh 1970.

Mamboga Blanco Rubiaceae

Origins:
From the Philippine local plant name, *mambóg* (Tagálog), for *Mitragyna speciosa* Korth.

Mamillaria F. Reichb. Cactaceae

Origins:
From the Latin *mammilla, mamilla, ae* "a nipple, breast."

Mamillopsis Morren ex Britton & Rose Cactaceae

Origins:
Resembling a *mammilla, mamilla, ae* "a nipple, breast."

Mammariella J. Shafter Cactaceae

Origins:
Latin *mammilla, mamilla, ae* "a nipple, breast," *mamma, ae* "a breast," Greek *mamma*.

Mammea L. Guttiferae

Origins:
Mammey is a West Indian vernacular name; see Carl Linnaeus, *Species Plantarum*. 512. 1753 and *Genera Plantarum*. Ed. 5. 228. 1754.

Species/Vernacular Names:
M. sp.
Gabon: ibeca
Nigeria: agba, bolo, ibeckaror, ibeca, ibecka, isong, odudu, odo, okakilo, ukutu, abot-zok, oboto
M. africana Sabine
Central Africa: bolele, ogbonodu, bompeggya, bletune, djimbo, mohia
Yoruba: ologbomodu
Nigeria: bolo; ologbomodu (Yoruba); otien ogiorio (Edo); urherame (Urhobo); ekpili (Igbo); edeng (Efik); okut (Boki)
Congo: m'bossi, oboli
Gabon: ebornzork, ibeka, muburo, banga, ebor, ibore, ilolo, mbaga, mbanga, oboto
Cameroon: aborzrk, abozok, abodzok, ebot, abot soc, boto, oboto
Zaire: bangwon, ipeki, bokoli, nkoli, bokodji, okodi, oliti, boye, boza, tshilunga tshikunze, lindale, mangwondi, mubuku, mukudi, mulira
Ivory Coast: anibe, animbe, aramo, bomoku, djimbo
M. americana L.
English: mammee, mammee apple, South American apricot, mammey apple, Santo Domingo apricot, wild apricot, mammee tree
The Antilles: mamey
Cuba: mamey rojo, mamey de S. Domingo
Panama: mamey de Cartagena
Peru: mamey, otere
Brazil: abricó-do-pará
M. siamensis (Miq.) Anderson
Malaya: belimbing Siam
M. touriga (C. White & Francis) L.S. Smith
English: brown touriga, alligator bark

Mammilaria Torrey & A. Gray Cactaceae

Origins:
From the Latin *mammilla, mamilla, ae* "a nipple, breast."

Mammillaria Haw. Cactaceae

Origins:
From the Latin *mammilla, mamilla, ae* "a nipple, breast," referring to the small tubercles.

Species/Vernacular Names:
M. spp.

Mexico: guechi betzi, guichi betzi

M. bocasana Poselg.

English: snowball cactus, powder puff cactus, fish hooks

M. camptotricha Dams.

English: bird's nest cactus, golden bird's nest cactus

M. carnea Zucc.

Japan: tenshû-maru (maru = round)

M. elongata DC.

English: lace cactus, golden star cactus, gold lace cactus, golden lace, ladyfinger

M. gracilis Pfeiff. (*Mammillaria fragilis* Salm-Dyck ex K. Brandg.)

English: thimble mammillaria, thimble cactus

M. hahniana Werderm.

English: old woman cactus, old lady cactus, old lady of Mexico

M. heyderi Muehlenpf. (*Mammillaria gummifera* Engelm.)

English: coral cactus

M. karwinskiana Mart. (*Mammillaria fischeri* Pfeiff.) (for the Bavarian naturalist Baron Wilhelm Friedrich Karwinski von Karwin, 1780-1855, field collector in Mexico; see Gordon Douglas Rowley, *A History of Succulent Plants*. 371. [b. 1799] Strawberry Press, Mill Valley, California 1997; Irving William Knobloch, compil., "A preliminary verified list of plant collectors in Mexico." *Phytologia Memoirs*. VI. 1983; Ida Kaplan Langman, *A Selected Guide to the Literature on the Flowering Plants of Mexico*. Philadelphia 1964)

English: royal cross

M. plumosa F.A. Weber

English: feather cactus

M. polythele Mart. (*Mammillaria tetracantha* Pfeiff.)

English: ruby dumpling

M. pringlei (J. Coulter) K. Brandg. (after the American botanist Cyrus Guernsey Pringle, 1838-1911, Quaker, plant collector; see John H. Barnhart, *Biographical Notes upon Botanists*. 3: 111. 1965; Ethelyn Maria Tucker, *Catalogue of the Library of the Arnold Arboretum of Harvard University*. Cambridge, Massachusetts 1917-1933; S. Lenley et al., *Catalog of the Manuscript and Archival Collections and Index to the Correspondence of John Torrey*. Library of the New York Botanical Garden. 335-336. 1973; Ira L. Wiggins, *Flora of Baja California*. 42. Stanford, California 1980; Ida Kaplan Langman, *A Selected Guide to the Literature on the Flowering Plants of Mexico*. 596. Philadelphia 1964; Gordon Douglas Rowley, *A History of Succulent Plants*. California 1997; Irving William Knobloch, compil., "A preliminary verified list of plant collectors in Mexico." *Phytologia Memoirs*. VI. 1983; Helen Burns Davis, *Life and Work of Cyrus Guernsey Pringle*. Burlington, Vermont 1936)

English: lemon ball

M. prolifera (Mill.) Haw.

English: little candles, silver cluster cactus

M. wildii A. Dietr.

English: fish hook pincushion cactus

M. zeilmanniana Boed.

English: rose pincushion

Manaosella J.C. Gómes Bignoniaceae

Origins:
Manaus, formerly Manáos, a city on the Rio Negro in northwest Brazil, capital of Amazonas.

Mancanilla Miller Euphorbiaceae

Origins:
French *mancinelle, mancenille*, English *manchineel* for *Hippomane mancinella*, Spanish *manzanilla* (from *manzana, mazana*, a kind of apple), Latin *Matiana mala, Matianum malum* "a kind of apple (so named from the *Matia gens*)"; see C.T. Onions, *The Oxford Dictionary of English Etymology*. Oxford University Press 1966.

Mancinella Tussac Euphorbiaceae

Origins:
French *mancinelle*, Spanish *manzanilla* (from *manzana, mazana*, a kind of apple), Latin *Matiana mala* "a kind of apple," see also *Mancanilla* Mill.

Mandelorna Steud. Gramineae

Origins:
An anagram of *Lenormandia* Steud., dedicated to the French botanist Sébastien-René Lenormand, 1796-1871, botanical collector for his herbarium; see A. Lasègue, *Musée botanique de Benjamin Delessert*. Paris 1845; John

H. Barnhart, *Biographical Notes upon Botanists*. 2: 369. 1965; T.W. Bossert, *Biographical Dictionary of Botanists Represented in the Hunt Institute Portrait Collection*. 234. 1972; S. Lenley et al., *Catalog of the Manuscript and Archival Collections and Index to the Correspondence of John Torrey*. Library of the New York Botanical Garden. 462. 1973; H.N. Clokie, *Account of the Herbaria of the Department of Botany in the University of Oxford*. 198. Oxford 1964.

Mandevilla Lindley Apocynaceae

Origins:

For the British Henry John Mandeville, 1773-1861 (d. Buenos Aires), introduced many plants into Europe; see John Lindley, in *Edwards's Botanical Register*. 26: sub t. 7. 1840; F. Boerner & G. Kunkel, *Taschenwörterbuch der botanischen Pflanzennamen*. 4. Aufl. 131. Berlin & Hamburg 1989; Ray Desmond, *Dictionary of British & Irish Botanists and Horticulturists*. 464. 1994.

Species/Vernacular Names:

M. illustris Woodson

Brazil: jalapa-do-campo

M. laxa (Ruíz and Pavón) Woodson

English: Chilean jasmine

China: wen teng

Mandioca Link Euphorbiaceae

Origins:

Brazilian vernacular name for *Manihot esculenta* Crantz; see F. Boerner & G. Kunkel, *Taschenwörterbuch der botanischen Pflanzennamen*. 4. Aufl. 302. 1989; R. Zander, F. Encke, G. Buchheim and S. Seybold, *Handwörterbuch der Pflanzennamen*. 14. Aufl. 110, 642. Stuttgart 1993; Helmut Genaust, *Etymologisches Wörterbuch der botanischen Pflanzennamen*. 365. 1996.

Mandonia Hasskarl Commelinaceae

Origins:

For the French traveler Gilbert Mandon, 1799-1866, plant collector in Bolivia; see John H. Barnhart, *Biographical Notes upon Botanists*. 2: 442. 1965; E.M. Tucker, *Catalogue of the Library of the Arnold Arboretum of Harvard University*. 1917-1933; Stafleu and Cowan, *Taxonomic Literature*. 3: 273. Utrecht 1981; I.C. Hedge and J.M. Lamond, *Index of Collectors in the Edinburgh Herbarium*. Edinburgh 1970.

Mandragora L. Solanaceae

Origins:

Greek and Latin *mandragoras* for a plant, mandrake, perhaps from the Persian *mardumgia* "the plant of the man (i.e., the plant being supposed to resemble the human form)"; see C.J.S. Thompson, *The Mystic Mandrake*. London 1934; Manlio Cortelazzo & Paolo Zolli, *Dizionario etimologico della lingua italiana*. 3: 710. Bologna 1983; E. Hoffmann-Krayer, *Handwörterbuch des deutschen Aberglaubens*. I: 312-324. Berlin-Leipzig 1927; G. Aquilecchia, "La favola Mandragola si chiama." in *Schede di italianistica*. 97-126. Torino 1976; Ernest Weekley, *An Etymological Dictionary of Modern English*. 2: 888. Dover Publications, New York 1967.

Species/Vernacular Names:

M. autumnalis Bertol.

Arabic: bidh el ghoul, beid el-ghoul, lufah

French: pomme d'amour, mandragore

M. caulescens C.B. Clarke (*Mandragora chinghaiensis* Kuang & A.M. Lu)

English: caulescent mandrake, Chinghai mandrake

China: qien shen

M. officinarum L.

English: mandrake, devil's apples

Manettia Boehmer Scrophulariaceae

Origins:

For the Italian botanist Saverio (Xaverio) Manetti, 1723-1785, physician; see John H. Barnhart, *Biographical Notes upon Botanists*. 2: 442. 1965.

Manettia Mutis ex L. Rubiaceae

Origins:

For the Italian botanist Saverio (Xaverio) Manetti, 1723-1785, physician, 1747-1782 prefect of the Giardino dei Semplici at Florence. His writings include *Viridarium florentinum*. Florentiae [Firenze] 1751, *Della Inoculazione del Vajuolo trattato*, etc. Firenze 1761, Lettera del Sig. Dotto S.M. che puo servire di *supplemento al suo trattato sull'inoculazione del vajuolo*. Firenze 1762 and *Delle specie diverse di frumento e di pane siccome della panizzazione*, memoria. Venezia 1766, with L. Lorenzo and V. Vanni wrote *Storia naturale degli uccelli* trattata con metodo ... *Ornithologia methodice digesta*, etc. Ital. and Latin. Firenze 1767-1776. See R. Zander, F. Encke, G. Buchheim and S. Seybold, *Handwörterbuch der Pflanzennamen*.

14. Aufl. 746. 1993; F. Boerner & G. Kunkel, *Taschenwörterbuch der botanischen Pflanzennamen*. 4. Aufl. 131. 1989; Alberto Chiarugi, "Le date di fondazione dei primi Orti Botanici del mondo: Pisa (Estate 1543); Padova (7 Luglio 1545); Firenze (1° Dicembre 1545)." *Nuovo Giorn. Bot. Ital.*, n.s. 60: 785-839. 1953; Oreste Mattirolo, "Cenni cronologici sugli Orti Botanici di Firenze." *Pubblicazioni R. Ist. Studi Superiori Pratici e di Perfezionamento*. Sez. Scienze Fis. Natur. Firenze 1899; P.A. Micheli, *Catalogus plantarum Horti caesarei florentini*. Florentiae 1748; J.H. Barnhart, *Biographical Notes upon Botanists*. 2: 442. 1965; T.W. Bossert, *Biographical Dictionary of Botanists Represented in the Hunt Institute Portrait Collection*. 257. 1972; Jonas C. Dryander, *Catalogus bibliothecae historico-naturalis Josephi Banks*. London 1796-1800; E.M. Tucker, *Catalogue of the Library of the Arnold Arboretum of Harvard University*. 1917-1933; Mariella Azzarello Di Misa, a cura di, *Il Fondo Antico della Biblioteca dell'Orto Botanico di Palermo*. 181. Regione Siciliana, Palermo 1988; Gustavo Bertoli, "Librai, cartolai e ambulanti immatricolati nell'Arte dei medici e speziali di Firenze dal 1490 al 1600." *La Bibliofilia*. Anno XCIV. 2: 125-164 and 3: 227-262. 1992.

Mangenotia Pichon Asclepiadaceae

Origins:

Dedicated to the French botanist Georges-Marie Mangenot, 1899-1985, professor of botany, plant collector at the Ivory Coast; see F.N. Hepper and F. Neate, *Plant Collectors in West Africa*. 53. 1971.

Mangifera L. Anacardiaceae

Origins:

Manga, mangai, man-kay, mankay, manghi, Malayalam and Tamil names, plus Latin *fero, fers, tuli, latum, ferre* "to bear, carry"; see Carl Linnaeus, *Species Plantarum*. 200. 1753 and *Genera Plantarum*. Ed. 5. 93. 1754; C.T. Onions, *The Oxford Dictionary of English Etymology*. Oxford University Press 1966; E. Zaccaria, *Raccolta di voci affatto sconosciute o mal note ai lessicografi ed ai filologi*. Marradi 1919; A.J.G.H. Kostermans & J.-M. Bompard, *The Mangoes. Their Botany, Nomenclature, Horticulture and Utilization*. London 1993.

Species/Vernacular Names:

M. caesia Jack

Malaya: binjai

M. foetida Lour.

English: horse mango

Malaya: bachang, machang, machai, membachang, batel, kembachang, machang batu, machang ketur, sepam, kuini

Borneo: bachang, mangga pau

M. indica L.

English: mango, Indian mango

Arabic: amba

French: manguier

Mali: mangoro, mankoro

Nigeria: mangoro

Yoruba: mangoro, seri

Congo: manga, mumanga

Tanzania: embe

Rodrigues Island: mangue, manguier

Peru: mango, maxchequexu

Brazil: manga

Bolivia: mango, manga, maca

Southern Laos: lloong diub' (Nya Hön)

Malaya: pauh, mempelan, mangga, ampelam, hampelam

The Philippines: manga, mangga, mangang-kalabaw, paho

India: amra, amva, mangga, sunda, chuta, am, amb, ambo, amba, thayet, mamidi, mampalam, mavu, mavina-hannu, mavu

Japan: mangô

China: mang guo, meng kuo, an lo kuo, hsiang kai

Nepal: amp, anp

Tibetan: a-bras

Hawaii: manako, manako meneke, meneke

M. x odorata Griff.

Malaya: kuwini, kuini, kwini

Mangium Rumph. ex Scop. Rhizophoraceae

Origins:

Spanish *mangle*, Portuguese *mangue*, Malay *manggimanggi, mangi-mangi* "mangrove."

Manglesia Endlicher & Fenzl Proteaceae

Origins:

Named for the English naturalist James Mangles, 1786-1867, traveler and patron of botanical collecting, Royal Navy 1800, 1825 Fellow of the Royal Society, 1831 W. Australia (Swan River Settlement), received specimens from James Drummond. Among his writings are *The Floral Calendar*, monthly and daily. London 1839, *Papers and Despatches Relating to the Arctic Searching Expeditions of*

1850-1851. Together with a few brief remarks as to the probable course pursued by Sir John Franklin, etc. [Compiled by J. Mangles.] London 1851, *Papers and Despatches Relating to the Arctic Searching Expeditions of 1850-1851-1852*. Collected and arranged by J.M. Second edition, with copious additions. London 1852, *Synopsis of a Complete Dictionary*, graphical, descriptive and identical of the illustrated Geography and Hydrography of ... England and Wales, Scotland and Ireland. London 1848 and *Thames Estuary*. Guide to the navigation of the Thames Mouth, and Key to the Model. London 1853, with Charles L. Irby wrote *Travels in Egypt and Nubia, Syria, and the Holy Land*, etc. 1868; see S.L. Endlicher (1804-1849) & Eduard Fenzl (1808-1879), *Novarum stirpium decades*. Decas 4: 25-26. Wien 30 Mai 1839; Ray Desmond, *Dictionary of British & Irish Botanists and Horticulturists*. 465. London 1994.

Manglietia Blume Magnoliaceae

Origins:

A vernacular name; see David Hunt, ed., *Magnolias and Their Allies*. Proceedings of an International Symposium, Royal Holloway, University of London, Egham, Surrey, U.K., 12-13 April 1996. International Dendrology Society and The Magnolia Society. 1998; David G. Frodin & Rafaël Govaerts, *World Checklist and Bibliography of Magnoliaceae*. Royal Botanic Gardens, Kew 1996.

Manglietiastrum Law Magnoliaceae

Origins:

Resembling *Manglietia* Blume; see David Hunt, ed., *Magnolias and Their Allies*. Proceedings of an International Symposium, Royal Holloway, University of London, Egham, Surrey, U.K., 12-13 April 1996. International Dendrology Society and The Magnolia Society. 1998.

Mangonia Schott Araceae

Origins:

Latin *mango, mangonis* "a dealer, monger," *mangonium* "a setting off, displaying of wares," *mangonico, avi, atum* "to adorn," Greek *manganon* "a dealer."

Manicaria Gaertner Palmae

Origins:

Latin *manicae, arum* "the long sleeve of a tunic, glove, handcuff," referring to the spathe (the fibrous inflorescence bracts used as a hat, as caps); see Joseph Gaertner (1732-1791), *De fructibus et seminibus plantarum*. 2(3): 468. Stuttgart, Tübingen 1791.

Species/Vernacular Names:

M. saccifera Gaertner

English: sleeve palm, monkey cap palm, bussu palm

Brazil: ubuçu, ubuçú, buçu, buçú, bussú, coqueiro buçu, obuçu, tururi, geruá

Venezuela: timiche, temiche, truli, ouasi

Manihot Miller Euphorbiaceae

Origins:

From the Brazilian vernacular name; see C.T. Onions, *The Oxford Dictionary of English Etymology*. 552. Oxford University Press 1966; E. Zaccaria, *Raccolta di voci affatto sconosciute o mal note ai lessicografi ed ai filologi*. Marradi 1919; F. D'Alberti di Villanuova, *Dizionario universale, critico, enciclopedico della lingua italiana*. Lucca 1797-1805; E. Zaccaria, *L'elemento iberico nella lingua italiana*. Bologna 1927; Ottaviano Targioni Tozzetti, *Dizionario botanico italiano*. Firenze 1809; T.E. Hope, *Lexical Borrowing in the Romance Languages*. Oxford 1971; Manlio Cortelazzo & Paolo Zolli, *Dizionario etimologico della lingua italiana*. 3: 713. Zanichelli, Bologna 1983; Luíz Caldas Tibiriçá, *Dicionário Guarani-Português*. Traço Editora, Liberdade 1989; Luíz Caldas Tibiriçá, *Dicionário Tupi-Português*. Traço Editora, Liberdade 1984.

Species/Vernacular Names:

M. sp.

English: cassava, manioca, tapioca

Southern Africa: antsumbula (Tsonga); mjumbula (Zulu)

Peru: accana, yoki-yoki

Mexico: cal-po-me-quec, yucuelina

Brazil: maniva, maniva-de-veado, maniva-do-campo, maniva-da-mata

M. dulcis (J.F. Gmelin) Pax

English: sweet cassava

Peru: sacharuma

Mexico: yuca dulce, guacamote, yuca, guacamote dulce, cuacamote, ch-uhuk-ts'iim, kiki-tsiim, ts'in, guh-yaga, guu yada, yaga-yeda

M. esculenta Crantz (*Manihot utilissima* Pohl; *Manihot manihot* (L.) Cockerell; *Jatropha manihot* L.)

English: bitter cassava, tapioca, tapioca plant, cassava, sweet potato tree, manioc

Congo: ayaka, saka saka

S. Rhodesia: muFarinya

Yoruba: ege funfun, ege oke, ege, ege karagba, ege gbokog-baala, ege olowokunbo, ege atu, ege gbokogbaala, gbajada, paki, gbaguuda dale joro, gbaguuda, gbaguuda funfun, gbaguuda pupa, lanase

Tanzania: muhoko

Paraguay: pejek

Peru: yuca, yuca amarilla, yuca blanca, abam, adtza, atza, atsa, cañiri, canri, cuabe, chimeca, chunopa, chunopa rumu, eequi, hatsa, jimeca, kaniri, kañiri, maam, máma, ohi, quimeca, rumu, sekachi, timeca, vazino, yawiri, ytuxe

Mexico: yuca, yuca brava, guacamote, yuca manioc, casava, yuca amarga, yuca mansa, yuca blanca, coshquehui, ko'chka'hui, cuauh-camotli, guu-yaga, gu-yaga, huacamote, huacamotl, tsiim, ts'im, ts'iin, tzin, tinché, cuacamojtli

Brazil: manioc, mandioca, manduba, mandiba, maniva, juca-amarga, mani-oca

South Laos (Nya Hön people): buem thay (= Thay tubercle), buem thay maat, buem thay book (= light Thay tubercle), buem thay cruay, buem thay dum (= red Thay tubercle), buem thay lowee (= Lové Thay tubercle; Lové, a tribe of Attopeu)

Japan: imo-no-ki, casaba, tapioka-no-ki, ki-imo

Malaya: ubi, ubi kayu

The Philippines: kamoteng-kahoy, kamoting-kahoy, kamuting dutung, kamuting-dutung, balangai, balinghoy, balangkoy, kahoy, balañgeg, kamote ni moro, kamote ti moro, pad-padi, pangi-kahui

M. glaziovii Müll.Arg.

English: Ceara rubber tree

Malaya: pokok chat, ceara rubber

Yoruba: pafuroba

Manihotoides D.J. Rogers & Appan Euphorbiaceae

Origins:
Resembling *Manihot* Miller.

Manilkara Adanson Sapotaceae

Origins:
Manil-kara is a Malayalam/Malabar/South Indian vernacular name cited by van Rheede (*Hort. Ind. Mal.* 4: t. 25. 1683) for the species *Manilkara kaukii* (L.) Dubard; see Helmut Genaust, *Etymologisches Wörterbuch der botanischen Pflanzennamen.* 366. Basel 1996.

Species/Vernacular Names:

M. sp.

Central Africa: monghinza, boango, gumbe

Cameroon: boang, boango, gumbe, nom adjap oswe

Gabon: adzacon-aboga

M. bidentata (A. DC.) A. Chev.

English: balata tree, bullet tree

Central America: balata

M. bidentata (A. DC.) Chev. ssp. *surinamensis* (Miq.) Penn. (*Mimusops surinamensis* Miq.)

Bolivia: masaranduba

M. chicle (Pittier) Gilly

English: crow gum

M. concolor (Harvey ex C.H. Wr.) Gerstner (*Mimusops concolor* Harv. ex C.H. Wr.)

English: Zulu milkberry

Southern Africa: Zoeloemelkbessie; umNqambo, umNcambu, amaSethole amhlophe (Zulu)

M. discolor (Sond.) J.H. Hemsl. (*Muriea discolor* (Sond.) Hartog; *Labourdonnaisia discolor* Sond.; *Labourdonnaisia sericea* Benth.; *Eichleria discolor* (Sond.) Hartog; *Mimusops discolor* (Sond.) Hartog; *Mimusops natalensis* (Pierre) Engl.; *Mahea natalensis* Pierre; *Muriea discolor* (Sond.) Hartog)

English: forest milkberry, red milkwood

Southern Africa: bosmelkbessie, umnweba, rooimelkhout; uNweba, umNweba, umNweba wentaba (= the umNwheba of the hills), umNqambo (Zulu); aNywebe (Thonga)

M. cf. *excelsa* (Ducke) Standley

Bolivia: sapito

M. fouilloyana Aubr. & Pellegrin

Cameroon: boang, boango, gumbe, nom adjap oswe

Central Africa: monghinza

Gabon: adzacon-aboga

M. hexandra (Roxb.) Dubard

English: sixstamens balata

India: palu

China: tie xian zi

M. kauki (L.) Dubard (a native name)

Australia: wongi

Malaya: sawai, sawah, sawoh, sau

M. mochisia (Bak.) Dubard (*Mimusops mochisia* Bak.; *Manilkara macaulayae* (Hutch. & Corb.) H.J. Lam.) (the specific name perhaps an African name)

English: lowveld milkberry

Southern Africa: laeveldmelkbessie, nwambu; umNqambo, umNcambu (Zulu); muSikanyati (Shona)

M. nicholsonii Van Wyk

English: south coast milkberry

South Africa: suidkusmelkbessie

M. obovata (Sabine & G. Don f.) J.H. Hemsley (*Manilkara multinervis* (Bak.) Dubard)

English: African pear

Yoruba: emido, ako emido

Nigeria: kadanyar rafi (Hausa); emido, osere (Yoruba); ukpi, ukpwi (Igbo); nchome (Ekoi); wono (Ijaw)

M. zapota (L.) Royen (*Manilkara zapotilla* (Jacq.) Gilly; *Achras zapota* L.; *Sapota achras* Mill.; *Manilkara achras* (Mill.) Fosb.) (Greek *achras* "a wild pear")

English: naseberry, heart balata, sapodilla plum, beef apple

Central America: sapodilla, nispero, chicle, chicozapote

Mexico: chicozapote, yá, tzicotzapotl, tzicte (= chicle), tzictli (=chicle), guela chiña, quela chiña, ya guelde, quela dau

The Philippines: tsiko, chico, chiku tree, sapodilla

Japan: sapojira

Malaya: chiku, sauh menila, sau manila, sawa

Manilkariopsis (Gilly) Lundell Sapotaceae

Origins:
Resembling *Manilkara* Adans.

Maniltoa R. Scheffer Caesalpiniaceae

Origins:
Possibly a Malesian/Papuasian native name; see Rudolph Herman Christiaan Carel Scheffer (1844-1880), *Annales du Jardin Botanique de Buitenzorg.* 1: 20. 1876.

Manisuris L. Gramineae

Origins:
Greek *manos* "loose, sparse, flaccid, rare" and *oura* "a tail," referring to the appearance of the spikes, resembling a string of minute beads; see C. Linnaeus, *Mantissa Plantarum.* 2: 164. 1771.

Mannagettaea H. Sm. Orobanchaceae (Scrophulariaceae)

Origins:
For the Austrian botanist Günther Beck Ritter von Mannagetta und Lërchenau, 1856-1931, professor of botany in Vienna and Prague. Among his numerous works are *Flora von Nieder-Österreich.* Wien 1890-1893 and *Die Algen Kärnstens.* Dresden 1931; see J.H. Barnhart, *Biographical Notes upon Botanists.* 1: 149. 1965; R. Zander, F. Encke, G. Buchheim and S. Seybold, *Handwörterbuch der Pflanzennamen.* 14. Aufl. Stuttgart 1993; T.W. Bossert, *Biographical Dictionary of Botanists Represented in the Hunt Institute Portrait Collection.* 31. 1972; Frans A. Stafleu and Erik A. Mennega, *Taxonomic Literature. Supplement II.* 19-27. 1993; H.N. Clokie, *Account of the Herbaria of the Department of Botany in the University of Oxford.* 129. Oxford 1964; E.M. Tucker, *Catalogue of the Library of the Arnold Arboretum of Harvard University.* 1917-1933; Ida Kaplan Langman, *A Selected Guide to the Literature on the Flowering Plants of Mexico.* 1964.

Mannaphorus Raf. Oleaceae

Origins:
From the Greek *manna* "frankincense powder or granules" and *phoros* "bearing, carrying," Latin *manna, ae* "a grain, a vegetable juice hardened into grains" (Plinius); *manna, man, manhu* and *manna, ae* for the manna of the Hebrews, food for the soul, divine support; see C.S. Rafinesque, *Am. Monthly Mag. Crit. Rev.* 2: 175. 1818; Ernest Weekley, *An Etymological Dictionary of Modern English.* 2: 890. New York 1967; Harold Norman Moldenke and Alma Lance Moldenke, *Plants of the Bible.* New York 1986; Manlio Cortelazzo & Paolo Zolli, *Dizionario etimologico della lingua italiana.* 3: 713. 1983; E. Monaci, *Crestomazia italiana dei primi secoli.* Nuova edizione riveduta e aumentata per cura di F. Arese. Roma-Napoli-Città di Castello 1955; P. Zürcher, *Der Einfluss der lateinischen Bibel auf den Wortschatz der italienischen Literatursprache vor 1300.* Bern 1970.

Mannia Hook.f. Simaroubaceae

Origins:
For the German botanist Gustav Mann, 1836-1916, Kew gardener 1859, plant collector, botanical explorer, traveler, 1859-1862 on William Balfour Baikie's Niger Expedition, 1863 India and Assam, 1863-1891 Indian Forest Service, sent plants and seeds to Kew, with the German botanist Hermann Wendland (1825-1903) wrote "On the palms of Western Tropical Africa." *Trans. Linn. Soc.* 24: 421-439. (Nov.) 1864; see J.H. Barnhart, *Biographical Notes upon Botanists.* 2: 443. 1965; H.N. Clokie, *Account of the Herbaria of the Department of Botany in the University of Oxford.* 206. Oxford 1964; E.M. Tucker, *Catalogue of the Library of the Arnold Arboretum of Harvard University.* 1917-1933; F.N. Hepper and F. Neate, *Plant Collectors in West Africa.* 53. 1971; René Letouzey, "Les botanistes au

Cameroun." in *Flore du Cameroun.* 7: 48. Paris 1968; Ernest Nelmes and William Cuthbertson, *Curtis's Botanical Magazine Dedications, 1827-1927.* 274-276. [1931]; Joseph Vallot, "Études sur la flore du Sénégal." in *Bull. Soc. Bot. de France.* 29: 184. Paris 1882; Ronald William John Keay, "Botanical collectors in West Africa prior to 1860." in *Comptes Rendus A.E.T.F.A.T.* 55-68. Lisbon 1962; I.C. Hedge and J.M. Lamond, *Index of Collectors in the Edinburgh Herbarium.* Edinburgh 1970; F. Nigel Hepper, "Botanical collectors in West Africa, except French territories, since 1860." in *Comptes Rendus de l'Association pour l'étude taxonomique de la flore d'Afrique,* (A.E.T.F.A.T.). 69-75. Lisbon 1962; Sir Clements Robert Markham (1830-1916), *Peruvian Bark.* A popular account of the introduction of *Cinchona* cultivation into British India ... London 1880; Claude Spencer et alii, "Survey of plants for antimalarial activity." *Lloydia.* 10(3): 145-174. [referring to *Mannia africana* Hook.] 1947.

Manniella Reichb.f. Orchidaceae

Origins:

For the German botanist Gustav Mann, 1836-1916, explorer; see John H. Barnhart, *Biographical Notes upon Botanists.* 2: 443. 1965.

Manniophyton Müll.Arg. Euphorbiaceae

Origins:

For the German botanist Gustav Mann, 1836-1916; see John H. Barnhart, *Biographical Notes upon Botanists.* 2: 443. 1965.

Manochlamys Aellen Chenopodiaceae

Origins:

From the Greek *manos* "loose, sparse, flaccid" and *chlamys, chlamydos* "cloak," referring to the fleshy bracts.

Manostachya Bremek. Rubiaceae

Origins:

From the Greek *manos* "loose, sparse, flaccid" and *stachys* "a spike."

Manotes Sol. ex Planch. Connaraceae

Origins:

From the Greek *manotes* "looseness of texture, rarity, separateness, porousness."

Manothrix Miers Apocynaceae

Origins:

Greek *manos* "loose, sparse" and *thrix, trichos* "hair."

Mansoa DC. Bignoniaceae

Origins:

For the Brazilian botanist António Luiz Patricio da Silva Manso, 1788-1848, physician, 1823 Matto Grosso, 1834-1837 in the Brazilian Parliament. See John H. Barnhart, *Biographical Notes upon Botanists.* 3: 278. 1965; A. Lasègue, *Musée botanique de Benjamin Delessert.* Paris 1845; Maria do Carmo Marques et al. "Levantamento dos Tipos do Herbário do Jardim Botânico do Rio de Janeiro — Bignoniaceae II." *Rodriguésia.* 41: 37-63. 1976 and "Levantamento dos Tipos do Herbário do Jardim Botânico do Rio de Janeiro — Bignoniaceae I." *Arq. Jard. Bot.* 20: 63-75. Rio de Janeiro 1977.

Species/Vernacular Names:

M. alliacea (Lam.) A. Gentry

Latin America: bejuco de ajo

M. hymenaea (DC.) A. Gentry (*Adenocalymma alboviolaceum* Loes.)

Mexico: looba beete

Mansonia J.R. Drumm. ex Prain Sterculiaceae

Origins:

After F.B. Manson, Conservator of Indian Forest Service during the early 20th century, plant collector in Tenasserim, Burma; see Sir George Watt, *Economic Products of India* exhibited in the Economic Court, Calcutta International Exhibition, 1883-1884. (Pt. 7. *Timbers.* Compiled by F.B.M.) Calcutta 1883; F.B. Manson and Henry Haselfoot Haines, *Tables for Use with Brandis' Hypsometer for Measuring the Height of Trees.* Calcutta 1892; M.P. Nayar, *Meaning of Indian Flowering Plant Names.* 217. Dehra Dun 1985. According to Helmut Genaust (*Etymologisches Wörterbuch der botanischen Pflanzennamen.* 367. 1996) and *Lexicon der Biologie, 1983-1987.* Freiburg im Breisgau, the genus was named for the English physician Sir Patrick Manson, 1844-1922, in 1898 founded the London School of Tropical Medicine. His writings include *Tropical Diseases.* London 1898 and *The Filaria sanguinis hominis and Certain New Forms of Parasitic Disease in India, China and Warm Countries.* London 1883; see Garrison and Morton, *Medical Bibliography.* 2266 and 2455. 1961.

M. altissima (A. Chev.) A. Chev.

Yoruba: odo, ofun, otutu

Nigeria: urodo, odo, odogi, orodo, otutu; ofun (Yoruba)

W. Africa: aprono, bété

Cameroon: koul, n'koul, bambanja

Ghana: aprono

Congo: guissepa

Ivory Coast: bete

Central Africa: koul

M. gagei J.R. Drumm. ex Prain

Burma: kalamet

Mantisalca Cass. Asteraceae

Origins:
Anagram of Salmantica, from the Spanish town of Salamanca; see Lorenzo Ruiz Fidalgo, *La Imprenta en Salamanca (1501-1600)*. Madrid 1904.

Mantisia Sims Zingiberaceae

Origins:
From the mantis insect, *Mantis religiosa*, the praying mantis, referring to the shape of the flowers; Greek *mantis* for a diviner, a prophet, a kind of grasshopper, Akkadian *manu* "to recount events, to consider."

Manulea L. Scrophulariaceae

Origins:
Latin *manus* "a hand," referring to the fingerlike divisions of the corolla.

Manuleopsis Thell. Scrophulariaceae

Origins:
Resembling *Manulea* L.

Manungala Blanco Simaroubaceae

Origins:
From the Philippine local plant name, *manunggál*, for *Samadera indica* Gaertner (syn. *Manungala pendula* Blanco).

Maoutia Wedd. Urticaceae

Origins:
Named for the French physician Jean Emmanuel Maurice Le Maout, 1799-1877, botanist, author of *Botanique*. Paris

[1851-] 1852; see John H. Barnhart, *Biographical Notes upon Botanists*. 2: 367. 1965; E.M. Tucker, *Catalogue of the Library of the Arnold Arboretum of Harvard University*. 1917-1933; R. Zander, F. Encke, G. Buchheim and S. Seybold, *Handwörterbuch der Pflanzennamen*. 14. Aufl. Stuttgart 1993; Ida Kaplan Langman, *A Selected Guide to the Literature on the Flowering Plants of Mexico*. 1964.

Mapania Aublet Cyperaceae

Origins:
Possibly the vernacular name for one species in French Guiana; see Jean Baptiste C. Fusée Aublet (1720-1778), *Histoire des Plantes de la Guiane Françoise*. 1: 47, t. 17. Paris 1775.

Species/Vernacular Names:
M. macrocephala (Gaudich.) Schumann

English: sedge

M. palustris (Steudel) Fernandez-Villar

Malaya: mengkuang lubok, mengkuang tedong

Mapaniopsis C.B. Clarke Cyperaceae

Origins:
Resembling *Mapania* Aublet.

Mapouria Aublet Rubiaceae

Origins:
A Guiana vernacular name.

Mappa Adr. Juss. Euphorbiaceae

Origins:
Based on the epithet of *Ricinus mappa* L. See Stafleu and Cowan, *Taxonomic Literature*. 3: 284. Utrecht 1981; Adrien Henri Laurent de Jussieu (1797-1853), *De Euphorbiacearum generibus*. 44. 1824.

Mappia Jacq. Icacinaceae

Origins:
After the French botanist Marcus Mappus, 1666-1736, M.D. Strasbourg 1694. He was the son of the botanist Marcus Mappus (1632-1701). See John H. Barnhart, *Biographical Notes upon Botanists*. 2: 445. 1965; Jonas C.

Dryander, *Catalogus bibliothecae historico-naturalis Josephi Banks*. 1796-1800; E.M. Tucker, *Catalogue of the Library of the Arnold Arboretum of Harvard University*. 1917-1933; Stafleu and Cowan, *Taxonomic Literature*. 3: 284. Utrecht 1981.

Mappianthus Hand.-Mazz. Icacinaceae

Origins:

The genus *Mappia* Jacq. and Greek *anthos* "flower."

Maprounea Aublet Euphorbiaceae

Origins:

A Guiana vernacular name.

Species/Vernacular Names:
M. africana Müll. Arg.

Eastern Africa: msoro, mtunguru (Tabora)

N. Rhodesia: kavulamume

Congo: mutsangula

M. guianensis Aublet

Suriname: awatie, bonnie-bonnie-hoedoe, dekie-hatti, gingepau, perabisi, pira pisi, pisie pira, tei-hatti

Marah Kellogg Cucurbitaceae

Origins:

Marah, bitter water, see *Exodus*. XV, 23. Hebrew *marah*, fem. of *mar* "bitter," Akkadian *maru* "bitter," Latin *amarus, a, um* "bitter," *maereo, -es, maerui, maestus, maerere* "to be sad," Akkadian *marasu* "to fall ill, to have a disease," *mararu* "to be bitter," *marsu* "sick, bitter, diseased"; referring to taste of all parts of the plants.

Species/Vernacular Names:
M. fabaceus (Naudin) E. Greene

English: California man-root

M. oreganus (Torrey & A. Gray) Howell

English: coast man-root

Marahuacaea Maguire Rapateaceae

Origins:

Cerro Marahuaca, Venezuela.

Maranta L. Marantaceae

Origins:

Named for the Venetian botanist Bartolo(m)meo Maranta, *circa* 1500-1571, physician. Among his publications are *De aquae Neapoli in Luculliano scaturientis (quam ferream vocant) metallica materia*. Neapoli [Naples] 1559, *Della Theriaca ed del Mithridato* libri due. Vinegia [Venice] 1572, *Methodi cognoscendorum simplicium* libri tres. Venetjis [Venice] 1559 and *Lucullianarum Quaestionum* libri quinque. Basileae 1564. See Giovanni Battista de Toni, *Nuovi documenti sulla vita e sul carteggio di Bartolomeo Maranta, medico e simplicista del secolo XVI*. [Reale Istituto Veneto di Scienze, Lettere ed Arti. Atti, etc. tom. 71. pt. 2] Venezia 1912; Carl Linnaeus, *Species Plantarum*. 2. 1753 and *Genera Plantarum*. Ed. 5. 2. 1754.

Species/Vernacular Names:
M. arundinacea L.

English: arrowroot, obedience plant, Bermuda arrowroot, St. Vincent arrowroot

Peru: araruta, shimi-pampana, yuquilla

Mexico: viuxita

Brazil (Amazonas): hore kiki

Japan: kuzu-ukon, arôrûtu

Okinawa: aramutu

M. bicolor Ker Gawl.

Japan: futa-iro-maranta

M. leuconeura E. Morr.

English: prayer plant, ten commandments

Maranthes Blume Chrysobalanaceae

Origins:

From the Greek *maraino* "to wither away, waste away," *maransis* "a withering, wasting" and *anthos* "flower," referring to the flowers; Hebrew *marah* "to rub, to rub in"; some suggest from the Latin *mare, maris* and *anthos*, flowers by the sea; see Karl Ludwig von Blume, *Bijdragen tot de flora van Nederlandsch Indië*. 89. Batavia 1825.

Species/Vernacular Names:
M. chrysophylla (Oliv.) Prance ex F. White

Central Africa: mébamené, akoa, dombali, evess move, minkoka, mobokola, otsâa, kioro

Cameroon: mombokola, kioro

Gabon: mebamene

M. corymbosa Blume

English: sea bean

M. glabra (Oliv.) Prance (*Parinari glabra* Oliv.)

Nigeria: abo-idofin (Yoruba); oghoye (Edo)

Cameroon: n'ko, asila oman, bokanga

Congo: essou

Ivory Coast: aramon

Central Africa: oku

Gabon: ekoulebang

M. kerstingii (Engler) Prance (*Parinari kerstingii* Engl.)

Central Africa: biri biri, difike, gongo, kele, kelongoi, mambimbi, nkote, oku, amalaroué, aramon, zéri zéri, bakanza

Gabon: ekoulebang

Cameroon: asila oman, bokanga, ekoulebang

Nigeria: kaikayi (Hausa)

M. polyandra (Benth.) Prance (*Parinari polyandra* Benth.)

French: toutou vert

Nigeria: gwanja kusa, kaikayi (Hausa); chiboli (Fula); abaddima (Nupe); ibua kuna (Tiv); abo-idofin, idofun (Yoruba)

Yoruba: ako idofin

W. Africa: tutu

Mali: tutu, tutufin

M. robusta (Oliv.) Prance (*Parinari robusta* Oliv.)

Nigeria: idofun, aweawe (Yoruba); daba dogun (Edo); ohaba-uji (Ibo; see G.T. Basden, *Niger Ibos*: A description of the primitive life, customs and animistic beliefs., etc., of the Ibo people of Nigeria, etc. London 1938; Geoffrey Parrinder, *West African Religion. A Study of the Beliefs and Practices of Akan, Ewe, Yoruba, Ibo, and Kindred Peoples.* London 1961)

Yoruba: aye, ayeni, agege

Marantochloa Brongn. ex Gris Marantaceae

Origins:
The genus *Maranta* and Greek *chloe, chloa* "a grass."

Marara Karst. Palmae

Origins:
A vernacular name; see Karsten, *Linnaea.* 28: 389. 1857.

Marasmodes DC. Asteraceae

Origins:
From the Greek *marasmos* "withering, a wasting away," *maraino* "to fade, wither, waste away" and *-odes* "resembling, of the nature of, like, having the form," referring to the appearance of the plant.

Marathrum Bonpl. Podostemaceae

Origins:
Latin *marathrus, marathros* or *marathrum* and Greek *marathron, marathon* "fennel."

Marattia Sw. Marattiaceae

Origins:
After the Italian botanist (Francesco Giovanni) (Gaetano) Giovanni Francesco Maratti (Joannes Franciscus Marattius), 1723-1777, clergyman, professor at Rome University. His works include *Descriptio de vera florum existentia vegetatione et forma in plantis dorsiferis.* Romae 1760, *Flora romana ... Opus postumum.* [Edited by M.B. Oliverius.] Romae 1822 and *De plantis zoophytis et lithophytis in Mari mediterraneo viventibus.* Romae 1776. See Pier Andrea Saccardo (1845-1920), *Di un'operetta sulla flora della Corsica di autore pseudonimo e plagiario.* Venezia 1908; Angelo Calogierà, *Nuova raccolta d'opuscoli scientifici e filologici.* [Botanophili Romani ad ... C. Amadutium epistola, qua J.F. Marattium ab Adansonii ... censuris vindicat. 1770] Venezia 1755-1784; R. Zander, F. Encke, G. Buchheim and S. Seybold, *Handwörterbuch der Pflanzennamen.* 14. Aufl. 747. 1993; F. Boerner & G. Kunkel, *Taschenwörterbuch der botanischen Pflanzennamen.* 4. Aufl. 131. 1989; Jonas C. Dryander, *Catalogus bibliothecae historiconaturalis Josephi Banks.* London 1796-1800; Mariella Azzarello Di Misa, a cura di, *Il Fondo Antico della Biblioteca dell'Orto Botanico di Palermo.* 181. Palermo 1988; E.M. Tucker, *Catalogue of the Library of the Arnold Arboretum of Harvard University.* 1917-1933; John H. Barnhart, *Biographical Notes upon Botanists.* 2: 445. 1965.

Species/Vernacular Names:
M. fraxinea Sm. (*Marattia salicina* Smith)

English: king fern, horseshoe fern

Maori names: para

M. oreades Domin

English: potato fern

Marcania J.B. Imlay Acanthaceae

Origins:
For the English (Yorks.) plant collector Alexander Marcan, 1883-1953, in Indochina and Thailand, wrote "The story of drugs with special reference to Siamese medicinal plants." *Jour. Siam. Soc., Nat. Hist. Suppl.* 7(2): 107-117. 1927.

Marcelliopsis Schinz Amaranthaceae

Origins:

Resembling *Marcellia* Baill.

Marcgravia L. Marcgraviaceae

Origins:

For the German (b. Liebstadt, Meissen) naturalist and traveler Georg (Georgius) Marcgrave (Markgraf, Marcgraf, Marggraff, Margraff, Margravius, Marcgravius, Marggravius), 1610-1644 (d. of a fever, Luanda, Angola), engineer, geographer, with the pioneer of tropical medicine Willem Piso (or Willem Pies, 1611-1678) wrote *Historia naturalis Brasiliae: De Medicina Brasiliensi* libri IV (Piso); *Historiae Rerum Naturalium Brasiliae* libri VIII (Marggravius). Lugdun. Batavorum (F. Hackius), Amstelodami (L. Elzevir) 1648 and (a second edition, much enlarged, with Piso and the Dutch physician in the East Jacobus Bontius, 1592 or 1599-1631) *De Indiae utriusque Re Naturali et Medica* libri XIV: libri VI (Piso); libri II (Marggravius); libri VI (Bontius), to which is appended *Mantissa Aromatica* (Piso). Amstelaedami (L. & D. Elzevir) 1658. See Carl Linnaeus, *Species Plantarum.* 1: 503. 1753; F. Markgraf, in *Dictionary of Scientific Biography* 9: 122-123. 1981; John H. Barnhart, *Biographical Notes upon Botanists.* 2: 89 and 447. 1965; Carl F.P. von Martius, *Versuch eines Commentars über die Pflanzen in den Werken von Marcgrav und Piso über Brasilien. Kryptogamen.* München 1853; Garrison and Morton, *Medical Bibliography.* 1825, 5303. New York 1961; Jonas C. Dryander, *Catalogus bibliothecae historico-naturalis Josephi Banks.* London 1796-1800; A. Lasègue, *Musée botanique de Benjamin Delessert.* 474, 505. Paris 1845; S. Lenley et al., *Catalog of the Manuscript and Archival Collections and Index to the Correspondence of John Torrey.* Library of the New York Botanical Garden. 280. 1973; E.M. Tucker, *Catalogue of the Library of the Arnold Arboretum of Harvard University.* 1917-1933; F. Boerner & G. Kunkel, *Taschenwörterbuch der botanischen Pflanzennamen.* 4. Aufl. 131. 1989; Georg Christian Wittstein, *Etymologisch-botanisches Handwörterbuch.* 556. Ansbach 1852.

Marenopuntia Backeb. Cactaceae

Origins:

After Maren B. Parsons; see Irving William Knobloch, compil., "A preliminary verified list of plant collectors in Mexico." *Phytologia Memoirs.* VI. 1983; Hans M. Fittkau, "The systematic position of *Opuntia marenae* S.H. Parsons." *J. Cactus and Succ. Soc. America.* 49: 261-262. 1977; R. Zander, F. Encke, G. Buchheim and S. Seybold, *Handwörterbuch der*

Pflanzennamen. 14. Aufl. 760. Stuttgart 1993; F. Boerner & G. Kunkel, *Taschenwörterbuch der botanischen Pflanzennamen.* 4. Aufl. 131. Berlin & Hamburg 1989.

Maresia Pomel Brassicaceae

Origins:

Dedicated to the French botanist Paul Marès, 1826-1900, explorer, in Algeria and the Balearic Islands, author of *Histoire des progrès de l'agriculture en Algérie.* Paris 1878, with Guillaume Vigineix wrote *Catalogue raisonné des plantes vasculaires des îles Baléares.* Paris 1880. See J.H. Barnhart, *Biographical Notes upon Botanists.* 2: 447. 1965; Miguel Colmeiro y Penido, *La Botánica y los Botánicos de la Peninsula Hispano-Lusitana.* 1858; E.M. Tucker, *Catalogue of the Library of the Arnold Arboretum of Harvard University.* 1917-1933; Ernest Saint-Charles Cosson, *Compendium florae atlanticae.* 1: 68-69. Paris 1881.

Mareya Baillon Euphorbiaceae

Origins:

After the French physician and naturalist Étienne-Jules Marey, 1830-1904 (d. Paris), physiologist, zoologist, invented the sphygmograph in its modern form and the use of graphical methods in scientific research. Among his writings are *Physiologie médicale de la circulation du sang.* [8vo, first edition; the graphical recording of the pulse.] Paris 1863, *Du mouvement dans les fonctions de la vie.* Leçons faites au Collége de France. [8vo, first edition, M.'s pioneer work on muscular contraction.] Paris 1868, *Mémoire sur le Vol des Insectes et des Oiseaux.* Paris 1869-1872, *La machine animale: locomotion terrestre et aérienne.* Paris 1873 [English translation, *Animal Mechanism: A Treatise on Terrestrial and Aerial Locomotion.* London 1874], *La méthode graphique dans les sciences expérimentales* et principalement en physiologie et en médecine. [8vo, first edition, first issue; the application of the graphical methods to physiology.] Paris [1878], *La circulation du sang à l'état physiologique et dans les maladies.* [A revised version of *Physiologie médicale de la circulation du sang.*] Paris 1881, *Physiologie du mouvement. Le vol des oiseaux.* Paris 1890 and *Le Mouvement.* Paris 1894. See Michael Gross, in *Dictionary of Scientific Biography* 9: 101-103. 1981; Coughtrie, *Aerial Locomotion. Pettigrew versus Marey.* [Reprinted from the *Quarterly Journal of Science*, April, 1875.] London 1875; James Bell Pettigrew, *La locomotion chez les animaux.* [Octavo, first edition in French; English 1872.] Paris 1874; Garrison and Morton, *Medical Bibliography.* 776. 1961.

Species/Vernacular Names:

M. micrantha (Benth.) Müll. Arg.

Nigeria: uhosa (Edo)

Mareyopsis Pax & K. Hoffm. Euphorbiaceae

Origins:

The genus *Mareya* and Greek *opsis* "resembling, aspect."

Margaranthus Schltdl. Solanaceae

Origins:

From the Greek *margarites, margaron* "a pearl" and *anthos* "flower," referring to the appearance of the flowers.

Margaris Griseb. Rubiaceae

Origins:

Greek *margarites, margaron* "a pearl, pearl-oyster," referring to the flowers; *margaris* is an ancient Greek plant name applied by Plinius to a kind of palm-tree.

Margaritaria L.f. Euphorbiaceae

Origins:

Latin *margarita, ae,* Greek *margarites,* Latin *margaritarius, a, um,* "pearl, of the pearls, pearly," referring to the flowers or to the white and shining glands or to the seeds of some species of these plants.

Species/Vernacular Names:

M. discoidea (Baill.) Webster (*Margaritaria discoidea* (Baill.) Webster subsp. *discoidea; Margaritaria discoidea* (Baill.) Webster subsp. *nitida* (Pax) Webster p.p.; *Phyllanthus discoideus* (Baill.) Müll. Arg.; *Phyllanthus amapondensis* Sim; *Phyllanthus flacourtioides* Hutch.)

English: common pheasant-berry

Yoruba: asasa, aweleso, ayiwe igi oko

Nigeria: ashasha, asiyin; asasa, awe (Yoruba); asiyin (Edo); ololo (Urhobo); isinkpi (Igbo)

Cameroon: ebegeng, ebebeng, be, kango

Ivory Coast: lie

Southern Africa: Egossa rooipeer, Egossa red pear, gewone fisantebessie; umPhunzito, umPhanzitha, umDlulamazembe (Xhosa); umDlulamazembe, umKhwangu, isiBangamlotha, uMadlozane, uMadlozini, uGugusakhethelo, uPata, umNandana (Zulu)

Central Africa: budela

W. Africa: badulafen, jula sungalani

M. dubium-traceyi Airy Shaw & B. Hyland (after John Geoffrey Tracey, born 1930, original collector)

English: Tracey's puzzle

M. nobilis L.f. (*Phyllanthus nobilis* (L.f.) Müll. Arg.; *Cicca antillana* Jussieu; *Phyllanthus antillanus* (Jussieu) Müll. Arg.; *Xylosma minutiflora* J.F. Macbride, Flacourtiaceae)

English: claw berries

Mexico: garbancillo (El Soconusco); k'ah-yuk, x-nabal-che (Maya l., Yucatan); mierda de loro (Tehuantepec, Oaxaca)

Guedeloupe: mille branches

Margaritolobium Harms Fabaceae

Origins:

Greek *margarites, margaron* "a pearl, pearl-oyster" and *lobion, lobos* "lobe, pod, small pod, fruit."

Margaritopsis Sauvalle Rubiaceae

Origins:

From the Greek *margarites, margaron* "a pearl, pearl-oyster" and *opsis* "like, likeness, appearance."

Margelliantha P.J. Cribb Orchidaceae

Origins:

Greek *margelis* "pearl" and *anthos* "flower," Latin *margella* "red coral."

Marginaria Bory Polypodiaceae

Origins:

Latin *margo, marginis* "edge, border, margin."

Marginariopsis C. Chr. Polypodiaceae

Origins:

From the genus *Marginaria* and *opsis* "like, likeness, appearance."

Marginatocereus (Backeb.) Backeb. Cactaceae

Origins:

Latin *margo, marginis* "edge, border, margin" plus *Cereus.*

Margotia Boiss. Umbelliferae

Origins:

For the Swiss botanist Henri Margot, 1807-1894, teacher, a pupil of A.P. de Candolle, 1834-1837 Ionian Isles, with Georges François Reuter (1805-1872) wrote *Essai d'une flore de l'île de Zante*. [Genève 1839-1840]. See J.H. Barnhart, *Biographical Notes upon Botanists*. 2: 448. 1965; A. Lasègue, *Musée botanique de Benjamin Delessert*. 346. Paris 1845; Ethelyn Maria Tucker, *Catalogue of the Library of the Arnold Arboretum of Harvard University*. Cambridge, Massachusetts 1917-1933.

Margyricarpus Ruíz & Pavón Rosaceae

Origins:

Greek *margarites* "a pearl, pearl-oyster," *argyros* "silver and *karpos* "fruit," referring to the white berries.

Species/Vernacular Names:

M. pinnatus (Lam.) Kuntze (*Margyricarpus setosus* Ruíz & Pav.)

English: pearl fruit

Peru: canglla, canlish, canlla, canlla queuña, canlli, china canlli, duraznillo, orccocanlli, perlilla, pique, perlillas, yurac tranca, yerba de perlilla

Marianthus Huegel ex Endl. Pittosporaceae

Origins:

After Princess Marie von Metternich, patroness of botany in Austria; see S.L. Endlicher et al., *Enumeratio plantarum quas in Novae Hollandiae ... collegit C. de Hügel*. 8. Wien 1837; F.A. Sharr, *Western Australian Plant Names and Their Meanings. A glossary*. 46. [named for Maria, the Virgin Mary.] University of Western Australia Press 1996; James A. Baines, *Australian Plant Genera. An Etymological Dictionary of Australian Plant Genera*. 231. Chipping Norton, N.S.W. 1981.

Mariarisqueta Guinea Orchidaceae

Origins:

For María Arisqueta de Guinea.

Marica Ker Gawl. Iridaceae

Origins:

Possibly a kind of an anagram of America; see Georg Christian Wittstein, *Etymologisch-botanisches Handwörterbuch*.

557. 1852; F. Boerner & G. Kunkel, *Taschenwörterbuch der botanischen Pflanzennamen*. 4. Aufl. 131. 1989; H. Genaust, *Etymologisches Wörterbuch der botanischen Pflanzennamen*. 415. Basel 1996.

Marila Sw. Guttiferae

Origins:

Greek *marile* "embers of charcoal, coal-dust," referring to the seeds.

Marina Liebm. Fabaceae

Origins:

After the name of an interpreter for Cortez, 16th century, see James C. Hickman, ed., *The Jepson Manual: Higher Plants of California*. 636. Berkeley 1993.

Species/Vernacular Names:

M. orcuttii (S. Watson) Barneby var. *orcuttii*

English: California marina

Mariposa (Alph. Wood) Hoover Calochortaceae (Liliaceae)

Origins:

Mariposa is a county in central California, Spanish *mariposa* "butterfly."

Mariscopsis Chermezon Cyperaceae

Origins:

Resembling Latin *mariscos* or *mariscus, i* "a rush."

Marisculus Goetgh. Cyperaceae

Origins:

Latin *mariscos* or *mariscus, i* "a rush.

Mariscus Vahl Cyperaceae

Origins:

Medieval Latin *mariscos* or *mariscus, i* (perhaps from *mare* "sea") "a rush" ("de junco, quem mariscon appellat", Plinius); Anglo-Saxon *mersc, merisc* "a marsh"; see Martin H.

Vahl (1749-1804), *M. Vahlii ... Enumeratio Plantarum*. 2: 372. 1805.

Species/Vernacular Names:

M. sp.

Hawaii: mau'u, kaluhaluha

M. alternifolius Vahl

Yoruba: ataponimomo, segi dudu, ewa okodo, ewa orisa, ewa sango

M. capensis (Steud.) Schrad.

English: monkey bulb

South Africa: aapuintjie, bobbejaanuintjie, qothoqothoane-e-nyenyane

M. hypochlorus (Hillebr.) C.B. Cl. (*Cyperus hypochlorus* Hillebr.)

Hawaii: 'ahu'awa

M. indecorus (Kunth.) Podlech

English: sedge

M. javanicus (Houtt.) Merr. & Metcalfe (*Cyperus javanicus* Houtt.)

Hawaii: 'ahu'awa, 'ehu'awa

M. rehmannianus C.B. Cl.

Southern Africa: makaladi (Tswana)

Maritimocereus Akers Cactaceae

Origins:

Latin *maritimus* "belonging to the sea" plus *Cereus*.

Markea Rich. Solanaceae

Origins:

After the French (b. Picardy) biologist Jean Baptiste Antoine Pierre de Monnet (Monet) de Lamarck, 1744-1829 (Paris), a great naturalist, botanist, zoologist, paleontologist, conchologist, from 1778 to 1793 botanist at the Jardin des Plantes in Paris (Jardin du Roi), naturalist and a forerunner of Darwin's theory of evolution. His writings include *Flore françoise* Paris 1778 and *Philosophie Zoologique*. [First edition, two volumes in one, 8vo.] Paris 1822. See Leslie J. Burlingame, in *Dictionary of Scientific Biography* 7: 584-594. 1981; Frans A. Stafleu, *Linnaeus and the Linnaeans. The spreading of their ideas in systematic botany, 1735-1789*. Utrecht 1971; Stafleu and Cowan, *Taxonomic Literature*. 2: 730-734. 1979; J.H. Barnhart, *Biographical Notes upon Botanists*. 2: 337. 1965; Ethelyn Maria Tucker, *Catalogue of the Library of the Arnold Arboretum of Harvard University*. Cambridge, Massachusetts 1917-1933; Denis I. Duveen, *Bibliotheca Alchemica et Chemica*. 334. London

1949; T.W. Bossert, *Biographical Dictionary of Botanists Represented in the Hunt Institute Portrait Collection*. 225. 1972; Mariella Azzarello Di Misa, a cura di, *Il Fondo Antico della Biblioteca dell'Orto Botanico di Palermo*. 143-145. Regione Siciliana, Palermo 1988; Emil Bretschneider, *History of European Botanical Discoveries in China*. 1981; R. Zander, F. Encke, G. Buchheim and S. Seybold, *Handwörterbuch der Pflanzennamen*. 14. Aufl. 739. Stuttgart 1993; G.C. Wittstein, *Etymologisch-botanisches Handwörterbuch*. 497. 1852.

Markhamia Seemann ex Baillon Bignoniaceae

Origins:

For the British (b. Yorkshire) traveler Sir Clements Robert Markham, 1830-1916 (d. London), botanist, geographer, explorer, plant collector, 1844-1851 in the Royal Navy, 1852-1854 in Peru, President of the Royal Geographical Society and introducer of *Cinchona* into India, 1858-1877 employed by the India Office, 1864 Fellow of the Linnean Society, 1873 Fellow of the Royal Society, he played an active role in preparations for R.F. Scott's *Discovery* voyage (1901-1904) and the expedition of 1910-1912. Among his valuable and very numerous works are *Antarctic Obsession. A Personal Narrative of the origins of the British National Antarctic Expedition 1901-1904*. Bluntisham Books, UK 1986, *Travels in Peru and India*. London 1862, *The Chinchona species of New Granada ...* London 1867, *A Memoir of the Lady Ana de Osorio, Countess of Chinchon and Vicequeen of Peru* (a.d. 1829-1839). London 1874, *Peruvian Bark. A popular account of the introduction of Cinchona cultivation into British India ...* London 1880, *Richard Hakluyt: His Life and Work*. London 1896, *Vocabularies of the General Language of the Incas of Peru or Runa Simi, Called Quichua by the Spanish Grammarians*. London 1908, *Lands of Silence. A History of Arctic and Antarctic exploration*. [Completed and edited by F.H.H. Guillemard.] Cambridge 1921. See Thomas F. Glick, in *Dictionary of Scientific Biography* 9: 123-124. 1981; Sir Albert Hastings Markham, *The Life of Sir Clements R. Markham*. London 1917; Harry Bernstein and Bailey Wallys Diffie, *Sir Clements R. Markham as a Translator ...* Reprinted from *The Hispanic American Historical Review*, etc. [1937]; Ethelyn Maria Tucker, *Catalogue of the Library of the Arnold Arboretum of Harvard University*. Cambridge, Massachusetts 1917-1933; John H. Barnhart, *Biographical Notes upon Botanists*. 2: 449. 1965; Sydney A. Spence, *Antarctic Miscellany. Books, Periodicals and Maps Relating to the Discovery and Exploration of Antarctica*. London 1980; J.J. von Tschudi, *Travels in Peru, During the Years 1838-1842, on the Coast, in the Sierra, across the Cordilleras and the Andes, into the Primeval Forests*. NY, Putnam 1849; M.E. and F.A. Markham, *The Life of Sir Albert Hastings*

Markham. London 1927; T.W. Bossert, *Biographical Dictionary of Botanists Represented in the Hunt Institute Portrait Collection.* 253. 1972; August Weberbauer, *Die Pflanzenwelt der peruanischen Andes in ihren Grundzügen dargestellt.* Leipzig 1911; Ida Kaplan Langman, *A Selected Guide to the Literature on the Flowering Plants of Mexico.* 1964; Isaac Henry Burkill, *Chapters on the History of Botany in India.* Delhi 1965; G. Murray, *History of the Collections Contained in the Natural History Departments of the British Museum.* 1: 166. London 1904; G.A. Doumani, ed., *Antarctic Bibliography.* Washington, Library of Congress 1965-1979; Peter H. Goldsmith, *A Brief Bibliography of Books in English, Spanish and Portuguese Relating to the Republics Commonly called Latin American,* with comments. New York 1915; R. Desmond, *The European Discovery of the Indian Flora.* Oxford 1992; F.D. Drewitt, *The Romance of the Apothecaries' Garden at Chelsea.* London 1924; Ray Desmond, *Dictionary of British & Irish Botanists and Horticulturists.* 467. 1994; I.C. Hedge and J.M. Lamond, *Index of Collectors in the Edinburgh Herbarium.* Edinburgh 1970; Mea Allan, *The Hookers of Kew.* London 1967.

Species/Vernacular Names:

M. sp.

Zaire: etaka, mushavu

N. Rhodesia: musase

M. lutea (Benth.) K. Schumann (*Markhamia platycalyx* Bak.; *Markhamia hildebrandtii* (Bak.) Sprague)

English: yello trumpet tree

Cameroon: maganga, osse, angossa, atag, gonja

M. obtusifolia (Bak.) Sprague (*Dolichandra obtusifolia* Bak.)

Southern Africa: mubfeya (Shona)

S. Rhodesia: muPfeya

M. tomentosa (Benth.) K. Schumann ex Engler

Yoruba: iru aaya

Nigeria: ogie-ikhimi, iru-aya, ognie-khimi; akoko, iru-aya (Yoruba); ogie ikhimi (Edo); onyiri akikara (Igbo)

Ivory Coast: tomboro

Congo: lubota

M. zanzibarica (Bojer ex DC.) K. Schum. (*Markhamia acuminata* (Klotzsch) K. Schum.; *Spathodea acuminata* Klotzsch)

English: bell bean tree, bean tree

Southern Africa: klokkiesboontjieboom, shidzanyi (Tsonga); muSikamyati, muSiramyati (Shona); mositsanyate (Kgatla dialect, Botswana); mositsanyate (Tawana dialect, Ngamiland); mupatalwala (Subya)

S. Rhodesia: muPetakwale, mu Siranyati

Marlea Roxb. Cornaceae

Origins:

The Bengali name for *Marlea begonifolia* Roxb., see William Roxburgh, *Plants of the Coast of Coromandel.* 3: 80, t. 283. 1795-1819; M.P. Nayar, *Meaning of Indian Flowering Plant Names.* 218. Dehra Dun 1985.

Marlieropsis Kiaersk. Myrtaceae

Origins:

Resembling *Marlierea* Cambess.

Marlothia Engl. Rhamnaceae

Origins:

For the South African botanist Hermann Wilhelm Rudolf Marloth, 1855-1931, pharmacist, chemist, botanical explorer, plant collector, from 1833 in South Africa. He is best known for "The historical development of the geographical botany of South Africa." *S. Afr. J. Sci.* 1: 251-257. 1903, "Notes on the vegetation of Southern Rhodesia." *S. Afr. J. Sci.* 2: 300-307. 1904, *The Chemistry of South African Plants and Plant Products.* Cape Town 1913, *The Flora of South Africa.* Capetown and London 1913-1932, *Dictionary of the Common Names of Plants* with a list of foreign plants cultivated in the open. Cape Town 1917, *Cape Flowers at Home.* Cape Town [1922]. See John H. Barnhart, *Biographical Notes upon Botanists.* 2: 449. 1965; R. Zander, F. Encke, G. Buchheim and S. Seybold, *Handwörterbuch der Pflanzennamen.* 14. Aufl. Stuttgart 1993; T.W. Bossert, *Biographical Dictionary of Botanists Represented in the Hunt Institute Portrait Collection.* 254. 1972; H.N. Clokie, *Account of the Herbaria of the Department of Botany in the University of Oxford.* 206. 1964; E.M. Tucker, *Catalogue of the Library of the Arnold Arboretum of Harvard University.* 1917-1933; Hans Herre (1895-1979), *The Genera of the Mesembryanthemaceae.* 49-50. Cape Town 1971; I.C. Hedge and J.M. Lamond, *Index of Collectors in the Edinburgh Herbarium.* Edinburgh 1970; A. White and B.L. Sloane, *The Stapelieae.* Pasadena 1937; A. Engler et al., "Plantae Marlothianae." in *Bot. Jahrb.* 10: 1-50, 242-285. 1889; Mary Gunn and Leslie E. Codd, *Botanical Exploration of Southern Africa.* Cape Town 1981; Gordon Douglas Rowley, *A History of Succulent Plants.* Strawberry Press, Mill Valley, California 1997.

Marlothiella H. Wolff Umbelliferae

Origins:

For the South African botanist Hermann Wilhelm Rudolf Marloth, 1855-1931; see John H. Barnhart, *Biographical Notes upon Botanists.* 2: 449. 1965.

Marlothistella Schwantes Aizoaceae

Origins:
For the South African botanist Hermann Wilhelm Rudolf Marloth, 1855-1931; see John H. Barnhart, *Biographical Notes upon Botanists*. 2: 449. 1965.

Marmaroxylon Killip Mimosaceae

Origins:
From the Greek *marmairo* "to flash, sparkle, glisten," *marmaros* "stone, rock, brightness, whiteness" and *xylon* "wood."

Marmoritis Bentham Labiatae

Origins:
Possibly from the Greek *marmaritis* "like marble," Plinius applied *marmaritis* to a plant that grows in marble quarries, *aglaophotis* or peony.

Species/Vernacular Names:
M. rotundifolia Bentham

English: roundleaf marmoritis

China: yuan ye niu lian qian

Marniera Backeb. Cactaceae

Origins:
For Julien Marnier-Lapostolle, 1902-1976 (d. Cap Ferrat, France); see Gordon Douglas Rowley, *A History of Succulent Plants*. Strawberry Press, Mill Valley, California 1997.

Marojejya Humbert Palmae

Origins:
Based on a Malagasy place name, the mountain of Marojejy, northeast of Madagascar; see H. Humbert, "Une merveille de la Nature à Madagascar. Première exploration botanique du massif du Marojeiy et ses satellites." *Mém. Inst. Rech. Sci. Madagascar*, sér. B. Biol. Vég. 6: 1-271. 1955; John Dransfield and Natalie W. Uhl, "A magnificent new palm from Madagascar." *Principes*. 28(4): 151-154. 1984.

Species/Vernacular Names:
M. darianii J. Dransfield & N.W. Uhl (for a Californian palm lover and palm collector, traveler, Dr. Mardy E. Darian)

English: big leaf palm

Madagascar: ravim-be, ravimbe (= big leaf) (Betsimisaraka)

M. insignis Humbert

Madagascar: maroalavehivavy, betefoka, besofina, hovotralanana, mandanzezika (Betsimisaraka); fohitanana, kona (Tanala); menamoso, beondroka (Tsimihety)

Marquisia A. Rich. Rubiaceae

Origins:
For the French botanist Alexandre Louis Marquis, 1777-1828, physician, professor of botany, author of *Essai sur l'histoire naturelle et médicale des Gentianes*. [Paris 1810]; see E.M. Tucker, *Catalogue of the Library of the Arnold Arboretum of Harvard University*. 1917-1933; Stafleu and Cowan, *Taxonomic Literature*. 3: 303-304. 1981.

Marrattia Sw. Marattiaceae

Origins:
Orthographic variant of *Marattia* Sw.

Marrubium L. Labiatae

Origins:
Latin *marrubium, ii* ("marrubium quod Graeci prasion vocant, alii linostrophon, nonnulli philopaeda, aut philochares", Plinius), Hebrew *marrob* "bitter, bitter juice"; see Carl Linnaeus, *Species Plantarum*. 582. 1753 and *Genera Plantarum*. Ed. 5. 254. 1754.

Species/Vernacular Names:
M. vulgare L.

English: white horehound, common hoarhound, hoarhound, white hoarhound, marvel, horehound

Peru: coronilla, nacnac, okce kcora

South Africa: houndsbene, koorsbossie

China: ou xia zhi cao

Arabic: morroubia, omerroubia, marriout, umm re-roubia, roubia, merriwa

Marsdenia R. Br. Asclepiadaceae

Origins:
Named for the Irish-born British traveler and plant collector William Marsden, 1754-1836 (d. Herts.), orientalist, numismatist, in 1771 he joined the service of British East India Company, Marsden spent some years in Sumatra as subsecretary and later principal secretary to the Government, from 1771 to 1779 collected in Sumatra, from 1779 to 1807

secretary to the British Board of Admiralty, 1783 Fellow of the Royal Society. Among his valuable writings are *The History of Sumatra*. London 1783, *A Catalogue of Dictionaries, Vocabularies, Grammars and Alphabets*. London 1796, *A Dictionary of the Malayan Language*. London 1812, *A Grammar of the Malayan Language*. London 1812, *Bibliotheca Marsdeniana Philologica et Orientalis*. London 1827, *Numismata Orientalia. The Oriental Coins, Ancient and Modern*. London 1823-1825 and *A Brief Memory of the Life and Writings of the Late W. Marsden. Written by Himself*. London 1838, in 1818 translated *The Travels of Marco Polo* into English. See Peter Marsden, *The Wreck of the Amsterdam*. New York 1975; Ray Desmond, *Dictionary of British & Irish Botanists and Horticulturists*. 468-469. London 1994; F. Boerner & G. Kunkel, *Taschenwörterbuch der botanischen Pflanzennamen*. 4. Aufl. 132. 1989; Georg Christian Wittstein, *Etymologisch-botanisches Handwörterbuch*. 558. Ansbach 1852; John H. Barnhart, *Biographical Notes upon Botanists*. 2: 450. Boston 1965; R. Brown, "On the Asclepiadeae." *Memoirs of the Wernerian Natural History Society*. 1: 28-29. Edinburgh 1811; M. Archer, *Natural History Drawings in the India Office Library*. London 1962; Warren R. Dawson, *The Banks Letters*, a Calendar of the Manuscript Correspondence of Sir Joseph Banks. London 1958; Jonas C. Dryander, *Catalogus bibliothecae historico-naturalis Josephi Banks*. London 1796-1800; A. Lasègue, *Musée botanique de Benjamin Delessert*. Paris 1845; Ethelyn Maria Tucker, *Catalogue of the Library of the Arnold Arboretum of Harvard University*. Cambridge, Massachusetts 1917-1933; G. Murray, *History of the Collections Contained in the Natural History Departments of the British Museum*. 1: 166. London 1904.

Species/Vernacular Names:

M. araujacea F. Muell.

Australia: milk vine

M. australis (R. Br.) Druce (*Leichhardtia australis* R. Br.)

Australia: native pear, Austral doubah, alunqua, cogola bush, bush banana, native banana

M. cinerascens R. Br.

Australia: milk vine

M. coronata Benth.

Australia: milk vine

M. cymulosa Benth.

Australia: milk vine

M. flavescens Cunn.

Australia: hairy milk vine, yellow milk vine

M. floribunda (brongn.) Schltr.

English: stephanotis, Madagascar jasmine

M. fraseri Benth.

Australia: narrow-leaved milk vine

M. glandulifera C. White

Australia: Fraser Island milk vine

M. griffithii Hook.f.

English: Griffith condor vine

China: bai yao niu nai cai

M. hainanensis Tsiang

English: Hainan condor vine

China: hai nan niu nai cai

M. hullsii Benth.

Australia: milk vine

M. koi Tsiang (*Marsdenia tsaiana* Tsiang)

English: Ko condor vine, Tsai condor vine

China: da ye niu nai cai

M. lachnostoma Bentham

English: hairy-throat condor vine

China: mao hou niu nai cai

M. leptophylla Benth.

Australia: wiry milk vine

M. longipes W.T. Wang ex Tsiang & P.T. Li

English: longstalk condor vine

China: bai ling cao

M. lloydii P. Forster (after Lloyd Bird)

Australia: corky milk vine

M. officinalis W.T. Wang ex Tsiang & P.T. Li

English: medicinal condor vine

China: hai feng teng

M. oreophila W.W. Smith

English: beakstyle condor vine

China: hui zhu niu nai cai

M. rostrata R. Br.

Australia: common milk vine, milk vine

M. stenantha Handel-Mazzetti

English: narrowflower condor vine

China: xia hua niu nai cai

M. suaveolens R. Br.

Australia: scented milk vine

M. tenacissima (Roxb.) Moon

English: tenacious condor vine

China: tong guang teng, tong guang san

M. tinctoria R. Br. (*Marsdenia globifera* Tsiang)

English: Java indigo, tinctorial condor vine, globose condor vine

China: lan ye teng

Malaya: akar tarum

M. tomentosa Morren & Decaisne

English: tomentose condor vine

China: jia fang ji

M. velutina R. Br.

Australia: milk vine

M. viridiflora R. Br.

Australia: green-flowered milk vine

Marshallfieldia J.F. Macbr. Melastomataceae

Origins:

Named after Captain Marshall Field, fl. 1920, a patron of botanical collecting.

Marshallia Schreber Asteraceae

Origins:

Dedicated to the American botanist Moses Marshall (1758-1813, West Bradford, USA), he was nephew of the American botanist Humphry Marshall, 1722-1801 (West Bradford, USA), dendrologist, Quaker, correspondent of John Fothergill and Sir J. Banks, cousin of John Bartram. His writings include *Arbustrum americanum*. Philadelphia 1785 [French translation by M. Lézermes, *Catalogue alphabétique des Arbres et Arbrisseaux*, etc. Paris 1788.] and *A Few Observations Concerning Christ, or the Eternal Word*. London 1755. See C.D. Beadle & F.E. Boynton, "Revision of the species of *Marshallia*." in *Biltmore Botanical Studies*. 1(1): 3-10. 1901; R.B. Channel, "A revisional study of the genus *Marshallia* (Compositae)." Contributions from the Gray Herbarium of Harvard University. 181: 41-130. 1957; John H. Barnhart, *Biographical Notes upon Botanists*. 2: 451. 1965; Warren R. Dawson, *The Banks Letters*, a Calendar of the Manuscript Correspondence of Sir Joseph Banks. 582-583. 1958; Jonas C. Dryander, *Catalogus bibliothecae historico-naturalis Josephi Banks*. 1796-1800; Ethelyn Maria Tucker, *Catalogue of the Library of the Arnold Arboretum of Harvard University*. Cambridge, Massachusetts 1917-1933; William Darlington (1782-1863), *Reliquiae Baldwinianae*. Philadelphia 1843; William Darlington, *Memorials of John Bartram and Humphry Marshall*. Philadelphia 1849; Ernest Earnest, *John and William Bartram, Botanists and Explorers 1699-1777, 1739-1823*. Philadelphia 1940; J. Ewan, ed., *A Short History of Botany in the United States*. New York and London 1969; Jeannette Elizabeth Graustein, *Thomas Nuttall, Naturalist. Explorations in America, 1808-1841*. Harvard University Press 1967; Howard Atwood Kelly and Walter Lincoln Burrage, *Dictionary of American Medical Biography*. New York 1928; J.W. Harshberger, *The Botanists of Philadelphia and Their Work*. 1899; Sarah Smith and Charles Shine,

"*Marshallia grandiflora*. Compositae." *Curtis's Botanical Magazine*. 15(3): 158-163. August 1998; R. Zander, F. Encke, G. Buchheim and S. Seybold, *Handwörterbuch der Pflanzennamen*. 14. Aufl. 747. Stuttgart 1993; F. Boerner & G. Kunkel, *Taschenwörterbuch der botanischen Pflanzennamen*. 4. Aufl. 132. Berlin & Hamburg 1989; Blanche Elizabeth Edith Henrey (1906-1983), *British Botanical and Horticultural Literature before 1800*. Oxford 1975.

Species/Vernacular Names:

M. grandiflora Beadle & F.E. Boynton

English: Barbara's button

Marshalljohnstonia Henrickson Asteraceae

Origins:

For the American botanist Marshall Conring Johnston (b. 1930), plant collector in Mexico, with the American botanist Donovan Stewart Correll (1908-1983) wrote *Manual of the Vascular Plants of Texas*. Texas Research Foundation, Renner, Texas 1970; see Irving William Knobloch, compil., "A preliminary verified list of plant collectors in Mexico." *Phytologia Memoirs*. VI. 1983; Ida Kaplan Langman, *A Selected Guide to the Literature on the Flowering Plants of Mexico*. 1964; M.N. Chaudhri, I.H. Vegter and C.M. De Wal, *Index Herbariorum*, Part II (3), *Collectors I-L*. Regnum Vegetabile vol. 86. 1972; Rogers McVaugh, "Botanical exploration in Nueva Galicia from 1790 to the present time." *Contr. Univ. Mich. Herb*. 9(3): 205-357. 1972; John H. Barnhart, *Biographical Notes upon Botanists*. 1: 383. 1965; R. Zander, F. Encke, G. Buchheim and S. Seybold, *Handwörterbuch der Pflanzennamen*. 14. Aufl. Stuttgart 1993.

Marshallocereus Backeb. Cactaceae

Origins:

For the American (b. Philadelphia) botanist William Taylor Marshall, 1886-1957, specialist in the Cactaceae, Director of the Desert Botanical Garden of Phoenix (Arizona). His writings include "Three Chilean species of cacti." *Cact. Succ. J.* (Los Angeles) 18: 171-173. 1946, *Arizona's Cactuses*. Desert Bot. Gard. Phoenix 1950, "The barrel cactus as a source of water." *Saguaroland Bull*. 10: 64-71. 1956, "A new genus of cactuses of Arizona." *Saguaroland Bull*. 10: 89-91, 107. 1956, with Thor Methven Bock wrote *Cactaceae* with illustrated keys of all tribes, sub-tribes and genera ... Supplementing the work of doctors Britton and Rose. Pasadena 1941 and "Cactaceae." Supplement no. 1. *Cact. Succ. J.* (Los Angeles) 17(8): 113-117. 1945, with R.S. Woods wrote *Glossary of Succulent Plant Terms*. Pasadena 1945; see Gordon Douglas Rowley, *A History of Succulent Plants*. Mill Valley, California 1997; Irving

William Knobloch, compil., "A preliminary verified list of plant collectors in Mexico." *Phytologia Memoirs*. VI. 1983; Ida Kaplan Langman, *A Selected Guide to the Literature on the Flowering Plants of Mexico*. 1964.

Marsilaea Necker Salviniaceae

Origins:

Dedicated to the Italian botanist Luigi Ferdinando Marsili (Marsigli), 1658 (but in Stafleu, 1656)-1730, mycologist, naturalist, scientist.

Marsilea L. Marsileaceae

Origins:

In honor of the celebrated Italian (b. Bologna) botanist and naturalist Luigi Ferdinando Marsili (Marsigli), 1658-1730 (Bologna), mycologist, scientist, traveler, London 1722 Fellow of the Royal Society. His works include *Dissertatio de generatione fungorum*. Roma 1714, *Histoire physique de la mer*. [The first scientific treatise on oceanography, with a preface by H. Boerhaave; first published in Italian in 1711.] Amsterdam 1725, *Osservationi intorno al Bosforo tracio overo canale di Constantinopoli*, rappresentate in Lettera alla ... Maestà di Cristina regina di Suezia. Roma 1681, *Danubius Panonico-Mysicus*. Hagae Comitum [The Hague], Amstelodami 1726 and *Stato militare dell'Imperio ottomano*. Haya 1732. See Sylvestre Dufour (Philippe) (1622-1687) [pseud. of Jacob Spon], Libellus primus sub titulo: *Jacobi Sponii Bevanda asiatica*, etc. [Lugduni] 1705; Icilio Guareschi, *Luigi Ferdinando Marsigli e la sua opera scientifica*. Torino 1915; Mario Longhena, *Il Conte L.F. Marsili*. Un uomo d'arme e di scienza. Milano 1930; Carlo Tagliavini, *Il "Lexicon Marsilianum," dizionario latino-rumeno-ungherese del sec. XVII*. Studio filologico e testo. Bucuresti 1930; Emilio Lovarini, ed., *La Schiavitù del Generale Marsigli sotto i Tartari e i Turchi*. Bologna 1931; Albano Sorbelli, ed., *Memorie intorno a Luigi Ferdinando Marsili*, pubblicate nel secondo centenario della morte per cura del Comitato Marsiliano. Bologna 1930; Abdulhak Adnan, *La Science chez les Turcs Ottomans*. Paris 1939; Giuseppe Gaetano Bolletti, *Dell'origine e de' progressi dell'Instituto delle Scienze di Bologna e di tutte le Accademie ad esso unite*. [8vo, first edition.] Bologna 1751; [Conte Francesco Algarotti, 1712-1764], *Memoria intorno alla vita ed agli scritti del Conte F. Algarotti*. [8vo, offprint.] Venice 1770; Paula Finden, "From Aldrovandi to Algarotti: the contours of science in early modern Italy." *BJHS*. 24: 353-360. 1991; Carl Linnaeus (1707-1778), *Species Plantarum*. 1099. 1753 and *Genera Plantarum*. Ed. 5. 485. 1754; Francesco Rodolico, in *Dictionary of Scientific Biography* 9: 134-136. 1981; A.A.V.V., *Celebrazione di Luigi Ferdinando*

Marsili nel secondo centenario della morte. Bologna 1930; Pericle Ducati, *Marsili. Libro e moschetto*. Milano 1930; Mariella Azzarello Di Misa, a cura di, *Il Fondo Antico della Biblioteca dell'Orto Botanico di Palermo*. 182. Regione Siciliana, Palermo 1988; Helmut Genaust, *Etymologisches Wörterbuch der botanischen Pflanzennamen*. 369. [genus named for the Italian botanist Giovanni Marsili, 1727-1794.] Basel 1996; R. Zander, F. Encke, G. Buchheim and S. Seybold, *Handwörterbuch der Pflanzennamen*. 14. Aufl. 747. Stuttgart 1993; F. Boerner & G. Kunkel, *Taschenwörterbuch der botanischen Pflanzennamen*. 4. Aufl. 132. Berlin & Hamburg 1989; Georg Christian Wittstein, *Etymologisch-botanisches Handwörterbuch*. 558. Ansbach 1852.

Species/Vernacular Names:

M. angustifolia R. Br.

Australia: common nardoo

M. crenata Presl

The Philippines: kaya-kayapuan, banig-usa

M. drummondii A. Braun

Australia: nardoo, common nardoo

M. exarata A. Braun

Australia: nardoo

M. hirsuta R. Br.

Australia: short-fruit nardoo

M. macrocarpa Presl

English: water clover, water fern

Southern Africa: inDlebe y ebhokhwe (Xhosa)

M. mutica Mett.

Australia: banded nardoo

M. quadrifolia L.

English: pepperwort, European water clover

Japan: den-ji-sô

Okinawa: tagusa

China: ping, p'in, ssu yeh tsai, tien tzu tsao

Marsippospermum Desv. Juncaceae

Origins:

From the Greek *marsippos, marsipos* "bag, pouch" and *sperma* "seed."

Marssonia Karsten Gesneriaceae

Origins:

For the German botanist Theodor Friedrich Marsson, 1816-1892, pharmacist, 1842-1867 at Wolgast, author of *Flora von Neu-Vorpommern und den Inseln Rügen und Usedom.* Leipzig 1869 and *Der Bryozoen der weissen Schreibkreide der Inseln Rügen.* 1887. See John H. Barnhart, *Biographical Notes upon Botanists.* 2: 452. 1965; E.M. Tucker, *Catalogue of the Library of the Arnold Arboretum of Harvard University.* 1917-1933; Ignatz Urban, *Geschichte des Königlichen Botanischen Museums zu Berlin-Dahlem (1815-1913). Nebst Aufzählung seiner Sammlungen.* Dresden 1916.

Marsupiaria Hoehne Orchidaceae

Origins:
Greek *marsippos* "bag, pouch," *marsippion, marsipion, marsypion, marsypeion* "small bag," Latin *marsupium,* referring to the base of the leaves.

Marsypianthes Martius ex Benth. Labiatae

Origins:
Greek *marsippos* "bag, pouch" and *anthos* "flower."

Species/Vernacular Names:
M. chamaedrys (Vahl) Kuntze
Peru: sacha albahaca, supi sacha

Marsypopetalum Scheffer Annonaceae

Origins:
From the Greek *marsippos, marsippion, marsyppion* "bag, pouch, purse" and *petalon* "petal."

Martensia Giseke Zingiberaceae

Origins:
For the German physician Friedrich Martens, explorer, 1671 Spitzbergen and Greenland. His writings include Friedrich Martens vom Hamburg *Spitzbergische oder Groenlandische Reise Beschreibung, gethan im Jahr 1671,* etc. [A work relating to whaling in the Arctic regions.] Hamburg 1675.

Marthella Urban Burmanniaceae

Origins:
To commemorate Urban's wife, Martha Urban *née* Kurtz.

Marticorenia Crisci Asteraceae

Origins:
After the Chilean botanist Clodomiro Marticorena, b. 1929, University of Concepción (Departamento de Botánica, Instituto Central de Biología, Concepción, Chile), author of *Bibliografía Botánica Taxonómica de Chile.* Missouri

Botanical Garden 1992, in cooperation with Jorge Victor Crisci (1945-) wrote "Sobre *Haplopappus scrobiculatus* (Compositae) de Chile y Argentina y su sinonimia." *Darwiniana.* 17: 467-472. 1972, with Max Quezada (1936-) wrote "Dos especies de *Nolana* (Nolanaceae) nuevas para Chile." *Bol. Soc. Biol.* 48: 91-97. Concepción 1974, "Catálogo de la flora vascular de Chile." *Gayana, Bot.* 42(1-2): 1-157. 1985, "Adiciones a la flora de Chile." *Gayana, Bot.* 44: 39-44. 1988, with Mary Therese Kalin Arroyo (b. 1944) wrote "El género *Bartsia* L. (Scrophulariaceae) en Chile." *Gayana, Bot.* 41(1-2): 47-51. 1984, with R. Rodriguez wrote "La presencia del género *Androsace* L. (Primulaceae) en Chile." *Bol. Soc. Biol.* 51(1): 303-304. Concepción 1978, with O. Parra wrote "Morfologia de los granos de polen de *Hesperomannia* Gray y *Moquinia* DC. (Compositae-Mutisieae). Estudio comparativo con generos afines." *Gayana, Bot.* 29: 1-22. 1975.

Martiella Tieghem Loranthaceae

Origins:
For the German (b. Erlangen) botanist Carl (Karl) Friedrich Philipp von Martius, 1794-1868 (d. Munich), physician, M.D. Erlangen 1814, ethnologist, traveler, botanical explorer, 1817-1820 Brazil, founder of the *Flora Brasiliensis.* [15 volumes, 1840-1906], plant collector, correspondent of Goethe and Miquel. His writings include *Beiträge zur Ethnographie und Sprachenkunde Amerikas zumal Brasilien.* Leipzig 1867 and *Die Kartoffelepidemie.* Munich 1842; see A.P.M. Sanders, in *Dictionary of Scientific Biography* 9: 148-149. 1981; Stafleu and Cowan, *Taxonomic Literature.* 3: 325-339. 1981; J.H. Barnhart, *Biographical Notes upon Botanists.* 2: 455. Boston 1965; T.W. Bossert, *Biographical Dictionary of Botanists Represented in the Hunt Institute Portrait Collection.* 255. 1972; A. Lasègue, *Musée botanique de Benjamin Delessert.* 1845; S. Lenley et al., *Catalog of the Manuscript and Archival Collections and Index to the Correspondence of John Torrey.* Library of the New York Botanical Garden. 281, 464. 1973; E.M. Tucker, *Catalogue of the Library of the Arnold Arboretum of Harvard University.* 1917-1933; João Barbosa Rodrigues (1842-1909), *A flora brasiliensis de Martius.* Rio de Janeiro 1907; Christian Friedrich Schwägrichen (1775-1853), "Bemerkungen über einige Stellen in der Flora brasiliensis von Endlicher und Martius." *Linnaea.* 14(5): 517-528. (Feb.) 1841.

Martinella Baillon Bignoniaceae

Origins:
For the French botanist Joseph Martin, flourished 1788-1826, plant collector, botanical explorer in French Guiana, Martinique and Mauritius.

Martinella H. Lév. Brassicaceae

Origins:

For the French missionary Léon François Martin, 1866-1919, botanical collector in China and Japan.

Martinezia Ruíz & Pavón Palmae

Origins:

Named for Baltazar (Baltasar) Jaime Martínez Compañon y Bujanda, 1737-1797, Archbishop of Santa Fé de Bogotá (1788-1797) and Bishop of Trujillo (1778-1788), sent many plants to Spain. See J.M. Pérez Ayala, *Baltasar Jaime Martínez Compañon y Bujanda, prelado español de Colombia y el Perú.* [Biblioteca de la Presidencia de Colombia.] Bogotá 1955; Eduardo Posada, *La imprenta en Santa Fé de Bogotá en el siglo XVIII.* Madrid 1917; Soledad Acosta de Samper, *Biografías de hombres ilustres o notables, relativas a Colombia.* Bogotá 1883; Fermin Ibañez, *Oracion funebre que en las solemnes exequias ... á la ... memoria del ... Señor D. B.J. Martinez Compañon, Arzobispo que fue de esta metropoli.* Santafe de Bogota 1798.

Martiodendron Gleason Caesalpiniaceae

Origins:

For Carl Friedrich Philipp von Martius, 1794-1868; see A.P.M. Sanders, in *Dictionary of Scientific Biography* 9: 148-149. 1981; Stafleu and Cowan, *Taxonomic Literature.* 3: 325-339. 1981; J.H. Barnhart, *Biographical Notes upon Botanists.* 2: 455. Boston 1965; T.W. Bossert, *Biographical Dictionary of Botanists Represented in the Hunt Institute Portrait Collection.* 255. 1972; A. Lasègue, *Musée botanique de Benjamin Delessert.* 1845; S. Lenley et al., *Catalog of the Manuscript and Archival Collections and Index to the Correspondence of John Torrey.* Library of the New York Botanical Garden. 281, 464. 1973; E.M. Tucker, *Catalogue of the Library of the Arnold Arboretum of Harvard University.* 1917-1933; Ida Kaplan Langman, *A Selected Guide to the Literature on the Flowering Plants of Mexico.* 1964.

Martiusella Pierre Sapotaceae

Origins:

For Carl Friedrich Philipp von Martius, 1794-1868.

Martiusia Benth. Caesalpiniaceae

Origins:

For Carl Friedrich Philipp von Martius, 1794-1868.

Martynia L. Pedaliaceae

Origins:

After the London physician John Martyn, 1699-1768 (d. Chelsea, London), botanist, from 1732 to 1761 professor of botany at Cambridge (on the death of the first professor, Richard Bradley; in 1762 Martyn was succeeded in the Cambridge chair by his son Rev. Thomas Martyn, 1735-1825, d. Beds.), translator of Tournefort, 1727 Fellow of the Royal Society, friend of Sherard, 1721 founded the Botanical Society of London. Among his works are *Historia plantarum rariorum.* Londini 1728 ("Isaaci Rand sermo de hocce libro habitus coram societate anglica reperitur." in *Philos. Transact.* no. 407), *Proposals for a Course of Botany in the University of Cambridge.* 1726-1727 and *The First Lecture of a Course of Botany; being an introduction to the rest.* [8vo, first edition; the first of a series of lectures which M. had read both in London and Cambridge.] London 1729. See [Royal Society of London], The *Philosophical Transactions* (from the year 1719, to the year 1733). Abridged ... by Mr. John Eames ... and John Martyn. In two volumes. [4to.] London 1734, The *Philosophical Transactions* (from the year 1732, to the year 1744). Abridged ... by John Martyn. In two volumes. [4to.] London 1747, The *Philosophical Transactions* (from the year 1743, to the year 1750). Abridged ... by John Martyn. [Two vols., 4to.] London 1756; Académie Royale des Sciences, *The Philosophical History and Memoirs of the Royal Academy of Sciences at Paris ...* The whole translated and abridged, by John Martyn ... and Ephraim Chambers. [Five volumes 8vo, first edition.] London 1742; G.C. Gorham, *Memoirs of John Martyn ... and Thomas Martyn.* London 1830; Richard Bradley (*circa* 1688-1732), *A Philosophical Account of the Works of Nature.* [4to, first edition, Newton was a subscriber.] London 1721; Carl Linnaeus, *Species Plantarum.* 618. 1753 and *Genera Plantarum.* Ed. 5. 270. 1754; R. Zander, F. Encke, G. Buchheim and S. Seybold, *Handwörterbuch der Pflanzennamen.* 14. Aufl. 1993; J.H. Barnhart, *Biographical Notes upon Botanists.* 2: 456. 1965; Blanche Elizabeth Edith Henrey, *British Botanical and Horticultural Literature before 1800.* Oxford 1975; H.N. Clokie, *Account of the Herbaria of the Department of Botany in the University of Oxford.* 207-208. Oxford 1964; Warren R. Dawson, *The Banks Letters*, a Calendar of the Manuscript Correspondence of Sir Joseph Banks. London 1958; Jonas C. Dryander, *Catalogus bibliothecae historico-naturalis Josephi Banks.* 5: 347. 1796-1800; E.M. Tucker, *Catalogue of the Library of the Arnold Arboretum of Harvard University.* 1917-1933; R. Pulteney, *Historical and Biographical Sketches of the Progress of Botany in England.* 2: 205-218. London 1790; R.T. Gunther, *Early Science in Cambridge.* Oxford 1937; Elisabeth Leedham-Green, *A Concise History of the University of Cambridge.* Cambridge, University Press 1996; Frans A. Stafleu, *Linnaeus and the Linnaeans.* The spreading of their ideas in systematic botany, 1735-1789. Utrecht 1971; Georg Christian Wittstein, *Etymologisch-botanisches Handwörterbuch.* 559f. Ansbach 1852; Ray Desmond, *Dictionary of British & Irish Botanists and Horticulturists.* 472. London 1994; Dawson Turner, *Extracts from the Literary and Scientific Correspondence of R. Richardson, of Bierly, Yorkshire: Illustrative of the State and Progress of Botany* [Edited by D. Turner. Extracted from the memoir of the Richardson family, by Mrs. D. Richardson] Yarmouth 1835; M. Hadfield et al., *British Gardeners: A Biographical Dictionary.* London 1980; J.D. Milner, *Catalogue of Portraits of Botanists Exhibited in the Museums of the Royal Botanic Gardens.* Royal Botanic Gardens, Kew, London 1906.

Species/Vernacular Names:
M. annua L.

Yoruba: aranbole

Marumia Blume Melastomataceae

Origins:

For the Dutch (b. Delft) scientist Martin (Martinus) van Marum, 1750-1837 (Haarlem, Netherlands), physician, 1773 received his medical degree, owner of a garden (in which he cultivated aloes especially), plant physiologist, naturalist, discovered carbon monoxide (with van Troostwijk), Director of Teyler's Cabinet of Physical and Natural Curiosities and Library, traveler. His writings include *Verhandelingen uitgeven door Teyler's tweede Genootschap.* 1785-1787-1795, Dissertatio … *de motu Fluidorum in plantis,* experimentis et observationibus indagato, etc. Groningae 1773, Dissertatio … *qua disquiritur, quousque motus fluidorum,* etc. Groningae 1773, *Lettre … à M. Berthollet,* … contenant la description d'un Gazomètre. [Haarlem 1791], *Seconde lettre à M. Berthollet,* contenant la description d'un Gazomètre très simple. Harlem 1792 and *Catalogus der Bibliotheek van Teyler's Stichting.* [Compiled by M. van Marum.] Haarlem 1832, worked with Gerhard Kuyper and C.H. Pfaff, in 1785 met Lavoisier in Paris, corresponded with C.P. Thunberg, Banks and A. Volta. See Alessandro G.A.A. Volta (1745-1827), *La Correspondance de A. Volta et M. van Marum,* publiée par J. Bosscha. Leiden 1905; John H. Barnhart, *Biographical Notes upon Botanists.* 2: 456. 1965; Warren R. Dawson, *The Banks Letters,* a Calendar of the Manuscript Correspondence of Sir Joseph Banks. London 1958; Jonas C. Dryander, *Catalogus bibliothecae historico-naturalis Josephi Banks.* London 1796-1800; Alida M. Muntendam, in *Dictionary of Scientific Biography* 9: 151-153. 1981.

Mascagnia (DC.) Colla Malpighiaceae

Origins:

After the Italian (b. near Volterra) physician Paolo Mascagni, 1752-1815 (d. Florence), anatomist, professor of anatomy at Siena, author of the famous *Vasorum lymphaticorum corporis humani historia et ichnographia.* Senis [Siena] 1787; see Garrison and Morton, *Medical Bibliography.* 1104. 1961; Federico Allodi, in *Dictionary of Scientific Biography* 9: 153-154. 1981.

Mascarena L.H. Bailey Palmae

Origins:

Named for the Mascarene Islands, in the southwest Indian Ocean; Mascarenhas was a Portuguese navigator of 16th century; see L.H. Bailey, *Gentes Herbarum.* 6: 71. 1942.

Species/Vernacular Names:
M. lagenicaulis L.H. Bailey (*Hyophorbe lagenicaulis* (L.H. Bail.) H.E. Moore; *Hyophorbe amaricaulis* sensu Lem.)

English: bottle palm

Japan: tokkuri-yashi

M. verschaffeltii (H.A. Wendl.) L.H. Bailey (*Hyophorbe verschaffeltii* H.A. Wendl.)

English: spindle palm

Japan: tokkuri-yashi-modoki

Mascarenhasia A. DC. Apocynaceae

Origins:

Mascarenhas was a Portuguese navigator of the 16th century; the Mascarene Islands are an island group in the western Indian Ocean, including La Réunion, Rodrigues and Mauritius.

Maschalanthe Blume Rubiaceae

Origins:

From the Greek *maschale* "axil, hollow at base of a shoot, arm pit" and *anthos* "flower."

Maschalanthus Nutt. Euphorbiaceae

Origins:

Greek *maschale* "axil, hollow at base of a shoot, arm pit" and *anthos* "flower."

Maschalocephalus Gilg & K. Schumann Rapateaceae

Origins:

From the Greek *maschale* "axil, hollow at base of a shoot, arm pit" and *kephale* "head."

Maschalocorymbus Bremek. Rubiaceae

Origins:

From the Greek *maschale* "axil, branch, arm pit" and *korymbos* "corymb, a cluster."

Maschalodesme K. Schumann & Lauterb. Rubiaceae

Origins:

From the Greek *maschale* "axil, branch, arm pit" and *desmis, desmos, desme* "a bond, band, bundle."

Maschalosorus Bosch Hymenophyllaceae

Origins:

From the Greek *maschale* "axil, branch, hollow at base of a shoot" and *soros* "a vessel, urn, a coffin, a heap."

Masdevallia Ruíz & Pavón Orchidaceae

Origins:

Named after the Spanish botanist José Masdevall, died 1801, physician; see F. Boerner & G. Kunkel, *Taschenwörterbuch der botanischen Pflanzennamen*. 4. Aufl. 132. Berlin & Hamburg 1989; Georg Christian Wittstein, *Etymologisch-botanisches Handwörterbuch*. 560. Ansbach 1852.

Masoala H. Jumelle Palmae

Origins:

Based on Malagasy place name, Cap Masoala or Masoala peninsula, northeastern Madagascar; see H. Jumelle, in *Annales de l'Institut Botanico-Géologique Colonial de Marseille*. sér. 5, 1(1): 8. 1933.

Species/Vernacular Names:

M. kona Beentje

Madagascar: kona, kogne (Tanala)

M. madagascariensis Jum.

Madagascar: kase, hovotralanana, mandanozezika (Betsimisaraka)

Massartina Maire Boraginaceae

Origins:

Dedicated to the Belgian botanist Jean Massart, 1865-1925, from 1902 to 1905 Curator of the Jardin Botanique de l'État, Bruxelles, 1906-1925 Director of the Botanical Institute Léo Errera. His works include *Un botaniste en Malasie*. Gand 1895, *La cicatrisation chez les végétaux*. Bruxelles 1898, *Comment les Belges résistent à la domination allemande*. Contributions au livre des douleurs de la Belgique. Lausanne, Paris 1916, *La presse clandestine dans la Belgique occupée*. Paris et Nancy 1917, *Nos arbres*. Bruxelles 1911 and *Un voyage botanique au Sahara*. Gand [Gent] 1898. See J.H. Barnhart, *Biographical Notes upon Botanists*. 2: 458. 1965; E.M. Tucker, *Catalogue of the Library of the Arnold Arboretum of Harvard University*. 1917-1933.

Massonia Thunb. ex L.f. Hyacinthaceae (Liliaceae)

Origins:

Named after the British (b. Aberdeen) gardener Francis Masson, 1741-1805 (d. Canada), traveler, plant collector for Kew, from 1772 to the Cape, from 1776 Canaries and Azores, 1779 West Indies, 1783 North Africa, from 1786 to 1795 to the Cape and interior with Thunberg, 1796 Fellow of the Linnean Society, 1798 North America, sent specimens to A.B. Lambert and Joseph Banks, wrote *Stapeliae novae; or a collection of several new species of that genus; discovered in the interior parts of Africa*. London 1796 [-1797] and "An account of three journeys from the Cape Town into the southern parts of Africa." *Phil. Trans. R. Soc.* 66: 268-317. 1776; see Karl Koenig (1774-1851) and John Sims, *Annals of Botany*. II: 592. London 1805-1806; A. White and B.L. Sloane, *The Stapelieae*. Pasadena 1937; H.N. Clokie, *Account of the Herbaria of the Department of Botany in the University of Oxford*. 208. Oxford 1964; G. Murray, *History of the Collections Contained in the Natural History Departments of the British Museum*. 1: 167. 1904; J.H. Barnhart, *Biographical Notes upon Botanists*. 2: 458. 1965; Henry C. Andrews, *The Botanist's Repository*. London 1803; Mary Gunn and Leslie E. Codd, *Botanical Exploration of Southern Africa*. 246-249. Cape Town 1981; T.W. Bossert, *Biographical Dictionary of Botanists Represented in the Hunt Institute Portrait Collection*. 256. 1972; Jonas C. Dryander, *Catalogus bibliothecae historico-naturalis Josephi Banks*. 1796-1800; Mea Allan, *The Hookers of Kew*. 64. 1967; Leonard Huxley, *Life and*

Letters of Sir J.D. Hooker. London 1918; P. MacOwan, *Trans. S. Afr. Philos. Soc.* 4: xxix-liii. 1887; J. Hutchinson, *A Botanist in Southern Africa.* 613, 617. London 1946; Blanche Elizabeth Edith Henrey (1906-1983), *British Botanical and Horticultural Literature before 1800.* Oxford 1975; Frans A. Stafleu, *Linnaeus and the Linnaeans.* The spreading of their ideas in systematic botany, 1735-1789. Utrecht 1971; Ray Desmond, *Dictionary of British & Irish Botanists and Horticulturists.* 474. London 1994; Jeannette Elizabeth Graustein, *Thomas Nuttall, Naturalist. Explorations in America, 1808-1841.* Harvard University Press 1967; Kenneth Lemmon, *The Golden Age of Plant Hunters.* London 1968; M. Hadfield et al., *British Gardeners: A Biographical Dictionary.* London 1980.

Species/Vernacular Names:

M. depressa Houtt. (*Massonia latifolia* L.f.)

South Africa: flattened massonia, suikerkannetjie, botterkannetjie

M. pustulata Jacq.

South Africa: skurweblaar

Massularia (K. Schumann) Hoyle Rubiaceae

Origins:

Latin *massula, ae* "a little lump, a little mass."

Mastersia Bentham Fabaceae

Origins:

Named after the English botanist John White Masters (*circa* 1792-1873, d. Kent), gardener, plant collector in British India; the genus was also dedicated to the English botanist Maxwell Tylden Masters, 1833-1907, physician, 1865-1907 editor of *Gardener's Chronicle*, author of *Vegetable Teratology*. [Ray Society.] London 1869.

Mastersiella Gilg-Benedict Restionaceae

Origins:

For the English botanist Maxwell Tylden Masters, 1833-1907 (Middx.), physician, 1860 Fellow of the Linnean Society, M.D. 1862, 1870 Fellow of the Royal Society. His writings include *Descriptive Catalogue of the Teratological Series in the Museum of the ... College ... Vegetable Malformations.* [Royal College of the Surgeons.] London 1893, *Botany for Beginners.* London 1872, *Life on the Farm. Plant Life.* 1881, *Plant Life on the Farm.* New York 1885, "Synopsis of the South African Restiaceae." *J. Linn. Soc. (Bot.)* 10: 209-279. 1867, "New garden plants." *Gard.*

Chron. 1873: 947-948. 1873, "A general view of the genus Pinus." *J. Linn. Soc. (Bot.)* 35: 560-659. 1904 and "Passifloraceae." in Endlicher and Martius, *Flora Brasiliensis*, etc. vol. XIII, pars I. 1875, he was son of the nurseryman William Masters (1796-1874). See G. Murray, *History of the Collections Contained in the Natural History Departments of the British Museum.* 1: 167. 1904; I. Urban, ed., *Symbolae Antillanae.* 2: 4. Berlin 1902; T.W. Bossert, *Biographical Dictionary of Botanists Represented in the Hunt Institute Portrait Collection.* 256. 1972; R. Zander, F. Encke, G. Buchheim and S. Seybold, *Handwörterbuch der Pflanzennamen.* 14. Aufl. Stuttgart 1993; H.N. Clokie, *Account of the Herbaria of the Department of Botany in the University of Oxford.* 208. Oxford 1964; E.M. Tucker, *Catalogue of the Library of the Arnold Arboretum of Harvard University.* 1917-1933; Leonard Huxley, *Life and Letters of Sir J.D. Hooker.* 1: 383. 1918; Ernest Nelmes and William Cuthbertson, *Curtis's Botanical Magazine Dedications, 1827-1927.* 191-192. 1932; Ida Kaplan Langman, *A Selected Guide to the Literature on the Flowering Plants of Mexico.* 1964; H.R. Fletcher, *Story of the Royal Horticultural Society, 1804-1968.* Oxford 1969.

Mastichodendron (Engl.) H.J. Lam. Sapotaceae

Origins:

From the Greek *mastiche* "mastic" and *dendron* "tree."

Species/Vernacular Names:
M. sloaneanum Box & Philipson
English: Barbados mastic

Mastigosciadium Rech.f. & Kuber Umbelliferae

Origins:

From the Greek *mastix, mastigos* "a whip" and *skiadion, skiadeion* "umbel, parasol."

Mastigostyla I.M. Johnston Iridaceae

Origins:

From the Greek *mastix, mastigos* "a whip" and *stylos* "a style, pillar."

Mastixia Blume Cornaceae (Mastixiaceae)

Origins:

Latin *mastiche, mastice, mastix, masticis* and Greek *mastixe* for mastic, an odoriferous gum from the mastic-tree (Plinius);

some suggest from the Greek *mastix, mastigos* "a whip, scourge"; see Helmut Genaust, *Etymologisches Wörterbuch der botanischen Pflanzennamen*. 370. Basel 1996.

Mastixiodendron Melch. Rubiaceae

Origins:

Latin *mastiche, mastice, mastix, masticis* and Greek *mastixe* for mastic, an odoriferous gum from the mastic-tree; some suggest from the Greek *mastix, mastigos* "a whip, scourge" and *dendron* "a tree."

Mastosuke Raf. Moraceae

Origins:

Greek *sykon* "fig"; see C.S. Rafinesque, *Sylva Telluriana*. 59. 1838.

Matayba Aublet Sapindaceae

Origins:

A vernacular name in French Guiana.

Mathewsia Hook. & Arn. Brassicaceae

Origins:

For the British gardener Andrew Mathews, d. 1841 (Chachapoyas, Peru), 1830-1841 plant collector in Chile and Peru. See A. Lasègue, *Musée botanique de Benjamin Delessert*. 255-257. Paris 1845; H.N. Clokie, *Account of the Herbaria of the Department of Botany in the University of Oxford*. 209. Oxford 1964; Alice Margaret Coats, *The Quest for Plants. A History of the Horticultural Explorers*. 373. London 1969.

Species/Vernacular Names:

M. sp.

Peru: ckori-huackack

Mathiasella Constance & C. Hitchcock Umbelliferae

Origins:

For the American botanist Mildred Esther Mathias, 1906-1995, professor of botany, specialist in Umbelliferae, conservationist, educator, 1929-1930 Missouri Botanical Garden, 1932-1936 New York Botanical Garden, founder of the University of California Natural Reserve System. Her writings include "The genus *Hydrocotyle* in northern South America." *Brittonia*. 2(3): 201-237. 1936, "Studies in the Umbelliferae. V." *Brittonia*. 2(3): 239-245. 1936 and "Umbelliferae." in "Plants collected in Ecuador by W.H. Camp." *Memoirs of the New York Botanical Garden*. Vol. 9, n. 2. Aug. 1955, with Lincoln Constance (b. 1909) wrote "A revision of the Andean genus *Niphogeton* (Umbelliferae)." *Univ. Calif. Publ. Bot*. 23(9): 405-425. 1951, "A revision of the genus *Bowlesia* Ruíz & Pav. (Umbelliferae-Hydrocotyloideae) and its relatives." *Univ. Calif. Publ. Bot*. 38: 1-73. 1965 and "Umbelliferae." in Harling and Sparre, *Flora of Ecuador*. 1976, in cooperation with L. Constance and D. Araujo wrote "Umbeliferas." *Flora Ilustrada Catarinense*. 1 (fasc. UMBE): 1-205. 1972, with W.L. Theobald wrote "A revision of the genus *Hyperbaena* (Menispermaceae)." *Brittonia*. 33(1): 81-104. 1981. See J.H. Barnhart, *Biographical Notes upon Botanists*. 2: 460. 1965; T.W. Bossert, *Biographical Dictionary of Botanists Represented in the Hunt Institute Portrait Collection*. 257. 1972; Ida Kaplan Langman, *A Selected Guide to the Literature on the Flowering Plants of Mexico*. 1964; J. Ewan, ed., *A Short History of Botany in the United States*. 141. 1969; Constance & Hitchcock, *Amer. J. Bot*. 41: 56. 1954; R. Zander, F. Encke, G. Buchheim and S. Seybold, *Handwörterbuch der Pflanzennamen*. 14. Aufl. Stuttgart 1993; Joseph Ewan, *Rocky Mountain Naturalists*. The University of Denver Press 1950.

Mathieua Klotzsch Amaryllidaceae (Liliaceae)

Origins:

After the German gardener Louis Mathieu, 1793-1867; see in *Allg. Gartenzeitung*. 21: 337. 22 Oct. 1853.

Mathiola R. Br. Brassicaceae

Origins:

An orthographic variant of *Matthiola* R. Br.

Mathiolaria Chevall. Brassicaceae

Origins:

Referring to *Matthiola* R. Br.; see F. F. Chevallier, *Fl. Gen. Env. Paris*. 2: 910. 5 Jan. 1828 [1827].

Matisia Bonpl. Bombacaceae

Origins:

For Francisco J. Matís, painter, with the *Real Expedición botánica del Nuevo Reino de Granada*; see Soledad Acosta de Samper, *Biografías de hombres ilustres o notables, relativas a Colombia*. Bogotá 1883.

Species/Vernacular Names:

M. sp.

Peru: sapote

Matonia R. Br. ex Wall. Matoniaceae

Origins:

Dedicated to the British botanist William George Maton, 1774-1835, physician, M.D. 1801, 1794 Fellow of the Linnean Society, 1800 Fellow of the Royal Society, author of *Observations Relative Chiefly to the Natural History ... of the Western Counties of England*, made in the years 1794 and 1796. Salisbury 1797, with Thomas Rackett wrote *An Historical Account of Testaceological Writers ...* From the transactions of the Linnean Society. London [1804], edited R. Pulteney, *A General View of the Writings of Linnaeus*. 1805. See J.H. Barnhart, *Biographical Notes upon Botanists*. 2: 461. 1965; William Munk, *The Roll of the Royal College of Physicians of London*. London 1878; Wallich, *Pl. Asiat. Rar.* 1(1): 16. Sep 1829; Warren R. Dawson, *The Banks Letters*, a Calendar of the Manuscript Correspondence of Sir Joseph Banks. London 1958; Ethelyn Maria Tucker, *Catalogue of the Library of the Arnold Arboretum of Harvard University*. Cambridge, Massachusetts 1917-1933; Lady Pleasance Smith, ed., *Memoir and Correspondence of ... Sir J.E. Smith*. London 1832; Jonas C. Dryander, *Catalogus bibliothecae historico-naturalis Josephi Banks*. London 1796-1800; A. Lasègue, *Musée botanique de Benjamin Delessert*. 138. Paris 1845; R. Zander, F. Encke, G. Buchheim and S. Seybold, *Handwörterbuch der Pflanzennamen*. 14. Aufl. 748. Stuttgart 1993.

Matonia Stephenson & Churchill Zingiberaceae

Origins:

For William George Maton, 1774-1835; see *Med. Bot.* 3: t. 106. 1831.

Matricaria L. Asteraceae

Origins:

Latin *matrix, tricis* (*mater, tris* "mother") "the womb"; *matricaria*, in Pseudo Apuleius Barbarus, *Herbarium*. 66, l. 9., in reference to its former medical use against diseases of the uterus. See Robert William Theodore Gunther, *The Herbal of Apuleius Barbarus*. London 1925; Carl Linnaeus, *Species Plantarum*. 890. 1753 and *Genera Plantarum*. Ed. 5. 380. 1754.

Species/Vernacular Names:

M. matricarioides (Less.) Porter (*Artemisia matricarioides* Less.; *Matricaria discoidea* DC.; *Santolina suaveolens* Pursh; *Chamomilla suaveolens* (Pursh) Rydb.)

English: pineapple weed, rounded chamomile, rayless chamomile

M. nigellifolia DC. var. *nigellifolia* (*Sphaeroclinium nigellifolium* (DC.) Sch.Bip.)

English: bovine staggers plant, steggersweed

Southern Africa: kerwel, rivierals, stootsiektebossie, waterkerwel; umSolo, umSolo womlambo, umHlonyane womlambo (river worm wood), umHlonyane omncinane, uKudliwa ngumLambo (bite of the river) (Xhosa)

M. recutita L. (*Chamomilla recutita* (L.) Rauschert) (Latin *recutitus, a, um* "circumcised," it looks as if its skin was taken away)

English: sweet false chamomile, wild chamomile, German chamomile, dog's chamomile

Arabic: babounig, babounag, babounej, bibounej, babnouj

Mexico: guia gueza, quije queza

China: mu ju

Matsumurella Makino Labiatae

Origins:

After the Japanese botanist Jinzô Matsumura, 1856-1928, Director of the Koishikawa Botanical Garden, professor of botany at the University of Tokyo. His writings include *A Classified Etymological Dictionary of the Japanese Language*, ancient and modern. Tokyo 1916 and *An Etymological Vocabulary of the Yamato Language*. Tokyo 1921, with Tokutarô Itô (1868-1941) wrote "Tentamen florae Lutchuensis. Sectio prima. Plantae Dicotyledoneae polypetalae." *Journ. Coll. Sci. Univ. Tokyo.* 12: 263-541. 1899. See J.H. Barnhart, *Biographical Notes upon Botanists*. 2: 462. 1965; T.W. Bossert, *Biographical Dictionary of Botanists Represented in the Hunt Institute Portrait Collection*. 257. 1972; Ethelyn Maria Tucker, *Catalogue of the Library of the Arnold Arboretum of Harvard University*. Cambridge, Massachusetts 1917-1933; R. Zander, F. Encke, G. Buchheim and S. Seybold, *Handwörterbuch der Pflanzennamen*. 14. Aufl. Stuttgart 1993.

Matsumuria Hemsl. Gesneriaceae

Origins:

After the Japanese botanist Jinzô Matsumura, 1856-1928.

Matteuccia Todaro Dryopteridaceae (Aspleniaceae, Woodsiaceae)

Origins:

After the Italian (b. Forlì) physiologist Carlo Matteucci, 1811-1868, studied at the University of Bologna and at Sorbonne, correspondent of Faraday, A. von Humboldt and James Clerk Maxwell (1831-1879), contributor of *Il Cimento. Giornale di Fisica*, in 1840 professor of physics at the University of Pisa, in 1842 discovered the induced twitch, in 1862 Minister of education (of the Kingdom of Italy), reorganized the Scuola Normale Superiore in Pisa. Among his writings are *Raccolta di scritti varii intorno all'Istruzione Pubblica*. Prato 1867, *Traité des phénomènes électro-physiologiques des Animaux*. Paris and Leipzig 1844, *Cours d'électro-physiologie*. Paris 1858 and *Lezioni di Fisica*. Terza edizione. Pisa 1847. See G. Moruzzi, *L'opera elettrofisiologica di Carlo Matteucci*. Ferrara 1973; Clelia Pighetti, *Carlo Matteucci e il Risorgimento scientifico*. Ferrara 1976; Nicomede Bianchi, *C. Matteucci e l'Italia del suo tempo*. Narrazione corredata di documenti inediti. Torino 1874; Giuseppe Moruzzi, in *Dictionary of Scientific Biography* 9: 176-177. 1981.

Species/Vernacular Names:

M. struthiopteris (L.) Tod.

English: ostrich fern

Mattfeldanthus H. Robinson & R.M. King Asteraceae

Origins:

For the German botanist Johannes Mattfeld, 1895-1951, from 1919 at the Botanical Museum, Berlin-Dahlem. His works include "Compositae novae Austro-Americanae I." *Repert. Spec. Nov. Regni Veg.* 17(1-3): 178-185. 1921, "Zwei neue Orobanchen aus Peru." *Notizbl. Bot. Gart. Mus. Berlin-Dahlem.* 8: 182-186. 1922, "Compositae novae sinenses." *Notizbl. Bot. Gart. Mus. Berlin-Dahlem.* 11(102): 107. 1931, "*Plettkea*, eine neue Gattung der Alsinoideae aus den Hochanden Perus." Schriften des Vereins für Naturkunde an der Unterweser. Wesermünde [Bremerhaven] 7: 5-27. 1934 and "Einige neue *Drymaria*-Arten aus Peru." *Notizbl. Bot. Gart. Mus. Berlin-Dahlem.* 13: 436-444. 1936. See J.H. Barnhart, *Biographical Notes upon Botanists.* 2: 462. 1965; T.W. Bossert, *Biographical Dictionary of Botanists Represented in the Hunt Institute Portrait Collection.* 257. 1972; R. Zander, F. Encke, G. Buchheim and S. Seybold, *Handwörterbuch der Pflanzennamen.* 14. Aufl. Stuttgart 1993.

Mattfeldia Urban Asteraceae

Origins:

For the German botanist Johannes Mattfeld, 1895-1951.

Matthiola L. Rubiaceae

Origins:

For the Italian botanist Pietro Andrea Gregorio Mattioli (Petrus Andreas Matthiolus), *circa* 1500-1577; see Carl Linnaeus, *Species Plantarum.* 1192. 1753 and *Genera Plantarum.* Ed. 5. 499. 1754.

Matthiola R. Br. Brassicaceae

Origins:

For the Italian (b. Siena) botanist Pietro Andrea Gregorio Mattioli (Petrus Andreas Matthiolus), *circa* 1500/1501-1577 (d. Trento), naturalist, herbalist, 1523 received an M.D. from the University of Padua, (in Prague) physician to Maximilian II and to Ferdinand I of Austria. Among his works are *Opusculum de simplicium medicamentorum facultatibus secundum locos, & genera*. [12mo, first edition, an abridgement of Mattioli's *Commentarii* on Dioscorides.] Venetiis 1569, *Apologia adversus Amathum Lusitanum*, cum censura in ejusdem enarrationes. Venetiis 1558, *Epistolarum medicinalium* libri quinque. Pragae 1561 and *Il magno palazzo del cardinale di Trento*. Venetia 1539. See Bruno Zanobio, in *Dictionary of Scientific Biography* 9: 178-180. 1981; Georg Christian Wittstein, *Etymologisch-botanisches Handwörterbuch.* 563. 1852; J.H. Barnhart, *Biographical Notes upon Botanists.* 2: 463. 1965; Ernst H.F. Meyer, *Geschichte der Botanik.* IV: 366-378. Königsberg 1854-1857; Wilfrid Blunt and W.T. Stearn, *The Art of Botanical Illustration.* London 1950; F. Ambrosi, in *Archivio Trentino.* 1: 49-61. 1882; Carlo Raimondi, ed., "Lettere di P.A. Mattioli ad Ulisse Aldrovandi." *Bollettino Senese di Storia Patria.* Anno 13. Fasc. 1, 2. Siena 1906; Josephus Tectander, editor, *Morbi Gallici curandi ratio ... a variis ... medicis conscripta: nempe P.A. Matthaeolo*, etc. [Included is the first work describing syphilis of the newborn by P.A. Mattioli; it contains a long section on guaiacum by the physician to Maximilian I, Nicholas Pol (ca. 1470-1532); Niccolò (Nicola) Massa (1489-1569) describes the neurological manifestations of syphilis, he speaks of Jamaican sarsaparilla; essays by Benedetto Vettori (1481-1561), Angelo Bolognini (fl. 1506-1517) and Juan Almenar appear in the present work.] Basel 1536; Niccolò Massa, *Liber de morbo gallico.* [Small 4to, second edition, one tract of the book is devoted to *Guaiacum*, or *lignum indicum*.] Venice 1532; Girolamo Fracastoro (1478-1553),

Syphilis sive morbus gallicus. [First edition of the most famous medical poem in literature, composed for Cardinal Pietro Bembo.] Verona 1530; Richard J. Durling, comp., *A Catalogue of Sixteenth Century Printed Books in the National Library of Medicine.* 3295, 2991, 3030. 1967; Garrison and Morton, *Medical Bibliography.* 2365, 2366. 1961; Mariella Azzarello Di Misa, a cura di, *Il Fondo Antico della Biblioteca dell'Orto Botanico di Palermo.* 183. Palermo 1988; Jonas C. Dryander, *Catalogus bibliothecae historico-naturalis Josephi Banks.* London 1796-1800; Ida Kaplan Langman, *A Selected Guide to the Literature on the Flowering Plants of Mexico.* 1964; Antoine Lasègue, *Musée botanique de M. Benjamin Delessert.* Paris, Leipzig 1845; Ethelyn Maria Tucker, *Catalogue of the Library of the Arnold Arboretum of Harvard University.* Cambridge, Massachusetts 1917-1933; J.D. Milner, *Catalogue of Portraits of Botanists Exhibited in the Museums of the Royal Botanic Gardens.* Royal Botanic Gardens, Kew, London 1906.

Species/Vernacular Names:

M. incana (L.) R.Br. (*Cheiranthus incanus* L.)

English: common stock, gill flower, stock, Brompton stock

Peru: alelí, alhelí

M. longipetala (Vent.) DC.

English: night-scented stock

Mattuschkaea Schreb. Rubiaceae

Origins:

After Heinrich Gottfried von Mattuschka, 1734-1779, naturalist, author of *Flora silesiaca.* Leipzig 1776-1777. See John H. Barnhart, *Biographical Notes upon Botanists.* 2: 463. 1965; Jonas C. Dryander, *Catalogus bibliothecae historico-naturalis Josephi Banks.* 3: 164. London 1796-1800; R. Zander, F. Encke, G. Buchheim and S. Seybold, *Handwörterbuch der Pflanzennamen.* 14. Aufl. Stuttgart 1993.

Mattuschkea Batsch Rubiaceae

Origins:

After Heinrich Gottfried von Mattuschka, 1734-1779, naturalist, author of *Flora silesiaca.* Leipzig 1776-1777.

Mattuskea Raf. Rubiaceae

Origins:

After Heinrich Gottfried von Mattuschka, 1734-1779; see C.S. Rafinesque, *Princ. Somiol.* 30. 1814.

Matucana Britton & Rose Cactaceae

Origins:

Referring to the town of Matucana, Lima, Peru; see Rob Bregman, *The Genus Matucana. Biology and Systematics of Fascinating Peruvian Cacti.* Rotterdam 1996.

Matudacalamus Maekawa Gramineae

Origins:

For the Mexican botanist Eizi Matuda, 1894-1978, from Japan to Mexico 1922, 1950-1978 University of Mexico. Among his many writings are "El género *Datura* en México." *Bol. Soc. Bot. Méx.* 14: 1-13. 1952, *Las Commelinaceas del Estado de México.* Toluca 1956 and *Las Ciperaceas del Estado de México.* Toluca 1959. See John H. Barnhart, *Biographical Notes upon Botanists.* 2: 463. 1965; T.W. Bossert, *Biographical Dictionary of Botanists Represented in the Hunt Institute Portrait Collection.* 258. 1972; Ida Kaplan Langman, *A Selected Guide to the Literature on the Flowering Plants of Mexico.* 1964; R. Zander, F. Encke, G. Buchheim and S. Seybold, *Handwörterbuch der Pflanzennamen.* 14. Aufl. Stuttgart 1993; Irving William Knobloch, compil., "A preliminary verified list of plant collectors in Mexico." *Phytologia Memoirs.* VI. Plainfield, N.J. 1983.

Matudaea Lundell Hamamelidaceae

Origins:

For the Mexican botanist Eizi Matuda, 1894-1978.

Matudanthus D.R. Hunt Commelinaceae

Origins:

For the Mexican (but Japanese-born) botanist Eizi Matuda, 1894-1978.

Matudina R.M. King & H. Robinson Asteraceae

Origins:

For the Mexican botanist Eizi Matuda, 1894-1978.

Maturna Raf. Orchidaceae

Origins:

Manturna, a Latin goddess of matrimony (invoked to render the marriage lasting), from Latin *maneo, nsi, nsum* "to stay, remain, last, endure"; see C.S. Rafinesque, *Flora Telluriana.* 2: 99. 1836 [1837].

Maughania J. St.-Hil. Fabaceae

Origins:

Probably dedicated to the Scottish botanist Robert Maughan, 1769-1844, in 1809 a Fellow of the Linnean Society, father of the botanist Edward James (1790-1868); see J.H.H. Saint-Hilaire, in *Bulletin des Sciences, par la Société Philomatique de Paris.* 3: 216. Paris 1813; Arthur D. Chapman, ed., *Australian Plant Name Index.* 1943. Canberra 1991.

Maughania N.E. Br. Aizoaceae

Origins:

After Dr. H. Maughan Brown.

Maughaniella L. Bolus Aizoaceae

Origins:

After Dr. H. Maughan Brown.

Maundia F. Muell. Juncaginaceae

Origins:

After Dr. John Maund, a medical doctor of Melbourne, 1823-1858; see F. von Mueller, *Fragmenta Phytographiae Australiae.* 1: 22. 1858.

Maurandella (A. Gray) Rothm. Scrophulariaceae

Origins:

The diminutive of *Maurandya.*

Maurandya Ortega Scrophulariaceae

Origins:

Named for the Spanish botanist Catalina (Catarina, Catherina) Pancratia Maurandy, fl. late 18th century, professor of botany, wife of A.J. Maurandy (Director of the Cartagena Botanic Garden).

Species/Vernacular Names:

M. petrophila Cov. & C. Morton

English: rock lady

Mauria Kunth Anacardiaceae

Origins:

Dedicated to the Italian botanist Ernesto Mauri, 1791-1836, professor of botany at Rome, 1820-1833 Director of the Second Botanical Garden (Palazzo Salviati alla Lungara, Roma). See John H. Barnhart, *Biographical Notes upon Botanists.* 2: 465. 1965; T.W. Bossert, *Biographical Dictionary of Botanists Represented in the Hunt Institute Portrait Collection.* 258. 1972; Ethelyn Maria Tucker, *Catalogue of the Library of the Arnold Arboretum of Harvard University.* Cambridge, Massachusetts 1917-1933; R. Zander, F. Encke, G. Buchheim and S. Seybold, *Handwörterbuch der Pflanzennamen.* 14. Aufl. Stuttgart 1993; Antoine Lasègue, *Musée botanique de M. Benjamin Delessert.* Paris, Leipzig 1845.

Mauritia L.f. Palmae

Origins:

From a vernacular name, or dedicated to Maurice, Prinz van Oranje, Graaf van Nassau (Prince of Orange, Count of Nassau), 1567-1625, Stadholder of the United Provinces of the Netherlands, the son of William the Silent; see Linnaeus filius, *Supplementum Plantarum.* 70: 454. 1782; Helmut Genaust, *Etymologisches Wörterbuch der botanischen Pflanzennamen.* 371. Basel 1996; William T. Stearn, *Stearn's Dictionary of Plant Names for Gardeners.* 202. Cassell, London 1993.

Species/Vernacular Names:

M. carana Wallace

English: Rio Negro fan palm

Brazil: caraná, caraná do mato, buritirana, miritirana, palmeira leque do Rio Negro

M. flexuosa L.f.

English: ita palm

Peru: achu, achua, achual, aguachi, aguaje, aguashi, ahuaque, ahuashi, banin, binón, buritisol, cananguacha, cananguche, mariti, miriti, moriche, muriti, kinema, wachori, xonuuña

Brazil: burití, mirití, murití, buritizeiro, carandá guassú, moriti

Mauritiella Burret Palmae

Origins:

The diminutive of *Mauritia*; see Burret, *Notizblatt des Botanischen Gartens und Museums zu Berlin-Dahlem.* 12: 609. 1935.

Species/Vernacular Names:

M. aculeata (Kunth) Burret

Brazil: buritirana, caranaí, caraná, carandai, carandaizinho, cariná canaiá, ripa

M. armata (Mart.) Burret

Brazil: buritirana, caranã, caraná, buriti-mirim, buriti-bravo, buriti Bahia

Maurocenia Miller Celastraceae

Origins:
Presumably honoring the Venetian Senator G.F. Morosini, 1658-1739, a patron of botany.

Species/Vernacular Names:
M. frangularia (L.) Mill. (*Cassine maurocenia* L.; *Maurocenia capensis* Sond.) (the specific name means "like Frangula," from the Latin *frango* "to break," the wood of some species is quite brittle, or it is supposed to be)

English: Hottentot cherry

South Africa: Hottentotskerbos, aasvoëlbessie

Maxburretia Furtado Palmae

Origins:
After the German botanist Karl Ewald Maximilian (Max) Burret, 1883-1964, worked in the herbarium at Berlin-Dahlem. His writings include "Eine neue Palmengattung von den Molukken." *Notizbl. Bot. Gart. Mus. Berlin-Dahlem.* 10: 198-201. 1927, "Die Palmengattung *Manicara* Gaertn." *Notizbl. Bot. Gart. Mus. Berlin-Dahlem.* 10: 389-394. 1928, "Die Palmengattung *Morenia* R. et P." *Notizbl. Bot. Gart. Mus. Berlin-Dahlem.* 13: 332-339. 1936, "Palmae chinenses." *Notizbl. Bot. Gart. Mus. Berlin-Dahlem.* 13: 582-606. 1937, "Myrtaceen-Studien." *Notizbl. Bot. Gart. Mus. Berlin-Dahlem.* 15: 479-550. 1941 and "Myrtaceenstudien. II." *Repert. Spec. Nov. Regni Veg.* 50: 50-60. 1941; see O. Beccari et R.E.G. Pichi Sermolli, "Subfamiliae Arecoidearum Palmae Gerontogeae. Tribuum et Generum Conspectus." in *Webbia.* 11: 1-187. 31 Mar. 1956; John H. Barnhart, *Biographical Notes upon Botanists.* 1: 288. 1965; Ida Kaplan Langman, *A Selected Guide to the Literature on the Flowering Plants of Mexico.* 1964; Elmer Drew Merrill, *Bernice P. Bishop Mus. Bull.* 144: 57. 1937; Eva Potztal, in *Principes.* Volume 2, no. 3: 87-91. 1958.

Maxillaria Ruíz & Pavón Orchidaceae

Origins:
Latin *maxilla* "jaw-bone, the jaw," referring to the flowers, to the column and lip inside the flowers, indicating the mandible formed by the column foot and the sepals seen in profile.

Species/Vernacular Names:
M. bicolor Ruíz & Pav. (*Dendrobium bicolor* (Ruíz & Pav.) Persoon; *Dicrypta bicolor* Bateman ex Loudon)

Peru: cacca-cacca

Maximiliana C. Martius Palmae

Origins:
Dedicated by Martius to Maximilian Joseph I, 1756-1825, King of Bavaria and sponsor of his travels in Brazil; according to W.T. Stearn and Joseph Ewan the genus was named for the German Prince Maximilian Alexander Philipp zu Wied-Neuwied, 1782-1867, traveler and plant collector in Brazil. See Claudio Urbano B. Pinheiro and Michael J. Balick, "Brazilian Palms. Notes on their uses and vernacular names, compiled and translated from Pio Corrêa's "Dicionário das Plantas Úteis do Brasil e das Exóticas Cultivadas," with updated nomenclature and added illustrations." in *Contributions from the New York Botanical Garden.* Volume 17. 1987; Joseph Ewan, *Rocky Mountain Naturalists.* 261. The University of Denver Press 1950; William T. Stearn, *Stearn's Dictionary of Plant Names for Gardeners.* 203. 1993; Helmut Genaust, *Etymologisches Wörterbuch der botanischen Pflanzennamen.* 372. Basel 1996; Stafleu and Cowan, *Taxonomic Literature.* 3: 381. 1981.

Species/Vernacular Names:
M. spp.

Peru: inajá, inayuga, maripá

M. maripa (Corrêa) Drude (*Maximiliana stenocarpa* Burret; *Maximiliana venatorum* (Poeppig ex C. Martius) H.A. Wendland ex Kerchove; *Palma maripa* Aublet)

English: inajá-palm, cucurit

Peru: anajá, anaju, anaú, kokerit, inayá, ina-yacu, ina-yuca, inayacu, inayuca, inija, inijá, juajá, juajuá, juayacu, ynayuca

northern South America: maripa, inajá, curcurita, kokerite

Guyana: kokerit palm

Venezuela: cucurito

Brazil: inajá, maripá, anajá, anajax, aritá, aritairé, coqueiro anajá, inajá da Guiana

Sri Lanka: coquirita palm

Maximowiczia Cogniaux Cucurbitaceae

Origins:
Dedicated to the Russian botanist Carl (Karl) Johann (Ivanovic, Ivanovich) Maximowicz (Maksimovich), 1827-

1891, explorer, traveler and plant collector, 1853-1857 Eastern Asia, 1859-1864 China and Japan. Among his valuable writings are "De *Coriaria, Ilice* et *Monochasmate* hujusque generibus proxime affinibus *Bungea* et *Cymbaria*." *Mém. Acad. Imp. Sci. Saint Pétersbourg.* 29(3): 1-70. 1881 and *Beiträge zur Kenntniss des Russischen Reiches.* Dritte Folge. Herausgegeben von L. von Schrenck und C.J. Maximovicz. 1839, etc. See John H. Barnhart, *Biographical Notes upon Botanists.* 2: 466. 1965; T.W. Bossert, *Biographical Dictionary of Botanists Represented in the Hunt Institute Portrait Collection.* 258. 1972; S. Lenley et al., *Catalog of the Manuscript and Archival Collections and Index to the Correspondence of John Torrey.* Library of the New York Botanical Garden. 464. 1973; Stafleu and Cowan, *Taxonomic Literature.* 3: 382-385. 1981; Ida Kaplan Langman, *A Selected Guide to the Literature on the Flowering Plants of Mexico.* University of Pennsylvania Press, Philadelphia 1964; R. Zander, F. Encke, G. Buchheim and S. Seybold, *Handwörterbuch der Pflanzennamen.* 14. Aufl. Stuttgart 1993; I.C. Hedge and J.M. Lamond, *Index of Collectors in the Edinburgh Herbarium.* Edinburgh 1970; E. Bretschneider, *History of European Botanical Discoveries in China.* 1981.

Maxonia C. Christensen Dryopteridaceae (Aspleniaceae)

Origins:

After the American botanist William Ralph Maxon, 1877-1948, pteridologist. His writings include *A Study of Certain Mexican and Guatemalan Species of Polypodium.* 1903, *Studies of Tropical American Ferns.* 1908, "Studies of tropical American ferns. No. 4." *Contr. U.S. Natl. Herb.* 17: 133-179. 1913, "Report upon a collection of ferns from western South America." *Smithsonian Misc. Collect.* 65(8): 1-12. 1915, "Studies of tropical American ferns. No. 6." *Contr. U.S. Natl. Herb.* 17: 541-608. 1916, "The lip-ferns of the southwestern United States related to *Cheilanthes myriophylla*." *Proc. Biol. Soc. Wash.* 31: 139-152. 1918, *New Selaginellas from the Western United States.* City of Washington 1920 and *Pteridophyta of Porto Rico and the Virgin Islands.* [New York] 1926, with C.A. Weatherby wrote "Some species of *Notholaena*, new and old. I. The group of *Notholaena nivea*." *Contr. Gray Herb.* 127: 3-15. 1939. See J.H. Barnhart, *Biographical Notes upon Botanists.* 2: 466. 1965; Ida Kaplan Langman, *A Selected Guide to the Literature on the Flowering Plants of Mexico.* University of Pennsylvania Press, Philadelphia 1964; T.W. Bossert, *Biographical Dictionary of Botanists Represented in the Hunt Institute Portrait Collection.* 258. 1972; S. Lenley et al., *Catalog of the Manuscript and Archival Collections and Index to the Correspondence of John Torrey.* Library of the New York Botanical Garden. 283-284. 1973; E.M.

Tucker, *Catalogue of the Library of the Arnold Arboretum of Harvard University.* 1917-1933; J. Ewan, ed., *A Short History of Botany in the United States.* 100. New York and London 1969; Thomas Henry Kearney (1874-1956), *Leafl. Western Bot.* 8: 277-278. 1958; R. Zander, F. Encke, G. Buchheim and S. Seybold, *Handwörterbuch der Pflanzennamen.* 14. Aufl. Stuttgart 1993.

Species/Vernacular Names:

M. apiifolia (Swartz) C. Christensen (*Dicksonia apiifolia* Swartz)

English: climbing wood fern

Maxwellia Baillon Sterculiaceae

Origins:

For the English botanist Maxwell Tylden Masters, 1833-1907 (Middx.), physician, 1860 Fellow of the Linnean Society, M.D. 1862, 1870 Fellow of the Royal Society. Among his writings are *Descriptive Catalogue of the Teratological Series in the Museum of the ... College ... Vegetable Malformations.* [Royal College of the Surgeons.] London 1893, *Botany for Beginners.* London 1872, *Life on the Farm. Plant Life.* 1881, *Plant Life on the Farm.* New York 1885, "Synopsis of the South African Restiaceae." *J. Linn. Soc. (Bot.)* 10: 209-279. 1867, "New garden plants." *Gard. Chron.* 1873: 947-948. 1873, "A general view of the genus *Pinus*." *J. Linn. Soc. (Bot.)* 35: 560-659. 1904 and "Passifloraceae." in Endlicher and Martius, *Flora Brasiliensis,* etc. vol. XIII, pars I. 1875, he was son of the nurseryman William Masters (1796-1874). See G. Murray, *History of the Collections Contained in the Natural History Departments of the British Museum.* 1: 167. 1904; I. Urban, ed., *Symbolae Antillanae.* 2: 4. Berlin 1902; T.W. Bossert, *Biographical Dictionary of Botanists Represented in the Hunt Institute Portrait Collection.* 256. 1972; H.N. Clokie, *Account of the Herbaria of the Department of Botany in the University of Oxford.* 208. Oxford 1964; Ethelyn Maria Tucker, *Catalogue of the Library of the Arnold Arboretum of Harvard University.* Cambridge, Massachusetts 1917-1933; Leonard Huxley, *Life and Letters of Sir J.D. Hooker.* 1: 383. 1918; Ernest Nelmes and William Cuthbertson, *Curtis's Botanical Magazine Dedications, 1827-1927.* 191-192. 1932.

Mayaca Aublet Mayacaceae

Origins:

A vernacular name in French Guiana, or named after the Mayas.

Mays Mill. Gramineae

Origins:
From the Mexican vernacular name for maize.

Maytenus Molina Celastraceae

Origins:
Maiten, mayten or *mayton*, a Chilean (Araucan) name for the type species *Maytenus boaria* Mol., Mapuche *mantun*; see Giovanni Ignazio Molina, *Saggio sulla storia naturale del Chili*. 177, 349. Bologna 1782 [impr. 1781].

Species/Vernacular Names:
M. sp.

Peru: apiranga, mayten

Mexico: agua bola, mangle, mangle agua bola, palo blanco

South America: cangorosa

M. abbottii Van Wyk

English: rock silky bark, Abbott's silky bark

South Africa: klipsybas

M. acuminata (L.f.) Loes var. *acuminata* (*Celastrus acuminatus* L.f.; *Gymnosporia acuminata* (L.f.) Szyszyl.)

English: silky bark, silkbark

Southern Africa: sybas; isiNama, umNama, umLulama, iNama elimhlophe (Zulu); umNama, umZungulwa (Xhosa); umNama (Swazi); tshikane (South Sotho)

M. apurimacensis Loesener

Peru: paltai-paltai, azar-azar

M. bachmannii (Loes.) Marais

English: willow koko tree

South Africa: wilgerkokoboom

M. bilocularis (F. Muell.) Loes.

English: orange bark

M. boaria Molina

Argentina: cancorosa, lena dura, maiten grande, mayten, naranjito, sal de inchias, yuki-ra

Chile: maiten, maiten grande, mayten

M. cordata (E. Mey. ex Sond.) Marais (*Celastrus cordatus* E. Mey. ex Sond.; *Gymnosporia cordata* (E. Mey. ex Sond.) Sim)

English: water silky bark

South Africa: watersybas

M. cunninghamii (Hook.) Loes.

English: yellow berry bush

M. cuzcoina Loesener

Peru: paltay-paltay, palltay-palltay

M. disperma (F. Muell.) Loes.

English: orange boxwood, orange bark

M. heterophylla (Ecklon & Zeyher) N.B.K. Robson (*Celastrus buxifolius* L.; *Celastrus heterophyllus* Eckl. & Zeyh.; *Celastrus lanceolatus* E. Mey. ex Sond.; *Celastrus linearis* L.f.; *Catha heterophylla* (Eckl. & Zeyh.) Presl; *Gymnosporia woodii* Szyszyl.; *Gymnosporia heterophylla* (Eckl. & Zeyh.) Loes.; *Gymnosporia condensata* Sprague; *Gymnosporia angularis* (Sond.) Sim; *Gymnosporia elliptica* (Thunb.) Schonl.; *Gymnosporia uniflora* Davison; *Gymnosporia buxifolia* (L.) Szyszyl.; *Gymnosporia crataegiflora* Davison; *Gymnosporia lanceolata* (E. Mey. ex Sond.) Loes.; *Gymnosporia linearis* (L.f.) Loes.; *Maytenus cymosa* (Soland.) Exell)

English: common spike-thorn, quickthorn, spikethorn

Southern Africa: gewone pendoring (= quill thorn), pendoring, lemoendoring, gifdoring; muKwokwoba, muTotova (Shona); isiHlangu (Swazi); shihlangwa (Thonga or Tsonga); uSala, uSolo, iNgqwangane, inGqwangane, inGqwangane yehlanze, isiBhubhu, isiHlangu (Zulu); mopasu (North Sotho); sefea-maeba (Sotho); umQaqoba (Xhosa); sefeamaeba se senyenyane (South Sotho); motlhono, mothono (Tswana: western Transvaal, northern Cape, Botswana); tshipandwa (Venda); murowanyero (Mbukushu: Okavango Swamps and western Caprivi)

M. ilicifolia Reissek ex Mart.

Argentina: sombra de toro

M. krukovii A.C. Smith

Peru: chuchu huasca, chuchu huasha, chuchuhuasca, chuchuhuasha

M. linearis (L.f.) Marais (*Celastrus linearis* L.f.; *Celastrus lanceolatus* E. Mey. ex Sond.; *Gymnosporia linearis* (L.f.) Loes.; *Gymnosporia lanceolata* (E. Mey. ex Sond.) Loes.)

South Africa: western pendoring

M. magellanica (Lam.) Hook.f.

Chile: maiten de magellanes

M. mossambicensis (Klotzsch) Blakelock var. *mossambicensis* (*Gymnosporia harveyana* Loes.; *Gymnosporia mossambicensis* (Klotzsch) Loes.; *Celastrus mossambicensis* Klotzsch; *Celastrus concinnus* N.E. Br.)

English: Mozambique maytenus, black forest spike-thorn, red forest spike-thorn, longspined maytenus

Southern Africa: swartbospendoring, rooibospendoring; iNgqwangane-yahlathi (Zulu); inGqwangane, umQaqoba (Xhosa); tshitongopfa (Venda)

M. nemorosa (Eckl. & Zeyh.) Marais (*Celastrus nemorosus* Eckl. & Zeyh.; *Gymnosporia nemorosa* (Eckl. & Zeyh.) Szyszyl.) (the specific name from the Latin *nemorosus, a, um* (*nemus, moris* "a tree, wood, grove") "woody, wooded, inhabiting woods, growing in groves")

English: white forest spikethorn, forest maytenus

Southern Africa: witbospendoring; iNgqwangane, inGqwangane (Zulu); umHlangwe (Xhosa)

M. octogona (L'Hér.) DC.

Peru: realengo, realingo

M. oleoides (Lam.) Loes. (*Celastrus laurinus* Thunb.; *Gymnosporia laurina* (Thunb.) Szyszyl.; *Scytophyllum angustifolium* Sond.; *Scytophyllum laurinum* (Thunb.) Eckl. & Zeyh.)

English: mountain maytenus, rock candlewood

South Africa: klipkersbos, klipkershout

M. peduncularis (Sond.) Loes. (*Celastrus peduncularis* Sond.; *Gymnosporia peduncularis* (Sond.) Loes.)

English: Cape blackwood, blackwood

Southern Africa: Kaapse swarthout, swarthout; umNqai, umNqayi omyama, iNqayi, iNqayi elimnyama (Zulu); umNqai, umNgqi (Xhosa); makhulu-wa-mukwatule (Venda)

M. phyllanthoides Benth.

Mexico: mangle, palo blanco; sak-ché (Maya l., Yucatan); mangle dulce (Baja California); mangle aguabola, agua bola (Sinaloa); granadilla (Guadalcazar, San Luis Potosí)

M. polyacantha (Sond.) Marais (*Celastrus polyacanthus* Sond.; *Gymnosporia polyacantha* (Sond.) Szyszyl.; *Gymnosporia vaccinifolia* Conrath)

South Africa: pendoring

M. procumbens (L.f.) Loes. (*Celastrus procumbens* L.f.; *Gymnosporia procumbens* (L.f.) Loes.)

English: dune koko tree, coast maytenus

Southern Africa: duinekokoboom; umPhophonono (Zulu); umPhono-phono (Xhosa)

M. senegalensis (Lam.) Exell (*Gymnosporia senegalensis* (Lam.) Loes.; *Gymnosporia crenulata* Engl.; *Gymnosporia dinteri* Loes.; *Celastrus senegalensis* Lam.)

English: red spike-thorn, confetti tree

Mali: ngege, nyenyele

Southern Africa: rooipendoring, bloupendoring, lemoendoring, pendoring, isihlangu; uBuhlangwe, isiHlangu, isiHlangwane (Zulu); isiHlangu (Swazi); mukutema tembuze (Subya: Botswana, eastern Caprivi); shihlangwa (Thonga); muGaranjewa, muGaranjua, chiVunabadza, chiZeza, muZhuzhu, chiZuzu (Shona); muthone, mothono, motlhonó (Tswana: western Transvaal, northern Cape, Botswana); tshibavhe, tshipandwa, tshiphandwa (Venda); gaú (Bushman)

Yoruba: isepolohun

Nigeria: kunkushewa, namijin tsada (Hausa); tultulde (Fula); momfofoji (Nupe); sepolohun (Yoruba)

M. umbellata (R. Br.) Mabb.

English: Madeira shrubby bittersweet

M. undata (Thunb.) Blakelock (*Celastrus undatus* Thunb.; *Celastrus zeyheri* Sond.; *Celastrus ilicinus* Burch.; *Celastrus albatus* N.E. Br.; *Catha fasciculata* Tul.; *Gymnosporia albata* (N. E. Br.) Sim; *Gymnosporia deflexa* Sprague; *Gymnosporia fasciculata* (Tul.) Loes.; *Gymnosporia peglerae* Davison; *Gymnosporia ilicina* Loes.; *Gymnosporia undata* (Thunb.) Szyszyl.; *Gymnosporia zeyheri* (Sond.) Szyszyl.; *Gymnosporia rehmannii* Szyszyl.; *Maytenus zeyheri* (Sond.) Loes.)

English: koko tree, South African holly

Southern Africa: kokoboom; iDohame, iGqwabali, iKhukhuze, iNqayielibomvu (Zulu); iNqayi-elibomvu (Zulu, Xhosa); umNqayi-mpofu, umGora (Xhosa)

M. verticillata (Ruíz & Pav.) DC.

Peru: duraznillo, picma, pigma, pilpus, rurama

Mayzea Raf. Gramineae

Origins:
For *Zea* L.; see C.S. Rafinesque, *Med. Fl.* 2: 241. 1830, *Flora Telluriana.* 1: 85, 86. 1836 [1837] and *Sylva Telluriana.* 1: 17. 1838; E.D. Merrill, *Index rafinesquianus.* 76. 1949.

Mazaea Krug & Urban Rubiaceae

Origins:
Dedicated to the Cuban botanist Manuel Goméz de la Maza y Jimenez, 1867-1916, Director of the Botanical Garden in Havana, author of *Catálogo de las periantiadas cubanas, espontaneas y cultivadas.* Sociedad Española de Historia Natural. Anales, etc. tom. 19. Madrid 1890.

Mazus Lour. Scrophulariaceae

Origins:
Greek *mazos* "a teat, papilla," referring to the tubercles or clavate hairs closing and blocking the corolla throat; see J. de Loureiro, *Flora cochinchinensis.* 385. Lisbon 1790.

Species/Vernacular Names:
M. pumilio R. Br. (*Lobelia pumila* Burm.f.)

Australia: swamp mazus

Japan: tokiwa-haze

Mazzettia Iljin Asteraceae

Origins:
For the Austrian botanist Heinrich Freiherr von Handel-Mazzetti, 1882-1940, explorer, moss collector, plant hunter,

from 1914 to 1919 in China, pupil of Richard Wettstein (1863-1931). His writings include *Monographie der Gattung Taraxacum*. Leipzig und Wien 1907, *Naturbilder aus Südwest China*. 1927 and *Symbolae sinicae*. Botanische Ergebnisse der Expedition der Akademie der Wissenschaften in Wien nach Südwest-China. 1914/1918. Wien 1929-1937, he was a contemporary of the great plant-hunters Forrest and Kingdon Ward. See John H. Barnhart, *Biographical Notes upon Botanists*. 2: 121. Boston 1965; Ida Kaplan Langman, *A Selected Guide to the Literature on the Flowering Plants of Mexico*. Philadelphia 1964; T.W. Bossert, *Biographical Dictionary of Botanists Represented in the Hunt Institute Portrait Collection*. 162. 1972; S. Lenley et al., *Catalog of the Manuscript and Archival Collections and Index to the Correspondence of John Torrey*. Library of the New York Botanical Garden. 210. 1973; E.M. Tucker, *Catalogue of the Library of the Arnold Arboretum of Harvard University*. 1917-1933; H.J. Noltje, in *The New Plantsman*. 5(1): 63-64. March 1998; [Anon.], in *Hortus*. no. 44: 121. [referring about *A Botanical Pioneer in South West China*, by Heinrich Handel-Mazzetti, translator and publisher David Winstanley, 1996.] Winter 1997; Stafleu and Cowan, *Taxonomic Literature*. 2: 43-44. 1979.

Mcvaughia W.R. Anderson Malpighiaceae

Origins:
For the American botanist Rogers McVaugh, b. 1909, professor of botany, traveler, plant collector and botanical historian. Among his writings are "A revision of *Laurentia* and allied genera in North America." *Bull. Torrey Bot. Club*. 67(9): 778-798. 1940, "A monograph of the genus *Downingia*." *Mem. Torrey Bot. Club*. 19(4): 1-57. 1941, "The genus *Triodanis* Rafinesque, and its relationships to *Specularia* and *Campanula*." *Wrightia*. 1(1): 13-52. 1945, "Generic status of *Triodanis* and *Specularia*." *Rhodora*. 50(590): 38-49. 1948, "Studies in the South American Lobelioideae (Campanulaceae) with special reference to Colombian species." *Brittonia*. 6(4): 450-493. 1949, "A revision of the North American black cherries (*Prunus serotina* Ehrh., and relatives)." *Brittonia*. 7(4): 279-315. 1951, "Tropical American Myrtaceae: notes on generic concepts and descriptions of previously unrecognized species." *Fieldiana, Bot*. 29(3): 145-228. 1956, "Tropical American Myrtaceae, II. Notes on generic concepts and descriptions of previously unrecognized species." *Fieldiana, Bot*. 29(8): 393-532. 1963, "The genera of American Myrtaceae — An interim report." *Taxon*. 17(4): 354-418. 1968, "Gramineae." *Fl. Novo-Galiciana*. 14: 1-436. 1983 and "Leguminosae." *Fl. Novo-Galiciana*. 5: 1-786. 1987. See John H. Barnhart, *Biographical Notes upon Botanists*. 2: 433. Boston 1965; T.W. Bossert, *Biographical Dictionary of Botanists Represented in the Hunt Institute Portrait Collection*. 261. 1972; S. Lenley et al., *Catalog of the Manuscript and Archival*

Collections and Index to the Correspondence of John Torrey. Library of the New York Botanical Garden. 278. 1973; Irving William Knobloch, compil., "A preliminary verified list of plant collectors in Mexico." *Phytologia Memoirs*. VI. 1983.

Mecardonia Ruíz & Pav. Scrophulariaceae

Origins:
After a Spanish Antonio de Meca y Cardona, a patron of botany.

Mecomischus Cosson ex Bentham Asteraceae

Origins:
Perhaps from the Greek *mekos* "length" and *mischos* "stalk."

Meconella Nutt. ex Torrey & A. Gray Papaveraceae

Origins:
Latin *mecon* and Greek *mekon* "poppy" plus diminutive.

Meconopsis R. Viguier Papaveraceae

Origins:
Greek *mekon* "poppy" and *opsis* "like, likeness, appearance," referring to the plant.

Species/Vernacular Names:
M. betonicifolia Franch. (*Cathcartia betonicifolia* (Franch.) Prain; *Meconopsis baileyi* Prain)

English: Tibetan blue poppy, Himalayan blue poppy

M. cambrica (L.) Vig. (*Papaver cambricum* L.; *Argemone cambrica* (L.) Desportes; *Cerastites cambrica* (L.) Gray; *Stylophorum cambricum* (L.) Spreng.)

English: Welsh poppy

M. grandis Prain

English: blue poppy

M. integrifolia (Maxim.) Franchet (*Cathcartia integrifolia* Maxim.; *Meconopsis integrifolia* var. *souliei* Fedde)

English: yellow poppywort

M. quintuplinervia Regel

English: harebell poppy

M. simplicifolia (D. Don) Walp. (*Papaver simplicifolium*
D. Don; *Meconopsis simplicifolia* var. *baileyi* Kingdon-
Ward; *Stylophorum simplicifolium* (D. Don) Spreng.)

English: blue poppy

Mecopus Benn. Fabaceae

Origins:
Greek *mekos* "length" and *pous, podos* "foot," referring to
the stem.

Mecosa Blume Orchidaceae

Origins:
Greek *mekos* "length," the long and linear lip.

Mecosorus Klotzsch Polypodiaceae

Origins:
Greek *mekos* "length" and *soros* "a vessel, heap."

Mecostylis Kurz ex Teijsm. & Binn. Euphorbiaceae

Origins:
From the Greek *mekos* "length" and *stylos* "a style, pillar,
a column."

Medea Klotzsch Euphorbiaceae

Origins:
From Medea, the sorceress, helped Jason get the golden
fleece.

Medeola L. Convallariaceae (Liliaceae, Trilliaceae, Medeolaceae)

Origins:
From Medea, the sorceress; Latin *medela, medella* for a
healing, a remedy, means of redress.

Species/Vernacular Names:
M. virginiana L.

English: cucumber root

Mediasia Pimenov Umbelliferae

Origins:
A genus endemic to southwest and central Asia.

Medica Cothen. Bignoniaceae

Origins:
Latin *medica, ae* for a kind of clover introduced from
Media, *Medicago sativa* L. (Plinius, Vergilius and Marcus
Terentius Varro).

Medicago L. Fabaceae

Origins:
The Greek *medike* (Media) "a grass, herba medica, lucerne,
a kind of clover" (Dioscorides); Latin *medica, ae* for a kind
of clover introduced from Media, *Medicago sativa* L. (Plinius, Vergilius and Marcus Terentius Varro); see Carl Linnaeus (Carl von Linnaeus, Carl von Linné) (1707-1778),
Species Plantarum. 778. 1753 and *Genera Plantarum.* Ed.
5. 339. 1754.

Species/Vernacular Names:
M. arabica (L.) Hudson (*Medicago polymorpha* L. var.
arabica L.)

English: spotted burclover, spotted medic, burclover

M. arborea L.

English: tree alfalfa, tree medic, moon trefoil, noon trefoil,
tree lucerne

M. falcata (L.) Arcang.

English: sickle medick

M. intertexta (L.) Miller (*Medicago echinus* DC.; *Medicago polymorpha* L. var. *intertexta* L.)

English: calvary clover, hedgehog medic

M. laciniata (L.) Miller (*Medicago aschersoniana* Urban;
Medicago laciniata (L.) Miller var. *brachyacantha* Boiss.;
Medicago polymorpha L. var. *laciniata* L.)

English: burclover, little burweed, veld shamrock, cut-leaf
medic

Southern Africa: karoolits, klawergras, klitsklawer;
bohomenyana (Sotho)

M. lupulina L. (*Medicago cupaniana* Guss.; *Medicago
lupulina* L. var. *cupaniana* (Guss.) Boiss.)

English: hop clover, black medic, black medick, nonesuch,
nonsuch, yellow trefoil, hop medic

Peru: trébol

Japan: kome-tsubu-uma-goyashi

China: lao wo sheng, niu yun tsao, huang hua

M. minima (L.) Bartal. (*Medicago minima* (L.) Bartal. var. *brachyodon* Reichb.; *Medicago minima* (L.) Bartal. var. *brevispina* Benth.; *Medicago polymorpha* L. var. *minima* L.; *Medicago sessilis* Peyron ex Post)

English: little burclover, small burclover, burclover

M. orbicularis (L.) Bartal. (*Medicago cuneata* J. Woods; *Medicago marginata* Willd.; *Medicago orbicularis* (L.) Bartal. var. *marginata* (Willd.) Benth.; *Medicago orbicularis* (L.) Bartal. var. *microcarpa* Rouy, nom. illeg.; *Medicago polymorpha* L. var. *orbicularis* L.)

English: buttonclover, buttonclover medic, button medic

M. polymorpha L. (*Medicago hispida* Gaertner; *Medicago denticulata* Willd.)

English: burclover, California burclover, toothed burclover, toothed medic, bur medic, rough medic, trefoil

Arabic: nafal

South Africa: growwe medicago, klawergras, klitsklawer, stekelklawer, wildeklawer

Japan: uma-goyashi

China: mu xu

M. rigidula (L.) All. (*Medicago agrestis* Ten.; *Medicago cinerascens* Jordan; *Medicago gerardii* Waldst. & Kit. ex Willd.; *Medicago morisiana* Jordan; *Medicago muricata* (L.) All.; *Medicago polymorpha* L. var. *rigidula* L.; *Medicago polymorpha* L. var. *muricata* L.; *Medicago rigidula* (L.) All. var. *agrestis* (Ten.) Burnat; *Medicago rigidula* (L.) All. var. *cinerascens* (Jordan) Rouy; *Medicago rigidula* (L.) All. var. *morisiana* (Jordan) Rouy)

English: Tifton burclover, Tifton medic

M. sativa L. (*Medicago caerulea* Less. ex Ledeb. var. *pauciflora* (Ledeb.) Grossh.; *Medicago karatschaica* Latsch.; *Medicago lavrenkoi* Vassilcz.; *Medicago pauciflora* Ledeb.; *Medicago sativa* L. var. *pilifera* Urban)

English: alfalfa, alfalfa U.S., lucerne

Arabic: qat, jat

Peru: alfar, alfalfa, omas

Bolivia: alp'alp'a

Mexico: goba bichinaxi xtilla

South Africa: klawer, lusering, lusern, makklawer

Japan: murasaki-uma-goyashi

China: mu xu, mu su

Hawaii: 'alapapa

M. sativa L. subsp. *falcata* (L.) Arcang. (*Medicago borealis* Grossh.; *Medicago falcata* L.; *Medicago quasifalcata* Sinsk.; *Medicago romanica* Prodan; *Medicago tenderiensis* Opperman ex Klokov)

English: sickle alfalfa, yellow-flowered alfalfa, yellow lucerne, sickle medic, sickle medick

M. sativa L. subsp. *sativa* (L.) Arcang. (*Medicago agropyretorum* Vassilcz.; *Medicago asiatica* Sinsk.; *Medicago mesopotamica* Vassilcz.; *Medicago praesativa* Sinsk.; *Medicago rivularis* Vassilcz.; *Medicago sogdiana* Vassilcz.; *Medicago transoxana* Vassilcz.)

English: alfalfa, lucerne

M. scutellata (L.) Miller (*Medicago polymorpha* L. var. *scutellata* L.)

English: snail medic

M. truncatula Gaertner (*Medicago tribuloides* Desr.; *Medicago tribuloides* Desr. var. *breviaculeata* Moris; *Medicago truncatula* Gaertner f. *tricycla* Negre; *Medicago truncatula* Gaertner var. *breviaculeata* (Moris) Urban; *Medicago truncatula* Gaertner var. *longiaculeata* Urban; *Medicago truncatula* Gaertner var. *tribuloides* (Desr.) Burnat; *Medicago truncatula* Gaertner var. *tricycla* (Negre) Heyn)

English: barrelclover, barrel medic, snail medic, caltrop medic

Medicosma Hook.f. Rutaceae

Origins:

Greek *medike* "a grass, medick, a kind of clover" and *osme* "smell, odor, perfume"; Latin *medica, ae* for a kind of clover introduced from Media.

Species/Vernacular Names:

M. cunninghamii (Hook.) Hook.f.

English: bonewood, pinkheart

Medinilla Gaudich. Melastomataceae

Origins:

Named for José de Medinilla y Pineda, Spanish Governor of the Mariana Islands in 1820; also called Northern Marianas, officially Commonwealth of the Northern Marianas Islands, self-governing commonwealth in political union with the United States, in the western Pacific Ocean, composed of 22 islands; Farallon de Medinilla, a southern island, is also dedicated to the former Governor; see Charles Gaudichaud-Beaupré (1789-1854), [Botany of the Voyage.] *Voyage autour du Monde ... sur ... l'Uranie et la Physicienne, pendant ... 1817-1820.* Paris 1826 [-1830].

Medinillopsis Cogn. Melastomataceae

Origins:

Resembling the genus *Medinilla.*

Mediocactus Britton & Rose Cactaceae

Origins:
From the Latin *medius* "middle, intermediate" plus *Cactus*, a genus intermediate between two others.

Mediocalcar J.J. Sm. Orchidaceae

Origins:
From the Latin *medius* "middle" and *calcar, calcaris* "the heel, spur," referring to the saccate middle part of the lip.

Mediocereus Fric & Kreuz. Cactaceae

Origins:
From the Latin *medius* "middle, intermediate" plus *Cereus*.

Mediolobivia Backeb. Cactaceae

Origins:
From the Latin *medius* "middle, intermediate" plus *Lobivia*.

Mediorebutia Fric Cactaceae

Origins:
From the Latin *medius* "middle, intermediate" plus *Rebutia* K. Schum.

Medusagyne Baker Medusagynaceae

Origins:
From *Medusa* and *gyne* "a woman, female," referring to the gynoecium superior with many locs and styles forming a crown, the pistil resembles the head of a gorgon and the flowers are fetid.

Species/Vernacular Names:
M. oppositifolia Baker
English: jellyfish tree
The Seychelles Islands: bois méduse

Medusandra Brenan Mudusandraceae

Origins:
From *Medusa* and *aner, andros* "a man, male, stamen," anthers opening by longitudinal recurving valves.

Medusanthera Seem. Icacinaceae

Origins:
From *Medusa* and *anthera* "anther," filaments free, anthers with longitudinal slits; in *Journal of Botany.* 2: 74. 1864.

Medusather Candargy Gramineae

Origins:
From *Medusa* and *ather* "an awn."

Medusea Haw. Euphorbiaceae

Origins:
From the daughter of Phorcys and Ceto, Medusa, in Hesiod one of the three Gorgones (with Stheno and Euryale), depicted as winged females, their heads covered with serpents in place of hair.

Meeboldia Pax & Hoffm. Capparidaceae (Capparaceae)

Origins:
After the German botanist Alfred Karl Meebold, 1863-1952, traveler and botanical collector, from 1928 to 1938 traveled widely in Australia, South Africa and the southwest USA. His works include *Indien.* München 1908 [1907]; see I.H. Vegter, *Index Herbariorum.* Part II (4), *Collectors M.* Regnum Vegetabile vol. 93. 1976; Mary Gunn and Leslie E. Codd, *Botanical Exploration of Southern Africa.* 249. Cape Town 1981.

Meeboldina Suess. Restionaceae

Origins:
After the German botanist Alfred Karl Meebold, 1863-1952, traveler and botanical collector, from 1928 to 1938 traveled widely in Australia, South Africa and the southwest USA; novelist and essayist, poet. His works include *Indien.* München 1908 [1907]; see I.H. Vegter, *Index Herbariorum.* Part II (4), *Collectors M.* Regnum Vegetabile vol. 93. 1976; Mary Gunn and Leslie E. Codd, *Botanical Exploration of Southern Africa.* 249. Cape Town 1981.

Meehania Britton Labiatae

Origins:
For the British nurseryman Thomas Meehan, 1826-1901, botanist, Kew gardener, at Germantown (Philadelphia), professor of vegetable physiology to the Pennsylvania State Board of Agriculture. He is known for *The American*

Handbook of Ornamental Trees. Philadelphia 1853 and *The Native Flowers and Ferns of the United States* in their botanical, horticultural, and popular aspects. Boston [Massachusetts] 1878[-1880]; see John H. Barnhart, *Biographical Notes upon Botanists*. 2: 470. 1965; Edith M. Allison, "Bibliography and History of Colorado Botany." *Univ. Colorado Studies*. 6: 51-76. 1908; Joseph Ewan, *Rocky Mountain Naturalists*. The University of Denver Press 1950; T.W. Bossert, *Biographical Dictionary of Botanists Represented in the Hunt Institute Portrait Collection*. 262. 1972; S. Lenley et al., *Catalog of the Manuscript and Archival Collections and Index to the Correspondence of John Torrey*. Library of the New York Botanical Garden. 284. 1973; E.M. Tucker, *Catalogue of the Library of the Arnold Arboretum of Harvard University*. 1917-1933; J. Ewan, ed., *A Short History of Botany in the United States*. 45, 136. New York and London 1969; Jeannette Elizabeth Graustein, *Thomas Nuttall, Naturalist. Explorations in America, 1808-1841*. Harvard University Press 1967; J.W. Harshberger, *The Botanists of Philadelphia and Their Work*. 249-256. 1899; Joseph William Blankinship (1862-1938), "A century of botanical exploration in Montana, 1805-1905: collectors, herbaria and bibliography." in *Montana Agric. Coll. Sci. Studies Bot*. 1: 1-31. 1904; Ida Kaplan Langman, *A Selected Guide to the Literature on the Flowering Plants of Mexico*. University of Pennsylvania Press, Philadelphia 1964.

Species/Vernacular Names:
M. fargesii (H. Léveillé) C.Y. Wu
English: Farges meehania
China: hua xi long tou cao
M. henryi (Hemsley) Sun ex C.Y. Wu
English: Henry meehania
China: long tou cao
M. urticifolia (Miquel) Makino
English: nettle-leaf meehania
China: qian ma ye long tou cao

Meehaniopsis Kudô Labiatae

Origins:
Resembling *Meehania*.

Megabaria Pierre ex Hutch. Euphorbiaceae

Origins:
From the Greek *megas* "big, large" and *baros* "weight," *barys* "heavy."

Megacarpaea DC. Brassicaceae

Origins:
Big-fruited, from Greek *megas* "big, large" and *karpos* "fruit," an allusion to the large pods.

Megacarpha Hochst. Rubiaceae

Origins:
Greek *megas* "big, large, great" and *karphos* "chip of straw, chip of wood."

Megacaryon Boiss. Boraginaceae

Origins:
Greek *megas* "big, large, great" and *karyon* "nut."

Megaclinium Lindley Orchidaceae

Origins:
Greek *megas* "big, large, wide" and *kline* "a bed," *klinion* "a little bed," alluding to the rachis and flattened inflorescence.

Megacodon (Hemsley) Harry Smith Gentianaceae

Origins:
Greek *megas* "big, large, great" and *kodon* "a bell," referring to the flowers.

Species/Vernacular Names:
M. stylophorus (C.B. Clarke) H. Smith
English: common megacodon
China: da zhong hua

Megadenia Maxim. Brassicaceae

Origins:
Greek *megas* "big, large, great" and *aden* "gland."

Megalachne Steud. Gramineae

Origins:
Greek *megas*, *megale* "big, large" and *achne* "husk, glume."

Megaleranthis Ohwi Ranunculaceae

Origins:

Greek *megas*, *megale* "big, large" plus the genus *Eranthis* Salisb.

Megalobivia Y. Ito Cactaceae

Origins:

Greek *megas*, *megale* "big, large" plus *Lobivia*.

Megalochlamys Lindau Acanthaceae

Origins:

Greek *megas*, *megale* "big, large" and *chlamys* "cloak."

Megalodonta Greene Asteraceae

Origins:

Greek *megas*, *megale* "big, large" and *odous*, *odontos* "tooth."

Megalonium (Berger) Kunkel Crassulaceae

Origins:

Greek *megas*, *megale* "big, large" plus *Aeonium*.

Megalopanax Ekman Araliaceae

Origins:

Greek *megas*, *megale* "big, large" plus *Panax*.

Megaloprotachne C.E. Hubb. Gramineae

Origins:

Greek *megas*, *megale* "big," *protos* "first" and *achne* "husk, glume."

Megalopus K. Schum. Rubiaceae

Origins:

Greek *megas*, *megale* "big" and *pous* "foot."

Megalorchis H. Perrier Orchidaceae

Origins:

Greek *megas*, *megale* "big" plus *Orchis*, referring to the flowers.

Megalostoma Leonard Acanthaceae

Origins:

Greek *megas*, *megale* "big" and *stoma* "mouth," *megalostomos* "with large mouth."

Megalostylis S. Moore Euphorbiaceae

Origins:

Greek *megas*, *megale* "big" and *stylos* "style."

Megalotheca F. Muell. Restionaceae

Origins:

Greek *megas*, *megale* "big" and *theke* "a box, case"; see F. von Mueller, *Fragmenta Phytographiae Australiae*. 8: 98. 1873.

Megalotus Garay Orchidaceae

Origins:

Greek *megas*, *megale* "big" and *ous*, *otos* "an ear."

Megaphrynium Milne-Redh. Marantaceae

Origins:

Greek *megas*, *megale* "big" and *phrynos* "a toad," the genus *Phrynium*.

Species/Vernacular Names:
M. macrostachyum (Benth.) Milne-Redh.
Yoruba: gbodogi

Megaphyllaea Hemsl. Meliaceae

Origins:

With big leaves, from the Greek *megas* "big, large" and *phyllon* "leaf."

Megaphyllum Spruce ex Baillon Rubiaceae

Origins:

Greek *megas* "big, large" and *phyllon* "leaf."

Megapleilis Raf. Gesneriaceae

Origins:

See Constantine Samuel Rafinesque, *Flora Telluriana*. 2: 57. 1836 [1837].

Megapterium Spach Onagraceae

Origins:

Greek *megas* "big, large" and *pterion*, the diminutive of *pteron* "wing, feathers," stipules more or less present.

Megarrhena Schrad. ex Nees Cyperaceae

Origins:

Greek *megas* "big, large" and *arrhen* "male," referring to the stamens.

Megarrhiza Torrey & A. Gray Cucurbitaceae

Origins:

Greek *megas* "big, large" and *rhiza* "a root," referring to the tuberous rootstock.

Megasea Haw. Saxifragaceae

Origins:

From the Greek *megas* "big, large."

Megaskepasma Lindau Acanthaceae

Origins:

Greek *megas* "big, large" and *skepasma* "covering," referring to the colored bracts.

Megastachya P. Beauv. Gramineae

Origins:

Greek *megas* "big, large" and *stachys* "spike," referring to the very large spikes.

Megastigma Hook.f. Rutaceae

Origins:

Greek *megas* "big, large" and *stigma* "stigma."

Megastoma (Benthem) Bonnet & Barratte Boraginaceae

Origins:

Greek *megas* "big, large" and *stoma* "mouth."

Megastylis (Schltr.) Schltr. Orchidaceae

Origins:

Greek *megas* "big, large" and *stylos* "style," the large column.

Megatritheca Cristóbal Sterculiaceae

Origins:

Greek *megas* "big," *treis*, *tria* "three" and *theke* "a box, case."

Megistostegium Hochr. Malvaceae

Origins:

Greek *megistos* "very big, very large" and *stege*, *stegos* "roof, cover."

Megistostigma Hook.f. Euphorbiaceae

Origins:

Greek *megistos* (superl. of *megas*) "very big, very large" and *stigma* "a stigma."

Megozipa Raf. Lentibulariaceae

Origins:

See C.S. Rafinesque, *Flora Telluriana*. 4: 110. 1836 [1837]; E.D. Merrill, *Index rafinesquianus*. 221-222. 1949.

Meialisa Raf. Euphorbiaceae

Origins:

See C.S. Rafinesque, *Sylva Telluriana*. 63. 1838; E.D. Merrill, *Index rafinesquianus*. 155. 1949.

Meiandra Markgr. Melastomataceae

Origins:

Greek *meion* "less, smaller, lesser" and *andros* "a man, male."

Meiena Raf. Loranthaceae

Origins:

See C.S. Rafinesque, *Sylva Telluriana*. 125. 1838; E.D. Merrill, *Index rafinesquianus*. 113. 1949.

Meiocarpidium Engl. & Diels Annonaceae

Origins:

Greek *meion* "less, smaller, lesser" and *karpos* "fruit," botanical Latin *carpidium* "carpel."

Meiogyne Miq. Annonaceae

Origins:

Greek *meion* "less, smaller" and *gyne* "a woman, female," an allusion to the small ovary.

Meioluma Baillon Sapotaceae

Origins:

Greek *meion* "less, smaller" plus *luma*.

Meiomeria Standley Chenopodiaceae

Origins:

Greek *meion* "less, smaller" and *meris* "a part, portion."

Meionandra Gauba Rubiaceae

Origins:

Greek *meion* "less, smaller" and *aner, andros* "a man, male, stamen."

Meionectes R. Br. Haloragidaceae (Haloragaceae)

Origins:

Perhaps from Greek *meionekteo* "to be poor, come short, fall short, to be short," *meionektikos* "disposed to take too little," *meion* "smaller, less" and *echo* "to hold, to sustain," the plant has only two petals and sepals, instead of three or four as in related genera; see Matthew Flinders (1774-1814), *A Voyage to Terra Australis*. App. 550. London 1814.

Meionula Raf. Lentibulariaceae

Origins:

See C.S. Rafinesque, *Flora Telluriana*. 4: 108. 1836 [1838]; E.D. Merrill, *Index rafinesquianus*. 222. 1949.

Meiostemon Exell & Stace Combretaceae

Origins:

Greek *meion* "less, smaller" and *stemon* "stamen," referring to the few stamens.

Meiota O.F. Cook Palmae

Origins:

Greek *meion* "less, smaller," *meiotes* "minimizing, minority."

Meiracyllium Reichb.f. Orchidaceae

Origins:

Greek *meirakyllion* "stripling," referring to the low habit.

Meisneria DC. Melastomataceae

Origins:

Dedicated to the Swiss botanist Carl Friedrich (Carolo Friderico) Meisner (Meissner), 1800-1874, physician, 1824 M.D. Göttingen, 1825-1828 with A.P. de Candolle, professor of physiology and professor of botany at the University of Basel from 1836 to 1867. Among his works are *Plantarum vascularium genera*. Lipsiae 1836-1843 and *De amphibiorum quorundam pupillis glandulisque femoralibus*. Basileae 1832. See J.H. Barnhart, *Biographical Notes upon Botanists*. 2: 472. 1965; Stafleu and Cowan, *Taxonomic Literature*. 3: 405-407. 1981; T.W. Bossert, *Biographical Dictionary of Botanists Represented in the Hunt Institute Portrait Collection*. 263. 1972; S. Lenley et al., *Catalog of the Manuscript and Archival Collections and Index to the Correspondence of John Torrey*. Library of the New York Botanical Garden. 464. 1973; Ida Kaplan Langman, *A Selected Guide to the Literature on the Flowering Plants of Mexico*. University of Pennsylvania Press, Philadelphia 1964; Mea Allan, *The Hookers of Kew*. London 1967; Leonard Huxley, *Life and Letters of Sir J.D. Hooker*.

London 1918; E.M. Tucker, *Catalogue of the Library of the Arnold Arboretum of Harvard University*. 1917-1933; R. Zander, F. Encke, G. Buchheim and S. Seybold, *Handwörterbuch der Pflanzennamen*. 14. Aufl. Stuttgart 1993; Antoine Lasègue, *Musée botanique de M. Benjamin Delessert*. Paris, Leipzig 1845; I.C. Hedge and J.M. Lamond, *Index of Collectors in the Edinburgh Herbarium*. Edinburgh 1970.

Melachone Gilli Rubiaceae

Origins:
From the Greek *melas* "black" and *chone* "funnel."

Meladendron St.-Lag. Myrtaceae

Origins:
From the Greek *melas* "black" and *dendron* "a tree."

Meladenia Turcz. Fabaceae

Origins:
From the Greek *melas* "black" and *aden* "gland."

Meladerma Kerr Asclepiadaceae (Periplocaceae)

Origins:
From the Greek *melas* "black" and *derma* "skin."

Melaleuca L. Myrtaceae

Origins:
From the Greek *melas* "black" and *leukos* "white," referring to the black trunks and the white branches of some species or to the colors of the bark; see Arthur D. Chapman, ed., *Australian Plant Name Index*. 1950-1972. Canberra 1991.

Species/Vernacular Names:
M. acuminata F. Muell.
Australia: mallee honey myrtle

M. armillaris (Sol. ex Gaertn.) Sm. (*Melaleuca alba* hort. ex Steud.)
English: bracelet

M. brevifolia Turcz.
Australia: mallee honey myrtle, D'Alton's melaleuca, white-flowered paperbark

M. cajuputi Powell
Vietnam: tram, che dong, che cay

M. cardiophylla F. Muell.
Australia: tangling melaleuca, umbrella bush

M. cuticularis Labill. (*Melaleuca abietina* Sm.)
English: salt-water paperbark

M. decussata R. Br. (*Melaleuca tetragona* Lodd ex Otto; *Melaleuca parviflora* Reichb.; *Melaleuca elegans* Hornsch.; *Melaleuca oligantha* F. Muell. ex Miq.; *Melaleuca decussata* R. Br. var. *ovoidea* J. Black)
Australia: totem poles, cross-leaved honey myrtle

M. eleutherostachya F. Muell.
Australia: hummock honey myrtle, swamp paperbark

M. ericifolia Sm.
English: swamp paperbark

M. glomerata F. Muell. (*Melaleuca hakeoides* F. Muell. ex Benth.)
Australia: white tea tree, desert paperbark, desert honey myrtle, inland paperbark

M. halmaturorum Miq.
Australia: kangaroo honey myrtle, kangaroo paperbark, salt paperbark, coastal paperbark, South Australian swamp paperbark

M. huegelii Endl.
English: honey myrtle, chenille honey myrtle

M. hypericifolia Smith
English: hillock bush, red honey myrtle

M. irbyana R. Baker (after Llewellyn G. Irby, forester and original collector)
English: weeping paperbark

M. lanceolata Otto (*Melaleuca pubescens* Schauer; *Melaleuca curvifolia* Schltdl.; *Cajuputi pubescens* (Schauer) Skeels; *Melaleuca seorsiflora* F. Muell.)
Australia: moonah honey myrtle, moonah, dryland tea tree, black tea tree

M. lateritia A. Dietr. (*Melaleuca callistemonea* Lindl.)
English: Robin red breast bush

M. leucadendra (L.) L.
English: cajeput, river tea tree, weeping tea tree, broad-leaved paperbark, paperbark, paperbark tree, broad-leaved tea tree, punk tree, cajeput tree, white wood
Indonesia: cajeput
China: bai qian ceng
Malaya: gelam, kayu puteh

M. linariifolia Smith
English: snow-in-summer, flax-leaf paperbark

M. nesophila F. Muell.

English: western tea myrtle

M. preissiana Schauer

English: Preiss's paperbark

M. rhaphiophylla Schauer

English: needle-leaved honey myrtle, swamp paperbark

M. suberosa (Schauer) C. Gardner

English: cork-barked honey myrtle

M. uncinata R. Br.

English: broom honey myrtle, broombush

M. undulata Benth.

English: hidden honey myrtle

M. wilsonii F. Muell.

Australia: Wilson's honey myrtle, violet honey myrtle

Melaleucon St.-Lag. Myrtaceae

Origins:

Greek *melas* "black" and *leukos* "white," see also *Melaleuca* L.

Melampodium L. Asteraceae

Origins:

Latin and Greek *melampodion* for the black hellebore (Plinius), Greek *melas* "black" and *pous*, *podion*, *podos* "foot, little foot," referring to the color of the base of the stem and roots; see Helmut Genaust, *Etymologisches Wörterbuch der botanischen Pflanzennamen.* 374-375. Basel 1996.

Melampyrum L. Scrophulariaceae

Origins:

Greek *melas* "black" and *pyros* "wheat," referring to the grains; Greek *melampyron* and *melampyros* used by Theophrastus (*HP.* 8.4.6, 8.8.3) for ball-mustard, a kind of *Neslia*, probably for *Neslia paniculata*; see Helmut Genaust, *Etymologisches Wörterbuch der botanischen Pflanzennamen.* 375. 1996.

Melananthus Walp. Solanaceae

Origins:

From the Greek *melas*, *melanos* "black" and *anthos* "flower."

Melandrium Roehl. Caryophyllaceae

Origins:

Melandryon, an ancient classical Greek name used by Theophrastus, from *melas* "black" and *drys* "oak"; Latin *melandryum* for a piece of salted tunny-fish (... like the black heart of oak, Plinius); see H. Genaust, *Etymologisches Wörterbuch der botanischen Pflanzennamen.* 375-376. 1996.

Melandryum Reichb. Caryophyllaceae

Origins:

Melandryon, an ancient classical Greek name used by Theophrastus, from *melas* "black" and *drys* "oak."

Melanea Pers. Rubiaceae

Origins:

From the Greek *melas*, *melanos* "black," *melania* "blackness" or referring to *Malanea* Aublet.

Melanium P. Browne Lythraceae

Origins:

From the Greek *melas*, *melanos* "black," *melania* "blackness," referring to the flowers.

Melanocarpum Hook.f. Amaranthaceae

Origins:

From the Greek *melas*, *melanos* "black" and *karpos* "fruit."

Melanocenchris Nees Gramineae

Origins:

From the Greek *melas*, *melanos* "black" and *kenchros* "millet."

Melanochyla Hook.f. Anacardiaceae

Origins:

From the Greek *melas*, *melanos* "black" and *chylos* "juice, sap."

Species/Vernacular Names:
M. auriculata Hook.f.

Malaya: swamp rengas, kerbau jalang, mempian, mempiang, rengas lanjut

Melanococca Blume Anacardiaceae

Origins:
From the Greek *melas*, *melanos* "black" and *kokkos* "berry, grain, seed."

Melanocommia Ridley Anacardiaceae

Origins:
From the Greek *melas*, *melanos* "black" and *kommi* "gum," sap drying as a black resin.

Melanodendron DC. Asteraceae

Origins:
From the Greek *melas*, *melanos* "black" and *dendron* "tree," an allusion to the color of the wood.

Melanodiscus Radlk. Sapindaceae

Origins:
From the Greek *melas*, *melanos* "black" and *diskos* "a disc."

Melanolepis Reichb.f. ex Zoll. Euphorbiaceae

Origins:
From the Greek *melas*, *melanos* "black" and *lepis* "scale, husk."

Species/Vernacular Names:
M. sp.
Malaya: chawan, jarak kayu

Melanoleuca St.-Lag. Myrtaceae

Origins:
From the Greek *melas*, *melanos* "black" and *leukos* "white," see *Melaleuca*.

Melanophylla Baker Cornaceae (Melanophyllaceae)

Origins:
From the Greek *melas*, *melanos* "black" and *phyllon* "leaf," sometimes accumulating aluminum and/or inulin.

Melanopsidium Cels ex Colla Rubiaceae

Origins:
Greek *melas*, *melanos* "black" and *psidion* "the pomegranate."

Melanopsidium Poit. ex DC. Rubiaceae

Origins:
From the Greek *melas*, *melanos* "black" and *psidion* "the pomegranate," referring to the edible fruits.

Melanorrhoea Wallich Anacardiaceae

Origins:
Greek *melas*, *melanos* "black" and *rheo* "to flow," referring to a black and waxy juice.

Species/Vernacular Names:
M. sp.
Malaya: rengas, mepoah

Melanosciadium Boissieu Umbelliferae

Origins:
From the Greek *melas*, *melanos* "black" and *skiadion, skiadeion* "umbel, parasol."

Melanoselinum Hoffm. Umbelliferae

Origins:
From the Greek *melas*, *melanos* "black" and *selinon* "parsley, celery."

Species/Vernacular Names:
M. decipiens (Schrad. & Wendl.) Hoffm.
English: black parsley

Melanospermum Hilliard Scrophulariaceae

Origins:
From the Greek *melas*, *melanos* "black" and *sperma* "seed."

Melanosticta DC. Caesalpiniaceae

Origins:
From the Greek *melas*, *melanos* "black" and *stiktos* "spotted, punctured, dotted, dappled," referring to the resinous dots.

Melanoxylon Schott Caesalpiniaceae

Origins:
Greek *melas, melanos* "black" and *xylon* "wood."

Melanthera J.P. Rohr Asteraceae

Origins:
From the Greek *melas* "black" and *anthera* "anther."

Species/Vernacular Names:
M. elliptica O. Hoffm.
Yoruba: abo yunriyun, yunriyun gbodo, ewe agbu igbo
M. scandens (Schumach. & Thonn.) Roberty
Yoruba: ajidari, ayaki, aboba dudu

Melanthes Blume Euphorbiaceae

Origins:
From the Greek *melas* "black" and *anthos* "flower," *anthesis* "flowering," *melanthes* "black, swarthy."

Melanthesa Blume Euphorbiaceae

Origins:
Greek *melas* "black" and *anthesis* "flowering," referring to the flowers; some suggest from Latin *thesion* and *thesium, ii,* an ancient name for a species of *Linaria,* toad flax, used by Plinius; see Karl Ludwig von Blume (1796-1862), *Bijdragen tot de flora van Nederlandsch Indië.* 590. Batavia 1825-1826.

Melanthesopsis Müll. Arg. Euphorbiaceae

Origins:
Resembling the genera *Melanthes* and *Melanthesa.*

Melanthium L. Melanthiaceae (Liliaceae)

Origins:
Latin *melanthium* and Greek *melanthion* for the plant *gith,* cultivated fennel-flower (Plinius); Greek *melas* "black" and *anthos* "a flower," the flower segments become dark after flowering; Greek *melanthion* was an herb whose seeds were used as spice, black cummin, originally applied to a kind of *Nigella;* see Carl Linnaeus, *Species Plantarum.* 339. 1753 and *Genera Plantarum.* Ed. 5. 157. 1754; Helmut

Genaust, *Etymologisches Wörterbuch der botanischen Pflanzennamen.* 376-377. Basel 1996.

Species/Vernacular Names:
M. virginicum L.
English: bunchflower

Melasma Bergius Scrophulariaceae

Origins:
From the Greek *melasma* "black or livid spot," *melasmos* "blackening, dyeing black, black spot," referring to a black spot when the plants dry.

Melasphaerula Ker Gawl. Iridaceae

Origins:
From the Greek *melas* "black" and *sphaira* "a globe, a ball," Latin *sphaerula* (*sphaera*) "a small ball," in reference to the small black corms.

Species/Vernacular Names:
M. ramosa (L.) N.E. Br. (*Melasphaerula graminea* (L.f.) Ker Gawl.)
South Africa: baardmannetjie, bokbaardjie, feëklokkies

Melastoma L. Melastomataceae

Origins:
From the Greek *melas* "black" and *stoma* "mouth," some fruits have staining effects, the fruit of some species has purple pulp; see Carl Linnaeus, *Species Plantarum.* 389. 1753 and *Genera Plantarum.* Ed. 5. 184. 1754.

Species/Vernacular Names:
M. sp.
Malaya: sendudok, kedudok, sekedudok
M. affine D. Don
English: native lasiandra
M. candidum D. Don (*Melastoma candidum* var. *nobotan* (Blume) Makino; *Melastoma septemnervium* Lour. non Jacq.)
English: melastoma
Japan: no-botan, hanki-tabu
China: ye mu dan
M. malabathricum L. (*Melastoma banksii* A. Cunn. ex Triana)

English: Indian rhododendron, Singapore rhododendron, Malabar laurel

Malaya: common sendudok, kedudok, sendudok, engkudu

M. sanguineum Sims (*Melastoma decemfidum* Roxb. ex W. Jack)

English: fox-tongued melastoma

Malaya: great sendudok, sendudok puteh, sendudok gajah

Melastomastrum Naudin Melastomataceae

Origins:

From the genus *Melastoma* L. and -*astrum* "incomplete resemblance, resembling in an inferior manner, wild, poor."

Species/Vernacular Names:

M. theifolium (G. Don) A. Fern. & R. Fern.

Yoruba: oju aguntan

Melchiora Kobuski Theaceae

Origins:

After the German botanist Hans Melchior, 1894-1984; see John H. Barnhart, *Biographical Notes upon Botanists*. 2: 472. 1965; T.W. Bossert, *Biographical Dictionary of Botanists Represented in the Hunt Institute Portrait Collection*. 263. 1972; Ida Kaplan Langman, *A Selected Guide to the Literature on the Flowering Plants of Mexico*. 1964; R. Zander, F. Encke, G. Buchheim and S. Seybold. *Handwörterbuch der Pflanzennamen*. 14. Aufl. Stuttgart 1993. Melchior was one of the three Magi and the genus *Balthasaria* Verdc. (Theaceae) was dedicated to Balthasar, another of the Magi.

Melchioria Penzig & Saccardo Theaceae

Origins:

After the Dutch botanist Melchior Treub, 1851-1910 (St. Raphael, France), traveler and botanical collector, from 1873 to 1880 collaborated in the botanical institute at Leiden with professor of botany Willem Frederik Reinier Suringar (1832-1898), from 1880 Director (he succeeded Rudolf Herman Christiaan Carel Scheffer, 1844-1880) of the Buitenzorg Botanical Gardens (now Bogor, West Java). Among his scientific works are "Notice sur la nouvelle flore de Krakatau." *Ann. Jard. Bot. Buitenzorg*. Leide [Leiden] 1888 and "Observations sur les Loranthacées." *Ann. Jard. Bot. Buitenzorg*. Leide [Leiden] 1881, from 1881 to 1910 editor of the *Annales du Jardin Botanique de Buitenzorg*. See Friedrich August Ferdinand Christian Went (1863-1935), "In Memoriam." *Ann. Jard. Bot. Buitenzorg*. 9: i-xxxii. 1911; C.G.G.J. van Steenis, in *Dictionary of Scientific Biography* 13: 458-460. 1981; G.J. Symons, ed., *The Eruption of Krakatoa and Subsequent Phenomena*. Report of the Krakatoa Committee of the Royal Society. London 1888; J.H. Barnhart, *Biographical Notes upon Botanists*. 3: 592. 1965; H.N. Clokie, *Account of the Herbaria of the Department of Botany in the University of Oxford*. 406. 1964; E.M. Tucker, *Catalogue of the Library of the Arnold Arboretum of Harvard University*. Cambridge, Massachusetts 1917-1933; P.M.W. Dakkus, *An Alphabetical List of Plants Cultivated in the Botanic Gardens, Buitenzorg*. Buitenzorg/Java, Dutch-East Indies 1930; Stafleu and Cowan, *Taxonomic Literature*. 6: 469-474. 1986; Gordon Douglas Rowley, *A History of Succulent Plants*. Strawberry Press, Mill Valley, California 1997.

Melhania Forssk. Sterculiaceae

Origins:

After Mount Melhan, in Yemen, [Djebbel Melhan, *Arabia felix*]; see P. Forsskål (1732-1763), *Flora aegyptiaco-arabica*. 64. Copenhagen 1775.

Species/Vernacular Names:

M. forbesii Planch. ex Mast.

South Africa: moulhwadambo

M. oblongifolia F. Muell. (*Melhania incana* sensu J. Black)

Australia: velvet hibiscus

Melia L. Meliaceae

Origins:

From *melia*, the classical Greek name used by Theophrastus (*HP*. 3.11.3) for the manna ash or flowering ash tree (*Fraxinus*), the leaves are quite similar; see Carl Linnaeus, *Species Plantarum*. 384. 1753 and *Genera Plantarum*. Ed. 5. 182. 1754.

Species/Vernacular Names:

M. azedarach L. (*Melia japonica* D. Don; *Melia sempervirens* Sw.; *Melia florida* Salisb.; *Melia azedarach* var. *glabrior* C. DC.; *Azedarach deleteria* Medik.; *Azedara speciosa* Raf.; *Azedaraca amena* Raf.; *Azedarach sempervirens* (L.) Kuntze var. *glabrior* Kuntze; *Azedarach vulgaris* Gomez de la Maza)

English: Persian lilac, Indian lilac, Cape lilac, berry tree, China berry, China tree, Chinaberry tree, pride of China, pride of Persia, pride of India, umbrella tree, Chinese umbrella, Texas umbrella tree, common bead tree, bead tree, Indian bead tree, Syrian bead tree, Japanese bead tree, syringa, syringa tree, red seringea, Cape seringa, Cape syringa, South African syringa, bastard cedar, white cedar, Ceylon cedar, night bloom, paradise tree, sycomore, azedarach, mahogany

French: acajou rose du Tonkin, margousier, lilac des Indes, lilas, lilas de la Chine, lilas de Chine, lilas de la Perse, lilas des Indes, lilas du Japon, paradisier, faux sycomore, arbre à chapelets, camphrier faux, sycomore faux

Portuguese: amargoseira

South Africa: maksering, mak-seringboom, seringboom, bessieboom, Kaapse sering, seringbessieboom

Portuguese Africa: bombolo ia n'buto

Madagascar: kandelaka

Yoruba: afoforo oyinbo, eke oyinbo, eke ile, afoforo igbalode

Nigeria: foreign kurna, kurna-na-sara, nassara, eke-oyinbo (= white man's rafter), itchin-kurdi; kurnan nasara (kurna = *Ziziphus*; nasara = Christian) (Hausa); chigban anasara (Nupe)

Indochina: hay san, lien, xoan, xoan dao

Sri Lanka: lunumidella, Malai vembu, kirikohomba

India: bukain, drek, dek, jek, bakain, betain, deikna, maha limbo, malla nim, muhli; chein, drek, heb bevu, kachein (Punjab); chik bevu, pejri padrai (Bombay); makan nim, mallay vembu, taraka vepa (Madras); male vimbou, turka bepa (Deccan); malla nim, darachik, deikna, dek, deknoi, denkan, hlerm, hlim, maha-limbo, maha nimb, muhli (East India)

Nepal: bakaino

China: ku lian pi, ku lien, lien, sen shu

Vietnam: chann mou, hou lien, kho luyen, san dan, sau dau, xun lien, yu mou, xoan, may rieu

Malaya: mindi kechil

Burma: tamaga

Japan: senn-dan, sen-yoo-si, shen lien, sendan, shindan

Venezuela: aleli

Brazil: cinamomo, lila da Percia, jasmim-de-cachorro, jasmim-de-soldado, saboeiro

Puerto Rico: alilaila

Dutch Indies: kakera, kikera, mindi

Peru: cinamono, jasmín de Arabia, flor de Paraíso, paraíso

Mexico: arbol paraiso, paraíso, lila, lilaila, lila de la Indias, paraíso morado, piocha, jacinto, pasilla; canelo (San Luis Potosí); cinamon (Guadalcazar, San Luis Potosí); granillo (Oaxaca); lila de China (San Luis Potosí and Nuevo Leon); paraíso chino (Chihuahua; see William B. Griffen, *Indian Assimilation in the Franciscan Area of Nueva Vizcaya*. Tucson 1979); paraíso (Michoacán, Veracruz, Yucatan, etc.); pioch (southeast San Luis Potosí)

Guadeloupe: arbre à chapelets, arbre saint, lilas du pays

Tahiti: tira

Hawaii: 'inia, 'ilinia

M. azedarach L. var. *australasica* (Adr. Juss.) DC. (*Melia australasica* Adr. Juss.)

Australia: white cedar, kooribill, tulip cedar

Meliandra Ducke Melastomataceae

Origins:
From the Greek *meli* "honey" and *aner, andros* "man, stamen."

Melianthus L. Melianthaceae

Origins:
From the Greek *meli* "honey" and *anthos* "a flower," the flowers are rich in nectar; see Carl Linnaeus, *Species Plantarum*. 639 (as '939). 1753 and *Genera Plantarum*. Ed. 5. 287. 1754.

Species/Vernacular Names:
M. comosus Vahl

English: tufted honey flower

South Africa: Kruidjie-roer-my-nie, Truitjie-roer-my-nie

M. dregeanus Sond. subsp. *dregeanus*

South Africa: Kruidjie-roer-my-nie

M. major L.

English: honey flower, large honey flower, Cape honey flower

Southern Africa: heuningblom, klappers, kruie, kriekiebos, krikkiebos, kruidjiebos, Kruidjie-roer-my-nie, Truitjie-roer-my-nie; ubuhlungubemamba (Xhosa)

Melica L. Gramineae

Origins:
From the Greek name *melike* (*meli* "honey") for a grass, or from Latin *herba(m) medica(m)* "grass from Media," *melicus, a, um* "Median"; or from Latin *melica, ae* for a kind of vessel (Marcus Terentius Varro). See C. Abegg-Mengold, *Die Bezeichnungsgeschichte von Mais, Kartoffel und Ananas in Italienischen*. Bern 1979; Carl Linnaeus, *Species Plantarum*. 66. 1753 and *Genera Plantarum*. Ed. 5. 76. 1754; Manlio Cortelazzo & Paolo Zolli, *Dizionario etimologico della lingua italiana*. 3: 738. Zanichelli, Bologna 1983.

Species/Vernacular Names:
M. altissima L.

English: Siberian melic

M. aristata Bolander

English: awned melic

M. bulbosa Geyer

English: onion grass

M. californica Scribner

English: California melic

M. ciliata L.

English: silky-spike melic

M. nutans L.

English: mountain melic, nodding mellic, onion grass, melic grass

M. racemosa Thunb. (*Melica bolusii* Stapf; *Melica brevifolia* Stapf; *Melica decumbens* sensu Gordon-Gray, non Thunb.; *Melica ovalis* Nees; *Melica pumila* Stapf)

English: steggersgrass, melic grass

Southern Africa: dronkgras, Kaapse dronkgras, haakgras; ntlo-ea-motintinyane, ntlo-ea-motintinyane-e-nyenyane (Sotho)

M. spectabilis Scribner

English: purple onion grass

M. uniflora Retz.

English: wood melic, wood melick

Melichrus R. Br. Epacridaceae

Origins:
Latin *melichros* for a precious stone yellow as honey (Plinius), Greek *melichros* "honey-sweet, honey-colored," *meli* "honey" and *chroa* "color," referring to the copious nectar exuding from the glands; see Robert Brown (1773-1858), *Prodromus florae Novae Hollandiae*. 539. London 1810.

Species/Vernacular Names:
M. erubescens DC.

English: western urn-heath

M. procumbens (Cav.) Druce

English: jam-tarts

M. urceolatus R. Br.

English: urn-heath

Meliclis Raf. Orchidaceae

Origins:
Greek *meli* "honey" and *kleis* "key," the lip is saccate and sometimes holds nectar; see C.S. Rafinesque, *Flora Telluriana*. 2: 99. 1836 [1837].

Melicocca L. Sapindaceae

Origins:
See *Melicoccus* P. Browne; see Carl Linnaeus, *Species Plantarum*. 495. 1762 and *Genera Plantarum*. Ed. 6. 188. 1762.

Melicoccus P. Browne Sapindaceae

Origins:
From the Greek *meli* "honey" and *kokkos* "berry, grain, seed," the fruits are sweet.

Melicope Forst. & Forst.f. Rutaceae

Origins:
Greek *meli* "honey" and *kope* (*kopto* "to cut off, to cut small, to pierce") "a division," referring to the glands at the base of the ovary; see J.R. Forster and J.G.A. Forster, *Characteres generum plantarum, quas in itinere ad insulas maris australis, etc.* 55, t. 28. London (Nov.) [1775].

Species/Vernacular Names:
M. elleryana (F. Muell.) T. Hartley

English: pink-flowered doughwood

M. erythrococca (F. Muell.) Benth.

English: tingletongue, clubwood

M. hayesii T. Hartley

English: small-leaved doughwood

M. melanophloia C. White

English: black-barked doughwood

M. micrococca (F. Muell.) T. Hartley

English: hairy-leaved doughwood

M. octandra (F. Muell.) Druce

English: doughwood, soapwood, silver birch

M. ternata Forst. & Forst.f.

Maori names: wharangi

Melicytus Forst. & Forst.f. Violaceae

Origins:
From the Greek *meli* "honey" and *kytos* "a hollow container, jar, cell," referring to the hollow staminal nectaries of the flowers; see J.R. Forster and J.G.A. Forster, *Characteres generum plantarum*. 123, t. 62. London (Nov.) [1775]; H.E. Connor and E. Edgar, "Name changes in the indigenous New Zealand flora, 1960-1986 and Nomina Nova IV, 1983-1986." *New Zealand Journal of Botany*. Vol. 25: 115-170.

1987; H.H. Allan, *Fl. New Z.* 1: 191-193. 1961; P.S. Green, *J. Arnold Arbor.* 51: 218-220. 1970; E.J. Beuzenberg, *New Zeal. J. Sci.* 4: 337-349. 1961.

Species/Vernacular Names:

M. micranthus Hook.f.

Maori name: manakura

M. ramiflorus Forst. & Forst.f.

Maori name: mahoe

Melilotus Miller Fabaceae

Origins:

Latin and Greek *melilotos* for a kind of clover, melilot, also called *sertula*; Greek *meli* "honey" and *lotos* "lotus, clover," alluding to the fragrant smell of the foliage; see Manlio Cortelazzo & Paolo Zolli, *Dizionario etimologico della lingua italiana.* 3: 738. Zanichelli, Bologna 1983.

Species/Vernacular Names:

M. alba Medikus (*Melilotus albus* Medikus var. *annuus* H.S. Coe; *Melilotus leucanthus* Koch ex DC.)

English: hubam, white melilot, white sweetclover, Bokhara clover, Bukhara clover, clover, hubam clover, sweet clover

Peru: alfalfa chilena

South Africa: Bokhaargras, Bokhaarklawer, witstinkklawer

M. altissimus Thuill. (*Melilotus macrorrhizus* (Waldst. & Kit.) Pers.; *Trifolium macrorrhizum* (Waldst. & Kit.) Pers.)

English: tall yellow sweetclover

M. indica (L.) All. (*Trifolium melilotus-indica* L.; *Trifolium indicum* L.; *Melilotus parviflorus* Desf.)

English: sourclover, Indian sweetclover, Indian melilot, annual yellow sweetclover, Hexham scent, melilot, stink clover, sweetclover, yellow sweetclover, white sweetclover, King Island melilot, sweet melilot

China: pi han cao

Arabic: handaqouq, handaquq murr, reqraq, qort

Peru: alfalfa macho, shacko-álfar, trébol macho

South Africa: bitterklawer, eenjarige geel stinkklawer, geel stinkklawer, steenklawer, stinkklawer

M. officinalis Lam. (*Melilotus arvensis* Wallr.; *Melilotus officinalis* Lam. var. *micranthus* O. Schulz; *Melilotus vulgaris* Hill; *Trifolium officinale* L.; *Trifolium Melilotus officinalis* L.)

English: yellow melilot, yellow sweetclover, common melilot

China: hsun tsao, ling ling hsiang

M. suaveolens Ledeb.

English: annual yellow sweetclover

Japan: Shinagawa-hagi

Melinis P. Beauv. Gramineae

Origins:

From the Greek *meline* "millet, Italian millet, *Panicum miliaceum*" (Herodotus) or a kind of *Setaria*; Latin *milium, ii* "millet" (Vergilius, Plinius and Marcus Terentius Varro); see Ambroise Marie François Joseph Palisot de Beauvois, *Essai d'une nouvelle Agrostographie*, ou nouveaux genres des Graminées. 54, t. 11, fig. 4. Paris 1812.

Species/Vernacular Names:

M. minutiflora Beauv. (*Melinis tenuinervis* (Stapf) Stapf)

English: Brazilian stink grass, dordura grass, efwatakala grass, gordura grass, honey grass, molasses grass

Spanish: yerba de gordura

South Africa: heuninggras, melassegras, stinkgras

The Philippines: gordura

Melio-schinzia K. Schumann Meliaceae

Origins:

For the Swiss botanist Hans Schinz, 1858-1941, traveler, professor of botany. Among his writings are *Plantae menyharthianae ein Beitrag zur Kenntniss der Flora des Unteren Sambesi ...* Wien, 1905 [Collector: Ladislav Menyharth, 1849-1897], "Durch Südwestafrika." *Verh. Ges. Erdk. Berl.* 14: 322-324. 1887, *Mein Lebenslauf.* Zürich 1940, *Observations sur une collection de plantes du Transvaal.* Genève 1891. See Albert Thellung (1881-1928), "Verzeichnis der Veröffentlichungen von Prof. Dr. Hans Schinz." *Beibl. Viertelj.-Schr. naturf. Ges. Zürich* 15 (Jahrg. 73): 773-783. 1928, the bibliography of works and papers by Schinz; Théophile Alexis Durand (1855-1912), *Conspectus florae Africae*, ou, Enumération des plantes d'Afrique, par T.D. et Hans Schinz. Bruxelles 1895-1898; John H. Barnhart, *Biographical Notes upon Botanists.* 3: 227. 1965; Mary Gunn and Leslie E. Codd, *Botanical Exploration of Southern Africa.* 311-313. Cape Town 1981; T.W. Bossert, *Biographical Dictionary of Botanists Represented in the Hunt Institute Portrait Collection.* 353. 1972; E.M. Tucker, *Catalogue of the Library of the Arnold Arboretum of Harvard University.* 1917-1933; Elmer Drew Merrill, *Bernice P. Bishop Mus. Bull.* 144: 163. 1937; R. Zander, F. Encke, G. Buchheim and S. Seybold, *Handwörterbuch der Pflanzennamen.* 14. Aufl. Stuttgart 1993; A. White and B.L. Sloane, *The Stapelieae.* Pasadena 1937; Stafleu and Cowan, *Taxonomic Literature.* 5: 175-181. 1985.

Meliosma Blume Sabiaceae (Meliosmaceae)

Origins:

Greek *meli* "honey" and *osme* "smell, odor, perfume," referring to the honey-scented flowers.

Species/Vernacular Names:

M. pinnata (Roxb.) Walp. (*Meliosma rhoifolia* Maxim.; *Meliosma oldhamii* Maxim.; *Meliosma sinensis* Nak.)

Japan: yanbaru-awabu-ki, nurude-awabuki, fushi-no-ki-awabuki, tisan

Melissa L. Labiatae

Origins:

From the Latin *melissophyllum* and Greek *melissophyllon*, Greek *melissa* "a honeybee, bee, honey," leaves in skeps alleged to attract bee-swarms, lemon-scented leaves, bees are supposed to be delighted with this herb; Melissa was a nymph who is said to have invented the art of keeping bees; see Carl Linnaeus (1707-1778), *Species Plantarum*. 592. 1753 and *Genera Plantarum*. Ed. 5. 257. 1754; Helmut Genaust, *Etymologisches Wörterbuch der botanischen Pflanzennamen*. 378-379. Basel 1996; Serapiom, *El libro agregà de Serapiom*. A cura di G. Ineichen. Venezia-Roma 1962-1966;[John Lemprière, 1765?-1824], *Lemprière's Classical Dictionary of Proper Names Mentioned in Ancient Authors*. Third edition. 369. London and New York 1984; Manlio Cortelazzo & Paolo Zolli, *Dizionario etimologico della lingua italiana*. 3: 738. 1983; P. Sella, *Glossario latino emiliano*. Città del Vaticano 1937.

Species/Vernacular Names:

M. axillaris (Bentham) Bakh.f. (*Melissa parviflora* Bentham; *Melissa hirsuta* Blume)

English: axillary balm

China: mi feng hua

M. flava Bentham ex Wallich

English: yellow balm

China: huang mi feng hua

M. officinalis L.

English: balm, common balm, bee balm, lemon balm, sweet balm, tea balm, balm leaf

Peru: toronjil

China: xiang feng hua

Arabic: louiza, merzizou

M. yunnanensis C.Y. Wu & Y.C. Huang

English: Yunnan balm

China: yun nan mi feng hua

Melittacanthus S. Moore Acanthaceae

Origins:

From the Greek *melissa, melitta* "a honeybee" plus *Acanthus*.

Melittis L. Labiatae

Origins:

From the Greek *melissa, melitta* "a honeybee," bees enjoy these plants.

Species/Vernacular Names:

M. melissophyllum L.

English: bastard balm

Mella Vand. Scrophulariaceae

Origins:

After Mr. Mello, a Portuguese Minister of marine and colonies.

Mellera S. Moore Acanthaceae

Origins:

For the English naturalist Charles James Meller, c. 1835-1869 (d. Sydney), botanist, physician, 1867 Fellow of the Linnean Society, 1860-1863 surgeon on Dr. Livingstone's African expedition, plant collector, 1866-1869 Mauritius; see Guy Rouillard and Joseph Guého, *Le Jardin des Pamplemousses 1729-1979. Histoire et Botanique*. 41, 68. Les Pailles 1983.

Melliniella Harms Fabaceae

Origins:

After the German botanist Adolf Mellin, b. 1910, plant collector (Togo); see John H. Barnhart, *Biographical Notes upon Botanists*. Boston 1965; F.N. Hepper and F. Neate, *Plant Collectors in West Africa*. 54. 1971.

Mellissia Hook.f. Solanaceae

Origins:

Named for the British botanist John Charles Melliss, Surveyor on St. Helena from *circa* 1860 to 1891, author of *St. Helena*. London 1875. See John H. Barnhart, *Biographical Notes upon Botanists*. 2: 473. 1965; H.N. Clokie, *Account of the Herbaria of the Department of Botany in the University of Oxford.* 209. Oxford 1964; E.M. Tucker, *Catalogue of the Library of the Arnold Arboretum of Harvard University.* 1917-1933; I.C. Hedge and J.M. Lamond, *Index of Collectors in the Edinburgh Herbarium.* Edinburgh 1970.

Melo Mill. Cucurbitaceae

Origins:
Latin *melo, melonis* for an apple-shaped melon.

Melocactus Link & Otto Cactaceae

Origins:
Greek *melon* "an apple" and *Cactus*, melon cactus.

Species/Vernacular Names:
M. intortus (Miller) Urban
English: Turk's cap cactus

Melocalamus Benth. Gramineae

Origins:
From the Greek *melon* "an apple" and *kalamos* "reed."

Melocanna Trin. Gramineae

Origins:
From the Greek *melon* "an apple" and *kanna* "a reed, cane," referring to the fruit, a berry size of an avocado.

Species/Vernacular Names:
M. baccifera (Roxb.) Kurz
English: Terai bamboo

Melochia L. Sterculiaceae

Origins:
Presumably from the Greek *meli* "honey" and *echo* "to hold, to sustain," or from *meli* and *locheia* "childbirth," ancient name for a plant supposed to ease parturition; *melochich* is an Arabic name for *Corchorus olitorius* L., tossa jute; see Carl Linnaeus, *Species Plantarum.* 674. 1753 and *Genera Plantarum.* Ed. 5. 304. 1754.

Species/Vernacular Names:
M. corchorifolia L. (*Melochia concatenata* L.)
English: juteleaf melochia

Sri Lanka: gal kura

Japan: noji-aoi

Malaya: lemak ketam, limah ketam, bunga padang, pulut-pulut

M. umbellata (Houtt.) Stapf (*Melochia velutina* (DC.) Wall. ex Bedd.; *Riedlia velutina* DC.; *Visenia umbellata* Houtt.)

English: meloch

Sri Lanka: malkenda

Malaya: tampu, tapu, chapah, singah, singa, chapak

Melodinus Forst. & Forst.f. Apocynaceae

Origins:
Greek *melon* "an apple" and *dineo* "I twist," referring to the shape of the fruit and the climbing and twining habit, the fruit is a large and pulpy berry; see J.R. Forster and J.G.A. Forster, *Characteres generum plantarum.* 37, t. 19. London (Nov.) [1775].

Species/Vernacular Names:
M. angustifolius Hayata
English: narrow-leaved melodinus
China: taiwan shan chen
M. australis Pierre
English: Harlequin fruit
M. baccellianus (F. Muell.) S.T. Blake
Australia: murpe (a name given to the fruit by the Russell River Aborigines)
M. cochinchinensis (Loureiro) Merrill (*Melodinus henryi* Craib)
English: Henry melodinus
China: si mao shan chen
M. fusiformis Champion ex Bentham (referring to the fusiform berries)
English: fusiform melodinus
China: jian shan chen
M. hemsleyanus Diels
English: Hemsley melodinus
China: chuan shan chen
M. khasianus Hook.f.
English: Khasia melodinus
China: jing dong shan chen
M. magnificus Tsiang
English: magnificent melodinus
China: cha teng
M. suaveolens (Hance) Champion ex Bentham

English: fragrant melodinus

China: shan chen

M. tenuicaudatus Tsiang & P.T. Li

English: thin-leaved melodinus

China: si mao shan chen

Melodorum Lour. Annonaceae

Origins:

From the Greek *melon* "an apple" and *doron* "gift," referring to the fruit; or from the Latin *mel, mellis* "honey" and *odor, odoris* "smell, perfume," referring to the honey-scented leaves; see J. de Loureiro, *Flora cochinchinensis.* 329, 351. 1790.

Melolobium Ecklon & Zeyher Fabaceae

Origins:

Greek *meli* "honey" and *lobos* "pod," referring to the sweets seeds; some suggest from *melos* "a limb" and *lobos* "pod, legume," referring to the constrictions between seeds.

Species/Vernacular Names:

M. candicans (E. Mey.) Eckl. & Zeyh. (*Dichilus candicans* E. Mey.)

English: honey bush

Southern Africa: heuningbossie, stroopbossie, voëltjie-kannie-sit-nie; sehlabane (Sotho)

M. humile Eckl. & Zeyh.

South Africa: heuningbossie

Meloneura Raf. Lentibulariaceae

Origins:

See C.S. Rafinesque, *Flora Telluriana.* 4: 109. 1836 [1838].

Melongena Mill. Solanaceae

Origins:

Arabic *badingian*, Greek *melizana*; see C.T. Onions, *The Oxford Dictionary of English Etymology.* Oxford University Press 1966; Helmut Genaust, *Etymologisches Wörterbuch der botanischen Pflanzennamen.* 380. Basel 1996; Yuhanna ibn Sarabiyun [Joannes Serapion], *Liber aggregatus in medicinis simplicibus.* Venetijs 1479; Serapiom, *El libro agregà de Serapiom.* A cura di G. Ineichen. Venezia-Roma 1962-1966; P. Sella, *Glossario latino italiano.* Stato della Chiesa–Veneto–Abruzzi. Città del Vaticano 1944; S. Battaglia, *Grande dizionario della lingua italiana.* X: 16. Torino 1978.

Melopepo Mill. Cucurbitaceae

Origins:

Greek *melopepon* "an apple shaped melon, cucumber-melon," Latin *melopepo, melopeponis*, Plinius.

Melosperma Benth. Scrophulariaceae

Origins:

From the Greek *melon* "an apple" and *sperma* "seed," large seeds.

Melothria L. Cucurbitaceae

Origins:

Latin *melothron* or *melotrum* for a plant, the white bryony, also called *vitis alba* (Plinius), Greek *melothron* (*melon* "an apple"), ancient name used by Theophrastus and Dioscorides for a kind of white grape or a plant related to the genus *Bryonia*; see Carl Linnaeus, *Species Plantarum.* 35. 1753 and *Genera Plantarum.* Ed. 5. 21. 1754.

Melothrianthus Mart. Crov. Cucurbitaceae

Origins:

From the genus *Melothria* L. and Greek *anthos* "flower."

Melpomene A.R. Sm. & R.C. Moran Grammitidaceae

Origins:

Melpomene, the songstress, the muse of tragedy, of tragic and lyric poetry.

Memecylanthus Gilg & Schltr. Alseuosmiaceae

Origins:

Latin *memecylon* and Greek *memekylon*, ancient name for the edible fruit of the strawberry tree plus *anthos* "flower."

Memecylon L. Melastomataceae (Memecylaceae)

Origins:

From the Greek *memekylon*, ancient name for the fruit of *Arbutus unedo*, the strawberry tree; Latin *memecylon, i* "the edible fruit of the strawberry-tree" (Plinius); see Carl Linnaeus, *Species Plantarum.* 349. 1753 and *Genera Plantarum.* Ed. 5. 166. 1754.

Species/Vernacular Names:

M. sp.

Malaya: nipis kulit, mangas, delek, kelat, kelat berdarah, ubah bukit

N. Rhodesia: muzele, muzelekavangu

M. bachmannii Engl. (*Memecylon grandiflorum* A. & R. Fernandes) (the specific name after the medical doctor Franz Ewald Bachmann, 1856- *circa* 1916, naturalist, collector in Pondoland in 1880s; see J. Lanjouw and F.A. Stafleu, *Index Herbariorum.* 2: 47. Utrecht 1954; Eve Palmer & Norah Pitman, *Trees of Southern Africa.* 1: 611. Cape Town 1972; Mary Gunn and Leslie E. Codd, *Botanical Exploration of Southern Africa.* 84. Cape Town 1981; Frans A. Stafleu and Erik A. Mennega, *Taxonomic Literature. Supplement I: A-Ba.* 232-235. Königstein 1992; John H. Barnhart, *Biographical Notes upon Botanists.* 1: 98. 1965; T.W. Bossert, *Biographical Dictionary of Botanists Represented in the Hunt Institute Portrait Collection.* 19. 1972)

English: Pondo rose-apple, Pondoland memecylon

Southern Africa: Pondoroosappel; umBande, umBondi (Xhosa)

M. natalense Markg. (*Memecylon australe* Gilg & Schltr.)

English: Natal rose-apple, Natal memecylon

Southern Africa: Natalroosappel, isiKwelamfene esikulu; uGalagala-oluncinci (Xhosa)

M. sousae A. & R. Fernandes

English: Tonga rose-apple

South Africa: Tongaroosappel

Memorialis Buch-Ham. ex Wedd. Urticaceae

Origins:

Latin *memorialis* "belonging to memory."

Menadena Raf. Orchidaceae

Origins:

Greek *mene* "the crescent moon, moon" and *aden* "gland," the shape of the glands; see C.S. Rafinesque, *Flora Telluriana.* 2: 98. 1836 [1837].

Menadenium Raf. Orchidaceae

Origins:

From the Greek *mene* "the moon, the crescent moon" and *aden* "gland," the shape of the callus at the base of the lip; see C.S. Rafinesque, *Flora Telluriana.* 4: 45. 1836 [1837].

Mendoncella A.D. Hawkes Orchidaceae

Origins:

After Luis Mendonça, editor of the orchid journal *Orquidea* [Sociedad Mexicana Amigos de las Orquideas]; see Hubert Mayr, *Orchid Names and Their Meanings.* 158. Vaduz 1998. Some suggest that the genus was named for the Portuguese botanist Francisco de Ascenção Mendonça, 1899-1982, traveler, naturalist of the Instituto Botânico da Universidade de Coimbra, with Arthur Wallis Exell, Eduardo José Santos Moreira Mendes and Abílio Fernandes wrote *Conspectus florae angolensis.* Lisboa [1937-] 1970, with A.W. Exell wrote "Novidades da flora de Angola." *Bol. Soc. Brot.* 25: 101-112. 1951; see [Missão Botânica de Moçambique], "Itinerário Fitogeográfico da Campanha de 1942 da Missão Botânica de Moçambique." in *Ann. Junt. Miss. Geogr. e Invest. Colon.* 3, 1. 1948; John H. Barnhart, *Biographical Notes upon Botanists.* 2: 475. 1965; T.W. Bossert, *Biographical Dictionary of Botanists Represented in the Hunt Institute Portrait Collection.* 264. 1972; António de Figueiredo Gomes e Sousa, *Dendrologia de Moçambique. Estudo geral.* Lourenço Marques 1966.

Mendoncia Vell. ex Vand. Mendonciaceae (Acanthaceae)

Origins:

Named for Cardinal Mendonca.

Species/Vernacular Names:

M. sp.

Peru: añallio-caspi

Menepetalum Loes. Celastraceae

Origins:

From the Greek *mene* "the moon" and *petalon* "petal."

Menephora Raf. Orchidaceae

Origins:

Greek *mene* "the moon" and *phoros* "bearing," the lunate staminode; see C.S. Rafinesque, *Flora Telluriana.* 4: 46. 1836 [1837].

Meniscium Schreber Thelypteridaceae

Origins:

From *meniskos*, diminutive of the Greek *meis, menos* "month, the lunar month, crescent moon, crescent-shaped figure," referring to the fruits.

Meniscogyne Gagnepain Urticaceae

Origins:

From *meniskos* "lunar crescent" and *gyne* "woman, female."

Menisorus Alston Thelypteridaceae

Origins:

From the Greek *mene* "the moon, the crescent moon" and *soros* "a vessel, urn, a heap, a spore case."

Menispermum L. Menispermaceae

Origins:

Moonseed, from the Greek *mene* "the moon, the crescent moon" and *sperma* "seed," referring to the shape of the seeds.

Species/Vernacular Names:

M. canadense L.

English: moonseed, yellow parilla

M. davuricum DC.

English: Siberian moonseed

China: bian fu ge

Menkea Lehm. Brassicaceae

Origins:

Named for the German malacologist Karl Theodor Menke, 1791-1861, physician, botanist. Among his works are Dissertatio inauguralis botanico-philologico-medica *de Leguminibus veterum*. Goettingae 1814, *Synopsis methodica Molluscorum generum omnium et specierum earum, quae in Museo Menkeano adservantur.* Pyrmonti [Pyrmont] 1828 and *Molluscorum Novae Hollandiae specimen,* quod ... praefuit ... J.G.C. Lehmann ... scripsit C.T. Menke. Hannoverae 1843; see J.G.C. Lehmann (1792-1860), *Delectus seminum quae in horto hamburgensium botanico e collectione* etc. Hamburgi 1849 [-1852].

Species/Vernacular Names:

M. australis Lehm. (*Stenopetalum procumbens* Hook.; *Menkea procumbens* (Hook.) F. Muell.; *Menkea coolgardiensis* S. Moore)

English: fairy spectacles

Menoceras (R. Br.) Lindley Goodeniaceae

Origins:

From the Greek *mene* "the moon" and *keras* "a horn."

Menodora Bonpl. Oleaceae

Origins:

Greek *mene* "the moon" and *doron* "gift," referring to the flowering period; or from *mene* "the moon" and *dory* "spear, shaft, pole," referring to the appearance of fruit, see James C. Hickman, ed., *The Jepson Manual: Higher Plants of California.* 776. University of California Press, Berkeley 1993.

Menodoropsis (A. Gray) Small Oleaceae

Origins:

Resembling the genus *Menodora.*

Menonvillea R. Br. ex DC. Brassicaceae

Origins:

For the French naturalist Nicolas (Nicholas) Joseph Thiéry de Menonville, 1739-1780, lawyer, in 1776 to Mexico and South America, see *Traité de la culture du nopal, et de l'éducation de la cochenille, dans les colonies françaises de l'Amerique*; précédé d'un voyage à Guaxaca. (Éloge de M. Thiéry de Menonville, par M. Arthaud.) Paris 1787; see Gordon Douglas Rowley, *A History of Succulent Plants.* Strawberry Press, Mill Valley, California 1997; Ida Kaplan Langman, *A Selected Guide to the Literature on the Flowering Plants of Mexico.* 1964; Irving William Knobloch, compil., "A preliminary verified list of plant collectors in Mexico." *Phytologia Memoirs.* VI. 1983.

Mentha L. Labiatae

Origins:

The Latin name for mint, *menta, mentha*; Greek *mintha, minthe, minthes*; Akkadian *mitum* "dead," Minthe was a daughter of Cocytus, loved by Pluto, Ovidius: "... in olentes mentas", *Met.* X, 729; see Carl Linnaeus, *Species Plantarum.* 576. 1753 and *Genera Plantarum.* Ed. 5. 250. 1754; M. Cortelazzo & P. Zolli, *Dizionario etimologico della lingua italiana.* 3: 741. Bologna 1983; Giovanni Semerano, *Le origini della cultura europea.* Dizionario della lingua Latina e di voci moderne. 2(2): 472. Leo S. Olschki Editore, Firenze 1994; G. Semerano, *Le origini della cultura europea.* Dizionari Etimologici. Basi semitiche delle lingue indeuropee. Dizionario della lingua Greca. 2(1): 184. Firenze 1994.

Species/Vernacular Names:

M. sp.

French: menthe

South Laos (people Nya Hön): phak a tom

Congo: nanaye

M. aquatica L. (*Mentha hirsuta* Huds.)

English: watermint

Brazil: hortelã

M. arvensis L. (*Mentha austriaca* Jacq.)

English: corn mint, field mint, Japanese mint, mint, peppermint, marsh mint

Brazil: hortelã-do-brasil, hortelã-japonesa, hortelã-menta

The Philippines: yerba buena, herba buena, ablebana, hierba buena

China: bo he, po ho, pa ho, fan ho

Vietnam: bac ha, bac ha nam

M. arvensis L. var. *haplocalyx* Briq.

English: corn mint, field mint, Japanese mint

China: bohe

M. arvensis L. var. *piperascens* Malinv.

English: Japanese mint

Japan: hakka

M. asiatica Boriss.

English: Asian mint

China: jia bo he

M. australis R. Br.

English: river mint, native mint, native peppermint, Australian mint

M. citrata Ehrhart (*M.* x *piperita* L. var. *citrata* (J.F. Ehrh.) Briq.; *Mentha aquatica* var. *citrata* (J.F. Ehrh.) Benth.)

English: bergamot mint, lemon mint

China: ning nemg liu lan xiang

M. crispata Schrader ex Willdenow

English: wrinkled-leaf mint

China: zhou ye liu lan xiang

M. dahurica Fischer ex Bentham

English: Dahurian thyme

China: xing an bo he

M. diemenica Sprengel (after Van Diemen's Land, now Tasmania)

English: slender mint

M. x *gracilis* Sole. (*Mentha* x *gentilis* L.; *Mentha cardiaca* (S.F. Gray) Bak.)

English: gingermint, redmint, Scotch mint

Brazil: hortelã

M. laxiflora Benth.

English: forest mint

M. longifolia (L.) Hudson (*Mentha sylvestris* L.; *Mentha tomentosa* D'Urv.; *Mentha incana* Willd.)

English: mint, pennyroyal, horsemint

Arabic: habaq, dabbab, nemdar

China: ou bo he

Brazil: levante, hortelã

Southern Africa: inXina, inZinziniba (Xhosa)

M. x *piperita* L.

English: peppermint

China: la bo he

Peru: hierba buena

Brazil: hortelã, hortelã-de-folha-longa, menta

Mexico: bete, biti, pete, piti, nocuana bete, nocuana beti, nocuana pete, nocuana piti

M. pulegium L.

English: pennyroyal, European pennyroyal, pennyroyal mint, pudding grass

Brazil: poejo, poejo-do-campo, dictamo-da-virgínia

China: chun e bo he

Arabic: habaq, flayou, fliou, fulayya, fulayha

M. requienii Benth.

English: Corsican royal, menthella, crème-de-menthe plant

M. sachalinensis (Briquet ex Miyabe & Miyake) Kudô

English: Sachalin mint

China: dong bei bo he

M. satureioides R. Br.

English: creeping mint, native pennyroyal

M. spicata L. (*Mentha viridis* (L.) L.; *Mentha crispa* L.)

English: spearmint, garden mint

Peru: hierba buena, khoa

Central America: hierba buena, alavina, arvino, menta dulce, pan sut, yerba buena

Brazil: hortelã-de-folha-minuda, hortelã, menta, hortelã comum

Japan: midori-hakka

China: liu lan xiang, xiang hua cai

Arabic: naanaa, na'na', nemdar, hana

Hawaii: kepemineka

M. suaveolens Ehrh. (*Mentha macrostachya* Ten.; *Mentha insularis* Req.)

English: applemint, woolly mint

China: yuan ye bo he

Arabic: mersit, mersita, timersidi, timersat

M. vagans Boriss.

English: grey mint

China: hui bo he

Mentodendron Lundell Myrtaceae

Origins:
Aromatic tree, from Greek *dendron* "tree" and Latin *menta* "mint."

Mentzelia L. Loasaceae

Origins:
Named for the German botanist Christian (Christianus) Mentzel, 1622-1701, physician. Among his writings are *Index nominum plantarum universalis multilinguis*. Berolini 1682 and *Sylloge Minutiarum Lexici Latino-Sinico-Characteristici*. Norimbergae 1685; see Mariella Azzarello Di Misa, a cura di, *Il Fondo Antico della Biblioteca dell'Orto Botanico di Palermo*. 185. Regione Siciliana, Palermo 1988.

Species/Vernacular Names:
M. laevicaulis (Douglas) Torrey

English: blazing star

Menyanthes L. Menyanthaceae

Origins:
Menyanthos, a Greek classical name for a water plant, possibly from the Greek *mene* "moon, crescent moon" and *anthos* "flower," or from *minyos* "small, tiny" and *anthos*; see Carl Linnaeus, *Species Plantarum*. 145. 1753 and *Genera Plantarum*. Ed. 5. 71. 1754.

Species/Vernacular Names:
M. trifoliata L.

English: bogbean, buckbean

Brazil: trevo-aquático, trifólio-librino, trevo-da-água, trevo-dos-charcos

China: shi cai, shui cai, shui tsai, ming tsai, cho tsai, tsui tsao

Menziesia Smith Ericaceae

Origins:
After the British (Scotsman) physician Archibald Menzies, 1754-1842 (London), pupil of Hope, zoologist, gardener and botanist, 1782 naval surgeon Royal Navy, collected Scottish plants for John Fothergill and William Pitcairn, plant collector in Western Australia, 1790 Fellow of the Linnean Society, from 1791 to 1795 with Captain George Vancouver (1757-1798) on the voyage of the *Discovery* and *Chatham*, with Sir Everard Home (1756-1832) wrote *A*

Description of the Anatomy of the Sea Otter. (Read before the Royal Society, May 26, 1796.) [London 1796]. His works include *Hawaii Nei* 128 years ago. [Edited by W.F. Wilson] Honolulu 1920 and *Menzies' Journal of Vancouver's Voyage, April to October, 1792*. Edited, with ... notes, by C.F. Newcombe ... and a biographical note by J. Forsyth. Victoria 1923 [Archives of British Columbia. Memoir no. 5]. See George Vancouver, *A Voyage of Discovery to the North Pacific Ocean and Round the World*. London 1798; G. Godwin, *Vancouver: A Life 1757-1798*. London 1930; B. Anderson, *Surveyor of the Sea: The Life and Voyages of George Vancouver*. Seattle 1960; E. Bell, ed., "The log of the *Chatham* (by Peter Puget)." *Honolulu Mercury*. 1, 4. 1929; W. Kaye Lamb, ed., *The Voyage of George Vancouver 1791-1795*. London 1984; H. Suzanne Maxwell and Martin F. Gardner, "The quest for Chilean green treasure: some notable British collectors before 1940." *The New Plantsman*. 4(4): 195-214. December 1997; R. Zander, F. Encke, G. Buchheim and S. Seybold, *Handwörterbuch der Pflanzennamen*. 14. Aufl. 750. Stuttgart 1993; Ray Desmond, *Dictionary of British & Irish Botanists and Horticulturists*. 482-483. 1994; John H. Barnhart, *Biographical Notes upon Botanists*. 2: 476. 1965; H.R. Fletcher, *Story of the Royal Horticultural Society, 1804-1968*. Oxford 1969; T.W. Bossert, *Biographical Dictionary of Botanists Represented in the Hunt Institute Portrait Collection*. 264. 1972; H.N. Clokie, *Account of the Herbaria of the Department of Botany in the University of Oxford*. 209-210. Oxford 1964; Jeannette Elizabeth Graustein, *Thomas Nuttall, Naturalist. Explorations in America, 1808-1841*. Harvard University Press 1967; Leonard Huxley, *Life and Letters of Sir J.D. Hooker*. London 1918; S. Lenley et al., *Catalog of the Manuscript and Archival Collections and Index to the Correspondence of John Torrey*. Library of the New York Botanical Garden. 285. 1973; Warren R. Dawson, *The Banks Letters*, a Calendar of the Manuscript Correspondence of Sir Joseph Banks. 604-607. London 1958; Jonas C. Dryander, *Catalogus bibliothecae historico-naturalis Josephi Banks*. London 1796-1800; Mary Gunn and Leslie E. Codd, *Botanical Exploration of Southern Africa*. 250. Cape Town 1981; A. Lasègue, *Musée botanique de Benjamin Delessert*. Paris 1845; J. Ewan, ed., *A Short History of Botany in the United States*. 76. New York and London 1969; Günther Schmid, *Chamisso als Naturforscher*. Eine Bibliographie. Leipzig 1942; John T. Walbran, *British Columbia Coast Names, 1592-1906. To Which are Added a Few Names in Adjacent United States Territory, Their Origin and History*. First edition. Ottawa: Government Printing Bureau, 1909; J.D. Milner, *Catalogue of Portraits of Botanists Exhibited in the Museums of the Royal Botanic Gardens*. Royal Botanic Gardens, Kew, London 1906; I.C. Hedge and J.M. Lamond, *Index of Collectors in the Edinburgh Herbarium*. Edinburgh 1970; M. Hadfield et al., *British Gardeners: A Biographical Dictionary*. London 1980; G.P.V. Akrigg & Helen B. Akrigg, *British Columbia Place*

Names. Sono Nis Press, Victoria 1986; Jonathan Wantrup, *Australian Rare Books, 1788-1900*. Sydney 1987.

Species/Vernacular Names:
M. ferruginea Smith
English: mock azalea

Mephitidia Reinw. ex Blume Rubiaceae

Origins:
Latin *mephitis, is* "a noxious exhalation."

Meranthera Tieghem Loranthaceae

Origins:
From the Greek *meris, meros* "part, portion, share," *merizo* "make a division" and *anthera* "anther."

Merathrepta Raf. Gramineae

Origins:
See Constantine Samuel Rafinesque (1783-1840), *Seringe Bull. Bot.* 1: 221. 1830.

Merciera A. DC. Campanulaceae

Origins:
For the French botanist Marie Philippe Mercier, 1781-1831, plant collector, traveler, a friend of Nicolas Charles Seringe (1776-1858).

Mercklinia Regel Proteaceae

Origins:
After the German botanist Carl (Karl) Eugen von Mercklin, 1821-1904, plant physiologist, professor of botany at St. Petersburg, author of *Zur Entwicklungsgeschichte der Blattgestalten*. Beobachtungen von C.E. von M. Jena 1846 and *Palaeodendrologikon Rossicum*. St. Petersburg 1855 [1856?]. See J.H. Barnhart, *Biographical Notes upon Botanists*. 2: 477. 1965; Ethelyn Maria Tucker, *Catalogue of the Library of the Arnold Arboretum of Harvard University*. Cambridge, Massachusetts 1917-1933; Arthur D. Chapman, ed., *Australian Plant Name Index*. 1981. Canberra 1991; Stafleu and Cowan, *Taxonomic Literature*. 3: 421-422. 1981.

Mercurialis L. Euphorbiaceae

Origins:
From the Latin *Mercurialis, e* "belonging to the god Mercury," *Mercurius, ii* "the son of Jupiter and Maia," originally *herba mercurialis*; see Carl Linnaeus, *Species Plantarum*. 1035. 1753 and *Genera Plantarum*. Ed. 5. 457. 1754.

Species/Vernacular Names:
M. spp.
English: mercury
M. annua L.
English: annual mercury, mercury, French mercury
Arabic: mourkeba, halbub, bou zenzir
South Africa: bingelkruie
M. perennis L.
English: dog's mercury

Mercuriastrum Fabr. Euphorbiaceae

Origins:
Resembling the genus *Mercurialis* L.

Merendera Ramond Colchicaceae (Liliaceae)

Origins:
Latin *merenda, ae* "an afternoon luncheon, a feed for the beast, midday meal."

Meriana Vent. Melastomataceae

Origins:
After a German-Dutch painter of plants and insects, Maria Sybilla Merian afterwards Graff, 1647-1717; see Mariella Azzarello Di Misa, a cura di, *Il Fondo Antico della Biblioteca dell'Orto Botanico di Palermo*. 186. Regione Siciliana, Palermo 1988; Kurt Wettengl, ed., *Maria Sybilla Merian 1647-1717*. Künstlerin und Naturforscherin. Gerd Hatje Verlag 1998; Asher Rare Books & Antiquariaat Forum, *Catalogue Natural History*. item no. 109. The Netherlands 1998.

Meriandra Benth. Labiatae

Origins:
From the Greek *meris, meros* "part, portion, share," *merizo* "divide, make a division" and *aner, andros* "man, stamen."

Meriania Sw. Melastomataceae

Origins:

After a German-Dutch painter of plants and insects, Maria Sybilla Merian afterwards Graff, 1647-1717, author of *Dissertatio de generatione et metamorphosibus insectorum Surinamensium*. [nomina latina adjecit Caspar Commelyn] The Hague 1726. See W. Blunt, *The Art of Botanical Illustration*. 127-29. London 1950; W. Blunt and S. Raphael, *The Illustrated Herbal*. London 1979; Bertus Aafjes, *Maria Sybilla Merian*. [First edition, 8vo.] Amsterdam 1952; Leonard de Vries, *De exotische kunst van Maria Sybilla Merian*. Amsterdam 1984; S. Sitwell and W. Blunt, *Great Flower Books 1700-1900: A Bibliographical Record of Two Centuries of Finely Illustrated Flower Books*. 5-6, 18. London 1956; Kurt Wettengl, ed., *Maria Sybilla Merian 1647-1717. Künstlerin und Naturforscherin*. Gerd Hatje Verlag 1998.

Merianthera Kuhlm. Melastomataceae

Origins:

From the Greek *meris, meros* "part, portion, share" and *anthera* "anther."

Mericarpaea Boiss. Rubiaceae

Origins:

From the Greek *meris, meros* "part, portion, share" and *karpos* "fruit."

Mericocalyx Bamps Rubiaceae

Origins:

From the Greek *merikos* "minutely subdivided" and *kalyx* "calyx."

Merinthopodium J. Donn. Sm. Solanaceae

Origins:

From the Greek *merinthos* "cord, line, string" and *podion* "a little foot."

Merinthosorus Copel. Polypodiaceae

Origins:

From the Greek *merinthos* "cord, line, string" and *soros* "a vessel, heap."

Meriolix Raf. ex Endl. Onagraceae

Origins:

See C.S. Rafinesque, *Am. Monthly Mag. Crit. Rev.* 4: 192. 1819 and *Jour. Phys. Chim. Hist. Nat.* 89: 259. 1819; E.D. Merrill, *Index rafinesquianus*. 177. 1949.

Merisachne Steud. Gramineae

Origins:

Greek *meris, meros* "part, portion, share" and *achne* "husk, glume."

Merismia Tieghem Loranthaceae

Origins:

From the Greek *merisma* "part."

Merismostigma S. Moore Rubiaceae

Origins:

From the Greek *merismos* "dividing, division" and *stigma* "stigma."

Meristotropis Fischer & C.A. Meyer Fabaceae

Origins:

From the Greek *meristos* "divided" and *tropis, tropidos* "a keel," referring to the flower.

Merleta Raf. ex Endl. Onagraceae

Origins:

See Constantine Samuel Rafinesque, *Autikon botanikon. Icones plantarum select. nov. vel rariorum, etc*. 49. Philadelphia 1840; E.D. Merrill, *Index rafinesquianus*. 155. 1949.

Merope M. Roemer Rutaceae

Origins:

Merope is one of the Atlantides, *vid*. in the constellation of the Pleiades.

Merostachys Sprengel Gramineae

Origins:

Greek *meris, meros* "part, portion, share" and *stachys* "spike."

Merostela Pierre Meliaceae

Origins:

Greek *meris, meros* "part, portion" and *stele* "a pillar, column."

Merremia Dennst. ex Endlicher Convolvulaceae

Origins:

For the German naturalist Blasius Merrem, 1761-1824, botanist, mathematician, 1804 professor of political economy and botany at Marburg. His works include *Avium rariorum et minus cognitarum icones et descriptiones ... e Germanicis Latinae factae. Lipsiae* 1786, *Reise nach Paris im August und September, 1798,* etc. 1800, *De Animalibus Scythicis apud Plinium ... disputabit B.M.* etc. Gottingae [1781], *Handbuch der Pflanzenkunde nach dem Linneischen System.* Marburg 1809 and *Index plantarum horti academici Marburgensis.* Marburg 1807; see J.H. Barnhart, *Biographical Notes upon Botanists.* 2: 478. 1965; Georg Christian Wittstein, *Etymologisch-botanisches Handwörterbuch.* 572f. Ansbach 1852.

Species/Vernacular Names:

M. spp.

Yoruba: olohun adunmo

M. aegyptia (L.) Urb. (*Ipomoea aegyptia* L.; *Ipomoea pentaphylla* (L.) Jacq.; *Operculina aegyptia* (L.) House)

English: hairy merremia

Hawaii: koali kua hulu, kuahulu

Yoruba: okoju orisa, moki

M. boisiana (Gagnep.) van Ooststr.

English: Bois merremia

China: jin zhong teng

M. cordata C.Y. Wu & R.C. Fang

English: heart-leaf merremia

China: xin ye shan tu gua

M. discoidesperma (J.D. Sm.) O'Don.

English: Mary's bean

M. dissecta (Jacq.) Hall.f.

English: dissected merremia

China: duo lie yu huang cao

M. emarginata (Burm.f.) Hall.f.

English: emarginate merremia

China: shen ye shan zhu cai

M. gemella (Burm.f.) Hall.f.

English: twin merremia

China: jin hua yu huang cao

M. hederacea (Burm.f.) Hall.f.

English: ivy-like merremia

China: li lan wang

Yoruba: adere eko, ata koko, irin wanjanwanjan

M. hirta (L.) Merrill

English: hairy merremia

China: mao shan zhu cai

M. hungaiensis (Lingelsh. & Borza) R.C. Fang

English: tuberous merremia

China: shan tu gua

M. longipedunculata (C.Y. Wu) R.C. Fang

English: long-peduncle merremia

China: chang sheng tu gua

M. quinata (R. Br.) van Ooststr.

English: finger-like merremia

China: zhi ye shan zhu cai

M. sibirica (L.) Hall.f.

English: Siberian merremia

China: bei yu huang cao

M. tridentata (L.) Hallier f.

Malaya: kangkong pasir

M. tridentata (L.) Hallier f. subsp. *angustifolia* (Jacq.) Van Ooststr. var. *angustifolia* (*Merremia angustifolia* (Jacq.) Hallier f.; *Merremia tridentata* (L.) Hallier f. subsp. *angustifolia* (Desr.) Van Ooststr.)

English: merremia

India: prasarini, talanili, tala nili, sendera-clandi, cantrakranti

M. tuberosa (L.) Rendle (*Ipomoea tuberosa* L.; *Operculina tuberosa* (L.) Meissn.)

English: wood rose, yellow morning glory, Spanish woodbine, Hawaiian wood rose, Ceylon merremia, Brazilian jalap

Japan: bara-asa-gao

Hawaii: pilikai

M. vitifolia (Burm.f.) Hall.f.

English: grapeleaf merremia

China: zhang ye yu huang cao

Malaya: lulang bulu

M. yunnanensis (Courchet & Gagnepain) R.C. Fang

English: Yunnan merremia

China: lan hua tu gua

Merrillanthus Chun & Tsiang Asclepiadaceae

Origins:

After the American botanist Elmer Drew Merrill, 1876-1956, plant collector. See J.H. Barnhart, *Biographical Notes upon Botanists.* 2: 479. 1965; T.W. Bossert, *Biographical Dictionary of Botanists Represented in the Hunt Institute Portrait Collection.* 264. 1972; S. Lenley et al., *Catalog of the Manuscript and Archival Collections and Index to the Correspondence of John Torrey.* Library of the New York Botanical Garden. 286-289. 1973; Ethelyn Maria Tucker, *Catalogue of the Library of the Arnold Arboretum of Harvard University.* Cambridge, Massachusetts 1917-1933.

Species/Vernacular Names:

M. hainanensis Chun & Tsiang

English: Hainan merrillanthus

China: tuo feng teng

Merrillia Swingle Rutaceae

Origins:

After the American (b. Maine) taxonomist Elmer Drew Merrill, 1876-1956 (d. Massachusetts), botanist and plant collector, from 1902 to 1923 with the Philippine Bureau of Agriculture, from 1906 to 1923 co-editor *The Philippine Journal of Science*, in 1919 became Director of the Bureau of Science, Manila, 1927-1929 Director of the California Botanic Garden, 1930-1935 Director of the New York Botanical Garden and professor of botany at Columbia University, in 1931 founder of *Brittonia*, Harvard University from 1935 to 1948 professor of botany, from 1937 Director of the Arnold Arboretum. Among his many and valuable works are *A Dictionary of the Plant Names of the Philippine Islands.* Manila 1903, *A Flora of Manila.* Manila 1912, *A Bibliography of Eastern Asiatic Botany.* 1938, "Loureiro and his botanical work." *Proc. Amer. Phil. Soc.* 72: 229-239. 1933, "Bibliography of Polynesian botany." *Bernice P. Bishop Mus. Bull.* 13: 1-68. 1924, *Plant Life of the Pacific World.* New York 1945, "A botanical bibliography of the islands of the Pacific." *Contr. U.S. Natl. Herb.* 30: 1-322. 1947 and "The botany of Cook's voyages." *Chron. Bot.* 14: 161-384. 1954; see J. Ewan, in *Dictionary of Scientific Biography* 15: 421-422. 1981; Stafleu and Cowan, *Taxonomic Literature.* 3: 425-429. 1981; J.H. Barnhart, *Biographical Notes upon Botanists.* 2: 479. 1965; R. Zander, F. Encke, G. Buchheim and S. Seybold, *Handwörterbuch der Pflanzennamen.* 14. Aufl. Stuttgart 1993; I.H. Vegter, *Index Herbariorum.* Part II (4), *Collectors M.* Regnum Vegetabile vol. 93. 1976; Joseph Ewan, *Rocky Mountain Naturalists.* The University of Denver Press 1950; T.W. Bossert, *Biographical Dictionary of Botanists Represented in the Hunt Institute Portrait Collection.* 264. 1972; S. Lenley et

al., *Catalog of the Manuscript and Archival Collections and Index to the Correspondence of John Torrey.* Library of the New York Botanical Garden. 286-289. 1973; Ethelyn Maria Tucker, *Catalogue of the Library of the Arnold Arboretum of Harvard University.* Cambridge, Massachusetts 1917-1933; J. Ewan, ed., *A Short History of Botany in the United States.* New York and London 1969; Gilbert Westacott Reynolds (1895-1967), *The Aloes of South Africa.* Balkema, Rotterdam 1982.

Species/Vernacular Names:

M. caloxylon (Ridley) Swingle

Malaya: kemuning gajah, kemuning limau, ketenggah

Merrilliodendron Kanehira Icacinaceae

Origins:

After the American taxonomist Elmer Drew Merrill, 1876-1956, botanist and plant collector; see J.H. Barnhart, *Biographical Notes upon Botanists.* 2: 479. 1965; I.H. Vegter, *Index Herbariorum.* Part II (4), *Collectors M.* Regnum Vegetabile vol. 93. 1976.

Merrilliopanax H.L. Li Araliaceae

Origins:

After the American taxonomist Elmer Drew Merrill, 1876-1956, botanist and plant collector; see J.H. Barnhart, *Biographical Notes upon Botanists.* 2: 479. 1965; I.H. Vegter, *Index Herbariorum.* Part II (4), *Collectors M.* Regnum Vegetabile vol. 93. 1976.

Mertensia Roth Boraginaceae

Origins:

Named for the German botanist Franz Karl (Carl) Mertens, 1764-1831, professor of botany at Bremen, with W.D.J. Koch (1771-1849) published ed. 3 of Johann Christoph Röhling (1757-1813), *Deutschlands Flora.* 1823, etc. See Jeannette Elizabeth Graustein, *Thomas Nuttall, Naturalist. Explorations in America, 1808-1841.* Harvard University Press 1967; S. Lenley et al., *Catalog of the Manuscript and Archival Collections and Index to the Correspondence of John Torrey.* Library of the New York Botanical Garden. 464. 1973; J.H. Barnhart, *Biographical Notes upon Botanists.* 2: 478. 1965; T.W. Bossert, *Biographical Dictionary of Botanists Represented in the Hunt Institute Portrait Collection.* 265. 1972; H.N. Clokie, *Account of the Herbaria of the Department of Botany in the University of Oxford.* 210. Oxford 1964; R. Zander, F. Encke, G. Buchheim and S. Seybold, *Handwörterbuch der Pflanzennamen.* 14. Aufl.

750. Stuttgart 1993; Stafleu and Cowan, *Taxonomic Literature*. 3: 430-431. 1981; Ida Kaplan Langman, *A Selected Guide to the Literature on the Flowering Plants of Mexico*. Philadelphia 1964; A. Lasègue, *Musée botanique de Benjamin Delessert*. Paris 1845; Günther Schmid, *Chamisso als Naturforscher*. Eine Bibliographie. Leipzig 1942.

Species/Vernacular Names:
M. bella Piper
English: Oregon lungwort
M. ciliata (Torrey) G. Don
English: streamside bluebells
M. cusickii Piper
English: Toiyabe bluebells
M. longiflora E. Greene
English: long bluebells
M. maritima (L.) Gray
English: gromwell, sea lungwort, oysterleaf
M. oblongifolia (Nutt.) G. Don
English: sagebrush bluebells
M. virginica (L.) Link
English: Virginian bluebell, Virginian cowslip

Mertensia Willd. Gleicheniaceae

Origins:
For the German botanist Franz Karl Mertens, 1764-1831; see Carl L. von Willdenow (1765-1812), in *Kongl. Vetenskaps Academiens Nya Handlingar*. 25: 166. Stockholm 1804.

Merxmuellera Conert Gramineae

Origins:
After the German botanist Hermann Merxmüller, 1920-1988. Among his writings are "Compositen-Studien I." *Mitt. Bot. Staatssamml. München* 1: 33-46. 1950 and *Prodromus einer Flora von Südwestafrika*. 1966-1972, with Wolfgang Engelhardt wrote [Was lebt in Tümpel, Bach und Weiher?] *The Young Specialist Looks at Pond-Life*, etc. [Translated by Heather J. Fisher ... Edited and adapted by Roderick C. Fisher.] London 1964, with A. Schreiber and Peter Frederick Yeo wrote "*Aster* L." in *Flora Europaea*. 4: 112-116. 1976; see Gustav Hegi, [*Alpenflora.*] *Flora alpina*, etc. [Translated from the German edition revised by H. Merxmüller.] Milano 1953; Mary Gunn and Leslie E. Codd, *Botanical Exploration of Southern Africa*. 251. Cape Town 1981.

Meryta Forst. & Forst.f. Araliaceae

Origins:
Greek *merytos* "glomerate," referring to the male flowers; see H. Genaust, *Etymologisches Wörterbuch der botanischen Pflanzennamen*. 382. 1996; R. Zander, F. Encke, G. Buchheim and S. Seybold, *Handwörterbuch der Pflanzennamen*. 14. Aufl. 378. 1993; Georg Christian Wittstein, *Etymologisch-botanisches Handwörterbuch*. 573. Ansbach 1852.

Species/Vernacular Names:
M. sinclairii (Hook.f.) Seem.
Maori name: puka

Mesadenella Pabst & Garay Orchidaceae

Origins:
The diminutive of *Mesadenus*, a genus of orchids.

Mesadenus Schltr. Orchidaceae

Origins:
From the Greek *mesos* "in the middle, middle" and *aden* "gland," gland lying between the two pollinia.

Mesandrinia Raf. Euphorbiaceae

Origins:
See Constantine S. Rafinesque, *Neogenyton, or Indication of Sixty-Six New Genera of Plants of North America*. 3. 1825; E.D. Merrill, *Index rafinesquianus*. 155. 1949.

Mesanthemum Koern. Eriocaulaceae

Origins:
From the Greek *mesos* "in the middle, middle" and *anthemon* "flower."

Mesanthophora H. Robinson Asteraceae

Origins:
From the Greek *mesos* "in the middle, middle," *anthos* "flower" and *phoreo* "to bear."

Mesanthus Nees Restionaceae

Origins:
From the Greek *mesos* "in the middle" and *anthos* "flower."

Mesaulosperma Slooten Flacourtiaceae

Origins:
Greek *mesos* "in the middle," *aulos* "a tube, flute," *aule* "the court-yard, court, hall," *mesaulos* "the inner court, inside the *aule*" and *sperma* "seed."

Mesechinopsis Y. Ito Cactaceae

Origins:
From the Greek *mesos* "in the middle" plus *Echinopsis* Zucc.

Mesechites Müll. Arg. Apocynaceae

Origins:
From the Greek *mesos* "in the middle" plus the genus *Echites*.

Mesembryanthemum L. Aizoaceae

Origins:
Latin *mesembrianthemum*, Greek *mesembria* "middle of the day, midday" and *anthemon* "flower," referring to the opening of the flowers in the sun; see Carl Linnaeus, *Species Plantarum*. 480. 1753 and *Genera Plantarum*. Ed. 5. 215. 1754; S. Battaglia, *Grande dizionario della lingua italiana*. X: 200. Torino 1978; some suggest from the Greek *mesos* "in the middle, middle," *embryon* "embryo" and *anthemon* "flower," referring to the position of the ovary.

Species/Vernacular Names:
M. aitonis Jacq. (*Cryophytum aitonis* (Jacq.) N.E. Br.; *Gasoul aitonis* (Jacq.) H. Eichler; *Mesembryanthemum angulatum* Thunb.; *Micropterum puberulum* (Haw.) Schwant.; *Micropterum sessiliflorum* (Ait.) Schwant. var. *album* (Haw.) Jacobsen)
English: sea spinach, angled iceplant
Southern Africa: brakslaai, soutslaai; mabone (Sotho)
M. crystallinum L.
English: iceplant, crystalline iceplant
M. guerichianum Pax (the specific name honors Georg Julius Ernst Gürich, 1859-1938, plant collector in southwest Africa, author of *Beiträge zur Geologie von West-Afrika*. Berlin 1887 and *Während des Krieges in Deutsch-Ostafrika und Südafrika ...* Berlin 1916; see Wm. Roger Louis, *Great Britain and Germany's Lost Colonies 1914-1919*. Oxford 1967; Hans Meyer, *Die Barundi*. Leipzig 1916; Ritchie Moore, *With Botha in the Field*. London 1915; Emmanuel Muller, *Les troupes du Katanga et les Campagnes d'Afrique* *1914-1918*. Bruxelles 1937; F.N. Hepper and F. Neate, *Plant Collectors in West Africa*. 34. 1971)
Namibia: brakslaai, volstruisslaai (Afrikaans)
M. nodiflorum L.
English: slender-leaved iceplant, Egyptian fig-marigold
Portuguese: barrilha

Mesicera Raf. Orchidaceae

Origins:
Greek *mesos* "middle" and *keras* "horn," referring to the stigmatic processes; see C.S. Rafinesque, *Neogenyton, or Indication of Sixty-Six New Genera of Plants of North America*. 4. 1825 and *Herb. Raf.* 72. 1833 and *Flora Telluriana*. 2: 39. 1836 [1837]; E.D. Merrill, *Index rafinesquianus*. 103. 1949.

Mesochlaena R. Br. ex J. Sm. Thelypteridaceae

Origins:
From the Greek *mesos* "in the middle, middle, half" and *chlaena* "cloak."

Mesoclastes Lindley Orchidaceae

Origins:
Greek *mesos* "in the middle" and *klastos* "broken in pieces," *mesoklastos* "broken off in the middle," referring to the lip.

Mesodactylis Wallich Orchidaceae

Origins:
From the Greek *mesodaktylon* "space between two fingers," *mesos* "in the middle" and *daktylos* "a finger," *mesodaktylos* "middle phalanx of a finger," the dorsal stamen.

Mesoglossum Halbinger Orchidaceae

Origins:
Greek *mesos* "in the middle" and *glossa* "tongue," referring to an intermediate position of this genus, between *Odontoglossum* and *Oncidium*.

Mesogyne Engl. Moraceae

Origins:
From the Greek *mesos* "in the middle" and *gyne* "female, woman."

Mesomelaena Nees Cyperaceae

Origins:

Greek *mesos* "in the middle, middle, half" and *melaina* "black" (the feminine form of *melas, melanos*), referring to the involucral bracts; see Johann Georg Christian Lehmann (1792-1860), *Plantae Preissianae ...* Plantarum quas in Australasia occidentali et meridionali-occidentali annis 1838-41 collegit L. Preiss. 2: 88. Hamburgi 1846.

Species/Vernacular Names:

M. tetragona (R. Br.) Benth.

Australia: semaphore sedge

Mesoneuron Ching Thelypteridaceae

Origins:

From the Greek *mesos* "in the middle, middle" and *neuron* "nerve."

Mesoneurum DC. Caesalpiniaceae

Origins:

Orthographic variant of *Mezonevron* Desf.

Mesophlebion Holttum Thelypteridaceae

Origins:

From the Greek *mesophlebion* "space between two veins."

Mesoptera Raf. Orchidaceae

Origins:

Greek *mesos* "middle" and *pteron* "wing," indicating the lateral wings on the column; see C.S. Rafinesque, *Herb. Raf.* 73. 1833 and *Flora Telluriana.* 4: 49. 1836 [1837].

Mesosorus Hassk. Gleicheniaceae

Origins:

From the Greek *mesos* "in the middle, middle" and *soros* "a heap."

Mesosphaerum P. Browne Labiatae

Origins:

Greek *mesosphairon*, Latin *mesosphaerum, i* "a kind of nard with middle-sized leaves."

Mesospinidium Reichb.f. Orchidaceae

Origins:

Greek *mesos* "middle" and *spinidion*, diminutive of *spinos* "chaffinch," referring to the shape of the anther, the column of the flower is similar to the beak of these birds.

Mesostemma Vved. Caryophyllaceae

Origins:

From the Greek *mesos* "in the middle, middle" and *stemma* "a crown."

Mesotriche Stschegl. Epacridaceae

Origins:

From the Greek *mesos* "in the middle, middle" and *thrix, trichos* "hair."

Mespilodaphne Nees Lauraceae

Origins:

Mespilus plus *Daphne.*

Mespilus L. Rosaceae

Origins:

Theophrastus (*HP.* 3.12.5) used the Greek *mespile* for the medlar-tree, the oriental thorn and the hawthorn; Latin *mespilum*, Greek *mespilon* "a medlar," Latin *mespilus, i* or *mespila, ae*, Greek *mespile* "a medlar-tree, a medlar."

Species/Vernacular Names:

M. germanica L.

English: medlar

Arabic: bousaa

Messermidia Raf. Boraginaceae

Origins:

Based on *Messerschmidia* L.; see Constantine Samuel Rafinesque (1783-1840), *Sylva Telluriana.* 167. 1838; E.D. Merrill, *Index rafinesquianus.* 203. 1949.

Messerschmidia Hebenstr. Boraginaceae

Origins:

After the German physician Daniel Gottlieb Messerschmidt (Messerschmid), 1685-1735, traveler, author of *Nachricht von D.G. Messerschmidt's siebenjähriger Reise in Sibirien.*

1781 and Dissertatio ... *de ratione praeside universae medicinae*. Praes. F. Hoffmanno. Halae Magdeburgicae [1713]. See [Aboul-Ghâzi, Husain Abu Al-Ghazi, Abu-l-Ghazi, Hossein Bakara], *Abulgasi Bagadur Chan's Geschlechtbuch der mungalisch-mogulischen oder mogorischen Chanen*. Aus einer türkischen Handschrift ins Teutsche übersetz von D. Dan. Gottlieb Messerschmid. Göttingen 1780; Johann C. Hebenstreit (1720-1795), in *Novi Commentarii Academiae Scientiarum Imperialis Petropolitanae*. 8: 315. 1763; Raymond Louis Specht (b. 1924), *Records of the American-Australian Scientific Expedition to Arnhem Land*. Melbourne 1958; Peter [Pyotr] Simon Pallas (1741-1811), ed., *Neue Nordische Beiträge*. St. Petersburg & Leipzig 1781-1796.

Messersmidia L. Boraginaceae

Origins:

For the German physician Daniel Gottlieb Messerschmidt (Messerschmid), 1685-1735.

Mestoklema N.E. Br. ex Glen Aizoaceae

Origins:

From the Greek *mestos* "full, filled" and *klema, klematos* "a shoot, twig."

Mesua L. Guttiferae

Origins:

To commemorate Yuhanna ibn Masawaih (Yahya ibn Musawi, Yuhanna Ibn Masawayh) (Joannes Mesuë, J. Damasceni, Joannis Mesue, Ioannis M., Iean de Damascene), a celebrated Arabian physician and botanist of Damascus. His works include *Liber de complexionibus proprietatibus simplicium medicinarum laxatiuarum* [and other works]. Padua 1471 and *Il libro della consolatione delle medicine*. [Modena] 1475. See A. Wagner, *La vie et l'oeuvre ophthalmologique de Jean Mésué*. Lyon 1932; Jean Tagault, *Commentariorum ... De purgantibus medicamentis simplicibus, libri II. In gratiam Pharmacopoeiae candidatorum, nuper in lucem editi*. Lyons 1549; Mesue, the Younger, *De re medica, libri tres*. Jacobo Sylvio [Jacques Dubois, Jacobus Sylvius] medico interprete. Paris 1542; Ernst H.F. Meyer, *Geschichte der Botanik*. III: 178-183. Königsberg 1854-1857; Carl Linnaeus, *Species Plantarum*. 515. 1753 and *Genera Plantarum*. Ed. 5. 231. 1754; Richard J. Durling, compil., *A Catalogue of Sixteenth Century Printed Books in the National Library of Medicine*. 3143. 1967; Konrad (Conrad) Gesner, *De chirurgia scriptores optimi quique veteres et recentiores*, plerique in Germania antehac non editi. Tiguri 1555; Garrison and Morton, *Medical Bibliography*.

5562. 1961; Giorgio Montecchi, *Aziende tipografiche, stampatori e librai a Modena dal Quattrocento al Settecento*. Mucchi, Modena 1988; Georg Christian Wittstein, *Etymologisch-botanisches Handwörterbuch*. 574. Ansbach 1852.

Species/Vernacular Names:

M. ferrea L.

English: ironwood, ironwood tree, Ceylon ironwood, Assam ironwood

Malaya: penaga, penaga lilin, penaga sabut, lenggapus, matopus, mentepus, nagasari, tapis

Nepal: nagesori

Metabolos Blume Rubiaceae

Origins:

Greek *metabolos* "changeable," Latin and Greek *metabole* for the transition to another key or set of tones.

Metabriggsia W.T. Wang Gesneriaceae

Origins:

Greek *meta* "other side, with, sharing, next to" plus *Briggsia* Craib.

Metachilum Lindley Orchidaceae

Origins:

Greek *meta* "near, other side, with, sharing, next to, between" and *cheilos* "lip," the union of the lip with the column foot.

Metaeritrichium W.T. Wang Boraginaceae

Origins:

From the Greek *meta* "other side, with, sharing, next to" plus the genus *Eritrichium* Schrad. ex Gaudin.

Species/Vernacular Names:

M. microuloides W.T. Wang

English: microula-like metaeritrichium

China: jing guo cao

Metalasia R. Br. Asteraceae

Origins:

Greek *meta* "changed in nature, with, sharing, next to" and *lasios* "hairy, woolly, shaggy," in allusion to the foliage of some species.

Species/Vernacular Names:

M. muricata (L.) D. Don (*Gnaphalium muricatum* L.; *Metalasia ericoides* Sieber ex DC.; *Metalasia umbellata* Cass.)

English: honey tree, white bristle bush

Southern Africa: witsteekbos, blombos, steekbos; lohlohlo, lehlohlo, sehalahala-se-seputsoa (South Sotho)

Metalepis Griseb. Asclepiadaceae

Origins:

From the Greek *meta* "changed in nature, next to" and *lepis* "scale."

Metalonicera M. Wang & A.G. Gu Caprifoliaceae

Origins:

From the Greek *meta* "changed in nature, next to, between, instead of" plus *Lonicera* L.

Metanarthecium Maxim. Melanthiaceae (Liliaceae)

Origins:

From the Greek *meta* "changed in nature, next to, between, instead of" plus *Narthecium* Hudson.

Metanemone W.T. Wang Ranunculaceae

Origins:

Greek *meta* "changed in nature, next to, between, instead of" plus *Anemone* L.

Metapetrocosmea W.T. Wang Gesneriaceae

Origins:

From the Greek *meta* "changed in nature, next to, between, instead of" plus *Petrocosmea* Oliver.

Metaplexis R. Br. Asclepiadaceae

Origins:

Greek *meta* "other side, change, sharing, next to" and *plektos* "twisted, plaited," *plexis* "plaiting, weaving," referring to stamens and corona or to the stems (for rope).

Species/Vernacular Names:

M. hemsleyana Oliver

English: Hemsley metaplexis

China: hua luo mo

M. japonica (Thunberg) Makino

English: Japanese metaplexis

China: luo mo, lo mo

Metapolypodium Ching Polypodiaceae

Origins:

From the Greek *meta* "changed in nature, next to, between, instead of" plus *Polypodium* L.

Metaporana N.E. Br. Convolvulaceae

Origins:

Greek *meta* "other side, sharing, next to, change" plus the genus *Porana* Burm.f.; see G.W. Staples, "The genus *Porana* (Convolvulaceae) in Australia." *Nuytsia*. 6(1): 51-59. 1987.

Metarungia Baden Acanthaceae

Origins:

From the Greek *meta* "other side, sharing, next to" and the genus *Rungia* Nees.

Metasasa Lin Gramineae

Origins:

From the Greek *meta* "change, sharing, next to" plus *Sasa* Makino & Shib.

Metasequoia Miki ex Hu & W.C. Cheng Taxodiaceae

Origins:

Greek *meta* and the genus *Sequoia*; see E. Charles Nelson, "*Metasequoia glyptostroboides*, the dawn redwood." in *Curtis's Botanical Magazine*. 15(1): 77-80. 1998; Earl of Rosse, "Dr. H.H. Hu's plant-hunting expeditions 1937-1940." *Yearbook of the International Dendrology Society*. 7-11. 1971.

Species/Vernacular Names:

M. glyptostroboides Hu & W.C. Cheng

English: dawn redwood, fossil tree

Japan: akebono-sugi

Metasocratea Dugand Palmae

Origins:
From the Greek *meta* "change, instead of, sharing, next to, after" and the genus *Socratea* Karsten.

Metastachydium Airy Shaw ex C.Y. Wu & H.W. Li Labiatae

Origins:
From the Greek *meta* "instead of, sharing, next to, after" and *stachys* "spike."

Species/Vernacular Names:
M. sagittatum (Regel) C.Y. Wu & H.W. Li (*Phlomis sagittata* Regel; *Metastachys sagittata* (Regel) Knorring)

English: arrow-leaf betony

China: jian ye shui su

Metastachys (Benth.) Tieghem Loranthaceae

Origins:
From the Greek *meta* "instead of, sharing, next to" and *stachys* "spike."

Metastachys Knorring Labiatae

Origins:
From the Greek *meta* "instead of, sharing, next to, after" and *stachys* "spike."

Metastelma R. Br. Asclepiadaceae

Origins:
From the Greek *meta* "other side, next to" and *stelma* "a girdle, belt," referring to the corolla.

Metastevia Grashoff Asteraceae

Origins:
From the Greek *meta* "other side, next to" plus the genus *Stevia* Cav.

Metathelypteris (H. Ito) Ching Thelypteridaceae

Origins:
From the Greek *meta* "instead of, sharing, next to, after" with the genus *Thelypteris* Schmidel.

Metatrophis F.B.H. Br. Moraceae

Origins:
Greek *meta* "instead of, sharing, next to, after" plus *Trophis* P. Browne.

Metaxya C. Presl Metaxyaceae

Origins:
Greek *metaxy* (late form *metoxy*) "between, in the midst," possibly referring to the sori; see R.E.G. Pichi Sermolli, "Fragmenta Pteridologiae." in *Webbia.* 24(2): 699-722. 28 Apr. 1970.

Metcalfia Conert Gramineae

Origins:
After the British botanist Charles Russell Metcalfe, 1904-1991, plant anatomist, traveler, botanical explorer, plant collector in W. Cameroon (1937 with Hutchinson), Keeper of the Jodrell Laboratory at Royal Botanic Gardens, Kew. His writings include "The wood structure of *Fokienia hodginsii* and certain related Coniferae." *Kew Bull.* 420-425. 1931, "The structure and botanical identity of some scented woods from the East." *Kew Bull. Misc. Inf.* 1933(1): 3-15, pl. 1-4. 1933, "The structure of some sandalwoods and their substitutes and of some other little known scented woods." *Bull. Misc. Inform.* 4: 165-195. 1935, "Some thoughts on the structure of bamboo leaves." *Bot. Mag. Tokyo.* 69: 391-400. 1956 and *Anatomy of the Monocotyledons. Gramineae.* London 1960, with Laurence Chalk (1896-1979) wrote *Anatomy of the Dicotyledons.* London 1950 [with the assistance of Mary Margaret Chattaway (b. 1899), Frederick Reginald Richardson (b. 1915), C.L. Hare and E.M. Slatter.]. See J.H. Barnhart, *Biographical Notes upon Botanists.* 2: 480. 1965; T.W. Bossert, *Biographical Dictionary of Botanists Represented in the Hunt Institute Portrait Collection.* 265. 1972; J. Ewan, ed., *A Short History of Botany in the United States.* 52. New York and London 1969; F.N. Hepper and F. Neate, *Plant Collectors in West Africa.* 55. 1971; Elmer Drew Merrill, *Contr. U.S. Natl. Herb.* 30(1): 211. 1947; Ida Kaplan Langman, *A Selected Guide to the Literature on the Flowering Plants of Mexico.* Philadelphia 1964.

Meteoromyrtus Gamble Myrtaceae

Origins:

Possibly referring to a myrtle of different climate, from the Greek *meteoros, meteoron* "raised up above the earth, off the ground, in air, natural phenomena, things in the air" and the genus *Myrtus* L.

Meterana Raf. Euphorbiaceae

Origins:

See C.S. Rafinesque, *Sylva Telluriana*. 65. 1838; E.D. Merrill, *Index rafinesquianus*. 155. 1949.

Methyscophyllum Ecklon & Zeyher Celastraceae

Origins:

Greek *methysko* "to make drunk with vine, to soak, to be drunken," *methyo, methyein* "to get drunk, to be full of, to be drunken with" and *phyllon* "leaf."

Methysticodendron R.E. Schultes Solanaceae

Origins:

From the Greek *methysko* "to make drunk with vine, to be drunken," *methystikos* "intoxicating, drunken" and *dendron* "tree," scopolamine leads to hallucinations after intoxication. See Richard Evans Schultes and Norman R. Farnworth, "Ethnomedical, botanical and phytochemical aspects of natural hallucinogeous." *Botanical Museum Leaflets*. 28(2): 123-214. Cambridge 1980; Richard Evans Schultes and Robert F. Raffauf, *Vine of the Soul. Medicine Men, their Plants and Rituals in the Colombian Amazonia*. Synergetic Press, Oracle, Arizona 1992 and *The Healing Forest. Medicinal and Toxic Plants of the Northwest Amazonia*. 420, 422, 430. Dioscorides Press, Portland, Oregon 1992; Holger Kalweit, *Shamans, Healers, and Medicine Men*. Shambhala, Boston and London 1992; Mark J. Plotkin, *Tales of a Shaman's Apprentice*. Viking 1993.

Species/Vernacular Names:
M. amesianum R.E. Schultes

Colombia: culebra borrachero

Metopium P. Browne Anacardiaceae

Origins:

Latin *metopion* or *metopium* for the gum of an African tree, also called *ammoniacum* (Plinius); Greek *metopon, metopion* "forehead, containing oil of bitter almonds," purging resins from *Metopium toxiferum* (L.) Krug & Urban.

Metrodorea St.-Hil. Rutaceae

Origins:

Probably from the Greek *metra* "core, heartwood or the heart of a tree" and *doron* "gift"; according to *Paxton's Botanical Dictionary* the genus was named after Metrodora Sabino, a botanical artist; Metrodorus was a physician of Chios, a disciple of Democritus.

Species/Vernacular Names:
M. pubescens St.-Hil. & Tul. (see Eurico Teixeira da Fonseca, "Plantas medicinales brasileñas." *R. Flora Medicinal*. Rio de Janeiro. 6(2): 95-110. 1939)

Brazil: caputuma, cataguá, laranjeira-do-mato, chupa-machado, limoeiro-do-mato

Metrosideros Banks ex Gaertner Myrtaceae

Origins:

Greek *metra* "core, heartwood, the heart of a tree" and *sideros* "iron," referring to the hardness of the wood; see Arthur D. Chapman, ed., *Australian Plant Name Index*. 1988-1993. Canberra 1991; J.W. Dawson, in *Blumea*. 23: 7-11. 1976; H.E. Connor and E. Edgar, "Name changes in the indigenous New Zealand flora, 1960-1986 and Nomina Nova IV, 1983-1986." in *New Zealand Journal of Botany*. Vol. 25: 115-170. 1987.

Species/Vernacular Names:
M. angustifolia Dum.-Cours. (*Myrtus angustifolia* L.)

South Africa: smalblad

M. excelsa Gaertn. (*Metrosideros tomentosa* A. Rich.)

English: Christmas tree

New Zealand: pohutukawa

M. macropus Hook. & Arnott (*Nania macropus* (Hook. & Arnott) Kuntze)

Hawaii: 'ohia'a, 'ohi'a lehua, lehua

M. polymorpha Gaudich. (*Nania glabrifolia* A. Heller; *Nania pumila* A. Heller)

Hawaii: 'ohia'a, 'ohi'a lehua, lehua

M. queenslandica L.S. Smith

Australia: pink myrtle, myrtle satin ash

M. robusta Cunn.

Maori names: rata

M. rugosa A. Gray (*Nania rugosa* (A. Gray) Kuntze)

Hawaii: lehua papa

M. scandens (Forst. & Forst.f.) Druce

Maori names: aka, akatawhiwhi, puatawhiwhi, aka kura

M. tremuloides (A. Heller) P. Knuth (*Nania tremuloides* A. Heller)

Hawaii: lehua 'ahihi, 'ahihi, 'ahihi ku ma kua, 'ahihi lehua, kumakua, 'ohi'a 'ahihi

M. umbellata Cav.

New Zealand: southern rata

Metroxylon Rottb. Palmae

Origins:

Greek *metra* "core, the heart of a tree" and *xylon* "wood," referring to the large pith, *Metroxylon* species sources of sago and materials for house construction, seeds furnish a form of vegetable ivory.

Species/Vernacular Names:

M. sagu Rottb.

English: sago palm, sago

The Philippines: lumbia

Malaya: rembia, rumbia, sagu

China: suo mu mian

Metteniusa Karsten Icacinaceae

Origins:

After the German (b. Frankfurt am Mein) botanist Georg Heinrich Mettenius, 1823-1866 (Leipzig, died of cholera), pteridologist, professor of botany, Director of the Leipzig Botanical Garden, botanical collector. Among his writings are "Ueber *Azolla*." *Linnaea*. 20: 259-282. 1847 and *Reise seine Majestät Fregatte Novara um die Erde*. Botanischer Theil. 1 Band, 4 Heft. Geffäss-Kryptogamen bearbeitet von Dr. Georg Mettenius. Ophioglossen und Equisetaceen bearbeitet von Dr. Julius Milde. Wien 1870. See William T. Stearn, in *Dictionary of Scientific Biography* 9: 340. 1981; J.H. Barnhart, *Biographical Notes upon Botanists*. 2: 481. 1965; T.W. Bossert, *Biographical Dictionary of Botanists Represented in the Hunt Institute Portrait Collection*. 265. 1972; Ethelyn Maria Tucker, *Catalogue of the Library of the Arnold Arboretum of Harvard University*. Cambridge, Massachusetts 1917-1933; E.D. Merrill, in *Bernice P. Bishop Mus. Bull*. 144: 134-135. 1937 and *Contr. U.S. Natl. Herb*. 30: 211. 1947; Stafleu and Cowan, *Taxonomic Literature*. 3: 432-435. 1981; R. Zander, F. Encke, G. Buchheim and S. Seybold, *Handwörterbuch der Pflanzennamen*. 14. Aufl. Stuttgart 1993.

Metternichia J.C. Mikan Solanaceae

Origins:

In honor of Prince Metternich, 1773-1859.

Metula Tieghem Loranthaceae

Origins:

Latin *metula, ae* "a small pyramid, obelisk."

Meum Mill. Umbelliferae

Origins:

Meon, the Greek name applied by Dioscorides and Plinius to an herb, possibly baldmoney, *Meum athamanticum*; Latin *meum, i* for an umbelliferous plant, bear-wort (Plinius).

Species/Vernacular Names:

M. athamanticum Jacq.

English: baldmoney

Mexacanthus T.F. Daniel Acanthaceae

Origins:

From Mexico and *Acanthus*.

Mexianthus B.L. Robinson Asteraceae

Origins:

Dedicated to the American botanist Ynés Enriquetta (Enriqueta) Julietta Mexia, 1870-1938, traveler, plant collector and botanical explorer (Andes, Mexico, Peru, Ecuador, Brazil, Alaska), author of "Botanical trails in old Mexico, the lure of the unknown." *Madroño*. 1: 227-238. 1929. See J.H. Barnhart, *Biographical Notes upon Botanists*. 2: 482. 1965; I.C. Hedge and J.M. Lamond, *Index of Collectors in the Edinburgh Herbarium*. Edinburgh 1970; H.P. Bracelin, "Ynes Mexia." *Madroño*. 4(8): 273-275. 1938; H.N. Clokie, *Account of the Herbaria of the Department of Botany in the University of Oxford*. 211. Oxford 1964; J. Ewan, ed., *A Short History of Botany in the United States*. 20. New York and London 1969; T. Harper Goodspeed, *Plant Hunters in the Andes*. 11-12. University of California Press, Berkeley and Los Angeles 1961; Irving William Knobloch, compiled by, "A preliminary verified list of plant collectors in Mexico." *Phytologia Memoirs*. VI. Plainfield, N.J. 1983; Ida Kaplan Langman, *A Selected Guide to the Literature on the Flowering Plants of Mexico*. Philadelphia 1964.

Mexicoa Garay Orchidaceae

Origins:
Named after Mexico.

Meximalva Fryxell Malvaceae

Origins:
From Mexico plus *Malva*.

Mexipedium V. Albert & M. Chase Orchidaceae

Origins:
A Mexican endemic species, *Mexipedium xerophyticum* (Soto Arenas, Salazar & Hágsater) V.A. Albert & Chase, Greek *pedilon* "a slipper"; see V.A. Albert and M.V. Chase, "*Mexipedium*: A new genus of slipper orchid (Cypripedioideae: Orchidaceae)." *Lindleyana*. 7: 172-176. 1992.

Meyenia Backeb. Cactaceae

Origins:
After the German botanist Franz Julius Ferdinand Meyen, 1804-1840, physician.

Meyenia Nees Acanthaceae

Origins:
After the German (Baltic German, Prussian, b. Tilsit) botanist Franz Julius Ferdinand Meyen, 1804-1840 (d. Berlin), physician, M.D. Berlin 1826, traveler, professor of botany at the University of Berlin, ship's physician. His writings include *Reise um die Erde ausgeführt auf dem Königlich Preussischen Seehandlungs-Schiffe Prinzess Louise, commandirt von Capitain W. Wendt, in den Jahren 1830, 1831 und 1832 ... Erster Theil. Historischer Bericht. 1834, Zweiter Theil ... 1835* and *Grundriss der Pflanzengeographie.* Berlin 1836; see Hans Querner, in *Dictionary of Scientific Biography* 9: 344-345. 1981; R. Zander, F. Encke, G. Buchheim and S. Seybold, *Handwörterbuch der Pflanzennamen.* 14. Aufl. Stuttgart 1993; J.H. Barnhart, *Biographical Notes upon Botanists.* 2: 482. 1965; J. Ewan, ed., *A Short History of Botany in the United States.* 116. New York and London 1969; T.W. Bossert, *Biographical Dictionary of Botanists Represented in the Hunt Institute Portrait Collection.* 265. 1972; E. Bretschneider, *History of European Botanical Discoveries in China.* 1981; A. Lasègue, *Musée botanique de Benjamin Delessert.* Paris 1845; E.M.

Tucker, *Catalogue of the Library of the Arnold Arboretum of Harvard University.* 1917-1933; Elmer Drew Merrill, *Contr. U.S. Natl. Herb.* 30: 211. 1947; G. Murray, *History of the Collections Contained in the Natural History Departments of the British Museum.* 1: 168. London 1904; August Weberbauer, *Die Pflanzenwelt der peruanischen Andes in ihren Grundzügen dargestellt.* 9-10. Leipzig 1911; Mariella Azzarello Di Misa, a cura di, *Il Fondo Antico della Biblioteca dell'Orto Botanico di Palermo.* 186-187. Palermo 1988; Gordon Douglas Rowley, *A History of Succulent Plants.* Strawberry Press, Mill Valley, California 1997; Ida Kaplan Langman, *A Selected Guide to the Literature on the Flowering Plants of Mexico.* Philadelphia 1964; Günther Schmid, *Chamisso als Naturforscher. Eine Bibliographie.* Leipzig 1942.

Meyenia Schltdl. Solanaceae

Origins:
Named after the German (Baltic German) botanist Franz Julius Ferdinand Meyen, 1804-1840, physician.

Meyerophytum Schwantes Aizoaceae

Origins:
For the German missionary G. Meyer, traveler, botanical explorer and plant collector, in Namibia; see Stafleu and Cowan, *Taxonomic Literature.* 3: 445. 1981.

Mezia Schwacke ex Niedenzu Malpighiaceae

Origins:
After the German botanist Carl Christian Mez, 1866-1944, professor of botany at Breslau, from 1899 to 1910 professor of botany at Halle, from 1910 to 1935 Director of the Königsberg Botanical Garden, contributor to C.F.P. von Martius *Flora Brasiliensis* (Bromeliaceae), contributor to Engler *Das Pflanzenreich* (Bromeliaceae, Myrsinaceae, Theophrastaceae, etc.), founder and editor of *Botanisches Archiv.* Königsberg 1922-1938. Among his writings are *Lauraceae americanae.* Berlin 1889, "Spicilegium Laureanum." *Arbeiten Königl. Bot. Gart. Breslau.* 1(1): 71-166. 1892, "Bromeliaceae." *Monogr. Phan.* 9: 1-990. 1896, "Novae species Panicearum." *Notizbl. Bot. Gart. Berlin-Dahlem.* 7: 45-78. 1917, "Additamenta monographica 1919. I. Bromeliaceae." *Repert. Spec. Nov. Regni Veg.* 16: 2-10. 1919, "Gramineae novae vel minus cognitae. IV. Stipeae cont." *Repert. Spec. Nov. Regni Veg.* 17: 204-214. 1921 and "*Stylagrostis*, novum graminearum genus." *Bot. Arch.* 1: 20. 1922. See J.H. Barnhart, *Biographical Notes upon Botanists.* 2: 484. 1965; Hermann Hager, *Das Mikroscop und*

seine Anwendung. Berlin 1908; E.M. Tucker, *Catalogue of the Library of the Arnold Arboretum of Harvard University.* 1917-1933; E.D. Merrill, in *Bernice P. Bishop Mus. Bull.* 144: 135. 1937 and *Contr. U.S. Natl. Herb.* 30: 212. 1947; T.W. Bossert, *Biographical Dictionary of Botanists Represented in the Hunt Institute Portrait Collection.* 266. 1972; S. Lenley et al., *Catalog of the Manuscript and Archival Collections and Index to the Correspondence of John Torrey.* Library of the New York Botanical Garden. 291. 1973; August Weberbauer (1871-1948), *Die Pflanzenwelt der peruanischen Andes in ihren Grundzügen dargestellt.* 33. Leipzig 1911; Ida Kaplan Langman, *A Selected Guide to the Literature on the Flowering Plants of Mexico.* Philadelphia 1964; R. Zander, F. Encke, G. Buchheim and S. Seybold, *Handwörterbuch der Pflanzennamen.* 14. Aufl. Stuttgart 1993.

Meziella Schindler Haloragidaceae (Haloragaceae)

Origins:
After the German botanist Carl Christian Mez, 1866-1944, professor of botany at Breslau.

Mezilaurus Kuntze ex Taubert Lauraceae

Origins:
After the German botanist Carl Christian Mez, 1866-1944.

Meziothamnus Harms Bromeliaceae

Origins:
After the German botanist Carl Christian Mez, 1866-1944; see J.H. Barnhart, *Biographical Notes upon Botanists.* 2: 484. 1965.

Mezobromelia L.B. Smith Bromeliaceae

Origins:
After the German botanist Carl Christian Mez, 1866-1944, professor of botany at Breslau.

Mezochloa Butzin Gramineae

Origins:
After the German botanist Carl Christian Mez, 1866-1944.

Mezoneuron Desf. Caesalpiniaceae

Origins:
Greek *meizon* "greater" (*megas* "big, large, great") and *neuron* "nerve," referring to the winged pod.

Mezoneurum DC. Caesalpiniaceae

Origins:
Orthographic variant of *Mezonevron* Desf.

Mezonevron Desf. Caesalpiniaceae

Origins:
From the Greek *meizon* "greater" (*megas* "big, large, great") and *neuron* "nerve," referring to the winged pod; see R.L. Desfontaines, in *Mémoires du Muséum National d'Histoire Naturelle.* 4: 245. 1818.

Mezzettiopsis Ridl. Annonaceae

Origins:
Resembling *Mezzettia* Becc.

Micagrostis Juss. Gramineae

Origins:
Latin *mica* "grain, a little bit" plus *agrostis.*

Michauxia L'Hérit. Campanulaceae

Origins:
Named for the French explorer André Michaux, 1746-1803 (died on an expedition to Madagascar), botanist and plantsman, plant collector, 1782-1785 Persia, 1785-1796 North America, author of *Flora Boreali-Americana.* 1: 51. Paris 1803 [Repr. New York 1973, intro by J. Ewan.], father of the French botanist François-André Michaux (1770-1855). See Stafleu and Cowan, *Taxonomic Literature.* 3: 456-464. 1981; J.H. Barnhart, *Biographical Notes upon Botanists.* 2: 485. 1965; Warren R. Dawson, *The Banks Letters.* London 1958; Jonas C. Dryander, *Catalogus bibliothecae historico-naturalis Josephi Banks.* 3: 185. London 1796-1800; A. Lasègue, *Musée botanique de Benjamin Delessert.* Paris 1845; E.M. Tucker, *Catalogue of the Library of the Arnold Arboretum of Harvard University.* 1917-1933; Frans A. Stafleu, *Linnaeus and the Linnaeans.* The spreading of their ideas in systematic botany, 1735-1789. Utrecht 1971; Jeannette E. Graustein, *Thomas Nuttall, Naturalist. Explorations in America, 1808-1841.* Harvard University Press 1967; J. Ewan, ed., *A Short History of Botany in the United*

States. 1969; F. Boerner & G. Kunkel, *Taschenwörterbuch der botanischen Pflanzennamen.* 4. Aufl. 134. Berlin & Hamburg 1989; R. Zander, F. Encke, G. Buchheim and S. Seybold, *Handwörterbuch der Pflanzennamen.* 14. Aufl. 751. Stuttgart 1993; Helmut Genaust, *Etymologisches Wörterbuch der botanischen Pflanzennamen.* 384. Basel 1996; Joseph Ewan, in *Dictionary of Scientific Biography* 9: 365-366. [d. 1802] 1981; Ida Kaplan Langman, *A Selected Guide to the Literature on the Flowering Plants of Mexico.* Philadelphia 1964; William Darlington, *Reliquiae Baldwinianae.* Philadelphia 1843; E. Earnest, *John and William Bartram, Botanists and Explorers 1699-1777, 1739-1823.* Philadelphia 1940; John Dunmore, *Who's Who in Pacific Navigation.* Honolulu 1991.

Michelia Kuntze Lecythidaceae

Origins:

After the Italian botanist Pier Antonio Micheli, 1679-1737.

Michelia L. Magnoliaceae

Origins:

After the Italian botanist Pier (Pietro) Antonio Micheli, 1679-1737, botanical collector throughout southern and central Europe, founder of the Società Botanica Fiorentina, from 1718 to 1737 Curator of the Botanical Garden of Florence. His works include *Relazione dell'erba detta da' botanici Orobanche* e volgarmente succiamele, fiamma, e mal d'occhio. Firenze 1723, *Catalogus plantarum Horti caesarei florentini.* Florentiae [Florence] 1748 and *Icones plantarum submarinarum.* [Florence?] 1748; see Francesco Rodolico, in *Dictionary of Scientific Biography* 9: 368-369. New York 1981; Giovanni Targioni-Tozzetti (1712-1783), *Notizie della vita e delle opere di Pier Antonio Micheli,* pubblicate per cura di Adolfo Targioni-Tozzetti. Firenze 1858; J.H. Barnhart, *Biographical Notes upon Botanists.* 2: 486. 1965; James Britten, *The Sloane Herbarium,* revised and edited by J.E. Dandy. London 1958; Antonio Targioni-Tozzetti (1785-1856), *Catalogo delle piante coltivate nell'Orto Botanico-Agrario detto dei Semplici.* Firenze 1841; T.W. Bossert, *Biographical Dictionary of Botanists Represented in the Hunt Institute Portrait Collection.* 266. 1972; E.M. Tucker, *Catalogue of the Library of the Arnold Arboretum of Harvard University.* 1917-1933; Frans A. Stafleu, *Linnaeus and the Linnaeans.* The spreading of their ideas in systematic botany, 1735-1789. Utrecht 1971; Jonas C. Dryander, *Catalogus bibliothecae historico-naturalis Josephi Banks.* London 1796-1800; H.N. Clokie, *Account of the Herbaria of the Department of Botany in the University of Oxford.* 1964; A. Lasègue, *Musée botanique de Benjamin Delessert.* Paris 1845; Mariella Azzarello Di Misa, a cura di, *Il Fondo Antico della Biblioteca dell'Orto Botanico di Palermo.* 189. Regione Siciliana, Palermo 1988; David Hunt, ed., *Magnolias and Their Allies.* Proceedings of an International Symposium, Royal Holloway, University of London, Egham, Surrey, U.K., 12-13 April 1996. International Dendrology Society and The Magnolia Society. 1998.

Species/Vernacular Names:

M. champaca L.

India: campakah, campakam, campakamu, campa, chamba, champa, campha, schampakam, champacam, cempakam, champaca, tsjampac, tchampaka

China: chen po, chen p'o, chen p'o ka

Malaya: orange chempaka, chempaka merah, chempa puteh, chempaka

Indonesia: cempaka

Bali: bungan capaka

Nepal: champ, chanp, aule chanp

M. compressa (Maxim.) Sarg. (*Michelia formosana* (Kaneh.) Masam.; *Michelia compressa* var. *formosana* Kaneh.; *Magnolia compressa* Maxim.)

Japan: oga-tama-no-ki

Okinawa: rusan

M. figo (Lour.) Spreng. (*Michelia fuscata* (Andrews) Wallich; *Liriodendron figo* Lour.)

English: banana shrub

Malaya: dwarf chempaka, chempaka ambon

China: han hsiao

M. kisopa DC.

Nepal: sirmoo

Micheliella Briquet Labiatae

Origins:

To commemorate the Swiss botanist Marc Micheli, 1844-1902, owner of a private botanical garden, wrote *Leguminosae langlasseanae.* Genève 1903; see J.H. Barnhart, *Biographical Notes upon Botanists.* 2: 486. 1965; T.W. Bossert, *Biographical Dictionary of Botanists Represented in the Hunt Institute Portrait Collection.* 266. 1972; Ethelyn Maria Tucker, *Catalogue of the Library of the Arnold Arboretum of Harvard University.* Cambridge, Massachusetts 1917-1933.

Michelsonia Hauman Caesalpiniaceae

Origins:

Named for the Russian explorer A. Michelson (Mikhel'son), flourished 1912, traveler; see Stafleu and Cowan, *Taxonomic Literature.* 3: 464. 1981.

Species/Vernacular Names:

M. microphylla (Troupin) Hauman

Zaire: musisi

Miconia Ruíz & Pav. Melastomataceae

Origins:

In honor of Francisco Micón (Micó), b. 1528, Spanish physician and botanist; see Georg Christian Wittstein, *Etymologisch-botanisches Handwörterbuch*. 576. Ansbach 1852; F. Boerner & G. Kunkel, *Taschenwörterbuch der botanischen Pflanzennamen*. 4. Aufl. 134. Berlin & Hamburg 1989.

Species/Vernacular Names:

M. sp.

Peru: purima-caspi, sanguillo, ananiroqui, botani tesseperini, canacan chuxúacěn, meta, palo tigre, patu ro, purma caspi

Mexico: laa chita, yo lila

M. albicans (Swartz) Triana (*Melastoma albicans* Swartz)

Peru: yúrac mullaca

M. amazonica Triana

Peru: dispera sacha, dispero blanco, níspero, níspero blanco, níspero sacha blanco, dispero sacha blanco

M. aurea (D. Don) Naudin (*Chitonia aurea* D. Don)

Peru: paoon vurovi, yausa quiru

M. latifolia (D. Don) Naudin (*Miconia andina* (Naudin) Naudin)

Peru: mote-mote

M. tomentosa (Richard) D. Don (*Miconia amplexans* (Crueger) Cogniaux)

Peru: carache caspi, chaita ida, muringa, pichirina

Micractis DC. Asteraceae

Origins:

From the Greek *mikros* "small" and *aktis* "a ray," referring to the flowers.

Micraeschynanthus Ridley Gesneriaceae

Origins:

From the Greek *mikros* "small" plus the related genus *Aeschynanthus* Jack.

Micraira F. Muell. Gramineae

Origins:

From the Greek *mikros* "small" and the genus *Aira* L., hair grass, in the same family; see Sir Ferdinand Jacob Heinrich von Mueller (1825-1896), *Fragmenta Phytographiae Australiae*. 5: 208. 1866.

Species/Vernacular Names:

M. subulifolia F. Muell.

English: mountain couch

Micrampelis Raf. Cucurbitaceae

Origins:

Greek *mikros* "small" and *ampelos* "a vine," referring to the appearance of the plant; see C.S. Rafinesque, *Med. Repos.* II. 5: 350. 1808.

Micrandra Benn. & R. Br. Euphorbiaceae

Origins:

Greek *mikros* "small" and *andros* "male, stamen."

Micrandra Benth. Euphorbiaceae

Origins:

Greek *mikros* "small" and *andros* "a man, male, stamen."

Species/Vernacular Names:

M. spruceana (Baillon) R. Schultes (*Cunuria spruceana* Baillon)

Peru: cunurí, hiringa masha

Micrandropsis Rodrigues Euphorbiaceae

Origins:

Resembling the genus *Micrandra*.

Micrantha Dvorak Brassicaceae

Origins:

From the Greek *mikros* "small" and *anthos* "flower," small-flowered.

Micranthella Naudin Melastomataceae

Origins:

From the Greek *mikros* "small" and *anthemon* "flower," *anthele* "a little flower."

Micranthemum Michx. Scrophulariaceae

Origins:

From the Greek *mikros* "small" and *anthemon* "flower."

Micrantheum Desf. Euphorbiaceae

Origins:
From the Greek *mikros* "small" and *anthos, anthemon* "flower," the flowers are tiny.

Species/Vernacular Names:
M. demissum F. Muell. (*Allenia blackiana* Ewart & Rees; *Allenia blackiana* Ewart & Rees var. *microphyllum* Ewart, J.R. White & Rees)
English: dwarf micrantheum
M. ericoides Desf.
English: heath micrantheum
M. hexandrum Hook.f.
English: box micrantheum

Micranthocereus Backeb. Cactaceae

Origins:
From the Greek *mikros* "small," *anthos* "flower" plus *Cereus*.

Micranthus (Pers.) Ecklon Iridaceae

Origins:
Greek *mikros* "small" and *anthos* "flower."

Micranthus J.C. Wendl. Acanthaceae

Origins:
Greek *mikros* and *anthos* "flower."

Micrargeria Benth. Scrophulariaceae

Origins:
From the Greek *mikros* "small" and *argyros* "silver," referring to the surface of the leaves.

Micrargeriella R.E. Fries Scrophulariaceae

Origins:
The diminutive of the genus *Micrargeria*.

Micrasepalum Urban Rubiaceae

Origins:
From the Greek *mikros* "small" and Latin *sepalum*.

Microchites Miq. Apocynaceae

Origins:
From the Greek *mikros* "small" plus *Echites*.

Microbahia Cockerell Asteraceae

Origins:
From the Greek *mikros* "small" plus the genus *Bahia*.

Microbambus K. Schumann Gramineae

Origins:
From the Greek *mikros* "small" plus *bambos, bambusa*, etc.

Microberlinia A. Chev. Caesalpiniaceae

Origins:
From the Greek *mikros* "small" plus *Berlinia* Sol. ex Hook.f. & Benth.

Species/Vernacular Names:
M. brazzavillensis A. Chev.
English: zebrawood (for the timber), zebrano, zingana
Gabon: izingani, zingana
Cameroon: anouk

Microbignonia Kraenzl. Bignoniaceae

Origins:
From the Greek *mikros* "small" plus *Bignonia* L.

Microbiota Komarov Cupressaceae

Origins:
Greek *mikros* "small" plus *bios* "life," genus *Biota* (D. Don) Endl.

Microblepharis (Wight & Arn.) M. Roemer Passifloraceae

Origins:
From the Greek *mikros* "small" plus *blepharis* "eyelash."

Microbriza Parodi ex Nicora & Rugolo Gramineae

Origins:
From the Greek *mikros* "small" plus *Briza* L.

Microbrochis C. Presl Dryopteridaceae

Origins:
Greek *mikros* and *brochos* "noose, snare for birds, mesh of a net," *brochis, idos* dim. of *brochos*.

Microcachrys Hook.f. Podocarpaceae

Origins:
From the Greek *mikros* "small" and *kachrys* "catkin, cone," referring to the small cones; in *Hooker's London Journal of Botany*. 4: 149. 1845.

Species/Vernacular Names:
M. tetragona (Hook.) Hook.f.
English: creeping pine, strawberry pine

Microcala Hoffsgg. & Link Gentianaceae

Origins:
Greek *mikros* "small" and *kalos* "beautiful" or *kalia* "wooden, dwelling, hut, barn"; see Johann C. von Hoffmannsegg and Heinrich Friedrich Link, *Flore portugaise*. 1: 359. Berlin 1813-1820.

Microcalamus Franchet Gramineae

Origins:
From the Greek *mikros* "small" and *kalamos* "reed," like a bamboo.

Microcalamus Gamble Gramineae

Origins:
Greek *mikros* and *kalamos* "reed."

Microcardamum O.E. Schulz Brassicaceae

Origins:
Greek *mikros* "small" and Latin *cardamomum, i* "a spice, cardamon," *cardamum, i* for a kind of cress.

Microcarpaea R. Br. Scrophulariaceae

Origins:
From the Greek *mikros* "small" and *karpos* "fruit," the fruits and the seeds are small; see Robert Brown (1773-1858), *Prodromus florae Novae Hollandiae*. 435. London 1810.

Microcaryum I.M. Johnston Boraginaceae

Origins:
From the Greek *mikros* "small" and *karyon* "a nut," the seeds are small.

Species/Vernacular Names:
M. pygmaeum (C.B. Clarke) I.M. Johnston
English: dwarf microcaryum
China: wei guo cao

Microcasia Becc. Araceae

Origins:
From the Greek *mikros* "small" and the genus *Colocasia* Schott.

Microcephala Pobed. Asteraceae

Origins:
From the Greek *mikros* "small" and *kephale* "head."

Microchaete Benth. Asteraceae

Origins:
From the Greek *mikros* "small" and *chaite* "bristle, long hair."

Microcharis Benth. Fabaceae

Origins:
From the Greek *mikros* "small" and *charis* "grace, beauty."

Microchilus C. Presl Orchidaceae

Origins:
From the Greek *mikros* "small" and *cheilos* "lip," the small size of the lip.

Microchlaena Ching Dryopteridaceae (Woodsiaceae)

Origins:
From the Greek *mikros* "small" and *chlaena* "cloak."

Microchlaena Kuntze Gramineae

Origins:
Greek *mikros* "small" and *chlaena* "cloak."

Microchloa R. Br. Gramineae

Origins:
From the Greek *mikros* "small" and *chloe, chloa* "a grass," an allusion to the small size of the plant; see Robert Brown, *Prodromus florae Novae Hollandiae.* 208. London 1810.

Species/Vernacular Names:
M. caffra Nees
English: pincushion grass
South Africa: elsgras

Microchonea Pierre Apocynaceae

Origins:
From the Greek *mikros* "small" and *chone* "funnel," referring to the flowers.

Microcitrus Swingle Rutaceae

Origins:
Greek *mikros* "small" and *kitron* "a lemon," the genus *Citrus*; see Walter Tennyson Swingle (1871-1952), *Journal of the Washington Academy of Sciences.* 5: 570. 1915.

Species/Vernacular Names:
M. australasica (F. Muell.) Swingle
English: finger lime, Australian finger lime
M. australasica (F. Muell.) Swingle var. *sanguinea* (Bailey) Swingle
English: red pulp finger lime
M. australis (Planchon) Swingle
English: wild lime, round lime, Australian round lime
M. garrawayi (Bailey) Swingle (after Mrs. R.W. Garraway, original collector)
Australia: Mount White lime
M. inodora (Bailey) Swingle
Australia: Russell River lime, large-leaf Australian wild lime

Microclisia Benth. Menispermaceae

Origins:
From the Greek *mikros* "small" and *kleis* "lock, key."

Microcnemum Ung.-Sternb. Chenopodiaceae

Origins:
From the Greek *mikros* "small" and *kneme* "limb, leg."

Micrococca Benth. Euphorbiaceae

Origins:
From the Greek *mikros* "small" and *kokkos* "berry, grain, seed," the fruits are very small.

Species/Vernacular Names:
M. capensis (Baill.) Prain (*Claoxylon capense* Baill.)
English: common bead-string
Southern Africa: gewone kralesnoer, ububu; uBubu (Zulu, Xhosa); iNkanga yehlathi, iNyothi (Xhosa)

Micrococos Phil. Palmae

Origins:
From the Greek *mikros* "small" plus *Cocos.*

Microcodon A. DC. Campanulaceae

Origins:
From the Greek *mikros* "small" and *kodon* "a bell," referring to the flowers.

Microcoelia Lindley Orchidaceae

Origins:
From the Greek *mikros* "small" and *koilos* "hollow," possibly referring to the sheathing and non-sheathing bracts or to the globose or inflated spur.

Microcoelum Burret & Potztal Palmae

Origins:
From the Greek *mikros* "small" and *koilos* "hollow," referring to the endosperm.

Microconomorpha (Mez) Lundell Myrsinaceae

Origins:
From the Greek *mikros* "small," *konos* "a cone" and *morphe* "a form, shape."

Microcorys R. Br. Labiatae

Origins:

From the Greek *mikros* "small" and *korys, korythos* "helmet," the upper corolla lobes are small and hooded; see Robert Brown, *Prodromus florae Novae Hollandiae.* 502. London 1810.

Microcos L. Tiliaceae

Origins:

Possibly from the Greek *mikros* "small" and *kos* "public prison, a prisoner," the leaves are used for wrapping.

Microculcas Peter Araceae

Origins:

From the Greek *mikros* "small" plus the genera *Colocasia* or *Culcasia*.

Microcybe Turcz. Rutaceae

Origins:

From the Greek *mikros* "small" and *kybe* "head," referring to the flowerheads; see Porphir K.N.S. Turczaninow, *Bulletin de la Société Impériale des Naturalistes de Moscou.* 25(2): 166. 1852.

Species/Vernacular Names:
M. albiflora Turcz.
English: white microcybe
M. multiflora Turcz.
English: red microcybe
M. pauciflora Turcz. (*Eriostemon capitatus* F. Muell.)
English: yellow microcybe

Microcycas A. DC. Zamiaceae

Origins:
From the Greek *mikros* "small" and the genus *Cycas* L.

Microdactylon Brandegee Asclepiadaceae

Origins:
From the Greek *mikros* "small" and *daktylos* "a finger."

Microderis A. DC. Asteraceae

Origins:

From the Greek *mikros* "small" and *deire, dere, dera, deris* "neck, throat, collar."

Microderis D. Don ex Gand. Asteraceae

Origins:
Orthographic error, see *Microseris* D. Don.

Microdesmis Hook.f. ex Hook. Pandaceae

Origins:

From the Greek *mikros* "small" and *desmis, desmos* "a bond, band, bundle," referring to the flowers.

Species/Vernacular Names:
M. haumaniana J. Léon.
Congo: nkuti
M. keayana J. Léonard
Nigeria: apata esunsun; erankpata (Edo); nkperi, akbalata (Igbo); kawa (Boki); ntanebit (Ibibio)
M. puberula Hook.f. ex Hook.
Yoruba: apata, idi apata, igi ori apata, arin igo, arin igo dudu, imeyinfun, akanju ile, esunsun
Nigeria: apata esunsun, apata, igi-apata, ubelu, esun-sun, ehran-apata, amomilla, amomilan, anamomilla; erankpata (Edo); nkperi, akbalata (Igbo); kawa (Boki); ntanebit (Ibibio)
Congo: dikota
M. yafungana J. Léonard
Zaire: me, bisamu

Microdon Choisy Globulariaceae (Scrophulariaceae)

Origins:

Small-toothed, from the Greek *mikros* "small" and *odous, odontos* "tooth."

Microepidendrum Brieger Orchidaceae

Origins:

From the Greek *mikros* "small" plus *Epidendrum* L.

Microglossa DC. Asteraceae

Origins:
From the Greek *mikros* "small" and *glossa* "a tongue," an allusion to the corolla segments, to the shortness of the ray florets.

Species/Vernacular Names:
M. pyrifolia (Lam.) Kuntze
Yoruba: anikan segbo torisa, anikan segbo
Congo: muntantali

Microgonium Presl Hymenophyllaceae

Origins:
From the Greek *mikros* "small" and *gonia* "an angle."

Species/Vernacular Names:
M. motleyi Bosch
English: tiny bristle fern

Microgramma C. Presl Polypodiaceae

Origins:
From the Greek *mikros* "small" and *gramma* "line, letter," referring to the sori.

Species/Vernacular Names:
M. heterophylla (L.) Wherry (*Polypodium heterophyllum* L.)
English: climbing vine fern

Microgyne Less. Asteraceae

Origins:
From the Greek *mikros* "small" and *gyne* "female," referring to the small pistils.

Microgynella Grau Asteraceae

Origins:
From the Greek *mikros* "small" and *gyne* "female," referring to the small pistils.

Microgynoecium Hook.f. Chenopodiaceae

Origins:
Greek *mikros* "small" and *gyne* "female," *gynoecium*, *gynaecium* "the female reproductive organs of a flower," referring to the small ovary.

Microholmesia Cribb Orchidaceae

Origins:
From the Greek *mikros* "small" and *Holmesia* Cribb, after Mr. W.D. Holmes, an English orchid collector.

Microjambosa Blume Myrtaceae

Origins:
Greek *mikros* "small" and *jambos, jambosa.*

Microkentia H.A. Wendland ex Hook.f. Palmae

Origins:
Greek *mikros* with *Kentia.*

Microlaena R. Br. Gramineae

Origins:
From the Greek *mikros* "small" and *chlaena, chlaenion* "cloak," referring to the outer glumes; see Robert Brown, *Prodromus florae Novae Hollandiae*. 210. London 1810.

Species/Vernacular Names:
M. stipoides (Labill.) R. Br.
English: weeping grass, meadow rice-grass, weeping rice-grass

Microlecane Sch. Bip. ex Benth. Asteraceae

Origins:
From the Greek *mikros* "small" and *lekane* "a dish, pot, saucer."

Microlepia Presl Dennstaedtiaceae

Origins:
From the Greek *mikros* "small" and *lepis* "scale," alluding to the indusia; see C. Presl (1794-1852), *Tentamen Pteridographiae, seu genera Filicacearum*. 124, t. IV, figs. 21-23. Prague 1836.

Species/Vernacular Names:
M. hookeriana (Wall.) Presl (*Davallia hookeriana* Wall. ex Hook.)
Japan: yanbaru-fumoto-shida

M. marginata (Panzer) C. Chr. (*Polypodium marginatum* Panzer)

Japan: fumoto-shida (= foot-of-the-mountain fern)

South Africa: bokhoring, bokhorinkies, heuningblommetjie, kannetjies, suikerkannetjies, melkblommetjie, skilpadkos

Microlepidium F. Muell. Brassicaceae

Origins:
Greek *mikros* "small" and the genus *Lepidium*; see F. von Mueller, in *Linnaea*. 25: 371. 1853.

Species/Vernacular Names:
M. pilosulum F. Muell. (*Capsella pilosula* (F. Muell.) F. Muell.; *Bursa pilosula* (F. Muell.) Kuntze)

English: hairy shepherd's purse

Microlepis (DC.) Miq. Melastomataceae

Origins:
From the Greek *mikros* "small" and *lepis* "scale."

Microliabum Cabrera Asteraceae

Origins:
From the Greek *mikros* "small" and *Liabum* Adans.

Microlobium Liebm. Fabaceae

Origins:
From the Greek *mikros* "small" and *lobion* "small pod, small lobe."

Microlobius C. Presl Fabaceae

Origins:
From the Greek *mikros* "small" and *lobion* "small pod, small lobe."

Microloma R. Br. Asclepiadaceae

Origins:
From the Greek *mikros* "small" and *loma* "border, margin, fringe," alluding to the hairy corolla tube.

Species/Vernacular Names:
M. tenuifolium (L.) K. Schum.

English: coral creeper, redwax creeper, wax twiner

Microlonchoides Candargy Asteraceae

Origins:
Resembling *Microlonchus*.

Microlonchus Cass. Asteraceae

Origins:
From the Greek *mikros* "small" and *lonche* "a lance, spear."

Microluma Baillon Sapotaceae

Origins:
Greek *mikros* "small" with *luma*.

Micromeles Decne. Rosaceae

Origins:
From the Greek *mikros* "small" and *melon* "an apple," the calyx falling away from the apex of the fruit.

Micromelum Blume Rutaceae

Origins:
From the Greek *mikros* "small, little" and *melon* "an apple" or *melos* "a limb, part, member," referring to the small and apple-like fruits or to the small branchlets; see Karl Ludwig von Blume, *Bijdragen tot de flora van Nederlandsch Indië*. 137. 1825.

Species/Vernacular Names:
M. minutum (Forst.) Wight & Arn.

English: red lime berry

Malaya: chama, chemama, chememar jantan, cherek, secherek, cherek puteh

Micromeria Bentham Labiatae

Origins:
Greek *mikros* "small, little" and *meris* "a part, portion," referring to leaves and flowers.

Species/Vernacular Names:

M. barosma (W. Smith) Handel-Mazzetti

English: small micromeria

China: xiao xiang ru

M. biflora (Buch.-Ham. ex D. Don) Bentham

English: two-flower micromeria

China: jiang wei cao

M. chamissonis (Bentham) E. Greene

Spanish: yerba buena

M. formosana Marquand

English: Taiwan micromeria

China: tai wan jiang wei cao

M. wardii Marquand & Airy Shaw

English: Ward micromeria

China: xi zang jiang wei cao

Micromonolepis Ulbr. Chenopodiaceae

Origins:

From the Greek *mikros* "small, little" plus *Monolepis* Schrad.

Micromyrtus Benth. Myrtaceae

Origins:

Greek *mikros* "small" and the genus *Myrtus*; see J.W. Green, "Taxonomy of *Micromyrtus ciliata* (Myrtaceae) and allied species including three new species of *Micromyrtus* from eastern Australia and lectotypification of *M. minutiflora*." *Nuytsia.* 4(3): 317-331. 1983.

Species/Vernacular Names:

M. ciliata (Smith) Druce (*Imbricaria ciliata* Smith; *Thryptomene ciliata* (Smith) F. Muell. ex Woolls; *Thryptomene plicata* (F. Muell.) F. Muell.; *Baeckea plicata* F. Muell.)

Australia: fringed honey-myrtle

M. leptocalyx (F. Muell.) Benth.

Australia: Queensland honey-myrtle

M. minutiflora Benth.

Australia: tiny honey-myrtle

Micromystria O.E. Schulz Brassicaceae

Origins:

From the Greek *mikros* "small" and *mystron* "a spoon," *mystrion* "a small spoon."

Micronoma H. Wendl. Palmae

Origins:

From the Greek *mikros* "small" and *nomos* "meadow, usage, pasture, custom."

Micronychia Oliver Anacardiaceae

Origins:

From the Greek *mikros* "small" and *onyx, onychos* "a claw, nail."

Micropapyrus Suesseng. Cyperaceae

Origins:

From the Greek *mikros* "small" plus *papyros* "linen, cord, papyrus," Latin *papyrus* and *papyrum* "the paper-reed, papyrus."

Microparacaryum (Riedl) Hilger & Podlech Boraginaceae

Origins:

From the Greek *mikros* "small" plus *Paracaryum* (A. DC.) Boiss.

Micropeplis Bunge Chenopodiaceae

Origins:

From the Greek *mikros* "small" and *peplos* "a robe."

Micropera Lindley Orchidaceae

Origins:

From the Greek *mikros* "small" and *pera* "a pouch," referring to the shape of the labellum; see J. Lindley, *The Botanical Register.* 18: sub t. 1522. 1832; B.A. Lewis and P.J. Cribb, *Orchids of the Solomon Islands and Bougainville.* Royal Botanic Gardens, Kew 1991.

Species/Vernacular Names:

M. fasciculata (Lindley) Garay

The Solomon Islands: gora goraha ubutu

Micropetalum Poit. ex Baillon Euphorbiaceae

Origins:

From the Greek *mikros* "small" and *petalon* "a petal," referring to the very small petals.

Microphacos Rydb. Fabaceae

Origins:

From the Greek *mikros* "small" and *phakos* "a lentil."

Microphlebodium L.D. Gómez Polypodiaceae

Origins:

From the Greek *mikros* and *phleps, phlebos* "vein," *phlebodes* "full of veins, with large veins," genus *Phlebodium* (R. Br.) J. Smith Polypodiaceae.

Micropholis (Griseb.) Pierre Sapotaceae

Origins:

From the Greek *mikros* "small" and *pholis* "scale, horny scale."

Species/Vernacular Names:

M. guyanensis (A. DC.) Pierre

Bolivia: colonion, coquino, tiororiba, toroquehua

Microphyes Philippi Caryophyllaceae

Origins:

Greek *mikros* "small" and *phye* "shape, nature, growth," *phyo* "to grow," *mikrophyes* "of low growth."

Microphysa Schrenk Rubiaceae

Origins:

From the Greek *mikros* "small" and *physa* "a bladder."

Microphysca Naudin Melastomataceae

Origins:

From the Greek *mikros* "small" and *physke* "sausage, blister, gall-bag."

Microphytanthe (Schltr.) Brieger Orchidaceae

Origins:

From the Greek *mikros* "small," *phyton* "plant" and *anthos* "flower."

Micropleura Lag. Umbelliferae

Origins:

From the Greek *mikros* "small" and *pleura, pleuron* "side, rib, lateral."

Microplumeria Baillon Apocynaceae

Origins:

From the Greek *mikros* "small" plus the genus *Plumeria*.

Micropodium Mett. Aspleniaceae

Origins:

From the Greek *mikros* "small" and *pous, podion, podos* "foot, small foot."

Micropogon Pfeiff. Gramineae

Origins:

From the Greek *mikros* "small" and *pogon* "a beard."

Micropolypodium Hayata Grammitidaceae

Origins:

From the Greek *mikros* "small" plus *Polypodium*.

Micropora Hook.f. Lauraceae

Origins:

From the Greek *mikros* "small" and *poros* "opening, pore."

Micropsis DC. Asteraceae

Origins:

Greek *mikros* and *opsis* "resembling, aspect," an allusion to the appearance of the plants.

Micropteris Desv. Grammitidaceae

Origins:

From the Greek *mikros* "small" and *pteris* "fern."

Micropteris J. Sm. Polypodiaceae

Origins:
Greek *mikros* and *pteris* "fern."

Micropterum Schwantes Aizoaceae

Origins:
From the Greek *mikros* "small" and *pteron* "wing," referring to the capsule locules.

Micropteryx Walp. Fabaceae

Origins:
From the Greek *mikros* "small" and *pteryx* "wing."

Micropuntia Daston Cactaceae

Origins:
From the Greek *mikros* "small" plus *Opuntia.*

Micropus L. Asteraceae

Origins:
From the Greek *mikros* "small" and *pous, podos* "foot," referring to the very thin and slender stem near the root.

Species/Vernacular Names:
M. amphibolus A. Gray
English: Mount Diablo cottonweed
M. californicus Fischer & C. Meyer
English: slender cottonweed

Micropyropsis Romero-Zarco & Cabezudo Gramineae

Origins:
Resembling the genus *Micropyrum.*

Micropyrum (Gaudin) Link Gramineae

Origins:
From the Greek *mikros* "small" and *pyros* "wheat."

Micropyxis Duby Primulaceae

Origins:
From the Greek *mikros* "small" and *pyxis* "a small box, a small container with lid."

Microrhamnus A. Gray Rhamnaceae

Origins:
From the Greek *mikros* "small" and *Rhamnus.*

Microrhynchus Sch.Bip. Asteraceae

Origins:
From the Greek *mikros* "small" and *rhynchos* "horn, beak," referring to the beak of the achenes.

Microrphium C.B. Clarke Gentianaceae

Origins:
From the Greek *mikros* "small" and *Orphium* E. Meyer.

Microsaccus Blume Orchidaceae

Origins:
From the Greek *mikros* "small" and *sakkos* "sack," the small and saccate lip.

Microschizaea C.F. Reed Schizaeaceae

Origins:
From the Greek *mikros* "small" plus *Schizaea* Sm.; in *Boletim da Sociedade Broteriana.* Ser. 2, 21: 133. 1948.

Microschoenus C.B. Clarke Juncaceae

Origins:
From the Greek *mikros* "small" and *schoinos* "rush, reed, cord," an allusion to the small size.

Microschwenkia Bentham ex Hemsley Solanaceae

Origins:
After the Dutch physician Martin Wilhelm (Martinus Wilhelmus) Schwencke, 1707-1785, botanist. His works include Dissertatio ... *de operatione inguinali.* Lugduni Batavorum 1731, *Novae plantae Schwenckia* dictae a celeberrimo Linnaeo in *Gen. plant.* ed. VI. p. 567 ex celeb. Davidiis van Rooijen Charact. mss. 1761 communicata brevis descriptio et delineatio cum notis characteristicis. Hagae Comitum [The Hague] [typ. van Karnebeek] 1766 and *Officinalium plantarum catalogus.* Hagae-Comitum 1752. See John H. Barnhart, *Biographical Notes upon Botanists.* 3: 250. 1965; Jonas C. Dryander, *Catalogus bibliothecae historico-naturalis Josephi Banks.* London 1800; H.

Heine, in *Kew Bulletin*. 16(3): 465-469. 1963; Frans A.
Stafleu and Cowan, *Taxonomic Literature*. 5: 442-443.
1985.

Microsciadium Boiss. Umbelliferae

Origins:
From the Greek *mikros* "small" and *skiadion, skiadeion*
"umbel, parasol."

Microsciadium Hook.f. Umbelliferae

Origins:
Greek *mikros* and *skiadion, skiadeion* "umbel, parasol."

Microsechium Naudin Cucurbitaceae

Origins:
From the Greek *mikros* "small" and *Sechium* P. Browne.

Microsemma Labill. Thymelaeaceae

Origins:
Greek *mikros* and *sema* "a sign, standard, mark"; see J.J.H.
de Labillardière, *Sertum austro-caledonicum*. 58, fig. 57.
Parisiis 1824-1825; James A. Baines, *Australian Plant Genera. An Etymological Dictionary of Australian Plant Genera*. 239. Chipping Norton, N.S.W. 1981; see also *Disemma*
Labill. (Passifloraceae), in *Paxton's Botanical Dictionary*.
196. 1980; some suggest from *stemma* "a crown."

Microsepala Miq. Euphorbiaceae

Origins:
Small sepals.

Microseris D. Don Asteraceae

Origins:
From the Greek *mikros* "small" and *seris, seridos* "chicory,
lettuce," referring to the tuberous roots and to the nature of
the plants.

Species/Vernacular Names:
M. borealis (Bong.) Sch.Bip.
English: northern microseris

M. lanceolata (Walp.) Sch.Bip. (*Phylopappus lanceolatus*
Walp.)
Australia: yam daisy, native yam, murrnong (an Aboriginal
vernacular name)
M. scapigera Sch.Bip. (*Phylopappus lanceolatus* Walp.)
Australia: murrnong, murnong

Microsideros Baum.-Bod. Myrtaceae

Origins:
From the Greek *mikros* "small" and *sideros* "iron."

Microsisymbrium O.E. Schulz Brassicaceae

Origins:
From *mikros* "small" and *sisymbrion*, the Greek name for
a fragrant herb, related genus *Sisymbrium* L.

Microsorum Link Polypodiaceae

Origins:
From the Greek *mikros* "small, tiny" and *soros* "a vessel
for holding anything, a cinerary urn, a coffin" but also "a
heap" from Akkadian *sarru, zarru* "heap of grain," *zaru* "to
winnow," *za'ru, zeru* "seed of cereals."

Species/Vernacular Names:
M. fortunei (T. Moore) Ching (*Drynaria fortunii* T. Moore;
Microsorum takedai (Nakai) H. Itô)
Japan: shina-noki-shinobu
M. nigrescens (Bl.) Copel.
English: pimple fern
M. punctatum (L.) Copel.
English: climbing bird's-nest fern
Yoruba: ida
M. scolopendria (Burm.f.) Copel. (*Phymatodes scolopen-
dria* (Burm.f.) Ching)
English: wart fern, water fern

Microsperma Hook. Loasaceae

Origins:
Greek *mikros* "small, tiny" and *sperma* "seed."

Microspermia Fric Cactaceae

Origins:
Greek *mikros* "small, tiny" and *sperma* "seed."

Microspermum Lag. Asteraceae

Origins:

From the Greek *mikros* "small, tiny" and *sperma* "seed."

Microsplenium Hook.f. Rubiaceae

Origins:

From the Greek *mikros* "small" and *splenion* "a canopy, umbel, pad or compress of linen."

Microstachys A. Juss. Euphorbiaceae

Origins:

From the Greek *mikros* "small, tiny" and *stachys* "spike," referring to the male flowers.

Microstaphyla C. Presl Lomariopsidaceae (Aspleniaceae)

Origins:

From the Greek *mikros* "small, tiny" and *staphyle* "a cluster, a grape."

Microstegia Pierre ex Harms Caesalpiniaceae

Origins:

From the Greek *mikros* "small, tiny" and *stege* "roof, cover."

Microstegia Presl Dryopteridaceae (Woodsiaceae)

Origins:

Greek *mikros* "small, tiny" and *stege* "roof, cover."

Microstegium Nees Gramineae

Origins:

From the Greek *mikros* "small, tiny" and *stege, stegos* "roof, cover," referring to the lemma and to small spikes.

Species/Vernacular Names:

M. ciliatum (Trin.) A. Camus (*Pollinia ciliata* Trin.)

Japan: ô-sasa-gaya (= large *Microstegium japonicum*)

M. nudum (Trin.) A. Camus (*Pollinia nuda* Trin.; *Microstegium mayebaranum* Honda)

Japan: miyama-sasa-gaya

Microstegnus C. Presl Cyatheaceae

Origins:

From the Greek *mikros* "small, tiny" and *stegnos* "covered, sheltered."

Microsteira Baker Malpighiaceae

Origins:

From the Greek *mikros* "small, tiny" and *steiros* "barren, sterile."

Microstelma Baill. Asclepiadaceae

Origins:

From the Greek *mikros* "small, tiny" and *stelma, stelmatos* "a girdle, belt."

Microstemma R. Br. Asclepiadaceae

Origins:

From the Greek *mikros* "small, tiny" and *stemma* "a crown," referring to the stamens; see P.I. Forster, "Studies of the Australasian Asclepiadaceae. I. *Brachystelma* Sims in Australia." *Nuytsia.* 6(3): 285-294. 1988; P.I. Forster, "Correction and further notes to Studies of the Australasian Asclepiadaceae. I. Brachystelma Sims in Australia." *Nuytsia.* 7(2): 123-124. 1990.

Microstemon Engl. Anacardiaceae

Origins:

From the Greek *mikros* "small, tiny" and *stemon* "thread, filament."

Microstephanus N.E. Br. Asclepiadaceae

Origins:

From the Greek *mikros* "small, tiny" and *stephanos* "a crown."

Microstephium Less. Asteraceae

Origins:

From the Greek *mikros* "small, tiny" and *stephos* "crown," referring to the flowers.

Microstigma Trautv. Brassicaceae

Origins:
From the Greek *mikros* "small, tiny" and *stigma* "stigma."

Microstrobilus Bremek. Acanthaceae

Origins:
From the Greek *mikros* "small, tiny" and *strobilos* "a cone."

Microstrobos J.G. Garden & L.A.S. Johnson Podocarpaceae

Origins:
Greek *mikros* "small, tiny" and *strobos* "a whirling round," *strobeo* "to spin a top spin," referring to the very small cones.

Species/Vernacular Names:
M. fitzgeraldii (F. Muell.) J. Garden & L. Johnson (after the Irish-born Australian botanical artist Robert Desmond (David or Douglas) Fitzgerald, 1830-1892 (Sydney, N.S.W.), orchidologist, in Australia from 1856, from 1873 Surveyor-General, in 1874 a Fellow of the Linnean Society, plant collector in Lord Howe Island, author of *Australian Orchids*. Sydney [1875-]-1882[-1894]; see J.H. Barnhart, *Biographical Notes upon Botanists*. 1: 547. 1965)

Australia: dwarf mountain pine

M. niphophilus J. Garden & L. Johnson (snow loving, Greek *niphas* "snow")

Australia: dwarf pine

Microstylis (Nuttall) Eaton Orchidaceae

Origins:
Greek *mikros* "small, tiny" and *stylos* "a style, pillar, a column," referring to the very small and slender column.

Microtatorchis Schltr. Orchidaceae

Origins:
Greek *mikrotes, mikrotetos* "smallness" and *orchis* "orchid," referring to the habit.

Microtea Sw. Phytolaccaceae

Origins:
Perhaps from the Greek *mikros* "small, tiny" and *ous, otos* "an ear" or *mikrotes* "smallness," referring to the tiny flowers, some suggest from *thea* "a seeing, looking at, view, a sight."

Microterangis (Schltr.) Senghas Orchidaceae

Origins:
From the Greek *mikros* "small, tiny" and the related genus *Aerangis* Reichb.f.

Microtheca Schltr. Orchidaceae

Origins:
Greek *mikros* "small, tiny" and *theke* "a box, case, chest," the anther cavities.

Microthelys Garay Orchidaceae

Origins:
From the Greek *mikros* "small, tiny" and *thele* "nipple," *thelys* "feminine, female."

Microthlaspi F.K. Mey. Brassicaceae

Origins:
From the Greek *mikros* "small, tiny" plus *Thlaspi* L.

Microthuareia Thouars Gramineae

Origins:
Greek *mikros* "small, tiny" plus *Thuarea* Pers.

Microtis R. Br. Orchidaceae

Origins:
From the Greek *mikros* "small" and *ous, otos* "an ear," referring to the small column wings or membranaceous auricles; see Robert Brown, *Prodromus florae Novae Hollandiae*. 320. London 1810.

Species/Vernacular Names:
M. alba R. Br.

English: white onion orchid, white mignonette orchid

M. atrata Lindley (*Microtis minutiflora* F. Muell.) (Latin *atratus, a, um* "blackened")

English: yellow onion orchid, tiny onion orchid, swamp mignonette orchid

M. orbicularis R. Rogers

English: rare onion orchid, dark mignonette orchid

M. parviflora R. Br. (*Microtis bipulvinaris* Nicholls)

English: slender onion orchid, granular onion orchid, slender mignonette orchid

M. rara R. Br. (*Microtis brownii* Reichb.f.; *Microtis oblonga* R. Rogers)

English: sweet onion orchid

M. unifolia (Forst.f.) Reichb.f. (*Microtis formosana* Schltr.; *Ophrys unifolia* Forst.f.)

English: common onion orchid, common mignonette orchid

Japan: nira-ba-ran

Microtoena Prain Labiatae

Origins:
From the Greek *mikros* "small" and *tainia* "fillet, band," an allusion to the small bands in the corolla.

Species/Vernacular Names:
M. delavayi Prain

English: Delavay microtoena, Yunnan microtoena

China: yun nan guan chun hua

M. insuavis (Hance) Prain ex Briquet

English: common microtoena

China: guan chun hua

M. maireana Handel-Mazzetti

English: Maire microtoena

China: shi shan guan chun hua

M. megacalyx C.Y. Wu

English: bigcalyx microtoena

China: da e guan chun hua

M. mollis H. Léveillé

English: pubescent microtoena

China: mao guan chun hua

M. moupinensis (Franchet) Prain

English: Moupin microtoena

China: bao xing guan chun hua

M. omeiensis C.Y. Wu & Hsuan

English: Omei Mountain microtoena

China: e mei guan chun hua

M. patchoulii (C.B. Clarke ex Hook.f.) C.Y. Wu & Hsuan

English: Chinese patchouly, patchoul microtoena

China: dian nan guan chun hua

M. prainiana Diels

English: Prain microtoena

China: nan chuan guan chun hua

M. robusta Hemsley

English: robust microtoena

China: cu zhuang guan chun hua

M. subspicata C.Y. Wu ex Hsuan

English: spike microtoena

China: jin sui zhuang guan chun hua

Microtrichia DC. Asteraceae

Origins:
Greek *mikros* "small" and *thrix, trichos* "hair," *mikrotrichos* "short-haired," with very short bracts.

Microtrichomanes (Mett.) Copel. Hymenophyllaceae

Origins:
From the Greek *mikros* "small" and the genus fern *Trichomanes* L.

Microtropis Wallich ex Meissner Celastraceae

Origins:
Greek *mikros* and *tropis, tropidos* "a keel," referring to the petals.

Species/Vernacular Names:
M. japonica (Fr. & Sav.) Haller f. (*Elaeodendron japonicum* Fr. & Sav.; *Otherodendron liukiuense* Nakai)

Japan: moku-reishi, marakuso-iku

Microula Bentham Boraginaceae

Origins:
Greek *mikros* "small" and *hyle* "a wood, forest, woodland," referring to the habit and habitat of these herbs, Bhutan, Nepal, Sikkim and northeast India.

Species/Vernacular Names:
M. blepharolepis (Maxim.) I.M. Johnston

English: sharp-leaf microula

China: jiang ye wei kong cao

M. diffusa (Maxim.) I.M. Johnston

English: narrow-leaf microula

China: shu san wei kong cao

M. sikkimensis (C.B. Clarke) Hemsley

English: Sikkim microula

China: wei kong cao

M. tibetica Bentham

English: Tibet microula

China: xi zang wei kong cao

Miersia Lindley Alliaceae (Liliaceae)

Origins:

For the British botanist John Miers, 1789-1879 (London), traveler, from 1819 to 1839 in South America, 1839 Fellow of the Linnean Society, 1843 Fellow of the Royal Society. Among his writings are *Travels in Chile and La Plata*. London 1826, "On a new genus of plants from Chile." *Trans. Linn. Soc. London*. 19: 95-98. 1842, "Volksnamen chilesischer Pflanzen." *Bonplandia*. 4: 201-205. 1856 and *On the Apocynaceae of South America*. London, Edinburgh 1878, collaborator of W.J. Hooker, *Niger Flora*. London 1849. See R. Zander, F. Encke, G. Buchheim and S. Seybold, *Handwörterbuch der Pflanzennamen*. 14. Aufl. Stuttgart 1993; Frans A. Stafleu and Cowan, *Taxonomic Literature*. 3: 471-474. 1981; A. Lasègue, *Musée botanique de Benjamin Delessert*. 1845; Ida Kaplan Langman, *A Selected Guide to the Literature on the Flowering Plants of Mexico*. Philadelphia 1964; S. Lenley et al., *Catalog of the Manuscript and Archival Collections and Index to the Correspondence of John Torrey*. Library of the New York Botanical Garden. 1973; J.H. Barnhart, *Biographical Notes upon Botanists*. 2: 487. 1965; G. Murray, *History of the Collections Contained in the Natural History Departments of the British Museum*. London 1904; Berthold Carl Seemann (1825-1871), *The Botany of the Voyage of H.M.S. Herald*, under the command of Captain Henry Kellett, R.N., C.B., during the years 1845-1851. Published under the authority of the Lords Commissionars of the Admiralty by Berthold Seemann ... naturalist of the expedition. London 1852-1857; I.C. Hedge and J.M. Lamond, *Index of Collectors in the Edinburgh Herbarium*. Edinburgh 1970; T.W. Bossert, *Biographical Dictionary of Botanists Represented in the Hunt Institute Portrait Collection*. 266. 1972; Ethelyn Maria Tucker, *Catalogue of the Library of the Arnold Arboretum of Harvard University*. Cambridge, Massachusetts 1917-1933; Elmer Drew Merrill, *Contr. U.S. Natl. Herb.* 30: 212-213. 1947.

Miersiella Urban Burmanniaceae

Origins:

For the British botanist John Miers, 1789-1879, metallurgist, from 1819 to 1839 traveled in South America. Among his writings are *Travels in Chile and La Plata*. London 1826; see J.H. Barnhart, *Biographical Notes upon Botanists*. 2: 487. 1965.

Miersiophyton Engler Menispermaceae

Origins:

For the British botanist John Miers, 1789-1879; see J.H. Barnhart, *Biographical Notes upon Botanists*. 2: 487. 1965.

Mikania Willd. Asteraceae

Origins:

For the Bohemian botanist Joseph Gottfried Mikan, 1743-1814. His works include *Catalogus plantarum omnium*. Pragae 1776 and *Dispensatorium oder Arzneiverzeichniss für Arme*, zusammengetragen von der medizinischen Fakultät ... Prag. Herausgegeben von J.G.M. Prag 1786; see J.H. Barnhart, *Biographical Notes upon Botanists*. 2: 488. 1965; Jonas C. Dryander, *Catalogus bibliothecae historico-naturalis Josephi Banks*. 1796-1800; E.M. Tucker, *Catalogue of the Library of the Arnold Arboretum of Harvard University*. 1917-1933; Frans A. Stafleu and Cowan, *Taxonomic Literature*. 3: 481-483. 1981. Some suggest the genus was dedicated to Johann Christian Mikan (1769-1844), son of Joseph Gottfried Mikan, botanist, entomologist, author of *Delectus florae et faunae brasiliensis*. Vindobonae 1820[-1825] and *Monographia Bombyliorum Bohemiae*, iconibus illustrata. Pragae 1796; see J.H. Barnhart, *Biographical Notes upon Botanists*. 2: 488. 1965; Jonas C. Dryander, *Catalogus bibliothecae historico-naturalis Josephi Banks*. 1796-1800; E.M. Tucker, *Catalogue of the Library of the Arnold Arboretum of Harvard University*. 1917-1933; A. Lasègue, *Musée botanique de Benjamin Delessert*. Paris 1845; R. Zander, F. Encke, G. Buchheim and S. Seybold, *Handwörterbuch der Pflanzennamen*. 14. Aufl. 751. Stuttgart 1993; Georg Christian Wittstein, *Etymologisch-botanisches Handwörterbuch*. 583. Ansbach 1852; H. Genaust, *Etymologisches Wörterbuch der botanischen Pflanzennamen*. 385-386. 1996.

Species/Vernacular Names:

M. cordata (Burm.f.) B.L. Robinson

Yoruba: oje dudu, akoeela, iyawa, okorowu

Congo: oyilu, oyiligui

Mikaniopsis Milne-Redhead Asteraceae

Origins:
Resembling *Mikania*.

Mila Britton & Rose Cactaceae

Origins:
An anagram of Lima, Peru.

Mildbraedia Pax Euphorbiaceae

Origins:
After the German botanist Gottfried Wilhelm Johannes Mildbraed, 1879-1954, traveler and explorer, plant collector, 1908 Duke of Mecklenburg's first expedition to Cam-

eroon and 1910-1911 second expedition from Congo to Lake Chad (also Cameroon and Fernando Po), 1913 Keeper of Berlin Bot. Mus., 1928 Likomba. Among his writings are "Plantae Tessmannianae Peruvianae II." *Notizbl. Bot. Gart. Berlin-Dahlem.* 9(84): 260-268. 1925, "Plantae Tessmannianae Peruvianae III." *Notizbl. Bot. Gart. Berlin-Dahlem.* 9(89): 964-997. 1926, "Acanthaceae novae." *Notizbl. Bot. Gart. Berlin-Dahlem.* 11(101): 63-71. 1930 and "Beiträge zur Flora von Deutsch-Südwestafrika." *Notizbl. Bot. Gart. Berlin.* 15: 448-471, 633-634. 1940, 15: 757-761. 1942, with Franz Otto Koch wrote *Die Banane ihre Kultur und Verbreitung.* Berlin 1926. See J.H. Barnhart, *Biographical Notes upon Botanists.* 2: 489. Boston 1965; F.N. Hepper and F. Neate, *Plant Collectors in West Africa.* 56. 1971; Ida Kaplan Langman, *A Selected Guide to the Literature on the Flowering Plants of Mexico.* Philadelphia 1964; Réné Letouzey (1918-1989), "Les botanistes au Cameroun." in *Flore du Cameroun.* 7: 1-110. Paris 1968; Jan Czekanovski, *Ethnographisch-Antropologischer Atlas*: Zwischenseen Bantu-Pygmaen und Pygmoiden-Urwaldstamme. Leipzig 1911; T.W. Bossert, *Biographical Dictionary of Botanists Represented in the Hunt Institute Portrait Collection.* 267. 1972; Adolf Friedrich Georg Ernst Albert Eduard, Duke of Mecklenburg (b. 1882), *Wissenschaftliche Ergebnisse der Deutschen ZentralAfrika-Expedition 1907-1908,* unter Führung Adolf Friedrichs ... Band ii. *Botanik.* Leipzig 1914 and *From the Congo to the Niger and the Nile. An Account of the German Central African Expedition of 1910-1911.* Duckworth 1913; Anthonius Josephus Maria Leeuwenberg, "Isotypes of which holotypes were destroyed in Berlin." in *Webbia.* 19(2): 862. 1965; R. Zander, F. Encke, G. Buchheim and S. Seybold, *Handwörterbuch der Pflanzennamen.* 14. Aufl. Stuttgart 1993.

Mildbraediochloa Butzin Gramineae

Origins:

After the German botanist Gottfried Wilhelm Johannes Mildbraed (1879-1954), traveler and explorer, plant collector.

Mildbraediodendron Harms Fabaceae (Caesalpiniaceae)

Origins:

After the German botanist Gottfried Wilhelm Johannes Mildbraed (1879-1954), traveler and explorer, plant collector, 1908 Duke of Mecklenburg's first expedition to Cameroon and 1910-1911 second expedition from Congo to Lake Chad (also Cameroon and Fernando Po), 1913 Keeper of Berlin Bot. Mus., 1928 at Likomba; see J.H. Barnhart, *Biographical Notes upon Botanists.* 2: 489. 1965.

Species/Vernacular Names:
M. excelsum Harms
Zaire: bolelembe
Cameroon: abies, ekela

Mildella Trevisan Pteridaceae (Adiantaceae)

Origins:

After the German botanist Carl August Julius Milde, 1824-1871. His writings include "Ueber exotische Equiseten." *Verh. K.K. Zool.-Bot. Ges. Wien.* 11: 345-364. 1861, "On the geographical distribution of the Equisetaceae." *J. Bot.* 1: 321-325. 1863, "Nachträge zu meinen Beschreibungen exotischer Equiseten." *Verh. K.K. Zool.-Bot. Ges. Wien.* 13: 225-232. 1863, "Index Equisetorum. Editio altera aucta et emendata." *Verh. K.K. Zool.-Bot. Ges. Wien.* 14: 525-550. 1864 and *Reise seine Majestät Fregatte Novara und die Erde.* Botanischer Theil. 1 Band, 4 Heft. Geffäss-Kryptogamen bearbeitet von Dr. Georg Mettenius. Ophioglossen und Equisetaceen bearbeitet von Dr. Julius Milde. 199-261. Wien 1870. See F.A. Stafleu and Cowan, *Taxonomic Literature.* 3: 485-488. 1981; J.H. Barnhart, *Biographical Notes upon Botanists.* 2: 489. 1965; T.W. Bossert, *Biographical Dictionary of Botanists Represented in the Hunt Institute Portrait Collection.* 267. 1972; S. Lenley et al., *Catalog of the Manuscript and Archival Collections and Index to the Correspondence of John Torrey.* Library of the New York Botanical Garden. 291. 1973; R. Zander, F. Encke, G. Buchheim and S. Seybold, *Handwörterbuch der Pflanzennamen.* 14. Aufl. Stuttgart 1993; I.C. Hedge and J.M. Lamond, *Index of Collectors in the Edinburgh Herbarium.* Edinburgh 1970.

Miliarium Moench Gramineae

Origins:
Latin *miliarius, a, um* or *milliarius* "belonging to millet," see also *Milium* L.

Miliastrum Fabr. Gramineae

Origins:
Resembling *milium.*

Milicia Sim Moraceae

Origins:
Named for Mr. Milicia, administrator in Mozambique; see Y. Tailfer, *La Forêt dense d'Afrique centrale.* CTA,

Ede/Wageningen 1989; J. Vivien & J.J. Faure, *Arbres des Forêts denses d'Afrique Centrale*. Agence de Coopération Culturelle et Technique. Paris 1985.

Species/Vernacular Names:

M. excelsa (Welw.) C.C. Berg (*Morus excelsa* Welw.; *Chlorophora excelsa* (Welw.) Benth. & Hook.f.; *Milicia africana* Sim)

English: African teak, E. African teak, rock elm, counter wood, iroko, cokewood, mulberry, African oak, Nigerian teak

French: chêne d'Afrique, teck d'Afrique

Angola: mercira, amoreira, moreira (Portuguese); mukuma, mukamba-kamba (Kimbundu); mova, sanga, makamba (Kioko); kamba (Umbundu); kambula (Kikongo)

Cameroon: abang, bang, bangi, abeng, beng, bing, eloum, emang, mokongo, ntong, adoum, menangi, momangi

Central African Republic: bangui

Congo: kambala

Dahomey: roco, rocco, rokko

Eastern Africa: mufula, mvule, mgonda, mbang

Gabon: abang, aban, abang heli, kambala, mandji, nkolo, eloun, kambala, nombo

Ghana: ala, iroko, odoum, edid, elunli, eluwi, erui, kusaba, odum

Ivory Coast: agui, akede, ala, iroko, odoum, odum, edoum, bakana, bang, bonzo, bouzu, di, n'di, diedie, dou, egouzi, elui, elwi, guenle, guenlo, guento, monangi, mui, muui, roko, sili, simme

Liberia: semli, ge-ay, kambalo

Nigeria: oroko, iroko, loko, rokko, oji, oloko, uloko, reko zhiko; loko (Hausa); roko (Nupe); iroko (Yoruba); uloko (Edo); olokpata (Ijaw); uno (Urhobo); uroko (Itsekiri); oji (Igbo); nsan (Ekoi); osan (Boki)

Portug. E. Africa: magundo, mgunde

Senegal: toumbohiro noir

Sierra Leone: sime

Togo: logo asagu, odum, ssare, ssere, ukloba

Tropical Africa: oroko

Uganda: mutumba, muvule, mvule

W. Africa: iroko, mbang

Zaire: bolondo, bolondu, kamba, kamba-kamba, mbara, molongo, molondu, mufula, nkamba, punga, sanga-sanga, ulundu; bangi (Kisangani); bokongo (Turumbu); bolondo, bolundu (Lokundu); bwagashanga (Bangende); dondo (Bumba); kalanda kakunze, lusanga (Tshiluba); kamba (Kisantu; see J. van Wing, *Études Bakongo*. Bruxelles 1959), kambala (Mayumbe); lebia (Mobwasa); mbara (Irumu); malamu (Kwilu); molongo (Lukolela); muvulu, mufula (muvula = rain) (Maniéma); mpunga, punga

(Kinande); mukobakoba mukunze (Bakuba; see Efraim Andersson, *Contribution à l'Ethnographie des Kuta*. Uppsala 1974; E. Torday and T.A. Joyce, *Notes ethnographiques sur les Peuples communement ... Bakuba, ainsi sur les ... Bushongo*. Bruxelles 1910); mukamba (Shaba); mulundu (Mai-Ndombe lake); musongwe, uposhu (Kilur); nagwande (Uele); sanga (Tshofa); ulundu (Bashobwa)

Milium Adans. Gramineae

Origins:

Latin *milium, ii* "millet."

Milium L. Gramineae

Origins:

Latin *milium, ii* "millet," Greek *meline*; see Carl Linnaeus, *Species Plantarum*. 61. 1753 and *Genera Plantarum*. Ed. 5. 30. 1754; S. Battaglia, *Grande dizionario della lingua italiana*. X: 383. Torino 1978; H. Genaust, *Etymologisches Wörterbuch der botanischen Pflanzennamen*. 386. 1996.

Species/Vernacular Names:

M. effusum L.

English: millet grass

Miliusa Leschen. ex A. DC. Annonaceae

Origins:

Possibly the genus was named after the Italian botanist Josephus Mylius, author of *De hortorum cultura*. Brixiae 1574, or dedicated to Joannes Mylius, 1585-1618, physician and philosopher in Hessen, author of *Antidotarium Medico-Chymicum Reformatum*. Francofurti 1620, or from Latin *milium, ii* "millet" or from a vernacular name; see L.W. Jessup, "The genus *Miliusa* Leschen. ex A. DC. (Annonaceae) in Australia." in *Austrobaileya*. 2: 517-523. 1988; M.P. Nayar, *Meaning of Indian Flowering Plant Names*. 227. Dehra Dun 1985; S.K. Jain & R.R. Rao, *An assessment of threatened plants of India*. Proceedings of the seminar held at Dehra Dun, 14-17 Sept., 1981. Howrah, Botanical Survey of India. 334 pp. 1983; D. Mitra & P. Chakraborty, "*Miliusa mukerjeeana* D. Mitra & Chakrab. (Annonaceae) — A new species from Andaman & Nicobar Islands." in *Bull. Bot. Surv. India*. 33: 326-328, ill. 1994 [1991]; B. Li et al., 1997. "Epidermal feature and venation of the leaves of Chinese *Miliusa* and *Saccopetalum* (Annonaceae) in relation to taxonomy." *J. S. China Agric. Univ*. 18: 41-44. 1997.

Milla Cav. Alliaceae (Liliaceae)

Origins:

Named for Julian (or Julio) Milla, 18th century gardener to the King of Spain; see Antonio José Cavanilles (1745-1804), *Icones et Descriptiones Plantarum*. 2: 76. 1793-1794; Georg Christian Wittstein, *Etymologisch-botanisches Handwörterbuch*. 583. Ansbach 1852.

Species/Vernacular Names:

M. biflora Cav.

Mexico: guia gana, quije cana

Milleria L. Asteraceae

Origins:

After the British botanist Philip Miller, 1691-1771 (London), gardener, 1722-1770 Superintendent at Chelsea Physick Garden, in 1729 Fellow of the Linnean Society, author of *The Gardeners Dictionary*. Abr. ed. 4. London 1754. See William T. Stearn, in *Dictionary of Scientific Biography* 9: 390-391. 1981; Ray Desmond, *Dictionary of British & Irish Botanists and Horticulturists*. 487-488. 1994; Blanche Elizabeth Edith Henrey, *British Botanical and Horticultural Literature before 1800*. Oxford 1975; Stafleu and Cowan, *Taxonomic Literature*. 3: 491-499. 1981; John H. Barnhart, *Biographical Notes upon Botanists*. 2: 492. 1965; R. Zander, F. Encke, G. Buchheim and S. Seybold, *Handwörterbuch der Pflanzennamen*. 14. Aufl. Stuttgart 1993; Jonas C. Dryander, *Catalogus bibliothecae historico-naturalis Josephi Banks*. London 1796-1800; T.W. Bossert, *Biographical Dictionary of Botanists Represented in the Hunt Institute Portrait Collection*. 268. 1972; E.M. Tucker, *Catalogue of the Library of the Arnold Arboretum of Harvard University*. 1917-1933; H.R. Fletcher, *Story of the Royal Horticultural Society, 1804-1968*. Oxford 1969; Ida Kaplan Langman, *A Selected Guide to the Literature on the Flowering Plants of Mexico*. Philadelphia 1964; A. Lasègue, *Musée botanique de Benjamin Delessert*. Paris 1845; H.R. Fletcher and W.H. Brown, *Royal Botanic Garden Edinburgh, 1670-1970*. Edinburgh 1970; G. Murray, *History of the Collections Contained in the Natural History Departments of the British Museum*. 1: 168. London 1904; R. Pulteney, *Historical and Biographical Sketches of the Progress of Botany in England*. 1790; E. Earnest, *John and William Bartram, Botanists and Explorers 1699-1777, 1739-1823*. Philadelphia 1940; E. Bretschneider, *History of European Botanical Discoveries in China*. 1981; J.D. Milner, *Catalogue of Portraits of Botanists Exhibited in the Museums of the Royal Botanic Gardens*. Kew, London 1906; Frans A. Stafleu, *Linnaeus and the Linnaeans*. The spreading of their ideas in systematic botany, 1735-1789. Utrecht 1971.

Millettia Wight & Arn. Fabaceae

Origins:

Named after the British plant collector Dr. Charles Millett of Canton, China, fl. 1825-34, probably a physician, in the 1830s (about 1825-1834) he was in the service of the Honourable East India Company, he lived in Macao and Canton, Sri Lanka and Malabar, friend and correspondent of W.J. Hooker, in 1830-1831 introduced some Chinese plants into the Glasgow Botanical Gardens; see Robert Wight and G. Arnott Walker Arnott, *Prodromus florae Peninsulae Indiae Orientalis*. 263. London 1834; James A. Baines, *Australian Plant Genera. An Etymological Dictionary of Australian Plant Genera*. 240. Chipping Norton, N.S.W. 1981. According to some authors the genus was named after the 18th century French botanist J.A. Millet; see M.P. Nayar, *Meaning of Indian Flowering Plant Names*. 227. Dehra Dun 1985.

Species/Vernacular Names:

M. sp.

Cameroon: bongongi

Yoruba: ogun malarere

Nigeria: bongoni, bongongi, ibakwi, ito, ekimi

Ivory Coast: ekimi, sohinta, vandague, sointa

Liberia: ju-ehn-jrah, tog-beli

Ghana: okuro-sante, pem, frafrah

French Guiana: missa amadan

Togo: sso abalu

Japan: fuji

Malaya: jenaris, tulang dain, mempari

M. aboensis (Hook.f.) Bak.

Nigeria: awo (Edo); kpukpumanya (Igbo)

M. grandis (E. Meyer) Skeels (*Millettia caffra* Meisn.; *Virgilia grandis* E. Meyer)

English: kaffir ironwood

Southern Africa: Umzimbeet, kafferysterhout, omsambeet; umSimbithi (= ironwood), umSimbithwa, umKunye (Xhosa); umSimbithi (Zulu)

M. griffoniana Baill. (*Lonchocarpus griffonianus* (Baill.) Dunn)

Nigeria: turburku (Hausa); ito (Yoruba); ehiengbo (Edo); njasi (Igbo); katep ashie (Boki)

M. japonica (Siebold & Zucc.) A. Gray (*Kraunhia japonica* (Siebold & Zucc.) Taubert; *Wisteria japonica* Siebold & Zucc.)

Japan: natsu-fuji

M. laurentii De Wild. (for the Belgian botanist Émile Laurent, 1861-1904, explorer, plant collector)

French: bois d'Ambam, bois de fer

Zaire: bokonghe, bokonge, mokonge, mbotu, mokonghe, mundambi, wenge

Equatorial Africa: kundu baenge

Congo: ntoko, ontoko, mutoko, mutoto, n'toko, ngondou

Gabon: awong, otogo, son-so

Cameroon: awong, nsonso

M. megasperma (F. Muell.) Benth.

English: Australian wisteria, native wisteria

M. oblata Dunn

East Africa: mhafa

M. pendula Benth.

Burma: thinwin

M. pilipes Bailey

English: northern wistaria

M. reticulata Benth.

Japan: murasaki-natsu-fuji

China: kun ming ji xue teng

M. rhodantha Baill.

Nigeria: nzachi (Igbo)

Sierra Leone: torlu gbélé

M. stuhlmannii Taubert

Southern Africa: panga panga; muSara, muSaru (Shona)

S. Rhodesia: muSaru

M. sutherlandii Harv. (the species honors Peter Cormack Sutherland, 1822-1900 (d. Natal), geologist and physician, author of *Journal of a Voyage in Baffin's Bay and Barrow Straits in the Years 1850-51 ... in Search of the Missing Crews of H.M. Ships Erebus and Terror.* London 1852; see Edward Augustus Inglefield, *A Summer Search for Sir J. Franklin ... With Short Notices ... by Dr. Sutherland on the Meteorology and Geology.* London 1853; Mary Gunn and Leslie E. Codd, *Botanical Exploration of Southern Africa.* 338. 1981; Ray Desmond, *Dictionary of British & Irish Botanists and Horticulturists.* 665. London 1994)

Southern Africa: giant umzimbeet, bastard umzimbeet, basteroemzimbiet, reuseomsambeet; umQunye, umGunye, umKhunye (Xhosa); umKhunye, umSimbatshani, umSimbishana (Zulu)

M. thonningii (Schum.) Baker (*Robinia thonningii* Schum.)

Nigeria: turburku (Hausa); ito (Yoruba)

Yoruba: itoo, asunlera, abe werewere ori ito, agbawi kowee, olukotun eye igbo

M. versicolor Welw. ex Baker

French: bois d'or

Angola: bobata

Congo: lubota, luboto, ombolo, omboro

Zaire: bobota, bota, boboto, bosoko, hoto, lubota, mbota, mumboto

M. zechiana Harms

Nigeria: katep oshie (Boki)

Milligania Hook.f. Asteliaceae (Liliaceae)

Origins:

After the Scottish-born Australian (Tasmanian) botanist Joseph Milligan, 1807-1883/1884, geologist, plant collector for W.J. Hooker, explorer, from 1829-1830 to 1860 worked in Tasmania, surgeon to the Van Diemen's Land Company (at Surrey Hills), naturalist, botanical collector, secretary of the Royal Society of Van Diemen's Land, in December 1843 became Superintendent of Aboriginals, in 1850 Fellow of the Linnean Society. His writings include *Vocabulary of the Dialects of Some of the Aboriginal Tribes of Tasmania.* (Appendix. List of words in use by the Oyster Bay Tribe ... by T. Scott.) Hobart Town 1866 and *Vocabulary of the Aborigines of Tasmania.* Tasmania 1857, acquainted with R.C. Gunn. See I.H. Vegter, *Index Herbariorum.* Part II (4), *Collectors M.* Regnum Vegetabile vol. 93. 1976; Douglas Pike, ed., *Australian Dictionary of Biography.* 2: 230-231. Melbourne 1967; W.J. Hooker, in *Hooker's Journal of Botany & Kew Garden Miscellany.* 5: 296. London 1853.

Species/Vernacular Names:

M. densiflora Hook.f.

English: short-leaf milligania

M. johnstonii F. Muell. ex Benth.

English: tiny milligania

M. longifolia Hook.f.

English: long-leaf milligania

M. stylosa (Hook.f.) Benth.

English: Huon milligania (from Huon Valley, SE. Tasmania)

Millingtonia L.f. Bignoniaceae

Origins:

Named to honor Sir Thomas Millington, 1628-1704 (London), 1659 M.D. Oxon, the discoverer of sexuality in plants, physician to William and Mary and to Queen Anne, in 1675 professor of natural philosophy at Oxford, in 1680 knighted. See *The Report of the Physicians and Surgeons, Commanded to Assist at the Dissecting the Body of his Late Majesty (William III) at Kensington.* From the original delivered to the Right Honorable the Privy Council. [The report drawn up by Sir T. Millington, Sir Richard Blackmore and Sir Edward Hannes.] 1702; Robert John Thornton (1768?-1837), *New Illustration of the Sexual System of C. von Linnaeus.* London [1799-] 1807; R. Pulteney, *Histori-*

cal and Biographical Sketches of the Progress of Botany in England. 1: 336-337. London 1790; William Munk, *The Roll of the Royal College of Physicians of London.* London 1878; Gordon Douglas Rowley, *A History of Succulent Plants.* Strawberry Press, Mill Valley, California 1997.

Species/Vernacular Names:
M. hortensis L.f. (*Millingtonia dubiosa* Span.)
English: Indian cork-tree

Millotia Cass. Asteraceae

Origins:
After the French historian Claude François Xavier Millot, 1726-1785, scientist. Among his numerous publications are *Elements of the History of England,* from the invasion of the Romans to the Reign of George the Second. Translated by Mrs. Brooke. London 1771, *Harangues choisies des Historiens Latins, traduites.* Lyon 1764, *Observations d'un citoyen sur le chapitre concernant l'administration de la justice, du projet de la nouvelle constitution.* [Paris 1791] and *Eléments de l'Histoire de France, depuis Clovis jusqu'à Louis XV.* Paris 1770; see Jean Baptiste de La Curne de Sainte-Palaye, *Histoire littéraire des Troubadours,* etc. [Arranged and published anonymously by C.F.X. Millot from materials collected by J.B. de La Curne de Sainte-Palaye, whose name is mentioned in the preface.] Paris 1774.

Species/Vernacular Names:
M. greevesii F. Muell. (after Augustus F.A. Greeves, 19th century medical practitioner and politician)
English: creeping millotia
M. macrocarpa (F. Muell. & Tate) Schodde
English: large-fruited millotia
M. myosotidifolia (Benth.) Steetz
English: broad-leaved millotia
M. tenuifolia Cass.
English: soft millotia

Millottia Stapf Asteraceae

Origins:
Orthographic variant of *Millotia* Cass.

Millspaughia B.L. Robinson Polygonaceae

Origins:
For the American botanist Charles Frederick Millspaugh, 1854-1923, physician, professor of botany, traveler and plant collector (four trips Yucatan). Among his works are *American Medicinal Plants.* New York [1974] and "Contributions to North American Euphorbiaceae-VI." *Publ. Field Mus. Nat. Hist., Bot. Ser.* 2: 401-420. 1916, with Earl Edward Sherff (1886-1966) wrote "Revision of the North American species of *Xanthium*." *Field Mus. Nat. Hist., Bot. Ser.* 4(2): 9-49. 1919. See John H. Barnhart, *Biographical Notes upon Botanists.* 2: 493. 1965; R. Zander, F. Encke, G. Buchheim and S. Seybold, *Handwörterbuch der Pflanzennamen.* 14. Aufl. Stuttgart 1993; I.C. Hedge and J.M. Lamond, *Index of Collectors in the Edinburgh Herbarium.* Edinburgh 1970; Ida Kaplan Langman, *A Selected Guide to the Literature on the Flowering Plants of Mexico.* Philadelphia 1964; T.W. Bossert, *Biographical Dictionary of Botanists Represented in the Hunt Institute Portrait Collection.* 268. 1972; S. Lenley et al., *Catalog of the Manuscript and Archival Collections and Index to the Correspondence of John Torrey.* Library of the New York Botanical Garden. 291-292. 1973; E.M. Tucker, *Catalogue of the Library of the Arnold Arboretum of Harvard University.* 1917-1933; J. Ewan, ed., *A Short History of Botany in the United States.* 148. New York and London 1969; Irving William Knobloch, compiled by, "A preliminary verified list of plant collectors in Mexico." *Phytologia Memoirs.* VI. Plainfield, N.J. 1983.

Milnea Roxb. Meliaceae

Origins:
Named for the British (b. Aberdeen) clergyman Rev. Colin Milne, c. 1743-1815 (d. Kent), botanist and botanical collector, Rector of North Chapel, near Petworth in Sussex, translated Linnaeus from the Latin, with Alexander Gordon (fl. 1790s, a son of James Gordon, the nurseryman of Mile-end) wrote *Indigenous Botany: or Habitations of English Plants.* London [1792-]1793. Among his writings are *A Botanical Dictionary.* London 1770, *A Sermon* [on Rom. xiv.16] *Preached before the Grand Lodge ... of Free and Accepted Masons.* London [1788], *Sermons.* London 1780, *The Boldness and Freedom of Apostolical Eloquence* recommended to the imitation of Ministers. A sermon [on Acts ii.4] occasioned by the death of the Reverend ... J. Bate. London 1775 and (presumably) *A Descriptive Catalogue of Rare and Curious Plants,* the seeds of which were lately received from the East-Indies. London [1773]; see John H. Barnhart, *Biographical Notes upon Botanists.* 2: 493. 1965; E.M. Tucker, *Catalogue of the Library of the Arnold Arboretum of Harvard University.* 1917-1933; Jonas C. Dryander, *Catalogus bibliothecae historico-naturalis Josephi Banks.* London 1796-1800; Blanche Henrey, *British Botanical and Horticultural Literature before 1800.* Oxford 1975; T.W. Bossert, *Biographical Dictionary of Botanists Represented in the Hunt Institute Portrait Collection.* 268. 1972; Warren R. Dawson, *The Banks Letters,* a

Calendar of the Manuscript Correspondence of Sir Joseph Banks. London 1958.

Miltitzia A. DC. Hydrophyllaceae

Origins:
After the German botanical bibliographer Friedrich Joseph Franx Xaver von Miltitz, died 1840, author of *Handbuch der botanischen Literatur.* [with the preface of Heinrich Gottlieb Ludwig Reichenbach, 1793-1879] Berlin 1829; see J.H. Barnhart, *Biographical Notes upon Botanists.* 2: 494. 1965.

Miltonia Lindley Orchidaceae

Origins:
For the British patron and benefactor of horticulture and orchid grower Charles William Wentworth Fitzwilliam, Viscount Milton, 1786-1857 (Wentworth Woodhouse, Yorks.), politician and ardent reformer, owner of a garden at Wentworth, received the garter in 1851, in 1857 was appointed Lord-lieutenant of the West Riding. Among his writings are *Address to the Landowners of England on the Corn Laws.* London 1832 and *A Letter to the Rev. J. Sargeaunt* [in vindication of the speech at Sheffield urging the necessity of assisting Ireland from the Imperial Treasury] London 1847, see *Miltonics: A Mock-Heroic Poem, Dedicated to the Freeholders of the County of York.* London 1826.

Miltonioides Brieger & Lückel Orchidaceae

Origins:
Resembling the genus *Miltonia.*

Miltoniopsis Godefroy-Leb. Orchidaceae

Origins:
Resembling *Miltonia.*

Milula Prain Alliaceae (Liliaceae)

Origins:
An anagram derived from the generic name of *Allium.*

Mimetanthe Greene Scrophulariaceae

Origins:
From the Greek *mimetes* "an imitator" and *anthos* "flower."

Mimetes Salisb. Proteaceae

Origins:
Greek *mimeomai* "to imitate," *mimetes* "an imitator," *mimo*, *mimos* "an ape, a mimic," referring to the complex compound inflorescence and the bird-pollination flowers, the Orange-breasted Sunbird, *Nectarinia violacea*, and the Cape Sugarbird, *Promerops caffer*, etc.; see J.P. Rourke, "A revision of *Mimetes* Salisb. (Proteaceae)." *Journal of South African Botany.* 50(2): 171-236. 1984.

Species/Vernacular Names:
M. fimbriifolius Salisb. ex Knight (*Mimetes cucullatus* (L.) R. Br. var. *hartogii* (R. Br.) Phill.; *Mimetes hartogii* R. Br.)
South Africa: fringe-leaved mimetes

Mimetophytum L. Bolus Aizoaceae

Origins:
From the Greek *mimetes* "an imitator" and *phyton* "plant."

Mimophytum Greenman Boraginaceae

Origins:
Greek *mimo*, *mimos* "an ape, a mimic" and *phyton* "a plant."

Mimosa L. Mimosaceae

Origins:
Latin *mimus, mimum* "a mime, actor," Greek *mimos* "a mimic, mime, imitator," *mimeomai* "to mimic, imitate"; see Carl Linnaeus, *Species Plantarum.* 516. 1753 and *Genera Plantarum.* Ed. 5. 233. 1754; S. Battaglia, *Grande dizionario della lingua italiana.* X: 418-419. Torino 1978.

Species/Vernacular Names:
M. bimucronata (DC.) Kuntze (*Mimosa sepiaria* Benth.; *Acacia bimucronata* DC.)
English: giant mimosa
M. pigra L.
English: mimosa, sensitive plant
Yoruba: patonmo, paidimo, oniwa agogo
N. Rhodesia: mungonga, chikwata
M. pudica L.
English: sensitive plant, touch-me-not, shame plant, live-and-die, humble plant, action plant, sleeping grass
Central America: ix mutz, corona de cristo, dormidillo
The Philippines: dicut malamarine, makahiya, damohia, damohiya, makahia, sipug-sipug, ambabaing, babain, bain-

bain, baeng-baeng, andi-baing, dilgansusu, huya-huya, tuyag-huyag, torog-torog, harupai, kiromkirom, kirom-kirom

Vietnam: co then, ham tu thao

Malaya: kemunchup, semalu, daun sopan, puteri malu

Hawaii: pua hilahila

Japan: ojigi-sô

Yoruba: patonmo, paidimo, pamamo aluro

M. pudica L. var. *hispida* Brenan

English: action plant, humble plant, live-and-die, sensitive plant, shame plant, sheme weed, touch-me-not

South Africa: Kruidjie-roer-my-nie

M. scabrella Benth. (*Mimosa bracaatinga* Hoehne)

Brazil: bracaatinga

M. sensitiva L.

English: sensitive plant

Mimosopsis Britton & Rose Mimosaceae

Origins:
Resembling the genus *Mimosa* L.

Mimozyganthus Burkart Mimosaceae

Origins:
From the Greek *mimos* "a mimic, mime, imitator," *zygon* "a yoke" and *anthos* "flower."

Mimulicalyx Tsoong Scrophulariaceae

Origins:
From the genus *Mimulus* and *kalyx* "calyx."

Mimulopsis Schweinf. Acanthaceae

Origins:
Resembling *Mimulus*.

Mimulus L. Scrophulariaceae

Origins:
Referring to the flowers, monkey-faced; from the Latin *mimulus, i* "a little mime," diminutive of *mimus, i* "a mimic actor," Greek *mimos*; see Carl Linnaeus, *Species Plantarum.* 634. 1753 and *Genera Plantarum.* Ed. 5. 283. 1754.

Species/Vernacular Names:
M. clevelandii Brandegee

English: Cleveland's bush monkey-flower

M. exiguus A. Gray

English: San Bernardino Mountains monkey-flower

M. filicaulis S. Watson

English: slender-stemmed monkey-flower

M. glaucescens E. Greene

English: shield-bracted monkey-flower

M. gracilipes Robinson

English: slender-stalked monkey-flower

M. gracilis R. Br.

Australia: slender monkey-flower

M. inconspicuus A. Gray

English: small-flowered monkey-flower

M. laciniatus A. Gray

English: cut-leaved monkey-flower

M. mohavensis Lemmon

English: Mojave monkey-flower

M. moschatus Douglas ex Lindley

Australia: musk monkey-flower, musk plant, musk-scented monkey-flower, musk mimulus, monkey musk

M. nudatus E. Greene

English: bare monkey-flower

M. pictus (E. Greene) A. Gray

English: Calico monkey-flower

M. prostratus Benth.

Australia: small monkey-flower, monkey face

M. pulchellus (E. Greene) A.L. Grant

English: pansy monkey-flower

M. purpureus A.L. Grant

English: purple monkey-flower

M. pygmaeus A.L. Grant

English: Egg Lake monkey-flower

M. repens R. Br.

English: creeping monkey-flower, Maori musk

M. rupicola Cov. & A.L. Grant

English: Death Valley monkey-flower

M. traskiae A.L. Grant

English: Santa Catalina Island monkey-flower

Mimusops L. Sapotaceae

Origins:
Greek *mimo, mimus* "an ape, a mimic" and *opsis* "resembling, aspect," *ops* "eye, face," the corolla and the shape of

flowers resemble the face of a monkey; Latin *mimus, i* "a mimic actor"; see Carl Linnaeus, *Species Plantarum.* 349. 1753 and *Genera Plantarum.* Ed. 5. 165. 1754.

Species/Vernacular Names:

M. caffra E. Mey. ex A. DC. (*Mimusops revoluta* Hochst.)

English: coastal red milkwood, red milkwood

Southern Africa: kusrooimelkhout, moepel, melkhout, rooimelkhout, shrope; umHayihayi, aMasethole, umHlalankwazi, umNole, umNole uMagayi, umNweba wasolwandle (= the umNweba of the sea), umThunzi (= shade tree), umKhakhayi, amaSethole-abomvu (Zulu); umThunzi, umHlophe, umHlope, umNweba (Xhosa)

M. elengi L. (a Malabar name)

English: Spanish cherry, medlar, tanjong tree

Malaya: tanjong, mengkulah, mengkula, mengkulang, bunga tanjong, pokok tanjong, pekola batu

India: bakulah, ilanni, bolasiri, elengi, elanni

M. kummel Bruce ex A. DC.

Nigeria: emido (Yoruba)

Tanzania: ghana

M. obovata Sond. (*Mimusops oleifolia* N.E. Br.; *Mimusops woodii* Engl.; *Mimusops rudatisii* Engl. & Krause)

English: red milkwood

Southern Africa: rooimelkhout, moepel, bosmelkhout; amaSethole-ehlathi, amaSethole-abomvu, umNolwe, umPhumbulu, umHlalankwazi (Zulu); aMasethole, umTunzi wehlathi, umNtunzi (Xhosa)

M. zeyheri Sond. (*Mimusops kirkii* Bak.) (the specific name after the German botanist and plant collector Carl Ludwig Philipp Zeyher, 1799-1858; see John H. Barnhart, *Biographical Notes upon Botanists.* 3: 540. 1965)

English: Transvaal red milkwood, red milkwood

Southern Africa: moepel; umPushane, mKele (Zulu); umPhushane (Swazi); moopudu, mmupudu (Western Transvaal, northern Cape, Botswana); mmupudu, mopupudu (North Sotho); mububulu (Venda); muChechefe, muChechete, uChirinsi, muTunzi (Shona); mobu (Subya)

S. Rhodesia: uTunzi, muCheningi, muChechete

Minasia H. Robinson Asteraceae

Origins:
Minas Geraes, Brazil.

Minicolumna Brieger Orchidaceae

Origins:
Referring to the tiny column.

Minquartia Aublet Olacaceae

Origins:

A native name for *Minquartia guianensis* Aublet, this tree is called *minquer* by the Creoles.

Species/Vernacular Names:

M. guianensis Aublet

Bolivia: caricuara amarilla, iscocharasi

Minthostachys (Benth.) Spach Labiatae

Origins:
From the Greek *minthe* "mentha" and *stachys* "spike."

Minuartia L. Caryophyllaceae

Origins:
For the Spanish botanist Juan Minuart, 1693-1768, chemist, apothecary, in Barcelona and then Madrid, a correspondent of Linnaeus, author of Illustrissimae regiae physico-medicae Matritensi Academiae ... *Cotyledon hispanica* seditereti folii. [Madrid 1739] and *Cerviana* sub auspiciis illustrissimi viri D.D. Josephi Cervi, ... feliciter edita. [Madrid 1739]. See Miguel Colmeiro y Penido (1816-1901), *La Botánica y los Botánicos de la Peninsula Hispano-Lusitana.* Madrid 1858; John H. Barnhart, *Biographical Notes upon Botanists.* 2: 495. 1965; F. Boerner & G. Kunkel, *Taschenwörterbuch der botanischen Pflanzennamen.* 4. Aufl. 135. Berlin & Hamburg 1989; Georg Christian Wittstein, *Etymologisch-botanisches Handwörterbuch.* 584. Ansbach 1852.

Species/Vernacular Names:

M. decumbens T.W. Nelson & J.P. Nelson

English: sandwort

M. howellii (S. Watson) Mattf.

English: Howell's sandwort

M. hybrida (Vill.) Schischkin (*Arenaria hybrida* Vill.; *Arenaria tenuifolia* L.; *Alsine tenuifolia* (L.) Crantz; *Minuartia tenuifolia* (L.) Hiern)

English: slender-leaved sandwort, slender sandwort, fine-leaved sandwort

M. obtusiloba (Rydb.) House

English: alpine sandwort

M. rosei (Maguire & Barneby) McNeill

English: peanut sandwort

M. stolonifera T.W. Nelson & J.P. Nelson

English: Scott Mountain sandwort

M. stricta (Sw.) Hiern

English: Teesdale sandwort

Minuopsis W.A. Weber Caryophyllaceae

Origins:
Possibly referring to *Minuartia*.

Minuria DC. Asteraceae

Origins:
Greek *minyros* "small, thin, weak," referring to the tiny flowers or to the stems and leaves of the type species, *Minuria leptophylla* DC.; see Nicholas S. Lander and Rhonda Barry, "A review of the genus *Minuria* DC. (Asteraceae, Astereae)." in *Nuytsia*. 3(2): 221-237. 1980.

Species/Vernacular Names:
M. annua (Tate) Tate ex J. Black (*Minuriella annua* Tate)
Australia: annual minuria

M. cunninghamii (DC.) Benth. (*Elachothamnos cunninghamii* DC.; *Eurybiopsis intricata* F. Muell.; *Therogeron tenuifolius* Sonder)
Australia: bush minuria

M. denticulata (DC.) Benth.
Australia: woolly minuria

M. integerrima (DC.) Benth.
Australia: smooth minuria

M. leptophylla DC.
Australia: Minnie daisy

Minuriella Tate Asteraceae

Origins:
The diminutive of the genus *Minuria*; see R. Tate, in *Transactions of the Royal Society of South Australia*. 23: 288. 1899.

Minurothamnus DC. Asteraceae

Origins:
Greek *minyros* "small, thin, weak" and *thamnos* "bush."

Miphragtes Nieuwl. Gramineae

Origins:
An anagram of the generic name *Phragmites*.

Miquelia Arnott & Nees Gramineae

Origins:
For the Dutch botanist Friedrich Anton Wilhelm Miquel, 1811-1871.

Miquelia Blume Gesneriaceae

Origins:
For the Dutch botanist Friedrich Anton Wilhelm Miquel, 1811-1871.

Miquelia Meissner Icacinaceae

Origins:
For the Dutch (b. in Germany) botanist Friedrich Anton Wilhelm Miquel, 1811-1871 (d. Utrecht), M.D. Gröningen 1833, professor of botany, Director of the Rotterdam Botanical Garden, Director of the Amsterdam Botanical Garden. Among his very numerous and valuable writings are "Species aliquot novas valdivianas, a Domino W. Lechler collectas." *Linnaea*. 25: 650-654. 1853 and "Animadversiones in Piperaceas Herbarii Hookeriani." *London J. Bot*. 4: 410-470. 1845. See Frans A. Stafleu and Richard S. Cowan, *Taxonomic Literature*. 3: 508-520. 1981; F.A. Stafleu, in *Wentia*. 16: 1-95. 1966; Frans A. Stafleu, in *Dictionary of Scientific Biography* 9: 417. 1981; J.H. Barnhart, *Biographical Notes upon Botanists*. 2: 495. 1965; E.D. Merrill, *Bernice P. Bishop Mus. Bull*. 144: 136-137. 1937; T.W. Bossert, *Biographical Dictionary of Botanists Represented in the Hunt Institute Portrait Collection*. 268. 1972; Ida Kaplan Langman, *A Selected Guide to the Literature on the Flowering Plants of Mexico*. Philadelphia 1964; A. Lasègue, *Musée botanique de Benjamin Delessert*. Paris 1845; E.M. Tucker, *Catalogue of the Library of the Arnold Arboretum of Harvard University*. 1917-1933; R. Zander, F. Encke, G. Buchheim and S. Seybold, *Handwörterbuch der Pflanzennamen*. 14. Aufl. Stuttgart 1993; Leonard Huxley, *Life and Letters of Sir J.D. Hooker*. London 1918.

Species/Vernacular Names:
M. caudata King
Malaya: pisang-pisang bulu, seluang

Miqueliopuntia Fric ex F. Ritter Cactaceae

Origins:
See also *Opuntia* Mill.

Mirabilis L. Nyctaginaceae

Origins:

Latin *mirabilis, e* "wonderful," *miror, atus, sum, ari* "to wonder"; see Carl Linnaeus, *Species Plantarum*. 1753 and *Genera Plantarum*. Ed. 5. 82. 1754.

Species/Vernacular Names:

M. californica A. Gray

English: wishbone bush

M. jalapa L. (*Mirabilis uniflora* Schrank)

English: beauty-of-the-night, false jalap, four-o'clock, marvel of Peru, four-o'clock flower

Peru: buenas tardes, clavanilla, clavenilla, Don Diego de la noche, trompetillas, flor de Panamá

Brazil: maravilha

Mexico: maravilla, Don Diego de noche, arrebolera; tlalquilín, tlaquilín (Aztec l.); tsutsuy-xiu, tutsuixiu (Maya l.); tzujoyó (Zoque l., Tapalapa)

Malay name: kembang pukul empat, bunga pechah empat

The Philippines: a las quatro, oraciones, gilala

China: zi mo li gen, tzu mo li, yen chih, huo tan mu tsao

Japan: oshiroi-bana

Okinawa: yasandi-bânâ

Hawaii: nani ahiahi, pua ahiahi, puahiahi

South Africa: vieruurblom, vieruurtjie

Madagascar: belakariva, folera, voampolera, nyctage, faux-jalape, belle-de-nuit

Yoruba: tannaposo, ododo elede, tannapaku, tannatanna, tannapowo, tanna pa oso

M. tenuiloba S. Watson

English: long-lobed four-o'clock

Miraglossum Kupicha Asclepiadaceae

Origins:

From the Greek *myria* "many, myriad, numberless" and *glossa* "a tongue."

Mirandaceltis Sharp Ulmaceae

Origins:

For the Mexican botanist Faustino Antonio Miranda González, 1905-1964, traveler, plant collector. His works include "Plantas nuevas del sur de México." *Bol. Soc. Bot. Méx.* 26: 120-132. 1961 and "Dos arbustos notáble del estado de Yucatan." *Bol. Soc. Bot. Méx.* 21: 9-21. 1957. See John H. Barnhart, *Biographical Notes upon Botanists*. 2: 495. 1965; T.W. Bossert, *Biographical Dictionary of Bota-*

nists Represented in the Hunt Institute Portrait Collection. 268. 1972; William R. Holland, *Medicina maya en los altos de Chiapas*. México 1963; Ida Kaplan Langman, *A Selected Guide to the Literature on the Flowering Plants of Mexico*. Philadelphia 1964; Irving William Knobloch, compiled by, "A preliminary verified list of plant collectors in Mexico." *Phytologia Memoirs*. VI. Plainfield, N.J. 1983.

Mirandea Rzed. Acanthaceae

Origins:

For the Mexican botanist González Faustino Miranda, 1905-1964.

Mirbelia Smith Fabaceae

Origins:

After the French (b. Paris) botanist Charles François Brisseau de Mirbel, 1776-1854 (d. Paris), plant anatomist, physiologist, plant collector, in 1796 forced into exile in the Pyrenées, from 1802 to 1806 head-gardener at Malmaison, from 1806 to 1810 in Holland with Louis Napoléon, after 1808 professor of botany at the Faculty of Sciences in Paris, in 1829 appointed administrator of the Jardin des Plantes. He is best known for his *Histoire naturelle ... des plantes*. Paris an x [1802] -1806, *Elements de botanique*. 1815 and *Recherches anatomiques et physiologyques sur le Marchantia polymorpha*. [Paris 1835]. See A. Nougarède, in *Dictionary of Scientific Biography* 9: 418-419. 1981; R. Zander, F. Encke, G. Buchheim and S. Seybold, *Handwörterbuch der Pflanzennamen*. 14. Aufl. Stuttgart 1993; John H. Barnhart, *Biographical Notes upon Botanists*. 2: 496. 1965; T.W. Bossert, *Biographical Dictionary of Botanists Represented in the Hunt Institute Portrait Collection*. 268. 1972; A. Lasègue, *Musée botanique de Benjamin Delessert*. 524. Paris 1845; S. Lenley et al., *Catalog of the Manuscript and Archival Collections and Index to the Correspondence of John Torrey*. Library of the New York Botanical Garden. 464. 1973; E.M. Tucker, *Catalogue of the Library of the Arnold Arboretum of Harvard University*. 1917-1933; Mea Allan, *The Hookers of Kew*. London 1967; Leonard Huxley, *Life and Letters of Sir J.D. Hooker*. London 1918; Ida Kaplan Langman, *A Selected Guide to the Literature on the Flowering Plants of Mexico*. Philadelphia 1964.

Species/Vernacular Names:

M. aotoides F. Muell.

Australia: northern mirbelia

M. baueri (Benth.) J. Thompson

Australia: Bauer's mirbelia

M. dilatata Dryander

English: holly-leaved mirbelia

M. oxylobioides F. Muell.

English: mountain mirbelia

M. platyloboides F. Muell.

English: tableland mirbelia

M. pungens G. Don

English: prickly mirbelia

M. rubiifolia (Andrews) G. Don

English: mirbelia, heathland mirbelia

M. seorsifolia (F. Muell.) C. Gardner

Australia: yilgara poison

M. speciosa DC.

Australia: purple poison

Miricacalia Kitam. Asteraceae

Origins:
A wonderful *Cacalia*.

Mirmecodia Gaudich. Rubiaceae

Origins:
See *Myrmecodia* Jack.

Mirobalanus Rumph. Euphorbiaceae

Origins:
From the Greek *myron* "a perfume, ointment, scent" and *balanos* "acorn."

Miscanthidium Stapf Gramineae

Origins:
Referring to the genus *Miscanthus*.

Miscanthus Andersson Gramineae

Origins:
Stalked flowers, Greek *mischos* "stalk" and *anthos* "flower," referring to the spikelets; in form *miskos* "shell, husk."

Species/Vernacular Names:
M. floridulus (Labill.) Warb. (*Saccharum floridulum* Labill.; *Miscanthus japonicus* Anderss.; *Miscanthus ryukyuensis* Honda), in Japan: tokiwa-susuki (= evergreen *Miscanthus*)

M. sinensis Anderss.

English: silver grasses, Japanese plume grass, miscanthus

Japan: susuki

China: mang jing

Mischarytera (Radlk.) H. Turner Sapindaceae

Origins:
From the Greek *mischos* "stalk" plus *Arytera* Blume.

Mischobulbon Schltr. Orchidaceae

Origins:
From the Greek *mischos* "stalk" and *bolbos* "a bulb," referring to the stipitate pseudobulbs.

Mischobulbum Schltr. Orchidaceae

Origins:
From the Greek *mischos* "stalk" and *bolbos* "a bulb," referring to the pseudobulbs.

Mischocarpus Blume Sapindaceae

Origins:
Greek *mischos* "a stalk, pedicel, leaf-stalk" and *karpos* "fruit," referring to the long stalked fruit; see Karl Ludwig von Blume (1796-1862), *Bijdragen tot de flora van Nederlandsch Indië*. 238. Batavia 1825.

Species/Vernacular Names:
M. sp.

Malaya: sugi

M. australis S. Reyn.

Australia: shiny brush apple

M. pyriformis (F. Muell.) Radlk.

Australia: pear-fruited tamarind

Mischocodon Radlk. Sapindaceae

Origins:
From the Greek *mischos* "a stalk, pedicel, leaf-stalk" and *kodon* "a bell."

Mischodon Thwaites Euphorbiaceae

Origins:
From the Greek *mischos* "a stalk, pedicel" and *odous, odontos* "tooth," referring to the calyx lobes.

Mischogyne Exell Annonaceae

Origins:
From the Greek *mischos* "a stalk, pedicel" and *gyne* "female."

Mischophloeus Scheffer Palmae

Origins:
From the Greek *mischos* "a stalk, pedicel" and *phloios* "bark of trees," referring to the pedicel-like base of the male flowers.

Mischopleura Wernham ex Ridl. Ericaceae

Origins:
From the Greek *mischos* "a stalk, pedicel" and *pleura, pleuron* "side, rib, lateral."

Misodendron G. Don f. Misodendraceae

Origins:
Greek *misos* "hate, hatred" and *dendron* "tree," see also *Misodendrum* Banks ex DC.

Misodendrum Banks ex DC. Misodendraceae

Origins:
Greek *misos* "hate, hatred" and *dendron* "tree," hemiparasitic shrublets.

Misopates Raf. Scrophulariaceae

Origins:
Misopathes, a classical Greek plant name used by Dioscorides; see Constantine Samuel Rafinesque, *Autikon botanikon. Icones plantarum select. nov. vel rariorum, etc.* 158. Philadelphia 1840.

Species/Vernacular Names:
M. orontium (L.) Raf. (*Antirrhinum orontium* L.)

English: lesser snapdragon, corn snapdragon, small snapdragon, weasel's snout

Mitchella L. Rubiaceae

Origins:
Named for John Mitchell, 1711-1768, physician, botanist, M.D. Leyden 1719, 1748 Fellow of the Royal Society, with the Duke of Argyll in North Scotland. His writings include *The Contest in America between Great Britain and France ... By an impartial hand.* [i.e., J. Mitchell] 1757 and *The Present State of Great Britain and North America.* [By J. Mitchell] 1767. See John H. Barnhart, *Biographical Notes upon Botanists.* 2: 496. 1965; H.N. Clokie, *Account of the Herbaria of the Department of Botany in the University of Oxford.* 213. Oxford 1964; A. Lasègue, *Musée botanique de Benjamin Delessert.* Paris 1845; William Darlington (1782-1863), *Reliquiae Baldwinianae.* Philadelphia 1843; W. Darlington, *Memorials of John Bartram and Humphry Marshall.* Philadelphia 1849; Jonas C. Dryander, *Catalogus bibliothecae historico-naturalis Josephi Banks.* London 1796-1800; Howard Atwood Kelly (1858-1943) and Walter Lincoln Burrage (1860-1935), *Dictionary of American Medical Biography.* Lives of eminent physicians of the United States and Canada, from the earliest times. New York 1928; E. Earnest, *John and William Bartram, Botanists and Explorers 1699-1777, 1739-1823.* Philadelphia 1940; J. Ewan, ed., *A Short History of Botany in the United States.* New York and London 1969; G. Murray, *History of the Collections Contained in the Natural History Departments of the British Museum.* 1: 168. London 1904; R. Pulteney, *Historical and Biographical Sketches of the Progress of Botany in England.* 2: 278-281. London 1790; Pehr Kalm (1716-1779), "Kalm's account of his visit to England on his way to America in 1748." [Extracted from *En Resa til Norra America.*] Translated by Joseph Lucas. London 1892.

Species/Vernacular Names:
M. repens L.

English: partridge berry

Mitella L. Saxifragaceae

Origins:
Referring to the fruits, from the Greek *mitra* "bishop's head-dress, cap," Latin *mitella, ae* the diminutive of *mitra, ae* "head-band, coif, a kind of turban."

Mitellastra Howell Saxifragaceae

Origins:
Resembling *Mitella.*

Mitellopsis Meisn. Saxifragaceae

Origins:
Resembling *Mitella.*

Mitolepis Balf.f. Asclepiadaceae (Periplocaceae)

Origins:

From the Greek *mitos* "thread, string, seed" and *lepis* "scale."

Mitopetalum Blume Orchidaceae

Origins:

From the Greek *mitos* "thread, string" and *petalon* "leaf, petal."

Mitophyllum E. Greene Brassicaceae

Origins:

From the Greek *mitos* "thread, string" and *phyllon* "leaf."

Mitostemma Mast. Passifloraceae

Origins:

Greek *mitos* "thread, string" and *stemma* "a garland, crown."

Mitostigma Blume Orchidaceae

Origins:

Greek *mitos* "thread, string" and *stigma* "stigma," the stigmatic processes are filiform.

Mitostigma Decne. Asclepiadaceae

Origins:

From the Greek *mitos* "thread, string" and *stigma* "stigma."

Mitozus Miers Apocynaceae

Origins:

From the Greek *mitos* "thread, string" and *ozo* "to smell."

Mitracarpus Zucc. ex Schultes & Schultes f. Rubiaceae

Origins:

From the Greek *mitra* "a turban, bishop's head-dress, cap, mitra, headband, girdle" and *karpos* "fruit," referring to the fruits.

Mitragyna Korth. Rubiaceae

Origins:

From the Greek *mitra* "a turban, bishop's head-dress" and *gyne* "a woman, female organ," referring to the stigma and to the cap-shaped ovary.

Species/Vernacular Names:

M. inermis (Willd.) Kuntze

Nigeria: giyeya (Hausa); koli, kwoli (Fula); okobo (Yoruba); akpatenyi (Igbo)

Mali: jun, ko baro

M. ledermannii (K. Krause) Ridsdale (*Adina ledermannii* K. Krause; *Mitragyna ciliata* Aubrév. & Pellegrin; *Hallea ciliata* (Aubrév. & Pellegrin) Leroy)

Central Africa: oro

Ghana: bahia, suhaba

Ivory Coast: bahia, suzo

Nigeria: abura (Yoruba); eben (Edo); uburu (Igbo); uwen (Efik)

Cameroon: ohombe, afop, afop zam, moukonia malamba, elolom, lwambo, langango, okobi

Gabon: elelome-nzame, ehoupou, epoukou, elelome, n'towo, nzamé, tobou

M. macrophylla Hiern

Zambia: munga, musangu

M. speciosa (Korth.) Haviland

Malaya: bia, biak, ketum, kutum

M. stipulosa (DC.) Kuntz (*Hallea stipulosa* (DC.) Leroy)

Central Africa: oro

Gabon: elelome-nzame, ehoupou, epoukou, elelome, n'towo, nzamé, tobou

Ivory Coast: bahia, suzo

Yoruba: abura, ewe obi, igbagbo

Nigeria: abura, abura eben, ba, bahia, ntowo, elilom, elobom, eti-ayip, eba, ebar, eben, idaba-korikor, igbei, igbeyi, ibi-ewoi, ketchi-lebet, subaha, ewe-obi, oboku, ovbebe, ubulu-milli, uburu, uburu mini, uwen, mukonia-malambo, ohombi; ganyen gori (Hausa); abura (Yoruba); eben (Edo); uburu (Igbo); uwen (Efik)

Cameroon: ohombe, afop, afop zam, moukonia malamba, elolom, lwambo, langango, okobi

Congo: vuku, mulongua, mulungua

Ghana: bahia, suhaba

Mitranthes O. Berg Myrtaceae

Origins:
From the Greek *mitra* "a turban, cap" and *anthos* "flower."

Mitrantia Peter G. Wilson & B. Hyland Myrtaceae

Origins:
From the Greek *mitra* "a turban, bishop's head-dress" and the suffix *antia* in reference to the genus *Ristantia* Peter G. Wilson & J.T. Waterhouse; see in *Telopea*. 3: 264. 1988.

Species/Vernacular Names:
M. bilocularis Peter G. Wilson & Hyland

Australia: Whyanbeel tristania (Timber Reserve 55, near Whyanbeel, in Queensland)

Mitraria Cav. Gesneriaceae

Origins:
From the Greek *mitra* "a turban, mitre, a cap," referring to the seed-pod.

Mitrasacme Labill. Loganiaceae

Origins:
From the Greek *mitra* "a mitre, bishop's head-dress" and *akme* "the top, summit," referring to the floral tube; see Jacques Julien Houtton de Labillardière (1755-1834), *Novae Hollandiae plantarum specimen*. 1: 35. Parisiis 1804-1806 [1807].

Species/Vernacular Names:
M. alsinoides R. Br.
English: annual mitrewort
M. archeri Hook.f. (after the Tasmanian botanist William H. Archer, 1820-1874)
English: cushion mitrewort
M. distylis F. Muell.
English: tiny mitrewort
M. montana Benth.
English: mountain mitrewort
M. paludosa R. Br.
English: swamp mitrewort
M. paradoxa R. Br.
English: wiry mitrewort
M. pilosa Labill.

English: hairy mitrewort
M. polymorpha R. Br.
English: varied mitrewort, mitre weed
M. pygmaea R. Brown
English: dwarf mitrasacme, dwarf mitrewort
China: shui tian bai
M. serpyllifolia R. Br.
English: thyme mitrewort

Mitrasacmopsis Jovet Rubiaceae

Origins:
Resembling *Mitrasacme*.

Mitrastemma Makino Rafflesiaceae (Mitrastemmataceae)

Origins:
From the Greek *mitra* "a mitre, cap" and *stemma, stemmatos* "a garland, crown" or *stemon* "thread, filament," see also *Mitrastemon* Makino.

Mitrastemon Makino Rafflesiaceae (Mitrastemmataceae)

Origins:
From the Greek *mitra* "a mitre, cap" and *stemon* "a stamen, thread, filament," referring to the nature of the stamens.

Mitrastigma Harv. Rubiaceae

Origins:
From the Greek *mitra* "a mitre, cap" and *stigma* "a stigma."

Mitrastylus Alm & T.C.E. Fries Ericaceae

Origins:
From the Greek *mitra* "a mitre, cap" and *stylos* "a style."

Mitratheca K. Schumann Rubiaceae

Origins:
From the Greek *mitra* "a mitre, cap" and *theke* "a case, capsule, box."

Mitrella Miq. Annonaceae

Origins:
The diminutive of the Greek *mitra* "a mitre, cap," possibly referring to the stigma.

Mitreola Boehm. Rubiaceae

Origins:
The diminutive of *mitra* "a mitre, cap."

Mitreola L. Loganiaceae

Origins:
The diminutive of the Greek *mitra* "a mitre, cap," referring to the fruits; see C. Linnaeus, *Opera varia* in quibus continentur Fundamenta botanica, sponsalia plantarum, et Systema naturae. 214. Lucae [Lucca] 1758.

Species/Vernacular Names:
M. pedicellata Bentham
English: hairy-leaf mitreola
China: da ye du liang cao
M. petiolata (J.F. Gmelin) Torrey & A. Gray
English: shiny-leaf mitreola
China: du liang cao

Mitrephora Hook.f. & Thomson Annonaceae

Origins:
From the Greek *mitra* "a mitre, cap" and *phoreo* "to bear," referring to the carpels.

Mitriostigma Hochst. Rubiaceae

Origins:
From the Greek *mitra* "a mitra, a cap, a turban" and *stigma*, in reference to the cap-like stigma, thick and bifid.

Species/Vernacular Names:
M. axillare Hochst. (*Gardenia citriodora* Hook.)
English: wild coffee
South Africa: kafferkoffie, wildekoffie

Mitrocereus (Backeb.) Backeb. Cactaceae

Origins:
Greek *mitra* "a mitre, cap" plus *Cereus*.

Mitrophyllum Schwantes Aizoaceae

Origins:
From the Greek *mitra* "a mitra, a cap, a turban" and *phyllon* "leaf," the second pair of leaves resembling a bishop's mitre.

Mitropsidium Burret Myrtaceae

Origins:
From the Greek *mitra* "a mitra, a cap, a turban" plus *Psidium*.

Mitrosicyos Maxim. Cucurbitaceae

Origins:
From the Greek *mitra* "a mitra, a cap, a turban" and *sikyos* "wild cucumber, gourd."

Mkilua B. Verdcourt Annonaceae

Origins:
Mkilua or *mkilua mwitu* (*mwitu* = of the woods, wild), the vernacular Swahili names for this shrub in the East Africa coastal area; see Bernard Verdcourt, "A new genus of Annonaceae from the East African coastal forests." *Kew Bulletin*. 24(3): 449-453. 1970.

Mnesithea Kunth Gramineae

Origins:
Greek *mnesis* "memory, remembrance," *mnemon* "mindful, remembering," *mnaomai* "to think on, to remember," referring to an aspect of another genus.

Mnianthus Walp. Podostemaceae

Origins:
From the Greek *mnion* "moss, seaweed" and *anthos* "flower."

Mniarum Forst. & Forst.f. Caryophyllaceae (Illecebraceae)

Origins:
From the Greek *mniaros* "mossy, soft as moss"; see Johann Reinhold Forster and his son Johann Georg Adam, *Characteres generum plantarum*. 1. 1776.

Mniochloa Chase Gramineae

Origins:

From the Greek *mnion* "moss, seaweed" and *chloe, chloa* "grass."

Mniodes (A. Gray) Benth. Asteraceae

Origins:

From the Greek *mniodes, mniaros* "mossy," cushion-formers.

Mniopsis Martius & Zucc. Podostemaceae

Origins:

From the Greek *mnion* "moss, seaweed" and *opsis* "like, resembling."

Mniothamnea (Oliver) Niedenzu Bruniaceae

Origins:

From the Greek *mnion* "moss, seaweed" and *thamnos* "shrub."

Moacroton Croizat Euphorbiaceae

Origins:

Resembling *Croton*.

Mobilabium Rupp Orchidaceae

Origins:

Latin *mobilis, mobile* "loose, movable" and *labium, labii* "lip," referring to the mobile or hinged labellum; see Herman Montague Rucker Rupp (1872-1956), in *The North Queensland Naturalist*. 13(78): 2. 1946.

Modecca Lam. Passifloraceae

Origins:

A native Malayalam name for *Modecca palmata* Lam., see van Rheede in *Hortus Indicus Malabaricus*. 8: t. 20. 1688.

Modiola Moench Malvaceae

Origins:

From the Latin *modiolus, i* (*modius, ii* "the Roman corn-measure") "a small measure, wheel hub, a nave of water-wheel," referring to the shape of the fruit; Greek *modiolos* "nave of a wheel"; see Conrad Moench, *Methodus plantas horti botanici et agri Marburgensis a staminum situ describendi*. 619. Marburgi Cattorum [Marburg] 1794.

Species/Vernacular Names:

M. carolinianum (L.) G. Don f. (*Modiola multifida* Moench; *Malva caroliniana* L.; *Malva hawaiensis* H. Lév.)

English: red flowered mallow, Carolina mallow

Modiolastrum K. Schumann Malvaceae

Origins:

The genus *Modiola* plus *astrum*, a Latin substantival suffix.

Moehringia L. Caryophyllaceae

Origins:

Named for the German botanist Paul Heinrich Gerhard (Paulus Henricus Gerardus) Möhring (Moehring), 1710-1792, physician, ornithologist. He is best known for *Avium Genera*. Auricae [Aurich] 1752, *Geslachten der Vogelen*. Amsteldam 1758, *Historiae Medicinales* junctis fere ubique corollariis, praxin medicam illustrantibus. Amstelodami 1739 and *Catalogus Bibliothecae Moehringianae* historiae praeprimis naturali atque arti medicae inservientis. Jeverae 1794. See John H. Barnhart, *Biographical Notes upon Botanists*. 2: 500. 1965; Jonas C. Dryander, *Catalogus bibliothecae historico-naturalis Josephi Banks*. London 1796-1800; G.C. Wittstein, *Etymologisch-botanisches Handwörterbuch*. 587. 1852.

Moenchia Ehrhart Caryophyllaceae

Origins:

Named after the German (b. Kassel) botanist Conrad Moench, 1744-1805 (d. Marburg), pharmacist, chemist, M.D. Marburg 1781, in 1781 professor of botany at Collegium Medicum Carolinianum at Kessel, founder of the Marburg Botanic Garden, an opponent of Linnaeus, a follower of Friedrich Kasimir Medikus (1736-1808). His works include *Enumeratio plantarum indigenarum Hassiae*. Cassellis [Kassel] 1777, *Methodus plantas horti botanici et agri Marburgensis a staminum situ describendi*. Marburgi Cattorum [Marburg] 1794-1802, *Systematische Lehre von denen gebräuchlichsten einfachen und zusammengesezten Arzney-Mitteln*. Marburg 1795, *Einleitung zur Pflanzen-Kunde*. Marburg 1798 and *Verzeichniss ausländischer Bäume und Stauden*. Frankfurt & Leipzig 1785; see William T. Stearn, in *Dictionary of Scientific Biography* 9: 434. 1981; John H. Barnhart, *Biographical Notes upon Bot-*

anists. 2: 500. 1965; Mariella Azzarello Di Misa, a cura di, *Il Fondo Antico della Biblioteca dell'Orto Botanico di Palermo.* 194. Regione Siciliana, Palermo 1988; R. Zander, F. Encke, G. Buchheim and S. Seybold, *Handwörterbuch der Pflanzennamen.* 14. Aufl. 752. Stuttgart 1993; Frans A. Stafleu, *Linnaeus and the Linnaeans.* The spreading of their ideas in systematic botany, 1735-1789. Utrecht 1971; Ida Kaplan Langman, *A Selected Guide to the Literature on the Flowering Plants of Mexico.* Philadelphia 1964.

Species/Vernacular Names:
M. erecta (L.) P. Gaertn., Meyer & Scherb.

English: cerastium

Moerenhoutia Blume Orchidaceae

Origins:
For the collector of the type species, in Tahiti.

Moeroris Raf. Euphorbiaceae

Origins:
See C.S. Rafinesque, *Sylva Telluriana.* 91. 1838.

Moghania J. St.-Hil. Fabaceae

Origins:
Orthographic variant of *Maughania* J. St.-Hil., probably in honor of the Scottish botanist Robert Maughan, 1769-1844, in 1809 a Fellow of the Linnean Society, father of the botanist Edward James (1790-1868); some suggest a Latinized form of an Indian name.

Mogiphanes Mart. Amaranthaceae

Origins:
From the Greek *mogis* "hardly, scarcely" and *phaneros* "evident, visible, distinct," *phanos* "light, bright."

Mohavea A. Gray Scrophulariaceae

Origins:
Mojave River, in southern California.

Species/Vernacular Names:
M. confertiflora (Benth.) A.A. Heller

English: ghost-flower

Mohlana Martius Phytolaccaceae

Origins:
To commemorate the German (b. Stuttgart) botanist Hugo von Mohl, 1805-1872 (d. Tübingen), plant anatomist, physician, professor of physiology and history. His works include *Beiträge zur Anatomie und Physiologie der Gewächse.* Bern 1834, *Grundzüge der Anatomie und Physiologie der vegetabilischen Zelle.* Braunschweig 1851, *Mikrographie.* Tübingen 1846 and *Ueber die Cuticula der Gewächse.* [Tübingen 1842]. See Marc Klein, in *Dictionary of Scientific Biography* 9: 441-442. 1981; Stafleu and Cowan, *Taxonomic Literature.* 3: 541-543. Utrecht 1981; John H. Barnhart, *Biographical Notes upon Botanists.* 2: 501. 1965; T.W. Bossert, *Biographical Dictionary of Botanists Represented in the Hunt Institute Portrait Collection.* 270. 1972; A. Lasègue, *Musée botanique de Benjamin Delessert.* 524. Paris 1845; Ethelyn Maria Tucker, *Catalogue of the Library of the Arnold Arboretum of Harvard University.* Cambridge, Massachusetts 1917-1933.

Mohria Swartz Schizaeaceae

Origins:
Named for the German botanist Daniel Matthias (Matthaeus) Heinrich Mohr, 1780-1808, professor of philosophy at Kiel, plant collector, author of *Observationes botanicae.* Kiliae [Kiel] 1803, with the German botanist Friedrich Weber (1781-1823) wrote *Handbuch der Enleitung in das Studium der kryptogamischen Gewächse.* 1807 and *Naturhistorische Reise durch einen Theil Schwedens.* 1804. See John H. Barnhart, *Biographical Notes upon Botanists.* 2: 502. Boston 1965; R. Zander, F. Encke, G. Buchheim and S. Seybold, *Handwörterbuch der Pflanzennamen.* 14. Aufl. 753. Stuttgart 1993; A. Lasègue, *Musée botanique de Benjamin Delessert.* Paris 1845; Ethelyn Maria Tucker, *Catalogue of the Library of the Arnold Arboretum of Harvard University.* Cambridge, Massachusetts 1917-1933; Stafleu and Cowan, *Taxonomic Literature.* 3: 544-545. 1981 and 7: 126-128. 1988.

Moldavica Fabr. Labiatae

Origins:
Moldava or Moldova, a historical province of eastern Rumania, southeastern Europe.

Moldenhauera Sprengel Icacinaceae

Origins:

After the German botanist Johann Jacob Paul Moldenhawer, 1766-1827, professor of botany at Kiel; see John H. Barnhart, *Biographical Notes upon Botanists*. 2: 502. 1965.

Moldenhawera Schrader Caesalpiniaceae

Origins:

After the German (b. Hamburg) botanist Johann Jacob Paul Moldenhawer, 1766-1827 (d. Kiel), one of the founders of plant anatomy, professor of botany at Kiel, wrote *Tentamen in historiam plantarum Theophrasti*. Hamburgi 1791 and *Beyträge zur Anatomie der Pflanzen*. Kiel 1812. See John H. Barnhart, *Biographical Notes upon Botanists*. 2: 502. Boston 1965; Jonas C. Dryander, *Catalogus bibliothecae historico-naturalis Josephi Banks*. 3: 52. London 1796-1800; E.M. Tucker, *Catalogue of the Library of the Arnold Arboretum of Harvard University*. 1917-1933; Jörn Henning Wolf, in *Dictionary of Scientific Biography* 9: 455-456. 1981.

Moldenkea Traub Amaryllidaceae

Origins:

For the American botanist Harold Norman Moldenke, traveler, studied algae under M.A. Howe (in 1929), Curator and Administrator of the Herbarium of the New York Botanical Garden, specialist in Avicenniaceae, Eriocaulaceae and Verbenaceae. His writings include "Contributions to the flora of extra-tropical South America. I." *Lilloa*. 5: 353-440. 1940, "Notes on new and noteworthy plants. I." *Phytologia*. 2(7): 213-242. 1947, "Notes on new and noteworthy plants. XIII." *Phytologia*. 4(1): 41-65. 1952, "Novelties from Brazil, Venezuela, and Thailand." *Phytologia*. 23(1): 180-181. 1972 and "Verbenaceae." in "Flora of Panama." *Ann. Missouri Bot. Gard.* 60(1): 41-148. 1973, with L.B. Smith wrote "Eriocauláceas." in P.R. Reitz, *Flora Ilustrada Catarinense*. Conselho Nacional de Pesquisas. Instituto Brasileiro de Desenvolvimento Florestal. Herbário "Barbosa Rodrigues". Itajaí, Brazil, with H.P. Traub wrote *Amaryllidaceae: Tribe Amarylleae*. Stanford, California 1949, with his wife Alma Lance Moldenke (*née* Ericson, b. 1908) wrote *Plants of the Bible*. New York 1986, he was son of Charles Edward Moldenke (1860-1935). See John H. Barnhart, *Biographical Notes upon Botanists*. 2: 502. Boston 1965; T.W. Bossert, *Biographical Dictionary of Botanists Represented in the Hunt Institute Portrait Collection*. 271. 1972; S. Lenley et al., *Catalog of the Manuscript and Archival Collections and Index to the Correspondence of John Torrey*. Library of the New York Botanical Garden. 296. 1973;

Elmer Drew Merrill, *Contr. U.S. Natl. Herb.* 30(1): 215. 1947; August Adriaan Pulle, ed., *Flora of Suriname* ("Avicenniaceae." by H.N. Moldenke, vol. 4, pt. 2: 322-325. 1940). Amsterdam 1932, etc.; Irving William Knobloch, compil., "A preliminary verified list of plant collectors in Mexico." *Phytologia Memoirs*. VI. Plainfield, N.J. 1983; Ida Kaplan Langman, *A Selected Guide to the Literature on the Flowering Plants of Mexico*. Philadelphia 1964; R. Zander, F. Encke, G. Buchheim and S. Seybold, *Handwörterbuch der Pflanzennamen*. 14. Aufl. Stuttgart 1993.

Moldenkeanthus Morat Eriocaulaceae

Origins:

For the American botanist Harold Norman Moldenke, b. 1909.

Molina Gay Euphorbiaceae

Origins:

Dedicated to the Chilean (b. Guaraculen, Talca) botanist Giovanni Ignazio (Juan Ignacio, Juan Ignatius) Molina, 1737-1829 (d. Bologna, Italy), Jesuit priest, missionary, plant collector, 1768 to Italy, professor of natural sciences. His writings include *Compendio della storia del regno del Chile*. Bologna 1776, *Saggio sulla storia naturale del Chili*. Bologna 1782, *Saggio sulla storia civile del Chili*. Bologna 1787 and *Memorie di storia naturale* lette in Bologna. Bologna 1821. See H. Gunckel, "Don Juan Ignacio Molina. Su vida, sus obras y su importancia cientifica." *Revista Univ.* 14(1-2): 195-216, 14(3-4): 320-341. [Santiago] 1929 and *Bibliografía Moliniana*. Santiago 1980; R. Zander, F. Encke, G. Buchheim and S. Seybold, *Handwörterbuch der Pflanzennamen*. 14. Aufl. 753. Stuttgart 1993; Miguel Colmeiro y Penido, *La Botánica y los Botánicos de la Peninsula Hispano-Lusitana*. 1858; John H. Barnhart, *Biographical Notes upon Botanists*. 2: 503. Boston 1965; T.W. Bossert, *Biographical Dictionary of Botanists Represented in the Hunt Institute Portrait Collection*. 271. 1972; Jonas C. Dryander, *Catalogus bibliothecae historico-naturalis Josephi Banks*. London 1796-1800; L. Polgar, *Bibliography of the History of the Society of Jesus*. Rome 1967; Francisco Guerra, in *Dictionary of Scientific Biography* 9: 458. 1981; E.M. Tucker, *Catalogue of the Library of the Arnold Arboretum of Harvard University*. 1917-1933.

Molinadendron P.K. Endress Hamamelidaceae

Origins:

For the Honduran botanist Antonio R. Molina, b. 1926.

Molinaea Commerson ex Jussieu Sapindaceae

Origins:

For the French botanist Jean (de) Desmoulins (Johannes or Joannes Molinaeus, Jean des Moulins), 1530-1620?, physician. See Alexander Trallianus, *Libri duodecim, Joanne Guinterio Andernaco interprete & emendatore*. Nunc demum Joannis Molinaei D.M. doctissimis annotationibus illustrati, multisque in locis suo nitori restituti. Lugduni 1575 [1576]; Pietro Andrea Mattioli, *Commentaires ... sur les six livres de Ped. Dioscoride Anazarbeen de la matiere medecinale ... mis en françois ... par M. Jean des Moulins.* Lyon 1572; Richard J. Durling, comp., *A Catalogue of Sixteenth Century Printed Books in the National Library of Medicine.* 3563-3564, 3017-3018, 153-154. 1967.

Species/Vernacular Names:

M. alternifolia Willd.

La Réunion Island: tan Georges, bois de gaulette blanc

Mauritius: bois de gaulette blanc

Molineria Colla Hypoxidaceae

Origins:

After the Italian botanist Ignazio Bernardo Molineri, 1741-1818. See Luigi (Aloysius) Colla (1766-1848), *Illustrationes et icones rariorum Stirpium* quae in ejus horto Ripulis florebant, Anno 1824 (-28), addita ad *Hortum Ripulensem*. Append. [Turin 1827-1831]; O. Mattirolo, *Cronistoria dell'Orto Botanico della Regia Università di Torino*. in *Studi sulla vegetazione nel Piemonte* pubblicati a ricordo del II Centenario della fondazione dell'Orto Botanico della R. Università di Torino. Torino 1929.

Species/Vernacular Names:

M. capitulata (Lour.) Herbert

English: weevil lily

Molineria Parlatore Gramineae

Origins:

After the Italian botanist Ignazio Bernardo Molineri, 1741-1818; see F. Parlatore, *Flora Italiana*. 1: 236. 1850.

Molineriella Rouy Gramineae

Origins:

After the Italian botanist Ignazio Bernardo Molineri, 1741-1818.

Species/Vernacular Names:

M. minuta (L.) Rouy (*Aira minuta* L.; *Molineria minuta* (L.) Parl.; *Periballia minuta* (L.) Asch. & Graebner)

English: hair grass, small hair grass

Molinia Schrank Gramineae

Origins:

After the Chilean botanist Giovanni Ignazio (Juan Ignacio) Molina, 1737-1829, Jesuit priest, missionary, plant collector, 1768 to Italy, professor of natural sciences, author of *Saggio sulla storia naturale del Chili*. Bologna 1782. See Miguel Colmeiro y Penido, *La Botánica y los Botánicos de la Peninsula Hispano-Lusitana*. 1858; John H. Barnhart, *Biographical Notes upon Botanists*. 2: 503. Boston 1965; T.W. Bossert, *Biographical Dictionary of Botanists Represented in the Hunt Institute Portrait Collection*. 271. 1972; Jonas C. Dryander, *Catalogus bibliothecae historico-naturalis Josephi Banks*. London 1796-1800.

Species/Vernacular Names:

M. caerulea (L.) Moench

English: purple moor grass

Moliniopsis Gand. Gramineae

Origins:

Resembling the genus *Molinia* Schrank.

Moliniopsis Hayata Gramineae

Origins:

Resembling *Molinia* Schrank.

Molkenboeria Vriese Goodeniaceae

Origins:

In honor of the Dutch botanist Julianus (Julius) Hendrik Molkenboer, 1816-1854, bryologist, M.D. Leiden 1840. See John H. Barnhart, *Biographical Notes upon Botanists*. 2: 503. 1965; T.W. Bossert, *Biographical Dictionary of Botanists Represented in the Hunt Institute Portrait Collection*. 271. 1972; E.M. Tucker, *Catalogue of the Library of the Arnold Arboretum of Harvard University*. 1917-1933; Friedrich Anton Wilhelm Miquel, *Plantae junghuhnianae*. Lugduni-Batavorum [Leiden], Parisiis [1851]- 1853-1855[-1857]; François Dozy and J.H. Molkenboer, *Prodromus florae bryologicae surinamensis*. Haarlem 1854; W.H. de Vriese, in *Natuurkundige Verhandelingen van de Holland-*

sche Maatschappij der Wetenschappen te Haarlem. [= Goodenovieae] Ser. 2, 10: 44. Amsterdam 1854.

Mollera O. Hoffmann Asteraceae

Origins:
Dedicated to the Portuguese naturalist Adolpho (Adolphe) Frederico Moller,1842-1920, botanist; see Yakov Vladimirovich Bedryaga, *Amphibiens et reptiles*, recueillis en Portugal par A.F. Moller. Coimbra 1889.

Mollia J.F. Gmelin Myrtaceae

Origins:
Named for L.B. von Moll.

Mollia Martius Tiliaceae

Origins:
For the German (Bavarian) politician L.B. von Moll; see Stafleu and Cowan, *Taxonomic Literature.* 3: 554. Utrecht 1981.

Mollinedia Ruíz & Pav. Monimiaceae

Origins:
After the Spanish chemist Francisco de Mollinedo; see H. Ruíz López and J.A. Pavón, *Flora peruvianae, et chilensis prodromus.* 83. Madrid 1794.

Molloya Meissner Proteaceae

Origins:
To commemorate the English-born (near Carlisle, Cumberland) Australian botanist Georgiana Molloy (*née* Kennedy), 1805-1843 (Busselton, Australia), 1829 married Captain John Molloy (1780-1867), 1830 to Western Australia, an early settler, gardener, plant collector in W. Australia, collected plants and seeds for Captain James Mangles; see C.F. Meisner (Meissner) (1800-1874), in *Hooker's Journal of Botany & Kew Garden Miscellany.* 7: 382. London 1855; D.J. Carr and S.G.M. Carr, eds., *People and Plants in Australia.* 334-338, 357-373. 1981; Douglas Pike, ed., *Australian Dictionary of Biography.* 2: 244-245. Melbourne 1967; George Bentham, *Flora Australiensis.* London 1863-1878; Alexandra Hasluck, *Portrait with Background: A Life of Georgiana Molloy.* Melbourne 1955.

Mollugo L. Molluginaceae

Origins:
From the Latin *mollugo, inis* (possibly from *mollis, e* "soft") used by Plinius for a variety of the plant *lappago*, referring to the tender leaves; see Carl Linnaeus, *Species Plantarum.* 89. 1753 and *Genera Plantarum.* Ed. 5. 39. 1754.

Species/Vernacular Names:
M. cerviana (L.) Ser. (*Pharnaceum cervianum* L.)
English: wire-stem chickweed, thread-stem carpetweed
M. pentaphylla L. (*Mollugo stricta* L.)
English: carpetweed
Japan: zakuro-sô
China: di ma huang
Malaya: rumput belangkas, tapak burong
M. verticillata (L.)
English: carpetweed

Molongum Pichon Apocynaceae

Origins:
From *molongó*, a Brazilian vernacular plant name for *Zschokkea arborescens* Müll. Arg., *Ambelania grandiflora* Hub., *Ambelania laxa* Müll. Arg. and *Zschokkea arborescens* Müll. Arg.; *Malouetia tamaquarina* A. DC. (Apocynaceae) is *molongó-de-colher.*

Molopanthera Turcz. Rubiaceae

Origins:
From the Greek *molops* "a stripe, weal, blood-clot" and *anthera* "anther."

Molopospermum W.D.J. Koch Umbelliferae

Origins:
Greek *molops* "a stripe, mark of a stripe" and *sperma* "seed," referring to the fruits.

Moltkia Lehm. Boraginaceae

Origins:
For the Danish noble Joachim Godske Moltke, 1746-1818; see Börge (Birgerus) Thorlacius [Rector of the University of Copenhagen], *Velgjøreren Grev Joachim Godske Moltkes Minde* etc. Copenhagen 1819; H. Genaust, *Etymologisches*

Wörterbuch der botanischen Pflanzennamen. 390. 1996; F. Boerner & G. Kunkel, *Taschenwörterbuch der botanischen Pflanzennamen.* 4. Aufl. 136. Berlin & Hamburg 1989; Georg Christian Wittstein, *Etymologisch-botanisches Handwörterbuch.* 589. Ansbach 1852.

Moltkiopsis I.M. Johnston Boraginaceae

Origins:

Resembling *Moltkia* Lehm.; see A.M. Rizk *et alii,* "Constituents of plants growing in Qatar." *Fitoterapia.* 57(1): 3-9. Milano 1986.

Moluccella L. Labiatae

Origins:

Presumably from an Arabic word meaning "king," or a diminutive of Molucca; see Carl Linnaeus, *Species Plantarum.* 587. 1753 and *Genera Plantarum.* Ed. 5. 255. 1754; H. Genaust, *Etymologisches Wörterbuch der botanischen Pflanzennamen.* 390. 1996; F. Boerner & G. Kunkel, *Taschenwörterbuch der botanischen Pflanzennamen.* 4. Aufl. 136. 1989; Georg Christian Wittstein, *Etymologisch-botanisches Handwörterbuch.* 589. 1852.

Species/Vernacular Names:
M. laevis L.

English: Molucca balm, shell flower, bells of Ireland

Momordica L. Cucurbitaceae

Origins:

Latin *mordeo, momordi, morsum, mordere* "to bite," refers to the jagged and chewed appearance of the seeds, the seeds appear if they have been bitten; see Carl Linnaeus, *Species Plantarum.* 1009. 1753 and *Genera Plantarum.* Ed. 5. 440. 1754; H. Genaust, *Etymologisches Wörterbuch der botanischen Pflanzennamen.* 391-392. 1996; F. Boerner & G. Kunkel, *Taschenwörterbuch der botanischen Pflanzennamen.* 4. Aufl. 136. 1989; Georg Christian Wittstein, *Etymologisch-botanisches Handwörterbuch.* 589. 1852.

Species/Vernacular Names:

M. balsamina L. (*Momordica involucrata* E. Meyer ex Sond.; *Momordica schinzii* Cogn.)

English: African cucumber, balsam apple, balsamina, balsam pear

Peru: balsamina

Southern Africa: masegasegane (Pedi); mmápuúpuú (Tswana); mohodu (Sotho); nkaka (Tonga); tsuúdáde (Bushmen)

S. Rhodesia: nGaka

Mozambique: bale, imbala, caca, kaka, kata, gaka, incaca, cacana, cácâna, incacana, mubabe, tia cana, zamba neluma

Yoruba: ejinrin

M. boivinii Baillon (first described in 1886 from a plant collected by the French botanist Louis Hyacinthe Boivin (1808-1852) in Mombasa, Kenya, see Mary Wilkins Ellert, in *Cactus and Succulent Journal.* vol. 68(2): 88-91. 1996; J. Lanjouw and F.A. Stafleu, *Index Herbariorum.* Part II, *Collectors A-D.* Regnum Vegetabile vol. 2. 1954; F.N. Hepper and Fiona Neate, *Plant Collectors in West Africa.* 11. Utrecht 1971; J.H. Barnhart, *Biographical Notes upon Botanists.* 1: 212. 1965; A. Lasègue, *Musée botanique de Benjamin Delessert.* Paris 1845; Mary Gunn and Leslie E. Codd, *Botanical Exploration of Southern Africa.* Cape Town 1981; I.C. Hedge and J.M. Lamond, *Index of Collectors in the Edinburgh Herbarium.* Edinburgh 1970; Frans A. Stafleu and Erik A. Mennega, *Taxonomic Literature.* Supplement II. 286-287. Königstein 1993)

Zimbabwe: umatukululu

M. charantia L. (*Momordica charantia* var. *abbreviata* Ser.; *Sicyos fauriei* H. Lév.)

English: balsam pear, bitter gourd, African cucumber, wild cucumber, bitter cucumber, bitter melon, bitter apple, carilla fruit, carilla plant, carilla seed, leprosy gourd, balsam apple

Peru: fu-kua, papayilla

Central America: cundeamor, balsamina, pepinillo, sorosi

Brazil: melão-de-são-caetano, melão-de-são-vicente, fruto-de-cobra (see Theodoro Peckolt, "Plantas medicinais e úteis do Brasil." *R. Flora Medicinal.* 3(4): 203-214. Rio de Janeiro 1937; Artur Lourenço Vienna, "Melão-de-são-caetano." *Tribuna Farmacêutica.* 7(5): 108-109. Curitiba 1939)

The Philippines: ampalaya, apalia, palia, paria, pulia, saligun, apape, apapet, amargoso, margoso

Malaya: peria laut, daun periok

China: ku gua, k'u kua, chin li chih, lai pu tao

Japan: naga-reishi, gôyâ

Tibetan: gser-gyi metog

Mozambique: nhadzumba

Yoruba: ejinrin wewe

Congo: lumbuzi, lubuzi-buzi, lumbuzi-busi

M. cochinchinensis (Lour.) Spreng. (*Muricia cochinchinensis* Lour.)

English: spiny bitter cucumber, Cochinchina gourd, Cochinchina balsam pear, Cochinchina balsam apple

Japan: nanban-karasu-uri

China: mu pieh tzu, mu bei zi, mu hsieh

Vietnam: moc miet

M. operculata L.

Brazil: cabacinha, buchinha, abobrinha-do-norte, bucha-dos-paulistas, purga-de-joão-paes (see Eurico Teixeira da Fonseca, "Plantas medicinales brasileñas." *R. Flora Medicinal.* Rio de Janeiro. 6(1): 37-49. 1939)

Monachanthus Lindley Orchidaceae

Origins:

From the Greek *monachos* "solitary, monk" and *anthos* "flower," referring to the labellum.

Monachather Steudel Gramineae

Origins:

From the Greek *monachos* "solitary, monk," *monos* "solitary" and *ather* "an awn."

Species/Vernacular Names:

M. paradoxa Steudel (*Danthonia bipartita* F. Muell.)

Australia: bandicoot grass

Monachne P. Beauv. Gramineae

Origins:

Greek *monos* "solitary, one" and *achne* "husk, glume"; see Ambroise Marie François Joseph Palisot de Beauvois (1752-1820), *Essai d'une nouvelle Agrostographie*, ou nouveaux genres des Graminées. 49, 168, t. 10, figs. 9, 10. Paris 1812.

Monachochlamys Baker Acanthaceae

Origins:

From the Greek *monachos* "solitary" and *chlamys* "cloak."

Monachosorella Hayata Monachosoraceae (Dennstaedtiaceae, Adiantaceae)

Origins:

The diminutive of *Monachosorum*.

Monachosorum Kunze Monachosoraceae (Dennstaedtiaceae, Adiantaceae)

Origins:

Greek *monachos* "solitary" and *soros* "a vessel, a spore case, a heap."

Monachyron Parl. Gramineae

Origins:

From the Greek *monos* "solitary, one" and *achyron* "chaff, husk."

Monactinocephalus Klatt Asteraceae

Origins:

From the Greek *monos* "solitary, one," *aktis* "a ray" and *kephale* "head."

Monactis Kunth Gramineae

Origins:

From the Greek *monos* "solitary, one" and *aktis* "a ray," referring to the flowers.

Monadelphanthus Karst. Rubiaceae

Origins:

Greek *monos* "one," *adelphos* "brother" and *anthos* "flower," *monadelphia* "possession of only one brother," the filaments of the stamens joined into one.

Monadenia Lindley Orchidaceae

Origins:

Greek *monos* "single, one, only" and *aden* "a gland," in form *monaden* "solitary-wise, only," referring to the single viscidium; see John Lindley, *The Genera and Species of Orchidaceous Plants.* 356. London 1830-1840.

Monadenium Pax Euphorbiaceae

Origins:

From the Greek *monos* and *aden* "a gland," having only one gland.

Monandraira Desv. Gramineae

Origins:
From the Greek *monos* "single, one," *aner, andros* "stamen" plus the genus *Aira* L.

Monandriella Engl. Podostemaceae

Origins:
From the Greek *monos* "single" and *aner, andros* "stamen."

Monanthella A. Berger Crassulaceae

Origins:
The genus *Monanthes* Haw. plus the diminutive.

Monanthes Haw. Crassulaceae

Origins:
From the Greek *monos* "only, one" and *anthos* "flower," but the flowers are not solitary.

Monanthochloe Engelm. Gramineae

Origins:
Greek *monos* "only, one," *anthos* "flower" and *chloe, chloa* "grass."

Species/Vernacular Names:
M. littoralis Engelm.
English: shore grass

Monanthocitrus Tanaka Rutaceae

Origins:
From the Greek *monos* "only, one," *anthos* "flower" plus *Citrus.*

Monanthos (Schltr.) Brieger Orchidaceae

Origins:
Greek *monos* and *anthos* "flower"; see Friedrich Richard Rudolf Schlechter (1872-1925), *Die Orchideen*; ihre Beschreibung, Kultur und Züchtung. 3: 659. 1981.

Monanthotaxis Baillon Annonaceae

Origins:
From the Greek *monos* "only, one," *anthos* "flower" and *taxis* "order," sometimes the flowers are solitary.

Species/Vernacular Names:
M. caffra (Sond.) Verdc. (*Guatteria caffra* Sond.; *Popowia caffra* (Sond.) Benth.)

English: dwaba-berry

Southern Africa: dwababessie; iDwabe, iDwaba (Xhosa); uMaluswembe, umaVumba, mKonjane, uMazwenda-omhlophe, umGogi-wezihlanya, iThunganhlanzi (Zulu)

Monarda L. Labiatae

Origins:
For the influential Spanish (b. Seville) physician Nicolás Bautista Monardes, *circa* 1493 (1512?)-1578 (1588?) (d. Seville), botanist. Among his writings are *Brief traité de la racine mechoacan*, venue de l'Espagne nouvelle. 1588 and *Simplicium medicamentorum ex novo orbe delatorum*, quorum in medicina usus est, historia, Hispanico sermone descripta. Antverpiae 1579, his first printer was Domingo de Robertis. See Francisco Guerra, in *Dictionary of Scientific Biography* 9: 466. 1981; Carl Linnaeus, *Species Plantarum*. 22. 1753 and *Genera Plantarum*. Ed. 5. 14. 1754; Richard J. Durling, *A Catalogue of Sixteenth Century Printed Books in the National Library of Medicine*. 3213, 3217, 3420. 1967; Garrison and Morton, *Medical Bibliography*. 1817. 1961; Johann David Schoepf (1752-1800), "Materia medica americana potissimum regni vegetabilis." *B. of the Lloyd Library of Botany, Pharmacy and Materia Medica*. [Reproduction Series no. 3] Cincinnati 1903; Garcia d'Orta, *Due libri dell'historia de i semplici aromati, et altre cose ... con brevi annotationi di Carlo Clusio ... Con un trattato della neve & del bever fresco* di Nicolo Monardes ... tradotti da Annibale Briganti. [Second edition of the Italian translation, first 1576.] Venice 1582; Charles Ralph Boxer, *Two Pioneers of Tropical Medicine: Garcia d'Orta and Nicolás Monardes*. London [1963]; F. Boerner & G. Kunkel, *Taschenwörterbuch der botanischen Pflanzennamen*. 4. Aufl. 136. Berlin & Hamburg 1989; G.C. Wittstein, *Etymologisch-botanisches Handwörterbuch*. 590. 1852; Francisco Guerra, *Nicolás Bautista Monardes. Su vida y su obra [ca. 1493-1588]*. Mexico 1961.

Species/Vernacular Names:
M. didyma L.

English: Oswego tea, bee balm, bergamot, fragrant balm

China: mei guo bo he

M. fistulosa L.

English: wild bergamot

China: ni mei guo bo he

Monardella Benth. Labiatae

Origins:
The diminutive of *Monarda* L.

Species/Vernacular Names:
M. beneolens J.R. Shevock, B. Ertter & Jokerst
English: sweet smelling monardella
M. cinerea Abrams
English: gray monardella
M. crispa Elmer
English: crisp monardella
M. frutescens (Hoover) Jokerst
English: San Luis Obispo monardella
M. lanceolata A. Gray
English: mustang mint
M. leucocephala A. Gray
English: merced monardella
M. palmeri A. Gray
English: Palmer's monardella
M. pringlei A. Gray
English: Pringle's monardella
M. purpurea Howell
English: Siskiyou monardella
M. robinsonii Epling
English: Robinson's monardella
M. stebbinsii Hardham & J. Bartel
English: Stebbin's monardella
M. villosa Benth.
English: coyote mint

Monarrhenus Cass. Asteraceae

Origins:
From the Greek *monos* "only, one" and *arrhen* "male," an allusion to the flowers.

Monarthrocarpus Merr. Fabaceae

Origins:
From the Greek *monos* "only, one," *arthron* "a joint" and *karpos* "fruit."

Monathera Raf. Gramineae

Origins:
From the Greek *monos* "only, one" and *ather* "stalk, barb, awn"; see C.S. Rafinesque, *Am. Monthly Mag. Crit. Rev.* 4: 190. 1819 and *Jour. Phys. Chim. Hist. Nat.* 89: 262. 1819.

Mondia Skeels Asclepiadaceae (Periplocaceae)

Origins:
From *uMondi*, the Zulu name for the plant.

Species/Vernacular Names:
M. whitei (Hook.f.) Skeels
Yoruba: ogba
Congo: mudiondo, mundiondo

Monechma Hochst. Acanthaceae

Origins:
From the Greek *monos* "one, single" and *echma* (*echo* "to hold, to sustain") "stoppage, a defense against, hold-fast."

Species/Vernacular Names:
M. debile (Forssk.) Nees (*Dianthera debilis* Forssk.; *Monechma bracteatum* Hochst.; *Monechma monechmoides* (S. Moore) Hutch.; *Monechma tettense* C.B. Cl.)
South Africa: perdebos

Monelytrum Hackel ex Schinz Gramineae

Origins:
From the Greek *monos* "one, single" and *elytron* "sheath, cover, scale, husk."

Monencyanthes A. Gray Asteraceae

Origins:
Presumably from the Greek *monos* "one," *enkymon, enkyos* "pregnant" and *anthos* "flower"; in *Hooker's Journal of Botany & Kew Garden Miscellany.* 4: 229. London 1852.

Monenteles Labill. Asteraceae

Origins:
Greek *monos* "one, single" and *enteles* "complete, perfect," referring to the single and perfect disk floret surrounded by

the ray florets; see J.J.H. de Labillardière, *Sertum austro-caledonicum*. 42, t. 43, 44. Parisiis 1824-1825.

Monerma P. Beauv. Gramineae

Origins:

From the Greek *monos* "alone, solitary" and *herma, hermatos* "support, prop," *moneres* "solitary," referring to the single and outer glume; see A.M.F.J. Palisot de Beauvois, *Essai d'une nouvelle Agrostographie*, ou nouveaux genres des Graminées. 116, 168, pl. XX, fig. X. Paris 1812.

Moneses Salisb. ex Gray Ericaceae (Pyrolaceae)

Origins:

From the Greek *monos* "one, single" and *esis* "a sending forth, delight," referring to the solitary flowers.

Species/Vernacular Names:

M. uniflora (L.) A. Gray

English: wood-nymph

Monetia L'Hérit. Salvadoraceae

Origins:

For the French biologist Jean Baptiste Antoine Pierre de Monnet (Monet) de Lamarck, 1744-1829 (d. Paris), a great naturalist, botanist, zoologist, paleontologist, conchologist, botanist at the Jardin des Plantes in Paris, naturalist and a forerunner of Darwin's theory of evolution, professor of zoology at the Museum d'Histoire Naturelle in Paris. His writings include *Flore françoise* Paris 1778 and *Philosophie Zoologique*. [First edition, two volumes in one, 8vo.] Paris 1822; see Leslie J. Burlingame, in *Dictionary of Scientific Biography* 7: 584-594. 1981; Frans A. Stafleu, *Linnaeus and the Linnaeans*. The spreading of their ideas in systematic botany, 1735-1789. Utrecht 1971; Stafleu and Cowan, *Taxonomic Literature*. 2: 730-734. 1979; Mariella Azzarello Di Misa, a cura di, *Il Fondo Antico della Biblioteca dell'Orto Botanico di Palermo*. 143-145. Palermo 1988; J.H. Barnhart, *Biographical Notes upon Botanists*. 2: 337. 1965; B. Glass et al., eds., *Forerunners of Darwin: 1745-1859*. Baltimore 1959; Emil Bretschneider, *History of European Botanical Discoveries in China*. 1981; Denis I. Duveen, *Bibliotheca Alchemica et Chemica*. 334. London 1949; W.M. Wheeler and T. Barbour, eds., *The Lamarck Manuscripts at Harvard*. Cambridge 1933; H.G. Cannon, *Lamarck and Modern Genetics*. London 1959; R.W. Burkhardt, *The Spirit of System: Lamarck and Evolutionary*

Biology. Cambridge Massachusetts 1977; Garrison and Morton, *Medical Bibliography*. 216. New York 1961; G.C. Wittstein, *Etymologisch-botanisches Handwörterbuch*. 497. 1852; R. Zander, F. Encke, G. Buchheim and S. Seybold, *Handwörterbuch der Pflanzennamen*. 14. Aufl. 739. Stuttgart 1993.

Moniera Loefling Rutaceae

Origins:

See the genus *Monniera* Loefl.

Moniera P. Browne Scrophulariaceae

Origins:

For the French botanist Louis Guillaume Le Monnier (Lemonnier), 1717-1799, physician, plant collector, 1758-1786 professor of botany at the Jardin du Roi, wrote *Observations d'histoire naturelle*, faites dans les provinces méridionales de la France. Paris 1744. See J.H. Barnhart, *Biographical Notes upon Botanists*. 2: 368. 1965; Georges Léopold Chrétien Frédéric Dagobert Cuvier, 1769-1832, *Notice historique sur L.G. Le Monnier*, lue à la séance publique de l'Institut national de France du 15 vendémiaire an 9. Paris, an ix [1801]; Patrick Browne, *The Civil and Natural History of Jamaica in Three Parts*. 269. London 1756; Jonas C. Dryander, *Catalogus bibliothecae historico-naturalis Josephi Banks*. London 1796-1800; A. Lasègue, *Musée botanique de Benjamin Delessert*. Paris 1845; Frans A. Stafleu, *Linnaeus and the Linnaeans*. The spreading of their ideas in systematic botany, 1735-1789. Utrecht 1971.

Monilaria (Schwantes) Schwantes Aizoaceae

Origins:

From the Latin *monile, monilis* "a necklace, collar."

Monilia Gray Gramineae

Origins:

An anagram of the generic name *Molinia* Schrank, or from the Latin *monile, monilis* "a necklace, collar," pl. *monilia*.

Monimia Thouars Monimiaceae

Origins:

From the Greek *monos, monimos* "one, single," referring to the fruits, each with a single ovule.

Monimiastrum Guého & A.J. Scott Myrtaceae

Origins:
From the genus *Monimia* and the Latin suffix *astrum*.

Monimopetalum Rehder Celastraceae

Origins:
From the Greek *monos, monimos* "one, single" and *petalon* "petal."

Monium Stapf Gramineae

Origins:
From the Greek *monios* "solitary, savage," *monia* "solitude, celibacy."

Monnieria Loefling Rutaceae

Origins:
For the French botanist Louis Guillaume Le Monnier (or Lemonnier), 1717-1799, physician, pupil of Bernard de Jussieu, plant collector, from 1758 to 1786 professor of botany at the Jardin du Roi, author of *Observations d'histoire naturelle*, faites dans les provinces méridionales de la France. Paris 1744. See John H. Barnhart, *Biographical Notes upon Botanists*. 2: 368. 1965; Georges Léopold Chrétien Frédéric Dagobert Cuvier, 1769-1832, *Notice historique sur L.G. Le Monnier*, lue à la séance publique de l'Institut national de France du 15 vendémiaire an 9. Paris, an ix [1801]; Frans A. Stafleu, *Linnaeus and the Linnaeans*. The spreading of their ideas in systematic botany, 1735-1789. Utrecht 1971.

Species/Vernacular Names:
M. trifolia L. (*Syst. Nat.* ed. 10. 2: 1153. Mai-Jun 1759)

Brazil: alfavaca-de-cobra, jaborandi-de-três-folhas, jaborandi, alfavaca-de-caboclo

Monnina Ruíz & Pav. Polygalaceae

Origins:
After the Spanish patron of botany José Moñino y Redondo (Josephus Monninus), Count de Florida-Blanca (Conde de Floridablanca), politician. See Cayetano Alcázar Molina, *El Conde de Floridablanca*. Madrid 1929, *El Conde de Floridablanca — Siglo XVIII*. Madrid [1935] and *Los Hombres del Despotismo Ilustrado en España*. El Conde de Floridablanca. Su vida y su obra. Murcia 1934; Buenaventura Carlos Aribau, ed., Biblioteca de autores españoles. *Obras originales del Conde de Floridablanca*. Madrid 1867; Paul Perès, *Relation historique de l'assassinat* commis en la personne du comte de Florida-Blanca, le 18 juin 1790, par P. Perès, etc. [Paris 1790]; Joaquim Mas-Guindal, "Las especies medicinales de Ruíz & Pavón." *Tribuna Farmacêutica*. 13(4): 65-69. Curitiba 1945.

Species/Vernacular Names:
M. conferta Ruíz & Pavón

Peru: muchuysa, tuta huiña

M. polystachya Ruíz & Pavón

Peru: quitaporquería, yallhoy

M. salicifolia Ruíz & Pavón

Peru: anca usa, condorpa usan, hacchiquis, muchi, muchuisa, muchuy, muchuysa, pahuata-huinac, sambo-ckorota, tuta huiña, urpay jacha

Monocardia Pennell Scrophulariaceae

Origins:
From the Greek *monos* "single" and *kardia* "heart."

Monocarpia Miq. Annonaceae

Origins:
From the Greek *monos* "alone, lonely, one" and *karpos* "fruit."

Monocelastrus F.T. Wang & T. Tang Celastraceae

Origins:
Greek *monos* "alone, lonely, one, single" plus *Celastrus* L.

Monocephalium S. Moore Icacinaceae

Origins:
From the Greek *monos* "alone, lonely, one" and *kephale* "head."

Monocera Elliott Gramineae

Origins:
Greek *monos* "alone, lonely, one" and *keras* "a horn," referring to the spike.

Monocera Jack Elaeocarpaceae

Origins:
Greek *monos* "alone, lonely, one" and *keras* "a horn."

Monochaete Döll Gramineae

Origins:

From the Greek *monos* "alone, lonely, one" and *chaite* "bristle, long hair."

Monochaetum (DC.) Naudin Melastomataceae

Origins:

Greek *monos* and *chaite* "bristle, long hair," referring to the connective of the stamen, androecium dimorphic.

Monochasma Maxim. ex Franchet & Sav. Scrophulariaceae

Origins:

From the Greek *monos* "alone, single" and *chasme* "gaping, yawning."

Monochilus Fischer & C.A. Meyer Labiatae (Verbenaceae)

Origins:

From the Greek *monos* "alone, lonely, one" and *cheilos* "lip," referring to the form of the flower.

Monochilus Wallich ex Lindley Orchidaceae

Origins:

Greek *monos* and *cheilos* "lip," referring to the partial union of the column and lip.

Monochlaena Gaudich. Dryopteridaceae (Aspleniaceae)

Origins:

From the Greek *monos* "alone, lonely" and *chlaena* "cloak, coat."

Monochoria C. Presl Pontederiaceae

Origins:

Greek *monos* "alone, lonely" and *chorion* "membrane" or *chora* "country, place, region" or *choris* "separate, asunder, apart," *chorizo* "to separate," one stamen is larger than the others. See C. Presl, *Reliquiae haenkeanae*. 1: 127. [Collector Thaddäus Peregrinus Xaverius Haenke, 1761-1816 or 1817] Pragae [Praha] 1827; Lorenzo Raimundo Parodi (1895-1966), "Thaddaeus Peregrinus Haenke a dos siglos de su nacimiento." *Anales Acad. Nac. Ci. Exact. Buenos Aires*. 17: 9-28. 1964; Georg Christian Wittstein, *Etymologisch-botanisches Handwörterbuch*. 592. Ansbach 1852; F. Boerner & G. Kunkel, *Taschenwörterbuch der botanischen Pflanzennamen*. 4. Aufl. 136. Berlin & Hamburg 1989; F.A. Sharr, *Western Australian Plant Names and Their Meanings. A Glossary*. 49. University of Western Australia Press, Nedlands, Western Australia 1996; James A. Baines, *Australian Plant Genera. An Etymological Dictionary of Australian Plant Genera*. 243. Chipping Norton, N.S.W. 1981.

Species/Vernacular Names:

M. hastata L.

China: tzu ku, shui ping

M. korsakowii Regel & Maack

China: ping

M. vaginalis (Burm.f.) Kunth (*Pontederia vaginalis* Burm.f.)

English: cordate monochoria, duck's-tongue monochoria

Malaya: rumput ayer

China: ya she cao, ya she tsao, fou shih

Monocladus L.C. Chia, H.L. Fung & Y.L. Yang Gramineae

Origins:

From the Greek *monos* "alone, single" and *klados* "a branch."

Monococcus F. Muell. Phytolaccaceae

Origins:

From the Greek *monos* "alone, lonely, one" and *kokkos* "a berry," the fruits have a single carpel; see F. von Mueller, *Fragmenta Phytographiae Australiae*. 1: 46. Melbourne 1858.

Monocosmia Fenzl Portulacaceae

Origins:

Greek *monos* "alone, lonely, one" and *kosmos* "ornament, decoration."

Monocostus K. Schumann Zingiberaceae (Costaceae)

Origins:

Greek *monos* "alone, lonely, one" with *Costus* L.

Monocyclanthus Keay Annonaceae

Origins:

From the Greek *monos* "alone, lonely," *kyklos* "circle" and *anthos* "flower," alluding to the arrangement of flowers.

Monocymbium Stapf Gramineae

Origins:

From the Greek *monos* "single" and *kymbe* "boat," *kymbos* "cavity," referring to the shape of the spathe of the racemes.

Species/Vernacular Names:

M. ceresiiforme (Nees) Stapf

English: oatgrass, wild oat grass, wild oats

Southern Africa: hawergras, wildehawergras; tshinwamul-wadze (Venda)

Monodiella Maire Gentianaceae

Origins:

For the French botanist M. Théodore Monod, b. 1902, plant collector (Senegal, Mali); see Auguste Jean Baptiste Chevalier, *Flore vivante de l'Afrique Occidentale Française.* Paris 1938; F.N. Hepper and F. Neate, *Plant Collectors in West Africa.* 57. 1971.

Monodora Dunal Annonaceae

Origins:

Greek *monos* "only, lonely, one" with *dora* "a skin, skin when taken off, hide," referring to the solitary flowers and to the single terminal carpel; some suggest from *doron* "gift"; see H. Genaust, *Etymologisches Wörterbuch der botanischen Pflanzennamen.* 393. 1996; Georg Christian Wittstein, *Etymologisch-botanisches Handwörterbuch.* 592. 1852; F. Boerner & G. Kunkel, *Taschenwörterbuch der botanischen Pflanzennamen.* 4. Aufl. 136. 1989; R. Zander, F. Encke, G. Buchheim and S. Seybold, *Handwörterbuch der Pflanzennamen.* 14. Aufl. 384. 1993.

Species/Vernacular Names:

M. sp.

Liberia: kray-bu

M. brevipes Benth.

Nigeria: yellow flowering nutmeg, lakoshin, alakoshin, ilakosin, osa, ukposa; lakosin (Yoruba); ukposa, iyoha (Edo)

M. junodii Engl. & Diels (the specific name after the Rev. Henri A. Junod, 1863-1934, a Swiss missionary at Lourenço Marques, an authority on Bantu life and also a botanical collector (in Northern Transvaal and Mozambique). Among his writings are *The Life of a South African Tribe (the Thonga).* London, Neuchâtel [printed] 1912, *Les Ba-Ronga. Étude ethnographique sur les indigènes de la Baie de Delagoa.* Neuchâtel 1898, *Manuel de conversation et Dictionnaire Ronga-Portugais-Français-Anglais.* Lausanne 1896 and *Grammaire Ronga.* Lausanne 1896; see John Wesley Haley, *Life in Mozambique and South Africa.* Free Methodist Publishing House, Chicago 1926; D.M. Goodfellow, *Principles of Economic Sociology. The Economics of Primitive Life as Illustrated from the Bantu Peoples of South and East Africa.* London 1939; António Augusto Pereira Cabral, *Vocabulario português, shironga, shitsua, guitonga, shishope, shisena, shinhungue, shishuabo, kikua, shi-yao e kissuahili.* Lourenço Marques 1924; Henri Philippe Junod, *NwaMpfundla-NwaSisana.* (The Romance of the Hare). [Written in the new Tsonga Orthography] Pretoria 1940; Henri A. Junod, *Moeurs et Coutumes des Bantous.* Paris 1936; Mary Gunn and Leslie E. Codd, *Botanical Exploration of Southern Africa.* 203. Cape Town 1981)

English: green apple

Southern Africa: groenappel; umKotshi (Zulu); ntsivila (eastern Transvaal); muShikoshiko (Shona)

S. Rhodesia: muShikoshiko

M. myristica (Gaertner) Dunal

English: calabash nutmeg, Jamaica nutmeg

French: muscadier d'Afrique, fausse noix de muscade

Cameroon: dzin, pebe

Central Africa: bende bende, mbende, mombendebende, mombende, bominingo, oniningo, pinguingu, ifuafua, kimbula, musahusa, mukasa, feup, ding, fep, ozek, annéhia, effoin, m'bo, moué

Congo: ntzinku, nzingu, dzingu

Ivory Coast: efuen, efueno, hane, mbong, moué

Gaboon: mpoussa, zing

Ghana: ayikui, ayirewamba, efuaba, kotokorowa, motukrodua, yikwi

Liberia: gboite

Tropical Africa: lakoshe

Tropical W. Africa: m'poussan

Uganda: musamwu

Yoruba: abalakose, ariwo, ilakosin igbo

Nigeria: ebenoyoba, ehuru, din, igbo, abo-lakoshin, ilakoshin-igbo, njimgene, gbosa, ukposa, wofiove, n'pokoson, efwen, efouen; lakosin (Yoruba); ehinawosin (Ikale); uyenghen (Edo); ehuru ofia (Igbo)

Togoland: uyu

M. tenuifolia Bentham

English: African nutmeg

Ivory Coast: pétimoué

Yoruba: lakosin, lakose, ilakosin, ilakose

Nigeria: ehinawosin, lakoshin, ihe-igbe, uyengen; abo lakoshe (Yoruba); ebenoyoba (Edo); ehuru (Igbo)

Monogereion G.M. Barroso & R.M. King Asteraceae

Origins:

From the Greek *monos* "lonely, single" and *gereion* "thistledown."

Monogonia C. Presl Pteridaceae (Adiantaceae)

Origins:

Greek *monos* "lonely, single" and *gonia* "an angle."

Monogramma Comm. ex Schkuhr Vittariaceae (Adiantaceae)

Origins:

From the Greek *monos* "only, lonely, one" and *gramma* "line, writing, letter," alluding to the single slender sorus; see Christian Schkuhr (1741-1811), *Vier und zwanzigste Klasse des Linnéischen Pflanzensystems oder kryptogamische Gewächse*. 82, t. 87. Wittenberg [1804-]1809.

Monographidium C. Presl Rosaceae

Origins:

From the Greek *monos* "only, lonely, one" and the diminutive of *graphis* "brush, a style for writing, a needle."

Monolena Triana Melastomataceae

Origins:

From the Greek *monos* "only, lonely, one" and *olene* "an arm, the elbow, bundle," referring to the base of the anthers.

Monolepis Schrader Chenopodiaceae

Origins:

From the Greek *monos* "single, one" and *lepis* "scale," in reference of the perianth-segment; see H.A. Schrader, in *Index seminum horti academici goettingensis* anno 1830 collecta. 4. [1830].

Species/Vernacular Names:
M. pusilla S. Watson
English: poverty weed

Monolopia DC. Asteraceae

Origins:

From the Greek *monos* "one" and *lopos* "a covering, husk, shell, bark, peel, hide, leather," an allusion to the single series of involucral bracts, referring to the phyllaries.

Monomeria Lindley Orchidaceae

Origins:

From the Greek *monos* "single, one" and *meris* "part," in reference of the single anther or to the abortive petals.

Monomesia Raf. Boraginaceae

Origins:

See Constantine Samuel Rafinesque (1783-1840), *Flora Telluriana*. 4: 87. 1836 [1838]; E.D. Merrill, *Index rafinesquianus*. 203. 1949.

Monoon Miq. Annonaceae

Origins:

From the Greek *monos* "single, one" and *oon* "egg."

Monopanax Regel Araliaceae

Origins:

From the Greek *monos* "single, one" plus *Panax*.

Monopera Barringer Scrophulariaceae

Origins:

Greek *monos* and *pera* "a pouch, bag."

Monopetalanthus Harms Caesalpiniaceae

Origins:

Greek *monos* "one, single," *petalon* "a petal" and *anthos* "a flower."

Species/Vernacular Names:
M. sp.

Ivory Coast: toubaouate

M. heitzii Pellegrin

Gabon: andouong

M. microphyllus Harms

Gabon: ngabe

Cameroon: ngang

Monopholis S.F. Blake Asteraceae

Origins:

From the Greek *monos* "one, single" and *pholis, pholidos* "scale, horny scale."

Monophrynium K. Schumann Marantaceae

Origins:

From the Greek *monos* "one, single" plus *Phrynium* Willd.

Monophyllaea R. Br. Gesneriaceae

Origins:

From the Greek *monos* "one, single" and *phyllon* "a leaf."

Monophyllanthe K. Schumann Marantaceae

Origins:

From the Greek *monos* "one, single," *phyllon* "a leaf," *monophyllos* "one-leaved" and *anthos* "flower."

Monophyllorchis Schltr. Orchidaceae

Origins:

From the Greek *monophyllos* "one-leaved" plus *Orchis*, referring to the habit.

Monoplegma Piper Fabaceae

Origins:

From the Greek *monos* "one, single" and *plegma* "twisted, twined."

Monoploca Bunge Brassicaceae

Origins:

Greek *monos* and *plokos* "folded, lock of hair, wreath, a braid"; see Johann G.C. Lehmann, *Plantae Preissianae ... Plantarum quas in Australasia occidentali et meridionali-occidentali annis 1838-41 collegit L. Preiss.* 1: 259. 1845.

Monopogon C. Presl Gramineae

Origins:

From the Greek *monos* "one, single" and *pogon* "beard."

Monoporandra Thwaites Dipterocarpaceae

Origins:

From the Greek *monos* "one, single," *poros* "opening, pore" and *aner, andros* "stamen, man."

Monoporus A. DC. Myrsinaceae

Origins:

From the Greek *monos* "one, single" and *poros* "opening, pore."

Monopsis Salisb. Campanulaceae

Origins:

Greek *monos* and *opsis* "aspect, resemblance, appearance, face," the corolla is almost regular, not bilabiate; see R.A. Salisbury, in *Transactions of the Horticultural Society of London.* 2: 37. 1817.

Species/Vernacular Names:
M. decipiens (Sond.) Thulin (*Lobelia decipiens* Sond.)

English: butterfly lobelia

M. lutea (L.) Urb. (*Lobelia lutea* L.; *Monopsis lutea* (L.) Urb. var. *ericoides* (Presl) Urb.; *Monopsis lutea* (L.) Urb. var. *euphrasioides* (Eckl. & Zeyh.) Urb.; *Monopsis lutea* (L.) Urb. var. *subcoerulea* Zahlbr.)

English: yellow lobelia

M. scabra (Thunb.) Urb. (*Lobelia scabra* Thunb.; *Dobrowskya serratifolia* (Thunb.) A. DC.)

English: brown lobelia

Monopteryx Spruce ex Benth. Fabaceae

Origins:
From the Greek *monos* "one, single" and *pteryx* "wing, feathery foliage."

Species/Vernacular Names:
M. uacu Spruce
Brazil: uacu

Monoptilon Torrey & A. Gray Asteraceae

Origins:
From the Greek *monos* "one, single" and *ptilon* "a feather, wing," referring to the pappus.

Species/Vernacular Names:
M. bellidiforme A. Gray
English: desert star

Monopyle Moritz ex Benth. Gesneriaceae

Origins:
From the Greek *monos* "one, single" and *pyle* "door, gate," referring to the capsule.

Monopyrena Speg. Verbenaceae

Origins:
From the Greek *monos* "one, single" and *pyren* "a kernel, a fruit stone."

Monorchis Séguier Orchidaceae

Origins:
From the Greek *monos* "one, single" and *orchis* "orchid," single tubers.

Monosalpinx N. Hallé Rubiaceae

Origins:
From the Greek *monos* "one, single" and *salpinx* "a trumpet, tube."

Monoschisma Brenan Mimosaceae

Origins:
From the Greek *monos* "one, single" and *schisma* "division, cleft, schism."

Monosemeion Raf. Fabaceae

Origins:
See Constantine Samuel Rafinesque, *Autikon botanikon. Icones plantarum select. nov. vel rariorum, etc.* 82. Philadelphia 1840; Elmer D. Merrill, *Index rafinesquianus.* 146. 1949.

Monosepalum Schltr. Orchidaceae

Origins:
From the Greek *monos* "one, single" and neolatin *sepalum* "sepal," the sepals are joined nearly to the apex.

Monosoma Griff. Meliaceae

Origins:
From the Greek *monos* "one, single" and *soma* "a body."

Monostachya Merr. Gramineae

Origins:
From the Greek *monos* "one, single" and *stachys* "spike," one-spiked.

Monostemon Henrard Gramineae

Origins:
From the Greek *monos* "one, single" and *stemon* "stamen."

Monostichanthus F. Muell. Annonaceae

Origins:
From the Greek *monos* "one, single," *stichos* "a row" and *anthos* "a flower," the petals are arranged in a single whorl; see F. von Mueller, in *The Victorian Naturalist.* 7: 180. 1891. Genus published as *Haplostichanthus*.

Monostylis Tul. Podostemaceae

Origins:

From the Greek *monos* "one, single" and *stylos* "style."

Monotagma K. Schumann Marantaceae

Origins:

Greek *monos* and *tagma, tagmatos* "a command, ordinance, order," *tasso* "to arrange, put in order."

Monotaxis Brongn. Euphorbiaceae

Origins:

From the Greek *monos* "single" and *taxis* "order, series," referring to the single row of stamens or to a genus apparently based on a single species.

Species/Vernacular Names:

M. grandiflora Endl.

English: diamond of the desert

Monotes A. DC. Dipterocarpaceae

Origins:

Greek *monos* "single," *monotes* "unity, uniqueness," at the time of publication it was the only genus of Dipterocarpaceae in Africa.

Species/Vernacular Names:

M. sp.

Nigeria: chimpampa, kipampa, mutembo

M. engleri Gilg (*Monotes tomentellus* Hutch.)

English: pink-fruited monotes

Southern Africa: muAra, muHarawashawa, chiNdharara, muNete, muNonye, muNunta, muNunywa, muNuya (Shona)

M. glaber Sprague

English: pale-fruited monotes, yellow wood

Southern Africa: muBarawashawa, baroshawa, muNinya, muNonye, muNyanyewa, muNyunyu, muShaba, muShawa, muVara, muWara (Shona)

S. Rhodesia: muBarawashava, umNonye, iNyunya

M. katangensis (De Wild.) De Wild.

English: red-fruited monotes

M. kerstingii Gilg

Nigeria: wasani, gasakura, farin-rura, wasane, gasu kura; hantso (Hausa)

Togo: kesang, kesau

W. Africa: kurunkurun, gandama, ngantama

Monotheca A. DC. Sapotaceae

Origins:

From the Greek *monos* "single" and *theke* "a case, capsule, box."

Monothecium Hochst. Acanthaceae

Origins:

Greek *monos* and *theke* "a case, capsule, box," *thekion* a diminutive, an allusion to the nature of the anther.

Monothrix Torrey Asteraceae

Origins:

From the Greek *monos* "single" and *thrix, trichos* "hair."

Monotoca R. Br. Epacridaceae

Origins:

Greek *monos* "single, one" and *tokos* "a birth," *monotokos* "bearing but one at a time," the ovary has only one ovule, the edible fruit has only one seed; see Robert Brown, *Prodromus florae Novae Hollandiae*. 546. London 1810.

Species/Vernacular Names:

M. elliptica R. Br.

English: tree broom heath

M. empetrifolia R. Br.

English: alpine broom heath

M. glauca (Labill.) Druce

English: currant-wood

M. ledifolia DC.

English: rare broom heath

M. linifolia (Rodway) W.M. Curtis

Australia: Longley currant-wood

M. rotundifolia J.H. Willis

English: trailing monotoca, slender broom heath

M. scoparia (Smith) R. Br. (*Styphelia scoparia* Smith)

English: broom heath, prickly broom, prickly broom heath

Monotrema Koern. Rapateaceae

Origins:

From the Greek *monos* "single, one" and *trema* "hole, aperture."

Monotris Lindley Orchidaceae

Origins:
Greek *monos* "single, one" and *tris* "three times," lip with three lobes.

Monotropa L. Ericaceae (Monotropaceae)

Origins:
Latin *monotropus* and Greek *monotropos* "of one kind, single," referring to the nodding flower and to stem, one-sided inflorescence.

Species/Vernacular Names:
M. hypopitys L.
English: yellow bird's nest, pinesap

M. uniflora L.
English: Indian pipe, corpse plant, convulsion root, pinesap, fitsroot

Monotropanthum Andres Ericaceae (Monotropaceae)

Origins:
From the genus *Monotropa* and *anthos* "flower."

Monotropastrum Andres Ericaceae (Monotropaceae)

Origins:
Resembling *Monotropa*, the genus *Monotropa* L. with the Latin suffix *astrum*.

Species/Vernacular Names:
M. macrocarpum H. Andres
English: false Indian pipe

Monotropsis Schwein. ex Elliott Ericaceae (Monotropaceae)

Origins:
Resembling the genus *Monotropa* L.

Monroa Torrey Gramineae

Origins:
See the genus *Munroa* Torr.

Monsonia L. Geraniaceae

Origins:
After Lady Ann Monson (*née* Vane), *circa* 1714-1776 (d. Calcutta), great-grandaughter of Charles II, botanical collector at the Cape of Good Hope and in Bengal, a correspondent of Linnaeus, at the Cape with Thunberg and F. Masson; see J. Britten, "Lady Ann Monson." *J. Bot., Lond.* 56: 147-149. 1918; J. Hutchinson, *A Botanist in Southern Africa.* 619. London 1946; Henry C. Andrews, *The Botanist's Repository.* t. 276. London 1803; Mary Gunn and Leslie E. Codd, *Botanical Exploration of Southern Africa.* 253. Cape Town 1981; E.J. Willson, *James Lee and the Vineyard Nursery, Hammersmith.* 39-40. London 1961.

Species/Vernacular Names:
M. angustifolia E. Mey. ex A. Rich.
English: crane's bill
Southern Africa: alsbos, malva naaldbossie, teebossie; malengoana (Sotho); phusana (Botswana)

M. burkeana Planch. ex Harv. (*Monsonia betschuanica* Knuth; *Monsonia biflora* DC.; *Monsonia glandulosissima* Schinz; *Monsonia malvaeflora* Schinz)
English: crane's bill, dysentery herb
Southern Africa: alsbossie, angelbossie, assegaaibossie, keitabossie (keita = disentery), naaldebossie; igqitha (Xhosa); khoara (Sotho); remarungana (Tswana)

M. emarginata (L.f.) L'Hérit. (*Monsonia ovata* Cav.; *Geranium emarginatum* L.f.)
English: dysentery herb
Southern Africa: igqitha, iGqita (Xhosa)

M. glauca Knuth (*Monsonia ovata* Cav. subsp. *glauca* (Knuth) Bowden & Muller; *Monsonia stricta* Knuth)
English: dysentery herb
Southern Africa: geitabossie, keitabossie, naaldebossie

Monstera Adans. Araceae

Origins:
The origins and derivation are quite obscure, possibly from the Latin *mons, montis* (*emineo, promineo*) "a mountain" and *teres, retis* (*tero*) "rounded off, smoothed, shapely," or from *monstrifer, era, erum* "monster-bearing," or from *monstrum, i* (*moneo*) "a warning, portent, wonder, monster, monstrosity, any unnatural person or thing," referring to the leaves; see D.H. Nicolson, "Derivation of aroid generic names." *Aroideana.* 10: 15-25. 1988.

Species/Vernacular Names:
M. acuminata K. Koch
English: shingle plant

M. adansonii Schott

Northwestern Amazonia: chupon khaki (Kofan); suso-iko (Shushufindi Siona)

Brazil (Amazonas): xaa a

M. deliciosa Liebm. (*Pothos pertusa* hort. non Roxb.)

English: ceriman, fruit salad plant, Swiss-cheese plant, hurricane palm, delicious monster, Mexican bread fruit

Japan: hôrai-shô

Brazil: chagas-de-São-Sebastião (= St. Sebastian's wounds), dragão-fedorento, folha-furada, folha-reta, imbé-farado, imbé-de-são-pedro

Mexico: guela gutzi

Cuba: cerimán de Méjico

Mauritius: banane Anglaise, taro vine

La Réunion Island: fruit délicieux

Latin America: camachillo, chirrivaca, costilla de adan

Montamans Dwyer Rubiaceae

Origins:

Possibly from the Latin *mons, montis* "a mountain" and *amans, amantis* "loving."

Montanoa Cervantes Asteraceae

Origins:

In honor of the Mexican (native of Puebla) Don Luis Montaña, b. 1755, physician and naturalist; see Pablo de La Llave (1773-1833) and Juan José Martinez de Lexarza (1785-1824), *Novorum Vegetabilium descriptiones.* 2: 11. Méxici 1824-1825; Arthur D. Chapman, ed., *Australian Plant Name Index.* 2032. 1991; H. Genaust, *Etymologisches Wörterbuch der botanischen Pflanzennamen.* 394. 1996; Georg Christian Wittstein, *Etymologisch-botanisches Handwörterbuch.* 594. 1852; F. Boerner & G. Kunkel, *Taschenwörterbuch der botanischen Pflanzennamen.* 4. Aufl. 136. 1989.

Species/Vernacular Names:

M. bipinnatifida (Kunth) C. Koch

English: daisy tree, tree daisy

M. hibiscifolia Benth.

English: tree daisy

M. tomentosa Cerv. (see Henry Laszlo and Paul S. Henshaw, "Plant materials used by primitive peoples to affect fertility." *Science.* 119: 626-631. 1954)

Mexico: zoapatle, nocuana titete xinini, cuana xana

Montbretia DC. Iridaceae

Origins:

Named for the French botanist Antoine François Ernest Coquebert de Montbret, 1781-1801, 1798 with Napoléon in Egypt.

Montbretiopsis L. Bolus Iridaceae

Origins:

Resembling *Montbretia* DC.

Montejacquia Roberty Convolvulaceae

Origins:

See *Jacquemontia* Choisy.

Montezuma Moçiño & Sessé ex DC. Bombacaceae (Malvaceae)

Origins:

Remembering the last Aztec Emperor Montezuma II, 1466-1520, warrior, legislator.

Montia L. Portulacaceae

Origins:

For the Italian botanist Giuseppe Monti, 1682-1760, professor of botany, from 1722 to 1760 Director of the Bologna Botanical Garden, father of Gaetano Lorenzo Monti (1712-1797), also professor and from 1760 to 1792 Director of the same garden and author of *Dizionario botanico Veronese.* Verona 1817; Giuseppe Monti wrote *Plantarum varii indices ad usum demonstrationum quae in Bononiensis Archigymnasii Publico Horto quotannis habentur.* Bononiae 1724, *De monumento diluviano nuper in agro Bononiensi detecto dissertatio.* Bononiae Studiorum 1719 and *Catalogi Stirpium agri Bononiensis prodromus, Gramina ac hujusmodi affinia complectens.* Bononiae 1719. See A. Bertoloni, *Sylloge plantarum horti bononiensis.* Bononiae 1827 and *Continuatio historiae horti botanici bononiensis.* Bononiae 1834; Carl Linnaeus, *Species Plantarum.* 87. 1753 and *Genera Plantarum.* Ed. 5. 38. 1754; Jonas C. Dryander, *Catalogus bibliothecae historico-naturalis Josephi Banks.* London 1796-1800; A. Lasègue, *Musée botanique de Benjamin Delessert.* 341. Paris 1845; Georg Christian Wittstein, *Etymologisch-botanisches Handwörterbuch.* 594. Ansbach 1852.

Species/Vernacular Names:

M. australasica (Hook.f.) Pax & Hoffm. (*Claytonia australasica* Hook.f.)

English: white purslane, montia

M. chamissoi (Sprengel) E. Greene

English: toad lily

M. fontana L.

English: water blinks, water chickweed

Montiastrum (A. Gray) Rydberg Portulacaceae

Origins:

Resembling *Montia* L.

Montinia Thunberg Grossulariaceae (Montiniaceae)

Origins:

After the Swedish botanist Lars (Laurentius) Jonasson Montin, 1723-1785, pupil of Linnaeus, physician, botanical collector. His works include Dissertatio botanica sistens *splachnum* ... Praes. ... C. Linnaeo. Upsaliae 1750, "Ericae tres novae species descriptae et delineatae." *Nova Acta R. Soc. Sci. Upsal.* 2: 291-296. 1775 and Dissertatio historico-medica *de medicina Lapponum Lulensium*. Praes. E. Rosen. Londini Gothorum [1751], he was uncle of Jonas Dryander. See John H. Barnhart, *Biographical Notes upon Botanists*. 2: 507. 1965; Warren R. Dawson, *The Banks Letters*, a Calendar of the Manuscript Correspondence of Sir Joseph Banks. London 1958; Jonas C. Dryander, *Catalogus bibliothecae historico-naturalis Josephi Banks*. London 1796-1800; A. Lasègue, *Musée botanique de Benjamin Delessert*. Paris 1845; E.D. Merrill, *Bernice P. Bishop Mus. Bull.* 144: 138. 1937; G. Murray, *History of the Collections Contained in the Natural History Departments of the British Museum*. 1: 27, 43. London 1904; I.C. Hedge and J.M. Lamond, *Index of Collectors in the Edinburgh Herbarium*. Edinburgh 1970.

Species/Vernacular Names:

M. caryophyllacea Thunb. (*Montinia acris* L.f.; *Montinia frutescens* Gaertn.)

English: pepper bush

South Africa: bergklapper, peperbos, perdebos

Montiopsis Kuntze Portulacaceae

Origins:

Resembling *Montia* L.

Montolivaea Reichb.f. Orchidaceae

Origins:

For Montolivo, Abyssinia.

Montrichardia Crueg. Araceae

Origins:

Dedicated to Gabriel de Montrichard; see D.H. Nicolson, "Derivation of aroid generic names." *Aroideana*. 10: 15-25. 1988.

Montrouziera Pancher ex Planchon & Triana Guttiferae

Origins:

Dedicated to the French botanist Rev. Xavier Montrouzier, 1820-1897, missionary, naturalist. Among his writings are *Essai sur la Faune de l'Ile de Woodlark ou Moiou*. Lyon 1857 and *Notice historique, ethnographique et physique sur la Nouvelle-Calédonie*. Paris 1860. See John H. Barnhart, *Biographical Notes upon Botanists*. 2: 507. 1965; Elmer Drew Merrill, *Contr. U.S. Natl. Herb.* 30(1): 216. 1947.

Monustes Raf. Orchidaceae

Origins:

One of the Danaides (the 50 daughters of Danaus King of Argos), who killed her husband Eurysthenes; see Constantine Samuel Rafinesque (1783-1840), *Flora Telluriana*. 2: 87. 1836 [1837]; E.D. Merrill, *Index rafinesquianus*. 103. 1949.

Monvillea Britton & Rose Cactaceae

Origins:

After Hyppolite Boissel, 1794-1863 (Paris, France), Baron de Monville, French amateur botanist, plant collector; see R. Zander, F. Encke, G. Buchheim and S. Seybold, *Handwörterbuch der Pflanzennamen*. 14. Aufl. 753. Stuttgart 1993; Gordon Douglas Rowley, *A History of Succulent Plants*. Strawberry Press, Mill Valley, California 1997.

Moonia Arnott Asteraceae

Origins:

After the Scottish botanist Alexander Moon, d. 1825 Ceylon, plant collector in North Africa, in 1815 Kew gardener, from 1817 to 1825 Superintendent of the Royal Botanical Gardens in Ceylon, wrote *A Catalogue of Indigenous and Exotic Plants Growing in Ceylon*. Colombo 1824. See John H. Barnhart, *Biographical Notes upon Botanists*. 2: 508. 1965; R. Zander, F. Encke, G. Buchheim and S. Seybold,

Handwörterbuch der Pflanzennamen. 14. Aufl. Stuttgart 1993; Warren R. Dawson, *The Banks Letters*. London 1958; E.M. Tucker, *Catalogue of the Library of the Arnold Arboretum of Harvard University*. 1917-1933; Isaac Henry Burkill (1870-1965), *Chapters on the History of Botany in India*. Delhi 1965; G. Murray, *History of the Collections Contained in the Natural History Departments of the British Museum*. London 1904; R. Desmond, *The European Discovery of the Indian Flora*. 162. Oxford 1992; *Nova Acta Physico-Medica Akademie Caesareae Leopoldino-Carolinae Naturae Curiosorum* [exhibentia ephemerides sive observationes historias et experimenta]. 18: 348. 1836.

Moorcroftia Choisy Convolvulaceae

Origins:

After William Moorcroft, *circa* 1765-1825 (Afghanistan), veterinary surgeon in East India Company, 1808 Bengal, plant collector with N. Wallich, traveler. His writings include *Cursory Account of the Various Methods of Shoeing Horses Hitherto Practised*; with incidental observations. London 1800, *Directions for Using the Contents of the Portable Horse Medicine Chest*, adapted for India. London 1795 and *Observations on the Breeding of Horses*, within the Provinces under the Bengal Establishment. Simla 1886, with George Trebeck wrote *Travels in the Himalayan Provinces of Hindustan and the Panjab*; etc. London 1841. See Nathaniel Wallich (1786-1854), *Plantae Asiaticae rariores*. 3: 7-9. London 1832; D.G. Crawford, *A History of the Indian Medical Service, 1600-1913*. 2: 141. London 1914; R. Desmond, *The European Discovery of the Indian Flora*. Oxford 1992; Ray Desmond, *Dictionary of British & Irish Botanists and Horticulturists*. 496. London 1994.

Moorea Lemaire Gramineae

Origins:

To commemorate the Scottish botanist David Moore (Muir, until 1828), 1808-1879 (Glasnevin, Dublin), botanical collector, from 1829 to 1834 at Trinity College Garden, in 1861 a Fellow of the Linnean Society, from 1838 to 1879 Curator of Glasnevin Garden (three kilometers northwest of the center of the city of Dublin), brother of Charles Moore (1820-1905) and father of Sir Frederick Moore (1857-1949). Among his works are *Guide to the Royal Botanic Gardens, Glasnevin*. [Revised and enlarged by Professor William Ramsay McNab, 1844-1889, Scottish botanist.] Dublin 1885 and *The Mosses of Ireland*. Dublin 1873, with the British botanist and entomologist Alexander Goodman More (1830-1895) wrote *Contributions Towards a Cybele Hibernica*, being outlines of the geographical distribution of plants in Ireland. Dublin and London 1866. See N. Colgan and Reginald William Scully (1858-1935), *Contribu-*

tions Towards a Cybele Hibernica. Ed. 2nd. Dublin 1898; J.H. Barnhart, *Biographical Notes upon Botanists*. 2: 508. 1965; E. Charles Nelson, "National Botanic Gardens, Glasnevin, Retrospect and Prospect." *Curtis's Botanical Magazine*. Volume 12. 4: 181-185. November 1995; E. Charles Nelson and E.M. McCracken, *The Brightest Jewel: A History of the National Botanic Gardens, Glasnevin, Dublin*. Kilkenny 1987; T.W. Bossert, *Biographical Dictionary of Botanists Represented in the Hunt Institute Portrait Collection*. 272. 1972; H.N. Clokie, *Account of the Herbaria of the Department of Botany in the University of Oxford*. 214. Oxford 1964; E.M. Tucker, *Catalogue of the Library of the Arnold Arboretum of Harvard University*. 1917-1933; G. Murray, *History of the Collections Contained in the Natural History Departments of the British Museum*. 1: 169. London 1904; Ernest Nelmes and William Cuthbertson, *Curtis's Botanical Magazine Dedications, 1827-1927*. 183-184. [1931]; Ray Desmond, *Dictionary of British & Irish Botanists and Horticulturists*. 496-497. 1994; I.C. Hedge and J.M. Lamond, *Index of Collectors in the Edinburgh Herbarium*. Edinburgh 1970.

Moorea Rolfe Orchidaceae

Origins:

Dedicated to the Irish botanist Sir Frederick William Moore, 1857-1949 (Dublin), son of the Scottish botanist David M. (1808-1879), between 1877 and 1879 Curator of the Trinity College Garden in Dublin, 1911 Fellow of the Linnean Society, knighted 1911, from 1879 to 1922 Director of the Glasnevin Botanic Garden; see Ernest Nelmes and William Cuthbertson, *Curtis's Botanical Magazine Dedications, 1827-1927*. 327-328. [1931]; M. Hadfield et al., *British Gardeners: A Biographical Dictionary*. London 1980; E.C. Nelson & Eileen May McCracken (1920-1988), *The Brightest Jewel: A History of the National Botanic Gardens, Glasnevin, Dublin*. Kilkenny 1987.

Mooria Montrouzier Myrtaceae

Origins:

For the Scottish-born (Dundee, Angus) Australian botanist Charles Moore (until 1828 Muir), 1820-1905 (Sydney), gardener, brother of David Moore (1808-1879), traveler and plant collector, from 1848 to 1896 Director of the Botanical Garden at Sydney, in 1863 a Fellow of the Linnean Society. His writings include *A Census of the Plants of New South Wales*. Sydney 1884 and *On the Woods of New South Wales*. Sydney 1871, co-author with Ern(e)st Betche (1851-1913) of *Handbook of the Flora of New South Wales*. Sydney 1893, with the Australian horticulturist Sir William Macarthur (1800-1882) of [New South Wales, Australia], *Cat-*

alogue des collections de bois indigènes des différents districts de cette colonie. Paris [1855]. See J.H. Barnhart, *Biographical Notes upon Botanists.* 2: 508. 1965; T.W. Bossert, *Biographical Dictionary of Botanists Represented in the Hunt Institute Portrait Collection.* 272. 1972; Ethelyn Maria Tucker, *Catalogue of the Library of the Arnold Arboretum of Harvard University.* Cambridge, Massachusetts 1917-1933; E.D. Merrill, "Bibliography of Polynesian botany." *Bernice P. Bishop Mus. Bull.* 144: 138. 1937 and *Contr. U.S. Natl. Herb.* 30(1): 216. 1947; E.C. Nelson & Eileen May McCracken, *The Brightest Jewel: A History of the National Botanic Gardens, Glasnevin, Dublin.* Kilkenny 1987; G. Murray, *History of the Collections Contained in the Natural History Departments of the British Museum.* 1: 169. London 1904; R. Zander, F. Encke, G. Buchheim and S. Seybold, *Handwörterbuch der Pflanzennamen.* 14. Aufl. Stuttgart 1993; Ray Desmond, *Dictionary of British & Irish Botanists and Horticulturists.* 496. London 1994.

Mopania Lundell Sapotaceae

Origins:

From an African vernacular name.

Moquinia DC. Asteraceae

Origins:

For the French botanist Christian Horace Bénédict Alfred Moquin-Tandon, 1804-1863, naturalist, from 1834 to 1853 Director of the Botanic Garden of Toulouse, one of the founders of the Société Botanique de France, professor of botany at the Faculté de Médecine at Paris, expert on the Provençal language, contributed to A.P. de Candolle, *Prodromus* (Phytolaccaceae, Salsolaceae, Basellaceae, Amaranthaceae). His publications include *Histoire naturelle des Mollusques ... de France.* Paris 1855, *Eléments de tératologie végetale.* Paris 1841, *Lettres inédites de Moquin-Tandon à Auguste de Saint-Hilaire.* Clermont l'Hérault 1893, *Eléments de Zoologie médicale.* Paris 1860 [1859], *Carya Magalonensis, ou Noyer de Maguelonne.* Montpellier 1844 and *Eléments de botanique médicale.* Paris 1861. See John H. Barnhart, *Biographical Notes upon Botanists.* 2: 510. 1965; Ida Kaplan Langman, *A Selected Guide to the Literature on the Flowering Plants of Mexico.* Philadelphia 1964; Stafleu and Cowan, *Taxonomic Literature.* 3: 573-575. 1981; R. Zander, F. Encke, G. Buchheim and S. Seybold, *Handwörterbuch der Pflanzennamen.* 14. Aufl. Stuttgart 1993; Ethelyn Maria Tucker, *Catalogue of the Library of the Arnold Arboretum of Harvard University.* Cambridge, Massachusetts 1917-1933; A. Lasègue, *Musée botanique de Benjamin Delessert.* Paris 1845; T.W. Bossert, *Biographical Dictionary of Botanists*

Represented in the Hunt Institute Portrait Collection. 273. 1972; Elmer Drew Merrill, *Contr. U.S. Natl. Herb.* 30(1): 220. 1947.

Moquinia Sprengel Loranthaceae

Origins:

For the French botanist Christian Horace Bénédict Alfred Moquin-Tandon (Alfred Frédol, pseudonym), 1804-1863, naturalist.

Moquiniella Balle Loranthaceae

Origins:

Genus named for the French botanist Christian Horace Bénédict Alfred Moquin-Tandon (Alfred Frédol, pseudonym), 1804-1863, naturalist, pupil of A.P. de Candolle at Montpellier, from 1834 to 1853 Director of the Botanic Garden of Toulouse, one of the founders of the Société Botanique de France, professor of botany at the Faculté de Médecine at Paris, expert on the Provençal language, contributed to A.P. de Candolle, *Prodromus* (Phytolaccaceae, Salsolaceae, Basellaceae, Amaranthaceae). His publications include *Histoire naturelle des Mollusques ... de France.* Paris 1855, *Eléments de tératologie végetale.* Paris 1841, *Lettres inédites de Moquin-Tandon à Auguste de Saint-Hilaire.* Clermont l'Hérault 1893, *Eléments de Zoologie médicale.* Paris 1860 [1859], *Carya Magalonensis, ou Noyer de Maguelonne.* Montpellier 1844 and *Eléments de botanique médicale.* Paris 1861; see John H. Barnhart, *Biographical Notes upon Botanists.* 2: 510. 1965; Stafleu and Cowan, *Taxonomic Literature.* 3: 573-575. 1981.

Species/Vernacular Names:

M. rubra (Spreng.f.) Balle (*Loranthus elegans* Cham. & Schltdl.; *Loranthus glaucus* Thunb. var. *burchellii* DC.; *Loranthus oleifolius* (Wendl.) Cham. & Schltdl. var. *elegans* (Cham. & Schltdl.) Harv.; *Loranthus oleifolius* sensu Eckl. & Zeyh., sensu Presl, sensu Marloth, non Cham. & Schltdl.; *Moquinia rubra* A. Spreng.; *Dendrophthoe elegans* (Cham. & Schltdl.) Mart.; *Scurrula elegans* (Cham. & Schltdl.) G. Don; *Lichtensteinia elegans* (Cham. & Schltdl.) Tiegh.)

English: lighting matches

South Africa: matchsticks

Mora Benth. Caesalpiniaceae

Origins:

From a South American vernacular name.

Moraea Miller Iridaceae

Origins:

Named in honor of the British (esquire of Shropshire) Robert More, 1703-1780, an amateur botanist and natural historian, in 1729 Fellow of the Royal Society of London, traveler, friend of Linnaeus; according to Georg Christian Wittstein (in *Etymologisch-botanisches Handwörterbuch.* 594. Ansbach 1852) and N.E. Brown (in *Journal of the Linn. Soc., Botany.* 48: 40-41. 1928) Miller changed *Morea* in *Moraea* in honor of Dr. Johan Moraeus, father of Sara Elisabeth Moraea, wife of Linnaeus; see George H.M. Lawrence, "Derivation of the generic name *Moraea* (Iridaceae)." *Baileya.* 3(3): 130. 1955.

Species/Vernacular Names:

M. sp.

English: fairy iris

Southern Africa: inCembu (Xhosa)

M. bipartita L. Bol. (*Moraea polyanthos* sensu Goldblatt, non L.f.)

English: blue tulp, Cape tulp

South Africa: bloutulp, Kaapse bloutulp, Kaapse-tulp

M. ciliata (L.f.) Ker Gawl.

South Africa: uintjie

M. fugax (Delaroche) Jacq. subsp. *fugax* (*Moraea edulis* (L.f.) Ker Gawl.)

South Africa: uintjie, soetuintjie

M. graminicola Oberm. subsp. *graminicola*

English: butterfly iris, moraea, Natal lily

M. insolens Goldbl.

South Africa: butterfly iris

M. lugubris (Salisb.) Goldbl. (*Moraea plumaria* (Thunb.) Ker Gawl.)

South Africa: kersblakertjie

M. neopavonia R.C. Fost.

South Africa: peacock flower, uiltjie, poublom

M. polyanthos L.f.

South Africa: blue tulp, bloutulp

M. polystachya (Thunb.) Ker Gawl.

English: blue moraea, blue tulp, Cape blue tulp

South Africa: bloutulp, Kaapse-bloutulp, blue tulp, kraaiuintjie, tulp, wildetulp

M. ramosissima (L.f.) Druce

South Africa: geeltulp

M. serpentina Bak. (*Moraea arenaria* Bak.)

South Africa: slanguintjie

M. spathulata (L.f.) Klatt (*Moraea spathulata* (L.f.) Klatt subsp. *autumnalis* Goldbl.; *Moraea spathulata* (L.f.) Klatt subsp. *saxosa* Goldbl.; *Moraea spathulata* (L.f.) Klatt subsp. *transvaalensis* Goldbl.)

English: large yellow moraea, mountain moraea, yellow tulip, yellow tulp, yellow iris, tulp

South Africa: geeltulp, grootgeeltulp, bergmoraea, bloutulp, kleintulp, tulp, nokha; leloele, teele-ea-noka (Sotho)

M. vegeta L. (*Moraea juncea* sensu N.E. Br., non L.; *Moraea sordescens* Jacq.)

South Africa: bruintulp

M. villosa (Ker Gawl.) Ker Gawl. subsp. *villosa*

South Africa: uiltjie, peacock flower

M. viscaria (L.f.) Ker Gawl.

English: yellow tulp

Morawetzia Backeb. Cactaceae

Origins:

For Viktor Morawetz, flourished in the mid-1930s; see Gordon Douglas Rowley, *A History of Succulent Plants.* Strawberry Press, Mill Valley, California 1997.

Morelia A. Rich. ex DC. Rubiaceae

Origins:

The name of the genus commemorates the French botanist Morel (d. *circa* 1825), a plant collector in Senegal 1824-1825; see Joseph Vallot, "Études sur la flore du Sénégal." in *Bull. Soc. Bot. de France.* 29: 168-238. Paris 1882; F.N. Hepper and F. Neate, *Plant Collectors in West Africa.* 57. 1971.

Species/Vernacular Names:

M. senegalensis A. Rich. ex DC.

Nigeria: osangodo; inuwar bauna (Hausa); dandojee, osangodo (Yoruba); akule (Igbo)

Yoruba: buje dudu, dandoje, osangodo, asogbodun, asogbodo, onipowoje

Morelotia Gaudich. Cyperaceae

Origins:

For the French pharmacist Simon Morelot, 1751-1809, author of *Histoire Naturelle* appliquée à la chimie, aux arts, aux différents genres de l'industrie, et aux besoins personnels de la vie. Paris 1809.

Morenia Ruíz & Pav. Palmae

Origins:

For the Peruvian physician Gabriel Moreno, 1735-1809, botanist.

Morettia DC. Brassicaceae

Origins:

Named after the Italian botanist Giuseppe Moretti, 1782-1853, professor of botany, from 1826 to 1833 and from 1833 to 1853 Director of the Botanical Garden of Pavia, author of *Prodromo di una monografia della specie del genere Morus.* [Milano 1842]. See V. Bianchi, E. Bruno and V. Giacomini, "Giuseppe Moretti." *Atti Ist. Bot. Lab. Critt. Univ. Pavia.* ser. 5, 16: 210-230. 1959; Gino Pollacci, "L'Orto Botanico di Pavia dalla fondazione al 1942." *Ticinum.* 6: 20-23. 1950; J.H. Barnhart, *Biographical Notes upon Botanists.* 2: 512. 1965; T.W. Bossert, *Biographical Dictionary of Botanists Represented in the Hunt Institute Portrait Collection.* 239. 1972; A. Lasègue, *Musée botanique de Benjamin Delessert.* 342. Paris 1845; Ethelyn Maria Tucker, *Catalogue of the Library of the Arnold Arboretum of Harvard University.* Cambridge, Massachusetts 1917-1933; Ida Kaplan Langman, *A Selected Guide to the Literature on the Flowering Plants of Mexico.* Philadelphia 1964; R. Zander, F. Encke, G. Buchheim and S. Seybold, *Handwörterbuch der Pflanzennamen.* 14. Aufl. Stuttgart 1993.

Morgania R. Br. Scrophulariaceae

Origins:

Named after the English apothecary Hugh Morgan, fl. 1540-1576 (d. 1613?), apothecary-in-ordinary to Queen Elizabeth I, owner of botanical gardens in London and at Battersea. See Charles Angell Bradford, *Hugh Morgan, Queen Elizabeth's Apothecary.* London 1939; John Gerard (1545-1612), *Herball, or Generall Historie of Plantes.* London 1597; Eleanour Sinclair Rohde, *The Old English Herbals.* London 1922; John Parkinson, 1567-1650, *Paradisi in Sole Paradisus Terrestris.* London 1629; Robert William Theodore Gunther, *Early British Botanists and Their Gardens* based on unpublished writings of Goodyer ... and others. 415. Oxford 1922; C.E. Raven, *English Naturalists from Neckam to Ray.* 116-117. Cambridge 1947; James Britten, *The Sloane Herbarium,* revised and edited by J.E. Dandy. 169. London 1958; Robert Brown, *Prodromus florae Novae Hollandiae.* 441. London 1810.

Species/Vernacular Names:

M. floribunda Benth.

Australia: bluerod, blue top, blue-flower, morgan flower, free-flowering morgania

M. glabra R. Br.

English: bluerod, smooth bluerod, morgan flower, blue top

South Australia: inmuda, inmorda (North of Innamincka)

Moricandia DC. Brassicaceae

Origins:

For the Swiss botanist Moïse Étienne (Stefano) Moricand, 1779-1854, author of *Plantes nouvelles d'Amérique.* Genève 1833-1846[-1847]. See John H. Barnhart, *Biographical Notes upon Botanists.* 2: 514. 1965; T.W. Bossert, *Biographical Dictionary of Botanists Represented in the Hunt Institute Portrait Collection.* 273. 1972; H.N. Clokie, *Account of the Herbaria of the Department of Botany in the University of Oxford.* 214. Oxford 1964; Ethelyn Maria Tucker, *Catalogue of the Library of the Arnold Arboretum of Harvard University.* Cambridge, Massachusetts 1917-1933; Ida Kaplan Langman, *A Selected Guide to the Literature on the Flowering Plants of Mexico.* Philadelphia 1964; R. Zander, F. Encke, G. Buchheim and S. Seybold, *Handwörterbuch der Pflanzennamen.* 14. Aufl. 754. 1993; Georg Christian Wittstein, *Etymologisch-botanisches Handwörterbuch.* 595. 1852; Irving William Knobloch, compil., "A preliminary verified list of plant collectors in Mexico." *Phytologia Memoirs.* VI. Plainfield, N.J. 1983.

Moriera Boissier Brassicaceae

Origins:

Named for the English (born of a Swiss family settled in England) traveler James (Jakub) Justinian Morier, about 1780-1849, novelist, diplomatist, Secretary of Embassy in Persia. Among his writings are *The Adventures of Hajji Baba of Ispahan.* Illustrated by H.R. Millar, with an introduction by E.G. Browne. London 1895, *Ayesha, the Maid of Kars.* Paris 1843, *A Journey through Persia, Armenia, and Asia Minor, to Constantinople, in the Years 1808 and 1809.* London 1812, *Martin Toutrond: or, Adventures of a Frenchman.* London 1852, *The Mirza.* London 1841 and *A Second Journey through Persia, Armenia, and Asia Minor to Constantinople between the years 1810 and 1816;* with a Journal of the voyage by the Brazils and Bombay to the Persian Gulph. London 1818; see W. Hauff, *The Banished* ... edited by J.M. 1839.

Morierina Vieillard Rubiaceae

Origins:

For the French botanist (Jules) Pierre Gilles Morière, 1817-1888, professor of botany at Caen, paleontologist, botanical collector. His writings include *Note sur une liliacée de la Californie.* Caen 1863, *De la Basse-Cour en général et des exportations d'oeufs* qui se sont faites par quelques ports du Calvados et de la Manche, etc. Caen 1868, *Excursion de la Société Linnéenne à Vivre*, le ... 8 Juillet 1866. Caen 1866 and *Note sur un cas de chorise dans le galanthus nivalis et de floriparité dans le cardamine pratensis.* Caen 1861. See J.H. Barnhart, *Biographical Notes upon Botanists.* 2: 514. 1965; Ethelyn Maria Tucker, *Catalogue of the Library of the Arnold Arboretum of Harvard University.* Cambridge, Massachusetts 1917-1933.

Morina L. Morinaceae

Origins:

After the French physician Louis Morin, 1636-1715, botanist, author of Quaestio medica, C. Guerin Praes. *An sit insita alicui homini naturaliter vis curandi morbos?* [Paris 1666] and Quaestio medica, C. Le Vasseur Praes. *An annus qui fructuum idem et morborum ferax.* [Paris 1665]; see H. Genaust, *Etymologisches Wörterbuch der botanischen Pflanzennamen.* 395. [b. 1635] 1996; F. Boerner & G. Kunkel, *Taschenwörterbuch der botanischen Pflanzennamen.* 4. Aufl. 136f. 1989; Georg Christian Wittstein, *Etymologisch-botanisches Handwörterbuch.* 595. Ansbach 1852.

Species/Vernacular Names:
M. sp.
Tibetan: spyang-tsher
M. alba Hand.-Mazz.
English: white flower morina
M. bulleyana Forr.
English: large-flower morina

Morinda L. Rubiaceae

Origins:

Indian mulberry, Latin *morus* and *indicus, indica*, because of the shape of the fruits; see Carl Linnaeus, *Species Plantarum.* 176. 1753 and *Genera Plantarum.* Ed. 5. 81. 1754.

Species/Vernacular Names:
M. sp.
Cameroon: atchek, n'keng
Yoruba: pawopawo

Malaya: mengkudu, bengkudu
M. acutifolia F. Muell.
English: veiny morinda
M. citrifolia L. (*Morinda quadrangularis* hort.)
English: Indian mulberry, awl tree, painkiller, great morinda, brimstone tree
Australia: koonjerung, tokoonja
Japan: yaeyama-aoki
Malaya: mengkudu besar, bengkudu, kemudu, mengkudu jantan
Indochina: dau, chau, ngao, rau
Vietnam: nhau, nhau rung
India: suranji
Hawaii: noni
The Cameroons: atchek, n'keng
M. jasminoides Cunn.
English: jasmine morinda, sweet morinda
M. lucida Benth.
Nigeria: owuru, eze-ogu; oruwo (Yoruba)
Yoruba: oruwo, oruwo funfun, apawoparun, iwo
Gabon: akian
Cameroon: kikengwe, kwakengue
West Africa: ake, atiati, sima
Congo: ossi, nsiki, ossika
M. morindoides (Bak.) Milne-Redhead
Sierra Leone: kojo logbo
Congo: kongobololo
M. officinalis How
English: medicinal Indian mulberry
China: ba ji tian
Vietnam: ba kich, day ruot ga
M. parvifolia Benth.
English: little-leaf Indian mulberry
M. reticulata Benth.
Australia: Ada-A, mapoon
M. trimera Hillebr. (*Morinda lanaiensis* St. John; *Morinda sandwicensis* Degener; *Morinda waikapuensis* St. John)
Hawaii: noni kuahiwi
M. umbellata L. (*Morinda royoc* hort.)
English: common Indian mulberry
Japan: hana-gasa-no-ki
Malaya: mengkudu hutan, mengkudu akar, mengkudu kechil, buah butang

Morindopsis Hook.f. Rubiaceae

Origins:
Resembling *Morinda* L.

Moringa Rheede ex Adans. Moringaceae

Origins:
Muringa or *murunga* or *moringo*, Malayalam and Tamil names for *Moringa oleifera* Lam.

Species/Vernacular Names:
M. oleifera Lam.

English: horse radish tree, oil of Ben tree, Ben, drumstick tree, radish tree, Bentree

Japan: wasabi-no-ki

India: sanjana, sigruh, murinna, murinkai, sajana, moringu, mouringou

Malaya: emmunggai, gemunggai, lemunggai, meringgai, merunggai, morunggai, remunggai, rembugai, lembugai, kelok, kachang kelur, kachang kelor, kelentang, germunga

The Philippines: malunggay, malungay, marunggay, kamalungai, arunggai, balungai, dool

Central America: paraiso blanco, arango, badumbo, brotón, caragua, caraño, marengo, moringa, perlas, sasafrás, teberindo, maranga calalu

Yoruba: ewe ile, ewe igbale, idagba manoye, idagba moloye

Nigeria: bagaruwa-makka, ewo-igbale, barambo, samarindanga, zogalagandi; zogallagandi (Hausa); ewe-igbale (Yoruba); okwe oyibo (Igbo)

W. Africa: masa yiri

M. ovalifolia Dinter & A. Berger (*Moringa oleifera* sensu Exell & Mendonça; *Moringa ovalifoliolata* Dinter & Berger)

English: African moringo

South Africa: moringa, meelsakboom

Morisia J. Gay Brassicaceae

Origins:
After the Italian botanist Giuseppe Giacinto (Joseph Hyacinthe, Josephus Hyacinthus) Moris, 1796-1869, professor of botany at Torino, from 1831 to 1869 Director of the Botanical Garden. Among his writings are "Caratteri di tre specie nuove di piante chilesi." *Ann. Storia Nat.* 4: 59-60. 1830, *Enumeratio seminum Regii horti botanici Taurinensis.* [in *Flora.* 16(1): 126. 1833] Torino 1831 and *Enumeratio seminum Regii horti botanici Taurinensis.* Torino 1835. See Arturo Ceruti, "L'Orto Botanico di Torino." *Agricoltura.* 7. 1963; Oreste Mattirolo, *Cronistoria dell'Orto Botanico della Regia Università di Torino.* in *Studi sulla vegetazione nel Piemonte* pubblicati a ricordo del II Centenario della fondazione dell'Orto Botanico della R. Università di Torino. Torino 1929; J.H. Barnhart, *Biographical Notes upon Botanists.* 2: 514. 1965; T.W. Bossert, *Biographical Dictionary of Botanists Represented in the Hunt Institute Portrait Collection.* 273. 1972; H.N. Clokie, *Account of the Herbaria of the Department of Botany in the University of Oxford.* 214. Oxford 1964; I.C. Hedge and J.M. Lamond, *Index of Collectors in the Edinburgh Herbarium.* Edinburgh 1970; Ethelyn Maria Tucker, *Catalogue of the Library of the Arnold Arboretum of Harvard University.* Cambridge, Massachusetts 1917-1933; S. Belli, "I *Hieracium* di Sardegna. Rivista critica delle specie note dalla 'Flora Sardoa' di Moris, etc." *Memorie della Reale Accademia delle Scienze di Torino.* Serie seconda. tom. 47. 1897; R. Zander, F. Encke, G. Buchheim and S. Seybold, *Handwörterbuch der Pflanzennamen.* 14. Aufl. 754. Stuttgart 1993.

Morisia Nees Cyperaceae

Origins:
After the Italian botanist Giuseppe Giacinto (Joseph Hyacinthe) Moris, 1796-1869. See J.H. Barnhart, *Biographical Notes upon Botanists.* 2: 514. 1965; G. Arnott Walker Arnott (1799-1868), *The Edinburgh New Philosophical Journal.* 17: 265. 1834.

Morisonia L. Capparidaceae (Capparaceae)

Origins:
For the Scottish botanist Robert Morison, 1620-1683 (London), physician, gardener, studied medicine and botany in Paris, professor of botany at Oxford 1669, author of *Plantarum historiae universalis oxoniensis.* Oxford 1680-1699. See J.H. Barnhart, *Biographical Notes upon Botanists.* 2: 514. 1965; Charles Webster, in *Dictionary of Scientific Biography* 9: 528-529. [b. Dundee] 1981; Ray Desmond, *Dictionary of British & Irish Botanists and Horticulturists.* 500. [b. Aberdeen] London 1994; Frans A. Stafleu, *Linnaeus and the Linnaeans. The spreading of their ideas in systematic botany, 1735-1789.* Utrecht 1971; T.W. Bossert, *Biographical Dictionary of Botanists Represented in the Hunt Institute Portrait Collection.* 274. 1972; H.N. Clokie, *Account of the Herbaria of the Department of Botany in the University of Oxford.* 215, 10-13. Oxford 1964; Jonas C. Dryander, *Catalogus bibliothecae historico-naturalis Josephi Banks.* London 1796-1800; E.M. Tucker, *Catalogue of the Library of the Arnold Arboretum of Harvard University.* 1917-1933; Richard Pulteney, *Historical and Biographical Sketches of the Progress of Botany in England.* London 1790; Francis Wall Oliver, ed., *Makers*

of British Botany. 8-43. Cambridge 1913; Blanche Elizabeth Edith Henrey (1906-1983), *British Botanical and Horticultural Literature before 1800.* Oxford 1975; Ida Kaplan Langman, *A Selected Guide to the Literature on the Flowering Plants of Mexico.* Philadelphia 1964; Asher Rare Books & Antiquariaat Forum, *Catalogue Natural History.* item no. 114. The Netherlands 1998.

Moritzia DC. ex Meissner Boraginaceae

Origins:
After the Swiss botanist Alexander Moritzi, 1806-1850, high school teacher. His works include *Systematisches Verzeichniss der von H. Zollinger in den Jahren 1842-1844 auf Java gesammelten Pflanzen,* etc. Solothurn 1845-1846. See J.H. Barnhart, *Biographical Notes upon Botanists.* 2: 515. 1965; E.M. Tucker, *Catalogue of the Library of the Arnold Arboretum of Harvard University.* 1917-1933; R. Zander, F. Encke, G. Buchheim and S. Seybold, *Handwörterbuch der Pflanzennamen.* 14. Aufl. Stuttgart 1993.

Mormodes Lindley Orchidaceae

Origins:
Greek *mormo* "phantom" and *-odes* "resembling, having the form," referring to the strange appearance of the flowers.

Mormolyca Fenzl Orchidaceae

Origins:
From the Greek *mormo* "hobgoblin, bogey," *mormolykeion* "bogey, hobgoblin," referring to the appearance of the flowers.

Morocarpus Siebold & Zucc. Urticaceae

Origins:
Greek *moron* "black mulberry, blackberry" and *karpos* "fruit."

Morolobium Kosterm. Mimosaceae

Origins:
From the Greek *moron* "black mulberry, blackberry" and *lobion, lobos* "pod, small pod," referring to the dark blue to black seeds.

Morongia Britton Mimosaceae

Origins:
For the American clergyman Thomas Morong, 1827-1894, botanist, plant collector in Paraguay and in Chile, from 1890 to 1894 Curator of the Columbia College Herbarium in New York. His writings include "An enumeration of the plants collected by Dr. Thomas Morong in Paraguay, 1888-1890." *Ann. New York Acad.* vol. 7, no. 2. 1892-1893 [by T. Morong and N.L. Britton, with the assistance of Miss A.M. Vail], "Studies in the Typhaceae. I. *Typha.*" *Bull. Torrey Bot. Club.* 15: 1-8. 1888 and "The flora of the desert of Atacama." *Bull. Torrey Bot. Club.* 18: 39-48. 1891; see J.H. Barnhart, *Biographical Notes upon Botanists.* 2: 515. 1965; T.W. Bossert, *Biographical Dictionary of Botanists Represented in the Hunt Institute Portrait Collection.* 274. 1972; S. Lenley et al., *Catalog of the Manuscript and Archival Collections and Index to the Correspondence of John Torrey.* Library of the New York Botanical Garden. 299. 1973; E.M. Tucker, *Catalogue of the Library of the Arnold Arboretum of Harvard University.* 1917-1933; G. Murray, *History of the Collections Contained in the Natural History Departments of the British Museum.* 1: 169. London 1904; Ida Kaplan Langman, *A Selected Guide to the Literature on the Flowering Plants of Mexico.* Philadelphia 1964; R. Zander, F. Encke, G. Buchheim and S. Seybold, *Handwörterbuch der Pflanzennamen.* 14. Aufl. 1993; I.C. Hedge and J.M. Lamond, *Index of Collectors in the Edinburgh Herbarium.* Edinburgh 1970.

Morrenia Lindley Asclepiadaceae

Origins:
After the Belgian botanist Charles François Antoine Morren, 1807-1858, horticulturist, naturalist, professor of botany and Director of the Botanical Garden at Liège. His works include *Descriptio Coralliorum fossilium in Belgio repertorum.* Gröningen 1827-1828, *Le Globe, le Temps, et la Vie.* Bruxelles 1850, *Revue systématique des nouvelles decouvertes d'ossemens fossiles,* faites dans le Brabant méridional. Gand 1828 and *Specimen academicum exhibens tentamen biozoogeniae generalis,* etc. Bruxelles 1829; Charles Jacques Édouard Morren (1833-1886) was the son of Charles François A. Morren, from 1857 to 1886 professor of botany and Director of the Botanical Garden at Liège, in 1839 with Louis Benoît Van Houtte (1810-1876) founded *L'Horticulteur belge.* Among his writings are *Correspondance botanique.* Liège 1874 and 1884, *Charles de l'Ecluse, sa vie et ses oeuvres.* Liège 1875, *Biographie de Auguste Grisebach 1814-1879.* Liège 1881 and *Mathias de L'Obel, sa Vie et ses Oeuvres. 1538-1616.* Liège 1875. See J.H. Barnhart, *Biographical Notes upon Botanists.* 2: 515. 1965; Juan A. Domínguez, "Contribuciones a la materia

medica argentina." *Trabajos del Instituto de Botánica y Farmacologia.* 44: 1-433. Buenos Aires 1928; Ida Kaplan Langman, *A Selected Guide to the Literature on the Flowering Plants of Mexico.* Philadelphia 1964; R. Zander, F. Encke, G. Buchheim and S. Seybold, *Handwörterbuch der Pflanzennamen.* 14. Aufl. 1993.

Morrisiella Aellen Chenopodiaceae

Origins:

After the Australian botanist Albert Morris, 1886-1939, see Stafleu and Cowan, *Taxonomic Literature.* 3: 595. 1981; or dedicated to the Australian botanist Patrick Francis Morris, 1896-1974, from 1913 to 1961 at the National Herbarium of Victoria at Melbourne, in 1943 President Field Naturalists' Club of Victoria, with James Wales Claredon Audas (1872-1959) wrote *Supplement* to Professor Ewart's *Weeds, Poison Plants and Naturalized Aliens of Victoria* ... Melbourne 1925; see Alfred James Ewart (1872-1937), *Flora of Victoria.* [Melbourne] 1930; James H. Willis, *Vict. Naturalist.* 91(7): 205-206. 1974; A.J. Ewart & James Richard Tovey (1873-1922), *The Weeds, Poison Plants and Naturalized Aliens of Victoria.* Melbourne 1909; John H. Barnhart, *Biographical Notes upon Botanists.* 2: 516. 1965; P. Aellen (1896-1973), *Botanische Jahrbücher.* 68: 422. 1938; James A. Baines, *Australian Plant Genera. An Etymological Dictionary of Australian Plant Genera.* 245. N.S.W. 1981.

Morsacanthus Rizzini Acanthaceae

Origins:

From the Latin *mordeo, momordi, morsum, mordere* "to bite" and *Acanthus.*

Mortonia A. Gray Celastraceae

Origins:

Dedicated to the North American (b. Philadelphia) naturalist Samuel George Morton, 1799-1851 (d. Philadelphia), anthropologist, physician, studied medicine at the University of Pennsylvania (graduating in 1800) and at Edinburgh University (M.D. degree in 1823), professor of anatomy, in 1820 elected to the Academy of Natural Sciences of Philadelphia, a founder of invertebrate paleontology in the United States. His principal writings are *Crania Americana.* Philadelphia 1839, *Some Observations on the Ethnography and Archaeology of the American Aborigines.* New Haven 1846, *Synopsis of the Organic Remains of the Cretaceous Group of the United States,* etc. Philadelphia 1834 and *Crania Aegyptiaca.* Philadelphia 1844. See Whitfield J. Bell, Jr., in *Dictionary of Scientific Biography* 9: 540-541.

1981; William Stanton, *The Leopard's Spots: Scientific Attitudes Toward Race in America, 1815-1859.* Chicago 1960; Charles D. Meigs, *A Memoir of Samuel George Morton, M.D.* [Philadelphia 1851].

Mortoniella Woodson Apocynaceae

Origins:

For the American botanist Conrad Vernon Morton, 1905-1972, specialist in Gesneriaceae and Solanaceae, from 1928 at the Smithsonian Institution. His works include "A revision of *Besleria.*" *Contr. U.S. Natl. Herb.* 26(9): 395-474. 1939, "Taxonomic studies of tropical American plants. Some South American species of *Solanum.*" *Contr. U.S. Natl. Herb.* 29(1): 41-72. 1944, "The classification of *Thelypteris.*" *Amer. Fern J.* 53(4): 149-154. 1963, "The Peruvian species of *Besleria* (Gesneriaceae)." *Bull. U.S. Natl. Mus.* 38(4): 125-151. 1968, "The genera, subgenera, and sections of the Hymenophyllaceae." *Contr. U.S. Natl. Herb.* 38(5): 153-214. 1968 and *A Revision of the Argentine Species of Solanum.* Academia Nacional de Ciencias, Córdoba, Argentina 1976. See John H. Barnhart, *Biographical Notes upon Botanists.* 2: 518. 1965; T.W. Bossert, *Biographical Dictionary of Botanists Represented in the Hunt Institute Portrait Collection.* 274. 1972; S. Lenley et al., *Catalog of the Manuscript and Archival Collections and Index to the Correspondence of John Torrey.* Library of the New York Botanical Garden. 300. 1973; Ida Kaplan Langman, *A Selected Guide to the Literature on the Flowering Plants of Mexico.* Philadelphia 1964; R. Zander, F. Encke, G. Buchheim and S. Seybold, *Handwörterbuch der Pflanzennamen.* 14. Aufl. Stuttgart 1993.

Mortoniodendron Standley & Steyermark Tiliaceae

Origins:

For the American botanist Conrad Vernon Morton, 1905-1972, specialist in Gesneriaceae and Solanaceae.

Mortoniopteris Pichi Sermolli Hymenophyllaceae

Origins:

For the American botanist Conrad Vernon Morton, 1905-1972, specialist in Gesneriaceae and Solanaceae; see R.E.G. Pichi Sermolli, "Fragmenta Pteridologiae — VI." in *Webbia.* 31(1): 237-259. Apr. 1977.

Morus L. Moraceae

Origins:

Latin *morum* and Greek *moron* for a mulberry, a blackberry, Latin *morus* and Greek *mora* or *morea* for a mulberry-tree; see Carl Linnaeus, *Species Plantarum.* 986. 1753 and *Genera Plantarum.* Ed. 5. 424. 1754; H. Genaust, *Etymologisches Wörterbuch der botanischen Pflanzennamen.* 396. 1996; P. Sella, *Glossario latino emiliano.* Città del Vaticano 1937; Y. Tailfer, *La Forêt dense d'Afrique centrale.* CTA, Ede/Wageningen 1989; J. Vivien & J.J. Faure, *Arbres des Forêts denses d'Afrique Centrale.* Agence de Coopération Culturelle et Technique. Paris 1985.

Species/Vernacular Names:

M. spp.

Mexico: yaga bey zaa, yaga peyo zaa

M. alba L. (*Morus japonica* Baillon; *Morus kagayamae* Koidz.; *Morus bombycis* Koidz.; *Morus mongolica* (Bur.) Schneid.; *Morus tatarica* L.)

English: white mulberry, white-fruited mulberry, silkworm mulberry, mulberry, white mulberry tree, Mongolian mulberry

Brazil: amoreira

Japan: yama-guwa (= mountain *Morus*), kuwa

Arabic: tout, touta, tout abyad, tout helw

China: sang bai pi, sang ye, sang, pai sang, lu sang, chi sang

Nepal: kumu, kimbu kaphal

Tibetan: srin-shing-'bru

Burma: labri, mawon, ngap-set-ting, posa

India: tunt, tuntri, tut, tutri

Hawaii: kilika

South Africa: moerbei

M. australis Poir.

English: Japanese mulberry

Vietnam: dau tam, dau tau, tang

M. cathayana Hemsl.

English: Chinese mulberry

M. macroura Miq.

English: yellow mulberry

Nepal: kimbu, nambyong

Sikkim: senta, singtok

India: tut

Burma: labri, malaing, posa, tawposa, tawpweesa

M. mesozygia Stapf (*Celtis lactea* Sim; *Morus lactea* (Sim) Mildbr.) (the specific name from the Greek *mesos* 'middle' and *zygon* "a yoke")

English: African mulberry, Tongaland mulberry

Southern Africa: Afrikaanse moerbei; inDuli, umDuli (Zulu); asiThondjwana (Thonga)

West Africa: wonton

Zaire: akwandia (Uele); ndeko ya mulungu (Iba); bongola (Basakata); bonkenyama, ntama (Mai-Ndombe lake); bonkese, monkese (d. Bangala); kamwefu (Kasai); kankate (Tumba); kebuni (Baboma); kesse (Mayumbe); mie (Batere); montela (Bolobo)

Congo: kesse

Cameroon: olape, osel, ossel

Central African Republic: bondé

Yoruba: aye, ita funfun

Nigeria: aye, ewe-aye, ewe-aiye (Yoruba)

Ivory Coast: apia, difou, ndongosan, sand

Ghana: wonter, wonton

Senegal: sanda

M. microphylla Buckl.

English: Texas mulberry

M. nigra L.

English: black mulberry, common mulberry

Brazil: amora

South Africa: swart moerbei

Arabic: tout arbi

M. rubra L.

English: red mulberry

Moscharia Ruíz & Pavón Asteraceae

Origins:

Greek *moschos* "musk," referring to the fragrance; see D.S. Brakuni *et alii*, "Screening of Chilean plants for anticancer activity. I." *Lloydia.* 39(4): 225-243. Cincinnati 1976.

Moschopsis Philippi Calyceraceae

Origins:

From the Greek *moschos* "musk" and *opsis* "resembling," herbs with inulin.

Moschosma Reichb. Labiatae

Origins:

From the Greek *moschos* "musk" and *osme* "smell, odor, perfume," the musky fragrance; see Heinrich Gottlieb Ludwig Reichenbach (1793-1879), *Conspectus Regni Vegetabilis.* 171. Lipsiae 1828.

Moschoxylum A. Juss. Meliaceae

Origins:

From the Greek *moschos* "musk" and *xylon* "wood."

Moseleya Hemsley Scrophulariaceae

Origins:

To commemorate the British traveler Henry Nottidge Moseley, 1844-1890 (Dorset), naturalist, plant collector, on the Challenger Expedition, 1877 American Northwest, in 1880 Fellow of the Linnean Society and in 1877 of the Royal Society of London, 1881-1887 professor of human and comparative anatomy at Oxford. He is best remembered for his *Notes by a Naturalist on the Challenger* ... during the voyage ... round the world ... 1872-1876. London 1879, "Notes on plants collected in the islands of the Tristan d'Acunha group." *J. Linn. Soc. Bot.* 14: 377-384. 1874, *Animal Life on the Ocean Surface.* London [1886] and *Oregon: Its Resources, Climate, People, and Productions.* London 1878. See Sir Wyville Thomas Charles Thomson (*olim* Wyville Thomson) (1830-1882) and John Murray, *Reports on the Scientific Results of the Voyage of H.M.S. Challenger, During the Years 1873-1876 under the Command of Captain George S. Nares R.N., F.R.S. and the Late Captain Frank Tourle Thomson R.N.* 50 vols. London 1880-1885; Stephen Ambrose, *Undaunted Courage: Meriwether Lewis, Thomas Jefferson, and the Opening of the American West.* Simon & Schuster 1996; J.H. Barnhart, *Biographical Notes upon Botanists.* 2: 519. 1965; E.M. Tucker, *Catalogue of the Library of the Arnold Arboretum of Harvard University.* 1917-1933; H.N. Clokie, *Account of the Herbaria of the Department of Botany in the University of Oxford.* 215. 1964; G. Murray, *History of the Collections Contained in the Natural History Departments of the British Museum.* 1: 43. 1904; Mary Gunn and Leslie E. Codd, *Botanical Exploration of Southern Africa.* 254. Cape Town 1981; John Dunmore, *Who's Who in Pacific Navigation.* 245-246. Honolulu 1991; I.C. Hedge and J.M. Lamond, *Index of Collectors in the Edinburgh Herbarium.* Edinburgh 1970.

Mosenodendron R.E. Fries Annonaceae

Origins:

For the Swedish botanist Carl Wilhelm Hjalmar Mosén, 1841-1887, on Regnell's Expedition 1873-1876, author of *Moss-studier på Kolmoren.* Stockholm 1873. See John H. Barnhart, *Biographical Notes upon Botanists.* 2: 519. 1965; Gustaf Oskar Andersson Malme (*né* Andersson) (1864-1937), *Ex herbario Regnelliano.* Stockholm 1898-1901 and "Die Compositen der zweiten Regnellschen Reise. III.

Puente del Inca und Las Cuevas (Mendoza)." *Ark. Bot.* 24A(8): 58-66. 1932.

Mosla (Bentham) Buch.-Ham. ex Maxim. Labiatae

Origins:

From the vernacular name in India.

Species/Vernacular Names:

M. cavaleriei J. Léveillé

English: Cavalerie mosla

China: xiao hua qi zhu

M. chinensis Maxim.

English: Chinese mosla

China: shi xiang ru, shi xiang rou

M. dianthera (Buch.-Ham. ex Roxburgh) Maxim.

English: two-anther mosla

China: xiao yu xian cao

M. formosana Maxim.

English: Taiwan mosla

China: tai wan qi zhu

M. grosseserrata Maxim.

English: largeserrate mosla

China: qi zhu, chou su, ching pai su, ji ning, chi ning

M. lanceolata (Benth.) Maxim.

Japan: nô-hanka

M. longibracteata (C.Y. Wu & Hsuan) C.Y. Wu & H.W. Li

English: longbract mosla

China: chao bao qi zhu

M. pauciflora (C.Y. Wu) C.Y. Wu & H.W. Li

English: fewflower mosla

China: shao hua qi zhu

M. scabra (Thunberg) C.Y. Wu & H.W. Li

English: scabrous mosla

China: shi qi zhu

M. soochowensis Matsuda

English: Suchow mosla

China: su zhou qi zhu

Mosquitoxylum Krug & Urban Anacardiaceae

Origins:

From the Spanish *mosquito* and Greek *xylon* "wood."

Species/Vernacular Names:

M. jamaicense Krug & Urban

English: mosquito wood, wild mahogany, redwood

British Honduras: chichimeca, nicta

Mossia N.E. Br. Aizoaceae

Origins:

Dedicated to the British (b. Cheshire) botanist Charles Edward Moss, 1870-1930 (d. Johannesburg, South Africa), editor of *Halifax Naturalist* 1899-1901, Curator of the Cambridge Herbarium, professor of botany, plant collector, traveler, botanical explorer, in 1912 Fellow of the Linnean Society. Among his writings are *The Cambridge British Flora.* [Alfred James Willmott, 1888-1950, the editor of vol. 3] Cambridge 1914-1920, "Some natural hybrids of *Clematis, Anemone* and *Gerbera* from the Transvaal." *Proc. Linn. Soc. Lond.* 141: 36-40. 1930 and "*Pleiosepalum,* a new genus of Caryophyllaceae." *J. Bot. Lond.* 69: 65-67. 1931, husband of the English botanist Margaret Heatley (1885-1953); see J.H. Barnhart, *Biographical Notes upon Botanists.* 2: 519. 1965; E.M. Tucker, *Catalogue of the Library of the Arnold Arboretum of Harvard University.* 1917-1933; Mary Gunn and Leslie E. Codd, *Botanical Exploration of Southern Africa.* 254-255. Cape Town 1981; R.T. Gunther, *Early Science in Cambridge.* Oxford 1937; Elisabeth Leedham-Green, *A Concise History of the University of Cambridge.* Cambridge, University Press 1996; Stafleu and Cowan, *Taxonomic Literature.* 3: 601-602. 1981.

Mostacillastrum O.E. Schulz Brassicaceae

Origins:

Perhaps from the Latin *motacilla, ae* "the white water-wagtail" and *-astrum,* a Latin substantival suffix indicating incomplete resemblance.

Mostuea Didrichsen Gelsemiaceae (Loganiaceae)

Origins:

For the Danish botanist Jens Laurentius (Lorenz) Moestue Vahl, 1796-1854, plant collector, traveler, librarian, son of the Norwegian-born Danish botanist Martin Vahl (1749-1804). See Paul Gaimard, *Voyages de la Commission Scientifique du Nord, en Scandinavie, en Laponie, au Spitzberg et aux Feröe, pendant les années 1838, 1839 et 1840, sur la Corvette La Recherche,* commandée par M. Fabvre ...

Géographie physique, Géographie botanique, Botanique et Physiologie, etc. Paris [1842-1848]; John Hendley Barnhart, *Biographical Notes upon Botanists.* 3: 419. 1965; A. Lasègue, *Musée botanique de Benjamin Delessert.* Paris 1845; Stafleu and Cowan, *Taxonomic Literature.* 3: 628. 1981; Carl Frederik Albert Christensen (1872-1942), *Den danske Botaniks Historie med tilhørende Bibliografi.* Copenhagen 1924-1926.

Motherwellia F. Muell. Araliaceae

Origins:

Named after Dr. J.B. Motherwell, physician and patron of science; see Sir Ferdinand Jacob Heinrich von Mueller (1825-1896), *Fragmenta Phytographiae Australiae.* 7: 107. 1870.

Motleyia Johannson Rubiaceae

Origins:

Probably for the British botanist James Motley, d. 1859 (murdered, Borneo), plant collector in Malaysia and Borneo, with Lewis Llewellyn Dillwyn wrote *Contributions to the Natural History of Labuan.* London 1855. See John H. Barnhart, *Biographical Notes upon Botanists.* 2: 522. 1965; E.M. Tucker, *Catalogue of the Library of the Arnold Arboretum of Harvard University.* 1917-1933; Isaac Henry Burkill, in Gard. Bull. Str. Settl. 4: 128. 1927; I.H. Vegter, *Index Herbariorum.* Part II (4), *Collectors M.* Regnum Vegetabile vol. 93. 1976; I.C. Hedge and J.M. Lamond, *Index of Collectors in the Edinburgh Herbarium.* Edinburgh 1970.

Moulinsia Raf. Gramineae

Origins:

See Constantine Samuel Rafinesque (1783-1840), *Seringe Bull. Bot.* 1: 221. 1830; Elmer D. Merrill, *Index rafinesquianus.* 76. 1949.

Moultonia Balf.f. & W.W. Sm. Gesneriaceae

Origins:

Named for the English botanist John Coney Moulton, 1886-1926, Curator Sarawak Museum 1905-1915, Director Raffles Museum at Singapore 1919-1923, collected insects and plants in Borneo (August 1913, Mount Kinabalu, with P. Skene Keith, the Assistant District Officer, Kota Belud).

Among his writings are *Cicadas of Malaysia*. Singapore 1923, "A collecting expedition to Mt. Kinabalu." *The Sarawak Gazette*. 248-250. November 1st 1913, "A collecting expedition to Mt. Kinabalu." *The Sarawak Gazette*. 262-264. November 17th 1913 and "An account of the various expeditions to Mt. Kinabalu." *Sarawak Museum J.* Vol. 6: 137-176. 1915.

Moultonianthus Merr. Euphorbiaceae

Origins:
Named for the English botanist John Coney Moulton, 1886-1926, Curator of the Sarawak Museum 1905-1915, Director Raffles Museum at Singapore 1919-1923, collected insects and plants in Borneo (August 1913, Mount Kinabalu, with P. Skene Keith, the Assistant District Officer, Kota Belud).

Mourera Aublet Podostemaceae

Origins:
From a vernacular South American name.

Mouretia Pitard Rubiaceae

Origins:
For the French botanist Marcellin Mouret, 1881-1915 (Argonne), soldier; see Stafleu and Cowan, *Taxonomic Literature*. 3: 609. 1981.

Mouriri Aublet Melastomataceae (Memecylaceae)

Origins:
The native South American name for *Mouriri guianensis* Poir.

Species/Vernacular Names:
M. sp.

English: deer's horn, half crown, jug

Brazil: apiranga, goyabarana, muriri

Cuba: palo torcido, torcido, yaya cimarrona, yaya mansa

Bolivia: yabe

Mexico: yo lila, yagalancito

M. apiranga Spruce

Brazil: apiranga, uapiranga

Mouriria Juss. Melastomataceae (Memecylaceae)

Origins:
The native South American name for *Mouriri guianensis* Poir.

Mouroucoa Aublet Convolvulaceae

Origins:
From a native South American name.

Moutabea Aublet Polygalaceae

Origins:
From a native South American name, *aymoutabou*.

Mozinna Ortega Euphorbiaceae

Origins:
For the Mexican botanist José Mariano Moçiño (Moziño Suarez de Figueroa), 1757-1820 (d. Barcelona, Spain), botanical explorer, 1795-1804 with Sessé on the Botanical Expedition to Nueva España. See John H. Barnhart, *Biographical Notes upon Botanists*. 2: 499. 1965; H.N. Clokie, *Account of the Herbaria of the Department of Botany in the University of Oxford*. 213. Oxford 1964; E.M. Tucker, *Catalogue of the Library of the Arnold Arboretum of Harvard University*. 1917-1933; Rogers McVaugh, in *Dictionary of Scientific Biography* 9: 432-434. 1981; F. Boerner & G. Kunkel, *Taschenwörterbuch der botanischen Pflanzennamen*. 4. Aufl. 309, 455. Berlin & Hamburg 1989; Irving William Knobloch, compil., "A preliminary verified list of plant collectors in Mexico." *Phytologia Memoirs*. VI. Plainfield, N.J. 1983; Ida Kaplan Langman, *A Selected Guide to the Literature on the Flowering Plants of Mexico*. Philadelphia 1964; R. Zander, F. Encke, G. Buchheim and S. Seybold, *Handwörterbuch der Pflanzennamen*. 14. Aufl. Stuttgart 1993; A. Lasègue, *Musée botanique de Benjamin Delessert*. Paris 1845; Miguel Colmeiro y Penido, *La Botánica y los Botánicos de la Peninsula Hispano-Lusitana*. 1858.

Mucronea Benth. Polygonaceae

Origins:
Latin *mucro, mucronis* "a sharp point," referring to the bracts.

Species/Vernacular Names:

M. californica Benth.

English: California spineflower

M. perfoliata (A. Gray) A.A. Heller

English: perfoliate spineflower

Mucuna Adans. Fabaceae

Origins:

From *mucunã*, Brazilian (Tupi-Guarani) vernacular name for these plants; *mucunã* is also *Dioclea malacocarpa*. See Antônio Geraldo da Cunha, *Dicionário Histórico das palavras portuguesas de origem tupi*. São Paulo 1978; Carlos Toledo Rizzini, *Árvores e Madeiras do Brasil*. Rio 1977; Edmundo Navarro de Andrade, *Les Bois Indigenes de São Paulo*. São Paulo 1916; Huascar Pereira, *Apontamentos sobre madeiras do Estado de São Paulo*. São Paulo 1905; Pierre Fatumbi Verger, *Ewé: The Use of Plants in Yoruba Society*. São Paulo 1995; Celia Blanco, *Santeria Yoruba*. Caracas 1995; Maria Helena Farelli, *Plantas que curam e cortam feitiços*. Rio de Janeiro 1988; William W. Megenney, *A Bahian Heritage*. University of North Carolina at Chapel Hill 1978.

Species/Vernacular Names:

M. sp.

Peru: añuje caspi, biik, huahuasencca, arari, crista de mutum, olho-de-boi, pucacuru sacha, 'unshin xama

Panama: kongi

Tibetan: gla-gor zho-sha, mkhal-ma zho-sha

M. spp.

Yoruba: yerepe odan, irepe odan

M. altissima DC. (see Eurico Teixeira da Fonseca, "Plantas medicinales brasileñas." *R. Flora Medicinal*. Rio de Janeiro. 6(2): 67-81. 1939; Paulo Occhiomi, "Estudos sobre plantas tóxicas do Brasil e a necessidade de sua sistematização." *R. Flora Medicinal*. 20(1/6): jan./jun. Rio de Janeiro 1953)

Brazil: mucunã

M. bennettii F. Muell.

English: New Guinea creeper

M. birdwoodiana Tutcher

English: white-flower mucuna

M. coriacea Baker

English: buffalo bean, fire bean, hell-fire bean

S. Rhodesia: hurukuru, uRiri, uReri

Southern Africa: brandboontjie, jeukpeul; chiriridzi, hurukuru, uliri, ureri (Shona)

M. elliptica (Ruíz & Pav.) DC.

Peru: llamapañani, llama pañaui

M. flagellipes Vogel ex Benth.

Congo: tsoko-mbele

M. gigantea (Willd.) DC. (*Dolichos giganteus* Willd.)

English: sea bean, burny bean

Hawaii: ka'e'e, ka'e'e'e

Malaya: kachang rimau

M. hainanensis Hayata

English: Hainan mucuna

M. nigricans (Lour.) Steudel (*Citta nigricans* Lour.)

Japan: Kashô-modama, hime-wani-guchi (wani = crocodile)

The Philippines: nipay, hipoi, ipa

M. novoguineensis R. Scheffer

English: New Guinea creeper

M. pruriens (L.) DC.

English: velvet bean, cow-itch, cowage

Yoruba: ejokun, yerebe, eesin, esinsin, esise, ewe ina, irepe, werepe

Nigeria: karara, yerepe, esisi, ighekpe

India: alkusa, itika, wakmi

Malaya: kachang babi, kekaras gatal

M. pruriens (L.) DC. var. *utilis* (Wight) Burck (*Carpopogon capitatum* Roxb.; *Carpopogon niveum* Roxb.; *Marcanthus cochinchinensis* Lour.; *Mucuna aterrima* (Piper & Tracy) Holland [Latin *ater, tra, trum* "black"]; *Mucuna cochinchinensis* (Lour.) A. Chev.; *Mucuna deeringiana* (Bort) Merr.; *Mucuna pachylobia* (Piper & Tracy) Rock; *Mucuna pruriens* (L.) DC. var. *biflora* Trimen; *Mucuna utilis* Wallich ex Wight; *Mucuna velutina* Hassk.; *Stizolobium aterrimum* Piper & Tracy; *Stizolobium capitatum* (Roxb.) Kuntze; *Stizolobium cinerium* Piper & Tracy; *Stizolobium deeringianum* Bort; *Stizolobium hassjoo* Piper & Tracy; *Stizolobium microspermum* Piper; *Stizolobium niveum* (Roxb.) Kuntze; *Stizolobium pachylobium* Piper & Tracy; *Stizolobium pruritum* (Hook.) Piper & Tracy subsp. *maculatum* Piper; *Stizolobium pruritum* (Hook.) Piper & Tracy subsp. *officinale* Piper; *Stizolobium pruritum* (Hook.) Piper & Tracy subsp. *biflorum* (Trimen) Piper; *Stizolobium utile* (Wallich ex Wight) Piper & Tracy; *Stizolobium velutinum* (Hassk.) Piper & Tracy)

English: Lyon-bean, velvet bean, Bengal velvet bean, cowage velvet bean, Florida velvet bean, Florida bean, Benghal bean, Bengal bean, Mauritius velvet bean, Yokohama velvet bean

China: li tou, hu tou

Japan: has-sho-mame, nabarume, Yokohama bean

M. rostrata Bentham (*Stizolobium rostratum* (Bentham) Kuntze)

Peru: aguacenqua, ahuacinca, ahuacincca, ancacjsillon, habilla, murcu huasca, llamacñahui, llamapa ñahui, ojo de llama, sachavaca ñahui

M. sempervirens Hemsl.

English: evergreen mucuna

M. urens (L.) DC. (*Dolichos urens* L.) (see Caminhoá, "Mucunan ou mucuná, comunicação feita pelo Conselheiro Caminhoá à Academia Imperial de Medicina do Rio de Janeiro." *R. Flora Medicinal.* 6(2): 67-81. Rio de Janeiro 1939 and "Mucunan ou mucuná." *R. Flora Medicinal.* 6(3): 143-149. Rio de Janeiro 1939; Joaquim Mas-Guindal, "Plantas medicinales y tintoreas." *R. da Associação Brasileira de Farmacêuticos.* 17(9): 394-397. Rio de Janeiro 1936)

English: sea bean, cow-itch plant, sheeps-eye

Brazil: mucunã

Muehlbergella Feer Campanulaceae

Origins:

After the Swiss botanist Friedrich Mühlberg (Muehlberg), 1840-1915, geologist, high school teacher, author of *Der Kreislauf der Stoffe auf der Erde.* 1886. See John H. Barnhart, *Biographical Notes upon Botanists.* 2: 522. 1965; E.M. Tucker, *Catalogue of the Library of the Arnold Arboretum of Harvard University.* 1917-1933.

Muehlenbeckia Meissner Polygonaceae

Origins:

Named for the Alsatian botanist Heinrich Gustav (Gustave) Muehlenbeck, 1798-1845, physician, author of *Dissertation sur la docimasie pulmonaire.* Strasbourg [1799], correspondent of the Alsatian physician and botanist Jean Baptiste Mougeot (1776-1858), and a friend of the Alsatian bryologist Wilhelm Philipp (Guillaume Philippe) Schimper (1808-1880). See J.H. Barnhart, *Biographical Notes upon Botanists.* 2: 522. 1965; Carl Friedrich Meissner (1800-1874), *Plantarum vascularium genera.* Lipsiae 1: 316. 1841; F. Boerner & G. Kunkel, *Taschenwörterbuch der botanischen Pflanzennamen.* 4. Aufl. 137. Berlin & Hamburg 1989; Georg Christian Wittstein, *Etymologisch-botanisches Handwörterbuch.* 597f. Ansbach 1852.

Species/Vernacular Names:

M. adpressa (Labill.) Meissner (*Muehlenbeckia adpressa* (Labill.) Meissn. var. *rotundifolia* Benth.; *Polygonum adpressum* Labill.)

Australia: climbing lignum, native sarsaparilla, Macquarie vine

M. axillaris (Hook.f.) Walp.

Australia: matted lignum

M. complexa (A. Cunn.) Meissner

Australia: maidenhair vine

M. cunninghamii (Meissner) F. Muell. (*Polygonum cunninghamii* Meissner)

Australia: lignum (a corruption of *Polygonum*), tangled lignum

M. diclina (F. Muell.) F. Muell.

Australia: twiggy lignum, slender lignum, weeping lignum

M. florulenta Meissner

Australia: lignum, tangled lignum

M. gracillima Meissner

Australia: slender lignum

M. gunnii (Hook.f.) Endl.

Australia: coastal lignum, Macquarie vine, native sarsaparilla

Muehlenbergia Hedwig Gramineae

Origins:

Orthographic variant of *Muhlenbergia* Schreber; see *Genera Plantarum.* 40. 1806.

Muellera L.f. Fabaceae

Origins:

After the Danish (b. Copenhagen) scientist Otto Friedrich (Fredrik, Frederik, Friderich, Fridrich) Müller, 1730-1784 (d. Copenhagen), naturalist, zoologist, linguist, botanist, microscopist, book collector, 1773-1782 one of the editors of *Flora danica.* His writings include *Flora Fridrichsdalina.* Argentorati 1767, *Animalcula infusoria fluviatilia et marina*, quae detexit, systematice descripsit et ad vivum delineari curavit O.F.M. Opus cura O. Fabricii. [First edition, posthumously published.] Copenhagen 1786, *Entomostraca seu Insecta Testacea;* quae in aquis Daniae et Norvegiae reperit, descripsit et iconibus illustravit O.F.M. Lipsiae et Havniae 1785, *Fauna Insectorum Fridrichsdalina.* Hafniae et Lipsiae 1764, *Vermium terrestrium et fluviatilium, seu animalium infusoriorum, helminthicorum et testaceorum, non marinorum, succincta historia.* Havniae et Lipsiae 1773, 1774 and *Zoologia Danica.* Havniae et Lipsiae 1779-1784. See E. Snorrason, in *Dictionary of Scientific Biography* 9: 574-576. 1981; John H. Barnhart, *Biographical Notes upon Botanists.* 2: 525. 1965; R. Zander, F. Encke, G. Buchheim and S. Seybold, *Handwörterbuch der Pflanzennamen.* 14. Aufl. Stuttgart 1993; Garrison and Morton, *Medical Bibliography.* 2466. 1961; T.W. Bossert,

Biographical Dictionary of Botanists Represented in the Hunt Institute Portrait Collection. 277. 1972; Jonas C. Dryander, *Catalogus bibliothecae historico-naturalis Josephi Banks.* 364. London 1796-1800; A. Lasègue, *Musée botanique de Benjamin Delessert.* 532. Paris 1845; Ethelyn Maria Tucker, *Catalogue of the Library of the Arnold Arboretum of Harvard University.* Cambridge, Massachusetts 1917-1933; C.F.A. Christensen, *Den danske Botaniks Historie med tilhørende Bibliografi.* Copenhagen 1924-1926; Frans A. Stafleu, *Linnaeus and the Linnaeans.* The spreading of their ideas in systematic botany, 1735-1789. Utrecht 1971.

Muelleranthus Hutchinson Fabaceae

Origins:

For the famous German-born (Rostock) Australian scientist Baron Sir Ferdinand Jacob (Jakob) Heinrich von Mueller (Müller), 1825-1896 (d. Melbourne), botanist, pharmacist, botanical explorer, plant collector, 1847 to Australia, 1856 collected with A.C. Gregory in Kimberley, from 1857-1873 Director of the Melbourne Botanical Garden, 1852-1896 Government botanist Victoria, 1859 Fellow of the Linnean Society, 1861 Fellow of the Royal Society, a prolific writer. Among his very numerous writings are *The Plants Indigenous to the Colony of Victoria.* Melbourne 1860-1865, *The Fate of Dr. Leichhardt,* and a proposal new search for his party. [Melbourne 1865] and *Key to the System of Victorian Plants.* Melbourne 1885-1888, contributed to George Bentham (1800-1884), *Flora Australiensis.* London 1863-1878. See J.H. Barnhart, *Biographical Notes upon Botanists.* 2: 524. 1965; Stafleu and Cowan, *Taxonomic Literature.* 3: 615-625. 1981; Ray Desmond, *Dictionary of British & Irish Botanists and Horticulturists.* 505. London 1994; R. Zander, F. Encke, G. Buchheim and S. Seybold, *Handwörterbuch der Pflanzennamen.* 14. Aufl. Stuttgart 1993; G. Murray, *History of the Collections Contained in the Natural History Departments of the British Museum.* 1: 170. 1904; Ida Kaplan Langman, *A Selected Guide to the Literature on the Flowering Plants of Mexico.* Philadelphia 1964; M. Willis, *By Their Fruits. A Life of Ferdinand von Mueller, Botanist and Explorer.* Sydney, London 1949; E.D. Merrill, in *Bernice P. Bishop Mus. Bull.* 144: 139-140. 1937; I.H. Vegter, *Index Herbariorum.* Part II (7), *Collectors T-Z.* Regnum Vegetabile vol. 117. 1988; T.W. Bossert, *Biographical Dictionary of Botanists Represented in the Hunt Institute Portrait Collection.* 276. 1972; Ernest Nelmes and William Cuthbertson, *Curtis's Botanical Magazine Dedications, 1827-1927.* 135-136. [1931]; H.N. Clokie, *Account of the Herbaria of the Department of Botany in the University of Oxford.* 215. Oxford 1964; I.C. Hedge and J.M. Lamond, *Index of Collectors in the Edinburgh Herbarium.* Edinburgh 1970; Merle A. Reinikka, *A History of the Orchid.* Timber

Press 1996; Jonathan Wantrup, *Australian Rare Books, 1788-1900.* Hordern House, Sydney 1987.

Species/Vernacular Names:

M. stipularis (J. Black) A. Lee (*Ptychosema stipulare* J. Black)

English: sand pea

M. trifoliolatus (F. Muell.) A. Lee (*Ptychosema trifoliolatum* F. Muell.)

English: spinifex pea

Muellerargia Cogn. Cucurbitaceae

Origins:

After the Swiss botanist Jean (Johannes) Mueller, called Argoviensis (of Aargau, Switzerland), 1828-1896, owner of a lichenological herbarium, from 1851 to 1869 Curator of the Candolle herbarium, from 1869 to 1896 Curator of the B. Delessert herbarium, from 1870 to 1874 Director of the Genève Botanic Garden, from 1871 to 1889 professor of botany. His works include *Monographie de la famille des Résédacées.* Zürich 1857, *Principes des classification des lichens.* Genève 1862, *Lichenologie Beiträge.* Regensburg and Bonn 1874-1891, contributed to C.F.P. von Martius *Flora Brasiliensis* (Apocynaceae, Rubiaceae, Euphorbiaceae). See John H. Barnhart, *Biographical Notes upon Botanists.* 2: 524. 1965; Stafleu and Cowan, *Taxonomic Literature.* 3: 628-635. 1981; E.D. Merrill, "Bibliography of Polynesian botany." *Bernice P. Bishop Mus. Bull.* 144: 140. 1937 and *Contr. U.S. Natl. Herb.* 30(1): 222. 1947; Sir Isaac Bayley Balfour (1853-1922), *Botany of Socotra.* Edinburgh, London 1888; T.W. Bossert, *Biographical Dictionary of Botanists Represented in the Hunt Institute Portrait Collection.* 277. 1972; S. Lenley et al., *Catalog of the Manuscript and Archival Collections and Index to the Correspondence of John Torrey.* Library of the New York Botanical Garden. 464. 1973; A. de Candolle and A.C.P. de Candolle, *Monographiae Phanerogamarum.* 3: 630. (Jun.) 1881; Ida Kaplan Langman, *A Selected Guide to the Literature on the Flowering Plants of Mexico.* Philadelphia 1964.

Muellerina Tieghem Loranthaceae

Origins:

Honoring Australia's greatest 19th century botanist Sir Ferdinand Jacob Heinrich von Mueller, 1825-1896, pharmacist, botanical explorer, plant collector, in 1847 to Australia, first Government botanist of Victoria, from 1857 to 1873 Director of the Melbourne Botanical Garden, in 1859 Fellow of the Linnean Society and in 1861 of the Royal Society of

London. Among his very numerous publications are *The Plants Indigenous to the Colony of Victoria*. Melbourne 1860-1865, *Fragmenta Phytographiae Australiae*. Melbourne 1858 to 1882, and *Key to the System of Victorian Plants*. Melbourne 1885-1888, contributed to George Bentham (1800-1884), *Flora Australiensis*. London 1863-1878. See John H. Barnhart, *Biographical Notes upon Botanists*. 2: 524. 1965; M. Willis, *By Their Fruits. A Life of Ferdinand von Mueller, Botanist and Explorer*. Sydney 1949; I.H. Vegter, *Index Herbariorum*. Part II (7), *Collectors T-Z*. Regnum Vegetabile vol. 117. 1988.

Species/Vernacular Names:

M. bidwillii (Benth.) Barlow

Australia: callitris mistletoe

M. celastroides (Schultes & J.H. Schultes) Tieghem

Australia: coast mistletoe

M. eucalyptoides (DC.) Barlow (*Loranthus eucalyptoides* DC.)

Australia: creeping mistletoe

M. myrtifolia (Benth.) Barlow

Australia: myrtle-leaf mistletoe

Muenteria Seemann Bignoniaceae

Origins:

For the German botanist Johann Andreas Heinrich August Julius Münter, 1815-1885; see John H. Barnhart, *Biographical Notes upon Botanists*. 2: 526. 1965; T.W. Bossert, *Biographical Dictionary of Botanists Represented in the Hunt Institute Portrait Collection*. 278. 1972; Ida Kaplan Langman, *A Selected Guide to the Literature on the Flowering Plants of Mexico*. Philadelphia 1964; Ethelyn (Daliaette) Maria Tucker, *Catalogue of the Library of the Arnold Arboretum of Harvard University*. Cambridge, Massachusetts 1917-1933.

Muhlenbergia Schreber Gramineae

Origins:

For the American Rev. Gotthilf Heinrich (Henry) Ernest (Ernst) Muhlenberg (Mühlenberg), 1753-1815 (Lancaster), Lutheran minister, amateur botanist. Among his works are *Catalogus plantarum Americae septentrionalis*. Lancaster 1813 and *Descriptio uberior graminum* et plantarum calamariarum Americae septentrionalis indigenarum et cicurum. Philadelphiae 1817. See John Christopher Schwab, *The Descendants of Henry Melchior Muhlenberg*. [New Haven, Conn. 1911]; R. Zander, F. Encke, G. Buchheim and S. Seybold, *Handwörterbuch der Pflanzennamen*.

14. Aufl. 755. Stuttgart 1993; William Jay Youmans, ed., *Pioneers of Science in America*. New York 1896; J.H. Barnhart, *Biographical Notes upon Botanists*. 2: 523. 1965; Jeannette E. Graustein, *Thomas Nuttall, Naturalist. Explorations in America, 1808-1841*. Harvard University Press 1967; T.W. Bossert, *Biographical Dictionary of Botanists Represented in the Hunt Institute Portrait Collection*. 276. 1972; A. Lasègue, *Musée botanique de Benjamin Delessert*. 319. Paris 1845; S. Lenley et al., *Catalog of the Manuscript and Archival Collections and Index to the Correspondence of John Torrey*. Library of the New York Botanical Garden. 301. 1973; E.M. Tucker, *Catalogue of the Library of the Arnold Arboretum of Harvard University*. 1917-1933; J. Ewan, ed., *A Short History of Botany in the United States*. New York and London 1969; J.W. Harshberger, *The Botanists of Philadelphia and Their Work*. 92-97. 1899; William Darlington, *Memorials of John Bartram and Humphry Marshall*. 466-474. Philadelphia 1849; Israel Smith Clare, *A Brief History of Lancaster County*. Argus 1892; Jonas C. Dryander, *Catalogus bibliothecae historico-naturalis Josephi Banks*. London 1796-1800; Ida Kaplan Langman, *A Selected Guide to the Literature on the Flowering Plants of Mexico*. Philadelphia 1964; William Darlington, *Reliquiae Baldwinianae*. Philadelphia 1843.

Species/Vernacular Names:

M. andina (Nutt.) Kunth

English: foxtail muhly

M. arsenei A. Hitchc.

English: tough muhly

M. asperifolia (Nees & Meyen) L. Parodi

English: scratch grass

M. californica Vasey

English: California muhly

M. filiformis (Thurber) Rydb.

English: pull-up muhly

M. fragilis Swallen

English: delicate muhly

M. macroura (Kunth) A. Hitchc.

Mexico: zakaton, zacate de carbonero, eb, malinalli, goba chita, coba chita

M. microsperma (DC.) Kunth

English: littleseed muhly

M. montana (Nutt.) A. Hitchc.

English: mountain muhly

M. pauciflora Buckley

English: few-flowered muhly

M. richardsonis (Trin.) Rydb.

English: mat muhly

M. rigens (Benth.) A. Hitchc.

English: deer grass

M. schreberi S. Gmelin

English: nimblewell

Muilla S. Watson ex Benth. Alliaceae (Liliaceae)

Origins:
Anagram of *Allium*.

Muiria C. Gardner Rutaceae

Origins:
After the Australian plant collector Sir Thomas Muir, born 1899, farmer at Warrungup (Borden, Western Australia), a companion of Charles A. Gardner (1896-1970); see Charles Austin Gardner, in *Journal of the Royal Society of Western Australia*. 19: 83. 1933.

Muiria N.E. Br. Aizoaceae

Origins:
For the Scottish (b. Castle Douglas) naturalist John Muir, 1874-1947 (d. Riversdale, C.P., South Africa), physician, plant collector at the Cape, his wife *née* Susanna Steyn; see Gordon Douglas Rowley, *A History of Succulent Plants*. Strawberry Press, Mill Valley, California 1997; Mary Gunn and Leslie E. Codd, *Botanical Exploration of Southern Africa*. 256. Cape Town 1981; Stafleu and Cowan, *Taxonomic Literature*. 3: 658. 1981; Ray Desmond, *Dictionary of British & Irish Botanists and Horticulturists*. 506. London 1994.

Muiriantha C. Gardner Rutaceae

Origins:
After the Australian plant collector Sir Thomas Muir, born 1899, farmer at Warrungup (Borden, Western Australia), a companion of Charles A. Gardner (1896-1970); see Charles Austin Gardner, in *Journal of the Royal Society of Western Australia*. 27: 181. (Aug.) 1942.

Mukia Arn. Cucurbitaceae

Origins:
From a Malayalam name, *mucca-piri* (*mucca* means three-fourth and *piri* spring, possibly referring to the curled tendrils), applied by van Rheede in his *Hortus Indicus Mala-* *baricus*. 8: t. 13. 1688 for *Mukia scabrella*; see Robert Wight (1796-1872), in *The Madras Journal of Literature and Science*. 12: 50. 1840; M.P. Nayar, *Meaning of Indian Flowering Plant Names*. 231. Dehra Dun 1985.

Species/Vernacular Names:
M. maderaspatana (L.) M. Roem. (*Cucumis maderaspatanus* L.; *Melothria maderaspatana* (L.) Cogn.; *Bryonia scabrella* L.; *Bryonia maderaspatana* L.; *Mukia scabrella* (L.f.) Arn.; *Melothria scabrella* (L.) Arn.)

English: mukia

Japan: sango-ju-suzume-uri

Yoruba: ori oka

M. micrantha (F. Muell.) F. Muell. (*Cucurbita micrantha* F. Muell.; *Zehneria micrantha* (F. Muell.) F. Muell.; *Melothria micrantha* (F. Muell.) Cogn.; *Cucumis muelleri* Naudin; *Melothria muelleri* (Naudin) Benth.)

Australia: desert cucumber, mallee cucumber

Mulgedium Cass. Asteraceae

Origins:
From the Latin *mulgeo* "to milk," Greek *amelgo*, referring to the closely related genus *Lactuca*.

Mullerina Tieghem Loranthaceae

Origins:
Orthographic variant of *Muellerina* Tieghem.

Multidentia Gilli Rubiaceae

Origins:
From the Latin *multi-* "many" and *dens, dentis* "a tooth."

Muluorchis J.J. Wood Orchidaceae

Origins:
From a mountain in Sarawak.

Munbya Boissier Boraginaceae

Origins:
For the British botanist Giles Munby, 1813-1876 (Surrey), physician, in Algeria, plant collector; see John H. Barnhart, *Biographical Notes upon Botanists*. 2: 528. 1965.

Munbya Pomel Fabaceae

Origins:

Named for the British botanist Giles Munby, 1813-1876 (Surrey), physician, M.D. Montpellier, 1839-1859 (or 1861) in Algeria, plant collector, author of *Catalogus plantarum in Algeria sponte nascentium*. Oran 1859. See John H. Barnhart, *Biographical Notes upon Botanists*. 2: 528. 1965; T.W. Bossert, *Biographical Dictionary of Botanists Represented in the Hunt Institute Portrait Collection*. 277. 1972; A. Lasègue, *Musée botanique de Benjamin Delessert*. 314. Paris 1845; J.D. Milner, *Catalogue of Portraits of Botanists Exhibited in the Museums of the Royal Botanic Gardens*. Royal Botanic Gardens, Kew, London 1906; I.C. Hedge and J.M. Lamond, *Index of Collectors in the Edinburgh Herbarium*. Edinburgh 1970; E.M. Tucker, *Catalogue of the Library of the Arnold Arboretum of Harvard University*. 1917-1933; Michel Charles Durieu de Maisonneuve, *Exploration scientifique de l'Algérie pendant les années 1840, 1841, 1842* publiée par ordre du gouvernement et avec le concours d'une commission académique. Sciences physiques. Botanique par MM. Bory de St.-Vincent et Durieu de Maisonneuve membres de la Commission scientifique d'Algérie. Paris 1846-1855[-1869]; A. White and B.L. Sloane, *The Stapelieae*. Pasadena 1937.

Mundia Kunth Polygalaceae

Origins:

The genus was named to honor the German (b. Berlin) pharmacist Johannes Ludwig Leopold Mund [often spelled Mundt], 1791-1831 (d. at the Cape), botanist, plant collector in South Africa from about 1816 onward; see Mary Gunn and Leslie E. Codd, *Botanical Exploration of Southern Africa*. 257. Cape Town 1981.

Mundulea (DC.) Benth. Fabaceae

Origins:

The meaning of the name seems obscure, it could originate from *Munduli*, an African name also used for *Arachis hypogaea* or *Apios tuberosa*, referring to the whole look of the plant, or from Latin *mundulus* "neat, trim," *mundule* "neatly, trimly"; according to Georg C. Wittstein it could be the diminutive of *Mundia* (Polygalaceae); see H. Genaust, *Etymologisches Wörterbuch der botanischen Pflanzennamen*. 398. 1996.

Species/Vernacular Names:

M. sericea (Willd.) A. Chev. (*Cytisus sericeus* Willd.; *Mundulea suberosa* (DC.) Benth.; *Robinia suberosa* Roxb.; *Tephrosia suberosa* DC.)

English: cork bush, silver bush, Rhodesia silver leaf, Rhodesian silver leaf

Southern Africa: kurkbos, blou-ertjieboom, olifantshout (meaning that which resists the elephants), visboontjie, visgif; umHlalantethe, uSekwane, umSindandlovu, umaMentabeni (Zulu); umSinndandlovana (Swazi); mosita-tlou (Hebron: central Transvaal); mukunda-ndou (Venda)

Nigeria: igun, lakuta

Angola: onkongo (Lunyaneka); ongeke (Tjiherero)

Namibia: omubanganyana or omumbanganyana (Kwanyama)

Zimbabwe: umSece, inKizaemaqaqa, umSindandhlovana, umPandula

Malawi: lusyunga (Chichewa, Nyanja, Yao; see J. Clyde Mitchell, *The Yao Village*: A study of the social structure of a Nyasaland tribe. Manchester 1956); nandolo (Ngoni; see Margaret Read, *Children of Their Fathers*. London 1959, *The Ngoni of Nyasaland*. London 1956 and *Native Standards of Living and African Culture Change*. London 1938); chiguluka (Yao)

Munnozia Ruíz & Pav. Asteraceae

Origins:

Named for Juan Bautista Muñoz, author of *Historia del Nuevo Mundo*. Madrid 1793. See A. Ballesteros y Beretta, "Don Juan Bautista Muñoz. La Historia del Nuevo Mundo." *Revista de Indias*. año 3. no. 10. 1942; Justo Pastor Fuster, *Copia de los manuscritos que recogió D. Juan Bautista Muñoz en sus viages y se entregaron en su muerte a Su Magestad*. [in his *Biblioteca valenciana*. II. 1827-1830].

Munroa Torrey Gramineae

Origins:

Named for Sir William Munro, 1818-1880 (Somerset), British botanist, plant collector, agrostologist, 1834-1838 India, 1847 Kashmir, 1870-1875 Barbados, in 1840 a Fellow of the Linnean Society, wrote "A monograph of the Bambusaceae, including descriptions of all the species." *Trans. Linn. Soc. London*. 26: 1-157. 1868. See T.W. Bossert, *Biographical Dictionary of Botanists Represented in the Hunt Institute Portrait Collection*. 278. 1972; E. Bretschneider, *History of European Botanical Discoveries in China*. 1981; H.N. Clokie, *Account of the Herbaria of the Department of Botany in the University of Oxford*. 216. Oxford 1964; Warren R. Dawson, *The Banks Letters*, a Calendar of the Manuscript Correspondence of Sir Joseph Banks. London 1958; E.M. Tucker, *Catalogue of the Library of the Arnold Arboretum of Harvard University*. 1917-1933; R. Zander, F. Encke, G. Buchheim and S. Seybold, *Handwörterbuch der Pflanzennamen*. 14. Aufl. Stuttgart 1993; Mea Allan, *The Hookers of Kew*. London 1967;

Leonard Huxley, *Life and Letters of Sir J.D. Hooker.* 199. London 1918; Isaac Henry Burkill, *Chapters on the History of Botany in India.* Delhi 1965; Ralph Randles Stewart, *An Annotated Catalogue of the Vascular Plants of West Pakistan and Kashmir.* Karachi 1972; Ignatz Urban, ed., *Symbolae Antillanae.* 3: 91. Berlin 1902; I.C. Hedge and J.M. Lamond, *Index of Collectors in the Edinburgh Herbarium.* Edinburgh 1970.

Species/Vernacular Names:

M. squarrosa (Nutt.) Torrey

English: false buffalo grass

Munroidendron Sherff Araliaceae

Origins:

After the naturalist George C. Munro, 1866-1963, a pioneer in Hawaiian ornithology, botany and horticulture, plant collector in the Hawaiian Islands. Among his principal writings are "Forest covers." *Hawaiian Forester Agric.* 19: 45-46. 1922, "Windbreaks for wind eroded lands." *Hawaiian Forester Agric.* 26: 124-125. 1929, "Birds of Hawaii and adventures in bird study. Bird islands off the coast of Oahu." *Elepaio.* 1: 46-49. 1941, "My first bird walks in Hawaii." *Elepaio.* 5: 6-7, 13-15. 1944 and "Hawaiian endemic flowering plants suitable for culture in gardens." *Elepaio.* 9: 14-15, 16-18, 23-25. 1948; see I.H. Vegter, *Index Herbariorum.* Part II (4), *Collectors M.* Regnum Vegetabile vol. 93. 1976.

Munronia Wight Meliaceae

Origins:

For Sir William Munro, 1818-1880, British botanist, plant collector, 1834-1838 India, 1847 Kashmir, 1870-1875 Barbados. See T.W. Bossert, *Biographical Dictionary of Botanists Represented in the Hunt Institute Portrait Collection.* 278. 1972; H.N. Clokie, *Account of the Herbaria of the Department of Botany in the University of Oxford.* 216. Oxford 1964; Warren R. Dawson, *The Banks Letters,* a Calendar of the Manuscript Correspondence of Sir Joseph Banks. London 1958; E.M. Tucker, *Catalogue of the Library of the Arnold Arboretum of Harvard University.* 1917-1933; Mea Allan, *The Hookers of Kew.* London 1967; Leonard Huxley, *Life and Letters of Sir J.D. Hooker.* 199. London 1918; Isaac Henry Burkill, *Chapters on the History of Botany in India.* Delhi 1965; Ralph Randles Stewart, *An Annotated Catalogue of the Vascular Plants of West Pakistan and Kashmir.* Karachi 1972; Ignatz Urban, ed., *Symbolae Antillanae.* 3: 91. Berlin 1902.

Muntingia L. Tiliaceae

Origins:

Named after the Dutch botanist Abraham Munting (Muntingius), 1626-1683, professor of medicine at Gröningen, attempted to identify a plant called *Britannica* by the ancient authors, which was used to cure scurvy. Among his many works are *Aloidarium.* [Amsterdam] 1680 and *De vera antiquorum Herba Britannica.* Amstelodami 1681. See *Rosa Leonina quam ... nuptiarum auspiciis ... A. Muntinck ... et E.A. Gabbema Sponsae fingebat ... & sacrabat Poetice Latina Anacreontis Umbra.* [Nuptial ode.] Leovardiae 1658; C.H. Andreas, *Hortus Muntingiorum.* [With special reference to Hindrick, Abraham and Albertus Munting.] Gröningen, Djakarta 1953; François [Franciscus] Kiggelaer (1648-1722), *Phytographia curiosa.* (J. Mensigae oratio funebris in obitum A. Muntingii.). Amsterdam & Leyden 1702-1713; Georg Christian Wittstein, *Etymologisch-botanisches Handwörterbuch.* 598f. Ansbach 1852; H.N. Clokie, *Account of the Herbaria of the Department of Botany in the University of Oxford.* 216. Oxford 1964; Jonas C. Dryander, *Catalogus bibliothecae historico-naturalis Josephi Banks.* London 1796-1800; E.M. Tucker, *Catalogue of the Library of the Arnold Arboretum of Harvard University.* 1917-1933; Mariella Azzarello Di Misa, a cura di, *Il Fondo Antico della Biblioteca dell'Orto Botanico di Palermo.* 199. Regione Siciliana, Palermo 1988; Ida Kaplan Langman, *A Selected Guide to the Literature on the Flowering Plants of Mexico.* Philadelphia 1964; Gilbert Westacott Reynolds (1895-1967), *The Aloes of South Africa.* 76, 77, 204. Balkema, Rotterdam 1982.

Species/Vernacular Names:

M. calabura L.

English: cherry tree, Jamaican cherry, jam tree

Peru: bolina, bolaina, ccoillor-ppanchu, guinda yumanasa, iumanasa, mullaca huayo, yumanasa, rupiña, tomaque

Tropical America: calabura, calabure, majaguillo, pacito, vijaguillo

Mexico: huiz lan

Guadeloupe: bois ramier, bois de soie

Malaya: buah cheri, kerukup siam

The Philippines: datiles, ratiles, cereza, zanitas, seresa

Sri Lanka: jam tree

Munzothamnus Raven Asteraceae

Origins:

For the American botanist Philip Alexander Munz, 1892-1974, Director of the Santa Ana Botanic Garden and Rancho Santa Ana Botanic Garden, professor of botany, specialist in Onagraceae. His writings include "Las Onagráceas de la Argentina." *Physis.* (Buenos Aires) 11: 266-292. 1933, "Las Onagráceas de Chile." *Farm. Chilena.* 8: 64-66, 82-83, 106-109, 123-125. 1934, "A revision of the genus *Fuchsia.*" *Proc. Calif. Acad. Sci.* 25(1): 1-138. 1943, *A California Flora.* [with David D. Keck] Berkeley, Los Angeles 1959 and *A Flora of Southern California.* Berkeley, Los Angeles, London 1974, with I.M. Johnston wrote "The *Oenotheras* of northwestern South America." *Contr. Gray Herb.* 75: 15-23. 1925. See S. Carlquist, "Philip A. Munz, botanist and friend." *Aliso.* 8(3): 211-220. 1975; J.H. Barnhart, *Biographical Notes upon Botanists.* 2: 530. 1965; Ida Kaplan Langman, *A Selected Guide to the Literature on the Flowering Plants of Mexico.* Philadelphia 1964; T.W. Bossert, *Biographical Dictionary of Botanists Represented in the Hunt Institute Portrait Collection.* 278. 1972; S. Lenley et al., *Catalog of the Manuscript and Archival Collections and Index to the Correspondence of John Torrey.* Library of the New York Botanical Garden. 301. 1973; R. Zander, F. Encke, G. Buchheim and S. Seybold, *Handwörterbuch der Pflanzennamen.* 14. Aufl. Stuttgart 1993; Joseph Ewan, *Rocky Mountain Naturalists.* The University of Denver Press 1950; Irving William Knobloch, compiled by, "A preliminary verified list of plant collectors in Mexico." *Phytologia Memoirs.* VI. Plainfield, N.J. 1983.

Muraltia DC. Polygalaceae

Origins:

After the Swiss surgeon Johannes von Muralt, 1645-1733, botanist, anatomist, in 1671 received an M.D. from the University of Basel, professor of medicine at Zürich, author of *Vade Mecum Anatomicum.* Amstelaedami 1785 and *Physicae specialis pars quarta. Botanologia, seu Helvetiae Paradisus.* Tiguri [Zürich] 1710; he often used the pseudonyms of Johannes Eutichus de Claromonte and Aretaeus [Aretaeus was a Greek physician of Cappadocia, 2nd century], using the last one he published *Academiae Naturae Curiosorum Aretaei ... Zoologia seu animalium contemplatio physica,* etc. Tiguri 1709. See J. Finsler, *Bemerkungen aus dem Leben des J. Muralt ...* nebst einem vollständingen Verzeichnisse der von ihm herausgegeben Schriften. Zürich 1833.

Species/Vernacular Names:

M. heisteria (L.) DC. (*Polygala heisteria* L.)
English: furze muraltia, African furze

Muratina Maire Chenopodiaceae

Origins:

Named for Marc Murat, 1909-1940, explorer and plant collector, 1935 from Chad to Tibesti, 1936-1938 Mauritania,

author of "Végétation de la zone prédésertique en Afrique centrale (Région du Tchad)." *Bull. Soc. Hist. Nat. Afr. Nord.* 1937 and "La Végétation du Sahara occidental en Mauritanie." *C.R. Acad. Sc.* 1937; see Auguste Jean Baptiste Chevalier (1873-1956), *Flore vivante de l'Afrique Occidentale Française.* Paris 1938; F.N. Hepper and F. Neate, *Plant Collectors in West Africa.* 58. 1971.

Murbeckiella Rothmaler Brassicaceae

Origins:

For Svante Samuel Murbeck, 1859-1946, Swedish botanist, professor of botany at Lund, plant collector, traveler. His works include "Monographie der Gattung *Verbascum.*" *Acta Univ. Lund.* n.s. 29(2): 1-630. 1933, *Tvenne Asplenier deras affiniteter och genesis.* Lund 1892, *Untersuchungen über das Androeceum der Rosaceen.* Lund, Leipzig 1941, *Untersuchungen über den Blütenbau der Papaveraceen.* Stockholm 1912, "Plantes du Sahara algérien, récoltées par Th. Orre." *Lunds Univ. Arsskrift.* N.F. Adv. 2, vol. 20, no. 11, 3-81. 1925, *Weitere Beobachtungen über Synaptospermie.* Lund, Leipzig 1943 and *Weitere Studien über die Gattungen Verbascum und Celsia.* Lund, Leipzig 1939. See J.H. Barnhart, *Biographical Notes upon Botanists.* 2: 530. 1965; T.W. Bossert, *Biographical Dictionary of Botanists Represented in the Hunt Institute Portrait Collection.* 278. 1972; H.N. Clokie, *Account of the Herbaria of the Department of Botany in the University of Oxford.* 216. Oxford 1964; E.M. Tucker, *Catalogue of the Library of the Arnold Arboretum of Harvard University.* 1917-1933; Ignatz Urban, *Geschichte des Königlichen Botanischen Museums zu Berlin-Dahlem (1815-1913). Nebst Aufzählung seiner Sammlungen.* Dresden 1916; R. Zander, F. Encke, G. Buchheim and S. Seybold, *Handwörterbuch der Pflanzennamen.* 14. Aufl. Stuttgart 1993; I.C. Hedge and J.M. Lamond, *Index of Collectors in the Edinburgh Herbarium.* Edinburgh 1970.

Murchisonia Brittan Anthericaceae (Liliaceae)

Origins:

After the Murchison district of Western Australia, named after Sir Roderick Impey Murchison, 1792-1871, British paleontologist and geologist, explorer, Director-General of the Geological Survey of the United Kingdom, Director of the Government School of Mines, Director of the Museum of Practical Geology, member of the Royal Geographical Society, member of the Geological Society of London, knighted in 1846, created a baronet in 1866, a patron of David Livingstone's expeditions. He is best known for his *The Silurian System.* London 1839, co-author (with E. de Verneuil and Count A. von Keyserling) of *The Geology of Russia in Europe and the Urals Mountains.* London, Paris

1845, acquainted with the English chemist and pioneer in the field of electrochemistry Sir Humphry Davy (1778-1829). See Sir Arch. Geikie, *Life of Sir Roderick I. Murchison*. London 1875; Sir Humphry Davy, *Researches, Chemical and Philosophical; Chiefly Concerning Nitrous Oxide, or Dephlogisticated Nitrous Air, and Its Respiration*. London 1800, *On the Safety Lamp for Coal Miners*; with some researches on flame. London 1818 and *Six Discourses Delivered before the Royal Society*. [Includes Davy's lectures between 1820-26 during his term as President of the Royal Society.] London 1827; Garrison and Morton, *Medical Bibliography*. By Leslie T. Morton. 5646. 1961; Norman Henry Brittan (b. 1920), in *Journal of the Royal Society of Western Australia*. 54: 95, figs 1, 2. 1972; Sir Francis Leopold Mac Clintock (M'Clintock), *The Voyage of the "Fox" in the Arctic Seas*. London 1859 and *A Narrative of the Discovery of the Fate of Sir John Franklin and His Companions*. Boston 1860; Elisha Kent Kane, *Arctic Explorations in Search of Sir John Franklin*. London 1882; Alexander B. Adams, *Eternal Quest. The Story of the Great Naturalists*. New York 1969; [George Stephenson], *Report upon the Claims of Mr. George Stephenson Relative to the Invention of His Safety Lamp*. [8vo, first edition.] Newcastle 1817; Andrew Crombie Ramsay (1814-1891), *Geological Map of England and Wales*. [R. was appointed to the Geological Survey by de la Beche in 1841, and thirty years later succeeded Murchison as Director-General.] London 1859.

Murdannia Royle Commelinaceae

Origins:
Named for the Indian plant collector Murdann Ali, Keeper of the Herbarium at Saharanpur Botanic Garden; the British surgeon and botanist John Forbes Royle (1798-1858) in 1833 was Curator of the same garden. See John H. Barnhart, *Biographical Notes upon Botanists*. 3: 187. 1965; John Forbes Royle, *Illustrations of the Botany and Other Branches of the Natural History of the Himalayan Mountains and the Flora of Cashmere*. London 1839-1840.

Species/Vernacular Names:
M. gigantea (Vahl) G. Brueckner

English: giant grass wool

M. graminea (R. Br.) G. Brueckner

English: grass lily, blue murdannia

Muricaria Desv. Brassicaceae

Origins:
Latin *muricatus, a, um* "roughened, with hard points, full of prickles, pointed, shaped like a purple-fish," referring to the pods.

Muricia Lour. Cucurbitaceae

Origins:
Latin *muriceus, a, um* "full of points, pointed, like a purple-fish, rough," *murex, icis* "purple, the purple dye, a caltrop."

Muricococcum Chun & How Euphorbiaceae

Origins:
Latin *muriceus, a, um* "pointed, like a purple-fish, rough," *murex, icis* "purple, the purple dye, a caltrop" and *kokkos* "a berry."

Murraea Koenig ex L. Rutaceae

Origins:
See *Murraya*.

Murraya Koenig ex L. Rutaceae

Origins:
After the Swedish botanist Johan (Johann) Andreas (Anders) Murray, 1740-1791, physician, botanical collector, 1763 graduated M.D. Göttingen, pupil and editor of Linnaeus, 1764-1791 professor of medicine and botany and Director of the Botanic Garden of Göttingen, 1774-1781 editor of *Medizinisch-praktische Bibliothek*. His works include *Enumeratio vocabulorum*. Holmiae [Stockholm] 1756, *Commentatio de Arbuto Uva Ursi*. Gottingae [Göttingen] [1764], *Enumeratio librorum praecipuorum medici argumenti*. Lipsiae 1773, *Apparatus medicaminum*. (Volumen sextum post mortem auctoris edidit Ludwig Christoph Althof) Gottingae 1776-1792, *De vermibus in lepra obviis, juncta leprosi historia*, etc. Gottingae 1769, *Index plantarum*. [Compiler of the index: Nicolaus Joseph von Jacquin, 1727-1817.] Viennae Austriae 1785 and *Prodromus Designationis Stirpium Göttingensium*. Gottingae 1770. See J.H. Barnhart, *Biographical Notes upon Botanists*. 2: 532. 1965; R. Zander, F. Encke, G. Buchheim and S. Seybold, *Handwörterbuch der Pflanzennamen*. 14. Aufl. 755. Stuttgart 1993; E. Bretschneider, *History of European Botanical Discoveries in China*. 1981; Georg Christian Wittstein, *Etymologisch-botanisches Handwörterbuch*. 599. Ansbach 1852; Pehr Kalm (1716-1779), *Beschreibung der Reise die er nach dem nördlichen Amerika ... unternommen hat. Eine Übersetzung* [translated from the Swedish, by J.A. Murray]. Göttingen 1754-1764; C. Linnaeus, *Systema Vegetabilium ... editio decima tertia*. [Edited by J.A. Murray] Göttingen Gotha 1774; T.W. Bossert, *Biographical Dictionary of Botanists Represented in the Hunt Institute Portrait Collection*. 279. 1972; Warren R. Dawson,

The Banks Letters, a Calendar of the Manuscript Correspondence of Sir Joseph Banks. London 1958; Jonas C. Dryander, *Catalogus bibliothecae historico-naturalis Josephi Banks*. 1796-1800; A. Lasègue, *Musée botanique de Benjamin Delessert*. Paris 1845; Blanche Henrey, *British Botanical and Horticultural Literature before 1800*. Oxford 1975; E.M. Tucker, *Catalogue of the Library of the Arnold Arboretum of Harvard University*. 1917-1933; C. Linnaeus, *Mantissa Plantarum*. 2: 554, 563. 1771; Mariella Azzarello Di Misa, a cura di, *Il Fondo Antico della Biblioteca dell'Orto Botanico di Palermo*. 199-200. Palermo 1988; Ida Kaplan Langman, *A Selected Guide to the Literature on the Flowering Plants of Mexico*. Philadelphia 1964; I.C. Hedge and J.M. Lamond, *Index of Collectors in the Edinburgh Herbarium*. Edinburgh 1970; Gilbert Westacott Reynolds, *The Aloes of South Africa*. Balkema, Rotterdam 1982; Frans A. Stafleu, *Linnaeus and the Linnaeans*. The spreading of their ideas in systematic botany, 1735-1789. Utrecht 1971.

Species/Vernacular Names:

M. koenigii (L.) Sprengel

English: curry leaf, curry bush, curry leaf tree

Malaya: karwa pale, garupillai, kerupulai

India: gandla, gandela, gandi, gant, gani, bowala, karapincha, surabhinimba, krishnanimba, karaypak, kadhee-nimba, katnim, kat-nim, barsunga, barsanga, harri, karrinim, karivaepamu, karuveppilai, karivaepu, karibevu

Nepal: mitho nim, mechia sag

M. paniculata (L.) Jack (*Murraya exotica* L.)

English: orange jessamine, satin wood, Andaman satin wood, Chinese box, Burmese box, Burmese boxwood, mock orange, common jasmine orange, Chinese myrtle, cosmetic bark tree

Peru: naranjillo

Venezuela: ayahar de la India

Japan: gek-kitsu, gikiji

China: jiu li xiang, chiu li hsiang tsao

Burma: mokson gayok, thanatka

The Philippines: banaasi, banasi, banati

Malaya: kemuning, kamoening, kamuning, kemuning kampong, kemuning lada

India: kamini, marchula, juti, bibzar, chaljuti, kada kongi cheddi, kamenee, kamini, kunli, kunti, marchi, marsan, marchula, marchulajuti, naga golunga, pandry

Nepal: bajardante

Indochina: cao ly yong, cay nguyet, keo, nguyet qui, nguyet qui tau

Murrinea Raf. Myrtaceae

Origins:

See Constantine Samuel Rafinesque, *Sylva Telluriana*. 104. 1838; E.D. Merrill, *Index rafinesquianus*. 203. 1949.

Murtekias Raf. Euphorbiaceae

Origins:

See Constantine Samuel Rafinesque, *Flora Telluriana*. 4: 116. 1836 [1838]; Elmer D. Merrill, *Index rafinesquianus*. 155. 1949.

Murtonia Craib Fabaceae

Origins:

For Henry James Murton, 1853-1881 (Bangkok), Kew gardener, 1875-1880 Singapore Botanic Gardens, see *Catalogue of Plants under Cultivation in Botanical Gardens*. Singapore 1879.

Murucuia Miller Passifloraceae

Origins:

A vernacular name, *maracuja* or *maracuya*, for some species of *Passiflora*.

Murucuja Medik. Passifloraceae

Origins:

A vernacular name, *maracuja* or *maracuya*, for some species of *Passiflora*.

Murucuja Persoon Passifloraceae

Origins:

Orthographic variant of *Murucuia* Miller; see C.H. Persoon, *Synopsis Plantarum*. 2: 222. 1870.

Musa L. Musaceae

Origins:

From Arabic *mauz, mouz* or *moz, muza*; see Carl Linnaeus, *Species Plantarum*. 1043. 1753 and *Genera Plantarum*. Ed. 5. 466. 1754; Salvatore Battaglia, *Grande dizionario della lingua italiana*. XI: 111. Torino 1981; M. Wis, in *Neuphilologische Mitteilungen*. LIX: 25. Helsingfors 1899-1958; H. Genaust, *Etymologisches Wörterbuch der botanischen Pflanzennamen*. 399-400. 1996; F. Boerner & G. Kunkel, *Taschenwörterbuch der botanischen Pflanzennamen*. 4.

Aufl. 137. Berlin & Hamburg 1989; Georg Christian Wittstein, *Etymologisch-botanisches Handwörterbuch.* 599. Ansbach 1852; Manlio Cortelazzo & Paolo Zolli, *Dizionario etimologico della lingua italiana.* 1: 110. 1979 and 3: 787. Zanichelli, Bologna 1983.

Species/Vernacular Names:

M. sp.

English: banana, plantain

Peru: arata, bahua paranta, banana, bari, bellaca, bellaco, beyaca, bissassa, bitiro, cantsin paranta, chapo, chihueron paranta, chila, chineo, chineo de isla, guayabo, guayaquil, guineo negro, harta bellaco, iná, iná platano, incárua, inguiri, isla, isleño, josho paranta, juxaacona, kinea, lacatan, llan guineo, manzano, mingado, muguichi, neamanihua, nónsi, paampa, pacoba, palillo, panara, parant, paranta, paranta chisó, parantacu, parianti, paruru, plantain, plantano, plátano domínico, poxi, sapo guineo, seda, shincun, shoiti paranta, sojete, soquete, tinti, yana guineo, yivara nava, zoquete

Brazil (Amazonas): aiamo si

Mexico: bitua, bituhua, pitohua castilla, edoo

Hawaii: mai'a

South Laos: (people Nya Hön) plää brüet, plää brüet bruun, plää brüet dab dab (= low banana), plää brüet glää (= egg banana), plää brüet llôông dum (= red tree banana or red banana tree), plää brüet mmuok, plää brüet ngiau (= dusky banana)

Tanzania: eligo

Yoruba: agbagba

M. acuminata Colla (*Musa cavendishi* Lamb. ex Paxt.)

English: banana, plantain, Chinese dwarf banana

Peru: musa enana, plátano enano

Japan: teikyaku-mi-bashô

Okinawa: Taiwan-basamai

Sierra Leone: manawa

Yoruba: ogede-ntiti, oyinbo

M. banksii F. Muell.

English: maroon-stemmed banana

M. basjoo Siebold & Zuccarini

English: Japanese banana, Ryukyu banana

China: ba jiao gen

M. fitzalanii F. Muell. (after Eugene Fitzalan, original collector)

English: native banana

M. jackeyi W. Hill (*Musa hillii* F. Muell.)

English: erect banana

M. nana Loureiro

Brazil: banana-anã, banana-da-china, banana-nanica, banana-de-italiano

M. ornata Roxb.

English: flowering banana

M. x *paradisiaca* L.

English: banana, edible banana, cooking banana, French plantain

French: bananier plantain

Paraguay: banana

Japan: ryôri-banana

Indonesia: kembang pisang

The Philippines: saging latundan, latundan, latunda, latundal, letondal, tordan, turdan, banangar, tukol, saguin a latondan

Bali: pusuh biu (pusuh = bud, biu = banana)

Italian: banana

Yoruba: ogede agbagba, ogedeapanta, ogede dudu, ogede weere

Congo: leka

M. sapientum L.

English: eating banana

French: bananier

Peru: huainama, huessesse, imama, jotete, maccocco, maninha, namidsadsa, omada, pirohua, potetera, ssapapa

Brazil: banana, banana-maçã, banana-de-são-tomé

Malaya: pisang, banana

China: hsiang chiao, kan chiao, pa chiao

Yoruba: ogede, ogede abo, ogede loboyo, ogede omini, agbagba

Congo: iko

M. textilis Née

English: Manila hemp

Japan: Manira-ito-basho

The Philippines: abaka

M. troglodytarum L. (*Musa fehi* Bertero ex Vieill.)

English: fe'i banana

Hawaii: mai'a he'i, mai'a Polapola

Musanga R. Br. Cecropiaceae

Origins:

A vernacular name in Zaire and Angola; see Y. Tailfer, *La Forêt dense d'Afrique centrale.* CTA, Ede/Wageningen 1989; J. Vivien & J.J. Faure, *Arbres des Forêts denses d'Afrique Centrale*. Agence de Coopération Culturelle et Technique. Paris 1985.

Species/Vernacular Names:

M. cecropioides R. Br. (*Musanga smithii* R. Br.)

English: umbrella tree, cork wood

French: parasolier

Angola: gofé; nsenga, nsanga, musenga, musanga (Kikongo); mulela (Kioko)

Cameroon: asseng, assan, bosengue, bosonge, bossengue, djseng, lisseng, leseng, kombo

Central African Republic: angope

Congo: senga, nsenga, nseenga, kombo-kombo

Gabon: assan, assang, asseng

Ghana: agyama, ajama

Ivory Coast: amonia, kodé, abome, agbome, agoumi, djuna, parasolier

Yoruba: aga, agbawo

Nigeria: agoken, aga, agbawo, asang, egbesu, agemanfuk, bosenge, bokumbu, egbu, ekombe, egimatuk, aju-eku, ajo-weku, lisenga, obonia, ofogo, oghohen, wosao, olo, oro, oru, aworo, ukporowi, uboniboni, ulu, uno, congo-congo, ukhurube, awunru, tako; aga (Yoruba); ogohen (Edo); ufogho (Etsako); egbesu (Itsekiri); ukhorube (Urhobo); ukporwe (Ijaw); oro (Igbo); uno (Efik); egimamfuk (Ekoi); bokuobe (Boki)

West Africa: bokombo, bosenge, combo combo

Zaire: senga; bokombo (Turumbu); bombambo (Lokundu); gombo (Dundusana); kitumbe (Kinande); kimbongo (Low Zaire); kombo, okombo, kimbu (Uele); mobambo (Lingala; see G. Malcolm Guthrie, ed., *Lingala grammar and dictionary*. Léopoldville-Ouest. 1935); mubena (Kihunde); musenga (Lukula); mosinki (Kwilu); mushake (Kirega); n'senga, senga (Mayumbe); tshilombalomba (Tshiluba); tumbe-tumbe (Kingwana)

Muscari Miller Hyacinthaceae (Liliaceae)

Origins:

A Turkish name recorded by Clusius in 1583; Latin *muscus, i* "moss, musk," Akkadian *musu* "outflow of water, land irrigated."

Species/Vernacular Names:

M. armeniacum Leichtlin ex Baker

English: grape hyacinth

M. comosum (L.) Miller (*Hyacinthus comosus* L.; *Leopoldia comosa* (L.) Parl.)

English: tufted grape hyacinth, tassel hyacinth

Muscarimia Kostel. Hyacinthaceae (Liliaceae)

Origins:

See *Muscari muscarimi* Medikus, original names *muscari, muschoromi* or *muscurimi*; a Turkish name recorded by Clusius in 1583; Latin *muscus, i* "moss, musk."

Muschleria S. Moore Asteraceae

Origins:

For the German botanist Reinhold (Reno) Conrad Muschler, 1883-1957, novelist, traveler, in Egypt, at the Botanical Museum at Berlin. Among his writings are *Bianca Maria*. Leipzig [1925], *Diana Beata*. Berlin 1939, *Énumération des algues marines et d'eau douce observées jusqu'à ce jour en Égypte*. Alexandria 1908, "Labiatas siamenses novae." *Repert. Sp. Nov. Fedde*. 4: 268-270. 1907, "Die Gattung *Coronopus* (L.) Gaertn." *Bot. Jahrb. Syst*. 41: 111-147. 1908, "Cruciferae andinae." *Bot. Jahrb. Syst*. 40: 267-277. 1908, "Caryophyllaceae andinae." *Bot. Jahrb. Syst*. 45: 441-461. 1911 and *Richard Strauss*. Hildesheim [1924]. See J.H. Barnhart, *Biographical Notes upon Botanists*. 2: 532. 1965; Christoph Bernhard Levin Schuecking, *Briefe von Levin Schücking und Louise von Gall*. Herausgegeben von Dr. R.C. Muschler. Leipzig 1928.

Musella (A.R. Franchet) H. Wu Li Musaceae

Origins:

The diminutive of *Musa* L., in *Acta Phytotax. Sin*. 16(3): 57. Aug. 1978.

Musgravea F. Muell. Proteaceae

Origins:

Named after Sir Anthony Musgrave, from 1873 to 1877 Governor of South Australia, from 1883 to 1889 Governor of Queensland, wrote *The Function of Money: Bimetallism*. Reprinted from the "Westminster Review." London 1886 and *Studies in Political Economy*. London 1875; see F. Mueller, in *Proceedings of the Linnean Society of New South Wales*. ser. 2, 5: 186. 1890.

Species/Vernacular Names:

M. heterophylla L.S. Smith

English: briar silky oak, briar oak, brown silky oak

M. stenostachya F. Muell.

English: crater silky oak, grey silky oak

Musineon Raf. Umbelliferae

Origins:

Referring to *Musenium* Nutt.; see Constantine Samuel Rafinesque (1783-1840), *Jour. Phys. Chim. Hist. Nat.* 91: 71. 1820; E.D. Merrill, *Index rafinesquianus.* 178, 181. 1949.

Mussaenda L. Rubiaceae

Origins:

A vernacular name for *Mussaenda frondosa* in Sri Lanka.

Species/Vernacular Names:

M. sp.

Indonesia: nusa indah (= beautiful island)

Bali: bungan nusa indah

M. arcuata Poir.

Southern Africa: muRidzameso, shingyarokasha (Shona)

S. Rhodesia: muRidzameso

Nigeria: gwawon-kare, tsada-kwuni

Yoruba: akekakara

M. erythrophylla Schumach. & Thonn.

English: red mussaenda, Ashanti blood, flame of the forest

The Philippines: Doña Trining (named in honor of Mrs. Trining Roxas)

M. glabra Vahl

English: common mussaenda

Malaya: adap-adap, balek hadap, cha padi, daun puteri, segoreh

M. landia Poiret var. *landia*

La Réunion Island: lingue en arbre, quinquina du pays

Mauritius: quinquina indigène

M. philippica A. Rich. (*Mussaenda acutifolia* Bartl.; *Mussaenda glabra* F. Vill.; *Mussaenda grandiflora* Rolfe; *Mussaenda frondosa* Blanco; *Calycophyllum grandiflorum* Meyen)

English: Philippine mussaenda, flag tree

Nepal: asari, dhobini

The Philippines: kahoy dalaga

M. philippica A. Rich. var. *aurorae* Sul. (named after a former first lady, Mrs. Aurora Quezon)

The Philippines: Doña Aurora

M. pubescens Ait.f. (*Mussaenda parviflora* Kanehira non Miq.)

English: Buddha's lamp

Mussaendopsis Baillon Rubiaceae

Origins:

Resembling *Mussaenda* L.

Musschia Dumort. Campanulaceae

Origins:

After the Belgian botanist Jean Henri Mussche, 1765-1834, Curator of the Botanical Garden at Gent, author of *Hortus gandavensis.* Gand [Gent] 1817. See John Hendley Barnhart, *Biographical Notes upon Botanists.* 2: 533. 1965; F. Boerner & G. Kunkel, *Taschenwörterbuch der botanischen Pflanzennamen.* 4. Aufl. 137. Berlin & Hamburg 1989; Georg Christian Wittstein, *Etymologisch-botanisches Handwörterbuch.* 599. Ansbach 1852; T.W. Bossert, *Biographical Dictionary of Botanists Represented in the Hunt Institute Portrait Collection.* 279. 1972; E.M. Tucker, *Catalogue of the Library of the Arnold Arboretum of Harvard University.* 1917-1933.

Species/Vernacular Names:

M. aurea (L.f.) Dumort.

English: golden musschia

Mustelia Steud. Gramineae

Origins:

Latin *mustela* "a weasel."

Mutisia L.f. Asteraceae

Origins:

Named for the Spanish (b. Cadiz) botanist José Célestino Bruno Mutis y Bosio (Bossio), 1732-1808 (d. Nueva Granada, now Bogotá, Colombia), physician, clergyman, in 1750 Nueva Granada, Director and leader of the *Real Expedición botánica del Nuevo Reino de Granada,* sent specimens to Linnaeus, professor of mathematics. See A.F. Gredilla, *Biografía de José Mutis.* Madrid 1911; J. Vernet, in *Dictionary of Scientific Biography* 15: 429-430. 1981; R. Zander, F. Encke, G. Buchheim and S. Seybold, *Handwörterbuch der Pflanzennamen.* 14. Aufl. 754f. 1993; Georg Christian Wittstein, *Etymologisch-botanisches Handwörterbuch.* 600. 1852; J.H. Barnhart, *Biographical Notes upon Botanists.* 2: 533. 1965; T.W. Bossert, *Biographical Dictionary of Botanists Represented in the Hunt Institute Portrait Collection.* 279. 1972; Warren R. Dawson, *The Banks Letters.* London 1958; Miguel Colmeiro y Penido, *La Botánica y los Botánicos de la Peninsula Hispano-Lusitana.* Madrid 1858; Jonas C. Dryander, *Catalogus bibliothecae historico-naturalis Josephi Banks.* London 1800; A. Lasègue, *Musée botanique de Benjamin Delessert.* Paris 1845; Ethelyn Maria Tucker, *Catalogue of the Library of the Arnold Arboretum of Harvard University.* Cambridge, Massachusetts 1917-1933; Friedrich Wilhelm Heinrich Alexander von Humboldt and Aimé Jacques Alexandre Bonpland, *Plantae aequinoctiales.* 1808.

Myagropsis Hort. ex O.E. Schulz Brassicaceae

Origins:
Resembling *Myagrum.*

Myagrum L. Brassicaceae

Origins:
Greek *myagros* (*mus, mys* "mouse" and *agra* "a prey, a catching"), the mouser, a kind of snake, but also a name for a plant used by Dioscorides (4.116) and Plinius, *melampyros*; Latin *myagros, i* for an unknown plant; *myiagros* "fly-catcher," referring to the stickiness of the plants; see Carl Linnaeus, *Species Plantarum.* 640. 1753 and *Genera Plantarum.* Ed. 5. 289. 1754.

Species/Vernacular Names:
M. perfoliatum L.
English: musk weed

Myanthus Lindley Orchidaceae

Origins:
Probably from the Greek *mya, myia* "fly" and *anthos* "flower," referring to sepals and petals.

Mycaranthes Blume Orchidaceae

Origins:
From the Greek *mykaris* "bat" and *anthos* "flower," the lip resembles a small bat.

Mycaridanthes Blume Orchidaceae

Origins:
Greek *mykaris* "bat" and *anthos* "flower."

Mycelis Cass. Asteraceae

Origins:
An ancient name; see H. Genaust, *Etymologisches Wörterbuch der botanischen Pflanzennamen.* 403. 1996; F.

Boerner & G. Kunkel, *Taschenwörterbuch der botanischen Pflanzennamen.* 4. Aufl. 137. Berlin & Hamburg 1989.

Species/Vernacular Names:
M. muralis (L.) Dumort.
English: wall lettuce

Mycerinus A.C. Sm. Ericaceae

Origins:
Mycerinus was a son of Cheops, king of Egypt, he reigned with great justice and moderation; *mykeros*, a Greek word for almond.

Mycetia Reinw. Rubiaceae

Origins:
Probably from Greek *mykes, myketos* "mushroom, fungus, any knobbed round body"; Latin *mycetias*, Greek *myketias seismos*, for *mycematias*, an earthquake, attendent with a rumbling noise, Greek *myketes* "bellower"; see M.P. Nayar, *Meaning of Indian Flowering Plant Names.* 233. Dehra Dun 1985.

Myconia Lapeyr. Gesneriaceae

Origins:
Possibly from *Myconius, a, um* "Myconian, belonging to Myconos."

Mygalurus Link Gramineae

Origins:
Greek *mygale* "shrew-mouse, field-mouse" and *oura* "a tail," Latin *mygale, es* "a small species of mouse."

Myginda Jacq. Celastraceae

Origins:
After the Austrian Franz Mygind, 1710-1789, a friend of N.J. von Jacquin. See John Hendley Barnhart, *Biographical Notes upon Botanists.* 2: 534. 1965; H.N. Clokie, *Account of the Herbaria of the Department of Botany in the University of Oxford.* 216. Oxford 1964; L. von Heufler, *Franz von Mygind, der Freund Jacquin's.* Wien 1870; I.C. Hedge and J.M. Lamond, *Index of Collectors in the Edinburgh Herbarium.* Edinburgh 1970; Frans A. Stafleu, *Linnaeus and the Linnaeans.* The spreading of their ideas in systematic botany, 1735-1789. Utrecht 1971.

Myladenia Airy Shaw Euphorbiaceae

Origins:

From the Greek *myle, mylos* "mill, knee-pan" and *aden* "gland."

Myllanthus R. Cowan Rutaceae

Origins:

From the Greek *myllos* "crooked" and *anthos* "flower."

Myoda Lindley Orchidaceae

Origins:

From the Greek *myodes* "mouse-like, a mouse-tail."

Myodium Salisb. Orchidaceae

Origins:

Greek *myodes* "mouse-like."

Myodocarpus Brongn. & Gris Araliaceae

Origins:

From the Greek *myodes* "mouse-like" and *karpos* "fruit."

Myopordon Boiss. Asteraceae

Origins:

From the Greek *myo* "mouse" and *pordon, pordos* "stinkard."

Myoporum Sol. ex Forster f. Myoporaceae

Origins:

Greek *myo, myein* "to close, shut" and *poros* "opening, pore," referring to the glands or transparent spots on the leaves; see J.G.A. Forster, *Florulae insularum australium prodromus.* Göttingen 1786; F. Boerner & G. Kunkel, *Taschenwörterbuch der botanischen Pflanzennamen.* 4. Aufl. 137. Berlin & Hamburg 1989; H. Genaust, *Etymologisches Wörterbuch der botanischen Pflanzennamen.* 403. 1996; Salvatore Battaglia, *Grande dizionario della lingua italiana.* X: 501. Torino 1978.

Species/Vernacular Names:

M. acuminatum R. Br. (*Myoporum montanum* R. Br.; *Myoporum cunninghamii* Benth.; *Myoporum dampieri* Cunn. ex A. DC.; *Myoporum acuminatum* R. Br. var. *angustifolium* Benth.)

English: waterbush, Australian blueberry tree

Australia: boobialla, western boobialla, northern boobialla, puturryu, mee mee, nymoo, pulkaru, tjurrkajurrka, walkapa, native myrtle, native daphne, waterbush, strychnine bush, dogwood, water bush

French: faux santal

M. betcheanum L.S. Smith (after the German botanist Ernst (Ernest) Betche, 1851-1913 (Sydney), botanical collector, New South Wales, gardener, from 1880-1881 collected plants in Samoa, from 1881 to 1897 collector for Sydney Botanic Garden, co-author with Charles Moore (1820-1905) of *Handbook of the Flora of New South Wales.* Sydney 1893 and with Joseph Henry Maiden (1859-1925) of *Census of New South Wales Plants.* 1916; see J. Lanjouw and F.A. Stafleu, *Index Herbariorum.* Part II, *Collectors A-D.* Regnum Vegetabile vol. 2. 1954; Friedrich Richard Rudolf Schlechter (1872-1925), in *Botanische Jahrbücher.* 52: 146. (Nov.) 1914)

Australia: mountain sugarwood

M. boninense Koidz.

Australia: boobialla

M. brevipes Benth.

Australia: djindijii

M. caprarioides Benth.

Australia: slender myoporum

M. floribundum Benth.

Australia: slender myoporum, shower tree

M. insulare R. Br. (*Myoporum adscendens* R. Br.; *Myoporum serratum* R. Br. var. *obovatum* Benth.)

Australia: boobyalla, boobialla, common boobialla, cookatoo bush, native juniper, water bush, native mangrove, blueberry, blueberry tree

Italian: mioporo

M. laetum Forster f.

Maori names: ngaio

M. montanum R. Br.

Australia: boobialla, western boobialla, boomeralla, meeme, nymoo, tjuruku tjuruku, native myrtle, native daphne, water bush

M. oppositifolium R. Br.

Australia: twin-leaf myoporum

M. parvifolium R. Br. (*Myoporum humile* R. Br.)

Australia: creeping myoporum, creeping boobialla, dwarf native myrtle

M. platycarpum R. Br.

Australia: boolgar, boolgarba, bulgara, miljiling, ngural, yumburra, sandalwood, false sandalwood, red sandalwood, sugarwood, bastard sandalwood, dogwood

M. sandwicense (A. DC.) A. Gray (*Myoporum degeneri* (Webster) Degener & I. Degener; *Myoporum fauriei* H. Lév.; *Myoporum lanaiense* (Webster) Degener & I. Degener)

English: bastard sandalwood

Pacific Islands: ngaio, naio

Hawaii: naio, naeo, naieo

M. tetrandrum (Labill.) Domin (*Myoporum serratum* R. Br.)

English: manitoka

Australia: boobialla

M. turbinatum Chinn.

Australia: salt myoporum

M. viscosum R. Br.

Australia: sticky boobialla

Myopsia C. Presl Campanulaceae

Origins:
From the Greek *myo* "mouse" and *opsis* "like."

Myoschilos Ruíz & Pavón Santalaceae

Origins:
From the Greek *myo, myos* "mouse" and *cheilos* "lip."

Myosotidium Hook. Boraginaceae

Origins:
Resembling *Myosotis.*

Species/Vernacular Names:
M. hortensia (Decne.) Baillon
English: Chatham Is. forget-me-not

Myosotis L. Boraginaceae

Origins:
Latin *myosota, ae* and *myosotis, idis* (Plinius), Greek *myosotis, myosotidos* (*mus, mys, myos* "mouse" and *ous, otos* "ear"), referring to the hairy leaves of some species of the genus; see Carl Linnaeus, *Species Plantarum.* 131. 1753 and *Genera Plantarum.* Ed. 5. 63. 1754; Manlio Cortelazzo & Paolo Zolli, *Dizionario etimologico della lingua italiana.* 3: 760. Bologna 1983; Georg Christian Wittstein, *Etymologisch-botanisches Handwörterbuch.* 601. 1852; Salvatore

Battaglia, *Grande dizionario della lingua italiana.* X: 502. 1978.

Species/Vernacular Names:
M. afropalustris C.H. Wr.
English: forget-me-not
M. arvensis (L.) Hill
English: forget-me-not, common forget-me-not
M. australis R. Br.
English: Austral forget-me-not, southern forget-me-not, native forget-me-not
M. caespitosa Schultz
English: water forget-me-not, tufted forget-me-not
China: shi di wu wang cao
M. discolor Pers. (*Myosotis collina* Hoffm.; *Myosotis versicolor* Sm.)
English: yellow and blue forget-me-not
M. exarrhena F. Muell.
English: sweet forget-me-not
M. scorpioides L.
English: water forget-me-not
M. sylvatica Ehrenb. ex Hoffm.
English: wood forget-me-not, garden forget-me-not, woodland forget-me-not

Myosoton Moench Caryophyllaceae

Origins:
From the Latin *myosoton* applied by Plinius to the plant *alsine.*

Myospyrum Lindl. Oleaceae

Origins:
From the Greek *mus, mys, myos* "mouse" and *pyros* "wheat," see also *Myxopyrum* Blume.

Myosurus L. Ranunculaceae

Origins:
From the Greek *mus, myos, mys* "mouse" and *oura* "a tail," referring to the fruiting elongated spikes; see Carl Linnaeus (Carl von Linnaeus, Carl von Linné), *Species Plantarum.* 284. 1753 and *Genera Plantarum.* Ed. 5. 137. 1754.

Species/Vernacular Names:
M. minimus L.

English: mousetail

Myoxanthus Poepp. & Endl. Orchidaceae

Origins:
Greek *myoxos* "dormouse" and *anthos* "flower," possibly referring to the flower of the type species; some suggest from *mys* "muscle" and *xanthos* "yellow"; see Richard Evans Schultes and Arthur Stanley Pease, *Generic Names of Orchids. Their Origin and Meaning*. 204. New York and London 1963; Hubert Mayr, *Orchid Names and Their Meanings*. 160. Vaduz 1998; H. Genaust, *Etymologisches Wörterbuch der botanischen Pflanzennamen*. 404. 1996.

Myrceugenella Kausel Myrtaceae

Origins:
The diminutive of *Myrceugenia* O. Berg.

Myrceugenia O. Berg Myrtaceae

Origins:
Genera *Myrcia* and *Eugenia*, Myrtaceae.

Myrcia DC. ex Guill. Myrtaceae

Origins:
Possibly from the Greek *myron* "a perfume, ointment, scent."

Species/Vernacular Names:
M. sp.
Peru: isayre, cambucá, folha de ouro
Mexico: yaga lan
M. amazonica DC.
Brazil: pedra-ume-caá
M. tomentosa (Aublet) DC.
Tropical America: cabelluda

Myrcialeucus Rojas Acosta Myrtaceae

Origins:
Genus *Myrcia* plus Greek *leukos* "white."

Myrcianthes O. Berg Myrtaceae

Origins:
Genus *Myrcia* plus Greek *anthos* "flower."

Species/Vernacular Names:
M. sp.
Peru: lanche, unca
M. pungens (Berg) Legr.
Brazil: guabiju, guabiju guaçu, guabira guaçu, guavira guaçu, guajaraí da várzea, iba viyu, guaviyú (see Luíz Caldas Tibiriçá, *Dicionário Guarani-Português*. Traço Editora, Liberdade 1989)
Argentina: mato
M. umbellulifera (Kunth) Alain
West Indies: monos plum

Myrciaria O. Berg Myrtaceae

Origins:
Referring to the genus *Myrcia*, or an anagram of *Myrcaria*.

Species/Vernacular Names:
M. cauliflora (C. Martius) O. Berg
Brazil: jaboticaba, jabuticaba
M. cuspidata O. Berg
Brazil: cambuí, camboim, camboinzinho
M. delicatula (DC.) O. Berg
Brazil: cambuí, camboim, camboinzinho, cambuim, camboí, araçazeiro, cambuim preto, camboim bala
M. tenella (DC.) O. Berg
Brazil: cambuí, cambuim, camboí, camboim, cambuí preto, camboizinho, Murta do campo

Myrciariopsis Kausel Myrtaceae

Origins:
Resembling the genus *Myrciaria*.

Myriachaeta Moritzi Gramineae

Origins:
From the Greek *myria* "many" and *chaite* "bristle, long hair."

Myriactis Less. Asteraceae

Origins:
Greek *myria* "many" and *aktis* "a ray, sunbeam," referring to the ray florets.

Myrialepis Becc. Palmae

Origins:

From the Greek *myria* "many" and *lepis* "scale," referring to the scales of the fruits.

Myrianthemum Gilg Melastomataceae

Origins:

From the Greek *myria* "many" and *anthemon* "a flower."

Myrianthus P. Beauv. Cecropiaceae

Origins:

Greek *myria, myrioi* "many" and *anthos* "a flower," referring to the many and tiny flowers; see Y. Tailfer, *La Forêt dense d'Afrique centrale.* CTA, Ede/Wageningen 1989; J. Vivien & J.J. Faure, *Arbres des Forêts denses d'Afrique Centrale.* Agence de Coopération Culturelle et Technique. Paris 1985.

Species/Vernacular Names:

M. arboreus P. Beauv.

Congo: ikamu, nsongoti, okamon

Zaire: bakombu, bonkowna mokili (Kundu); balandu (Mayogo); bokamu, bonkomu (Lingala); bokomu (lake Mai-Ndombe); bongunguna, ongunguna (Turumbu); buba, m'buba (Mayumbe); dikomokomo, mokamu a mukomo, mukomu, munkala (Tshiluba); ekama, ekamu (Musa); gwolo (Likimi); mamala (Kinande); mutuse (Kisantu); tshikala kala (Kaniama)

Yoruba: alade

Nigeria: ibishere, obishere, ebisheghe, ihi-egghi, bekeku, charaka, ekokom, ekokwan, isasa, ntinsek, wakeku; tsakpachi (Nupe); ibisere (Yoruba); ihieghe (Edo); oseghe (Etsako); ujuju (Igbo); ndisok (Efik); kekeku (Boki)

Angola: muxibiri, pernanbuco, muzubidi (Kimbundu); umbukusu (Umbundu); dijikala-kala, mupalapanga (Kioko); mbuba, vuiba (Kikongo)

Malawi: chiwere, mkwakwa (Chichewa); mukwajo (Chichewa, Yao); mufwisa (Sukwa); mwa(n)ja (Yao)

Cameroon: angom, bokekou, engakom, mongo

Ivory Coast: anianahia, doba, niangama, grand wounian

M. holstii Engl.

Zaire: mwamba, mufe, tshefu

Southern Africa: muDenya, guvu, muTeswa (Shona)

S. Rhodesia: muDenya, guvu

Myriaspora DC. Melastomataceae

Origins:

From the Greek *myria* "many" and *sporos* "a seed."

Myrica L. Myricaceae

Origins:

From the Greek name for tamarisk, *myrike*, from *myron* "a perfume, ointment, scent"; Latin *myrice, es*, and *myrica, ae* "tamarisk." C. Plinius Secundus, *Naturalis historia* 13, 116; "pinguia corticibus sudent electra myricae", Publius Vergilius Maro, *Eclogae* 8, 54; Salvatore Battaglia, *Grande dizionario della lingua italiana.* X: 525-526. 1978; Publius Ovidius Naso, *De arte amandi.* 1, 747.

Species/Vernacular Names:

M. sp.

Peru: búccu

M. californica Cham.

English: California wax myrtle, California bayberry, bayberry, wax myrtle

M. cerifera L.

English: wax myrtle, candleberry

M. conifera Burm.f.

Southern Africa: chiNdzere, chiNdzewere (Shona)

S. Rhodesia: chiNzere

M. cordifolia L. (*Myrica cordifolia* L. var. *microphylla* A. Chev.)

English: waxberry, candleberry, waxbush

South Africa: glashout, wasbes

M. esculenta Buch.-Ham. ex D. Don

English: Malay gale, box myrtle

Nepal: kafal, kaphal

Malaya: telur chichak

M. faya Aiton

English: candleberry myrtle, firetree, Azorean candleberry tree, wax myrtle

Portuguese: samouco

M. gale L.

English: sweet gale, bog myrtle, meadow fern, gale

M. integra (A. Chev.) Killick (*Myrica conifera* Burm.f. var. *integra* A. Chev.; *Myrica linearis* sensu A. Chev., Hutch.)

South Africa: waxberry, myrica-with-untoothed-leaves

M. pensylvanica Lois.

English: bayberry, candleberry, swamp candleberry

M. pilulifera Rendle (*Myrica pilulifera* Rendle var. *puberula* Rendle; *Myrica rogersii* Burtt Davy) (Latin *pilulifera* "bearing hairs," *pilus, i* "hair" and *fero, tuli, latum, ferre*)

English: broad-leaved waxberry, waxberry tree, waxberry bush

Southern Africa: breëblaarwasbessie, wasbes; iNyamvura (Shona)

S. Rhodesia: nyamvura

M. quercifolia L. (*Myrica incisa* A. Chev.; *Myrica zeyheri* C. DC.)

South Africa: wasbes, maagpyne bossie

M. rubra Siebold & Zucc.

English: strawberry tree, edible bayberry

Japan: yama-momo

Okinawa: mumu

China: yang mei, chiu tzu

M. serrata Lam. (*Myrica conifera* sensu Hutch., Adamson, non Burm.f.; *Myrica mossii* Burtt Davy)

English: lance-leaved waxberry, waxberry

Southern Africa: smalblaarwasbessie, wasbessie, gammabos; uMakhuthula, uLethi, iLethi, iYethi (Zulu); isiBhara, uMakhuthula (Xhosa); monnamotho, maloleha (South Sotho); murandela (Venda)

Myricanthe Airy Shaw Euphorbiaceae

Origins:
From the Greek *myrike* "tamarisk" and *anthos* "flower."

Myricaria Desv. Tamaricaceae

Origins:
From the Greek *myrike* "tamarisk," false tamarisk.

Myrinia Lilja Onagraceae

Origins:
After the Swedish botanist Claës Gustaf Myrin, 1803-1835, economist, wrote *Corollarium florae upsaliensis*. Upsaliae [1833]; see John Hendley Barnhart, *Biographical Notes upon Botanists*. 2: 534. 1965.

Myriocarpa Benth. Urticaceae

Origins:
From the Greek *myrios* "many, myriad, numberless" and *karpos* "fruit."

Myriocephalus Benth. Asteraceae

Origins:
Greek *myrios* "many, myriad, numberless" and *kephale* "head," referring to the flowers or to the numerous capitula.

Species/Vernacular Names:
M. appendiculatus Benth.

English: white-tip myriocephalus

M. rhizocephalus (DC.) Benth. (*Hyalolepis rhizocephala* DC.; *Hyalolepis occidentalis* F. Muell.)

English: woolly-heads

M. stuartii (Sonder) Benth. (*Polycalymma stuartii* Sonder)

English: poached-egg daisy, ham-and-eggs daisy, white billybutton

Myriocladus Swallen Gramineae

Origins:
From the Greek *myrios* "many, myriad, numberless" and *kladion* "a branchlet, a small branch," *klados* "a branch."

Myriodon (Copel.) Copel. Hymenophyllaceae

Origins:
From the Greek *myrios* "many, numberless" and *odous, odontos* "tooth."

Myriogomphus Didr. Euphorbiaceae

Origins:
From the Greek *myrios* "many, myriad" and *gomphos* "a nail, pin, peg, club."

Myriogyne Less. Asteraceae

Origins:
From the Greek *myrios* "many, myriad" and *gyne* "female, woman, female organs"; see *Linnaea*. 6: 219. 1831.

Myrioneuron R. Br. ex Hook.f. Rubiaceae

Origins:
From the Greek *myrios* "many, myriad" and *neuron* "nerve," referring to the many-nerved leaves.

Myriophyllum L. Haloragidaceae (Haloragaceae)

Origins:
Latin and Greek *myriophyllon* the plant milfoil, yarrow, Greek *myrios* "many, myriad, numberless, a great many, countless" and *phyllon* "a leaf," the leaves are very finely divided; see Carl Linnaeus, *Species Plantarum.* 992. 1753 and *Genera Plantarum.* Ed. 5. 429. 1754.

Species/Vernacular Names:
M. alpinum Orch.
English: water-milfoil

M. amphibium Labill.
English: broad water-milfoil

M. aquaticum (Vell. Conc.) Verdc. (*Enydria aquatica* Vell.; *Myriophyllum brasiliense* Cambess.; *Myriophyllum proserpinacoides* Gill. ex Hook. & Arnott)
English: parrot's feather, diamond milfoil, milfoil, water feather, water-milfoil, Brazilian water-milfoil
South Africa: duisendblaar, waterduisendblaar

M. austropygmaeum Orch.
English: water-milfoil

M. caput-medusae Orch.
English: coarse water-milfoil, cat tail

M. crispatum Orch.
English: water-milfoil

M. dicoccum F. Muell.
English: water-milfoil

M. gracile Benth.
English: water-milfoil, slender water-milfoil

M. hippuroides Torrey & A. Gray
English: western milfoil

M. integrifolium (Hook.f.) Hook.f.
English: small water-milfoil

M. latifolium F. Muell.
English: water-milfoil

M. lophatum Orch.
English: water-milfoil

M. muelleri Sonder
English: hooded water-milfoil, slender milfoil, Mueller's water-milfoil

M. muricatum Orch.
English: water-milfoil

M. papillosum Orch.
English: water-milfoil

M. pedunculatum Hook.f.
English: mat water-milfoil, stalked water-milfoil

M. salsugineum Orch.
English: saline water-milfoil

M. simulans Orch.
English: water-milfoil, common water-milfoil

M. spicatum L.
English: spiked water-milfoil, water-milfoil, Eurasian milfoil
Japan: kin-gyo-mo, hozaki-no-fusa-mo
China: shui tsao

M. striatum Orch.
English: water-milfoil

M. trachycarpum F. Muell.
English: water-milfoil

M. variifolium Hook.f.
English: variable water-milfoil

M. verrucosum Lindley
English: red water-milfoil

M. verticillatum L.
English: myriad leaf

Myriopteris Fée Pteridaceae (Adiantaceae)

Origins:
From the Greek *myrios* "many, countless" and *pteris* "fern"

Myriopteron Griffith Asclepiadaceae

Origins:
From the Greek *myrios* "many, numberless, a great many, countless" and *pteron* "wing," alluding to the papery and longitudinal wings of the follicles.

Species/Vernacular Names:
M. extensum (Wight & Arnott) K. Schumann
English: extensed wing-fruitvine
China: chi guo teng

Myriostachya (Benth.) Hook.f. Gramineae

Origins:
From the Greek *myrios* "many, numberless, a great many, countless" and *stachys* "spike," alluding to the spikes.

Myriotheca Comm. ex Juss. Marattiaceae

Origins:
From the Greek *myrios* and *theke* "box," referring to the capsules.

Myristica Gronov. Myristicaceae

Origins:
From the Greek *myristikos* "fragrant, fit for anointing," *myron* "a perfume, ointment, scent, sweet smelling, sweet oil," *myrizo, myrizein* "to rub with ointment"; see Johan Frederik Gronovius, *Flora orientalis*. 141. Lugduni Batavorum [Leiden] 1755.

Species/Vernacular Names:
M. fragrans Houtt.
English: nutmeg, mace, common nutmeg, nutmeg tree
Arabic: gouz bouwa
India: jati, jai
Malaya: pala
Tibetan: dzati
China: rou dou kou, jou tou kou, jou kuo
Brazil: noz-moscada, macis
M. insipida R. Br.
English: native nutmeg
M. simiarum A. DC.
English: Antao nutmeg
China: anuping

Myrmechis (Lindley) Blume Orchidaceae

Origins:
From the Greek *myrmex, myrmekos* "ant," referring to the flowers.

Myrmecodendron Britton & Rose Mimosaceae

Origins:
Greek *myrmex, myrmekos* "ant" and *dendron* "tree."

Myrmecodia Jack Rubiaceae

Origins:
From the Greek *myrmex, myrmekos* "ant" and *eidos, oides* "resemblance" or *kodeia* "head, bulb, capsule," referring to the friendship and symbiosis between ants and plants; see W. Jack, in *Transactions of the Linnean Society of London*. 14: 122. 1823.

Species/Vernacular Names:
M. beccarii Hook.f. (after the Italian botanist Odoardo Beccari, 1843-1920, botanical explorer, traveler)
English: prickly ant plant
M. muelleri Becc.
English: ant plant
M. platyrea Becc.
English: ridged ant plant

Myrmecoides Elmer Rubiaceae

Origins:
Resembling *Myrmecodia* Jack.

Myrmeconauclea Merr. Rubiaceae

Origins:
From the Greek *myrmex, myrmekos* "ant" plus the genus *Nauclea* L., myrmecophilous.

Myrmecophila (Christ) Nakai Polypodiaceae

Origins:
Greek *myrmex, myrmekos* "ant" and *philos* "lover, loving," in *Bot. Mag.* (Tokyo) 43: 6. Jan. 1929 ("Myrmechophila"); (non Rolfe 1917).

Myrmecophila Rolfe Orchidaceae

Origins:
Greek *myrmex, myrmekos* "ant" and *philos* "lover, loving," the hollow pseudobulbs give shelter to ants.

Myrmecopteris Pichi Sermolli Polypodiaceae

Origins:
From the Greek *myrmex, myrmekos* "ant" and *pteris* "a fern," referring to rhizomes and ants; see R.E.G. Pichi Sermolli, "Fragmenta Pteridologiae — VI." in *Webbia*. 31(1): 237-259. Apr. 1977.

Myrmecosicyos C. Jeffrey Cucurbitaceae

Origins:
From the Greek *myrmex, myrmekos* "ant" and *sikyos* "wild cucumber, gourd," originally found around holes of harvester ants.

Myrmecostylum C. Presl Hymenophyllaceae

Origins:
From the Greek *myrmex, myrmekos* "ant" and *stylos* "a column."

Myrmedoma Becc. Rubiaceae

Origins:
From the Greek *myrmex, myrmekos* "ant" and *doma, domatos* "house," see also *Myrmephytum* Becc.

Myrmephytum Becc. Rubiaceae

Origins:
From the Greek *myrmex, myrmekos* "ant" and *phyton* "plant."

Myrmidone Mart. Melastomataceae

Origins:
Greek *myrmedon* "ant's nest"; Myrmidone was one of the 50 daughters of Danaüs, who killed her husband Mineus; Myrmidones, a people on the southern borders of Thessaly.

Myrobalanifera Houtt. Combretaceae

Origins:
From the Greek *myron* "a sweet-smelling oil, perfume," *balanos* "acorn" and *phoros* "bearing, carrying."

Myrobalanus Gaertner Combretaceae

Origins:
From the Greek *myron* "a sweet-smelling oil" and *balanos* "acorn"; Latin *myrobalanum* applied by Plinius to a balsam and to the fruit of a palm tree, from which the balsam itself was made.

Myrobroma Salisb. Orchidaceae

Origins:
Greek *myron* "a sweet-smelling oil, perfume" and *broma* "food"; see R.E. Schultes and A.S. Pease, *Generic Names of Orchids. Their Origin and Meaning.* 204. Academic Press, New York and London 1963.

Myrocarpus Allemão Fabaceae

Origins:
From the Greek *myron* "a sweet-smelling oil, perfume" and *karpos* "fruit."

Myrosma L.f. Marantaceae

Origins:
From the Greek *myron* "a sweet-smelling oil, perfume" and *osme* "smell, odor, perfume."

Species/Vernacular Names:
M. cannifolia L.f.
English: marble arrowroot

Myrosmodes Reichb.f. Orchidaceae

Origins:
From the Greek *myron* "a sweet-smelling oil" and *osmodes* "fragrant."

Myrospermum Jacq. Fabaceae

Origins:
From the Greek *myron* "a sweet-smelling oil" and *sperma* "seed."

Myrothamnus Welw. Myrothamnaceae

Origins:
From the Greek *myron* "perfume" and *thamnos* "bush, shrub," aromatic bush, resinous, from cells in the leaf epidermis.

Species/Vernacular Names:
M. flabellifolia Welw.
English: resurrection plant
Southern Africa: muFanichimuka, muForichimuka (Shona)
S. Rhodesia: iKalimela, uFazimuke

Myroxylon Forst. & Forst.f. Flacourtiaceae

Origins:
From the Greek *myron* "perfume" and *xylon* "wood"; see J.R. Forster and J.G.A. Forster, *Characteres generum plantarum.* 125, t. 63. (Nov.) 1775.

Myroxylon L.f. Fabaceae

Origins:

Greek *myron* "perfume, a sweet-smelling oil" and *xylon* "wood," referring to the resinous heartwood of these trees, a source of balsam.

Species/Vernacular Names:

M. balsamum (L.) Harms

Brazil: bálsamo-de-tolu

M. peruiferum L.f.

Brazil: bálsamo, óleo vermelho, árvore-de-bálsamo

Myrrhidendron J.M. Coulter & Rose Umbelliferae

Origins:

From the Greek *myrrhis, myrr(h)a*, a name used by Dioscorides for a plant, probably sweet cicely, *Myrrhis odorata*, and *dendron* "tree."

Myrrhis Miller Umbelliferae

Origins:

Latin *myrrha, murra, murrha* for the myrrh-tree and myrrh (Plinius), Greek (of Semitic origin) *myrrhis, myrrhidos, myrr(h)a*, a name used by Dioscorides for a plant, probably sweet cicely, *Myrrhis odorata*; see E. Masson, *Recherches sur les plus anciens emprunts sémitiques en grec.* Paris 1967; Giovanni Semerano, *Le origini della cultura europea. Dizionari Etimologici. Basi semitiche delle lingue indeuropee. Dizionario della lingua Greca.* 2(1): 189. Leo S. Olschki Editore, Firenze 1994; Ernest Weekley, *An Etymological Dictionary of Modern English.* 2: 969. Dover Publications, New York 1967; S. Battaglia, *Grande dizionario della lingua italiana.* X: 529. Torino 1978.

Species/Vernacular Names:

M. odorata (L.) Scop.

English: sweet cicely, garden myrrh

Myrrhoides Heister ex Fabr. Umbelliferae

Origins:

Resembling *Myrrhis*.

Myrsine L. Myrsinaceae

Origins:

Myrsine, myrrhine, ancient Greek names for the myrtle; Akkadian *murdudu*, Sumerian *mur-du-du* "a plant," Akka-dian *murdinnu, amurdinnu* "bramble"; see Carl Linnaeus, *Species Plantarum.* 196. 1753 and *Genera Plantarum.* Ed. 5. 90. 1754.

Species/Vernacular Names:

M. africana L.

English: Cape myrtle, African boxwood, Cape beech, wild myrtle, African myrsine

South Africa: mirting, vliegebos, wildemirt, wildemirting; chikuma, muDongera (Shona); thakxisa (Sotho); thlare-samadi (Tswana); tshilalantsa (Venda)

China: tie zai

M. australis (A. Rich) Allan

English: red matipo

New Zealand: mapau (Maori name)

M. gillianus Sond. (*Rapanea gilliana* (Sond.) Mez) (after the surgeon and plant collector William Gill, 1792-1863, d. Somerset East, Cape Province; see Mary Gunn and Leslie E. Codd, *Botanical Exploration of Southern Africa.* 168. Cape Town 1981)

South Africa: Gill's rapanea

M. helleri (Degener & I. Degener) St. John (*Rapanea helleri* Degener & I. Degener; *Suttonia angustifolia* Mez)

Hawaii: 'oliko, kolea

M. lessertiana A. DC. (*Myrsine fauriei* H. Lév.; *Suttonia cuneata* H. Lév. & Faurie; *Suttonia fauriei* (H. Lév.) H. Lév.)

Hawaii: kolea lau nui

M. melanophloeos (L.) Sweet

English: Cape beech

M. pillansii Adamson (for the South African botanist Neville Stuart Pillans, 1884-1964, who collected it in the Cape, author of "The genus *Phylica* Linn." *J. S. Afr. Bot.* 8: 1-164. 1942 and "Destruction of indigenous vegetation by burning on the Cape Peninsula." *S. Afr. J. Sci.* 21: 348-350. 1924; see John Hendley Barnhart, *Biographical Notes upon Botanists.* 3: 86. 1965; T.W. Bossert, *Biographical Dictionary of Botanists Represented in the Hunt Institute Portrait Collection.* 311. 1972; Mary Gunn and Leslie Edward W. Codd, *Botanical Exploration of Southern Africa.* 281-282. Cape Town 1981; Gordon Douglas Rowley, *A History of Succulent Plants.* Strawberry Press, Mill Valley, California 1997; John Hutchinson (1884-1972), *A Botanist in Southern Africa.* 53-54. London 1946; A. White and B.L. Sloane, *The Stapelieae.* Pasadena 1937)

South Africa: grootmirting

English: large Cape myrtle

M. sandwicensis A. DC. (*Myrsine vanioti* H. Lév.; *Suttonia mauiensis* (H. Lév.) H. Lév.; *Suttonia sandwicensis* (A.

DC.) Mez; *Suttonia vanioti* (H. Lév.) H. Lév.; *Rapanea sandwicensis* (A. DC.) Degener & Hosaka)

Hawaii: kolea lau li'i

M. semiserrata Wallich

English: needletooth myrsine

China: zhen chi tie zai

Myrsiphyllum Willd. Asparagaceae (Liliaceae)

Origins:

Myrsine, the ancient Greek name for myrtle and myrtle branch, and *phyllon* "leaf," or from the Greek *myrsos* "a basket" and *phyllon*, referring to the form of the leaves.

Species/Vernacular Names:

M. asparagoides (L.) Willd. (*Medeola asparagoides* L.; *Dracaena medeoloides* L.f.; *Asparagus asparagoides* (L.) Wight; *Asparagus medeoloides* (L.f.) Thunb.)

English: bridle creeper, smilax asparagus, smilax, florists smilax

M. scandens (Thunb.) Oberm. (*Asparagus scandens* Thunb.)

English: asparagus fern

Myrstiphylla Raf. Rubiaceae

Origins:

Referring to *Myrstiphyllum* P. Browne (1756); see Constantine Samuel Rafinesque (1783-1840), *Sylva Telluriana*. 148. 1838; E.D. Merrill, *Index rafinesquianus*. 226. 1949.

Myrtama Ovcz. & Kinsik. Tamaricaceae

Origins:

Myricaria and *Tamarix*.

Myrtastrum Burret Myrtaceae

Origins:

Referring to the genus *Myrtus*.

Myrtella F. Muell. Myrtaceae

Origins:

A diminutive of *myrtus*, myrtle; see F. von Mueller, *Descriptive Notes on Papuan Plants*. App. 105. Melbourne 1875[-1890].

Myrteola O. Berg Myrtaceae

Origins:

From the Latin *myrteolus*, of the color of the myrtle-blossoms.

Myrthoides Wolf Myrtaceae

Origins:

Resembling *myrtus*, myrtle.

Myrthus Scop. Myrtaceae

Origins:

Latin *myrtus*, myrtle.

Myrtillocactus Console Cactaceae

Origins:

From the Latin *myrtillus* "a small myrtle" plus *Cactus*, referring to the small, edible fruits, like *Vaccinium myrtillus*.

Myrtillocereus Fric & Kreuz. Cactaceae

Origins:

From the Latin *myrtillus* "a small myrtle" plus *Cereus*, referring to the fruits.

Myrtiluma Baillon Sapotaceae

Origins:

Latin *myrtus* with *luma*.

Myrtobium Miq. Eremolepidaceae

Origins:

From the Latin *myrtus* and Greek *bios* "life," hemiparasitic.

Myrtoleucodendron Kuntze Myrtaceae

Origins:

Myrtus plus *leukos* "white" and *dendron* "tree."

Myrtopsis Engl. Rutaceae

Origins:
Resembling *myrtus*.

Myrtus L. Myrtaceae

Origins:
Latin *myrtum*, *murtum* for the fruit of the myrtle, a myrtle-berry (Plinius), Latin *myrtus*, *murtus* for a myrtle, myrtle-tree (Plinius) and for a spear of myrtle-wood, Greek *myrtos*, *myrsine*; Akkadian *murdudû*, Sumerian *mur-dù-dù* "a plant," Akkadian *murdinnu*, *amurdinnu* "bramble"; see Carl Linnaeus, *Species Plantarum*. 471. 1753 and *Genera Plantarum*. Ed. 5. 212. 1754.

Species/Vernacular Names:
M. communis L.
English: myrtle
Italian: mirto
Bolivia: arrayan, mirto, chequen, arayana
Arabic: rihan, guemmam, adhera, mersin, aas, shalmun

Mystacidium Lindley Orchidaceae

Origins:
From the Greek *mystax*, *mystakos* "moustache," probably referring to the barbate lobes, or to the wiry peduncle.

Species/Vernacular Names:
M. capense (L.f.) Schltr.
English: tree orchid

Mystropetalon Harvey Balanophoraceae

Origins:
From the Greek *mystron* "a spoon" and *petalon* "a petal," referring to the perianth segments.

Mystroxylon Ecklon & Zeyher Celastraceae

Origins:
From the Greek *mystron* "a spoon" and *xylon* "wood."

Mytilaria Lecomte Hamamelidaceae

Origins:
Referring to the Latin *mitulus* and *mytulus*, *mytilus, ii* "a kind of mussel, sea-mussel," Greek *mytylos*.

Myuropteris C. Chr. Polypodiaceae

Origins:
From the Greek *myouros* for a plant, mouse-tail, plus *pteris* "a fern."

Myxopappus Källersjö Asteraceae

Origins:
From the Greek *myxa* "slime" and *pappos* "fluff, pappus"; Latin *myxa* for a kind of plum-tree, *Cordia myxa*, Plinius.

Myxopyrum Blume Oleaceae

Origins:
Greek *myxa* "slime" and *pyros* "wheat," slimy or pulpy seeds.

Species/Vernacular Names:
M. pierrei Gagnep. (*Myxopyrum hainanense* L.C. Chia)
English: Pierre myxopyrum
China: hai nan jiao he mu

Myxos Raf. Boraginaceae

Origins:
From the Latin and Greek *myxa* for a kind of plum-tree, Latin *myxum* is the fruit of the *myxa*, type *Cordia myxa* L.; see Constantine Samuel Rafinesque (1783-1840), *Sylva Telluriana*. 37. 1838; E.D. Merrill, *Index rafinesquianus*. 203. 1949.

Myzodendron Banks & Sol. ex R. Br. Misodendraceae

Origins:
From the Greek *myzo* "to suck" and *dendron* "tree," hemi-parasitic shrublets, see *Misodendrum* Banks ex DC.

Myzorrhiza Philippi Orobanchaceae

Origins:
From the Greek *myzo* "to suck" and *rhiza* 'a root, see *Orobanche* L.

N

Nabaluia Ames Orchidaceae

Origins:
Derived from Mount Kinabalu, Borneo.

Nabea Lehm. ex Klotzsch Ericaceae

Origins:
Probably named for the Scottish gardener William McNab, 1780-1848 (d. Edinburgh), Kew 1801-1810, Curator Royal Botanic Gardens Edinburgh 1810-1848. His writings include *Hints on the Planting and General Treatment of Hardy Evergreens, in the Climate of Scotland.* Edinburgh 1831 and *A Treatise on the Cultivation and General Treatment of Cape Heaths,* in a climate where they require protection during the winter months. Edinburgh 1832; see M. Hadfield et al., *British Gardeners: A Biographical Dictionary.* London 1980; H.R. Fletcher and W.H. Brown, *Royal Botanic Garden Edinburgh, 1670-1970.* 80-90. Edinburgh 1970.

Nabelekia Roshevitz Gramineae

Origins:
For Frantisek Nabelek, 1884-1965, Moravian professor of botany, author of *Iter- turcico-persicum.* Brno 1923-1929. See J.H. Barnhart, *Biographical Notes upon Botanists.* 2: 534. 1965; T.W. Bossert, *Biographical Dictionary of Botanists Represented in the Hunt Institute Portrait Collection.* 280. 1972; Stafleu and Cowan, *Taxonomic Literature.* 3: 678-679. 1981.

Nablonium Cass. Asteraceae

Origins:
From the Greek *nabla, nablas* "a musical instrument of ten or twelve strings, a Phoenician lyre," referring to the shape of the fruit; Hebrew *nebel* "harp," Akkadian *naba'um* "to invoke a deity, a lament."

Naegelia Regel Gesneriaceae

Origins:
For the Swiss (b. near Zürich) naturalist and botanist Carl Wilhelm von Nägeli (Naegeli), 1817-1891 (d. Munich), microscopist, a friend of Albert von Koelliker (1817-1905), at Geneva studied botany under Alphonse de Candolle, in Jena worked with the morphologist and anatomist Matthias Jacob Schleiden (1804-1881. He was the first to recognize the unique importance of the cell in biology; he was the cofounder, with Schwann, of the cell theory), 1850 in Zürich collaborated with Carl Cramer (1831-1901) in plant physiology research, in 1844 discovered the antherozoids of ferns and in 1850 those of the Rhizocarps, from 1857 professor of botany in the University of Munich, he may be considered to have anticipated the concept of mutation. His works include *Mechanisch-physiologische Theorie der Abstammungslehre.* [This contribution formed the basis of the future work in cytology by biologists such as Weismann and Nussbaum.] Munich & Leipzig 1884, *Botanische Mittheilungen.* München 1863-1881, *Die Cirsien der Schweiz.* Neuchatel 1840, with Simon Schwendener (1829-1919) wrote *Das Mikroskop.* Leipzig 1865-1867. See Robert Olby, in *Dictionary of Scientific Biography* 9: 600-602. 1981; Stafleu and Cowan, *Taxonomic Literature.* 3: 681-685. 1981; Carl E. Cramer, *Leben und Wirken von Carl Wilhelm Nägeli.* Zürich 1896; C. Correns, *Gregor Mendels Briefe an Carl Nägeli 1866-1873.* Leipzig 1903; B. Glass et al., eds., *Forerunners of Darwin: 1745-1859.* Baltimore 1959; J.H. Barnhart, *Biographical Notes upon Botanists.* 2: 534. 1965; E. Wunschmann, *Carl Wilhelm von Naegeli.* Berlin 1893; Gordon Douglas Rowley, *A History of Succulent Plants.* Strawberry Press, Mill Valley, California 1997; I.C. Hedge and J.M. Lamond, *Index of Collectors in the Edinburgh Herbarium.* Edinburgh 1970.

Nageia Gaertner Podocarpaceae

Origins:
Nagi, a vernacular name; see Jisaburo Ohwi, *Flora of Japan.* 110. [*Podocarpus nagi* (Thunb.) Zoll. & Moritzi ex Makino (*Myrica nagi* Thunb., *Nageia japonica* Gaertn., *Nageia nagi* (Thunb.) O. Kuntze)] Smithsonian Institution, Washington, D.C. 1965.

Nageliella L.O. Williams Orchidaceae

Origins:

For the plant collector Otto Nagel, traveled and collected orchids in Mexico; see Irving William Knobloch, compiled by, "A preliminary verified list of plant collectors in Mexico." *Phytologia Memoirs*. VI. Plainfield, N.J. 1983; Ida Kaplan Langman, *A Selected Guide to the Literature on the Flowering Plants of Mexico*. 1964; F. Verdoorn, "Botanical collectors in Latin American countries (prelim. list)." in *Chronica Botanica*. 6: 171-172. 1941.

Nagelocarpus Bullock Ericaceae

Origins:

Anagram of the generic name *Lagenocarpus*, from Greek *lagenos* "a flask, bottle, flagon" and *karpos* "fruit."

Nahusia Schneev. Onagraceae

Origins:

After Alexander Peter Nahuys, *circa* 1736-1794, author of *Oratio inauguralis de religiosa plantarum contemplatione, acerrimo ad divini numinis amorem et cultum stimulo*, etc. Trajecti ad Rhenum 1775.

Naias Adans. Najadaceae (Hydrocharitaceae)

Origins:
See *Najas* L.

Najas L. Najadaceae (Hydrocharitaceae)

Origins:

From the Greek *naias, naiados* (*nao* "to flow"), a Naiad, a nymph of fresh water; Akkadian *zinu* "rain"; see Carl Linnaeus, *Species Plantarum*. 1015. 1753 and *Genera Plantarum*. Ed. 5. 445. 1754.

Species/Vernacular Names:
N. flexilis (Willd.) Rostkov & Schmidt
English: slender water-nymph
N. gracillima (A. Braun) Magnus
English: thread-leaved water-nymph
N. graminea Delile
English: naiad, rice-field, water-nymph
Japan: hossu-mo

N. guadalupensis (Sprengel) Magnus
English: common water-nymph
N. horrida A. Br. (*Najas pectinata* (Parl.) Magnus)
English: sawgrass, saw-weed
South Africa: watersaagblaar
N. marina L.
English: holly-leaved water-nymph
N. minor Allioni
Japan: torige-mo
N. tenuifolia R. Br.
English: water nymph, Australian najad

Nama L. Hydrophyllaceae

Origins:

Greek *nao* "spring, to flow," *nama, namatos* "spring, running water, stream," Latin *nama, namatis* "a fluid, liquid," referring to the habitat; Akkadian *naba'um* "to rise, said of flood," *namba'u* "a large spring," *naqûm* "to pour out," *niqûm* "libation"; Hebrew *naba'* "to pour out," *nebeh* "spring."

Species/Vernacular Names:
N. sandwicensis A. Gray (*Conanthus sandwicensis* (A. Gray) A. Heller; *Marilaunidium sandwicense* (A. Gray) Kuntze)
Hawaii: hinahina kahakai

Namacodon Thulin Campanulaceae

Origins:
Bell from the Namib, Greek *kodon* "a bell."

Namaquanthus L. Bolus Aizoaceae

Origins:
Flower from Namaqualand, Greek *anthos* "flower."

Namaquanula D. Mueller-Doblies & U. Mueller-Doblies Amaryllidaceae (Liliaceae)

Origins:
From Namaqualand.

Namation Brand Scrophulariaceae

Origins:
From *namation*, the diminutive of the Greek *nama, namatos* "spring, running water, stream."

Namibia (Schwantes) Schwantes Aizoaceae

Origins:
From the Namib Desert, Southwest Africa — Namibia.

Nananthea DC. Asteraceae

Origins:
From the Greek *nanos* "dwarf" and *anthos* "flower."

Nananthus N.E. Br. Aizoaceae

Origins:
Greek *nanos* "dwarf" and *anthos* "flower."

Nanarepenta Matuda Dioscoreaceae

Origins:
Latin *nanus* "dwarf" and *repens* "creeping, prostrate."

Nandina Thunb. Berberidaceae (Nandinaceae)

Origins:
The Japanese name *nanten*.

Species/Vernacular Names:
N. domestica Thunb.
English: heavenly bamboo
Japan: nanten
Okinawa: nantin
China: nan tian zhu zi, nan tien chu, nan chu

Nandiroba Adans. Cucurbitaceae

Origins:
A vernacular name.

Nannoglottis Maxim. Asteraceae

Origins:
From the Greek *nannos* "dwarf" and *glossa, glotta* "tongue."

Nannorrhops H.A. Wendl. Palmae

Origins:
From the Greek *nannos* "dwarf" and *rhops* "a bush," bushy palms.

Nannoseris Hedb. Asteraceae

Origins:
From the Greek *nannos* "dwarf" and *seris, seridos*, a species of chicory or endive.

Nannothelypteris Holttum Thelypteridaceae

Origins:
Greek *nannos* and the genus *Thelypteris* Schmidel.

Nanochilus K. Schumann Zingiberaceae

Origins:
From the Greek *nanos* "dwarf" and *cheilos* "lip."

Nanocnide Blume Urticaceae

Origins:
From the Greek *nanos* "dwarf" and *knide* "nettle."

Species/Vernacular Names:
N. lobata Wedd.
Japan: yaeyama-katen-sô, shima-katen-sô

Nanodea Banks ex Gaertner f. Santalaceae

Origins:
From the Greek *nanodes* "dwarfish."

Nanodes Lindley Orchidaceae

Origins:

From the Greek *nanodes* "dwarfish," plant and flowers are very small.

Nanolirion Benth. Anthericaceae (Liliaceae)

Origins:

From the Greek *nanos* "dwarf" and *leirion* "a lily."

Nanopetalum Hassk. Euphorbiaceae

Origins:

From the Greek *nanos* "dwarf" and *petalon* "a petal."

Nanophyton Less. Chenopodiaceae

Origins:

From the Greek *nanos* "a dwarf" and *phyton* "plant."

Nanorrhinum Betsche Scrophulariaceae

Origins:

Greek *nanos* and *rhis, rhinos* "snout, nose," referring to the flower.

Nanostelma Baillon Asclepiadaceae

Origins:

From the Greek *nanos* "a dwarf" and *stelma, stelmatos* "a girdle, belt."

Nanothamnus Thomson Asteraceae

Origins:

From the Greek *nanos* "a dwarf" and *thamnos* "bush."

Napaea L. Malvaceae

Origins:

Latin *napaeus* "belonging to a wooded vale," *nymphae napaeae* "dell-nymphs," Greek Napaeae (-paiai) "nymphs of glens."

Napeanthus Gardner Gesneriaceae

Origins:

Greek Napaeae (-paiai) "nymphs of glens," *nape* "woodland vale, dell, glen" and *anthos* "flower."

Napeodendron Ridl. Meliaceae

Origins:

From the Greek *nape* "woodland vale, dell, glen" and *dendron* "tree."

Napina Fric Cactaceae

Origins:

From the Latin *napina, ae* "a turnip-field," *napus* "a kind of turnip."

Napoleona P. Beauv. Lecythidaceae (Napoleonaeaceae)

Origins:

Named in honor of the Emperor Napoléon Bonaparte, 1769-1821.

Napoleonaea P. Beauv. Lecythidaceae (Napoleonaeaceae)

Origins:

Named in honor of the Emperor Napoléon Bonaparte, 1769-1821.

Species/Vernacular Names:

N. imperialis P. Beauv.

Nigeria: irosun-igbo (Yoruba); ukpakonrisa (Edo); akbodo (Igbo); mabungi (Hausa)

Yoruba: boiboi, boribori, bongibongi

N. vogelii Hooker & Planchon (*Napoleonaea parviflora* Bak.f.)

Nigeria: boiboi, bobori, bori-bori, ukpagongbaragia, afo, itor; gbogbori (Yoruba); ukpagberajia (Edo)

Yoruba: igo, boiboi, boribori

Naravelia Adans. Ranunculaceae

Origins:

From *narawael*, the Sinhalese name for *Naravelia zeylanica* (L.) DC., the Malayalam name *naru-walli* refers to a trailing habit.

Narcissus L. Amaryllidaceae (Liliaceae)

Origins:

Classical ancient Greek name, from Akkadian *narum* "river" and *gissu* "a thorny bush or tree," some suggest from *narke* "dullness of sense, numbness" or from *naros* "wet" (Sanskrit *nira*) or from Sanskrit *nara* "very perfumed plant" and *kirros* "yellow"; Narcissus (-kissos), handsome youth, was son of the river-god Cephissus and Liriope; see Carl Linnaeus, *Species Plantarum.* 289. 1753 and *Genera Plantarum.* Ed. 5. 141. 1754; [Crusca], *Vocabolario degli Accademici della Crusca.* Firenze 1691, 1729-1738 and 1863-1923; Ernest Weekley, *An Etymological Dictionary of Modern English.* 2: 972. 1967; Manlio Cortelazzo and Paolo Zolli, *Dizionario etimologico della lingua italiana.* 3: 792. Bologna 1983; N. Tommaseo & B. Bellini, *Dizionario della lingua italiana.* Torino 1865-1879; Helmut Genaust, *Etymologisches Wörterbuch der botanischen Pflanzennamen.* 409-410. Basel 1996; G. Volpi, "Le falsificazioni di Francesco Redi nel Vocabolario della Crusca." in *Atti della R. Accademia della Crusca per la lingua d'Italia.* 33-136. 1915-1916; Giovanni Semerano, *Le origini della cultura europea.* Dizionari Etimologici. Basi semitiche delle lingue indeuropee. Dizionario della lingua Greca. 2(1): 192. Firenze 1994.

Species/Vernacular Names:

N. poeticus L.

English: poet's narcissus, pheasant's eye

N. pseudonarcissus L.

English: daffodil, wild daffodil, Lent lily, Tenby daffodil

N. tazetta L.

English: angel's tears, narcissus, polyanthus narcissus, tazetta

Arabic: nargis, behar, berengat

China: shui hsien, chin chan yin tai

Nardophyllum (Hook. & Arn.) Hook. & Arn. Asteraceae

Origins:

From the Greek *nardos* "spikenard" and *phyllon* "leaf."

Nardosmia Cass. Asteraceae

Origins:

From the Greek *nardos* "spikenard" and *osme* "smell, odor, perfume."

Nardostachys DC. Valerianaceae

Origins:

From the Greek *nardos* "spikenard" and *stachys* "a spike," used by Theophrastus (*HP.* 9.7.2); Latin *nardostachyon, nardostatius* "spikenard."

Species/Vernacular Names:

N. grandiflora DC.

English: spikenard, nard

Narduroides Rouy Gramineae

Origins:

Resembling *Nardurus.*

Nardurus (Bluff, Nees & Schauer) Reichb. Gramineae

Origins:

From the Greek *nardos* "spikenard, a fragrant shrub."

Nardus L. Gramineae

Origins:

Greek *nardos* "spikenard," Latin *nardus, i* and *nardum* "nard," Hebrew *nerd*, Akkadian *nadû* "to pour, to let water flow, to scatter"; see Carl Linnaeus, *Species Plantarum.* 53. 1753 and *Genera Plantarum.* Ed. 5. 27. 1754.

Species/Vernacular Names:

N. stricta L.

English: mat grass, nard grass

Narega Raf. Rubiaceae

Origins:

From the Malayalam *naregam* "citrus," based on *Catunaregam*; see C.S. Rafinesque, *Sylva Telluriana.* 98. 1838.

Naregamia Wight & Arn. Meliaceae

Origins:

From the Malayalam *nila-naregam*, from *nila* "ground" and *naregam* "citrus," used by van Rheede in *Hortus Indicus Malabaricus.* 10: t. 22. 1690.

Species/Vernacular Names:

N. alata Wight & Arn.

English: Goa ipecacuanha

Narica Raf. Orchidaceae

Origins:

Possibly dedicated to a nymph; see Constantine Samuel Rafinesque, *Flora Telluriana*. 2: 87. 1836 [1837].

Naringi Adans. Rutaceae

Origins:

From *narinjin*, a vernacular name for *Citrus maxima*.

Narthecium Huds. Melanthiaceae (Liliaceae)

Origins:

Latin *narthecium, ii* "an ointment-box, a medicine-chest," Greek *narthex, narthekos* "rod, giant fennel, casket," used by Theophrastus (*HP*. 1.2.7) and Plinius for *Ferula communis, narthekion* "small splint, small rod."

Species/Vernacular Names:

N. ossifragum (L.) Hudson

English: bog asphodel

Narthex Falc. Umbelliferae

Origins:

From the Greek *narthex, narthekos* "giant fennel, casket, a plant with a hollow stalk," Latin *narthex, ecis* "the shrub ferula," applied by Plinius and Theophrastus (*HP*. 1.2.7) to *Ferula communis*.

Narum Adans. Annonaceae

Origins:

A vernacular name, *Uvaria narum* (Dunal) Wall. (*Unona narum* Dunal), Rheede, *Hort. Mal.* 2: t. 9.

Naruma Raf. Annonaceae

Origins:

A vernacular name; see C.S. Rafinesque, *Anal. Nat. Tabl. Univ.* 175. 1815.

Nashia Millspaugh Verbenaceae

Origins:

Named for the American botanist George Valentine Nash, 1864-1921, horticulturist, plant collector, gardener, in Florida and the West Indies, author of "Poaceae (pars.)" in *North Amer. Fl.* 17(1): 77-98. 1909, 17(2): 99-196. 1912, 17(3): 197-198. 1915 and "The Kaffir-bread plants." *Journal of the N.Y. Bot. Gard.* 10: 275-277. 1909. See J.H. Barnhart, *Biographical Notes upon Botanists*. 2: 538. 1965; T.W. Bossert, *Biographical Dictionary of Botanists Represented in the Hunt Institute Portrait Collection*. 282. 1972; S. Lenley et al., *Catalog of the Manuscript and Archival Collections and Index to the Correspondence of John Torrey*. Library of the New York Botanical Garden. 365. 1973; Ethelyn Maria Tucker, *Catalogue of the Library of the Arnold Arboretum of Harvard University*. Cambridge, Massachusetts 1917-1933; Ida Kaplan Langman, *A Selected Guide to the Literature on the Flowering Plants of Mexico*. Philadelphia 1964; R. Zander, F. Encke, G. Buchheim and S. Seybold, *Handwörterbuch der Pflanzennamen*. 14. Aufl. Stuttgart 1993; I.C. Hedge and J.M. Lamond, *Index of Collectors in the Edinburgh Herbarium*. Edinburgh 1970.

Nasonia Lindley Orchidaceae

Origins:

From the Latin *nasus* "the nose," referring to the column, to the form of the anthers.

Nassella (Trin.) Desv. Gramineae

Origins:

Latin *nassa, ae* "a fish basket"; see Claude Gay (1800-1873), *Historia física y política de Chile ... Botánica [Flora chilena]*. 263. Paris 1846.

Species/Vernacular Names:

N. cernua (Stebb. & Löve) Barkworth

English: nodding needle grass

N. lepida (A. Hitchc.) Barkworth

English: foothill needle grass

N. pulchra (A. Hitchc.) Barkworth

English: purple needle grass

N. tenuissima (Trin.) Barkworth (*Stipa tenuissima* Trin.)

English: white tussock

South Africa: katdoogras, saagtandpolgras, witgras, witpolgras

N. trichotoma (Nees) Arechav (*Stipa trichotoma* Nees)

English: Australian serrated tussock, nasella tussock, nasella tussock-grass, New Zealand tussock-grass, serrated tussock, serrated tussock-grass, tumbleweed, Yass river tussock

South Africa: nassella-polgras, Nu Zeelandse polgras, saagtand polgras

Nastanthus Miers Calyceraceae

Origins:
From the Greek *nastos, naston* "close-pressed, firm" and *anthos* "flower," *nasso, natto* "to press, squeeze close," *nastos* is the Greek name for a kind of reed.

Nasturtiastrum (Gren. & Godr.) Gillet & Magne Brassicaceae

Origins:
Resembling *Nasturtium.*

Nasturtiicarpa Gilli Brassicaceae

Origins:
From the genus *Nasturtium* and *karpos* "fruit."

Nasturtiopsis Boiss. Brassicaceae

Origins:
Resembling *Nasturtium.*

Nasturtium R. Br. Brassicaceae

Origins:
The etymology of the generic name quite uncertain, possibly from the Latin *nasturtium, ii (nasi tortium)* ("quod nasum torqueat") or *nasturcium* or *nasturcum* for a kind of cress, referring to the pungent taste or to the acidity of some species; according to Plinius "nasturtium nomen accepit a narium tormento"; see F. Boerner & G. Kunkel, *Taschenwörterbuch der botanischen Pflanzennamen.* 4. Aufl. 139. 1989; H. Genaust, *Etymologisches Wörterbuch der botanischen Pflanzennamen.* 411-412. 1996; Georg Christian Wittstein, *Etymologisch-botanisches Handwörterbuch.* 608. 1852.

Species/Vernacular Names:
N. microphyllum Reichb. (*Nasturtium officinale* sensu Hawaiian botanists, non R. Br.)

Hawaii: leko

N. officinale R. Br. (*Nasturtium fontanum* (Lam.) Asch.)

English: common watercress, watercress

Arabic: guernech, karsun, rashad, harriqa

Maori name: kowhitiwhiti

Brazil: agrião-saúde-do-corpo, agrião oficial, agrião-da-fonte, agrião

China: xi yang cai gan

N. sarmentosum (DC.) Schinz & Guillaumin (*Cardamine sarmentosa* DC.; *Arabis o-waihiensis* Cham. & Schlechtend.; *Rorippa elstonii* Hochr.)

Hawaii: pa'ihi, 'ihi ku kepau, pa'ihi'ihi

Nastus Juss. Gramineae

Origins:
From the Greek *nastos, naston* "close-pressed, firm," *nastos* is the Greek name for a kind of reed, referring to the stem, tree-like.

Nastus Lunell Gramineae

Origins:
Nastos is the Greek name for a kind of reed.

Natalanthe Sond. Rubiaceae

Origins:
Flower from Natal.

Natsiatopsis Kurz Icacinaceae

Origins:
Resembling the genus *Natsiatum* Buch.-Ham ex Arn.

Natsiatum Buch.-Ham. ex Arn. Icacinaceae

Origins:
From *natsiat*, a Javanese plant name.

Nauclea L. Rubiaceae (Naucleaceae)

Origins:
Greek *naus, neos* "ship" and *kleos* "glory," referring to wood not suitable for the building of a ship, or from *naus*

and *kleio* "to close, shut, confine," the half capsule is boat-shaped; or from Latin *naucula, ae* "a little ship," because of the shape of the fruits; Latin *nauclerus*, Greek *naukleros* "skipper"; see Carl Linnaeus, *Species Plantarum*. 243. 1762; F.A. Sharr, *Western Australian Plant Names and Their Meanings*. A glossary. 50. University of Western Australia Press 1996; James A. Baines, *Australian Plant Genera. An Etymological Dictionary of Australian Plant Genera.* 250-251. Chipping Norton, N.S.W. 1981.

Species/Vernacular Names:

N. diderrichii (De Wild.) Merrill

Central Africa: n'gulu, bilinga, kilu, ntombo

Cameroon: lingui, lingi, ntoma, ntomba, hipen-lip-an, akondok, moukonia mamundi, mosayuri

Congo: n'gulu

Gabon: aloma, bilinga

Ivory Coast: badi

Yoruba: opepe

Nigeria: obiakhe; opepe (Yoruba); obliakhe (Edo); awesu (Itsekiri); urherekor (Urhobo); owoso (Ijaw); uburu (Igbo); ochi kanerung (Boki)

West Africa: badi, bilinga, opepe

N. junghuhnii (Miq.) Merrill (*Sarcocephalus horsfieldii* Elm.; *Sarcocephalus junghuhnii* Miq.) (after (Friedrich) Franz Wilhelm Junghuhn, 1809-1864, German botanist in Java, traveler, author of "Nova genera et species plantarum florae Javanicae." *Tydschr. nat. Gesch. Physiol.* 7: 285-317. 1840 and *Java*. Amsterdam 1850-1854. See J.H. Barnhart, *Biographical Notes upon Botanists*. 2: 263. 1965; M.C.P. Schmidt, ed., *Franz Junghuhn*. Leipzig 1909; M. Schmidt et al., *Gedenkboek Franz Junghuhn 1809-1909*. 's Gravenhage 1910; C.W. Wormser, *Frans Junghuhn*. Deventzer [1941]; M.N. Chaudhri, I.H. Vegter and C.M. De Wal, *Index Herbariorum*, Part II (3), *Collectors I-L*. Regnum Vegetabile vol. 86. 1972; Friedrich Anton Wilhelm Miquel (1811-1871), *Plantae junghuhnianae*. Enumeratio plantarum, quas in insulis Java et Sumatra, detexit Fr. Junghuhn. Lugduni-Batavorum [Leiden], Parisiis [1851]- 1853-1855[-1857]; Ida Kaplan Langman, *A Selected Guide to the Literature on the Flowering Plants of Mexico*. Philadelphia 1964; A. Lasègue, *Musée botanique de Benjamin Delessert*. Paris 1845; T.W. Bossert, *Biographical Dictionary of Botanists Represented in the Hunt Institute Portrait Collection*. 201. 1972; E.M. Tucker, *Catalogue of the Library of the Arnold Arboretum of Harvard University*. 1917-1933)

English: lesser bur-flower tree

Malaya: bengkai, bengkal, mengkal

N. latifolia Smith

French: liane-fraise

Sierra Leone: yumbuyambe

Congo: tsienga, kienga

Mali: bari, badi, ayugu

N. officinalis (Pierre ex Pitard) Merrill ex Chun

English: medicinal fatheadtree

N. orientalis (L.) L. (*Nauclea cordata* Roxb.; *Nauclea undulata* Roxb.; *Cephalanthus orientalis* L.)

English: cheesewood, Leichhardt's pine, soft Leichhardt, Leichhardt tree, Canary cheesewood

East Asia: kauluang, kanluang

N. pobeguinii (Pellegr.) Petit (after the French botanist Charles Henri Oliver Pobéguin, 1856-1951, colonial administrator in French Africa, plant collector, author of *Essai sur la flore de la Guinée française* produits forestiers, agricoles et industriels. Paris 1906 and of *Les plantes médicinales de la Guinée*. Paris 1912; see Auguste Jean Baptiste Chevalier (1873-1956), *Flore vivante de l'Afrique Occidentale Française*. 1938; J.H. Barnhart, *Biographical Notes upon Botanists*. 3: 93. 1965; E.M. Tucker, *Catalogue of the Library of the Arnold Arboretum of Harvard University*. 1917-1933; F.N. Hepper and Fiona Neate, *Plant Collectors in West Africa*. 65. 1971)

Cameroon: andinding, monse

Gabon: ntoma biliba

Ivory Coast: sibo

Nigeria: opepe ira, ira, opepe; opepe ira (Yoruba)

N. vanderguchtii (De Wild.) Petit

Nigeria: opepe ira (Yoruba)

Yoruba: opepe ira

Naucleopsis Miq. Moraceae

Origins:
Resembling *Nauclea*.

Naudinia Planchon & Linden Rutaceae

Origins:
After the French (b. Autun) botanist Charles Victor Naudin, 1815-1899 (d. near Antibes), horticulturist and arboriculturist, plant hybridizer, pioneer of contemporary biology, friend of Joseph Decaisne (1807-1882), Director of the Botanical Garden of the Villa Thuret at Antibes (France), established a private experimental garden. Among his writings are "Carioffleas." in Gay, *Fl. Chil.* 1: 250-284. 1846, "Elatineas." in Gay, *Fl. Chil.* 1: 284-286. 1846, *Le jardin du cultivateur*. Paris 1857, *Melastomacearum*. Parisiis 1849-1853, *Mémoire sur les eucalyptus introduits dans la région méditerranéenne*. Paris 1883, "De l'hybridité considérée comme cause de la variabilité dans les végétaux."

in *Comptes rendus ... de l'Académie des Sciences*. 59: 837-845. 1864 and "Espèces et variétés nouvelles de Cucurbitacées." *Ann. Sci. nat. Bot.* ser. 4. 16: 154-192 and 193-199. Aug.-Sep. 1862, with Ferdinand Jacob Heinrich von Mueller (1825-1896) wrote *Manuel de l'Acclimateur*. Paris 1887, with Joseph Decaisne wrote *Manuel de l'amateur des jardins*. Paris 1862-1871. See Pierre Eugène Marcelin Berthelot (1827-1907), *Notice historique sur la vie et les travaux de M. Naudin*. Paris 1900; William Coleman, in *Dictionary of Scientific Biography* 9: 618-619. 1981; R. Zander, F. Encke, G. Buchheim and S. Seybold, *Handwörterbuch der Pflanzennamen*. 14. Aufl. Stuttgart 1993; R.C. Olby, *Origins of Mendelism*. 62-66. London 1966; J.H. Barnhart, *Biographical Notes upon Botanists*. 2: 539. 1965; H.F. Roberts, *Plant Hybridization Before Mendel*. Princeton 1929; Ethelyn Maria Tucker, *Catalogue of the Library of the Arnold Arboretum of Harvard University*. Cambridge, Massachusetts 1917-1933; T.W. Bossert, *Biographical Dictionary of Botanists Represented in the Hunt Institute Portrait Collection*. 282. 1972; A. Lasègue, *Musée botanique de Benjamin Delessert*. Paris 1845; S. Lenley et al., *Catalog of the Manuscript and Archival Collections and Index to the Correspondence of John Torrey*. Library of the New York Botanical Garden. 1973; E.D. Merrill, in *Bernice P. Bishop Mus. Bull.* 144: 142. 1937 and *Contr. U.S. Natl. Herb.* 30(1): 225-226. 1947; E. Bretschneider, *History of European Botanical Discoveries in China*. 1981; Ida Kaplan Langman, *A Selected Guide to the Literature on the Flowering Plants of Mexico*. Philadelphia 1964; C.F.A. Christensen, *Den danske Botaniks Historie med tilhørende Bibliografi*. Copenhagen 1924-1926.

Naudiniella Krasser Melastomataceae

Origins:
After the French botanist Charles Victor Naudin, 1815-1899.

Nauenia Klotzsch Orchidaceae

Origins:
For a German orchid grower whose name was Nauen.

Naufraga Constance & Cannon Umbelliferae

Origins:
From the Latin *naufragus* "shipwrecked, wrecked," *navis frango*.

Naumburgia Moench Primulaceae

Origins:
For the German botanist Samuel Johann Naumburg, 1768-1799, professor of botany at Erfurt, author of *Dissertatio inauguralis botanica sistens delineationes Veronicae chamaedryos*. Erfordiae [Erfurt] [1792] and D.J.S.Naumburg's ... *Abhandlung von der Beinkrümmung*, nebst einer Beschreibung der Ehrenmannischen Fussmaschine ... Mit ... Kupfertafeln. Leipzig 1796; see Jonas C. Dryander, *Catalogus bibliothecae historico-naturalis Josephi Banks*. London 1796-1800.

Nauplius (Cass.) Cass. Asteraceae

Origins:
Nauplios was a son of Neptune and Amymone, a legendary king of Euboea, Greek *nauplios* "a shell fish," *naus* "ship, boat" and *pleo* "to sail, go by sea," Latin *nauplius, ii* "a kind of shell-fish"; *nauplius* is also a larval stage of certain crustaceans.

Nautilocalyx Linden Gesneriaceae

Origins:
From the Greek *nautilos* "of a ship, sailor, seaman, the nautilus" and *kalyx* "calyx," Latin *nautilus* for the nautilus, a cephalopod mollusk.

Nautochilus Bremek. Labiatae

Origins:
Greek *nautes* "sailor," *naus* "boat" and *cheilos* "lip," referring to the shape of the lower corolla lip.

Nautophylla Guillaumin Loganiaceae

Origins:
From the Greek *nautes* "sailor" and *phyllon* "leaf."

Navajoa Croizat Cactaceae

Origins:
Navaho or Navajo, a tribe of North American Indians, Athapascan stock; Navajo Mountain, in southern Utah; see Fritz Hochstätter, *In the Shadow of the Rocky Mountains: The Genera Pediocactus-Navajoa-Toumeya*. Mannheim, Germany 1995.

Navarretia Ruíz & Pavón Polemoniaceae

Origins:
After the Spanish physician Francisco Fernandez de Navarrete, professor of medicine at the University of Granada, Spain, author of *Ephemerides barometrico-medicas Matritenses.* 1737. See Ruíz & Pavón, *Flora peruvianae, et chilensis prodromus.* 20. Madrid 1794; Sir William J. Hooker and G.A.W. Arnott, *The Botany of Capt. Beechey's Voyage;* comprising an account of the plants collected by Messrs. Lay and Collie ... during the voyage to the Pacific and Bering's Strait, performed in H.M.S. *Blossom* ... 1825-1828. London [1830-] 1841.

Species/Vernacular Names:
N. divaricata (A. Gray) E. Greene
English: mountain navarretia
N. eriocephala H. Mason
English: hoary navarretia
N. heterandra H. Mason
English: Tehama navarretia
N. jepsonii Jepson
English: Jepson's navarretia
N. myersii P.S. Allen & Day
English: pincushion navarretia
N. rosulata Brand
English: Marin County navarretia
N. squarrosa (Eschsch.) Hook. & Arnott (*Hoitzia squarrosa* Eschsch.; *Gilia squarrosa* (Eschsch.) Hook. & Arn.)
English: skunkweed, Californian stinkweed
N. subuligera E. Greene
English: awl-leaved navarretia
N. tagetina E. Greene
English: marigold navarretia

Navia Schult.f. Bromeliaceae

Origins:
From the Latin *navia, ae,* a corruption of *navis* "a ship."

Navicularia Raddi Gramineae

Origins:
Latin *navicula* "a small vessel," *navicularius, a* "belonging to a small ship or boat."

Naxiandra (Baill.) Krasser Crypteroniaceae

Origins:
An anagram of the generic name *Axinandra* Thwaites.

Nealchornea Huber Euphorbiaceae

Origins:
Referring to *Alchornea,* a related genus; Greek *neos* "new" plus the genus.

Species/Vernacular Names:
N. yapurensis Huber
Amazon: botóka, na-nee-wa-kê-ta (see Richard E. Schultes and Robert F. Raffauf, *The Healing Forest. Medicinal and Toxic Plants of the Northwest Amazonia.* 177, 183. Portland, Oregon 1992)

Neanotis W.H. Lewis Rubiaceae

Origins:
Referring to the genus *Anotis* DC.

Neanthe O.F. Cook Palmae

Origins:
From the Greek *neos* "new" and *anthos* "flower."

Neatostema I.M. Johnston Boraginaceae

Origins:
Greek *neatos* "the last, uttermost, lowest" and *stema* "stamen," referring to the position of the stamens; see I.M. Johnston, in *Journal of the Arnold Arboretum.* 34: 2, 5. 1953.

Species/Vernacular Names:
N. apulum (L.) I.M. Johnston (*Lithospermum apulum* (L.) M. Vahl; *Myosotis apula* L.)
English: hairy sheepweed, blackweed

Nebelia Necker ex Sweet Bruniaceae

Origins:
Presumably from the Greek *nabla, nablas* "a musical instrument of ten or twelve strings, a Phoenician lyre"; Hebrew *nebel* "harp," Akkadian *naba'um* "to invoke a deity, a lament."

Neblinaea Maguire & Wurdack Asteraceae

Origins:

Pico da Neblina, Brazil (the country's highest elevation, at 3,014 m), in the Amazon Plains, in the Guiana Highlands north of the border with Venezuela.

Neblinantha Maguire Gentianaceae

Origins:

Pico da Neblina, Brazil, plus Greek *anthos* "flower."

Neblinanthera Wurdack Melastomataceae

Origins:

Pico da Neblina, Brazil, and Greek *anthera* "anther."

Neblinaria Maguire Guttiferae (Bonnetiaceae)

Origins:

Pico da Neblina, Brazil.

Neblinathamnus Steyerm. Rubiaceae

Origins:

Pico da Neblina, Brazil, plus Greek *thamnos* "shrub."

Nebrownia O. Kuntze Araceae

Origins:

After the Kew botanist Dr. Nicholas Edward Brown, 1849-1934 (d. Kew, Surrey). Among his writings are "Notes on the genera *Cordyline, Dracaena, Pleomele, Sansevieria* and *Taetsia*." *Bull. Misc. Inform.* 1914: 273-279. 1914, "The *Stapeliae* of Thunberg's herbarium." *J. Linn. Soc. Bot.* 17: 162-172. 1878, "*Acicarpha rosulata*, N.E. Brown (Calyceraceae)." *Hooker's Icon. Pl.* 1900, "The genera *Aloe* and *Mesembryanthemum* as represented in Thunberg's herbarium." *Bothalia* 1: 139-169. 1923, "The South African *Iridaceae* of Thunberg's herbarium." *J. Linn. Soc. Bot.* 48: 15-55. 1928 and "*Mesembryanthemum* and allied genera." *J. Bot.* 66(785): 138-145. 1928, contributor to Daniel Oliver (1830-1916), *Flora of Tropical Africa*. See J.H. Barnhart, *Biographical Notes upon Botanists.* 1: 263. 1965; T.W. Bossert, *Biographical Dictionary of Botanists Represented in the Hunt Institute Portrait Collection.* 55. 1972; Mary Gunn and Leslie E. Codd, *Botanical Exploration of Southern Africa.* 105. Cape Town 1981; Alain Campbell White

and Boyd Lincoln Sloane, *The Stapelieae.* Pasadena 1937; R. Zander, F. Encke, G. Buchheim and S. Seybold, *Handwörterbuch der Pflanzennamen.* 14. Aufl. Stuttgart 1993; Ray Desmond, *Dictionary of British & Irish Botanists and Horticulturists.* 107-108. London 1994; Adolar Gottlieb Julius Hans Herre (1895-1979), *The Genera of the Mesembryanthemaceae.* Cape Town 1971; Frans A. Stafleu and Erik A. Mennega, *Taxonomic Literature.* Supplement III. 136-138. Königstein 1995.

Necalistis Raf. Moraceae

Origins:

See Constantine Samuel Rafinesque, *Sylva Telluriana.* 58. 1838; Elmer D. Merrill, *Index rafinesquianus.* 112. 1949.

Necepsia Prain Euphorbiaceae

Origins:

Necepsus (Necepso), an astrologer in Egypt, a disciple of Aesculapius and Anubis.

Neckia Korthals Ochnaceae

Origins:

In honor of Jacob van Neck (Jacobo Neccio, I.C. Necq, J. Corneliszen Nek), 1564-1638, a Dutch Admiral, navigator. See Isaac Commelin, (ed), *Begin ende Voortgangh vande ... Oost-Indische Compagnie*, etc. [*Voyage to the East Indies* by Jac. Cornelisz van Neck and Wybrand van Warwijck (1598-1600). 56pp. Twenty-four plates and two maps; *Second Voyage to the East Indies* by Jac. van Neck (1600-2) (description by Roelof Roelofsz). 51pp. Four plates.] Amsterdam. 1646; J. Keuning, ed., *De tweede schipvaart der Nederlanders naar Oost-Indië onder Jacob Cornelisz. van Neck en Wybrant Warwijck, 1598-1600.* Journalen, documenten en andere bescheiden. 's-Grv., Nijhoff 1938-1951.

Necranthus Gilli Scrophulariaceae (Orobanchaceae)

Origins:

From the Greek *nekros* "corpse, dying person, the dead" and *anthos* "flower."

Nectandra Rol. ex Rottb. Lauraceae

Origins:

Greek *nektar* "nectar" and *andros* "man, male, stamen."

Species/Vernacular Names:

N. spp.

Peru: canela moena, carunje, dacoco

Mexico: ya yexru yoo

N. elaiophora Barb. Rodr.

Brazil: inhame, nhamu, inhamu

N. pichurim (Kunth) Mez

English: purchury bean

Peru: ambi caspi, huarme tashango, huarmi tashango, isula micuna, isula micuna muina, moena, muena, moena negra, pishco nahui muina, loro pucheri, pucherí, pucheri roble

Brazil: pichurim, pixuri, peixuri, pechurim, puchuri

N. rigida (Kunth) Nees (*Ocotea rigida* Kunth; *Ocotea incana* Schott ex Meissn.; *Ocotea ramentacea* Kunth; *Nectandra oppositifolia* Nees; *Nectandra rigida* Nees)

Brazil: canela amarela, canela fedorenta, canela garuva, canela ferrugem, canela ceiba, canela seiva, canela dura, canela inuçara, catinga de negro, louro da mata virgem

Nectaropetalum Engl. Erythroxylaceae

Origins:
Petals with scales, *nektar* and *petalon*, nectar scales grow on the petals.

Species/Vernacular Names:
N. capense (H. Bol.) Stapf & Boodle (*Erythroxylum capense* (H. Bol.) Stapf; *Peglera capensis* H. Bol.) Stapf (the genus *Peglera* after the botanical collector Miss Alice Pegler, author of "On the flora of Kentani." *Ann. Bolus Herbarium* 1916-1918)

Southern Africa: Keikokaboom; isiGqalaba (Zulu); iQande (Xhosa)

English: kei coca tree, Cape nectaropetalum

N. zuluense (Schonl.) Corbishley (*Erythroxylum zuluense* Schonl.)

English: Zulu nectaropetalum, Natal coca tree

Southern Africa: Natalkokaboom; iDweleba, isiGqalaba, iXweleba, iNgqalaba (Zulu); iQande

Nectaroscordum Lindley Alliaceae (Liliaceae)

Origins:
From the Greek *nektar* "nectar" and *skorodon, skordon* "garlic," referring to the nectaries on the ovary or to the scent.

Species/Vernacular Names:
N. siculum (Ucria) Lindl.

English: Sicilian honey-garlic

Nectouxia Kunth Solanaceae

Origins:
For Hippolyte (Hipolyte) Nectoux, a French botanist, with Napoléon to Egypt, author of *Voyage dans la Haute Égypte.* Paris 1808; see J.H. Barnhart, *Biographical Notes upon Botanists.* 2: 542. 1965; Jonas C. Dryander, *Catalogus bibliothecae historico-naturalis Josephi Banks.* 3: 613. London 1800; Ethelyn Maria Tucker, *Catalogue of the Library of the Arnold Arboretum of Harvard University.* Cambridge, Massachusetts 1917-1933.

Neea Ruíz & Pavón Nyctaginaceae

Origins:
For the French botanist Luis Née, from 1789 to 1794 on the Malaspina Expedition (the complement included the noted botanist Thaddeus Haenke); the first Italian to visit the Vancouver area was probably Captain Alessandro Malaspina, who explored the British Columbia coast for Spain in 1791. See J.H. Barnhart, *Biographical Notes upon Botanists.* 2: 542. Boston 1965; R. Zander, F. Encke, G. Buchheim and S. Seybold, *Handwörterbuch der Pflanzennamen.* 14. Aufl. Stuttgart 1993; Ida Kaplan Langman, *A Selected Guide to the Literature on the Flowering Plants of Mexico.* Philadelphia 1964; Miguel Colmeiro y Penido, *La Botánica y los Botánicos de la Peninsula Hispano-Lusitana.* Madrid 1858; A. Lasègue, *Musée botanique de Benjamin Delessert.* Paris 1845; Antonio José Cavanilles, *Icones et Descriptiones Plantarum.* 5: i-iii. 1799; August Weberbauer, *Die Pflanzenwelt der peruanischen Andes in ihren Grundzügen dargestellt.* 4-5. Leipzig 1911; John Dunmore, *Who's Who in Pacific Navigation.* 2, 170-171. Honolulu 1991; Günther Schmid, *Chamisso als Naturforscher. Eine Bibliographie.* Leipzig 1942; Alessandro (Alejandro) Malaspina (1754-1810, d. Pontremoli, Italy), *Viaje político-científico alrededor del mundo por las corbetas Descubierta y Atrevida* al mando de los capitanes de navío D. Alejandro Malaspina y Don José de Bustamante y Guerra desde 1789 á 1794. Publicado con una introducción por Don Pedro de Novo y Colson ... Segunda edición. Madrid 1885; Carlo Caselli, *Alessandro Malaspina e la sua spedizione scientifica intorno al mondo,* etc. [Lettere inedite di A. Malaspina a D. Paolo Greppi.] Alpes, Milano 1929; Emma Bona, *Alessandro Malaspina. Sue navigazioni ed esplorazioni.* Roma 1935; A. Giordano, "Alessandro Malaspina." in *Il vetro.* XVII. 1973; Alessandro D'Ancona, *Viaggiatori ed avventurieri.* Firenze, Sansoni [1911].

Species/Vernacular Names:

N. spp.

Peru: djírai, palo ceniza, palometa micuna

N. divaricata Poeppig & Endlicher

Peru: cumala, shula, palometa huayo, sacha cumala

N. floribunda Poeppig & Endlicher

Peru: mullo caspi, mullo huayo

N. laxa Poeppig & Endlicher

Peru: puca huayo

N. parviflora Poeppig & Endlicher

Peru: yana muco, yano muco

N. spruceana Heimerl

Peru: topa maca blanca

N. subpubescens Heimerl

Peru: intuto caspi, yututo caspi, yntutu caspi

Neeania Raf. Nyctaginaceae

Origins:

For the French botanist Luis Née, from 1789 to 1794 on the Malaspina Expedition (the first Italian to visit the Vancouver area was probably Captain Alessandro Malaspina, who explored the British Columbia coast for Spain in 1791); see C.S. Rafinesque, *Princ. Somiol.* 30. 1814 and *Flora Telluriana.* 1: 86. 1836 [1837].

Needhamia R. Br. Epacridaceae

Origins:

For the English Rev. John Turberville Needham (Needhamus Turbervillus), 1713-1781; see Robert Brown, *Prodromus florae Novae Hollandiae.* 549. London 1810.

Needhamiella L. Watson Epacridaceae

Origins:

Named for the English (b. London) Rev. John Turberville Needham (Needhamus Turbervillus), 1713-1781 (d. Belgium), microscopist, scientist, ordained a priest in 1738, Fellow of the Royal Society 1747, settled in Bruxelles 1768. Among his well-known works are *Observations upon the Generation, Composition, and Decomposition of Animal and Vegetable Substances.* London 1749, *An Account of Some New Microscopical Discoveries Founded on an Examination of the Calamary.* London 1745, *A Letter from Paris, Concerning Some New Electrical Experiments Made There.* 1746 and *Observations des hauteurs faites avec le baromètre au mois d'Aoust, 1751, sur une partie des Alpes.* Berne 1760. See Pierre Martial Cibot, *Lettre de Pékin.* London 1773; Herbert Kenneth Airy Shaw (1902-1985), in *Kew Bulletin.* 18: 272. 1965; Lazzaro Spallanzani (1729-1799), *Nouvelles recherches sur les découvertes microscopiques, et la génération des corps organisés.* Londres et Paris 1769; Rachel Horwitz Westbrook, in *Dictionary of Scientific Biography* 10: 9-11. 1981.

Neeopsis Lundell Nyctaginaceae

Origins:

Resembling *Neea.*

Neesenbeckia Levyns Cyperaceae

Origins:

Named for the German (b. near Erbach, Hesse) botanist Christian Gottfried (Daniel) Nees von Esenbeck, 1776-1858 (d. Breslau, Wroclaw), physician, 1800 Dr. med. Giessen, editor of Robert Brown, *Vermischte botanische Schriften.* Nürnberg 1825-1835, from 1818 to 1858 President of the Deutschen Akademie der Naturforscher *Leopoldina*, professor of botany, botanical collector. Among his works are *Handbuch der Botanik.* Nürnberg 1820-1821, *Hornschuchia*, novum plantarum brasiliensium genus. Ratisbonae [Regensburg] [1882], *Systema laurinarum.* Berolini [Berlin] 1836 and *Genera et species Asterearum.* Vratislaviae [Breslau] 1832, brother of the German botanist Theodor Friedrich Ludwig Nees von Esenbeck (1787-1837); see J.H. Barnhart, *Biographical Notes upon Botanists.* 2: 542. 1965; T.W. Bossert, *Biographical Dictionary of Botanists Represented in the Hunt Institute Portrait Collection.* 283. 1972; Johannes Proskauer, in *Dictionary of Scientific Biography* 10: 11-14. 1981; Günther Schmid, *Chamisso als Naturforscher. Eine Bibliographie.* Leipzig 1942; Ida Kaplan Langman, *A Selected Guide to the Literature on the Flowering Plants of Mexico.* Philadelphia 1964; A. Lasègue, *Musée botanique de Benjamin Delessert.* Paris 1845; S. Lenley et al., *Catalog of the Manuscript and Archival Collections and Index to the Correspondence of John Torrey.* Library of the New York Botanical Garden. 464. 1973; Ethelyn (Daliaette) Maria Tucker, *Catalogue of the Library of the Arnold Arboretum of Harvard University.* Cambridge, Massachusetts 1917-1933; E.D. Merrill, *Contr. U.S. Natl. Herb.* 30: 227. 1947; Stafleu and Cowan, *Taxonomic Literature.* 3: 705-712. 1981; R. Zander, F. Encke, G. Buchheim and S. Seybold, *Handwörterbuch der Pflanzennamen.* 14. Aufl. Stuttgart 1993; I.C. Hedge and J.M. Lamond, *Index of Collectors in the Edinburgh Herbarium.* Edinburgh 1970; Leonard Huxley, *Life and Letters of Sir J.D. Hooker.* London 1918.

Neesia Blume Bombacaceae

Origins:

For the German botanist Theodor Friedrich Ludwig Nees von Esenbeck, 1787-1837, professor of pharmacy at Bonn University and co-director of the Botanic Garden, author of *De muscorum propagatione*. Erlangae 1818 and *Radix plantarum mycetoidearum*. Bonnae 1820, joint author with Fridolin Carl Leopold Spenner (1798-1841), Alois (Aloys) Putterlick (1810-1845), Stephan Friedrich Ladislaus Endlicher, Carl Wilhelm Bischof (b. 1825), Johann Xaver Robert Caspary (1818-1887), Adalbert Carl Friedrich Hellwig Conrad Schnizlein (1814-1868) and Dietrich Brandis (1824-1907) of *Genera plantarum Florae germanicae*. Bonnae [1833-] 1835-1860, co-author with Carl Heinrich Ebermaier (1802-1870) of *Handbuch der medicinisch-pharmaceutischen Botanik*. Düsseldorf 1830-1832, with his brother the German botanist Christian Gottfried Daniel Nees von Esenbeck (1776-1858) wrote *De Cinnamomo disputatio*. Bonnae 1823 and *Plantarum in Horto Medico Bonnensi nutritarum icones selectae*. Bonnae 1824; see J.H. Barnhart, *Biographical Notes upon Botanists*. 2: 542. 1965; T.W. Bossert, *Biographical Dictionary of Botanists Represented in the Hunt Institute Portrait Collection*. 283. 1972; Ida Kaplan Langman, *A Selected Guide to the Literature on the Flowering Plants of Mexico*. 1964; E.M. Tucker, *Catalogue of the Library of the Arnold Arboretum of Harvard University*. 1917-1933; Stafleu and Cowan, *Taxonomic Literature*. 3: 712-715. 1981; R. Zander, F. Encke, G. Buchheim and S. Seybold, *Handwörterbuch der Pflanzennamen*. 14. Aufl. Stuttgart 1993; J.D. Milner, *Catalogue of Portraits of Botanists Exhibited in the Museums of the Royal Botanic Gardens*. Royal Botanic Gardens, Kew, London 1906.

Species/Vernacular Names:

N. sp.

Malaya: apa apa, ha ha

Neesiochloa Pilger Gramineae

Origins:

Named after the German botanist Christian Gottfried Daniel Nees von Esenbeck, 1776-1858, physician, philosopher, botanical collector, 1800 Dr. med. Giessen, from 1818 to 1858 President of the Deutschen Akademie der Naturforscher *Leopoldina*, professor of botany. Among his works are *Handbuch der Botanik*. Nürnberg 1820-1821, *Hornschuchia*, novum plantarum brasiliensium genus. Ratisbonae [Regensburg] [1882] and *Die Algen des süssen Wassers*. Bamberg 1814; see J.H. Barnhart, *Biographical Notes upon Botanists*. 2: 542. 1965.

Negria Chiov. Gramineae

Origins:

For the Italian botanist Giovanni Negri, 1877-1960, professor of botany, from 1925 to 1949 Director of the Istituto Botanico and Orto Botanico at Firenze. His works include *Erbario figurato*. Milano 1904 and *Atlante dei principali funghi commestibili e velenosi*. Torino 1908; see J.H. Barnhart, *Biographical Notes upon Botanists*. 2: 542. 1965; T.W. Bossert, *Biographical Dictionary of Botanists Represented in the Hunt Institute Portrait Collection*. 283. 1972; P. Luzzi, "L'Orto Botanico *Giardino dei Semplici*." in *Museo Nazionale di Storia Naturale* a Firenze: ipotesi di insediamento. Alinea Editrice, Firenze 1987; Guido Moggi, "Il Museo Botanico ed il Giardino dei Semplici dell'Università di Firenze." *Boll. Soc. It. Iris*. 26-39. 1975.

Negria F. Muell. Gesneriaceae

Origins:

For the Italian geographer Cristoforo (Christophorus) Negri, 1809-1896 (Firenze), politician and patriot. Among his works are *Le memorie di G. Pallavicino*. Considerazioni di Negri. Torino 1882, *Memorie storico-politiche sugli antichi Greci e Romani*. Torino 1864, *La storia antica restituita a verità, e raffrontata alla moderna*. Torino 1865 and *La storia politica dell'antichità paragonata alla moderna*. Venezia 1866, edited Giovanni Berchet (1783-1851), *La Repubblica di Venezia e la Persia*. 1865. See M. Carazzi, *La Società Geografica Italiana e l'esplorazione coloniale in Africa 1867-1900*. Firenze 1972; Ferdinand J.H. von Mueller, *Fragmenta Phytographiae Australiae*. 7: 151. 1871.

Species/Vernacular Names:

N. rhabdothamnoides F. Muell.

English: pumpkin flower

Negripteris Pic. Serm. Pteridaceae (Adiantaceae, Negripteridaceae)

Origins:

For the Italian botanist Giovanni Negri, 1877-1960, professor of botany, from 1925 to 1949 Director of the Orto Botanico at Firenze. His works include *Erbario figurato*. Milano 1904 and *Atlante dei principali funghi commestibili e velenosi*. Torino 1908; see J.H. Barnhart, *Biographical Notes upon Botanists*. 2: 542. 1965; R.E.G. Pichi Sermolli, "Negripteridaceae e Negripteris, nuova famiglia e nuovo genere delle Filicales." in *Nuovo Giorn. Bot. Ital*. ser. 2. 53(1-2): 129-169. 1947 [1946].

Negundium Raf. Aceraceae

Origins:

From Sanskrit and Bengali *nirgundi*, referring to *Acer* L.; see Constantine Samuel Rafinesque, *Med. Repos.* II. 5: 352. 1808; H. Genaust, *Etymologisches Wörterbuch der botanischen Pflanzennamen.* 413. 1996; R. Zander, F. Encke, G. Buchheim and S. Seybold, *Handwörterbuch der Pflanzennamen.* 14. Aufl. 647. Stuttgart 1993; F. Boerner & G. Kunkel, *Taschenwörterbuch der botanischen Pflanzennamen.* 4. Aufl. 139. Berlin & Hamburg 1989; Georg Christian Wittstein, *Etymologisch-botanisches Handwörterbuch.* 611. Ansbach 1852.

Neillia D. Don Rosaceae

Origins:

After the Scottish (b. Edinburgh) botanist Patrick Neill, 1776-1851 (d. Canonmills, Edinburgh), plant collector, printer, (founder and) secretary of the Caledonian Horticultural Society and Wernerian Society, in 1813 a Fellow of the Linnean Society, friend of G. Don. His writings include *Journal of a Horticultural Tour.* Edinburgh, London 1823, *An Address to the Members of the Wernerian Natural History Society.* Edinburgh 1830, *A Tour Through Some of the Islands of Orkney and Shetland.* Edinburgh 1806, "Remarks made in a *Tour thro' some of the Shetland Islands in 1804.*" *The Scots Magazine.* 67: 347-352 and 431-435. 1805, "Remarkable shower of hail in Orkney." *Transactions, Royal Society of Edinburgh.* 1818, *Fruit, Flower and Kitchen Garden.* Edinburgh 1840 and *Canonmills Loch and Meadow.* [Edinburgh 1832]. See J.H. Barnhart, *Biographical Notes upon Botanists.* 2: 543. 1965; E.M. Tucker, *Catalogue of the Library of the Arnold Arboretum of Harvard University.* 1917-1933; H.R. Fletcher and W.H. Brown, *Royal Botanic Garden Edinburgh, 1670-1970.* Edinburgh 1970; I.C. Hedge and J.M. Lamond, *Index of Collectors in the Edinburgh Herbarium.* Edinburgh 1970; F. Boerner & G. Kunkel, *Taschenwörterbuch der botanischen Pflanzennamen.* 4. Aufl. 139. Berlin & Hamburg 1989; Georg Christian Wittstein, *Etymologisch-botanisches Handwörterbuch.* 611. Ansbach 1852.

Neippergia C. Morren Orchidaceae

Origins:

Dedicated to the German patron of sciences Graf Alfred Neipperg, Württemberg, naturalist.

Neisosperma Raf. Apocynaceae

Origins:

Possibly from the Greek *nesos, nasos* "island" and *sperma* "seed," about twenty species from the Seychelles to the western Pacific; see C.S. Rafinesque, *Sylva Telluriana.* 162. Philadelphia 1838.

Species/Vernacular Names:

N. poweri (F.M. Bailey) Fosberg & Sachet

English: milkbush

Nelanaregam Adans. Meliaceae

Origins:

From the Malayalam *nila-naregam*, from *nila* "ground" and *naregam* "citrus," see also *Naregamia* Wight & Arnott.

Neleixa Raf. Rubiaceae

Origins:

See Constantine Samuel Rafinesque, *New Fl. N. Am.* 4: 56. 1836 [1838]; E.D. Merrill, *Index rafinesquianus.* 226. 1949.

Nelia Schwantes Aizoaceae

Origins:

After the South African (b. Natal) botanist Gert Cornelius Nel, 1885-1950 (d. Stellenbosch, South Africa), plant collector, cactus specialist, from 1921 to 1950 professor of botany at Stellenbosch University. Among his works are *Lithops* plantae succulentae, rarissimae, in terra obscuratae, etc. [1946], "Some South African succulents." *Cactus Succ. J.* 11: 4-5. Los Angeles 1949, *The Gibbaeum Handbook.* [Posth. edited by P.G. Jordan and Ernest William Shurly, 1888-1963] London 1953 and "Studien über die Amaryllidaceae-Hypoxideae, unter besonderer Berücksichtigung der afrikanischen Arten." *Bot. Jb.* 51: 234-340. 1914. See Hendrik Wijbrand De Boer (1885-1970), "In memoriam Professor Dr. G.C. Nel." *Succulenta, Amst.* 1950: 37-38. 1950; R. Zander, F. Encke, G. Buchheim and S. Seybold, *Handwörterbuch der Pflanzennamen.* 14. Aufl. 756. Stuttgart 1993; F. Boerner & G. Kunkel, *Taschenwörterbuch der botanischen Pflanzennamen.* 4. Aufl. 139. Berlin & Hamburg 1989; John H. Barnhart, *Biographical Notes upon Botanists.* 2: 543. 1965; Adolar Gottlieb Julius Hans Herre (1895-1979), *The Genera of the Mesembryanthemaceae.* Cape Town 1971; A. White and B.L. Sloane, *The Stapelieae.* Pasadena 1937; Gordon Douglas Rowley, *A History of Succulent Plants.* Strawberry Press, Mill Valley, California 1997; Mary Gunn and Leslie Edward W. Codd, *Botanical Exploration of Southern Africa.* 258-259. Cape Town 1981.

Nelipus Raf. Lentibulariaceae

Origins:

Greek *nelipous* "unshod, barefooted"; see Constantine Samuel Rafinesque (1783-1840), *Flora Telluriana.* 4: 109. Philadelphia 1838.

Nellica Raf. Euphorbiaceae

Origins:

See Constantine Samuel Rafinesque, *Sylva Telluriana.* 92. 1838; E.D. Merrill, *Index rafinesquianus.* 155. 1949.

Nelmesia Veken Cyperaceae

Origins:

Named after the British botanist Ernest Nelmes, 1895-1959 (d. Kew, Surrey), gardener, librarian, 1921-1958 Kew. His writings include "The genus *Carex* in Malaysia." *Reinwardtia.* 1(3): 221-450. 1951, "The genus *Carex* in Indo-Chine, incl. Thailand and Lower Burma." *Mém. Mus. Nat. Hist. Paris,* sér. B., 4(2): 83-182. 1955, "A key to the Carices of Malaysia and Polynesia." *Kew Bull.* 1946(1): 5-29. 1946 and "Notes on Cyperaceae: XXIV." *Kew Bull.* 1950: 189-208. 1950, with the Scottish William Cuthbertson (c. 1859-1934) compiled *Curtis's Botanical Magazine Dedications, 1827-1927.* [1931]. See John H. Barnhart, *Biographical Notes upon Botanists.* 2: 543. 1965; T.W. Bossert, *Biographical Dictionary of Botanists Represented in the Hunt Institute Portrait Collection.* 284. 1972; S. Lenley et al., *Catalog of the Manuscript and Archival Collections and Index to the Correspondence of John Torrey.* Library of the New York Botanical Garden. 307. 1973; Ida Kaplan Langman, *A Selected Guide to the Literature on the Flowering Plants of Mexico.* 1964.

Nelsonia R. Br. Acanthaceae

Origins:

After the Kew gardener David Nelson, died 1789 (Koepang, Timor), plant and seed collector, explorer, from 1776 to 1780 collected for Banks on Cook's third and last voyage, in 1787 with Captain (subsequently Rear Admiral) William Bligh on HMS *Bounty* (collected breadfruit trees); see Charles Rathbone Low, Lt., ed., *Captain Cook's Three Voyages Round the World, with a Sketch of His Life.* London 1906 (probably); John H. Barnhart, *Biographical Notes upon Botanists.* 2: 544. 1965; Eric Swenson, *The South Sea Shilling: Voyages of Captain Cook, R.N.* Ill. by Charles Michael Daugherty. NY 1952; Charles Louis L'Héritier de Brutelle (1746-1800), *Sertum anglicum seu plantae rariores quae in hortis juxta Londinum ... ab anno 1786 ad annum 1787 observatae.* Parisiis, Londini, Argentorati [Strasbourg] 1788; Robert Brown, *Prodromus florae Novae Hollandiae.* 480-481. London 1810; J. Britten, "Some early Cape botanists and collectors." *J. Linn. Soc. Bot.* 45: 29-51. 1920; Joseph Henry Maiden, *Sir Joseph Banks, the Father of Australia.* 124-125. Sydney and London 1909; Ida Lee (afterwards Marriott), *Early Explorers in Australia.* 76-77.

London 1925; Kenneth Lemmon, *The Golden Age of Plant Hunters.* 79-106. London 1968; Harold B. Carter, *Sir Joseph Banks (1743-1820). A Guide to Biographical and Bibliographical Sources.* Winchester 1987; E. Bretschneider, *History of European Botanical Discoveries in China.* 1981; J. Ewan, ed., *A Short History of Botany in the United States.* New York and London 1969; R. Glenn, *The Botanical Explorers of New Zealand.* Wellington 1950; I.C. Hedge and J.M. Lamond, *Index of Collectors in the Edinburgh Herbarium.* Edinburgh 1970; G. Murray, *History of the Collections Contained in the Natural History Departments of the British Museum.* London 1904; Frans A. Stafleu, *Linnaeus and the Linnaeans.* The spreading of their ideas in systematic botany, 1735-1789. Utrecht 1971; [John Barrow], *A Description of Pitcairn's Island and Its Inhabitants, with an Authentic Account of the Mutiny of the Ship Bounty and of the Subsequent Fortunes of the Mutineers.* J. & J. Harper, New York 1833; Captain Josiah N. Knowles, *Crusoes of Pitcairn Island. The Shipwreck Diary of ..., Master of the California Clipper, Wild Wave, 1858.* Edited by Richard S. Dillon 1957; Mary Gunn and Leslie Edward W. Codd, *Botanical Exploration of Southern Africa.* 259. Cape Town 1981.

Nelsonianthus H. Robinson & Brettell Asteraceae

Origins:

For the American naturalist Edward William Nelson, 1855-1934, explorer, in Mexico and Guatemala, plant collector, from 1890 to 1929 with the USDA. His writings include *Descriptions of New Genera, Species and Subspecies of Birds from Panama, Colombia and Ecuador.* Washington 1912, *The Eskimo about Bering Strait.* Washington 1881, "Lower California and its natural resources." *Mem. Natl. Acad. Sci.* 16: 1-194. Washington 1922, "A winter expedition in southwestern Mexico." *Natl. Geog. Mag.* 15(9): 339-356. 1904 and *Wild Animals of North America.* Washington 1930. See [edited by H.W. Henshaw] *Report upon Natural History Collections made in Alaska between the years 1877 and 1881 by E.W. Nelson.* 1887; E.A. Goldman, "Edward William Nelson, naturalist, 1855-1934." *Auk.* 52: 135-148. 1935; John H. Barnhart, *Biographical Notes upon Botanists.* 2: 544. 1965; H. Robinson & R.D. Brettell, in *Phytologia.* 27(1): 54. 1973; Irving William Knobloch, compiled by, "A preliminary verified list of plant collectors in Mexico." *Phytologia Memoirs.* VI. Plainfield, N.J. 1983; Ida Kaplan Langman, *A Selected Guide to the Literature on the Flowering Plants of Mexico.* 1964; Joseph Ewan, *Rocky Mountain Naturalists.* The University of Denver Press 1950.

Nelumbicum Raf. Nelumbonaceae

Origins:
See C.S. Rafinesque, *Med. Repos.* II. 5: 351. 1808; E.D. Merrill, *Index rafinesquianus.* 123. 1949.

Nelumbium Juss. Nelumbonaceae

Origins:
Orthographic variant of *Nelumbo* Adans.

Nelumbo Adans. Nelumbonaceae

Origins:
Sinhalese name for lotus plant, *nelumbu, nelun.*

Species/Vernacular Names:
N. lutea (Willd.) Pers.
English: water chinquapin, American lotus
N. nucifera Gaertner
English: sacred lotus, Hindu lotus, sacred lotus of India, lotus, pink water lily, red lily, pink lotus lily, Egyptian bean
French: nénuphar de Chine, lotus sacré, fève d'Egypte
Japan: Hasu (= the Throne of Hasu = The Lotus Throne), renkon (for the roots, rhizomes)
Okinawa: rin, din
China: he ye, ho yeh, lian zi, ho, fu chu
Vietnam: sen, lien
The Philippines: baino, balbalino, sukao
Indonesia: padma (padma is a Sanskrit word that means lotus, and the lotus is the seat of God), teratai
Balinese name: bungan tunjung
Malaya: seroja, telepok, teratai

Nemacaulis Nutt. Polygonaceae

Origins:
Greek *nema, nematos* "thread" and *kaulos* "stem," Latin *caulis.*

Species/Vernacular Names:
N. denudata Nutt.
Australia: woolly-heads

Nemacladus Nutt. Campanulaceae

Origins:
From the Greek *nema* "thread" and *klados* "a branch."

Species/Vernacular Names:
N. gracilis Eastw.
English: slender nemacladus
N. twisselmannii J. Howell
English: Twisselmann's nemacladus

Nemaconia Knowles & Westcott Orchidaceae

Origins:
From the Greek *nema* "thread" and *akon* "dart, javelin," referring to the leaves.

Nemaluma Baill. Sapotaceae

Origins:
Greek *nema* "thread" and *luma.*

Nemastachys Steud. Gramineae

Origins:
From the Greek *nema* "thread" and *stachys* "a spike."

Nemastylis Nutt. Iridaceae

Origins:
From the Greek *nema* "thread" and *stylos* "a column," referring to the slender style.

Nematanthus Schrader Gesneriaceae

Origins:
From the Greek *nema, nematos* "thread" and *anthos* "flower," alluding to the flower-stalks or pedicels.

Nematoceras Hook.f. Orchidaceae

Origins:
From the Greek *nema, nematos* "thread" and *keras* "horn," sepals and petals.

Nematolepis Turcz. Rutaceae

Origins:
Greek *nema* and *lepis* "scale," referring to the base of the stamens, to the long hairs at the base of the style; see Porphir

Kiril N.S. Turczaninow (1796-1863), *Bulletin de la Société Impériale des Naturalistes de Moscou.* 25(2): 158. 1852.

Nematophyllum F. Muell. Fabaceae

Origins:

Greek *nema* "thread" and *phyllon* "a leaf"; in *Hooker's Journal of Botany & Kew Garden Miscellany.* 9: 20. London 1857.

Nematopoa C.E. Hubb. Gramineae

Origins:

From the Greek *nema, nematos* "thread" and *poa* "grass, pasture grass."

Nematopogon Bureau & K. Schumann Bignoniaceae

Origins:

From the Greek *nema, nematos* "thread" and *pogon* "a beard, hair."

Nematopteris Alderw. Grammitidaceae

Origins:

From the Greek *nema, nematos* "thread" and *pteris* "a fern."

Nematopus A. Gray Asteraceae

Origins:

Greek *nema, nematos* "thread" and *pous, podos* "foot"; in *Hooker's Journal of Botany & Kew Garden Miscellany.* 3: 98 and 3: 150. London 1851.

Nematosciadium H. Wolff Umbelliferae

Origins:

From the Greek *nema, nematos* "thread" and *skiadion, skiadeion* "umbel, parasol."

Nematostemma Choux Asclepiadaceae

Origins:

From the Greek *nema, nematos* "thread" and *stemma* "crown, a garland."

Nematostigma A. Dietrich Iridaceae

Origins:

From the Greek *nema, nematos* "thread" and *stigma.*

Nematostigma Planchon Ulmaceae

Origins:

From the Greek *nema, nematos* "thread" and *stigma.*

Nematostylis Hook.f. Rubiaceae

Origins:

From the Greek *nema, nematos* "thread" and *stylps* "column, style."

Nematuris Turcz. Asclepiadaceae

Origins:

From the Greek *nema, nematos* "thread" and *oura* "tail."

Nemcia Domin Fabaceae

Origins:

After the Czech botanist Dr. Bohumil Rehor Nemec, 1873-1966, mycologist and plant physiologist, from 1904 professor of plant anatomy and physiology at Charles University at Prague, author of *Die Reizleitung und die reizleitenden Strukturen bei den Pflanzen.* Jena 1901, *Dejiny ovocnictvi.* Praha 1955 and *Studien über die Regeneration.* Berlin 1905, with L. Pastyrik wrote *Vseobecná botanika.* Bratislava 1948, with Otakar Matousek wrote *Jan Evangelista Purkyne,* badatel — národni buditel. Praha 1955. See John H. Barnhart, *Biographical Notes upon Botanists.* 2: 545. 1965; Jan Kapras, the Younger, *Idea Ceskoslovenského státu.* 1936; T.W. Bossert, *Biographical Dictionary of Botanists Represented in the Hunt Institute Portrait Collection.* 284. 1972; S. Lenley et al., *Catalog of the Manuscript and Archival Collections and Index to the Correspondence of John Torrey.* Library of the New York Botanical Garden. 307. 1973; E.M. Tucker, *Catalogue of the Library of the Arnold Arboretum of Harvard University.* 1917-1933; Karel Domin (1882-1953), in *Preslia.* 2: 27. 1923; Ida Kaplan Langman, *A Selected Guide to the Literature on the Flowering Plants of Mexico.* 1964.

Species/Vernacular Names:
N. capitata (Benth.) Domin

English: bacon and eggs

N. leakeana (Drumm.) Crisp

English: mountain pea

N. obovata (Benth.) Crisp

English: boat-leaved poison

Nemesia Vent. Scrophulariaceae

Origins:

From the ancient Greek name *nemesion, nemeseion* (*nemo* "to distribute, to enjoy, to pasture, to feed," *nemos* "wooded pasture, glade, a grove") used by Dioscorides for *Silene* or a plant similar to these herbs; see W.P.U. Jackson, *Origins and Meanings of Names of South African Plant Genera.* 130. Rondebosch 1990.

Species/Vernacular Names:

N. fruticans (Thunb.) Benth. (*Antirrhinum fruticans* Thunb.; *Nemesia capensis* (Thunb.) Kuntze; *Nemesia divergens* Benth.; *Nemesia foetens* Vent.; *Nemesia gracillima* Dinter)

South Africa: leeubekkie

N. strumosa Benth.

English: nemesia

South Africa: leeubekkie, rooileeubekkie

Nemodon Griff. Convolvulaceae

Origins:

From the Greek *nema* "thread" and *odous, odontos* "tooth."

Nemopanthes Raf. Aquifoliaceae

Origins:

See Constantine Samuel Rafinesque, *Jour. Phys. Chim. Hist. Nat.* 89: 96. 1819 and *Sylva Telluriana.* 51. 1838.

Nemopanthus Raf. Aquifoliaceae

Origins:

Greek *nema* "thread," *pous* "foot" and *anthos* "flower," referring to the peduncles, or from *nema, ops, opsis* "aspect, sight, appearance, resemblance" and *anthos*; see Constantine Samuel Rafinesque, *Florula ludoviciana.* 167. New York 1817; E.D. Merrill, *Index rafinesquianus.* 159-160. 1949.

Nemophila Nutt. Hydrophyllaceae

Origins:

Greek *nemos* "wooded pasture, glade, a grove" and *philos* "loving," referring to the habitat.

Species/Vernacular Names:

N. maculata Lindley

English: fivespot

N. menziesii Hook. & Arn.

English: baby blue-eyes

Nemopogon Raf. Liliaceae

Origins:

Probably from the Greek *nema* "thread" and *pogon* "a beard, hair"; see C.S. Rafinesque, *Flora Telluriana.* 2: 27. 1836 [1837].

Nemosenecio (Kitam.) B. Nord. Asteraceae

Origins:

Greek *nemos* "wooded pasture, a grove" plus *Senecio*.

Nemoseris Greene Asteraceae

Origins:

Greek *nemos* "wooded pasture, a grove" or *nema* "thread" and *seris, seridos*, a species of chicory or endive.

Nemostylis Steven Rubiaceae

Origins:

Presumably from the Greek *nema* "thread" and *stylos* "pillar, style."

Nemum Desv. Cyperaceae

Origins:

Nemus was the sacred grove of Diana, poet. transf. a tree, also wood; Latin *nemus* and Greek *nemos* "wooded pasture, a grove."

Nemuranthes Raf. Orchidaceae

Origins:

From the Greek *nema* "thread," *oura* "tail" and *anthos* "flower," referring to the spur; see C.S. Rafinesque, *Flora Telluriana.* 2: 61. 1836 [1837].

Nenga H.A. Wendland & Drude Palmae

Origins:

A Javanese vernacular name, *nenge*, type species *Pinanga nenga* (Blume ex Martius) Blume (*Areca nenga* Blume ex Martius); see Wendland & O. Drude, "Palmae Australasicae." in *Linnaea*. 39: 155-238. 1875; Edwino S. Fernando, "A revision of the genus *Nenga*." *Principes*. 27(2): 55-70. 1983; O. Beccari & R.E.G. Pichi Sermolli, "Subfamiliae Arecoidearum Palmae Gerontogeae. Tribuum et Generum Conspectus." 25 Mar. 1955, seors. impr. ex *Webbia*. 11: 1-187. 31 Mar. 1956.

Species/Vernacular Names:

N. pumila (Martius) H.A. Wendland (*Areca pumila* Martius)

English: pinang palms

Nengella Beccari Palmae

Origins:
The diminutive of *Nenga*.

Neo-urbania Fawcett & Rendle Orchidaceae

Origins:
After the German botanist Ignatz Urban, 1848-1931, specialized in the flora of the West Indies, from 1889 professor Botanical Garden and Museum at Berlin-Dahlem, wrote on botanical biography. Among his most valuable writings are *Plantae novae Antillanae*. Berlin 1895 and *Zur Flora Südamerikas*. Halle a.S. 1882, edited *Symbolae Antillanae seu fundamenta florae Indiae occidentalis*. Berolini etc. 1898-1928; see J.H. Barnhart, *Biographical Notes upon Botanists*. 3: 417. 1965; T.W. Bossert, *Biographical Dictionary of Botanists Represented in the Hunt Institute Portrait Collection*. 411. 1972; E.M. Tucker, *Catalogue of the Library of the Arnold Arboretum of Harvard University*. 1917-1933; S. Lenley et al., *Catalog of the Manuscript and Archival Collections and Index to the Correspondence of John Torrey*. Library of the New York Botanical Garden. 418. 1973; R. Zander, F. Encke, G. Buchheim and S. Seybold, *Handwörterbuch der Pflanzennamen*. 14. Aufl. Stuttgart 1993; Stafleu and Cowan, *Taxonomic Literature*. 6: 606-619. 1986.

Neo-uvaria Airy Shaw Annonaceae

Origins:
Resembling *Uvaria* L.

Neoabbottia N.L. Britton & J.N. Rose Cactaceae

Origins:
After the American William Louis Abbott, b. 1860, patron of botany and natural history.

Neoacanthophora Bennet Araliaceae

Origins:
Greek *neos* "new" and the genus *Acanthophora* Merr.

Neoalsomitra Hutch. Cucurbitaceae

Origins:
Greek *neos* "new" and *Alsomitra* (Blume) Roem. (Greek *alsos* "a grove" and *mitra* "a turban, bishop's head-dress, cap, mitra").

Neoancistrophyllum Rauschert Palmae

Origins:
Greek *neos* "new" and *Ancistrophyllum* (G. Mann & H. Wendland) H. Wendland.

Neoapaloxylon Rauschert Caesalpiniaceae

Origins:
Referring to the genus *Apaloxylon* Drake, Greek *hapalos* "soft, tender, delicate, weak" and *xylon* "wood."

Neoastelia J.B. Williams Asteliaceae (Liliaceae)

Origins:
From the Greek *neos* "new" and the genus *Astelia*, Greek *a* "without" and *stele* "a pillar, column, trunk."

Neoathyrium Ching & Z.R. Wang Dryopteridaceae (Woodsiaceae)

Origins:
Greek *neos* "new" and the genus *Athyrium* Roth.

Neoaulacolepis Rauschert Gramineae

Origins:
Greek *neos* "new" and the genus *Aulacolepis* Hackel, Greek *aulax, aulakos* "a furrow" and *lepis* "a scale."

Neobaclea Hochr. Malvaceae

Origins:

After the Swiss naturalist César Hippolyte Bacle, 1794-1838, plant collector; see F.N. Hepper and Fiona Neate, *Plant Collectors in West Africa*. 6. Utrecht 1971; J.H. Barnhart, *Biographical Notes upon Botanists*. Boston 1965; Joseph Vallot, "Études sur la flore du Sénégal." in *Bull. Soc. Bot. de France*. 29: 168-238. Paris 1882; J. Lanjouw and F.A. Stafleu, *Index Herbariorum*. Part II, *Collectors A-D*. Regnum Vegetabile vol. 2. 1954.

Neobakeria Schlechter Hyacinthaceae (Liliaceae)

Origins:

For the British botanist John Gilbert Baker, 1834-1920 (d. Kew, Surrey), Herbarium of Kew, botanical collector, 1866 Fellow of the Linnean Society, in 1878 a Fellow of the Royal Society. See R.G.C. Desmond, in *Dictionary of Scientific Biography* 1: 412-413. 1981; J.H. Barnhart, *Biographical Notes upon Botanists*. 1: 108. 1965; T.W. Bossert, *Biographical Dictionary of Botanists Represented in the Hunt Institute Portrait Collection*. 1972; Ida Kaplan Langman, *A Selected Guide to the Literature on the Flowering Plants of Mexico*. 1964; Mea Allan, *The Hookers of Kew*. 1967; Ernest Nelmes and William Cuthbertson, *Curtis's Botanical Magazine Dedications, 1827-1927*. [1931]; S. Lenley et al., *Catalog of the Manuscript and Archival Collections and Index to the Correspondence of John Torrey*. 17. 1973; E. Bretschneider, *History of European Botanical Discoveries in China*. 1981; H.N. Clokie, *Account of the Herbaria of the Department of Botany in the University of Oxford*. 124. Oxford 1964; E.D. Merrill, in *Bernice P. Bishop Mus. Bull*. 144: 39-41. 1937; Clyde F. Reed, *Bibliography to Floras of Southeast Asia*. Baltimore, Maryland 1969.

Neobalanocarpus Ashton Dipterocarpaceae

Origins:

From the Greek *neos* "new" plus the genus *Balanocarpus*.

Species/Vernacular Names:

N. heimii (King) Ashton

Southeast Asia: chengal

Neobambus Keng f. Gramineae

Origins:

Greek *neos* "new" plus *bambus, bambusa*.

Neobaronia Baker Fabaceae

Origins:

For the English missionary Rev. Richard Baron, 1847-1907 (d. Lancs.), botanist, plant collector in Madagascar, 1882 Fellow of the Linnean Society.

Neobartlettia R. King & H. Robinson Asteraceae

Origins:

After the American botanist Harley Harris Bartlett, 1886-1960, professor of botany, biologist, traveler and botanical explorer (Southeast Asia, Mexico, Central America and Argentina), plant collector. See John H. Barnhart, *Biographical Notes upon Botanists*. 1: 131. 1965; T.W. Bossert, *Biographical Dictionary of Botanists Represented in the Hunt Institute Portrait Collection*. 27. 1972; Ida Kaplan Langman, *A Selected Guide to the Literature on the Flowering Plants of Mexico*. 115. Philadelphia 1964; S. Lenley et al., *Catalog of the Manuscript and Archival Collections and Index to the Correspondence of John Torrey*. Library of the New York Botanical Garden. 24. 1973; Ethelyn Maria Tucker, *Catalogue of the Library of the Arnold Arboretum of Harvard University*. Cambridge, Massachusetts 1917-1933; J. Ewan, ed., *A Short History of Botany in the United States*. New York and London 1969; R. Zander, F. Encke, G. Buchheim and S. Seybold, *Handwörterbuch der Pflanzennamen*. 14. Aufl. Stuttgart 1993; Irving William Knobloch, compiled by, "A preliminary verified list of plant collectors in Mexico." *Phytologia Memoirs*. VI. Plainfield, N.J. 1983.

Neobartlettia Schlechter Orchidaceae

Origins:

For the English botanist Albert William Bartlett, 1875-1943, plant collector, mycologist, 1903 Fellow of the Linnean Society, Government Botanist and Superintendent of the Botanic Garden in Georgetown, British Guiana.

Neobassia A.J. Scott Chenopodiaceae

Origins:

From the Greek *neos* "new" and the genus *Bassia*, for the Italian botanist Ferdinando Bassi, 1710-1774, naturalist, author of *Ambrosina, novum plantae genus*. Bononiae 1763 and *Delle terme porrettane*. 1768. See Antonio Bertoloni (1775-1869), *Sylloge plantarum horti bononiensis*. Bononiae 1827; John H. Barnhart, *Biographical Notes upon Botanists*. 1: 135. 1965; Carlo Allioni (1728-1804), *Mélanges de Philosophie et de Mathématique de la Société Royale de Turin* pour les années 1762-1765. 177, t. IV, fig. 2. 1766; R. Zander, F. Encke, G. Buchheim and S. Seybold, *Handwörterbuch der Pflanzennamen*. 14. Aufl. Stuttgart 1993; Jonas C. Dryander, *Catalogus bibliothecae historico-naturalis Josephi Banks*. London 1800; A. Lasègue, *Musée botanique de Benjamin Delessert*. Paris 1845.

Species/Vernacular Names:
N. proceriflora (F. Muell.) A.J. Scott (*Threlkeldia proceriflora* F. Muell.)
Australia: soda bush, desert glasswort

Neobathiea Schltr. Orchidaceae

Origins:
After the French botanist Joseph Marie Henri Alfred Perrier de la Bâthie, 1873-1958, plant collector, from 1896 to 1933 in Madagascar. Among his works are *La végétation malgache*. Marseille, Paris 1921, contributed to Henri Humbert (1887-1967), *Flore de Madagascar et des Comores*. 1936-1967. See Alfred Lacroix, *Notice historique sur quatre botanistes*, etc. [On P.H. Lecomte, H. Perrier de la Bâthie, E.M. Heckel and H. Jumelle.] 1938; J.H. Barnhart, *Biographical Notes upon Botanists*. 3: 70. 1965; R. Zander, F. Encke, G. Buchheim and S. Seybold, *Handwörterbuch der Pflanzennamen*. 14. Aufl. Stuttgart 1993.

Neobaumannia Hutch. & Dalziel Rubiaceae

Origins:
For the German administrator Ernst Baumann, b. 1893 (1895 or 1897?); see J. Lanjouw and F.A. Stafleu, *Index Herbariorum*. Part II, *Collectors A-D*. Regnum Vegetabile vol. 2. 1954; F.N. Hepper and Fiona Neate, *Plant Collectors in West Africa*. 9. Utrecht 1971.

Neobeckia Greene Brassicaceae

Origins:
For the American naturalist Lewis (Louis) Caleb Beck, 1798-1853, physician, wrote *Botany of the Northern and Middle States*. Albany 1833; see J.H. Barnhart, *Biographical Notes upon Botanists*. 1: 149. 1965; Howard Atwood Kelly and Walter Lincoln Burrage, *Dictionary of American Medical Biography*. New York 1928; J. Ewan, ed., *A Short History of Botany in the United States*. New York and London 1969; S. Lenley et al., *Catalog of the Manuscript and Archival Collections and Index to the Correspondence of John Torrey*. Library of the New York Botanical Garden. 1973; H.N. Clokie, *Account of the Herbaria of the Depart-*

ment of Botany in the University of Oxford. 129. Oxford 1964; E.M. Tucker, *Catalogue of the Library of the Arnold Arboretum of Harvard University*. 1917-1933.

Neobeguea J.-F. Leroy Meliaceae

Origins:
After the French botanist Louis Henry Bégué, b. 1906, plant collector in West Africa; see J. Lanjouw and F.A. Stafleu, *Index Herbariorum*. Part II, *Collectors A-D*. Regnum Vegetabile vol. 2. 1954; F.N. Hepper and Fiona Neate, *Plant Collectors in West Africa*. 9. 1971.

Neobenthamia Rolfe Orchidaceae

Origins:
Named after the English botanist George Bentham, 1800-1884 (London), nephew of Jeremy Bentham, taxonomist, from 1829 to 1840 Secretary of the Horticultural Society, 1862 Fellow of the Royal Society of London and in 1826 of the Linnean Society, from 1861 to 1874 President of the Linnean Society. Among his most valuable writings are *Handbook of the British Flora*. London 1858, *Flora hongkongensis*. London 1861, *Labiatarum genera et species*. London 1832-1836, *The Botany of the Voyage of H.M.S. Sulphur*, under the command of captain Sir Edward Belcher ... during the years 1836-1842. London 1844 [-1846], author of most of the *Genera Plantarum* (London 1862-1883) of Bentham and Joseph Dalton Hooker (1817-1911) and in collaboration with Ferdinand Mueller of *Flora Australiensis*. London 1863-1878, his herbarium amounted to over 100,000 specimens; see J.H. Barnhart, *Biographical Notes upon Botanists*. 1: 165. 1965; S. Lenley et al., *Catalog of the Manuscript and Archival Collections and Index to the Correspondence of John Torrey*. Library of the New York Botanical Garden. 447. 1973; Emil Bretschneider, *History of European Botanical Discoveries in China*. [Reprint of the original edition 1898.] Leipzig 1981; Mea Allan, *The Hookers of Kew*. London 1967; J.D. Milner, *Catalogue of Portraits of Botanists Exhibited in the Museums of the Royal Botanic Gardens*. Royal Botanic Gardens, Kew, London 1906; H.R. Fletcher, *Story of the Royal Horticultural Society, 1804-1968*. Oxford 1969; R. Zander, F. Encke, G. Buchheim and S. Seybold, *Handwörterbuch der Pflanzennamen*. 14. Aufl. Stuttgart 1993; T.W. Bossert, *Biographical Dictionary of Botanists Represented in the Hunt Institute Portrait Collection*. 34. 1972; Ida Kaplan Langman, *A Selected Guide to the Literature on the Flowering Plants of Mexico*. University of Pennsylvania Press, Philadelphia 1964; A. Lasègue, *Musée botanique de Benjamin Delessert*. Paris 1845.

Neobinghamia Backeb. Cactaceae

Origins:

Greek *neos* and genus *Binghamia* Britton & Rose, for the American explorer Hiram Bingham, 1875-1956.

Neoblakea Standley Rubiaceae

Origins:

After the American botanist Sidney Fay Blake, 1892-1959, bibliographer, traveler, correspondent of Degener and Fogg. Among his writings are *New Plants from Guatemala and Honduras.* Washington 1922, *Native Names and Uses of Some Plants of Eastern Guatemala and Honduras.* Washington 1922 and *New Plants from Oaxaca.* Cambridge, Massachusetts 1918, with Alice Cary Atwood (1876-1947) wrote *Geographical Guide to Floras of the World.* Washington 1942; see John H. Barnhart, *Biographical Notes upon Botanists.* 1: 196. 1965; T.W. Bossert, *Biographical Dictionary of Botanists Represented in the Hunt Institute Portrait Collection.* 40. 1972; Ida Kaplan Langman, *A Selected Guide to the Literature on the Flowering Plants of Mexico.* University of Pennsylvania Press, Philadelphia 1964; S. Lenley et al., *Catalog of the Manuscript and Archival Collections and Index to the Correspondence of John Torrey.* Library of the New York Botanical Garden. 33. 1973; J. Ewan, ed., *A Short History of Botany in the United States.* New York and London 1969; Frans A. Stafleu and Erik A. Mennega, *Taxonomic Literature. Supplement II.* 190-195. Königstein 1993; R. Zander, F. Encke, G. Buchheim and S. Seybold, *Handwörterbuch der Pflanzennamen.* 14. Aufl. Stuttgart 1993.

Neoboivinella Aubrév. & Pellegr. Sapotaceae

Origins:

For the French botanist Louis Hyacinthe Boivin, 1808-1852 (Brest), traveler and plant collector (islands of the Indian Ocean and coasts of Africa, Canary Islands). See F.N. Hepper and Fiona Neate, *Plant Collectors in West Africa.* 11. 1971; J.H. Barnhart, *Biographical Notes upon Botanists.* 1: 212. 1965; A. Lasègue, *Musée botanique de Benjamin Delessert.* Paris 1845; Mary Gunn and Leslie Edward W. Codd, *Botanical Exploration of Southern Africa.* Cape Town 1981; J. Lanjouw and F.A. Stafleu, *Index Herbariorum.* Part II, *Collectors A-D.* Regnum Vegetabile vol. 2. 1954; I.C. Hedge and J.M. Lamond, *Index of Collectors in the Edinburgh Herbarium.* Edinburgh 1970.

Neobolusia Schltr. Orchidaceae

Origins:

For the South African botanist Harry Bolus, 1834-1911 (Surrey), banker, plant collector in South Africa, 1850 to Cape, Fellow of the Linnean Society 1873, in 1902 founded the Chair of Botany at South African College, wrote *Icones orchidearum Austro-Africanarum extra-tropicarum.* London 1893-1913 and *The Orchids of the Cape Peninsula.* Cape Town 1888, with D. Barclay (paintings) and E.J. Steer (photos) wrote *A Book of the South African Flowers.* Capetown, Wijnberg, Johannesburg 1925. See John H. Barnhart, *Biographical Notes upon Botanists.* 1: 214. 1965; T.W. Bossert, *Biographical Dictionary of Botanists Represented in the Hunt Institute Portrait Collection.* 44. 1972; Merle A. Reinikka, *A History of the Orchid.* Timber Press 1996; Ernest Nelmes and William Cuthbertson, *Curtis's Botanical Magazine Dedications, 1827-1927.* 270-272. [1931]; John Hutchinson, *A Botanist in Southern Africa.* 645. London 1946; W.J. Lütjeharms, *Tribute to Harriet Margaret Louisa Bolus.* Cape Town 1970; Alain White (1880-1951) and Boyd Lincoln Sloane (1886-1955), *The Stapelieae.* Pasadena 1937; Mary Gunn and Leslie E. Codd, *Botanical Exploration of Southern Africa.* 97-99. Cape Town 1981; Leonard Huxley, *Life and Letters of Sir J.D. Hooker.* London 1918.

Neobouteloua Gould Gramineae

Origins:

After the Spanish botanist Estéban Boutelou y Soldevilla, 1776-1813 (Madrid), professor in Madrid; see Yahya Ibn Muhammad called Ibn Al-'Auwan, *Libro de Agricultura ...* Arreglo hecho ... por D. C. Boutelou, precedido de una introduccion de D. E. Boutelou, etc. 1878; Miguel Colmeiro y Penido, *La Botánica y los Botánicos de la Peninsula Hispano-Lusitana.* Madrid 1858; *Genera et Species Plantarum.* 1816.

Neoboutonia Müll. Arg. Euphorbiaceae

Origins:

For the botanist Louis Sulpice Bouton (1799-1878), naturalist, secretary of the Société d'Histoire Naturelle de l'Ile Maurice, secretary of the Société Royale des Arts et Sciences de l'Ile Maurice, Curator of the Muséum d'Histoire Naturelle, Port Louis, plant collector, a contemporary of the amateur botanist and artist Malcy de Chazal (1803-1880), in 1831 elected member of the Muséum Royal d'Histoire Naturelle, Paris. Among his writings are *Sur le décroissement des forêts à Maurice.* Maurice 1838, "Plantes médicinales de Maurice. Indigènes ou cultivées dans les jardins." *Société Royale des Arts & Sciences de Maurice.* Transactions, vol. 1, part 1. Port Louis 1857 [Engl. transl. "Medicinal plants growing or cultivated in the island of Mauritius." *Trans. Roy. Soc. Arts Sci. Mauritius,* New Ser. 1: 1-177. 1857], *Rapport présenté à la Chambre d'Agriculture sur les diverses espèces de cannes à sucre cultivées à Maurice.*

1863 and *Plantes médicinales de Maurice*. Port Louis 1864. See J.H. Barnhart, *Biographical Notes upon Botanists*. 1: 232. 1965; John Gilbert Baker (1834-1920), *Flora of Mauritius [Rodriguez] and the Seychelles*. London 1877; Auguste Toussaint and H. Adolphe, *Bibliography of Mauritius, 1502-1954*. Port Louis 1956; Auguste Toussaint, ed., *Dictionnaire de biographie Mauricienne*. [Société de l'Histoire de l'Ile Maurice. Publication no. 2] [Port Louis] 1941-; N. Cantley, *Catalogue of Plants in the Royal Botanic Gardens at Pamplemousses, Mauritius*. Port Louis 1880; John Vaughan Thompson (1779-1847), *A Catalogue of the Exotic Plants Cultivated in the Mauritius ... to which are added the english and french names....* Compiled under the auspices of R.T. Farquhar Esq. Governor of Mauritius. Mauritius 1816; Frans A. Stafleu and Erik A. Mennega, *Taxonomic Literature. Supplement II*. 393-394. ["1800-1879, or 1878"] 1993; E.M. Tucker, *Catalogue of the Library of the Arnold Arboretum of Harvard University*. Cambridge, Massachusetts 1917-1933.

Species/Vernacular Names:

N. glabrescens Prain

Nigeria: keperu-mbang (Boki); ejuraip (Ekoi)

Neobrittonia Hochr. Malvaceae

Origins:

Named for the American botanist Nathaniel Lord Britton, 1859-1934 (d. New York), founder and first Director of The New York Botanical Garden; see J.H. Barnhart, *Biographical Notes upon Botanists*. 1: 253. 1965; H.N. Clokie, *Account of the Herbaria of the Department of Botany in the University of Oxford*. 139. Oxford 1964; William T. Stearn, in *Dictionary of Scientific Biography* (Editor in Chief Charles Coulston Gillispie.) 2: 476-477. New York 1981; J. Ewan, ed., *A Short History of Botany in the United States*. New York and London 1969; S. Lenley et al., *Catalog of the Manuscript and Archival Collections and Index to the Correspondence of John Torrey*. Library of the New York Botanical Garden. 77-102. 1973; Stafleu and Cowan, *Taxonomic Literature*. 1: 332-348. 1976; Frans A. Stafleu and Erik A. Mennega, *Taxonomic Literature. Supplement III: Br-Ca*. 97-107. 1995; Ida Kaplan Langman, *A Selected Guide to the Literature on the Flowering Plants of Mexico*. University of Pennsylvania Press, Philadelphia 1964; R. Zander, F. Encke, G. Buchheim and S. Seybold, *Handwörterbuch der Pflanzennamen*. 14. Aufl. Stuttgart 1993.

Neobuchia Urban Bombacaceae

Origins:

After the German botanist Wilhelm Buch, b. 1862, plant collector in Haiti.

Neobuxbamia Backeb. Cactaceae

Origins:

For the Austrian botanist Franz Buxbaum, 1900-1979; see Frans A. Stafleu and Erik A. Mennega, *Taxonomic Literature. Supplement III*. 289-290. 1995; Gordon Douglas Rowley, *A History of Succulent Plants*. California 1997; John H. Barnhart, *Biographical Notes upon Botanists*. 1: 293. 1965; R. Zander, F. Encke, G. Buchheim and S. Seybold, *Handwörterbuch der Pflanzennamen*. 14. Aufl. Stuttgart 1993.

Neobyrnesia J.A. Armstr. Rutaceae

Origins:

After Norman Brice Byrnes (b. 1922), Australian botanist and collector; see J.A. Armstrong and J.M. Powell, in *Telopea*. 1: 399, t. XV, XVI. 1980.

Neocabreria R.M. King & H. Robinson Asteraceae

Origins:

For the Argentine botanist Angel Lulio Cabrera, b. 1908; see John H. Barnhart, *Biographical Notes upon Botanists*. 1: 295. 1965.

Neocardenasia Backeb. Cactaceae

Origins:

After the Bolivian (born of Indian parents) botanist Martín Cárdenas Hermosa, 1899-1973 (d. Cochabamba, Bolivia), professor of botany, Quechua scholar, historian, Rector of the Universidad Autónoma de Cochabamba, traveler, plant collector, author of *Manual de Plantas Económicas de Bolivia*. 1969; see Gordon Douglas Rowley, *A History of Succulent Plants*. Strawberry Press, California 1997; National Research Council, *Lost Crops of the Incas: Little-Known Plants of the Andes with Promise for Worldwide Cultivation*. National Academy Press, Washington, D.C. 1989.

Neocarya (DC.) Prance ex F. White Chrysobalanaceae

Origins:

From the Greek *neos* "new" and *karyon* "walnut, nut."

Species/Vernacular Names:

N. macrophylla (Sabine) Prance ex F. White (*Parinari macrophylla* Sabine)

English: gingerbread plum

Nigeria: gawasa (Hausa); putu (Nupe); nawarre (Fula)

Neocastela Small Simaroubaceae

Origins:

From the Greek *neos* "new" and *Castela* Turpin.

Neocentema Schinz Amaranthaceae

Origins:

From the Greek *neos* "new" and the genus *Centema* Hook.f.

Neocheiropteris Christ Polypodiaceae

Origins:

From the Greek *neos* "new, recent" and the genus *Cheiropteris*, *cheir* "hand" and *pteris* "fern."

Species/Vernacular Names:

N. ensata (Thunb.) Ching (*Polypodium ensatum* Thunb.; *Microsorium ensatum* (Thunb.) H. Itô; *Polypodium phyllomanes* Christ)

Japan: kuri-ha-ran

Neochevaliera A. Chev. & Beille Euphorbiaceae

Origins:

Dedicated to the French botanist Auguste Jean Baptiste Chevalier, 1873-1956 (Paris), explorer, father of applied tropical botany, 1905 established Botanical Garden at Dalaba (Guinea). His works include *Travaux bryologiques dédiés à la mémoire de Pierre-Tranquille Husnot*. Paris 1942, *Michel Adanson, voyageur, naturaliste et philosophe*. Paris 1934, *Nos connaissances actuelles sur la géographie botanique et la flore économique du Sénégal et du Soudan*. [Exposition Universelle Internationale de 1900. Colonies françaises.] Paris 1900, *L'Afrique Centrale Française*. Mission Chari-Lac Tchad, 1902-1904. Paris 1907 and *Monographie des Myricacées*. Cherbourg 1901. See John H. Barnhart, *Biographical Notes upon Botanists*. 1: 340. 1965; François Gagnepain (1866-1952), in Paul Henri Lecomte's *Flore générale de l'Indo-Chine*. Paris 1944; Clyde F. Reed, *Bibliography to Floras of Southeast Asia*. Baltimore, Maryland 1969; R. Zander, F. Encke, G. Buchheim and S. Seybold,

Handwörterbuch der Pflanzennamen. 14. Aufl. Stuttgart 1993; A. White and B.L. Sloane, *The Stapelieae*. Pasadena 1937.

Neochevalierodendron J. Léonard Caesalpiniaceae

Origins:

Named after the French botanist Auguste Jean Baptiste Chevalier, 1873-1956, explorer. Among his works *L'Afrique Centrale Française*. Mission Chari-Lac Tchad, 1902-1904. Paris 1907 and *Monographie des Myricacées*. Cherbourg 1901; see John H. Barnhart, *Biographical Notes upon Botanists*. 1: 340. 1965.

Neocinnamomum H. Liou Lauraceae

Origins:
Referring to *Cinnamomum*.

Neoclemensia Carr Orchidaceae

Origins:

For the Reverend Joseph Clemens, (b. Cornwall) 1862-1932 (d. New Guinea), Chaplain in U.S. Army (but retired around 1925), with his wife Mary Strong Clemens collected orchids on Mount Kinabalu. His writings include "The cleft mountain." *Brit. N. Born. Herald.* vol. L, 14: 143-144. 1932, "Mount Kinabalu: A naturalist's description." *Brit. N. Born. Herald.* vol. LI, 1: 7. 1933 and "Mount Kinabalu: the roaring falls of Pinokok." *Brit. N. Born. Herald.* vol. LI, no. 23. 1933. See Jeffrey J. Wood, Reed S. Beaman and John H. Beaman, *The Plants of Mount Kinabalu. Orchids.* Royal Botanic Gardens, Kew 1993; Oakes Ames and Charles Schweinfurth, *The Orchids of Mount Kinabalu*, British North Borneo. Merrymount Press, Boston 1920; A.M. Carter, "The itinerary of Mary Strong Clemens in Queensland, Australia." *Contr. Univ. Michigan Herb.* 15: 163-169. 1982; B.J. Conn, "Mary Strong Clemens: A botanical collector in New Guinea (1935-1941)." in P.S. Short, ed. *History of Systematic Botany in Australasia*. Australia Systematic Botany Society Inc. 217-229. 1990; R.F. Langdon, "The remarkable Mrs Clemens." in D.J. Carr and S.G.M. Carr, eds., *People and Plants in Australia*. 374-383. Academic Press 1981; W.B. Turrill, "J. Clemens." XXXI-Miscell. Notes. *Kew Bull.* 1936: 287-289. 1936; Datin Margaret Luping, Chin Wen and E. Richard Dingley, eds., *Kinabalu — Summit of Borneo*. The Sabah Society, Kota Kinabalu, Sabah, Malaysia 1978.

Neocogniauxia Schltr. Orchidaceae

Origins:

Dedicated to the Belgian botanist Célestin Alfred Cogniaux, 1841-1916, with A. Goossens wrote *Dictionnaire Iconographique des orchidées*. [plates printed by Goffart, after watercolors by A. Goossens, the Belgian orchid painter.] Brussels 1896-1907, contributor to C.F.P. von Martius *Flora Brasiliensis* (Orchidaceae) and to Engler *Das Pflanzenreich*; see John H. Barnhart, *Biographical Notes upon Botanists*. 1: 363. 1965; R. Zander, F. Encke, G. Buchheim and S. Seybold, *Handwörterbuch der Pflanzennamen*. 14. Aufl. Stuttgart 1993; Frans A. Stafleu and Erik A. Mennega, *Taxonomic Literature. Supplement IV: Ce-Cz*. 248-254. Königstein 1997.

Neocollettia Hemsley Fabaceae

Origins:

Dedicated to the British botanist Sir Henry Collett, 1836-1901 (Kew, Surrey), plant collector, 1855 Bengal Army, in 1879 a Fellow of the Linnean Society, author of *Flora simlensis*. Calcutta, Simla and London 1902; see John H. Barnhart, *Biographical Notes upon Botanists*. 1: 367. 1965; R. Zander, F. Encke, G. Buchheim and S. Seybold, *Handwörterbuch der Pflanzennamen*. 14. Aufl. Stuttgart 1993; Frans A. Stafleu and Erik A. Mennega, *Taxonomic Literature. Supplement IV*. 269-270. Königstein 1997; I.C. Hedge and J.M. Lamond, *Index of Collectors in the Edinburgh Herbarium*. Edinburgh 1970; A. White and B.L. Sloane, *The Stapelieae*. Pasadena 1937.

Neoconopodium (Koso-Pol.) Pimenov & Kljuykov Umbelliferae

Origins:

Greek *neos* plus the genus *Conopodium* Koch.

Neocouma Pierre Apocynaceae

Origins:

Greek *neos* plus the genus *Couma* Aublet.

Neocracca Kuntze Fabaceae

Origins:

Greek *neos* plus the genus *Cracca* Benth.

Neocryptodiscus Hedge & Lamond Umbelliferae

Origins:

Greek *neos* with the genus *Cryptodiscus* Schrenk ex Fischer & C. Meyer.

Neocuatrecasia R.M. King & H. Robinson Asteraceae

Origins:

Dedicated to the Spanish (b. Gerona) botanist José Cuatrecasas Arumi, 1903-1906, professor of systematic botany, professor of botany, a specialist in Andean Compositae. His writings include "Estudios sobre plantas andinas. V." *Caldasia*. 2(8): 209-240. [*Diplostephium*] 1943, "Estudios sobre plantas andinas. X." *Caldasia*. 10(46): 3-26. [*Baccharis*] 1943, "Taxonomic notes on neotropical trees." *Trop. Woods*. 101: 10-28. 1955, "A taxonomic revision of the Humiriaceae." *Contr. U.S. Natl. Herb*. 35(2): 25-214. 1961 and "Cacao and its allies: A taxonomic revision of the genus *Theobroma*." *Contr. U.S. Natl. Herb*. 35(6): 379-614. 1964; see J.H. Barnhart, *Biographical Notes upon Botanists*. 1: 401. 1965; J. Ewan, ed., *A Short History of Botany in the United States*. 1969; H.E. Robinson, V.A. Funk, J.F. Pruski and R.M. King, "José Cuatrecasas Arumí." in *Comp. Newsl*. 29: 1-30. 1996.

Neocussonia Hutch. Araliaceae

Origins:

After the French Jesuit and botanist Pierre Cusson, 1727-1783.

Neodawsonia Backeb. Cactaceae

Origins:

After Elmer Yale Dawson, 1920-1966 (drowned, Egypt), at the Smithsonian Institution (Washington), traveler and botanical explorer, collector of algae; see Gordon Douglas Rowley, *A History of Succulent Plants*. California 1997.

Neodeutzia (Engl.) Small Hydrangeaceae

Origins:

After the Dutch Johan van der Deutz, 1743-1788, friend and patron of the Swedish botanist Carl Peter Thunberg (1743-1828); see F. Boerner & G. Kunkel, *Taschenwörter-*

buch der botanischen Pflanzennamen. 4. Aufl. 92. Berlin & Hamburg 1989.

Neodielsia Harms Fabaceae

Origins:

For the German botanist Friedrich Ludwig Emil Diels, 1874-1945; see John H. Barnhart, *Biographical Notes upon Botanists.* 1: 454. 1965; T.W. Bossert, *Biographical Dictionary of Botanists Represented in the Hunt Institute Portrait Collection.* 102. 1972; Ida Kaplan Langman, *A Selected Guide to the Literature on the Flowering Plants of Mexico.* 241. University of Pennsylvania Press, Philadelphia 1964.

Neodissochaeta Bakh.f. Melastomataceae

Origins:

Greek *neos* plus *Dissochaeta* Blume.

Neodistemon Babu & A.N. Henry Urticaceae

Origins:

From the Greek *neos* with the genus *Distemon* Wedd.

Neodregea C.H. Wright Colchicaceae (Liliaceae)

Origins:
For J.L. Drège.

Neodriessenia M.P. Nayar Melastomataceae

Origins:

Greek *neos* plus *Driessenia* Korth.

Neodryas Reichb.f. Orchidaceae

Origins:

Greek *neos* "new" and *dryas* "a wood nymph," referring to the habitat.

Neodunnia R. Vig. Fabaceae

Origins:

After the British botanist Stephen Troyte Dunn, 1868-1938 (d. Surrey), Kew Herbarium, traveler, plant collector (China, Korea, Taiwan, Japan and Hong Kong), 1895 Fellow of the Linnean Society, with William James Tutcher (1867-1920) wrote *Flora of Kwantgtung and Hongkong.*

London 1912; see John H. Barnhart, *Biographical Notes upon Botanists.* 1: 483. 1965; Ida Kaplan Langman, *A Selected Guide to the Literature on the Flowering Plants of Mexico.* 255. Philadelphia 1964; Stafleu and Cowan, *Taxonomic Literature.* 1: 702-703. 1976; Ray Desmond, *Dictionary of British & Irish Botanists and Horticulturists.* 222. 1994.

Neodypsis Baillon Palmae

Origins:

Greek *neos* "new" plus the genus *Dypsis* Noronha ex Martius.

Neoescobaria Garay Orchidaceae

Origins:

Presumably for a Gilberto Escobar, Colombian botanist and orchid grower.

Neofabricia J. Thompson Myrtaceae

Origins:

From the Greek *neos* "new, recent" plus the genus *Fabricia* Gaertner; see J. Thompson, in *Telopea.* 2: 380. 1983.

Species/Vernacular Names:

N. myrtifolia (Gaertn.) J. Thomps.

English: black tea-tree

Australia: untarra

Neofinetia H.H. Hu Orchidaceae

Origins:

For the French botanist Eugène Achille Finet, 1863-1913, orchidologist, botanical artist, (with the French botanist François Gagnepain, 1866-1952, a specialist in the flora of southeastern Asia) participated to the publication of Paul Henri Lecomte's (1856-1934) *Flore générale de l'Indo-Chine.* (Dilleniaceae, Anonaceae, Ranunculaceae and Magnoliaceae) Paris 1907-1950; see J.H. Barnhart, *Biographical Notes upon Botanists.* 1: 541. 1965; Ida Kaplan Langman, *A Selected Guide to the Literature on the Flowering Plants of Mexico.* Philadelphia 1964; Clyde F. Reed, *Bibliography to Floras of Southeast Asia.* Baltimore, Maryland 1969; David G. Frodin & Rafaël Govaerts, *World Checklist and Bibliography of Magnoliaceae.* Royal Botanic Gardens, Kew 1996.

Neogaillonia Lincz. Rubiaceae

Origins:

Referring to the genus *Gaillonia* A. Rich. ex DC., dedicated to the French algologist François Benjamin Gaillon, 1782-1839; see J.H. Barnhart, *Biographical Notes upon Botanists*. 2: 23. 1965; J.D. Milner, *Catalogue of Portraits of Botanists Exhibited in the Museums of the Royal Botanic Gardens*. Royal Botanic Gardens, Kew, London 1906.

Neogardneria Schltr. ex Garay Orchidaceae

Origins:

After the English botanist and explorer George Gardner, 1812-1849 (Sri Lanka), plant collector, M.D. at Glasgow 1835, pupil of W.J. Hooker at Glasgow, traveler in southern India and (from 1836 to 1841) Brazil, in 1842 Fellow of the Linnean Society, from September 1843 (on the recommendation of Sir William Hooker) Superintendent of the Peradeniya Botanic Garden in Sri Lanka, author of *Travels in the Interior of Brazil ... during the years 1836-1841*. London 1846; see J.H. Barnhart, *Biographical Notes upon Botanists*. 2: 29. Boston 1965; Sir James Emerson Tennent (1804-1889), *Ceylon, an Account of the Island ... with Notices of its Natural History*. London 1859; Isaac Henry Burkill, *Chapters on the History of Botany in India*. Delhi 1965; Alice Margaret Coats, *The Quest for Plants. A History of the Horticultural Explorers*. London 1969; R. Desmond, *The European Discovery of the Indian Flora*. 163. Oxford 1992; A. Lasègue, *Musée botanique de Benjamin Delessert*. Paris 1845; H.N. Clokie, *Account of the Herbaria of the Department of Botany in the University of Oxford*. 169-170. Oxford 1964; Ray Desmond, *Dictionary of British & Irish Botanists and Horticulturists*. 270. London 1994; R. Zander, F. Encke, G. Buchheim and S. Seybold, *Handwörterbuch der Pflanzennamen*. 14. Aufl. 714. Stuttgart 1993; Antoine Lasègue, *Musée botanique de M. Benjamin Delessert*. Paris, Leipzig 1845; Richard Evans Schultes and Arthur Stanley Pease, *Generic Names of Orchids. Their Origin and Meaning*. 208. New York and London 1963.

Neoglaziovia Mez Bromeliaceae

Origins:

The genus name commemorates the French botanist Auguste François Marie Glaziou, 1828-1906, traveler, garden designer, plant collector in Brazil; see J.H. Barnhart, *Biographical Notes upon Botanists*. 2: 54. 1965; D. de Andrade Lima, "Plantas das Caatingas." *Academia Brasileira de Ciências*, Rio de Janeiro. 218-219. 1989; Raymond M. Harley, "Plant diversity: Kew's role in north-east Brazil." *The Kew Magazine*. Volume 9. 3: 103-116. August 1992; Simon Mayo, "*Neoglaziovia variegata* (Bromeliaceae)." *The Kew Magazine*. Volume 9. 3: 124-127. August 1992; R. Zander, F. Encke, G. Buchheim and S. Seybold, *Handwörterbuch der Pflanzennamen*. 14. Aufl. 715. Stuttgart 1993.

Species/Vernacular Names:

N. variegata (Arruda da Câmara) Mez

Brazil: caroá

Neogoodenia C. Gardner & A.S. George Goodeniaceae

Origins:

Greek *neos* "new, recent" and the closely related genus *Goodenia* Smith, in *Journal of the Royal Society of Western Australia*. 46: 138, fig. 6. 1963; for the Rev. Samuel Goodenough, 1743-1827 (Worthing, Sussex), amateur botanist, in 1788 a founder (with the entomologist Thomas Marsham, d. 1819, and the physician and botanist Sir James Edward Smith, 1759-1828) of the Linnean Society of London (and initial Treasurer), 1789 Fellow of the Royal Society, with Thomas Jenkinson Woodward (1745-1820) wrote *Observations on the British Fuci*. [London 1797]; see J.E. Smith, *A Specimen of the Botany of New Holland*. 15, t. 5. London 1793[-1795]; Georg Christian Wittstein, *Etymologisch-botanisches Handwörterbuch*. 398. Ansbach 1852; *Memoir and Correspondence of ... Sir J.E. Smith ...* Edited by Lady Pleasance Smith. London 1832; Andrew Thomas Gage (1871-1945), *A History of the Linnean Society of London*. London 1938.

Neogunnia Pax & K. Hoffm. Aizoaceae

Origins:

Greek *neos* "new, recent" plus *Gunnia* F. Muell., dedicated to the South African botanist Ronald Campbell Gunn, 1808-1881, traveler and naturalist, scientist, land owner, plant collector, migrated to Tasmania 1829-1830, Superintendent of Convicts for Northern Tasmania, Police Magistrate at Stanley and Hobart, correspondent of William Jackson Hooker (1785-1865) and John Lindley, a Fellow of the Linnean Society (1850) and of the Royal Society (1854); see F. von Mueller, "Report on the plants collected during Mr. Babbage's expedition into the north-western interior of South Australia in 1858 [by F. Mueller]." *Victorian Parliamentary Papers*. 9. 1859-1860; T.E. Burns and John Rowland Skemp, *Van Diemen's Land Correspondents. Letters from R.C. Gunn [and others] ... to Sir William J. Hooker, 1827-1849*. Queen Victoria Museum. [Launceston] 1961; Antoine Lasègue, *Musée botanique de M. Benjamin Delessert*. 283. Paris, Leipzig 1845.

Neogymnantha Y. Ito Cactaceae

Origins:

Greek *neos, gymnos* "naked" and *anthos* "flower," the genus *Gymnantha* Y. Ito.

Neogyna Reichb.f. Orchidaceae

Origins:

From the orchid generic name *Coleogyne*, Greek *neos* "new, recent" plus *gyne* "female, woman, female organs"; see F. Boerner & G. Kunkel, *Taschenwörterbuch der botanischen Pflanzennamen.* 4. Aufl. 139. Berlin & Hamburg 1989; F. Boerner, *Taschenwörterbuch der botanischen Pflanzennamen.* 2. Aufl. 145. Berlin & Hamburg 1966; Helmut Genaust, *Etymologisches Wörterbuch der botanischen Pflanzennamen.* 414. Basel 1996.

Neogyne Reichb.f. Orchidaceae

Origins:

See the genus *Neogyna*.

Neohallia Hemsley Acanthaceae

Origins:

Dedicated to a daughter of William Borrer (1781-1862), Mrs. Eardley Hall, of Barrow Hill, Henfield, Sussex.

Neoharmsia R. Viguier Fabaceae

Origins:

For the German botanist Hermann August Theodor Harms, 1870-1942, professor of botany, from 1900 editor of *Pflanzenreich*, joint author with Karl (Carl) Wilhelm von Dalla Torre von Thurnberg-Sternhoff (1850-1928) of *Genera Siphonogamarum*. Leipzig 1907-1908; see J.H. Barnhart, *Biographical Notes upon Botanists.* 2: 128. 1965; T.W. Bossert, *Biographical Dictionary of Botanists Represented in the Hunt Institute Portrait Collection.* 164. 1972; Ida Kaplan Langman, *A Selected Guide to the Literature on the Flowering Plants of Mexico*. Philadelphia 1964; Ethelyn Maria Tucker, *Catalogue of the Library of the Arnold Arboretum of Harvard University*. Cambridge, Massachusetts 1917-1933; R. Zander, F. Encke, G. Buchheim and S. Seybold, *Handwörterbuch der Pflanzennamen.* 14. Aufl. Stuttgart 1993; Elmer Drew Merrill, *Contr. U.S. Natl. Herb.* 30(1): 143-144. 1947.

Neohenricia L. Bolus Aizoaceae

Origins:

For the Swiss plant physiologist Dr. Marguerite (Margaret) Gertrude Anna Henrici, 1892-1971, plant collector, of Fauresmith, O.F.S.; see F. Boerner & G. Kunkel, *Taschenwörterbuch der botanischen Pflanzennamen.* 4. Aufl. 139. Berlin & Hamburg 1989; W.P.U. Jackson, *Origins and Meanings of Names of South African Plant Genera.* 131. Rondebosch 1990; Mary Gunn and Leslie E. Codd, *Botanical Exploration of Southern Africa.* 182. Cape Town 1981.

Neohenrya Hemsley Asclepiadaceae

Origins:

Dedicated to the Irish botanist Mrs. Augustine Henry (1857-1930), and the plant collector in China Rev. Benjamin Couch Henry (1850-1901).

Neohickenia Fric Cactaceae

Origins:

For the Argentine botanist Cristóbal Maria Hicken, 1875-1933; see John H. Barnhart, *Biographical Notes upon Botanists.* 2: 171. 1965; Fortunato Luciano Herrera y Garmendia (1875-1945), *Plantarum cuzcorum herrerarianum.* Estudios sobre la flora del Departamento del Cuzco. 15. Lima [1930]; T.W. Bossert, *Biographical Dictionary of Botanists Represented in the Hunt Institute Portrait Collection.* 174. 1972; S. Lenley et al., *Catalog of the Manuscript and Archival Collections and Index to the Correspondence of John Torrey.* Library of the New York Botanical Garden. 222. Boston, Massachusetts 1973; Ethelyn Maria Tucker, *Catalogue of the Library of the Arnold Arboretum of Harvard University.* Cambridge, Massachusetts 1917-1933; Stafleu and Cowan, *Taxonomic Literature.* 2: 190-191. 1979.

Neoholstia Rauschert Euphorbiaceae

Origins:

For the German gardener C.H.E.W. Holst, 1865-1894, plant collector and traveler in East Africa.

Neohumbertiella Hochr. Malvaceae

Origins:

Referring to the related genus *Humbertiella*, named for the French botanist Henri Humbert, 1887-1967, botanical traveler, explorer of Madagascar. His writings include *Les composées de Madagascar*. Caen 1923, *La destruction d'une flore insulaire par le feu. Principaux aspects de la végétation à Madagascar*. [in *Mémoires de l'Académie Malgache*. fasc. 5.] Tananarive 1927 and "François Gagnepain, in memoriam." *Not. Syst*. 14(4): 221-229. 1952, editor and joint author of *Flore de Madagascar et des Comores*. 1936-1967, with F. Gagnepain wrote *Supplement* à la *Flore générale de l'Indo-Chine*. Vol. 1: 1-1027, f. 1-131. Paris 1938-1951. See John H. Barnhart, *Biographical Notes upon Botanists*. 2: 218. 1965; T.W. Bossert, *Biographical Dictionary of Botanists Represented in the Hunt Institute Portrait Collection*. 186. 1972; Ida Kaplan Langman, *A Selected Guide to the Literature on the Flowering Plants of Mexico*. Philadelphia 1964; R. Zander, F. Encke, G. Buchheim and S. Seybold, *Handwörterbuch der Pflanzennamen*. 14. Aufl. 729. Stuttgart 1993.

Neohusnotia A. Camus Gramineae

Origins:

For the French botanist Pierre Tranquille Husnot, 1840-1929, bryologist, agrostologist, botanical collector, traveler. His works include *Catalogue des cryptogames recueillis aux Antilles françaises en 1868*. Caen 1870, *Flore analytique et descriptive des mousses du Nord-Ouest*. Cahan (Orne) [Caen] [1873] and *Muscologia gallica*. Cahan [Caen] 1884-1894, editor of the *Revue bryologique*. vols. 1-53. 1874-1926. See Ernest Roussel, *Énumération des champignons récoltés par M. T. Husnot aux Antilles françaises en 1868*. [Extrait du Bulletin de la Société Linnéenne de Normandie, etc.] Caen 1870; T.W. Bossert, *Biographical Dictionary of Botanists Represented in the Hunt Institute Portrait Collection*. 187. 1972; John H. Barnhart, *Biographical Notes upon Botanists*. 2: 222. 1965; Auguste Jean Baptiste Chevalier (1873-1956) et al., *Travaux bryologiques dédiés à la mémoire de Pierre-Tranquille Husnot*. Paris 1942; H.N. Clokie, *Account of the Herbaria of the Department of Botany in the University of Oxford*. 188. Oxford 1964; Ignatz Urban, ed., *Symbolae Antillanae*. 3: 65. Berlin 1902; S. Lenley et al., *Catalog of the Manuscript and Archival Collections and Index to the Correspondence of John Torrey*. Library of the New York Botanical Garden. 234. Boston, Massachusetts 1973; E.M. Tucker, *Catalogue of the Library of the Arnold Arboretum of Harvard University*. Cambridge, Massachusetts 1917-1933.

Neohymenopogon Bennet Rubiaceae

Origins:

Genus *Hymenopogon* Wallich, Greek *hymen* "a membrane" and *pogon* "beard."

Neohyptis J.K. Morton Labiatae

Origins:

Greek *neos* and genus *Hyptis* Jacq.

Neojatropha Pax Euphorbiaceae

Origins:

Greek *neos* and genus *Jatropha* L.

Neojeffreya Cabrera Asteraceae

Origins:

For Charles Jeffrey, born 1934, botanist at Kew; see Stafleu and Cowan, *Taxonomic Literature*. 2: 433. 1979.

Neojunghuhnia Koorders Ericaceae

Origins:

For the German botanist in Java Franz Wilhelm Junghuhn, 1809-1864, traveler, plant collector. Among his writings are "Nova genera et species plantarum florae Javanicae." *Tydschr. nat. Gesch. Physiol*. 7: 285-317. 1840 and *Topographische und naturwissenschaftliche Reisen durch Java*. Magdeburg 1845. See C.W. Wormser, *Frans Junghuhn*. Deventzer [1941]; M.C.P. Schmidt, ed., *Franz Junghuhn*. Beiträge zur 100. Leipzig 1909; M. Schmidt et al., *Gedenkboek Franz Junghuhn 1809-1909*. 's Gravenhage 1910; M.N. Chaudhri, I.H. Vegter and C.M. De Wal, *Index Herbariorum*, Part II (3), *Collectors I-L*. Regnum Vegetabile vol. 86. 1972; Friedrich Anton Wilhelm Miquel (1811-1871), *Plantae junghuhnianae*. Enumeratio plantarum, quas in insulis Java et Sumatra, detexit Fr. Junghuhn. Lugduni-Batavorum [Leiden], Parisiis [1851]- 1853-1855[-1857]; Ida Kaplan Langman, *A Selected Guide to the Literature on the Flowering Plants of Mexico*. Philadelphia 1964; J.H. Barnhart, *Biographical Notes upon Botanists*. 2: 263. 1965; A. Lasègue, *Musée botanique de Benjamin Delessert*. Paris 1845; T.W. Bossert, *Biographical Dictionary of Botanists Represented in the Hunt Institute Portrait Collection*. 201. 1972; E.M. Tucker, *Catalogue of the Library of the Arnold Arboretum of Harvard University*. 1917-1933; Ignatz Urban, *Geschichte des Königlichen Botanischen Museums zu Berlin-Dahlem (1815-1913). Nebst Aufzählung seiner Sammlungen*. Dresden 1916.

Neokoehleria Schlechter Orchidaceae

Origins:

For E. (or H.) Koehler, flourished 1906-1913, collector of Peruvian plants and orchids; see Stafleu and Cowan, *Taxonomic Literature*. 2: 595. 1979; Richard Evans Schultes and

Arthur Stanley Pease, *Generic Names of Orchids. Their Origin and Meaning*. 208. 1963; Hubert Mayr, *Orchid Names and Their Meanings*. 162. [for the German botanist J.Ch.G. Köhler] Vaduz 1998.

Neolabatia Aubrév. Sapotaceae

Origins:

Greek *neos* and the genus *Labatia* Sw.; for the French naturalist Jean Baptiste Labat, 1663-1738 (Paris), botanist, traveler, Dominican missionary. His writings include *Nouveau voyage aux isles de l'Amérique*. Paris 1722 and *Nouvelle relation de l'Afrique occidentale* (rédigée d'après les Mémoires d'André Brue.) [Five volumes in-duodecimo.] Paris 1728; see Johann Joachim Schwabe, *Allgemeine Historie der Reisen zu Wasser und (zu) Lande*. Leipzig 1747-1774; Laurent d'Arvieux, *Mémoires du Chevalier d'Arvieux*, ... mis en ordre ... par J.B.L. [Six volumes in-duodecimo.] Paris 1735.

Neolacis Wedd. Podostemaceae

Origins:

The genus *Lacis* Lindley, Greek *lakis* "a rent," *lakizo* "to tear," *lakistos* "torn, rent, split."

Neolauchea Kraenzlin Orchidaceae

Origins:

For the Director of the garden of Prince John of Liechtenstein, the German horticulturist Lauche; see Richard Evans Schultes and Arthur Stanley Pease, *Generic Names of Orchids. Their Origin and Meaning*. 208. 1963; Hubert Mayr, *Orchid Names and Their Meanings*. 162. Vaduz 1998.

Neolaugeria Nicolson Rubiaceae

Origins:

Greek *neos* and the genus *Laugeria* L.

Neolehmannia Kraenzlin Orchidaceae

Origins:

Dedicated to the German botanist Friedrich Carl Lehmann, 1850-1903, plant and orchid collector in Central and South America (Guatemala, Costa Rica, Colombia, Ecuador), German Consul in Colombia.

Neolemaireocereus Backeb. Cactaceae

Origins:

For the French botanist Antoine Charles Lemaire, 1801-1871, naturalist, specialist in Cactaceae. His works include *Cactearum aliquot novarum ac insuetarum in horto monvilliano cultarum accurata descriptio*. Lutetiae Parisiorum [Paris] et Argentorati [Strasbourg] 1838, *Les Cactées. Histoire, ... culture*, etc. Paris [1868] and *Iconographie descriptive des cactées*. Paris [1841-1850?]; see John H. Barnhart, *Biographical Notes upon Botanists*. 2: 366. 1965; E.M. Tucker, *Catalogue of the Library of the Arnold Arboretum of Harvard University*. Cambridge, Massachusetts 1917-1933; Elmer Drew Merrill, *Contr. U.S. Natl. Herb.* 30(1): 187. 1947; Will Tjaden, "Charles Lemaire (1801-1871) and the Christmas Cactus." in *The Plantsman*. 15(3): 161-167. December 1993; Jonas C. Dryander, *Catalogus bibliothecae historico-naturalis Josephi Banks*. London 1796-1800; Gordon Douglas Rowley, *A History of Succulent Plants*. 1997; Ida Kaplan Langman, *A Selected Guide to the Literature on the Flowering Plants of Mexico*. Philadelphia 1964; R. Zander, F. Encke, G. Buchheim and S. Seybold, *Handwörterbuch der Pflanzennamen*. 14. Aufl. Stuttgart 1993; F.A. Stafleu and Cowan, *Taxonomic Literature*. 2: 834-837. 1979.

Neolemonniera H. Heine Sapotaceae

Origins:

For the French botanist G. Le Monnier, born 1843.

Species/Vernacular Names:
N. clitandrifolia (A. Chev.) Heine
Zaire: muyabi, muiabi

Neolepisorus Ching Polypodiaceae

Origins:

Greek *neos* with the genus *Lepisorus* (J. Sm.) Ching.

Neoleptopyrum Hutch. Ranunculaceae

Origins:

Greek *neos* with the genus *Leptopyrum* Reichb.

Neoleroya A. Cavaco Rubiaceae

Origins:

For J.-F. Leroy, b. 1915; see A. Cavaco, "*Neoleroya*, nouveau genre de Rubiaceae-Vanguerieae." *Adansonia*. 11: 119-123. 1971.

Neolindenia Baillon Acanthaceae

Origins:

Named after the Luxemburg botanist Jean Jules Linden, 1817-1898, botanical explorer, traveler in America, co-editor of *Pescatorea*. Iconographie des Orchidées. Bruxelles [1854-] 1860. His writings include *Hortus lindenianus*. Bruxelles 1859-1860 and *Catalogue des plantes exotiques, nouvelles et rares*. Bruxelles 1855 etc. He was father of Auguste Charles Joseph Linden (1852-1894) and Charles Lucien Linden (1851-1940). See John H. Barnhart, *Biographical Notes upon Botanists*. 2: 385. 1965; Stafleu and Cowan, *Taxonomic Literature*. 3: 41-46. Utrecht 1981; H.N. Clokie, *Account of the Herbaria of the Department of Botany in the University of Oxford*. 200. Oxford 1964; Ida Kaplan Langman, *A Selected Guide to the Literature on the Flowering Plants of Mexico*. Philadelphia 1964; A. Lasègue, *Musée botanique de Benjamin Delessert*. 213-215. Paris 1845; R. Zander, F. Encke, G. Buchheim and S. Seybold, *Handwörterbuch der Pflanzennamen*. 14. Aufl. Stuttgart 1993; S. Lenley et al., *Catalog of the Manuscript and Archival Collections and Index to the Correspondence of John Torrey*. Library of the New York Botanical Garden. 268. 1973; Ignatz Urban, *Geschichte des Königlichen Botanischen Museums zu Berlin-Dahlem (1815-1913). Nebst Aufzählung seiner Sammlungen*. Dresden 19162; Frederico Carlos Hoehne, M. Kuhlmann and Oswaldo Handro, *O jardim botânico de São Paulo*. 1941; Ignatz Urban, ed., *Symbolae Antillanae*. 3: 76-78. Berlin 1902; J. Lindley, *Orchideae lindenianae*. London 1846; Merle A. Reinikka, *A History of the Orchid*. Timber Press 1996.

Neolindleya Kraenzlin Orchidaceae

Origins:

For the English (b. near Norwich) botanist John Lindley, 1799-1865 (d. Middlesex), horticulturist, in 1820 Fellow of the Linnean Society, 1828 Fellow of the Royal Society, taxonomist, botanical artist, from 1829 to 1860 professor of botany at University College London, author of numerous and valuable botanical and horticultural works. His writings include *The Genera and Species of Orchideaceous Plants*. London 1830-1840 and *Sertum orchideaceum*. London [1837-] 1838 [-1841], he was the son of George Lindley (*circa* 1769-1835). See William T. Stearn, in *Dictionary of Scientific Biography* 8: 371-373. 1981; R. Zander, F. Encke, G. Buchheim and S. Seybold, *Handwörterbuch der Pflanzennamen*. 14. Aufl. Stuttgart 1993; Mariella Azzarello Di Misa, a cura di, *Il Fondo Antico della Biblioteca dell'Orto Botanico di Palermo*. 158. Regione Siciliana, Palermo 1988; John H. Barnhart, *Biographical Notes upon Botanists*. 2: 386. 1965; T.W. Bossert, *Biographical Dictionary of Botanists Represented in the Hunt Institute Portrait*

Collection. 239. 1972; Blanche Henrey, *British Botanical and Horticultural Literature before 1800*. Oxford 1975; H.N. Clokie, *Account of the Herbaria of the Department of Botany in the University of Oxford*. 200. Oxford 1964; Warren R. Dawson, *The Banks Letters*, a Calendar of the Manuscript Correspondence of Sir Joseph Banks. London 1958; S. Lenley et al., *Catalog of the Manuscript and Archival Collections and Index to the Correspondence of John Torrey*. Library of the New York Botanical Garden. 463. 1973; Emil Bretschneider, *History of European Botanical Discoveries in China*. [Reprint of the original edition 1898.] Leipzig 1981; H.R. Fletcher, *Story of the Royal Horticultural Society, 1804-1968*. Oxford 1969; M. Hadfield et al., *British Gardeners: A Biographical Dictionary*. London 1980; Mea Allan, *The Hookers of Kew*. London 1967; J.D. Milner, *Catalogue of Portraits of Botanists Exhibited in the Museums of the Royal Botanic Gardens*. Royal Botanic Gardens, Kew, London 1906; Ray Desmond, *Dictionary of British & Irish Botanists and Horticulturists*. 429. London 1994; Stafleu and Cowan, *Taxonomic Literature*. 3: 49-60. Utrecht 1981; J. Ewan, ed., *A Short History of Botany in the United States*. 1969.

Neolindleyella Fedde Rosaceae

Origins:

After the British botanist John Lindley, 1799-1865.

Neolitsea Merr. Lauraceae

Origins:

From the Greek *neo* "new" and the genus *Litsea* Lam.

Species/Vernacular Names:

N. australiensis Kosterm.

English: grey bollywood

N. dealbata (R. Br.) Merr.

English: hairy-leaved bolly gum

N. sericea (Blume) Koidz. (*Laurus sericea* Blume; *Neolitsea glauca* (Sieb.) Koidz.; *Litsea glauca* Sieb.; *Neolitsea latifolia* Koidz. non S. Moore; *Neolitsea sieboldii* (Kuntze) Nakai)

Japan: shiro-damo

China: yueh kuei, pu shih hua

N. zeylanica (Nees) Merr. (*Litsea zeylanica* Nees & T. Nees)

English: shore laurel

Malaya: tejur, tejur pasir, teja pasir, medang pasir

Neolloydia Britton & Rose Cactaceae

Origins:
For the American botanist Francis Ernest Lloyd, 1868-1947.

Neolobivia Y. Ito Cactaceae

Origins:
Greek *neos* with the genus *Lobivia* Britton & Rose.

Neolophocarpus Camus Cyperaceae

Origins:
Greek *neos* plus the genus *Lophocarpus* Boeck.

Neoluederitzia Schinz Zygophyllaceae

Origins:
The generic name after the brother of the German merchant and botanical collector August Lüderitz (1838-1922), Franz Adolph Eduard Lüderitz (1834-1896) was one of the chief protagonists of German colonial politics and perished in Namibia; see Hubert Henoch, *Adolph Lüderitz. Eine biographische Skizze*, etc. [1910] [Koloniale Abhandlungen. Hft. 25.]; Wilhelm Schuessler, *Adolf Lüderitz. Ein deutscher Kampf um Südafrika 1883-1886*. Geschichte des ersten Kolonialpioniers im Zeitalter Bismarcks. Bremen [1936]; *Die Erschliessung von Deutsch-Südwest-Afrika durch Adolf Lüderitz*. Oldenburg 1945; Mary Gunn and Leslie E. Codd, *Botanical Exploration of Southern Africa*. 233. Cape Town 1981.

Neoluffa Chakrav. Cucurbitaceae

Origins:
Greek *neos* and the genus *Luffa* Miller.

Neomacfadya Baillon Bignoniaceae

Origins:
Named for the Scottish botanist Dr. James Macfadyen, 1798 (or 1800)-1850 (in Jamaica), physician, M.D. Glasgow 1821-1822, from 1826 in Jamaica, 1838 Fellow Linnean Society, author of the incomplete *Flora of Jamaica*. London, Edinburgh, Glasgow 1837. See J.H. Barnhart, *Biographical Notes upon Botanists*. 2: 423. 1965; Georg Christian Wittstein, *Etymologisch-botanisches Handwörterbuch*. 361. Ansbach 1852; R. Zander, F. Encke, G.

Buchheim and S. Seybold, *Handwörterbuch der Pflanzennamen*. 14. Aufl. 363, 746. Stuttgart 1993; H.N. Clokie, *Account of the Herbaria of the Department of Botany in the University of Oxford*. 204. Oxford 1964; A. Lasègue, *Musée botanique de Benjamin Delessert*. Paris 1845; E.M. Tucker, *Catalogue of the Library of the Arnold Arboretum of Harvard University*. 1917-1933; Mea Allan, *The Hookers of Kew*. London 1967; Ignatz Urban, ed., *Symbolae Antillanae*. 1904; I.C. Hedge and J.M. Lamond, *Index of Collectors in the Edinburgh Herbarium*. Edinburgh 1970.

Neomammillaria Britton & Rose Cactaceae

Origins:
Greek *neos* plus *Mammillaria* Haw.

Neomandonia Hutchinson Commelinaceae

Origins:
For the French traveler Gilbert Mandon, 1799-1866, plant collector in Bolivia; see John H. Barnhart, *Biographical Notes upon Botanists*. 2: 442. 1965; E.M. Tucker, *Catalogue of the Library of the Arnold Arboretum of Harvard University*. 1917-1933; Stafleu and Cowan, *Taxonomic Literature*. 3: 273. Utrecht 1981; I.C. Hedge and J.M. Lamond, *Index of Collectors in the Edinburgh Herbarium*. Edinburgh 1970.

Neomanniophyton Pax & K. Hoffmann Euphorbiaceae

Origins:
Greek *neos* plus the genus *Manniophyton*.

Neomarica Sprague Iridaceae

Origins:
Referring to the genus *Marica* Ker Gawl.; see Georg Christian Wittstein, *Etymologisch-botanisches Handwörterbuch*. 557. Ansbach 1852; F. Boerner & G. Kunkel, *Taschenwörterbuch der botanischen Pflanzennamen*. 4. Aufl. 131. Berlin & Hamburg 1989; Helmut Genaust, *Etymologisches Wörterbuch der botanischen Pflanzennamen*. 415. Basel 1996.

Neomartinella Pilger Brassicaceae

Origins:
For the French missionary Léon François Martin, 1866-1919, botanical collector in China and Japan.

Neomazaea Krug & Urban Rubiaceae

Origins:

Dedicated to the Cuban botanist Manuel Goméz de la Maza y Jimenez, 1867-1916, Director of the Botanical Garden in Havana; see also the genus *Mazaea* Krug & Urban.

Neomezia Votsch Theophrastaceae

Origins:

After the German botanist Carl Christian Mez, 1866-1944, professor of botany at Breslau, from 1899 to 1910 professor of botany at Halle, from 1910 to 1935 Director of the Königsberg Botanical Garden, contributor to C.F.P. von Martius *Flora Brasiliensis* (Bromeliaceae), contributor to Engler *Das Pflanzenreich* (Bromeliaceae, Myrsinaceae, Theophrastaceae, etc.), author of *Lauraceae americanae.* Berlin 1889, founder and editor of *Botanisches Archiv.* Königsberg 1922-1938. See J.H. Barnhart, *Biographical Notes upon Botanists.* 2: 484. 1965; Hermann Hager, *Das Mikroscop und seine Anwendung.* Berlin 1908; Ida Kaplan Langman, *A Selected Guide to the Literature on the Flowering Plants of Mexico.* Philadelphia 1964; R. Zander, F. Encke, G. Buchheim and S. Seybold, *Handwörterbuch der Pflanzennamen.* 14. Aufl. Stuttgart 1993.

Neomicrocalamus Keng f. Gramineae

Origins:

Greek *neos* plus the generic name *Microcalamus* Franchet.

Neomillspaughia S.F. Blake Polygonaceae

Origins:

Greek *neos* plus *Millspaughia* B.L. Robinson, dedicated to the American botanist Charles Frederick Millspaugh, 1854-1923, physician, professor of botany, traveler and plant collector (four trips to the Yucatan); see John H. Barnhart, *Biographical Notes upon Botanists.* 2: 493. 1965; R. Zander, F. Encke, G. Buchheim and S. Seybold, *Handwörterbuch der Pflanzennamen.* 14. Aufl. Stuttgart 1993; I.C. Hedge and J.M. Lamond, *Index of Collectors in the Edinburgh Herbarium.* Edinburgh 1970; Ida Kaplan Langman, *A Selected Guide to the Literature on the Flowering Plants of Mexico.* Philadelphia 1964; T.W. Bossert, *Biographical Dictionary of Botanists Represented in the Hunt Institute Portrait Collection.* 268. 1972; S. Lenley et al., *Catalog of the Manuscript and Archival Collections and Index to the Correspondence of John Torrey.* Library of the New York Botanical Garden. 291-292. 1973; E.M. Tucker, *Catalogue of the Library of the Arnold Arboretum of Harvard*

University. 1917-1933; J. Ewan, ed., *A Short History of Botany in the United States.* 148. 1969; Irving William Knobloch, compiled by, "A preliminary verified list of plant collectors in Mexico." *Phytologia Memoirs.* VI. Plainfield, N.J. 1983.

Neomimosa Britton & Rose Mimosaceae

Origins:

Greek *neos* plus *Mimosa* L.

Neomirandea R.M. King & H. Robinson Asteraceae

Origins:

For the Mexican botanist Faustino Antonio Miranda González, 1905-1964, traveler, plant collector. His works include "Plantas nuevas del sur de México." *Bol. Soc. Bot. Méx.* 26: 120-132. 1961 and "Dos arbustos notáble del estado de Yucatan." *Bol. Soc. Bot. Méx.* 21: 9-21. 1957. See John H. Barnhart, *Biographical Notes upon Botanists.* 2: 495. 1965; T.W. Bossert, *Biographical Dictionary of Botanists Represented in the Hunt Institute Portrait Collection.* 268. 1972; William R. Holland, *Medicina maya en los altos de Chiapas.* México 1963; Ida Kaplan Langman, *A Selected Guide to the Literature on the Flowering Plants of Mexico.* Philadelphia 1964; Irving William Knobloch, compiled by, "A preliminary verified list of plant collectors in Mexico." *Phytologia Memoirs.* VI. Plainfield, N.J. 1983.

Neomitranthes Legrand Myrtaceae

Origins:

Greek *neos* and the genus *Mitranthes* O. Berg.

Neomolinia Honda & Sakis. Gramineae

Origins:

For Giovanni Ignazio Molina (1737-1829), author of *Saggio sulla storia naturale del Chili.* Bologna 1782.

Neomoorea Rolfe Orchidaceae

Origins:

Dedicated to the Irish botanist Sir Frederick William Moore, 1857-1949 (Dublin), son of the Scottish botanist David M. (1808-1879), between 1877 and 1879 Curator of the Trinity College Garden in Dublin, in 1911 a Fellow of the Linnean

Society, knighted 1911, from 1879 to 1922 Director of the Glasnevin Botanic Garden.

Neomortonia Wiehler Gesneriaceae

Origins:

For the American botanist Conrad Vernon Morton, 1905-1972, specialist in Gesneriaceae and Solanaceae, from 1928 Smithsonian Institution, he was the author of "A revision of *Besleria*." *Contr. U.S. Natl. Herb.* 26(9): 395-474. 1939, "Taxonomic studies of tropical American plants. Some South American species of *Solanum*." *Contr. U.S. Natl. Herb.* 29(1): 41-72. 1944, "The classification of *Thelypteris*." *Amer. Fern J.* 53(4): 149-154. 1963, "The Peruvian species of *Besleria* (Gesneriaceae)." *Bull. U.S. Natl. Mus.* 38(4): 125-151. 1968, "The genera, subgenera, and sections of the Hymenophyllaceae." *Contr. U.S. Natl. Herb.* 38(5): 153-214. 1968 and *A Revision of the Argentine Species of Solanum*. Academia Nacional de Ciencias, Córdoba, Argentina 1976. See John H. Barnhart, *Biographical Notes upon Botanists*. 2: 518. 1965; T.W. Bossert, *Biographical Dictionary of Botanists Represented in the Hunt Institute Portrait Collection*. 274. 1972; S. Lenley et al., *Catalog of the Manuscript and Archival Collections and Index to the Correspondence of John Torrey*. Library of the New York Botanical Garden. 300. 1973; Ida Kaplan Langman, *A Selected Guide to the Literature on the Flowering Plants of Mexico*. Philadelphia 1964; R. Zander, F. Encke, G. Buchheim and S. Seybold, *Handwörterbuch der Pflanzennamen*. 14. Aufl. 754. Stuttgart 1993.

Neomuellera Briquet Labiatae

Origins:

Named after the Swiss botanist Jean (Johannes) Mueller, called Argoviensis (Aargau or Argovie, Switzerland), 1828-1896, owner of a lichenological herbarium, from 1851 to 1869 Curator of the Candolle herbarium, from 1869 to 1896 Curator of the B. Delessert herbarium, from 1870 to 1874 Director of the Genève Botanic Garden, from 1871 to 1889 professor of botany. His works include *Monographie de la famille des Résédacées*. Zürich 1857, *Principes des classification des lichens*. Genève 1862, *Lichenologie Beiträge*. Regensburg and Bonn 1874-1891, contributed to C.F.P. von Martius *Flora Brasiliensis* (Apocynaceae, Rubiaceae, Euphorbiaceae). See John H. Barnhart, *Biographical Notes upon Botanists*. 2: 524. 1965; Stafleu and Cowan, *Taxonomic Literature*. 3: 628-635. 1981; E.D. Merrill, "Bibliography of Polynesian botany." *Bernice P. Bishop Mus. Bull.* 144: 140. 1937 and *Contr. U.S. Natl. Herb.* 30(1): 222. 1947; Sir Isaac Bayley Balfour (1853-1922), *Botany of Socotra*. Edinburgh, London 1888; T.W. Bossert, *Biographical Dictionary of Botanists Represented in the Hunt Institute Portrait Collection*. 277. 1972; S. Lenley et al.,

Catalog of the Manuscript and Archival Collections and Index to the Correspondence of John Torrey. Library of the New York Botanical Garden. 464. 1973; Ida Kaplan Langman, *A Selected Guide to the Literature on the Flowering Plants of Mexico*. 529. University of Pennsylvania Press, Philadelphia 1964; A. de Candolle and A.C.P. de Candolle, *Monographiae Phanerogamarum*. 3: 630. (Jun.) 1881; R. Zander, F. Encke, G. Buchheim and S. Seybold, *Handwörterbuch der Pflanzennamen*. 14. Aufl. Stuttgart 1993.

Neomyrtus Burret Myrtaceae

Origins:

From the Greek *neos* "new" and the genus *Myrtus*.

Neonauclea Merr. Rubiaceae

Origins:

From the Greek *neos* "new" and the genus *Nauclea*.

Species/Vernacular Names:
N. gordoniana (F.M. Bailey) Ridsdale
English: hard Leichhardt tree

Neonelsonia J.M. Coulter & Rose Umbelliferae

Origins:

For the American naturalist Edward William Nelson, 1855-1934, explorer, in Mexico and Guatemala, plant collector, from 1890 to 1929 with the USDA. His writings include *Descriptions of New Genera, Species and Subspecies of Birds from Panama, Colombia and Ecuador*. Washington 1912, *The Eskimo about Bering Strait*. Washington 1881, *Lower California and Its Natural Resources*. Washington 1922 and *Wild Animals of North America*. Washington 1930. See [edited by H.W. Henshaw] *Report upon Natural History Collections Made in Alaska Between the Years 1877 and 1881 by E.W. Nelson*. 1887; John H. Barnhart, *Biographical Notes upon Botanists*. 2: 544. 1965; Joseph Ewan, *Rocky Mountain Naturalists*. The University of Denver Press 1950; Ida Kaplan Langman, *A Selected Guide to the Literature on the Flowering Plants of Mexico*. University of Pennsylvania Press, Philadelphia 1964; Irving William Knobloch, compil., "A preliminary verified list of plant collectors in Mexico." *Phytologia Memoirs*. VI. 1983.

Neonicholsonia Dammer Palmae

Origins:

After the British gardener George Nicholson, 1847-1908 (d. Richmond, Surrey), at Kew, in 1898 Fellow of the Linnean Society; see John H. Barnhart, *Biographical Notes upon Botanists.* 3: 2. 1965; Ida Kaplan Langman, *A Selected Guide to the Literature on the Flowering Plants of Mexico.* 540. Philadelphia 1964; E.M. Tucker, *Catalogue of the Library of the Arnold Arboretum of Harvard University.* 1917-1933; T.W. Bossert, *Biographical Dictionary of Botanists Represented in the Hunt Institute Portrait Collection.* 286. 1972; see Ray Desmond, *Dictionary of British & Irish Botanists and Horticulturists.* 517. London 1994; H.N. Clokie, *Account of the Herbaria of the Department of Botany in the University of Oxford.* 217. Oxford 1964; S. Lenley et al., *Catalog of the Manuscript and Archival Collections and Index to the Correspondence of John Torrey.* Library of the New York Botanical Garden. 314. 1973; R. Zander, F. Encke, G. Buchheim and S. Seybold, *Handwörterbuch der Pflanzennamen.* 14. Aufl. 757. Stuttgart 1993; I.C. Hedge and J.M. Lamond, *Index of Collectors in the Edinburgh Herbarium.* Edinburgh 1970; Ernest Nelmes and William Cuthbertson, *Curtis's Botanical Magazine Dedications, 1827-1927.* 302-304. [1931]; Stafleu and Cowan, *Taxonomic Literature.* 3: 740-742. 1981; M. Hadfield et al., *British Gardeners: A Biographical Dictionary.* London 1980.

Neoniphopsis Nakai Polypodiaceae

Origins:
The genus *Niphopsis* J. Sm., Greek *nipha* "snow" and *opsis* "appearance."

Neonotonia Lackey Fabaceae

Origins:
From the Greek *notos, noton* "the south, the southwest wind, the south wind, back" (cf. Latin *notus* and *notos, i* "the south wind, auster, wind"; Akkadian *nadûm* "to pour," *nataku* "to drip"; Armenian *nay*), or after Benjamin Noton, 1812-1835 botanical collector in Peninsular India, Nilgiri Hills; see Ray Desmond, *Dictionary of British & Irish Botanists and Horticulturists.* 522. London 1994.

Species/Vernacular Names:
N. wightii (Wight & Arn.) Verdc. (*Glycine wightii* (Arn.) Verdc. subsp. *wightii* var. *longicauda* (Schweinf.) Verdc.)

English: soya bean, perennial soybean

Southern Africa: olieboontjie, sooiboontjie, soyaboontjie; dinawá tsá nága (Tswana)

Neopallasia Poljakov Asteraceae

Origins:
For the German botanist Pyotr (Peter) Simon Pallas, 1741-1811, physician, explorer, naturalist, traveler, scientist, professor of natural history (St. Petersburg). His works include *Reise durch verschiedene Provinzen des russischen Reichs ...* St. Petersburg 1771-1776 and *A Naturalist in Russia. Letters from Peter Simon Pallas to Thomas Pennant.* Edited by Carol Urness. Minneapolis [1967], editor of *Neue Nordische Beiträge.* St. Petersburg & Leipzig 1781-1796. See John H. Barnhart, *Biographical Notes upon Botanists.* 3: 43. 1965; Norman Douglas, *Looking Back. An Autobiographical Excursion.* Chatto and Windus, London 1938; E.M. Tucker, *Catalogue of the Library of the Arnold Arboretum of Harvard University.* 1917-1933; T.W. Bossert, *Biographical Dictionary of Botanists Represented in the Hunt Institute Portrait Collection.* 298. 1972; Jonas C. Dryander, *Catalogus bibliothecae historico-naturalis Josephi Banks.* London 1800; H.N. Clokie, *Account of the Herbaria of the Department of Botany in the University of Oxford.* 220. Oxford 1964; A. Lasègue, *Musée botanique de Benjamin Delessert.* Paris 1845; Vasiliy A. Esakov, in *Dictionary of Scientific Biography* (Editor in Chief Charles Coulston Gillispie.) 10: 283-285. New York 1981; Emil Bretschneider (1833-1901), *History of European Botanical Discoveries in China.* [Reprint of the original edition, St. Petersburg 1898.] Leipzig 1981; Blanche Elizabeth Edith Henrey, *British Botanical and Horticultural Literature before 1800.* Oxford 1975; R. Zander, F. Encke, G. Buchheim and S. Seybold, *Handwörterbuch der Pflanzennamen.* 14. Aufl. Stuttgart 1993; G. Murray, *History of the Collections Contained in the Natural History Departments of the British Museum.* 1: 172. 1904; Stafleu and Cowan, *Taxonomic Literature.* 4: 20-27. 1983; Blanche Henrey, *No Ordinary Gardener — Thomas Knowlton, 1691-1781.* Edited by A.O. Chater. British Museum (Natural History). London 1986.

Neopanax Allan Araliaceae

Origins:
Greek *neos* plus *Panax.*

Neoparrya Mathias Umbelliferae

Origins:
For the American botanist Charles Christopher Parry, 1823-1890 (Davenport, Iowa, USA), plant collector, explorer (Oregon, western Wyoming, Rocky Mountains, Utah, Nevada, California, Mexico), physician, M.D. Columbia College 1846, 1861 Colorado Expedition, 1862 the Parry, Hall and Harbour Expedition, expedition of 1864 (Hot Sulphur Springs, excursion to Long's Peak); see John H. Barnhart, *Biographical Notes upon Botanists.* 3: 52. 1965; T.W. Bossert, *Biographical Dictionary of Botanists Represented in the Hunt Institute Portrait Collection.* 301. 1972; H.N.

Clokie, *Account of the Herbaria of the Department of Botany in the University of Oxford.* 221. Oxford 1964; Ida Kaplan Langman, *A Selected Guide to the Literature on the Flowering Plants of Mexico.* 566-567. Philadelphia 1964; S. Lenley et al., *Catalog of the Manuscript and Archival Collections and Index to the Correspondence of John Torrey.* Library of the New York Botanical Garden. 465-466. 1973; Joseph Ewan, *Rocky Mountain Naturalists.* The University of Denver Press 1950; E.M. Tucker, *Catalogue of the Library of the Arnold Arboretum of Harvard University.* 1917-1933; J. Ewan, ed., *A Short History of Botany in the United States.* 1969; Charles Francis Saunders, *Western Wild Flowers.* 73. New York 1933; Howard Atwood Kelly and Walter Lincoln Burrage, *Dictionary of American Medical Biography.* New York 1928; G. Murray, *History of the Collections Contained in the Natural History Departments of the British Museum.* 1: 172. London 1904; Margaret Miller Rocq, ed., *California Local History. A Bibliography and Union List of Library Holdings.* Second edition. Stanford, California 1970; A.E. Weber, *King of Colorado Botany: Charles Christopher Parry, 1823-1890.* 1997; Gordon Douglas Rowley, *A History of Succulent Plants.* Strawberry Press, Mill Valley, California 1997; R. Zander, F. Encke, G. Buchheim and S. Seybold, *Handwörterbuch der Pflanzennamen.* 14. Aufl. Stuttgart 1993; Stafleu and Cowan, *Taxonomic Literature.* 4: 79-82. 1983.

Neopatersonia Schönland Hyacinthaceae (Liliaceae)

Origins:

Named in honor of Mrs. Florence Mary Paterson (*née* Hallack), 1869-1936, botanical collector, married Mr. T.V. Paterson of Redhouse, South Africa; see Mary Gunn and Leslie E. Codd, *Botanical Exploration of Southern Africa.* 272-273. Cape Town 1981.

Neopaxia O.E.G. Nilsson Portulacaceae

Origins:

Named after the German botanist Ferdinand Albin Pax, 1858-1942, collaborator of A. Engler, in 1893 succeeded Karl Anton Eugen Prantl (1849-1893) at Breslau, from 1893 to 1925 professor of botany and Director of the Botanical Garden at Breslau. Among his many works are *Beitrag zur Kenntnis des Ovulums* von *Primula elatior* Jacq. und *officinalis* Jacq. Breslau 1882 and *Pflanzengeographie von Rumänien.* Halle 1919, contributed to H.G.A. Engler & K.A.E. Prantl *Die Natürlichen Pflanzenfamilien* and to Engler *Das Pflanzenreich.* See Adolf Friedrich Georg Ernst Albert Eduard, Duke of Mecklenburg (b. 1882), *Wissenschaftliche Ergebnisse der Deutschen ZentralAfrika-Expedition 1907-*

1908, unter Führung Adolf Friedrichs ... Band ii. *Botanik.* Leipzig 1914; Charles Jacques Édouard Morren (1833-1886), *Correspondance botanique.* Liège 1884; T.W. Bossert, *Biographical Dictionary of Botanists Represented in the Hunt Institute Portrait Collection.* 303. 1972; Orjan Eric Gustaf Nilsson (1933-), in *Botaniska Notiser.* 119: 274. 1966; Ida Kaplan Langman, *A Selected Guide to the Literature on the Flowering Plants of Mexico.* 571. Philadelphia 1964; E.M. Tucker, *Catalogue of the Library of the Arnold Arboretum of Harvard University.* 1917-1933; R. Zander, F. Encke, G. Buchheim and S. Seybold, *Handwörterbuch der Pflanzennamen.* 14. Aufl. Stuttgart 1993.

Species/Vernacular Names:

N. australasica (Hook.f.) O. Nilsson

English: white purslane

Neopeltandra Gamble Euphorbiaceae

Origins:

Greek *neos* plus the genus *Peltandra.*

Neopentanisia Verdc. Rubiaceae

Origins:

Greek *neos* with the generic name *Pentanisia* Harv.

Neopetalonema Brenan Melastomataceae

Origins:

Greek *neos* plus the genus *Petalonema* Gilg.

Neophloga Baillon Palmae

Origins:

From the Greek *neos* "new" and the genus *Phloga.*

Neophylum Tieghem Loranthaceae

Origins:

From the Greek *neos* "new" and *phylon* "tribe, family."

Neopicrorhiza D.Y. Hong Scrophulariaceae

Origins:

From the Greek *neos* "new" and *Picrorhiza* Royle ex Benth.

Neopieris Britton Ericaceae

Origins:
From the Greek *neos* "new" with *Pieris.*

Neopilea Leandri Urticaceae

Origins:
Greek *neos* "new" with *Pilea* Lindl.

Neoplatytaenia Geld. Umbelliferae

Origins:
Greek *neos* and the genus *Platytaenia* Nevski & Vved., Greek *platys* "broad" and *tainia* "fillet."

Neoplingia Ramam., Hiriart & Medran Labiatae

Origins:
For Carl Clawson Epling (Eppling), 1894-1968, American naturalist, botanist, specialist in Labiatae, with the Russian-born American scientist, evolutionist and naturalist Theodosy Grigorievich Dobrzhansky (1900-1975, also Theodosius Dobzhansky) wrote *Contributions to the Genetics, Taxonomy, and Ecology of Drosophila pseudoobscura and Its Relatives.* Washington 1944; see Bentley Glass, ed., *The Roving Naturalist. Travel Letters of Theodosius Dobzhansky.* Philadelphia 1980; Joseph Ewan, *Rocky Mountain Naturalists.* The University of Denver Press 1950.

Neopometia Aubrév. Sapotaceae

Origins:
Greek *neos* and the genus *Pometia* Vell.

Neoporteria Backeb. Cactaceae

Origins:
Probably named after the Chilean naturalist Carlos Emilio Porter, 1868-1942; see Adriana E. Hoffmann J., *Cactaceas en la flora silvestre de Chile.* Ediciones Fundacion Claudio Gay. Santiago de Chile 1989.

Species/Vernacular Names:
N. castanea Ritter
Chile: castañita
N. clavata (Soehr.) Werdermann
Chile: cacto maza
N. nidus (Soehr.) Britton & Rose

Chile: viejito, nidito
N. subgibbosa (Haw.) Werdermann
Chile: cacto rosado, quisquito
N. villosa (Monv.) Berg.
Chile: quisco peludo
N. wagenknechtii Ritter
Chile: quisquito don Rodolfo

Neoporteria Britton & Rose Cactaceae

Origins:
Named after the Chilean naturalist Carlos Emilio Porter, 1868-1942, zoologist, entomologist, Director of the Zoological Museum of the Instituto Agronómico de Chile 1914-1927, from 1897 to 1942 editor of the *Revista Chilena de Historia Natural.* Among his writings are "Don Claudio Gay. Notas biográficas i bibliográficas." in *Revista Chilena Hist. Nat.* 6(3): 110-132. 1902, *Bibliografía Chilena de Antropología y Etnología.* [Museo Nacional de Historia Natural. Antropología, Etnología y Arqueología. Publicacion 30.] Buenos Aires 1910 and "Don Federico Philippi. Notas biográficas i bibliográficas." in *Revista Chilena Hist. Nat.* 7: 106-107. 1903. See John H. Barnhart, *Biographical Notes upon Botanists.* 3: 100. 1965; S. Lenley et al., *Catalog of the Manuscript and Archival Collections and Index to the Correspondence of John Torrey.* Library of the New York Botanical Garden. 333. 1973.

Species/Vernacular Names:
N. aricensis (Ritter) Don. & Rowl. (Arica, Región de Tarapacá, Chile)
Chile: ariqueño
N. carrizalensis (Ritter) A. Hoffmann (Carrizal, Huasco, Región de Atacama, Chile)
Chile: quisquito de Carrizal
N. castanea Ritter
Chile: castañita
N. chilensis (Hidmann) Britton & Rose
Chile: chilenito
N. clavata (Soehr.) Werdermann
Chile: cacto maza
N. curvispina (Bert.) Don. & Rowl.
Chile: cacto rojo
N. eriosyzoides (Ritter) Don. & Rowl. (resembling *Eriosyce*)
Chile: quisco de Huanta (= Huanta, Elqui, Chile)
N. esmeraldana (Ritter) Don. & Rowl. (Esmeralda, Región de Antofagasta, Chile)
Chile: esmeraldano

N. horrida (Remy ex Gay) Hunt

Chile: hórrido

N. intermedia (Ritter) Don. & Rowl.

Chile: quisquito de Chañaral

N. jussieuii (Monville) Britton & Rose

Chile: quisquito del Elqui

N. krainziana (Ritter) Don. & Rowl. (after Hans Krainz, 1906-1980, Director of the State Succulent Collection in Zürich; see Gordon Douglas Rowley, *A History of Succulent Plants*. Strawberry Press, Mill Valley, California 1997)

Chile: erizo de Krainz

N. kunzei (Foerster) Back.

Chile: cunze

N. napina (Phil.) Back.

Chile: napín

N. nidus (Soehr.) Britton & Rose

Chile: viejito, nidito

N. occulta (Schumann) Britton & Rose

Chile: cacto oculto

N. odieri (Salm-Dyck) Berger (for James Odier)

Chile: odieri

N. paucicostata (Ritter) Don. & Rowl.

Chile: pocas costillas, peludín

N. recondita (Ritter) Don. & Rowl.

Chile: escondido

N. simulans (Ritter) Don. & Rowl. (resembling *Copiapoa*)

Chile: simulador

N. subgibbosa (Haw.) Britton & Rose

Chile: quisquito, cacto rosado

N. taltalensis Hutchinson (Taltal, Región de Antofagasta, Chile)

Chile: quisquito de Taltal

N. vallenarensis (Ritter) Hoffmann (Vallenar, Región de Atacama, Chile)

Chile: quisquito de Vallenar

N. villosa (Monv.) Berg.

Chile: quisco peludo

N. wagenknechtii Ritter (for Rodolfo Wagenknecht)

Chile: quisquito de don Rodolfo

Neopreissia Ulbrich Chenopodiaceae

Origins:
After the German botanist Johann August Ludwig Preiss, 1811-1883, traveler, plant collector, from 1838 to 1842 in Western Australia; see Johann G.C. Lehmann (1792-1860), *Plantae Preissianae* ... Plantarum quas in Australasia occidentali et meridionali-occidentali annis 1838-41 collegit L. Preiss. Hamburgi 1844-1847[-1848]; John H. Barnhart, *Biographical Notes upon Botanists*. 3: 107. 1965; D.J. Carr and S.G.M. Carr, eds., *People and Plants in Australia*. 1981; Mary Gunn and Leslie E. Codd, *Botanical Exploration of Southern Africa*. 287. Cape Town 1981; A. Lasègue, *Musée botanique de Benjamin Delessert*. Paris 1845; F. Ludwig Emil Diels (1874-1945), *Die Pflanzenwelt von West-Australien südlich des Wendekreises*. 1906, in H.G.A. Engler, *Die Vegetation der Erde*. No. VII. 1906; I.C. Hedge and J.M. Lamond, *Index of Collectors in the Edinburgh Herbarium*. Edinburgh 1970; N.S. Lander, "Asteraceae specimens collected by Johann August Ludwig Preiss," in *Kingia*. 1(1): 9-19. 1987.

Neopringlea S. Watson Flacourtiaceae

Origins:
Named after the American botanist Cyrus Guernsey Pringle, 1838-1911, Quaker, plant collector (Pacific States and Mexico) and plant breeder, wrote *The Record of a Quaker Conscience. C. Pringle's Diary*. New York 1918. See John H. Barnhart, *Biographical Notes upon Botanists*. 3: 111. 1965; T.W. Bossert, *Biographical Dictionary of Botanists Represented in the Hunt Institute Portrait Collection*. 318. 1972; E.M. Tucker, *Catalogue of the Library of the Arnold Arboretum of Harvard University*. 1917-1933; S. Lenley et al., *Catalog of the Manuscript and Archival Collections and Index to the Correspondence of John Torrey*. Library of the New York Botanical Garden. 335-336. 1973; Ira L. Wiggins, *Flora of Baja California*. 42. Stanford, California 1980; Ida Kaplan Langman, *A Selected Guide to the Literature on the Flowering Plants of Mexico*. 596. University of Pennsylvania Press, Philadelphia 1964; Gordon Douglas Rowley, *A History of Succulent Plants*. California 1997; Irving William Knobloch, compil., "A preliminary verified list of plant collectors in Mexico." *Phytologia Memoirs*. VI. 1983; Helen Burns Davis, *Life and Work of Cyrus Guernsey Pringle*. Burlington, Vt. 1936; I.C. Hedge and J.M. Lamond, *Index of Collectors in the Edinburgh Herbarium*. Edinburgh 1970.

Neoptychocarpus Buchheim Flacourtiaceae

Origins:
From the Greek *neos* "new" plus *Ptychocarpus* Kuhlm.

Neopycnocoma Pax Euphorbiaceae

Origins:
Greek *neos* "new" plus *Pycnocoma* Benth.

Neoraimondia Britton & Rose Cactaceae

Origins:

For the Italian botanist Antonio Raimondi, 1826-1890, naturalist and geologist, from 1850 in Peru, professor of botany, botanical collector. His writings include *Apuntes sobra la provincia litoral de Loreto*. Lima 1862, *El Departamento de Ancachs y sus riquezas minerales*. Lima 1873, *Elementos de Botanica aplicada á la medicina y á la industria*. Lima 1857, *Minerales del Perú*. Lima 1878, *El Perú. Estudios mineralógicos y geológicos*. Lima 1874-1902, *El Perú. Itinerarios de viajes*. 1929. See Manuel Rouaud y Paz-Soldan, *Dos ilustres Sabios* [S. Lorente and A. Raimondi] *vindicados* [from the criticisms of E. Desjardins]. Lima 1868; J.H. Barnhart, *Biographical Notes upon Botanists*. 3: 124. 1965; August Weberbauer, *Die Pflanzenwelt der peruanischen Andes in ihren Grundzügen dargestellt*. 13-14, 35. Leipzig 1911; Ettore Janni, [Vita di Antonio Raimondi] *Vida de Antonio Raimondi*. Lima 1942; R. Zander, F. Encke, G. Buchheim and S. Seybold, *Handwörterbuch der Pflanzennamen*. 14. Aufl. Stuttgart 1993.

Neorapinia Mold. Labiatae (Verbenaceae)

Origins:

The genus *Rapinia* Montrouzier, for the French botanist René Rapin, 1621-1687. His writings include *Carmina*. Parisiis 1723, *Christus patiens*, carmen heroicum. Londini 1713, Rapini *hortorum lib. IV, et disputatio de cultura hortensi*. Paris 1665, *Hortorum* libri IIII. Lugduni-Batav. 1668 and 1672, *Hortorum libri IV, et cultura hortensis*. Parisiis 1780 and *Oeuvres diverses* de P. Rapin, nouv. édition, augmentée du poëme des Jardins. La Haye 1725.

Neoraputia Emmerich Rutaceae

Origins:

The Greek *neos* plus the genus *Raputia* Aublet.

Neorautanenia Schinz Fabaceae

Origins:

After a Finnish missionary, the Rev. Martti (Martin) Rautanen, 1845-1926, collected 1886-91 southwest Africa (Ovamboland); see I.H. Vegter, *Index Herbariorum*. Part II (5), *Collectors N-R*. Regnum Vegetabile vol. 109. 1983; Mary Gunn and Leslie E. Codd, *Botanical Exploration of Southern Africa*. 290-291. Cape Town 1981.

Neoregelia L.B. Smith Bromeliaceae

Origins:

Named for the Russian-born botanist Constantin Andreas von Regel, 1890-1970, from 1922 to 1940 professor of botany at Kaunus University, Lithuania, author of *Fontes florae Lituaniae-Lietuvos floros* ... Kaunas 1931-1939 [Mémoires de la Faculté des sciences de l'Université de Vytautus le Grand — Scripta Horti botanici Universitatis Vytauti magni] and *Pflanzen in Europa liefern Rohstoffe*. Stuttgart 1944; see John H. Barnhart, *Biographical Notes upon Botanists*. 3: 138. 1965; T.W. Bossert, *Biographical Dictionary of Botanists Represented in the Hunt Institute Portrait Collection*. 327. 1972; Stafleu and Cowan, *Taxonomic Literature*. 4: 637-638. 1983.

Species/Vernacular Names:

N. carolinae (Beer) L.B. Sm.

English: blushing bromeliad

N. marmorata (Bak.) L.B. Sm.

English: marble plant

N. spectabilis (Moore) L.B. Sm.

English: painted fingernail, fingernail plant

Neoregnellia Urban Sterculiaceae

Origins:

After the Swedish botanist Anders Fredrik (André Frederick) Regnell, 1807-1884, plant collector (with Gustaf Anders Lindberg (1832-1900) and Salomon Eberhard Henschen, 1847-1930), physician, lichenologist, 1837 M.D. Uppsala, 1840 in Brazil. See John H. Barnhart, *Biographical Notes upon Botanists*. 3: 138. 1965; T.W. Bossert, *Biographical Dictionary of Botanists Represented in the Hunt Institute Portrait Collection*. 327. 1972; A. Lasègue, *Musée botanique de Benjamin Delessert*. Paris 1845.

Neorhine Schwantes Aizoaceae

Origins:

See the genus *Rhinephyllum* N.E. Br.

Neorites L.S. Smith Proteaceae

Origins:

From the Greek *neos* "new" and the genus *Orites* R. Br.; see Lindsay Stewart Smith (1917-1970), in *Contributions from the Queensland Herbarium*. 6: 15. 1969.

Species/Vernacular Names:

N. kevediana L.S. Smith (after the officers of Queensland Forestry Dept. Kevin White and Edgar Volck, original collectors of the type in 1952, NNW. of Kuranda, Forest Res. 315, Australia)

English: fishtail silky oak

Neoroepera Müll.Arg. & F. Muell. Euphorbiaceae

Origins:

From the Greek *neos* "new" and the related genus *Roeperia* Spreng., named in honor of the German botanist Johannes August Christian Roeper (also Röper), 1801-1885, physician, M.D. Göttingen 1823, studied with A.P. de Candolle, professor of botany at Basel and Rostock, University librarian at Rostock, correspondent of the German botanist Diederich Franz Leonhard von Schlechtendal (1794-1866), contributor to C.F.P. von Martius *Flora Brasiliensis* (Euphorbiaceae). His publications include *Zur Flora Mecklenburgs*. Rostock 1843 [-1844], *Enumeratio Euphorbiarum quae in Germania et Pannonia gignuntur*. Gottingae 1824, *Vorgefasste botanische Meinungen*. Rostock 1860, and *De floribus et affinitatibus Balsaminearum*. Basiliae [Basel] 1830; see John H. Barnhart, *Biographical Notes upon Botanists*. 3: 170. 1965; Charles J. Édouard Morren, *Correspondance botanique*. Liège 1874 and 1884; A.H.L. de Jussieu, in *Mémoires du Muséum d'Histoire Naturelle*. 12: 454, t. 15, no. 3. [Mémoires sur les Rutacées.] Paris 1825; T.W. Bossert, *Biographical Dictionary of Botanists Represented in the Hunt Institute Portrait Collection*. 335. 1972; Antoine Lasègue, *Musée botanique de M. Benjamin Delessert*. 335. Paris 1845; E.M. Tucker, *Catalogue of the Library of the Arnold Arboretum of Harvard University*. 1917-1933; G. Schmid, *Goethe und die Naturwissenschaften*. Halle 1940.

Neorosea N. Hallé Rubiaceae

Origins:

After the German apothecary Valentin Rose, 19th century. See Berlin. *Berlinisches Jahrbuch für die Pharmacie, etc.* [Edited by V.R., etc.] 1818 etc.; A. Pabst, in *Dictionary of Scientific Biography* 11: 539-540. 1981; Stuart Pierson, in *Dictionary of Scientific Biography* 11: 540-542. 1981.

Neorudolphia Britton Fabaceae

Origins:

For the Swedish naturalist Israel Karl Asmund Rudolphi, 1771-1832, physician, professor of medicine and anatomy; see J.H. Barnhart, *Biographical Notes upon Botanists*. 3:

189. 1965; T.W. Bossert, *Biographical Dictionary of Botanists Represented in the Hunt Institute Portrait Collection*. 342. 1972; Stafleu and Cowan, *Taxonomic Literature*. 4: 974-975. 1983; R. Zander, F. Encke, G. Buchheim and S. Seybold, *Handwörterbuch der Pflanzennamen*. 14. Aufl. Stuttgart 1993; Günther Schmid, *Chamisso als Naturforscher*. Eine Bibliographie. Leipzig 1942; G. Schmid, *Goethe und die Naturwissenschaften*. Halle 1940.

Neosabicea Wernham Rubiaceae

Origins:

Greek *neos* plus the genus *Sabicea* Aublet.

Neosasamorpha Tatew. Gramineae

Origins:

Greek *neos* plus the genus *Sasamorpha* Nakai.

Neoschimpera Hemsley Rubiaceae

Origins:

For the German botanist Andreas Franz Wilhelm Schimper, (b. Strasbourg, France) 1856-1901 (d. Basel, Switzerland), son of Wilhelm Philipp Schimper (1808-1880, d. Strasbourg, France [then part of Germany], professor of natural history and geology at the University of Strasbourg), interested in plant geography, worked with Julius von Sachs (1832-1897) at Würzburg, 1880-1881 traveled in the eastern United States, 1882-1883 with the German botanist Friedrich Richard Adalbert Johow (1859-1933) to West Indies (Barbados, Trinidad, Venezuela and Dominica), 1886 Brazil, Jena 1888-1901 editor of *Botanische Mittheilungen aus den Tropen*, 1889-1890 Ceylon and Java, succeeded the German botanist Georg Albrecht Klebs (1857-1918) at Basel, 1898 professor of botany at the University of Basel, 1898-1899 with the German *Valdivia* Deep Sea Expedition ("Deutsche Tiefsee-Expedition") for study of plankton, visited Cameroon. Among his works are *Untersuchungen über die Proteinkrystalloide der Pflanzen*. Strasbourg 1878, *Taschenbuch der medicinisch-pharmaceutischen Botanik und pflanzlichen Drogenkunde*. Strassburg 1886, *Die epiphytische Vegetation Amerikas*. Jena 1888, *Die indo-malaysche Strandflora ...* Jena 1891 and "Rhizophoraceae." in *Nat. Pflanzenfam*. 3(7): 42-48. 1892 and 49-56. 1893, until 1898 worked in the laboratory of Eduard Adolf Strasburger (1844-1912) at Bonn. See A.P.M. Sanders, in *Dictionary of Scientific Biography* 12: 165-167. 1981; J.H. Barnhart, *Biographical Notes upon Botanists*. 3: 225. 1965; H. Schenk, "A.F.W. Schimper." in *Berichte der Deutschen botanischen Gesellschaft, XIX, Generalversammlungsheft*. 1: 54-70.

1901; Mary Gunn and Leslie E. Codd, *Botanical Exploration of Southern Africa*. 311. Cape Town 1981; Heinz Tobien, in *Dictionary of Scientific Biography* 12: 167-168. 1981; P.W. Richards, in *Dictionary of Scientific Biography* 12: 168-169. 1981; Ignatz Urban, ed., *Symbolae Antillanae*. 1: 150-151. 1898 and 3: 120. 1902; E.M. Tucker, *Catalogue of the Library of the Arnold Arboretum of Harvard University*. 1917-1933; T.W. Bossert, *Biographical Dictionary of Botanists Represented in the Hunt Institute Portrait Collection*. 352. 1972; Ida Kaplan Langman, *A Selected Guide to the Literature on the Flowering Plants of Mexico*. 679. University of Pennsylvania Press, Philadelphia 1964; F.N. Hepper and Fiona Neate, *Plant Collectors in West Africa*. 72. Utrecht 1971.

Neoschischkinia Tzvelev Asteraceae

Origins:
For the Russian botanist Boris Konstantinovich Schischkin (Shishkin), 1886-1963, traveler, botanical explorer, professor of botany. See John H. Barnhart, *Biographical Notes upon Botanists*. 3: 228. 1965; T.W. Bossert, *Biographical Dictionary of Botanists Represented in the Hunt Institute Portrait Collection*. 353. 1972; R. Zander, F. Encke, G. Buchheim and S. Seybold, *Handwörterbuch der Pflanzennamen*. 14. Aufl. Stuttgart 1993.

Neoschroetera Briquet Zygophyllaceae

Origins:
For the Swiss botanist Carl (Karl) Joseph Schröter, 1855-1939, traveler, professor of systematic botany. See John H. Barnhart, *Biographical Notes upon Botanists*. 3: 520. 1965; T.W. Bossert, *Biographical Dictionary of Botanists Represented in the Hunt Institute Portrait Collection*. 355. 1972; E.M. Tucker, *Catalogue of the Library of the Arnold Arboretum of Harvard University*. Cambridge, Massachusetts 1917-1933; Mary Gunn and Leslie E. Codd, *Botanical Exploration of Southern Africa*. 318. Cape Town 1981.

Neoschumannia Schlechter Asclepiadaceae

Origins:
For the German botanist Karl Moritz Schumann, 1851-1904, taxonomist, botanical collector. See John H. Barnhart, *Biographical Notes upon Botanists*. 3: 247. 1965; T.W. Bossert, *Biographical Dictionary of Botanists Represented in the Hunt Institute Portrait Collection*. 356. 1972; Ida Kaplan Langman, *A Selected Guide to the Literature on the Flowering Plants of Mexico*. 687-688. University of Pennsylvania Press, Philadelphia 1964; E.M. Tucker, *Catalogue*

of the Library of the Arnold Arboretum of Harvard University*. Cambridge, Massachusetts 1917-1933; A. White and B.L. Sloane, *The Stapelieae*. Pasadena 1937; Stafleu and Cowan, *Taxonomic Literature*. 5: 400-408. 1985; R. Zander, F. Encke, G. Buchheim and S. Seybold, *Handwörterbuch der Pflanzennamen*. 14. Aufl. Stuttgart 1993.

Neosciadium Domin Umbelliferae

Origins:
From the Greek *neos* "new" and *skias* "a canopy, umbel," *skiadion*, *skiadeion* "umbel, parasol," possibly referring to the new genus of umbelliferous plants.

Neoscortechinia Pax Euphorbiaceae

Origins:
Named after the Italian botanist Benedetto Scortechini, 1845-1886 (Calcutta), botanical explorer, clergyman, traveler, in Australia and the Malay Peninsula, collected ferns, in 1881 a Fellow of the Linnean Society. See John H. Barnhart, *Biographical Notes upon Botanists*. 3: 251. 1965; T.W. Bossert, *Biographical Dictionary of Botanists Represented in the Hunt Institute Portrait Collection*. 358. 1972; D.J. Carr and S.G.M. Carr, eds., *People and Plants in Australia*. 1981; Henry Nicholas Ridley (1855-1956), *The Flora of the Malay Peninsula*. London 1922-1925; I.H. Vegter, *Index Herbariorum*. Part II (6), *Collectors S*. Regnum Vegetabile vol. 114. 1986.

Neosepicaea Diels Bignoniaceae

Origins:
After Sepik River, Papua New Guinea; in *Botanische Jahrbücher*. 57: 500. (May) 1972.

Species/Vernacular Names:
N. jucunda (F. Muell.) Steenis
English: jungle vine

Neosieversia Bolle Rosaceae

Origins:
See *Novosieversia* F. Bolle.

Neosinocalamus Keng f. Gramineae

Origins:
Greek *neos* plus *Sinocalamus* McClure.

Neosloetiopsis Engl. Moraceae

Origins:
Greek *neos* plus the genus *Sloetiopsis* Engl.

Neosparton Griseb. Verbenaceae

Origins:
Greek *neos* and *sparton* "a rope, bond."

Neosprucea Sleumer Tiliaceae

Origins:
For the British botanist Richard Spruce, 1817-1893 (near Malton, Yorkshire), cartographer, traveler, botanical explorer and collector, from 1849 to 1864 in South America (Venezuela, Brazil, Peru, Ecuador, Bolivia, the Amazon Valley and Andes). His writings include "Palmae amazonicae." *J. Linn. Soc., Bot.* 11(50-51): 65-185. 1869, *Notes on the Valleys of Piura and Chira, in Northern Peru, and on the Cultivation of Cotton Therein.* London 1864, *Hepaticae of the Amazon and of the Andes of Peru and Ecuador.* London [1884-] 1885 and *Report on the Expedition to Procure Seeds and Plants of the Cinchona Succirubra, or Red Bark Tree.* London 1861. See Alfred Russell Wallace, editor, *Notes of a Botanist on the Amazon & Andes* ... during the years 1849-1864 by Richard Spruce. London 1908; J.H. Barnhart, *Biographical Notes upon Botanists.* 3: 312. 1965; T.W. Bossert, *Biographical Dictionary of Botanists Represented in the Hunt Institute Portrait Collection.* 379. 1972; G. Murray, *History of the Collections Contained in the Natural History Departments of the British Museum.* 1: 184. 1904; Ethelyn Maria Tucker, *Catalogue of the Library of the Arnold Arboretum of Harvard University.* Cambridge, Massachusetts 1917-1933; H.N. Clokie, *Account of the Herbaria of the Department of Botany in the University of Oxford.* 247. 1964; S. Lenley et al., *Catalog of the Manuscript and Archival Collections and Index to the Correspondence of John Torrey.* Library of the New York Botanical Garden. 381. 1973; Charles Lyte, *The Plant Hunters.* London 1983; Henri Pittier, *Manual de las Plantas Usuales de Venezuela* y su Suplemento. Caracas 1978; August Weberbauer, *Die Pflanzenwelt der peruanischen Andes in ihren Grundzügen dargestellt.* 14-15. Leipzig 1911; R. Desmond, *The European Discovery of the Indian Flora.* 1992; R.G.C. Desmond, in *Dictionary of Scientific Biography* 12: 594. 1981.

Neostachyanthus Exell & Mendonça Icacinaceae

Origins:
Greek *neos* plus *Stachyanthus* Engl.

Neostapfia Davy Gramineae

Origins:
After the Austrian botanist Otto Stapf, 1857-1933, traveler, 1900-1922 Keeper of the Herbarium of the Royal Botanic Gardens, Kew, 1922-1933 editor of the *Botanical Magazine.* Among his numerous and valuable publications are *On the Flora of Mount Kinabalu in North Borneo.* London 1894 and *The Aconites of India.* Calcutta 1905; see John H. Barnhart, *Biographical Notes upon Botanists.* 3: 317. 1965; Mia (Maria) Caroline Karsten, *The Old Company's Garden at the Cape and Its Superintendents.* Cape Town 1951; James Edgar Dandy (1903-1976), in *The Journal of Botany.* 69: 54. (Feb.) 1931; T.W. Bossert, *Biographical Dictionary of Botanists Represented in the Hunt Institute Portrait Collection.* 380. 1972; Ida Kaplan Langman, *A Selected Guide to the Literature on the Flowering Plants of Mexico.* Philadelphia 1964; S. Lenley et al., *Catalog of the Manuscript and Archival Collections and Index to the Correspondence of John Torrey.* Library of the New York Botanical Garden. 382. 1973; Ethelyn Maria Tucker, *Catalogue of the Library of the Arnold Arboretum of Harvard University.* Cambridge, Massachusetts 1917-1933; Leonard Huxley, *Life and Letters of Sir Joseph Dalton Hooker.* London 1918; Stafleu and Cowan, *Taxonomic Literature.* 5: 839-843. 1985; R. Zander, F. Encke, G. Buchheim and S. Seybold, *Handwörterbuch der Pflanzennamen.* 14. Aufl. Stuttgart 1993; Emil Bretschneider, *History of European Botanical Discoveries in China.* [Reprint of the original edition, St. Petersburg 1898.] Leipzig 1981; Ray Desmond, *Dictionary of British & Irish Botanists and Horticulturists.* 650. London 1994.

Species/Vernacular Names:
N. colusana (Burtt Davy) Burtt Davy
English: Colusa grass

Neostapfiella A. Camus Gramineae

Origins:
The diminutive of *Neostapfia.*

Neostenanthera Exell Annonaceae

Origins:
From the Greek *neos* "new" and the genus *Stenanthera* Engl. & Diels ("narrow anthers," from the Greek *stenos* "narrow" and *anthera*).

Species/Vernacular Names:
N. myristicifolia (Oliv.) Exell
Nigeria: uyenghen eze (Edo)

Neostrearia L.S. Smith Hamamelidaceae

Origins:

From the Greek *neos* "new" and the genus *Ostrearia* Baill.

Neotainiopsis Bennet & Raizada Orchidaceae

Origins:

Greek *neos* "new" and the genus *Tainiopsis* Schlechter.

Neotatea Maguire Guttiferae

Origins:

After the American (born in Great Britain) naturalist George Henry Hamilton Tate, 1894-1953 (Morristown, New Jersey), zoologist, plant and botanical collector in South America, 1922 Field Assistant American Museum of Natural History, author of *Mammals of Eastern Asia*. New York 1947, with Thomas D. Carter wrote *Mammals of the Pacific World*. 1945. See Douglas C. McMurtrie, *A Bibliography of Morristown Imprints 1798-1820*. [from the *Proceedings of the New Jersey Historical Society*, April 1936] Newark 1936; J.H. Barnhart, *Biographical Notes upon Botanists*. 3: 361. 1965; E.M. Tucker, *Catalogue of the Library of the Arnold Arboretum of Harvard University*. Cambridge, Massachusetts 1917-1933.

Neotchihatchewia Rauschert Brassicaceae

Origins:

After the Russian botanist Pierre de Tchihatcheff (Petr Aleksandrovich Tchichatscheff, Chikhachef, Tschihatcheff), 1812-1890, traveler, geographer; see John H. Barnhart, *Biographical Notes upon Botanists*. 3: 365. 196; T.W. Bossert, *Biographical Dictionary of Botanists Represented in the Hunt Institute Portrait Collection*. 396. 1972; E.M. Tucker, *Catalogue of the Library of the Arnold Arboretum of Harvard University*. 1917-1933.

Neotessmannia Burret Tiliaceae

Origins:

For the German ethnographer Günther (Guenther) Tessmann, explorer and plant collector in Africa and Peru, missionary. Among his writings are *Die Bubi von Fernando Poo*. Herausgegeben von Prof. Dr. O. Reche. Hagen & Darmstadt 1923, *Menschen Ohne Gott*. Stuttgart 1928 and *Die Pangwe*. ... Ergebnisse der Lübecker Pangwe-Expedition 1907-1909 und früherer Forschungen 1904-1907. Berlin 1913; see John H. Barnhart, *Biographical Notes upon Botanists*. 3: 369. 1965; Stafleu and Cowan, *Taxonomic Literature*. 6: 228-229. 1986; Réné Letouzey, "Les botanistes au Cameroun." in *Flore du Cameroun*. 7: 58. Paris 1968; F.N. Hepper and Fiona Neate, *Plant Collectors in West Africa*. 78-79. 1971; Anthonius Josephus Maria Leeuwenberg, "Isotypes of which holotypes were destroyed in Berlin." *Webbia*. 19(2): 861-863. 1965; Gottfried Wilhelm Johannes Mildbraed (1879-1954), "Plantae Tessmannianae Peruvianae I." *Notizbl. Bot. Gart. Berlin-Dahlem*. 9: 136-144. 1924, "Plantae Tessmannianae Peruvianae II." *Notizbl. Bot. Gart. Berlin-Dahlem*. 9(84): 260-268. 1925 and "Plantae Tessmannianae Peruvianae III." *Notizbl. Bot. Gart. Berlin-Dahlem*. 9(89): 964-997. 1926, etc.

Neothorelia Gagnepain Capparidaceae (Capparaceae)

Origins:

After the French botanist Clovis Thorel, 1833-1911, physician, plant collector; see John H. Barnhart, *Biographical Notes upon Botanists*. 3: 380. 1965; Emil Bretschneider, *History of European Botanical Discoveries in China*. Leipzig 1981; E.M. Tucker, *Catalogue of the Library of the Arnold Arboretum of Harvard University*. 1917-1933.

Neotinea Reichb.f. Orchidaceae

Origins:

After the Italian botanist Vincenzo (Vincentius) Tineo, 1791-1856, professor of botany, from 1814 to 1856 Director of the Botanical Garden of Palermo. His works include *Catalogus plantarum Horti Regii Panormitani*. Panormi 1827, *Plantarum rariorum Siciliae minus cognitarum* pugillus primus. Palermo 1817 and *Plantarum rariorum Siciliae minus cognitarum* ... Panormi 1846, he was son of the Italian botanist Giuseppe Tineo (1757-1812). See John H. Barnhart, *Biographical Notes upon Botanists*. 3: 386. 1965; T.W. Bossert, *Biographical Dictionary of Botanists Represented in the Hunt Institute Portrait Collection*. 402. 1972; A. Lasègue, *Musée botanique de Benjamin Delessert*. 314. Paris 1845; E.M. Tucker, *Catalogue of the Library of the Arnold Arboretum of Harvard University*. 1917-1933; F. Tornabene, "Elogio accademico del cav. Vincenzo Tineo." *Atti Accad. Gioenia Sci. Nat*. Catania 1856; Giuseppe M. Mira, *Bibliografia Siciliana*. 2: 411. Palermo 1881; Mariella Azzarello Di Misa, a cura di, *Il Fondo Antico della Biblioteca dell'Orto Botanico di Palermo*. 271. Palermo 1988; R. Zander, F. Encke, G. Buchheim and S. Seybold, *Handwörterbuch der Pflanzennamen*. 14. Aufl. 790. Stuttgart 1993.

Species/Vernacular Names:

N. maculata (Desf.) Stearn

English: neotinea

Neotorularia Hedge & Léonard Brassicaceae

Origins:
Greek *neos* plus the genus *Torularia* O. Schulz.

Neotreleasea Rose Commelinaceae

Origins:
For William Trelease, 1857-1945, American botanist, professor of botany, Director of the Missouri Botanical Garden, plant collector; see J.H. Barnhart, *Biographical Notes upon Botanists.* 3: 399. 1965; S. Lenley et al., *Catalog of the Manuscript and Archival Collections and Index to the Correspondence of John Torrey.* Library of the New York Botanical Garden. 1973; Ida Kaplan Langman, *A Selected Guide to the Literature on the Flowering Plants of Mexico.* Philadelphia 1964; E.M. Tucker, *Catalogue of the Library of the Arnold Arboretum of Harvard University.* Cambridge, Massachusetts 1917-1933; Joseph Ewan, *Rocky Mountain Naturalists.* The University of Denver Press 1950; Joseph Ewan, in *Dictionary of Scientific Biography* 13: 456. 1981; R. Zander, F. Encke, G. Buchheim and S. Seybold, *Handwörterbuch der Pflanzennamen.* 14. Aufl. Stuttgart 1993; T.W. Bossert, *Biographical Dictionary of Botanists Represented in the Hunt Institute Portrait Collection.* 406. 1972; Stafleu and Cowan, *Taxonomic Literature.* 6: 458-468. Utrecht 1986; Irving William Knobloch, compil., "A preliminary verified list of plant collectors in Mexico." *Phytologia Memoirs.* VI. 1983; J.T. Buchholz, "William Trelease 1857-1945." *Science.* n.s. 101: 192-193. 1945.

Neotrewia Pax & K. Hoffmann Euphorbiaceae

Origins:
For the German botanist Christoph Jakob Trew, 1695-1769, physician, traveler, correspondent of the botanical artist Georg Dionysius Ehret (1708-1770), author of *Plantae selectae, quarum imagines ad exemplaria naturalia Londini in hortis Curiosorum nutrita manu artificiosa doctaque pinxit Georgius Dionysius Ehret ...* [Norimbergae] 1750-1773. See John H. Barnhart, *Biographical Notes upon Botanists.* 3: 400. 1965; T.W. Bossert, *Biographical Dictionary of Botanists Represented in the Hunt Institute Portrait Collection.* 406. 1972; Jonas C. Dryander, *Catalogus bibliothecae historico-naturalis Josephi Banks.* London 1800; J.D. Milner, *Catalogue of Portraits of Botanists Exhibited in the Museums of the Royal Botanic Gardens.* Royal Botanic Gardens, Kew, London 1906; Blanche Elizabeth Edith Henrey (1906-1983), *British Botanical and Horticultural Literature before 1800.* Oxford 1975; Gordon Douglas Rowley, *A History of Succulent Plants.* Strawberry Press, Mill Valley, California 1997; R. Zander, F. Encke, G. Buchheim and S. Seybold, *Handwörterbuch der Pflanzennamen.* 14. Aufl. Stuttgart 1993.

Neotrigonostemon Pax & K. Hoffmann Euphorbiaceae

Origins:
Greek *neos* plus *Trigonostemon* Blume.

Neottia Guett. Orchidaceae

Origins:
Greek *neossia, neottia, nossia* "nest, a bird's nest, nest of young birds, beehive," *neotteia, neosseia* "nest-building," referring to the fibers of the roots.

Species/Vernacular Names:
N. nidus-avis (L.) Rich.
English: bird's nest orchid

Neottianthe (Reichb.) Schltr. Orchidaceae

Origins:
Genus *Neottia* Guett. and *anthos* "flower," referring to the flowers.

Neottidium Schltdl. Orchidaceae

Origins:
Diminutive of *Neottia*.

Neottopteris J. Sm. Aspleniaceae

Origins:
From the Greek *neottia* "nest, a bird's nest" and *pteris* "fern."

Neotuerckheimia Donn. Sm. Bignoniaceae

Origins:
For the German plant collector Hans von Türckheim, 1853-1920, traveler, in Guatemala and Santo Domingo; see Theodore W. Bossert, *Biographical Dictionary of Botanists Represented in the Hunt Institute Portrait Collection.* 408. Boston, Massachusetts 1972.

Neoturczaninovia Koso-Pol. Umbelliferae

Origins:

For the Russian botanist Porphir Kiril Nicolai Stepanowitsch Turczaninow, 1796-1863 (or 1864), traveler, botanical explorer, administrator; see J.H. Barnhart, *Biographical Notes upon Botanists*. 3: 408. 1965; Emil Bretschneider, *History of European Botanical Discoveries in China*. [Reprint of the original edition 1898.] Leipzig 1981; H.N. Clokie, *Account of the Herbaria of the Department of Botany in the University of Oxford*. 257. Oxford 1964; E.M. Tucker, *Catalogue of the Library of the Arnold Arboretum of Harvard University*. Cambridge, Massachusetts 1917-1933; Ida Kaplan Langman, *A Selected Guide to the Literature on the Flowering Plants of Mexico*. Philadelphia 1964; Antoine Lasègue, *Musée botanique de M. Benjamin Delessert*. 1845; Stafleu and Cowan, *Taxonomic Literature*. 6: 537-541. 1986; R. Zander, F. Encke, G. Buchheim and S. Seybold, *Handwörterbuch der Pflanzennamen*. 14. Aufl. Stuttgart 1993.

Neotysonia Dalla Torre & Harms Asteraceae

Origins:

For Isaac Tyson, c. 1859-1942, original collector of the plant near the Upper Murchison River, Western Australia; see Karl (Carl) Wilhelm von Dalla Torre von Thurnberg-Sternhoff (1850-1928) and Hermann August Theodor Harms, *Genera Siphonogamarum*. Leipzig 1907-1908; I.H. Vegter, *Index Herbariorum*. Part II (7), *Collectors T-Z*. Regnum Vegetabile vol. 117. 1988; James A. Baines, *Australian Plant Genera. An Etymological Dictionary of Australian Plant Genera*. 385. Chipping Norton, N.S.W. 1981.

Neoveitchia Beccari Palmae

Origins:

Greek *neos* "new" and the generic name *Veitchia* H. Wendland.

Neowashingtonia Sudw. Palmae

Origins:

Greek *neos* "new" and the genus *Washingtonia* H.A. Wendland.

Neowawraea Rock Euphorbiaceae

Origins:

For the Austrian (b. Brünn) botanist Dr. Heinrich Ritter Wawra von Fernsee (Ritter von Fernsee), 1831-1887 (d.

near Vienna), traveler, plant collector (Brazil, Luanda and South Africa), ship's surgeon, ennobled 1873; see John H. Barnhart, *Biographical Notes upon Botanists*. 3: 466. 1965; T.W. Bossert, *Biographical Dictionary of Botanists Represented in the Hunt Institute Portrait Collection*. 428. 1972; E. Bretschneider, *History of European Botanical Discoveries in China*. Leipzig 1981; Ethelyn Maria Tucker, *Catalogue of the Library of the Arnold Arboretum of Harvard University*. Cambridge, Massachusetts 1917-1933; Stafleu and Cowan, *Taxonomic Literature*. 7: 111-113. 1988; R. Zander, F. Encke, G. Buchheim and S. Seybold, *Handwörterbuch der Pflanzennamen*. 14. Aufl. Stuttgart 1993; Mary Gunn and Leslie E. Codd, *Botanical Exploration of Southern Africa*. 371. 1981; Irving William Knobloch, compil., "A preliminary verified list of plant collectors in Mexico." *Phytologia Memoirs*. VI. 1983; H. Dolezal, *Portug. Acta Biol*. 6: 257-323. 1959 and 7: 324-551. 1961.

Neowerdermannia Fric Cactaceae

Origins:

Dedicated to the German (b. Berlin) botanist Erich Werdermann, 1892-1959 (Bremen), specialist in Cactaceae and fungi, plant geographer, traveler (Latin America, South Africa, Namibia, Chile, Bolivia, Brazil, Mexico), explorer, plant collector, anatomist, professor of botany, plant physiologist; see John H. Barnhart, *Biographical Notes upon Botanists*. 3: 477. 1965; Mary Gunn and L.E. Codd, *Botanical Exploration of Southern Africa*. 374. 1981; T.W. Bossert, *Biographical Dictionary of Botanists Represented in the Hunt Institute Portrait Collection*. 431. 1972; Ida Kaplan Langman, *A Selected Guide to the Literature on the Flowering Plants of Mexico*. 798-799. Philadelphia 1964; Gordon Douglas Rowley, *A History of Succulent Plants*. Mill Valley, California 1997; R. Zander, F. Encke, G. Buchheim and S. Seybold, *Handwörterbuch der Pflanzennamen*. 14. Aufl. Stuttgart 1993.

Species/Vernacular Names:

N. chilensis Back.

Chile: macso, achacana

Neowilliamsia Garay Orchidaceae

Origins:

Dedicated to the American botanist Louis Otho (Otto) Williams, 1908-1991, traveler, orchidologist, botanical collector. His works include *Orchid Studies*. Harvard University [1937-1947], with P.H. Allen wrote "Orchids of Panama." *Monogr. Syst. Bot. Missouri Botanical Garden*. vol. 4. 1980. See John H. Barnhart, *Biographical Notes upon Botanists*. 3: 499. 1965; T.W. Bossert, *Biographical Dictionary of Botanists Represented in the Hunt Institute Portrait Collec-*

tion. 437. 1972; Ida Kaplan Langman, *A Selected Guide to the Literature on the Flowering Plants of Mexico*. 805-806. Philadelphia 1964; S. Lenley et al., *Catalog of the Manuscript and Archival Collections and Index to the Correspondence of John Torrey*. Library of the New York Botanical Garden. 436. 1973; R. Zander, F. Encke, G. Buchheim and S. Seybold, *Handwörterbuch der Pflanzennamen*. 14. Aufl. Stuttgart 1993; Irving William Knobloch, compil., "A preliminary verified list of plant collectors in Mexico." *Phytologia Memoirs*. VI. 1983.

Neowimmeria O. Degener & I. Degener Campanulaceae

Origins:

For the Austrian botanist Franz Elfried Wimmer, 1881-1961, clergyman, naturalist; see J.H. Barnhart, *Biographical Notes upon Botanists*. 3: 505. 1965; T.W. Bossert, *Biographical Dictionary of Botanists Represented in the Hunt Institute Portrait Collection*. 439. 1972; Ida Kaplan Langman, *A Selected Guide to the Literature on the Flowering Plants of Mexico*. 808. Philadelphia 1964; R. Zander, F. Encke, G. Buchheim and S. Seybold, *Handwörterbuch der Pflanzennamen*. 14. Aufl. Stuttgart 1993; Stafleu and Cowan, *Taxonomic Literature*. 7: 358-359. 1988.

Neowollastonia Wernham ex Ridley Apocynaceae

Origins:

After the English naturalist Alexander Frederick Richmond Wollaston, 1875-1930 (d. Cambridge), traveler, plant collector; see Ray Desmond, *Dictionary of British & Irish Botanists and Horticulturists*. 752. London 1994.

Neowormia Hutch. & Summerh. Dilleniaceae

Origins:

Greek *neos* plus the genus *Wormia* Rottb.

Neozenkerina Mildbr. Acanthaceae

Origins:

After the German plant collector and botanist George August Zenker, 1855-1922, 1896-1916 professional plant collector for Woermann Trading Co. in East and West Cameroon, in West Africa, with the German botanist and botanical collector Alois Staudt in East Cameroon; see G.W.J. Mildbraed (1879-1954), *Notizbl. Bot. Gart. Berl.* 8: 317-

324. 1923; H. Walter, "Afrikanische pflanzen in Hamburg." in *Mitt. Geog. Gesell. Hamburg*. 56: 92-93. Hamburg 1965; Réné Letouzey (1918-1989), "Les botanistes au Cameroun." in *Flore du Cameroun*. 7: 1-110. Paris 1968; Frank Nigel Hepper, "Botanical collectors in West Africa, except French territories, since 1860." in *Comptes Rendus de l'Association pour l'étude taxonomique de la flore d'Afrique*, (A.E.T.F.A.T.). 69-75. Lisbon 1962; Anthonius Josephus Maria Leeuwenberg, "Isotypes of which holotypes were destroyed in Berlin." in *Webbia*. 19: 861-863. 1965; F.N. Hepper and Fiona Neate, *Plant Collectors in West Africa*. 76 and 87. 1971.

Nepa Webb Fabaceae

Origins:

Latin *nepa* for a scorpion, a crab.

Nepenthandra S. Moore Euphorbiaceae

Origins:

Greek *nepenthes* and *aner, andros* "man, stamen, male."

Nepenthes L. Nepenthaceae

Origins:

Greek *nepenthes*, from *ne, ni* "not" and *penthes, penthos* "mourning," a plant or drug that drives away sadness, see Homer, *Odyssey*. 4.221, and Theophrastus, *HP*. 9.15.1 (Loeb Classical Library 1916), Latin *nepenthes* for a plant which mingled with wine had an exhilarating effect (Plinius); see Carl Linnaeus, *Species Plantarum*. 955. 1753 and *Genera Plantarum*. Ed. 5. 909. 1754; Helmut Genaust, *Etymologisches Wörterbuch der botanischen Pflanzennamen*. 416. Basel 1996; A. Bonavilla, *Dizionario etimologico di tutti i vocaboli usati nelle scienze, arti e mestieri, che traggono origine dal greco*. Milano 1819-1821; C.T. Onions, *The Oxford Dictionary of English Etymology*. Oxford University Press 1966.

Species/Vernacular Names:

N. madagascariensis Poir.

Madagascar: ampangandrano, mohara, oranitoko, ponga, rakotra, rapaoranjanahary, anramitaco (recorded by Flacourt in 1661, *Hist. de la Grande Isle de Madagascar* 130)

N. mirabilis (Lour.) Druce (*Phyllamphora mirabilis* Lour.)

English: monkey cup, tropical pitcher plant, pitcher plant

Japan: utsubo-kazura

China: zhu long cao

N. rajah Hook.f.

English: king monkey cup

Nepeta L. Labiatae

Origins:

Nepeta, ae "Italian catnip, *Nepeta italica* Willd.," Latin *nepos, otis* "a descendant, a sucker," Hebrew *nib* "to produce," Arabic *'nb* "offspring, child"; see Carl Linnaeus, *Species Plantarum.* 570. 1753 and *Genera Plantarum.* Ed. 5. 249. 1754; Giovanni Semerano, *Le origini della cultura europea.* Dizionario della lingua Latina e di voci moderne. 2(2): 484-485. Firenze 1994.

Species/Vernacular Names:

N. cataria L.

English: catmint, catnip

China: jing jie, jia jing jie

N. coerulescens Maxim.

English: blueflower nepeta

China: lan hua jing jie

N. densiflora Karelin & Kirilov

English: denseflower nepeta

China: mi hua jing jie

N. discolor Royle ex Bentham

English: diverse-color nepeta

China: yi se jing jie

N. fordii Hemsley

English: Ford nepeta

China: xin ye jing jie

N. hemsleyana Oliver ex Prain

English: Hemsley nepeta

China: zang jing jie

N. jomdaensis H.W. Li

English: Jomda nepeta

China: jiang da jing jie

N. laevigata (D. Don) Handel-Mazzetti

English: smooth nepeta

China: si hua jing jie

N. manchuriensis S. Moore

English: Manchurian nepeta

China: hei long jiang jing jie

N. prattii H. Léveillé

English: Pratt nepeta

China: kang zang jing jie

N. sibirica L.

English: Siberian nepeta

China: da hua jing jie

N. stewartiana Diels

English: Stewart nepeta

China: duo hua jing jie

N. sungpanensis C.Y. Wu

English: Songpan nepeta

China: song pan jing jie

N. taxkorganica Y.F. Chang

English: Taxkorgan nepeta

China: ka shi jing jie

N. veitchii Duthie

English: Veitch nepeta

China: chuan xi jing jie

Nephelaphyllum Blume Orchidaceae

Origins:

From the Greek *nephele* "a cloud" and *phyllon* "a leaf," referring to the upper surface of the leaves.

Nephelea R.M. Tryon Cyatheaceae

Origins:

From the Greek *nephele* "a cloud," Latin *nephela, nefela* "a kind of thin cake," in Greek legend Nephele was the wife of Atamas and mother of Phrixus and Helle.

Nephelium L. Sapindaceae

Origins:

Greek *nephele* "a cloud"; Pseudo Apuleius Barbarus in his *Herbarium* applied Latin *nephelion* to a plant also called *personata* (Latin *personata, ae,* is a kind of large burdock (Plinius), said to be synonymous with *persolata* or *persollata,* the brown mullein); see Carl Linnaeus, *Systema Naturae.* Ed. 12. 2: 623. 1767 and *Mantissa Plantarum.* 125. 1767.

Species/Vernacular Names:

N. lappaceum L.

English: rambutan, rambootan

Malaya: rambutan

China: shao zi, shao tzu

N. lappaceum L. var. *topengii* (Merr.) How & Ho

English: Hainan rambutan

N. ramboutan-ake (Labill.) Leenh.

Malaya: pulasan, gerak, paik, sangga lotong

Nephelochloa Boiss. Gramineae

Origins:
From the Greek *nephele* "a cloud" and *chloe, chloa* "grass."

Nephopteris Lellinger Pteridaceae (Adiantaceae, Hemionitidaceae)

Origins:
From the Greek *nephos, nepheos* "a cloud, mass of clouds" and *pteris* "fern."

Nephradenia Decne. Asclepiadaceae

Origins:
From the Greek *nephros* "a kidney" and *aden* "a gland."

Nephrangis (Schltr.) Summerh. Orchidaceae

Origins:
From the Greek *nephros* "a kidney" and *angeion, aggeion* "a vessel, cup," the shape of the lip.

Nephranthera Hassk. Orchidaceae

Origins:
From the Greek *nephros* "a kidney" and *anthera* "anther," see also *Renanthera* Lour.

Nephrocarpus Dammer Palmae

Origins:
From the Greek *nephros* "a kidney" and *karpos* "fruit," referring to the kidney-shaped seed.

Nephrocarya Candargy Boraginaceae

Origins:
From the Greek *nephros* "a kidney" and *karyon* "nut."

Nephrodesmus Schindler Fabaceae

Origins:
Greek *nephros* "a kidney" and *desmos* "a bond, band, bundle."

Nephrodium Michaux Dryopteridaceae (Aspleniaceae)

Origins:
Greek *nephrodes* "kidney-like," referring to the indusium; see André Michaux (1746-1803), *Flora Boreali-Americana.* 2: 266. Paris 1803.

Nephrolepis Schott Nephrolepidaceae (Oleandraceae, Davalliaceae)

Origins:
Greek *nephros* "kidney" and *lepis* "scale," referring to the shape of the indusia.

Species/Vernacular Names:
N. sp.

Yoruba: omu ebe

N. auriculata (L.) Trimen

English: giant fishbone fern, large sword fern, broad sword fern, giant sword fern

Japan: hô-bi-kanju (bi = tail)

Niue Island: kohuku tane

N. biserrata (Sw.) Schott

English: broad sword fern

Congo: tseyobia

N. cordifolia (L.) Presl (*Polypodium cordifolium* L.; *Aspidium cordifolium* (L.) Swartz)

English: sword fern, erect sword fern, ladder fern, fishbone fern, tuberous sword fern, tuber sword fern, common sword fern

The Philippines: bayabang

N. exaltata (L.) Schott (*Polypodium exaltatum* L.)

English: Boston fern, wild Boston fern, sword fern

N. hirsutula (Forst.f.) Presl (*Polypodium hirsutulum* G. Forst.)

English: hairy sword fern

Japan: ô-tama-shida

The Philippines: alolokdo

Niue Island: kohuku

N. multiflora (Roxb.) Jarrett ex Morton (*Davallia multiflora* Roxb.)

English: Asian sword fern

N. pectinata (Willd.) Schott

English: basket fern, toothed sword fern

Peru: plumaje

Nephromeria (Benth.) Schindler Fabaceae

Origins:
From the Greek *nephros* "kidney, kidneys" and *meros* "part."

Nephropetalum B.L. Robinson & Greenman Sterculiaceae

Origins:
From the Greek *nephros* "kidney, kidneys" and *petalon* "petal."

Nephrophyllidium Gilg Menyanthaceae

Origins:
From the Greek *nephros* "kidney" and *phyllon* "leaf," *phyllas, phyllados* "bed, foliage, heap."

Nephrophyllum A. Rich. Convolvulaceae

Origins:
From the Greek *nephros* "kidney" and *phyllon* "leaf."

Nephrosperma Balf.f. Palmae

Origins:
From the Greek *nephros* "kidney" and *sperma* "seed," seed more or less globose, somewhat kidney shaped.

Species/Vernacular Names:
N. vanhoutteanum (H.A. Wendl. ex van Houtte) Balf. f.
French: latanier millepattes

Nephrostylus Gagnep. Euphorbiaceae

Origins:
From the Greek *nephros* "kidney" and *stylos* "pillar, style."

Nephthytis Schott Araceae

Origins:
Daughter of Geb and Nut, depicted in human form wearing a crown in the form of the hieroglyph for house; in Egyptian religion Nut, a goddess of the sky, gave birth successively to the deities Osiris, Horus, Seth, Isis (Egyptian Aset, or Eset), and Nephthys (Egyptian Neb-hut, Nebthet); according to one tradition Nephthys was the mother of Anubis by Osiris and wife of Typhonis (Typhon), with the goddesses Isis, Neith, and Selket, she protected the dead.

Nepsera Naudin Melastomataceae

Origins:
For the German botanist Fridolin Carl Leopold Spenner, 1798-1841, physician, professor of medical botany; see J.H. Barnhart, *Biographical Notes upon Botanists.* 3: 309. 1965; Ethelyn Maria Tucker, *Catalogue of the Library of the Arnold Arboretum of Harvard University.* Cambridge, Massachusetts 1917-1933; R. Zander, F. Encke, G. Buchheim and S. Seybold, *Handwörterbuch der Pflanzennamen.* 14. Aufl. Stuttgart 1993. Some suggest from the Greek *nepsis* "sobriety," *nepho, napho* "to be sober, drink no wine."

Neptunia Lour. Mimosaceae

Origins:
From Neptunus (Neptune), the Roman god of water, he became god of the sea after his identification with the Greek Poseidon; see J. de Loureiro, *Flora cochinchinensis: sistens plantas in regno Cochinchina nascentes.* 2: 641, 653. Ulyssipone [Lisboa] 1790.

Species/Vernacular Names:
N. dimorphantha Domin
English: sensitive plant
N. gracilis Benth.
English: native sensitive plant

Neraudia Gaudich. Urticaceae

Origins:
Named to commemorate J. Néraud, 1794-1855, French lawyer and amateur botanist of Madagascar.

Species/Vernacular Names:
N. melastomifolia Gaud. (*Boehmeria melastomaefolia* (Gaud.) Hook. & Arnott; *Neraudia pyramidalis* St. John)
Hawaii: ma'aloa, ma'aloa, 'oloa

Neriacanthus Benth. Acanthaceae

Origins:

Greek *neros, nearos* "wet, fresh," *nero, neros* "water" plus *Acanthus*, or from *Nerium* and *Acanthus*.

Nerine Herbert Amaryllidaceae (Liliaceae)

Origins:

Nerine was a Nereid, a sea nymph, daughter of Doris and Nereus; the nymphs (in Greek *numphai*, in Latin *nymphae*) were female demigods with whom the ancients peopled all parts of nature, the Nereids (in Greek *Nereides*) were daughters of Nereus (son of Pontus and Ge) and nymphs of Ocean.

Species/Vernacular Names:

N. angustifolia (Bak.) Bak.

South Africa: berglelie

N. filifolia Bak. (*Nerine filamentosa* W.F. Barker; *Nerine filifolia* Bak. var. *parviflora* W.F. Barker)

Southern Africa: iTswele lenyoka (Xhosa)

N. humilis (Jacq.) Herb. (*Nerine breachiae* W.F. Barker; *Nerine flexuosa* (Jacq.) Herb., p.p.; *Nerine tulbaghensis* W.F Barker; *Nerine persii* W.F Barker; *Nerine pulchella* Herb.)

South Africa: berglelie

N. laticoma (Ker Gawl.) Dur. & Schinz (*Nerine duparquetiana* Bak.; *Nerine falcata* W.F Barker; *Nerine lucida* (Herb.) Herb.)

South Africa: vleilelie, gifbol

N. sarniensis (L.) Herbert

English: Guernsey lily

South Africa: berglelie

N. undulata (L.) Herb. (*Nerine alta* W.F Barker)

English: nerine

Nerissa Raf. Orchidaceae

Origins:

Possibly dedicated to a nymph; see Constantine Samuel Rafinesque (1783-1840), *Flora Telluriana*. 2: 89. 1836 [1837].

Nerisyrenia Greene Brassicaceae

Origins:

Probably from the Greek *neros* "wet, fresh" and *syreon* for a plant, called also *tordylion* or *tordylon* by Plinius.

Nerium L. Apocynaceae

Origins:

Nerion is the ancient classical Greek name used by Dioscorides for the oleander, *neros* "wet, fresh," *nero, neros* "water"; Latin *nerion* or *nerium* for the oleander or rosebay (Plinius).

Species/Vernacular Names:

N. oleander L. (*Nerium indicum* Miller; *Nerium odorum* Solander)

English: Ceylon rose, rose of Ceylon, oleander, rosebay, common oleander, dog-bane, double oleander, laurier rose, South Sea rose, sweet-scented oleander, scented oleander

French: laurier rose

Italian: oleandro

Arabic: defla, khadhraya, haban, ward el-homar

Peru: adelfa laurel, naranjillo, oleander

South Africa: oleander, rose of Ceylon, Selonsroos

Japan: kyô-chiku-tô, chochikutô

China: jia zhu tao

Vietnam: truc dao

Malaya: bunga jepun, bunga anis, pedendang

Bali: bungan kenyeri

India: kamili ba, rajbaka

The Philippines: adelfa, baladre

Nerophila Naudin Melastomataceae

Origins:

From the Greek *neros* "wet, fresh," *nero, neros* "water" and *philos* "loving."

Nertera Banks & Sol. ex Gaertner Rubiaceae

Origins:

Greek *nerteros, enerteros* "lowly, lower, nether," *nerthe, nerthen, enerthe* "below, beneath," referring to the prostrate or creeping habit of growth.

Species/Vernacular Names:

N. granadensis (L.f.) Druce (*Gomozia granadensis* Mutis)

English: matted nertera, bead plant

Hawaii: makole

N. reptans (F. Muell.) Benth. (*Diodia reptans* F. Muell.)

English: dwarf nertera

Nervilia Comm. ex Gaudich. Orchidaceae

Origins:

Latin *nervus, i* "nerve," referring to the veined leaves or to the pseudobulbs; see Richard Evans Schultes and Arthur Stanley Pease, *Generic Names of Orchids. Their Origin and Meaning.* 210. Academic Press, New York and London 1963; Hubert Mayr, *Orchid Names and Their Meanings.* 162. Vaduz 1998.

Species/Vernacular Names:

N. aragoana Gaudich. (*Nervilia yaeyamensis* Hayata)

Japan: Yaeyama-kuma-sô, aoi-bokuro

Nesaea Comm. ex Kunth Lythraceae

Origins:

According to Greek mythology, Nesaea or Nesaie was a name given to a sea nymph, one of the Nereids; Greek *nesos* "an island"; *Nesaea triflora* (L.f.) Kunth was found on the island of Mauritius.

Nesiota Hook.f. Rhamnaceae

Origins:

Greek *nesos* "island," *nesiotes* "islander, insular," one species living in St. Helena.

Neslia Desv. Brassicaceae

Origins:

For the French botanist J.A.N. de Nesle (Denesle), 1784-1856, author of *Introduction à la botanique.* Poitiers; see Léon Faye (d. 1855), *Notice sur J.A. Denesle.* Poitiers 1844; Nicaise Auguste Desvaux (1784-1856), in *Journal de Botanique*, appliquée à l'Agriculture, à la Pharmacie, à la Médicine et aux Arts. 3: 162, t. 24, fig. 1. Paris 1815; Georg Christian Wittstein (1810-1887), *Etymologisch-botanisches Handwörterbuch.* 614. Ansbach 1852.

Species/Vernacular Names:

N. paniculata (L.) Desv. (*Myagrum paniculatum* L.)

English: ball mustard

Nesobium R. Phil. ex Fuentes Urticaceae

Origins:

From the Greek *nesos, nasos* "an island, land flooded" and *bios* "life."

Nesocaryum I.M. Johnston Boraginaceae

Origins:

Greek *nesos, nasos* "island" and *karyon* "nut, fruit," a genus from Desventuradas Island, Chile.

Nesocodon Thulin Campanulaceae

Origins:

From the Greek *nesos, nasos* "island" and *kodon* "a bell," a genus from Mauritius.

Nesodaphne Hook.f. Lauraceae

Origins:

Greek *nesos, nasos* "island" and *daphne* "the laurel, the bay tree"; see J.D. Hooker, *Flora Novae-Zelandiae.* 1: 217. 1853.

Nesodoxa Calest. Araliaceae

Origins:

From the Greek *nesos, nasos* "island" and *doxa* "glory."

Nesodraba Greene Brassicaceae

Origins:

From the Greek *nesos, nasos* plus *Draba* L.

Nesogenes A. DC. Nesogenaceae (Verbenaceae)

Origins:

Greek *nesos* and *genos* "people, nation," *gennao* "to generate," the plants occur on islands of the Indian and Pacific Oceans.

Nesogordonia Baillon Sterculiaceae

Origins:

Presumably from the Greek *nesos* "an island" and the genus *Gordonia.*

Species/Vernacular Names:

N. papaverifera (A. Chév.) R. Capuron

Gabon: aborbora

Yoruba: oro, otutu, alele, oronla

Nigeria: otutu, opepe-ira (Yoruba); urhuaro (Edo); otalo (Igbo)

W. Africa: danta, kotibé

Central Africa: aborbora, kotibe, naouya, kotibe

Ghana: danta

Cameroon: ovoé, tekeleke

Ivory Coast: kotibé

Nesohedyotis (Hook.f.) Bremek. Rubiaceae

Origins:

Greek *nesos* "an island" and the genus *Hedyotis* L., from St. Helena; see A.B. Ellis, *West African Islands*. London 1885; J.C. Melliss, *St. Helena*. London 1875.

Nesoluma Baillon Sapotaceae

Origins:

Greek *nesos* "an island" plus *luma*, a Sapotacea from Polynesia.

Species/Vernacular Names:

N. polynesicum (Hillebr.) Baillon

Hawaii: keahi

Nesopanax Seem. Araliaceae

Origins:

From the Greek *nesos* "an island" plus *Panax*.

Nesopteris Copel. Hymenophyllaceae

Origins:

From the Greek *nesos* "an island" and *pteris* "a fern," growing in Samoa.

Nesoris Raf. Pteridaceae (Adiantaceae)

Origins:

Greek *soros* "a spore case"; see C.S. Rafinesque, *New Fl. N. Am.* 4: 104. 1836 [1838] and *The Good Book*. 47. Philadelphia 1840; E.D. Merrill (1876-1956), *Index rafinesquianus*. The plant names published by C.S. Rafinesque, etc. 71. Jamaica Plain, Massachusetts, USA 1949.

Nesothamnus Rydb. Asteraceae

Origins:

From the Greek *nesos* "an island" and *thamnos* "a bush, shrub."

Nestegis Raf. Oleaceae

Origins:

Of unknown origin (Constantine Samuel Rafinesque, 1783-1840, often invented generic names!, see E.D. Merrill (1876-1956), *Index rafinesquianus*. The plant names published by C.S. Rafinesque, etc. Jamaica Plain, Massachusetts, USA 1949), perhaps has no meaning at all or possibly named from the Greek *stegos* "cover," referring to the lack of a corolla in the type species; see C.S. Rafinesque, *Sylva Telluriana*. 10. Philadelphia 1838.

Species/Vernacular Names:

N. ligustrina (Vent.) L. Johnson (*Notelaea ligustrina* Vent.)

English: privet mock-olive, silkwood

N. sandwicensis (A. Gray) Degener, I. Degener & L. Johnson (*Olea sandwicensis* A. Gray; *Gymnelaea sandwicensis* (A. Gray) L. Johnson; *Osmanthus sandwicensis* (A. Gray) Knobl.)

Hawaii: olopua, pua, ulupua

Nestlera Sprengel Asteraceae

Origins:

Named for the Alsatian botanist Chrétien Géofroy (Christian Gottfried) Nestler, 1778-1832, professor of botany and pharmacy, studied with L.C. Richard. See John H. Barnhart, *Biographical Notes upon Botanists*. 2: 545. 1965; Ida Kaplan Langman, *A Selected Guide to the Literature on the Flowering Plants of Mexico*. University of Pennsylvania Press, Philadelphia 1964; A. Lasègue, *Musée botanique de Benjamin Delessert*. Paris 1845; E.M. Tucker, *Catalogue of the Library of the Arnold Arboretum of Harvard University*. 1917-1933; R. Zander, F. Encke, G. Buchheim and S. Seybold, *Handwörterbuch der Pflanzennamen*. 14. Aufl. Stuttgart 1993; I.C. Hedge and J.M. Lamond, *Index of Collectors in the Edinburgh Herbarium*. Edinburgh 1970; Stafleu and Cowan, *Taxonomic Literature*. 3: 728-729. 1981.

Nestoria Urban Bignoniaceae

Origins:

From the Greek *nestoris, nestoridos* "a kind of cup."

Nestronia Raf. Santalaceae

Origins:

See Constantin S. Rafinesque, *New Fl. N. Am.* 3: 12. 1836 [1838] and *Flora Telluriana*. 4: 106. 1836 [1838], *The Good Book*. 45. Philadelphia 1840, *Autikon botanikon*. Icones plantarum select. nov. vel rariorum, etc. 6. Philadelphia 1840; E.D. Merrill (1876-1956), *Index rafinesquianus*. The plant names published by C.S. Rafinesque, etc. 114. Massachusetts, USA 1949.

Nettlera Raf. Rubiaceae

Origins:
See Constantin Samuel Rafinesque, *Sylva Telluriana*. 147. 1838; E.D. Merrill, *Index rafinesquianus*. The plant names published by C.S. Rafinesque, etc. 226. Massachusetts, USA 1949.

Nettoa Baillon Tiliaceae

Origins:
After the Brazilian botanist Ladislau de Souza Mello e Netto, 1837-1893, Director of the Rio de Janeiro National Museum, from 1876 to 1887 editor of the *Archivos Museu Nacional do Rio de Janeiro*, author of *Itinéraire botanique dans la province de Minas Geraes*. Paris 1866, *Apontamentos relativos á botanica applicada no Brasil*. Rio de Janeiro. 1871 and *Le Muséum national de Rio-de-Janeiro*. Paris 1889. See Emmanuel Liais, *Explorations scientifiques au Brésil*. Paris 1865; Paris — Exposition Universelle de 1867. [Brazil], *Breve noticio sobre a coleção das madeiras do Brasil*. Paris. 1867; Rio de Janeiro — Exposição Anthropologica Brazileira, *Revista da Exposição Anthropologica Brazileira*. 1882; Rio de Janeiro — Museu Nacional, *Archivos do Museu Nacional do Rio de Janeiro*. 1876; Sylvio Romero, *Ethnographia Brazileira*. Estudos criticos sobre ... L. Netto. Rio de Janeiro 1888; E.M. Tucker, *Catalogue of the Library of the Arnold Arboretum of Harvard University*. 1917-1933.

Neumannia A. Rich. Flacourtiaceae (Neumanniaceae)

Origins:
For the French gardener J.H.F. Neumann, 1800-1858.

Neuracanthus Nees Acanthaceae

Origins:
From the Greek *neuron* "nerve, tendon" and *akantha* "thorn."

Neurachne R. Br. Gramineae

Origins:
Greek *neuron* "nerve" and *achne* "husk, glume," referring to the glumes; see Robert Brown, *Prodromus florae Novae Hollandiae*. 196. London 1810.

Species/Vernacular Names:
N. alopecuroides R. Br. (as *Alopecurus*)
English: foxtail mulga-grass
N. munroi (F. Muell.) F. Muell.
English: slender-headed mulga-grass, window mulga-grass

Neuractis Cass. Asteraceae

Origins:
From the Greek *neuron* "nerve" and *aktis, aktin* "a ray," referring to the flowers.

Neurada L. Neuradaceae

Origins:
Possibly from the Greek *neuron* "nerve, sinew, tendon, string" and *aden* "gland," referring to the leaves or to the mucilage-ducts in pith; Latin *neuras, nevras, neuradis* "the plant *manicon*, which excites the nerves"; *neuras, neurados* was a Greek plant name used by Dioscorides and Plinius for *potirrion* or *poterion*, goat's thorn, a species of *Astragalus*.

Neuradopsis Bremek. & Oberm. Neuradaceae

Origins:
Resembling the genus *Neurada* L.

Neurocallis auctt. Pteridaceae

Origins:
From the Greek *neuron* "nerve" and *kallos* "beauty," see *Nevrocallis* Fée.

Neurocalyx Hook. Rubiaceae

Origins:
Greek *neuron* "nerve" and *kalyx* "calyx," referring to the calyx lobes.

Neurocarpaea K. Schumann Rubiaceae

Origins:
From the Greek *neuron* "nerve" and *karpos* "fruit," see also *Nodocarpaea* A. Gray.

Neurocarpaea R. Br. Rubiaceae

Origins:
Greek *neuron* "nerve" and *karpos* "fruit," see also *Pentas* Benth.

Neurodium Fée Polypodiaceae

Origins:
From the Greek *neuron* "nerve," referring to the veinlets.

Species/Vernacular Names:
N. lanceolatum (L.) Fée (*Pteris lanceolata* L.; *Paltonium lanceolatum* (L.) C. Presl)
English: ribbon fern

Neurogramma Link Pteridaceae (Adiantaceae)

Origins:
From the Greek *neuron* "nerve" and *gramma* "letter, thread, marking."

Neurolaena R. Br. Asteraceae

Origins:
Greek *neuron* "nerve" and *chlaena, laina* "cloak, blanket, coat."

Species/Vernacular Names:
N. lobata (L.) R. Br.
Central America: tres puntas, capitana, contragavilana, gavilana, hierba amarga, mano de lagarto, quina
Mexico: arnica, cola de faisán, rabo de faisán, tabaco cimarrón

Neurolakis Mattf. Asteraceae

Origins:
From the Greek *neuron* "nerve" and *lakis, lakidos* "a rent, gap, rending," *lakizo* "to tear."

Neurolepis Meissner Gramineae

Origins:
From the Greek *neuron* "nerve" and *lepis, lepidos* "scale."

Neurolobium Baillon Apocynaceae

Origins:
Greek *neuron* and *lobion, lobos* "lobe, pod, small pod, fruit."

Neuromanes Trevis. Hymenophyllaceae

Origins:
Greek *neuron* "nerve" and *manes* "a cup, a kind of cup."

Neuronia D. Don Oleandraceae

Origins:
Greek *neuron* "nerve, bowstring, fibers of plants," the fronds are simple, the sori are near midrib.

Neuropeltis Wallich Convolvulaceae

Origins:
From the Greek *neuron* "nerve, bowstring" and *peltis* "a shield," alluding to the leaves.

Species/Vernacular Names:
N. racemosa Wallich
English: racemose neuropeltis
China: dun bao teng
Malaya: akar china puteh, jengkal, oran merah

Neuropeltopsis Ooststr. Convolvulaceae

Origins:
Resembling *Neuropeltis* Wall.

Neurophyllodes (A. Gray) O. Degener Geraniaceae

Origins:
From the Greek *neuron* "nerve, bowstring" and *phyllodes* "like leaves, having petalled flowers."

Neurophyllum C. Presl Hymenophyllaceae

Origins:
From the Greek *neura, neuron* "nerve" and *phyllon* "leaf."

Neuroplatyceros Fée Polypodiaceae

Origins:
Greek *neura, neuron* "nerve," *platys* "broad" and *keras* "horn," see *Platycerium* Desv.

Neuropoa Clayton Gramineae

Origins:
From the Greek *neura, neuron* "nerve" and *poa* "grass, pasture grass"; see W.D. Clayton, in *Kew Bulletin.* 40: 728. 1985.

Neuropteris Desv. Dennstaedtiaceae

Origins:
From the Greek *neuron* "nerve" and *pteris* "fern," *pteron* "wing."

Neurosoria Mettenius ex Kuhn Pteridaceae (Adiantaceae, Sinopteridaceae)

Origins:
From the Greek *neuron* "nerve" and *soros* "a vessel, a heap, a spore case."

Neurosperma Raf. Cucurbitaceae

Origins:
From the Greek *neuron* "nerve" and *sperma* "seed"; see Constantine Samuel Rafinesque, *Am. Monthly Mag. Crit. Rev.* 4: 207. 1819; E.D. Merrill, *Index rafinesquianus.* The plant names published by C.S. Rafinesque, etc. 230. 1949.

Neurospermum Bartl. Cucurbitaceae

Origins:
Greek *neuron* "nerve" and *sperma* "seed."

Neurotecoma K. Schumann Bignoniaceae

Origins:
From the Greek *neuron* "nerve" with the generic name *Tecoma.*

Neurotheca Salisb. ex Benth. Gentianaceae

Origins:
From the Greek *neuron* "nerve" and *theke* "a box, case," referring to the reticulated capsule.

Neustanthus Benth. Fabaceae

Origins:
Greek *neustazo* "nod," *neuster, neusteros* "swimmer," *neo* "to swim" and *anthos* "flower."

Neuwiedia Blume Orchidaceae (Apostasiaceae)

Origins:
Dedicated to the German botanist Maximilian Alexander Philipp zu Wied-Neuwied, 1782-1867, naturalist, traveler and plant collector in Brazil, 1832-1834 North America, author of *Reise in das innere Nord America in den Jahren 1832 bis 1834.* Coblenz 1839-1841 and *Reise nach Brasilien in den Jahren 1815 bis 1817 ...* Mit Kupfern [by C. Bodmer] Frankfurt a.M. 1820-1821; see T.W. Bossert, *Biographical Dictionary of Botanists Represented in the Hunt Institute Portrait Collection.* 258. 1972; E.M. Tucker, *Catalogue of the Library of the Arnold Arboretum of Harvard University.* 1917-1933; A. Lasègue, *Musée botanique de Benjamin Delessert.* Paris 1845; Stafleu and Cowan, *Taxonomic Literature.* 3: 381. 1981; Joseph Ewan, *Rocky Mountain Naturalists.* The University of Denver Press 1950; John H. Barnhart, *Biographical Notes upon Botanists.* Boston 1965.

Neves-armondia Schumann Bignoniaceae

Origins:
Possibly named for Amaro Ferreira das Neves-Armond, 1854-1944.

Neviusia A. Gray Rosaceae

Origins:
After Rev. Ruben Denton Nevius, 1827-1913, plant collector.

Nevrilis Raf. Bignoniaceae

Origins:

From the Greek *neuron* "nerve"; see C.S. Rafinesque, *Sylva Telluriana*. 138. 1838.

Nevrocallis Fée Pteridaceae

Origins:

From the Greek *neuron* "nerve" and *kallos* "beauty."

Nevroctola Raf. Gramineae

Origins:

Greek *neuron* "nerve"; see C.S. Rafinesque, *Neogenyton, or Indication of Sixty-Six New Genera of Plants of North America.* 4. 1825 and *Seringe Bull. Bot.* 1: 221. 1830; E.D. Merrill, *Index rafinesquianus.* 76. 1949.

Nevroloma Raf. Gramineae

Origins:

From the Greek *neuron* "nerve" and *loma* "border, margin, fringe, edge"; see C.S. Rafinesque, *Jour. Phys. Chim. Hist. Nat.* 89: 106. 1819; E.D. Merrill, *Index rafinesquianus.* 76. 1949.

Nevrosperma Raf. Cucurbitaceae

Origins:

From the Greek *neuron* "nerve" and *sperma* "a seed"; see Constantine Samuel Rafinesque, *Am. Monthly Mag. Crit. Rev.* 4: 40. 1818 and *Jour. Phys. Chim. Hist. Nat.* 89: 101. 1819; E.D. Merrill, *Index rafinesquianus. The plant names published by C.S. Rafinesque, etc.* 230. 1949.

Nevskiella Krecz. & Vved. Gramineae

Origins:

Named for Sergei Arsenjevic Nevski, 1908-1938, author of "Bromeae tribus Graminearum naturalis." *Trudy Sredne-Aziatsk. Gosud. Univ.* Ser. 8b, Bot. 17: 14-15. 1934, "Beiträge zur Flora des Kuhitang-Tau und seiner Vorgebirge. Resumé." 1937 and "Beiträge zur Kenntnis der Wildwachsenden Gersten in Zusammenhang mit der Frage über den Ursprung von *Hordeum vulgare* L. und *Hordeum distichum* L. (Versuch einer Monographie der Gattung *Hordeum* L.)." *Trudy Bot. Inst. Akad. Nauk SSSR*, Ser. 1, Fl. Sist. Vyss. Rast. 5: 64-255. 1941.

Newberrya Torrey Ericaceae

Origins:

For the American botanist John Strong Newberry, 1822-1892, surgeon-naturalist, paleontologist, physician, M.D. 1848, professor of geology. Among his many writings are *The Paleozoic Fishes of North America.* Washington 1889 and *Report of the Exploring Expedition from Santa Fé, New Mexico, to the Junction of the Grand and Green Rivers of the Great Colorado of the West, in 1859 ...* Geological Report. 1876. See John H. Barnhart, *Biographical Notes upon Botanists.* 2: 548. 1965; Joseph Ewan, *Rocky Mountain Naturalists.* The University of Denver Press 1950; T.W. Bossert, *Biographical Dictionary of Botanists Represented in the Hunt Institute Portrait Collection.* 285. 1972; S. Lenley et al., *Catalog of the Manuscript and Archival Collections and Index to the Correspondence of John Torrey.* Library of the New York Botanical Garden. 313-314. 1973; E.M. Tucker, *Catalogue of the Library of the Arnold Arboretum of Harvard University.* 1917-1933; Irving William Knobloch, compil., "A preliminary verified list of plant collectors in Mexico." *Phytologia Memoirs.* VI. 1983; N.L. Britton "John Strong Newberry." *Bull. Torr. Bot. Club.* 20: 89-98. 1893; A.E. Waller, "The breadth of vision of Dr. John Strong Newberry." *Ohio State Arch. & Hist. Quart.*, Oct., 324-346. 1943; Ida Kaplan Langman, *A Selected Guide to the Literature on the Flowering Plants of Mexico.* University of Pennsylvania Press, Philadelphia 1964; Stafleu and Cowan, *Taxonomic Literature.* 3: 733-735. 1981; J. Ewan, ed., *A Short History of Botany in the United States.* New York and London 1969.

Newbouldia Seemann ex Bureau Bignoniaceae

Origins:

After the British botanist Rev. William Williamson Newbould, 1819-1886 (Kew, Surrey), Curate at Bluntisham, Hunts. and Comberton, Cambr., a close friend of Rev. Churchill Babington (1821-1889) and H.C. Watson, a member of Ray Society, 1863 a Fellow of the Linnean Society, editor (with John Gilbert Baker, 1834-1920) of Hewett Cottrell Watson (1804-1881), *Topographical Botany.* Ed. 2. London 1883. See James Sowerby (1757-1822), *English Botany ... Supplement ...* The descriptions, synonyms, and places of growth by ... W.W. Newbould. Vol. V. London 1863; John H. Barnhart, *Biographical Notes upon Botanists.* 2: 548. 1965; T.W. Bossert, *Biographical Dictionary of Botanists Represented in the Hunt Institute Portrait Collection.* 285. 1972; H.N. Clokie, *Account of the Herbaria of the Department of Botany in the University of Oxford.* 217. Oxford 1964; I.C. Hedge and J.M. Lamond, *Index of Collectors in the Edinburgh Herbarium.* Edinburgh 1970.

Species/Vernacular Names:

N. laevis (P. Beauv.) Bureau

Congo: mumeni

Nigeria: akoko, akoka, ikhimi, okurimi, ogirishi, oji-karisi, bareshi, aduruku, oririsi; aduruku (Hausa); kontor (Tiv); akoko (Yoruba); ikhimi (Edo); ogirisi (Igbo); obot (Efik)

Yoruba: akoko

Newcastelia F. Muell. Labiatae (Chloanthaceae, Verbenaceae)

Origins:

Named after Henry Pelham Fiennes Pelham-Clinton, 5th Duke of Newcastle, 1811-1864, a patron of natural sciences, from 1852 to 1854 Secretary of State for the Colonies, in 1855 sponsor of the A.C. Gregory's Expedition to Northern Australia. His publications include *Archaeological Association. Inaugural discourse delivered at the opening of the Newark congress.* [London? 1852] and *The Duke of Newcastle and his Tenantry.* Speeches delivered by ... the Duke of Newcastle, to his tenantry, at the rent-audit for Michaelmas, 1851. London 1852. See John Martineau, *The Life of Henry Pelham, fifth Duke of Newcastle, 1811-1864.* London 1908; Thomas Chisholm Anstey, *Case of Hong Kong.* Correspondence between T.C. Anstey, late Attorney-General of Hong Kong, and ... the Duke of Newcastle, Secretary of State for the Colonies. 3rd to 29th July 1859. London 1859; Alfred Gatty, *In piam memoriam. A sermon* (dedicated to the memory of the late Duke of Newcastle). London 1865; F. von Mueller, in *Hooker's Journal of Botany & Kew Garden Miscellany.* 9: 22. London 1857.

Species/Vernacular Names:

N. hexarrhena F. Muell.

English: lamb's tail

Newcastlia F. Muell. Labiatae (Chloanthaceae, Verbenaceae)

Origins:

Orthographic variant of *Newcastelia* F. Muell.

Newtonia Baillon Mimosaceae

Origins:

According to Stafleu and Cowan (see their *Taxonomic Literature.* 3: 738. 1981) the name of the genus honors the great English (b. Woolsthorpe) mathematician and scientist Sir Isaac Newton, 1642-1727 (d. London). Among his writings are *Philosophiae Naturalis Principia Mathematica.* London 1687 and *Observations Upon the Profecies of Daniel, and the Apocalypse of St. John.* [Edited by Benjamin Smith, Newton's half brother, first edition.] London 1733, pupil of Isaac Barrow (1630-1677) the first Lucasian professor of mathematics at Cambridge; see G.J. Gray, *A Bibliography of the Works of Sir Isaac Newton.* [Reprint of the second edition.] London 1966; Bernard le Bovier (Bouyer, Bouvier) de Fontenelle (1657-1757, Paris), *Éloges historiques des académiciens morts depuis le renouvellement* etc. [Two vols. 12mo, fourth, but first complete; the first biography of Newton.] Paris 1742; [Sotheby's], *Catalogue of the Newton Papers.* Sold by Order of The Viscount Lymington to whom they have descended from Catherine Conduitt, Viscountess Lymington, Great-Niece of Sir Isaac Newton. London Sale of 13-14 July 1936; Jeremiah Horrocks (1618-1641), *Opera posthuma;* viz. *Astronomia Kepleriana,* defensa & promota ... In calce adjiciuntur Johannis Flamstedii, Derbiensis, *De temporis aequatione diatriba.* Numeri ad lunae theoriam Horroccianam. [The foundation of British astronomy, a bridge between Kepler and Newton; pages 465-470 contain a letter to William Crabtree dated 20 December 1638.] London 1673 [1672]; John Keill (1671-1721), *Trigonometriae planae & sphericae elementa.* [First edition, K. was a pupil of David Gregory (1659-1708) and an important early exponent of Newtonian philosophy, Savilian professor of astronomy in 1712.] Oxford 1715 and *Introductio ad veram physicam.* [The first textbook of Newtonian physics.] Oxoniae 1702; Thomas Birch, *The History of the Royal Society of London for Improving of Natural Knowledge, from Its First Rise.* London, for A. Millar 1756 [1757]; Charles Coulston Gillispie, editor, *Dictionary of Scientific Biography* 10: 42-103. New York 1981; Benjamin Daydon Jackson (1846-1927), "A list of the contributors to the herbarium of the Royal Botanic Gardens, Kew, brought down to 31st December 1899." *Bull. Misc. Inf. Kew.* 169-171. 1901; R.T. Gunther, *Early Science in Cambridge.* Oxford 1937; Elisabeth Leedham-Green, *A Concise History of the University of Cambridge.* Cambridge, University Press 1996; Richard S. Westfall (1924-1996), *The Richard Westfall Library of Newton and Newtoniana.* Jeff Weber Rare Books (1923 Foothill Drive, Glendale, California 91201-1242), Catalogue 57. 1998; Grace K. Babson, *A Descriptive Catalogue of the Grace K. Babson Collection of the Works of Sir Isaac Newton* and the material relating to him in the Babson Institute Library, Babson Park, Mass. New York 1950; Alexander B. Adams, *Eternal Quest. The Story of the Great Naturalists.* New York 1969.

Species/Vernacular Names:

N. sp.

Gabon: ensale

N. buchananii (Baker) Gilbert & Boutique (*Piptadenia buchananii* Baker)

English: forest newtonia

Southern Africa: muFumiti, muFumoti, muJairaiya, mupfumboti (Shona)

N. glandulifera (Pellegrin) Gilbert & Boutique

Zaire: kamalufu

N. hildebrandtii (Vatke) Torre var. *hildebrandtii* (*Piptadenia hildebrandtii* Vatke) (the specific name honors a German traveler and collector in East Africa, Johann Maria Hildebrandt, 1847-1881)

English: Lebombo wattle, lowveld newtonia

Southern Africa: Lebombowattel; umFomothi, uDongolokamadilika (Zulu)

Gabon: ensale

N. leucocarpa (Harms) Gilbert & Boutique

Zaire: pangi ia nsinga

Gabon: ekango, mossinga, ossimiale

Neyraudia Hook.f. Gramineae

Origins:
The anagram of *Reynaudia* Kunth.

Nhandiroba Adans. Cucurbitaceae

Origins:
A Brazilian vernacular name. See Eurico Teixeira da Fonseca, "Plantas medicinales brasileñas." *R. Flora Medicinal.* 6(6): 357-367. Rio de Janeiro 1940; Carlos Stellfeld, "As drogas da farmacopéia paulista." *Tribuna Farmacêutica.* 8(7): 152-166. Curitiba 1940; Gustavo Edwall, "Ensayo para uma synonimia dos nomes populares das plantas indigenas do estado de São Paulo, 2a parte." *B. da Commissão Geographica e Geologica do estado de São Paulo.* São Paulo 16: 3-63. 1906.

Nicandra Adans. Solanaceae

Origins:
After the Greek botanist Nikander of Colophon (Nikandros Kolophonios) (c. 100-150 A.D.), physician, poet, medical writer, author of *Alexipharmaca.* Halae [Halle an der Saale] 1792 and *Theriaca* [and other works]. Venetiis 1522; see Ernst H.F. Meyer (1791-1858), *Geschichte der Botanik.* I: 244-250. Königsberg 1854-1857.

Species/Vernacular Names:
N. physalodes (L.) Gaertner

English: shoo fly, shoo fly plant, apple of Peru, Chinese lantern

Peru: anrreshuailla, capuli cimarrón, ccarapamacmam, corneta jacha, joto-joto, jarrito, orzita de pellejo, toccoro

East Africa: chemogong'it-cheptitet

South Africa: basterappelliefie, bloubitterappelliefie, bloubitter, wildebitter

Madagascar: boreda, gaboroda, tsipokipoky, tsitsipoky

China: jia suan jiang

Nepal: isamgoli

Nicarago Britton & Rose Caesalpiniaceae

Origins:
One of the popular names of *Caesalpinia* is Nicaragua wood.

Nichallea Bridson Rubiaceae

Origins:
For Nicolas Hallé, born 1927, botanist, specialist in Rubiaceae.

Nicobariodendron Vasudeva Rao & Chakrab. Celastraceae

Origins:
One species, from Nicobar Island.

Nicodemia Ten. Buddlejaceae (Loganiaceae)

Origins:
Named for the Italian botanist Gaetano Nicodemo, died 1803, from 1799 Curator of the Lyons Botanical Garden; see Michele Tenore (1780-1861), *Catalogo delle piante che si coltivano nel Regio Orto Botanico di Napoli.* Napoli 1845.

Nicolaia Horan. Zingiberaceae

Origins:
After Nicholas I, 1796-1855 (died during the Crimean War), the Czar (Tsar) of Russia (1825-1855), the third son of Paul I, an absolute despot.

Nicolasia S. Moore Asteraceae

Origins:

After the Kew botanist Dr. Nicholas Edward Brown, 1849-1934. Among his writings are "Notes on the genera *Cordyline, Dracaena, Pleomele, Sansevieria* and *Taetsia.*" *Bull. Misc. Inform.* 1914: 273-279. 1914, "The *Stapeliae* of Thunberg's herbarium." *J. Linn. Soc. Bot.* 17: 162-172. 1878, "*Acicarpha rosulata,* N.E. Brown (Calyceraceae)." *Hooker's Icon. Pl.* 1900, "The genera *Aloe* and *Mesembryanthemum* as represented in Thunberg's herbarium." *Bothalia* 1: 139-169. 1923, "The South African *Iridaceae* of Thunberg's herbarium." *J. Linn. Soc. Bot.* 48: 15-55. 1928 and "*Mesembryanthemum* and allied genera." *J. Bot.* 66(785): 138-145. 1928, contributor to Daniel Oliver (1830-1916), *Flora of Tropical Africa.* See J.H. Barnhart, *Biographical Notes upon Botanists.* 1: 263. 1965; T.W. Bossert, *Biographical Dictionary of Botanists Represented in the Hunt Institute Portrait Collection.* 55. 1972; Mary Gunn and Leslie E. Codd, *Botanical Exploration of Southern Africa.* 105. Cape Town 1981; Alain Campbell White and Boyd Lincoln Sloane, *The Stapelieae.* Pasadena 1937.

Nicolletia A. Gray Asteraceae

Origins:

For the French astronomer Joseph Nicolas Nicollet, 1786-1843, explorer, he is remembered for *Observations astronomiques,* faites ... (par MM. Bouvard, ... Nicollet, etc.). tom. 1. 1825, etc. and *Essay on Meteorological Observations.* [Washington 1839].

Nicolsonia DC. Fabaceae

Origins:

In honor of a Père Nicolson, a Dominican priest, author of *Essai sur l'histoire naturelle de l'isle de Saint-Domingue.* Paris 1776; see John H. Barnhart, *Biographical Notes upon Botanists.* 3: 4. 1965; A. Lasègue, *Musée botanique de Benjamin Delessert.* Paris 1845; E.M. Tucker, *Catalogue of the Library of the Arnold Arboretum of Harvard University.* 1917-1933; Jonas C. Dryander, *Catalogus bibliothecae historico-naturalis Josephi Banks.* London 1800.

Nicotiana L. Solanaceae

Origins:

To commemorate the French diplomat Jean Nicot, 1530-1600, ambassador to Portugal, introduced tobacco into France (about 1560) and Portugal, author of *Dictionnaire francois-latin* ... recuilli des obseruations de plusieurs hommes doctes, entre autres de *M. Nicot* conseiller du roy ... Paris 1573; see Albert Puech, *Un Homme de Lettres au XVIe siècle* (J. Nicot). Nîmes 1892; *Jean Nicot, ambassadeur de France en Portugal au XVIe siècle. Sa correspondance diplomatique inédite.* Par E. Falgairolle. Paris 1897; F. André Thevet (1502-1592), *Les singularitez de la France Antarctique, autrement nommée Amérique, et de plusieurs terres et isles découvertes de notre temps.* Paris 1558 (Italian translation: *Historia dell'India America, detta altramente Francia antartica...* Vinegia 1561); Pierre Borel, *Dictionnaire des termes du vieux françois* ... Augmenté de tout ce qui s'est trouvé de plus dans les Dictionnaires de Nicot, etc. 1882; Carl Linnaeus, *Species Plantarum.* 180. 1753 and *Genera Plantarum.* Ed. 5. 84. 1754; R. Gordon Wasson, "Notes on the present status of Ololiuhqui and the other hallucinogens of Mexico." from *Botanical Museum Leaflets, Harvard University.* Vol. 20(6): 161-212. Nov. 22, 1963; Blas Pablo Reko, *Mitobotánica Zapoteca.* [Appended by an analysis of "Lienzo de Santiago Guevea"] Tacubaya 1945.

Species/Vernacular Names:

N. spp.

Peru: pacháró, rumë

N. alata Link & Otto

English: jasmine tobacco, flowering tobacco, winged tobacco

N. clevelandii A. Gray

Mexico: tabaco de perro

N. glauca R.C. Graham

English: wild tobacco, shrub tobacco, Mexican tobacco, tree tobacco, coneton, mustard tree, San Juan tree, tobacco plant, tobacco bush, grey blue tobacco

Arabic: dokkhane

China: guang yan cao

Bolivia: helado, karalawa, karallanta

Mexico: árbol de tabaco, cornetón, buena moza, lengua de buey, cornetón, Don Juan, maraquiana, gigante, gretaño, mostaza montés, hierba del gigante, hoja de cera, levántate Don Juan, me-he-kek, tabaco, palo virgen, tacote, tabaco amarillo, tabaco cimarrón, tabaquillo, tronadora de España, tzinyacua, Virginio, xiutecuitlanextli, Alamo loco

Peru: ccjamachu, cjamata, ccjamata, supai ccarcco

Portuguese: charuto do rei, charuteira

Southern Africa: tabakboom, tabakbos, Jan Twak, volstruisgifboom, wildetabak, wildetwak; mohlafotha (Sotho); tabaka bume (South Sotho)

Hawaii: makahala, paka

N. goodspeedii H. Wheeler (after Thomas Harper Goodspeed (1887-1966), botanist, monographed the genus

Nicotiana in 1954, see T.H. Goodspeed et al., *Chronica Botanica.* 16: 1-536 (323-536). 1954)

English: small-flower tobacco, small-flowered tobacco, smooth-flowered tobacco

N. gossei Domin (after the explorer who discovered Ayers Rock, William Christie Gosse, 1842-1881, see *W.C. Gosse's ... Report and Diary of ... Central and Western Exploring Expedition, 1873.* [Adelaide 1874]; Karel Domin, *Beiträge zur Flora und Pflanzengeographie Australiens.* [= Bibliotheca Botanica Heft 89 (Dec. 1929) 592] 1929)

English: Gosse's tobacco, native tobacco

N. knightiana Goodspeed

Peru: tabaquilla

N. longiflora Cav.

English: long-flowered tobacco

Japan: naga-bana-tabako

N. maritima H. Wheeler

English: coast tobacco

N. megalosiphon Van Heurck & Müll. Arg.

English: long-flowered tobacco

N. otophora Griseb.

Bolivia: tabacachi, tabaquilla

N. paniculata L.

Peru: ccamasayri, kkuru, koles-koles, sairi, tabaco cimarrón

N. plumbaginifolia Viviani

Mexico: tabaquillo

N. rustica L.

English: wild tobacco, tobacco, Aztec tobacco

Arabic: dokhan akhdar, dokhan, dokhan soufi

China: huang hua yan cao

Mexico: tabaco, tabaquillo, andumucua, macuche, nohol-x'i-k'uts, picietl, yetl, k'uts, tabaco macuche, teneshil, tenapete, tabaco pequeño

Peru: petúm, piciete

N. suaveolens Lehm.

English: Austral tobacco

N. tabacum L.

English: tobacco, common tobacco

French: tabac

Italian: tabacco

Hawaii: paka

South Laos: iyaa (people Nya Hön)

China: yan cao, yen tsao, jen tsao, yu yen tsao

India: tamrakuta, tamak, tamaku, tamakhu, tambaku, tanbak, pogaku, pugaiyilai, pugaielai, pugere, pukayila, pukayil, hogesoppu, pokala, dhurapan

Malaya: tembakau

The Philippines: tabako, tabaco, tobacco

Mexico: tabaco, tabaco bobo, apuga, a'xcu't, ayic, cuauhyetl, cuayetl, cutz, may, k'uts, kuutz, gueeza, gueza, guexa, huepá, huepaca, huipá, iyátl, ju'uikill, me-e, otzi, picietl, ro-hú, ro-u, uipa, ya, yaná, hapis copxot

Paraguay: sidí

Peru: chiri, chiri tseri, iri, yiri, shahuano, pori, petima, tsaang, rume, tabaco, romu, sairi, seri, sheri, shiña, ssina, tsiña, yemats

Brazil: tabaco, fumo, petum

Brazil (Amazonas): pee nahe

Yoruba: taba, taba esu

Congo: fumu, mbuli

N. tomentosa Ruíz & Pav.

Bolivia: saipi saipi

Peru: jarato

N. undulata Ruíz & Pav.

Peru: asnak-kcora, ccjama-sairi, ccjamasairi

N. velutina H. Wheeler

English: velvet tobacco

N. wigandioides Koch & Fintelm. (*Nicotiana herzogii* Dammer; *Nicotiana rusbyi* Britton)

Bolivia: kura kura, kuro kuro

Nidema Britton & Millspaugh Orchidaceae

Origins:
Apparently an anagram of the generic name *Dinema.*

Nidorella Cass. Asteraceae

Origins:
Latin *nidor, oris* "steam, smell"; Greek *knisa* "the steam and odor of fat which comes from meat roasting or burning"; Akkadian *nisu* "raising," *nidu, nadu* "to scatter incense, to pour oil for divination and other purposes; Hebrew *nasi* "raising vapor, mist, cloud," *naza* "to sprinkle."

Species/Vernacular Names:
N. anomala Steetz (*Nidorella angustifolia* O. Hoffm.; *Nidorella depauperata* Harv.; *Nidorella solidaginea* auct. non DC., Wild)

Southern Africa: mokoteli (Sotho)

N. resedifolia DC. subsp. *resedifolia* (*Nidorella densifolia* O. Hoffm; *Nidorella hirta* DC.; *Nidorella krookii* O.

Hoffm.; *Nidorella pinnatilobata* DC.; *Nidorella rapunculoides* DC.; *Nidorella resedifolia* DC. var. *subvillosa* Merxm.; *Nidorella solidaginea* DC.)

Southern Africa: stinkkruid, wurmbossie; kgôtôdúa (Tswana)

Nidularium Lemaire Bromeliaceae

Origins:

Latin *nidus* "nest," referring to the inflorescence enclosed within the rosette.

Nidus Rivinus Orchidaceae

Origins:

From the Latin *nidus* "nest," from the specific epithet of *Ophrys Nidus avis*, or referred to the genus *Neottia*.

Niebuhria DC. Capparidaceae (Capparaceae)

Origins:

Named for the German-born (b. Holstein) Danish botanist Carsten Niebuhr, 1733-1815 (d. Holstein), explorer, studied mathematics and astronomy at the University of Göttingen, traveler, member of the Royal Society of Göttingen, in 1802 *associé* de l'Institut de France, knight of the Danebrog, joined Pehr Forsskål's expedition to Arabia as a geographer 1761-1763 (he was the sole survivor of the expedition!). Among his writings are *Beschreibung von Arabien.* Kopenhagen 1772 and *Reisebeschreibung nach Arabien und andern umliegenden Ländern.* Kopenhagen 1774-1778, edited Pehr (Peter or Petrus) Forsskål (1732-1763, on 11 July), *Flora aegyptiaco-arabica.* Copenhagen 1775, *Descriptiones animalium.* Copenhagen 1775 and *Icones rerum naturalium.* Copenhagen 1776; among the other members of the expedition were the philologist van Haven (died on 25 May, 1763), the Dane physician Christian Carl Cramer (1732-1764), the Swedish ex-dragoon Berggren (died 1763, on 31 August) and the German artist Georg Wilhelm Baurenfeind (1728-1763, on 29 August). See Nils Spjeldnaes, in *Dictionary of Scientific Biography* 10: 117. 1981; Frans A. Stafleu, *Linnaeus and the Linnaeans.* The spreading of their ideas in systematic botany, 1735-1789. Utrecht 1971; John H. Barnhart, *Biographical Notes upon Botanists.* 3: 4. 1965; Barthold Georg Niebuhr, *Carsten Niebuhr's Leben.* Kiel 1817; Frank Nigel Hepper (b. 1929) and Ib Friis (b. 1945), *The Plants of Pehr Forsskål's "Flora Aegyptiaco-Arabica." Collected on the Royal Danish Expedition to Egypt and the Yemen 1761-1763.* Royal Botanic Gardens, Kew 1994; Pehr Forsskål, *Resa til Lyck-* *lige Arabien. Petrus Forsskål's dagbok 1761-1763. Med anmärkningar utgiven av Svenska Linné-Sällskapet.* Uppsala 1950; Anne Fox Maule, "Danish botanical expeditions and collections in foreign countries." *Botanisk Tidsskrift.* 69: 167-205. 1974; T.W. Bossert, *Biographical Dictionary of Botanists Represented in the Hunt Institute Portrait Collection.* 286. 1972; A. Lasègue, *Musée botanique de Benjamin Delessert.* 426. Paris 1845; E.M. Tucker, *Catalogue of the Library of the Arnold Arboretum of Harvard University.* 1917-1933; C.F.A. Christensen, *Den danske Botaniks Historie med tilhørende Bibliografi.* Copenhagen 1924-1926.

Niedenzua Pax Euphorbiaceae

Origins:

After the German botanist Franz Joseph Niedenzu, 1857-1937, from 1892 to 1926 professor of mathematics and natural history at the Braunsberg Lyceum Hosianum, contributor to Engler and K.A.E. Prantl *Die Natürlichen Pflanzenfamilien,* to Engler *Das Pflanzenreich* (Malpighiaceae), etc. His works include *Handbuch für botanische Bestimmunsübungen.* Leipzig 1895 and "Myrtaceae." *Nat. Pflanzenfam.* III. 7: 57-105. 1893. See John H. Barnhart, *Biographical Notes upon Botanists.* 3: 4. 1965; T.W. Bossert, *Biographical Dictionary of Botanists Represented in the Hunt Institute Portrait Collection.* 286. 1972; Ida Kaplan Langman, *A Selected Guide to the Literature on the Flowering Plants of Mexico.* University of Pennsylvania Press, Philadelphia 1964; E.M. Tucker, *Catalogue of the Library of the Arnold Arboretum of Harvard University.* 1917-1933; R. Zander, F. Encke, G. Buchheim and S. Seybold, *Handwörterbuch der Pflanzennamen.* 14. Aufl. Stuttgart 1993; Stafleu and Cowan, *Taxonomic Literature.* 3: 746-749. 1981.

Niederleinia Hieronymus Frankeniaceae

Origins:

After the German botanist Gustavo Niederlein, 1858-1924, traveler and plant collector, in Brazil and Argentina 1879-1886. Among his publications are *Ressources végétales des Colonies françaises.* Paris 1902, *The Republic of Costa Rica.* Philadelphia 1905, *Herbario Bettfreund.* Enumeración sistemática de las plantas recogidas en Buenos Ayres y sus alrededores por el Señor Don Carlos Bettfreund. Buenos Ayres 1898, *Resultados botanicos de exploraciones hechas en Misiones.* Buenos Ayres 1890 and*Mis exploraciones en el territorio de Misiones.* Buenos Ayres 1891. See *Comision cientifica de la Expedicion al Rio negro, 1879. Botánica,* por P.G. Lorentz ... y G. Niederlein. 1881; John H. Barnhart, *Biographical Notes upon Botanists.* 3: 4.

1965; E.M. Tucker, *Catalogue of the Library of the Arnold Arboretum of Harvard University*. 1917-1933; Ida Kaplan Langman, *A Selected Guide to the Literature on the Flowering Plants of Mexico*. University of Pennsylvania Press, Philadelphia 1964; I.C. Hedge and J.M. Lamond, *Index of Collectors in the Edinburgh Herbarium*. Edinburgh 1970.

Niemeyera F. Muell. Orchidaceae

Origins:

After the German physician Felix von Niemeyer, pathologist, professor of medicine at the University of Tübingen. His works include *Lehrbuch der speciellen Pathologie und Therapie*. Berlin 1858-1861, *Die epidemische Cerebro-Spinal Meningitis*. Berlin 1865 and *On the Symptomatic Treatment of Cholera ...* Translated from the German by P.W. Latham. Cambridge 1872. See Achille Casanova, *La Critica ... del nichilismo antiflogistico di Niemeyer*. Milano 1876; William Banting, *Letter on Corpulence*. London 1869; F. von Mueller, *Fragmenta Phytographiae Australiae*. 6: 96. 1867.

Niemeyera F. Muell. Sapotaceae

Origins:

After the German physician Felix von Niemeyer, professor of medicine at the University of Tübingen, author of *Lehrbuch der speciellen Pathologie und Therapie*. Berlin 1858-1861, *Die epidemische Cerebro-Spinal Meningitis*. Berlin 1865 and *On the Symptomatic Treatment of Cholera ...* Translated from the German by P.W. Latham. Cambridge 1872. See Achille Casanova, *La Critica ... del nichilismo antiflogistico di Niemeyer*. Milano 1876; William Banting, *Letter on Corpulence*. London 1869; F. von Mueller, *Fragmenta Phytographiae Australiae*. 7: 114. Melbourne 1870.

Species/Vernacular Names:

N. chartacea (Bailey) C. White (*Lucuna chartacea* Bailey; *Chrysophyllum chartacea* (Bailey) Vink)

English: plum boxwood

Nienokuea A. Chev. Orchidaceae

Origins:

Ivory Coast, a mountain near Cavally or Cavalla, possibly Nékaounié; some suggest from Nienokué.

Nierembergia Ruíz & Pavón Solanaceae

Origins:

Named for the Spanish Jesuit Juan Eusebio (Jean Eusèbe de Nieremberg) Nieremberg, 1595-1658. Among his numerous publications are *Historia naturae, maxime peregrinae, libris XVI distinctae*. Antwerpiae, Plantin-Moretus 1635, *Honor de S. Ignacio de Loyola*, en que se propone su vida y la de su discipulo S. Fr. Xavier, con las noticias de gran multitud de hijos del mismo S. Ignacio. Madrid 1645-1647, *Obras Christianas*. Sevilla 1686, *Epistolas*. Publicadas por M. de Faria y Sousa. Madrid 1649, *Obras y Dias: manual de Señores y Principes*. Madrid 1629 and *Causa y Remedio de los Males publicos*. Madrid 1642. See Ruíz & Pavón, *Flora peruvianae, et chilensis prodromus*. Madrid 1794; Alonso de Andrade, *Vida del ... padre Francisco Aguado ...* Inserta ... una breue relacion de la vida del padre J.E. Nieremberg. Madrid 1658; L. Polgar, *Bibliography of the History of the Society of Jesus*. Rome 1967; Georg Christian Wittstein, *Etymologisch-botanisches Handwörterbuch*. 617. Ansbach 1852; F. Boerner & G. Kunkel, *Taschenwörterbuch der botanischen Pflanzennamen*. 4. Aufl. 140. Berlin & Hamburg 1989; Asher Rare Books & Antiquariaat Forum, *Catalogue Natural History*. item no. 118. The Netherlands 1998.

Nietneria Klotzsch ex Benth. Melanthiaceae (Liliaceae)

Origins:

Probably named for the plant collector Johannes Nietner, see Karl Müller (1818-1899), "De muscorum Ceylonensium." in *Linnaea*. 36: 1-40. 1869.

Nigella L. Ranunculaceae

Origins:

The classic Latin name for the plant, *nigella, ae* (in Theodorus Priscianus), *nigellus, a, um* "somewhat black, dark," the diminutive of the Latin *niger, nigra, nigrum* "black," referring to the color of the seeds; see Carl Linnaeus, *Species Plantarum*. 534. 1753 and *Genera Plantarum*. Ed. 5. 238. 1754.

Species/Vernacular Names:

N. damascena L.

English: love-in-a-mist

Arabic: habba souda, sinouj

N. sativa L.

English: fennel flower, black cumin, common fennel flower

Arabic: habba sooda, kammun aswad

Tibetan: zira nagpo

Malaya: jintan hitam

Nigrina L. Scrophulariaceae

Origins:

Latin *niger, nigra, nigrum* "black," see also *Melasma* Bergius (Greek *melasma* "black or livid spot," *melasmos* "blackening, dyeing black," referring to a black spot when the plants dry).

Nigritella L.C. Richard Orchidaceae

Origins:

Latin *niger, nigra, nigrum* "black," *Nigritae, arum* "the people living near the Niger," *nigritia* and *nigrities* "blackness," referring to the flowers.

Nigromnia Carolin Goodeniaceae

Origins:

The generic name *nigromnia* is the Latin translation (*niger, nigra, nigrum* "black" and *omnis* "all") of the name of the original collector (Australia, between Yuna and Dartmoor, 20 Sept. 1940) of *Nigromnia globosa* Carolin, the British botanist Dr. William Edward Blackall, d. 1941, physician and surgeon, botanical artist; see Roger Charles Carolin, "*Nigromnia*, a new genus of Goodeniaceae." in *Nuytsia*. 1(4): 292-293. 1974 and "Floral structure and anatomy in the family Goodeniaceae Dumort." *Proc. Linn. Soc. N.S.W.* 84: 242. 1959.

Nilgirianthus Bremek. Acanthaceae

Origins:

Nilgiris, mountains in W. Ghats, Tamil Nadu State, India.

Nimmoia Wight Meliaceae

Origins:

Named for Joseph Nimmo, flourished 1830s-1854, at Surat 1819, plant collector in Socotra 1834-1839, sent plants to R. Wight; see Ray Desmond, *Dictionary of British & Irish Botanists and Horticulturists*. 519. London 1994.

Niopa (Benth.) Britton & Rose Mimosaceae

Origins:

From a vernacular name for a species of *Anadenanthera* Speg.

Nipa Thunberg Palmae

Origins:

From the Moluccan vernacular name for the plant, see *Nypa*; see C.P. Thunberg, in *Kongl. Vetenskaps Academiens Nya Handlingar*. 3: 231. Stockholm (Jul.-Sep.) 1782.

Niphaea Lindley Gesneriaceae

Origins:

From the Greek *nipha* "snow," *niphas* "snowflake," alluding to the flowers.

Niphidium J. Sm. Polypodiaceae

Origins:

From the Greek *nipha* "snow."

Niphobolus Kaulf. Polypodiaceae

Origins:

From the Greek *niphobolos* "snowclad."

Niphogeton Schltdl. Umbelliferae

Origins:

From the Greek *nipha* "snow" and *geiton* "a neighbor."

Niphopsis J. Sm. Polypodiaceae

Origins:

From the Greek *nipha* "snow" and *opsis* "appearance."

Nipponanthemum Kitam. Asteraceae

Origins:

Japanese flower, Greek *anthemon* "flower."

Nipponobambusa Muroi Gramineae

Origins:

From Japan and the genus *Bambusa*.

Nipponocalamus Nakai Gramineae

Origins:

From Japan and Greek *kalamos* "reed."

Nipponorchis Masam. Orchidaceae

Origins:
From Japan and *Orchis*, the native country.

Niruri Adans. Euphorbiaceae

Origins:
A vernacular name, see *Phyllanthus niruri* L.

Niruris Raf. Euphorbiaceae

Origins:
A vernacular name; see C.S. Rafinesque, *Sylva Telluriana.* 91. 1838; E.D. Merrill, *Index rafinesquianus.* The plant names published by C.S. Rafinesque, etc. 155. Massachusetts, USA 1949.

Nisomenes Raf. Euphorbiaceae

Origins:
Perhaps from Greek *nissomai, nisomai, nisomenos* "to go, to come"; see C.S. Rafinesque, *Flora Telluriana.* 4: 116. 1836 [1838]; E.D. Merrill, *Index rafinesquianus.* 155. Massachusetts, USA 1949.

Nisoralis Raf. Sterculiaceae

Origins:
See C.S. Rafinesque, *Sylva Telluriana.* 74. 1838; E.D. Merrill, *Index rafinesquianus.* 167. Massachusetts, USA 1949.

Nispero Aubrév. Sapotaceae

Origins:
A Spanish word for medlar.

Nissolia Jacq. Fabaceae

Origins:
After the French botanist Guillaume Nissole (Nissolle), 1647-1735, physician. See John H. Barnhart, *Biographical Notes upon Botanists.* 3: 7. 1965; T.W. Bossert, *Biographical Dictionary of Botanists Represented in the Hunt Institute Portrait Collection.* 287. 1972; James Britten, *The Sloane Herbarium,* revised and edited by J.E. Dandy. 172. London 1958; H.N. Clokie, *Account of the Herbaria of the*

Department of Botany in the University of Oxford. 218. Oxford 1964; Jonas C. Dryander, *Catalogus bibliothecae historico-naturalis Josephi Banks.* London 1800; E.M. Tucker, *Catalogue of the Library of the Arnold Arboretum of Harvard University.* 1917-1933; Helmut Genaust, *Etymologisches Wörterbuch der botanischen Pflanzennamen.* 420. [d. 1734] Basel 1996.

Nitraria L. Nitrariaceae (Zygophyllaceae)

Origins:
From the Latin *nitrum, i* "natron, native soda," Greek *nitron,* soda sources, the plant was first found on the saline plains in Siberia; see Carl Linnaeus, *Systema Naturae.* Ed. 10. 1044. 1759.

Species/Vernacular Names:
N. billardierei DC. (*Nitraria schoberi* sensu J. Black)

English: dillon bush, nitre bush, wild grape

China: kou chi

Nitrophila S. Watson Chenopodiaceae

Origins:
From the Greek *nitron* "natron, native soda" and *philos* "loving."

Species/Vernacular Names:
N. mohavensis Munz & Roos

English: Amargosa nitrophila (Amargosa Desert)

Nivenia R. Br. Proteaceae

Origins:
For the Scottish (b. near Edinburgh) gardener James (David) Niven, *circa* 1774-1826/1827 (d. near Edinburgh), gardener at the Royal Botanic Garden of Edinburgh and at Syon House (Middx.), from 1798 to 1812 plant collector in South Africa (at the Cape), collected plants in South Africa for the garden of George Hibbert (1757-1837); see I.H. Vegter, *Index Herbariorum.* Part II (5), *Collectors N-R.* Regnum Vegetabile vol. 109. 1983; J. Britten, "Some early Cape botanists and collectors." *J. Linn. Soc. (Bot.)* 45: 29-51. 1920; Henry C. Andrews (1794-1830), *The Botanist's Repository.* t. 193. London 1801; Mary Gunn and Leslie E. Codd, *Botanical Exploration of Southern Africa.* 262. [d. 1828] Cape Town 1981; Ray Desmond, *Dictionary of British & Irish Botanists and Horticulturists.* 519. [d. 1827] 1994; Christopher Hibbert, *Africa Explored — Europeans in the Dark Continent 1769-1889.* W.W. Norton Co., New

York 1982; J.H. Barnhart, *Biographical Notes upon Botanists*. 3: 8. 1965; A. Lasègue, *Musée botanique de Benjamin Delessert*. 447. Paris 1845; G. Murray, *History of the Collections Contained in the Natural History Departments of the British Museum*. 1: 170. London 1904; Alice Margaret Coats, *The Quest for Plants. A History of the Horticultural Explorers*. 260. London 1969.

Nivenia Vent. Iridaceae

Origins:
After the Scottish gardener James (David) Niven, c. 1774-1826/1827.

Niveophyllum Matuda Bromeliaceae

Origins:
From *niveus* "snowy, snow-" and *phyllon* "leaf."

Noaea Moq. Chenopodiaceae

Origins:
For the German botanist Friedrich Wilhelm Noë, d. 1858, pharmacist, plant collector in the Near East and the Balkan Peninsula; see J.H. Barnhart, *Biographical Notes upon Botanists*. 3: 9. Boston 1965; H.N. Clokie, *Account of the Herbaria of the Department of Botany in the University of Oxford*. 218. Oxford 1964; A. Lasègue, *Musée botanique de Benjamin Delessert*. Paris 1845; R. Zander, F. Encke, G. Buchheim and S. Seybold, *Handwörterbuch der Pflanzennamen*. 14. Aufl. Stuttgart 1993; I.C. Hedge and J.M. Lamond, *Index of Collectors in the Edinburgh Herbarium*. Edinburgh 1970.

Noahdendron Endress, Hyland & Tracey Hamamelidaceae

Origins:
From the Australian locality Noah's Creek near Cape Tribulation, Northern Queensland, and the Greek *dendron* "a tree."

Nocca Cavanilles Asteraceae

Origins:
For the Italian botanist Domenico Nocca, 1758-1841, clergyman.

Noccaea Moench Brassicaceae

Origins:
Dedicated to the Italian botanist Domenico Nocca, 1758-1841, clergyman, pupil of Giovanni Antonio Scopoli (1723-1788), professor of botany, Praefectus of the Botanic Garden at Pavia (in 1788, 1797-1820 and from 1820 to 1826). His writings include *Onomatologia*. Papiae [Pavia] [1813], *Istituzioni di botanica pratica*. [Pavia] 1801, *Historia atque Ichnographia Horti Botanici Ticinensis*. Ticini Regii [Pavia] 1818 and *Elementi di botanica*. Pavia 1801, with Giovanni Battista Balbis (1765-1831) wrote *Flora ticinensis*. Ticini [Pavia] 1816-1821. See Gino Pollacci (1872-1963), "L'Orto Botanico di Pavia dalla fondazione al 1942." *Ticinum*. 6:20-23. 1950; J.H. Barnhart, *Biographical Notes upon Botanists*. 3: 9. 1965; E.M. Tucker, *Catalogue of the Library of the Arnold Arboretum of Harvard University*. 1917-1933; Jonas C. Dryander, *Catalogus bibliothecae historico-naturalis Josephi Banks*. London 1800; A. Lasègue, *Musée botanique de Benjamin Delessert*. 342. Paris 1845.

Noccidium F.K. Meyer Brassicaceae

Origins:
Referring to *Noccaea*.

Nodocarpaea A. Gray Rubiaceae

Origins:
Latin *nodus* "a knot" and Greek *karpos* "fruit."

Nodonema B.L. Burtt Gesneriaceae

Origins:
Latin *nodus* "a knot" and Greek *nema* "thread."

Nogalia Verdc. Boraginaceae

Origins:
Greek *nogala* "dainties, dessert."

Nogo Baehni Sapotaceae

Origins:
A West African vernacular name.

Nogra Merr. Fabaceae

Origins:
From the generic name *Grona* Lour., an anagram.

Noisettia Kunth Violaceae

Origins:
Dedicated to the French horticulturist Louis Claude Noisette, 1772-1849. His works include *Le jardin fuitier*. Paris 1821, *Manuel complet du Jardinier*. Paris 1825, 1826 and *Traité complet de la greffe*. Paris 1852, with Pierre Boitard wrote *Manuel du jardinier des primeurs*, ou l'art de forcer les plantes à donner leurs fruits ou leurs fleurs dans toutes les saisons. Paris 1832.

Nolana L. ex L.f. Nolanaceae (Solanaceae)

Origins:
Latin *nola* "small bell, a little bell."

Species/Vernacular Names:
N. sp.
Peru: palo-palo

Nolina Michx. Dracaenaceae (Agavaceae, Nolinaceae)

Origins:
For P.C. Nolin, a French botanist and writer, see Abbé Nolin and J.L. Blavet, *Essai sur l'Agriculture moderne*. 1755.

Species/Vernacular Names:
N. interrata H. Gentry
English: Dehesa nolina
N. longifolia (Schultes & Schultes f.) Hemsley
Mexico: zacate
N. microcarpa S. Wats.
English: beargrass
Mexico: sacahuista
N. recurvata (Lem.) Hemsley
English: bottle palm, elephant-foot tree, pony tail, nolina
N. stricta (Lem.) Cif. & Giac.
Mexico: gobiila, cobijla

Nolitangere Raf. Balsaminaceae

Origins:
Touch-me-not, for *Impatiens* L., Latin *impatiens, entis*, referring to the sudden and violent dehiscence of the capsules; see Constantine Samuel Rafinesque, *Herb. Raf.* 68. 1833; E.D. Merrill, *Index rafinesquianus*. The plant names published by C.S. Rafinesque, etc. 162. Jamaica Plain, Massachusetts, USA 1949.

Nolletia Cass. Asteraceae

Origins:
For the French physicist Jean Antoine Nollet, 1700-1770, author of *L'Art des Expériences*. Paris 1770 and *Essai sur l'électricité des corps*. Paris 1754; see Joseph-Aignan Sigaud de La Fond (1730-1810), *Description et usage d'un cabinet de physique expérimentale*. [Two volumes, first edition; Sigaud de La Fond was a follower of the Abbé Nollet, whom he succeeded as professor of experimental physics at the Collège Louis-le-Grand in 1760.] Paris 1755 and *Précis historique et expérimental des phénomènes électriques*. [First edition; electricity and magnetism.] Paris 1781.

Species/Vernacular Names:
N. ciliaris (DC.) Steetz (*Leptothamnus ciliaris* DC.)
South Africa: kaalriekte, moloka

Noltea H.G.L. Reichenbach Rhamnaceae

Origins:
Named after the German botanist Ernst Ferdinand Nolte, 1791-1875, physician, Dr. med. Göttingen 1817, from 1824 to 1826 in Denmark with Jens Wilken Hornemann, from 1826 to 1873 professor of botany at the University of Kiel, author of *Novitiae florae holsaticae* ... Kilonii [Kiel] 1826 and *Botanische Bemerkungen über Stratiotes und Sagittaria*. Kopenhagen 1825. See J.W. Hornemann, *Naturh. Tidsskr.* 1: 587-588. 1837; R. Zander, F. Encke, G. Buchheim and S. Seybold, *Handwörterbuch der Pflanzennamen*. 14. Aufl. 757. Stuttgart 1993; Heinrich Gustav Reichenbach (1824-1889), *E.F. Nolte, ein hamburger Botaniker*. Eine Skizze. Hamburg 1881; C.F.A. Christensen, *Den danske Botaniks Historie med tilhørende Bibliografi*. Copenhagen 1924-1926; Günther Schmid, *Chamisso als Naturforscher*. Eine Bibliographie. Leipzig 1942.

Species/Vernacular Names:
N. africana (L.) W.H. Harvey et Sonder (*Ceanothus africana* L.)

English: soap dogwood, soap bush

Southern Africa: seepblinkblaar, noltia, seepbos; uMahlahl-akwa (Zulu); umaLuleka, umKuthuhla, umGlindi, iPhalode, maKutula, amaLuleka (Xhosa)

Noltia Schumach. & Thonn. Ebenaceae

Origins:
For the German botanist Ernst Ferdinand Nolte, 1791-1875, physician, Dr. med. Göttingen 1817, 1824-1826 in Denmark with Jens Wilken Hornemann (1770-1841), 1826-1873 professor of botany at the University of Kiel, author of *Novitiae florae holsaticae* ... Kilonii [Kiel] 1826 and *Botanische Bemerkungen über Stratiotes und Sagittaria*. Kopenhagen 1825; see J.W. Hornemann, *Naturh. Tidsskr.* 1: 587-588. 1837.

Nomaphila Blume Acanthaceae

Origins:
From the Greek *nomos* "meadow, pasture" and *philos* "loving," indicating the habitat.

Nomismia Wight & Arn. Fabaceae

Origins:
Probably from the Latin *nomisma, atis* "a coin, a piece of money, a medal," Greek *nomisma*, fruits orbicular and flattened.

Nomocharis Franchet Liliaceae

Origins:
Greek *nomos* "meadow, pasture" and *charis* "grace, loveliness, charm, beauty," growing in pasture lands.

Nomosa I.M. Johnston Boraginaceae

Origins:
From the Greek *nomos* "meadow, pasture."

Nonatelia Aublet Rubiaceae

Origins:
Derivation uncertain, probably from *non* "not" and Greek *ateleia, atelea* "imperfection," *ateles* "imperfect, without end, incomplete."

Nonatelia Kuntze Rubiaceae

Origins:
Probably from *non* "not" and Greek *ateleia, atelea* "imperfection," *ateles* "imperfect, without end, incomplete."

Nonea Medikus Boraginaceae

Origins:
Possibly named after the German botanist Johann Philipp Nonne, 1729-1772. His works include *Flora in territorio erfordensi indigena*. Erfordiae [Erfurt] 1763, *J.P. Nonne ... botanophilis S.P.D. et publicas scholas aperit*, quaedam de botanicae usu, et ratione, qua studium hoc rite ingrediendum sit, commentans. Erfordiae 1763 and *Quaedam de plantis nothis*. 1790, with Johann Jacob Planer (1743-1789) wrote *Index plantarum quas in agro erfurtensi sponte provenientes*. Gothae 1788. See Paul Usteri (1768-1831), *Delectus opusculorum botanicorum*. Argentorati [Strasbourg] 1790; Jonas C. Dryander, *Catalogus bibliothecae historico-naturalis Josephi Banks*. London 1800; F. Boerner & G. Kunkel, *Taschenwörterbuch der botanischen Pflanzennamen*. 4. Aufl. 141. Berlin & Hamburg 1989; Georg Christian Wittstein, *Etymologisch-botanisches Handwörterbuch*. 619. 1852.

Species/Vernacular Names:
N. capsica (Willdenow) G. Don
English: common nonea
China: jia lang zi cao

Nopalea Salm-Dyck Cactaceae

Origins:
The Náhuatl name *nopalli*; see Alicja Iwanska, *Purgatory and Utopia: A Mazahua Indian Village of Mexico*. Cambridge 1971; Santiago da Cruz e Gonçalves, *Nova instrucção sobre a cultura dos nopales, e criação da cochinilha d'America*, para uso dos lavradores das Canarias. Lisbon 1837.

Species/Vernacular Names:
N. cochenillifera (L.) Salm-Dyck
English: cochineal cactus
The Philippines: dilang-baka, dila-dila, nopal, palad, dapal
Mexico: biaa, piaa

Nopalxochia Britton & Rose Cactaceae

Origins:
Possibly from an Aztec name, *nopalli*; Náhuatl, the language of the Aztecs, was the chief imperial tongue of

pre-conquest Mexico. See the first dictionary in the New World written by Fray Alonso de Molina, *Vocabulario en lengua castellana y mexicana.* Mexico 1571; Cyrus Thomas and John R. Swanton, *Indian Languages of Mexico and Central America and Their Geographical Distribution.* Washington DC 1911; Faustino Chimalpopoca[tl Galicia], *Epítome ó modo fácil de aprender el idioma nahuatl ó lengua mexicana.* México. 1869; Faustino Chimalpopocatl Galicia, *Silabario de idioma mexicano.* Mexico. 1849; Muriel Porter Weaver, *The Aztec, Maya, and Their Predecessors: Archaeology of Mesoamerica.* New York 1972; E.L. Hewett, *Ancient Life in Mexico and Central America.* Indianapolis 1936; F. Peterseon, *Ancient Mexico: An Introduction to the Pre-hispanic Cultures.* NY 1959; Carmen Aguilera, *Flora y fauna Mexicana.* Mitología y tradiciones. 116-117. México s.d. [1985].

Norantea Aublet Marcgraviaceae

Origins:
Probably after a vernacular name (*gonora-antegri*? see *Paxton's Botanical Dictionary.* 394. 1868).

Species/Vernacular Names:
N. adamantium Camb.

Brazil: planta-dos-diamantes, parreira-da-pedra

N. guianensis Aubl.

Brazil: rabo-de-arara

Normanbokea Klad. & Buxb. Cactaceae

Origins:
For the American botanist Norman Hill Boke, 1913-1996, cactologist, traveler, 1945-1981 professor of botany and microbiology at the University of Oklahoma, 1970-1975 editor of the *American Journal of Botany.* Among his writings are "Histogenesis of the vegetative shoot in *Echinocereus.*" *Amer. J. Bot.* 38: 23-38. 1951, "The genus *Pereskia* in Mexico." *Cact. Succ. J. Amer.* 35: 3-10. 1963, "A botanist looks at Mexico." *Bios.* 33(4): 187-199. 1962 and "Developmental morphology and anatomy in Cactaceae." *BioScience.* 30: 605-610. 1980; see Larry W. Mitich, in *Cactus and Succulent Journal.* 69(1): 15-16. 1997; Irving William Knobloch, compil., "A preliminary verified list of plant collectors in Mexico." *Phytologia Memoirs.* VI. 1983; Ida Kaplan Langman, *A Selected Guide to the Literature on the Flowering Plants of Mexico.* University of Pennsylvania Press, Philadelphia 1964.

Normanboria Butzin Gramineae

Origins:
For the Irish (b. at Tramore, Co. Waterford) botanist Norman Loftus Bor, 1893-1972 (London), agrostologist, plant collector, 1921-1948 Indian Forest Service, Forest Botanist of the Forest Research Institute (Dehra Dun), 1945 President of the Indian Botanical Society, 1948-1959 Assistant Director Kew, 1931 Fellow of the Linnean Society, author of *Manual of Indian Forest Botany.* Oxford University Press, London 1953, with Mukat Behari Raizada wrote *Some Beautiful Indian Climbers and Shrubs.* Bombay Natural History Society–Oxford University Press, Bombay 1982, contrib. to *Indian Forester.* See John H. Barnhart, *Biographical Notes upon Botanists.* 1: 220. 1965; C.E. Hubbard, "Norman Loftus Bor (1893-1972)." in *Kew Bulletin.* 30(1): 1-4. 1975; Carolyn M.K. Pope, "A bibliography of the work of Dr. N.L. Bor." in *Kew Bulletin.* 30(1): 4-10. 1975.

Normanbya F. Muell. ex Becc. Palmae

Origins:
After George Augustus Constantine Phipps, 1819-1890, 2nd Marquis of Normanby, from 1871 to 1874 Governor of Queensland, then Governor of New Zealand and from 1879 to 1884 Governor of Victoria, see *Tenure of Office.* [On the controversy between Lord Normanby, the Lieutenant Governor, and the Government of Nova Scotia on the dismissal of permanent officials for political reasons.] [1864], *Governor Normanby's Visit to Gympie*; with descriptive accounts of the Gympie Goldfield and the Newsa District. Gympie 1873, *The Visit of His Excellency the Governor of the North, 1876.* Auckland [1876]; see also O. Beccari, *Annales du Jardin Botanique de Buitenzorg.* 2: 91, 170, 171. 1885.

Species/Vernacular Names:
N. normanbyi (W. Hill) L. Bailey

Australia: dowar, black palm

Normania Lowe Solanaceae

Origins:
After Commander Francis M. Norman, plant collector and botanical discoverer in Madeira 1865-1866.

Norna Wahlenb. Orchidaceae

Origins:
Old Norse Norn, one of the three Fates.

Noronhaea Post & Kuntze Oleaceae

Origins:
Orthographic variant of *Noronhia*.

Noronhia Stadman ex Thouars Oleaceae

Origins:
For the Spanish physician Francisco (François, Fernando) Noroña (Noronha), c. 1748-1788, traveled to the Philippines and Java, wrote *Prodromus Phytologicus Vegetabilia exhibens nuperrime insula Madagascar detecta*, etc. in Louis-Marie Aubert Aubert du Petit-Thouars (1758-1831), *Mélanges de Botanique et de Voyages ...* Paris 1811. See Jonas C. Dryander, *Catalogus bibliothecae historico-naturalis Josephi Banks*. London 1800; A. Lasègue, *Musée botanique de Benjamin Delessert*. Paris 1845; E.M. Tucker, *Catalogue of the Library of the Arnold Arboretum of Harvard University*. 1917-1933.

Norrisia Gardner Loganiaceae

Origins:
Named after Sir William Norris, 1793-1859 (d. Berks.), to India 1829, traveler and plant collector, sent plants to G. Gardner and W. Griffith; see Ray Desmond, *Dictionary of British & Irish Botanists and Horticulturists*. 521. London 1994.

Nortenia Thou. Scrophulariaceae

Origins:
A derivation from the generic name *Torenia* L., an anagram or near; for the Swedish clergyman Rev. Olof Torén, 1718-1753, traveler, botanist and plant collector, ship chaplain with the Swedish East India Company at Surat, India (1750-1752) and in China (1748-1749); see J.H. Barnhart, *Biographical Notes upon Botanists*. 3: 391. 1965; Pehr Osbeck (1723-1805), *Dagbok öfwer en Ostindisk Resa åren 1750, 1751, 1752 ...* Stockholm 1757, English translation by Johann Reinhold Forster (1729-1798), (in English) *A Voyage to China and the East Indies*, by P.O., ... together with a voyage to Suratte, by Olof Toreen, etc. London 1771; Emil Bretschneider, *History of European Botanical Discoveries in China*. Leipzig 1981; A. Lasègue, *Musée botanique de Benjamin Delessert*. 1845; E.M. Tucker, *Catalogue of the Library of the Arnold Arboretum of Harvard University*. Cambridge, Massachusetts 1917-1933.

Northea Hook.f. Sapotaceae

Origins:
Named for the English flower painter Miss Marianne North, 1830-1890, traveler.

Northia Hook.f. Sapotaceae

Origins:
Named for the extraordinary botanical painter Miss Marianne North, 1830-1890 (d. Alderley, Gloucestershire), traveled widely. Her work is of great botanical interest, because she aimed to depict nature as she saw it, without attempting to sanitize and prettify her work. In 1871 she undertook the first of her many journeys, visiting the United States, Canada and Jamaica. She stayed for 8 months in Brazil and completed over 100 paintings; in 1875 she crossed the American continent on her way to Japan, returning home in 1877 via Sarawak, Java and Sri Lanka. Six months later she traveled to India where she stayed for 15 months and produced over 200 paintings. Just 2 months after the opening of her Gallery (June 1882) she traveled to South Africa; in 1883 she was in the Seychelles and in 1884, despite ill-health, she was painting plants in Chile; she wrote *Recollections of a Happy Life*. London 1892, *Further Recollections of a Happy Life*. Macmillan, London 1893 and *A Vision of Eden*. Second edition. Royal Botanic Gardens, Kew and Webb and Bower 1986; see Shirley Heriz-Smith, "Western collectors on eastern islands." *Hortus*. 7: 83-92. 1988; Ray Desmond, *Dictionary of British & Irish Botanists and Horticulturists*. 521. London 1994; M. Hadfield et al., *British Gardeners: A Biographical Dictionary*. London 1980; Mary Gunn and Leslie Edward W. Codd, *Botanical Exploration of Southern Africa*. 263. Cape Town 1981.

Species/Vernacular Names:
N. hornei (Hartog) Pierre
French: capucin

Northiopsis Kaneh. Sapotaceae

Origins:
Resembling *Northia*.

Nosema Prain Labiatae

Origins:
Greek *sema* "sign, standard," or an anagram of the generic name *Mesona* Blume.

Species/Vernacular Names:

N. cochinchinensis (Lour.) Merrill

English: Cochinchina nosema

China: long chuan cao

Nostelis Raf. Labiatae

Origins:

See C.S. Rafinesque, *Sylva Telluriana*. 76. 1838; E.D. Merrill (1876-1956), *Index rafinesquianus*. The plant names published by C.S. Rafinesque, etc. 208. Jamaica Plain, Massachusetts, USA 1949.

Nostolachma T. Durand Rubiaceae

Origins:

See also *Lachnastoma* Korth.

Notanthera (DC.) G. Don f. Loranthaceae

Origins:

From the Greek *notos* "the south, back" and *anthera* "anther."

Notaphoebe Griseb. Lauraceae

Origins:

See *Nothaphoebe* Blume.

Notechidnopsis Lavranos & Bleck Asclepiadaceae

Origins:

From the Greek *notos* "the south, back" and the genus *Echidnopsis* Hook.f.

Notelaea Vent. Oleaceae

Origins:

Greek *notos* "the south" and *elaia* "the olive tree, olive," referring to these Oleaceae of the Southern Hemisphere.

Species/Vernacular Names:

N. johnsonii P. Green

English: veinless mock-olive

N. ligustrina Vent.

English: privet mock-olive

N. linearis Benth.

English: yellow mock-olive

N. longifolia Vent.

English: large mock-olive, large-leaved olive

N. microcarpa R. Br.

English: small-fruit mock-olive, native olive, velvet mock-olive

N. ovata R. Br. (*Notelaea longifolia* var. *ovata* (R. Br.) Domin)

English: netted mock-olive

N. venosa F. Muell.

English: veined mock-olive, smooth mock-olive

Notelea Raf. Oleaceae

Origins:

Greek *notos* "the south" and *elaia* "the olive tree, olive"; see Constantine Samuel Rafinesque (1783-1840), *Sylva Telluriana*. 10. 1838; E.D. Merrill (1876-1956), *Index rafinesquianus*. The plant names published by C.S. Rafinesque, etc. 190. Jamaica Plain, Massachusetts, USA 1949.

Noterias Raf. Oleaceae

Origins:

See Constantine Samuel Rafinesque, *Sylva Telluriana*. 76. 1838; E.D. Merrill, *Index rafinesquianus*. The plant names published by C.S. Rafinesque, etc. 208. 1949.

Noterophila Mart. Melastomataceae

Origins:

From the Greek *noteros* "damp, moist" and *philos* "lover, loving."

Nothaphoebe Blume Lauraceae

Origins:

From the Greek *nothos* "false" and the related genus *Phoebe* Nees.

Nothapodytes Blume Icacinaceae

Origins:

Greek *nothos* "false" and *apodyo*, *apodutos* "to strip off, undressed," referring to the calyx or to the petals.

Species/Vernacular Names:
N. foetida (Wight) Sleumer (*Stemonurus foetidus* Wight; *Mappia ovata* Miers; *Mappia insularis* (Matsum.) Hatusima; *Mappia ovata* Miers var. *insularis* Matsum.)

Japan: kusa-mizu-ki

India: kalgur, ghanera, arali, chorla, kodsa, hedare

Nothoalsomitra Telford Cucurbitaceae

Origins:
From the Greek *nothos* "false" and the genus *Alsomitra* (Blume) M. Roemer.

Species/Vernacular Names:
N. suberosa (F.M. Bailey) Telford

English: corky cucumber

Nothobaccharis R.M. King & H. Robinson Asteraceae

Origins:
From the Greek *nothos* "false, bastard" and the genus *Baccharis* L.

Nothocalais Greene Asteraceae

Origins:
Greek *nothos* "false, spurious, not genuine" plus the genus *Calais*.

Species/Vernacular Names:
N. alpestris (A. Gray) Chambers

English: false-agoseris

Nothocarpus Post & Kuntze Rubiaceae

Origins:
From the Greek *nothos* "false, bastard" and *karpos* "fruit."

Nothocestrum A. Gray Solanaceae

Origins:
From the Greek *nothos* "false" and the genus *Cestrum*.

Species/Vernacular Names:
N. sp.

Hawaii: 'aiea, hâlena

Nothochelone (A. Gray) Straw Scrophulariaceae

Origins:
From the Greek *nothos* "false" and the genus *Chelone* L.

Nothochilus Radlk. Scrophulariaceae

Origins:
From the Greek *nothos* "false" and *cheilos* "lip."

Nothochlaena Kaulf. Pteridaceae

Origins:
Orthographic variant of *Notholaena* R. Br.

Nothocissus (Miq.) Latiff Vitaceae

Origins:
From the Greek *nothos* "false, bastard, baseborn" plus *Cissus*.

Nothocnestis Miq. Celastraceae

Origins:
From the Greek *nothos* "false, bastard, baseborn" plus *Cnestis* Juss.

Nothocnide Blume ex Chew Urticaceae

Origins:
From the Greek *nothos* "false" and *knide* "nettle"; see Karl Ludwig von Blume, *Museum Botanicum Lugduno-Batavum*. Lugduni-Batavorum 1856.

Nothodoritis Tsi Orchidaceae

Origins:
From the Greek *nothos* "false, spurious" plus the genus *Doritis* Lindl.

Nothofagus Blume Fagaceae

Origins:
From the Greek *nothos* "false" and the genus *Fagus* L., false or mongrel beech, in Southern Hemisphere; see Karl Ludwig von Blume, *Museum Botanicum Lugduno-Batavum*. Lugduni-Batavorum 1850.

Species/Vernacular Names:

N. antarctica (Forster f.) Oersted

English: Antarctic beech

N. cunninghamii (Hook.) Oersted (*Fagus cunninghamii* Hook.)

English: myrtle beech, myrtle, Southern beech, beech

N. dombeyi (Mirbel) Oersted

Chile: coigue

N. fusca (Hook.f.) Oersted

English: red beech

N. gunnii (Hook.f.) Oersted (*Fagus gunnii* Hook.f.)

English: tanglefoot, fagus, deciduous beech, tanglefoot beech

N. menziesii (Hook.f.) Oersted

English: silver beech, Southland beech

N. moorei (F. Muell.) Krasser (*Fagus moorei* F. Muell.)

English: negrohead beech, niggerhead, Antarctic beech, niggerhead beech

N. solanderi (Hook.f.) Oersted

English: black beech

Nothoholcus Nash Gramineae

Origins:

From the Greek *nothos* "false, bastard" and the genus *Holcus* L.

Notholaena R. Br. Pteridaceae (Polypodiaceae, Adiantaceae)

Origins:

Cloak ferns, from the Greek *nothos* "false" and *chlaena, laina* "cloak, blanket, coat," referring to the leaf margins and to the incomplete indusium; see Robert Brown (1773-1858), *Prodromus florae Novae Hollandiae et Insulae van-Diemen.* 145. London 1810.

Notholcus Hitchc. Gramineae

Origins:

From the Greek *nothos* "false, bastard" and the genus *Holcus* L.

Notholex Raf. Aquifoliaceae

Origins:

Greek *nothos* "false, bastard" and the genus *Ilex* L.; see C.S. Rafinesque, *Sylva Telluriana.* 47. 1838; E.D. Merrill, *Index rafinesquianus.* 160. 1949.

Notholirion Wallich ex Boiss. Liliaceae

Origins:

Greek *nothos* "false" and *leirion* "a lily," lily-like plants.

Nothomyrcia Kausel Myrtaceae

Origins:

From the Greek *nothos* "false" and the genus *Myrcia* DC. ex Guill.

Nothopanax Miq. Araliaceae

Origins:

From the Greek *nothos* "false" and the genus *Panax* L., *panakes* "all healing, a panacea," *pan* "all" and *akos* "a remedy"; see Friedrich A.W. Miquel (1811-1871), *Flora van Nederlandsch Indië.* [= *Flora Indiae batavae.*] 1(1): 765. Amsterdam, Utrecht and Leipzig 1856.

Nothopegia Blume Anacardiaceae

Origins:

From the Greek *nothos* "false" and the genus *Pegia* Colebr.

Nothopegiopsis Lauterb. Anacardiaceae

Origins:

Resembling *Notopegia.*

Nothoperanema (Tagawa) Ching Dryopteridaceae (Aspleniaceae)

Origins:

From the Greek *nothos* "false, bastard" and the genus *Peranema* D. Don.

Nothophlebia Standley Rubiaceae

Origins:

From the Greek *nothos* "false, bastard" and *phleps, phlebos* "vein."

Nothoprotium Miq. Anacardiaceae

Origins:

From the Greek *nothos* "false, bastard" plus *Protium* Burm.f., Burseraceae, the stem is a source of oil.

Nothoruellia Bremek. & Nannenga-Bremek. Acanthaceae

Origins:
From the Greek *nothos* "false, bastard, spurious" plus the related *Ruellia* Bremek.

Nothosaerva Wight Amaranthaceae

Origins:
From the Greek *nothos* "false, bastard, spurious" plus *Aerva* Forsskål.

Nothoscordum Kunth Alliaceae (Liliaceae)

Origins:
Greek *nothos* "false" and *skordon, skorodon* "garlic," *Allium*-like and odorless; see Carl Sigismund Kunth (1788-1850), *Enumeratio Plantarum.* 4: 457. Stutgardia & Tubingae 1843.

Species/Vernacular Names:
N. gracile (Aiton) Stearn (*Nothoscordum fragrans* (Vent.) Kunth; *Allium gracile* Ait.; *Allium fragrans* Vent.)

English: fragrant false garlic, Gowie's curse, Gowie weed, onion weed, fragrant onion, wild onion, false garlic

South Africa: basterknoffel, wilde tjienkerientjie

Nothosmyrnium Miq. Umbelliferae

Origins:
From the Greek *nothos* "false" and the genus *Smyrnium* L.

Nothospondias Engl. Simaroubaceae

Origins:
Greek *nothos* "false" and the genus *Spondias*.

Nothostele Garay Orchidaceae

Origins:
From the Greek *nothos* "false" and *stele* "a pillar, column."

Nothotaxus Florin Taxaceae

Origins:
From the Greek *nothos* "bastard, false" plus *Taxus* L.

Nothotsuga H.H. Hu ex C.N. Page Pinaceae

Origins:
From the Greek *nothos* "bastard, false" with the genus *Tsuga*.

Notiophrys Lindley Orchidaceae

Origins:
Greek *notios* "southern" plus the genus *Ophrys*.

Notiosciadium Speg. Umbelliferae

Origins:
Greek *notios* "southern" plus *skiadion, skiadeion* "umbel, parasol," a species from Argentina.

Notobasis (Cass.) Cass. Asteraceae

Origins:
Greek *notis* "moisture," *notios, notos, noton* "moist, damp, rainy" and *basis* "base, pedestal," referring to the habitat.

Notobuxus Oliver Buxaceae

Origins:
From the Greek *notos* "south, back" and the genus *Buxus* L.

Notocactus (K. Schumann) Fric Cactaceae

Origins:
From the Greek *notos* "south, back" plus *Cactus*, Latin *notus, auster*.

Notocentrum Naudin Melastomataceae

Origins:
From the Greek *notos* "south, back" and *kentron* "a spur, point," connective spurred.

Notoceras R. Br. Brassicaceae

Origins:
Greek *notos* "back" and *keras* "a horn," referring to the fruits.

Notochaete Benth. Labiatae

Origins:
Greek *notos* "back" and *chaite* "bristle, long hair," stems stellate or hirsute, lips villous outside, nutlets glabrous or stellate at the apex.

Species/Vernacular Names:
N. hamosa Bentham
English: hookedsepal
China: gou e cao
N. longiaristata C.Y. Wu & H.W. Li
English: longspiny hookedsepal
China: chang ci gou e cao

Notochloe Domin Gramineae

Origins:
From the Greek *notos* "south" and *chloe, chloa* "grass," a southern grass; see Karel Domin, *Repertorium specierum novarum regni vegetabilis*. 1911.

Notodanthonia Zotov Gramineae

Origins:
From the Greek *notos* "south" and the genus *Danthonia* DC.; see Victor Dmitrievich Zotov (1908-1977), in *New Zealand Journal of Botany*. 1: 104, 122. 1963.

Notodon Urban Fabaceae

Origins:
From the Greek *notos* "south, back" and *odous, odontos* "tooth."

Notodontia Pierre ex Pitard Rubiaceae

Origins:
From the Greek *notos* "south, back" and *odous, odontos* "tooth."

Notogramme C. Presl Pteridaceae (Adiantaceae)

Origins:
From the Greek *notos* "south, back" and *gramma* "letter, thread, marking."

Notonema Raf. Gramineae

Origins:
Greek *notos* "south, back" and *nema* "thread"; see C.S. Rafinesque, *Neogenyton, or Indication of Sixty-Six New Genera of Plants of North America*. 4. 1825; E.D. Merrill, *Index rafinesquianus*. 76. 1949.

Notonerium Benth. Apocynaceae

Origins:
From the Greek *notos* "south" and the genus *Nerium* L., in reference to the resemblance of the leaves to those of the oleander.

Notonia DC. Asteraceae

Origins:
Greek *notos, noton* "the south, the southwest wind, the south wind, back," or after Benjamin Noton, plant collector in India; see Ray Desmond, *Dictionary of British & Irish Botanists and Horticulturists*. 522. London 1994.

Notoniopsis B. Nord. Asteraceae

Origins:
Resembling the genus *Notonia* DC.

Notophaena Miers Rhamnaceae

Origins:
From the Greek *notos, noton* "the south-west wind, the south wind, south" and *phaino* "to shine, to appear."

Notopleura (Hook.f.) Bremek. Rubiaceae

Origins:
Greek *notos, noton* "back, south" and *pleura, pleuron* "side, rib, lateral."

Notopora Hook.f. Ericaceae

Origins:
From the Greek *notos, noton* and *poros* "opening, pore."

Notoptera Urban Asteraceae

Origins:

From the Greek *notos* "back" and *pteron* "a wing."

Notopterygium Boissieu Umbelliferae

Origins:

From the Greek *notos* "back" and *pterygion* "a small wing."

Notosceptrum Benth. Asphodelaceae (Aloaceae, Liliaceae)

Origins:

From the Greek *notos* "back" and *skeptron* "a sceptre, royal staff."

Notoseris C. Shih Asteraceae

Origins:

From the Greek *notos* "back, south" and *seris* "a species of chicory or a kind of endive."

Notospartium Hook.f. Fabaceae

Origins:

From the Greek *notos* "southern, south" and *Spartium* L., *spartion* "a small cord, broom," growing in New Zealand.

Notothixos Oliver Viscaceae

Origins:

Greek *notos, noton* "the south, the southwest wind, the south wind, back" (cf. Latin *notus* and *notos, i* "the south wind, auster, wind"; Akkadian *nadûm* "to pour," *nataku* "to drip"; Armenian *nay*) and *ixia, ixos* "mistletoe, bird-lime," some suggest from *thixis* "touching, a touch"; see Daniel Oliver (1830-1916), in *Journal of the Linnean Society. Botany*. 7: 92, 103. 1863.

Species/Vernacular Names:
N. cornifolius Oliver
English: tiny-flowered mistletoe
N. incanus (Hook.) Oliver (*Viscum incanum* Hook.)
English: downy mistletoe
N. subaureus Oliver (*Viscum incanum* Hook. var. *aureum* Ettingsh.)

English: golden mistletoe

Notothlaspi Hook.f. Brassicaceae

Origins:

From the Greek *notos, noton* "the south, southern" plus the genus *Thlaspi* L., from New Zealand.

Species/Vernacular Names:
N. rosulatum Hook.f.
English: penwiper plant

Nototriche Turcz. Malvaceae

Origins:

From the Greek *notos, noton* "back" and *thrix, trichos* "hair."

Species/Vernacular Names:
N. sp.
Peru: shiric-yalckoy, tupucpa-ashuan

Nototrichium (A. Gray) W.F. Hillebrand Amaranthaceae

Origins:

From the Greek *notos, noton* "the south, back, southern" and *thrix, trichos* "hair."

Notoxylinon Lewton Malvaceae

Origins:

From the Greek *notos, noton* "the south, southern" and *xylinos* "of wood, wooden," *xylon* "wood," woody plants of the Southern Hemisphere; see Frederick Lewis Lewton (1874-1959), in *Journal of the Washington Academy of Sciences*. 1915.

Notylia Lindley Orchidaceae

Origins:

Derivation from the Greek *noton* "back" and *tylos, tyle* "lump, any swelling, hump," referring to the position of the pollen and cap on the column, an allusion to the callosity on the stigma.

Nouletia Endlicher Bignoniaceae

Origins:

After the French botanist Jean Baptiste Noulet, 1802-1890, anthropologist, M.D. Toulouse 1826. Among his writings are *Essai sur l'histoire littéraire des patois du midi de la France au XVIe et XVIIe siècles*. Paris 1859, *Essai sur l'histoire littéraire des patois du midi de la France au VXIIIe siècle*. Paris 1877, *Flore du bassin sous-pyrénéen*. Toulouse 1837, *Las nonpareilhas Receptas*. Montpellier 1880 and *Flore analytique de Toulouse et de ses environs*. Toulouse 1855. See John H. Barnhart, *Biographical Notes upon Botanists*. 3: 15. 1965; E.M. Tucker, *Catalogue of the Library of the Arnold Arboretum of Harvard University*. 1917-1933.

Novosieversia F. Bolle Rosaceae

Origins:

Dedicated to the German pharmacist Johann August Carl Sievers, d. 1795, botanist, explorer, 1790-1795 traveled in southern Siberia and Mongolia, author of *J. Sievers ... Briefe aus Sibirien an seine Lehrer ... Herrn Brande, ... Herrn Ehrhart, und ... Herrn Westrumb*. St. Petersburg 1796; see Stafleu and Cowan, *Taxonomic Literature*. 5: 596. [for Johann Erasmus Sievers] Utrecht 1985; Jonas C. Dryander, *Catalogus bibliothecae historico-naturalis Josephi Banks*. 3: 274. London 1800; A. Lasègue, *Musée botanique de Benjamin Delessert*. Paris 1845; Emil Bretschneider, *History of European Botanical Discoveries in China*. Leipzig 1981. According to other authors the genus was named after the German explorer Wilhelm Sievers, 1860-1921, geographer, professor of geography at the Universities of Würzburg and Giessen, traveler. His works include *Die Cordillere von Mérida*. Wien und Olmütz. 1888, *Zweite Reise in Venezuela in den Jahren 1892-1893*. [Hamburg 1896] and *Reise in Peru und Ecuador*. München 1914; see F. Oliver Brachfeld, *Sievers en Mérida*. [Universidad de los Andes. Direccion de Cultura. Publicaciones. no. 15.] Mérida 1951; José Toribio Medina, *La imprenta en Mérida de Yucatán*. Amsterdam 1964; Ida Kaplan Langman, *A Selected Guide to the Literature on the Flowering Plants of Mexico*. University of Pennsylvania Press, Philadelphia 1964.

Nowickea J. Martínez & J.A. McDonald Phytolaccaceae

Origins:

Named for Joan W. Nowicke, b. 1938, B.A. (1958) Washington University, M.A. (1962) University of Missouri, Ph.D. (1968) Washington University, now at the Department of Botany, Smithsonian Institution (Staff of the United States National Herbarium) and National Museum of Natural History, Smithsonian Institution, Curator Research Training Program, research specialties: palynology, palynological analysis of families and orders, structure, function and pattern analysis of exines, adaptive significance of exine structure, pollen morphology and higher order systematics. Among his works are "Palynotaxonomic study of the Phytolaccaceae." *Ann. Missouri Bot. Gard.* 55(3): 294-363. 1968 [1969], "Apocynaceae." *Fl. Panama*. in *Ann. Missouri Bot. Gard.* 57(1): 59-130. 1970, "Pollen morphology and exine structure." in *Caryophyllales, Evolution and Systematics*. Eds. H.-D. Behnke and T. Mabry. Berlin 1994, "Rhamnaceae." *Fl. Panama*. in *Ann. Missouri Bot. Gard.* 58(3): 267-283. 1971 and "A palynological study of Crotonoideae (Euphorbiaceae)." in *Annals of the Missouri Botanical Garden*. 81: 245-269. 1994, with John J. Skvarla wrote "Pollen morphology." in Engler and Prantl, *Die Natürlichen Pflanzenfamilien*, Angiospermae: Ordung Ranunculales, Fam. Ranunculaceae. P. Hiepko, ed. Band 17 a IV. 129-159. Duncker and Humblot, Berlin 1995, with Thomas D. Seeley and Matthew Meselson wrote "Yellow rain." in *Scientific American*. 253: 128-137. (September) 1985, with P. S. Ashton, J. P. Robinson, T. D. Seeley and Matthew Meselson wrote "Comparison of yellow rain and bee excrement." presented at the Annual Meeting of the American Association for the Advancement of Science, Detroit 31 May 1983, with Matthew Meselson wrote "Yellow rain: A palynological analysis." in *Nature*. 309: 205-206. 17 May 1984.

Nowodworskya Presl Gramineae

Origins:

After Johann Nowodworsky, d. 1811, professor of botany.

Nubigena Raf. Fabaceae

Origins:

Referring to *Spartium nubigenum* Ait. and *Cytisus nubigenus* Link; see C.S. Rafinesque, *Sylva Telluriana*. 23. 1838; E.D. Merrill, *Index rafinesquianus*. The plant names published by C.S. Rafinesque, etc. 147. 1949.

Nucularia Batt. Chenopodiaceae

Origins:

From the Latin *nucula* "a small nut," *nux, nucis* "a nut."

Nudilus Raf. Oleaceae

Origins:
See C.S. Rafinesque, *Atl. Jour.* 1: 176. 1833, *Herb. Raf.* 20. 1833, *New Fl. N. Am.* 1: 62. 1836 and *The Good Book.* 46. Philadelphia 1840; E.D. Merrill, *Index rafinesquianus.* The plant names published by C.S. Rafinesque, etc. 190. 1949.

Numisaureum Raf. Linaceae

Origins:
See C.S. Rafinesque, *Flora Telluriana.* 3: 32. 1836 [1837]; E.D. Merrill, *Index rafinesquianus.* The plant names published by C.S. Rafinesque, etc. 150. 1949.

Nunnezharia Ruíz & Pav. Palmae

Origins:
See Hipólito Ruíz López and José Antonio Pavón, *Flora peruvianae, et chilensis prodromus.* 147. Madrid 1794.

Nunnezia Willd. Palmae

Origins:
See Willdenow, *Species plantarum.* 4: 890, 1154. 1806.

Nuphar Sm. Nymphaeaceae

Origins:
An Arabic or Persian name; see Auguste Adolphe Lucien Trécul (1818-1896), *Recherches sur la structure et le développement du Nuphar lutea.* Paris 1845; H. Genaust, *Etymologisches Wörterbuch der botanischen Pflanzennamen.* 425. 1996.

Species/Vernacular Names:
N. japonicum DC.
China: ping peng tsao, shui su (= water millet), shui li tzu
N. luteum (L.) Sm.
English: brandy bottle, yellow waterlily
N. pumilum (Timm) DC.
English: yellow pond lily
China: ping peng cao zi

Nutalla Raf. Loasaceae

Origins:
For Thomas Nuttall, 1786-1859; see C.S. Rafinesque, *Am. Monthly Mag. Crit. Rev.* 2: 266. 1818.

Nuttalia Raf. Loasaceae

Origins:
For Thomas Nuttall, 1786-1859; see C.S. Rafinesque, *Flora Telluriana.* 1: 89. 1836 [1837].

Nuttalla Raf. Loasaceae

Origins:
For Thomas Nuttall, 1786-1859; see C.S. Rafinesque, *Am. Monthly Mag. Crit. Rev.* 2: 175-176. 1818.

Nuttallanthus D.A. Sutton Scrophulariaceae

Origins:
For Thomas Nuttall, 1786-1859 (d. Lancashire), eminent English (b. Yorkshire) botanist, explorer, traveler, botanical and plant collector, printer, 1813 Fellow of the Linnean Society, naturalist, ornithologist, celebrated author of *The Genera of North American Plants*, and catalogue of the species, to the year 1817. Philadelphia 1818. See Jeannette Elizabeth Graustein, *Thomas Nuttall, Naturalist. Explorations in America, 1808-1841.* Cambridge, Harvard University Press 1967; Phillip Drennon Thomas, in *Dictionary of Scientific Biography* 10: 163-165. 1981; Stafleu and Cowan, *Taxonomic Literature.* 3: 781-787. 1981; Ray Desmond, *Dictionary of British & Irish Botanists and Horticulturists.* 523. London 1994; John H. Barnhart, *Biographical Notes upon Botanists.* 3: 17. 1965; William Jay Youmans, ed., *Pioneers of Science in America.* New York 1896; T.W. Bossert, *Biographical Dictionary of Botanists Represented in the Hunt Institute Portrait Collection.* 290. 1972; Joseph Ewan, *Rocky Mountain Naturalists.* The University of Denver Press 1950; J. Ewan, ed., *A Short History of Botany in the United States.* New York and London 1969; S. Lenley et al., *Catalog of the Manuscript and Archival Collections and Index to the Correspondence of John Torrey.* Library of the New York Botanical Garden. 465. 1973; Ethelyn Maria Tucker, *Catalogue of the Library of the Arnold Arboretum of Harvard University.* Cambridge, Massachusetts 1917-1933; J.W. Harshberger, *The Botanists of Philadelphia and Their Work.* 1899; R. Zander, F. Encke, G. Buchheim and S. Seybold, *Handwörterbuch der Pflanzennamen.* 14. Aufl. Stuttgart 1993; Ida Kaplan Langman, *A Selected Guide to the Literature on the Flowering Plants of Mexico.* University of Pennsylvania Press, Philadelphia 1964; A. Lasègue, *Musée botanique de Benjamin Delessert.* 1845; G. Murray, *History of the Collections Contained in the Natural History Departments of the British Museum.* London 1904; William Darlington, *Reliquiae Baldwinianae.* Philadelphia 1843; I.C. Hedge and J.M. Lamond, *Index of Collectors in the Edinburgh Herbarium.* Edinburgh 1970;

E.D. Merrill, *Bernice P. Bishop Mus. Bull.* 144: 143-144.
1937 and *Contr. U.S. Natl. Herb.* 30(1): 229. 1947; J.D.
Milner, *Catalogue of Portraits of Botanists Exhibited in the
Museums of the Royal Botanic Gardens*. Royal Botanic
Gardens, Kew, London 1906.

Nuttallia Raf. Loasaceae

Origins:

For Thomas Nuttall, 1786-1859; see C.S. Rafinesque, *Am.
Monthly Mag. Crit. Rev.* 1: 358. 1817.

Nuttallia Torrey & A. Gray Rosaceae

Origins:

For Thomas Nuttall, 1786-1859, English botanist, explorer,
traveler in USA, plant collector.

Nuxia Comm. ex Lam. Buddlejaceae (Loganiaceae)

Origins:

The genus was named after a French botanist on La Réunion
Island, M. de la Nux.

Species/Vernacular Names:

N. congesta R. Br. ex Fresen. (*Lachnopylis congesta* (R.
Br. ex Fresen.) C.A. Sm.; *Lachnopylis ternifolia* Hochst.;
Lachnopylis saxatilis C.A. Sm.; *Lachnopylis viscidulosa*
C.A. Sm.; *Lachnopylis heterotricha* C.A. Sm.; *Lachnopylis
speciosa* C.A. Sm.; *Lachnopylis schistotricha* C.A. Sm.;
Lachnopylis montana C.A. Sm.; *Lachnopylis emarginata*
(Sond.) C.A. Sm.; *Lachnopylis pubescens* (Sond.) C.A.
Sm.; *Lachnopylis breviflora* (S. Moore) C.A. Sm.; *Lach-
nopylis tomentosa* (Sond.) C.A. Sm.; *Nuxia congesta* var.
brevifolia Sond.; *Nuxia congesta* var. *tomentosa* (Sond.)
Cummins; *Nuxia congesta* var. *emarginata* (Sond.) Prain;
Nuxia emarginata Sond.; *Nuxia pubescens* Sond.; *Nuxia
tomentosa* Sond.; *Nuxia breviflora* S. Moore; *Nuxia dentata*
var. *transvaalensis* S. Moore; *Nuxia viscosa* Gibbs)

English: common wild elder, brittlewood, bogwood

Cameroon: evoun

Southern Africa: gewone wildevlier, bergsalie, broshout,
witblomsalie; mohatantswe, mokwerekwere (Ngwaketse
dialect, Botswana); umKhobeza, isiPhofane (Zulu);
umKhobeza (Xhosa)

N. floribunda Benth. (*Lachnopylis floribunda* (Benth.) C.A.
Sm.)

English: forest elder, white elder, kite tree, wild peach

Southern Africa: bosvlier, vlier, wildevlier; motlhabare
(North Sotho); umHlambandlazi (= mousebird washer),
umDlambandlaze, iThambo, inGobese, umSunuwembuzi,
umGwaqu, umKhobeza, umKhombeza (Zulu); iNgqota,
isiKhali, isiKali (Xhosa); mula-notshi, mpupumwa (Venda)

N. glomerulata (C.A. Sm.) Verdoorn (*Lachnopylis glomerulata* C.A. Sm.; *Lachnopylis suaveolens* C.A. Sm.)

South Africa: rock nuxia

N. oppositifolia (Hochst.) Benth. (*Lachnopylis oppositifolia*
Hochst.; *Nuxia dentata* R. Br.; *Nuxia schlechteri* Gilg;
Nuxia antunesii Gilg)

English: water elder, bushveld nuxia

Southern Africa: watervlier; iNkhweza (Zulu)

Nuytsia R. Br. ex G. Don f. Loranthaceae

Origins:

After the Dutch explorer Pieter Nuyts (Nuijts), member of
the Council of Dutch Indies, explored almost 1,000 miles
of the southern coast of Australia in 1626-27. On
26.01.1627 the Dutch East India Company vessel *Guilden
Zeepaard* (= Golden Sea-horse), captained by Francois
Thyssen, conveyed Pieter Nuyts as far east as Fowler's Bay
in the Great Australian Bight, in the neighborhood of the
Head of the Great Australian Bight; on his arrival in Batavia,
he was sent as ambassador to Japan and afterwards was
made Governor of Formosa; for more than one century the
region from Cape Leewin to Nuyts Archipelago was known
as *Nuyts Land*. See Matthew Flinders (1774-1814), *A Voyage to Terra Australis*. London 1814; C. Halls, "The voyage
of the *Gulden Zeepaard*." *J. R. Geog. Soc. Aust.* 72: 19-32.
1971; George Collingridge [de Tourcey], *The Discovery of
Australia. A Critical, Documentary and Historic Investiga-
tion Concerning the Priority of Discovery in Australia by
Europeans before the arrival of Lieut. James Cook*, in the
"Endeavour," in the year 1770. [The original date of publication was 1895. The facsimile edition 1983]; Robert
Brown, in *Journal of the Royal Geographical Society of
London*. 1: 17. 1831; John Landwehr, *VOC: A Bibliography
of Publications Relating to the Dutch East India Company,
1602-1800*. Ed. Peter van der Krogt. HES Publisher, Utrecht
1991.

Species/Vernacular Names:

N. floribunda (Labill.) R. Br. ex G. Don f. (*Loranthus
floribundus* Labill.)

English: Western Australia Christmas tree, mistletoe tree,
Swan River blaze tree, fire tree, flame tree, Christmas tree

Australia: mudja

Nyctaginia Choisy Nyctaginaceae

Origins:

Greek *nyx, nyktos* "night," Latin *nyctago, nyctaginis*, the flowers open at night; see A. Bonavilla, *Dizionario etimologico di tutti i vocaboli usati nelle scienze, arti e mestieri, che traggono origine dal greco.* Milano 1819-1821.

Nyctanthes L. Oleaceae (Verbenaceae)

Origins:

From the Greek *nyx, nyktos* "night" and *anthos* "flower," the flowers open at night and fall off at the break of the day, at dawn.

Species/Vernacular Names:

N. arbor-tristis L.

English: tree of sadness, night jasmine, coral jasmine, night-blooming jasmine, musk flower

India: manjapumaram, manja pumeram,, mannappumaram, parijata, harsinghar, sephalika, pavizamalli, pavalamallikai, siharu

Malaya: seri gading

China: nai hua, hung mo li

Nycteranthus Rothm. Aizoaceae

Origins:

From the Greek *nyx, nyktos* "night," *nykteros* "nightly" and *anthos* "flower."

Nycterianthemum Haw. Aizoaceae

Origins:

From the Greek *nykteros* "nightly" and *anthemon* "flower."

Nycterinia D. Don Scrophulariaceae

Origins:

From the Greek *nykteris* "bat, a night bird," *nyx, nyktos* "night," *nykteros* "nightly."

Nycterisition Ruíz & Pavón Sapotaceae

Origins:

From the Greek *nykteris* "bat, a night bird" and *sition* "food," food for bats.

Nycterium Vent. Solanaceae

Origins:

From the Greek *nykteros, nykterinos* "nightly."

Nycticalanthus Ducke Rutaceae

Origins:

From the Greek *nyx, nyktos* "night," *nykteros* "nightly," *kallos* "beauty" and *anthos* "flower."

Nyctocalos Teijsm. & Binnend. Bignoniaceae

Origins:

From the Greek *nyx, nyktos* "night" and *kallos* "beauty," *kalos, kalli* "beautiful," the flowers of some species open at night; see F.A.W. Miquel, in *Journal de Botanique Néerlandaise.* Fasc. 1. Amsterdam, Utrecht, Leipzig, Paris, Londres 1861.

Nyctocereus (A. Berger) Britton & Rose Cactaceae

Origins:

Greek *nyx, nyktos* plus the genus *Cereus*; see Nathaniel Lord Britton (1859-1934) and Joseph Nelson Rose (1862-1928), in *Contributions from the United States National Herbarium.* 12: 423. (July) 1909.

Nyctosma Raf. Orchidaceae

Origins:

Greek *nyx, nyktos* "night" and *osme* "smell, odor, perfume," the nocturnal fragrance; see C.S. Rafinesque, *Flora Telluriana.* 2: 9. 1836 [1837].

Nylandtia Dumort. Polygalaceae

Origins:

For the Dutch botanist Petrus Nylandt (Peter Nyland), physician. His works include *Den Ervaren Huys-Houder.* Amsterdam 1669 (in *Het vermakelyck landt-leven.* Amsterdam 1669, joint author the 17th century Dutch gardener Jan van der Groen), *De Nederlandtse Herbarius of Kruydt-Boeck,* etc. Amsterdam 1670 [the German translation: *Neues Medicinalisches Kräuterbuch.* Osnabrück 1678], with Jan van Hextor wrote *Het Schouw-toneel der aertsche schepselen,* etc. Amsterdam 1672 and *Schauplatz irdischer*

Geschöpfe ... in unser Hochteutsche Sprache übergesetzet. Osnabrück 1678.

Species/Vernacular Names:

N. spinosa (L.) Dumort. (*Mundia spinosa* (L.) DC.)

South Africa: buazifiber, duinebessie, bokbessie, skilpadbessie

Nymania K. Schum. Euphorbiaceae

Origins:

For the Swedish botanist Carl Fredrik Nyman, 1820-1893; see John H. Barnhart, *Biographical Notes upon Botanists.* 3: 18. 1965.

Nymania Lindberg Meliaceae

Origins:

Probably after the Swedish botanist Carl Fredrik Nyman, 1820-1893, from 1855 to 1889 Curator of the Botanical Dept. of Naturhistorika Riksmuseum, traveled in Sicily. Among his publications are *Sylloge florae Europeae...* Oerebroae 1854-1855, *Praktisk handbok i botanik.* Stockholm 1858 and *Conspectus florae europeae...* Örebro Sueciae 1878-1890, with C.G. Theodor Kotschy (1813-1866) contributed to Heinrich Wilhelm Schott (1794-1865), *Analecta botanica.* Scripta a H. Schott, adjutoribus C.F. Nyman et T. Kotschy. Vindobonae [Wien] 1854. See John H. Barnhart, *Biographical Notes upon Botanists.* 3: 18. 1965; Miguel Colmeiro y Penido, *La Botánica y los Botánicos de la Peninsula Hispano-Lusitana.* Madrid 1858; G. Murray, *History of the Collections Contained in the Natural History Departments of the British Museum.* 1: 171. London 1904; I.C. Hedge and J.M. Lamond, *Index of Collectors in the Edinburgh Herbarium.* Edinburgh 1970; T.W. Bossert, *Biographical Dictionary of Botanists Represented in the Hunt Institute Portrait Collection.* 290. 1972; S. Lenley et al., *Catalog of the Manuscript and Archival Collections and Index to the Correspondence of John Torrey.* Library of the New York Botanical Garden. 1973; H.N. Clokie, *Account of the Herbaria of the Department of Botany in the University of Oxford.* 218. Oxford 1964; Ethelyn Maria Tucker, *Catalogue of the Library of the Arnold Arboretum of Harvard University.* Cambridge, Massachusetts 1917-1933; R. Zander, F. Encke, G. Buchheim and S. Seybold, *Handwörterbuch der Pflanzennamen.* 14. Aufl. Stuttgart 1993. Latin *nyma, ae* used by Plinius for a plant otherwise unknown.

Species/Vernacular Names:

N. capensis (Thunb.) Lindb. (*Aitonia capensis* Thunb. nom. illegit.; *Aitonia capensis* var. *microphylla* Schinz)

English: Chinese lantern

South Africa: klapper, klapperbos, brosdoring

Nymphaea L. Nymphaeaceae

Origins:

Greek *nymphaia* "goddess of springs, water nymph," Latin *nymphaea* "water lily," Akkadian *nib'u* "growth," *naba'u* "to spring," *namba'u* "spring," Hebrew *nub* "to bud, to sprout, to grow, to thrive"; see Carl Linnaeus, *Species Plantarum.* 510. 1753 and *Genera Plantarum.* Ed. 5. 227. 1754; [Crusca], *Vocabolario degli Accademici della Crusca.* Firenze 1691, 1729-1738; N. Tommaseo & B. Bellini, *Dizionario della lingua italiana.* Torino 1865-1879; G. Volpi, "Le falsificazioni di Francesco Redi nel Vocabolario della Crusca." in *Atti della R. Accademia della Crusca per la lingua d'Italia.* 1915-1916; Giovanni Semerano, *Le origini della cultura europea.* Dizionari Etimologici. Basi semitiche delle lingue indeuropee. Dizionario della lingua Greca. 2(1): 198. Leo S. Olschki Editore, Firenze 1994; Ernest Weekley, *An Etymological Dictionary of Modern English.* 2: 998. Dover Publications, New York 1967.

Species/Vernacular Names:

N. spp.

Mexico: exta ga benne, quie benne, quie penne

N. alba L.

English: European waterlily

N. atrans S.W.L. Jacobs

English: waterlily

N. capensis Thunb.

English: blue waterlily, Cape blue waterlily, lotus lily, waterlily

South Africa: bloublom, blouwaterblom, blouwaterlelie, kaaimanblom, paddapreekstoel, waterlelie

N. elleniae S.W.L. Jacobs

English: waterlily

N. gigantea Hook.

English: Australian waterlily, blue waterlily, giant waterlily

N. hastifolia Domin

English: waterlily

N. immutabilis S.W.L. Jacobs

English: waterlily

N. lotus L.

English: Egyptian lily, Egyptian waterlily, lotus, sacred lotus, waterlily, white waterlily, white lily

South Africa: lotus lily, waterlelie

Nigeria: osipata, bado

Yoruba: osibata

N. macrosperma Merr. & Perry

English: waterlily

N. mexicana Zucc.

English: yellow waterlily, Florida waterlily, banana waterlily

N. minima F.M. Bailey

English: waterlily

N. nouchali Burm.f.

The Philippines: lawas, lauasa, pulau, tunas, talailo

N. nouchali Burm.f. var. *caerulea* (Savigny) Verdc. (*Nymphaea caerulea* Sav.; *Nymphaea calliantha* Conard; *Nymphaea capensis* Thunb. var. *alba* K. Landon; *Nymphaea capensis* Thunb. var. *capensis*; *Nymphaea mildbraedii* Gilg; *Nymphaea nelsonii* Burtt Davy; *Nymphaea spectabilis* Gilg; *Nymphaea stellata* sensu Harv. non Willd. sensu stricto; *Nymphaea stellata* sensu Oliv. p.p. non Willd. sensu stricto)

English: blue lotus, blue waterlily, Egyptian lotus, waterlily

South Africa: blue waterlily, waterlelie, blouwaterlelie, blouwaterblom, paddapreekstoel

N. odorata Aiton

English: fragrant waterlily, pond lily, white waterlily

N. pubescens Willd.

English: waterlily

N. rubra Roxb.

English: Indian red waterlily

Japan: aka-bana-hitsuji-gusa (= red flowered *Nymphaea tetragona*)

Okinawa: suirin

N. tetragona Georgi

English: pigmy waterlily

Japan: hitsuji-gusa

China: shui lien

N. violacea Lam.

English: waterlily

Nymphanthus Lour. Euphorbiaceae

Origins:

Latin *nymphaea* "water lily," Greek *nymphaia* "water nymph" and *anthos* "flower."

Nymphoides Hill Menyanthaceae

Origins:

Resembling the genus *Nymphaea*. See John Hill (1716-1775), *The British Herbal*. London 1756; Arthur D. Chapman, ed., *Australian Plant Name Index*. 2098-2099. Canberra 1991.

Nymphoides Séguier Menyanthaceae (Gentianaceae)

Origins:

Resembling the genus *Nymphaea*.

Species/Vernacular Names:

N. aquatica (Walter) Kuntze (*Nymphoides lacunosa* (Vent.) O. Kuntze; *Limnanthemum aquaticum* (Walt.) Britt.)

English: banana plant, fairy water lily

N. aurantiaca (Dalzell) Kuntze

English: orange fringe, marshwort

N. beaglensis Aston (from Beagle Bay Mission, Dampierland Peninsula, Kimberley Region, Western Australia)

English: marshwort

N. crenata (F. Muell.) Kuntze (*Limnanthemum crenatum* F. Muell.)

English: wavy marshwort, yellow fringe

N. cristata (Roxburgh) Kuntze (*Menyanthes cristata* Roxb.)

English: cristate floating heart

China: shui pi lian

N. disperma Aston

English: marshwort

N. elliptica Aston

English: marshwort

N. exigua (F. Muell.) Kuntze (*Limnanthemum exiguum* F. Muell.)

English: small marshwort, marshwort

N. exiliflora (F. Muell.) Kuntze

English: marshwort

N. furculifolia Specht

English: marshwort

N. geminata (R. Br.) Kuntze (*Limnanthemum geminatum* (R. Br.) Griseb.; *Villarsia geminata* R. Br)

English: large marshwort, star fringe, entire marshwort

N. indica (L.) Kuntze (*Limnanthemum indicum* (L.) Thwaites)

English: water snowflake, Indian marshwort, fringed water lily, Indian floating heart, white fringe

China: jin yin lian hua

N. indica (L.) Kuntze subsp. *occidentalis* A. Raynal

South Africa: geelwateruintjie, drywende hart

English: floating heart, water snowflake, yellow pond lily

N. peltata (S.G. Gmelin) Kuntze (*Nymphoides nymphae-oides* Britt.; *Menyanthes nymphoides* L.; *Limnanthemum nymphoides* (L.) Hoffm. & Link; *Limnanthemum peltatum* S.G. Gmel.)

English: yellow floating heart, water fringe, shield floating heart

China: xing cai, hui tiao, hui tiao tsai, ching so tien

N. thunbergiana (Griseb.) Kuntze

Southern Africa: marombodane (Venda)

Nypa Steck Palmae

Origins:

From Malayan or Moluccan word *nipah*.

Species/Vernacular Names:

N. fruticans Wurmb (*Nipa fruticans* Thunberg)

English: water coconut, mangrove palm, nipa palm, nypa palm, water palm

Brazil: palmeira ripa, palmeira do mangue

Indonesia: nipah

Sri Lanka: gin-pol

Japan: nippa-yashi

The Philippines: nipa, sasa, saga, lasa, pawid, anipa, tata, pinok, pinog

Nyssa L. Cornaceae (Nyssaceae)

Origins:

Greek *nysso* "to prick, to pierce," Nyssa or Nysa was the name of one of the water nymphs, some species love moist or swampy habitat; Dionysus was the Greek god of wine and revelry, he was the son of Zeus and the youngest of the twelve Olympians, the nymphs of Nysa raised him, and were later rewarded by being changed into a constellation; Nysa was the name of several mountains sacred to Dionysus; in the Homeric "Hymn to Demeter," the story is told of how Persephone was gathering flowers in the Vale of Nysa when she was seized by Hades and removed to the underworld.

Species/Vernacular Names:

N. aquatica L.

English: cotton gum

N. ogeche Bartram ex Marshall

English: ogeechee lime

N. sinensis Oliver

English: Chinese tupelo

N. sylvatica Marshall

English: black gum, cotton gum, pepperidge

Nyssanthes R. Br. Amaranthaceae

Origins:

Greek *nysso* "to prick, to pierce" and *anthos* "a flower," referring to the prickly flowers; see Robert Brown, *Prodromus florae Novae Hollandiae et Insulae van-Diemen*. 418. London 1810.

Species/Vernacular Names:

N. diffusa R. Br.

English: spreading nyssanthes

N. erecta R. Br.

English: erect nyssanthes

Oakes-amesia C. Schweinfurth & P.H. Allen Orchidaceae

Origins:

For the American botanist Oakes Ames, 1874-1950, orchidologist, plant collector, botanical explorer, Fellow of the Linnean Society, in 1895 began his long association with Harvard University, in 1941 donated his orchid herbarium to Harvard University, collaborated with Eduardo Quisumbing (1895-1986). His writings include *Orchidaceae*. New York, Boston 1905-1922 and *Schedulae Orchideanae*. Boston, Massachusetts 1922-1930, with Donovan Stewart Correll (1908-1983) wrote *Orchids of Guatemala*. Chicago 1952-1953. See J.H. Barnhart, *Biographical Notes upon Botanists*. 1: 50. 1965; Frans A. Stafleu and Erik A. Mennega, *Taxonomic Literature. Supplement I: A-Ba*. 98-102. 1992; Donovan S. Correll, *Supplement to Orchids of Guatemala, and British Honduras*. Chicago 1965; T.W. Bossert, *Biographical Dictionary of Botanists Represented in the Hunt Institute Portrait Collection*. 10. 1972; S. Lenley et al., *Catalog of the Manuscript and Archival Collections and Index to the Correspondence of John Torrey*. Library of the New York Botanical Garden. 6. 1973; R. Zander, F. Encke, G. Buchheim and S. Seybold, *Handwörterbuch der Pflanzennamen*. 14. Aufl. Stuttgart 1993; Ida Kaplan Langman, *A Selected Guide to the Literature on the Flowering Plants of Mexico*. 85-86. University of Pennsylvania Press, Philadelphia 1964.

Oakesiella Small Colchicaceae (Uvulariaceae, Liliaceae)

Origins:

For William Oakes, 1799-1848, American botanist, lawyer, naturalist, wrote *Catalogue of Vermont Plants*. [Burlington 1842]; see John H. Barnhart, *Biographical Notes upon Botanists*. 3: 18. 1965; R. Zander, F. Encke, G. Buchheim and S. Seybold, *Handwörterbuch der Pflanzennamen*. 14. Aufl. Stuttgart 1993; T.W. Bossert, *Biographical Dictionary of Botanists Represented in the Hunt Institute Portrait Collection*. 291. 1972; H.N. Clokie, *Account of the Herbaria of the Department of Botany in the University of Oxford*. 218. Oxford 1964; S. Lenley et al., *Catalog of the Manuscript and Archival Collections and Index to the Correspondence of John Torrey*. Library of the New York Botanical Garden. 465. 1973; Ethelyn Maria Tucker, *Catalogue of the Library of the Arnold Arboretum of Harvard University*. Cambridge, Massachusetts 1917-1933; J. Ewan, ed., *A Short History of Botany in the United States*. 92, 115. New York and London 1969; Jeannette Elizabeth Graustein, *Thomas Nuttall, Naturalist. Explorations in America, 1808-1841*. 1967.

Oaxacania B.L. Robinson & Greenman Asteraceae

Origins:

From Oaxaca, a city and a state in Mexico. See Martin Diskin and Scott Cook, eds., *Mercados de Oaxaca*. Inst. Nacional Indigenista. Mexico 1975; John K. Chance, *Race and Class in Colonial Oaxaca*. Stanford 1978; José Toribio Medina, *La imprenta en Oaxaca*. Amsterdam 1964.

Obbea Hook.f. Rubiaceae

Origins:

Anagram of *Bobea* Gaudich.

Oberonia Lindley Orchidaceae

Origins:

Named after Oberon, the mythical King of the Fairies, husband of Titania, in reference to the variable forms; see John Lindley (1799-1865), *The Genera and Species of Orchidaceous Plants*. 15. London 1830-1840.

Species/Vernacular Names:

O. japonica (Maxim.) Makino (*Malaxis japonica* Maxim.; *Oberonia makinoi* Masam.)

Japan: yôraku-ran

O. titania (Endl.) Lindl.

English: soldier's crest orchid

Obesia Haw. Asclepiadaceae

Origins:

From the Latin *obesus* "fat, succulent," *obedo, edi, esum* "to eat away, devour," *edo, edi, esum* "to eat."

Obetia Gaudich. Urticaceae

Origins:

The derivation of the generic name is not clear; see Charles Gaudichaud-Beaupré (1789-1854), [Botany of the Voyage.] *Voyage autour du Monde ... sur ... l'Uranie et la Physicienne, pendant ... 1817-1820*. Paris 1826 [-1830].

Species/Vernacular Names:

O. carruthersiana (Hiern) Rendle (the specific name after the British botanist William Carruthers, 1830-1922 (d. Surrey), a former Keeper of the Department of Botany of the British Museum, paleobiologist, 1861 Fellow of the Linnean Society, 1871 Fellow of the Royal Society, President of the Linnean Society, author of "Lebanon plants." in Sir Richard Francis Burton (1821-1890) and Charles Frederick Tyrwhitt Drake (1846-1874), *Unexplored Syria*, etc. Vol. II, Appendix III. London 1872; see John H. Barnhart, *Biographical Notes upon Botanists*. 1: 316. 1965; Joseph Ewan, *Rocky Mountain Naturalists*. The University of Denver Press 1950)

South Africa: stinging nettle tree

O. ficifolia (Poiret) Gaudich.

Rodrigues Island: figue marron

O. tenax (N.E. Br.) Friis (*Urera tenax* N.E. Br.)

English: mountain nettle, stinging nettle tree, giant nettle, tree nettle

South Africa: bergbrandnetel, bergbrandnekel, brandneukel, brandnetel, uluzi; bogozimbe, iMpongozembe, umDadi-omkhulu, imBati, imBadi enkulu, uluZi (Zulu); mmabi, lebabi, lebadi (North Sotho); mmabi (Tswana: western Transvaal, northern Cape, Botswana); umBabazane (Xhosa); muvhazwi (= stinging) (Venda)

Obione Gaertner Chenopodiaceae

Origins:

Probably after the river Ob; see Helmut Genaust, *Etymologisches Wörterbuch der botanischen Pflanzennamen*. 427. Basel 1996.

Oblivia Strother Asteraceae

Origins:

From the Latin *oblivio, onis* "oblivion, forgetfulness," some suggest an anagram of Bolivia.

Obolaria L. Gentianaceae

Origins:

From the Greek *obolos* "obol," Latin *obolus* "a small Greek coin, an obol," referring to the shape of the fruits.

Obregonia Fric Cactaceae

Origins:

For Álvaro Obregon, 1880-1928 (assassinated), President of Mexico 1920-1924.

Ocampoa A. Rich. & Galeotti Orchidaceae

Origins:

For the Mexican botanist Melchor Ocampo, 1810-1861, a patriot, author of *Esposicion que el C.M. Ocampo dirigió al Exmo. Sr. Presidente de la República Lic. D.B. Juarez sobre las circulares que llevan el nombre del mismo Ocampo*. México 1861; see Jesús Romero Flores, *Melchor Ocampo, el filósofo de la Reforma*. México 1944.

Oceanopapaver Guillaumin Capparidaceae (Capparaceae)

Origins:

Greek *Okeanos*, Latin *Oceanus* "the ocean" and *papaver, eris* "poppy, a kernel, seed," one species growing in New Caledonia.

Ochagavia Philippi Bromeliaceae

Origins:

For Sylvestre Ochagavia, between 1853 and 1854 Minister of Education in Chile; see R.A. Philippi, "Observaciones sobre la flora de Juan Fernández." in *Anales Univ. Chile*. 13: 157-169. (May) 1856 and "Bemerkungen über die Flora der Insel Juan Fernandez." in *Bot. Zeitung*. 14: 625-636, 641-650. 1856.

Ochanostachys Mast. Olacaceae

Origins:

From the Greek *ochanon* "holder of a shield, a bar or band fastened crosswise" and *stachys* "a spike."

Species/Vernacular Names:

O. amentacea Mast.

Malaya: petaling, petaling bukit, petaling misu, mentatai, petikal

Ochetocarpus Meyen Loasaceae

Origins:

From the Greek *ochetos* "water-pipe, streams, channel" and *karpos* "fruit."

Ochetophila Poepp. ex Reissek. Rhamnaceae

Origins:

Greek *ochetos* "water-pipe, streams, currents" and *philos* "loving."

Ochlandra Thwaites Gramineae

Origins:

From the Greek *ochlos* "crowd, mass" and *aner, andros* "man, stamen, male."

Ochlogramma C. Presl Dryopteridaceae (Aspleniaceae, Woodsiaceae)

Origins:

From the Greek *ochlos* "crowd, mass" and *gramma* "letter, thread, marking."

Ochna L. Ochnaceae

Origins:

Greek *ochne* "wild pear, a pear tree, a pear," Homer used this name for the wild pear tree, see also Theophrastus (*HP.* 2.5.6); see Carl Linnaeus, *Species Plantarum.* 513. 1753 and *Genera Plantarum.* Ed. 5. 229. 1754.

Species/Vernacular Names:

O. arborea Burch. ex DC. var. *arborea* (*Ochna arborea* Burch. ex DC.)

English: Cape plane, Transvaal plane, African boxwood, redwood, Cape redwood

Southern Africa: Kaapse rooihout, Transvaalrooihout, rooihout, pleinhout; umBovane, umThelelo, umShelele, umBomvane, umHlazane (Zulu); umThentsema, umTensema, inTansema, umVithi (Xhosa)

O. arborea Burch. ex DC. var. *oconnorii* (Phill.) Du Toit (*Ochna oconnorii* Phill.) (the name of the variety after A.J. O'Connor, Director of Forestry in South Africa)

English: Transvaal boxwood, African boxwood

Southern Africa: rooihout, rooiysterhout, morelle; murambo, murambo-thavha (Venda)

O. barbosae N.K.B. Robson

English: sand plane

South Africa: sandrooihout

O. calodendron Gilg & Mildbr.

Congo: mbane noir

Cameroon: mulébengoye, tonso

Central Africa: tobolassongué

O. glauca Verdoorn

Southern Africa: mubovu, chinyanu, chinyanyu, musanku, muwezebanga

O. holstii Engl. (the specific name honors the German gardener and plant collector Carl H.E.W. Holst, 1865-1894, traveler in East Africa)

English: red ironwood, real red pear

E. Africa: m'muaga, mtakula, takula

Southern Africa: rooiysterhout, regterrooipeer; isiBhanku (Zulu); tshipfure (Venda)

O. inermis (Forssk.) Schweinf.

South Africa: unarmed ochna

O. leptoclada Oliv. (*Ochnella leptoclada* (Oliv.) Tiegh.; *Ochna debeerstii* De Wild.; *Ochnella debeerstii* (De Wild.) Tiegh.)

Southern Africa: tswatswa (Shona)

O. membranacea Oliv.

Nigeria: iso gbedu (Yoruba; see Daryll Forde, *The Yoruba speaking peoples of south-western Nigeria.* London 1951)

O. natalitia (Meisn.) Walp. (*Ochna atropurpurea* sensu Harv. var. *natalitia* (Meisn.) Harv.; *Ochna chilversii* Phill.; *Diporidium natalitium* Meisn.)

English: Natal plane, Mickey Mouse bush, coast boxwood, coast redwood, showy ochna

Southern Africa: Natalrooihout, coast rooihout, rodehout; umBovu, umBhovane-ongcingci, umBhovane, umBovane, isiThundu, umNandi, umShelele, iSendengulube, ummilamatsheni, uMilamatsheni (= grow among rocks) (Zulu); isiBomvu, umTensema (Xhosa)

O. pulchra Hook. (*Ochna rehmannii* Szyszyl.; *Ochna aschersoniana* Schinz; *Ochna pulchra* forma *integra* Suess.; *Ochna fuscescens* Heine; *Polythecium pulchrum* (Hook.) Tiegh.; *Polythecium rehmannii* (Szyszyl.) Tiegh.)

English: beautiful ochna, peeling-bark ochna, peeling plane, wild pear, wild plum

N. Rhodesia: kachale, musang'u

Southern Africa: lekkerbreek (this name refers to the brittleness of the branches), skilferbas, pypsteel, pypsteelhout, morsaf, mansaf, Barnardsgif, seerbas, seerbos, seermak (probably from siermaak = bringing good cheer), slegbreek, vervelbas; nzololo (Tsonga); monyelenyele (Tswana: western Transvaal, northern Cape, Botswana); monamane, mopha (North Sotho); musuma, tshithothonya (Venda); mozwe (Mbukushu: Okavango Swamps and western Caprivi); ujue (Samui); omu (Ovambo); muChedza, muChoa, muMinu, muNino, muNinu, muNzeremanga, muParamoswa, muParamota, murezeremanga, muSonzoa, muSwaswari (Shona)

O. schweinfurthiana F. Hoffm. (*Ochna cyanophylla* N.K.B. Robson)

English: brick-red ochna

French: ochna jaune

W. Africa: manakeni, manakenyi

Southern Africa: muminu, mupalamosa, muParamoswa, muRampambari, nSiwi (Shona)

O. serrulata (Hochst.) Walp. (*Ochna atropurpurea* sensu Harv. non DC.; *Diporidium serrulata* Hochst.)

English: small-leaved plane, Mickey Mouse plant, bird's eye bush, carnival bush

Southern Africa: fynblaarrooihout, rooihout; umBovu (Zulu); iLitiye (= stone) (Xhosa)

Ochoterenaea F.A. Barkley Anacardiaceae

Origins:

Named for the Mexican botanist Isaac Ochoterena, 1885-1950, he is best remembered for *Las cactáceas de México*. México 1922 and "Las regiones geográfico-botánicas de México." *Rev. Esc. Nac. Preparatoria*. 1: 261-331. 1923. See John H. Barnhart, *Biographical Notes upon Botanists*. 3: 20. 1965; Julio Riquelme Inda, "Los naturalistas desaparecidos de 1936-1961." *Rev. Soc. Mex. Hist. Nat*. 22: 242-276. 1961; Fred Alexander Barkley (1908-1989), "A key to the genera of the Anacardiaceae." *Amer. Midl. Naturalist*. 28(2): 465-474. 1942 and "A new genus of the Anacardiaceae from Colombia." *Bull. Torrey Bot. Club*. 69(6): 442-444. 1942; George Neville Jones, *An Annotated Bibliography of Mexican Ferns*. University of Illinois Press 1966; Irving William Knobloch, compil., "A preliminary verified list of plant collectors in Mexico." *Phytologia Memoirs*. VI. 1983; Ida Kaplan Langman, *A Selected Guide to the Literature on the Flowering Plants of Mexico*. Philadelphia 1964.

Ochradenus Delile Resedaceae

Origins:

Greek *ochros, ochra* "pale yellow, yellow ochre" and *aden* "gland."

Ochreinauclea Ridsdale & Bakh.f. Rubiaceae

Origins:

From the Greek *ochros* "pale yellow, pale" plus the genus *Nauclea* L.

Ochrocarpos Thouars Guttiferae

Origins:

From the Greek *ochros* "pale yellow, pale, wan, paleness" and *karpos* "fruit," referring to the color of the fruits.

Ochrocephala Dittrich Asteraceae

Origins:

From the Greek *ochros* "pale yellow, pale, wan" and *kephale* "head."

Ochrolasia Turcz. Dilleniaceae

Origins:

Greek *ochros, ochra* and *lasios* "shaggy, woolly, velvety"; see Porphir Kiril N.S. Turczaninow (1796-1863), *Bulletin de la Société Impériale des Naturalistes de Moscou*. 22(2): 3. 1849.

Ochroluma Baillon Sapotaceae

Origins:

Greek *ochros* "pale yellow, pale" plus *luma*.

Ochroma Sw. Bombacaceae

Origins:

From the Greek *ochros* "pale yellow," referring to the color of the flowers.

Species/Vernacular Names:

O. sp.

Mexico: diyum (Oaxaca)

O. pyramidale (Lam.) Urban (*Ochroma lagopus* Swartz; *Bombax pyramidale* Cav. ex Lam.; *Ochroma boliviana* Rowlee; *Ochroma peruviana* I.M. Johnston)

English: balsa, down tree, cork tree

Peru: apiwa, balsa, chinchipá, huambu caspi, huambuna, huambuna caspi, huampo, jellma, uampu, kina-kina, mapalo, musso, musso-jihui, musu, muxu, palo de balsa, pañaañuro, pumacuchu, shintipa, topa, wawa

Mexico: ha-ma, mo-hó, pepe balsa, arbol del algodón, balsa, corcho, jonote real, jopi, jubiguy, mo-ma-ah, pata de liebre, pomoy, pomay

Ochronerium Baillon Apocynaceae

Origins:

From the Greek *ochros* "pale yellow" and the genus *Nerium* L.

Ochropteris J. Sm. Pteridaceae (Adiantaceae)

Origins:

From the Greek *ochros* "pale yellow" and *pteris* "fern."

Ochrosia A.L. Juss. Apocynaceae

Origins:

From the Greek *ochros, ochra* "pale yellow, yellow ochre," referring to the flowers or to the fruit color or to the yellow color of the timber; see A.L. de Jussieu, *Genera Plantarum.* pl. 144. 1789.

Species/Vernacular Names:

O. borbonica J.F. Gmelin

English: Bourbon ochrosia

China: mei gui shu

O. elliptica Labill.

English: elliptical ochrosia, mangrove ochrosia

China: gu cheng mei gui shu

O. moorei (F. Muell.) F. Muell. ex Benth. (for Charles Moore, Government botanist, NSW, Australia)

English: southern ochrosia

O. oppositifolia (Lam.) K. Schumann (*Cerbera oppositifolia* Lam.; *Lactaria iwasakiana* Koidz.; *Ochrosia iwasakiana* (Koidz.) Koidz.)

English: twin-apple

Japan: shima-sokei, yama-fukum

Ochrosperma Trudgen Myrtaceae

Origins:

From the Greek *ochros* "pale yellow" and *sperma* "a seed"; see Malcolm Eric Trudgen "*Ochrosperma*, a new genus of Myrtaceae (Leptospermeae, Baeckeinae) from New South Wales and Queensland)." in *Nuytsia.* 6(1): 9-17. 1987.

Species/Vernacular Names:

O. lineare (C.T. White) Trudgen

English: straggly baeckea

Ochrothallus Pierre ex Baillon Sapotaceae

Origins:

From the Greek *ochros* "pale yellow" and *thallos* "blossom, branch."

Ochthephilus Wurdack Melastomataceae

Origins:

From the Greek *ochthos* "hill, tubercle, bank," *ochthe* "any height ground, raised banks" and *philos* "loving."

Ochthocharis Blume Melastomataceae

Origins:

From the Greek *ochthos* "hill, tubercle, bank" and *charis* "grace, beauty."

Ochthochloa Edgew. Gramineae

Origins:

From the Greek *ochthos* "hill, tubercle, bank" and *chloe, chloa* "grass."

Ochthocosmus Benth. Ixonanthaceae

Origins:

Greek *ochthos* "hill, tubercle, bank" and *kosmos* "ornament, decoration."

Species/Vernacular Names:

O. africanus Hook.f.

Gabon: alane-afane

Ivory Coast: abrahassa

Sierra Leone: twanyé

Ochthodium DC. Brassicaceae

Origins:

From the Greek *ochthos* "hill, tubercle, bank," *ochthodes* "hilly, tuberous."

Ocimastrum Rupr. Onagraceae

Origins:

Resembling *okimon*, Latin *ocimum* "basil."

Ocimum L. Labiatae

Origins:

From the ancient Greek name *okimon* used by Theophrastus (*HP*. 1.6.6) and Dioscorides and Galenus for an aromatic herb, basil, Latin *ocimum* "basil" (Plinius), *ocinum* also *ocimum*, *ocymum* and *ozymum* for an herb which serves for fodder, perhaps a sort of clover (Plinius); see Carl Linnaeus, *Species Plantarum*. 597. 1753 and *Genera Plantarum*. Ed. 5. 259. 1754; Yuhanna ibn Sarabiyun [Joannes Serapion], *Liber aggregatus in medicinis simplicibus*. Venetijs 1479; M. Cortelazzo, *L'influsso linguistico greco a Venezia*. Bologna 1970; Serapiom, *El libro agregà de Serapiom*. A cura di G. Ineichen. Venezia-Roma 1962-1966; V. Bertoldi, in *Archivio glottologico italiano*. XXI: 140-142. Torino and Firenze 1927; Franco Montanari, *Vocabolario della Lingua Greca*. 2284. Loescher Editore, Torino 1995.

Species/Vernacular Names:

O. sp.

South Laos: kloyh (people Nya Hön)

Peru: albahacal, allbahaca, pichana alabaca, sanisisa alabaca

Brazil: manjericão, alfavaca-de-vaqueiro

Yoruba: akeroro

O. americanum L. (*Ocimum canum* Sims; *Ocimum africanum* Loureiro; *Ocimum dinteri* Briq.; *Ocimum fruticulosum* Burchell; *Ocimum simile* N.E. Br.)

English: American basil, common basil, sweet basil, hoary basil

Peru: garawa, shara mashan, shara mashu

China: hui luo he

India: tulasi, thulasi, kala tulshi, kala tulsi, tulsi, kuppatulasi, kukka-tulasi, nai tulasi, nayitulasi, kattu ram tulasi, kattu tulasi, thiksnamanu, gunjamkorai, bharbhari, gramya, ajaka, gambhira, kuthera, mamri, runhmui

Brazil: esturaque, alfavaca-do-campo, alfavaca-campestre, alfavaca-de-cheiro, remédio-dos-vaqueiros, segurelha Santa Maria

Southern Africa: kinuka (Swahili); manhuwe (Shona); mniaywatwane (South Sotho)

Nigeria: efinrin

Yoruba: eruyanntefe, efinrin otu, eye obale efinrin, efinrin wewe

O. basilicum L.

English: common basil, sweet basil, basil, holy basil, monk's basil, sacred basil, lemon basil

Arabic: rehan, h'baq, habaq, rayhan, hamahim

Italian: basilico (Latin *basilicum*, Greek *basilikos okimon*)

Peru: albahaca, ashra mashán, basil, pichana blanca, salvaca, nooro, moro, wurolo

Guatemala: albahaca, albahaca cimarrona, basen, cacaltun

Mexico: albaca, albacarón, albahaca, guiestia, albacar

Brazil: manjericão, alfavaca

India: niazbo, niyazbo, panr, tulsi, Krishna tulsi, karpura tulasi, bisva tulasi, dhala-tulasi, bhu-tulasibabui-tulsi, bhutulasi, kukkatulasi, ram-tulasi, gulal tulsi, kali tulsi, sabzah, sabza, sabajhi, sajjagida, hazbo, tiru nitru, tirunitri, tirunirupachai, tirnirupachai, tirnut-patchi, shashasfaram, pharanjamuskh, firanj-mushk, furrunj-mushk, bharbari, baburi, babui, varvara, varavara, marua, marva, manjariki, munjariki, vebudipatri, vepudupachha, karandai, kam kasturi, kama kasturi, kala pingain, surasa, damaro, nasabo, rudrajada, kapur kanti

Malaya: ruku, selaseh antan

The Philippines: ruku-ruku, balanoi, valanoi, bouak, samilig, samirig, solasi, albanaka, bauing, bidai, kalu-ui, kamañgi

Japan: me-bôki

China: luo le, lo le, hsiang tsai, ai kang, i tzu tsao

Hawaii: ki 'a'ala, ki paoa

Yoruba: efinrin ata, efinrin wewe, efinrin aya, efinrin marugbosanyan, aruntantan

O. basilicum L. var. *pilosum* (Willdenow) Bentham

English: sweet basil

China: shu rou mao bian zhong

O. gratissimum L.

Yoruba: efinrin nla, efinrin oso, efinrin ogaja, efinfin nla, efinrin, amowokuro aye, woromoba

Congo: dumaduma

Vietnam: huong nhu trang

O. gratissimum L. var. *suave* (Willdenow) Hook.f.

English: sweet-scented basil

China: wu mao ding xiang luo le

O. micranthum Willdenow

Peru: abaca, albaca silvestre, alvaca silvestre, iroro, fweroro, pichana alvaca, pichana blanca, salvaca, shara mashari, shara mashan, vii roro

Mexico: albahaca, albahaca de monte, kakaltum, xkakaltum, guie, guie huece, quije huece, guia belaga, quije pelaga, guie'stia, quie-nacuana, quije nocuana, guia nocuana, nocuana

O. tashiroi Hayata

English: Taiwan basil

China: tai wan luo le

O. tenuiflorum L. (*Ocimum sanctum* L.)

English: holy basil, Thai basil, sacred basil, mosquito plant of South Africa

French: basilic sacré

The Philippines: solasi, balanoi, albahaca, colocogo, camange, bidai, kamangkau, loko-loko

China: sheng luo le

Vietnam: e do, e tia, huong nhu tia

India: Vishnu-priya, Krishnamul, tulasi, tulasa, tulashi, tulsi, thulasi, tulasi chajadha, Krishna-tulasi, Krishna-tulsi, Krushna tulasi, Shiva-tulasi, suvasa tulasi, Vishnu tulasi, kala-tulasi, kala tulsi, kari-tulasi, nalla tulasi, sri tulasi, tunrusi, divya, bharati, baranda, brinda, brynda, jiyal, jiuli, choiharr, gaggera, gaggerachettu, oddhi, gumpina, maduru-tulla, lun, ajaka, manjari, parnasa, patrapuspha, trittavu

Ocotea Aublet Lauraceae

Origins:
Based on the native name in French Guiana, the Garipons called *Ocotea guianensis* Aublet *aiou-hou-ha*.

Species/Vernacular Names:
O. sp.

Peru: itahuba, negrillo, carunje, roble corriente, yono

O. aciphylla (Nees) Mez

Peru: muena negra, roble blanco

O. architectorum Mez

Peru: roble blanco

O. aurantiodora (Ruíz & Pav.) Mez

Peru: negrito

O. bullata (Burchell) Baillon (*Laurus bullata* Burch.)

English: stinkwood, blank stinkwood, Cape laurel, laurel wood, Cape olive, African acorn

Southern Africa: swartstinkhout, laurelhout, stinkhout, knobbed ocotea, bubbled ocotea; umHlungulu, umNukani, umNukane (Xhosa); umNukani, umNukane (Zulu)

O. cernua (Nees) Mez (*Oreodaphne cernua* Nees)

Mexico: aguacatillo, laurel, laurel amarillo, laurel de bajo

O. cymbarum Kunth

Brazil: inamui

O. dielsiana O. Schmidt

Peru: muena, moena, jojaavuhxo, moena blanca

O. guianensis Aublet

Peru: macaa, quinilla amarilla

O. jelskii Mez

Peru: ishpingo

O. kenyensis (Chiov.) Robyns (*Ocotea viridis* Kosterm.; *Ocotea gardneri* Hutch. & M.M. Moss; *Tylostemon kenyensis* Chiov.)

English: Transvaal stinkwood, bastard stinkwood

Southern Africa: basterstinkhout; umNukani (Xhosa); motshega, modula-tshwene (north and northeast Transvaal)

O. licanioides A.C. Smith

Peru: muena negra, moena negra

O. longifolia Kunth (*Ocotea grandifolia* (Nees) Mez; *Oreodaphne grandifolia* Nees)

Peru: muena blanca, maraco-fuina, sipra muena, maraco fuima, moena blanca, sipra moena

O. pauciflora (Nees) Mez (*Ocotea laxiflora* (Meissner) Mez; *Oreodaphne pauciflora* Nees; *Mespilodaphne laxiflora* Meissner)

Peru: canela muena, canela moena, muena

O. pretiosa (Nees) Mez (*Mespilodaphne pretiosa* Nees; *Laurus odorifera* Vell.; *Aydendron suaveolens* Nees)

Brazil: canela sassafrás, canela funcho, canela parda, sassafrás, sassafrás amarelo, sassafrás preto, sassafrás rajado, sassafrazinho

O. rubra Mez

Peru: determa, louro vermelho, wana

O. rubrinervis Mez

Peru: muena blanca, yúrac moena, yurac muena

O. sassafras Mez

Brazil: anhu-iba-pea-bia, couro-sassafrás, canela-sassafrás, canela-funcho, pão-funcho, sassafrás-amarelo, sassafrás-de-cantagalo, sassafrás-do-brasil, sassafrás-do-pará

O. tarapotana (Meissner) Mez (*Oreodaphne tarapotana* Meissner)

Peru: canela, muena aguarás, turpentina muena, canela muena, muena aguarrás

O. trianae Rusby

Peru: muena blanca, pampa muena

O. usambarensis Engl.

English: East Africa camphorwood

Octadesmia Benth. Orchidaceae

Origins:
From the Greek *okto* "eight" and *desmos* "a bond, band, bundle," referring to the pollinia, to the eight pollen masses.

Octamyrtus Diels Myrtaceae

Origins:
Probably from the Greek *okto* "eight" plus *Myrtus*.

Octandrorchis Brieger Orchidaceae

Origins:
From the Greek *okto* "eight," *aner, andros* "man, stamen" plus *Orchis*, see also *Octomeria* R. Br.

Octarrhena Thwaites Orchidaceae

Origins:
From the Greek *okto* "eight" and *arrhen* "male, stamen," indicating the eight free pollinia.

Octoceras Bunge Brassicaceae

Origins:
From the Greek *okto* "eight" and *keras* "a horn."

Octoclinis F. Muell. Cupressaceae

Origins:
From the Greek *okto* "eight" and *kline* "a bed, couch, receptacle"; see F. von Mueller, *Transactions and Proceedings of the Philosophical Institute of Victoria.* 2: 21. Melbourne (Sep.) 1857 [from 1860 superseded by *Transactions and Proceedings of the Royal Society of Victoria.* Melbourne].

Octodon Thonn. Rubiaceae

Origins:
From the Greek *okto* "eight" and *odous, odontos* "tooth."

Octoknema Pierre Olacaceae

Origins:
From the Greek *okto* "eight" and *kneme* "limb, leg" or *nema* "thread," referring to the seeds.

Species/Vernacular Names:
O. affinis Pierre (*Octoknema winkleri* Engl.)
Nigeria: ofan ebu (Boki)
O. borealis Huch. & Dalz.

Ivory Coast: bagba

Octolepis Oliver Thymelaeaceae

Origins:
Greek *okto* "eight" and *lepis, lepidos* "scale," in reference to the petals.

Octolobus Welw. Sterculiaceae

Origins:
Greek *okto* "eight" and *lobos* "a pod," Latin *octo* "eight" and *lobus, i* "berry, pod."

Species/Vernacular Names:
O. spectabilis Welw. (*Octolobus angustatus* Hutch.)
Nigeria: tutugboro (Yoruba)

Octomeles Miq. Datiscaceae (Tetramelaceae)

Origins:
From the Greek *okto* "eight" and *melon* "an apple," flowers 5-8-merous, styles 5-8, capsules urceolate.

Species/Vernacular Names:
O. sumatrana Miq.
Malaya: minuang

Octomeria Pfeiff. Rubiaceae

Origins:
Greek *okto* "eight" and *meros* "part," see also *Otomeria* Benth., Greek *ous, otos* "an ear" and *meris* "part."

Octomeria R. Br. Orchidaceae

Origins:
Greek *okto* "eight" and *meros* "part," referring to the number of pollinia, to the pollen-bearing masses.

Octomeron Robyns Labiatae

Origins:
From the Greek *okto* "eight" and *meros, meris* "part, portion."

Octonum Raf. Melastomataceae

Origins:

Probably derived from *Melastoma octonum* Humb. & Bonpl., Latin *octoni, ae, a* "eight each, by eights," late Latin *octonus*; see Constantine Samuel Rafinesque (1783-1840), *Sylva Telluriana*. 95. 1838; E.D. Merrill (1876-1956), *Index rafinesquianus*. The plant names published by C.S. Rafinesque, etc. 176. Jamaica Plain, Massachusetts, USA 1949.

Octopleura Griseb. Melastomataceae

Origins:

From the Greek *okto* "eight" and *pleura, pleuron* "side, rib, lateral."

Octopoma N.E. Br. Aizoaceae

Origins:

From the Greek *okto* "eight" and *poma, pomatos* "a lid, cover," eight species, the capsules are 7- to 10-locular.

Octosomatium Gagnep. Boraginaceae

Origins:

From the Greek *okto* "eight" and *somation* "small body, corpse," *somateion* "a corporate body," *soma* "body."

Octospermum Airy Shaw Euphorbiaceae

Origins:

From the Greek *okto* "eight" and *sperma* "seed."

Octotheca R. Vig. Araliaceae

Origins:

From the Greek *okto* "eight" and *theke* "a box, case."

Octotropis Beddome Rubiaceae

Origins:

From the Greek *okto* "eight" and *tropis* "keel," referring to the ovary.

Ocyroe Philippi Asteraceae

Origins:

Ocyrrhoe, Ocyrhoe or Ocyroe was a daughter of Chiron by Chariclo, Greek *okys* "quick" and *rhoe, rhoa* "river, stream."

Odicardis Raf. Scrophulariaceae

Origins:

See C.S. Rafinesque, *Flora Telluriana*. 4: 55. 1836 [1838]; E.D. Merrill, *Index rafinesquianus*. 217. 1949.

Odisca Raf. Bignoniaceae

Origins:

See E.D. Merrill, *Index rafinesquianus*. 220. 1949; C.S. Rafinesque, *Sylva Telluriana*. 80. 1838.

Odixia Orchard Asteraceae

Origins:

An anagram of the genus *Ixodia* R. Br.; see Anthony Edward Orchard (1946-), in *Brunonia*. (Feb.) 1982.

Ododeca Raf. Lythraceae

Origins:

See Elmer D. Merrill, *Index rafinesquianus*. 173. 1949; C.S. Rafinesque, *Neogenyton, or Indication of Sixty-Six New Genera of Plants of North America*. 3. 1825.

Odollamia Raf. Apocynaceae

Origins:

Based on *Cerbera manghas* L. (*Cerbera odollam* Gaertner) and *Odollam* Adans.; see C.S. Rafinesque, *Sylva Telluriana*. 162. 1838.

Odonectis Raf. Orchidaceae

Origins:

Greek *odous, odontos* "tooth, anything pointed or sharp" and Latin *necto* "to bind, join," probably referring to the leaves; see E.D. Merrill, *Index rafinesquianus*. 103. 1949; C.S. Rafinesque, *Med. Repos*. II. 5: 357. 1808.

Odonellia K.R. Robertson Convolvulaceae

Origins:
Dedicated to the Argentine botanist Carlos Alberto O'Donell, 1912-1954. Among his writings are "Convolvuloideas chilenas." *Bol. Soc. Argent. Bot.* 6(3-4): 143-184. 1957, "Convolvuláceas Argentinas." *Lilloa.* 29: 87-364. 1959 and "Las especies de *Jacquemontia* de Perú." *Lilloa.* 30: 71-105. 1960, with Horacio Raul Descole wrote "El *Crinodendron tucumanum* Lillo y su relación con las especies chilenas del género." *Lilloa.* 2(2): 341-352. 1938, with H.R. Descole and Alicia Lourteig wrote "*Wendtia y Balbisia* en Argentina." *Lilloa.* 4(2): 197-216. 1939, "Revisión de las Zigofiláceas argentinas." *Lilloa.* 5: 257-352. 1940 and "Zygophyllaceae." in Descole, *Genera et species plantarum argentinarum.* 1: 1-46. Buenos Aires 1943, with A. Lourteig wrote "Acalypheae Argentinae (Euphorbiaceae)." *Lilloa.* 8: 273-333. 1942, "Euphorbiaceae." in Descole, *Genera et species plantarum argentinarum.* 1: 143-317. 1943 and "Las Celastráceas de Argentina y Chile." *Natura.* 1(2): 181-233. Buenos Aires 1955. See A. Lourteig, "Carlos Alberto O'Donell." *Revista Argent. Agron.* 21(2): 105-110. 1954; H.R. Descole, "Carlos Alberto O'Donell." *Bol. Soc. Argent. Bot.* 5(3): 160-165. 1954; John H. Barnhart, *Biographical Notes upon Botanists.* 3: 21. 1965; Ida Kaplan Langman, *A Selected Guide to the Literature on the Flowering Plants of Mexico.* Philadelphia 1964.

Odonostephana Alexander Asclepiadaceae

Origins:
From the Greek *odous, odontos* "tooth, anything pointed or sharp" and *stephanos* "a crown."

Odontadenia Benth. Apocynaceae

Origins:
Greek *odous, odontos* "tooth, anything pointed or sharp" and *aden* "gland," referring to the pistil.

Odontandra Willd. ex Roemer & Schultes Meliaceae

Origins:
Greek *odous, odontos* and *aner, andros* "male, man, stamen."

Odontanthera Wight Asclepiadaceae

Origins:
From the Greek *odous, odontos* "tooth" and *anthera* "anther."

Odonteilema Turcz. Euphorbiaceae

Origins:
From the Greek *odous, odontos* "tooth" and *eilema* "a veil, covering, involucre."

Odontella Tieghem Loranthaceae

Origins:
From the Greek *odous, odontos* "tooth," a diminutive.

Odontelytrum Hackel Gramineae

Origins:
From the Greek *odous, odontos* "tooth" and *elytron* "sheath, cover, scale, husk."

Odontitella Rothm. Scrophulariaceae

Origins:
The diminutive of *Odontites*.

Odontites Ludwig Scrophulariaceae

Origins:
Greek *odous, odontos* "tooth," Latin *odontitis* used by Plinius for a plant good for toothache, tooth-wort.

Species/Vernacular Names:
O. vernus (Bellardi) Dumort.
English: red bartsia

Odontocarya Miers Menispermaceae

Origins:
From the Greek *odous, odontos* "tooth" and *karyon* "nut."

Odontochilus Blume Orchidaceae

Origins:
Having a toothed lip, from the Greek *odous, odontos* "tooth" and *cheilos* "lip."

Odontocline B. Nord. Asteraceae

Origins:
From the Greek *odous, odontos* "tooth" and *kline* "a bed."

Odontocyclus Turcz. Brassicaceae

Origins:
From the Greek *odous, odontos* "tooth" and *kyklos* "a circle, ring."

Odontoglossum Kunth Orchidaceae

Origins:
From the Greek *odous, odontos* and *glossa* "a tongue," referring to the callus on the lip.

Odontoloma J. Sm. Dennstaedtiaceae

Origins:
From the Greek *odous, odontos* "tooth" and *loma* "border, margin, fringe, edge," referring to the achenes.

Odontomanes C. Presl Hymenophyllaceae

Origins:
From the Greek *odous, odontos* "tooth" and *manes* "a cup, a kind of cup."

Odontonema Nees Acanthaceae

Origins:
From the Greek *odous, odontos* "tooth" and *nema* "a thread, filament," referring to the corolla or to the stamens.

Odontonemella Lindau Acanthaceae

Origins:
The diminutive of the genus *Odontonema* Nees.

Odontophorus N.E. Br. Aizoaceae

Origins:
From the Greek *odous, odontos* "tooth" and *phoros* "bearing, carrying."

Odontophyllum Sreem. Acanthaceae

Origins:
From the Greek *odous, odontos* "tooth" and *phyllon* "leaf."

Odontopteris Bernh. Schizaeaceae

Origins:
From the Greek *odous, odontos* and *pteris* "a fern."

Odontorrhynchus M.N. Corrêa Orchidaceae

Origins:
From the Greek *odous, odontos* "tooth" and *rhynchos* "horn, beak," the shape of the rostellum.

Odontosiphon M. Roem. Meliaceae

Origins:
From the Greek *odous, odontos* "tooth" and *siphon* "tube."

Odontosoria Fée Dennstaedtiaceae (Lindsaeaceae, Lindsaeoideae)

Origins:
Greek *odous, odontos* and *sorus* "a spore case," leaves forming spiny thickets; see Antoine Laurent Apollinaire Fée (1789-1874), *Mémoires sur la famille des Fougères.* V. Genera Filicum. Strasbourg & Paris 1850-1852.

Species/Vernacular Names:
O. chinensis (L.) J. Sm. (*Sphenomeris chinensis* (L.) Maxon; *Trichomanes chinense* L.; *Adiantum chusanum* L.; *Sphenomeris chusana* (L.) Copel.)

Japan: hora-shinobu (= cave *Davallia*)

China: kwei li, wu chiu, shih hsu, shih i, shih tai, shih hua, shih ma tsung

O. clavata (L.) J. Smith (*Adiantum clavatum* L.; *Sphenomeris clavata* (L.) Maxon; *Stenoloma clavatum* (L.) Fée)

English: wedgelet fern

Odontospermum Necker ex Sch. Bip. Asteraceae

Origins:
Greek *odous, odontos* "tooth" and *sperma* "seed."

Odontostelma Rendle Asclepiadaceae

Origins:
From the Greek *odous, odontos* "tooth" and *stelma, stelmatos* (*stello* "to bring together, to bind, to set") "a girdle, belt."

Odontostemum Baker Tecophilaeaceae (Liliaceae)

Origins:

See *Odontostomum* Torr.

Odontostigma Zoll. & Moritzi Acanthaceae

Origins:

From the Greek *odous, odontos* "tooth" and *stigma* "stigma."

Odontostomum Torrey Tecophilaeaceae (Liliaceae)

Origins:

Greek *odous, odontos* "tooth" and *stoma* "mouth," alluding to the staminodes.

Odontostyles Breda Orchidaceae

Origins:

From the Greek *odous, odontos* "tooth" and *stylos* "pillar, style," the shape of the appendages of the column.

Odontotecoma Bureau & K. Schumann Bignoniaceae

Origins:

From the Greek *odous, odontos* "tooth" plus *Tecoma*.

Odontotrichum Zucc. Asteraceae

Origins:

From the Greek *odous, odontos* "tooth" and *thrix, trichos* "hair."

Odontychium K. Schumann Zingiberaceae

Origins:

Possibly from the Greek *odous, odontos* "tooth" and *chion* "snow," referring to the flowers.

Odosicyos Keraudren Cucurbitaceae

Origins:

Possibly from the Greek *odo-* "swollen" and *sikyos* "wild cucumber, gourd."

Odostelma Raf. Passifloraceae

Origins:

Possibly from the Greek *odo-* "swollen" and *stelma* "a girdle, belt"; see C.S. Rafinesque, *Flora Telluriana.* 4: 104. 1836 [1838].

Odostemon Raf. Berberidaceae

Origins:

Possibly from the Greek *odo-* "swollen" and *stemon* "a stamen, a thread"; see C.S. Rafinesque, *Am. Monthly Mag. Crit. Rev.* 2: 265. 1818, *Jour. Phys. Chim. Hist. Nat.* 89: 259. 1819 and *Sylva Telluriana.* 69. 18385; Elmer D. Merrill, *Index rafinesquianus.* 126. 1949.

Odotalon Raf. Euphorbiaceae

Origins:

See C.S. Rafinesque, *Sylva Telluriana.* 66, 67. 1838.

Odyendea Pierre ex Engl. Simaroubaceae

Origins:

An African vernacular name.

Species/Vernacular Names:

O. gabonensis (Pierre) Engl.

Cameroon: onzam, onzan, ozek, kokop, bwadendu, dondagueengo, kohope, njain

Gabon: onzan, nzan, onzeng, onzang, onzon, osendje, ozendje, ozenje, benzeng, bondjengi, odiendle, odieneje, odjenge, odyendie, odzense, dibindi, disengo, lebvola, moussiguiri, musigiri, musigiti, noka

Odyssea Stapf Gramineae

Origins:

The genus was named after Odysseus (Odusseus), in Greek legend, son of Laertes and Anticleia, husband of Penelope (daughter of Icarius), and father by her of Telemachus; in Latin Ulysses or Ulixes.

Species/Vernacular Names:

O. paucinervis (Nees) Stapf (*Diplachne cinerea* Hack.)

South Africa: brakkweek, steekgras

German: brackquecke

Oeceoclades Lindley Orchidaceae

Origins:
Greek *oikeios* "private, related, relative" and *klados* "a branch," possibly referring to a taxonomic separation from *Angraecum*.

Oecopetalum Greenman & C.H. Thompson Icacinaceae

Origins:
From the Greek *oikeios* "private, related," *oikos* "house, any dwelling-place" and *petalon* "leaf, petal."

Oedematopus Planchon & Triana Guttiferae

Origins:
From the Greek *oideo, oidao* "swell, become swollen," *oidema, oidematos* "swelling, tumor" and *pous* "foot."

Oedera L. Asteraceae

Origins:
After the German botanist George Christian Edler von Old-enburg Oeder, 1728-1791, physician, M.D. 1749 Göttingen, the leader of the so-called "Botanical Institutions," between 1754 and 1755 botanical journey in Europe, from 1754 to 1770 professor of botany at Copenhagen, ennobled in 1788. Among his works are *Elementa botanicae*. Hafniae [Copenhagen] 1764-1766, *Nomenclator botanicus*. Hafniae 1769 and *Enumeratio plantarum florae danicae*. Hafniae 1770, co-author and editor of *Flora Danica. Icones plantarum sponte nascentium in regnis Daniae et Norvegiae*, etc. Copenhagen [1761-] 1764-1883. See Gerhard Anton von Halem, *Andenken an Oeder*. Altona 1793; John H. Barnhart, *Biographical Notes upon Botanists*. 3: 21. 1965; R. Zander, F. Encke, G. Buchheim and S. Seybold, *Handwörterbuch der Pflanzennamen*. 14. Aufl. Stuttgart 1993; A. Lasègue, *Musée botanique de Benjamin Delessert*. 532. 1845; Jens Wilken Hornemann, *Naturh. Tidsskr*. 1: 568. 1837; Ethelyn Maria Tucker, *Catalogue of the Library of the Arnold Arboretum of Harvard University*. Cambridge, Massachusetts 1917-1933; T.W. Bossert, *Biographical Dictionary of Botanists Represented in the Hunt Institute Portrait Collection*. 291. 1972; Jonas C. Dryander, *Catalogus bibliothecae historico-naturalis Josephi Banks*. London 1800; C.F.A. Christensen, *Den danske Botaniks Historie med tilhørende Bibliografi*. Copenhagen 1924-1926.

Oedibasis Koso-Pol. Umbelliferae

Origins:
From the Greek *oideo, oidao* "swell, become swollen" and *basis* "base, pedestal."

Oedicephalus Nevski Fabaceae

Origins:
From the Greek *oideo, oidao* "swell, become swollen" and *kephale* "head."

Oedina Tieghem Loranthaceae

Origins:
Greek *oideo, oidao* "swell, become swollen," *oidos* "swelling," *oidaino* "swell."

Oedipachne Link Gramineae

Origins:
From the Greek *oidipous* "the swollen-footed," *oideo* and *pous*, and *achne* "chaff, glume."

Oehmea Buxb. Cactaceae

Origins:
For Hans Oehme, author of "*Thelocactus krainzianus* Oeh. spec. nov." *Beitr. Sukk*. 1: 1-3. 1940 and "Entwicklungstendenzen innerhalb der gattung *Thelocactus* B. and R." *Beitr. Sukk*. 2: 40-41. 1941; see Ida Kaplan Langman, *A Selected Guide to the Literature on the Flowering Plants of Mexico*. 549. University of Pennsylvania Press, Philadelphia 1964.

Oemleria Reichb. Rosaceae

Origins:
After Augustus Gottlieb Oemler, pharmacist and naturalist.

Species/Vernacular Names:
O. cerasiformis (Hook. & Arn.) J.W. Landon
English: Oso berry

Oenanthe L. Umbelliferae

Origins:
Greek *oinos, oenos* "wine" and *anthos* "flower," a plant smelling of wine; see Carl Linnaeus, *Species Plantarum*. 254. 1753 and *Genera Plantarum*. Ed. 5. 122. 1754.

Species/Vernacular Names:

O. aquatica (L.) Poiret

English: water fennel

O. crocata L.

English: water dropwort

O. divaricata (R. Br.) Mabb.

English: Madeira water dropwort

Portuguese: salsa brava

O. javanica (Blume) DC. (*Oenanthe japonica* Miq.; *Oenanthe stolonifera* (Roxb.) DC.; *Sium javanicum* Blume)

English: water dropwort

Japan: seri, shiiriba

China: shui qin, shui chin, ku chin, chin tsai, shui ying, chu kuei

O. pimpinelloides L.

English: water dropwort, corky-fruited water dropwort

Oenocarpus Martius Palmae

Origins:

Greek *oinos, oenos* "wine" and *karpos* "fruit," wine palm, fruit an oil source, mesocarp fleshy and oily, from the mesocarp, a creamy drink; see C. von Martius, *Historia Naturalis Palmarum*. 2: 21. Munich 1823; M.J. Balick, "Systematics and Economic Botany of the *Oenocarpus-Jessenia* (Palmae) Complex." *Advances in Economic Botany*. Volume 3. The New York Botanical Garden. 1986.

Species/Vernacular Names:

O. spp.

Peru: batauá, ciami patauá

O. bacaba C. Martius

Peru: bacaba, obaanco, milpesos, ungurany, bacabá, lu, manaka, turu, ungurahui, ungurauy

Brazil: bacaba, bacaba-açú, bacabão, bacaba verdadeira, bacaba de azeite, bacaba vermelha

O. bataua C. Martius (*Jessenia bataua* (C. Martius) Burret subsp. *bataua*; *Jessenia weberbaueri* Burret)

Peru: chocolatera, obaanjoro, shega, ingurabe, ungurauy, ungurahuy, sacumana, hungurahui, sinami, hizaan

Brazil: patauá

O. distichus C. Martius

Brazil: bacaba, bacaba de azeite, palmeira, yandy-bacaba, bacaba de leque, bacaba do pará, palmeira norte sul

O. mapora Karsten subsp. *mapora* (*Oenocarpus multicaulis* Spruce)

Peru: ciamba, ciana, siama, siamba, sinami, sinamillo, ciama

Brazil: bacabinha, bacaba, coqueiro bacaba

O. minor C. Martius

Peru: ciamba, manaqui, ciama

Brazil: bacabinha, bacaba-mirim, bacabí

Oenone Tul. Podostemaceae

Origins:

Oenone was a nymph of Mount Ida.

Oenosciadium Pomel Umbelliferae

Origins:

From the Greek *oinos, oenos* "wine" and *skiadion, skiadeion* "umbel, parasol."

Oenostachys Bullock Iridaceae

Origins:

From the Greek *oinos, oenos* "wine" and *stachys* "spike."

Oenothera L. Onagraceae

Origins:

Greek name *oinotheris, oinotheras* (*oinos* "wine" and *thera* "booty") "wine catcher," or from *onotheras* (*onos* "an ass," *thera* "hunting" or *ther* "wild beast, wild animal"); Latin *oenothera, ae* "a plant, whose juice may cause sleep" ("onothera, sive onear, hilaritatem afferens in vino", Plinius); see Carl Linnaeus, *Species Plantarum*. 346. 1753 and *Genera Plantarum*. Ed. 5. 163. 1754.

Species/Vernacular Names:

O. sp.

Peru: antañahui

O. biennis L.

English: common evening primrose, evening primrose, German rampion

South Africa: gewone nagblom, nagblom

O. caespitosa Nutt.

English: fragrant evening primrose

O. deltoides Torr. & Frém.

English: desert evening primrose, devil's lantern, lion-in-a-cage, basket evening primrose

O. elata subsp. *hookeri* (Torr. & Gray) W. Dietr. & W.L. Wagner

English: Hooker's evening primrose

O. fruticosa L.

English: sundrops

O. glazoviana M. Micheli (*Oenothera erythrosepala* Borbás)

English: large-flowered evening primrose

O. grandiflora Aiton

English: evening primrose, godetia

O. indecora Cambess. subsp. *indecora*

English: evening primrose

South Africa: nagblom

O. jamesii Torrey & Gray

English: giant evening primrose

O. laciniata Hill

English: cutleaf evening primrose, cut-leaved evening primrose

Southern Africa: aandblom, boer-in-die-nag; lesoma (Sotho)

O. macrocarpa Nutt.

English: Ozark sundrops

O. multicaulis Ruíz & Pav.

Peru: huailla-cajetilla, saya-saya, yahuar chchuncca

O. perennis L.

English: sundrops

O. pilosella Raf.

English: sundrops

O. rosea Aiton

English: evening primrose, rose evening primrose

Peru: chupa sangre, San Juan, yahuar chchunga, yahuar chonca, yahuar chchunca

Southern Africa: aandblom, rooskleurige nagblom

O. rubida Rusby

Peru: saya-saya

O. speciosa Nutt.

English: white evening primrose, rose evening primrose, Mexican evening primrose

Japan: hiru-zaki-tsuki-mi-sô

O. stricta Ledeb. ex Link [Ort. var., see *O. striata* Link]

English: evening primrose, common evening primrose, sweet-scented evening primrose

Southern Africa: aandblom, nagblom

Japan: matsuyoi-gusa

O. tetraptera Cav.

English: evening primrose, white evening primrose

Southern Africa: aandblom, witnagblom

Japan: tsuki-mi-sô (tsuki = moon)

O. verrucosa I.M. Johnston

Peru: choclillo

O. versicolor Lehmann (*Onagra fusca* Krause)

Peru: alto-yahuar-chchuncca, huailla-yahuar-chchuncca

Oenotheridium Reiche Onagraceae

Origins:
Referring to *Oenothera*.

Oenotrichia Copel. Dennstaedtiaceae

Origins:
From the Greek *oinos, oenos* "wine" and *thrix, trichos* "hair."

Species/Vernacular Names:
O. dissecta (C.T. White & Goy) S.B. Andrews

English: lace fern

O. tripinnata (F. Muell. ex Benth.) Copel.

English: hairy lace fern

Oeonia Lindley Orchidaceae

Origins:
Greek *oionos* "a bird of omen, a large bird, bird of prey, birds, presage," possibly in allusion to the flower, to the perianth parts.

Oeoniella Schltr. Orchidaceae

Origins:
The diminutive of the orchid genus *Oeonia*.

Oerstedella Reichb.f. Orchidaceae

Origins:
Dedicated to the Danish botanist Anders Sandoe Oersted (A. Sandøe Ørsted, A. Sandö Örsted), 1816-1872, zoologist, traveled in Central America, plant collector, 1851-1862 professor of botany at the University of Copenhagen, collected the holotype. Among his numerous publications are *Centralamerikas Rubiaceer.* [Copenhagen 1852] and *L'Amérique centrale.* Copenhague 1863. See J.H. Barnhart, *Biographical Notes upon Botanists.* 3: 22. 1965; Stafleu and

Cowan, *Taxonomic Literature*. 3: 808-812. 1981; R. Zander, F. Encke, G. Buchheim and S. Seybold, *Handwörterbuch der Pflanzennamen*. 14. Aufl. Stuttgart 1993; C.F.A. Christensen, *Den danske Botaniks Historie med tilhørende Bibliografi*. Copenhagen 1924-1926; G. Murray, *History of the Collections Contained in the Natural History Departments of the British Museum*. 1: 171. London 1904; M.N. Chaudhri, I.H. Vegter and C.M. De Wal, *Index Herbariorum*, Part II (3), *Collectors I-L*. Regnum Vegetabile vol. 86. 1972; T.W. Bossert, *Biographical Dictionary of Botanists Represented in the Hunt Institute Portrait Collection*. 292. 1972; Irving William Knobloch, compil., "A preliminary verified list of plant collectors in Mexico." *Phytologia Memoirs*. VI. 1983; E.M. Tucker, *Catalogue of the Library of the Arnold Arboretum of Harvard University*. 1917-1933; Ida Kaplan Langman, *A Selected Guide to the Literature on the Flowering Plants of Mexico*. Philadelphia 1964; Benjamin Daydon Jackson (1846-1927), "A list of the contributors to the herbarium of the Royal Botanic Gardens, Kew, brought down to 31st December 1899." *Bull. Misc. Inf. Kew*. 1901 and "A list of the collectors whose plants are in the herbarium of the Royal Botanic Gardens, Kew, to 31st December 1899." in *Kew Bulletin*. 1-80. 1901.

Oerstedianthus Lundell Myrsinaceae

Origins:

For the Danish botanist Anders Sandoe Oersted, 1816-1872, zoologist, traveler and plant collector, professor of botany at the University of Copenhagen, author of *Myrsinaceae centroamericanae et mexicanae*. [Copenhagen] 1862.

Oerstedina Wiehler Gesneriaceae

Origins:

For the Danish botanist Anders Sandoe Oersted (A. Sandøe Ørsted, A. Sandö Örsted), 1816-1872, zoologist, traveled in Central America, plant collector, professor of botany at the University of Copenhagen; see J.H. Barnhart, *Biographical Notes upon Botanists*. 3: 22. 1965.

Oeschynomene Raf. Fabaceae

Origins:

Referring to *Aeschynomene* L.; see C.S. Rafinesque, *Florula ludoviciana*. 137. New York 1817; E.D. Merrill, *Index rafinesquianus*. 147. 1949.

Oestlundorchis Szlach. Orchidaceae

Origins:

Karl Erik Magnus Östlund (Oestlund), orchid collector in Mexico; see Irving William Knobloch, compil., "A preliminary verified list of plant collectors in Mexico." *Phytologia Memoirs*. VI. 1983.

Ofaiston Raf. Chenopodiaceae

Origins:

See C.S. Rafinesque, *Flora Telluriana*. 3: 46. 1836 [1837]; E.D. Merrill, *Index rafinesquianus*. 118. 1949.

Ogastemma Brummitt Boraginaceae

Origins:

An anagram of *Megastoma*.

Ogcerostylus Cass. Asteraceae

Origins:

Greek *ogkeros, onkeros* "swollen" and *stylos* "a column, pillar, style," nom. illeg. superfluous, see *Siloxerus* Labill.

Ogcodeia Bureau Moraceae

Origins:

Greek *onkodes* "swelling, rounded, bulky."

Oianthus Benth. Asclepiadaceae

Origins:

Greek *oios* "alone" and *anthos* "flower."

Oiospermum Less. Asteraceae

Origins:

Possibly from the Greek *oios* "alone" and *sperma* "seed."

Oistanthera Markgr. Apocynaceae

Origins:

Greek *oistos* "arrow, endurable" and *anthera* "anther."

Oistonema Schltr. Asclepiadaceae

Origins:

Probably from the Greek *oistos* "arrow, endurable" and *nema* "a thread, filament."

Okenia Schlechtendal & Chamisso Nyctaginaceae

Origins:

For the German naturalist Lorenz Oken (Okenfuss, until 1802), 1779-1851 (Zurich, Switzerland), physician, M.D. Freiburg 1804, 1817-1842 founder and editor of *Isis*. Among his writings are *Okens Lehrbuch der Naturgeschichte*. Jena 1825-1826, *Idee sulla classificazione filosofica dei tre regni della natura esposte dal professore Oken alla riunione dei naturalisti in Pisa etc.* Milano 1840, *Lehrbuch der Naturphilosophie*. Jena 1809-1811 and *Elements of Physiophilosophy*. London 1847. See A. Ecker, *Lorenz Oken, A Biographical Sketch*. London 1883; Marc Klein, in *Dictionary of Scientific Biography* 10: 194-196. 1981; Günther Schmid, *Chamisso als Naturforscher*. Eine Bibliographie. Leipzig 1942; Karl Gustav Carus (1789-1869), *Vorlesungen über Psychologie, gehalten im Winter 1829/1830 zu Dresden*. Leipzig 1831 [In the journal *Isis* L.O. commented that this work has delivered to the world the embryo of psychology.]; T.W. Bossert, *Biographical Dictionary of Botanists Represented in the Hunt Institute Portrait Collection*. 293. 1972; R. Zander, F. Encke, G. Buchheim and S. Seybold, *Handwörterbuch der Pflanzennamen*. 14. Aufl. Stuttgart 1993; Bentley Glass et al., eds., *Forerunners of Darwin: 1745-1859*. Edited by ... O. Temkin, William Strauss, Jr. First edition. John Hopkins Press, Baltimore 1959.

Okoubaka Pellegrin & Normand Santalaceae

Origins:

Okoubaka is a vernacular name in Agni (Anyin or Anyi) language (Ivory Coast), from *okou* "death" and *baka* "tree," no plants can survive too close to our tree.

Species/Vernacular Names:

O. aubrevillei Pellegrin & Normand (*Octoknema okoubaka* Aubréville & Pellegrin)

Nigeria: igi-nha (Yoruba); akoebisi (Edo); akoebilisi (Igbo)

Zaire: etale (Turumbu); dibwe mutshi ditoke (Tshiluba)

Ivory Coast: borolo, gbagba, okoubaka

Olax L. Olacaceae

Origins:

Probably from the Latin *olax, olacis* "odorous, smelling" (Latin *oleo, es, ui, ere* "to smell"; Akkadian *alu, elu*, Hebrew *ala* "to move up, to rise"; Akkadian *elesu* "to rejoice," *ullusu* "to cause rejoice"), referring to the scent of some species, see Martianus Minneius [Mineus] Felix Capella, 4th-5th century a.C., author of *De nuptiis Philologie et Mercurii*. Vicentiae 1499; or, according to other authors the genus was named after the Doric *olax, olakos*, Greek *aulax, aulakos* "a furrow," referring to the ridged bark and branches; see Carl Linnaeus, *Species Plantarum*. 34. 1753 and *Genera Plantarum*. Ed. 5. 20. 1754.

Species/Vernacular Names:

O. benthamiana Miq.

Australia: western olax

O. dissitiflora Oliv. (*Olax stuhlmannii* Engl.) (Latin *dissitus, a, um* "well spaced, scattered, lying apart, disperse, remote, apart," referring to the well-spaced flowers)

English: small sourplum, olax

Southern Africa: kleinsuurpruim, small sourplum; uMaphunzana, mapunzana (Zulu); mutandavha (Venda: Soutpansberg, northern Transvaal); mudowe (Mbukushu: Okavango Swamps and western Caprivi)

O. stricta Miq.

Australia: eastern olax

O. subscorpioidea Oliver

Nigeria: gwanon rafi (Hausa); ifon (Yoruba); ukpakon (Edo); igbulu (Igbo)

Yoruba: ifon, awefin

Congo: ombiena-mbiena, otsotsolo

Ivory Coast: acagnikaba

Oldenburgia Less. Asteraceae

Origins:

After the Swedish soldier and traveler Franz Pehr Oldenburg, 1740-1774 (d. Madagascar), plant collector for Kew 1772-73, companion of Thunberg and Francis Masson (1741-1805) on their trips in South Africa; see J. Lanjouw and F.A. Stafleu, *Index Herbariorum*. 2(5): 619. Utrecht 1983; Mary Gunn and Leslie Edward W. Codd, *Botanical Exploration of Southern Africa*. 265. Cape Town 1981; P. MacOwan, "Personalia of botanical collectors at the Cape." *Trans. S. Afr. Philos. Soc.* 4(1): xxx-liii. 1884-1886.

Species/Vernacular Names:

O. grandis (Thunb.) Baill. (*Arnica grandis* Thunb.; *Oldenburgia arbuscula* DC.)

English: rabbit's ears

South Africa: bastersuikerbos, lamb's ears, donkey's ears, oldenburgia

Oldenlandia L. Rubiaceae

Origins:

After the Danish botanist Henrik (Hendrik) Bernard Oldenland (Henricus Bernardus Oldenlandus), c.1663-1699, physician, naturalist, pupil of Paul Hermann (1646-1695) at Leyden, plant collector at the Cape of the Good Hope and Curator-Superintendent of the Botanical Garden of the Dutch East Indian Company; he probably executed the drawings for Herman Boerhaave, *Index alter plantarum*. Lugduni Batavorum. 1720. See J. Burman (1707-1779), *Catalogi duo plantarum Africanarum, quorum prior complectitur plantas ab Hermanno observatas, posterior vero quas Oldenlandus et Hartogius indagarunt*. [Contains lists of nearly 1,000 plants collected at the Cape by Paul Hermann (1646-1695) and of about 400 collected by H.B. Oldenland and J. Hartog.] Amstelaedami 1737; Frans A. Stafleu, *Linnaeus and the Linnaeans*. The spreading of their ideas in systematic botany, 1735-1789. Utrecht 1971; N.L. Burman, *Flora Indica*: cui accedit series Zoophytorum Indicorum, nec non Prodromus Florae Capensis. [The *Florae Capensis Prodromus* is based on Oldenland's collections.] Lugduni Batavorum & Amstelaedami 1768; Peter Kolbe (1675-1726), *Caput Bonae Spei hodiernum. Beschreibung des Africanischen Vorgebürges der Guten Hofnung*. Nürnberg 1719; Carl Linnaeus, *Flora capensis*. Upsaliae [1759] (based on the collections of J. Burman, J. Hartog, P. Hermann and Oldenland); François Valentijn (1666-1725), *Oud en nieuw Oost-Indiën ... Beschryvinge van de Kaap der Goede Hoope*, etc. Dordrecht & Amsterdam 1724-1726; P. MacOwan, "Personalia of botanical collectors at the Cape." *Trans. S. Afr. Philos. Soc.* 4(1): xxx-liii. 1884-1886; J.H. Barnhart, *Biographical Notes upon Botanists*. 3: 26. 1965; John Hutchinson, *A Botanist in Southern Africa*. 610. London 1946; J. Britten, "Some early Cape botanists and collectors." *J. Linn. Soc. (Bot.)* 45: 34-36. 1920; J. Hoge, *Africana Notes*. (On H.B. Oldenland) 3: 125. 1946; Mia C. Karsten, *The Old Company's Garden at the Cape and Its Superintendents*: involving an historical account of early Cape botany. Cape Town 1951; Jens Wilken Hornemann, *Naturh. Tidsskr*. 1: 562-563. 1837; G. Murray, *History of the Collections Contained in the Natural History Departments of the British Museum*. London 1904; Carl Linnaeus, *Species Plantarum*. 119. 1753 and *Genera Plantarum*. Ed. 5. 55. 1754; H.N. Clokie, *Account of the Herbaria of the Department of Botany in the University of Oxford*. 219. Oxford 1964; A. Lasègue, *Musée botanique de Benjamin Delessert*. 66. 1845; John Landwehr, *VOC: A Bibliography of Publications Relating to the Dutch East India Company,*

1602-1800. Ed. Peter van der Krogt. HES Publisher, Utrecht 1991; James Britten, *The Sloane Herbarium*, revised and edited by J.E. Dandy. 1958; A. White and B.L. Sloane, *The Stapelieae*. Pasadena 1937; Adolar Gottlieb Julius Hans Herre (1895-1979), *The Genera of the Mesembryanthemaceae*, ... illustrations by Harry Bolus, Beatrice Carter, Mary Page and Maisie Walgate. Cape Town 1971; Mary Gunn and Leslie E. Codd, *Botanical Exploration of Southern Africa*. 265-266. [1663-1697] 1981; Gilbert Westacott Reynolds (1895-1967), *The Aloes of South Africa*. Balkema, Rotterdam 1982.

Species/Vernacular Names:

O. corymbosa L.

Congo: mundumbu

China: shui xian cao

O. herbacea (L.) Roxb. var. *herbacea* (*Hedyotis herbacea* L.; *Hedyotis heynii* (G. Don) Sond.)

English: false spurry

Southern Africa: seobi (Sotho)

O. umbellata L. (*Hedyotis puberula* (G. Don f.) Arn.)

English: Indian madder

Oldenlandiopsis Terrell & W.H. Lewis Rubiaceae

Origins:

Resembling *Oldenlandia* L.

Oldfieldia Benth. & Hook. Euphorbiaceae

Origins:

After the British physician Richard Albert K. Oldfield, plant collector in Nigeria and Sierra Leone, 1832-1834 on 1st Niger Expedition led by Laird and Lander; see Macgregor Laird and R.A.K. Oldfield, *Narrative of an Expedition into the Interior of Africa by the River Niger*. London 1837; F.N. Hepper and F. Neate, *Plant Collectors in West Africa*. 62. 1971; Ray Desmond, *Dictionary of British & Irish Botanists and Horticulturists*. 526. London 1994; Benjamin Daydon Jackson (1846-1927), "A list of the contributors to the herbarium of the Royal Botanic Gardens, Kew, brought down to 31st December 1899." *Bull. Misc. Inf. Kew*. 1901 and "A list of the collectors whose plants are in the herbarium of the Royal Botanic Gardens, Kew, to 31st December 1899." in *Kew Bulletin*. 1-80. 1901; R.W.J. Keay, "Botanical collectors in West Africa prior to 1860." *Comptes Rendus A.E.T.F.A.T.* Lisbon 1962.

Species/Vernacular Names:

O. africana Benth. & Hook.

English: African oak, African teak

Nigeria: fou

Cameroon: bobindo, alenile

Congo: vesambata

Ivory Coast: angouran, esson, esson angouaran, esui, etui, fou, fu, dantoue

Liberia: pau-lau, saye

Tropical West Africa: paulati

Olea L. Oleaceae

Origins:

Greek *elaio*, *elaia*, Latin *olea* "an olive, olive-berry, an olive-tree"; see Carl Linnaeus, *Species Plantarum*. 8. 1753 and *Genera Plantarum*. Ed. 5. 8. 1754; Salvatore Battaglia, *Grande dizionario della lingua italiana*. Torino 1961-1989 etc.; Manlio Cortelazzo & Paolo Zolli, *Dizionario etimologico della lingua italiana*. 4: 827-828. Bologna 1985; Giovanni Semerano, *Le origini della cultura europea*. Dizionario della lingua Latina e di voci moderne. 2(2): 493. 1994; Ernest Weekley, *An Etymological Dictionary of Modern English*. 2: 1007. New York 1967; G. Semerano, *Le origini della cultura europea*. Dizionari Etimologici. Basi semitiche delle lingue indeuropee. Dizionario della lingua Greca. 2(1): 88. Firenze 1994.

Species/Vernacular Names:

O. capensis L. subsp. *capensis* (*Olea capensis* L.; *Olea laurifolia* Lam.; *Olea undulata* Jacq.; *Olea undulata* Jacq. var. *planifolia* E. Mey.; *Olea concolor* E. Meyer)

English: false ironwood, bastard black ironwood, black ironwood

Southern Africa: basterysterhout, baster swartysterhout; umSishane, umSinjane, isiNhletshe, iGwanxi (Zulu)

O. capensis L. subsp. *enervis* (Harv. ex C.H. Wr.) Verdoorn (*Olea enervis* Harv. ex C.H. Wr.)

English: bushveld ironwood

Southern Africa: bosveldysterhout, ysterhout; umSishane, unSinjane, isiNhletshe, iGwanxi (Zulu); motshere (North Sotho)

O. capensis L. subsp. *hochstetteri* (Bak.) Friis & P.S. Green (*Olea hochstetteri* Bak.)

Nigeria: zitum (Hausa)

O. capensis L. subsp. *macrocarpa* (C.H. Wr.) Verdoorn (*Olea macrocarpa* C.H. Wr.)

English: ironwood, forest ironwood, black ironwood

Southern Africa: ysterhout, regte swartysterhout; isiTimane (Swazi); umZimane, umSishane, umSinjane, iGwanxi (Zulu); uGqwangxe, umHlebe (Xhosa)

O. europaea L.

English: olive tree, common olive, edible olive, European olive tree, wild olive tree

Japan: oribu-no-ki

Arabic: zitoun, zaytun, zebbour

China: mu xi lan

Hawaii: 'oliwa, 'oliwa haole

Mexico: biache riche zaa castilla, piache castilla nititi zaa niza, yaga biache

O. europaea L. subsp. *africana* (Mill.) P.S. Green (*Olea africana* Mill.; *Olea buxifolia* Mill.; *Olea chrysophylla* Lam.; *Olea europaea* var. *verrucosa* Willd.; *Olea sativa* Hoffm. var. *verrucosa* R. & S.; *Olea verrucosa* (Willd.) Link)

English: wild olive, South African olive

Southern Africa: olienhout, olyfboom, swartolienhout; umNquma (Swazi); muGuma, muToba (Shona); mohlware (South Sotho); motlhware (North Sotho); mokgware (Ngwaketse dialect, Botswana); motlhware (Western Transvaal, northern Cape, Botswana); umNqumo, umHlwathi, isAdlulambazo (Zulu); umNquma (Xhosa); khérob, khéraheis (Nama: southern Southwest Africa)

O. europaea L. var. *oleaster* (Hoffm. & Link) DC.

English: wild olive

O. exasperata Jacq. (*Olea humilis* Eckl.)

English: dune olive

South Africa: slanghout (= snake-wood), glashoutolien, swarthout

O. lancea Lam.

Rodrigues Island: bois malaya

O. paniculata R. Brown

Australia: Australian olive, native olive, pigeonberry ash, clove berry

China: xian ye mu xi lan

O. welwitschii (Knobl.) Gilg & Schellenb.

Zaire: ndobo

O. woodiana Knobl. (*Olea mackenii* Harv. ex C.H. Wr.; *Olea listeriana* Sim ex Lister) (after the Natal (b. England, Notts.) botanist John Medley Wood, 1827-1915 (d. Durban), 1882-1903 Curator of the Natal Botanic Garden and Director of the Natal Herbarium (1903-1913). His writings include *Handbook to the Flora of Natal*. Cape Town 1907 and *Catalogue of Plants in Natal Botanic Gardens*. Durban 1890; see Alain White and Boyd Lincoln Sloane, *The Stapelieae*. Pasadena 1937; Ray Desmond, *Dictionary of British & Irish Botanists and Horticulturists*. 753. 1994; John H.

Barnhart, *Biographical Notes upon Botanists*. 3: 516. 1965; T.W. Bossert, *Biographical Dictionary of Botanists Represented in the Hunt Institute Portrait Collection*. 442. 1972; E.M. Tucker, *Catalogue of the Library of the Arnold Arboretum of Harvard University*. 1917-1933; Mary Gunn and Leslie E. Codd, *Botanical Exploration of Southern Africa*. 379-381. Cape Town 1981; Ernest Nelmes and William Cuthbertson, *Curtis's Botanical Magazine Dedications, 1827-1927*. 338-340. [1931]; Gordon Douglas Rowley, *A History of Succulent Plants*. Mill Valley, California 1997; Gilbert Westacott Reynolds, *The Aloes of South Africa*. 65-66, 96. Rotterdam 1982)

English: forest olive

Southern Africa: bosolienhout; maNyatsi (Swazi); iSahlulambhazo, isaDlulambazo (= the tree defying the axe), umNqumo, umNqugunya (Zulu); umGqukunqa, uSintlwa (Xhosa)

Oleandra Cav. Oleandraceae (Davalliaceae)

Origins:
Referring to the simple and oleander-like fronds, laminae simple, entire; see Antonio José Cavanilles, in *Anales de Historia Natural Madrid*. 1(2): 115. 1799.

Species/Vernacular Names:
O. maquilingensis Copel. (habitat on Mount Makiling)
The Philippines: kaliskis-ahas

O. neriiformis Cav.
English: stilt fern

O. pistillaris (Sw.) C. Chr.
The Philippines: kaliskis-ahas

Oleandropsis Copel. Polypodiaceae

Origins:
Resembling *Oleandra* Cav.

Olearia Moench Asteraceae

Origins:
Probably named after the German horticulturist Johann Gottfried Ölschläger (Olearius), 1635-1711, author of *Specimen Florae Hallensis*. Halle an der Saale 1666-1668 and of *Hyacinth-betrachtung*. Leipzig 1665; or for the German mathematician Adam Ölschläger (Olearius, Olivarius, Oelschläger) (1599/1603-1671), poet, traveler and librarian in Holstein, (author and) editor of Johann Albrecht von Mandelslo (1616-1644), *Morgenländische Reyse-*

Beschreibung. Schleswig 1658 [*The Voyages and Travels of the Ambassadors sent by Frederick Duke of Holstein, to the Great Duke of Muscovy, and the King of Persia. Begun in the year 1633, and finished in 1639*. London 1662], or for Johann Christoph Olearius (1668-1747) who wrote *Aloedarium historicum*. Arnstadt 1713, or from the Latin *olea, ae* "an olive." See Samuel H. Baron, transl. and ed., *The Travels of Olearius in Seventeenth-Century Russia*. Stanford University Press [n.d.]; James A. Baines, *Australian Plant Genera. An Etymological Dictionary of Australian Plant Genera*. 261-262. Chipping Norton, N.S.W. 1981; Francis Aubie Sharr, *Western Australian Plant Names and Their Meanings*. 52. University of Western Australia Press 1996; Helmut Genaust, *Etymologisches Wörterbuch der botanischen Pflanzennamen*. 434. Basel 1996; Georg Christian Wittstein, *Etymologisch-botanisches Handwörterbuch*. 629. Ansbach 1852; Arthur D. Chapman, ed., *Australian Plant Name Index*. 2107-2115. Canberra 1991.

Species/Vernacular Names:
O. adenolasia (F. Muell.) Benth. (*Aster adenolasius* F. Muell.)
Australia: woolly-glandular daisy-bush

O. adenophora (F. Muell.) F. Muell. ex Benth.
English: scented daisy-bush, forest daisy-bush

O. algida Wakef.
Australia: mountain daisy-bush

O. allenderae J.H. Willis (after Marie Allender, the original collector, Wilson's Promontory National Park, South Victoria)
Australia: Promontory daisy-bush

O. alpicola (F. Muell.) Benth.
Australia: alpine daisy-bush

O. argophylla (Labill.) Benth.
Australia: native musk, silver shrub, musk daisy-bush

O. asterotricha (F. Muell.) Benth.
Australia: rough daisy-bush, starhair daisy-bush

O. avicenniifolia (Raoul) Hook.f.
New Zealand: akeake

O. axillaris (DC.) Benth. (*Eurybia axillaris* DC.)
Australia: coast daisy-bush, sandhill daisy

O. ballii (F. Muell.) Hemsl. (for the discoverer of Lord Howe Island in 1788, Henry Lidgbird Ball)
English: mountain daisy

O. calcarea F. Muell. ex Benth.
English: limestone daisy-bush

O. canescens (Benth.) Hutch.
Australia: New England daisy-bush

O. ciliata (Benth.) F. Muell. ex Benth.

English: fringed daisy-bush

O. decurrens (DC.) Benth.

English: clammy daisy-bush

O. elliptica DC.

Australia: sticky daisy-bush, shining daisy-bush

O. erubescens (DC.) Dippel

Australia: silky daisy-bush, moth daisy-bush

O. floctoniae Maiden & E. Betche

Australia: Dorrigo daisy-bush

O. floribunda (Hook.f.) Benth.

Australia: heath daisy-bush

O. glandulosa (Labill.) Benth. (*Aster glandulosus* Labill.)

Australia: swamp daisy-bush

O. glutinosa (Lindley) Benth.

Australia: sticky daisy-bush

O. grandiflora Hook. (*Aster sonderi* F. Muell.)

Australia: Mount Lofty daisy-bush

O. gravis (F. Muell.) Benth.

Australia: Tableland daisy-bush

O. hookeri (Sonder) Benth.

Australia: Hooker's daisy-bush

O. hygrophila (DC.) Benth.

Australia: Island daisy-bush

O. iodochroa (F. Muell.) Benth.

Australia: violet daisy-bush

O. ledifolia (DC.) Benth.

Australia: grooved daisy-bush

O. lepidophylla (Pers.) Benth.

Australia: clubmoss daisy-bush

O. lirata (Sims) Hutch.

Australia: snow daisy-bush, snowy daisy-bush

O. megalophylla (F. Muell.) Benth.

Australia: large-leaf daisy-bush

O. microdisca J. Black

Australia: small-flowered daisy-bush

O. microphylla (Vent.) Maiden & E. Betche

Australia: bridal daisy-bush

O. muelleri (Sonder) Benth.

Australia: dusky daisy-bush

O. myrsinoides (Labill.) Benth.

Australia: blush daisy-bush, silky daisy-bush

O. nernstii (F. Muell.) Benth.

Australia: jagged daisy-bush

O. obcordata (Hook.f.) Benth.

Australia: fan daisy-bush

O. passerinoides (Turcz.) Benth.

Australia: slender daisy-bush

O. persoonioides (DC.) Benth.

Australia: lacquered daisy-bush

O. phlogopappa (Labill.) DC.

Australia: dusty daisy-bush

O. picridifolia (F. Muell.) Benth.

Australia: rasp daisy-bush

O. pimeleoides (DC.) Benth. (*Eurybia pimeleoides* DC.)

Australia: showy daisy-bush

O. pinifolia (Hook.f.) Benth.

Australia: needle daisy-bush

O. quercifolia DC.

Australia: oak-leaf daisy-bush

O. ramosissima (DC.) Benth.

Australia: moss daisy-bush

O. ramulosa (Labill.) Benth.

Australia: twiggy daisy-bush

O. rosmarinifolia (DC.) Benth.

Australia: rosemary daisy-bush

O. rudis (Benth.) Benth.

Australia: azure daisy-bush

O. rugosa (W. Archer) Hutch.

Australia: wrinkled daisy-bush

O. speciosa Hutch.

Australia: netted daisy-bush

O. stellulata (Labill.) DC.

Australia: starry daisy-bush

O. viscidula (F. Muell.) Benth.

Australia: wallaby weed, brush daisy-bush

Oleicarpon Airy Shaw Fabaceae

Origins:
Greek *elaio, elaia*; Latin *olea, ae* for the olive and *karpos* "fruit."

Oleiocarpon Dwyer Fabaceae

Origins:
From Greek *elaio, elaia*; Latin *olea, ae* for the olive and *karpos* "fruit."

Olgasis Raf. Orchidaceae

Origins:

According to Rafinesque possibly named for a Greek nymph; see C.S. Rafinesque, *Flora Telluriana*. 2: 51. 1836 [1837].

Oligactis (Kunth) Cass. Asteraceae

Origins:

From the Greek *oligos* "few, little" and *aktin* "ray."

Oligactis Raf. Asteraceae

Origins:

See C.S. Rafinesque, *Flora Telluriana*. 2: 44. 1836 [1837].

Oligandra Less. Asteraceae

Origins:

From the Greek *oligos* "few, little, small" and *andros* "male," referring to the small number of stamens.

Oliganthemum F. Muell. Asteraceae

Origins:

Greek *oligos* "few" and *anthos, anthemon* "flower"; see F. von Mueller, *Transactions and Proceedings of the Philosophical Institute of Victoria*. 3: 56. Melbourne 1859.

Oliganthes Cass. Asteraceae

Origins:

From the Greek *oligos* "few" and *anthos, anthemon* "flower."

Oligarrhena R. Br. Epacridaceae

Origins:

Greek *oligos* and *arrhen* "male," referring to the small number of stamens; see Robert Brown (1773-1858), *Prodromus florae Novae Hollandiae et Insulae van-Diemen*. 549. London 1810.

Oligobotrya Baker Convallariaceae (Liliaceae)

Origins:

From the Greek *oligos* "few" and *botrys* "cluster," referring to the inflorescence.

Oligocarpus Less. Asteraceae

Origins:

With few fruits, from the Greek *oligos* "few" and *karpos* "fruit."

Oligoceras Gagnepain Euphorbiaceae

Origins:

From the Greek *oligos* "few, small, little" and *keras* "a horn."

Oligochaeta (DC.) K. Koch Asteraceae

Origins:

From the Greek *oligos* "few, small, little" and *chaite* "bristle, long hair," small bristles.

Oligocladus Chodat & Wilczek Umbelliferae

Origins:

From the Greek *oligos* "few, small" and *klados* "branch."

Oligocodon Keay Rubiaceae

Origins:
Greek *oligos* "few" and *kodon* "a bell."

Oligodora DC. Asteraceae

Origins:

From the Greek *oligos* "few" and *dora* "a skin, hide."

Oligogynium Engl. Araceae

Origins:

From the Greek *oligos* "few" and *gyne* "female, woman, female organ."

Oligolobos Gagnepain Hydrocharitaceae

Origins:
From the Greek *oligos* "few" and *lobos* "a pod, fruit, lobe."

Oligomeris Cambess. Resedaceae

Origins:
From the Greek *oligos* "few" and *meris* "part," referring to the petals or to the small flowers with lobed calyx and capsule.

Oligoneuron Small Asteraceae

Origins:
From the Greek *oligos* "few" and *neuron* "nerve."

Oligophyton Linder Orchidaceae

Origins:
From the Greek *oligos* "few, small, little" and *phyton* "plant."

Oligoscias Seem. Araliaceae

Origins:
From the Greek *oligos* "few, small, little" and *skias* "a canopy, pavilion."

Oligospermum D.Y. Hong Scrophulariaceae

Origins:
From the Greek *oligos* "few, small, little" and *sperma* "seed."

Oligostachyum Z.P. Wang & G.H. Ye Gramineae

Origins:
From the Greek *oligos* "few, little" and *stachys* "a spike."

Oligostemon Benth. Caesalpiniaceae

Origins:
Greek *oligos* and *stemon* "a stamen, a thread."

Oligothrix DC. Asteraceae

Origins:
From the Greek *oligos* "few" and *thrix, trichos* "hair," alluding to the bristles and the pappus.

Olinia Thunberg Oliniaceae

Origins:
After the Swedish botanist Johan Henrik (Henric) Olin, 1769-1824, medical man and student of Thunberg, Dr. med. Uppsala 1797, author of *Plantae svecanae*. Upsaliae [Uppsala] 1797 and *Dissertatio de arnica*. Upsaliae 1799; see John Hendley Barnhart, *Biographical Notes upon Botanists*. 3: 27. 1965.

Species/Vernacular Names:
O. emarginata Burtt Davy

English: mountain hard pear, Transvaal hard pear

Southern Africa: berghardepeer, rooibessie; umNganalahla, iQudu (Xhosa); uQudu (Zulu); mmasephaletsi (North Sotho)

O. radiata J. Hofmeyr. & Phill.

English: Natal hard pear, Pondoland hard pear

Southern Africa: Natalhardepeer, rooibessieboom; umZaneno, umPhanzi (Zulu); umBovana, umBomvane, umPhanzi (Xhosa)

O. rochetiana Juss. (*Olinia usambarensis* Gilg)

English: sando tree

Southern Africa: sando (Venda)

O. ventosa (L.) Cufod. (*Olinia cymosa* (L.) Thunb.; *Olinia capensis* Klotzsch; *Olinia acuminata* Klotzsch; *Sideroxylon cymosum* L.f.)

English: hard pear

Southern Africa: hardepeer, rooibessieboom; iNgobamakhosi, umNgenelahla, ongenalahle (= tree that has no embers or charcoal), iNqudu (Xhosa)

Olivaea Sch. Bip. ex Benth. Asteraceae

Origins:
For the Mexican botanist Leonardo Oliva, 1805-1873, pharmacologist, plant collector, wrote *Lecciones de farmacologia*. Guadalajara 1853-1854, "El copal." *Mexicano* 1: 439-440. 1866 and "Flórula del Departamento de Jalisco." *Naturaleza* 5: 88-99, 127-133. 1880. See J.H. Barnhart, *Biographical Notes upon Botanists*. 3: 27. 1965; Eric Van Young, *Hacienda and Market in Eighteenth-Century Mexico: The Rural Economy of the Guadalajara Region, 1675-*

1820. Berkeley, University of California 1981; José Toribio Medina, *La imprenta en Guadalajara de México*. Amsterdam 1966; Ida Kaplan Langman, *A Selected Guide to the Literature on the Flowering Plants of Mexico*. Philadelphia 1964; Irving William Knobloch, compil., "A preliminary verified list of plant collectors in Mexico." *Phytologia Memoirs*. VI. 1983.

Oliveranthus Rose Crassulaceae

Origins:

For George W. Oliver, 1857-1923, American horticulturist, author of *New Methods of Plant Breeding*. [United States of America. Department of Agriculture. Bureau of Plant Industry. Bulletin no. 167.] 1910.

Oliverella Tieghem Loranthaceae

Origins:

After the British botanist Daniel Oliver, 1830-1916 (Kew, Surrey), at the Kew Herbarium, professor of botany at London (University College), 1853 Fellow of the Linnean Society, 1863 Fellow of the Royal Society, editor and joint author of *Flora of Tropical Africa* (vols. 1-3). Among his very numerous publications are "Notes on the Loranthaceae, with a synopsis of the genera." *J. Proc. Linn. Soc, Bot*. 7: 90-106. 1864, *Official Guide to the Kew Museum. A Handbook to the Museums of Economic Botany of the Royal Gardens, Kew*. [London] 1861 and *First Book of Indian Botany*. London 1869. See John H. Barnhart, *Biographical Notes upon Botanists*. 3: 27. 1965; R. Zander, F. Encke, G. Buchheim and S. Seybold, *Handwörterbuch der Pflanzennamen*. 14. Aufl. Stuttgart 1993; Stafleu and Cowan, *Taxonomic Literature*. 3: 819-828. 1981; T.W. Bossert, *Biographical Dictionary of Botanists Represented in the Hunt Institute Portrait Collection*. 293. 1972; H.N. Clokie, *Account of the Herbaria of the Department of Botany in the University of Oxford*. 219. Oxford 1964; E.M. Tucker, *Catalogue of the Library of the Arnold Arboretum of Harvard University*. 1917-1933; Mea Allan, *The Hookers of Kew*. London 1967; Leonard Huxley, *Life and Letters of Sir J.D. Hooker*. London 1918; E.D. Merrill, *Bernice P. Bishop Mus. Bull*. 144: 145. 1937 and *Contr. U.S. Natl. Herb*. 30(1): 231. 1947; Ernest Nelmes and William Cuthbertson, *Curtis's Botanical Magazine Dedications, 1827-1927*. 159-160. [1931]; M. Hadfield et al., *British Gardeners: A Biographical Dictionary*. London 1980; Emil Bretschneider (1833-1901), *History of European Botanical Discoveries in China*. [Reprint of the original edition, St. Petersburg 1898.] Leipzig 1981; Ida Kaplan Langman, *A Selected Guide to the Literature on the Flowering Plants of Mexico*. 557. Philadelphia 1964; J.D. Milner, *Catalogue*

of Portraits of Botanists Exhibited in the Museums of the Royal Botanic Gardens. Royal Botanic Gardens, Kew, London 1906.

Oliveria Ventenat Umbelliferae

Origins:

After the French botanist Guillaume Antoine Olivier, 1756-1814, zoologist, traveler, entomologist. His writings include *Voyage dans l'empire Othoman*. Paris [1801]-1807 and *Entomologie ou histoire naturelle des insectes ... Coléoptères*. Paris 1798-1808. See Ernest Olivier, *1756-1814. G.A. Olivier ... sa vie, ses travaux, ses voyages*. Documents inédits. Moulins 1880; John H. Barnhart, *Biographical Notes upon Botanists*. 3: 28. 1965; T.W. Bossert, *Biographical Dictionary of Botanists Represented in the Hunt Institute Portrait Collection*. 293. 1972; Jonas C. Dryander, *Catalogus bibliothecae historico-naturalis Josephi Banks*. London 1800; A. Lasègue, *Musée botanique de Benjamin Delessert*. Paris 1845; E.M. Tucker, *Catalogue of the Library of the Arnold Arboretum of Harvard University*. 1917-1933; R. Zander, F. Encke, G. Buchheim and S. Seybold, *Handwörterbuch der Pflanzennamen*. 14. Aufl. Stuttgart 1993.

Oliveriana Reichb.f. Orchidaceae

Origins:

After the British botanist Daniel Oliver, 1830-1916 (Kew, Surrey), at the Kew Herbarium, professor of botany at London (University College), 1853 Fellow of the Linnean Society, 1863 Fellow of the Royal Society, editor and joint author of *Flora of Tropical Africa* (vols. 1-3). Among his very numerous publications are "Notes on the Loranthaceae, with a synopsis of the genera." *J. Proc. Linn. Soc, Bot*. 7: 90-106. 1864, *Official Guide to the Kew Museum. A Handbook to the Museums of Economic Botany of the Royal Gardens, Kew*. [London] 1861 and *First Book of Indian Botany*. London 1869.

Olmeca Söderstr. Gramineae

Origins:

The Olmec people lived in hot, humid lowlands along the Gulf Coast in what is now southern Veracruz and Tabasco states in southern Mexico; the Olmec domain extends from the Tuxtlas Mountains in the west to the lowlands of the Chontalpa in the east; the first elaborate pre-Columbian culture of Mesoamerica; see Carmen Aguilera, *Flora y fauna Mexicana*. Mitología y tradiciones. 137-138. Editorial Everest Mexicana México s.d. [1985].

Olmedia Ruíz & Pav. Moraceae

Origins:
Dedicated to Vicente de Olmedo, in 1790 in Ecuador.

Olmediella Baillon Flacourtiaceae

Origins:
Dedicated to Vicente de Olmedo, in 1790 in Ecuador.

Olmedioperebea Ducke Moraceae

Origins:
The genera *Olmedia* Ruíz & Pav. and *Perebea* Aublet.

Olmediophaena Karst. Moraceae

Origins:
The genus *Olmedia* Ruíz & Pav. and Greek *phaino, phaeino* "bring to light, reveal."

Olmediopsis Karst. Moraceae

Origins:
Resembling *Olmedia*.

Olneya A. Gray Fabaceae

Origins:
For the American botanist Stephen Thayer Olney, 1812-1878, author of *Catalogue of Plants*. Providence 1845. See John H. Barnhart, *Biographical Notes upon Botanists*. 3: 28. 1965; T.W. Bossert, *Biographical Dictionary of Botanists Represented in the Hunt Institute Portrait Collection*. 294. 1972; S. Lenley et al., *Catalog of the Manuscript and Archival Collections and Index to the Correspondence of John Torrey*. Library of the New York Botanical Garden. 465. 1973; E.M. Tucker, *Catalogue of the Library of the Arnold Arboretum of Harvard University*. 1917-1933; J. Ewan, ed., *A Short History of Botany in the United States*. 92. New York and London 1969.

Species/Vernacular Names:
O. tesota A. Gray

English: desert ironwood, tesota, ironwood

Olofuton Raf. Capparidaceae (Capparaceae)

Origins:
Greek *holos* "entire, whole" and *phyton* "plant"; see C.S. Rafinesque, *Sylva Telluriana*. 108. 1838.

Olostyla DC. Rubiaceae

Origins:
From the Greek *holos* "entire, whole" and *stylos* "pillar, style."

Olotrema Raf. Cyperaceae

Origins:
Greek *holos* "entire, whole" and *trema* "hole, aperture"; see C.S. Rafinesque, *The Good Book*. 25. 1840.

Olsynium Raf. Iridaceae

Origins:
See C.S. Rafinesque, *New Fl. N. Am.* 1: 72, 4: 104. 1836 and *The Good Book*. 45. 1840.

Oluntos Raf. Moraceae

Origins:
Greek *olynthos, olonthos* "the edible fruit of the wild fig"; see C.S. Rafinesque, *Sylva Telluriana*. 58. Philadelphia 1838.

Olympia Spach Guttiferae

Origins:
A sacred region.

Olymposciadium H. Wolff Umbelliferae

Origins:
From Olympus and Greek *skiadion, skiadeion* "umbel, parasol."

Olyra L. Gramineae

Origins:
Greek *olyra* "rice-wheat," Theophrastus (*HP.* 8.9.2) and Dioscorides, Latin *olyra, ae* applied by Plinius to a kind of grain, called also *arinca*, which resembles spelt.

Species/Vernacular Names:
O. sp.
Peru: 'insun chëxëti ro?
O. caudata Trinius
Peru: shashap
O. fasciculata Trinius (*Olyra heliconia* Lindman)

Peru: pinguil shucush

O. latifolia L. (*Olyra cordifolia* Kunth)

Peru: carrizo

Yoruba: ofa etu, fodun, eto igbo pe laye

Omalanthus A. Juss. Euphorbiaceae

Origins:
Greek *homalos* "smooth, flat" and *anthos* "a flower," an allusion to the shape and the appearance of the flower, see also *Homalanthus* A. Juss.; see Adrien Henri Laurent de Jussieu (1797-1853), *De Euphorbiacearum generibus.* 50. Parisiis 1824.

Species/Vernacular Names:
O. nutans (Forst.) Guill
English: native poplar, bleeding heart

Omalocaldos Hook.f. Rubiaceae

Origins:
Greek *homalos* "smooth, flat."

Omalocarpus Choux Sapindaceae

Origins:
From the Greek *homalos* "smooth, flat" and *karpos* "a fruit."

Omalotheca Cass. Asteraceae

Origins:
From the Greek *homalos* "smooth, flat" and *theke* "a box, case"; see Alexandre Henri Gabriel Comte de Cassini (1781-1832), *Dictionnaire des Sciences Naturelles.* Paris 1828.

Omania S. Moore Scrophulariaceae

Origins:
From the independent Sultanate of Oman, Arabic Saltanat 'Uman, country occupying the southeastern coast of the Arabian Peninsula, facing the Persian Gulf, the Gulf of Oman, and the Arabian Sea.

Ombrocharis Handel-Mazzetti Labiatae

Origins:
From the Greek *ombros* "storm of rain, thunderstorm, water" and *charis* "grace, beauty."

Species/Vernacular Names:
O. dulcis Handel-Mazzetti
English: sweet ombrocharis
China: xi yu cao

Ombrophytum Poeppig ex Endl. Balanophoraceae

Origins:
Greek *ombros* "rain, water" and *phyton* "plant."

Species/Vernacular Names:
O. peruvianum Poeppig & Endlicher (*Ombrophytum zamioides* Weddell)
Peru: maíz del monte

Omeiocalamus Keng f. Gramineae

Origins:
From Mount Omei, Emei Shan (*mei* "devil" and *shan* "mount"), Sichuan (Szechwan) province, central China and *kalamos* "reed."

Omiltemia Standley Rubiaceae

Origins:
Omiltemi Ecological State Park (17°31′–17°35′N and 99°30′–99°44′W) is 27 km west of Chilpancingo, including 36 km² of the western portion of the Sierra de Igualatlaco, Mexico. In Omiltemi Ecological State Park there are 37 species of amphibians and reptiles, with 13 endemics (Muñoz-Alonso 1988); see L.A. Muñoz-Alonso, *Estudio herpetofaunístico del Parque Ecológico Estatal de Omiltemi.* Mpio. de Chilpancingo de los Bravo, Guerrero. Thesis. UNAM, Mexico, D.F. 111 pp. 1988.

Ommatodium Lindley Orchidaceae

Origins:
From the Greek *omma, ommatos* "eye, light," *ommato* "give sight to," spots on the labellum.

Omoea Blume Orchidaceae

Origins:

From the Greek *homoios* "like, resembling, the same, similar," referring to the petals or to the similarity to *Ceratochilus*.

Omphacomeria (Endl.) A. DC. Santalaceae

Origins:

From the Greek *omphax, omphakos* "unripe fruit, sour, bitter" and *meris* "a portion, part," referring to the very sour fruits.

Species/Vernacular Names:

O. acerba (R. Br.) A. DC. (*Leptomeria acerba* R. Br.)

English: leafless sourbush

Omphalandria P. Browne Euphorbiaceae

Origins:

From the Greek *omphalos* "umbilicus, a navel" and *andros* "male, man, male parts."

Omphalea L. Euphorbiaceae

Origins:

Greek *omphalos* "umbilicus, a navel," the anthers are umbilicate, also referring to the flower carpels; see Carl Linnaeus, *Systema Naturae.* Ed. 10. 1254, 1264, 1378. 1759.

Species/Vernacular Names:

O. triandra L.

English: Jamaican cobnut

Peru: comadre de azeite

Omphalobium Gaertner Connaraceae

Origins:

From the Greek *omphalos* "umbilicus, a navel" and *lobion, lobos* "lobe, pod, small pod, fruit," referring to the shape of the fruits.

Omphalocarpum P. Beauv. Sapotaceae

Origins:

From the Greek *omphalos* "umbilicus, navel" and *karpos* "fruit."

Species/Vernacular Names:

O. elatum Miers

Nigeria: usha (Igbo)

Cameroon: bele, mebemengono, pinbi, abo, aboc, mbate

Ivory Coast: aguia

Gabon: olong

O. lecomteanum Pierre

Zaire: diwala, kakesi

O. mortehanii De Wild.

Zaire: bodimba, bolula, diwala

O. procerum P. Beauv.

Yoruba: akisapo

Nigeria: akisapo, akishapo, ikassa, mantor, ipasi, finsika, akissa (Yoruba); ikassa (Edo); mantor (Kiaka); ipase (Ijaw)

Cameroon: mebemengono, mebemengon, m'pounde, abo, abok

Congo: eoualab

Gabon: abok

Omphalodes Miller Boraginaceae

Origins:

Greek *omphalos* "umbilicus, navel," *omphalodes, omphaloeides* "navel-like, like a navel or boss," in allusion to the hollowed nutlets.

Species/Vernacular Names:

O. verna Moench

English: blue-eyed Mary

Omphalogonus Baillon Asclepiadaceae

Origins:

From the Greek *omphalos* "umbilicus, navel" and *gonia* "an angle."

Omphalogramma (Franchet) Franchet Primulaceae

Origins:

From the Greek *omphalos* "umbilicus, navel" and *gramma* "letter, thread, marking," referring to the winged and flattened seeds.

Species/Vernacular Names:

O. vinciflorum (Franchet) Franchet

English: hairy oneflower-primrose

China: du hua bao chun

Omphalolappula Brand Boraginaceae

Origins:

The mericarps are intermediate between those of *Omphalodes* Mill. and *Lappula* Moench; see August Brand (1863-1930), *Das Pflanzenreich*. Heft 97. (February) 1931.

Species/Vernacular Names:

O. concava (F. Muell.) Brand (*Echinospermum concavum* F. Muell.; *Lappula concava* (F. Muell.) F. Muell.; *Cynoglossum concavum* (F. Muell.) Kuntze)

English: burr stickseed

Omphalopappus O. Hoffm. Asteraceae

Origins:

From the Greek *omphalos* "umbilicus, navel" and *pappos* "fluff, pappus."

Omphalophthalma Karsten Asclepiadaceae

Origins:

From the Greek *omphalos* "navel" and *ophthalmos* "eye."

Omphalopus Naudin Melastomataceae

Origins:

From the Greek *omphalos* "navel, anything like a navel, knob" and *pous* "a foot."

Omphalotrigonotis W.T. Wang Boraginaceae

Origins:

Greek *omphalos* "navel, knob" plus *Trigonotis* Steven.

Species/Vernacular Names:

O. cupulifera (I.M. Johnston) W.T. Wang

English: cup omphalotrigonotis

China: min guo cao

Onagra Miller Onagraceae

Origins:

Greek *onagros* "the wild ass," *onos* "ass" and *agrios* (*agra*) "wild," referring to the shape of the calyx; Dioscorides applied the name *onagra* to the oleander.

Oncella Tieghem Loranthaceae

Origins:

The diminutive of the Greek *onkos* "tumor, tubercle."

Oncidium Swartz Orchidaceae

Origins:

The diminutive of the Greek *onkos* "tumor, tubercle, swelling," alluding to the shape of the calluses on the lip.

Oncinema Arn. Asclepiadaceae

Origins:

From the Greek *onkinos* "hook, hooked," *onkos* "tumor, tubercle, bulk" and *nema* "a thread, filament," referring to the anthers with appendages.

Oncinocalyx F. Muell. Labiatae (Verbenaceae)

Origins:

From the Greek *onkinos* "hook, hooked" and *kalyx* "a calyx"; see F. von Mueller, in *Southern Science Record*. 3: 69. Melbourne (March) 1883.

Oncinotis Benth. Apocynaceae

Origins:

Possibly from the Greek *onkinos* "hook, hooked" and *ous, otos* "an ear" or *notos* "back," referring to the anthers.

Species/Vernacular Names:

O. tenuiloba Stapf (*Oncinotis inandensis* Wood & Evans; *Oncinotis natalensis* Stapf)

English: magic rope

Southern Africa: towertou, phamba plant; umZongazonga, iPhamba, uBhuku, uMathunzi wehlathi (Zulu); uBuka, ubuKa (Xhosa)

Oncoba Forssk. Flacourtiaceae

Origins:

From the Arabic name *onkob*; see Pehr (Peter) Forsskål (1732-1763), *Flora aegyptiaco-arabica*. 103. Copenhagen 1775.

Species/Vernacular Names:

O. spinosa Forssk.

English: snuff-box tree, African dog-rose

French: tabatière

Arabic: onkob, korkor

Nigeria: kakandika, kokociko, okongul, okonkoul, bilau, wofongo, wufongo, ngumi, njora, njore, njori, naki, gbomishere, parisa, wosao; kokochiko (Hausa); amurikpa (Junkun); gbonsere, kakandika (Yoruba); okpoko (Igbo)

Yoruba: kakandika, ponse, ponsere, ajisabere, gamugamusu

W. Africa: sirabara, babara, kongobarani

Southern Africa: snuifkalbassie, kafferklapper; umThongwane (Swazi); mutuzwu (Venda); umThongwane, umShungu, iShungu elikhulu, umThongwana, isiNgongongo (Zulu)

Oncocalamus (G. Mann & H.A. Wendland) Hook.f. Palmae

Origins:
Greek *onkos* "bulk, mass" and *kalamos* "reed," rattan palms, hapaxanthic, spiny, stem with long internodes, stem eventually becomes bare and is used as a source of canes.

Oncocalyx Tieghem Loranthaceae

Origins:
From the Greek *onkos* "bulk, mass, tumor" and *kalyx* "calyx."

Oncocarpus A. Gray Anacardiaceae

Origins:
From the Greek *onkos* "bulk, tumor" and *karpos* "fruit."

Oncodia Lindley Orchidaceae

Origins:
From the Greek *onkodes* "swelling, rounded, bulky," dorsal sepal cucullate, lip very fleshy and saccate at the base.

Oncodostigma Diels Annonaceae

Origins:
From the Greek *onkodes* "swelling, rounded, bulky" and *stigma* "stigma."

Oncosiphon Källersjö Asteraceae

Origins:
From the Greek *onkos* "bulk, mass" and *siphon* "tube," referring to the corolla tube.

Species/Vernacular Names:
O. grandiflorum (Thunb.) Källersjö (*Matricaria grandiflora* (Thunb.) Fenzl ex Harv.; *Pentzia grandiflora* (Thunb.) Hutch.; *Tanacetum grandiflorum* Thunb.)

English: matricaria, stinkweed

South Africa: stinkkruid

O. piluliferum (L.f.) Källersjö (*Cotula globifera* Thunb.; *Cotula pilulifera* L.f.; *Matricaria globifera* (Thunb.) Fenzl ex Harv.; *Matricaria pilulifera* (L.f.) Druce; *Pentzia globifera* (Thunb.) Hutch.; *Pentzia pilulifera* (L.f.) Fourc.)

English: stink-net, cattle bush

South Africa: beesbossie, beesterkruid, miskruie, stinkkruid, stinknet

O. suffruticosum (L.) Källersjö (*Cotula tanacetifolia* L.; *Pentzia suffruticosa* (L.) Hutch. ex Merxm.; *Pentzia tanacetifolia* (L.) Hutch.; *Tanacetum suffruticosum* L.)

English: karoobush

South Africa: karoobos, stinkkruid, wormkruie, wurmbossie, wurmkruid

Oncosperma Blume Palmae

Origins:
From the Greek *onkos* "bulk, mass" and *sperma* "seed," fruit spherical.

Species/Vernacular Names:
O. gracilipes Becc.

The Philippines: anibong

O. platyphyllum Becc.

The Philippines: anibong-laparan

O. tigillarium (Jack) Ridley (*Oncosperma filamentosum* Blume)

Brazil: palmeira nibung

Malaya: nibong, linau

Oncosporum Putterl. Pittosporaceae

Origins:
Greek *onkos* "bulk, mass" and *sporos, spora* "a seed"; see Stephan F. Ladislaus Endlicher and Eduard Fenzl, *Novarum stirpium decades*. 2: 9. Vienna 1839.

Oncostemma K. Schumann Asclepiadaceae

Origins:

From the Greek *onkos* "bulk, mass" and *stemma* "crown, a garland."

Oncostemum A. Juss. Myrsinaceae

Origins:

From the Greek *onkos* "bulk, mass" and *stemon* "a stamen, a thread."

Oncostylus (Schltdl.) Bolle Rosaceae

Origins:

From the Greek *onkos* "bulk, mass" and *stylos* "pillar, style."

Oncotheca Baill. Oncothecaceae

Origins:

From the Greek *onkos* "bulk, mass" and *theke* "a box, case."

Ondetia Benth. Asteraceae

Origins:

The Damaras called Ondetu the locality (Elephantskloof) in Damaraland where the type-specimen was collected.

Ondinea Hartog Nymphaeaceae

Origins:

From ondine, variant of undine, Latin *undina* (Paracelsus, *De Nymphis* etc., Works. 1658), a water nymph, a female being, a nymph, Latin *unda, undae* "a wave"; see Cornelis de Hartog (1931-), in *Blumea*. 18: 413-417. 1970; Kevin F. Kenneally and Edward L. Schneider, "The genus *Ondinea* (Nymphaeaceae) including a new subspecies from the Kimberley region, Western Australia." *Nuytsia*. 4(3): 359-365. 1983.

Species/Vernacular Names:

O. purpurea Hartog

English: ondinea

Ongokea Pierre Olacaceae

Origins:

An African vernacular name; see Y. Tailfer, *La Forêt dense d'Afrique centrale*. CTA, Ede/Wageningen 1989; J. Vivien

& J.J. Faure, *Arbres des Forêts denses d'Afrique Centrale*. Agence de Coopération Culturelle et Technique. Paris 1985.

Species/Vernacular Names:

O. gore (Hua) Pierre

French: noix de gore

Zaire: isanu, sanou; boleko (lake Mai-Ndombe); bosiko (Mbelo); buleka (Bakuba); dileka, dimuma (Tshiluba); ibeka (Batwa); muketu, muleku (Bakete, Tshofa); ndake (Kisangani); oleke (Turumbu); oleko, olekwa (Bashobwa); sanu (Kiombe); tshifulufulu (Lulua; see Walter Henry Stapleton, *Comparative Handbook of Congo Languages. Being a Comparative Grammar of the Eight Principal Languages [Kongo, Bangui, Lolo, Ngale, Poto, Ngombe, Soko and Kele] spoken along the banks of the Congo River from the West Coast of Africa to Stanley Falls*, etc. Compiled and prepared for the Baptist Missionary Society, London, etc. 1903)

Sierra Leone: gbui

Gabon: angueuk, moungéké, nziéké, okéka

Cameroon: angok, angueuk, busolo, oniek, njek, bwelabako, nké

Central African Republic: mobenge

Congo: sanou, nsanou, ongueke, onguegue, mungueke

Nigeria: elede, ekuso, igbepo, wenren-wenren; elede (Yoruba); ekuso (Edo); okufuo (Boki)

Ivory Coast: komonti, kouero, kouéro, ndrouhia

Onira Ravenna Iridaceae

Origins:

Latin *oniros* for the wild poppy.

Onix Medik. Fabaceae

Origins:

From the Greek *onyx, onychos* "a claw, nail."

Onixotis Raf. Colchicaceae (Liliaceae)

Origins:

Greek *onyx, onychos* "a claw, nail" and *ous, otos* "an ear," *otion* "small ear, auricle"; see C.S. Rafinesque, *Flora Telluriana*. 2: 32. 1836 [1837].

Species/Vernacular Names:

O. stricta (Burm.f.) Wijnands (*Onixotis triquetra* (L.f.) Mabberley; *Dipidax triquetra* (L.f.) Bak.)

South Africa; hanekom, vleiblom, waterblom

Onkeripus Raf. Orchidaceae

Origins:
Greek *ogkeros, onkeros* "swollen" and *pous* "foot," the perianth parts; see C.S. Rafinesque, *Flora Telluriana*. 4: 42. 1836 [1838].

Onobrychis Miller Fabaceae

Origins:
From the Greek *onos* "an ass" and *bryche* "gnashing, bellowing," *bryko, brycho* "bite, gobble, eat greedily"; see Philip Miller, *The Gardeners Dictionary*. Abr. ed. 4. London (28 Jan.) 1754.

Species/Vernacular Names:
O. caput-galli (L.) Lam. (*Hedysarum caput-galli* L.)

English: cockshead sainfoin

O. crista-galli Lam. (*Hedysarum caput-galli* L. var. *crista-galli* L.; *Hedysarum crista-galli* (L.) Murray; *Onobrychis squarrosa* Viv.)

English: cockscomb sainfoin

O. viciifolia Scop. (*Hedysarum onobrychis* L.; *Onobrychis sativa* Lam.; *Onobrychis vulgaris* Hill)

English: esparcet, holy clover, sainfoin, common sainfoin, saintfoin, sanfoin

Onochiles Bubani Boraginaceae

Origins:
Greek *onocheiles, onocheilis*, Latin *onochiles, is* and *onochelis, is* and *onochilon* for a plant, a kind of bugloss (Plinius).

Onoclea L. Dryopteridaceae (Onocleaceae, Aspleniaceae, Woodsiaceae)

Origins:
From the Greek *onos* "a vessel" and *kleio* "to close, shut," referring to the sori or the rolled fertile fronds; *onokleia* was an ancient name for another plant (Dioscorides, Galenus); see Carl Linnaeus, *Species Plantarum*. 1062. 1753 and *Genera Plantarum*. Ed. 5. 484. 1754.

Species/Vernacular Names:
O. sensibilis L. (*Onoclea sensibilis* forma *hemiphyllodes* (Kiss & Kümmerle) Gilbert; *Onoclea sensibilis* forma *obtusilobata* (Schkuhr) Gilbert; *Onoclea sensibilis* var. *obtusilobata* (Schkuhr) Torrey)

English: sensitive fern

Onocleopsis Ballard Dryopteridaceae (Onocleaceae, Aspleniaceae, Woodsiaceae)

Origins:
Resembling *Onoclea*.

Onohualcoa Lundell Bignoniaceae

Origins:
The state of Tabasco is located to the southeast of the Mexican Republic, has 17 municipalities in 4 zones (The Chontalpa, The Center, The Saw and The Rivers); the nahuas called this Mayan territory Onohualco, when it was governed by Taabz Cobb and the power resided in its capital Comalcalco.

Ononis L. Fabaceae

Origins:
The classical Greek name *ononis* for the rest-harrow, *Ononis antiquorum*; Latin *ononis, idis, anonis* used by Plinius for a plant, the tall rest-harrow; see Carl Linnaeus, *Species Plantarum*. 716. 1753 and *Genera Plantarum*. Ed. 5. 321. 1754.

Species/Vernacular Names:
O. natrix L.
Arabic: mellita

Onopordum L. Asteraceae

Origins:
Greek *onos* "an ass" and *porde* "fart, crepitus ventris," *pordon, pordos* "stinkard," Latin *onopordon, i* (al. *onopradon*) "a plant, St. Mary's thistle," Greek *onopordon* for the pellitory, *Parietaria cretica*; see Carl Linnaeus, *Species Plantarum*. 827. 1753 and *Genera Plantarum*. Ed. 5. 359. 1754.

Species/Vernacular Names:
O. acanthium L.

English: cotton thistle, Scotch thistle

Maori name: kotimana

O. acaule L.

English: stemless thistle, stemless onopordum, horse thistle

O. illyricum L.

English: Illyrian thistle

O. tauricum Willd.

English: Taurian thistle

Onoseris Willd. Asteraceae

Origins:
From the Greek *onos* "an ass" and *seris, seridos* "chicory, lettuce."

Onosma L. Boraginaceae

Origins:
Greek *onosma* for stone bugloss, *Onosma echioides*, Latin *onosma, atis* for a kind of *anchusa* (Plinius).

Species/Vernacular Names:
O. cingulatum W.W. Smith & Jeffrey (*Onosma tsiangii* I.M. Johnston)

English: Tsiang onosma

China: zhao tong dian zi cao

O. confertum W.W. Smith

English: denseflower onosma

China: mi hua dian zi cao

O. exsertum Hemsley

English: exserted onosma

China: lu rui dian zi cao

O. farrerii I.M. Johnston

English: Farrer onosma

China: xiao hua dian zi cao

O. fistulosum I.M. Johnston

English: fistular onosma

China: guang zhuang dian zi cao

O. frutescens Lam.

English: golden drop

O. hookeri C.B. Clarke var. *longiflorum* (Duthie) Duthie ex Stapf

English: longflower onosma

China: chang hua xi hua dian zi cao

O. mertensioides I.M. Johnston

English: Chedo Mountain onosma

China: chuan xi dian zi cao

O. paniculatum Bureau & Franchet

English: paniculate onosma

China: dian zi cao, zi cao

O. sinicum Diels

English: Chinese onosma

China: xiao ye dian zi cao

O. simplicissimum L.

English: yellow-flower onosma

China: dan jing dian zi cao

Onosmodium Michx. Boraginaceae

Origins:
Referring to the genus *Onosma* L.

Onosuris Raf. Onagraceae

Origins:
Some suggest from the Greek *onos* "an ass" and *oura* "tail"; see C.S. Rafinesque, *Florula ludoviciana*. 95. New York 1817 and *Am. Monthly Mag. Crit. Rev.* 4: 188, 192. 1819; Elmer D. Merrill, *Index rafinesquianus*. 177. 1949.

Onosurus G. Don Onagraceae

Origins:
Greek *onos* "an ass" and *oura* "tail," see also *Oenothera* L.

Onuris Philippi Brassicaceae

Origins:
Latin *onuris* and Greek *onouris* for a plant, called also *oenothera*; Greek *onouris, onotheras, onothouris* for oleander, *Nerium oleander*; see Franco Montanari, *Vocabolario della Lingua Greca*. 1396, 1398. Loescher Editore, Torino 1995.

Onus Gilli Acanthaceae

Origins:
Greek *onos* "an ass," Latin *onus, eris* "a load, burden."

Onychacanthus Nees Acanthaceae

Origins:
From the Greek *onyx, onychos* "a claw, nail" plus *Acanthus*.

Onychium Blume Orchidaceae

Origins:
Greek *onyx, onychos* "a claw, nail," the dim. *onychion*, referring to the nature of the bracteoles.

Onychium Kaulfuss Pteridiaceae (Adiantaceae, Cryptogrammaceae)

Origins:
Greek *onyx, onychos* "a claw, nail," referring to the shape of the lobes of the fronds; in *Berlin. Jahrb. Pharm. Verbundenen Wiss.* 21: 45. 1820; Pichi Sermolli, *Webbia.* 17: 308. 20 Apr. 1963.

Species/Vernacular Names:
O. japonicum (Thunb.) Kunze (*Trichomanes japonicum* Thunb.)
Japan: tachi-shinobu (= erect *Davallia*)
O. siliculosum (Desv.) C. Chr.
The Philippines: pakong anuang

Onychium Reinw. Polypodiaceae

Origins:
Greek *onyx, onychos* "a claw, nail."

Onychopetalum R.E. Fr. Annonaceae

Origins:
From the Greek *onyx, onychos* "a claw, nail" and *petalon* "petal."

Onychosepalum Steudel Restionaceae

Origins:
From the Greek *onyx, onychos* "a claw, nail" and Latin *sepalum*, referring to the perianth segments; see Ernst Gottlieb von Steudel (1783-1856), *Synopsis plantarum glumacearum.* Fasc. 2: 249. Stuttgartiae 1855.

Oocarpon Micheli Onagraceae

Origins:
From the Greek *oon* "egg" and *karpos* "fruit."

Oochlamys Fée Thelypteridaceae

Origins:
From the Greek *oon* "egg" and *chlamys* "cloak."

Oonopsis (Nutt.) Greene Asteraceae

Origins:
From the Greek *oon* "egg" and *opsis* "like, resembling."

Oophytum N.E. Br. Aizoaceae

Origins:
From the Greek *oon* "egg" and *phyton* "plant," an allusion to the shape of the leaves.

Oothrinax O.F. Cook Palmae

Origins:
Greek *oon, oion* "egg" plus genus *Thrinax* Sw.

Opanea Raf. Myrtaceae

Origins:
Opa Lour. as a synonym of *Opanea*; see C.S. Rafinesque, *Sylva Telluriana.* 106. 1838; Elmer D. Merrill, *Index rafinesquianus.* 174. 1949.

Opercularia Gaertner Rubiaceae

Origins:
From the Latin *operculum, i* "a cover, lid," *operio, perui, pertum, ire* "to cover, hide," referring to the partial dehiscence of the fruits; see Joseph Gaertner (1732-1791), *De fructibus et seminibus plantarum.* 1: 111. Stuttgart, Tübingen 1788.

Species/Vernacular Names:
O. aspera Gaertner
English: thin stinkweed, coarse stinkweed, common stinkweed
O. diphylla Gaertner
English: twin-leaf stinkweed
O. hispida Sprengel
English: hairy stinkweed
O. ovata Hook.f.
English: broad-leaved stinkweed, broad stinkweed
O. scabrida Schltdl.
English: rough stinkweed
O. turpis Miq.
English: twiggy stinkweed, grey stinkweed
O. varia Hook.f.
English: variable stinkweed

Operculicarya H. Perrier Anacardiaceae

Origins:
From *operculum, i* "a cover, lid" and *karyon* "nut."

Operculina Silva Manso Convolvulaceae

Origins:
The diminutive of the Latin *operculum, i* "a cover, lid," referring to the capsules.

Species/Vernacular Names:
O. macrocarpa Urban
Brazil: jalapa-do-brasil, jalapão, batata-da-purga
O. turpethum (L.) Silva Manso
English: turpeth root, turpethum, boxfruit vine
China: he guo teng

Ophelia D. Don ex G. Don Gentianaceae

Origins:
From the Greek *ophelos* "advantage, help, furtherance." See David Don (1799-1841), in *Transactions of the Linnean Society of London*. 17: 503-532. 8 July-8 August 1837; George Don (1798-1856), *A General History of the Dichlamydeous Plants*. 4: 173, 178. London 1837.

Ophellantha Standley Euphorbiaceae

Origins:
Greek *ophelos* "advantage, help, furtherance" and *anthos* "flower."

Ophiala Desv. Ophioglossaceae

Origins:
See also *Helminthostachys* Kaulf.

Ophidion Luer Orchidaceae

Origins:
From *ophidion*, the diminutive of the Greek *ophis* "serpent, a creeping plant."

Ophiobotrys Gilg Flacourtiaceae

Origins:
From the Greek *ophis* "a snake, serpent" and *botrys* "cluster, a bunch of grapes," referring to the branches of the racemes.

Species/Vernacular Names:
O. zenkeri Gilg

Nigeria: bofan (Boki)
Cameroon: uolobo

Ophiocarpus (Bunge) Ikonn. Fabaceae

Origins:
From the Greek *ophis* "a serpent" and *karpos* "fruit."

Ophiocaryon Endl. Sabiaceae (Meliosmaceae)

Origins:
Greek *ophis* and *karyon* "nut," embryo like snake.

Species/Vernacular Names:
O. paradoxum Schomb. ex Hook.
English: snakenut

Ophiocaulon Hook.f. Passifloraceae

Origins:
From the Greek *ophis* "a snake, serpent" and *kaulos* "stem," referring to the climbing habit.

Ophiocaulon Raf. Caesalpiniaceae

Origins:
Greek *ophis* "a snake, serpent" and *kaulos* "stem"; see C.S. Rafinesque, *Sylva Telluriana*. 129. Philadelphia 1838.

Ophiocephalus Wiggins Scrophulariaceae

Origins:
From the Greek *ophis* "a snake, serpent" and *kephale* "head."

Ophiochloa Filgueiras, Davidse & Zulonga Gramineae

Origins:
From the Greek *ophis* "a snake, serpent" and *chloe, chloa* "grass."

Ophiocolea H. Perrier Bignoniaceae

Origins:
From the Greek *ophis* "a snake, serpent" and *koleos* "a sheath."

Ophioderma (Blume) Endl. Ophioglossaceae

Origins:

Greek *ophis* "a snake, serpent" and *derma, dermatos* "skin"; see Stephan F. Ladislaus Endlicher, *Genera Plantarum.* 66. 1836.

Ophioglossum L. Ophioglossaceae

Origins:

From the shape of the fructification, from the Greek *ophis* "a snake, serpent" and *glossa* "a tongue," snake-tongue, referring to the sometimes bifid apex above the fertile spike; see Carl Linnaeus, *Species Plantarum.* 1062. 1753 and *Genera Plantarum.* Ed. 5. 484. 1754.

Species/Vernacular Names:

O. azoricum Presl

Portuguese: língua de cobra

O. costatum R. Br.

English: large adder's tongue

O. gramineum Willd.

English: narrow adder's tongue

O. lusitanicum L.

English: common adder's tongue

Portuguese: língua de cobra, língua de cobra menor

O. pendulum L. (*Ophioderma pendulum* (L.) Presl)

English: hanging adder's tongue fern, ribbon fern

Japan: kobu-ran (= knobby orchid)

O. polyphyllum A. Br.

English: adder's tongue, large adder's tongue, inland adder's tongue

Namibia: adder's tongue fern

German: natternzunge, natternfarn

O. reticulatum L.

English: ground adder's tongue fern, adder's tongue

O. thermale Komarov

English: adder's tongue fern

Japan: hama-hana-yasuri (= beach *Ophioglossum*)

O. valdivianum Phil.

Chile: huentru-lahuén

O. vulgatum L.

Portuguese: língua de cobra maior, língua de serpente

Spain: llengua de serp, llança de Crist, lengua de serpiente, lengua serpentina, lanza de Cristo

Ophiomeris Miers Burmanniaceae

Origins:

From the Greek *ophis* "a snake, serpent" and *meris* "part," possibly referring to the leaves, spiral and scale-like.

Ophione Schott Araceae

Origins:

Greek *ophis* "a snake, serpent, a creeping plant," *ophioneos, ophione* "like a serpent," see also *Dracontium* L.

Ophionella Bruyns Asclepiadaceae

Origins:

From the Greek *ophione* "like a serpent," *ophis* "a snake, a creeping plant."

Ophiopogon Ker Gawl. Convallariaceae (Liliaceae)

Origins:

Greek *ophis* "a serpent, snake" and *pogon* "a beard, hair," tuft forming; see John Bellenden Ker Gawler, 1764-1842, in *The Botanical Magazine.* 27: t. 1063. 1807.

Species/Vernacular Names:

O. intermedius D. Don (*Ophiopogon spicatus* D. Don)

China: mai men tung

O. jaburan (Kunth) Lodd. (*Flueggea jaburan* Kunth)

English: white lily turf

Japan: noshi-ran

O. japonicus (L.f.) Ker Gawl. (*Convallaria japonica* Thunb.)

English: dwarf lily turf, Japanese snake's beard, lily turf

Japan: ryu-no-hige (= dragon's beard)

Okinawa: habukusa

China: mai dong, mai men dong

Vietnam: lan tien, mach mon

Ophiopteris Reinw. Oleandraceae

Origins:

From the Greek *ophis* "a serpent, snake" and *pteris* "fern."

Ophiorrhiza L. Rubiaceae

Origins:
From the Greek *ophis* "a snake, serpent" and *rhiza* "a root," referring to the serpentine roots; see Carl Linnaeus, *Species Plantarum*. 150. 1753 and *Genera Plantarum*. Ed. 5. 74. 1754.

Species/Vernacular Names:
O. australiana Benth.

English: Australian snake-root

Ophiorrhiziphyllon Kurz Acanthaceae

Origins:
From the Greek *ophis* "a snake, serpent," *rhiza* "a root" and *phyllon* "leaf."

Ophioseris Raf. Asteraceae

Origins:
Greek *ophis* and *seris, seridos*, the Greek name for a species of chicory or endive; see C.S. Rafinesque, *New Fl. N. Am.* 4: 87. 1836 [1838].

Ophioxylon L. Apocynaceae

Origins:
Greek *ophis* "a snake, serpent" and *xylon* "wood."

Ophiria Beccari Palmae

Origins:
The genus is named after Mount Ophir (Gunung Ledang), Johor, Malaysia; Latin Ophir for a region in southern Arabia.

Ophismenus Poir. Gramineae

Origins:
See also *Oplismenus* P. Beauv.

Ophiurinella Desv. Gramineae

Origins:
The diminutive of the genus *Ophiuros*.

Ophiuros Gaertner f. Gramineae

Origins:
From the Greek *ophis* "a snake, serpent" and *oura* "tail," referring to the spikes; see Joseph Gaertner (1732-1791), *De fructibus et seminibus plantarum*. Stuttgart, Tübingen 1788.

Species/Vernacular Names:
O. exaltatus (L.) O. Kuntze

The Philippines: girum, talangiu

Ophiurus R. Br. Gramineae

Origins:
Orthographic variant of *Ophiuros* Gaertner; see Robert Brown, *Prodromus florae Novae Hollandiae*. 206. London 1810.

Ophrestia H.M.L. Forbes Fabaceae

Origins:
An anagram of the generic name *Tephrosia* Pers.

Ophryococcus Oerst. Rubiaceae

Origins:
From the Greek *ophrys* "brow, eyebrow" and *kokkos* "a berry."

Ophryosporus Meyen Asteraceae

Origins:
Greek *ophrys* and *sporos* "a seed."

Ophrypetalum Diels Annonaceae

Origins:
From the Greek *ophrys* "brow, eyebrow" and *petalon* "leaf, petal."

Ophrys L. Orchidaceae

Origins:
From the Greek *ophrys* "brow, eyebrow," lip often very hairy, particularly around margins, some species of *Ophrys* are known as spider orchids because their flower lips resemble

the bodies of spiders; Latin *ophrys, yos* for a plant with two leaves, bifoil (Plinius).

Species/Vernacular Names:
O. sp.
Arabic: el haya ouel miyta
O. apifera Hudson
English: bee orchid
O. insectifera L.
English: fly orchid

Ophthalmoblapton Allemão Euphorbiaceae

Origins:
From the Greek *ophthalmos* "eye" and *blapto* "to blind, to harm, to damage."

Ophthalmophyllum Dinter & Schwantes Aizoaceae

Origins:
Greek *ophthalmos* "eye" and *phyllon* "leaf," referring to the lens-like epidermis.

Ophyoxylon Raf. Apocynaceae

Origins:
Greek *ophis* "a snake, serpent" and *xylon* "wood," see also *Ophioxylon* L.; see C.S. Rafinesque, *First Cat. Bot. Gard. Transylv. Univ.* 22. 1824; E.D. Merrill, *Index rafinesquianus.* 194. 1949.

Opicrina Raf. Asteraceae

Origins:
See C.S. Rafinesque, *New Fl. N. Am.* 4: 85. 1836 [1838]; E.D. Merrill, *Index rafinesquianus.* 240. 1949.

Opilia Roxb. Opiliaceae

Origins:
Derivation not known, possibly from the Greek *ope* "a hole, opening" and *eilo* "to be shut, to assemble," referring to the imbricate bracts concealing the flowers before opening; or from the Latin *opilio, onis* "a shepherd," or from an Indian name; see William Roxburgh (1751-1815), *Plants of the Coast of Coromandel.* 2: 31, t. 158. London (Apr.) 1802.

Opisthiolepis L.S. Smith Proteaceae

Origins:
From the Greek *opisthen* "behind, at the back," *opisthios* "hinder, belonging to the hinder part" and *lepis* "a scale"; see Lindsay Stewart Smith (1917-1970), "A new genus of Proteaceae from Queensland." *Proceedings of the Royal Society of Queensland.* 62: 79, t. IV. 1952.

Species/Vernacular Names:
O. heterophylla L.S. Smith
English: blush silky oak

Opisthocentra Hook.f. Melastomataceae

Origins:
From the Greek *opisthen* "behind, at the back" and *kentron* "a spur, point."

Opisthopappus C. Shih Asteraceae

Origins:
From the Greek *opisthen* "behind, at the back" and *pappos* "fluff, pappus."

Opithandra B.L. Burtt Gesneriaceae

Origins:
From the Greek *opisthen, opithe, opithen* "behind, at the back" and *aner, andros* "man, stamen."

Opizia Presl Gramineae

Origins:
For Philipp (Filip) Maximilian Opiz, 1787-1858, Czech botanist. See John H. Barnhart, *Biographical Notes upon Botanists.* 3: 31. 1965; T.W. Bossert, *Biographical Dictionary of Botanists Represented in the Hunt Institute Portrait Collection.* 294. 1972; H.N. Clokie, *Account of the Herbaria of the Department of Botany in the University of Oxford.* 219. Oxford 1964; E.M. Tucker, *Catalogue of the Library of the Arnold Arboretum of Harvard University.* 1917-1933; Stafleu and Cowan, *Taxonomic Literature.* 3: 839-841. 1981; Ida Kaplan Langman, *A Selected Guide to the Literature on the Flowering Plants of Mexico.* Philadelphia 1964; R. Zander, F. Encke, G. Buchheim and S. Seybold, *Handwörterbuch der Pflanzennamen.* 14. Aufl. Stuttgart 1993.

Opizia Raf. Brassicaceae

Origins:

See C.S. Rafinesque, *New Fl. N. Am.* 2: 29. 1836 [1837]; E.D. Merrill, *Index rafinesquianus.* 130. 1949.

Oplax Raf. Smilacaceae (Liliaceae)

Origins:

See C.S. Rafinesque, *Autikon botanikon.* Icones plantarum select. nov. vel rariorum, etc. 125. Philadelphia 1840; E.D. Merrill, *Index rafinesquianus.* 92. 1949.

Oplexion Raf. Boraginaceae

Origins:

See C.S. Rafinesque, *Flora Telluriana.* 4: 86. 1836 [1838]; E.D. Merrill, *Index rafinesquianus.* 203. 1949.

Oplismenopsis Parodi Gramineae

Origins:

Resembling the genus *Oplismenus.*

Oplismenus P. Beauv. Gramineae

Origins:

From the Greek *hoplismos, hoplisis* "a weapon, equipment for war, arming," referring to the awned spikelets; see Ambroise Palisot de Beauvois (1752-1820), *Flore d'Oware et de Benin en Afrique.* 2: 14. Paris 1810.

Species/Vernacular Names:

O. aemulus (R. Br.) Roemer & Schultes

English: Australian basketgrass, creeping beard-grass, wavy beard-grass

O. compositus (L.) P. Beauv.

The Philippines: balibatong, balibis, balisibis, banig-usa, kauakauayan, kawakawayan, huphuplit, litlitum, malakauayan, marikauayan, yamong-yamong, yamog-yamog, bailituganalu

O. hirtellus (L.) P. Beauv.

English: basketgrass

Hawaii: honohono kukui, honohono, honohono maoli

O. imbecillis (R. Br.) Roemer & Schultes

English: creeping beard-grass

Oplonia Raf. Acanthaceae

Origins:

From the Greek *hoplon* "a tool, implement, large shield, weapon"; see C.S. Rafinesque, *Flora Telluriana.* 4: 64. 1836 [1838]; E.D. Merrill (1876-1956), *Index rafinesquianus.* The plant names published by C.S. Rafinesque, etc. 224. Jamaica Plain, Massachusetts, USA 1949.

Oplopanax (Torrey & A. Gray) Miq. Araliaceae

Origins:

Greek *hoplon* "a tool, implement, weapon" plus the related genus *Panax*, referring to the spiny habit of these prickly treelets.

Species/Vernacular Names:

O. horridus (Sm.) Miq.

English: devil's club

Oploteca Raf. Amaranthaceae

Origins:

Referring to *Oplotheca* Nutt.; see C.S. Rafinesque, *New Fl. N. Am.* 4: 46. 1836 [1838]; E.D. Merrill, *Index rafinesquianus.* The plant names published by C.S. Rafinesque, etc. 119. Jamaica Plain, Massachusetts, USA 1949.

Oplukion Raf. Solanaceae

Origins:

Greek *hoplon* "a tool, implement, weapon" and *Lykion*, a Greek name used by Dioscorides and Plinius for a thorny shrub; see C.S. Rafinesque, *Sylva Telluriana.* 53. Philadelphia 1838.

Opocunonia Schltr. Cunoniaceae

Origins:

From the Greek *opos* "juice, sap, gum, resin" plus *Cunonia.*

Opodix Raf. Salicaceae

Origins:

See C.S. Rafinesque, *Am. Monthly Mag. Crit. Rev.* 1: 439. 1817; E.D. Merrill (1876-1956), *Index rafinesquianus.* The plant names published by C.S. Rafinesque, etc. 107. Jamaica Plain, Massachusetts, USA 1949.

Opoidea Lindley Umbelliferae

Origins:
Greek *opos* "juice, sap, gum, resin."

Opopanax W.D.J. Koch Umbelliferae

Origins:
Latin *opopanax, acis* for the juice of the herb *panax*, Greek *opopanax, akos* for the gum of *Opopanax hispidus*, Hercules' woundwort, Greek *opos* "juice, sap, gum, milky juice, resin" plus *Panax, panax, panakos* "a remedy," referring to the sap, supposed to cure all disorders, the gum used in scent making.

Opophytum N.E. Br. Aizoaceae

Origins:
From *opos* "juice, sap, gum, milky juice, resin" and *phyton*, some suggest from the Greek *opora* "autumn, the end of summer, the fruit-time, the fruit, tree-fruit" and *phyton* "plant," see also *Mesembryanthemum* L.

Opsago Raf. Solanaceae

Origins:
See Constantine Samuel Rafinesque (1783-1840), *Sylva Telluriana*. 54. 1838; E.D. Merrill, *Index rafinesquianus*. 212. 1949.

Opsiandra O.F. Cook Palmae

Origins:
Greek *opson* "food" and *aner, andros* "man, stamen"; see Orator Fuller Cook (1867-1949), in *Journal of the Washington Academy of Sciences*. 13: 182. 1923.

Opsianthes Lilja Onagraceae

Origins:
From the Greek *opson* "food" and *anthos* "flower."

Opuntia Miller Cactaceae

Origins:
Latin *herba Opuntia*, from *Opus, Opuntis* "a town of Locris, in Greece," *Opuntius, a, um* "Opuntian," Greek *Opous,*

Opountos; some suggest from the Papago Indian name *opun*; see Philip Miller, *The Gardeners Dictionary*. Abr. ed. 4. London (28 Jan.) 1754; Helmut Genaust, *Etymologisches Wörterbuch der botanischen Pflanzennamen*. 440. Basel 1996; F. Boerner & G. Kunkel, *Taschenwörterbuch der botanischen Pflanzennamen*. 4. Aufl. 143. 1989; G.C. Wittstein, *Etymologisch-botanisches Handwörterbuch*. 636. Ansbach 1852; Salvatore Battaglia, *Grande dizionario della lingua italiana*. XI: 1089. UTET, Torino 1981.

Species/Vernacular Names:
O. sp.

Peru: cajaruro, lalo, airampo, tchai

Mexico: a'xitl, biaa-gueta, guichi, biichi, bitzu, bironi, bitoni, pitoni, chog, irá, la-po-nécaiualá, lo, nacá, nacari-té, ndi-tu, paré, shodo, shoto, naabo

Malaya: lidah jin

O. atacamensis Philippi

Chile: chuchampe

O. aurantiaca Lindley

English: jointed cactus, jointed prickly pear, tiger pear

Southern Africa: kaatjie, lidjiesturksvy, litjiesturksvy, platturksvy, rankturksvy, litjies kaktus, taaietjie; makonde (Venda)

O. basilaris Engelm. & J. Bigelow

English: beavertail cactus

O. berteri (Colla) A. Hoffmann (for the Italian botanist Carlo Luigi Giuseppe Bertero, 1789-1831, plant collector in Chile; see Frans A. Stafleu and Erik A. Mennega, *Taxonomic Literature. Supplement II*. 118-119. 1993; John H. Barnhart, *Biographical Notes upon Botanists*. 1: 175. Boston 1965; A. Lasègue, *Musée botanique de Benjamin Delessert*. Paris 1845; E.M. Tucker, *Catalogue of the Library of the Arnold Arboretum of Harvard University*. 1917-1933; T.W. Bossert, *Biographical Dictionary of Botanists Represented in the Hunt Institute Portrait Collection*. 36. 1972; H.N. Clokie, *Account of the Herbaria of the Department of Botany in the University of Oxford*. 131. Oxford 1964; R. Zander, F. Encke, G. Buchheim and S. Seybold, *Handwörterbuch der Pflanzennamen*. 14. Aufl. Stuttgart 1993; Irving William Knobloch, compil., "A preliminary verified list of plant collectors in Mexico." *Phytologia Memoirs*. VI. 1983)

Chile: perrito, gatito, puskaye

O. bigelovii Engelm. (or *bigelowii*) (for the American botanist John Milton Bigelow, 1804-1878, botanical collector, U.S. Army surgeon, accompanied Lieut. A.W. Whipple's Exploring Expedition; see Gordon Douglas Rowley, *A History of Succulent Plants*. 1997; Irving William Knobloch, compil., "A preliminary verified list of plant collectors in Mexico." *Phytologia Memoirs*. VI. 1983; J. Lanjouw and F.A. Stafleu, *Index Herbariorum*. Part II, *Collectors A-D*.

Regnum Vegetabile vol. 2. 1954; Frans A. Stafleu and Erik A. Mennega, *Taxonomic Literature. Supplement II: Be-Bo.* 163-164. 1993; A.E. Waller, "Dr. John Milton Bigelow, 1804-1878, an early Ohio physician-botanist." *Pap. Dept. Bot. Ohio State Univ.* 449: 313-331. 1942; John H. Barnhart, *Biographical Notes upon Botanists.* 1: 185. Boston 1965; E.M. Tucker, *Catalogue of the Library of the Arnold Arboretum of Harvard University.* 1917-1933; J. Ewan, ed., *A Short History of Botany in the United States.* 1969; Joseph Ewan, *Rocky Mountain Naturalists.* The University of Denver Press 1950)

English: teddy-bear cholla

O. brasiliensis (Willdenow) A. Berger (*Brasiliopuntia brasiliensis* (Willdenow) A. Berger)

Peru: sincuc-casha, shucuru casha, supuai manchachi, shucucu casha, siucu casha, siucus casha

O. chlorotica Engelm. & J. Bigelow

English: pancake prickly pear

O. cochenillifera (L.) Mill. (*Cactus cochenillifer* L.; *Nopalea cochenillifera* (L.) Salm-Dyck)

English: cochineal cactus

O. conoidea (Back.) A. Hoffmann

Chile: conofdeo

O. dillenii (Ker Gawler) Haw. (*Cactus dillenii* Ker Gawl.)

English: Dillen's prickly pear, pipestem prickly pear, prickly pear

Mexico: cuija, chaparra, pak'an, ya-ak-pak'an, y'ax-pak'an

Southern Africa: geeldoringturksvy, pypsteelturksvy, pypturksvy; makonde (Venda)

Japan: sen-nin-saboten

China: xian ren zhang

O. echinacea (Ritter) A. Hoffmann

Chile: puscaya, espina

O. echinocarpa Engelm. & J. Bigelow

English: silver cholla, golden cholla

O. engelmannii Engelm. var. *engelmannii*

English: Engelmann prickly pear

Mexico: héel

O. erinacea Engelm. var. *erinacea*

English: Mojave prickly pear

O. exaltata Berg.

English: long spine cactus

Southern Africa: langdoringkaktus; makonde (Venda)

O. ficus-indica (L.) Miller (*Cactus ficus-indica* L.; *Opuntia engelmannii* Salm-Dyck; *Opuntia occidentalis* Engelm. & Bigelow)

English: Indian fig, mission prickly pear, prickly pear, spineless cactus, sweet prickly pear, barbary fig, Burbank's spineless cactus

French: figuier d'Inde

Arabic: hendi, seurti, nowara hindia

Italian: ficodindia (see M. Cortelazzo & P. Zolli, *Dizionario etimologico della lingua italiana.* 2: 430. Zanichelli, Bologna 1980)

Peru: pupa, tuna

Mexico: nopal de Castilla, nopal sin espinas, tuna de Castilla

Venezuela: tuna de España, tuna real, tuna mansa

Hawaii: panini, papipi

Southern Africa: struksvy, stuksvy, turksvy, wyfieturksvy-bobbejaansturksvy, boereturksvy, doringblad, Indiaansche vij, Indiaanse turksvy, kaalblad, mannetjiesturksvy; makonde (Venda)

O. fragilis (Nutt.) Haw.

English: brittle prickly pear, little prickly pear

O. humifusa (Raf.) Schltr.

English: creeping pear

O. ignescens Vaupel

Chile: jala-jala, puskayo

O. imbricata (Haw.) DC. (*Cereus imbricatus* Haw.; *Opuntia decipiens* DC.; *Opuntia arborescens* Engelm.)

English: chain-link cactus, devil's rope, devil's rope pear, imbricate cactus, imbricate prickly pear

Mexico: abrojo, cardenche, cardón, cojonostle, coyonostle, coyonostli, coyonoxtle, joconostle, tasajo, tuna joconoxtli, velas de coyote, xoconochtli

Southern Africa: kabelturksvy, toukaktus; makonde (Venda)

O. leoncito Werdermann

Chile: leoncito

O. lindheimeri Engelm.

English: small roundleaved prickly pear, Lindheimer prickly pear, Texas prickly pear

South Africa: klein rondeblaarturksvy

O. microdasys (Lehm.) Pfeiffer

English: golden bristle cactus

Mexico: cegador, ciega borrego, nopal cegador, nopal real, nopalillo, tlacota-nochtli, tlatoc-nochtli

O. miquelii Monv.

Chile: tunilla, tuna de Miguel

O. ovata Pfeiffer

Chile: gatito, perrito

O. parishii Orc.

English: club cholla, mat cholla

O. parryi Engelm.

English: cane cholla, snake cholla

O. prolifera Engelm.

English: cholla

O. pubescens Wendland (*Cactus nanus* Kunth; *Cactus pubescens* (Wendland) Lemaire; *Cereus nanus* (Kunth) DC.; *Opuntia pascoensis* Britton & Rose; *Opuntia pestifer* Britton & Rose; *Platyopuntia nana* (Kunth) F. Ritter)

Peru: upa casha

O. pulchella Engelm.

English: sand cholla

O. ramosissima Engelm.

English: diamond cholla, pencil cactus

O. robusta Pfeiffer

English: wheel cactus

Mexico: cochinera, tuna camuesa, tuna tapona, nopal, k'oh, nopalli

O. rosea DC.

English: Douglas pest, rosea cactus

Southern Africa: roseakaktus; makonde (Venda)

O. schumannii Weber

English: Schumann prickly pear

Southern Africa: Schumann-turksvy; makonde (Venda)

O. soehrensii Britton & Rose (*Platyopuntia soehrensii* (Britton & Rose) F. Ritter)

Peru: airampo, hairampu, ayrampo

Chile: ayrampu

O. spinulifera Salm-Dyck

English: large round-leaved prickly pear, saucepan cactus

Southern Africa: blouturksvy, grootrondeblaarturksvy, rondbladturksvy; makonde (Venda)

O. streptacantha Lemaire

English: Cardona pear, white-spined pear

Mexico: nopal cardón, tuna cardona

O. stricta (Haw.) Haw.

Southern Africa: suurturksvy; makonde (Venda)

O. stricta (Haw.) Haw. var. *stricta* (*Cactus strictus* Haw.; *Opuntia inermis* DC.; *Opuntia stricta* (Haw.) Haw.)

Australia: common prickly pear, smooth pest pear, spiny pest pear

O. tomentosa Salm-Dyck

English: velvet tree pear

O. vulgaris Mill. (*Cactus monacanthus* Willd.; *Opuntia monacantha* (Willd.) Haw.)

English: barberry fig, Barbary fig, cochineal fig, Indian fig, prickly pear, prickly pear cactus, sour prickly pear, cochineal prickly pear, English prickly pear, drooping prickly pear, drooping tree pear, drooping pear, smooth tree pear, Irish-mittens

Southern Africa: Engelseturksvy, gewone turksvy, suurturksvy; makonde (Venda)

Japan: hira-uchiwa

O. wolfii (L. Benson) M. Baker

English: Wolf's cholla

Opuntiopsis Knebel Cactaceae

Origins:
Resembling the genus *Opuntia*.

Orania Zipp. Palmae

Origins:
Named after the Prince of Orange (Oranje), 1792-1849, Crown Prince of the Netherlands; see Karl Ludwig von Blume (1796-1862), in *Alg. Konst- en Letterbode*. 1829(19): 297. 1829.

Species/Vernacular Names:
O. palindan (Blanco) Merrill

The Philippines: palindan

O. rubiginosa Becc.

The Philippines: palindan-pula

Oraniopsis (Beccari) J. Dransfield, A.K. Irvine & N. Uhl Palmae

Origins:
Resembling the genus *Orania* Zipp.; see John Dransfield (1945-), Anthony Kyle Irvine (1937-) and Natalie Whitford Uhl (1919-), "*Oraniopsis appendiculata*, a previously misunderstood Queensland palm." in *Principes*. 29(2): 56-63. 1985.

Species/Vernacular Names:
O. appendiculata (F.M. Bailey) J. Dransfield, Irvine and Uhl

Brazil: orânia

Orbea Haw. Asclepiadaceae

Origins:
From the Latin *orbis* "a disc, a circular shape, circle."

Orbeanthus L.C. Leach Asclepiadaceae

Origins:
From the genus *Orbea* Haw. and *anthos* "flower."

Orbeopsis L.C. Leach Asclepiadaceae

Origins:
Resembling the genus *Orbea* Haw.

Orbexilum Raf. Fabaceae

Origins:
See C.S. Rafinesque, *Atl. Jour.* 1: 145. 1832; Elmer D. Merrill, *Index rafinesquianus.* 147. 1949.

Orbicularia Baillon Euphorbiaceae

Origins:
From the Latin *orbicularis* "circular," *orbiculus* "a small disk, a sheave."

Orbignia Martius Palmae

Origins:
Orthographic variation of *Orbignya* Martius ex Endlicher.

Orbignya Martius ex Endlicher Palmae

Origins:
Named for the French (b. Loire-Atlantique) naturalist Alcide-Charles-Victor Dessalines d'Orbigny, 1802-1857 (d. near Sant-Denis), traveler, zoologist, paleontologist, explorer, 1826-1834 in South America, palm collector, studied in Paris under Pierre-Louis-Antoine Cordier (1777-1861), professor of paleontology at the Muséum d'Histoire Naturelle de Paris. Among his numerous writings are *Voyage dans l'Amerique méridionale.* Paris, Strasbourg 1834[-1847], *Tableau méthodique de la classe des Céphalopodes.* Paris 1826 and *Paléontologie française. Description zoologique et géologique de tous les animaux mollusques et rayonnés fossiles de France.* Paris 1840-1856. See P. Fischer, "Notice sur la vie et sur les travaux d'Alcide d'Orbigny." *Bulletin de la Société géologique de France.* sér. 3. 6: 434-453. Paris 1878; A. Gaudry, "Alcide d'Orbigny, ses voyages et ses travaux." *Revue des Deux Mondes.* Paris 1859; A. Lasègue, *Musée botanique de Benjamin Delessert.* Paris 1845; B. Glass et al., eds.,

Forerunners of Darwin: 1745-1859. Baltimore 1959; August Weberbauer (1871-1948), *Die Pflanzenwelt der peruanischen Andes in ihren Grundzügen dargestellt.* 8-9. Leipzig 1911; Claudio Urbano B. Pinheiro and Michael J. Balick, "Brazilian Palms. Notes on their uses and vernacular names, compiled and translated from Pio Corrêa's 'Dicionário das Plantas Úteis do Brasil e das Exóticas Cultivadas,' with updated nomenclature and added illustrations." in *Contributions from the New York Botanical Garden.* Volume 17: 39-40. 1987; Heinz Tobien, in *Dictionary of Scientific Biography* 10: 221-222. 1981; F. Boerner & G. Kunkel, *Taschenwörterbuch der botanischen Pflanzennamen.* 4. Aufl. 143. Berlin & Hamburg 1989; H. Genaust, *Etymologisches Wörterbuch der botanischen Pflanzennamen.* 440. [the genus is dedicated to Alcide-Charles-Victor Dessalines d'Orbigny, 1802-1857] 1996; R. Zander, F. Encke, G. Buchheim and S. Seybold, *Handwörterbuch der Pflanzennamen.* 14. Aufl. 407, 785f. Stuttgart 1993; Georg Christian Wittstein, *Etymologisch-botanisches Handwörterbuch.* 636. Ansbach 1852; Stafleu and Cowan, *Taxonomic Literature.* 3: 842-844. [the genus is dedicated to Alcide Dessalines d'Orbigny, 1802-1857] 1981.

Species/Vernacular Names:
O. agrestis (Barb. Rodr.) Burr.
Brazil: curuaí

O. campestris Barb. Rodr.
Brazil: indaiá redondo, indaiá verdadeiro

O. cohune (C. Martius) Standley
English: cohune nut, cohune, cohune palm
Central America: cohune, corozo
Mexico: bega rago, biga rago

O. cuatrecasana Dugand
Colombia: táparos

O. eichleri Dr.
Brazil: piaçaba do Piauí, Pindoba (State of Goiás)

O. longibracteata Barb. Rodr.
Brazil: indaiá-mirim, indaiá crespo

O. macrocarpa Barb. Rodr.
Brazil: indaiá-açu

O. phalerata C. Martius
English: babassu palm
Brazil: coco de macaco, babassu, babaçu, Aguassú, Auassú, baguassú, bauassú, coco de palmeira, coco nayá, coco pindoba, guaguassú, oauassú, palha branca, uassú

O. pixuna Barb. Rodr.
Brazil: curuá pixúna, curuá preto, palha preta

O. racemosa (Spr.) Dr.

Brazil: piaçaba verdadeira, piaçaba do Amazonas

O. sabulosa Barb. Rodr.

Brazil: inaiaſ, Caruaru

O. spectabilis (C. Martius) Burret

Brazil: curua

Orchadocarpa Ridley Gesneriaceae

Origins:
Greek *orchas, orchados* "enclosing, a kind of olive" and *karpos* "a fruit," Latin *orchas, adis* for a kind of edible olive of an oblong shape.

Orchiastrum Séguier Orchidaceae

Origins:
From the Greek *orchis, orchidos* "a testicle" and *-astrum*, a Latin substantival suffix indicating incomplete resemblance.

Orchidantha N.E. Br. Lowiaceae

Origins:
From *Orchis* and the Greek *anthos* "flower."

Orchidium Swartz Orchidaceae

Origins:
From the Greek *orchis, orchidos* "a testicle," the diminutive *orchidion*, referring to the small pseudobulbs.

Orchidocarpum Michx. Annonaceae

Origins:
From the Greek *orchis, orchidos* "a testicle" and *karpos* "fruit."

Orchidofunckia A. Rich. & Galeotti Orchidaceae

Origins:
Named presumably for the Luxemburg artist and naturalist Nicolas Funck, 1816-1896, plant collector in Central and South America 1835-1847, collaborated with Linden.

Orchidotypus Kraenzlin Orchidaceae

Origins:
Greek *orchis* "orchid" and *typos* "model," a new type of orchid.

Orchiodes Kuntze Orchidaceae

Origins:
Greek *orchis, orchidos* "a testicle" and *-odes* "resembling, of the nature of, like, having the form," resembling the genus *Orchis*; see Otto Kuntze (1843-1907), *Revisio generum Plantarum Vascularium.* 2: 675. Leipzig 1891.

Orchipeda Blume Apocynaceae

Origins:
Presumably from the Greek *orchipedon* "testicles," *orchipede* "restraint of the testicles."

Orchipedum Breda Orchidaceae

Origins:
From the Greek *orchipedon* "testicles."

Orchis L. Orchidaceae

Origins:
Greek *orchis* "a testicle," referring to the shape of the tuberoids; Latin *orchis, is* for a plant, a kind of olive (Columella), or for another plant, so called from the shape of its roots, the *orchis* (Plinius); see Georg Christian Wittstein, *Etymologisch-botanisches Handwörterbuch.* 636. Ansbach 1852; A. Bonavilla, *Dizionario etimologico di tutti i vocaboli usati nelle scienze, arti e mestieri, che traggono origine dal greco.* Milano 1819-1821; Ernest Weekley, *An Etymological Dictionary of Modern English.* 2: 1013. New York 1967; Manlio Cortelazzo & Paolo Zolli, *Dizionario etimologico della lingua italiana.* 4: 840. Zanichelli, Bologna 1985; Salvatore Battaglia, *Grande dizionario della lingua italiana.* XII: 17-18. Torino 1984.

Species/Vernacular Names:
O. sp.

Arabic: el haya ouel miyta

O. coriophora L.

English: bug orchid

O. laxiflora Lam.

English: Jersey orchid

O. mascula (L.) L.

English: early purple orchid, dead man's finger

O. militaris L.

English: military orchid, soldier orchid

O. morio L.

English: green-winged orchid

O. papilionacea L.

English: butterfly orchid

O. provincialis Balb.

English: Provence orchid

O. purpurea Huds.

English: lady orchid

O. sancta L.

English: holy orchid

O. simia L.

English: monkey orchid

O. tridentata Scop.

English: toothed orchid

O. ustulata L.

English: burnt orchid

Orcuttia Vasey Gramineae

Origins:
Dedicated to the American (California) botanist Charles Russell Orcutt, 1864-1929, plant collector and naturalist, 1882-1886 Baja California, editor of *The West American Scientist*. vol. 1-21. 1881-1919. He is best known for his *Flora of Southern and Lower California*. San Diego 1885, *Botany of Southern California*. San Diego 1901 and *American Plants*. San Diego 1907-1912. See H. Du Shane, *The Baja California Travels of Charles Russel Orcutt*. Los Angeles 1971; G.P. Trujillo, *Bibliografía de Baja California*. Tijuana 1967; Ira L. Wiggins, *Flora of Baja California*. 42. Stanford, California 1980; G. Murray, *History of the Collections Contained in the Natural History Departments of the British Museum*. 1: 172. London 1904; R. Zander, F. Encke, G. Buchheim and S. Seybold, *Handwörterbuch der Pflanzennamen*. 14. Aufl. Stuttgart 1993; T.W. Bossert, *Biographical Dictionary of Botanists Represented in the Hunt Institute Portrait Collection*. 295. 1972; S. Lenley et al., *Catalog of the Manuscript and Archival Collections and Index to the Correspondence of John Torrey*. Library of the New York Botanical Garden. 318. 1973; E.M. Tucker, *Catalogue of the Library of the Arnold Arboretum of Harvard University*. 1917-1933.

Species/Vernacular Names:

O. inaequalis Hoover

English: San Joaquin Valley Orcutt grass

O. pilosa Hoover

English: hairy Orcutt grass

O. tenuis A. Hitchc.

English: slender Orcutt grass

O. viscida (Hoover) Reeder

English: Sacramento Orcutt grass

Oreacanthus Benth. Acanthaceae

Origins:
From the Greek *oros* "mountain" and *Acanthus*.

Oreanthes Benth. Ericaceae

Origins:
From the Greek *oros* "mountain" and *anthos* "flower," growing in the Andes of Ecuador.

Oreas Cham. & Schltdl. Brassicaceae

Origins:
From the Greek *oros* "mountain," *oreias, oreiados* "belonging to mountains, mountain," Latin *Oreas, Oreadis* and Greek *Oreias* "a mountain-nymph, Oread," Plinius used *oreion, oreon, orion* for a mountain-plant.

Orectanthe Maguire Xyridaceae (Abolbodaceae)

Origins:
From the Greek *orektos* "stretched out, desired" and *anthos* "flower."

Oregandra Standley Rubiaceae

Origins:
Greek *orego* "reach, stretch, hold out" and *aner, andros* "man, stamen."

Oreiostachys Gamble Gramineae

Origins:
From the Greek *oreios* "from the mountains" and *stachys* "spike."

Oreithales Schltdl. Ranunculaceae

Origins:
From the Greek *oreithales* "blooming on the hills," Andes.

Oreobambos K. Schumann Gramineae

Origins:
From the Greek *oros* "mountain" plus *Bambos*.

Oreoblastus Susl. Brassicaceae

Origins:
From the Greek *oros* "mountain" and *blastos* "bud, sprout, germ, ovary."

Oreobliton Durieu Chenopodiaceae

Origins:
From the Greek *oros* "mountain" and *bliton* "blite."

Oreobolopsis Koyama & Guagl. Cyperaceae

Origins:
Resembling the genus *Oreobolus* R. Br.

Oreobolus R. Br. Cyperaceae

Origins:
Greek *oros* "mountain" and *bolos* "a lump of earth, lump," referring to the habit of growth and to the mountain habitat; see Robert Brown, *Prodromus florae Novae Hollandiae et Insulae van-Diemen*. 235. London 1810.

Species/Vernacular Names:
O. sp.
Hawaii: mau'u, kaluhaluha
O. distichus F. Muell.
English: fan tuft-rush
O. pumilio R. Br.
English: alpine tuft-rush

Oreobroma Howell Portulacaceae

Origins:
From the Greek *oros* "mountain" and *broma* "food."

Oreobulus Boeckeler Cyperaceae

Origins:
Orthographic variant of *Oreobolus* R. Br.

Oreocalamus Keng Gramineae

Origins:
From the Greek *oros* "mountain" and *kalamos* "a reed, cane."

Oreocallis R. Br. Proteaceae

Origins:
Greek *oros* "mountain" and *kallos* "beauty," referring to the mountain habitat and to the beauty of the flowers; see Robert Brown, in *Transactions of the Linnean Society of London*. 10: 196. 1810.

Oreocallis Small Ericaceae

Origins:
Greek *oros* "mountain" and *kallos* "beauty."

Oreocarya Greene Boraginaceae

Origins:
From the Greek *oros* "mountain" and *karyon* "nut."

Oreocaryon Kuntze ex K. Schumann Rubiaceae

Origins:
From the Greek *oros* "mountain" and *karyon* "nut."

Oreocereus (A. Berger) Riccobono Cactaceae

Origins:
Greek *oros* "mountain" plus *cereus*.

Species/Vernacular Names:
O. australis (Ritter) Hoffmann
Chile: arequipa
O. hempelianus (Gürke) Hunt
Chile: achacaño

O. leucotrichus (Phil.) Wagen.

Chile: viejito, chastudo

O. varricolor Backeberg

Chile: chastudo

Oreocharis (Decne.) Lindley Boraginaceae

Origins:

Greek *oros* "mountain" and *charis* "grace, beauty."

Oreocharis Benth. Gesneriaceae

Origins:

From the Greek *oros* "mountain" and *charis* "grace, beauty," referring to the habitat.

Oreochloa Link Gramineae

Origins:

From the Greek *oros* "mountain" and *chloe, chloa* "grass."

Oreochorte Koso-Pol. Umbelliferae

Origins:

From the Greek *oros* "mountain" and *chortos* "green herbage, grass."

Oreochrysum Rydb. Asteraceae

Origins:

Greek *oros* "mountain" and *chrysos* "gold."

Oreocnide Miquel Urticaceae

Origins:

From the Greek *oros* "mountain" and *knide* "nettle."

Species/Vernacular Names:

O. fruticosa (Gaudich.) Hand.-Mazz. (*Boehmeria fruticosa* Gaudich.)

Japan: iwa-ga-ne

Oreocome Edgew. Umbelliferae

Origins:

From the Greek *oros* "mountain" and *kome* "a tuft of hairs."

Oreocosmus Naudin Melastomataceae

Origins:

From the Greek *oros* and *kosmos* "ornament, decoration."

Oreodaphne Nees Lauraceae

Origins:

Greek *oros* "mountain" and *daphne* "the laurel, the bay tree"; see in *Linnaea*. 8: 39. 1833.

Oreodendron C.T. White Thymelaeaceae

Origins:

From the Greek *oros* "mountain" (*oreos*, genitive of *oros*) and *dendron* "a tree"; see Cyril Tenison White (1890-1950), in *Contributions from the Arnold Arboretum of Harvard University*. 4: 74, t. IX. 1933.

Oreodoxa Willd. Palmae

Origins:

From the Greek *oros* "mountain" and *doxa* "glory," referring to the mountain habitat and to the growth of some species.

Oreogenia I.M. Johnston Boraginaceae

Origins:

From the Greek *oros* "mountain" and *genea* "race, family, tribe," *genos* "race, offspring, generation."

Oreogeum (Ser.) Golubkova Rosaceae

Origins:

From the Greek *oros* "mountain" plus the genus *Geum* L.

Oreogrammitis Copel. Grammitidaceae

Origins:

From the Greek *oros* "mountain" plus *Grammitis* Sw.

Oreograstis K. Schumann Cyperaceae

Origins:

From the Greek *oros* "mountain" and *grastis* "grass, green fodder."

Oreoherzogia W. Vent Rhamnaceae

Origins:

For the German botanist Theodor Carl (Karl) Julius Herzog, 1880-1961, bryologist, plant collector in Sri Lanka and Bolivia (flora of the Bolivian Andes), professor of botany. His writings include "Siphonogamae novae Bolivienses in itinere per Boliviam orientalem ab auctore lectae." *Repert. Spec. Nov. Regni Veg.* 7: 49-69. 1909, "Die von Dr. Th. Herzog auf seiner zweiten Reisen durch Bolivien in den Jahren 1910 und 1911 gesammelten Pflanzen. Teil II." *Meded. Rijks-Herb.* 27: 1-90. 1915, "Die von Dr. Th. Herzog auf seiner zweiten Reisen durch Bolivien in den Jahren 1910 und 1911 gesammelten Pflanzen. Teil III." *Meded. Rijks-Herb.* 29: 1-94. 1916, *Die Pflanzenwelt der bolivischen Anden und ihres östlichen Vorlandes.* Leipzig 1923, "Studien über kritische und neue Lejeuneaceae der Indomalaya." *Svensk Bot. Tidskr.* 42: 230-241. 1948 and "Hepaticae ecuadorienses a cl. d.re Gunnar Harling annis 1946-1947 lectae." *Svensk Bot. Tidskr.* 46: 62-108. 1952, with R. Pilger wrote "Rhamnaceae." in "Die von Dr. Th. Herzog auf seiner zweiten Reisen durch Bolivien in den Jahren 1910 und 1911 gesammelten Pflanzen." *Meded. Rijks-Herb.* 1921. See J.H. Barnhart, *Biographical Notes upon Botanists.* 2: 167. 1965; T.W. Bossert, *Biographical Dictionary of Botanists Represented in the Hunt Institute Portrait Collection.* 173. 1972; S. Lenley et al., *Catalog of the Manuscript and Archival Collections and Index to the Correspondence of John Torrey.* 222. 1973; Elmer Drew Merrill, *Contr. U.S. Natl. Herb.* 30(1): 149. 1947; Gordon Douglas Rowley, *A History of Succulent Plants.* Strawberry Press, Mill Valley, California 1997; Ida Kaplan Langman, *A Selected Guide to the Literature on the Flowering Plants of Mexico.* 367. Philadelphia 1964; Stafleu and Cowan, *Taxonomic Literature.* 2: 178-181. 1979; Ethelyn Maria Tucker, *Catalogue of the Library of the Arnold Arboretum of Harvard University.* Cambridge, Massachusetts 1917-1933.

Oreoleysera Bremer Asteraceae

Origins:

Greek *oros* "mountain" and the genus *Leysera* L., dedicated to the German botanist Friedrich Wilhelm von Leysser (Fridericus Wilhelmus a Leyser), 1731-1815, author of *Flora halensis.* Halae Salicae [Halle a. Saale] 1761 and [Editio altera, aucta et reformata.] 1783; see John Hendley Barnhart (1871-1949), *Biographical Notes upon Botanists.* 2: 377. Boston 1965; Frans A. Stafleu (1921-1997), *Linnaeus and the Linnaeans. The spreading of their ideas in systematic botany, 1735-1789.* Utrecht 1971; Jonas Carlsson Dryander (1748-1810), *Catalogus bibliothecae historico-naturalis Josephi Banks.* London 1796-1800; Ethelyn (Daliaette) Maria Tucker, *Catalogue of the Library*

of the Arnold Arboretum of Harvard University. Cambridge, Massachusetts 1917-1933.

Oreoloma Botsch. Brassicaceae

Origins:

From the Greek *oros* "mountain" and *loma* "border, margin, fringe, edge."

Oreomitra Diels Annonaceae

Origins:

From the Greek *oros* "mountain" and *mitra* "a head-band."

Oreomyrrhis Endl. Umbelliferae

Origins:

From the Greek *oros* "mountain" and the genus *Myrrhis* Mill.; see Stephan F. Ladislaus Endlicher, *Genera Plantarum.* 787. 1839.

Species/Vernacular Names:

O. argentea (Hook.f.) Hook.f.

English: silvery carraway

O. brevipes E.M. Mathias & Constance

English: rock carraway

O. ciliata Hook.f.

English: bog carraway

O. eriopoda (DC.) Hook.f. (*Caldasia eriopoda* DC.)

Australia: Australian caraway, Australian carraway

O. pulvinifica F. Muell.

English: cushion carraway

Oreonana Jepson Umbelliferae

Origins:

Mountain dwarf, from the Greek *oros* "mountain" and *nanos* "dwarf"; see Shevock and Norris, in *Fremontia.* 9: 22-25. 1981.

Species/Vernacular Names:

O. purpurascens J.R. Shevock & Constance

English: purple mountain-parsley

O. vestita (S. Watson) Jepson

English: woolly mountain-parsley

Oreonesion J. Raynal Gentianaceae

Origins:
Greek *oros* "mountain" and *nesos* "island," *nesion* "islet."

Oreopanax Decne. & Planchon Araliaceae

Origins:
Greek *oros* "mountain" with *Panax.*

Oreophylax Endl. Gentianaceae

Origins:
From the Greek *oros* "mountain" and *phylax, phylakos* "a guardian, protector," *oreophylax* "saltuarius, desert-guard"; see H.H. Allan, *Fl. New Z.* 1: 766. 1961; H.E. Connor and E. Edgar, "Name changes in the indigenous New Zealand flora, 1960-1986 and Nomina Nova IV, 1983-1986." *New Zealand Journal of Botany.* Vol. 25: 115-170. 1987; A. Löve, in *Taxon.* 32: 511. 1983.

Oreophysa (Boiss.) Bornm. Fabaceae

Origins:
Greek *oros* "mountain" and *physa* "a bladder," from mountain near Teheran in Persia (Iran).

Oreophyton O.E. Schulz Brassicaceae

Origins:
From the Greek *oros* "mountain" and *phyton* "plant."

Oreopoa Gand. Gramineae

Origins:
From the Greek *oros* "mountain" and *poa* "grass, pasture grass," genus *Poa* L.

Oreopolus Schltdl. Rubiaceae

Origins:
From the Greek *oreopolos* "haunting mountains," *oreopoleo* "haunt mountains."

Oreoporanthera Hutch. Euphorbiaceae

Origins:
Greek *oros* "mountain" plus the genus *Poranthera* Rudge; see J. Hutchinson, in *American Journal of Botany.* 56: 747 adnot. 1969.

Oreopteris Holub Thelypteridaceae

Origins:
Greek *oros* "mountain, hill" and *pteris* "fern," referring to the habitat.

Oreorchis Lindley Orchidaceae

Origins:
From the Greek *oros* "mountain" and *orchis* "orchid," referring to the habitat.

Oreorhamnus Ridl. Rhamnaceae

Origins:
From the Greek *oros* "mountain" and *Rhamnus.*

Oreoschimperella Rauschert Umbelliferae

Origins:
From the Greek *oros* "mountain" and *Schimperella* H. Wolff, named for the German botanist Georg Heinrich Wilhelm Schimper, 1804-1878, plant collector, brother of Karl Friedrich Schimper (1803-1867), cousin of the Alsatian bryologist and paleontologist Wilhelm Philipp (Guillaume Philippe) Schimper (1808-1880); see John H. Barnhart, *Biographical Notes upon Botanists.* 3: 226. 1965; A.P.M. Sanders, *Dictionary of Scientific Biography* 12: 165-167. 1981; Heinz Tobien, *Dictionary of Scientific Biography* 12: 167-168. 1981; P.W. Richards, *Dictionary of Scientific Biography* 12: 168-169. 1981.

Oreosciadium Wedd. Umbelliferae

Origins:
From the Greek *oros* "mountain" and *skiadion, skiadeion* "umbel, parasol."

Oreosedum Grulich Crassulaceae

Origins:
From the Greek *oros* "mountain" plus *Sedum* L.

Oreoselinon Raf. Umbelliferae

Origins:
Latin *oreoselinon, oreoselinum* and Greek *oreoselinon* "mountain-parsley"; see C.S. Rafinesque, *The Good Book.* 56. 1840; E.D. Merrill, *Index rafinesquianus.* 181. 1949.

Oreoselinum Miller Umbelliferae

Origins:

Latin *oreoselinon, oreoselinum* and Greek *oreoselinon* "mountain-parsley."

Oreoselis Raf. Umbelliferae

Origins:

Latin *oreoselinon, oreoselinum* and Greek *oreoselinon* "mountain-parsley"; see C.S. Rafinesque, *The Good Book.* 56. 1840; E.D. Merrill, *Index rafinesquianus.* 181. 1949.

Oreosolen Hook.f. Scrophulariaceae

Origins:

From the Greek *oros* "mountain" and *solen* "a tube, channel, pipe."

Oreosphacus Leybold Labiatae

Origins:

From the Greek *oros* "mountain" and *sphakos* "sage."

Oreostemma Greene Asteraceae

Origins:

From the Greek *oros* "mountain" and *stemma* "crown, a garland."

Oreostylidium Berggren Stylidiaceae

Origins:

From the Greek *oros* "mountain" and the genus *Stylidium* Sw. ex Willd., *stylidion* "a small pillar."

Oreosyce Hook.f. Cucurbitaceae

Origins:

From the Greek *oros* "mountain" and *sykon* "fig" or *sikyos* "wild cucumber, gourd."

Oreotelia Raf. Umbelliferae

Origins:

Referring to *Seseli* L.; see C.S. Rafinesque, *The Good Book.* 50. 1840; E.D. Merrill, *Index rafinesquianus.* 181. 1949.

Oreothyrsus Lindau Acanthaceae

Origins:

From the Greek *oros* "mountain" and *thyrsos* "a panicle."

Oreoxis Raf. Umbelliferae

Origins:

See C.S. Rafinesque, *Seringe Bull. Bot.* 1: 217. 1830; Elmer D. Merrill, *Index rafinesquianus.* 181. 1949.

Oresitrophe Bunge Saxifragaceae

Origins:

Greek *oresitrophos, oreitrophos, oreitrephes* "mountain-bred, mountain-fed," a genus native to the mountains of Northern China; Oresitrophos was one of Actaeon's hounds.

Orestias Ridl. Orchidaceae

Origins:

From the Greek *orestias* "of the mountains."

Orestion Raf. Asteraceae

Origins:

Latin *orestion, ii* for a plant, called also *helenion* and *nectarea* (Plinius); see Constantine Samuel Rafinesque (1783-1840), *Flora Telluriana.* 2: 48. Philadelphia 1837.

Orfilea Baillon Euphorbiaceae

Origins:

Named for the chemist Mathieu Joseph Bonaventure Orfila, (he was born in Minorca) 1787-1853 (d. Paris), toxicologist, one of the founders of the Académie de Médecine, professor of medical jurisprudence and of chemistry. Among his works are *Traité de toxicologie générale.* Paris 1813, *Traité des poisons.* 2me éd. Paris 1818, *Élemens de Chimie médicale.* Paris 1817, *Traité de Médecine légale.* Paris 1836, *Mémoire sur la nicotine et sur la conicine.* Paris 1851 and *Nouvelles recherches sur l'urine des ictériques.* Paris 1811; see Garrison and Morton, *Medical Bibliography.* 2072. New York 1961; J. Barse, *Manuel de la Cour d'Assises dans les questions d'empoisonnement.* Paris 1845; Amédé Fayol, *La vie et l'oeuvre d'Orfila.* Paris [1930].

Orias Dode Lythraceae

Origins:
From the Greek *orestias, orias* "of the mountains, mountain-wind."

Oribasia Schreb. Rubiaceae

Origins:
Oribasius (or Oreibasius) was a celebrated physician, esteemed by the emperor Julian, he abridged the works of Galenus; see Fridolf Kudlien, in *Dictionary of Scientific Biography* 10: 230-231. 1981. Some suggest an origin from the Greek *ouribatas, oreibates, oribates* "walking the mountains," Latin *oribata* and Greek *oreibates* "a mountain-climber," Oribasus or Oreibasos (mountain-climber) was one of Actaeon's hounds.

Oricia Pierre Rutaceae

Origins:
Origins obscure, possibly from the Greek *oros* "mountain" or Latin *Oricius, a, um* "belonging to Oricum, Orician"; Oricum or Oricus, a town of Epirus, on the Ionian Sea, the tree which produces turpentine grew there in abundance, see Vergilius, *The Aeneid.* 10, v. 136.

Species/Vernacular Names:
O. bachmannii (Engl.) Verdoorn (*Teclea bachmannii* Engl.; *Teclea swynnertonii* Bak.f.; *Oricia swynnertonii* (Bak.f.) Verdoorn; *Oricia transvaalensis* Verdoorn) (the specific name honors Charles Francis Massey Swynnerton, 1877-1938, botanical collector in Gazaland, b. India 1877-d. Tanganyika 1938, farmer, botanical collector in Rhodesia and Mozambique, 1907 Fellow of the Linnean Society; see Mary Gunn and Leslie E. Codd, *Botanical Exploration of Southern Africa.* 339. Cape Town 1981; Ray Desmond, *Dictionary of British & Irish Botanists and Horticulturists.* 668. 1994)

English: twin-berry tree

Southern Africa: tweelingbessieboom, oricia; ruAnzili, ruAnziti, chiRgwanzili (Shona); uMozane (Zulu); iNzanyane (Xhosa)

O. suaveolens (Engl.) Verdoorn

Nigeria: ain-adie (Yoruba)

Oriciopsis Engl. Rutaceae

Origins:
Resembling *Oricia.*

Origanum L. Labiatae

Origins:
Ancient classical Greek name, *origanon, oreiganon, origanos, oreiganos,* possibly from the Greek *oros* "mountain" and *ganos* "beauty, brightness, ornament, delight," Latin *origanum* and *origanon* and *origanus* for the plant wild-marjoram, origan (Plinius); see Carl Linnaeus, *Species Plantarum.* 588. 1753 and *Genera Plantarum.* Ed. 5. 256. 1754; Manlio Cortelazzo & Paolo Zolli, *Dizionario etimologico della lingua italiana.* 4: 844. Bologna 1985; V. Bertoldi, in *Revue de linguistique romane.* II: 140. Paris 1926; R. Strömberg, *Griechische Pflanzennamen.* 24-26, 117. Göteborg 1940; *Testi fiorentini del Dugento e dei primi del Trecento,* con introduzione, annotazioni linguistiche e glossario a cura di Alfredo Schiaffini. Firenze 1926; Ernest Weekley, *An Etymological Dictionary of Modern English.* 2: 1015. Dover Publications, New York 1967; Salvatore Battaglia, *Grande dizionario della lingua italiana.* XII: 101. Torino 1984.

Species/Vernacular Names:
O. dictamnus L.

English: dittany

O. majorana L.

English: sweet marjoram, marjoram, knotted marjoram

Arabic: mardqouche, mardaddoush, mardaqoush, bardaqoush

O. onites L.

English: pot marjoram

O. vulgare L.

English: wild marjoram, common origanum, pot marjoram, oregano

Italian: origano, origamo, regamo, oregano

China: niu zhi, tu xiang ru

Orinocoa Raf. Solanaceae

Origins:
Orinoco, a river in Venezuela; see C.S. Rafinesque, *Sylva Telluriana.* 57. Philadelphia 1838.

Orinus Hitchc. Gramineae

Origins:
Greek *orino* "move, excite."

Orites R. Br. Proteaceae

Origins:
Greek *oros* "mountain," referring to the alpine habitat; Greek Orites "Ruler of the Seasons," of Apollo; see Robert

Brown, in *Transactions of the Linnean Society of London*. 10: 189. 1810.

Species/Vernacular Names:
O. acicularis (R. Br.) Roemer & Schultes (*Oritina acicularis* R. Br.)
English: prickly orites, yellow bush
O. diversifolia R. Br.
English: varied orites
O. excelsa R. Br.
English: northern orites, prickly ash, mountain silky oak
O. fragrans Bailey
English: fragrant orites, fragrant silky oak
O. lancifolia F. Muell.
English: alpine orites, lance-leaf orites
O. revoluta R. Br.
English: rolled orites

Orithalia Blume Gesneriaceae

Origins:
From the Greek *oros* "mountain" and *thaleia* "full of bloom, blooming, luxuriant," *thalia* "bloom."

Oritina R. Br. Proteaceae

Origins:
From the genus *Orites* R. Br., see Robert Brown, in *Transactions of the Linnean Society of London*. Botany. 10: 224. 1810.

Oritrephes Ridl. Melastomataceae

Origins:
From the Greek *oritrephes*, *oritrophos* "mountain-bred."

Oritrophium (Kunth) Cuatrec. Asteraceae

Origins:
From the Greek *oritrephes*, *oritrophos* "mountain-bred," growing in the Andes.

Orixa Thunb. Rutaceae

Origins:
From a Japanese name.

Species/Vernacular Names:
O. japonica Thunb.

English: Chinese quinine
China: chou yang, chang shan, hen shan, hu tsao, shu chi

Orlaya Hoffm. Umbelliferae

Origins:
For Johann Orlay, physician in Moscow, *circa* 1770-1829.

Orleanesia Barb. Rodr. Orchidaceae

Origins:
After Gaston d'Orléans, 1608-1660, a patron of botany; see Abel Brunyer, *Hortus Regius Blesensis*. Paris, Antoine Vitré 1653.

Ormenis (Cass.) Cass. Asteraceae

Origins:
From the Greek *oro* "raise" and *men, menos* "month," reputed to aid menstruation, or from *ormenos, hormenos* "shoot, sprout, stem, stalk," *ormenoeis* "having a long stalk," or named for Ormenis, the female descendant of Ormenus, a King of Thessaly.

Ormiastis Raf. Labiatae

Origins:
See C.S. Rafinesque, *Flora Telluriana*. 3: 93. 1836 [1837].

Ormilis Raf. Labiatae

Origins:
See C.S. Rafinesque, *Flora Telluriana*. 3: 94. 1836 [1837].

Ormiscus DC. Brassicaceae

Origins:
Greek *hormos* "a necklace, a chain," a diminutive, a small necklace, referring to the fruits; Latin *hormiscion, ii* and Greek *hormiskos* "a precious stone."

Ormocarpopsis R. Vig. Fabaceae

Origins:
Resembling *Ormocarpum* P. Beauv.

Ormocarpum P. Beauv. Fabaceae

Origins:
Greek *hormos* "a necklace, a chain" and *karpos* "fruit," the pods are moniliform or necklace-like; see Ambroise Palisot de Beauvois (1752-1820), *Flore d'Oware et de Benin en Afrique*. 1: 95. Paris 1807.

Species/Vernacular Names:
O. bibracteatum (A. Rich.) Bak.
Nigeria: faskara giwa (Hausa)

O. kirkii S. Moore (*Diphaca discolor* (Vatke) Chiov.; *Diphaca kirkii* (S. Moore) Taub.; *Ormocarpum affine* De Wild; *Ormocarpum discolor* Vatke; *Ormocarpum mimosoides* S. Moore)
Southern Africa: mupotonzoa, purupuru, musankanakutcha, muswutaderere (Shona)

O. trichocarpum (Taub.) Engl. (*Diphaca trichocarpa* Taub.; *Ormocarpum setosum* Burtt Davy)
English: caterpillar bush, jackal-tail tree, caterpillar tree
Southern Africa: ormocarpum, rusperboontjie; umSindandlovu, umSindandlovana (= even an elephant will recover), isiThibane (Zulu); isiTsibane, isiTibane (Swazi)

Ormoloma Maxon Dennstaedtiaceae (Lindsaeoideae)

Origins:
From the Greek *hormos* "a necklace, a chain" and *loma* "border, margin, fringe, edge."

Ormopteris J. Sm. Pteridaceae (Adiantaceae)

Origins:
From the Greek *hormos* "a necklace, a cord, wreath" and *pteris* "a fern."

Ormopterum Schischkin Umbelliferae

Origins:
Greek *hormos* "a necklace, a cord, chain" and *pteron* "a wing."

Ormosciadium Boiss. Umbelliferae

Origins:
From the Greek *hormos* "a necklace, a chain" and *skias* "a canopy, pavilion," *skiadion, skiadeion* "umbel, parasol."

Ormosia G. Jackson Fabaceae

Origins:
From the Greek *hormos* "a necklace, a chain," referring to the seeds of *Ormosia coccinea* (Aublet) Jackson; see George Jackson, (1790-1811), in *Transactions of the Linnean Society of London*. 10: 360. 1811.

Species/Vernacular Names:
O. ormondii (F. Muell.) Merr.
English: yellow bean

Ormosiopsis Ducke Fabaceae

Origins:
Resembling *Ormosia*.

Ormosolenia Tausch Umbelliferae

Origins:
From the Greek *hormos* "a necklace, a chain" and *solen* "a tube."

Ormostema Raf. Orchidaceae

Origins:
Greek *hormos* "a chain" and *stema* "stamen," referring to the rhizome; see C.S. Rafinesque, *Flora Telluriana*. 4: 38. 1836 [1838].

Ornanthes Raf. Oleaceae

Origins:
Greek *ornis, orneon* "a bird" and *anthos*; see C.S. Rafinesque, *New Flora and Botany of North America*. 3: 93. Philadelphia 1836 [1838], *Alsographia americana*. 39. Philadelphia 1838, *Sylva Telluriana*. 10. 1838 and *The Good Book*. 46, 49. 1840.

Ornanthus Raf. Oleaceae

Origins:
See *Ornanthes* Raf.

Ornichia Klack. Gentianaceae

Origins:
An anagram of *Chironia*.

Ornitharium Lindley & Paxton Orchidaceae

Origins:

From the Greek *ornitharion* "a small bird," indicating the shape of the lip.

Ornithidium R. Br. Orchidaceae

Origins:

Greek *ornis, ornithos* "a bird," *ornithion* "small bird," alluding to the upper lip of the stigma or the appearance of the flowers.

Ornithoboea Parish ex C.B. Clarke Gesneriaceae

Origins:

From the Greek *ornis, ornithos* "a bird" and the genus *Boea* Comm. ex Lam.

Ornithocarpa Rose Brassicaceae

Origins:

From the Greek *ornis, ornithos* "a bird" and *karpos* "fruit."

Ornithocephalochloa Kurz Gramineae

Origins:

From the Greek *ornis, ornithos* "a bird" plus the genus *Cephalochloa* Coss. & Durieu.

Ornithocephalus Hooker Orchidaceae

Origins:

Greek *ornis, ornithos* "a bird" and *kephale* "head," referring to the form of the column-apex and anther.

Ornithochilus (Lindley) Wallich ex Bentham Orchidaceae

Origins:

From the Greek *ornis, ornithos* "a bird" and *cheilos* "lip," referring to the shape and appearance of the bilobed lip.

Ornithogalon Raf. Hyacinthaceae (Liliaceae)

Origins:

See C.S. Rafinesque, *Flora Telluriana.* 2: 22. 1836 [1837], see also *Ornithogalum* L.

Ornithogalum L. Hyacinthaceae (Liliaceae)

Origins:

From the Greek *ornis, ornithos* "a bird" and *gala* "milk," in reference to the very fleshy bulbs or to the very white flowers; name used by Dioscorides and Plinius; Latin *ornithogale, es* and Greek *ornithogale* for a plant, the star of Bethlehem. See Jacques Julien Houtton de Labillardière (1755-1834), *Novae Hollandiae plantarum specimen.* Parisiis 1804-1806 [1807]; Carl Linnaeus, *Species Plantarum.* 306. 1753 and *Genera Plantarum.* Ed. 5. 145. 1754.

Species/Vernacular Names:

O. angustifolium Bor

English: star of Bethlehem

O. arabicum L.

English: lesser Cape lily

O. conicum Jacq.

English: chink, chinkerinchee

South Africa: tjienkerientjie

O. dubium Houtt. (*Ornithogalum flavescens* Jacq.; *Ornithogalum flavissimum* Jacq.; *Ornithogalum miniatum* Jacq.)

Southern Africa: geeltjienkerientjie, yellow chinkerinchee; iTsweletswele lasethafeni (Xhosa)

O. longibracteatum Jacq.

English: wild onion, sea onion, false sea onion, German onion

Southern Africa: masxabana (Xhosa)

O. maculatum Jacq. (*Ornithogalum thunbergianum* Bak.)

English: snake flower

South Africa: viooltjie, slangblom, snake flower, nagslangblom, geelviooltjie, pampoentjie, chinkerinchee, rooi Dirkie

O. multifolium Bak. (*Ornithogalum aurantiacum* Bak.)

South Africa: slangblom, kliptjienk, geelviooltjie

O. ornithogaloides (Kunth) Oberm. (*Ornithogalum zeyheri* Bak.; *Bulbinella ornithogaloides* Kunth)

English: grass chinkerinchee

South Africa: vlei chinkerinchee, gras-tjienkerientjiee, vlei-tjienkerientjiee

O. prasinum Lindl.

South Africa: gifbol, maerman

O. pruinosum Leighton (having a waxy bloom, from the Latin *pruinosus, a, um (pruina)* "frosty, rimy")

South Africa: chinkerinchee, wit Dirkie

O. pyrenaicum L.

English: bath asparagus, Prussian asparagus, star of Bethlehem

O. saundersiae Bak.

English: chincherinchee, giant chincherinchee, giant chinkerinchee, Transvaal chinkerinchee

South Africa: tjienkerientjiee, Transvaalse tjienkerientjiee

O. suaveolens Jacq. (*Ornithogalum breviscapum* Leighton; *Ornithogalum roodiae* Phill.)

South Africa: geelviooltjie

O. thyrsoides Jacq. (*Ornithogalum revolutum* sensu Ker Gawl., non Jacq.; *Ornithogalum gilgianum* Schltr. ex Poelln.; *Ornithogalum ceresianum* Leighton; *Ornithogalum hermannii* Leighton)

English: African wonder flower, wonder flower, star-of-Bethlehem, Cape lily, chincherinchee, chinkerinchee, chinckerinchee, common chinkerinchee

Portuguese: torrões de açúcar, pinhas

South Africa: gewone tjienkerientjie, tjienkerientjie, viooltjie, witviooltjie

O. umbellatum L.

English: star of Bethlehem

Ornithogloson Raf. Colchicaceae (Liliaceae)

Origins:
See C.S. Rafinesque, *Flora Telluriana*. 2: 32. 1836 [1837], see also *Ornithoglossum* Salisb.

Ornithoglossum Salisb. Colchicaceae (Liliaceae)

Origins:
Bird's tongue, from the Greek *ornis, ornithos* "a bird" and *glossa* "a tongue," the tepals are very narrow.

Species/Vernacular Names:
O. viride (L.f.) Ait. (*Melanthium viride* L.f.; *Ornithoglossum glaucum* Salisb.)

English: poison onion, Cape poison onion

South Africa: slangkop, geelslangkop, Karoo-slangkop, Kaapse slangkop, eendjies, Cape slangkop, yellow slangkop

Ornithophora Barb. Rodr. Orchidaceae

Origins:
From the Greek *ornis, ornithos* "a bird" and *phoros* "bearing," referring to the column.

Ornithopteris Bernh. Schizaeaceae

Origins:
From the Greek *ornis, ornithos* "a bird" and *pteris* "fern."

Ornithopteris (J.G. Agardh) J. Sm. Dennstaedtiaceae

Origins:
Greek *ornis, ornithos* "a bird" and *pteris* "fern."

Ornithopus L. Fabaceae

Origins:
Resembling a bird's foot, from the Greek *ornis, ornithos* "a bird" and *pous* "foot"; see Carl Linnaeus, *Species Plantarum*. 743-744. 1753 and *Genera Plantarum*. Ed. 5. 331. 1754.

Species/Vernacular Names:
O. sativus Brot.

English: common seradella, serradella

Ornithospermum Dumoulin Gramineae

Origins:
From the Greek *ornis, ornithos* "a bird" and *sperma* "seed."

Ornithostaphylos Small Ericaceae

Origins:
From the Greek *ornis, ornithos* "a bird" and *staphyle* "a cluster, a grape."

Species/Vernacular Names:
O. oppositifolia (C. Parry) Small

English: Baja California birdbush

Ornithrophus Bojer ex Engl. Cunoniaceae

Origins:
From the Greek *ornis, ornithos* "a bird" and *trophe* "food."

Ornus Boehm. Oleaceae

Origins:
Latin *ornus* for the wild mountain ash, a lance (Plinius), Akkadian *eranu, elanu, eranum, esum* "a tree"; see Helmut Genaust, *Etymologisches Wörterbuch der botanischen Pflanzennamen*. 442. 1996; Manlio Cortelazzo & Paolo Zolli, *Dizionario etimologico della lingua italiana*. 4: 846. 1985; Salvatore Battaglia, *Grande dizionario della lingua italiana*. XII: 136. Torino 1984; Giovanni Semerano, *Le origini della cultura europea*. Dizionario della lingua Latina e di voci moderne. 2(2): 497. Firenze 1994.

Orobanche L. Orobanchaceae (Scrophulariaceae)

Origins:
Greek *orobagche, orobanche, orobos* "a kind of vetch" and *anchein* "to strangle," Akkadian *arawu (aramu), erewu, eremu* "to cover," *erwum, ermu* "covering, cover," referring to the parasitic habit of the plant; Latin *orobanche, es* for the broom-rape, choke-weed (Plinius); see Carl Linnaeus, *Species Plantarum*. 632. 1753 and *Genera Plantarum*. Ed. 5. 281. 1754; H. Genaust, *Etymologisches Wörterbuch der botanischen Pflanzennamen*. 442. Basel 1996; Manlio Cortelazzo & Paolo Zolli, *Dizionario etimologico della lingua italiana*. 4: 847. 1985; Salvatore Battaglia, *Grande dizionario della lingua italiana*. XII: 140. Torino 1984; Giovanni Semerano, *Le origini della cultura europea*. Dizionari Etimologici. Basi semitiche delle lingue indeuropee. Dizionario della lingua Greca. 2(1): 212. Leo S. Olschki Editore, Firenze 1994.

Species/Vernacular Names:
O. spp.
China: lie dang
O. alba Stephan ex Willd. (*Orobanche epithymum* DC.; *Orobanche rubra* Sm.)
English: thyme broom rape
O. australiana F. Muell.
English: Australian broom rape, broom rape
O. crenata Forssk.
English: scalloped broom rape
Arabic: halouk metabi, diker el foul
O. fasciculata Nutt.

English: clustered broom rape
O. flava Mart.
English: yellow broom rape
O. minor Smith
English: clover broom rape, lesser broom rape
South Africa: klawerbesemraap
O. purpurea Jacq. (*Orobanche arenaria* auct. ex Wallr.; *Orobanche caerulea* Vill.)
English: purple broom rape
O. ramosa L.
English: blue broom rape, branched broom rape
Arabic: halouk
South Africa: blouduiwel, geelpop
O. uniflora L.
English: naked broom rape

Orobus L. Fabaceae

Origins:
Greek *orobos* "a kind of vetch," Latin *orobus* for the bitter vetch (Plinius Valerianus), Latin *ervum*, Akkadian *arawu (aramu), erewu (eremu* "to cover"), *erwum, ermu* "covering"; see G. Semerano, *Le origini della cultura europea*. Dizionario della lingua Greca. 2(1): 212. Firenze 1994.

Orochaenactis Coville Asteraceae

Origins:
From the Greek *oros* "mountain" plus the genus *Chaenactis* DC.

Orogenia S. Watson Umbelliferae

Origins:
Mountain race, from the Greek *oros* "mountain" and *genia* "born," *genea* "race, family, tribe."

Orontium L. Araceae

Origins:
From the ancient Greek *orontion*, name of a water-plant, a remedy for jaundice (Galenus), possibly from Orontes, a river in Syria, Latin *Oronteus*, poet. for Syrian.

Species/Vernacular Names:
O. aquaticum L.
English: golden club

Oropetium Trin. Gramineae

Origins:
Possibly from the Greek *oros* "mountain" and *pedion* "a plain, flat, open country, a field."

Orophaca Nutt. Fabaceae

Origins:
From the Greek *oros* "mountain" and *phakos* "a lentil."

Orophea Blume Annonaceae

Origins:
From the Greek *orophe* "roof, ceiling of a room," referring to the petals.

Orophochilus Lindau Acanthaceae

Origins:
From the Greek *orophe* "roof, ceiling of a room" and *cheilos* "lip."

Orophoma Spruce Palmae

Origins:
From the Greek *orophe* "roof," *orophos* "reed used for thatching houses, cover," a source of fiber for weaving and tying, leaves are used for thatching.

Orostachys (DC.) Sweet Crassulaceae

Origins:
From the Greek *oros* "mountain" and *stachys* "a spike," referring to the habitat and flowering.

Species/Vernacular Names:
O. fimbriata (Turcz.) A. Berger
China: wa song

Orostachys Fischer ex A. Berger Crassulaceae

Origins:
Greek *oros* "mountain" and *stachys* "a spike."

Orostachys Steudel Gramineae

Origins:
Greek *oros* "mountain" and *stachys* "a spike."

Orothamnus Pappe ex Hook. Proteaceae

Origins:
From the Greek *oros* "mountain" and *thamnos* "bush."

Species/Vernacular Names:
O. zeyheri Pappe ex Hook.
English: marsh rose

Oroxylum Vent. Bignoniaceae

Origins:
From the Greek *oros* "mountain" and *xylon* "wood," a pachycaul tree.

Species/Vernacular Names:
O. indicum (L.) Kurz
English: tree of Damocles, midnight horror, India trumpet flower
Malaya: beka, beka kampung, bonglai, bongloi, bonglai kayu, boli, boloi, bulai, bulai kayu, merlai
India: sheonak
Nepal: tatelo
Vietnam: may ca, nuc nac, moc ho diep
China: mu hu die

Oroya Britton & Rose Cactaceae

Origins:
From the name of a village in Peru, in the Andes, Junin; see also *Oreocereus* (A. Berger) Riccobono.

Species/Vernacular Names:
O. peruviana (Schumann) Britton & Rose (*Echinocactus peruvianus* Schumann)
Peru: uman-casha, uman-cusan

Orphanidesia Boiss. & Balansa Ericaceae

Origins:
Named after the Greek botanist Theodoros Georgios Orphanides, 1817-1886, poet, plant collector, studied in Paris, from 1848 professor of botany in Greece. Among his

works are *Prospectus flora graeca exsiccata*. Athènes 1850, *Chloridis Hellenicae classium, ordinum, specierumque enumeratio*. [in Greek] Athens 1868 and *Chios esclave*. Poème épique. [Translated into French prose by Charles Schaub.] Genève 1864. See John H. Barnhart, *Biographical Notes upon Botanists*. 3: 32. 1965; Theodor von Heldreich (1822-1902), *Catalogus systematicus herbarii Theodori G. Orphanidis* ... nunc ... in Museo Botanico Universitatis Athenarum. Fasciculus primus Leguminosae ... Florentiae 1877; H.N. Clokie, *Account of the Herbaria of the Department of Botany in the University of Oxford*. 219. Oxford 1964; G. Murray, *History of the Collections Contained in the Natural History Departments of the British Museum*. 1: 172. London 1904; T.W. Bossert, *Biographical Dictionary of Botanists Represented in the Hunt Institute Portrait Collection*. 295. 1972; R. Zander, F. Encke, G. Buchheim and S. Seybold, *Handwörterbuch der Pflanzennamen*. 14. Aufl. 759. Stuttgart 1993.

Orphanodendron Barneby & J.W. Grimes Caesalpiniaceae

Origins:

Greek *orphanos* "fatherless, without parents" and *dendron* "tree," only one species, Colombia.

Orphium E. Meyer Gentianaceae

Origins:

From the legendary Orpheus, a Greek poet and musician, son of the King Oeagrus (Oiagros), husband of Eurydice and one of the Argonauts.

Orrhopygium A. Löve Gramineae

Origins:

From the Greek *orropygion, orsopygion, orthopygion* "rump of birds, tail, rump of any animal," *orros* "rump, end of the *os sacrum*," Latin *orrhopygium* (a false read. for *orthopygium*), *orthopygium* "the rump and tail feathers of birds."

Orsidice Reichb.f. Orchidaceae

Origins:

Orsedice was a daughter of Cinyras and Metharme.

Ortegia L. Caryophyllaceae

Origins:

The name honors the Spanish botanist Casimiro Gómez de Ortega, 1740-1818, Madrid Botanical Garden 1771-1801, sent plants to Banks. His writings include *Tratado de la naturaleza y virtudes de la cicuta*. Madrid 1763, *De laudibus Caroli III. Hispaniarum Regis carmina*. Bononiae 1759, *Tabulae Botanicae*. Matriti 1773 and *Resumen historico del primer viage hecho al rededor del mundo*. Madrid 1769. See J.H. Barnhart, *Biographical Notes upon Botanists*. 3: 33 and 2: 62. 1965; Miguel Colmeiro y Penido, *La Botánica y los Botánicos de la Peninsula Hispano-Lusitana*. Madrid 1858; Ida Kaplan Langman, *A Selected Guide to the Literature on the Flowering Plants of Mexico*. Philadelphia 1964; Jonas C. Dryander, *Catalogus bibliothecae historico-naturalis Josephi Banks*. London 1800; Ethelyn Maria Tucker, *Catalogue of the Library of the Arnold Arboretum of Harvard University*. Cambridge, Massachusetts 1917-1933; G. Murray, *History of the Collections Contained in the Natural History Departments of the British Museum*. 1: 172. London 1904; R. Zander, F. Encke, G. Buchheim and S. Seybold, *Handwörterbuch der Pflanzennamen*. 14. Aufl. Stuttgart 1993.

Ortegocactus Alexander Cactaceae

Origins:

For the original discoverer of the species, the Mexican plant collector Francisco Ortega, Tehuantepec, Oaxaca, collaborator of the American (but Scottish-born) anthropologist and plant collector Thomas Baillie MacDougall (1895-1973, d. Oaxaca, Mexico); see Gordon Douglas Rowley, *A History of Succulent Plants*. Strawberry Press, Mill Valley, California 1997; Judith S. Stix, "Yo soy botánico." *J. Cac. & Succ. Soc. Mex.* 46: 117-119. 1974; Stafleu and Cowan, *Taxonomic Literature*. 3: 848. [genus dedicated to the Ortega family of San José Lachiguirí, Mexico] 1981; Irving William Knobloch, compil., "A preliminary verified list of plant collectors in Mexico." *Phytologia Memoirs*. VI. 1983.

Ortgiesia Regel Bromeliaceae

Origins:

After Karl Eduard Ortgies, 1829-1916, German botanist, contributed to *Gartenflora* [vols. 1-33, 1852-1884, edited by Eduard A. von Regel, 1815-1892.]; see Ida Kaplan Langman, *A Selected Guide to the Literature on the Flowering Plants of Mexico*. 557. Philadelphia 1964.

Orthaea Klotzsch Ericaceae

Origins:
Orthaea was a daughter of Hyacinthus.

Orthandra Burret Tiliaceae

Origins:
From the Greek *orthos* "upright, straight" and *aner, andros* "stamen, man, male."

Orthantha (Bernth.) Wettst. Scrophulariaceae

Origins:
From the Greek *orthos* "upright, straight" and *anthos* "flower."

Orthanthe Lem. Gesneriaceae

Origins:
Greek *orthos* and *anthos* "flower."

Orthanthella Rauschert Scrophulariaceae

Origins:
The diminutive of the genus *Orthantha*.

Orthanthera Wight Asclepiadaceae

Origins:
From the Greek *orthos* "upright, straight" and *anthera* "anther," referring to the pollen masses.

Orthechites Urban Apocynaceae

Origins:
From the Greek *orthos* "upright, straight" plus the genus *Echites* P. Browne.

Orthilia Raf. Ericaceae (Pyrolaceae)

Origins:
Greek *orthos, orthelos* "straight, tall," referring to the one-sided raceme; see Constantine Samuel Rafinesque, *Autikon botanikon*. Icones plantarum select. nov. vel rariorum, etc.

103. Philadelphia 1840; E.D. Merrill, *Index rafinesquianus*. 185. 1949.

Species/Vernacular Names:
O. secunda (L.) House
English: one-sided wintergreen

Orthion Standley & Steyerm. Violaceae

Origins:
From the Greek *orthios* "straight up, uphill," Latin *orthius* "high, lofty," Orthios was also an epithet of Asclepius (Asklepios, Aesculapius), the Greek god of medicine.

Orthiopteris Copel. Dennstaedtiaceae

Origins:
From the Greek *orthios* "straight up, uphill" and *pteris* "a fern."

Orthocarpus Nutt. Scrophulariaceae

Origins:
From the Greek *orthos* "upright, straight" and *karpos* "fruit," the upright pods; see Thomas Nuttall, *The Genera of North American Plants*, and catalogue of the species, to the year 1817. 2: 56. Philadelphia 1818.

Species/Vernacular Names:
O. lithospermoides Benth.
English: cream sacs
O. pachystachyus A. Gray
English: Shasta orthocarpus
O. purpurascens Benth.
Mexico: escobita

Orthoceras R. Br. Orchidaceae

Origins:
Greek *orthos* and *keras* "a horn," referring to the stiff and erect horn-like bracts; see Robert Brown (1773-1858), *Prodromus florae Novae Hollandiae*. 316. London 1810.

Species/Vernacular Names:
O. strictum R. Br. (*Orthoceras strictum* R. Br. forma *viride* Hatch)
English: horned orchid, bird's mouth

Orthochilus Hochst. ex A. Rich. Orchidaceae

Origins:

From the Greek *orthos* "upright, straight" and *cheilos* "a lip," the long claw.

Orthoclada P. Beauv. Gramineae

Origins:

With straight shoots, from the Greek *orthos* "upright, straight" and *klados* "branch."

Orthodon Bentham Labiatae

Origins:

From the Greek *orthos* "upright, straight" and *odous, odontos* "a tooth," referring to the calyx teeth.

Orthogoneuron Gilg Melastomataceae

Origins:

From the Greek *orthos* "upright, straight," *gonia* "an angle" and *neuron* "nerve."

Orthogramma Presl Blechnaceae

Origins:

From the Greek *orthos* "upright" and *gramma* "a line, writing, letter, thread, mark," referring to the position of the sori; see Karl (Carl, Carel, Carolus) Boriwog (Boriwag, Borivoj) Presl (1794-1852), *Epimeliae botanicae.* 121. Pragae 1849 [reprinted from *Abhandlungen der Königlichen Böhmischen Gesellschaft der Wissenschaften.* 6: 481. 1851].

Orthogynium Baillon Menispermaceae

Origins:

From the Greek *orthos* "upright" and *gyne* "female, woman, female organs."

Ortholobium Gagnep. Mimosaceae

Origins:

From the Greek *orthos* "upright" and *lobion* "small pod."

Ortholoma Hanst. Gesneriaceae

Origins:

From the Greek *orthos* "upright" and *loma* "border, margin, fringe."

Orthomene Barneby & Krukoff Menispermaceae

Origins:

From the Greek *orthos* "upright" and *mene* "the moon," possibly referring to the endocarp.

Orthopappus Gleason Asteraceae

Origins:

Greek *orthos* "upright" and *pappos* "fluff, pappus."

Orthopenthea Rolfe Orchidaceae

Origins:

From the Greek *orthos* "upright" plus the genus of orchids *Penthea* Lindl.

Orthophytum Beer Bromeliaceae

Origins:

From the Greek *orthos* "upright, erect, straight" and *phyton* "plant," referring to the flowers, to the scape with its inflorescence.

Orthopichonia H. Huber Apocynaceae

Origins:

From the Greek *orthos* "upright" plus the genus *Pichonia* Pierre, Sapotaceae.

Orthopogon R. Br. Gramineae

Origins:

From the Greek *orthos* "upright" and *pogon* "a beard"; see Robert Brown (1773-1858), *Prodromus florae Novae Hollandiae et Insulae van-Diemen.* 194. London 1810.

Orthopterum L. Bolus Aizoaceae

Origins:

Greek *orthos* and *pteron* "a wing," an allusion to the erect wings on the cell lid of the capsule.

Orthopterygium Hemsley Anacardiaceae (Julianiaceae)

Origins:

From the Greek *orthos* "upright" and *pteron* "a wing," *pterygion* "a small wing," referring to the winged pedicels.

Orthoraphium Nees Gramineae

Origins:

From the Greek *orthos* "upright" and *rhaphis, rhaphidos, raphis* "a needle, raphe."

Orthosia Decne. Asclepiadaceae

Origins:

From the Greek *orthos* "upright."

Orthosiphon Bentham Labiatae

Origins:

From the Greek *orthos* "upright" and *siphon* "a tube," referring to the tube of the corolla; see John Lindley (1799-1865), *The Botanical Register.* [Continued as *Edwards's Botanical Register.* London 1829-1837.] 15, sub t. 1300. London 1830.

Species/Vernacular Names:

O. aristatus (Blume) Miquel

English: cat's moustache, cat's whiskers

French: moustache de chat, thé de Java

Japan: mulisu-kuchin

The Philippines: balbas-pusa, kabling-gubat, kabling-parang

Malaya: kumis kuching

Vietnam: rau meo, cay bong bac

O. marmoritis (Hance) Dunn

English: stone-living Java tea

China: shi sheng ji jiao shen

O. wulfenioides (Diels) Handel-Mazzetti

English: common Java tea

China: ji jiao shen

Orthosphenia Standley Celastraceae

Origins:

From the Greek *orthos* "upright" and *sphen* "wedge."

Orthosporum (R. Br.) T. Nees Chenopodiaceae

Origins:

From the Greek *orthos* "upright" and *sporos* "a seed"; see Theodor F. L. Nees von Esenbeck (1787-1837), *Genera plantarum Florae germanicae.* Plantarum Dicotyledonearum. Bonnae 1835; Arthur D. Chapman, ed., *Australian Plant Name Index.* 2130. Canberra 1991.

Orthostemma Wall. ex Voigt Rubiaceae

Origins:

From the Greek *orthos* "upright" and *stemma* "crown, a garland."

Orthostemon O. Berg Myrtaceae

Origins:

From the Greek *orthos* "upright" and *stemon* "a stamen, a thread."

Orthostemon R. Br. Gentianaceae

Origins:

Greek *orthos* and *stemon* "a stamen, a thread"; see Robert Brown, *Prodromus florae Novae Hollandiae et Insulae van-Diemen.* 451. London 1810.

Orthotactus Nees Acanthaceae

Origins:

From the Greek *orthos* "straight, true, correct" and *tasso* "to arrange, put in order," *taxis* "a series, order, arrangement."

Orthotheca Pichon Bignoniaceae

Origins:

From the Greek *orthos* "upright, straight" and *theke* "a box, case."

Orthothylax (Hook.f.) Skottsb. Philydraceae

Origins:

Greek *orthos* "upright" and *thylakos* "a bag, sack, pouch," referring to the fruits; see the Swedish botanist Carl Johan Fredrik Skottsberg (1880-1963), in *Botanische Jahrbücher.* 65: 264. 1932.

Orthotropis Lindley Fabaceae

Origins:

From the Greek *orthos* "upright" and *trope* "turning, defeat," *trepein* "to turn"; see John Lindley (1799-1865), *Edwards's Botanical Register.* Appendix to Vols. 1-23: *A Sketch of the Vegetation of the Swan River Colony.* xxi. London (November) 1839.

Orthrosanthes Sweet ex Raf. Iridaceae

Origins:

Orthographic variant of *Orthrosanthus* Sweet; see Constantine Samuel Rafinesque (1783-1840), *Flora Telluriana.* 4: 30. Philadelphia 1838.

Orthrosanthus Sweet Iridaceae

Origins:

Flower opens early in the day, from the Greek *orthros* "morning" and *anthos* "flower," referring to the short-living flowers, fading before noon; see the British horticulturist and botanist Robert Sweet (1783-1835), *Flora Australasica.* t. 11. London 1828.

Species/Vernacular Names:

O. laxus (Endl.) Benth.

English: morning iris

O. multiflorus Sweet (*Eveltria multiflora* (Sweet) Raf.; *Sisyrinchium multiflorum* (Sweet) Steudel; *Sisyrinchium cyaneum* Lindley)

English: morning flag, morning iris

Orthurus Juz. Rosaceae

Origins:

From the Greek *orthos* "upright" and *oura* "tail," with erect flowers.

Orthylia Raf. Ericaceae (Pyrolaceae)

Origins:

See also *Orthilia* Raf.; see Constantine Samuel Rafinesque, *Med. Fl.* 2: 70. 1830; E.D. Merrill, *Index rafinesquianus.* 185. 1949.

Ortiga Neck. Loasaceae

Origins:

From the Spanish *ortiga* "nettle," referring to the stinging hairs of these species.

Ortmannia Opiz Orchidaceae

Origins:

For the German-Bohemian botanist Anton Ortmann, 1801-1861, pharmacist, wrote *Die Flora Karlsbads und seiner Umgegend.* Stuttgart 1838; see J.H. Barnhart, *Biographical Notes upon Botanists.* 3: 33. 1965; L. Schlesinger, *Deutsche Chroniken. Die Chronik der Stadt Elbogen. 1471-1504.* Prague 1879; Ethelyn Maria Tucker, *Catalogue of the Library of the Arnold Arboretum of Harvard University.* Cambridge, Massachusetts 1917-1933.

Orumbella J.M. Coulter & Rose Umbelliferae

Origins:

Possibly a derivation from Latin *umbella, ae* "a sunshade, umbrella, parasol," a dim. of *umbra, ae* "shadow, a shade."

Orxera Raf. Orchidaceae

Origins:

See C.S. Rafinesque, *Flora Telluriana.* 4: 37. 1836 [1838]; E.D. Merrill, *Index rafinesquianus.* 104. 1949.

Orychophragmus Bunge Brassicaceae

Origins:

Greek *orycho, orysso* "dig up, dig" and *phragma* "fence, partition," referring to the septum of the pod, the septum is pitted.

Oryctanthus (Griseb.) Eichler Loranthaceae

Origins:

Greek *orykte, orygma, orygmatos* "excavation, trench, pit" and *anthos* "flower," photosynthetic hemiparasites on tree-branches.

Oryctes S. Watson Solanaceae

Origins:

Greek *oryktes* "digger, implement for digger," *orykte, orygma, orygmatos* "excavation, trench, pit."

Species/Vernacular Names:

O. nevadensis S. Watson

English: Nevada oryctes

Oryctina Tieghem Loranthaceae

Origins:

From the Greek *orykte, orygma, orygmatos* "excavation, trench, pit," the diminutive.

Orygia Forssk. Molluginaceae

Origins:

An Arabic plant name.

Oryza L. Gramineae

Origins:

Latin and Greek *oryza* "rice," Arabic *eruz*, Tamil *arisi* or *erisi*, Malayalam *ari*; see Carl Linnaeus, *Species Plantarum*. 333. 1753 and *Genera Plantarum*. Ed. 5. 155. 1754; P. Sella, *Glossario latino emiliano*. Città del Vaticano 1937; Salvatore Battaglia, *Grande dizionario della lingua italiana*. Torino 1961-1989 etc.; Helmut Genaust, *Etymologisches Wörterbuch der botanischen Pflanzennamen*. 444-445. Basel 1996; Manlio Cortelazzo & Paolo Zolli, *Dizionario etimologico della lingua italiana*. 4: 1091. Bologna 1985; Ernest Weekley, *An Etymological Dictionary of Modern English*. 2: 1236. New York 1967.

Species/Vernacular Names:

O. sativa L.

English: rice, rice plant, wild rice, paddy

Australia: anboa, kwangan, jikan, mokomurdo (all Aboriginals names)

Mexico: xoba nagati xtilla, xoopa nagati castilla

The Philippines: palai, palay, pale, ammai, humay, pagai, pagay, pagei, parai, paroy, pai

India: tandula, tandul, arishi, arshi, dhanya, dhanyamu, dhan, dangar, paral, pari, ari, akki, bhat, bhatta, bhattadahullu, hal, chan, bras, biyyam, nellu, chaval, chaul, thomul, chokha, pendha, saryun, pulut, ketanshali, vrihi

Tibetan: bras

China: jing mi, tao, tu, hsien

Malaya: padi, paddy

South Laos: (people Nya Hön) cäh duan (= holiday rice), cäh maat (= everyday rice), cäh mûön (= late rice), cäh ngiau, cäh ddak dông, cäh gye' (= early rice), cäh gleet,

cäh roh (= washed rice), cäh hlak (= Alak rice), cäh dang (= bitter rice), cäh da'ôôn, cäh kuan dean, cäh llôông lang (= rice tree lang), cäh dum (= red rice), cäh boh (= salted rice)

Japan: ine (meaning rice plant), kome (meaning rice grain), gohan (meaning cooked rice)

Okinawa: mai

Italian: riso

Tanzania: mshele

Yoruba: resi, iresi

Oryzidium C.E. Hubb. & Schweick. Gramineae

Origins:

Referring to the generic name *Oryza* L.

Oryzopsis Michaux Gramineae

Origins:

Greek *oryza* "rice" and *opsis* "appearance"; see André Michaux (1746-1803), *Flora Boreali-Americana*. 1: 51. Paris 1803.

Species/Vernacular Names:

O. hymenoides (Roemer & Schultes) Ricker

English: silkgrass, Indian millet

O. miliacea (L.) Asch. & Schweinf.

English: smilo grass, smilo rice grass

South Africa: managras, smilo rys gras, wilderys

O. obtusa Stapf

Japan: ine-gaya

Osbeckia L. Melastomataceae

Origins:

For the Swedish clergyman Pehr Osbeck, 1723-1805, naturalist and botanist, a student of Linnaeus, a member of the Academy of Stockholm and of the Society of Sciences (Uppsala), traveler and plant collector in Southeast Asia, Java and China, chaplain with the Svenska Ostindiska Kompaniet, author of *Dagbok öfwer en Ostindisk Resa åren 1750, 1751, 1752 ... Stockholm 1757*, father of the Swedish surgeon and plant collector Carl Gustaf Osbeck (1766-1841); see John H. Barnhart, *Biographical Notes upon Botanists*. 3: 33. 1965; Miguel Colmeiro y Penido, *La Botánica y los Botánicos de la Peninsula Hispano-Lusitana*. Madrid 1858; Georg Christian Wittstein, *Etymologisch-botanisches*

Handwörterbuch. 643. 1852; Carl Linnaeus, *Species Plantarum.* 345. 1753 and *Genera Plantarum.* Ed. 5. 162. 1754; T.W. Bossert, *Biographical Dictionary of Botanists Represented in the Hunt Institute Portrait Collection.* 295. 1972; S. Lenley et al., *Catalog of the Manuscript and Archival Collections and Index to the Correspondence of John Torrey.* Library of the New York Botanical Garden. 318. 1973; Mary Gunn and Leslie E. Codd, *Botanical Exploration of Southern Africa.* 267-268. 1981; Ethelyn Maria Tucker, *Catalogue of the Library of the Arnold Arboretum of Harvard University.* Cambridge, Massachusetts 1917-1933; Jonas C. Dryander, *Catalogus bibliothecae historico-naturalis Josephi Banks.* London 1800; A. Lasègue, *Musée botanique de Benjamin Delessert.* Paris 1845; Emil Bretschneider (1833-1901), *History of European Botanical Discoveries in China.* [Reprint of the original edition, St. Petersburg 1898.] Leipzig 1981; R. Zander, F. Encke, G. Buchheim and S. Seybold, *Handwörterbuch der Pflanzennamen.* 14. Aufl. 759. 1993; Frans A. Stafleu, *Linnaeus and the Linnaeans.* The spreading of their ideas in systematic botany, 1735-1789. Utrecht 1971.

Species/Vernacular Names:

O. chinensis L.

English: Chinese osbeckia

Japan: hime-no-botan

China: tin xiang lu

Osbeckiastrum Naudin Melastomataceae

Origins:

The genus *Osbeckia* L. and the Latin substantival suffix *astrum*, indicating inferiority or incomplete resemblance.

Osbornia F. Muell. Myrtaceae

Origins:

After John Walter Osborne, chemist, invented a new method of photolithography; see Ferdinand von Mueller, *Fragmenta Phytographiae Australiae.* 3: 30. Melbourne 1862.

Species/Vernacular Names:

O. octodonta F. Muell. (with eight calyx teeth)

English: myrtle mangrove

Oschatzia Walpers Umbelliferae

Origins:

Dedicated to the German botanist Adolph Oschatz, 1812-1857, physician, the inventor of microtomy, wrote *De Phalli impudici germinatione.* D. Vratislaviae 1842; see the German botanist Wilhelm Gerhard Walpers (1816-1853), in *Annales botanices systematicae.* 1: 340. Lipsiae 1848.

Species/Vernacular Names:

O. cuneifolia (F. Muell.) Drude

English: wedge oschatzia

Oscularia Schwantes Aizoaceae

Origins:

From the Latin *osculum* "little mouth, kiss, pretty mouth," diminutive of *os, oris* "the mouth," an allusion to the beauty of the leaves.

Osmadenia Nutt. Asteraceae

Origins:

Greek *osme* "smell, odor, perfume" and *aden* "gland."

Species/Vernacular Names:

O. tenella Nutt.

English: osmadenia

Osmanthus Lour. Oleaceae

Origins:

From the Greek *osme* "smell, odor, perfume" and *anthos* "flower," with fragrant flowers.

Species/Vernacular Names:

O. americanus (L.) A. Gray

English: devilwood, American olive, wild olive

O. fragrans Lour. (*Osmanthus asiaticus* Nakai; *Osmanthus asiaticus* var. *latifolius* Makino; *Olea fragrans* Thunb.)

English: fragrant olive, sweet tea, sweet olive, tea olive, sweet osmanthus

Japan: usu-gin-mokusei, gin-mokusei

China: mu xi, mu hsi, gui hua, yen kuei

O. fragrans Lour. f. *aurantiacus* (Mak.) P. Green (*Osmanthus aurantiacus* (Mak.) Nak.)

Japan: kin-mokusei

O. heterophyllus (G. Don) P. Green (*Osmanthus aquifolium* Siebold & Zucc.; *Olea aquifolium* Sieb. & Zucc.; *Olea ilicifolia* Hassk.; *Ilex heterophyllus* G. Don)

English: holly olive, Chinese holly, false holly, diversifolius osmanthus

Japan: hi-ira-gi

China: zhong shu

Osmaronia Greene Rosaceae

Origins:
From the Greek *osme* "smell, odor, perfume" and the genus *Aronia* Medik.

Osmelia Thwaites Flacourtiaceae

Origins:
Possibly from the Greek *ozo*, *osme* "smell, odor, to smell" and *meli* "honey."

Osmhydrophora Barb. Rodr. Bignoniaceae

Origins:
From the Greek *osme* "odor, perfume," *hydor* "water" and *phoros* "bearing, carrying."

Osmiopsis R.M. King & H. Robinson Asteraceae

Origins:
From the Greek *osme* "odor, perfume" and *opsis* "like, appearance, resemblance."

Osmites L. Asteraceae

Origins:
Greek *osme* "scent, odor, perfume," *osmites* "mentastrum," *osmitis* "kalaminthe."

Osmitopsis Cass. Asteraceae

Origins:
Resembling *Osmites*.

Osmoglossum (Schltr.) Schltr. Orchidaceae

Origins:
Greek *osme* "smell, scent, perfume" and *glossa* "a tongue," referring to the fragrance of the flowers, referring also to the genus *Odontoglossum*.

Osmorhiza Raf. Umbelliferae

Origins:
Greek *osme* "smell, odor, perfume" and *rhiza* "a root"; see C.S. Rafinesque, *Am. Monthly Mag. Crit. Rev.* 2: 176. 1818, *Jour. Phys. Chim. Hist. Nat.* 89: 257. 1819, *Med. Fl.* 2: 249. 1830, *New Fl. N. Am.* 4: 34. 1836 [1838] and *The Good Book.* 53. 1840; Elmer D. Merrill, *Index rafinesquianus.* 181. 1949.

Osmoshiza Raf. Umbelliferae

Origins:
See *Osmorhiza* Raf.; see C.S. Rafinesque, *Med. Fl.* 2: 249. 1830.

Osmoxylon Miq. Araliaceae

Origins:
From the Greek *osme* "smell, odor, perfume" and *xylon* "wood."

Osmunda L. Osmundaceae

Origins:
Uncertain attribution, French *osmunde*, English *osmund*, of unknown origin, possibly after the Saxon Osmunder, a name for Thor, the god of war, or for Osmundus, c. 1025, a Scandinavian writer of runes, or after Osmun, Bishop of Salisbury, d. 1099, etc.; see Carl Linnaeus, *Species Plantarum.* 1063. 1753 and *Genera Plantarum.* Ed. 5. 484. 1754; F. Boerner & G. Kunkel, *Taschenwörterbuch der botanischen Pflanzennamen.* 4. Aufl. 145. Berlin & Hamburg 1989; Helmut Genaust, *Etymologisches Wörterbuch der botanischen Pflanzennamen.* 445. Basel 1996; Georg Christian Wittstein, *Etymologisch-botanisches Handwörterbuch.* 643. Ansbach 1852; Salvatore Battaglia, *Grande dizionario della lingua italiana.* XII: 201. Torino 1984; Ernest Weekley, *An Etymological Dictionary of Modern English.* 2: 1017. Dover Publications, New York 1967.

Species/Vernacular Names:
O. banksiifolia (Presl) Kuhn (*Nephrodium banksiaefolium* Presl)

English: flowering fern

Japan: Shiroyama-zenmai (= Shiroyama *Osmunda japonica* Thunb.)

O. cinnamomea L.

English: cinnamon fern, fiddleheads

O. regalis L.

English: royal fern, flowering fern

China: wei

Osmundastrum C. Presl Osmundaceae

Origins:
Referring to the genus *Osmunda* L.

Osmundopteris (J. Milde) Small Ophioglossaceae

Origins:
From the genus *Osmunda* and Greek *pteris* "fern."

Ossiculum P.J. Cribb & van der Laan Orchidaceae

Origins:
Latin *ossiculum* "a small bone, ossicle."

Ossifraga Rumph. Euphorbiaceae

Origins:
From the Latin subst. *ossifraga* "the sea-eagle, osprey," adj. *ossifragus, a, um* "bone-breaking," *os-frango.*

Ostenia Buchenau Limnocharitaceae

Origins:
For the German botanist Cornelius Osten, 1863-1936, botanical collector in Uruguay; see John H. Barnhart, *Biographical Notes upon Botanists.* 3: 35. 1965.

Osteocarpum F. Muell. Chenopodiaceae

Origins:
From the Greek *osteon* "bone" and *karpos* "fruit," referring to the hard fruiting perianth; see F. von Mueller, in *Transactions of the Philosophical Institute of Victoria.* 2: 77. (September) 1857.

Species/Vernacular Names:
O. acropterum (F. Muell. & Tate) Volkens (*Babbagia acroptera* F. Muell. & Tate)
Australia: water weed, babbagia

O. salsuginosum F. Muell. (*Threlkeldia salsuginosa* (F. Muell.) F. Muell. ex Benth.; *Chenolea salsuginosa* (F. Muell.) F. Muell.; *Bassia salsuginosa* (F. Muell.) F. Muell.)
Australia: bonefruit

Osteocarpus Kuntze Chenopodiaceae

Origins:
Orthographic variant of *Osteocarpum* Mueller.

Osteocarpus Phil. Nolanaceae (Solanaceae)

Origins:
From the Greek *osteon* "bone" and *karpos* "fruit."

Osteomeles Lindley Rosaceae

Origins:
From the Greek *osteon* "a bone" and *melon* "an apple," referring to the hard nutlets.

Species/Vernacular Names:
O. anthyllidifolia (Sm.) Lindl. (*Pyrus anthyllidifolia* Sm.)
Hawaii: 'ulei, eluehe, u'ulei
O. subrotunda K. Koch
English: bonyberry
Japan: ten-no-ume

Osteophloeum Warb. Myristicaceae

Origins:
From the Greek *osteon* "a bone" and *phloios* "bark of trees, smooth bark, husk."

Osteospermum L. Asteraceae

Origins:
From the Greek *osteon* "bone" and *sperma* "seed," the achenes are hard; see Carl Linnaeus, *Species Plantarum.* 923. 1753 and *Genera Plantarum.* Ed. 5. 395. 1754.

Species/Vernacular Names:
O. barberae (Harvey) T. Norlindh (*Dimorphotheca barberae* Harv.; *Dimorphotheca lilacina* Regel & V. Herd.)
English: daisy
O. clandestinum (Less.) T. Norlindh (*Tripteris clandestina* Less.)

English: tripteris, stinking Roger

O. ecklonis (DC.) T. Norl. (*Dimorphotheca ecklonis* DC.) (after the Danish botanist Christian Friedrich (Frederik) Ecklon, 1795-1868, apothecary and botanical collector, traveler, sent plants to Bentham (1835), author of *Topographisches Verzeichniss der Pflanzensammlung von C.F. Ecklon.* Esslingen 1827 and "A list of plants found in the district of Uitenhage between the months of July 1829 and February 1830." *S. Afr. Quart. J.* 1: 358-380. 1830, with Karl Ludwig Philipp Zeyher wrote *Enumeratio plantarum africae australis extratropicae.* Hamburg [1834-] 1835-1836[-1837]; see John H. Barnhart, *Biographical Notes upon Botanists.* 1: 494. 1965; Karl Boriwog Presl, *Botanische Bemerkungen.* Prague 1844; Peter MacOwan, "Personalia of botanical collectors at the Cape." *Trans. S. Afr. Philos. Soc.* 4(1): xliii-xlvi. 1884-1886; John Hutchinson, *A Botanist in Southern Africa.* 641-642. London 1946; Gordon Douglas Rowley, *A History of Succulent Plants.* 1997; Mary Gunn and Leslie E. Codd, *Botanical Exploration of Southern Africa.* Cape Town 1981; H.N. Clokie, *Account of the Herbaria of the Department of Botany in the University of Oxford.* Oxford 1964; Günther Schmid, *Chamisso als Naturforscher.* Eine Bibliographie. Leipzig 1942)

English: African daisy, blue-and-white daisy bush, white daisy bush, marigold, sunday's river daisy, Vanstaden's daisy, Van Staden osteospermum

South Africa: bietou, blue and white daisy bush, bergbietou, jakkalsbos, Vanstaden-osteospermum

O. muricatum E. Mey. ex DC. subsp. *muricatum* (*Oligocarpus calendulaceus* auct. non Less., Dinter; *Oligocarpus hamiltonei* S. Moore)

Southern Africa: bietou, boegoebossie; motlapa-tsunyana (Sotho)

Ostodes Blume Euphorbiaceae

Origins:

Greek *osteodes* "bony," *ostoeides* "like bones," *ostodes* "bony, like bone."

Ostrearia Baillon Hamamelidaceae

Origins:

Latin *ostrea, ae* "an oyster," Greek *ostreion, ostreon*, referring to the form of the fruits; see the French botanist Henri Ernest Baillon (1827-1895), in *Adansonia.* 10: 131. 1873.

Species/Vernacular Names:
O. australiana Baill.

English: hard pink alder

Ostrowskia Regel Campanulaceae

Origins:

For the Russian patron of botany Michail Nikolaevic (Michael Nicholazewitsch von O.) Ostrowski (Ostrovskij, Ostrowsky), Minister of Imperial Domains.

Ostrya Scop. Betulaceae (Carpinaceae, Corylaceae)

Origins:

Latin *ostrya, ae* and *ostrys, yos* for a tree with hard wood, perhaps the common hornbeam (Plinius), Greek *ostrys, ostrya, ostrye, ostryis*, hop hornbeam, *Ostrya carpinifolia* Scop., Theophrastus (*HP.* 3.10.3) and Plinius, presumably referring to the hardwood.

Species/Vernacular Names:
O. virginiana (Miller) K. Koch

English: ironwood, leverwood

Ostryocarpus Hook.f. Fabaceae

Origins:

Presumably from the Greek *ostreion, ostreon* "an oyster" and *karpos* "fruit," some suggest from the Greek *ostrya*.

Ostryoderris Dunn Fabaceae

Origins:

Possibly from the Greek *ostreion, ostreon* "an oyster" and the genus *Derris*, referring to the fruit.

Species/Vernacular Names:
O. sp.

Nigeria: owo, erumaki, akirankhiri, erchun, ori, durbi

Ostryopsis Decne. Betulaceae (Carpinaceae, Corylaceae)

Origins:

From the genus *Ostrya* and *opsis* "like, appearance."

Osyricera Blume Orchidaceae

Origins:

From the Egyptian god Osyris and the Greek word *keras* "horn," origin and meaning obscure; see Richard Evans

Schultes and Arthur Stanley Pease, *Generic Names of Orchids. Their Origin and Meaning.* 225. Academic Press, New York and London 1963.

Osyridicarpos A. DC. Santalaceae

Origins:
From the genus *Osyris* L. and *karpos* "fruit," fruit like that of *Osyris*.

Osyris L. Santalaceae

Origins:
Possibly from the Greek *ozos* "branch, knot," the small tree is much branched. Plinius and Dioscorides used *osyris* or *osiris, osiridos* as a plant name for poet's cassia, *Osyris alba*; Latin *osyris* applied by Plinius to a plant, probably the broom-like goose-foot or summer cypress.

Species/Vernacular Names:
O. lanceolata Hochst. & Steud. (*Osyrys abyssinica* Hochst. & A. Rich.)

English: Transvaal sumach, barkbush

Southern Africa: bergbas, basbessie, pruimbos, looibos; mpere (Tsonga or Shangaan: eastern Transvaal); mofetela (South Sotho: Lesotho, Orange Free State, southeast Transvaal); muritho, mpeta (Venda: Soutpansberg, northern Transvaal); iNtshakasa, inGondotha-mpete, uMbulunyathi (Zulu); inTekeza, uMbulunyathi (Xhosa)

O. tenuifolia Engl.

English: E. Africa sandalwood

Otacanthus Lindley Scrophulariaceae

Origins:
Greek *ous, otos* "an ear" plus *akantha*, the genus was originally classified in the Acanthaceae.

Species/Vernacular Names:
O. coeruleus Lindley

English: little boy blue

Otachyrium Nees Gramineae

Origins:
From the Greek *ous, otos* "an ear" and *achyron* "chaff, husk."

Otandra Salisb. Orchidaceae

Origins:
Greek *ous, otos* "an ear" and *aner, andros* "stamen, male, man," referring to the appendages on the anther.

Otanthera Blume Melastomataceae

Origins:
From the Greek *ous, otos* "an ear" and *anthera* "anther" referring to the stamen appendages; see Karl Ludwig von Blume, *Flora oder allgemeine Botanische Zeitung.* 14: 488. (July-September) 1831.

Otanthus Hoffmanns. & Link Asteraceae

Origins:
From the Greek *ous, otos* "an ear" and *anthos* "flower," referring to the corolla.

Species/Vernacular Names:
O. maritimus (L.) Hoffmanns. & Link

English: cottonweed, sea cudweed

Arabic: shiba

Otatea (McClure & E.W. Sm.) Calderón & Söderstr. Gramineae

Origins:
From the Mexican vernacular name *otate*; in some areas *otate* is applied to the types of cane or bamboo with solid stalks, see Louise C. Schoenhals, *A Spanish-English Glossary of Mexican Flora and Fauna.* 80. Hidalgo, México 1988.

Species/Vernacular Names:
O. spp.

Mexico: gui yaa, qui yaa

Othake Raf. Asteraceae

Origins:
See C.S. Rafinesque, *New Fl. N. Am.* 4: 73. 1836 [1838].

Othonna L. Asteraceae

Origins:
Greek *othonna*, used by Dioscorides for the greater celandine, *Chelidonium majus*, Latin *othonna, ae* used by Plinius for a Syrian plant; see Carl Linnaeus, *Species Plantarum.*

924. 1753 and *Genera Plantarum*. Ed. 5. 396. 1754; Gordon Rowley, *Succulent Compositae (Senecio* and *Othonna).* Strawberry Press 1994; Gordon Douglas Rowley, *A History of Succulent Plants.* Strawberry Press, Mill Valley, California 1997.

Othonnopsis Jaub. & Spach Asteraceae

Origins:
Resembling *Othonna.*

Otilix Raf. Solanaceae

Origins:
Possibly from the Greek *oteile* "wound," referring to medicinal properties, the fruits are rich in vitamin C; see C.S. Rafinesque, *Med. Fl.* 2: 87. 1830; E.D. Merrill, *Index rafinesquianus.* 212. 1949.

Otiophora Zucc. Rubiaceae

Origins:
Greek *otion* "auricle, a little handle," diminutive of *ous, otos* "ear" and *phoros* "bearing, carrying."

Otoba (DC.) Karst. Myristicaceae

Origins:
A vernacular name.

Otocalyx Brandegee Rubiaceae

Origins:
From the Greek *ous, otos* "an ear" and *kalyx* "a calyx."

Otocarpus Durieu Brassicaceae

Origins:
From the Greek *ous, otos* "an ear" and *karpos* "fruit."

Otocephalus Chiov. Rubiaceae

Origins:
From the Greek *ous, otos* "an ear" and *kephale* "head."

Otochilus Lindley Orchidaceae

Origins:
Greek *ous, otos* and *cheilos* "lip," referring to the appendages at the base of the lip.

Otochlamys DC. Asteraceae

Origins:
From the Greek *ous, otos* "an ear" and *chlamys* "cloak," alluding to the ear-like appendage to the corolla.

Otoglossum (Schltr.) Garay & Dunsterville Orchidaceae

Origins:
Greek *ous, otos* "an ear" and *glossa* "tongue," referring to the side lobes of the lip.

Otomeria Benth. Rubiaceae

Origins:
From the Greek *ous, otos* and *meris* "part."

Otonephelium Radlk. Sapindaceae

Origins:
From the Greek *ous, otos* "an ear" and *nephele* "a cloud," *nephelion* "a small cloud, a cloud-like spot (on the eye)."

Otopappus Benth. Asteraceae

Origins:
From the Greek *ous, otos* "an ear" and *pappos* "fluff, pappus."

Otopetalum Lehm. & Kraenzlin Orchidaceae

Origins:
From the Greek *ous, otos* "an ear" and *petalon* "leaf, petal," indicating the auriculate petals.

Otophora Blume Sapindaceae

Origins:
From the Greek *ous, otos* "an ear" and *phoros* "bearing."

Otoptera DC. Fabaceae

Origins:

From the Greek *ous*, *otos* "an ear" and *pteron* "wing."

Otospermum Willk. Asteraceae

Origins:

From the Greek *ous*, *otos* "an ear" and *sperma* "seed."

Otostegia Benth. Labiatae

Origins:

From the Greek *ous*, *otos* "an ear" and *stege* "roof, cover, covering," referring to the petals.

Otostemma Blume Asclepiadaceae

Origins:

From the Greek *ous*, *otos* "an ear" and *stemma* "crown, a garland."

Otostylis Schltr. Orchidaceae

Origins:

From the Greek *ous*, *otos* "an ear" and *stylos* "pillar, style," the auriculate wings of the column.

Otoxalis Small Oxalidaceae

Origins:

From the Greek *ous*, *otos* "an ear" plus the genus *Oxalis* L., *oxalis* (*oxys* "acid, sour, sharp").

Ottelia Pers. Hydrocharitaceae

Origins:

Ottel-ambel, the native name for an Indian aquatic species, *Ottelia alismoides* (L.) Pers., used by van Rheede in *Hortus Indicus Malabaricus*. 11: t. 46. 1692; see Christiaan Hendrik Persoon (1761/1762-1836), *Synopsis plantarum*. 1: 400. Paris et Tubingae 1805-1807; Georg Christian Wittstein, *Etymologisch-botanisches Handwörterbuch*. 645. Ansbach 1852; Helmut Genaust, *Etymologisches Wörterbuch der botanischen Pflanzennamen*. 447. Basel 1996.

Species/Vernacular Names:

O. alismoides (L.) Pers.

English: tropical swamp lily

Japan: mizu-ôba-ko (= water *Plantago*)

Okinawa: karanazu, takubu

O. ovalifolia (R. Br.) Rich. (*Damasonium ovalifolium* R. Br.)

English: swamp lily

Ottoa Kunth Umbelliferae

Origins:

For the German botanist Christoph Friedrich Otto, 1783-1856, gardener, from 1801 to 1843 at the Berlin Botanical Garden, with Johann Heinrich Friedrich Link (1767-1851) wrote *Icones plantarum selectarum*. Berolini [Berlin] 1820-1828, *Über die Gattungen Melocactus und Echinocactus*. Berlin 1827 and *Icones plantarum rariorum*. Berolini 1828[-1831], with Link and Johann Friedrich Klotzsch (1805-1860) *Icones plantarum rariorum horti regii bot. Berol.* Berlin [1840-]1841-1844; see A. Lasègue, *Musée botanique de Benjamin Delessert*. 334. Paris 1845; Ethelyn Maria Tucker, *Catalogue of the Library of the Arnold Arboretum of Harvard University*. Cambridge, Massachusetts 1917-1933; Günther Schmid, *Chamisso als Naturforscher. Eine Bibliographie*. Leipzig 1942; R. Zander, F. Encke, G. Buchheim and S. Seybold, *Handwörterbuch der Pflanzennamen*. 14. Aufl. Stuttgart 1993.

Ottochloa Dandy Gramineae

Origins:

Named for the Austrian botanist Otto Stapf, 1857-1933, traveler, from 1882 to 1889 assistant with Kerner von Marilaun in Wien, 1900-1922 Keeper of the Herbarium of the Royal Botanic Gardens, Kew, 1908-1916 botanical secretary of the Linnean Society, 1922-1933 editor of the *Botanical Magazine*, contributor to Daniel Oliver (1830-1916), *Flora of Tropical Africa* (Apocynaceae, Verbenaceae, Myristicaceae, Gramineae, etc.), contributor to Harvey and Sonder, *Flora Capensis* (Apocynaceae, Laurineae, Proteaceae, etc.), contributed to Hooker's *Icones Plantarum*. Among his numerous and valuable publications are *On the Flora of Mount Kinabalu in North Borneo*. London 1894 and *The Aconites of India*. Calcutta 1905; see John H. Barnhart, *Biographical Notes upon Botanists*. 3: 317. 1965; Mia C. Karsten, *The Old Company's Garden at the Cape and Its Superintendents*. Cape Town 1951; James Edgar Dandy (1903-1976), in *The Journal of Botany*. 69: 54. (Feb.) 1931; T.W. Bossert, *Biographical Dictionary of Botanists Represented in the Hunt Institute Portrait Collection*. 380.

1972; Ida Kaplan Langman, *A Selected Guide to the Literature on the Flowering Plants of Mexico.* Philadelphia 1964; S. Lenley et al., *Catalog of the Manuscript and Archival Collections and Index to the Correspondence of John Torrey.* Library of the New York Botanical Garden. 382. 1973; Ethelyn Maria Tucker, *Catalogue of the Library of the Arnold Arboretum of Harvard University.* Cambridge, Massachusetts 1917-1933; Leonard Huxley, *Life and Letters of Sir Joseph Dalton Hooker.* London 1918; Stafleu and Cowan, *Taxonomic Literature.* 5: 839-843. 1985; R. Zander, F. Encke, G. Buchheim and S. Seybold, *Handwörterbuch der Pflanzennamen.* 14. Aufl. Stuttgart 1993; Emil Bretschneider, *History of European Botanical Discoveries in China.* [Reprint of the original edition, St. Petersburg 1898.] Leipzig 1981; Ray Desmond, *Dictionary of British & Irish Botanists and Horticulturists.* 650. London 1994.

Species/Vernacular Names:

O. nodosa (Kunth) Dandy

The Philippines: banig-usa, kauakauayanan, kumut-usa

Ottonia Sprengel Piperaceae

Origins:
Dedicated to the German botanist Christoph Friedrich Otto, 1783-1856, from 1801 to 1843 at the Berlin Botanical Garden, with Johann Heinrich Friedrich Link (1767-1851) wrote *Icones plantarum selectarum.* Berolini [Berlin] 1820-1828, *Über die Gattungen Melocactus und Echinocactus.* Berlin 1827 and *Icones plantarum rariorum.* Berolini 1828[-1831], with Link and Johann Friedrich Klotzsch (1805-1860) *Icones plantarum rariorum.* Berlin [1840-]1841-1844.

Ottoschmidtia Urban Rubiaceae

Origins:
After the German botanist Otto Christian Schmidt, 1900-1951, professor of pharmacognosy (Universities of Berlin and Münster), author of *Die marine Vegetation der Azoren.* Stuttgart 1931; see John H. Barnhart, *Biographical Notes upon Botanists.* 3: 658. 1965; Elmer Drew Merrill, *Contr. U.S. Natl. Herb.* 30(1): 268. 1947; Stafleu and Cowan, *Taxonomic Literature.* 5: 259-260. 1985.

Ottoschulzia Urban Icacinaceae

Origins:
After the German botanist Otto Eugen Schulz, 1874-1936, collaborated with I. Urban and A. Engler; see John H.

Barnhart, *Biographical Notes upon Botanists.* 3: 246. 1965; T.W. Bossert, *Biographical Dictionary of Botanists Represented in the Hunt Institute Portrait Collection.* 356. 1972; Ida Kaplan Langman, *A Selected Guide to the Literature on the Flowering Plants of Mexico.* 687. Philadelphia 1964; R. Zander, F. Encke, G. Buchheim and S. Seybold, *Handwörterbuch der Pflanzennamen.* 14. Aufl. Stuttgart 1993.

Ottosonderia L. Bolus Aizoaceae

Origins:
After the German botanist Otto Wilhelm Sonder, 1812-1881, pharmacist, botanical explorer and plant collector, joint author with William H. Harvey (1811-1866) of the first three volumes of *Flora capensis.* Dublin, Capetown 1860-1865, co-editor of Johann G.C. Lehmann, *Plantae Preissianae.* Hamburgi 1844-1848. Among his many works are *Flora hamburgensis.* Hamburg 1851 [1850] and *Die Algen des tropischen Australiens.* [Hamburg 1871]; see John H. Barnhart, *Biographical Notes upon Botanists.* 3: 303. Boston 1965; Dennis John Carr (1915-) and S.G.M. Carr (1912-1988), eds., *People and Plants in Australia.* 1981; Mary Gunn and Leslie E. Codd, *Botanical Exploration of Southern Africa.* 328-329. Cape Town 1981; Michel Gandoger (1850-1926), "L'herbier africain de Sonder." *Bull. Soc. bot. France.* 60: 414-422, 445-462. 1913; T.W. Bossert, *Biographical Dictionary of Botanists Represented in the Hunt Institute Portrait Collection.* 376. 1972; Ethelyn Maria Tucker, *Catalogue of the Library of the Arnold Arboretum of Harvard University.* Cambridge, Massachusetts 1917-1933; R. Zander, F. Encke, G. Buchheim and S. Seybold, *Handwörterbuch der Pflanzennamen.* 14. Aufl. Stuttgart 1993; Leonard Huxley, *Life and Letters of Sir Joseph Dalton Hooker.* London 1918; Gordon Douglas Rowley, *A History of Succulent Plants.* Strawberry Press, Mill Valley, California 1997.

Oubanguia Baillon Scytopetalaceae

Origins:
Named after the Oubangui or Ubangi River, largest right-bank tributary of the Congo River, forming the border between Zaire and the People's Republic of the Congo; see Auguste Jean Baptiste Chevalier, 1873-1956, *L'Afrique Centrale Française.* Mission Chari-Lac Tchad, 1902-1904. Paris 1907; Franz Thonner, *Vom Kongo zum Ubangi.* Berlin 1910.

Species/Vernacular Names:

O. africana Baillon

Gabon: akok, mupapambu

Oudneya R. Br. Brassicaceae

Origins:

For the Scottish (b. Edinburgh) physician Walter Oudney, 1790-1824 (Nigeria), traveler, naval surgeon, 1817 M.D. Edinburgh, plant collector, explorer, 1822-1824 (after crossing the Sahara to Lake Chad) expedition to Northern Provinces of Nigeria with Dixon Denham (1786-1828, d. Freetown, Sierra Leone) and Hugh Clapperton (1788-1827); see Rev. Thomas Nelson, *A Biographical Memoir of the Late Dr. W. Oudney*, Captain H. Clapperton, and Major A.G. Laing. Edinburgh 1830; Denham, Clapperton and Oudney, *Narrative of Travels and Discoveries in Northern and Central Africa. 1822-1824.* [Botanical appendix by R. Brown.] London 1826; F.N. Hepper and Fiona Neate, *Plant Collectors in West Africa.* 63. 1971; Joseph Vallot, "Études sur la flore du Sénégal." in *Bull. Soc. Bot. de France.* 29: 168-238. Paris 1882; R.W.J. Keay, "Botanical collectors in West Africa prior to 1860." *Comptes Rendus A.E.T.F.A.T.* Lisbon 1962; Auguste J.B. Chevalier, *Flore vivante de l'Afrique Occidentale Française.* 1938.

Ouratea Aublet Ochnaceae

Origins:

From *ourati*, a vernacular name from Guiana.

Species/Vernacular Names:

O. sp.

Peru: loro micuna

Nigeria: kemegbohun

Yoruba: akoodo

O. calantha Gilg

Nigeria: okoribobo; okoribobodo (Ijaw)

O. calophylla (Hook.f.) Engl.

Nigeria: ntene (Boki)

O. parviflora (A. St.-Hil.) Engl.

Brazil: batiputa

O. staudtii (Tiegh.) Keay

Nigeria: lasigba (Yoruba)

Ourisia Comm. ex Juss. Scrophulariaceae

Origins:

Named for Ouris, Governor of the Falkland Islands, where the French botanist and traveler Philibert Commerson (1727-1773), Bougainville's naturalist, collected the species; see A.L. de Jussieu, *Genera Plantarum.* 1789; Antoine Joseph Pernety, *The History of a Voyage to the Malouine Islands made in 1763 and 1764 ... and of Two Voyages to the Streights of Magellan with an Account of the Patagonians.* London 1771; L.A. de Bougainville, *Voyage autour du monde par la frégate du Roi "La Boudeuse" et la flute L'Etoile" en 1766-1769.* Paris 1771; Robert Brown, *Prodromus florae Novae Hollandiae et Insulae van-Diemen.* London 1810; Margaret Patricia Henwood Laver, *An Annotated Bibliography of the Falkland Islands and the Falkland Island Dependencies* (as delimited on 3rd March, 1962). Cape Town 1977.

Ourisianthus Bonati Scrophulariaceae

Origins:

From the genus *Ourisia* and *anthos* "flower."

Ourouparia Aublet Rubiaceae

Origins:

A vernacular name.

Ouvirandra Thouars Aponogetonaceae

Origins:

Orthographic variant of *Urirandra* Mirbel.

Ovaria Fabr. Solanaceae

Origins:

Latin *ovum* "an egg," referring to the fruits.

Owenia F. Muell. Meliaceae

Origins:

For the English (b. Lancaster) comparative anatomist Sir Richard Owen, 1804-1892 (d. London), surgeon, biologist, naturalist, zoologist, geologist, physiologist, paleontologist, studied in Edinburgh and at St. Bartholomew's Hospital in London, assistant and successor to William Clift (1775-1849), 1836 Hunterian professor, Curator of the Hunterian Museum at the Royal College of Surgeons of England, London, from 1856 Superintendent of the Natural History departments in the British Museum. Among his very numerous publications are *Odontography: or, a Treatise on the Comparative Anatomy of the Teeth ... in the Vertebrate Animals.* [Two vols.] London 1840-1845, *On Parthenogenesis.* [8vo, first edition.] London 1849, *On the Anatomy and Physiology of the Vertebrates.* [Three vols.] London 1866-1868 and *A*

History of British Fossil Reptiles. London 1849-1884, described Darwin's fossils from South America (in *The Zoology of the Voyage of H.M.S. Beagle.* Part I. Fossil Mammalia by R.Owen. 1840), he was Thomas Henry Huxley's antagonist in the debates over Darwinism. See *The Life of Richard Owen, by His Grandson*, with the scientific portions revised by C.D. Sherborn, with "An Essay on Owen's Position in Anatomical Science," by T.H. Huxley. London 1895; J.R. Norman, *Squire*, memories of Charles Davies Sherborn. London 1944; Wesley C. Williams, in *Dictionary of Scientific Biography* (Editor in Chief Charles Coulston Gillispie.) 10: 260-263. 1981; Georges Léopold Chrétien Frédéric Dagobert Cuvier (1769-1832), *G. Cuvier's Briefe an C.H. Pfaff aus den Jahren 1788 bis 1792* ... Herausgegeben von ... W.F.G. Behn. Kiel 1845 and *Prof. R. Owen's Osteologie der Dronte. (Didus ineptus L.)* [Dresden 1868]; Casey A. Wood, *An Introduction to the Literature of Vertebrate Zoology.* Based chiefly on titles in the Blacker Library of Zoology ... and other libraries of McGill University, Montreal. London 1931; Garrison and Morton, *Medical Bibliography.* 329, 336. New York 1961; Alexander B. Adams, *Eternal Quest. The Story of the Great Naturalists.* New York 1969.

Species/Vernacular Names:
O. acidula F. Muell.

Australia: emu apple, sour plum, sour apple, gooya, gruie, colane, mooley plum, dillie boolen

O. cepiodora F. Muell.

Australia: onionwood, onion cedar

O. reticulata F. Muell.

Australia: desert walnut

O. venosa F. Muell.

Australia: crow's apple, rose almond, sour plum

O. vernicosa F. Muell.

Australia: emu apple

Oxalis L. Oxalidaceae

Origins:
Greek *oxalis* (*oxys* "acid, sour, sharp"), referring to the taste of the leaves and stem; Plinius used Latin *oxalis, idis*, for some species of *Rumex*. See T.M. Salter, "The genus *Oxalis* in South Africa: A taxonomic revision." *The Journal of South African Botany.* Supplementary Volume no. 1: 238-242. 1944; Carl Linnaeus, *Species Plantarum.* 433. 1753 and *Genera Plantarum.* Ed. 5. 198. 1754.

Species/Vernacular Names:
O. spp.

South Africa: suring, sorrel

Mexico: nocuana bee

O. acetosella L.

English: wood sorrel, cuckoo bread, alleluia

O. alstonii Lourteig

English: fire fern, red flame

O. caprina L.

English: goat's-foot, wood sorrel

O. corniculata L. (*Oxalis repens* Thunb.; *Oxalis lupulina* Kunth)

English: creeping lady's sorrel, creeping wood sorrel, creeping oxalis, creeping sorrel, yellow oxalis, creeping yellow oxalis, yellow sorrel, yellow wood sorrel, procumbent yellow sorrel, sour weed, sour grass

Arabic: hamd

India: carngeri, puliyaral, puliyarai, amrul, amboti

Nepal: chariamilo

China: cu jiang cao, tsao chiang, suan chiang, hsiao suan tsai

The Philippines: taingang daga, marasiksik, daraisig, kungi, iayo, kanapa, piknik, salmagi

Malaya: sikap dada

Southern Africa: ranksuring, steenboksuring, tuinranksuring

Congo: lopeto, ngongua

Hawaii: 'ihi 'ai, 'ihi 'awa, 'ihi maka 'ula, 'ihi makole

O. corymbosa DC. (*Ionoxalis martiana* (Zucc.) Small; *Oxalis martiana* Zucc.)

English: pink wood sorrel

Japan: murasaki-katabami, yafata

Hawaii: 'ihi pehu

O. depressa Eckl. & Zeyh. (*Oxalis commutata* Sond. var. *pusilla* Knuth; *Oxalis convexula* Jacq. var. *dilatata* Eckl. & Zeyh.; *Oxalis dammeriana* Schltr.; *Oxalis otaviensis* Knuth)

Southern Africa: suring; bolila (Sotho)

O. enneaphylla Cav.

English: scurvy grass

O. exilis A. Cunn.

English: slender wood sorrel

O. flava L. (*Oxalis flabellifolia* Jacq.; *Oxalis lupinifolia* Jacq.; *Oxalis pectinata* Jacq.)

English: finger-leaved oxalis

South Africa: bobbejaanuintjie, vingersuring

O. latifolia Kunth

English: oxalis, garden sorrel, red garden sorrel, large-leaved wood-sorrel, fish-tail oxalis

East Africa: akanyuunya-mbuzi, enyonyo, kajampuni, kanyobwa, ndabibi, obwekijunga

South Africa: rooisuring, rooituinsuring, suring, tuinsuring

O. luteola Jacq. (*Oxalis balsamifera* E. Mey. ex Sond.)

English: pink sorrel

South Africa: pienksuring

O. magellanica G. Forst.

English: white wood sorrel

O. obliquifolia Steud. ex Rich.

Southern Africa: bolila (Sotho)

O. oregana Nutt. ex Torr. & A. Gray

English: red wood-sorrel

O. ortgiesii Reg.

English: tree oxalis

O. perennans Haw.

English: scour wood sorrel

O. pes-caprae L.

English: Bermuda buttercup, sorrel, soursop, soursob, wild sorrel, wood sorrel, Cape sorrel, yellow sorrel, English weed, goat's foot

O. pes-caprae L. var. *pes-caprae* (*Oxalis cernua* Thunb.; *Oxalis mairei* Knuth ex Engler)

South Africa: yellow sorrel, geelsuring, kalwersoring, klawersoring, klawersuring, pypsoring, suring, tuinsuring, varksuring, wildesuring

O. polyphylla Jacq. var. *polyphylla* (*Oxalis amoena* Jacq.; *Oxalis filifolia* Jacq.)

English: finger sorrel

South Africa: vingersuring

O. purpurea L. (*Oxalis aemula* Schltr. ex Knuth; *Oxalis breviscapa* Jacq.; *Oxalis decipiens* Schltr.; *Oxalis humilis* Thunb.; *Oxalis laburnifolia* Jacq.; *Oxalis sanguinea* Jacq.; *Oxalis strictophylla* Sond.; *Oxalis variabilis* Jacq.)

English: one o'clock, large-flowered wood-sorrel, red flowering sorrel

South Africa: sorrel, suring

O. semiloba Sond.

English: Transvaal sorrel

Southern Africa: Transvaalse suring; bolila (Sotho); isinungu (Zulu)

O. smithiana Eckl. & Zeyh. (*Oxalis galpinii* Schltr.)

Southern Africa: klawersuring, rooisuring; umMuncwane (Xhosa); (for the bulbs) inKolowane, iZotho (Xhosa)

O. tetraphylla Cav.

English: lucky clover, good luck leaf, good luck plant

O. violacea L.

English: violet wood sorrel

Oxalistylis Baillon Euphorbiaceae

Origins:
Greek *oxys* "sharp" and *stylos* "style, pillar."

Oxandra A. Rich. Annonaceae

Origins:
From the Greek *oxys* "sharp" and *andros* "male, man."

Species/Vernacular Names:
O. lanceolata (Sw.) Baillon
English: lancewood

Oxanthera Montr. Rutaceae

Origins:
From the Greek *oxys* "sharp" and *anthera* "anther," indicating the nature of the anthers.

Oxera Labill. Labiatae (Verbenaceae)

Origins:
From the Greek *oxys* "acid, sour, sharp," *oxeros, oxera* "of vinegar, acid," referring to the sour and acrid sap.

Oxipolis Raf. Umbelliferae

Origins:
See *Oxypolis* Raf.; see C.S. Rafinesque, *The Good Book.* 59. 1840; E.D. Merrill, *Index rafinesquianus.* 182. 1949.

Oxleya Hook. Rutaceae

Origins:
After the English (b. near Westow, Yorkshire) explorer John Oxley, 1783/1785-1828 (d. Sydney), traveler, discovered the Brisbane River, Australia. His works include *Journals of Two Expeditions into the Interior of New South Wales,* undertaken by order of the British Government, in the years 1817-1818. London 1820, *An Historical Account of the Colony of New South Wales ...* To which is subjoined an accurate map of Port Macquarie, 1821 and *Report of an Expedition to Survey Port Curtis, Moreton Bay, and Port Bowen,* in Barron Field (1786-1846), *Geographical Memoirs on New South Wales.* By various hands ... Edited by B.F. London 1825; see Douglas Pike, ed., *Australian Dictionary of Biography.* 2: 305-307. Melbourne 1967.

Oxyanthe Steud. Gramineae

Origins:

From the Greek *oxys* "acid, sour, sharp" and *anthos* "flower."

Oxyanthera Brongniart Orchidaceae

Origins:

From the Greek *oxys* "sharp" and *anthera* "anther," the pointed anther.

Oxyanthus DC. Rubiaceae

Origins:

The generic name is based on the Greek words *oxys* "sharp" and *anthos* "flower," referring to the sharp teeth of the calyx and acute segments of the corolla.

Species/Vernacular Names:

O. sp.

Nigeria: nkpo-azu

O. latifolius Sond. (*Oxyanthus schlechteri* K. Schumann)

English: Zulu loquat, broad-leaved oxyanthus

Southern Africa: Zoeloelukwart; uMaphekemoyeni-omn-yama, isiBinda esingakhali-amasi, umDlankawu, umVilowehlathi (Zulu)

O. pyriformis (Hochst.) Skeels subsp. *pyriformis* (*Oxyanthus natalensis* Sond.; *Megacarpa pyriformis* Hochst.)

English: Natal loquat, Natal oxyanthus

Southern Africa: Natallukwart; isiBinda esingakhali-amasi (Zulu)

O. speciosus DC.

Congo: buko ba cafe, ekie

O. speciosus DC. subsp. *gerrardii* (Sond.) Bridson (*Oxyanthus gerrardii* Sond.)

English: wild loquat, whipstick tree

Southern Africa: wildelukwart, sweepstokboom, wilde-koffieboom; umKhulu-omncane, umKhuluomncane, uPhondo lwembabala (= the horns of the bushbuck), isiBinda esingakhali amasi (Zulu); iMpekana, umBindi (Xhosa)

Oxybaphus L'Hérit. ex Willd. Nyctaginaceae

Origins:

Greek *oxybaphon* "saucer, shallow earthen vessel, small vinegar saucer," Latin *oxybaphus* "a vinegar cup."

Oxycarpha S.F. Blake Asteraceae

Origins:

From the Greek *oxys* "sharp" and *karphos* "chip of straw, chip of wood, mote, splinter, nail."

Oxycarpus Lour. Guttiferae

Origins:

Greek *oxys* "sharp, acid, sour" and *karpos* "fruit."

Oxycaryum Nees Cyperaceae

Origins:

From the Greek *oxys* "sharp, acid, sour" and *karyon* "nut."

Oxyceros Lour. Rubiaceae

Origins:

Greek *oxys* "sharp, acid, sour" and *keras* "horn" or *keros* "wax."

Oxychlamys Schltr. Gesneriaceae

Origins:

From the Greek *oxys* "sharp, acid, sour" and *chlamys, chlamydos* "cloak."

Oxychloe Philippi Juncaceae

Origins:

From the Greek *oxys* "sharp, acid, sour" and *chloe, chloa* "grass."

Oxychloris Lazarides Gramineae

Origins:

Greek *oxys* "sharp" and the genus *Chloris* Swartz; see Michael Lazarides, "New taxa of tropical Australian grasses (Poaceae)." in *Nuytsia*. 5(2): 273-303. 1985.

Oxycladium F. Muell. Fabaceae

Origins:

Greek *oxys* "sharp" and *kladion* "a branchlet, a small branch."

Oxycoca Raf. Ericaceae

Origins:
Referring to *Oxycoccus* Adans.; see C.S. Rafinesque, *Med. Fl.* 2: 48. 1830; E.D. Merrill, *Index rafinesquianus*. 186. 1949.

Oxycoccus Hill Ericaceae

Origins:
From the Greek *oxys* "sharp" and *kokkos* "a berry."

Oxydectes Kuntze Euphorbiaceae

Origins:
From the Greek *oxys* "sharp, acid, sour" and *dektes* "biter."

Oxydendrum DC. Ericaceae

Origins:
From the Greek *oxys* "sharp, sour" and *dendron* "tree," referring to the bitter and acid-tasting leaves or to the shape of the trees.

Species/Vernacular Names:
O. arboreum (L.) DC.

English: sourwood, tree sorrel

Oxydenia Nutt. Gramineae

Origins:
From the Greek *oxys* "sharp, sour" and *aden* "gland."

Oxyglottis (Bunge) Nevski Fabaceae

Origins:
From the Greek *oxys* "sharp, sour" and *glotta* "tongue."

Oxygonium C. Presl Aspleniaceae (Woodsiaceae)

Origins:
Greek *oxygonios* "acute-angled."

Oxygonum Burch. ex Campdera Polygonaceae

Origins:
Greek *oxys* "sharp" and *gonia* "an angle," with sharp angles, with spiny fruits.

Species/Vernacular Names:
O. sinuatum (Meissner) Dammer

English: double thorn

East Africa: akitikemiria, awayo, kafumita bagenda, karinga, kuru, mbigiri, nyatiend-gweno, obucumita-mbogo, obwita-mbogo, okuro

Oxygraphis Bunge Ranunculaceae

Origins:
From the Greek *oxys* "sharp, acid" and *graphis* "a drawing, pencil."

Oxygyne Schltr. Burmanniaceae

Origins:
From the Greek *oxys* "sharp" and *gyne* "female, woman," referring to the style.

Oxylaena Benth. ex Anderb. Asteraceae

Origins:
From the Greek *oxys* "sharp" and *chlaena, chlaenion, laina* "cloak, blanket."

Oxylobium Andrews Fabaceae

Origins:
Greek *oxys* and *lobos* "a pod," *lobion* "small pod," the seedpods have sharp appendages; see Henry C. Andrews (1794-1830), *The Botanist's Repository*. London (November) 1807.

Species/Vernacular Names:
O. aciculiferum (F. Muell.) Benth.

English: needle shaggy pea

O. alpestre F. Muell.

English: alpine shaggy pea

O. arborescens R. Br.

English: tall shaggy pea

O. cordifolium Andrews

English: heart-leaved shaggy pea

O. ellipticum (Vent.) R. Br.

English: common shaggy pea

O. ilicifolium (Andrews) Domin

English: prickly shaggy pea

O. procumbens F. Muell.

English: trailing shaggy pea

O. pulteneae DC.

English: wiry shaggy pea

O. robustum J. Thompson

English: tree shaggy pea

O. scandens (Smith) Benth.

English: netted shaggy pea

Oxylobus (DC.) A. Gray Asteraceae

Origins:

From the Greek *oxys* "sharp" and *lobos* "a pod," referring to the pointed involucre.

Oxymeris DC. Melastomataceae

Origins:

From the Greek *oxys* "sharp" and *meris* "part."

Oxymitra (Blume) Hook.f. & Thomson Annonaceae

Origins:

From the Greek *oxys* "sharp, acid" and *mitra* "a head-band."

Oxymyrrhine Schauer Myrtaceae

Origins:

Greek *myrrhine, myrsine* for *Myrtus communis*, Latin *oxymyrsine, es* applied by Plinius to the plant prickly-myrtle, butcher's broom.

Oxyosmyles Speg. Boraginaceae

Origins:

Probably from the Greek *oxys* "acid, sour" and *osmyle* "a strong-smelling musky octopus," probably referring to the smell.

Oxypappus Benth. Asteraceae

Origins:

From the Greek *oxys* "sharp, acid" and *pappos* "fluff, pappus."

Oxypetalum R. Br. Asclepiadaceae

Origins:

From the Greek *oxys* "sharp" and *petalon* "a petal"; see R. Brown, "On the Asclepiadeae." *Memoirs of the Wernerian Natural History Society.* 1: 41. Edinburgh 1811.

Oxyphyllum Philippi Asteraceae

Origins:

From the Greek *oxys* "sharp" and *phyllon* "leaf," with sharp-pointed leaves.

Oxypogon Raf. Fabaceae

Origins:

See C.S. Rafinesque, *Jour. Phys. Chim. Hist. Nat.* 89: 98. 1819; E.D. Merrill, *Index rafinesquianus.* 147. 1949.

Oxypolis Raf. Umbelliferae

Origins:

Greek *oxys* "sharp" and *polos* "axis, pole," referring to the leaves; see C.S. Rafinesque, *Neogenyton, or Indication of Sixty-Six New Genera of Plants of North America.* 2. 1825, *Med. Fl.* 2: 250. 1830 and *Seringe Bull. Bot.* 1: 217-218. 1830; E.D. Merrill, *Index rafinesquianus.* 182. 1949.

Oxypteryx Greene Asclepiadaceae

Origins:

From the Greek *oxys* "sharp" and *pteryx, pterygos* "small wing."

Oxyrhachis Pilger Gramineae

Origins:

From the Greek *oxys* "sharp" and *rhachis* "rachis, axis, midrib of a leaf."

Oxyrhynchus Brandegee Fabaceae

Origins:
From the Greek *oxys* "sharp" and *rhynchos* "horn, beak, snout," sharp-snouted, sharp-pointed.

Oxyria Hill Polygonaceae

Origins:
From the Greek *oxys* "sharp, sour," indicating the acidity of the leaves.

Oxys Miller Oxalidaceae

Origins:
Greek *oxys* "acid, sour," referring to the leaves, see also *Oxalis* L.

Oxysepala Wight Orchidaceae

Origins:
From the Greek *oxys* "sharp" and Latin *sepalum* "sepal," acuminate sepals.

Oxyspermum Ecklon & Zeyher Rubiaceae

Origins:
From the Greek *oxys* "sharp" and *sperma* "seed."

Oxyspora DC. Melastomataceae

Origins:
From the Greek *oxys* "sharp, pointed" and *sporos* "a seed," referring to the awned and pointed seeds.

Oxystelma R. Br. Asclepiadaceae

Origins:
Greek *oxys* "sharp, pointed" and *stelma, stelmatos* (*stello* "to bring together, to bind, to set") "a girdle, belt," indicating the sharp corona segments; see R. Brown, "On the Asclepiadeae." *Memoirs of the Wernerian Natural History Society.* 1: 40. Edinburgh 1811.

Species/Vernacular Names:
O. esculentum (L.f.) Smith
English: edible oxystelma

China: jian huai teng

Oxystemon Planchon & Triana Guttiferae

Origins:
From the Greek *oxys* "sharp, sour" and *stemon* "a stamen, a thread."

Oxystigma Harms Caesalpiniaceae

Origins:
Greek *oxys* "sharp" and *stigma.*

Species/Vernacular Names:
O. mannii (Baill.) Harms
Nigeria: bosipi; ntufiak (Efik)
Cameroon: bidou, bosipi, bossipi, bussipi, elongo
O. oxyphyllum (Harms) J. Léonard
Nigeria: logagola; lolagbola (Yoruba)
West Africa: lolagbola, tchitola
Cameroon: gondo, tchitola
Gabon: emoli, emolo, m'babu
Congo: kitola, tchitola
Zaire: akwakwa, kalakati, kitola, waka

Oxystophyllum Blume Orchidaceae

Origins:
From the Greek *oxystos*, the superlative of *oxys* "sharp" and *phyllon* "leaf."

Oxystylis Torrey & Frém. Capparidaceae (Capparaceae)

Origins:
From the Greek *oxys* "sharp" and *stylos* "pillar, style."

Oxytenanthera Munro Gramineae

Origins:
Greek *oxytes* "sharpness, of acute angles, acidity," *oxytenes* "pointed" and *anthera* "anther," an allusion to the nature of the anthers.

Species/Vernacular Names:
O. abyssinica (A. Rich.) Munro

English: Bindura bamboo

Southern Africa: mushenjerere, musengere (Shona)

Yoruba: apako, pako, aparun, oparun, opa

Oxytenia Nutt. Asteraceae

Origins:
From the Greek *oxytenes* "pointed," or from *oxys* "sharp, sour" and *tainia* "fillet."

Oxytheca Nutt. Polygonaceae

Origins:
Greek *oxys* "sharp, sour" and *theke* "a box, case, capsule," referring to the involucre.

Species/Vernacular Names:
O. caryophylloides C. Parry

Australia: chickweed oxytheca

O. emarginata H.M. Hall

Australia: white-margined oxytheca

O. watsonii Torrey & A. Gray

Australia: Watson's oxytheca

Oxythece Miq. Sapotaceae

Origins:
Greek *oxys* "sharp, sour" and *theke* "a box, case, capsule."

Oxytria Raf. Hyacinthaceae (Liliaceae)

Origins:
See C.S. Rafinesque, *Flora Telluriana*. 2: 26. 1836 [1837]; E.D. Merrill, *Index rafinesquianus*. 92. 1949.

Oxytropis DC. Fabaceae

Origins:
From the Greek *oxys* "sharp" and *tropis* "keel," alluding to the pointed or beaked keels of the flowers.

Species/Vernacular Names:
O. deflexa (Pall.) DC. var. *sericea* Torrey & A. Gray

English: blue pendent pod oxytrope

Ozandra Raf. Myrtaceae

Origins:
Greek *ozo*, *osme* "smell, odor" (or *ozos* "branch, knot") and *aner*, *andros* "male, stamen"; see Constantine Samuel Rafinesque, *Autikon botanikon*. Icones plantarum select. nov. vel rariorum, etc. Philadelphia 1840.

Ozanthes Raf. Lauraceae

Origins:
Greek *ozo*, *osme* "smell, odor, to smell" and *anthos* "flower"; see Constantine Samuel Rafinesque, *Sylva Telluriana*. 133. 1838.

Ozodycus Raf. Cucurbitaceae

Origins:
See Constantine Samuel Rafinesque, *Atl. Jour*. 1: 145. 1832; E.D. Merrill, *Index rafinesquianus*. 230. 1949.

Ozomelis Raf. Saxifragaceae

Origins:
From the Greek *ozo*, *osme* "smell, odor, to smell" and *meli* "honey"; see C.S. Rafinesque, *Flora Telluriana*. 2: 73. 1836 [1837]; E.D. Merrill, *Index rafinesquianus*. 135. 1949.

Ozoroa Del. Anacardiaceae

Origins:
Probably after an Arabian name, or an Ethiopian name for "queen."

Species/Vernacular Names:
O. concolor (Presl ex Sond.) De Winter (*Heeria concolor* (C. Presl) O. Kuntze; *Rhus concolor* C. Presl) (Latin *concolor, oris* "of one color")

South Africa: Richtersveld Ozoroa, harpuisboom

O. crassinervia (Engl.) R. & A. Fernandes (*Anaphrenium crassinervium* (Engl.) Engl.; *Heeria aromatica* Dinter; *Heeria crassinervia* Engl.; *Heeria dinteri* Schinz)

South Africa: harpuisboom

O. dispar (Presl) R. & A. Fernandes (*Heeria dispar* (Presl) O. Kuntze; *Heeria rangeana* Engl.; *Ozoroa rangeana* (Engl.) R. & A. Fernandes; *Rhus dispar* Presl) (Latin *dispar, aris* "dissimilar")

South Africa: Namaqualand ozoroa, harpuisboom

O. engleri R. & A. Fernandes (the species was named after the German botanist Heinrich Gustav Adolf Engler, 1844-1930, author of *Das Pflanzenreich. Regni vegetabilis conspectus.* Leipzig 1900; see J.H. Barnhart, *Biographical Notes upon Botanists.* 1: 510. Boston 1965; Simon Mayo, Josef Bogner and Peter Boyce, "The acolytes of the Araceae." *Curtis's Botanical Magazine.* Volume 12. 3: 153-168. August 1995; Stafleu and Cowan, *Taxonomic Literature.* 1: 757-797. Utrecht 1976; T.W. Bossert, *Biographical Dictionary of Botanists Represented in the Hunt Institute Portrait Collection.* 117. 1972; Ida Kaplan Langman, *A Selected Guide to the Literature on the Flowering Plants of Mexico.* Philadelphia 1964; S. Lenley et al., *Catalog of the Manuscript and Archival Collections and Index to the Correspondence of John Torrey.* Library of the New York Botanical Garden. 154-155. 1973)

English: white resin tree, Makatini ozoroa

Southern Africa: witharpuisboom; isiFice, isiFica, isiFico (Zulu); shinungumafi (Thonga or Tsonga)

O. insignis Del. (*Heeria insignis* (Del.) O. Kuntze)

Southern Africa: bukati (Shona)

Nigeria: hawayen zaki, kasheshe (Hausa)

Mali: kalakari, sisan, cifaama

O. insignis Del. subsp. *latifolia* (Engl.) R. Fernandes (*Anaphrenium abyssinicum* var. *latifolium* Engl.)

South Africa: harpuisboom

O. insignis Del. subsp. *reticulata* (Bak.f.) J.B. Gillett (*Heeria insignis* (Del.) O. Kuntze var. *reticulata* Bak. f.; *Heeria reticulata* (Bak.f.) Engl.; *Ozoroa reticulata* (Bak.f.) R. & A. Fernandes)

Southern Africa: harpuisboom, isifico; isiFico (Zulu); shinungu (Tsonga); muacha, muBedu, chafitcha, maDsikavakadzi, muHacha, iHlanshwachipini, muRingu, muRungu, sukavu (Shona)

O. longipes (Engl. & Gilg) R. & A. Fernandes (*Heeria longipes* Engl. & Gilg)

South Africa: harpuisboom

O. mucronata (Bernh. ex Krauss) R. & A. Fernandes (*Heeria mucronata* Bernh. ex Krauss)

English: Eastern Cape resin tree, Cape ozoroa

South Africa: Oos-Kaapse harpuisboom

O. namaensis (Schinz & Dinter) R. Fernandes (*Heeria namaensis* Schinz & Dinter)

Southern Africa: harpuisboom, Nama Ozoroa (from southern Southwest Africa)

O. obovata (Oliv.) R. & A. Fernandes

English: broad-leaved resin tree

Southern Africa: breëblaarharpuisboom; isiFice, isiFica, isiFico (Zulu); ashiFisu, shinungumafi (Thonga or Tsonga); mochudi (Western Transvaal, northern Cape, Botswana); munungu-mahfi (Venda)

O. paniculosa (Sond.) R. & A. Fernandes var. *paniculosa* (*Rhus paniculosa* Sond.)

English: common resin tree, bushveld ozoroa

Southern Africa: gewone harpuisboom; isiFice, isiFica, isiFico sehlanze (Zulu); monokane (Hebron dialect, central Transvaal); monoko (North Sotho)

O. sphaerocarpa R. & A. Fernandes

English: currant resin tree

Southern Africa: korenteharpuisboom; isiFice (Zulu)

Ozothamnus R. Br. Asteraceae

Origins:

Greek *ozo, ozein* "to smell" and *thamnos* "shrub," many of the species are fragrant; some suggest from *ozos* "branch, knot"; see R. Brown, "Observations on the natural family of plants called Compositae." *Transactions of the Linnean Society of London.* 12: 125. 1818.

Species/Vernacular Names:

O. alpinus (N.A. Wakenf.) Anderb.

English: alpine everlasting

O. argophyllus (A. Cunn. ex DC.) Anderb.

English: spicy everlasting

O. blackallii N.T. Burb. (for William E. Blackall, botanist, Western Asutralia)

English: milky everlasting

O. conditus (N.A. Wakenf.) Anderb.

English: pepper everlasting

O. cuneifolius (Benth.) Anderb.

English: wedge everlasting

O. decurrens F. Muell.

English: ridged everlasting

O. diosmifolius (Vent.) DC.

English: ball everlasting, rice flower, sago flower, pill flower

O. diotophyllus (F. Muell.) Anderb.

English: heath everlasting

O. ferrugineus (Labill.) DC.

English: tree everlasting

O. hookeri Sond.

English: scaly everlasting, kerosene bush

O. ledifolius (DC.) Hook.f.

English: kerosene bush

O. lepidophyllus Steetz

English: scaly-leaved everlasting

O. obcordatus DC.

English: grey everlasting

O. occidentalis (N.T. Burb.) Anderb.

English: rough-leaved everlasting

O. retusus Sond. & F. Muell.

English: rough everlasting

O. rosmarinifolius (Labill.) Sweet

English: rosemary everlasting

O. secundiflorus (N.A. Wakenf.) C. Jeffrey

English: cascade everlasting

O. stirlingii (F. Muell.) Anderb.

English: Ovens everlasting

O. thyrsoideus DC.

English: sticky everlasting

O. turbinatus DC.

English: coast everlasting

P

Pabellonia Quezada & Martic. Alliaceae (Liliaceae)

Origins:
Spanish *pabellon* "pavillion, canopy, banner, flag, shelter, covering, external ear."

Pabstia Garay Orchidaceae

Origins:
For the botanist Guido Frederico João Pabst, 1914-1980, orchidologist, plant collector, Director of the Herbarium Bradeanum, Rio de Janeiro, Brazil (location: Rua São Francisco Xavier 524, Pavilhão João Lira Filho, Bloco F, II° andar), with Fritz Dungs (1915-1977) wrote *Orchidaceae Brasilienses*. Hildesheim 1975-1977; see R. Zander, F. Encke, G. Buchheim and S. Seybold, *Handwörterbuch der Pflanzennamen*. 14. Aufl. 759. Stuttgart 1993.

Pabstiella Brieger & Sanghas Orchidaceae

Origins:
The diminutive of *Pabstia*.

Pachira Aublet Bombacaceae

Origins:
Pachira is a native name in Guiana; see R. Zander, F. Encke, G. Buchheim and S. Seybold, *Handwörterbuch der Pflanzennamen*. 14. Aufl. 676. 1993; Georg Christian Wittstein, *Etymologisch-botanisches Handwörterbuch*. 650. 1852.

Species/Vernacular Names:
P. aquatica Aublet (*Bombax aquaticum* (Aublet) Schumann)

Peru: bellaco caspi, huimba, pasharo, punga, wimba

Latin America: zapote de bobo

P. insignis (Swartz) Swartz ex Savigny (*Bombax spruceanum* (Decaisne) Ducke)

Peru: mamorana grande

Pachistima Raf. Celastraceae

Origins:
Genus *Paxistima* Raf., Greek *pachys* "thick, stout" and *stigma* "stigma"; see C.S. Rafinesque, *Am. Monthly Mag. Crit. Rev.* 2: 176. 1818 and *Jour. Phys. Chim. Hist. Nat.* 89: 257. 1819.

Pachites Lindley Orchidaceae

Origins:
From the Greek *pachys* "thick, stout," referring to the rostellum.

Pachyanthus A. Rich. Melastomataceae

Origins:
Thick-flowered, from the Greek *pachys* "thick, stout" and *anthos* "a flower."

Pachycarpus E. Meyer Asclepiadaceae

Origins:
From the Greek *pachys* "thick, stout" and *karpos* "a fruit," with a thick pericarp.

Pachycentria Blume Melastomataceae

Origins:
From the Greek *pachys* "thick, stout" and *kentron* "a spur, point," referring to the thorns.

Pachycereus (A. Berger) Britton & Rose Cactaceae

Origins:
From the Greek *pachys* "thick, stout" plus *Cereus*, referring to the stems of these tree-like cacti.

Pachychilus Blume Orchidaceae

Origins:

Fleshy lip, Greek *pachys* "thick, stout" and *cheilos* "lip," see also *Pachystoma* Blume.

Pachychlamys Dyer ex Ridl. Dipterocarpaceae

Origins:

From the Greek *pachys* "thick, stout" and *chlamys* "cloak."

Pachycladon Hook.f. Brassicaceae

Origins:

From the Greek *pachys* "thick, stout" and *klados* "branch."

Pachycormus Coville ex Standley Anacardiaceae

Origins:

Thick stump (trunk or stem), from the Greek *pachys* "thick, stout" and *kormos* "trunk of a tree, logs of timber," referring to the basal trunk and main branches, thick.

Species/Vernacular Names:

P. discolor (Benth.) Cov. ex Standley (see Robert R. Humphrey, "Baja's Sacred Resin Tree." *Cactus and Succulent Journal.* vol. 63. 1: 35-39. 1991)

English: elephant tree

Baja California: copalquín, torote blanco

Pachycornia Hook.f. Chenopodiaceae

Origins:

Greek *pachys* "thick, stout" and Latin *cornu, us* "a horn," resembling the genus *Salicornia*, referring to the branches and spikes.

Species/Vernacular Names:

P. triandra (F. Muell.) J. Black (*Arthrocnemum triandrum* F. Muell.; *Salicornia triandra* (F. Muell.) Druce)

Australia: desert glasswort, desert samphire

Pachyctenium Maire & Pamp. Umbelliferae

Origins:

From the Greek *pachys* "thick, stout" and *ktenion* "a little comb."

Pachycymbium L.C. Leach Asclepiadaceae

Origins:

From the Greek *pachys* "thick, stout" and *kymbe* "boat," *kymbos* "cavity."

Pachyderma Blume Oleaceae

Origins:

From the Greek *pachys* "thick, stout" and *derma, dermatos* "skin."

Pachydesmia Gleason Melastomataceae

Origins:

From the Greek *pachys* "thick, stout" and *desmos* "a bond, band, bundle."

Pachydiscus Gilg & Schltr. Alseuosmiaceae

Origins:

From the Greek *pachys* "thick, stout" and *diskos* "disc."

Pachyelasma Harms Caesalpiniaceae

Origins:

Greek *pachys* and *elasma, elasmos* "a metal plate," referring to the fruits.

Species/Vernacular Names:

P. tessmannii (Harms) Harms

Yoruba: eru

Nigeria: eru (Yoruba); ogiesegheseghe (Edo); ire (Ijaw); mbaghe (Ekoi)

Congo: ediouk

Gabon: mekogbo, mekogho

Central Africa: dogabela

Zaire: bolubo, boliko, bomboli, dula

Cameroon: eyec, eyek, ndiai, lec, mbo

Pachygone Miers Menispermaceae

Origins:

Greek *pachys* "thick, stout" and *gonos* "seed," referring to the thick seeds, or from *gony* "joint, knee," indicating thick joints or nodes; see John Miers (1789-1879), in *Annals and*

Magazine of Natural History. Ser. II, 7: 37, 43. London 1851.

Pachylaena D. Don ex Hooker & Arnott Asteraceae

Origins:

From the Greek *pachys* "thick, stout" and *chlaena, chlaenion* "a cloak, blanket."

Pachylarnax Dandy Magnoliaceae

Origins:

Greek *pachys* and *larnax, larnakos* "a box, chest, vessel, boat," referring to the thick fruits, to the woody capsule; see David Hunt, ed., *Magnolias and Their Allies.* Proceedings of an International Symposium, Royal Holloway, University of London, Egham, Surrey, U.K., 12-13 April 1996. International Dendrology Society and The Magnolia Society. 1998; David G. Frodin & Rafaël Govaerts, *World Checklist and Bibliography of Magnoliaceae.* Royal Botanic Gardens, Kew 1996.

Pachylecythis Ledoux Lecythidaceae

Origins:

From the Greek *pachys* "thick, stout" plus the genus *Lecythis* Loefl.

Pachylobus G. Don Burseraceae

Origins:

From the Greek *pachys* "thick, stout" and *lobos* "a lobe, pod."

Pachyloma Bosch Hymenophyllaceae

Origins:

From the Greek *pachys* "thick, stout" and *loma* "border, margin, fringe, edge"; see Roelof Benjamin van den Bosch (1810-1862), in *Versl. Meded. Akad. Wetensch.* 11: 318. Amsterdam 1861.

Pachyloma DC. Melastomataceae

Origins:

Greek *pachys* "thick, stout" and *loma* "border, margin, fringe, edge," indicating the connective.

Pachylophus Spach Onagraceae

Origins:

From the Greek *pachys* "thick, stout" and *lophos* "a crest."

Pachymeria Benth. Melastomataceae

Origins:

From the Greek *pachys* "thick, stout" and *meris* "part," presumably referring to the connective or to the genus *Meriania* Sw.

Pachymitus O.E. Schulz Brassicaceae

Origins:

Greek *pachys* "thick, stout" and *mitos* "a thread, web, string," referring to the pedicels of the fruits; see Otto Eugen Schulz (1874-1936), *Das Pflanzenreich.* Heft 86. (July) 1924.

Species/Vernacular Names:

P. cardaminoides (F. Muell.) O. Schulz (*Sisymbrium cardaminoides* F. Muell.; *Blennodia cardaminoides* (F. Muell.) Benth.; *Erysimum cardaminoides* (F. Muell.) F. Muell.; *Erysimum lucae* F. Muell.)

English: sand cress

Pachyne Salisb. Orchidaceae

Origins:

From the Greek *pachys* "thick, stout," probably referring to the fleshy column, sepals and petals, to the flowers and leaves.

Pachynema R. Br. ex DC. Dilleniaceae

Origins:

Greek *pachys* "thick, stout" and *nema* "thread," referring to the thickened stamens; see Augustin Pyramus de Candolle, *Regni vegetabilis systema naturale.* Parisiis [1817] 1818-1821.

Pachyneurum Bunge Brassicaceae

Origins:

From the Greek *pachys* "thick, stout" and *neuron* "nerve."

Pachynocarpus Hook.f. Dipterocarpaceae

Origins:

From the Greek *pachyno* "fatten, grow fat, thicken" and *karpos* "fruit."

Pachypharynx Aellen Chenopodiaceae

Origins:

Greek *pachys* "thick" and *pharynx, pharyngos* "pharynx"; see Paul Aellen (1896-1973), in *Botanische Jahrbücher*. 68: 429. 1938.

Pachyphragma (DC.) Reichb. Brassicaceae

Origins:

From the Greek *pachys* "thick" and *phragma* "a hedge, a fence, screen," referring to the septum of the pod.

Pachyphyllum Kunth Orchidaceae

Origins:

Greek *pachys* "thick" and *phyllon* "leaf."

Pachyphytum Link, Klotzsch & Otto Crassulaceae

Origins:

From the Greek *pachys* "thick" and *phyton* "plant," alluding to the thickened stems and leaves.

Pachyplectron Schltr. Orchidaceae

Origins:

Greek *pachys* and *plektron* "spur," the fleshy spur.

Pachypleuria (C. Presl) C. Presl Davalliaceae

Origins:

From the Greek *pachys* "thick" and *pleura, pleuron* "side, rib, lateral," referring to the side of the indusium, separate and forming a cupped flap.

Pachypleurum Ledeb. Umbelliferae

Origins:

Greek *pachys* and *pleura, pleuron* "side, rib, lateral," referring to the ridges of mericarps.

Pachypodanthium Engl. & Diels Annonaceae

Origins:

Greek *pachys* "thick, stout," *pous, podion, podos* "foot" and *anthos* "flower," referring to the carpels.

Species/Vernacular Names:

P. sp.

Nigeria: ntom, bokingo, olon

P. barteri (Benth.) Hutch. & Dalz.

Nigeria: ntokon eto (Efik)

P. staudtii (Engl. & Diels) Engl. & Diels

Cameroon: ntombe, ntom, ntoma, molombo

Congo: touom

Gabon: ntom

Nigeria: osoko (Igbirra)

Central Africa: mabongila, okianga, touom, metoma, ndondongo, ntom, molombo, ntombe, ntouma, mbalinga, anioukéti, miedzo, vahé

Ivory Coast: aniokouety, anionketi, anioukéti, anioukeli, emiengre, miedzo, niango

Liberia: zree-chu

Pachypodium Lindley Apocynaceae

Origins:

Greek *pachys* "thick, stout" and *podion* "a small foot," alluding to the fleshy and thick roots.

Species/Vernacular Names:

P. densiflorum Bak. var. *densiflorum*

Madagascar: songosongo, somo, somoy (Betsileo dialect); vontaka (Bara dialect); veloarivatana (Merina dialect)

P. horombense Pichon (the specific epithet from Horombe Plateau, in south-central Madagascar)

Madagascar: somo (Betsileo dialect); vontake, vontakakely (Bara dialect)

P. lealii Welw. (*Pachypodium giganteum* Engl.) (the specific name honors the Portuguese cartographer Lt. Col. Fernando Da Costa Leal whose map of Angola assisted Welwitsch in his travels; Friedrich Welwitsch discovered this species in southern Angola and described it in 1869; see the historic plate of two *Welwitschia* plants made by Leal and Thomas Baines and which appeared in *The Transactions of the Linnean Society, 1863-1864.*)

English: bottle tree

South Africa: bottelboom

P. succulentum (L.f.) Sweet

South Africa: krachtman

Pachypodium Nutt. Brassicaceae

Origins:
Greek *pachys* and *podion* "a small foot."

Pachypodium Webb & Berthel. Brassicaceae

Origins:
Greek *pachys* "thick, stout" and *podion* "a small foot."

Pachyptera DC. ex Meissner Bignoniaceae

Origins:
From the Greek *pachys* "thick, stout" and *pteron* "wing, feather."

Pachypteris Kar. & Kir. Brassicaceae

Origins:
From the Greek *pachys* "thick, stout" and *pteron* "wing, feather."

Pachypterygium Bunge Brassicaceae

Origins:
Greek *pachys* and *pteron* "wing, feather," *pterygion* "a small wing."

Pachyraphea Presl Fabaceae

Origins:
From the Greek *pachys* "thick, stout" and *rhaphis, rhaphidos* "a needle."

Pachyrhizanthe (Schltr.) Nakai Orchidaceae

Origins:
Greek *pachys* "thick," *rhiza* "root" and *anthos* "flower," roots tufted, thickened rhizomes.

Pachyrhizus Rich. ex DC. Fabaceae

Origins:
Greek *pachys* "thick" and *rhiza* "a root," referring to the edible and tuberous roots; see A.P. de Candolle, *Prodromus.* 2: 402. 1825; National Research Council, *Lost Crops of the Incas: Little-Known Plants of the Andes with Promise for*

Worldwide Cultivation. National Academy Press, Washington, D.C. 1989.

Species/Vernacular Names:
P. ahipa (Wedd.) L. Parodi (*Dolichos ahipa* Wedd.)

Latin America: ahipa

P. erosus (L.) Urban (*Cacara erosa* (L.) Kuntze; *Cacara palmatiloba* (Moçiño & Sessé ex DC.) Kuntze; *Dolichos bulbosus* L.; *Dolichos erosus* L.; *Dolichos palmatilobus* Moçiño & Sessé ex DC.; *Pachyrhizus angulatus* Rich. ex DC., nom. illeg.; *Pachyrhizus bulbosus* (L.) Kurz, nom. illeg.; *Pachyrhizus erosus* (L.) Urban var. *palmatilobus* (Moçiño & Sessé ex DC.) R.T. Clausen; *Pachyrhizus strigosus* R.T. Clausen) (Latin *strigosus* "covered with strigae, with stiff bristles")

English: yam bean, chopsui potato

Latin America: jicama

Japan: kuso-imo

China: di gua

The Philippines: sinkamas, hinkamas, kamas, lakamas, sikamas, kaman

P. tuberosus (Lam.) Sprengel (*Dolichos tuberosus* Lam.)

English: yam bean, potato bean

Peru: achipa, ajipa, jiquima, goseo-o, namou, wuiso, yaspo

Latin America: ahipa

Japan: oo-kuzu-imo

Pachyrhynchus DC. Asteraceae

Origins:
From the Greek *pachys* "thick" and *rhynchos* "horn, beak."

Pachysandra Michx. Buxaceae

Origins:
Greek *pachys* "thick, stout" and *aner, andros* "a man," referring to the thick stamens.

Pachysanthus C. Presl Rubiaceae

Origins:
From the Greek *pachys* "thick, stout" and *anthos* "flower."

Pachysolen Phil. Nolanaceae (Solanaceae)

Origins:
From the Greek *pachys* "thick, stout" and *solen* "a tube."

Pachystachys Nees Acanthaceae

Origins:

From the Greek *pachys* "thick, stout" and *stachys* "a spike," referring to the nature of the inflorescence.

Species/Vernacular Names:

P. coccinea (Aubl.) Nees (*Jacobinia coccinea* (Aubl.) Hiern; *Justicia coccinea* Aubl.)

English: cardinal's guard, cardinal flower

Pachystegia Cheeseman Asteraceae

Origins:

From the Greek *pachys* "thick, stout" and *stege, stegos* "roof, cover," referring to the flowers.

Pachystela Pierre ex Radlk. Sapotaceae

Origins:

From the Greek *pachys* "thick, stout" and *stele* "a pillar, column," Latin *stela, ae*, referring to the style.

Species/Vernacular Names:

P. sp.

Nigeria: azimomo, osan-igbo, osan-odo, otiemmi

P. bequaertii De Wild.

Zaire: bombili, mbili, lilo

P. brevipes (Bak.) Baill.

Nigeria: itoki (Igbirra); osan igbo, osan odo, osan oye (Yoruba); otiemme (Edo)

Yoruba: osan igbo, osan odo

P. msolo Engl.

Cameroon: poki

Pachystele Schltr. Orchidaceae

Origins:

From the Greek *pachys* "thick, stout" and *stele* "a pillar, column," fleshy column.

Pachystelis Rauschert Orchidaceae

Origins:

Greek *pachys* and *stele* "a pillar, column."

Pachystelma Brandegee Asclepiadaceae

Origins:

From the Greek *pachys* "thick" and *stelma, stelmatos* "a girdle, belt."

Pachystemon Blume Euphorbiaceae

Origins:

From the Greek *pachys* "thick" and *stemon* "stamen, thread."

Pachystigma Hochst. Rubiaceae

Origins:

From the Greek *pachys* "thick, stout" and *stigma*, referring to the cylindrical stigma.

Species/Vernacular Names:

P. macrocalyx (Sond.) Robyns (*Vangueria macrocalyx* Sond.)

English: crowned medlar

Southern Africa: kroonmispel, isibuthe tree; isiPhuthe, isi-Buthe (Zulu)

P. pygmaeum (Schltr.) Robyns (*Vangueria pygmaea* Schltr.; *Vangueria setosa* Conrath)

South Africa: Transvaal gousiektebossie, Western Transvaal gousiektebossie, Wes-Transvaalse of harige gousiekte-bossie, Wes-Transvaalse gousiektebossie, hairy gousiekte-bossie, gousiektebossie, gousiekte bush, goubos, gougoubossie, grysappel, witappeltjie; umkukuzela (Nde-bele)

P. thamnus Robyns

South Africa: Natal gousiektebossie, smooth gousiekte-bossie, gladde gousiektebossie, Natalse gousiektebossie

Pachystima Raf. Celastraceae

Origins:

Genus *Paxistima* Raf., Greek *pachys* "thick, stout" and *stigma* "stigma."

Pachystoma Blume Orchidaceae

Origins:

From the Greek *pachys* and *stoma* "mouth," an allusion to the thick lip; see Karl Ludwig von Blume, *Bijdragen tot de flora van Nederlandsch Indië*. 376. Batavia 1825.

Pachystrobilus Bremek. Acanthaceae

Origins:
From the Greek *pachys* "thick, stout" and *strobilos* "a cone."

Pachystroma Müll. Arg. Euphorbiaceae

Origins:
From the Greek *pachys* "thick, stout" and *stroma* "a bed."

Pachystylidium Pax & K. Hoffmann Euphorbiaceae

Origins:
From the Greek *pachys* "thick, stout" and *stylidion* "a small pillar."

Pachystylis Blume Orchidaceae

Origins:
From the Greek *pachys* "thick" and *stylos* "pillar, style."

Pachystylus K. Schumann Rubiaceae

Origins:
From the Greek *pachys* "thick" and *stylos* "pillar, style."

Pachysurus Steetz Asteraceae

Origins:
Having a thick tail, from the Greek *pachys* "thick, stout" and *oura* "tail"; see Johann G.C. Lehmann (1792-1860), *Plantae Preissianae*. 1: 441. Hamburgi 1845.

Pachythamnus (R. King & H. Robinson) R. King & H. Robinson Asteraceae

Origins:
Fat stems, Greek *pachys* "thick" and *thamnos* "shrub, bush."

Pachytrophe Bureau Moraceae

Origins:
From the Greek *pachys* "thick" and *trophe* "food."

Pacouria Aublet Apocynaceae

Origins:
A native name, French Guiana.

Pacourina Aubl. Asteraceae

Origins:
A native name, French Guiana.

Padellus Vassilcz. Rosaceae

Origins:
The diminutive of Greek *pados, pedos*.

Padus Miller Rosaceae

Origins:
Greek *pados, pedos*, used by Theophrastus (*HP.* 4.1.3 and 5.7.6) for a species of *Prunus* or for a tree whose timber was used for axles.

Paederia L. Rubiaceae

Origins:
Latin *paedor* (*pedor*), *paedoris* "filth, stench, an offensive smell," Akkadian *padu*, Hebrew *pada* "to dismiss, to free," possibly referring to an unpleasant (fecal) smell of some species when bruised.

Species/Vernacular Names:
P. foetida L.
Madagascar: laingomaimbo, lengomena, liane caca, lingue caca
Malaya: dangdangking, daun kuntut, sekuntut
China: nu ching, chiao piao (= sparrow calabash)
P. scandens (Lour.) Merr. (*Gentiana scandens* Lour.)
English: chicken-dung creeper
Hawaii: maile pilau, maile ka kahiki
Vietnam: day dam cho, mo tam the
China: ji shen teng, ji shi teng

Paederota L. Scrophulariaceae

Origins:
Latin *paederos, otis*, a classical name applied to a number of different plants: holm-oak, *Quercus ilex*, or chervil, *Anthriscus cerefolium* (L.) G.F. Hoffm., or bear's-foot, etc., Greek *paideros, otos*; see Helmut Genaust, *Etymologisches*

Wörterbuch der botanischen Pflanzennamen. 450. Basel 1996.

Paederotella (Wulff) Kem.-Nath. Scrophulariaceae

Origins:
Referring to the genus *Paederota* L., the diminutive.

Paeonia L. Paeoniaceae

Origins:
Greek *paionia* "the peony," Theophrastus (*HP*. 9.8.6), Latin *paeonia*; Greek *paionios*, *paionikos* "healing"; another name for paeonia is Latin *fatuina rosa*; Paeon or Paion (Paean, Paian) was the physician of the immortal gods, subsequently the name was applied to Apollo; see [Crusca], *Vocabolario degli Accademici della Crusca.* Firenze 1691 and 1729-1738; N. Tommaseo & B. Bellini, *Dizionario della lingua italiana.* Torino 1865-1879; G. Volpi, "Le falsificazioni di Francesco Redi nel Vocabolario della Crusca." in *Atti della R. Accademia della Crusca per la lingua d'Italia.* 33-136. 1915-1916; Salvatore Battaglia, *Grande dizionario della lingua italiana.* XII: 1071-1072. Torino 1984; Manlio Cortelazzo & Paolo Zolli, *Dizionario etimologico della lingua italiana.* 4: 904. Bologna 1985.

Species/Vernacular Names:
P. lactiflora Pallas (*Paeonia fragrans* (Sab.) Redouté; *Paeonia edulis* Salisb.; *Paeonia albiflora* Pall.; *Paeonia chinensis* hort.; *Paeonia reevesiana* (Paxt.) Loud.)
English: white-flowered peony, Chinese white peony, Chinese peony
China: bai shao, shao yao, bai shao yao, chin shao yao
P. officinalis L.
China: mu shao yao
P. potaninii Komar.
English: tree-peony
P. suffruticosa Andrews (*Paeonia moutan* Sims; *Paeonia arborea* Donn)
English: moutan peony, tree paeony, tree peony, moutan
Japan: botan
China: mu dan pi, mou tan, hua wang (= the king of flowers), pai liang chin (= a hundred ounces of gold)

Paepalanthus Kunth Eriocaulaceae

Origins:
From the Greek *paipale* "the finest flour or meal" and *anthos* "flower."

Paesia St.-Hil. Dennstaedtiaceae

Origins:
After Fernando Dias Paes Leme, Portuguese administrator in Minas Geraes, c. 1600, Brazil. See *Descripção geographica, topographica, historica e politica da Capitania das Minas Geraes.* Se descobrimento, estado civil, politico e das rendas reaes (1781). 1909; Auguste François César Prouvençal de Saint-Hilaire (1779-1853), *Voyage dans le district des diamans* et sur le littoral du Brésil. 1: 385. Paris 1833.

Pagella Schönland Crassulaceae

Origins:
Dedicated to the English (b. London) botanical artist Mary Maud Page, 1867-1925 (d. South Africa; she was buried in the cemetery at Plumstead), botanical explorer, plant collector, Bolus Herbarium (University of Cape Town), worked at the School of Art (Caldrons), in 1911 sailed for South Africa (to Dealesville, in the Orange Free State), early in 1912 moved to Bloemfontein, August 1912 Palapye in Bechuanaland, November 1915 to Cape Point, collaborated with the *Journal of the Botanical Society,* illustrated *Elementary Lessons in Systematic Botany.* Cape Town [1919] by Harriet Margaret Louisa Bolus (1877-1970). See Mary Gunn and Leslie E. Codd, *Botanical Exploration of Southern Africa.* 269. Cape Town 1981; H.M.L. Bolus, "In Memoriam — M.M. Page." *The Annals of the Bolus Herbarium.* Vol. IV, Part II: 56-61. January 1926; Gordon Douglas Rowley, *A History of Succulent Plants.* Strawberry Press, Mill Valley, California 1997; Ray Desmond, *Dictionary of British & Irish Botanists and Horticulturists.* London 1994.

Pageria Raf. Scrophulariaceae

Origins:
See C.S. Rafinesque, *The Good Book.* 45. 1840; E.D. Merrill, *Index rafinesquianus.* 217. 1949.

Pagesia Raf. Scrophulariaceae

Origins:
See Constantine Samuel Rafinesque, *Florula ludoviciana.* 48, 49. New York 1817 and *New Fl. N. Am.* 2: 69. 1836 [1837]; E.D. Merrill, *Index rafinesquianus.* 214, 215, 217, 218. 1949.

Pagetia F. Muell. Rutaceae

Origins:

After the English surgeon and pathologist Sir James Paget, 1814-1899 (London), botanist, was considered the best diagnostician of the time, studied at St. Bartholomew's Hospital in London and in 1847 was appointed assistant-surgeon to the hospital, a member of the Royal College of Surgeons, 1851 Fellow of the Royal Society, 1871 Baronet, 1872 Fellow of the Linnean Society, professor of anatomy and surgery to the Royal College of Surgeons, edited *A Descriptive Catalogue of the Anatomical Museum of St. Bartholomew's Hospital.* London 1846. Among his publications are *Theology and Science.* An address, etc. London 1881, *Lectures on Surgical Pathology*, delivered at the Royal College of Surgeons of England. London 1853, *The Hunterian Oration* delivered ... on the 13th of February, 1877. London 1877 and *Sketch of the Natural History of Yarmouth.* [Co-author for zoology was Charles John Paget, 1814-1899, brother of James Paget] London 1834. See *Memoirs and Letters of Sir James Paget.* Edited by Stephen Paget, one of his sons. London 1901; Howard Marsh, *In Memoriam Sir James Paget.* London 1901; Helen Cordelia Putnam, *Sir James Paget in His Writings. Bibliography ...* Read ... at the ... meeting of the Rhode Island Medical Society, June 5, 1902 ... and reprinted from the Transactions. [1903]; John H. Barnhart, *Biographical Notes upon Botanists.* 3: 41. 1965; George C. Turner, *The Paget Tradition.* [Essays on Sir James and Stephen Paget.] 1938; Leonard Huxley, *Life and Letters of Sir Joseph Dalton Hooker.* London 1918; Ferdinand von Mueller, *Fragmenta Phytographiae Australiae.* 5: 178. Melbourne 1866; Mea Allan, *The Hookers of Kew.* London 1967; Garrison and Morton, *Medical Bibliography.* 2996, 4343, 5337, 5772. New York 1961.

Pahudia Miq. Caesalpiniaceae

Origins:

A synonym of *Afzelia*; see Jean Joseph Gustave Léonard, "Note sur les genres paléotropicaux *Afzelia, Intsia* et *Pahudia*." in *Reinwardtia.* 1: 61-66. 1950.

Pajanelia DC. Bignoniaceae

Origins:

From *pajaneli*, the Malabar/Malayalam name for *Pajanella longifolia*; see van Rheede tot Draakestein, *Hortus Indicus Malabaricus.* 1: t. 44. 1678.

Species/Vernacular Names:

P. longifolia (Willd.) Schumann

English: dagger-tree

Malaya: beka, beka gerian, beka utan

Pakaraimaea Maguire & P.S. Ashton Dipterocarpaceae

Origins:

Possibly from a vernacular name.

Palafoxia Lagasca Asteraceae

Origins:

Possibly named after the Spanish general José de Rebolledo Palafox y Melci (or Melzi), 1775/1776-1847, author of *Exhortacion del Señor Palafox, despues de la última Victoria conseguita por los Zaragozanos.* Sevilla [1808], of *Proclama hecha á los Aragoneses ... con motivo de la Batalla de las Heras de Zaragoza.* Cadiz [1808], among the defenders of Zaragoza (during the protracted siege (1808-09) by the French) was María Augustín, the "Maid of Saragossa," whose exploits are described in Lord Byron's poem *Childe Harold*; or named for the prelate Juan de Palafox y Mendoza (1600-1659), visitor general of New Spain (*visitador*), on December 27, 1639, consecrated Bishop of Puebla de Los Angeles (founded as Puebla de los Angeles in 1532, now Puebla of Zaragoza, capital of Puebla State, Central Mexico) and in 1655 Bishop of Osma (Soria, Spain), served briefly as Archbishop of Mexico (1642-1643), was very involved in affairs of government and held the posts of fiscal of the Consejo de Guerra, fiscal of the Consejo de Indias, a promoter of education and book collector, *juez de residencia*, and viceroy of New Spain, he was an administrative reformer who came into conflict with the Jesuits, with whom he was involved in litigation from 1647 to 1655. Among his many works and letters are *Obras.* Madrid 1762, *Oeuvres spirituelles ...* Marseille 1775, *Virtudes del Indio.* [1650?] and *Vida interior del ... Señor D.J. de Palafox y Mendoza ...* Copiada fielmente por la que el mismo escrivio con titulo de Confessiones y Confusiones ... Sacala a luz Don M. de Vergara. Sevilla 1691. See Carlos E. Castañeda and Jack Autrey Dabbs, eds., *Guide to the Latin American Manuscripts in the University of Texas Library.* Cambridge, Massachusetts 1939; Genaro García, *Colección de documentos inéditos o muy raros para la historia de México* (volume 7), Mexico, Vda. de C. Bouret. 1906; Genaro García, *Don Juan de Palafox y Mendoza.* Mexico 1918; *Juan de Palafox y Mendoza Collection, 1563-1750.* Benson Latin American Collection, General Libraries, University of Texas at Austin; José Toribio Medina, *La imprenta en la Puebla de los Angeles (1640-1821).* Amsterdam 1964; Georg Christian Wittstein, *Etymologisch-botanisches Handwörterbuch.* 653. Ansbach 1852; H. Genaust,

Etymologisches Wörterbuch der botanischen Pflanzenna-men. 451. 1996; Ana María Huerta Jaramillo, *El jardin de Cal. Antonio de la Cal y Bracho, la botánica y las ciencias de la salud en Puebla, 1766-1833.* Puebla 1996.

Species/Vernacular Names:

P. arida B. Turner & M. Morris var. *gigantea* (M.E. Jones) B. Turner & M. Morris

English: giant Spanish-needle

Palaquium Blanco Sapotaceae

Origins:

A Philippine native name, *palak-palak* or *palac*, for *Palaquium ellipticum* (Dalz.) Engl., in Tagalog language *palakihin* means to let grow, increase in size; see F.M. Blanco, *Flora de Filipinas.* Manila 1837; Herman Johannes Lam (1892-1977), in *Bull. Jard. Bot. Buitenzorg.* Ser. 3, 7: 107. 1925.

Species/Vernacular Names:

P. formosanum Hayata (*Palaquium hayatae* Lam)

English: Hayata nato tree

China: Taiwan jiao mu

P. galactoxylum (F. Muell.) Lam

Australia: red silkwood, Cairns pencil cedar, Daintree maple

P. gutta (Hook.f.) Baillon (*Isonandra gutta* Hook.; *Dichopsis gutta* (Hook.) Bentley & Trimen; *Palaquium oblongifolia* (de Vriese) Burck)

English: Malay guttapercha, nyato tree, nato tree, guttapercha tree, guttapercha

Malaya: taban, taban merah, getah percha, nyatoh barak, getah durian, getah rian

Palaua Cav. Malvaceae

Origins:

For the Spanish physician Antonio Paláu y Verdéra, d. 1793, naturalist, botanist, author of *Explicacion de la filosofia, y fundamentos botanicos de Linneo.* Madrid 1778, with Casimiro Gómez de Ortega (1740-1818) wrote *Curso elemental de botánica,* dispuesto para la enseñanza del Real Jardin de Madrid Madrid 1785. See John H. Barnhart, *Biographical Notes upon Botanists.* 3: 43. 1965; E.M. Tucker, *Catalogue of the Library of the Arnold Arboretum of Harvard University.* 1917-1933; Miguel Colmeiro y Penido, *La Botánica y los Botánicos de la Peninsula Hispano-Lusitana.* Madrid 1858; Jonas C. Dryander, *Catalogus bibliothecae historico-naturalis Josephi Banks.* London 1800; Ida

Kaplan Langman, *A Selected Guide to the Literature on the Flowering Plants of Mexico.* University of Pennsylvania Press, Philadelphia 1964; R. Zander, F. Encke, G. Buchheim and S. Seybold, *Handwörterbuch der Pflanzennamen.* 14. Aufl. Stuttgart 1993; F. Boerner & G. Kunkel, *Taschenwörterbuch der botanischen Pflanzennamen.* 4. Aufl. 146. 1989.

Species/Vernacular Names:

P. dissecta Bentham (*Palaua flexuosa* Masters; *Palaua geranioides* Ulbrich)

Peru: corilla

P. weberbaueri Ulbrich

Peru: corilla

Palava Juss. Malvaceae

Origins:

For the Spanish physician Antonio Paláu y Verdéra, d. 1793; see also *Palaua* Cav.

Paleaepappus Cabrera Asteraceae

Origins:

Latin *palea, ae* "chaff, straw, pollen" and *pappus*, Hebrew *palah* "to cleave, to cut in pieces, to let break forth"; see Giovanni Semerano, *Le origini della cultura europea.* Dizionario della lingua Latina e di voci moderne. 2(2): 500. Firenze 1994

Paleista Raf. Asteraceae

Origins:

See C.S. Rafinesque, *New Fl. N. Am.* 2: 40, 43. 1836 [1837] and *The Good Book.* 45. 1840; E.D. Merrill, *Index rafinesquianus.* 240. 1949.

Paletuviera Thouars ex DC. Rhizophoraceae

Origins:

A French word, *palétuvier* (in 1614 *appariturier*), for the mangroves *Rhizophora, Avicennia* and *Bruguiera,* from the Tupi *apareiba, apara* "curved" and *iba* "tree"; see Luíz Caldas Tibiriçá, *Dicionário Guarani-Português.* Traço Editora, Liberdade 1989 and *Dicionário Tupi-Português.* Traço Editora, Liberdade 1984; Salvatore Battaglia, *Grande dizionario della lingua italiana.* XII: 402. UTET, Torino 1984.

Palgianthus Baillon Malvaceae

Origins:

The orthographic variant of *Plagianthus* Forster.

Palhinhaea Franco & Vasconcellos Lycopodiaceae

Origins:

For the Portuguese (Azores-born) botanist Ruy Telles Palhinha, 1871-1957, from 1921-1941 Director of the Botanical Institute of the University of Lisbon, editor of Antonio Xavier Pereira Coutinho (1851-1939), *Flora de Portugal.* Ed. 2. Lisbon 1939; see Francisco de Mello, *Memoria sobre a malagueta* ... 2a edição prefaciada e revista por R.T. Palhinha. Lisboa 1945; John H. Barnhart, *Biographical Notes upon Botanists.* 3: 43. 1965.

Species/Vernacular Names:

P. cernua (L.) Vasconcellos & Franco (*Lycopodium cernuum* L.)

English: nodding club moss

Paliavana Vell. ex Vand. Gesneriaceae

Origins:

From Palhava.

Palicourea Aublet Rubiaceae

Origins:

The Palicour Indians of the Arcucua River in Brazil.

Species/Vernacular Names:

P. brachyloba (Müll. Arg.) B. Boom

Bolivia: ahuaramacha

P. grandifolia (Kunth) Standley

Bolivia: ahuaramacha

P. macrobotrys (Ruíz & Pav.) Roemer & Schultes

Bolivia: clavo rojo

P. quadrifolia (Rudge) DC.

Bolivia: bisatamane

P. rigida Kunth

Brazil: douradinha, gritadeira, douradinha-do-campo

Palicuria Raf. Rubiaceae

Origins:

For *Palicourea* Aublet; see C.S. Rafinesque, *Ann. Gén. Sci. Phys.* 6: 86. 1830.

Paliris Dumortier Orchidaceae

Origins:

An anagram of *Liparis* Rich.

Palisota Reichb. Commelinaceae

Origins:

For the French botanist Ambroise Marie François Joseph Palisot de Beauvois (Pallisat de Beauvois), 1752-1820, traveler (Africa, Haiti, United States), explorer (Ghana, Dahomey, S. Nigeria, Warri/Oware and Bénin), 1786-1788 Landolphe Expedition to the Gulf of Guinea (Niger Delta). His writings include *Essai d'une nouvelle Agrostographie, ou nouveaux genres des Graminées.* Paris 1812 and *Flore d'Oware et de Bénin, en Afrique.* Paris 1804-1807 [1803-1820]; see John H. Barnhart, *Biographical Notes upon Botanists.* 3: 43. 1965; E.M. Tucker, *Catalogue of the Library of the Arnold Arboretum of Harvard University.* 1917-1933; Jonas C. Dryander, *Catalogus bibliothecae historico-naturalis Josephi Banks.* London 1800; A. Lasègue, *Musée botanique de Benjamin Delessert.* Paris 1845; William Darlington (1782-1863), *Reliquiae Baldwinianae.* 160. Philadelphia 1843; F.N. Hepper and Fiona Neate, *Plant Collectors in West Africa.* 63. 1971; Auguste J.B. Chevalier, *Flore vivante de l'Afrique Occidentale Française.* 1938; Stafleu and Cowan, *Taxonomic Literature.* 4: 15-19. 1983; Jean François Landolphe (1765-1825), *Mémoires du Capitaine Landolphe,* contenant l'histoire de ses voyages pendant trente-six ans, aux côtes d'Afrique, et aux deux Amériques; rédigées sur son manuscrit, par J.S. Quesné. Paris 1823; E.D. Merrill, in *Proc. Amer. Phil. Soc.* 76: 899-920. 1936; H. Heine, in *Adansonia.* Sér. 2, 7: 115-140. 1967; Ida Kaplan Langman, *A Selected Guide to the Literature on the Flowering Plants of Mexico.* 1964; R. Zander, F. Encke, G. Buchheim and S. Seybold, *Handwörterbuch der Pflanzennamen.* 14. Aufl. 761. 1993; Georg Christian Wittstein, *Etymologisch-botanisches Handwörterbuch.* 653. Ansbach 1852; J. Ewan, ed., *A Short History of Botany in the United States.* 1969; J.D. Milner, *Catalogue of Portraits of Botanists Exhibited in the Museums of the Royal Botanic Gardens.* Royal Botanic Gardens, Kew, London 1906; Joseph Lanjouw (1902-1984) and Frans Antonie Stafleu (1921-1997), *Index Herbariorum. Part II, Collectors A-D.* Regnum Vegetabile vol. 2. 1954.

Species/Vernacular Names:

P. hirsuta (Thunb.) K. Schum.

Sierra Leone: ndumui

Yoruba: ojo, rogbo aguntan, jangborokun

Congo: abomo

Paliurus Mill. Rhamnaceae

Origins:

From *paliouros*, the ancient Greek name for *Paliurus spina-christi* Mill., Christ's thorn, or the great jujube, *Ziziphus spina-christi*, Latin *paliurus* for a plant, Christ's-thorn (Plinius), *paliuraeus* "covered with Christ's-thorn" (Plinius); see Pietro Bubani, *Flora Virgiliana*. 87-88. [Ristampa dell'edizione di Bologna 1870] Bologna 1978.

Species/Vernacular Names:

P. ramosissimus (Lour.) Poir. (*Paliurus aubletii* Benth.; *Aubletia ramosissima* Lour.)

English: thorny wingnut

Japan: hama-natsu-me

China: ma jia zi ye, pai chi

P. spina-christi Mill. (*Paliurus aculeatus* Lam.; *Paliurus australis* Gaertn.)

English: Christ's thorn, Jerusalem thorn

Pallasia Klotzsch Rubiaceae

Origins:

For the German (b. Berlin) botanist Pyotr (Peter) Simon Pallas, 1741-1811 (d. Berlin), physician, explorer, naturalist, traveler, scientist, professor of natural history (St. Petersburg). His works include *Reise durch verschiedene Provinzen des russischen Reichs* ... St. Petersburg 1771-1776 and *A Naturalist in Russia. Letters from Peter Simon Pallas to Thomas Pennant*. Edited by Carol Urness. Minneapolis [1967], editor of *Neue Nordische Beiträge*. St. Petersburg & Leipzig 1781-1796. See John H. Barnhart, *Biographical Notes upon Botanists*. 3: 43. 1965; Norman Douglas, *Looking Back. An Autobiographical Excursion*. Chatto and Windus, London 1938; E.M. Tucker, *Catalogue of the Library of the Arnold Arboretum of Harvard University*. 1917-1933; T.W. Bossert, *Biographical Dictionary of Botanists Represented in the Hunt Institute Portrait Collection*. 298. 1972; Jonas C. Dryander, *Catalogus bibliothecae historico-naturalis Josephi Banks*. London 1800; H.N. Clokie, *Account of the Herbaria of the Department of Botany in the University of Oxford*. 220. Oxford 1964; A. Lasègue, *Musée botanique de Benjamin Delessert*. Paris

1845; Emil Bretschneider (1833-1901), *History of European Botanical Discoveries in China*. [Reprint of the original edition, St. Petersburg 1898.] Leipzig 1981; Blanche Elizabeth Edith Henrey, *British Botanical and Horticultural Literature before 1800*. Oxford 1975; R. Zander, F. Encke, G. Buchheim and S. Seybold, *Handwörterbuch der Pflanzennamen*. 14. Aufl. Stuttgart 1993; G. Murray, *History of the Collections Contained in the Natural History Departments of the British Museum*. 1: 172. 1904; Vasiliy A. Esakov, in *Dictionary of Scientific Biography* (Editor in Chief Charles Coulston Gillispie.) 10: 283-285. New York 1981; Stafleu and Cowan, *Taxonomic Literature*. 4: 20-27. 1983; Blanche Henrey, *No Ordinary Gardener — Thomas Knowlton, 1691-1781*. Edited by A.O. Chater. British Museum (Natural History). London 1986.

Pallasia Scopoli Gramineae

Origins:

For the German botanist Pyotr (Peter) Simon Pallas, 1741-1811.

Pallenis (Cass.) Cass. Asteraceae

Origins:

From *Pallene, es*, a peninsula and town of Macedonia; see Alexandre H.G. Comte de Cassini, *Dictionnaire des Sciences Naturelles*. 23: 566. Paris 1822.

Species/Vernacular Names:

P. spinosa (L.) Cass. (*Buphthalmum spinosum* L.; *Asteriscus spinosus* (L.) Sch.Bip.)

English: golden pallenis

Palma Mill. Palmae

Origins:

Latin *palma, ae* "a palm-tree, a palm-wreath, the fruit of the palm-tree, a date, a palm-branch, a broom made of palm-twigs, the palm of the hand," Akkadian *palkû* "wide, open"; see H. Genaust, *Etymologisches Wörterbuch der botanischen Pflanzennamen*. 451-452. 1996; Manlio Cortelazzo & Paolo Zolli, *Dizionario etimologico della lingua italiana*. 4: 866-867. Bologna 1985; Giovanni Semerano, *Le origini della cultura europea*. Dizionario della lingua Latina e di voci moderne. 2(2): 501. Firenze 1994.

Palmerella A. Gray Campanulaceae

Origins:

For the American (b. England) botanist Edward Palmer, 1831-1911 (Washington, D.C.), ethnologist, ethnobotanist, to USA 1849, botanical and zoological collector, traveler, naturalist, physician, 1853-1855 La Plata Expedition, botanical explorer, described the ethnobotany and aboriginal medical practice of the American Southwest. His writings include "Food products of the North American Indians." *U.S. Dept. Agr. Rpt.* 1870: 404-428. 1871 and "Plants used by the Indians of the United States." *Amer. Nat.* 12: 593-606, 646-655. 1878. See Virgil J. Vogel, *American Indian Medicine.* University of Oklahoma Press 1977; Joseph Ewan, *Rocky Mountain Naturalists.* The University of Denver Press 1950; Rogers McVaugh, *Edward Palmer, Plant Explorer of the American West.* University of Oklahoma Press, Norman 1956; J.H. Barnhart, *Biographical Notes upon Botanists.* 3: 44. 1965; T.W. Bossert, *Biographical Dictionary of Botanists Represented in the Hunt Institute Portrait Collection.* 298. 1972; S. Lenley et al., *Catalog of the Manuscript and Archival Collections and Index to the Correspondence of John Torrey.* Library of the New York Botanical Garden. 320. 1973; Ida Kaplan Langman, *A Selected Guide to the Literature on the Flowering Plants of Mexico.* Philadelphia 1964; G. Murray, *History of the Collections Contained in the Natural History Departments of the British Museum.* 1: 172. London 1904; Ira L. Wiggins, *Flora of Baja California.* 42. Stanford, California 1980; Janice J. Beaty, *Plants on His Back: A Life of Edward Palmer, Adventurous Botanist and Collector.* Pantheon Books, New York 1964; Irving William Knobloch, compil., "A preliminary verified list of plant collectors in Mexico." *Phytologia Memoirs.* VI. 1983; Ray Desmond, *Dictionary of British & Irish Botanists and Horticulturists.* 532. London 1994; Rogers McVaugh, in *Dictionary of Scientific Biography* 10: 285-286. 1981; http://www.herbaria.harvard.edu/Libraries/archives/PALMER.html.

Palmeria F. Muell. Monimiaceae

Origins:

After the English-born Australian politician Sir John Frederick Palmer, 1804-1871, physician, medical practitioner, pastoralist, in 1846 Major of Melbourne; see Ferdinand von Mueller, *Fragmenta Phytographiae Australiae.* 4: 151. Melbourne 1864.

Species/Vernacular Names:

P. scandens F. Muell.

English: anchor plant, anchor vine, pomegranate vine

Palmerocassia Britton Caesalpiniaceae

Origins:

For the American botanist Edward Palmer, 1831-1911, ethnologist, to USA 1849, botanical and zoological collector, traveler, naturalist, physician, 1853-1855 La Plata Expedition, botanical explorer, described the ethnobotany and aboriginal medical practice of the American Southwest.

Palmijuncus Kuntze Palmae

Origins:

O. Kuntze, *Revisio Generum Plantarum.* 2: 731. 1891, *Palma* and *Juncus,* see also *Calamus* L. in Natalie W. Uhl & John Dransfield, *Genera Palmarum. A Classification of Palms Based on the Work of Harold E. Moore, Jr.* The L.H. Bailey Hortorium and The International Palm Society. 255-258. Allen Press, Lawrence, Kansas 1987.

Palmoglossum Klotzsch ex Reichb.f. Orchidaceae

Origins:

From *palma* and *glossa* "tongue."

Palmolmedia Ducke Moraceae

Origins:

Referring to *Olmedia* Ruíz and Pav., see also *Naucleopsis* Miq.

Palmorchis Barb. Rodr. Orchidaceae

Origins:

Greek *palme* "palm, date" and *orchis* "orchid," Latin *palma* "a palm tree, the fruit of palm tree."

Palmstruckia Sonder Brassicaceae

Origins:

After the Swedish botanical artist Johan Wilhelm Palmstruch, 1770-1811; see John H. Barnhart, *Biographical Notes upon Botanists.* 3: 45. 1965; T.W. Bossert, *Biographical Dictionary of Botanists Represented in the Hunt Institute Portrait Collection.* 299. 1972.

Paloue Aublet Caesalpiniaceae

Origins:
A vernacular name.

Palovea Juss. Caesalpiniaceae

Origins:
A vernacular name.

Palovea Raf. Rubiaceae

Origins:
See C.S. Rafinesque, *Flora Telluriana*. 1: 85. 1836 [1837].

Paloveopsis R.S. Cowan Caesalpiniaceae

Origins:
Resembling *Palovea* Juss.

Paltonium C. Presl Polypodiaceae

Origins:
From the Greek *palton* "missile, dart," referring to the form of the frond.

Palumbina Reichb.f. Orchidaceae

Origins:
Latin *palumbinus, a, um* "of wood-pigeons," *palumbes, palumbis, palumbus, palumba* "a wood-pigeon, ring-dove," referring to the perianth or to the white flowers.

Pamburus Swingle Rutaceae

Origins:
Presumably from the Greek *panboros* "all-devouring" or *pan* and *purros* "red," Latin *burrus, a, um* (= *rufus, rubens*); see M.P. Nayar, *Meaning of Indian Flowering Plant Names*. 253. Dehra Dun 1985.

Pamianthe Stapf Amaryllidaceae (Liliaceae)

Origins:
For Major Albert Pam, 1875-1955 (d. Herts.), in 1939 a Fellow of the Linnean Society, grew South American plants, wrote *Adventures and Recollections*. Oxford 1945, "Checklist of amaryllid coloured plates." *Herbertia*. 9: 85-98. 1942

and "*Jasminum primulinum*." *Gard. Chron.* III. 84: 326. 1928; see Ray Desmond, *Dictionary of British & Irish Botanists and Horticulturists*. 533. London 1994.

Pamphalea Lag. Asteraceae

Origins:
From the Greek *pan* "all" and *phalos* "shining, bright," referring to the leaves.

Panamanthus Kuijt Loranthaceae

Origins:
From the region of Panama and *anthos* "flower."

Panax L. Araliaceae

Origins:
Latin *panacea, ae, panaces, is* and also *panax, acis* for an herb to which was ascribed the power of healing all diseases, all-heal, panacea, catholicon (Plinius), Greek *panakes, panakeia, panax* "all healing, a panacea," *pan* "all" and *akos* "a remedy"; see Carl Linnaeus, *Species Plantarum*. 1058. 1753 and *Genera Plantarum*. Ed. 5. 481. 1754.

Species/Vernacular Names:
P. ginseng C. Meyer
English: ginseng
China: ren shen, jen shen, shen tsao
Vietnam: tho sam, tam that
P. quinquefolius L.
English: American ginseng
P. trifolius L.
English: dwarf ginseng, groundnut

Pancheria Brongn. & Gris Cunoniaceae

Origins:
For the French botanist Jean Armand Isidore Pancher, 1814-1877, plant collector (Tahiti and New Caledonia); see John H. Barnhart, *Biographical Notes upon Botanists*. 3: 46. Boston 1965; Elmer Drew Merrill, *Contr. U.S. Natl. Herb.* 30(1): 233. 1947; G. Murray, *History of the Collections Contained in the Natural History Departments of the British Museum*. 1: 173. London 1904; in *Bull. Soc. Bot. France*. 9: 74. 1862.

Panchezia Montrouz. Rubiaceae

Origins:

For the French botanist Jean Armand Isidore Pancher, 1814-1877.

Pancicia Visiani Umbelliferae

Origins:

Named for the Croatian botanist Josef (Giuseppe, Josif) Pancic (Pancio, Panchic), 1814-1888, physician; see John H. Barnhart, *Biographical Notes upon Botanists*. 3: 46. Boston 1965; Stafleu and Cowan, *Taxonomic Literature*. 4: 42-44. 1983; E.M. Tucker, *Catalogue of the Library of the Arnold Arboretum of Harvard University*. 1917-1933; T.W. Bossert, *Biographical Dictionary of Botanists Represented in the Hunt Institute Portrait Collection*. 299. 1972; R. Zander, F. Encke, G. Buchheim and S. Seybold, *Handwörterbuch der Pflanzennamen*. 14. Aufl. Stuttgart 1993.

Pancovia Willd. Sapindaceae

Origins:

The generic name after Thomas Panckow, 1622-1665, author of *Herbarium portatile*. Berlin [1654]; see Carl L. von Willdenow (1765-1812), *Species Plantarum*. Ed. 4, 2: 285. 1799.

Species/Vernacular Names:

P. golungensis (Hiern) Exell & Mendonça (*Aphania golungensis* Hiern)

English: false soap-berry

Southern Africa: basterseepbessie, pancovia; uMuthinzima, umuThionzima, umPofana (Zulu)

Pancratium L. Amaryllidaceae (Liliaceae)

Origins:

An old Greek name for a bulbous plant, *pankration*, from *pan* "all" and *kratus* "strong, mighty," *kratos* "strength, might, power," referring to its supposed medicinal properties, Latin *pancratium, pancration* used by Plinius for the herb succory or for a plant, called also *scilla pusilla*; see Carl Linnaeus, *Species Plantarum*. 290. 1753 and *Genera Plantarum*. Ed. 5. 141. 1754.

Species/Vernacular Names:

P. maritimum L.

English: sea daffodil

Panda Pierre Pandaceae

Origins:

A vernacular name in Douala (Nigeria) and in Cameroon; see Y. Tailfer, *La Forêt dense d'Afrique centrale*. CTA, Ede/Wageningen 1989; J. Vivien & J.J. Faure, *Arbres des Forêts denses d'Afrique Centrale*. Agence de Coopération Culturelle et Technique. Paris 1985.

Species/Vernacular Names:

P. oleosa Pierre

Yoruba: ipade

Nigeria: mpanda, obirijia, afam, iku, otiemme, otieme, uku; iyoku (Yoruba); otieme (Edo); obirijia (Ijaw); ojifo ewum (Boki)

Congo: pad, okana

Gabon: elongolongo, ewawa, afane, afam, afame, afan, afann, mubaka, muguba, muvaga, muvamba, nkuba, ouando, ovaga, ovanda, bepanda, m'panda, upando, uvando

Cameroon: afam, afan, afane, afann, abfan, bokol, mbanda, mfanda, mwanda, panda, pendo, ovanda, pate, nkana

Ivory Coast: gere, aukua, aoukoua

Ghana: apurokuma, krakun, tana

Pandanus S. Parkinson Pandanaceae

Origins:

From a Malayan name, *pandan* or *pandang*, meaning conspicuous; see Sydney C. Parkinson, *circa* 1745-1771, *A Journal of a Voyage to the South Seas, in His Majesty's Ship, the Endeavour*. London 1773; R. Zander, F. Encke, G. Buchheim and S. Seybold, *Handwörterbuch der Pflanzennamen*. 14. Aufl. 415. 1993; G.C. Wittstein, *Etymologisch-botanisches Handwörterbuch*. 655. Ansbach 1852.

Species/Vernacular Names:

P. sp.

Hawaii: hala, hala 'ula, hala lihilihi 'ula, hala pia

Malaya: pandan, bengkuang, mengkuang

P. aquaticus F. Muell.

English: water pandan, river pandan

P. basedowii C.H. Wright

English: sandstone screw pine

P. candelabrum P. Beauv.

English: screw pine

Nigeria: ekunkun (Yoruba); ebo (Edo); kokoro (Ijaw); akpale (Igbo); nkpoto (Efik)

Yoruba: ekunkun

P. conicus H. St. John

English: mace screw pine

P. cookii Martelli (for Cooktown, Queensland, Australia)

English: screw pine

P. forsteri C. Moore & F. Muell.

English: screw pine

P. heterocarpus Balf.f.

Rodrigues Island: gros vacoa, vacoa chevron, vacoa calé rouge, vacoa parasol

P. monticola F. Muell.

English: rainforest screw pine, scrub bread fruit

P. solms-laubachii F. Muell.

English: screw pine

P. spiralis R. Br.

English: spiral screw palm, spiral screw pine

P. tectorius Sol. ex Parkinson (*Pandanus chamissonis* Gaud.)

English: (fragrant) screw pine, coastal bread fruit, coastal screw pine

Hawaii: hala, pu hala

India: kea, ketaka, ketaki, kaitha, kedage, kedagi, bondayi, mundige, keora, keya, keur, keura, kevda, talhai, thazhai, kedagai, tala, talum, talamchedi, dhuli puspika, mogili, gajangi

P. tectorius Sol. ex Parkinson var. *australianus* Martelli

English: screw pine

P. tectorius Sol. ex Parkinson var. *pedunculatus* (R. Br.) Domin (*Pandanus pedunculatus* R. Br.)

English: screw pine

Australia: wynnum (so called by Moreton Bay Aborigines), Moreton Bay bread-fruit

P. whitei Martelli (after C.T. White, botanist, Queensland, Australia)

Australia: White's screw palm, screw pine

P. yalna R. Tucker

Australia: yalna

P. zea H. St. John

English: corncob screw pine

Panderia Fischer & C.A. Meyer Chenopodiaceae

Origins:

For the Latvian (b. Riga) anatomist Christian Heinrich Pander, 1794-1865 (d. St. Petersburg), embryologist, zoologist and zoological collector, naturalist, discovered the trilaminar structure of the chick blastoderm (Greek *blastos* "germ"

and *derma* "skin"), traveler, 1820 accompanied a Russian mission to Bokhara as a naturalist. His writings include *Der vergleichende Osteologie*. [Written with E. J. d'Alton.] Bonn 1821-1831, Dissertatio inauguralis sistens *historiam metamorphoseos*, quam ovum incubatum prioribus quinque diebus subit. Wirceburgi 1817 and *Über die Ctenodipterinen des Devonischen Systems*. St. Petersburg 1858; see Garrison and Morton, *Medical Bibliography*. 474. New York 1961; Vern L. Bullough, in *Dictionary of Scientific Biography* 10: 286-288. 1981.

Pandorea Spach Bignoniaceae

Origins:

Named for Pandora (Greek *pan* "all" and *doron* "gift"), according to Greek mythology, the first woman sent to earth, the first mortal woman. See Hesiod, *Opera & Dies & Theogonia & Clypeus* [and other works]. Florentiae 1540; Édouard Spach (1801-1879), *Histoire naturelle des Végétaux*. 9: 136. Paris 1838.

Species/Vernacular Names:

P. austrocaledonica (Bureau) Seem.

English: boat vine

P. baileyana (Maiden & Betche) Steenis

Australia: large-leaved wonga vine

P. doratoxylon (J. Black) J. Black (*Tecoma doratoxylon* J. Black)

English: spearwood bush

P. jasminoides (Lindley) Schumann (*Tecoma jasminoides* Lindley)

English: bower plant, pink trumpet-flower, bower climber, bower vine, bower of beauty

Japan: sokei-nôzen

P. pandorana (Andrews) Steenis (*Bignonia pandorana* Andrews; *Tecoma pandorana* (Andrews) Skeels; *Tecoma oxleyi* Cunn. ex A. DC.; *Tecoma doratoxylon* J. Black; *Pandorea doratoxylon* (J. Black) J. Black; *Pandorea australis* Spach)

English: spearwood, spearwood bush

Australia: wonga-wonga vine, wonga vine (Eastern Australia); ooratan (Everard Range); urtjunpa (Pitjantjatjara)

Paneguia Raf. Iridaceae

Origins:

See C.S. Rafinesque, *Flora Telluriana*. 4: 34. 1836 [1838]; E.D. Merrill, *Index rafinesquianus*. 99. 1949.

Panemata Raf. Lentibulariaceae

Origins:

See C.S. Rafinesque, *Flora Telluriana*. 4: 64. 1836 [1838]; E.D. Merrill, *Index rafinesquianus*. 224. 1949.

Panetos Raf. Rubiaceae

Origins:

See C.S. Rafinesque, *Ann. Gén. Sci. Phys.* 5: 225, 227 and 6: 81. 1820; E.D. Merrill, *Index rafinesquianus*. 226. 1949.

Pangium Reinw. Flacourtiaceae

Origins:

From the Malay vernacular name, *pangi*.

Species/Vernacular Names:

P. edule Reinw.

Malaya: payang, kepayang, buah keluak

Panicastrella Moench Gramineae

Origins:

Resembling *Panicum*.

Panicularia Colla Dicksoniaceae (Thyrsopteridoideae)

Origins:

Latin *panicula, panucula, panucla, paniculus* "a tuft, panicle."

Panicularia Fabr. Gramineae

Origins:

From the Latin *panicula, panucula, panucla, paniculus* "a tuft, panicle."

Paniculum Ard. Gramineae

Origins:

Latin *panicula, panucula, panucla, paniculus* "a tuft, panicle," see also *Panicum* L.

Panicum L. Gramineae

Origins:

From a classical Latin name for millet, *panicum, i* (*panus, i* "the thread, a tumor, an ear of millet," Akkadian *panu* "to turn"), Italian panic grass, *Panicum italicum* L.; see Carl Linnaeus, *Species Plantarum*. 55. 1753 and *Genera Plantarum*. Ed. 5. 29. 1754.

Species/Vernacular Names:

P. sp.

Peru: canarana, penacho, shukushina, shukushkina, taboquinha, toro urcu

Mexico: camalote, zacate de loma

Yoruba: ite aparo, motisan, kereiyale, kase, esin, ketuketu igbo

P. aequinerve Nees

South Africa: bosbuffelgras

P. antidotale Retz.

English: giant panic grass, blue panic, blue panic grass

P. australiense Domin

English: bunch panic grass

P. bisulcatum Thunb.

English: black-seeded panic, black-seed panic grass

P. bulbosum Kunth

English: bulbous panic, Texas grass

P. buncei Benth. (after Daniel Bunce, 1813-1872, first Curator Geelong Botanic Gardens, Australia; see Jonathan Wantrup, *Australian Rare Books, 1788-1900*. Hordern House, Sydney 1987)

English: Bunce's panic

P. capillare L.

English: old witch grass, witch grass

P. caudiglume Hack.

The Philippines: kumut-palaka

P. coloratum L.

English: blue panic grass

Arabic: qosseiba

Australia: Coolah grass

P. coloratum L. var. *coloratum* (*Panicum coloratum* L. var. *makarikariense* Goossens)

English: small buffalo grass, small panicum, white buffalo grass

South Africa: bamboeskweek, blousaadgras, buffelgras, kleinbuffelsgras, miergras, osgras, witbuffelgras

P. decompositum R. Br.

Australia: Australian millet, native millet, windmill grass, tindil, papa grass

P. deustum Thunb.

English: buffalo grass, reed panicum

South Africa: buffelsgras, rietbuffelsgras, breëblaarwintergras

P. effusum R. Br.

English: hairy panic, branched panic, hairy panic grass

P. lachnophyllum Benth.

Australia: woolly-leaved panic, don't panic

P. laevinode Lindley (*Panicum whitei* J. Black)

English: pepper grass, pigeon grass

P. luzonense Presl

The Philippines: niknikan

P. maximum Jacq.

English: barbe grass, rainbow grass, ubabe grass, bush buffalo grass, common buffalo grass, Guinea grass, purple top buffalo grass

Peru: pasto de Guinea, pasto Guinea, yerba Abadía, zaina

Mexico: coloniao, guineo, hoja fina, pasto guineo, privilegio, rabo de mula, zacate privilegio, zacate guinea, zacatón, panizo de guinea

East Africa: achuku, odunyo

Southern Africa: blousaad, blousaadsoetgras, buffelsgras, gewone buffelsgras, bush buffel grass, brown top buffelgrass, purple top buffelgrass, soetgras; lehola, mofantsoe (Sotho); mphaga (Tswana); ubabe (Zulu); umhatji (Ndebele)

Yoruba: ikin, ikin iruke, kooko, eru oparun

P. miliaceum L.

English: broom corn, broom corn millet, broom millet, common millet, hog millet, millet, proso millet, millet panic, French panic, French millet, Russian millet

Spanish: mijo

South Africa: kaffermanna, prosomanna

Japan: kibi (= millet)

Tibetan: khre

China: shu mi, chi

P. molle Sw.

English: water grass

P. natalense Hochst. (*Panicum fulgens* auct. non Stapf)

English: Natal buffalo grass

South Africa: Natalbuffelsgras

P. nephelophilum Gaud. (*Panicum havaiense* Reichardt)

Hawaii: konakona

P. niihauense St. John

Hawaii: lau'ehu

P. obseptum Trin.

English: white water panic

P. paludosum Roxb.

English: swamp panic, swamp panic grass

P. pellitum Trin.

Hawaii: kai'oi'o

P. pygmaeum R. Br.

English: dwarf panic, pigmy panic

P. queenslandicum Domin

Australia: Yadbila grass, Yabila grass, Coolibah grass, Coolabah grass

P. repens L. (*Panicum paludosum* Roxb.)

English: couch panicum, creeping panic grass, torpedo grass, quack grass

Arabic: zommar

South Africa: bamboeskweek, grootblousaadgras, kruipgras, kweekbuffelsgras, varkgras

Japan: hai-kibi (= creeping *Panicum*)

Okinawa: najichu

The Philippines: kayana, luya-luyahan

P. sarmentosum Roxb.

The Philippines: kanubsuban, kauakauya, kauayansauak, kauayan-kauyan

Malaya: janggut ali, kelubong, tongkat ali

P. schinzii Schinz (*Panicum laevifolium* Hack. var. *laevifolium*)

English: bluegrass, buffalograss, landgrass, sweetgrass, sweet buffalo grass

Southern Africa: blousaad, blousaadbuffelsgras, blousaadsoetgras, buffelsgras, oulandegras, soet buffelsgras, soetgras, vleibuffelsgras, vleigras, old land grassvlei panicum; mofantsoe-o-moholo (Sotho)

P. simile Domin (*Panicum effusum* var. *simile* (Domin) B. Simon)

English: two color panic

P. subalbidum Kunth (*Panicum glabrescens* Steud.)

English: elbow buffalo grass

South Africa: elmboogbuffelsgras

P. subxerophilum Domin

Australia: Gilgai grass, cane panic grass

P. sumatrense Roth ex Roemer & Schultes

English: little millet

P. tenuifolium Hook. & Arnott

Hawaii: mountain pili

P. texanum Buckley

English: Colorado grass, Texas millet

P. torridum Gaud.

Hawaii: kakonakona, hakonakona

P. umbellatum Trin.

Rodrigues Island: gazon chinois

P. virgatum L.

English: switch grass

P. volutans J.G. Anders.

English: rolling grass, tumbleweed

South Africa: rolgras, waaigras

P. xerophilum (Hillebr.) Hitchc.

Hawaii: kakonakona, he'upueo

Panisea (Lindley) Lindley Orchidaceae

Origins:

From the Greek *pan* "all" and *isos* "equal," referring to the similarity of the sepals and petals.

Panopia Noronha ex Thouars Euphorbiaceae

Origins:

Referring possibly to a sea nymph, Panope, Panopea.

Panopsis Salisb. Proteaceae

Origins:

From the Greek *pan* "all" and *ops, opos, opsis* "aspect, sight" referring to the appearance of the trees.

Panphalea Lag. Asteraceae

Origins:

From the Greek *pan* "all" and *phalos* "shining, bright," referring to the leaves, see also *Pamphalea* Lag.

Panstrepis Raf. Orchidaceae

Origins:

From the Greek *pan* "all" and *streptos* "twisted," referring to the perianth segments; see C.S. Rafinesque, *Flora Telluriana*. 4: 41. Philadelphia 1836 [1838].

Pantacantha Speg. Solanaceae

Origins:

From the Greek *pas, pantos* "all, the whole" and *akantha* "thorn, spine."

Pantadenia Gagnepain Euphorbiaceae

Origins:

From the Greek *pas, pantos* "all, the whole" and *aden* "gland."

Pantathera Philippi Gramineae

Origins:

From the Greek *pas, pantos* "all, the whole" and *ather* "stalk, barb, awn."

Panterpa Miers Bignoniaceae

Origins:

Greek *pas, pantos* "all, the whole" and *erpo, herpo* "to creep, crawl," big lianes, terminal leaflets often replaced by tendrils.

Pantlingia Prain Orchidaceae

Origins:

For the British botanist Robert Pantling, 1857-1910 (Egypt), Kew gardener, Curator Royal Botanic Gardens Calcutta and Deputy Superintendent of the Government Cinchona Plantation in Bengal, illustrated Sir George King's "The Orchids of the Sikkim-Himalaya." in *Annals of the Royal Botanic Garden, Calcutta*. 8(1-4). 1898; see John H. Barnhart, *Biographical Notes upon Botanists*. 3: 47. 1965; E.M. Tucker, *Catalogue of the Library of the Arnold Arboretum of Harvard University*. 1917-1933.

Pantorrhynchus Murb. Brassicaceae

Origins:

From the Greek *pas, pantos* "all, the whole" and *rhynchos* "horn, beak."

Panurea Spruce ex Benth. Fabaceae

Origins:

From the Greek *pan* "all" and *oura* "tail."

Panzeria J.F. Gmelin Solanaceae

Origins:

Dedicated to the German botanist Georg Wolfgang Franz Panzer, 1755-1829, physician, entomologist, 1777 M.D. Erlangen, traveler, plant collector. His writings include *Observationum botanicarum specimen*. Norimbergae et

Lipsiae 1781, *De dolore*. Altorfii [1785], *Symbolae Entomologicae*. Erlangae 1802, *Faunae Insectorum Americes Borealis prodromus*. Norimbergae 1794 and *Ideen zu einer künftigen Revision der Gattungen der Gräser*. München 1813. See John H. Barnhart, *Biographical Notes upon Botanists*. 3: 47. 1965; T.W. Bossert, *Biographical Dictionary of Botanists Represented in the Hunt Institute Portrait Collection*. 300. 1972; H.N. Clokie, *Account of the Herbaria of the Department of Botany in the University of Oxford*. 221. Oxford 1964; A. Lasègue, *Musée botanique de Benjamin Delessert*. 337. Paris 1845; Jonas C. Dryander, *Catalogus bibliothecae historico-naturalis Josephi Banks*. London 1800; Ida Kaplan Langman, *A Selected Guide to the Literature on the Flowering Plants of Mexico*. 563. Philadelphia 1964; I.C. Hedge and J.M. Lamond, *Index of Collectors in the Edinburgh Herbarium*. Edinburgh 1970; Frans A. Stafleu, *Linnaeus and the Linnaeans*. The spreading of their ideas in systematic botany, 1735-1789. Utrecht 1971.

Panzeria Moench Labiatae

Origins:

For the German botanist Georg Wolfgang Franz Panzer, 1755-1829, physician, entomologist.

Panzerina Soják Labiatae

Origins:

Referring to *Panzeria* Moench, the diminutive, for the German botanist Georg Wolfgang Franz Panzer, 1755-1829.

Species/Vernacular Names:

P. canescens (Bunge) Soják (*Panzeria canescens* Bunge; *Leonurus canescens* (Bunge) Bentham; *Leonurus bungeanus* Schischkin)

China: hui bai nong chuang cao

Papaver L. Papaveraceae

Origins:

The old Latin name *papaver, papaveris*, Akkadian *papallu*, Sumerian *pa-pal* "bud, sprout," Akkadian (*bir*)*birru* "to flame, to blaze," Hebrew *bera* "fire, burning," *ba'ar* "to burn," Latin *buro, uro, -is, ussi, ustum, urere* "to burn"; see Felix de Avellar Brotero (1744-1828), *Noções geraes das dormideiras, da sua cultura, e da extracção do verdadeiro opio, que ellas conte'm*. [Small 8vo, first edition.] Lisbon 1824; Carl Linnaeus, *Species Plantarum*. 506. 1753 and *Genera Plantarum*. Ed. 5. 224. 1754; Giovanni Semerano, *Le origini della cultura europea*. Dizionario della lingua Latina e di voci moderne. 2(2): 503. Leo S. Olschki Editore, Firenze 1994; Ernest Weekley, *An Etymological Dictionary of Modern English*. 2: 1037, 1125. New York 1967.

Species/Vernacular Names:

P. aculeatum Thunb. (*Papaver gariepinum* Burch.; *Papaver horridum* DC.) (Gariep River or Eijn River, South Africa; see Gilbert Westacott Reynolds, *The Aloes of South Africa*. 7, 18, 41, 402. Balkema, Rotterdam 1982)

English: poppy, Californian poppy, red poppy, thorny poppy, wild poppy, bristle poppy

Southern Africa: doringpapawer, koringpapawer, koringroos, rooipapawer, rooipoppie, wilde papawer, wildepoppie; sehloahloa (Sotho)

P. argemone L.

English: prickly poppy, long prickly-headed poppy, long pale poppy

P. californicum A. Gray

English: western poppy, fire poppy

P. nudicaule L.

English: Icelandic poppy, Arctic poppy

P. orientale L.

English: oriental poppy

P. rhoeas L.

English: corn poppy, field poppy, Flanders poppy, red poppy, common red poppy, red field poppy, Shirley poppy

Spanish: amapola

Mexico: abapol, amapola

South Africa: wilde papawer, koring papawer

Japan: hina-geshi

China: li chun hua, li chun tsao

Arabic: bougaroun, khad bougaroun, chkaik ennoman, ch'qayeq, ben na'aman, khushkhash manthur

P. somniferum L.

English: opium poppy

French: pavot somnifère, pavot blanc, pavot à opium

Spanish: adormidera

Mexico: adormidera, amapola de opio, guia-guiña, nocuana-bizuono-huceochoga-becale, nocuana bizoono huecochaga becala, quie-guiña, quije guiña

Arabic: khoch khache, bou en-noum, khashkhash aswad, boundi

India: afim, kasa kasa, aphin, afium, postaka, koso-kosa

Japan: keshi, chishi

China: ying su ke, ying su, ying tzu shu

Vietnam: A phu dung, a phien, anh tuc, lao fen, chu gia dinh

Paphia Seemann Ericaceae

Origins:

Greek *Paphia* was an epithet of Aphrodite, Latin *Paphie* "Venus, the Paphian," Columella applied *paphie* to a sort of lettuce that grew on the island of Cyprus. Paphos, on the island of Cyprus, was a city sacred to Venus (or Aphrodite); see Berthold Carl Seemann (1825-1871), in *The Journal of Botany*. 2: 77. 1864.

Paphinia Lindley Orchidaceae

Origins:

Paphia, an epithet of Aphrodite or Venus.

Paphiopedilum Pfitzer Orchidaceae

Origins:

Paphia, an epithet of Aphrodite and Greek *pedilon* "slipper, sandal," referring to the slipper-shaped lip.

Species/Vernacular Names:

P. bellatulum (Reichb.f.) Pfitzer

Thailand: rongtao naree fa hoi

P. callosum (Reichb.f.) Pfitzer

Thailand: rongtao naree kang gob

P. concolor (Batem.) Pfitzer

Thailand: rongtao naree lueng prachin

P. hirsutissimum (Lindl.) Pfitzer

Thailand: rongtao naree lueang loei

P. insigne (Wallich ex Lindley) Pfitz.

English: lady's slipper, Venus' slipper

P. niveum (Reichb.f.) Pfitzer

Thailand: rongtao naree dok kao

P. parishii (Reichb.f.) Pfitzer

Thailand: rongtao naree nuad rue-sii, rongtao naree muang-gaan

P. villosum (Lindl.) Pfitzer

Thailand: rongtao naree inthanon

Papilionanthe Schltr. Orchidaceae

Origins:

Latin *papilio, papilionis* "a butterfly, moth, tent" and Greek *anthos* "flower," an allusion to the beautiful flowers; Greek *papylion* "tent."

Papilionopsis Steenis Fabaceae

Origins:

Resembling *papilio, papilionis* "a butterfly, moth, tent."

Papiliopsis E. Morren Orchidaceae

Origins:

From the Latin *papilio, papilionis* "a butterfly" and Greek *opsis* "like," referring to the floral parts.

Papillilabium Dockrill Orchidaceae

Origins:

From the Latin *papilla, ae* "a nipple, teat, soft protuberance on a surface" and *labium, ii* "a lip"; see Alick William Dockrill (1915-), *Australasian Sarcanthinae*. 1967, and *Australian Indigenous Orchids*. Sydney 1969.

Pappagrostis Roshev. Gramineae

Origins:

Greek *pappos* "fluff, pappus, the woolly and hairy seed of certain plants" plus *agrostis, agrostidos* "grass, weed, couch grass."

Pappea Ecklon & Zeyher Sapindaceae

Origins:

To honor the German (b. Hamburg) physician and botanist Karl (Carl) Wilhelm Ludwig Pappe, 1803-1862 (d. Cape Town), M.D. Leipzig 1827, professor of botany at the South African College in Cape Town, published *A List of South African Indigenous Plants Used as Remedies by the Colonists of the Cape of Good Hope*. Cape Town 1847, *Florae capensis medicae prodromus*. Cape Town 1850 and *Silva capensis*. Cape Town 1854; see P. MacOwan, "Personalia of botanical collectors at the Cape." *Trans. S. Afr. Philos. Soc.* 4(1): xxx-liii. 1884-1886; John H. Barnhart, *Biographical Notes upon Botanists*. 3: 47. 1965; E.M. Tucker, *Catalogue of the Library of the Arnold Arboretum of Harvard University*. 1917-1933; Mary Gunn and Leslie E. Codd, *Botanical Exploration of Southern Africa*. 270-272. Cape Town 1981; R. Zander, F. Encke, G. Buchheim and S. Seybold, *Handwörterbuch der Pflanzennamen*. 14. Aufl. Stuttgart 1993; I.C. Hedge and J.M. Lamond, *Index of Collectors in the Edinburgh Herbarium*. Edinburgh 1970; Ray Desmond, *Dictionary of British & Irish Botanists and Horticulturists*. 533-534. London 1994.

Species/Vernacular Names:

P. capensis Eckl. & Zeyh. (*Pappea capensis* Eckl. & Zeyh. var. *radlkoferi* (Schweinf. ex Radlk.) Schinz; *Pappea schumanniana* Schinz; *Pappea fulva* Conrath; *Pappea ugandensis* Bak.f.; *Sapindus pappea* Sond., nom. illegit.; *Sapindus capensis* (Eckl. & Zeyh.) Hochst.)

English: jacket-plum, bushveld cherry, wild plum, wild plum tree, indaba tree, kaffir plum, wild amandel, wild cherry

Eastern Africa: mfunuguru (Usambara)

Tanzania: mboboyo, kiboboyo

Southern Africa: doppruim, oliepit, oliepitboom, wildepruim, bergpruim, pruimbessie, pruimboom, pruimbos, tkaambessie boom, kaambessie boom, kaambos-bessiepitte, noupit; umQhokwane, umKhokhwane, umQhoqho, umGqogqo, umVuma, iNdaba (Zulu); iliTye, umGqalutye, ilitye (Xhosa); kambeje (Ndebele); liLatsa (Swazi); gulaswimbi (Thonga); xikwakwaxu (Tsonga); sikwakwashe (Shangaan); mothata (Tswana: Western Transvaal, northern Cape, Botswana); mopennweng (Kgatla and Kwena dialects, Botswana); morobaliepe; mongatane, moroba-diepe (= axe-breaker) (North Sotho); chitununu (Shona)

Papperitzia Reichb.f. Orchidaceae

Origins:
For William Papperitz, a friend of Reichenbach.

Pappobolus S.F. Blake Asteraceae

Origins:
From the Greek *pappos* "fluff, pappus, woolly seeds" and *ballo, bolis, bolos* "casting."

Pappochroma Raf. Asteraceae

Origins:
Greek *pappos* "fluff, pappus" and *chroma* "color"; see C.S. Rafinesque, *Flora Telluriana*. 2: 48. Philadelphia 1836 [1838].

Pappophorum Schreber Gramineae

Origins:
From the Greek *pappos* "fluff, pappus, down, beard" and *phoros* "bearing, carrying," *phero, phoreo* "to bear," referring to the awns on the flowering glume; see Johann Christian Daniel von Schreber (1739-1810), *Genera Plantarum*. 2: 787. 1791.

Pappostyles Pierre Rubiaceae

Origins:
From the Greek *pappos* "fluff, pappus" and *stylos* "pillar, style."

Pappostylum Pierre Rubiaceae

Origins:
Greek *pappos* and *stylos* "pillar, style."

Pappothrix (A. Gray) Rydb. Asteraceae

Origins:
Greek *pappos* "fluff, pappus" and *thrix, trichos* "hair."

Papuacedrus H.L. Li Cupressaceae

Origins:
From Papua New Guinea, plus *Cedrus*, the genus *Papuacedrus* is endemic to eastern Malesia; see H.-L. Li, "A reclassification of *Libocedrus* and Cupressaceae." *Journal of the Arnold Arboretum*. 34: 17-26. 1953; F. von Mueller, "Records of observations on Sir William MacGregor's highland plants from New Guinea." *Transactions of the Royal Society of Victoria*. 1(2): 32. 1889.

Papuaea Schltr. Orchidaceae

Origins:
From Papua New Guinea.

Papuanthes Danser Loranthaceae

Origins:
From Papua New Guinea and *anthos* "flower."

Papuapteris C. Chr. Dryopteridaceae (Aspidiaceae)

Origins:
From Papua New Guinea and *pteris* "fern."

Papuastelma Bullock Asclepiadaceae

Origins:

From Papua New Guinea and *stelma, stelmatos* "a girdle, belt."

Papuechites Markgr. Apocynaceae

Origins:

From Papua New Guinea plus *Echites*.

Papuodendron C.T. White Malvaceae

Origins:

From Papua New Guinea and *dendron* "tree."

Papyrius Lam. Moraceae

Origins:

Greek *papyros* "linen, cord, papyrus," Latin *papyrus* and *papyrum* "the paper-reed, papyrus," *papyrius, a, um* "of papyrus, of paper," referring to the paper derived from the inner bark of the tree, see also *Broussonetia* L'Hérit. ex Vent.

Parabaena Miers Menispermaceae

Origins:

From the Greek *para* "near, similar to, beside" and *bao, baino* "to go, walk," alluding to the climbing or spreading nature.

Parabarium Pierre ex C. Spire Apocynaceae

Origins:

Possibly from the Greek *para* "beside" and *baros* "weight," *barys* "heavy."

Parabarleria Baillon Acanthaceae

Origins:

Greek *para* "beside, near" plus the genus *Barleria* L.

Parabeaumontia Pichon Apocynaceae

Origins:

From the Greek *para* "beside, near" plus *Beaumontia* Wall.

Parabenzoin Nakai Lauraceae

Origins:

Greek *para* "beside, near" plus the genus *Benzoin* Boerhaave ex J. C. Schaeffer, aromatic trees or shrubs, seed oil used as illuminant, leaves as a tea; see Nakai, *Bull. Soc. Bot. France.* 71: 180. 1925.

Paraberlinia Pellegr. Caesalpiniaceae

Origins:

From the Greek *para* "beside, near" plus the genus *Berlinia* Sol. ex Hook.f. & Benth.

Parabesleria Oerst. Gesneriaceae

Origins:

From the Greek *para* "beside, near" plus the genus *Besleria* L.

Parabignonia Bureau ex K. Schumann Bignoniaceae

Origins:

From the Greek *para* "beside, near" plus the genus *Bignonia* L.

Parablechnum Presl Blechnaceae

Origins:

Greek *para* "near, similar to" and the genus *Blechnum* L.; see Karl Boriwog (Boriwag, Borivoj) Presl, *Epimeliae botanicae.* Pragae 1849 [reprinted from *Abhandlungen der Königlichen Böhmischen Gesellschaft der Wissenschaften.* 1851].

Paraboea (C.B. Clarke) Ridley Gesneriaceae

Origins:

Greek *para* "near, similar to" and the genus *Boea* Comm. ex Lam.

Parabotrys J.C. Muell. Annonaceae

Origins:

From the Greek *para* "near, similar to" and *botrys* "cluster, a bunch of grapes, a cluster of grapes."

Parabouchetia Baill. Solanaceae

Origins:

From the Greek *para* "near, similar to" and *Bouchetia* Dunal.

Paracalanthe Kudô Orchidaceae

Origins:

From the Greek *para* "near, similar to, beside" and *Calanthe*.

Paracaleana Blaxell Orchidaceae

Origins:

Greek *para* plus the genus *Caleana* R. Br.; see Donald Frederick Blaxell (1934-), in *Contributions from the New South Wales National Herbarium*. 4: 280. 1972.

Species/Vernacular Names:

P. minor (R. Br.) Blaxell (*Caleana minor* R. Br.; *Paracaleana sullivanii* (F. Muell.) Blaxell)

English: small duck orchid

Paracalyx Ali Fabaceae

Origins:

From the Greek *para* "beside, near" and *kalyx* "a calyx."

Paracarpaea (Schumann) Pichon Bignoniaceae

Origins:

From the Greek *para* "beside, near" and *karpos* "fruit."

Paracaryopsis (H. Riedl) R. Mill Boraginaceae

Origins:

Resembling *Paracaryum* (A. DC.) Boiss.

Paracaryum (A. DC.) Boiss. Boraginaceae

Origins:

Greek *para* "beside, near" and *karyon* "nut," referring to the nutlets.

Paracautleya R.M. Sm. Zingiberaceae

Origins:

From the Greek *para* "beside, near" and the genus *Cautleya* (Benth.) Hook.f.

Paracelastrus Miq. Celastraceae

Origins:

From the Greek *para* "beside, near" and the genus *Celastrus* L.

Paracephaelis Baillon Rubiaceae

Origins:

From the Greek *para* "beside, near" plus *Cephaelis* Sw.

Paraceterach (F. Muell.) Copel. Pteridaceae (Adiantaceae, Hemionitidaceae)

Origins:

Greek *para* "near, similar to" and the fern genus *Ceterach* Willd.; see Edwin Bingham Copeland (1873-1964), *Genera Filicum*. 1947.

Parachampionella Bremek. Acanthaceae

Origins:

Greek *para* "near, similar to" plus *Championella* Bremek.

Parachimarrhis Ducke Rubiaceae

Origins:

Greek *para* "near, similar to" plus *Chimarrhis* Jacq.

Paraclarisia Ducke Moraceae

Origins:

Greek *para* "near, similar to" plus the genus *Clarisia* Ruíz & Pav.

Paracleisthus Gagnep. Euphorbiaceae

Origins:

Greek *para* "near, similar to" plus the genus *Cleistanthus* Hook.f. ex Planchon.

Paracoffea (Miq.) J. Leroy Rubiaceae

Origins:
Greek *para* "near, similar to" plus *Coffea* L.

Paracolea Baillon Bignoniaceae

Origins:
Greek *para* "near, similar to" plus *Colea* Bojer ex Meissner.

Paracolpodium (Tzvelev) Tzvelev Gramineae

Origins:
Greek *para* plus *Colpodium* Trin.

Paracorynanthe Capuron Rubiaceae

Origins:
Greek *para* "near, similar to" plus *Corynanthe* Welw.

Paracroton Miq. Euphorbiaceae

Origins:
Greek *para* "near, similar to" with *Croton* L.

Paracryphia Baker f. Paracryphiaceae (Theales)

Origins:
Greek *para* "near, similar to" and *kryphios* "covered, concealed, hidden."

Paractaenum P. Beauv. Gramineae

Origins:
Greek *para* "near, similar to, like, alongside" and *ktenion* "a little comb," referring to the spikes; see Ambroise Marie François Joseph Palisot de Beauvois (1752-1820), *Essai d'une nouvelle Agrostographie*, ou nouveaux genres des Graminées. Paris 1812.

Species/Vernacular Names:
P. novae-hollandiae P. Beauv. (*Panicum paractaenum* Kunth; *Panicum reversum* F. Muell.)

English: barbed-wire grass, reverse grass, reflexed panic

Australia: Wandie grass

Paracynoglossum Popov Boraginaceae

Origins:
From the Greek *para* "near, similar to, like, alongside, beside" and the genus *Cynoglossum* L.

Paradaniellia Rolfe Caesalpiniaceae

Origins:
Greek *para* "near, similar to, like, alongside, beside" and the genus *Daniellia* Benn., named for the British (b. Lancs.) botanist William Freeman Daniell, 1818 (1817?)-1865 (d. Hants.), plant collector in Sierra Leone and Senegal 1841-1853, surgeon in West Africa 1841-1853, later in West Indies and China, Fellow of the Linnean Society 1855; see J. Lanjouw and F.A. Stafleu, *Index Herbariorum. Collectors A-D.* Utrecht 1954; F.N. Hepper and Fiona Neate, *Plant Collectors in West Africa.* 23. 1971; G. Murray, *History of the Collections Contained in the Natural History Departments of the British Museum.* London 1904; Ray Desmond, *Dictionary of British & Irish Botanists and Horticulturists.* 192. [b. 1817] London 1994; Emil Bretschneider, *History of European Botanical Discoveries in China.* Leipzig 1981.

Paradavallodes Ching Davalliaceae

Origins:
Greek *para* "near, similar to, like, alongside, beside" plus *Davallodes* (Copel.) Copel., for the Swiss botanist (of English origin, b. near London) Edmund Davall, 1763-1798 (d. Switzerland), 1788 Fellow of the Linnean Society, plant collector; see Karl Koenig (Charles Konig) (1774-1851) and John Sims (1749-1831), in *Annals of Botany.* I: 576-577. London 1805; in *Ann. Bot.* (Oxford) 5(18): 201. 1891; *Memoir and Correspondence of ... Sir J.E. Smith ...* Edited by Lady Pleasance Smith. London 1832; J.H. Barnhart, *Biographical Notes upon Botanists.* 1: 421. 1965; J. Lanjouw and F.A. Stafleu, *Index Herbariorum.* Part II, *Collectors A-D.* Regnum Vegetabile vol. 2. 1954.

Paradennstaedtia Tagawa Dennstaedtiaceae

Origins:
Greek *para* "near, similar to, like" and *Dennstaedtia* Bernh.

Paradenocline Müll. Arg. Euphorbiaceae

Origins:
Greek *para* "near, similar to, like" and *Adenocline* Turcz.

Paraderris (Miq.) R. Geesink Fabaceae

Origins:

Greek *para* "near, similar to, like" and *Derris* Lour.

Paradina Pierre ex Pitard Rubiaceae

Origins:

Greek *para* "near, similar to, like" plus the genus *Adina* Salisb.

Paradisanthus Reichb.f. Orchidaceae

Origins:

Greek *paradeisos, paradisos* "paradise" and *anthos* "flower," referring to the beauty of the flowers.

Paradisea Mazzucc. Asphodelaceae (Liliaceae)

Origins:

From the Greek *paradeisos, paradisos* "paradise."

Species/Vernacular Names:
P. liliastrum (L.) Bertol.
English: St. Bruno's lily

Paradolichandra Hassler Bignoniaceae

Origins:

Greek *para* "near, similar to, like" plus the genus *Dolichandra* Cham.

Paradombeya Stapf Bombacaceae (Sterculiaceae)

Origins:
Greek *para* "near, similar to, like" plus the genus *Dombeya* Cav., for the French botanist Joseph Dombey, 1742-1796, plant collector, physician, naturalist, explorer and traveler, between 1777-1788 in Chile and Peru with H. Ruíz López (1754-1815) and José Antonio Pavón (1754-1844); see Francisco Guerra, in *Dictionary of Scientific Biography* 4: 156-157. [d. 1794] 1981; J.H. Barnhart, *Biographical Notes upon Botanists*. 1: 463. 1965; J. Lanjouw and F.A. Stafleu, *Index Herbariorum*. Part II, *Collectors A-D*. Regnum Vegetabile vol. 2. 1954; F. Boerner & G. Kunkel, *Taschenwörterbuch der botanischen Pflanzennamen*. 4. Aufl. 94. 1989; Frans A. Stafleu, *Linnaeus and the Linnaeans*. The spreading of their ideas in systematic botany, 1735-1789. Utrecht 1971; E. Alvarez López, "Dombey y la expedición al Perú y Chile." *Anales Inst. Bot. Cavanilles*. 14: 31-129. 1956.

Paradrymonia Hanst. Gesneriaceae

Origins:

From the Greek *para* "near" and the genus *Drymonia* Mart.

Paradrypetes Kuhlm. Euphorbiaceae

Origins:

From the Greek *para* "near" and the genus *Drypetes* Vahl.

Paraeremostachys Adylov, Kamelin & Machm. Labiatae

Origins:

Greek *para* "near" plus *Eremostachys* Bunge.

Parafaujasia C. Jeffrey Asteraceae

Origins:

Greek *para* "near" plus *Faujasia* Cass.

Parafestuca Alexeev Gramineae

Origins:

Greek *para* "near" and the genus *Festuca* L.

Paragelonium Leandri Euphorbiaceae

Origins:

From the Greek *para* "near" and *Gelonium* Roxb. ex Willd.

Paragenipa Baillon Rubiaceae

Origins:

From the Greek *para* "near" and the genus *Genipa* L.

Parageum Nakai & H. Hara Rosaceae

Origins:

Greek *para* "near" and the genus *Geum* L.

Paraglycine F.J. Herm. Fabaceae

Origins:

Greek *para* "near" plus the genus *Glycine*.

Paragnathis Sprengel Orchidaceae

Origins:

Greek *paragnathis* "cheekpiece of a helmet, tiara, cheek-muscle," referring to petals.

Paragoldfussia Bremek. Acanthaceae

Origins:

Greek *para* "near" and the genus *Goldfussia* Nees, for the German zoologist Georg August Goldfuss, 1782-1848; see John H. Barnhart, *Biographical Notes upon Botanists*. 2: 60. 1965.

Paragonia Bureau Bignoniaceae

Origins:

Possibly from the Greek *paragonios* "adjacent to an angle."

Paragramma (Blume) T. Moore Polypodiaceae

Origins:

Greek *para* "near, similar to, like, alongside, beside" and *gramma* "line, writing, letter."

Paragrewia Gagnepain ex Rao Sterculiaceae

Origins:

Greek *para* "near, similar to, like, beside" and the genus *Grewia* L., for the English botanist and physiologist Nehemiah Grew, 1641-1712, physician and microscopist, M.D. Leyden 1671, Fellow of the Royal Society 1671; see William LeFanu, *Nehemiah Grew*. 1990; William Munk, *The Roll of the Royal College of Physicians of London*. 1: 406-409. London 1878; Francis Wall Oliver, ed., *Makers of British Botany*. Cambridge 1913; Garrison and Morton, *Medical Bibliography*. 297. New York 1961; M. Hadfield et al., *British Gardeners: A Biographical Dictionary*. London 1980; Charles R. Metcalfe, in *Dictionary of Scientific Biography* 5: 534-536. 1981; H. Genaust, *Etymologisches Wörterbuch der botanischen Pflanzennamen*. 274. [d. 1711] Basel 1996; Georg Christian Wittstein, *Etymologisch-botanisches Handwörterbuch*. 402. 1852; Blanche Henrey, *British Botanical and Horticultural Literature before 1800*. Oxford 1975.

Paragulubia Burret Palmae

Origins:

From the Greek *para* "near, similar to, like" and the genus *Gulubia* Becc.

Paragutzlaffia H.P. Tsui Acanthaceae

Origins:

Greek *para* "near, similar to, like" plus the genus *Gutzlaffia* Hance.

Paragynoxys (Cuatrec.) Cuatrec. Asteraceae

Origins:

Greek *para* "near, similar to, like" and the genus *Gynoxys* Cass.

Parahancornia Ducke Apocynaceae

Origins:

Greek *para* "near, similar to, like" and *Hancornia* B.A. Gomes.

Parahebe W. Oliver Scrophulariaceae

Origins:

Greek *para* "near, beside" and the genus *Hebe* Comm. ex Juss., like *Hebe*; see Walter Reginald Brook Oliver (1883-1957), in *Records of the Dominion Museum*. 1: 229. Wellington 1944.

Species/Vernacular Names:

P. derwentiana (Andrews) B. Briggs & Ehrend. (*Veronica derwentiana* Andrews; *Veronica labiata* R. Br.)

English: Derwent speedwell, white veronica

Parahemionitis Panigr. Pteridaceae

Origins:

Greek *para* "near, beside, near by" and the genus *Hemionitis* L.

Parahyparrhenia A. Camus Gramineae

Origins:

Greek *para* "near, beside" and the genus *Hyparrhenia* Andersson ex Fourn.

Paraia Rohwer, Richter & van der Werff Lauraceae

Origins:

Pará, an estuary of the Amazon, a state in northeast Brazil.

Paraixeris Nakai Asteraceae

Origins:

Greek *para* "near, beside, along with" plus the genus *Ixeris* Cass.

Parajaeschkea Burkill Gentianaceae

Origins:

From the Greek *para* "near, beside, alongside" plus *Jaeschkea* Kurz, after Rev. Heinrich August Jaeschke (Jäschke), plant collector from Kashmir and the western Himalayas, author of *Romanized Tibetan and English Dictionary*. Kyelang in British Lahoul 1866; see Theodor Bechler, *Heinrich August Jäschke, der geniale Sprachforscher der Mission der Brüdergemeine unter den Tibetern*. Herrnhut 1930; M.N. Chaudhri, I.H. Vegter and C.M. De Wal, *Index Herbariorum*, Part II (3), *Collectors I-L*. Regnum Vegetabile vol. 86. 1972.

Parajubaea Burret Palmae

Origins:

From the Greek *para* "near, beside, alongside" and the genus *Jubaea* Kunth.

Parajusticia Benoist Acanthaceae

Origins:

From the Greek *para* "near, beside, alongside" and the genus *Justicia* L.

Parakaempferia A. Rao & Verma Zingiberaceae

Origins:

Greek *para* "near, beside, alongside" and the genus *Kaempferia* L.

Parakibara Philipson Monimiaceae

Origins:

Greek *para* "near, beside, alongside" and the genus *Kibara* Endl.

Parakmeria Hu & Cheng Magnoliaceae

Origins:

Greek *para* "near, beside, alongside" and the genus *Kmeria* (Pierre) Dandy; see David Hunt, ed., *Magnolias and Their Allies*. Proceedings of an International Symposium, Royal Holloway, University of London, Egham, Surrey, U.K., 12-13 April 1996. International Dendrology Society and The Magnolia Society. 1998.

Paraknoxia Bremek. Rubiaceae

Origins:

Greek *para* "near, beside, alongside" and the genus *Knoxia* L.

Parakohleria Wiehler Gesneriaceae

Origins:

Greek *para* and *Kohleria* Regel.

Paralabatia Pierre Sapotaceae

Origins:

Greek *para* "near, beside, alongside" and *Labatia* Sw.

Paralamium Dunn Labiatae

Origins:

Greek *para* "near, beside, alongside" and *Lamium* L.

Species/Vernacular Names:

P. gracile Dunn

English: false deadnettle

China: jia yue zhi ma

Paralbizzia Kosterm. Mimosaceae

Origins:

Greek *para* "near, beside, alongside" and *Albizzia* Benth. or *Albizia* Durazz.

Paralepistemon Lejoly & Lisowki Convolvulaceae

Origins:
Greek *para* "near, beside, alongside" plus *Lepistemon* Blume.

Paraleptochilus Copel. Polypodiaceae

Origins:
Greek *para* "near, beside, alongside" and *Leptochilus* Kaulf.

Paraligusticum Tichom. Umbelliferae

Origins:
Greek *para* "near, beside, alongside" and *Ligusticum* L.

Paralinospadix Burret Palmae

Origins:
From the Greek *para* "near, beside, alongside" and the genus *Linospadix* H.A. Wendland.

Paralstonia Baill. Apocynaceae

Origins:
From the Greek *para* "near, beside, alongside" and the genus *Alstonia* R. Br.

Paralychnophora MacLeish Asteraceae

Origins:
From the Greek *para* "near, beside, alongside" plus *Lychnophora* C. Martius.

Paralyxia Baillon Apocynaceae

Origins:
Greek *para* "near, beside, alongside" plus *Alyxia* Banks ex R. Br.

Paramachaerium Ducke Fabaceae

Origins:
From the Greek *para* "near, beside, alongside" plus *Machaerium* Pers.

Paramacrolobium Léonard Caesalpiniaceae

Origins:
Greek *para* "near, beside, like" plus *Macrolobium* Schreber.

Species/Vernacular Names:
P. coeruleum (Taub.) J. Léonard
Cameroon: libogol, nkop
Zaire: muhonga

Paramammea Leroy Guttiferae

Origins:
Greek *para* "near, beside, like" plus the genus *Mammea* L.

Paramansoa Baillon Bignoniaceae

Origins:
From the Greek *para* "near, beside, alongside" plus the genus *Mansoa* DC.

Paramapania Uittien Cyperaceae

Origins:
Greek *para* and the genus *Mapania* Aublet; see Hendrik Uittien (1898-1944), *Recueil des Travaux botaniques néerlandais*. 32: 186. 1935.

Paramelhania Arènes Sterculiaceae

Origins:
Greek *para* "near, beside, like" plus *Melhania* Forssk.

Parameria Benth. Apocynaceae

Origins:
From the Greek *para* "near, beside" and *meris* "part," an allusion to the same floral parts.

Species/Vernacular Names:
P. laevigata (Jussieu) Moldenke
English: laevigate parameria
China: chang jie zhu, jin si teng zhong

Parameriopsis Pichon Apocynaceae

Origins:
Resembling *Parameria* Benth.

Paramichelia Hu Magnoliaceae

Origins:

From the Greek *para* "near, beside" and the genus *Michelia* L.; see David Hunt, ed., *Magnolias and Their Allies.* Proceedings of an International Symposium, Royal Holloway, University of London, Egham, Surrey, U.K., 12-13 April 1996. International Dendrology Society and The Magnolia Society. 1998; David G. Frodin & Rafaël Govaerts, *World Checklist and Bibliography of Magnoliaceae.* Royal Botanic Gardens, Kew 1996.

Paramicropholis Aubrév. & Pellegrin Sapotaceae

Origins:

Greek *para* "near, beside" and the genus *Micropholis* (Griseb.) Pierre.

Paramignya Wight Rutaceae

Origins:

Greek *para* "near, similar to" and *mignymi* "to mix," *migas* "mixed," referring to the ovary surrounded by a rim or to the close alliance with the genus *Atalantia* Corr.; see Robert Wight (1796-1872), *Illustrations of Indian Botany.* 1: 108, t. 42. Madras 1840-1850.

Species/Vernacular Names:

P. confertifolia Swing.

English: denseleaf vinelime

P. monophylla Wight

Sri Lanka: wellangirya

Malaya: akar merlimau

Paramitranthes Burret Myrtaceae

Origins:

From the Greek *para* "near, beside" and the genus *Mitranthes* O. Berg.

Paramoltkia Greuter Boraginaceae

Origins:

From the Greek *para* "near, beside" and the genus *Moltkia* Lehm.

Paramomum S.Q. Tong Zingiberaceae

Origins:

Greek *para* "near, beside" and the genus *Amomum* Roxb.

Paramongaia Velarde Amaryllidaceae (Liliaceae)

Origins:

Paramonga, Peru, an area to the north of Lima; see Octavio Velarde-Núñez, "*Paramongaia*, un nuevo genero Peruano de Amarilidoideas." *Lilloa.* 17: 491-501. 1949.

Species/Vernacular Names:

P. weberbaueri Velarde (for the German botanist August Weberbauer, 1871-1948, phytogeographer, plant taxonomist, plant collector in Peru, explorer. His works include *Plantas tóxicas que sirven para la pesca en el Perú.* [Lima 1933], *Die Pflanzenwelt der peruanischen Andes in ihren Grundzügen dargestellt.* Leipzig 1911; see John H. Barnhart, *Biographical Notes upon Botanists.* 3: 469. 1965; Réné Letouzey, "Les botanistes au Cameroun." in *Flore du Cameroun.* 7: 60. 1968; Stafleu and Cowan, *Taxonomic Literature.* 7: 134-136. 1988; Theodore W. Bossert, *Biographical Dictionary of Botanists Represented in the Hunt Institute Portrait Collection.* 429. Boston, Massachusetts 1972; Ethelyn Maria Tucker, *Catalogue of the Library of the Arnold Arboretum of Harvard University.* Cambridge, Massachusetts 1917-1933; T. Harper Goodspeed, *Plant Hunters in the Andes.* University of California Press, Berkeley and Los Angeles 1961; F.N. Hepper and Fiona Neate, *Plant Collectors in West Africa.* 84. 1971)

English: Peruvian daffodil

Peru: cojomaria (Andean Indian vernacular name)

Paramyrciaria Kausel Myrtaceae

Origins:

Greek *para* "near, beside" plus *Myrciaria* O. Berg.

Paranecepsia R.-Sm. Euphorbiaceae

Origins:

Greek *para* "near, beside" plus *Necepsia* Prain.

Paranephelium Miq. Sapindaceae

Origins:

From the Greek *para* "near, alongside, near to" and the genus *Nephelium* L.

Species/Vernacular Names:

P. macrophyllum King

Malaya: gohor, gaha

Paraneurachne S.T. Blake Gramineae

Origins:

Greek *para* and the genus *Neurachne* R. Br.; see Stanley Thatcher Blake (1910-1973), in *Contributions from the Queensland Herbarium*. 13: 20. 1972.

Species/Vernacular Names:

P. muelleri (Hackel) S.T. Blake (*Neurachne muelleri* Hackel; *Neurachne clementii* Domin)

English: northern mulga-grass

Paranneslea Gagnep. Theaceae

Origins:

Greek *para* "near" and the genus *Anneslea* Wallich.

Paranomus Salisb. Proteaceae

Origins:

From the Greek *para* "contrary to" and *nomos* "law," referring to the different types of leaves on the same shoot.

Parantennaria Beauverd Asteraceae

Origins:

From the Greek *para* "near" and the genus *Antennaria* Gaertn., referring to the pappus of the flowers; see Gustave Beauverd (1867-1942), in *Bull. Soc. Bot. Genève*. 3: 255. 1911.

Paranthe O.F. Cook Palmae

Origins:

See Natalie W. Uhl & John Dransfield, *Genera Palmarum. A Classification of Palms Based on the Work of Harold E. Moore, Jr.* The L.H. Bailey Hortorium and The International Palm Society. 309. Allen Press, Lawrence, Kansas 1987.

Parapachygone Forman Menispermaceae

Origins:

Greek *para* "near" plus the genus *Pachygone* Miers.

Parapactis Zimmermann Orchidaceae

Origins:

From the Greek *para* "near" and the genus *Epipactis* Zinn.

Parapanax Miq. Araliaceae

Origins:

Greek *para* "near" plus *Panax*.

Parapantadenia Capuron Euphorbiaceae

Origins:

Greek *para* "near" and the genus *Pantadenia* Gagnep.

Parapentapanax Hutch. Araliaceae

Origins:

Greek *para* "near" and the genus *Pentapanax* Seem.

Parapentas Bremek. Rubiaceae

Origins:

From the Greek *para* "near" and the genus *Pentas* Benth.

Paraphalaenopsis A.D. Hawkes Orchidaceae

Origins:

From the Greek *para* "near" and the genus *Phalaenopsis* Blume.

Paraphlomis Prain Labiatae

Origins:

Greek *para* "near, beside, near to" plus genus *Phlomis* L.

Species/Vernacular Names:

P. albida Handel-Mazzetti

English: white-hairy paraphlomis

China: bai mao jia cao su

P. albiflora (Hemsley) Handel-Mazzetti

English: white-flower paraphlomis

China: bai hua jia cao su

P. foliata (Dunn) C.Y. Wu & H.W. Li

English: foliate paraphlomis

China: qu jing jia cao su

P. gracilis (Hemsley) Kudô

English: slender paraphlomis

China: xian xi mao jia cao su

P. hispida C.Y. Wu

English: hispid paraphlomis

China: gang mao jia cao su

P. intermedia C.Y. Wu & H.W. Li

English: intermediate paraphlomis

China: zhong jian jia cao su

P. javanica (Blume) Prain

English: Java paraphlomis

Japan: kuraru-odori-ko-sô

China: jia cao su

P. kwangtungensis C.Y. Wu & H.W. Li

English: Kwangtung paraphlomis, Guangdong paraphlomis

China: ba jiao hua

P. lanceolata Handel-Mazzetti

English: setulose paraphlomis

China: chang ye jia cao su

P. membranacea C.Y. Wu & H.W. Li

English: membranaceous paraphlomis

China: bo e jia cao su

P. setulosa C.Y. Wu & H.W. Li

English: setulose paraphlomis

China: xiao ci mao jia cao su

Parapholis C.E. Hubb. Gramineae

Origins:
Greek *para* "near, beside, near to" and the genus *Pholiurus* Trin., *pholis, pholidos* "scale, horny scale"; see Charles Edward Hubbard (1900-1980), in *Blumea*. Suppl. 3: 14. 1946.

Species/Vernacular Names:
P. incurva (L.) C.E. Hubb. (*Lepturus incurvatus* Trin.; *Pholiurus incurvus* (L.) Schinz. & Thell.; *Aegilops incurva* L.)

English: coast barb grass, curly ryegrass, sickle grass

Paraphyadanthe Mildbr. Flacourtiaceae

Origins:
The diminutive of the Greek *paraphyas, ados* "off-shoot, side-growth, sucker," Greek *paraphyadion* and *anthos* "flower."

Parapiptadenia Brenan Mimosaceae

Origins:
Greek *para* "near, beside, near to" and the genus *Piptadenia* Benth.

Parapiqueria R. King & H. Robinson Asteraceae

Origins:
Greek *para* "near, beside, near to" and *Piqueria* Cav.

Parapodium E. Meyer Asclepiadaceae

Origins:
From the Greek *para* "near, beside, near to" and *pous, podos* "a foot," referring to the rootstock.

Parapolystichum (Keyserling) Ching Dryopteridaceae (Aspidiaceae)

Origins:
Greek *para* and the genus *Polystichum* Roth; see Ren Chang Ching (1898-1986), in *Sunyatsenia*. 5: 239. 1940.

Paraprenanthes Chang ex Shih Asteraceae

Origins:
Greek *para* "near, beside, near to" and the genus *Prenanthes* L.

Paraprotium Cuatrec. Burseraceae

Origins:
Greek *para* "near, beside, near to" plus the genus *Protium* Burm.f.

Parapteroceras Averyanov Orchidaceae

Origins:
Greek *para* "near, beside, near to" and *Pteroceras* Hasselt ex Hassk.

Parapteropyrum A.J. Li Polygonaceae

Origins:
Greek *para* "near, beside, near to" and *Pteropyrum* Jaub. & Spach.

Parapyrenaria H.T. Chang Theaceae

Origins:
Greek *para* "near, beside, near to" with *Pyrenaria* Blume.

Paraquilegia J.R. Drumm. & Hutch. Ranunculaceae

Origins:
From the Greek *para* "near, beside, near to" and the genus *Aquilegia* L.

Pararchidendron I. Nielsen Mimosaceae

Origins:
Greek *para* "near, beside, near to" and the genus *Archidendron* F. Muell.

Species/Vernacular Names:
P. pruinosum (Benth.) Nielsen
English: snow wood

Parardisia Nayar & Giri Myrsinaceae

Origins:
Greek *para* "near, beside, near to" plus *Ardisia* Sw.

Pararistolochia Hutch. & Dalziel Aristolochiaceae

Origins:
Greek *para* "near, beside, near to" plus *Aristolochia* L.

Parartabotrys Miq. Annonaceae

Origins:
Greek *para* "near, beside, near to" plus *Artabotrys* R. Br.

Parartocarpus Baillon Moraceae

Origins:
Greek *para* plus *Artocarpus* Forster & Forster f.

Pararuellia Bremek. Acanthaceae

Origins:
Greek *para* "near, beside, near to" plus *Ruellia* L.

Parasamanea Kosterm. Mimosaceae

Origins:
Greek *para* "near, beside, near to" plus the genus *Samanea* (Benth.) Merr.

Parasarcochilus Dockrill Orchidaceae

Origins:
From the Greek *para* "near, beside, near to" and the genus *Sarcochilus* R. Br.; see Alick William Dockrill (1915-), *Australasian Sarcanthinae.* 22, t. 8. 1967, and *Australian Indigenous Orchids.* Sydney 1969.

Parasassafras Long Lauraceae

Origins:
Greek *para* "near, beside, near to" plus *Sassafras* Nees & Eberm.

Parascheelea Dugand Palmae

Origins:
For Wilhelm Scheele, 1742-1786, chemist, genus *Scheelea* Karsten.

Parascopolia Baillon Solanaceae

Origins:
For Giovanni Antonio (Joannes Antonius) Scopoli, 1723-1788, botanist and physician, chemist, professor of chemistry and botany, plant collector, naturalist; see John H. Barnhart, *Biographical Notes upon Botanists.* 3: 250. 1965; T.W. Bossert, *Biographical Dictionary of Botanists Represented in the Hunt Institute Portrait Collection.* 358. 1972; Jonas C. Dryander, *Catalogus bibliothecae historico-naturalis Josephi Banks.* 1800; A. Lasègue, *Musée botanique de Benjamin Delessert.* 1845; E.M. Tucker, *Catalogue of the Library of the Arnold Arboretum of Harvard University.* Cambridge, Massachusetts 1917-1933.

Paraselinum H. Wolff Umbelliferae

Origins:
Greek *para* "near, beside, near to" plus *Selinum* L.

Parasenecio W.W. Sm. & Small Asteraceae

Origins:
Greek *para* "near, beside, near to" plus *Senecio* L.

Paraserianthes I. Nielsen Mimosaceae

Origins:

From the Greek *para* "near, alongside, near to" and the genus *Serianthes* Benth.; *seris*, *seridos*, was the Greek name for a species of chicory or endive.

Species/Vernacular Names:

P. falcataria (L.) I. Nielsen (*Adenanthera falcataria* L.; *Albizia falcataria* (L.) Fosb.; *Albizia moluccana* Miq.)

English: batai-wood, Moluccan sau, albizia

Malaya: batai

P. lophantha (Willd.) I. Nielsen subsp. *lophantha* (*Acacia lophantha* Willd.; *Albizia distachya* (Vent.) J.F. Macbride; *Albizia lophantha* (Willd.) Benth.; *Mimosa distachya* Vent.)

English: Australian albizia, Cape Leeuwin wattle, Cape wattle, crested wattle, plume albizia, silk tree, sirus, stinkbean

South Africa: Australiese albizia, stinkboon

P. toona (Bailey) I. Nielsen

English: red siris, cedar acacia, Mackay cedar

Parashorea Kurz Dipterocarpaceae

Origins:

Greek *para* "near, alongside, near to" and *Shorea* Roxb. ex Gaertn.

Species/Vernacular Names:

P. malaanonan (Blanco) Merr.

SE Asia: bagtikan

P. stellata Kurz

Southeast Asia: thingadu

Malaya: damar laut

Parasicyos Dieterle Cucurbitaceae

Origins:

Greek *para* "near, alongside, near to" and *Sicyos* L.

Parasilaus Leute Umbelliferae

Origins:

Greek *para* "near, alongside, near to" and *Silaum silaus* (L.) Schinz & Thell.

Parasitaxus Laubenf. Podocarpaceae

Origins:

Latin *parasitus* "a guest" and *Taxus* or *taxon* or *axis*, etc., referring to a parasitic gymnosperm.

Parasitipomoea Hayata Convolvulaceae

Origins:

Latin *parasitus* "a guest, parasite" and *Ipomoea* L.

Parasorus Alderw. Davalliaceae

Origins:

Greek *para* "near, alongside, near to" and *soros* "a vessel, mound, heap."

Paraspalathus Presl Fabaceae

Origins:

From the Greek *para* "near, near to" and the genus *Aspalathus* L.

Parasponia Miq. Ulmaceae

Origins:

Greek *para* "near, near to" and the genus *Sponia* Comm. ex Decne.

Parastemon A. DC. Chrysobalanaceae

Origins:

From the Greek *para* "near, similar to, beside" and *stemon* "a stamen, a thread."

Species/Vernacular Names:

P. urophyllus (A. DC.) A. DC.

English: pink beam

Malaya: kelat putej, malas, nyalas, ngilas, kelat pasir, kelat puteh, siangus betina, sempalawan

Parastranthus G. Don Campanulaceae

Origins:

From the Greek *parastrepho* "to twist aside" and *anthos* "a flower," referring to the corolla.

Parastrephia Nutt. Asteraceae

Origins:
From the Greek *parastrepho* "to twist aside."

Parastriga Mildbr. Scrophulariaceae

Origins:
Greek *para* "near, similar to, beside" and *Striga* Lour.

Parastrobilanthes Bremek. Acanthaceae

Origins:
Greek *para* "near, similar to, beside" and *Strobilanthes* Blume.

Parastyrax W.W. Smith Styracaceae

Origins:
From the Greek *para* "near, similar to, beside" and the genus *Styrax* L.

Species/Vernacular Names:
P. lacei (W.W. Smith) W.W. Smith
English: common parastyrax
China: mo li guo
P. macrophyllus C.Y. Wu & K.M. Feng
English: large-leaf parastyrax
China: da ye mo li guo

Parasympagis Bremek. Acanthaceae

Origins:
Greek *para* "near, similar to, beside" plus *Sympagis* (Nees) Bremek.

Parasyringa W.W. Sm. Oleaceae

Origins:
From the Greek *para* "near, similar to, beside" and the genus *Syringa* L.

Parasystasia Baillon Acanthaceae

Origins:
Greek *para* "near, similar to, beside" and the genus *Asystasia* Blume.

Paratecoma Kuhlm. Bignoniaceae

Origins:
Greek *para* plus the genus *Tecoma* Juss.

Species/Vernacular Names:
P. peroba (Rec.) Kuhlm.
Brazil: ipê peroba, peroba amarela, peroba manchada, peroba tigrina, peroba remida

Paratephrosia Domin Fabaceae

Origins:
From the Greek *para* "near, similar to" and the genus *Tephrosia* Pers.; see Karel Domin, *Repertorium specierum novarum regni vegetabilis*. 11: 261. 1912.

Parathelypteris (H. Ito) Ching Thelypteridaceae

Origins:
From the Greek *para* "near, similar to" and the genus *Thelypteris*.

Paratheria Griseb. Gramineae

Origins:
From the Greek *para* "near, similar to" and *ather* "stalk, barb, awn."

Parathesis (A. DC.) Hook.f. Myrsinaceae

Origins:
Greek *parathesis* "juxtaposition, neighborhood, storing up, offering of food to a god."

Parathyrium Holttum Dryopteridaceae (Aspleniaceae, Woodsiaceae)

Origins:
Greek *para* "near, similar to" and *Athyrium* Roth.

Paratriaina Bremek. Rubiaceae

Origins:
Greek *para* "near, similar to" and *triaina* "trident."

Paratrophis Blume Moraceae

Origins:
Greek *para* "near, similar to" and *Trophis* P. Browne; see Karl Ludwig von Blume, *Museum Botanicum Lugduno-Batavum*. 2: 81. Lugduni-Batavorum 1856.

Paratropia (Blume) DC. Araliaceae

Origins:
From the Greek *paratropos* "turned aside," *paratrope* "turning away."

Paravallaris Pierre Apocynaceae

Origins:
Greek *para* and the genus *Vallaris* Burm.f.

Paravinia Hassk. Rubiaceae

Origins:
Referring to the genus *Praravinia* Korth.

Paravitex H.R. Fletcher Labiatae (Verbenaceae)

Origins:
Greek *para* and the genus *Vitex* L.

Pardanthopsis (Hance) Lenz Iridaceae

Origins:
Greek *pardos, pardalis, pordalis* "leopard, male panther," *anthos* "flower" and *opsis* "like," spotted flowers.

Pardoglossum E. Barbier & Mathez Boraginaceae

Origins:
From the Greek *pardos* "leopard, male panther" and *glossa* "tongue."

Parenterolobium Kosterm. Mimosaceae

Origins:
Greek *para* "near, similar to" and the genus *Enterolobium* Mart.

Parentucellia Viv. Scrophulariaceae

Origins:
Named for the Renaissance Pope Nicholas V, Pontifex Maximus (formerly Tomaso Parentucelli, 1397-1455), reigned from 1447 to 1455, founder of the Vatican Library, a patron of many artists and scholars. He transferred the *Viridarium* (created by Nicholas III, original name Giovanni Gaetano Orsini, Pope from 1277 to 1280) inside the Vatican walls, over the Belvedere hill. See *Indulgentia*. [Mainz] 1455, for contributions to the war against the Turks; E. Müntz and P. Fabre (eds.), *La Bibliothèque du Vatican au XV siècle*. 1887; A. Demski, *Papst N. III*. Münster 1903; C.W. Westfall, *In This Most Perfect Paradise. Alberti, Nicolas V and the Invention of Conscious Urban Planning in Rome 1447-1455*. University Park, Pa. 1975; Domenico Viviani (1772-1840), *Florae Libycae Specimen*. Genuae 1824.

Species/Vernacular Names:
P. latifolia (L.) Caruel (*Euphrasia latifolia* L.; *Bartsia latifolia* (L.) Sibth. & Smith)
English: red bartsia, common bartsia
P. viscosa (L.) Caruel (*Bartsia viscosa* L.)
English: yellow bartsia, sticky bartsia

Parepigynum Tsiang & P.T. Li Apocynaceae

Origins:
Greek *para* and the genus *Epigynum* Wight.

Species/Vernacular Names:
P. funingense Tsiang & P.T. Li
English: Funing parepigynum
China: fu ning teng

Parestia C. Presl Davalliaceae

Origins:
From the Greek *parestios* "by the earth," *hestia, histis* "altar, house, hearth of a house, home."

Pareugenia Turrill Myrtaceae

Origins:
From the Greek *para* "near, similar to" and the genus *Eugenia* L.

Parhabenaria Gagnepain Orchidaceae

Origins:
From the Greek *para* "near, similar to" and the genus *Habenaria*.

Parietaria L. Urticaceae

Origins:

Latin *parietaria, ae* "the herb pellitory, parietary," *parietarius, a, um* "belonging to walls," *paries, etis* "a wall"; see Carl Linnaeus, *Species Plantarum.* 1052. 1753 and *Genera Plantarum.* Ed. 5. 471. 1754.

Species/Vernacular Names:

P. debilis Forster f. (*Freirea australis* Nees; *Parietaria debilis* Forster f. var. *australis* (Nees) J. Black; *Parietaria australis* (Nees) Blume)

English: shade pellitory, forest pellitory, smooth nettle, soft nettle, native pellitory

P. judaica L.

English: wall pellitory, pellitory-of-the-wall, pellitory

Parinari Aublet Chrysobalanaceae

Origins:

From a popular plant name used in Guiana or in Brazil (Tupi); see J.B.C. Fusée Aublet (1720-1778), *Histoire des Plantes de la Guiane Françoise.* 1: 514. Paris 1775; Luíz Caldas Tibiriçá, *Dicionário Guarani-Português.* 140. Traço Editora, Liberdade 1989; Luíz Caldas Tibiriçá, *Dicionário Tupi-Português.* 154. Traço Editora, Liberdade 1984.

Species/Vernacular Names:

P. sp.

English: bow wood

The Philippines: aningat, balayau, barit, binggas, bingsau, bongog, botabon, dungon-dungonan, gimaimai, ginaiang, pasak, kagemkem, kamulitingan, kangkangkan, kapgangan, karatakat, kulatingan, laiusin, malaigang, malapiga, malapuyau, maluklik, salifungan, salutin, sarangan, sigaadan, tabon-tabon

Malaya: ubah ngilas, nyalin, dak ballow

Neth. Indies: kolakka

N. Rhodesia: muhanyi

Zaire: mupundu blanc, mupundu gris, musesji, mutontwe

Liberia: ve ay dweh

Nigeria: awewe, abere, gawasa, mobola, otin, oyat, rura, gwanja-kusa, soncui, songui, udari, ugu, dabadogun, oghegke, kaikai

P. capensis Harv.

English: sand apple

Congo: mututulu

P. capensis Harv. subsp. *capensis* (*Parinari capensis* Harv. subsp. *latifolia* (Oliv.) R. Graham; *Parinari capensis* Harv. var. *latifolia* Oliv.; *Parinari curatellifolia* Planch. ex Benth.

var. *fruticulosa* R.E. Fr.; *Parinari latifolia* (Oliv.) Exell; *Parinari pumila* Mildbr.)

English: bosapple, dwarf mobola, sand apple

Southern Africa: mmola (Tswana); mobolo-oa-fatsi (Sotho); muHacha kwa pasi, mushakata kwa pasi, muJakata kwa pasi (Shona); sagogwane (Matabele)

P. congensis F. Didr.

Nigeria: papaa (Yoruba); unga (Ijaw); ahada, akong (Igbo); ekamufit (Ekoi)

Zaire: bobombi, bogongo, ditschia, mondjukuba

Ivory Coast: kotosoma

P. corymbosum (Blume) Miq. (*Maranthes corymbosa* Blume)

English: sea beam

Malaya: batu, membatu, merbatu, merbatu laut

P. curatellifolia Planchon ex Bentham (*Parinari chapelieri* Baillon; *Parinari curatellifolia* Planchon ex Benth. subsp. *mobola* (Oliv.) R. Graham; *Parinari gardineri* Hemsl.; *Parinari mobola* Oliv.)

English: mobola plum, hissing tree, sandapple, cork tree

French: toutou blanc

Tropical Africa: mbura, mobola

West Africa: kura tamba, tutu tamba

N. Rhodesia: mucha, mpundu

Nigeria: abo, abo-idofun, rura, gutabo, masao, gwanja-kusa, mobola, odara; rura (Hausa); putu (Nupe); nawarre-badi (Fula); ibua (Tiv); idofun (Yoruba)

Yoruba: afun lehin, idofun, abo idofun, igiabo, abo

Southern Africa: grysappel, mobolapruim; amaBuye, umBulwa (Zulu); mbulwa (Tsonga); muBula, muUura, muJakata, muShakata, muCha, muChakata, muHacha, muMbhuni (Shona); mobola (Tswana: western Transvaal, northern Cape, Botswana); mmola (North Sotho); muvhula (Venda); nxa (Sambui: Okavango Native Territory)

P. excelsa Sabine

Nigeria: esagho, esgho, yinrin-yinrin, dee, egin-ato, ako-idofun, sougue, sougui, aroba; yinrinyinrin (Yoruba); esagho (Edo); ohehe (Ikale); dee (Ijaw)

Cameroon: asila akung, nombokola, fo

Ivory Coast: sougué, sougué à grandes feuilles

Gabon: eto

P. excelsa Sabine subsp. *holstii* (Engl.) Graham

Gabon: ossang-eli, eto

Cameroon: asila akung, nombokola

Central Africa: esgho, assain, piolo, sougué, welia, loona, bofale, bongongo

P. nonda F. Muell. ex Benth. (see James A. Baines, *Australian Plant Genera. An Etymological Dictionary of Australian Plant Genera*. 273. Chipping Norton, N.S.W. 1981)

Australia: nonda

Parinarium Juss. Chrysobalanaceae

Origins:
Orthographic variant of *Parinari* Aublet.

Paris L. Trilliaceae (Liliaceae)

Origins:
Possibly from the Latin *par, paris* "equal," referring to the parts of the plant; see G.C. Wittstein, *Etymologisch-botanisches Handwörterbuch*. 658. Ansbach 1852; Helmut Genaust, *Etymologisches Wörterbuch der botanischen Pflanzennamen*. 460. 1996; William T. Stearn, *Stearn's Dictionary of Plant Names for Gardeners*. 231. Cassell, London 1993.

Species/Vernacular Names:
P. polyphylla Sm.
Nepal: dai sua, daiswa
China: tsao hsiu
P. quadrifolia L.
English: herb paris
China: wang sun

Parishella A. Gray Campanulaceae

Origins:
Dedicated to the brothers Samuel Bonsall Parish (1838-1928) and William F. Parish, American botanists and botanical collectors; Samuel Bonsall Parish wrote *A Catalogue of Plants Collected in the Salton Sink*. Washington, D.C. 1913. See George Neville Jones (1903-1970), *An Annotated Bibliography of Mexican Ferns*. Univ. Illinois Press 1966; John H. Barnhart, *Biographical Notes upon Botanists*. 3: 48. 1965; Ida Kaplan Langman, *A Selected Guide to the Literature on the Flowering Plants of Mexico*. 566. University of Pennsylvania Press, Philadelphia 1964; T.W. Bossert, *Biographical Dictionary of Botanists Represented in the Hunt Institute Portrait Collection*. 301. 1972; S. Lenley et al., *Catalog of the Manuscript and Archival Collections and Index to the Correspondence of John Torrey*. Library of the New York Botanical Garden. 321. 1973; E.M. Tucker, *Catalogue of the Library of the Arnold Arboretum of Harvard University*. 1917-1933; G. Murray, *History of the Collections Contained in the Natural History Departments of the*

British Museum. 1: 172. London 1904; I.C. Hedge and J.M. Lamond, *Index of Collectors in the Edinburgh Herbarium*. Edinburgh 1970; Irving William Knobloch, compil., "A preliminary verified list of plant collectors in Mexico." *Phytologia Memoirs*. VI. 1983; George Edmund Lindsay (b. 1916), *Notes Concerning the Botanical Explorers and Exploration of Lower California, Mexico*. San Francisco 1955.

Parishia Hook.f. Anacardiaceae

Origins:
For the English (b. Calcutta) botanist Rev. Charles Samuel Pollock Parish, 1822-1897 (d. Somerset), collector of plants in Burma and the Andaman Island, from 1852 to 1878 chaplain at Maulmain (or Moulmein), Burma (Myanmar), an authority on Burmese orchids, contributed to *Journ. As. Soc. Bengal*. See Pahpoo, ed., *The Instructor and Morning Star*. Maulmain 1853-1857, [continued as:] *The Morning Star*. Maulmain 1858; Ernest Nelmes and William Cuthbertson, *Curtis's Botanical Magazine Dedications, 1827-1927*. 171-172. [1931]; Merle A. Reinikka, *A History of the Orchid*. Timber Press 1996; Ray Desmond, *Dictionary of British & Irish Botanists and Horticulturists*. 534. 1994; I.H. Vegter, *Index Herbariorum*. Part II (5), *Collectors N-R*. Regnum Vegetabile vol. 109. 1983.

Species/Vernacular Names:
P. insignis Hook.f.
English: red dhup
Malaya: sepul, surian, sapoi

Pariti Adans. Malvaceae

Origins:
Pariti is the Malayalam name for cotton, see van Rheede tot Draakestein (1637-1691), *Hortus Indicus Malabaricus*. 1: t. 30. 1678; Michel Adanson, *Familles des Plantes*. 2: 401. Paris 1763.

Paritium Juss. Malvaceae

Origins:
Orthographic variant of *Pariti* Adanson.

Parkeria W.J. Hooker Parkeriaceae

Origins:
For the British (b. Glasgow) botanist Charles Sandbach Parker, (d. 1869), plant collector in North America, Guiana and the West Indies; see Ray Desmond, *Dictionary of Brit-*

ish & Irish Botanists and Horticulturists. 535. London 1994; A. Lasègue, *Musée botanique de Benjamin Delessert.* Paris 1845; R. Zander, F. Encke, G. Buchheim and S. Seybold, *Handwörterbuch der Pflanzennamen.* 14. Aufl. 760. Stuttgart 1993.

Parkia R. Br. Mimosaceae

Origins:

After the Scottish (b. near Selkirk) explorer Mungo Park, 1771-1806 (d. Niger, he was drowned), surgeon and traveler in Africa. He was the first modern European to reach the Niger, friend and *protégé* of Sir Joseph Banks, brother-in-law of James Dickson, to India and Sumatra (he sailed in February 1792), leader of the African Association's Exploring Expedition, 1795-1797 from the Gambia River to Sansanding on the Upper Niger, 1805 sailed for Africa and Northern Nigeria (Park and his companions perished in a fight with the natives). His writings include *Travels in the Interior Districts of Africa*: performed under the direction ... of the African Association, in ... 1795, 1796, and 1797. London 1799 and *The Journal of a Mission to the Interior of Africa in ... 1805 ...* To which is prefixed an account of the life of Mr. Park. London 1815. See Thomas Edward Bowdich (1791-1824), *Mission from Cape Coast Castle to Ashantee.* London 1819; Peter Hudson, *Two Rivers — Travels in West Africa on the Trail of Mungo Park.* London 1991; Kenneth Lupton, *Mungo Park: The African Traveler.* Oxford, London 1979; Lewis Grassic Gibbon, *Niger: The Life of Mungo Park.* Edinburgh 1934; F.N. Williams, "Collectors of Gambian plants." in *Bull. Herb. Boiss.* sér. 2, 7: 82-85. Geneva 1907; R.W.J. Keay, "Botanical collectors in West Africa prior to 1860." in *Comptes Rendus A.E.T.F.A.T.* 55-68. Lisbon 1962; F.N. Hepper and Fiona Neate, *Plant Collectors in West Africa.* 63. 1971; William Henry Giles Kingston, *Travels of Mungo Park, Denham, and Clapperton.* London [1886]; Robert Huish, *The Travels of Richard and John Lander ... for the Discovery of the Course ... of the Niger, ...* with a prefatory analysis of the previous travels of Park, Denham, Clapperton, Adams, G.F. Lyon, J. Ritchie, etc. into the hitherto unexplored countries of Africa. London 1836; D.G. Crawford, *A History of the Indian Medical Service, 1600-1913.* London 1914; M. Hadfield et al., *British Gardeners: A Biographical Dictionary.* London 1980; Harold B. Carter, *Sir Joseph Banks (1743-1820). A Guide to Biographical and Bibliographical Sources.* Winchester 1987; Ray Desmond, *Dictionary of British & Irish Botanists and Horticulturists.* 535. London 1994; Georg Christian Wittstein, *Etymologisch-botanisches Handwörterbuch.* 658. Ansbach 1852.

Species/Vernacular Names:

P. sp.

Peru: shimbillo pashaco

Malaya: kedawong, kerayong, alai

Cameroon: bolondo, ekombolo

P. bicolor A. Chev. (see Y. Tailfer, *La Forêt dense d'Afrique centrale.* CTA, Ede/Wageningen 1989; J. Vivien & J.J. Faure, *Arbres des Forêts denses d'Afrique Centrale.* Agence de Coopération Culturelle et Technique. Paris 1985)

English: locust bean

Congo: ezieb

Liberia: boe, gumni

Yoruba: irugba abata, aridan abata, oso

Nigeria: igbado, igba odo, ibibia, ogbokowo, ogrili-okpi, shanago, akwukwo-kaucha, uba, ugboro; igba odo (Yoruba); ugboro (Edo); ogirili okpi (Igbo); etediuku (Efik); kakpaja (Boki)

Gabon: essang

Cameroon: agnian, atoul, ekombolo, ndembé, tsoumbou, essang, eseng

Ivory Coast: ananjui, lo, pouopo

Zaire: bolele, wamba, wambamba, luboko

P. biglandulosa Wight & Arn. (*Parkia pedunculata* J.F. Macbr., nom. illeg.)

India: sivalinga maram

Malaya: nering, nenering, neri, neneri, petai

P. biglobosa (Jacq.) R. Br. ex G. Don (*Inga biglobosa* (Jacq.) Willd.; *Mimosa biglobosa* Jacq; *Parkia africana* R. Br.; *Parkia clappertoniana* Keay)

English: African locust bean, cainda-wood

Tropical Africa: afiti

W. Africa: nere sun, netige, niri

Yoruba: irugba, igbaru, atawere iru, iru, igi-iru, iruworo, igba iru, ayunbo, igba, agbanire, igba iyere, woro

Nigeria: dorowa (Hausa); narehi (Fula); runo (Kanuri); maito (Shuwa Arabic); lonchi (Nupe); nune (Tiv); igba (Yoruba); ugba (Etsako); eyiniwan (Edo); ogirili (Igbo)

Togo: owati, sorono, ssulo, wo

P. filicoidea Welw. ex Oliv.

English: locust bean

Nigeria: dorowa, ogba, ogirili-okpi

Sudan: mudus, umdus

Zaire: ofiloli, mombo, mpalina, mupaku

Cameroon: dorouwa, eseng, edzin, mondous, ziya

Central Africa: zinya

Ivory Coast: pipigbalé, pipigbale, pipigpale

P. igneiflora Ducke

Peru: goma pashaco, goma guayo, goma huayo

P. multijuga Bentham

Peru: tankam, guarango

P. nitida Miquel (*Parkia oppositifolia* Spruce ex Bentham)

Peru: goma pashaco, goma guayo, goma huayo

P. pendula (Willdenow) Bentham ex Walpers

Peru: esponja, fava de bolotas, japacanim, manopé de praia, paricá, visgueiro, visquero, zarcillo

Suriname: apa kanilan, apa akaniran, ipana, kwata kama, kwatta, kwatta kama

P. speciosa Hassk.

Malaya: nyiring, petai

P. velutina Benoist

Peru: pashaco curtidor, pashaco colorado

Parkinsonia L. Caesalpiniaceae

Origins:
After the British apothecary John Parkinson, 1567-1650 (London), botanist, *Botanicus Regius Primarius*, appointed apothecary to James I, gardener in Long Acre, London, author of *Paradisi in Sole Paradisus Terrestris*. London 1629 and *Theatrum Botanicum: The Theatre of Plants*. London 1640. See Mathias de Lobel [de Lobelius] (1538-1616), *Stirpium illustrationes*. Londini 1655; James Newton (1639-1718), *A Compleat Herbal*. London 1752; J.H. Barnhart, *Biographical Notes upon Botanists*. 3: 50. 1965; Robert William Theodore Gunther, *Early British Botanists and their Gardens* based on unpublished writings of Goodyer ... and others. Oxford 1922; T.W. Bossert, *Biographical Dictionary of Botanists Represented in the Hunt Institute Portrait Collection*. 301. 1972; H.N. Clokie, *Account of the Herbaria of the Department of Botany in the University of Oxford*. 221. Oxford 1964; Jonas C. Dryander, *Catalogus bibliothecae historico-naturalis Josephi Banks*. London 1800; E.M. Tucker, *Catalogue of the Library of the Arnold Arboretum of Harvard University*. 1917-1933; R. Pulteney, *Historical and Biographical Sketches of the Progress of Botany in England*. 1: 138-154. London 1790; Blanche Henrey, *British Botanical and Horticultural Literature before 1800*. Oxford 1975; M. Hadfield et al., *British Gardeners: A Biographical Dictionary*. London 1980; Carl Linnaeus, *Species Plantarum*. 375. 1753 and *Genera Plantarum*. Ed. 5. 177. 1754; Ray Desmond, *Dictionary of British & Irish Botanists and Horticulturists*. 536. 1994; F. Boerner & G. Kunkel, *Taschenwörterbuch der botanischen Pflanzennamen*. 4. Aufl. 146. Berlin & Hamburg 1989.

Species/Vernacular Names:
P. aculeata L.

English: horsebean, Jerusalem thorn, Mexican paloverde

Mexico: guechi belle, quechi pelle, quechi pella

Peru: mataburro, azote de Cristo, espinillo, junco marino, palo verde

Cuba: espinillo

North Colombia: sauce

West Indies: flor de rayo

Argentina: cinacina

Nigeria: sasabami (Hausa); sharan labbi (Fula); ogbo-okuye (Yoruba)

Arabia: sesaban

P. africana Sond.

Southern Africa: lemoendoring, lemoenboom, lemoenhout, wildelemoenhout (lemoen = pale yellow), thaboom, waterboom, wildelemoen (Afrikaans); khas (Nama)

Namibia: thaboom, waterboom (Afrikaans); omuumbamenye (Herero); lkhab (Nama/Damara)

Parlatorea Barb. Rodr. Orchidaceae

Origins:
For the Italian botanist Filippo (Philippus, Philippo) Parlatore, 1816-1877, physician, M.D. Palermo 1837, botanical explorer, plant collector, professor of botany (Istituto degli Studi Superiori di Firenze), founded the Istituto Botanico (in Museo di via Romana, Florence). His writings include *Memoria su di una membrana sierosa dell'occhio*. Palermo 1834, *Trattato teorico-pratico del chòlera asiatico*, osservato in Palermo nel 1837. Palermo 1837, *Breve cenno sulla vita e sulle opere del barone A. Bivona Bernardi*. Palermo 1837, *Prospetto dello stato della botanica in Sicilia nel principio del secolo XIX*. Palermo 1838, *Biografia di F.G.V. Broussais*. Palermo 1839, *Flora Panormitana*. Panormi 1839, "Dubbi sui limiti assegnati da Cuvier alle diverse rivoluzioni del Globo." *Gazzetta Toscana delle Scienze Mediche Fisiche*. anno III. 1845, *Flora palermitana*. Firenze 1845, *Elogio di Jacopo Gräber de Hemso*. Firenze 1849, *Elogio storico di Luigi Colla*. Firenze 1850, *Elogio di Filippo Barker Webb*. Firenze 1856, *Necrologia di Roberto Brown*. Firenze 1859, *Elogio di Alessandro Humboldt*. Firenze 1860, *Coniferas novas* nonnullas descripsit Philippus Parlatore. Florentiae [1863], *Rimedio popolare per la cura del chòlera*. Palermo 1867, *Cenno cronologico di Adolfo Brongniart*. Firenze 1870 and *Flora italiana*. [vols. 6-10 ed. by T. Caruel, 1830-1898] Firenze 1848-1893, with Philip Barker Webb wrote *Florula aethiopico-aegyptiaca*. Florentiae 1851. See Giuseppe M. Mira, *Bibliografia Siciliana*. 2: 181-185. Palermo 1881; Stafleu and Cowan, *Taxonomic Literature*. 4: 66-72. 1983; Philip Barker Webb (1793-1854) and Sabin Berthelot (1794-1880), *Histoire naturelle des Iles Canaries*. Paris [1835-] 1836-1850; John H. Barnhart, *Biographical Notes upon Botanists*. 3: 51. Boston 1965; Ethelyn Maria Tucker, *Catalogue of the*

Library of the Arnold Arboretum of Harvard University.
Cambridge, Massachusetts 1917-1933; T.W. Bossert, *Bio-graphical Dictionary of Botanists Represented in the Hunt Institute Portrait Collection.* 301. 1972; H.N. Clokie, *Account of the Herbaria of the Department of Botany in the University of Oxford.* 221. Oxford 1964; A. Lasègue, *Musée botanique de Benjamin Delessert.* Paris 1845; R. Zander, F. Encke, G. Buchheim and S. Seybold, *Handwörterbuch der Pflanzennamen.* 14. Aufl. Stuttgart 1993; Ida Kaplan Langman, *A Selected Guide to the Literature on the Flowering Plants of Mexico.* University of Pennsylvania Press, Philadelphia 1964; I.C. Hedge and J.M. Lamond, *Index of Collectors in the Edinburgh Herbarium.* Edinburgh 1970.

Parlatoria Boissier Brassicaceae

Origins:
For the Italian botanist Filippo Parlatore, 1816-1877.

Parmentiera DC. Bignoniaceae

Origins:
After the French (b. Montdidier) botanist Antoine Augustin Parmentier, 1737-1813 (d. Paris), pharmacist (army pharmacist). His works include *L'Art de faire les Eaux-de-Vie,* d'après la doctrine de M. Chaptal. Paris 1819, *Dissertation sur la nature des Eaux de la Seine.* Paris 1787, *Manière de faire le pain de pommes de terres,* sans mélange de farine. Paris 1779, *Méthode facile de conserver à peu de frais les grains et les farines.* Londres 1784, *Le parfait Boulanger.* Paris 1778, *Les pommes de terre.* Paris 1781 and *Traité sur la culture et les usages des Pommes de terre, de la Patate e du Topinambour.* Paris 1789. See Alex Berman, in *Dictionary of Scientific Biography* (Editor in Chief Charles Coulston Gillispie.) 10: 325-326. 1981; John H. Barnhart, *Biographical Notes upon Botanists.* 3: 51. 1965; E.M. Tucker, *Catalogue of the Library of the Arnold Arboretum of Harvard University.* 1917-1933; T.W. Bossert, *Biograph-ical Dictionary of Botanists Represented in the Hunt Institute Portrait Collection.* 301. 1972; Ida Kaplan Langman, *A Selected Guide to the Literature on the Flowering Plants of Mexico.* 566. University of Pennsylvania Press, Philadelphia 1964; Jonas C. Dryander, *Catalogus bibliothecae historico-naturalis Josephi Banks.* London 1800; Blanche Henrey, *British Botanical and Horticultural Literature before 1800.* Oxford 1975; R. Zander, F. Encke, G. Buchheim and S. Seybold, *Handwörterbuch der Pflanzennamen.* 14. Aufl. 760. Stuttgart 1993; Georg Christian Wittstein, *Etymologisch-botanisches Handwörterbuch.* 659. 1852.

Species/Vernacular Names:
P. cerifera Semann
English: candle tree

Parmentiera Raf. Solanaceae

Origins:
For the French botanist Antoine Augustin Parmentier, 1737-1813; see Constantine Samuel Rafinesque, *Autikon botanikon.* Icones plantarum select. nov. vel rariorum, etc. 108, 109. Philadelphia 1840; E.D. Merrill, *Index rafinesquianus.* 212. 1949.

Parnassia L. Parnassiaceae (Saxifragaceae)

Origins:
Mount Parnassus, Greece.

Species/Vernacular Names:
P. palustris L.
English: grass of Parnassus

Parochetus Buch.-Ham. ex D. Don Fabaceae

Origins:
From the Greek *para* "near" and *ochetos* "water-pipe, streams, channel," referring to the habitat, along streams and brooks.

Parodia Spegazzini Cactaceae

Origins:
After the Italian botanist Domingo Parodi, 1823-1890, pharmacist, physician, in Argentina and Paraguay; see John H. Barnhart, *Biographical Notes upon Botanists.* 3: 51. 1965; Ethelyn Maria Tucker, *Catalogue of the Library of the Arnold Arboretum of Harvard University.* Cambridge, Massachusetts 1917-1933; Stafleu and Cowan, *Taxonomic Literature.* 4: 75-76. 1983.

Parodianthus Troncoso Verbenaceae

Origins:
After the Argentine botanist Lorenzo Raimundo Parodi, 1895-1966, agrostologist, professor of botany in Argentina, 1934-1962 editor and director of the *Revista Argentina de Agronomía.* His writings include "Nota sobre las especies de *Briza* de la flora argentina." *Revista Fac. Agron. Veterin.*

3: 113-138. 1920, "Los arroces de la flora argentina." *Physis* (Buenos Aires). 11: 238-252. 1933, "Resumen bibliográfico. Looser, Gualterio. Las Proteáceas chilenas." *Revista Argent. Agron.* 1(2): 151-153. 1934, "Albert Spear Hitchcock." *Revista Argent. Agron.* 3(2): 113-119. 1936, "El origen geográfico de algunas gramíneas coleccionadas por don Luis Née en su viaje alrededor del mundo." *Revista Argent. Agron.* 14(1): 61-69. 1947, "Robert Pilger." *Revista Argent. Agron.* 20(2): 107-114. 1953 and "Thaddaeus Peregrinus Haenke a dos siglos de su nacimiento." *Anales Acad. Nac. Ci. Exact. Buenos Aires.* 17: 9-28. 1964, with J. Camara wrote "El mango, cereal extinguido en cultivo, sobrevive en estado salvaje." *Ci. & Invest.* 20(12): 543-549. 1964. See Arturo E. Burkart (1906-1975), "Bibliografía del botánico argentino Lorenzo R. Parodi (1895-1966)." *Bol. Soc. Argent. Bot.* 12: 7-16. 1968; H. Augustín Garaventa (1911-1981), "El botánico argentino Lorenzo R. Parodi." *Revista Univ.* (Santiago) 52: 167-175. 1967 [1968]; J.H. Barnhart, *Biographical Notes upon Botanists.* 3: 51. 1965; T.W. Bossert, *Biographical Dictionary of Botanists Represented in the Hunt Institute Portrait Collection.* 301. 1972; S. Lenley et al., *Catalog of the Manuscript and Archival Collections and Index to the Correspondence of John Torrey.* Library of the New York Botanical Garden. 321. 1973; Ida Kaplan Langman, *A Selected Guide to the Literature on the Flowering Plants of Mexico.* Philadelphia 1964; R. Zander, F. Encke, G. Buchheim and S. Seybold, *Handwörterbuch der Pflanzennamen.* 14. Aufl. 760. Stuttgart 1993; Stafleu and Cowan, *Taxonomic Literature.* 4: 76-77. 1983.

Parodiella J.R. Reeder and C.G. Reeder Gramineae

Origins:

After the Argentine botanist Lorenzo Raimundo Parodi, 1895-1966.

Parodiochloa A.M. Molina Gramineae

Origins:

After the Argentine botanist Lorenzo Raimundo Parodi, 1895-1966.

Parodiochloa C.E. Hubb. Gramineae

Origins:

After the Argentine botanist Lorenzo Raimundo Parodi, 1895-1966.

Parodiodendron Hunz. Euphorbiaceae

Origins:

After the Argentine botanist Lorenzo Raimundo Parodi, 1895-1966.

Parodiodoxa O.E. Schulz Brassicaceae

Origins:

After the Argentine botanist Lorenzo Raimundo Parodi, 1895-1966.

Parodiolyra Söderstr. & Zuloaga Gramineae

Origins:

After the agrostologist Lorenzo Raimundo Parodi, 1895-1966, and the genus *Olyra* L.

Paronychia Miller Caryophyllaceae (Illecebraceae)

Origins:

Greek *paronychia* "a whitlow," *para* "beside, near" and *onyx, onychos* "nail, claw," the plant was supposed to cure a kind of whitlow under the nails; see Philip Miller, *The Gardeners Dictionary.* Abr. ed. 4. London (28 Jan.) 1754.

Species/Vernacular Names:

P. ahartii B. Ertter

English: Ahart's paronychia

P. brasiliana Poiret

English: Brazilian paronychia, Brazilian whitlow wort, Chilean whitlow wort, chikweed nailwort, Chile nailwort

South Africa: Brasiliaanse paronychia

Paropsia Noronha ex Thouars Passifloraceae

Origins:

Greek *paropsis, paropsidos, para* "beside, near, together" and *opson* "food," a dish on which the food is served, referring to the fruit.

Species/Vernacular Names:

P. guineensis Oliv.

Nigeria: koropo (Yoruba); akapue (Edo)

Yoruba: sansan ona

Paropsiopsis Engl. Passifloraceae

Origins:
Resembling the genus *Paropsia*.

Paropyrum Ulbr. Ranunculaceae

Origins:
Probably from the Latin *paro, are* "to make equal" or *paro, onis* "a small ship" and Greek *pyros* "grain, wheat," referring to the fruit, see also *Isopyrum* L.

Parosela Cav. Fabaceae

Origins:
An anagram of *Psoralea*.

Paroxygraphis W.W. Sm. Ranunculaceae

Origins:
Greek *para* "beside, near, together" plus the genus *Oxygraphis* Bunge, *oxys* "sharp" and *graphis* "drawing."

Parrotia C.A. Meyer Hamamelidaceae

Origins:
For the German botanist Johann Jacob Friedrich Wilhelm Parrot, 1792-1841, physician, professor of medicine, traveler. His works include *Reise zum Ararat.* Berlin 1834 and *Ueber Gasometrie.* Dorpat [1814]; see John H. Barnhart, *Biographical Notes upon Botanists.* 3: 51. 1965; A. Lasègue, *Musée botanique de Benjamin Delessert.* Paris 1845; Ethelyn Maria Tucker, *Catalogue of the Library of the Arnold Arboretum of Harvard University.* Cambridge, Massachusetts 1917-1933; Helmut Genaust, *Etymologisches Wörterbuch der botanischen Pflanzennamen.* 461. [dedicated to a Georg Friedrich Parrot, b. 1791] Basel 1996; Georg Christian Wittstein, *Etymologisch-botanisches Handwörterbuch.* 659. Ansbach 1852.

Species/Vernacular Names:
P. persica (DC.) Meyer
English: ironwood

Parrotiopsis (Niedenzu) C. Schneider Hamamelidaceae

Origins:
Resembling *Parrotia*.

Parrya R. Br. Brassicaceae

Origins:
For the British (b. Bath, Somerset) explorer Sir William Edward Parry, 1790-1855 (d. Germany), Arctic navigator, in 1821 elected a Fellow of the Royal Society, knighted 1829, 1853 Governor of Greenwich Hospital. His writings include *Journal of a Voyage for the Discovery of a North-West Passage.* London 1821 and *Journal of a Second Voyage for the Discovery of a North-West Passage.* London 1824. See A. Hervé & F. de Lanoye, *Voyages dans les glaces du Pole Arctique à la recherche du Passage Nord-Ouest.* Paris 1854; John H. Barnhart, *Biographical Notes upon Botanists.* 3: 52. 1965; H.N. Clokie, *Account of the Herbaria of the Department of Botany in the University of Oxford.* 221-222. Oxford 1964; A. Lasègue, *Musée botanique de Benjamin Delessert.* Paris 1845; Mea Allan, *The Hookers of Kew.* London 1967; G. Murray, *History of the Collections Contained in the Natural History Departments of the British Museum.* 1: 172. London 1904; M. Hadfield et al., *British Gardeners: A Biographical Dictionary.* London 1980; John Dunmore, *Who's Who in Pacific Navigation.* 18, 73, 214. University of Hawaii Press, Honolulu 1991; Ida Kaplan Langman, *A Selected Guide to the Literature on the Flowering Plants of Mexico.* 566-567. Philadelphia 1964; I.C. Hedge and J.M. Lamond, *Index of Collectors in the Edinburgh Herbarium.* 1970; Leonard Huxley, *Life and Letters of Sir Joseph Dalton Hooker.* London 1918; R. Zander, F. Encke, G. Buchheim and S. Seybold, *Handwörterbuch der Pflanzennamen.* 14. Aufl. 1993; Ray Desmond, *Dictionary of British & Irish Botanists and Horticulturists.* 537. 1994.

Parryella Torrey & A. Gray Fabaceae

Origins:
For the English-born American botanist Charles Christopher Parry, 1823-1890 (Davenport, Iowa, USA), 1832 to America, M.D. Columbia College 1846, plant collector, explorer (Colorado, Oregon, western Wyoming, Rocky Mountains, Utah, Nevada, California, Mexico), physician, from 1849 to 1861 with the Mexican Boundary Survey, 1861 Colorado Expedition, 1862 the Parry, Hall and Harbour Expedition, expedition of 1864 (Hot Sulphur Springs, excursion to Long's Peak), 1869-1872 USDA. His writings include *Botanical Observations in Western Wyoming.* Salem 1874 and "California manzanitas." *Bull. Calif. Acad. Sci.* 2: 483-496. 1886-1887. See John H. Barnhart, *Biographical Notes upon Botanists.* 3: 52. 1965; Ray Desmond, *Dictionary of British & Irish Botanists and Horticulturists.* 537. London 1994; T.W. Bossert, *Biographical Dictionary of Botanists Represented in the Hunt Institute Portrait Collection.* 301. 1972; H.N. Clokie, *Account of the Herbaria of*

the Department of Botany in the University of Oxford. 221. Oxford 1964; Ida Kaplan Langman, *A Selected Guide to the Literature on the Flowering Plants of Mexico*. 566-567. Philadelphia 1964; S. Lenley et al., *Catalog of the Manuscript and Archival Collections and Index to the Correspondence of John Torrey*. Library of the New York Botanical Garden. 465-466. 1973; Ira L. Wiggins, *Flora of Baja California*. 42. Stanford, California 1980; J.D. Milner, *Catalogue of Portraits of Botanists Exhibited in the Museums of the Royal Botanic Gardens*. Royal Botanic Gardens, Kew, London 1906; Irving William Knobloch, compil., "A preliminary verified list of plant collectors in Mexico." *Phytologia Memoirs*. VI. 1983; Joseph William Blankinship, "A century of botanical exploration in Montana, 1805-1905: collectors, herbaria and bibliography." in *Montana Agric. Coll. Sci. Studies Bot*. 1: 1-31. 1904; R. Zander, F. Encke, G. Buchheim and S. Seybold, *Handwörterbuch der Pflanzennamen*. 14. Aufl. Stuttgart 1993; Joseph Ewan, *Rocky Mountain Naturalists*. The University of Denver Press 1950; E.M. Tucker, *Catalogue of the Library of the Arnold Arboretum of Harvard University*. 1917-1933; J. Ewan, ed., *A Short History of Botany in the United States*. 1969; Charles Francis Saunders, *Western Wild Flowers*. 73. New York 1933; Howard Atwood Kelly and Walter Lincoln Burrage, *Dictionary of American Medical Biography*. New York 1928; G. Murray, *History of the Collections Contained in the Natural History Departments of the British Museum*. 1: 172. 1904; Margaret Miller Rocq, ed., *California Local History. A Bibliography and Union List of Library Holdings*. Second edition. Stanford, California 1970; A.E. Weber, *King of Colorado Botany: Charles Christopher Parry, 1823-1890*. 1997; Gordon Douglas Rowley, *A History of Succulent Plants*. Strawberry Press, Mill Valley, California 1997; I.C. Hedge and J.M. Lamond, *Index of Collectors in the Edinburgh Herbarium*. Edinburgh 1970.

Parryodes Jafri Brassicaceae

Origins:

For the British explorer William Edward Parry, 1790-1855 (Germany), Arctic navigator.

Parryopsis Botsch. Brassicaceae

Origins:

Resembling *Parrya*, for the British explorer William Edward Parry, 1790-1855 (Germany), Arctic navigator.

Parsana Parsa & Maleki Urticaceae

Origins:

For Ahmed Parsa.

Parsonsia P. Browne Lythraceae

Origins:

For the English botanist James Parsons, 1705-1770, physician, anatomist.

Parsonsia R. Br. Apocynaceae

Origins:

In memory of the distinguished English (b. Devon) botanist James Parsons, 1705-1770 (d. London), physician (specialized in obstetrics), antiquary, anatomist, he went to Paris to study medicine, pupil of the French botanist Bernard de Jussieu (1699-1777), M.D. Rheims in 1736, practiced in London, (in 1740 or 1741) a Fellow of the Royal Society, a Fellow of the Society of Antiquaries. His works include *The Microscopical Theatre of Seeds*. London [1744-] 1745, *Philosophical Observations on the Analogy Between the Propagation of Animals and that of Vegetables*. London 1752 and *Remains of Japhet: Being Historical Enquiries into the Affinity and Origin of the European Languages*. London 1767. See John Nichols, *Literary Anecdotes of the Eighteenth Century*. London 1812-1815; R. Brown, "On the Asclepiadeae." *Memoirs of the Wernerian Natural History Society*. 1: 64. Edinburgh 1811; William Munk, *The Roll of the Royal College of Physicians of London*. 2: 175-177. London 1878; Blanche Henrey, *British Botanical and Horticultural Literature before 1800*. 2: 41-42. Oxford 1975.

Species/Vernacular Names:

P. alboflavescens (Dennstedt) Mabberley (*Parsonsia laevigata* (Moon) Alston; *Parsonsia howii* Tsiang; *Echites laevigata* Moon)

English: smooth parsonsia, Kwangsi parsonsia

Japan: Paruson-kazura, hôrai-kagami

China: hai nan tong xin jie

P. brownii (Britten) Pichon (after Robert Brown, British botanist)

English: twining silkpod, mountain silkpod

P. densivestita C.T. White

English: silkpod

P. dorrigoensis J.B. Williams

English: milky silkpod

P. eucalyptophylla F. Muell.

Australia: gargaloo, gargalou, woodbine

P. fulva S.T. Blake

English: furry silkpod

P. goniostemon Hand.-Mazz.

English: Kwangsi parsonsia, Guangxi parsonsia

China: guang xi tong xin jie

P. induplicata F. Muell.

English: thin-leaved silkpod

P. lanceolata R. Br.

English: northern silkpod, rough silkpod

P. lenticellata C. White

English: narrow-leaf silkpod

P. lilacina F. Muell.

English: delicate silkpod, crisped silkpod

P. purpurascens J.B. Williams

English: black silkpod

P. rotata Maiden & E. Betche

English: veinless silkpod, corky silkpod

P. straminea (R. Br.) F. Muell.

English: common silkpod

P. tenuis S.T. Blake

English: slender silkpod

P. velutina R. Br.

English: velvet silkpod

P. ventricosa F. Muell.

English: pointed silkpod, acuminate silkpod

Parthenice A. Gray Asteraceae

Origins:
From the Greek *parthenos, parthenike* "virgin, maiden, girl," Latin *parthenice* used by Catullus for a plant, also called *parthenium*.

Parthenium L. Asteraceae

Origins:
Parthenion, used by Plinius and Dioscorides for American feverfew, from the Greek *parthenos* "virgin," referring to the white rays or to the shape of the ovary or in allusion to its supposed medicinal properties, or possibly because the fruits are produced only by female florets; Latin *parthenium*, the name of several plants, i.e. *perdicium, leucanthes, tamnacus, linozostis, hermupoa, mercurialis, chrysocollis,* etc.; see Carl Linnaeus, *Species Plantarum.* 998. 1753 and *Genera Plantarum.* Ed. 5. 426. 1754.

Species/Vernacular Names:
P. argentatum A. Gray

New Mexico: guayule

P. hysterophorus L.

English: false ragweed, Santa Maria

India: bish-gach

Latin America: altamisa

P. integrifolium L.

New Mexico: wild quinine, American feverfew, prairie dock

Parthenocissus Planchon Vitaceae

Origins:
Greek *parthenos* "virgin" and *kissos* "ivy," possibly referring to its English vernacular name, Virginia creeper, or to the unisexual flowers.

Species/Vernacular Names:
P. quinquefolia (L.) Planchon (*Hedera quinquefolia* L.)

English: Virginia creeper, woodbine, true Virginia creeper

Japan: Amerika-zuta

P. tricuspidata (Siebold & Zucc.) Planchon (*Ampelopsis tricuspidata* Sieb. & Zucc.; *Vitis inconstans* Miq.; *Cissus thunbergii* Sieb. & Zucc.; *Parthenocissus thunbergii* (Sieb. & Zucc.) Nakai)

English: Japanese creeper, Boston ivy, Virginia creeper

Japan: tsuta

China: di jin, chang chun teng, lung lin pi li

P. vitacea (Knerr) Hitchc.

English: woodbine

Parvatia Decne. Lardizabalaceae

Origins:
The mountaineer, the goddess Parvati, Maha-devi, the great goddess, a name of the wife of God Shiva (Siva).

Parviopuntia Soulaire & Marn.-Lap. Cactaceae

Origins:
From the Latin *parvus* "small" plus *Opuntia* Mill.

Parvisedum R.T. Clausen Crassulaceae

Origins:
Latin *parvus* "small" plus *Sedum* L. or *Sedella* Britton & Rose.

Species/Vernacular Names:
P. leiocarpum (H. Sharsm.) R.T. Clausen

English: Lake County stonecrop

Parvotrisetum Chrtek Gramineae

Origins:

From the Latin *parvus* "small" plus the genus *Trisetum* Pers.

Paryphantha Schauer Myrtaceae

Origins:

From the Greek *paryphes* "with a border, bordered robe" and *anthos* "flower"; see *Linnaea.* 17: 235. 1843.

Paryphosphaera Karst. Mimosaceae

Origins:

From the Greek *paryphes* "with a border" and *sphaira* "a globe, ball."

Pasaccardoa Kuntze Asteraceae

Origins:

Named for the Italian botanist Pier Andrea Saccardo, 1845-1920, mycologist, professor of natural sciences, professor of botany, from 1879 to 1915 Director of the Botanical Garden of the University of Padova, botanical collector. His writings include *Di un'operetta sulla flora della Corsica di autore pseudonimo e plagiario.* Venezia 1908, "Fungilli novi Europaei et Asiatici." *Grevillea.* 21: 65-69. 1893 and *Fungi italici.* Patavii 1877-1886, with Roberto de Visiani (1800-1878) wrote *Catalogo delle piante vascolari del Veneto.* Venezia 1869. See John H. Barnhart, *Biographical Notes upon Botanists.* 3: 197. 1965; Ethelyn Maria Tucker, *Catalogue of the Library of the Arnold Arboretum of Harvard University.* Cambridge, Massachusetts 1917-1933; T.W. Bossert, *Biographical Dictionary of Botanists Represented in the Hunt Institute Portrait Collection.* 344. 1972; Ida Kaplan Langman, *A Selected Guide to the Literature on the Flowering Plants of Mexico.* 661. University of Pennsylvania Press, Philadelphia 1964; S. Lenley et al., *Catalog of the Manuscript and Archival Collections and Index to the Correspondence of John Torrey.* Library of the New York Botanical Garden. 357. 1973; Stafleu and Cowan, *Taxonomic Literature.* 4: 1024-1040. 1983.

Pasania Oerst. Fagaceae

Origins:

See Helmut Genaust, *Etymologisches Wörterbuch der botanischen Pflanzennamen.* 461. Basel 1996; F. Boerner, *Taschenwörterbuch der botanischen Pflanzennamen.* 2. Aufl. 152. Berlin & Hamburg 1966.

Pasaniopsis Kudô Fagaceae

Origins:

Resembling *Pasania.*

Pascalia Ortega Asteraceae

Origins:

After the Italian botanist Diego Baldassa(r)re Pascal, 1768-1812, a pupil of Giovanni Battista Guatteri (from 1770 to 1793 Director of the Botanical Garden of Parma), from 1793 to 1802 Director of the Botanical Garden of Parma, professor of botany. See Francesco Lanzoni, "L'Orto Botanico e i suoi dirigenti dal 1600 ad oggi." *Aurea Parma.* Nuova Serie, n. 68. Parma 1933; Casimiro Gómez Ortega (1740-1818), *Novarum, aut rariorum Plantarum Horti Reg. Botan. Matrit. descriptionum decades*, etc. 39, t. 4. Matriti [Madrid] 1797.

Pascalium Cass. Asteraceae

Origins:

See *Psacalium* Cass.

Paschalococos Dransf. Palmae

Origins:

Easter Island, Latin *Pascha, ae* "the fest of the Passover, Easter."

Paschanthus Burch. Passifloraceae

Origins:

Latin *Pascha, ae* "the fest of the Passover, Easter" and Greek *anthos* "flower."

Pascopyrum Á. Löve Gramineae

Origins:

Pasture wheat, Latin *pasco, pavi, pastum* "to feed, pasture" and Greek *pyros* "grain, wheat."

Pasithea D. Don Anthericaceae (Liliaceae)

Origins:

Named after Pasithea or Pasithee, one of the Graces; *pasithea* was also the Greek name for a magical plant.

Paspalanthium Desv. Gramineae

Origins:

The genus *Paspalum* L. and *anthos* "flower."

Paspalidium Stapf Gramineae

Origins:

A diminutive of the generic name *Paspalum* L.; see Sir David Prain (1857-1944), in *Flora of Tropical Africa*. 9: 15, 582. 1920.

Species/Vernacular Names:

P. albovillosum S.T. Blake

English: panic grass

P. aversum Vickery

English: panic grass

P. caespitosum C.E. Hubb.

Australia: Brigalow grass

P. constrictum (Domin) C.E. Hubb. (*Panicum constrictum* Domin; *Paspalidium gracile* (R. Br.) Hughes var. *rugosum* Hughes)

English: knotty-butt paspalidium, box grass, slender grass, knotty-butt grass, slender panic

P. flavidum (Retz.) Camus

The Philippines: sabung-sabung, giling, baili ixao

P. globoideum (Domin) Hughes

English: shot grass, sago grass

P. gracile (R. Br.) Hughes

English: slender panic grass, slender grass, graceful panic grass

P. jubiflorum (Trin.) Hughes

Australia: Warrego grass, Warrego summer grass (the common names refer to the Warrego River, southwest Queensland), Vandyke grass, yellow-flowered panic grass

P. punctatum (Burm.f.) Camus

The Philippines: baririan, lalabok

Paspalum L. Gramineae

Origins:

From the Greek name *paspalos* for millet; see Carl Linnaeus, *Systema Naturae*. Ed. 10. 1359. 1759.

Species/Vernacular Names:

P. spp.

Mexico: guixi yaci, quixi yaci, yaza, yaci, quique piquiñi

P. ciliatifolium Michx.

English: one-spiked paspalum

P. commersonii Lam.

Rodrigues Island: herbe à épée

P. conjugatum Bergius

English: sour paspalum, sour grass, yellow grass, Hilo grass, Johnston River grass

Hawaii: Hilo grass, mau'u Hilo

The Philippines: laau-laau, kulape, bantotan, kauat-kauat, kauatkauat

Malaya: rumput kerbau

Congo: likele, kedigui

P. dilatatum Poiret

English: Dallis grass, paspalum, bastard millet grass, common paspalum, golden crown grass, hairy-flowered paspalum, large water grass, large waterseed paspalum, water grass

Southern Africa: bankrotkweek, breësaadgras, breësaadvleigras, gewone paspalum, paspalatum gras, paspalum gras, watergras; mupunganini (Tonga)

Japan: shima-suzume-no-hie (= island *Paspalum*)

P. distichum L. (*Paspalum paspalodes* (Michx.) Scribner; *Digitaria paspalodes* Michx.)

English: buffalo quick paspalum, couch paspalum, water couch, swamp couch, silt grass, knot grass, saltwater couch

South Africa: bankrotkweek, buffelgras, buffelskweek paspalum, bulkweek, kakiekweek, knopgras, kweek paspalum, militêrekweek, rooikweek, tweevingergras

Japan: Kishû-suzume-hie

P. longifolium Roxb.

The Philippines: tal-tal likod

P. notatum Fluegge

English: Bahia grass, notatum grass

South Africa: Bahia paspalum

P. paniculatum L.

English: Russell River grass

Mexico: guixi betaa, quixi petaa

P. quadrifarium Lam.

English: tussock paspalum

P. scrobiculatum L. (*Paspalum auriculatum* Presl; *Paspalum commersonii* Lam.; *Paspalum orbiculare* Forst.f.; *Paspalum polystachyum* R. Br.) (Latin *scrobiculus* "a little ditch")

English: wild paspalum, creeping paspalum, ditchgrass, ditch millet, ricegrass, koda millet, kodo millet, native millet, scrobic, water couch grass

Southern Africa: dronkgras, slootgras; isiamuyisane (Zulu)

Japan: hai-suzume-no-hie (= creeping *Paspalum*)

Hawaii: mau'u laiki

P. urvillei Steudel (for the French traveler and explorer Jules Sébastian César Dumont d'Urville, 1790-1842, plant collector, a member of the Linnean Society and the Société de Géographie, he took part in the voyage of the *Coquille* (commanded by L.I. Duperrey), from 1825 commander of the *Astrolabe* (former *Coquille*). Among his works are *Enumeratio Plantarum quas in insulis archipelagi aut littoribus Ponti-Euxinii, annis 1819 et 1820, collegit atque detexit J. Dumont D'Urville*. Parisiis 1822, *Voyage au Pôle Sud and dans l'Océanie sur les corvettes L'Astrolabe et La Zélée*. Paris 1841-1846 and *Voyage de Découvertes autour du Monde ... sur la corvette L'Astrolabe pendant les Années 1826-1829*. Paris 1832-1848; see E.S. Dodge, *Islands and Empire: Western Impact on the Pacific and East Asia*. Minneapolis 1976; Gaston Meissas, *Les grands voyageurs de notre siècle*. Paris 1889; J.H. Barnhart, *Biographical Notes upon Botanists*. 1: 480. 1965; Theodore W. Bossert, *Biographical Dictionary of Botanists Represented in the Hunt Institute Portrait Collection*. 109. Boston, Massachusetts 1972; R. Glenn, *The Botanical Explorers of New Zealand*. Wellington 1950; John Dunmore, *Who's Who in Pacific Navigation*. 89-91, 93, 135, 184, 216. Honolulu 1991; Stafleu and Cowan, *Taxonomic Literature*. 1: 696-698. Utrecht 1976)

English: giant paspalum, upright paspalum, Vasey grass, tall paspalum

South Africa: langbeenwatergras, langbeen paspalum, regop paspalum, reuse paspalum, Vasey-paspalum

Japan: tachi-suzume-no-hie (= erect *Paspalum*)

P. vaginatum Swartz

English: sea shore paspalum, swamp couch, salt water couch

South Africa: brakpaspalum

Japan: sawa-suzume-no-hie

P. wettsteinii Hack.

English: Warrel grass, broad-leaved paspalum

Paspalus Fluegge Gramineae

Origins:

Possibly an orthographic variant of *Paspalum* L.

Passacardoa Wild Asteraceae

Origins:

See *Pasaccardoa*.

Passerina L. Thymelaeaceae

Origins:

Probably from the Latin *passer, eris* "a sparrow," *passerinus, a, um* "relating to the sparrow," the black seeds are beaked; see Carl Linnaeus, *Species Plantarum*. 559. 1753 and *Genera Plantarum*. Ed. 5. 168. 1754.

Species/Vernacular Names:

P. spp.

English: gonna

Southern Africa: ganna soorte, kannabas, kannabos, koordehaar, taaibos; lekhapu (Sotho)

P. drakensbergensis Hilliard & Burtt

South Africa: Drakensberg gonna

P. falcifolia C.H. Wr.

English: passerina with the sickle-shaped leaves

South Africa: forest gonna

P. filiformis L.

English: brown gonna

Southern Africa: gonnabas, gonnabos, windmakersbessie, kaalgaarbos (from the Dutch kabelgaren= tarred rope), kordhaar, bakkerbos, bakbos, bruingonna; unWele oluncane (Zulu)

P. montana Thoday

English: mountain gonna

Southern Africa: bakkerbos, bakbos; lithaba (Sotho)

P. rigida Wikstr.

English: dune gonna

Southern Africa: seekoppiesgonna; uNyenyevu, iShoba (Zulu)

Passiflora L. Passifloraceae

Origins:

Latin *passio, inis* (*patior, passus sum, pati* "to suffer") "passion" and *flos, floris* "a flower," the flowers symbolize the passion and crucifixion of Jesus Christ; see Carl Linnaeus, *Species Plantarum*. 955. 1753 and *Genera Plantarum*. Ed. 5. 410. 1754; National Research Council, *Lost Crops of the Incas: Little-Known Plants of the Andes with Promise for Worldwide Cultivation*. National Academy Press, Washington, D.C. 1989; John Vanderplank, *Passion Flowers and Passion Fruit*. London 1991.

Species/Vernacular Names:

P. sp.

English: passion flaxoer

Peru: caxoori, puche-puche, puro-purillo, purush, caxori, curubo, estrela purpurú, gipetu chlma-jixi?, inti sisa, maracujá, mashu sisa, millua caspi, naa mutaa, ñorbito, ñorbo, purpurillo, tacso

Mexico: jujo verde, pushulucuate, yaga-igue-lau-bille, pux-lucú

P. antioquiensis Karst.

English: banana passion fruit, red passion flower

P. aurantia Forst.f.

English: golden passion flower, red passion flower, blunt-leaved passion flower

P. aurantia Forst.f. var. *aurantia*

English: blunt-leaved passion fruit

P. caerulea L.

English: passion flower, blue passion flower, blue crown passion flower

Spanish: flor de la pasión, pasionaria

Japan: tokei-sô

Hawaii: palikea

P. cinnabarina Lindley

English: red passion flower

P. coccinea Aublet

English: red passion flower, red granadilla

Peru: granadilla

P. coriacea Juss.

English: bat-leaf passion flower

Peru: costado-sacha, uchu-anquirisi

Mexico: ala de chinaca, ala de murciélago, bazo de venado, hoja de murciélago, laga-guidi, murcielago, pachauatuán, xik-sots, xik-zots, granada de ratón, ocobithut

P. edulis Sims

English: granadilla, purple granadilla, passion fruit, purple water lemon, common passion fruit

Spanish: granadilla

South Africa: grenadella, wildegrenadella

Japan: kudamono-tokei-sô

Hawaii: liliko'i

P. foetida L.

English: running pop, love-in-a-mist, wild water lemon, passion flower, lover-in-a-mist passion flower

Peru: bedoca, ñorbo hediondo, puru-puru, ñorbo cimarrón, puru purillo

Hawaii: lani wai, pohapoha

Vietnam: chum bao, lac tien

Malaya: timun dendang, timun padang, timun hutan, kapas bulan, kerang kerut, letup

Yoruba: abiirunpo

Congo: bimpfii, okuma, mumpolompolo

P. foetida L. var. *hispida* (Triana & Planchon) Killip

English: stinking passion flower

P. herbertiana Ker Gawler

English: native passion fruit

P. herbertiana Ker Gawler subsp. *herbertiana* (after Lady Carnarvon, *née* Herbert)

English: native passion fruit

P. incarnata L.

English: wild passion flower, may pops, apricot vine, may apple

P. laurifolia L.

English: yellow granadilla, water lemon, yellow water lemon, Jamaica honeysuckle, bell apple, vinegar pear

French: pomme de liane

Spanish: granadilla

Malaya: buah susu

P. ligularis Juss.

English: sweet granadilla, granadilla

Peru: apicoya, granadilla, hutu, tintin

Mexico: granadita, granadita china, pelul, peñul

Hawaii: lemi wai, lani wai, lemona

P. maliformis L.

English: sweet calabash, sweetcup, conch apple

P. manicata (Juss.) Pers. (*Tacsonia manicata* Juss.)

English: red passion flower

P. mollissima (Kunth) L. Bailey (*Tacsonia mollissima* Kunth)

English: banana passion fruit

Peru: purocksha, tacso, tin-tin, trompos, tumbo, tumbo del monte

Hawaii: banana poka

P. quadrangularis L.

English: granadilla, grenadilla, giant granadilla, passion flower

Peru: apincoya, barbadina, badea, granadilla, tumbo, uxu-bëru

Japan: ô-mi-tokei-sô

The Philippines: granadilla, parola, kasaflora, granada

Malaya: timun hutan, timun belanda

P. racemosa Brot.

English: red passion flower

Japan: hozaki-no-tokei-sô

P. suberosa L. (*Passiflora minima* L.)

English: cork passion flower, small passion fruit

Peru: ñorbo, ñorbo marron

Mexico: granadita de ratón, zak-kansel-ak, kansel-ak

Hawaii: huehue haole

P. subpeltata Ortega

English: wild grenadella, granadina, white passion flower

Spanish: granadina

Mexico: granada de zorra

Pastinaca L. Umbelliferae

Origins:
Latin *pastinaca, ae* used by A. Cornelius Celsus and Plinius for a parsnip, also the carrot, Latin *pastino, avi, atum* "to prepare the ground," *pastinum, i* "a dribble, the prepared ground," Akkadian *pastum, pasum* "adze, ax cleaver"; see Carl Linnaeus, *Species Plantarum*. 262. 1753 and *Genera Plantarum*. Ed. 5. 126. 1754.

Species/Vernacular Names:
P. sativa L.

English: parsnip

Pastinacopsis Golosk. Umbelliferae

Origins:
Resembling *Pastinaca* L.

Patagonula L. Boraginaceae (Ehretiaceae)

Origins:
Of Patagonia, a South American region, in Chile and Argentina.

Patellaria J.T. Williams & Ford-Lloyd Chenopodiaceae

Origins:
The diminutive of the Latin *patina* "a broad dish," *patella, ae*, "a small dish, a plate."

Patellifolia A.J. Scott, Ford-Lloyd & J.T. Williams Chenopodiaceae

Origins:
Latin *patina* "a broad dish," *patella, ae*, "a small dish, a plate."

Patellocalamus W.T. Lin Gramineae

Origins:
Latin *patella, ae*, "a small dish, a plate" plus *Calamus*.

Patersonia R. Br. Iridaceae

Origins:
After the British (b. Angus) naturalist William Paterson, 1755-1810 (died on a voyage from Australia), gardener, traveler, botanical collector in South Africa (from 1777 to 1781) and Australia, in India from 1781 to 1785, to Australia 1791, in 1797 Fellow of the Linnean Society and in 1799 of the Royal Society, from 1800 to 1810 Lieutenant-Governor of New South Wales, correspondent of W. Forsyth (1737-1804), collected for Banks and J. Lee, author of *A Narrative of Four Journeys into the Country of the Hottentots and Caffraria*. London 1789. See V.S. Forbes and J. Rourke, *Paterson's Cape Travels, 1777 to 1779*. Johannesburg 1980; J. Britten, "Some early Cape botanists and collectors." *J. Linn. Soc. (Bot.)* 45: 45-46. 1920; I.H. Vegter, *Index Herbariorum*. Part II (5), *Collectors N-R*. Regnum Vegetabile vol. 109. 1983; J.H. Barnhart, *Biographical Notes upon Botanists*. 3: 54. Boston 1965; Ray Desmond, *Dictionary of British & Irish Botanists and Horticulturists*. 539. London 1994; J. Hutchinson, *A Botanist in Southern Africa*. 620-623. London 1946; Mary Gunn and Leslie E. Codd, *Botanical Exploration of Southern Africa*. 273-275. Cape Town 1981; John Sims (1749-1831), in *The Botanical Magazine*. 26, t. 41. 1807; Robert Brown, *Prodromus florae Novae Hollandiae et Insulae van-Diemen*. 303-304. London 1810; A. Lasègue, *Musée botanique de Benjamin Delessert*. Paris 1845; E.M. Tucker, *Catalogue of the Library of the Arnold Arboretum of Harvard University*. Cambridge, Massachusetts 1917-1933; A. White and B.L. Sloane, *The Stapelieae*. Pasadena 1937; I.C. Hedge and J.M. Lamond, *Index of Collectors in the Edinburgh Herbarium*. Edinburgh 1970.

Species/Vernacular Names:
P. fragilis (Labill.) Asch. & Graebner (*Genosiris fragilis* Labill.; *Patersonia glauca* R. Br.)

English: swamp iris, short purple flag, blue iris

P. glabrata R. Br.

English: leafy purple flag

P. juncea Lindley

English: purple flag

P. lanata R. Br.

English: woolly patersonia

P. limbata Endl.

English: bordered purple flag

P. longifolia R. Br.

English: dwarf purple flag, purple flag

P. macrantha Benth.

English: tropical purple flag

P. maxwellii (F. Muell.) F. Muell. ex Benth.

English: purple flag

P. occidentalis R. Br.

English: long purple flag

P. pygmaea Lindley

English: small purple flag

P. rudis Endl.

English: hairy flag

P. sericea R. Br.

English: silky purple flag

P. umbrosa Endl.

English: purple flag

P. xanthina Oldfield & F. Muell. ex F. Muell.

English: yellow flag

Patinoa Cuatrec. Bombacaceae

Origins:
Named for Victor Manuel Patiño, agronomist.

Patis Ohwi Gramineae

Origins:
An anagram of the generic name *Stipa* L.

Patonia Wight Annonaceae

Origins:
Presumably named for Mrs. A.W. Walker (*née* Paton), wife of General George Warren Walker (d. 1844), plant collectors in Ceylon, she illustrated the *Flora of Ceylon*. See *Companion to the Botanical Magazine*. 2: 194-200. 1837; Robert Wight, *Icones plantarum Indiae orientalis*, or figures of Indian plants. Madras [1838-] 1840-1853; Isaac Henry

Burkill, *Chapters on the History of Botany in India*. 50. Delhi 1965; Ray Desmond, *Dictionary of British & Irish Botanists and Horticulturists*. 710. London 1994.

Patrinia Juss. Valerianaceae

Origins:
For Eugène Louis Melchior Patrin, 1742-1815, French naturalist and mineralogist, wrote *Histoire Naturelle des Minéraux*. Paris [1801]; see N.A. Desvaux, *Tableau synoptique des minéraux*. Paris 1805.

Species/Vernacular Names:
P. scabiosaefolia Fisch.
English: Dahurian patrinia
China: bai jiang cao, bai jiang, pai chiang

Patrisia Rich. Flacourtiaceae

Origins:
After J.B. Patris.

Pattalias S. Watson Asclepiadaceae

Origins:
Probably from the Greek *pattalias* "two-year-old stag, pricket."

Pattonia Wight Orchidaceae

Origins:
Presumably named for Mrs. A.W. Walker (*née* Paton), wife of General George Warren Walker (d. 1844), plant collectors in Ceylon, she illustrated the *Flora of Ceylon*. See *Companion to the Botanical Magazine*. 2: 194-200. 1837; Robert Wight, *Icones plantarum Indiae orientalis*, or figures of Indian plants. Madras [1838-] 1840-1853; Isaac Henry Burkill, *Chapters on the History of Botany in India*. 50. Delhi 1965; Ray Desmond, *Dictionary of British & Irish Botanists and Horticulturists*. 710. London 1994.

Paua Caballero Asteraceae

Origins:
After the Spanish botanist Carlos Pau, 1857-1937, pharmacist. His works include *Plantas del Norte de Yebala, Marruecos*. Madrid 1924, *Plantas de Persia y de Mesopotamia recogidas por D. Fernando Martínez de la Escalera*. [with

Carlos Vicioso Martínez, 1886/7-1968] Madrid 1918 and *Plantas de Almería*. Barcelona 1925. See E.M. Tucker, *Catalogue of the Library of the Arnold Arboretum of Harvard University*. Cambridge, Massachusetts 1917-1933; T.W. Bossert, *Biographical Dictionary of Botanists Represented in the Hunt Institute Portrait Collection*. 303. 1972; R. Zander, F. Encke, G. Buchheim and S. Seybold, *Handwörterbuch der Pflanzennamen*. 14. Aufl. Stuttgart 1993; I.C. Hedge and J.M. Lamond, *Index of Collectors in the Edinburgh Herbarium*. Edinburgh 1970.

Pauella Ramam. & Sebastine Araceae

Origins:

For the Spanish botanist Hermenegild Santapau, 1903-1970, clergyman, 1928 to India, Director of the Botanical Survey of India, professor of botany, author of *The Flora of Khandala on the Western Ghats of India*. Delhi 1953, with A.N. Henry wrote *A Dictionary of the Flowering Plants in India*. New Delhi 1975.

Pauia Deb & Dutta Solanaceae

Origins:

For Hermenegild Santapau, 1903-1970, Director of the Botanical Survey of India.

Pauldopia Steenis Bignoniaceae

Origins:

Dedicated to the French botanist Paul Louis Amans Dop, 1876-1954. His writings include "La végétation de l'Indo-Chine." *Trav. Lab. For. Toulouse*. 1(Art.9): 1-16. 1931 and "Les *Gmelina* arborescents de l'Indochine." *Rev. Bot. Appl.* 13: 893-897. 1933. See John H. Barnhart, *Biographical Notes upon Botanists*. 1: 465. 1965.

Pauletia Cavanilles Caesalpiniaceae

Origins:

For the French botanist Jean Jacques Paulet, 1740-1826, physician, mycologist, M.D. Montpellier 1764, author of *Traité des champignons*. Paris [1790-]1793, *Tabula Plantarum Fungosarum*. Parisiis 1791 and *Flore et faune de Virgile*. Paris 1824. See J.H. Barnhart, *Biographical Notes upon Botanists*. 3: 56. 1965; E.M. Tucker, *Catalogue of the Library of the Arnold Arboretum of Harvard University*. Cambridge, Massachusetts 1917-1933; Jonas C. Dryander, *Catalogus bibliothecae historico-naturalis Josephi Banks*. London 1800.

Paullinia L. Sapindaceae

Origins:

For Simon Paulli, 1603-1680 (d. Copenhagen), professor of botany, physician to the King of Denmark Christian V. His writings include *Quadripartitum botanicum de Simplicium medicamentorum facultatibus*. Rostochii 1639, *Flora Danica det er: Dansk Urteborg*. Copenhagen [1647-]1648 and *Commentarius de abusu tabaci americanorum veteri, et herbae thee Asiaticorum in Europa novo*. Argentorati [Strasbourg] 1665; see Georg Christian Wittstein, *Etymologisch-botanisches Handwörterbuch*. 661f. Ansbach 1852; Helmut Genaust, *Etymologisches Wörterbuch der botanischen Pflanzennamen*. 463. [genus dedicated to the German physician Christian Franz Paullini, 1643-1712] Basel 1996; F. Boerner & G. Kunkel, *Taschenwörterbuch der botanischen Pflanzennamen*. 4. Aufl. 147. Berlin & Hamburg 1989; Karin Figala, in *Dictionary of Scientific Biography* 10: 426-427. 1981.

Species/Vernacular Names:

P. sp.

Peru: lúcuma del norte, confite del monte

P. alata (Ruíz & Pav.) G. Don (*Paullinia rhizantha* Poeppig; *Semarillaria alata* Ruíz & Pav.)

Peru: macote, macota, yurac macote, cumba huasca

P. bidentata Radlkofer

Peru: sapo huasca, acero huasca, sapo wasca

P. caloptera Radlkofer (*Paullinia williamsii* J.F. Macbride)

Peru: sapo huasca

P. capreolata (Aublet) Radlkofer (*Enourea capreolata* Aublet)

Peru: tingui

P. cupana Kunth

Peru: cupana, guaraná

P. grandifolia Bentham

Peru: cururú, yurari

P. imberbis Radlkofer

Peru: cumba huasca, guaraná, iurari, sapo huasca, timbó, uchu huasca

P. laeta Radlkofer

Peru: imino-o

P. pinnata L.

Peru: cruape, cururu ape, matto porco, timbó, timbó cipo, vermelho

Yoruba: kakasenla, ogbe okuje, kakasemi sola awomi, lago-lago

P. tenera Poeppig

Peru: ampi huasca colorada, paujil chaqui

P. uchocacha J.F. Macbride

Peru: cacha, ucho huasco, uchu huasca

P. yoco R. Schultes & Killip

Peru: huarmi yoco, yoco, yoco blanco, yoco colorado

Paulownia Siebold & Zucc. Scrophulariaceae

Origins:

Named after Princess Anna Paulowna (1795-1865), daughter of Paul I (1754-1801), Tsar of Russia (1796-1801); see Philipp Franz (Balthasar) von Siebold (1796-1866) and Joseph Gerhard Zuccarini (1797-1848), *Flora japonica*. 1: 25. Lugduni Batavorum [Leyden] 1835-1836; H. Genaust, *Etymologisches Wörterbuch der botanischen Pflanzennamen*. 463. 1996; G.C. Wittstein, *Etymologisch-botanisches Handwörterbuch*. 662. 1852; F. Boerner & G. Kunkel, *Taschenwörterbuch der botanischen Pflanzennamen*. 4. Aufl. 147. 1989.

Species/Vernacular Names:

P. kawakamii Itô

Japan: Taiwan-giri

P. tomentosa (Thunberg) Steudel (*Paulownia imperialis* Siebold & Zucc.; *Bignonia tomentosa* Thunb.)

English: royal paulownia, princess tree, karri tree, paulownia

Japan: kiri, chiri

China: tong pi, tung, pai tung, huang tung, pao tung, jung tung

Paulseniella Briquet Labiatae

Origins:

For the Danish botanist Ove Vilhelm Paulsen, 1874-1947, explorer, traveler, naturalist, in the Danish West Indies, professor of botany. His works include *Plankton and Other Biological Investigations in the Sea around the Faeroes in 1913*. Copenhagen 1918, *The Second Danish Pamir Expedition* conducted by O. Olufsen ... *Studies in the Vegetation of Pamir*. Copenhagen 1920, *The Peridiniales of the Danish Waters*. Copenhagen 1907, with C.H. Ostenfeld wrote "A list of flowering plants from inner Asia, collected by Dr. Sven Hedin, determined by various authors and compiled by C.H. Ostenfeld and O. Paulsen." in Sven Hedin, *Southern Tibet: Discoveries in former times compared with my own researches 1906-1908*. 1922. See John H. Barnhart, *Biographical Notes upon Botanists*. 3: 57. 1965; C.F.A. Christensen, *Den danske Botaniks Historie med tilhørende Bibliografi*. Copenhagen 1924-1926; E.M. Tucker, *Catalogue of the Library of the Arnold Arboretum of Harvard University*. Cambridge, Massachusetts 1917-1933; J. Ewan, ed., *A Short History of Botany in the United States*. 130. 1969.

Pauridia Harvey Hypoxidaceae (Liliaceae)

Origins:

From the Greek *pauros, pauron* "small" and diminutive, in allusion to the very small size of some species.

Species/Vernacular Names:

P. minuta (L.f.) Dur & Schinz

South Africa: koringsterretjie, koringblommetjie

Pauridiantha Hook.f. Rubiaceae

Origins:

From the Greek *pauros* "small" and *anthos* "flower."

Species/Vernacular Names:

P. symplocoides (S. Moore) Bremek. (*Urophyllum symplocoides* S. Moore)

Southern Africa: muriyashoko, ru tenja (Shona)

Paurolepis S. Moore Asteraceae

Origins:

Greek *pauros* "small" and *lepis* "scale."

Paurotis O.F. Cook Palmae

Origins:

From the Greek *pauros* "small, little" and *ous, otos* "ear."

Pausandra Radlk. Euphorbiaceae

Origins:

Probably from the Greek *pausis* "to stop, stopping, ceasing" and *aner, andros* "man, stamen."

Pausia Raf. Oleaceae

Origins:

Latin *pausea, posea, posia* and *pausia* for a kind of olive which yielded an excellent oil (Plinius); see C.S. Rafinesque, *Sylva Telluriana*. 9. 1838; E.D. Merrill, *Index rafinesquianus*. 190. 1949.

Pausia Raf. Thymelaeaceae

Origins:
See C.S. Rafinesque, *Flora Telluriana*. 4: 105. 1836 [1838]; E.D. Merrill, *Index rafinesquianus*. 172. 1949.

Pausinystalia Pierre ex Beille Rubiaceae

Origins:
Greek *pausinystalos* "stopping drowsiness," *pausis* "to stop" and *nystalus* "drowsy" (perhaps, implying excitement in the activity); see Y. Tailfer, *La Forêt dense d'Afrique centrale*. CTA, Ede/Wageningen 1989; J. Vivien & J.J. Faure, *Arbres des Forêts denses d'Afrique Centrale*. Agence de Coopération Culturelle et Technique. Paris 1985.

Species/Vernacular Names:
P. johimbe (K. Schumann) Beille

Yoruba: idagbon

Nigeria: idagbon (Yoruba)

Congo: gabo, lubanga, ompopo

Gabon: belemi, endone

Cameroon: akalan, djombe, adjeck, adjadjo, toboli

P. macroceras (K. Schum.) Beille

Cameroon: bitok, akela, djombe, wasara

Yoruba: abo idagbon

Nigeria: abo-idagbon (Yoruba); nikiba (Edo)

Congo: lokodiolo

Gabon: akeul

Zaire: tsanya

P. talbotii Wernham

Yoruba: idagbon, dake, nwerewere, wenrenwenren

Nigeria: ako-idagbon (Yoruba)

Pavate Adans. Rubiaceae

Origins:
See *Pavetta* L.

Pavetta L. Rubiaceae

Origins:
A Malayalam (Sinhalese) vernacular plant name for *Pavetta indica*, see van Rheede tot Draakestein (1637-1691), *Hortus Indicus Malabaricus*. 5: t. 10. 1685; see Carl Linnaeus, *Species Plantarum*. 110. 1753 and *Genera Plantarum*. Ed. 5. 49. 1754.

Species/Vernacular Names:
P. arenosa Lour.

English: sand pavetta

P. crassipes K. Schum.

Mali: kumbafura, kumu

P. edentula Sond.

English: gland-leaf tree, large-leaved bride's bush

Southern Africa: kliertjiesboom, isimuncwane; isiMuncwane, isaMunyane, uMafayindlala, umFayindlala (Zulu)

P. eylesii S. Moore (for the Rhodesian (b. Bristol) botanist Frederick Eyles, 1864-1937 (d. Salisbury, Rhodesia), author of "A record of plants collected in Southern Rhodesia. Arranged on Engler's system." *Trans. R. Soc. S. Afr.* 5: 273-564. 1916; see Ray Desmond, *Dictionary of British & Irish Botanists and Horticulturists*. 238. 1994)

South Africa: bushveld pavetta

P. gardeniifolia A. Rich. var. **gardeniifolia** (*Pavetta assimilis* Sond. var. *assimilis*; *Pavetta assimilis* Sond. var. *glabra* Brem.; *Pavetta assimilis* Sond. var. *glabra-brevituba* Brem.)

English: common bride's bush, kaffir bride

Southern Africa: gewone bruidsbos; isiNyombolo, isAnyane, isaNywane (Zulu); mmilorotswans (Hebron dialect, central Transvaal)

P. gerstneri Brem. (the specific epithet after the German Rev. Jacob Gerstner, 1888-1948, Roman Catholic missionary and botanist, plant collector, from 1928 to 1942 he was Superior of Mission Farms in Zululand, author of "A preliminary check list of Zulu names of plants." *Bantu Stud.* 12: 251-236, 321-342. 1938; *loc.cit.* 13: 49-64, 131-149, 307-326. 1939; *loc.cit.* 15: 277-301, 369-383. 1941; see John Wesley Haley, *Life in Mozambique and South Africa*. Free Methodist Publishing House, Chicago 1926; Samuel Drew, *The Life of the Rev. Thomas Coke, LL.D.* London 1817; Filipe Gastão de Almeida de Eça, *Subsídio para uma bibliografia missionária moçambicana (católica)*. N.p. 1969; Gilbert Westacott Reynolds, *The Aloes of South Africa*. Balkema, Rotterdam 1982)

English: Zulu bride's bush

South Africa: Zoeloebruidsbos

P. harborii S. Moore (*Pavetta marlothii* Brem.)

English: pavetta

South Africa: pavetta-bossie, gousiekte pavetta, tonnabossie

P. inandensis Brem. (*Pavetta rattrayi* Brem.)

English: forest bride's bush

Southern Africa: bosbruidsbos; umHleza (Xhosa)

P. indica L.

English: white pavetta

Malaya: bunga jarum, jarum jarum, jarum jarum padang, jarum paya, jarum puteh, menjarum, nyarum nyarum, nyarong, angsoka, gading hutan, gading galoh, gading gading, pechah periok puteh, serau lipis, serungkok

P. kotzei Brem. (after J.J. Kotze, 1892-1967, author of "Afforestation in arid and semi-arid regions." *Bull. S. Afr. Dept. Agric.* (Pan-Afr. Agric. and Vet. Conf. Papers, Agric. Sect. Pretoria): 112-114. 1929)

English: glossy bride's bush

South Africa: blinkbruidsbos

P. lanceolata Eckl. (*Pavetta alexandrae* Brem.; *Pavetta tristis* Brem.)

English: weeping bride's bush, kaffir bride, Christmas tree, Christmas bush

Southern Africa: treurbruidsbos, kafferbruid, kermisboom; mufhanza, tshituku (Venda); umDleza, umDlezi, umHleza, umSunu wembuzi, umSunumbuzi, iGololembuzi (Zulu); umPhonyana, umPonyane, umHleza, umDlesa (Xhosa)

P. natalensis Sond. (*Pavetta bowkeri* Harv. var. *glabra* Brem.; *Pavetta suluensis* Brem.)

English: Natal bride's bush, Natal pavetta

Southern Africa: Natalbruidsbos; isiCeza (Xhosa)

P. oblongifolia (Hiern) Bremek.

Mali: warasakuman

P. revoluta Hochst. (*Pavetta obovata* E. Meyer ex Sonder)

English: dune bride's bush, coastal pavetta

Southern Africa: duinebruidsbos; umHlabambazo, umHlabambaza (Zulu); umCilikishe, usiKolpati (Xhosa)

P. schumanniana F. Hoffm. ex K. Schum. (for the German botanist Karl Moritz Schumann, 1851-1904, botanical collector, taxonomist, contributor to C.F.P. von Martius *Flora Brasiliensis*, to H.G.A. Engler *Plantae Marlothianae*, to Engler and K.A.E. Prantl *Die Natürlichen Pflanzenfamilien*, to Engler *Das Pflanzenreich*, etc. See Hermann Baum, *Kunene-Sambesi-Expedition H. Baum.* 1903 ... Herausgegeben von ... O. Warburg. Berlin 1903; John H. Barnhart, *Biographical Notes upon Botanists.* 3: 247. 1965; T.W. Bossert, *Biographical Dictionary of Botanists Represented in the Hunt Institute Portrait Collection.* 356. 1972; Ida Kaplan Langman, *A Selected Guide to the Literature on the Flowering Plants of Mexico.* 687-688. University of Pennsylvania Press, Philadelphia 1964; E.M. Tucker, *Catalogue of the Library of the Arnold Arboretum of Harvard University.* Cambridge, Massachusetts 1917-1933; Stafleu and Cowan, *Taxonomic Literature.* 5: 400-408. 1985; Gordon Douglas Rowley, *A History of Succulent Plants.* Strawberry Press, Mill Valley, California 1997; R. Zander, F. Encke, G. Buchheim and S. Seybold, *Handwörterbuch der Pflanzennamen.* 14. Aufl. Stuttgart 1993)

English: poison bride's bush, poison pavetta

Southern Africa: gifbruidsbos, gousiekte tree, tree gousiekte, gousiekteboom, boom gousiekte; mugaramondoro, chiFikau, chiFukawi, Nyapuntu, chiSwimbovarisi, muTandarombo, chiTunguru (Shona); isiMbuzana, uSawoti (Zulu); uSawoti (Swazi); tshituku, mukhobekwa, mukhobigwa (Venda)

P. zeyheri Sond. (*Pavetta dissimilis* Brem.; *Pavetta inconspicua* Dinter ex Brem.; *Pavetta lasiopeplus* K. Schum.; *Pavetta microlancea* K. Schumann; *Pavetta middelburgensis* Brem.; *Pavetta pseudo-zeyheri* Brem.) (for the German botanist and botanical collector Carl (Karl) Ludwig Philipp Zeyher, 1799-1858 (d. Cape Town); see Stafleu and Cowan, *Taxonomic Literature.* 7: 534-535. 1988; Mary Gunn and Leslie E. Codd, *Botanical Exploration of Southern Africa.* Cape Town 1981; J.H. Barnhart, *Biographical Notes upon Botanists.* 3: 540. 1965; C.F. Ecklon, "Nachricht über die von Ecklon und Zeyher unternommenen Reisen und deren Ausbeute in botanischer Hinsicht." *Linnaea.* 8: 390-400. 1833; H.N. Clokie, *Account of the Herbaria of the Department of Botany in the University of Oxford.* Oxford 1964; Antoine Lasègue, *Musée botanique de M. Benjamin Delessert.* 1845; Ethelyn Maria Tucker, *Catalogue of the Library of the Arnold Arboretum of Harvard University.* Cambridge, Massachusetts 1917-1933; Alain White and Boyd Lincoln Sloane, *The Stapelieae.* Pasadena 1937; R. Zander, F. Encke, G. Buchheim and S. Seybold, *Handwörterbuch der Pflanzennamen.* 14. Aufl. Stuttgart 1993)

English: tshitabanna tree

Southern Africa: tshitabanna (= too much for men), matiadule (western Transvaal, northern Cape, Botswana)

Pavieasia Pierre Sapindaceae

Origins:

A genus from Southeast Asia.

Pavonia Cav. Malvaceae

Origins:

For the Spanish botanist José Antonio Pavón y Jiménez, 1754-1844, traveler, explorer, between 1777-1788 he traveled with Hipolito Ruíz Lopez (1754-1815) and Joseph Dombey in Chile and Peru. His works include *Disertacion botanica sobre los generos Tovaria, Actinophyllum, Araucaria y Salmia.* [Madrid 1797], with H. Ruíz wrote *Flora peruvianae, et chilensis prodromus.* Madrid 1794, *Systema vegetabilium florae peruviana et chilensis.* [Madrid] 1798 and *Flora peruviana, et chilensis.* Madrid 1798-1802; see John H. Barnhart, *Biographical Notes upon Botanists.* 3: 57. 1965; Stafleu and Cowan, *Taxonomic Literature.* 4: 117-118, 981-986. 1983; S. Lenley et al., *Catalog of the Manu-*

script and Archival Collections and Index to the Correspondence of John Torrey. Library of the New York Botanical Garden. 466. 1973; E.M. Tucker, *Catalogue of the Library of the Arnold Arboretum of Harvard University.* 1917-1933; Miguel Colmeiro y Penido (1816-1901), *La Botánica y los Botánicos de la Peninsula Hispano-Lusitana.* Madrid 1858; Antonio José Cavanilles (1745-1804), *Monadelphiae classis dissertationes decem.* Matriti 1786-1787; G. Murray, *History of the Collections Contained in the Natural History Departments of the British Museum.* 1: 173. London 1904; Frans A. Stafleu, *Linnaeus and the Linnaeans.* The spreading of their ideas in systematic botany, 1735-1789. Utrecht 1971; August Weberbauer, *Die Pflanzenwelt der peruanischen Andes in ihren Grundzügen dargestellt.* 2-4. Leipzig 1911; H.N. Clokie, *Account of the Herbaria of the Department of Botany in the University of Oxford.* 222. Oxford 1964; Blanche Elizabeth Edith Henrey (1906-1983), *British Botanical and Horticultural Literature before 1800.* Oxford 1975; A. Lasègue, *Musée botanique de Benjamin Delessert.* Paris 1845; R.E.G. Pichi Sermolli, "Le collezioni cedute da J. Pavon a F.B. Webb e conservate nell'Herbarium Webbianum." *Nuovo Giorn. Bot. Ital.* ser. 2. 56(4): 699-701. 1950 [1949]; R. Zander, F. Encke, G. Buchheim and S. Seybold, *Handwörterbuch der Pflanzennamen.* 14. Aufl. Stuttgart 1993; Paul A. Fryxell, "The genus *Pavonia* Cav. (Malvaceae: Malvavisceae) in Australia." *Nuytsia.* 6(3): 305-308. 1988.

Species/Vernacular Names:
P. sp.
Peru: algodãorana, malva rosada
P. hastata Cav.
English: pink pavonia
P. leucantha Garcke
Peru: charapilla huatana, mashu shillo, mushisillo, mushu sillo, yerba del monte, yopixrisewatapitsa
P. mollis Kunth
Peru: nihi xova
P. paniculata Cav.
Peru: malva-malva, malva masha
P. spinifex (L.) Cavanilles (*Hibiscus spinifex* L.)
Peru: anguia, cuerno de venado, taroca-asta

Paxia Gilg Connaraceae

Origins:
Named for the German botanist Ferdinand Albin Pax, 1858-1942.

Paxia O.E.G. Nilsson Portulacaceae

Origins:
After the German botanist Ferdinand Albin Pax, 1858-1942, collaborator with A. Engler, in 1893 succeeded Karl Anton Eugen Prantl (1849-1893) at Breslau, 1893-1925 professor of botany and Director of the Botanical Garden at Breslau. Among his many works are *Beitrag zur Kenntnis des Ovulums* von *Primula elatior* Jacq. und *officinalis* Jacq. Breslau 1882 and *Pflanzengeographie von Rumänien.* Halle 1919, contributed to H.G.A. Engler & K.A.E. Prantl *Die Natürlichen Pflanzenfamilien* and to Engler *Das Pflanzenreich.* See T.W. Bossert, *Biographical Dictionary of Botanists Represented in the Hunt Institute Portrait Collection.* 303. 1972; R. Zander, F. Encke, G. Buchheim and S. Seybold, *Handwörterbuch der Pflanzennamen.* 14. Aufl. Stuttgart 1993; Adolf Friedrich Georg Ernst Albert Eduard, Duke of Mecklenburg (b. 1882), *Wissenschaftliche Ergebnisse der Deutschen ZentralAfrika-Expedition 1907-1908,* unter Führung Adolf Friedrichs ... Band ii. *Botanik.* Leipzig 1914; Charles Jacques Édouard Morren (1833-1886), *Correspondance botanique.* Liège 1884; Orjan Eric Gustaf Nilsson (1933-), in *Botaniska Notiser.* 119: 274. 1966; Ida Kaplan Langman, *A Selected Guide to the Literature on the Flowering Plants of Mexico.* 571. Philadelphia 1964; E.M. Tucker, *Catalogue of the Library of the Arnold Arboretum of Harvard University.* 1917-1933.

Paxiactes Raf. Umbelliferae

Origins:
See C.S. Rafinesque, *The Good Book.* 60. 1840; E.D. Merrill, *Index rafinesquianus.* 182. 1949.

Paxiodendron Engler Monimiaceae

Origins:
After the German botanist Ferdinand Albin Pax, 1858-1942, collaborator of A. Engler.

Paxistima Raf. Celastraceae

Origins:
Greek *pachys* "thick, stout" and *stigma* "stigma"; see C.S. Rafinesque, *Sylva Telluriana.* 42. 1838.

Species/Vernacular Names:
P. myrsinites (Pursh) Raf.
English: Oregon boxwood

Paxiuscula Herter Euphorbiaceae

Origins:

After the German botanist Ferdinand Albin Pax, 1858-1942.

Paxtonia Lindley Orchidaceae

Origins:

For the British (b. Beds.) horticulturist Sir Joseph Paxton, 1803-1865 (Kent), gardener, architect, designer, landscape gardener and horticultural architect, publisher of *Paxton's Magazine of Botany*. London [1833] 1834-1849, he was chosen a fellow of the Horticultural Society, he became a Fellow of the Linnean Society in 1831, knighted 1851, with John Lindley (1799-1865) wrote *Paxton's Flower Garden*. London [1850]-1853. See M. Hadfield et al., *British Gardeners: A Biographical Dictionary*. London 1980; Mea Allan, *The Hookers of Kew*. London 1967; J.H. Barnhart, *Biographical Notes upon Botanists*. 3: 58. 1965; Ida Kaplan Langman, *A Selected Guide to the Literature on the Flowering Plants of Mexico*. University of Pennsylvania Press, Philadelphia 1964; Ethelyn Maria Tucker, *Catalogue of the Library of the Arnold Arboretum of Harvard University*. Cambridge, Massachusetts 1917-1933; T.W. Bossert, *Biographical Dictionary of Botanists Represented in the Hunt Institute Portrait Collection*. 303. 1972; A. Lasègue, *Musée botanique de Benjamin Delessert*. 529. Paris 1845; H.R. Fletcher, *Story of the Royal Horticultural Society, 1804-1968*. Oxford 1969; Kenneth Lemmon, *The Golden Age of Plant Hunters*. London 1968; R. Zander, F. Encke, G. Buchheim and S. Seybold, *Handwörterbuch der Pflanzennamen*. 14. Aufl. Stuttgart 1993; Merle A. Reinikka, *A History of the Orchid*. Timber Press 1996; Ray Desmond, *Dictionary of British & Irish Botanists and Horticulturists*. 540-541. London 1994.

Payena A. DC. Sapotaceae

Origins:

After the French (b. Paris) chemist Anselme Payen, 1795-1871 (Paris), author of *Manuel de cours de chimie organique appliquée aux arts industriels et agricoles*. Paris 1842-1843; see J.-A. Barral, in *Mémoires publiés par la Société centrale d'agriculture de France*. 67-87. 1873; W.V. Farrar, in *Dictionary of Scientific Biography* 10: 436. 1981.

Species/Vernacular Names:

P. leerii (Teijsm. & Binn.) Kurz

Malaya: semaram, sundek, balam sundek, nyatoh burong, sundek burong

Payera Baillon Rubiaceae

Origins:

After the French botanist Jean-Baptiste Payer, 1818-1860, naturalist, physician, M.D. Paris 1852, bryologist, wrote *Familles naturelles des plantes*. Paris 1848. See J.H. Barnhart, *Biographical Notes upon Botanists*. 3: 58. 1965; Ethelyn Maria Tucker, *Catalogue of the Library of the Arnold Arboretum of Harvard University*. Cambridge, Massachusetts 1917-1933; Stafleu and Cowan, *Taxonomic Literature*. 4: 125-126. 1983.

Payeria Baill. Meliaceae

Origins:

After the French botanist Jean-Baptiste Payer, 1818-1860, physician, M.D. Paris 1852.

Pearsonia Dümmer Fabaceae

Origins:

For the English (b. Lincolnshire) botanist Henry Harold Welch Pearson, 1870-1916 (d. Wynberg, Cape Town, South Africa), Cambridge Herbarium, professor of botany (South African College, Capetown), plant collector and botanical explorer, 1901 Fellow of the Linnean Society, 1916 Fellow of the Royal Society, founder and Hon. Director of the National Botanic Gardens (Kirstenbosch, South Africa), edited *The Annals of the Bolus Herbarium*, specialist in *Welwitschia*. His writings include "Percy Sladen Memorial Expedition in South-West Africa, 1908-1909." *Nature*. vol. LXXXI. 1909 and "Itinerary of the Percy Sladen Memorial Expedition to the Orange River, 1910-1911." *Ann. S. Afr. Mus*. vol. IX. 1912; see A.C. Seward, "H.H.W. Pearson, F.R.S., Sc.D. (Cambridge)." in *The Annals of the Bolus Herbarium*. Vol. II, Part III: 131-147. July 1917; T.W. Bossert, *Biographical Dictionary of Botanists Represented in the Hunt Institute Portrait Collection*. 304. 1972; Ida Kaplan Langman, *A Selected Guide to the Literature on the Flowering Plants of Mexico*. 573. University of Pennsylvania Press, Philadelphia 1964; E.M. Tucker, *Catalogue of the Library of the Arnold Arboretum of Harvard University*. Cambridge, Massachusetts 1917-1933; Mary Gunn and Leslie Edward W. Codd, *Botanical Exploration of Southern Africa*. 275-276. A.A. Balkema, Cape Town 1981; Leonard Huxley, *Life and Letters of Sir Joseph Dalton Hooker*. London 1918; Ernest Nelmes and William Cuthbertson, *Curtis's Botanical Magazine Dedications, 1827-1927*. [1931]; A. White and B.L. Sloane, *The Stapelieae*. Pasadena 1937; Gordon Douglas Rowley, *A History of Succulent Plants*. Strawberry Press, Mill Valley, California 1997; Ray Desmond, *Dictionary of British & Irish Botanists and Horti-*

culturists. 542. London 1994; R. Zander, F. Encke, G. Buchheim and S. Seybold, *Handwörterbuch der Pflanzennamen.* 14. Aufl. 1993.

Peccana Raf. Euphorbiaceae

Origins:

See C.S. Rafinesque, *Sylva Telluriana.* 114, 115. 1838 and *Autikon botanikon.* Icones plantarum select. nov. vel rariorum, etc. 138. Philadelphia 1840; E.D. Merrill (1876-1956), *Index rafinesquianus.* The plant names published by C.S. Rafinesque, etc. 156. Jamaica Plain, Massachusetts, USA 1949.

Pecheya Scop. Rubiaceae

Origins:

For the English (b. Sussex) physician John Pechey (Pechy), 1654-1717 (Chichester, Sussex). His works include *A Collection of Chronical Diseases,* viz. the colick ...the hysteric diseases, the gout, etc. [From Sydenham, Riverius, etc.] London 1692, *The Compleat Herbal of Physical Plants.* London 1694, *Some Observations Made upon the Calumba Wood.* 1694, *The London Dispensatory.* London 1694, *Some Observations Made upon the Wood called Lignum Nephriticum.* London 1694, *A General Treatise of the Diseases of Infants and Children.* Collected from the best practical authors. London 1697 and *A Plain Introduction to the Art of Physick.* London 1697; see R. Pulteney, *Historical and Biographical Sketches of the Progress of Botany in England.* London 1790; William Munk, *The Roll of the Royal College of Physicians of London.* London 1878.

Pechuel-Loeschea O. Hoffm. Asteraceae

Origins:

After the German naturalist and geographer M. Eduard Pechuël-Loesche, 1840-1913 (d. Munich), plant collector in the Cape, Hereroland and Congo. Among his works are *Herrn Stanley's Partisane und meine offiziellen Berichte von Kongolande.* Leipzig 1886 and *Kongoland.* Jena 1887. See Richard P. Wilhelm Guessfeldt (b. 1840), *Die Loango-Expedition ausgesandt von der Deutschen Gesellschaft zur Erfoschung Aequatorial-Africas 1873-76 etc.* Abth. I, von P. Guessfeldt; II, von J. Falkenstein; III, von E. Pechuël-Loesche. Leipzig 1879-82; Otto Kuntze (1843-1907), "Plantae pechuelianae hereroenses." *Jb. K. Bot. Gart. Mus. Berl.* 4: 260-275. 1886; Alfred Edmund Brehm (1829-1884), *Dritte ... Auflage von ... Pechuel-Loesche.* 10 Bd. (The Mammalia, Aves, and Pisces edited by Pechuel-Loesche) Leipzig & Wien 1890-93; Hugo von Wobeser, *H.M. Stanley und Dr. Pechuël-Loesche.* Leipzig 1886; J.

Kiessling, *Untersuchungen über Dämmerungserscheinen* ... Mit neun Farbendrucktafeln nach Aquarellen von Prof. Dr. Pechuël-Loesche, etc. Hamburg und Leipzig 1888; M. Gunn and L.E.W. Codd, *Botanical Exploration of Southern Africa.* 276. Cape Town 1981.

Species/Vernacular Names:

P. leubnitziae (Kuntze) O. Hoffm. (*Pluchea leubnitziae* (Kuntze) N.E. Br.) (in honor of the maiden name of the wife of Pechuël-Loesche)

English: stinkbush

South Africa: bitterbossie, stinkbossie

Namibia: bitterbos (Afrikaans); omundumba (Herero); oshizimba (Ndonga); edimba (Kwanyama); autsi!khanneb (Nama/Damara)

German: bitterbusch, grauer stinkbusch

Peckeya Raf. Rubiaceae

Origins:

For the English physician John Pechey; see C.S. Rafinesque, *Ann. Gén. Sci. Phys.* 6: 87. 1820; E.D. Merrill, *Index rafinesquianus.* 226. 1949.

Peckia Vellozo Myrsinaceae

Origins:

Possibly dedicated to the American botanist William Dandridge Peck, 1763-1822, naturalist, entomologist, professor of natural history (Harvard). Among his writings are *Natural History of the Slug Worm.* Boston 1799, *A Catalogue of American and Foreign Plants Cultivated in the Botanic Garden, Cambridge, Massachusetts.* Cambridge 1818 and *Observationes Carpologicae in Kamelliam et Theam.* [Boston 1817]. See J.H. Barnhart, *Biographical Notes upon Botanists.* 3: 62. 1965; E.M. Tucker, *Catalogue of the Library of the Arnold Arboretum of Harvard University.* Cambridge, Massachusetts 1917-1933; T.W. Bossert, *Biographical Dictionary of Botanists Represented in the Hunt Institute Portrait Collection.* 304. 1972; William Darlington (1782-1863), *Reliquiae Baldwinianae.* Philadelphia 1843; Jeannette E. Graustein, *Thomas Nuttall, Naturalist. Explorations in America, 1808-1841.* Harvard University Press 1967; J. Ewan, ed., *A Short History of Botany in the United States.* 1969; H.A. Kelly & W.L. Burrage, *Dictionary of American Medical Biography.* New York 1928.

Peckoltia Fourn. Asclepiadaceae

Origins:

After the German botanist Theodor (Theodoro) Peckolt, 1822-1912, in Brazil 1847, botanical explorer, plant collector, pharmacist. His writings include "Plantas medicinais e úteis do Brasil." *R. Flora Medicinal*. 3(4): 203-214. Rio de Janeiro (Jan.) 1937 and *Historia das plantas alimentares*. Rio de Janeiro 1871-1884. See J.H. Barnhart, *Biographical Notes upon Botanists*. 3: 62. 1965; E.M. Tucker, *Catalogue of the Library of the Arnold Arboretum of Harvard University*. Cambridge, Massachusetts 1917-1933; Frederico Carlos Hoehne, M. Kuhlmann and Oswaldo Handro, *O jardim botânico de São Paulo*. 1941.

Pecluma M.G. Price Polypodiaceae

Origins:
From the Latin *pectinatus, a, um* "combed" and *plumula, ae* "a little feather," possibly referring to the leaf blades.

Species/Vernacular Names:
P. dispersa (A.M. Evans) M.G. Price (*Polypodium dispersum* A.M. Evans)

English: widespread polypody

P. plumula (Humboldt & Bonpland ex Willdenow) M.G. Price (*Polypodium plumula* Humboldt & Bonpland ex Willdenow)

English: plume polypody

Pectanisia Raf. Resedaceae

Origins:
See C.S. Rafinesque, *Flora Telluriana*. 3: 72, 73. 1836 [1837]; E.D. Merrill, *Index rafinesquianus*. 132. 1949.

Pectantia Raf. Saxifragaceae

Origins:
See C.S. Rafinesque, *Flora Telluriana*. 2: 73. 1836 [1837]; E.D. Merrill, *Index rafinesquianus*. 135. 1949.

Pecteilis Raf. Orchidaceae

Origins:
From the Latin *pecten, pectinis* "a comb," Greek *pekteo, peko* "to comb," referring to the side lobes of the lip; see C.S. Rafinesque, *Flora Telluriana*. 2: 37. 1836 [1837]; E.D. Merrill, *Index rafinesquianus*. 104. 1949.

Pectiantia Raf. Saxifragaceae

Origins:
See C.S. Rafinesque, *Flora Telluriana*. 2: 72. 1836 [1837]; E.D. Merrill, *Index rafinesquianus*. 135. 1949.

Pectinaria (Benth.) Hack. Gramineae

Origins:
Latin *pecten, pectinis* "a comb," *pectinarius, a, um* "belonging to combs."

Pectinaria Cordemoy Orchidaceae

Origins:
From *Mystacidium pectinatum*, the type species.

Pectinaria Haw. Asclepiadaceae

Origins:
Latin *pecten, pectinis* "a comb," *pectinarius, a, um* "belonging to combs," referring to the appearance of the outer corona edge.

Pectinella J.M. Black Cymodoceaceae (Potamogetonaceae)

Origins:
Diminutive of the Latin *pecten, pectinis* (*pecto* "to comb") "a comb"; see John McConnell Black (1855-1951), *Transactions and Proceedings of the Royal Society of South Australia*. 37: 1. 1913.

Pectis L. Asteraceae

Origins:
Latin *pecten, pectinis* (*pecto* "to comb") "a comb," referring to the marginally bristled leaves or to the form of the pappus.

Species/Vernacular Names:
P. papposa Harvey & A. Gray

English: chinch-weed

Pectocarya DC. ex Meissner Boraginaceae

Origins:
Greek *pekteo, pektein, peko* "to comb" and *karyon* "nut," in some species nutlets with dentate margins.

Pedaliodiscus Ihlenf. Pedaliaceae

Origins:
From the Greek *pedalion, pedon* "oar-blade, rudder" and *diskos* "disc."

Pedaliophyton Engl. Pedaliaceae

Origins:

From the Greek *pedalion* "rudder, steering-paddle" and *phyton* "plant."

Pedalium Royen ex L. Pedaliaceae

Origins:

Greek *pedalion* "rudder, steering-paddle," referring to the angles of the fruits, Latin *pedalion, ii* for a plant, called also *proserpinaca.*

Peddiea Harvey Thymelaeaceae

Origins:

Named after the plant collector John Peddie (d. 1840, Ceylon, Newara), soldier, he sent South African plants to William H. Harvey at Dublin; the German Adolphus Kummer, d. 1817 in Guinea (Rio Nunez), in 1815 was a naturalist on Major Peddie's expedition into Guinea and Senegal. See *Bonplandia* 10: 353. 1862; William Gray and Dochard [Staff surgeon], *Travels in Western Africa in the Years 1819-21, from the River Gambia ... to the River Niger...* London 1825; Mary Gunn and Leslie Edward W. Codd, *Botanical Exploration of Southern Africa.* 276-277. Cape Town 1981; F.N. Hepper and Fiona Neate, *Plant Collectors in West Africa.* 46. 1971; James Britten (1846-1924) and George E. Simonds Boulger (1853-1922), *A Biographical Index of Deceased British and Irish Botanists.* London 1931.

Species/Vernacular Names:

P. africana Harv.

English: poison olive, fiber-bark, green flower, fiber-bark tree

Southern Africa: gifolyf, sterkbas, sterkbos, trekbas; inTozane, inTozani, isiFufu (Xhosa); chiBurawaringu, chiBurawuronga, chiRinga, ruSitu (Shona); uSanginde, inTozwana-yehlathi, isiFufufu, inTozane, inTozwane, uSinga-olusalugazi (Zulu); mukalakata (Venda)

Pederia Raf. Rubiaceae

Origins:

For *Paederia* L.; see C.S. Rafinesque, *Ann. Gén. Sci. Phys.* 6: 86. 1820.

Pederlea Raf. Solanaceae

Origins:

See Constantine Samuel Rafinesque, *Sylva Telluriana.* 54. 1838; E.D. Merrill, *Index rafinesquianus.* 212. Jamaica Plain, Massachusetts, USA 1949.

Pedicellaria Schrank Capparidaceae (Capparaceae)

Origins:

Latin *pedicellus* "a little louse," *pediculus* "a little foot, the foot-stalk or pedicle of a fruit or leaf," *pediculus, pedis* "a louse."

Pedicellarum M. Hotta Araceae

Origins:

Latin *pediculus* "a little foot, the foot-stalk or pedicle of a fruit or leaf," flowers pedicellate; see D.H. Nicolson, "Derivation of aroid generic names." *Aroideana.* 10: 15-25. 1988.

Pediculariopsis Á. Löve & D. Löve Scrophulariaceae

Origins:

Resembling *Pedicularis* L.

Pedicularis L. Scrophulariaceae

Origins:

From the Latin *pediculus* "a louse," *pedicularis* "relating to lice," the plants were supposed to become lice when sheep contacted them, referring to the belief that ingestion by stock promoted lice infestation, to produce lice in sheep; Latin *herba pedicularis,* lousewort, so called because it kills lice.

Species/Vernacular Names:

P. attollens A. Gray

English: little elephant's head

P. canadensis L.

English: forest lousewort, lousewort, wood-betony, common lousewort

P. centranthera A. Gray

English: dwarf lousewort

P. crenulata Benth.

English: scallop-leaved lousewort

P. densiflora Hook.

English: Indian warrior

P. dudleyi Elmer

English: Dudley's lousewort

P. groenlandica Retz.

English: elephant's head

P. howellii A. Gray

English: Howell's lousewort

P. lanceolata Michx.

English: swamp lousewort

P. racemosa Hook.

English: leafy lousewort

P. sylvatica L.

English: small lousewort

Pedilanthus Necker ex Poit. Euphorbiaceae

Origins:

From the Greek *pedilon* "slipper, sandal" and *anthos* "flower," referring to the shape and appearance of the flowers.

Species/Vernacular Names:

P. tithymaloides (L.) Poit. (*Euphorbia tithymaloides* L.) (resembling the genus *Tithymalus*)

English: bird cactus, Japanese poinsettia, redbird flower, redbird cactus, ribbon cactus, slipper flower, slipper plant, Jewbush, devil's backbone, zigzag plant

Japan: gin-ryu

Malaya: pokok lipan, penawar lipan

India: airi, baire, agia

Yoruba: aperejo

Pedilea Lindley Orchidaceae

Origins:

Greek *pedilon* "slipper, sandal, shoe," the shape of the lip.

Pedilochilus Schltr. Orchidaceae

Origins:

Greek *pedilon* "slipper, sandal" and *cheilos* "lip," slipper-shaped lip.

Pedilonum Blume Orchidaceae

Origins:

Greek *pedilon* "slipper, sandal," referring to the form of the perianth, to the lateral sepals.

Pedinogyne Brand Boraginaceae

Origins:

Greek *pedinos* "flat, level," *pedion* "a plain, flat, open country, a field" and *gyne* "woman, pistil," referring to the nature of the pistil.

Pedinopetalum Urban & H. Wolff Umbelliferae

Origins:

From the Greek *pedinos* "flat, level" and *petalon* "petal, leaf."

Pediocactus Britton & Rose Cactaceae

Origins:

Greek *pedion* "flat, plain, field" plus *cactus*, referring to the habitat, Great Plains, U.S.A.

Pediomellum Rydb. Fabaceae

Origins:

Greek *pedion* "flat, plain, field" and *melon* "an apple, fruit."

Pediomelum Rydb. Fabaceae

Origins:

From the Greek *pedion* "flat, plain, field" and *melon* "an apple, fruit."

Species/Vernacular Names:

P. californicum (S. Watson) Rydb.

English: Indian breadroot

P. esculentum (Pursh) Rydb.

English: breadnut

Pedistylis Wiens Loranthaceae

Origins:

Greek *pedion* (*pedon* "the ground, earth") "flat, plain, field" and *stylos* "pillar, style."

Species/Vernacular Names:

P. galpinii (Schinz ex Sprague) Wiens (*Loranthus galpinii* Schinz ex Sprague; *Emelianthe galpinii* (Schinz ex Sprague) Danser) (the type specimen was collected near

Barberton in the eastern Transvaal in April 1890 by Ernest Edward Galpin, naturalist and botanist)

English: mistletoe

Peekelia Harms Fabaceae

Origins:
For the German Catholic priest Gerhard Peekel, 1876-1949, botanist, plant collector, for many years missionary in New Ireland (Bismarck Archipelago, a chain of islands off the northeast coast of New Guinea), author of *Religion und Zauberei auf dem mittleren Neu-Mecklenburg Bismarck-Archipel, Südsee.* Münster 1910 and *Flora of the Bismarck Archipelago for Naturalists.* English translation by E.E. Henty. Office of Forests, Division of Botany, Lae, Papua New Guinea. 1984; see Frederick B. Essig, "A checklist and analysis of the palms of the Bismarck Archipelago." *Principes.* Volume 39, no. 3: 123-129. 1995.

Peekeliodendron Sleumer Icacinaceae

Origins:
For the German botanist Gerhard Peekel, 1876-1949.

Peekeliopanax Harms Araliaceae

Origins:
For the German botanist Gerhard Peekel, 1876-1949.

Peersia L. Bolus Aizoaceae

Origins:
After the Australian (b. NSW) Victor Stanley Peers, 1874-1940 (d. South Africa, Cape Town), collected succulents plants and bulbs; see Gordon Douglas Rowley, *A History of Succulent Plants.* Strawberry Press, Mill Valley, California 1997; Mary Gunn and Leslie Edward W. Codd, *Botanical Exploration of Southern Africa.* 277. Cape Town 1981.

Pegaeophyton Hayek & Hand.-Mazz. Brassicaceae

Origins:
Possibly from the Greek *pegas* "thick, hard, solid, a rock" and *phyton* "plant," some suggest from Pegae (Pegai), a fountain in Bithynia.

Pegamea Vitman Rubiaceae

Origins:
Referring to the genus *Pagamea* Aublet.

Peganum L. Zygophyllaceae

Origins:
Ancient Greek name *peganon* for rue, *Ruta graveolens* (Theophrastus), Latin *peganon, i* for garden-rue or wild-rue (Greek *peganon oreinon*); see Carl Linnaeus, *Species Plantarum.* 444. 1753 and *Genera Plantarum.* Ed. 5. 204. 1754.

Species/Vernacular Names:
P. harmala L.
English: harmal, African rue, Syrian rue
Arabic: harmel, harmal, harmel sahari

Pegia Colebr. Anacardiaceae

Origins:
Greek and Latin *pege* "a source, spring, origin, fountain, stream."

Peglera Bolus Erythroxylaceae

Origins:
For Alice Marguerite Pegler, 1861-1929 (d. Umtata, Transkei), plant collector; see Mary Gunn and Leslie Edward W. Codd, *Botanical Exploration of Southern Africa.* 277-278. Cape Town 1981; Gilbert Westacott Reynolds, *The Aloes of South Africa.* 161. Balkema, Rotterdam 1982.

Peirescia Zucc. Cactaceae

Origins:
After the French numismatist Nicholas (Nicolas) Claude Fabry (Fabri) de Peiresc, 1580-1637, physician, antiquary, patron of botany and letters, naturalist and archaeologist; see also *Pereskia* Mill.

Peireskia Steud. Cactaceae

Origins:
After the French numismatist Nicholas (Nicolas) Claude Fabry (Fabri) de Peiresc, 1580-1637, physician, antiquary, patron of botany and letters, naturalist and archaeologist; see Requier, *Vie de Nicolas Claude Peiresc*, Conseiller au

Parlement de Provence. Paris 1770; [Nicolas Claude de Fabri de Peiresc], *Lettres inédites.* Paris, Sajou, Aix, Pontier 1815-1816; Jacqueline Hellin & Andrée Willems, *Nicolas-Claude Fabri de Peiresc. 1580-1637.* Bruxelles 1980.

Peireskiopsis Vaupel Cactaceae

Origins:
Resembling *Peireskia.*

Peixotoa A. Juss. Malpighiaceae

Origins:
Possibly after the Brazilian physician Domingos Ribeiro Dos Guimaraens Peixoto, author of *Dissertation ... sur les médicamens brésiliens.* Paris 1830 and *Projecto de Estatutos para a Escola de Medicina* do Rio de Janeiro. Rio de Janeiro 1836.

Pelagodendron Seemann Rubiaceae

Origins:
Greek *pelagos* "the sea" and *dendron* "tree," New Caledonia and Fiji.

Pelagodoxa Beccari Palmae

Origins:
Greek *pelagos* "the sea" and *doxa* "glory," only one species, confined to the Marquesas Islands, Nuku Hiva.

Pelargonium L'Hérit. Geraniaceae

Origins:
Greek *pelargos* "a stork," referring to the mericarp (derivation perhaps from the Greek *pelios* "black, dark" and *argos* "white, whitish," probably from the Akkadian *bel-*, *pel-* plus *arhu*: *belu* (*pe-lu*) "lord" and *arhu* "road," Hebrew *orho* "way, wanderer, caravan"); see William Aiton (1731-1793), *Hortus Kewensis.* 2: 417. London 1789; Giovanni Semerano, *Le origini della cultura europea.* Dizionario della lingua Greca. 2(1): 225. Leo S. Olschki Editore, Firenze 1994.

Species/Vernacular Names:
P. abrotanifolium (L.f.) Jacq.

English: southernwood geranium

P. x *asperum* Willd.

English: rose geranium

P. australe Willd.

English: Austral stork's bill, native stork's bill

Maori name: kopata

P. capitatum (L.) L'Hérit.

English: rose-scented geranium, wild pelargonium

P. caylae Humbert (the species was named after Mr. Léon Cayla, Governor-General of Madagascar from 1930 to 1939, a patron of scientific research)

P. chelidonium (Houtt.) DC. (*Geranium chelidonium* Houtt.; *Geranium trilobum* Thunb.; *Geraniospermum chelidonium* (Houtt.) Kuntze; *Pelargonium ficaria* Willd.; *Pelargonium trilobum* (Thunb.) DC.; *Pelargonium revolutum* (Andr.) Pers.; *Pelargonium meyeri* Harv.) (from the Greek *chelidon* "a swallow")

South Africa: speenkruid (= the Dutch name for pilewort, *Ranunculus ficaria* L.)

P. crispum (Berg.) L'Hérit.

South Africa: crisped-leaf pelargonium, lemon geranium

P. denticulatum Jacq.

English: pine geranium, fern-leaf geranium

P. x *domesticum* L. Bailey (*Pelargonium cucullatum* sensu J. Black)

English: regal pelargonium, garden geranium, pelargonium

P. drummondii Turcz.

English: Drummond's stork's bill

P. echinatum Curtis

English: cactus geranium, sweetheart geranium

P. ellaphieae E.M. Marais (in honor of the well-known South African botanical artist Johanna Ellaphie Ward-Hilhorst, 1920-1994 (d. Cape Town), interested in the genus *Pelargonium,* illustrator of Johannes Jacobus Adriaan Van der Walt (b. 1938) & Pieter Johannes Vorster (b. 1945), *Pelargoniums of Southern Africa.* Cape Town 1981, and of Deirdré Anne Snijman (b. 1949), *A Revision of the Genus Haemanthus L.* (Amaryllidaceae). Claremont 1984, in 1990 she was awarded of the Royal Horticultural Society's Gold Medal for her paintings of *Haemanthus*; see Gordon Douglas Rowley, *A History of Succulent Plants.* Strawberry Press, Mill Valley, California 1997; Mary Gunn and Leslie Edward W. Codd, *Botanical Exploration of Southern Africa.* 371. Cape Town 1981)

English: geranium

P. fragrans Willd.

English: nutmeg geranium

P. gibbosum (L.) L'Hérit.

English: gouty geranium, knotted geranium

P. glutinosum (Jacq.) L'Hérit.

English: pheasant's foot geranium

P. graveolens L'Hérit.

English: rose geranium, sweet-scented geranium

Arabic: attirchia

China: xiang ye

P. x *hortorum* L. Bailey

English: geranium, bedding geranium, zonal pelargonium, fish geranium

Japan: Tenjiku-aoi (= Indian *Althaea rosa*)

P. incrassatum (Andr.) Sims

South Africa: Namaqualand beauty, t'neitjie

P. inodorum Willd.

Maori name: kopata

P. odoratissimum (L.) L'Hérit.

English: apple geranium

P. panduriforme Ecklon & Zeyher

English: oak-leafed geranium

P. peltatum (L.) L'Hérit. (*Pelargonium lateripes* L'Hérit.)

English: ivy geranium, hanging geranium, ivy-leafed pelargonium

P. punctatum (Andrews) Willd. (*Geranium punctatum* Andrews; *Dimacria punctata* (Andrews) Sweet; *Geraniospermum punctatum* (Andrews) Kuntze) (the ligulate posterior petals have wine-red small dots)

English: spotted geranium

P. quercifolium (L.) L'Hérit.

English: oak-leaved geranium, almond geranium, village oak geranium

P. rapaceum (L.) L'Hérit.

South Africa: bergaartappel, norretjie

P. reniforme Curtis

Southern Africa: roois rabas; iYeza lezikhali, iKhubalo (Xhosa)

P. schizopetalum Sweet (*Pelargonium amatymbicum* (Eckl. & Zeyh.) Harv.) (from the Greek *schizo, schizein* "to split, to divide" and *petalum* or *petalon* "a petal," this species has split or fimbriate petals)

English: divided-petalled pelargonium, orchid pelargonium

South Africa: muishondbos, muishondbossie (= Afrikaans for skunkbush)

P. tetragonum (L.) L'Hérit.

English: square-stack cranesbill

P. triste (L.) L'Hérit. (*Pelargonium flavum* (L.) Ait.)

South Africa: aandblom, kaneelblom, rasmusbas

P. zonale (L.) L'Hérit.

English: horseshoe geranium

Japan: mon-tenjiku-aoi

Pelatantheria Ridley Orchidaceae

Origins:
From the Greek *pelates* "neighboring, a neighbor, approaching" and *anthera* "anther," referring to the column and the anther cap.

Pelea A. Gray Rutaceae

Origins:
Named in honor of Pele, the Hawaiian volcano goddess.

Species/Vernacular Names:
P. sp.

Hawaii: alani, alani kuahiwi

P. anisata H. Mann (*Evodia anisata* (H. Mann) Drake; *Pelea nodosa* H. Lév.; *Pelea subpeltata* H. Lév.)

Hawaii: mokihana, mokehana

P. barbigera (A. Gray) Hillebr. (*Melicope barbigera* A. Gray; *Evodia barbigera* (A. Gray) Drake)

Hawaii: uahiapele

P. clusiifolia A. Gray (*Evodia sapotaefolia* (H. Mann) Drake; *Evodia clusiaefolia* (A. Gray) Drake)

Hawaii: kukaemoa, kolokolo mokihana

P. cruciata A. Heller

Hawaii: pilo 'ula

P. elliptica (A. Gray) Hillebr. (*Evodia elliptica* (A. Gray) Drake)

Hawaii: leiohi'iaka, kaleiohi'iaka

P. hawaiensis Wawra (*Pelea brighamii* St. John)

Hawaii: mokihana kukae moa, manema

P. kavaiensis H. Mann (*Evodia kavaiensis* (H. Mann) Drake; *Pelea recurvata* Rock)

Hawaii: pilo 'ula

P. waialealae Wawra (*Evodia waialealae* (Wawra) Drake)

Hawaii: alani wai, 'anonia

Pelecostemon Leonard Acanthaceae

Origins:
From the Greek *pelekys* "an axe, two-edged axe" and *stemon* "stamen."

Pelecyphora Ehrenb. Cactaceae

Origins:
Greek *pelekys* "an axe, hatchet" and *phoros* "bearing," referring to the tubercles.

Pelexia Poit. ex Lindley Orchidaceae

Origins:
Greek *pelex* "helmet, crest," referring to the adnate dorsal sepal and petals.

Pelidnia Barnhart Lentibulariaceae

Origins:
From the Greek *pelidnos* "livid," *pelios* "a dark color, blackish, discolored"; see John H. Barnhart, in *Memoirs of the New York Botanical Garden*. 6: 50. 1916.

Peliosanthes Andrews Convallariaceae (Liliaceae)

Origins:
From the Greek *pelios* "livid, a dark color, blackish, discolored, purple" and *anthos* "flower."

Species/Vernacular Names:
P. sp.
Malaya: lumbah bukit, pinang lumbah

Peliostomum E. Meyer ex Benth. Scrophulariaceae

Origins:
Greek *pelios* "livid, dull, discolored" and *stoma* "mouth."

Peliotis E. Meyer Bruniaceae

Origins:
From the Greek *pelios* "livid, dull, dark" and *ous, otos* "ear."

Peliotus E. Meyer Bruniaceae

Origins:
Greek *pelios* and *ous, otos* "ear."

Pella Gaertner Moraceae

Origins:
Greek *pella, pelle* "a hide, a wooden bowl, a drinking cup"; Pella and Pelle "a city in Macedonia."

Pellacalyx Korth. Rhizophoraceae

Origins:
Greek *pella* "a hide, a wooden bowl, a drinking cup" and *kalyx, kalykos* "a calyx," referring to the shape of the flowers or to the hairy calyx.

Species/Vernacular Names:
P. axillaris Korth.
English: axilflower pellacalyx
Malaya: membuloh, bebuloh, buloh buloh, buloh, pianggu jantan

Pellaea Link Pteridaceae (Adiantaceae)

Origins:
Greek *pellos, pellaios* "dark, dusky," referring to the leaves or to the stalks.

Species/Vernacular Names:
P. falcata (R. Br.) Fée
English: sickle fern
P. myrtillifolia Mettenius ex Kuhn
Chile: coca, hierba coca, lendo del cerro
P. ternifolia (Cav.) Link
Chile: cusapi, yuquelahuén

Pellaeopsis J. Sm. Pteridaceae (Adiantaceae)

Origins:
Resembling the genus *Pellaea* Link.

Pellegrinia Sleumer Ericaceae

Origins:
Named for the French botanist François Pellegrin, 1881-1965. His writings include "*Walsura* nouveau du Tonkin." *Not. Syst.* 1: 227-229. 1910, "Notes sur les *Aglaia, Amoora* et *Lansium*." *Not. Syst.* 1: 284-290. 1910, "*Munronia* nouveau de l'Annam." *Not. Syst.* 2: 135-136. 1911, *La Flore du Mayombe* d'après les Récoltes de M. Georges Le Testu. [in *Mémoires de la Société Linnéenne de Normandie*. XXVI

volume. Two parts; collector Georges M.P.C. Le Testu, 1877-1967] Caen 1924-1928 and "*Perantha* Craib et *Oreocharis* Benth., Gesneracées du Yunnan." *Bull. Soc. Bot. France*. 72: 872-873. 1925. See J.H. Barnhart, *Biographical Notes upon Botanists*. 3: 64. 1965; E.M. Tucker, *Catalogue of the Library of the Arnold Arboretum of Harvard University*. Cambridge, Massachusetts 1917-1933; Paul Henri Lecomte, *Flore générale de l'Indo-Chine*. Paris 1944.

Species/Vernacular Names:
P. grandiflora (Ruíz & Pav.) Sleumer
Peru: uchu-uchu

Pellegriniodendron Léonard Caesalpiniaceae

Origins:
For the French botanist François Pellegrin, 1881-1965.

Pelleteria Poiret Primulaceae

Origins:
Orthographic variant of *Pelletiera*.

Pelletiera A. St.-Hil. Primulaceae

Origins:
For the French (b. Paris) botanist Pierre Joseph Pelletier, 1788-1842 (d. Paris), pharmacist, wrote *Essai sur la valeur des caractères physiques employés en minéralogie*. Thèse, etc. Paris 1812 and *Essai sur la nature des substances connues sous le nom de gommes résines*. Thèse, etc. Paris 1812, son of the French pharmacist Bertrand Pelletier (1761-1797); see John H. Barnhart, *Biographical Notes upon Botanists*. 3: 64. 1965; Alex Berman, in *Dictionary of Scientific Biography* (Editor in Chief Charles Coulston Gillispie.) 10: 497-499. 1981; R. Zander, F. Encke, G. Buchheim and S. Seybold, *Handwörterbuch der Pflanzennamen*. 14. Aufl. Stuttgart 1993; T.W. Bossert, *Biographical Dictionary of Botanists Represented in the Hunt Institute Portrait Collection*. 305. 1972; S. Lenley et al., *Catalog of the Manuscript and Archival Collections and Index to the Correspondence of John Torrey*. Library of the New York Botanical Garden. 321. 1973; E.M. Tucker, *Catalogue of the Library of the Arnold Arboretum of Harvard University*. 1917-1933.

Pelliceria Planchon & Triana Pellicieraceae

Origins:
See *Pelliciera*.

Pelliciera Planchon & Triana ex Benth. Pellicieraceae

Origins:
Probably from the Latin *pellicius* "made of skins," referring to the leathery leaves and dry and leathery fruits, or *pellicio, lexi, lectum* "to allure, decoy, to draw away the fruits of another's land to one's own," indicating the fruit.

Species/Vernacular Names:
P. rhizophorae Planchon & Triana
Colombia: piñuelo, mangle de buenaventura

Pellionia Gaudich. Urticaceae

Origins:
Named for the French Admiral Alphonse Odet Pellion, 1796-1868, a companion of the French navigator Louis de Freycinet on his second voyage around the world in 1817 to 1820; see Louis de Freycinet, *Voyage autour du Monde entrepris par ordre du Roi ... sur les corvettes de S.M. L'Uranie et La Physicienne, pendant ... 1817-1820*. (Atlas historique par ... A. Pellion, etc.) Paris 1826[-1830].

Species/Vernacular Names:
P. pulchra N.E. Br.
English: rainbow vine
P. repens (Lour.) Merrill (*Pellionia daveauana* (Carr.) N.E. Br.)
English: trailing watermelon begonia
Malaya: sisek naga

Pelma Finet Orchidaceae

Origins:
Greek *pelma, pelmatos* "sole of the foot, stalk, sole of sandal," the form of the column-foot.

Pelonastes Hook.f. Haloragaceae

Origins:
Greek *pelos, palos* "mud, clay, earth" and *naster, nastes* "inhabitant," referring to the habitat.

Pelozia Rose Onagraceae

Origins:
Anagram of *Lopezia* Cav.

Peltactila Raf. Umbelliferae

Origins:
See C.S. Rafinesque, *New Fl. N. Am.* 4: 28. 1836 [1838] and *The Good Book.* 55. 1840; E.D. Merrill, *Index rafinesquianus.* 182. 1949.

Peltaea (C. Presl) Standley Malvaceae

Origins:
From the Greek *pelte* "a shield, target," Latin *pelta, ae* "a small shield, a light shield in the shape of a half-moon."

Peltandra Raf. Araceae

Origins:
Greek *pelte* "a shield, target" and *aner, andros* "stamen, male," Latin *pelta, ae* for a small, light shield in the shape of a half-moon; see C.S. Rafinesque, *Florula ludoviciana.* 167. New York 1817, *Jour. Phys. Chim. Hist. Nat.* 89: 103. 1819, *New Fl. N. Am.* 1: 6, 86. 1836, *Flora Telluriana.* 3: 65. 1836 [1837] and *The Good Book.* 45. 1840; E.D. Merrill, *Index rafinesquianus.* 81. 1949; D.H. Nicolson, "Derivation of aroid generic names." *Aroideana.* 10: 15-25. 1988.

Species/Vernacular Names:
P. virginica (L.) Kunth
English: green arrow arum, tuckahoe

Peltandra Wight Euphorbiaceae

Origins:
Greek *pelte* "a shield, target" and *aner, andros* "stamen, male."

Peltanthera Benth. Buddlejaceae (Loganiaceae)

Origins:
From the Greek *pelte* "a shield" and *anthera* "anther."

Peltapteris Link Lomariopsidaceae (Aspleniaceae)

Origins:
From the Greek *pelte* "a shield" and *pteris* "a fern."

Peltaria Jacq. Brassicaceae

Origins:
Greek *pelte* "a shield," alluding to the leaf.

Peltariopsis (Boiss.) N. Busch Brassicaceae

Origins:
Resembling *Peltaria* Jacq.

Peltastes Woodson Apocynaceae

Origins:
Greek *peltastes* "one who bears a light shield," Latin *peltasta* "a soldier armed with the *pelta*, a peltast."

Pelticalyx Griff. Annonaceae

Origins:
Greek *pelte* "a shield" and *kalyx* "calyx."

Peltiphyllum (Engl.) Engl. Saxifragaceae

Origins:
Greek *pelte* "a shield" and *phyllon* "leaf," alluding to the form of the leaves.

Peltoboykinia (Engl.) H. Hara Saxifragaceae

Origins:
From the Greek *pelte* "a shield" plus the genus *Boykinia* Nutt., alluding to the peltate leaves.

Peltobractea Rusby Malvaceae

Origins:
Greek *pelte* "a shield" and Latin *bractea* "a thin plate of metal, gold-leaf, bract," referring to the form of the bracts.

Peltocalathos Tamura Ranunculaceae

Origins:
Greek *pelte* "a shield" and *kalathos* "a basket."

Peltodon Pohl Labiatae

Origins:

From the Greek *pelte* "a shield" and *odous, odontos* "a tooth."

Peltogyne Vogel Caesalpiniaceae

Origins:

Greek *pelte* and *gyne* "a woman, female," referring to the peltate pistil.

Peltophoropsis Chiov. Caesalpiniaceae

Origins:

Resembling the genus *Peltophorum*.

Peltophorum (Vogel) Benth. Caesalpiniaceae

Origins:

Referring to the shape of the stigma, from the Greek *pelte* "a shield" and *phoros* "bearing."

Species/Vernacular Names:

P. sp.

Indochina: hoang-linh, lem, lim set, lim xet

P. africanum Sonder (*Baryxylum africanum* (Sonder) Pierre; *Brasilettia africana* (Sond.) Kuntze)

English: African-wattle, weeping wattle, Natal wattle, Rhodesian black wattle, Rhodesian wattle

Southern Africa: huilboom, huilbos, dopperkiaat; mugija, Nyakambariro, muNyamashawa, muOra, muPangara, muPangasa, muSabanyoka, muZaze, muZeze, iZeze (Shona); umThobo, isiKhabamkhombe, umSehle (Zulu); isiKhabakhombe (Swazi); nhlanhlanhu, ndzedze (eastern Transvaal); mosetlha (western Transvaal, northern Cape, Botswana); mosehla (North Sotho: north and northeast Transvaal); musese (Venda); mosiru (Botswana, eastern Caprivi); nzeze (Kalanga: northern Botswana); movevi (Mbukushu); omuparara (Herero: central southwest Africa)

P. dasyrachis (Miq.) Kurz (*Baryxylum dasyrachis* (Miq.) Pierre; *Caesalpinia dasyrachis* Miq.)

Vietnam: hoan linh, lim xet

Indochina: hoang-linh, lim vang, lim xet, mun si

Cambodia: tram kang, tramkan, trasec

Malaya: yellow batai, batai, jemerelang, alai

P. dubium (Sprengel) Taubert (*Baryxylum dubium* (Sprengel) Pierre; *Caesalpinia dubia* Sprengel; *Peltophorum vogelianum* Benth., nom. illeg.)

Brazil: angico

Argentina: ibira pita, ibira puita, ibira puita guassu, ivira pita, ivira pita guazu, virapita, ybira pyita, ybira pyita guazu, yuira pita, cana fistula

Puerto Rico: cana fistula

P. pterocarpum (DC.) K. Heyne (*Baryxylum inerme* (Roxb.) Pierre; *Caesalpinia ferruginea* Decne.; *Caesalpinia inermis* Roxb.; *Inga pterocarpa* DC.; *Peltophorum ferrugineum* (Decne.) Benth.; *Peltophorum inerme* (Roxb.) Naves ex Fern.-Vill.)

English: copperpod, yellow-flamboyant, yellow-flame, yellow-poinciana, peltophorum, yellow flame tree

Indochina: hoang-linh, lim vang, lim xet

Malaya: jemerelang, batai, batai laut

Vietnam: lim vangh

Peltophorus Desv. Gramineae

Origins:

Greek *pelte* "a shield" and *phoros* "bearing," alluding to the shape of the flowers.

Peltophyllum Gardner Triuridaceae

Origins:

From the Greek *pelte* "a shield" and *phyllon* "a leaf."

Peltospermum Benth. Rubiaceae

Origins:

From the Greek *pelte* and *sperma* "seed."

Peltostegia Turcz. Malvaceae

Origins:

From the Greek *pelte* "a shield" and *stege, stegos* "roof, cover."

Peltostigma Walp. Rutaceae

Origins:

From the Greek *pelte* "a shield" and *stigma* "stigma," from the form of the stigma.

Pemphis Forster & Forster f. Lythraceae

Origins:

Greek *pemphis, pemphidos* "bladder, blister, a bubble, a swelling," referring to the ovary or to the globular or swollen capsule; see Johann Reinhold Forster (1729-1798) and his son Johann Georg Adam (1754-1794), *Characteres generum plantarum.* 67, t. 34. 1775.

Species/Vernacular Names:

P. acidula Forster & Forster f.

English: digging stick tree

Rodrigues Island: bois matelot

Japan: mizu-ganpi

Malaya: mentigi, keremak batu, mentagu

Penaea L. Penaeaceae

Origins:

For the French physician Pierre Pena, fl. 1520/1535-1600/1605, botanist, assistant to Mathias de Lobel [de Lobelius, L'Obel, etc.] (1538-1616), physician of Henri III, with L'Obel wrote *Stirpium adversaria nova.* [Londini 1570] 1571. See Ludovic Legré (1838-1904), *La Botanique en Provence au XVIe siècle, Pierre Pena et Mathias de Lobel.* Marseille 1899; Ethelyn Maria Tucker, *Catalogue of the Library of the Arnold Arboretum of Harvard University.* Cambridge, Massachusetts 1917-1933; Jules Émile Planchon (1823-1888), *Rondelet et ses disciples.* Montpellier 1866; John H. Barnhart, *Biographical Notes upon Botanists.* 3: 65. 1965; Jonas C. Dryander, *Catalogus bibliothecae historico-naturalis Josephi Banks.* London 1800; Edward Lee Greene, *Landmarks of Botanical History.* Edited by Frank N. Egerton. 879, 884, 1027. Stanford University Press, Stanford, California 1983; Richard J. Durling, comp., *A Catalogue of Sixteenth Century Printed Books in the National Library of Medicine.* 1967.

Penianthus Miers Menispermaceae

Origins:

Greek *pene, penion* "thread, spool" and *anthos* "a flower."

Species/Vernacular Names:

P. longifolius Miers

Congo: kuluku

Penicillaria Willd. Gramineae

Origins:

Latin *penicillum, i* "a pencil, a painter's brush," see also *Pennisetum* Rich.

Peniculifera Ridl. Euphorbiaceae

Origins:

Latin *peniculus, i* "a little tail, brush, pencil" and *fero, fers, tuli, latum, ferre* "to bear, carry."

Peniculus Swallen Gramineae

Origins:

Latin *peniculus, i* "a little tail, brush, pencil."

Peniocereus (A. Berger) Britton & Rose Cactaceae

Origins:

Greek *pene, penion* "thread, spool" plus *Cereus*, referring to the very slender stems.

Peniophyllum Pennell Onagraceae

Origins:

From the Greek *pene, penion* "thread, spool" and *phyllon* "leaf."

Pennantia Forst. & Forst.f. Icacinaceae

Origins:

After the Welsh (b. Flintshire, Wales) zoologist Thomas Pennant, 1726-1798 (d. Downing, Wales), traveler and antiquary, amateur natural scientist, naturalist, 1767 Fellow of the Royal Society, correspondent of Banks and Linnaeus, natural history collector. His works include *Indian Zoology.* London 1790 [1791], *The View of the Malayan Isles.* London 1800, *The History of the Parishes of Whiteford and Holywell.* [London] 1796, *Of the Patagonians. Formed with the Relation of Father Falkener, a Jesuit who had Resided Among Them Thirty Eight Years*; and from ... different voyagers. Darlington 1788, *Tour on the Continent 1765.* London 1948 and *A Tour in Scotland and Voyage to the Hebrides.* London 1774-76. See *Downing Hall, near Holywell. Catalogue of Sale of the Remainder of the Downing Library* (formed by Thomas Pennant ... 1726-98, and augmented by his son, David Pennant), consisting of about 5000 volumes ... Sale by public auction ... May 26th, 27th & 28th, 1913, at Downing Hall. Bangor [1913]; Peter [Pyotr] Simon Pallas (1741-1811), *A Naturalist in Russia. Letters from Peter Simon Pallas to Thomas Pennant.* Edited by Carol Urness. Minneapolis [1967]; Rev. Charles Cordiner, *Antiquities & Scenery of the North of Scotland,* in a series of letters to Thomas Pennant, Esq. by the Rev. Chas. Cordiner, minister of St. Andrew's Chapel, Banff. London 1780; John Lightfoot (1735-1788), *Flora scotica*. London 1777; J.R. Forster & J.G. Forster, *Characteres generum plantarum*. 133. 1775; Harold B. Carter, *Sir Joseph Banks (1743-1820). A Guide to Biographical and Bibliographical Sources*. Winchester 1987; *Memoir and Correspondence of ... Sir J.E. Smith ...* Edited by Lady Pleasance Smith. London 1832; J.M. Chalmers-Hunt, *Natural History Auctions 1700-1972, a Register of Sales in the British Isles*. London 1976; T.W. Bossert, *Biographical Dictionary of Botanists Represented in the Hunt Institute Portrait Collection*. 305. 1972; Blanche Henrey, *British Botanical and Horticultural Literature before 1800*. Oxford 1975; Ethelyn Maria Tucker, *Catalogue of the Library of the Arnold Arboretum of Harvard University*. Cambridge, Massachusetts 1917-1933; Blanche Henrey, *No Ordinary Gardener — Thomas Knowlton, 1691-1781*. Edited by A.O. Chater. British Museum (Natural History). London 1986; Edmund Berkeley and D.S. Berkeley, *Dr. Alexander Garden of Charles Town*. University of North Carolina Press [1969]; H.R. Fletcher and W.H. Brown, *Royal Botanic Garden Edinburgh, 1670-1970*. 94. Edinburgh 1970; Carol Urness, in *Dictionary of Scientific Biography* (Editor in Chief Charles Coulston Gillispie.) 10: 509-510. 1981; Ray Desmond, *Dictionary of British & Irish Botanists and Horticulturists*. 544-545. London 1994.

Species/Vernacular Names:

P. cunninghamii Miers

English: brown beech

Pennellia Nieuwl. Brassicaceae

Origins:
For the American botanist Francis Whittier Pennell, 1886-1952, bibliographer, traveler, plant collector, specialist in Scrophulariaceae, 1924-1952 edited *Bartonia*. His works include "What is *Commelina communis*?" *Proc. Acad. Nat. Sci. Philadelphia*. 90: 31-39. 1939 and "*Veronica* in North and South America." *Rhodora*. 23(265): 1-22, 23(266): 29-41. 1921. See John H. Barnhart, *Biographical Notes upon Botanists*. 3: 66. Boston 1965; Walter Mackinett Benner (1888-1970), "Francis Whittier Pennell." *Bartonia*. 26: 16-17. 1952; Joseph Ewan, *Rocky Mountain Naturalists*. The University of Denver Press 1950; J. Ewan, ed., *A Short History of Botany in the United States*. 1969; J. Ewan and Nesta Dunn Ewan, "Biographical dictionary of Rocky Mountain naturalists." *Reg. Veget*. 107: 1-253. 1981; Irving William Knobloch, compil., "A preliminary verified list of plant collectors in Mexico." *Phytologia Memoirs*. VI. 1983; R. Zander, F. Encke, G. Buchheim and S. Seybold, *Handwörterbuch der Pflanzennamen*. 14. Aufl. Stuttgart 1993; George Neville Jones, *An Annotated Bibliography of*

Mexican Ferns. University of Illinois Press 1966; Ida Kaplan Langman, *A Selected Guide to the Literature on the Flowering Plants of Mexico*. Philadelphia 1964; T.W. Bossert, *Biographical Dictionary of Botanists Represented in the Hunt Institute Portrait Collection*. 305. 1972; S. Lenley et al., *Catalog of the Manuscript and Archival Collections and Index to the Correspondence of John Torrey*. Library of the New York Botanical Garden. 324. 1973; E.M. Tucker, *Catalogue of the Library of the Arnold Arboretum of Harvard University*. 1917-1933.

Pennellianthus Crosswh. Scrophulariaceae

Origins:
For the American botanist Francis Whittier Pennell, 1886-1952, specialist in Scrophulariaceae, bibliographer, traveler, plant collector.

Pennilabium J.J. Smith Orchidaceae

Origins:
Latin *penna, ae* "a feather" plus *labia* and *labium* "a lip," referring to the nature of the labellum, to the margins of the lateral lobes of the lip.

Pennisetum Rich. Gramineae

Origins:
From the Latin *penna, ae* "a feather, plume" and *saeta* (*seta*), *ae* "a bristle, hair," referring to the bristly spikes; see Christiaan Hendrik Persoon (1761/1762-1836), *Synopsis plantarum*. 1: 72. Paris et Tubingae 1805.

Species/Vernacular Names:

P. alopecuroides (L.) Sprengel

English: Chinese pennisetum, swamp foxtail grass, fountain grass, swamp foxtail

P. alopecuroides (L.) Sprengel var. *viridescens* (Miq.) Ohwi (*Gymnothrix japonica* (Trin.) Kunth)

Japan: ao-chikara-shiba

P. basedowii Summerh. & C.E. Hubb.

English: spotter's grass

P. clandestinum Hochst. ex Chiov.

English: Kikuyu, Kikuyu grass (after the Kikuyu tribe, Kenya; see W. Scoreby Routledge and K. Routledge, *With a Prehistoric People. The Akikuyu of British East Africa*. London 1910; Fr. C. Cagnolo, *The Akikuyu*. Catholic Mission of the Consolata Fathers. Nyeri, Kenya 1933)

East Africa: chikoko, esereti, kigombe, lindadongo, olobobo

Southern Africa: Kikoejoe, Kikoejoegras, Kikujugras, Kikuyugras; tajoe (Sotho)

P. glaucum (L.) R. Br. (*Pennisetum americanum* (L.) Leeke subsp. *americanum*; *Pennisetum albicauda* Stapf & C.E. Hubb.; *Pennisetum echinurus* (K. Schum.) Stapf & C.E. Hubb.; *Pennisetum nigritarum* (Schlecht.) Dur. & Schinz; *Pennisetum typhoides* (Burm.f.) Stapf & C.E. Hubb.; *Setaria lutescens* (Wiegel) F.T. Hubb.)

English: African millet, babala, Indian millet, kaffir millet, pearl millet, poko grass, pussy grass

South Africa: babalagras, kaffermanna koring, manna koring, pokogras

Yoruba: emeye, mayi

P. latifolium Sprengel

English: Uruguay pennisetum

P. polystachyon (L.) Schultes

English: foxtail, mission grass

The Philippines: buntot-pusa

Yoruba: ilosun, inasua, irunmunu efon

P. purpureum Schumacher

English: elephant grass, Napier's fodder, Napier grass, Merkerr grass

Southern Africa: olifantsgras; mufufu (Shona)

Yoruba: eesun, eesun funfun, eesun pupa, iken, esisun, eesu

The Philippines: darai, gulalay, handalaui, lagoli

P. setaceum (Forsskål) Chiov.

English: fountain grass

South Africa: pronkgras

P. unisetum (Nees) Benth. (*Beckeropsis uniseta* (Nees) K. Schum.)

English: Duncan grass, Natal grass, silky grass

Southern Africa: sygras; nsipi (Zulu)

P. villosum Fresenius

English: feathertop, long style feather grass

Penstemon Schmidel Scrophulariaceae

Origins:
Greek *pente* "five" and *stemon* "stamen," alluding to the four fertile stamens and one staminode.

Species/Vernacular Names:
P. spp.

Mexico: jarritos, cebadilla falsa, lengua de arista; mocuepanixóchtl (Aztec l.)

P. albomarginatus M.E. Jones

English: white-margined beardtongue

P. barnebyi N. Holmgren

English: Barneby's beardtongue

P. calcareus Brandegee

English: limestone beardtongue

P. californicus (Munz & I.M. Johnston) Keck

English: California penstemon

P. canescens (Britton) Britton

English: Appalachian beardtongue, gray beardtongue

P. centranthifolius (Benth.) Benth.

English: scarlet bugler

P. cinicola Keck (Latin *cinisculus* "a little ashes," *cinis* "ashes")

English: ash beardtongue

P. eatonii A. Gray

English: Eaton's firecracker, Eaton's penstemon

P. filiformis (Keck) Keck

English: thread-leaved beardtongue

P. grandiflorus Nutt.

English: large beardtongue, northeastern beardtongue

P. neotericus Keck

English: Plumas County beardtongue

P. newberryi A. Gray

English: mountain-pride

P. pahutensis N. Holmgren

English: Pahute beardtongue

P. papillatus J. Howell

English: Inyo beardtongue

P. personatus Keck

English: close-throated beardtongue

P. purpusii Brandegee

English: snow mountain beardtongue

P. stephensii Brandegee

English: Stephen's beardtongue

P. thurberi Torrey

English: Thurber's beardtongue

P. tracyi Keck

English: Tracy's beardtongue

Pentabothra Hook.f. Asclepiadaceae

Origins:
From the Greek *pente* "five" and *bothros* "a pit," alluding to five pits.

Pentabrachion Müll. Arg. Euphorbiaceae

Origins:

From the Greek *pente* "five" and *brachion* "the arm, the forearm."

Pentacaena Bartl. Caryophyllaceae (Illecebraceae)

Origins:

Probably from the Greek *pente* "five" and *akaina* "thorn," presumably referring to the valves or apical teeth of the capsules.

Pentacalia Cass. Asteraceae

Origins:

From the Greek *pente* "five" and *kalia* "wooden, dwelling, hut, barn."

Pentacarpaea Hiern Rubiaceae

Origins:

From the Greek *pente* "five" and *karpos* "fruit."

Pentacarpus Post & Kuntze Rubiaceae

Origins:

From the Greek *pente* "five" and *karpos* "fruit."

Pentace Hassk. Tiliaceae

Origins:

Greek *pentakis* "five times," or from *pente* and *ake, akis* "tip, thorn, a sharp point."

Species/Vernacular Names:
P. burmannica Kurz

English: Burmese mahogany

Pentaceras Hook.f. Rutaceae

Origins:

From the Greek *pente* "five" and *keras* "horn," alluding to the carpels.

Species/Vernacular Names:
P. australis (F. Muell;) Hook.f. ex Benth.

English: bastard crow's ash, black tea, penta ash

Pentachaeta Nutt. Asteraceae

Origins:

Greek *pente* "five" and *chaite* "bristle, long hair," slender bristles at the base of the pappus.

Species/Vernacular Names:
P. aurea Nutt.

English: golden-rayed pentachaeta

P. bellidiflora E. Greene

English: white-rayed pentachaeta

P. lyonii A. Gray

English: Lyon's pentachaeta

Pentachlaena H. Perrier Sarcolaenaceae

Origins:

From the Greek *pente* "five" and *chlaena* "a cloak, blanket."

Pentachondra R. Br. Epacridaceae

Origins:

From the Greek *pente* "five" and *chondros* "cartilage," referring to the base of the ovary; see R. Brown, *Prodromus florae Novae Hollandiae et Insulae van-Diemen*. 549. London 1810.

Species/Vernacular Names:
P. pumila (Forst. & Forst.f.) R. Br.

English: carpet heath, dwarf heath

Pentaclathra Endl. Cucurbitaceae

Origins:

Greek *pente* "five" and *klathron, klethron, kleithron* "gate, a lock," see also *Polyclathra* Bertol.

Pentaclethra Benth. Mimosaceae

Origins:

Greek *pente* "five" and *kleis* "lock, key" (*kleio* "to close, to shut"); Greek *klethra* "the alder"; see Y. Tailfer, *La Forêt dense d'Afrique centrale*. CTA, Ede/Wageningen 1989; J.

Vivien & J.J. Faure, *Arbres des Forêts denses d'Afrique Centrale*. Agence de Coopération Culturelle et Technique. Paris 1985.

Species/Vernacular Names:
P. eetveldeana De Wild. & Th. Dur.

Congo: nabo

Gabon: engona, mpasse

Cameroon: ebaye, ebaye bekwe

Zaire: elai

P. macroloba (Willd.) Kuntze (*Acacia macroloba* Willd.; *Pentaclethra filamentosa* Benth.)

Brazil: pracaxi, pao choca

P. macrophylla Benth.

English: oil-bean tree, owala oiltree, atta bean, wild locust

Central Africa: apara, ogba, ukana

Ivory Coast: alta, dio, ovada, ovala, seredieu

Ghana: akuamma, ekuama, ata, ataa, atawa, odenya, tsaklo

Yoruba: apara, apaha, pala, kako

Nigeria: apara (Yoruba); okpagha (Edo); ukpaghan (Itsekiri); okpaghan (Urhobo); ukpakara (Ijaw); ugba (Igbo); ukana (Efik); nkpa (Boki)

Zaire: mabula, mubala, mvanza, panza, vaanza, tshibamba, tshibambamba, tshibambabamba

Congo: essiri, kihanzi, nvandza, mukandzi, obala

Gabon: ébé, ebe, obaa, obada, obala, ovala, owala, bembada, bombaha, dimbalo, ombala, mbala, moulla-panza, mpanza, mupandji, muvandji, muvemdji, mvala, m'vans, ouala

Cameroon: balé, ba, bali, bemba, ebal, ebaye, ebe, kombolo, kommott, mbalaka, mbara, mba, owala

Liberia: blayhu, fai

Pentacme A. DC. Dipterocarpaceae

Origins:
From the Greek *pente* "five" and *akme* "the top, highest point," referring to the calyx members.

Pentacraspedon Steudel Gramineae

Origins:
Greek *pente* "five" and *kraspedon* "a fringe, border, hem"; see Ernst Gottlieb von Steudel (1783-1856), *Synopsis plantarum glumacearum*. 1: 151. 1854.

Pentacrostigma K. Afzel. Convolvulaceae

Origins:
From the Greek *pente* "five," *akros* "terminal, tip" and *stigma* "stigma."

Pentactina Nakai Rosaceae

Origins:
From the Greek *pente* "five" and *aktin* "ray."

Pentacyphus Schltr. Asclepiadaceae

Origins:
Greek *pente* and *kyphos* "bent, curved, humped."

Pentadactylon C.F. Gaertner Proteaceae

Origins:
From the Greek *pente* "five" and *daktylos* "a finger"; see Carl (Karl) Friedrich von Gaertner (1772-1850), *Supplementum carpologicae* seu continuati operis Josephi Gaertner de fructibus et seminibus plantarum. 3: 219. Leipzig 1807.

Pentadenia (Planchon) Hanst. Gesneriaceae

Origins:
From the Greek *pente* "five" and *aden* "gland."

Pentadesma Sabine Guttiferae

Origins:
Greek *pente* "five" and *desmos* "a bond, band, bundle," referring to the stamens in bundles of five.

Species/Vernacular Names:
P. sp.
French: arbre à beurre
Cameroon: avoum
Ivory Coast: b'bere
Gabon: agnuhe
Ghana: agyapa
P. butyracea Sabine
English: tallow tree, butter tree
Fr. Guiana: lami
Cameroon: avoum, nom onie, gambe
Central Africa: pangi la mundagila, n'kandika, agnihé, agnouhé, avoum, gambé, orogbo erin, lami ovotera

Congo: n'kandika

Gabon: agnuhe

Ghana: bromabine

Ivory Coast: lami

Liberia: waye, waye-kpay

Yoruba: orogbo, ekuso, kuro

Nigeria: izeni; orogbo-erin, iriro, uroro (Yoruba); aghe (Kwale Igbo); akanti (Ijaw); oze (Igbo)

Zaire: kiasose

P. exelliana Staner

Central Africa: bonzo, kampangi pangi, koma mufike, mukomo, mombele, nzibu, usudi

Pentadiplandra Baillon Pentadiplandraceae (Capparidaceae)

Origins:

Greek *pente* "five," *diploos* "double" and *aner, andros* "man, stamen, male," referring to the number and nature of the stamens.

Species/Vernacular Names:

P. brazzeana Baillon

Congo: ngama, kikamu, kikuolo, nguza

Pentadynamis R. Br. Fabaceae

Origins:

Greek *pente* "five" and *dynamis* "strength, power, might," possibly referring to the stamens; see Charles Sturt (1795-1869), *Narrative of an Expedition into Central Australia, Performed ... During the Years 1844, 1845 and 1846.* 2. App. 76. London 1849.

Pentaglottis Tausch Boraginaceae

Origins:

From the Greek *pente* "five" and *glotta* "tongue," referring to the scales of the corolla; see Ignaz Friedrich Tausch (1793-1848), in *Flora oder allgemeine Botanische Zeitung.* 12: 643. (Nov.) 1829.

Pentagonanthus Bullock Asclepiadaceae (Periplocaceae)

Origins:

From the Greek *pente* "five," *gonia* "an angle, corner" and *anthos* "a flower."

Pentagonaster Klotzsch Myrtaceae

Origins:

Greek *pente* "five," *gonia* "an angle, corner" and *aster* "a star"; see Christoph Friedrich Otto (1783-1856) and Albert Gottfried Dietrich (1795-1856), in *Allgemeine Gartenzeitung.* 4: 113. (Apr.) 1836.

Pentagonia Benth. Rubiaceae

Origins:

Greek *pente* "five" and *gonia* "an angle, corner," referring to the divisions of the corolla.

Pentagonia Fabr. Solanaceae

Origins:

Greek *pente* "five" and *gonia* "an angle, corner."

Pentagramma Yatskievych, Windham & E. Wollenweber Pteridaceae (Adiantaceae)

Origins:

Greek *pente* "five" and *gramma* "line, writing, letter," referring to the leaf blades.

Pentalepis F. Muell. Asteraceae

Origins:

Greek *pente* "five" and *lepis* "scale"; see F. von Mueller, in *Edinburgh New Philosophical Journal.* Ser. 2, 17: 230. 1863, and *Transactions of the Botanical Society.* Edinburgh. 7: 496. 1863.

Pentalinon Voigt Apocynaceae

Origins:

From the Greek *pente* "five" and *linon* "rope, flax, thread, a net, anything made of flax."

Pentaloncha Hook.f. Rubiaceae

Origins:

Greek *pente* "five" and *lonche* "a lance, spear."

Pentameria Klotzsch ex Baillon Euphorbiaceae

Origins:
From the Greek *pente* "five" and *meris* "part," having parts in five.

Pentameris P. Beauv. Gramineae

Origins:
From the Greek *pente* "five" and *meris* "part."

Pentamerista Maguire Tetrameristaceae

Origins:
From the Greek *pente* "five" and *meristos* "divided, individual."

Pentanema Cass. Asteraceae

Origins:
From the Greek *pente* "five" and *nema* "filament, thread."

Pentanisia Harvey Rubiaceae

Origins:
From the Greek *pente* "five" and *anisos* (*a* and *isos* "equal") "unequal," referring to the lobes of the calyx.

Species/Vernacular Names:
P. prunelloides (Klotzsch ex Eckl. & Zeyh.) Walp. subsp. *prunelloides* (*Pentanisia variabilis* Harv. var. *intermedia* Sond.)
Southern Africa: sooibrandbossie; isiCimamlilo, iRubuxa (Xhosa)

Pentanura Blume Asclepiadaceae (Periplocaceae)

Origins:
From the Greek *pente* "five" and *oura* "tail."

Pentapanax Seemann Araliaceae

Origins:
Greek *pente* "five" and the genus *Panax* in the same family, an allusion to the pistil; see Berthold Carl Seemann, in *The Journal of Botany*. 2: 290. (Oct.) 1864.

Pentapeltis (Endl.) Bunge Umbelliferae

Origins:
Greek *pente* "five" and *pelte* "a shield"; see Johann G.C. Lehmann, *Plantae Preissianae* ... 1: 292. Hamburgi (Feb.) 1845.

Pentapera Klotzsch Ericaceae

Origins:
Greek *pente* "five" and *pera* "a pouch, bag," referring to the ovary.

Pentapetes L. Sterculiaceae

Origins:
From the Greek *pentapetes* (see *pentaphyllon*, Theophrastus in *HP*. 9.13.5 and Dioscorides) "cinque-foil," Latin *quinquefolium, ii* "a plant, cinque-foil"; see Carl Linnaeus, *Species Plantarum*. 698. 1753 and *Genera Plantarum*. Ed. 5. 310. 1754.

Species/Vernacular Names:
P. phoenicea L.
Japan: goji-ka

Pentaphalangium Warb. Guttiferae

Origins:
From the Greek *pente* "five" and *phalanx, phalangos* "phalanx."

Pentaphitrum Reichb. Solanaceae

Origins:
From the Greek *pente* "five" and *phitros* "log, block of wood, a trunk of a tree."

Pentaphragma Wallich ex G. Don f. Pentaphragmataceae

Origins:
Greek *pente* "five" and *phragma* "a hedge, a fence," five members of the calyx, gynoecium separated from hypanthium by nectariferous pits.

Pentaphylax Gardner & Champ. Pentaphylacaceae (Theaceae)

Origins:

Greek *pente* "five" and *phylax, phylakos* "a guardian, protector," *pentaphylakos* "divided into five watches," flowers pentamerous, gynoecium five-loculated.

Pentaphylloides Duhamel Rosaceae

Origins:

From the Greek *pente* "five," *phyllon* "leaf" and *eidos, oides* "resemblance."

Pentapleura Hand.-Mazz. Labiatae

Origins:

From the Greek *pente* "five" and *pleura, pleuron* "side, rib, lateral," *pentapleuron* "figure with five sides."

Pentapogon R. Br. Gramineae

Origins:

From the Greek *pente* "five" and *pogon* "a beard," referring to the lemma; see Robert Brown (1773-1858), *Prodromus florae Novae Hollandiae et Insulae van-Diemen*. 173. London (Mar.) 1810.

Species/Vernacular Names:

P. quadrifidus (Labill.) Baillon (*Agrostis quadrifida* Labill.; *Pentapogon billardieri* R. Br.)

English: five-awned spear-grass, five-awn spear grass

Pentaptera Roxb. Combretaceae

Origins:

Greek *pente* "five" and *pteron* "a wing."

Pentapterygium Klotzsch Ericaceae

Origins:

Greek *pente* and *pterygion* "a small wing," referring to the small wings in the calyx.

Pentaptilon E. Pritzel Goodeniaceae

Origins:

Greek *pente* "five" and *ptilon* "a feather, wing," referring to the winged fruits; see Friedrich Ludwig Emil Diels (1874-1945) and Ernst Georg Pritzel (1875-1946), in *Botanische Jahrbücher*. 35: 564. (Feb.) 1905.

Pentarhaphia Lindley Gesneriaceae

Origins:

From the Greek *pente* "five" and *raphis* "a needle," referring to the calyx.

Pentarhizidium Hayata Dryopteridaceae (Woodsiaceae)

Origins:

Greek *pente* "five" and *rhiza* "a root."

Pentarhopalopilia (Engl.) Hiepko Opiliaceae

Origins:

Greek *pente* "five" plus the genus *Rhopalopilia* Pierre.

Pentaria M. Roemer Passifloraceae

Origins:

Greek *pente* "five," referring to the parts of the flower.

Pentarrhaphis Kunth Gramineae

Origins:

From the Greek *pente* "five" and *raphis* "a needle," referring to the flowers.

Pentarrhinum E. Meyer Asclepiadaceae

Origins:

From the Greek *pente* "five" and *rhis, rhinos* "snout, nose," referring to the corona lobes.

Pentas Benth. Rubiaceae

Origins:

From the Greek *pentas* "a series of five," *pente* "five," referring to the pentamerous flowers.

Species/Vernacular Names:

P. lanceolata (Forssk.) Deflers (*Pentas carnea* Benth.)

English: star-cluster, Egyptian star-cluster

Pentasachme Wallich ex Wight Asclepiadaceae

Origins:
See *Pentasacme.*

Pentasacme Wallich ex Wight Asclepiadaceae

Origins:
From the Greek *pentas* "a series of five," *pente* "five" and *akme* "the top, highest point," referring to the nature of the flowers.

Species/Vernacular Names:
P. caudatum Wallich ex Wight (*Pentasacme championii* Bentham)
English: Champion pentasacme
Malaya: chermin hantu, serai ayer

Pentaschistis (Nees) Spach Gramineae

Origins:
Greek *pente* "five" and *schistos* "cut," *schizo, schizein* "to divide," the lemma has five divisions; Édouard Spach (1801-1879), *Histoire naturelle des Végétaux.* 13: 164. Paris (Jun.) 1841.

Species/Vernacular Names:
P. rosea Linder subsp. *rosea*
South Africa: rooigras
P. thunbergii (Kunth) Stapf (*Danthonia thunbergii* Kunth)
English: pussy tail
P. triseta (Thunb.) Stapf
English: dune grass
South Africa: duinegras, haasgras

Pentascyphus Radlk. Sapindaceae

Origins:
From the Greek *pente* "five" and *skyphos* "a cup, goblet, a jug."

Pentaspadon Hook.f. Anacardiaceae

Origins:
From the Greek *pente* "five" and *spadon* "eunuch," referring to the five sterile stamens.

Species/Vernacular Names:
P. motleyi Hook.f. (after the British plant collector James Motley, d. 1859, civil engineer and botanist in Malaysia and Borneo, co-author with Lewis Llewellyn Dillwyn (fl. 1855) of *Contributions to the Natural History of Labuan, and the Adjacent Coasts of Borneo.* London 1855. See John H. Barnhart, *Biographical Notes upon Botanists.* 2: 520. 1965; I.H. Vegter, *Index Herbariorum.* Part II (4), *Collectors M.* Regnum Vegetabile vol. 93. 1976)

Malaya: white pelong-tree, pelong, pelajau, pelajoh

Pentaspatella Gleason Ochnaceae

Origins:
Greek *pente* "five" and *spathe* "spathe."

Pentastachya Steud. Gramineae

Origins:
From the Greek *pente* "five" and *stachys* "spike."

Pentastelma Tsiang & P.T. Li Asclepiadaceae

Origins:
Greek *pente* "five" and *stelma, stelmatos* "a girdle, belt," referring to the five corona lobes.

Species/Vernacular Names:
P. auritum Tsiang & P.T. Li
English: longeared pentastelma
China: bai shu teng

Pentastemon L'Hérit. Scrophulariaceae

Origins:
Greek *pente* "five" and *stemon* "stamen."

Pentastemona Steenis Pentastemonaceae (Stemonaceae)

Origins:
Greek *pente* "five" and *stemon* "stamen," pentamerous, 5-mery.

Pentastemonodiscus Rech.f. Caryophyllaceae

Origins:
From the Greek *pente* "five," *stemon* "stamen" and *diskos* "disc."

Pentasticha Turcz. Cyperaceae

Origins:
From the Greek *pente* "five" and *stichos* "a row."

Pentataphrus Schltdl. Epacridaceae

Origins:
Greek *pente* "five" and *taphros* "a trench, ditch"; see Diederich Franz Leonhard von Schlechtendal (1794–1866), in *Linnaea*. 20: 618. (Oct.) 1847.

Pentatherum Náb. Gramineae

Origins:
From the Greek *pente* "five" and *ather* "stalk, barb, awn."

Pentathymelaea Lecomte Thymelaeaceae

Origins:
Greek *pente* "five" and *Thymelaea* Miller, Latin *thymelaea, ae* for a plant, the flax-leaved daphne.

Pentatrichia Klatt Asteraceae

Origins:
From the Greek *pente* "five" and *thrix, trichos* "hair."

Pentatropis R. Br. ex Wight & Arn. Asclepiadaceae

Origins:
Greek *pente* "five" and *tropis, tropidos* "a keel," referring to the corona lobes, the flowers having five keels; see Robert Wight, *Contributions to the Botany of India*. London 1834.

Penteca Raf. Euphorbiaceae

Origins:
See Constantine S. Rafinesque, *Sylva Telluriana*. 62. 1838; E.D. Merrill, *Index rafinesquianus*. The plant names published by C.S. Rafinesque, etc. 156. Jamaica Plain, Massachusetts, USA 1949.

Pentelesia Raf. Bignoniaceae

Origins:
See Constantine S. Rafinesque, *Sylva Telluriana*. 146. 1838; E.D. Merrill, *Index rafinesquianus*. 220. 1949.

Penthea Lindley Orchidaceae

Origins:
Greek *penthos, pentheia* "grief, sorrow, mourning," flowers variously colored, often with darker veins and spots.

Pentheriella O. Hoffm. & Muschl. Asteraceae

Origins:
For the Austrian (b. Rome, Italy) botanical collector and naturalist Arnold Penther, 1865-1931 (d. Vienna), zoologist, traveler, plant collector in South Africa and Rhodesia, he published (with the Austrian botanist Emmerich Zederbauer, 1877-1950) *Ergebnisse einer naturwissenschaftlichen Reise zum Erdschias-Dag (Kleinasien)*. Vienna 1905-1906; see Mary Gunn and L.E.W. Codd, *Botanical Exploration of Southern Africa*. 278. 1981; Alexander Zahlbruckner (1860-1938), *Plantae Pentherianae. Aufzählung der von Dr. A. Penther ... in SüdAfrika gesammelten Pflanzen*. Wien [Annalen der K.K. Naturhistorischen Hofmuseums] 1900 [-1905].

Penthorum L. Saxifragaceae (Penthoraceae)

Origins:
From the Greek *pente* "five" and *horos* "a column, pillar, stone, border," referring to the angled and beaked capsule.

Penthysa Raf. Boraginaceae

Origins:
Greek *pente* "five" and *thysanos* "a fringe, tassel"; see Constantine S. Rafinesque, *Flora Telluriana*. 4: 86. 1836 [1838]; Elmer D. Merrill, *Index rafinesquianus*. 203. 1949.

Pentodon Hochst. Rubiaceae

Origins:
From the Greek *pente* "five" and *odous, odontos* "a tooth."

Pentopetiopsis Costantin & Gallaud Asclepiadaceae

Origins:
Resembling *Pentopetia*.

Pentossaea W. Judd Melastomataceae

Origins:

Greek *pente* and *Ossaea* DC.

Pentstemon Aiton Scrophulariaceae

Origins:

From the Greek *pente* "five" and *stemon* "a stamen," see also *Penstemon* Schmidel.

Pentstemonacanthus Nees Acanthaceae

Origins:

Greek *pente* "five," *stemon* "a stamen" and *akantha* "thorn."

Pentulops Raf. Orchidaceae

Origins:

Possibly from the Greek *pente* "five," *tylos* "lump, knob" and *opsis* "like, sight, appearance," the calluses on the lip; see Constantine S. Rafinesque, *Flora Telluriana*. 4: 42. 1836 [1838]; E.D. Merrill, *Index rafinesquianus*. 104. 1949.

Pentzia Thunb. Asteraceae

Origins:

After Carolus Johannes Pentz, author of *De Diosma*. Praeside Carolo Petro Thunberg dissertationes academicae. Upsaliae [1797]; see Carl Peter Thunberg (1743-1828), *Prodromus plantarum Capensium*: quas in promontorio Bonae Spei Africes, annis 1772-1775 collegit... Upsaliae 1800.

Species/Vernacular Names:

P. cooperi Harv.

English: pentzia

P. globosa Less. (*Pentzia globifera* Licht. ex Less.)

English: bitter karoo bush, hair karroo

South Africa: bitterbultkaroo, bitterkaroo, bitterkaroobossie, goedkaro, langbeenkaroo, rooikarobos, skaapbossie, vaalkaroowitkarobos

P. incana (Thunb.) Kuntze (*Pentzia virgata* Less.; *Chrysanthemum incanum* Thunb.)

English: common karoo, good karroo bush, karroo bush, sheep bush, African sheep bush, sweet karoo

Southern Africa: ankerkaroo, gansie, goedkaro, goeiekaroo, grootgansie, karobos, karoobossie, kleingansie, kortbeenka-

roo, rooikarobos, soetkaroo, skaapbossie, skaapkaroo, vaalkarroo, witkarobos; mohantsoana (Sotho)

P. sphaerocephala DC. (*Pentzia cinerascens* DC.; *Pentzia grisea* Muschl. ex Dinter)

English: large karroo bush

South Africa: berggansiekaroo, gansiekaroo, grootberggansiekaroo, karoobossie, langsteelkaroo

P. suffruticosa (L.) J.B. Hutch. ex Merxm. (*Tanacetum suffruticosum* L.; *Matricaria suffruticosa* (L.) Druce)

English: Calomba daisy

Peperomia Ruíz & Pavón Piperaceae (Peperomiaceae)

Origins:

Greek *peperi* "pepper" and *homoios, homios* "resembling"; see Ruíz & Pavón, *Flora peruvianae, et chilensis prodromus*. Madrid 1794.

Species/Vernacular Names:

P. sp.

Hawaii: 'ala'ala wai nui, 'ala'ala wai nui kane, 'ala'ala wai nui kupa li'i, 'ala'ala wai nui pehu, 'ala'ala wai nui pohina, 'awalauakane, kupaoa, kupali'i

Peru: ahui ccobori, erav de jabotí, lancetilla, lancetilla de hoja ancha, lancetilla hembra, lansetilla

P. acutifolia C. DC.

Peru: afasiquihua

P. argyreia Morr. (*Peperomia arifolia* Miq. var. *argyreia* Miq.)

English: watermelon begonia, watermelon pepper

Japan: biyô-sada-sô (= beautifully-faced *Peperomia japonica*), shima-aoi-sô

P. caperata Yunck.

English: emerald-ripple pepper, green-ripple pepper, little fantasy pepper

P. dolabriformis Kunth

English: prayer pepper

P. dindygulensis Miq.

Japan: birôdo-goshô

P. elegantifolia C. DC.

Peru: sacha chullco, sacha chullcu

P. fraseri C. DC.

English: flowering pepper

P. griseoargentea Yunck.

English: ivy-leaf pepper, silver-leaf pepper, platinum pepper

P. inaequalifolia Ruíz & Pav. (*Piper aromaticum* Willldenow)

Peru: congona, huinay quilla

P. japonica Mak.

Japan: Sada-sô

P. obtusifolia (L.) Dietr.

English: baby rubber plant, American rubber plant, pepper face

Peru: came

P. pellucida (L.) Kunth

The Philippines: olasiman-ihalas, ikmong-bata, ikmo-ikmo-han

Yoruba: rinrin

P. urvilleana A. Rich.

English: two-leaved peperomia

Peplidium Delile Scrophulariaceae

Origins:

Similar to the genus *Peplis*, from the Greek *peplis, peplidos*, Latin *peplis, peplidis* applied to two plants, one of which was also called *porcilaca* and the other *syce meconion*, or *mecon aphrodes*; Latin *peplium* for a medicinal plant, a species of spurge (Plinius); see Alire Delile (Raffeneau-Delile) (1778-1850), *Florae Aegyptiacae illustratio*. 2: 50, 148. 1813.

Peplis L. Lythraceae

Origins:

From the Greek *peplis, peplidos*, ancient name used by Dioscorides for the wild purslane, a species of *Euphorbia*; Latin *peplis, peplidis* Plinius applied to two plants, one of which also called *porcilaca*; see Carl Linnaeus, *Species Plantarum*. 332. 1753 and *Genera Plantarum*. Ed. 5. 154. 1754.

Pepo Miller Cucurbitaceae

Origins:

From the Greek *pepon, peponos* "ripe, mild," Latin *pepo, peponis* "a species of large melon."

Peponia Naudin Cucurbitaceae

Origins:

From the Latin *pepo, peponis* "a species of large melon, a pumpkin."

Peponidium Baillon ex Arènes Rubiaceae

Origins:

Resembling a *pepo, peponis* "a species of large melon, a pumpkin."

Peponiella Kuntze Cucurbitaceae

Origins:

The diminutive of *pepo, peponis* "a species of large melon, a pumpkin."

Peponium Engl. Cucurbitaceae

Origins:

From the Latin *pepo, peponis* "a species of large melon, a pumpkin."

Species/Vernacular Names:
P. pageanum C. Jeffrey
Mozambique: matanhoca

Peponopsis Naudin Cucurbitaceae

Origins:

Resembling *pepo, peponis* "a species of large melon, a pumpkin."

Pera Mutis Euphorbiaceae (Peraceae)

Origins:

Greek *pera, pere, pare* "a pouch, wallet," possibly referring to the leaves with petiolar glands.

Species/Vernacular Names:
P. sp.

Peru: machusacha, mapiche, mapichi, pereiro, tatacáa

Peracarpa Hook.f. & Thomson Campanulaceae

Origins:

Greek *pera, pere, pare* "a pouch, wallet" and *karpos* "fruit."

Perakanthus F. Robyns Rubiaceae

Origins:

From Perak and Greek *anthos* "flower."

Peramium Salisbury ex Coulter Orchidaceae

Origins:
Greek *peramion* "a pouch, little wallet," the saccate basal part of the lip.

Peranema D. Don Dryopteridaceae (Aspleniaceae, Peranemataceae)

Origins:
From the Greek *pera* "further, much, beyond" and *nema* "thread."

Peraphyllum Nutt. Rosaceae

Origins:
Greek *pera* "further, much" and *phyllon* "leaf," very leafy.

Species/Vernacular Names:
P. ramosissimum Nutt.

English: wild crab apple

Perdicium L. Asteraceae

Origins:
Greek *perdikion*, the name of a plant, applied by Theophrastus (*HP.* 1.6.11) to a species of *Polygonum*, *perdikion* is the diminutive of *perdix* "partridge"; Latin *perdicium, ii* for the pellitory, the plant *parthenium* (Plinius).

Peregrina W.R. Anderson Malpighiaceae

Origins:
Latin *peregrinus, a, um* "foreign, exotic," *peregrina, ae* "a foreign woman."

Perescia Lem. Cactaceae

Origins:
After the French numismatist Nicholas Claude Fabry (Fabri) de Peiresc, 1580-1637, patron of botany and letters, naturalist and archaeologist.

Pereskia Miller Cactaceae

Origins:
After the French (b. Var) numismatist Nicholas Claude Fabry (Fabri) de Peiresc, 1580-1637 (d. Aix-en-Provence), patron of botany and letters, naturalist and archaeologist, eminent antiquarian, a correspondent of the supporter of Copernicanism Marin Mersenne (1588-1648) (Mersenne and Gassendi united in their controversy over hermeticism with Robert Fludd, the Rosicrucian and Oxford physician). See [N.C.F. de Peiresc], *Correspondance de Peiresc avec plusieurs missionaires et religieux de l'ordre des Capucins, 1631-1637* ... Publiée par le P. Apollinaire de Valence. Paris 1891; [N.C.F. de Peiresc], *Lettres de Peiresc à Jacques Gaffarel (1627-1637)*. Digne 1909; Francis West Gravit, *The Peiresc Papers.* [University of Michigan. Contributions in Modern Philology. no. 14.] Ann Arbor 1950; Pierre Gassendi (1592-1655), *Viri illustris N.C. Fabricii de Peiresc ... vita.* (Peireskii laudatio habita in concione funebri Academicorum Romanorum ... J.J. Buccardo ... perorante). Parisiis 1641 and *Epistolica exercitatio, in qua Principia Philosophiae Robert Fluddi Medici retiguntur.* Paris 1630; [Théophraste (1586-1653) and] Eusèbe Renaudot (1613-1679), *A General Collection of Discourses of the Virtuosi of France.* London 1664 and *Another Collection of Philosophical Conferences of the French Virtuosi.* London 1665 [Two volumes, first and only English edition.]; Philip Miller, *The Gardeners Dictionary.* Abr. ed. 4. London (28 Jan.) 1754; Joseph Bougarel (1680-1753), *Vie de Pierre Gassendi.* Paris 1737; Harcourt Brown, in *Dictionary of Scientific Biography* 10: 488-492. 1981.

Species/Vernacular Names:
P. aculeata Mill.

English: Barbados gooseberry, blade apple, Cape gooseberry, leafy cactus, lemon vine, pereskia creeper, primitive cactus, Spanish gooseberry

South Africa: Barbados stekelbessie

P. grandifolia Haw.

English: large-flowered Barbados gooseberry, large-leaved Barbados gooseberry, rose cactus

South Africa: grootblaar Barbadosstekelbessie, grootblom, Barbadosstekelbessie

P. horrida DC.

Peru: ataquisca

Pereskiopsis Britton & Rose Cactaceae

Origins:
Resembling the genus *Pereskia* Mill.

Perezia Lagasca Asteraceae

Origins:
For the Spanish apothecary Lázaro (in Stafleu: Laurentio) Pérez, botanist; see Georg Christian Wittstein (1810-1887), *Etymologisch-botanisches Handwörterbuch.* 670. Ansbach

1852; Stafleu and Cowan, *Taxonomic Literature*. 4: 164. 1983.

Species/Vernacular Names:
P. coerulescens Weddell (*Perezia cirsiifolia* Weddell; *Perezia nivalis* Weddell)

Peru: contrayerba, sotoma, sutuma, valeriana, intipa sapran

P. multiflora (Humboldt & Bonpl.) Lessing (*Chaetanthera multiflora* Humboldt & Bonpl.)

Peru: chancoruma, escorzonera, chancorma

P. pungens (Humboldt & Bonpl.) Lessing (*Perezia conaicaensis* Tovar)

Peru: ocksha huayta, azul corpus

Pereziopsis J. Coulter Asteraceae

Origins:
Resembling *Perezia*.

Pergamena Finet Orchidaceae

Origins:
From the Latin *pergamena, pergamina* "parchment," referring to the leaves.

Pergularia L. Asclepiadaceae

Origins:
Latin *pergula, ae* "a vine-arbor, a stall, a projection," referring to the twining habit of the plants; see C. Linnaeus, *Mantissa Plantarum*. Holmiae [Stockholm] 1767 [-1771].

Species/Vernacular Names:
P. tomentosa L.
French: pergulaire
Arabic: ghelga

Periandra C. Martius ex Benth. Fabaceae

Origins:
From the Greek *peri* "around" and *aner, andros* "stamen, man, male."

Perianthomega Bureau ex Baillon Bignoniaceae

Origins:
From the Greek *peri* "around, round about, all round," *anthos* "flower," *perianthion* "perianth" and *megas* "big, large."

Perianthopodus Silva Manso Cucurbitaceae

Origins:
From the Greek *perianthion* "perianth" and *pous, podos* "foot."

Perianthostelma Baill. Asclepiadaceae

Origins:
From the Greek *perianthion* "perianth" and *stelma, stelmatos* "a girdle, belt."

Periarrabidaea A. Samp. Bignoniaceae

Origins:
From the Greek *peri* "around, round about" plus *Arrabidaea* DC.

Periballia Trin. Gramineae

Origins:
Greek *periballo* "throw round, put on, put round," possibly in reference to the dispersion of the seeds; see Carl Bernhard von Trinius (1778-1844), *Fundamenta Agrostographiae*. 133. Viennae 1820.

Periblema DC. Acanthaceae

Origins:
From the Greek *periblema, periblematos* "garment."

Pericallis D. Don Asteraceae

Origins:
From the Greek *perikalles* "very beautiful."

Pericalymma (Endl.) Endl. Myrtaceae

Origins:
Greek *perikalymma* "a covering, garment"; see Stephan F. Ladislaus Endlicher, *Genera Plantarum*. 2: 1230. 1840.

Species/Vernacular Names:
P. ellipticum (Endl.) Schauer
English: swamp tea-tree

Pericalymna Meissn. Myrtaceae

Origins:
See *Pericalymma*.

Pericalypta Benoist Acanthaceae

Origins:
From the Greek *perikalypto* "cover all round."

Pericampylus Miers Menispermaceae

Origins:
Greek *peri* "around" and *kampylos* "curved," referring to the fruits; see John Miers (1789-1879), in *Annals and Magazine of Natural History*. Ser. II, 7: 36. London (Jan.) 1851.

Species/Vernacular Names:
P. formosanus Diels
Japan: hôrai-tsuzura-fuji

Perichasma Miers Menispermaceae

Origins:
Greek *peri* "around" and *chasme* "gaping, yawning," *perichasko* "open the mouth wide."

Perichlaena Baill. Bignoniaceae

Origins:
From the Greek *peri* "around" and *chlaena* "a cloak, blanket," *perichlainizomai* "wrap oneself in a cloak."

Periclesia A.C. Sm. Ericaceae

Origins:
Greek *peri* "around" and *kleis* "lock, key," *perikleisis* "enclosing all round"; some suggest from Pericles, the son of Xanthippus and Agariste.

Pericome A. Gray Asteraceae

Origins:
From the Greek *peri* "around" and *kome* "a tuft of hairs," referring to the tufted achenes.

Pericopsis Thwaites Fabaceae

Origins:
Greek *perikope* "cutting all round, mutilation, a section," *perikopto* "to cut all round," referring to the calyx.

Species/Vernacular Names:
P. elata (Harms) van Meeuwen (*Afrormosia elata* Harms)
English: afrormosia
Ghana: kokrodua
Ivory Coast: asamela, assamela
Cameroon: mobay, obang, nom eyen
Central Africa: obang
Nigeria: egbin (Yoruba)
Congo: obang
P. laxiflora (Baker) van Meeuwen (*Afrormosia laxiflora* Benth. ex Bak.)
English: false dalbergia
Nigeria: sedun (Yoruba); makarfo (Hausa); kpankangichi (Nupe); tserama (Tiv); abua ocha (Igbo)
Yoruba: ayan, sedun, awin, amuyin, sedun, egbi
W. Africa: kolo kolo, cincime

Perictenia Miers Apocynaceae

Origins:
From the Greek *peri* "around" and *kteis, ktenos* "a comb."

Pericycla Blume Palmae

Origins:
From the Greek *perikyklas* "revolving."

Perideridia Reichb. Umbelliferae

Origins:
From the Greek *perideris* "necklace."

Peridictyon Seberg, Frederiksen & Baden Gramineae

Origins:
Greek *peri* "around" and *diktyon* "a net."

Peridiscus Benth. Peridiscaceae (Bixaceae)

Origins:
Greek *peri* "around" and *diskos* "disc," stamens around outside of the disk.

Peridium Schott Euphorbiaceae

Origins:
From the Greek *pera* "pouch, bag."

Periestes Baillon Acanthaceae

Origins:
From the Greek *peri* "all round, around" and *esthes, esthos* "garment, dress," see also *Hypoestes* Sol. ex R. Br.

Perieteris Raf. Solanaceae

Origins:
See Constantine Samuel Rafinesque (1783-1840), *Sylva Telluriana.* 3: 74. 1836 [1838]; E.D. Merrill (1876-1956), *Index rafinesquianus.* 212. Jamaica Plain, Massachusetts, USA 1949.

Periglossum Decne. Asclepiadaceae

Origins:
From the Greek *peri* "around" and *glossa* "tongue."

Perilepta Bremek. Acanthaceae

Origins:
Greek *perileptos* "embraced, to be embraced," *perilambano* "clasp, embrace, surround," referring to the stem-clasping leaf bases.

Perilimnastes Ridl. Melastomataceae

Origins:
Greek *perilimnazo* "surround with water," *limne* "pool of standing water," *limnasia* "marshy ground."

Perilla L. Labiatae

Origins:
Derivation obscure, possibly a diminutive of the Latin *pera, ae* "a bag, wallet, pocket," Greek *pera* "a pouch," in reference to the form of the fruiting calyx, or from the Hindu name; Latin Perilla is a female proper name.

Species/Vernacular Names:
P. frutescens (L.) Britton (*Perilla ocimoides* L.)

English: purple common perilla, acute common perilla, perilla, beefsteak plant, common perilla

China: bai su zi, pai su, zi su, zi su zi, tzu su

Vietnam: tia to, tu to

Japan: shiso, aka-jiso (the red variety), ao-jiso (the green variety)

P. frutescens (L.) Britton var. *crispa* (Bentham) Deane ex Bailey

English: crisped common perilla

China: hui hui su

P. frutescens (L.) Britton var. *purpurascens* (Hayata) H.W. Li (*Perilla frutescens* (L.) Britton var. *acuta* (Thunberg) Kudô)

English: purple perilla, acute common perilla

China: ye sheng zi su

Perillula Maxim. Labiatae

Origins:
The diminutive of the genus *Perilla* L.

Periloba Raf. Nolanaceae (Solanaceae)

Origins:
Greek *peri* "around" and *lobos* "a lobe"; see C.S. Rafinesque, *Flora Telluriana.* 4: 87. 1836 [1838].

Perilomia Kunth Labiatae

Origins:
From the Greek *peri* "around" and *loma* "fringe," referring to the winged nutlets.

Perima Raf. Mimosaceae

Origins:
Possibly referring to *Perimkakuvalli*; see Constantine S. Rafinesque, *Sylva Telluriana.* 118. 1838; E.D. Merrill (1876-1956), *Index rafinesquianus.* The plant names published by C.S. Rafinesque, etc. 147. 1949.

Periomphale Baill. Alseuosmiaceae

Origins:
From the Greek *peri* "around" and *omphalos* "umbilicus, a navel."

Peripentadenia L.S. Smith Elaeocarpaceae

Origins:
From the Greek *peri* "around, near," *pente* "five" and *aden* "gland"; see Lindsay Stewart Smith (1917-1970), in *Proceedings of the Royal Society of Queensland.* 68: 45. 1957.

Species/Vernacular Names:
P. mearsii (C. White) L.S. Smith
Australia: buff quandong, grey quandong, boonjie

Peripeplus Pierre Rubiaceae

Origins:
From the Greek *peri* "around, all round" and *peplos* "a robe."

Periphanes Salisb. Amaryllidaceae (Liliaceae)

Origins:
Greek *peri* "around, all round" and *phaeinos, phanos* "bright, light," *phane, phanos* "torch."

Periphragmos Ruíz & Pavón Polemoniaceae

Origins:
Greek *peri* "around, all round" and *phragma* "a hedge, a fence, screen," *phragmon* "a thorn-hedge," *periphragma* "fence round a place, enclosure," referring to the habitat.

Periplexis Wall. Euphorbiaceae

Origins:
From the Greek *peripleko* "intertwining, twine round with something," *periploke* "twining round," *plexis* "plaiting, weaving."

Periploca L. Asclepiadaceae (Periplocaceae)

Origins:
Greek *periploke* "twining round, interlacing, entanglement," *periplokos* "entwined, twined about."

Species/Vernacular Names:
P. calophylla (Wight) Falconer
English: pretty leaf silk vine
China: qing she teng
P. forrestii Schlechter
English: Forrest silk vine
China: hei long gu
P. graeca L.
English: silk vine
P. laevigata L.
Tunisia: halleb
P. sepium Bunge
English: Chinese silk vine
China: gang liu, xiang jia pi

Periptera DC. Malvaceae

Origins:
From the Greek *peri* "around, near" and *pteron* "wing," *peripteros* "flying round about, surrounded by a gallery."

Peripteris Raf. Pteridaceae (Adiantaceae)

Origins:
For *Pteris* L., Greek *peri* "around, near, all round" and *pteris* "fern"; see C.S. Rafinesque, *Anal. Nat. Tabl. Univ.* 205. 1818.

Peripterygia (Baillon) Loes. Celastraceae

Origins:
From the Greek *peri* "around, all round" and *pteryx, pterygos* "small wing."

Peripterygium Hassk. Cardiopteridaceae (Rosidae, Celastrales)

Origins:
From the Greek *peri* "around, near" and *pteryx, pterygos* "small wing," referring to the base of the ovary, or referring to the genus *Pterygium*; see Justus Carl (Karl) Hasskarl (1811-1894), in *Tijdschrift voor natuurlijke Geschiedenis en Physiologie.* 10: 142. Amsterdam 1843.

Perispermum O. Degener Convolvulaceae

Origins:

From the Greek *peri* "around, near" and *sperma* "seed."

Perissocarpa J.A. Steyermark & B. Maguire Ochnaceae

Origins:

Greek *perissos, perittos* "prodigious, out of common, strange, beyond the regular number" and *karpos* "fruit," Plinius applied Latin *perisson* to a plant, also called *dorycnion.*

Perissocoeleum Mathias & Constance Umbelliferae

Origins:

Greek *perissos* "out of common, strange, beyond the regular number" and *koilos* "hollow."

Perissolobus N.E. Br. Aizoaceae

Origins:

Greek *perissos* "out of common, beyond the regular number" and *lobos* "lobe, pod."

Peristeranthus Hunt Orchidaceae

Origins:

Greek *peristera* "a pigeon, dove" and *anthos* "a flower," referring to the dove-like appearance of the flowers; see Trevor Edgar Hunt (1913-), in *The Queensland Naturalist.* 15: 17. (Sep.) 1954.

Peristeria Hooker Orchidaceae

Origins:

Peristerion, the diminutive of the Greek *peristera* "a pigeon, dove," referring to the appearance of the column-apex and anther, when seen from the front the flowers bear a resemblance to a dove.

Species/Vernacular Names:
P. elata Hook.

English: dove plant

Peristethium Tieghem Loranthaceae

Origins:

From the Greek *peristethion* "breastband," *peri* "around" and *stethos* "breast."

Peristrophe Nees Acanthaceae

Origins:

Greek *peristrophe* "a turning around," *peri* "around" and *strophe* "turning, twist," *strophos* "twisted cord, belt, band," referring to the twisted corolla tube or to the bracts surrounding and enclosing the calyx or to the involucre; see Nathaniel Wallich (1786-1854), *Plantae Asiaticae rariores.* 3: 112. London 1832.

Peristylus Blume Orchidaceae

Origins:

From the Greek *peri* "around" and *stylos* "a column," referring to the shape of the column; see Karl Ludwig von Blume, *Bijdragen tot de flora van Nederlandsch Indië.* 404. Batavia (Sep.-Dec.) 1825.

Species/Vernacular Names:
P. papuanus (Kraenzlin) J.J. Sm.

English: green habenaria

Peritassa Miers Celastraceae

Origins:

Greek *peri* and *tasso* "to arrange, put in order."

Perithrix Pierre Asclepiadaceae

Origins:

From the Greek *peri* "around" and *thrix, trichos* "hair," *perithrix, peritrikos* "the first growth of hair before it is cut."

Peritoma DC. Capparidaceae

Origins:

Greek *peri* "around" and *tome, tomos, temno* "division, section, to slice."

Perittostemma I.M. Johnston Boraginaceae

Origins:
Greek *perittos* "prodigious, out of common, strange" and *stemma* "a crown."

Perityle Benth. Asteraceae

Origins:
Greek *peri* "around" and *tyle* "lump, knob, any swelling," referring to the fruit margin.

Species/Vernacular Names:
P. inyoensis (Ferris) A. Powell
California: Inyo laphamia (Inyo Mountains)
P. villosa (S.F. Blake) Shinn.
California: Hanaupah laphamia

Perizoma (Miers) Lindl. Solanaceae

Origins:
From the Greek *perizoma, perizomatos* "girdle worn round the loins," *perizonnimi* "gird upon a person."

Perlaria Fabr. Gramineae

Origins:
From the Italian *perla* "a pearl."

Pernettya Gaudich. Ericaceae

Origins:
For the French academician Antoine Joseph Pernetty (Pernety), 1716-1801, with Louis-Antoine de Bougainville (1729-1811) visited the Falkland Islands and South America. His works include *The History of a Voyage to the Malouine Islands Made in 1763 and 1764 ... and of Two Voyages to the Streights of Magellan with an Account of the Patagonians.* London 1771, *La Connoissance de l'Homme moral par celle de l'Homme physique.* Berlin 1776 and 1777, *Dictionnaire mytho-hermétique.* Paris 1787, *Dissertation sur l'Amérique et les Américains.* Berlin 1770, *Les fables égyptiennes et grecques dévoilées.* Paris 1758 and *Des Herrn Pernety Handlexicon der bildenden Künste ... Aus dem Französischen übersetzt.* Berlin 1764. See Joanny Bricaud, *Les Illuminés d'Avignon. Étude de dom Pernety et son groupe.* Paris 1927; Charles François Brisseau de Mirbel (1776-1854), in *Annales des Sciences Naturelles.* 5: 102. 1825; L.A. de Bougainville, *Voyage autour du monde par la frégate du Roi "La Boudeuse" et la flute L'Etoile" en 1766-1769.* Paris 1771; Robert O. Cunningham, *Notes on the Natural History of the Strait of Magellan and West Coast of Patagonia Made During the Voyage of H.M.S. Nassau in the Years 1866, 67, 68, & 69.* Edinburgh 1871; R. Zander, F. Encke, G. Buchheim and S. Seybold, *Handwörterbuch der Pflanzennamen.* 14. Aufl. 427. Stuttgart 1993; John Dunmore, *Who's Who in Pacific Navigation.* University of Hawaii Press, Honolulu 1991; Margaret Patricia Henwood Laver, *An Annotated Bibliography of the Falkland Islands and the Falkland Island Dependencies* (as delimited on 3rd March, 1962). Cape Town 1977.

Pernettyopsis King & Gamble Ericaceae

Origins:
Resembling *Pernettya.*

Peronema Jack Labiatae (Verbenaceae)

Origins:
From the Greek *peroo, paroo* "to mutilate," *peros* "deficient" and *nema* "thread," referring to the missing stamens.

Species/Vernacular Names:
P. canescens Jack
English: false elder
Malaya: sungkai, sukai, cherek

Perotis Aiton Gramineae

Origins:
Greek *peroo* "to mutilate," *peros* "deficient," *perosis* "disabling," in reference to the nature of the small spikelets (often dangerous to the eyes), the glumes are awned; see William Aiton, *Hortus Kewensis.* 1: 85 and 3: 506. London 1781.

Species/Vernacular Names:
P. patens Gand.
English: bottlebrush grass, purplespike perotis
South Africa: katstertgras, per-aar-perotis
Yoruba: ero yewa
P. rara R. Br.
Australia: Comet grass, Comet River grass (Comet River, a tributary of Mackenzie River, Queensland)

Perotriche Cass. Asteraceae

Origins:

From the Greek *peroo* "to mutilate," *peros* "deficient" and *thrix, trichos* "hair."

Perovskia Karelin Labiatae

Origins:

For the Russian General Leo A. Perovski (Perovskij), 1792-1856, Governor of Orenburg; see Georg Christian Wittstein, *Etymologisch-botanisches Handwörterbuch.* 673. Ansbach 1852; F. Boerner & G. Kunkel, *Taschenwörterbuch der botanischen Pflanzennamen.* 4. Aufl. 149. Berlin & Hamburg 1989.

Species/Vernacular Names:

P. abrotanoides Karelin

English: Caspian sage

China: fen yao hua

P. atriplicifolia Bentham (*Perovskia pamirica* C.Y. Yang & B. Wang)

English: Pamir sage

China: bin li ye fen yao hua

Perralderia Cosson Asteraceae

Origins:

Named for the French botanist Henri René Letourneux de la Perraudière, 1831-1861, see also the genus *Tourneuxia* Cosson (Asteraceae).

Perralderiopsis Rauschert Asteraceae

Origins:

Resembling *Perralderia* Coss.

Perriera Courchet Simaroubaceae

Origins:

For the French botanist Joseph Marie Henri Alfred Perrier de la Bâthie, 1873-1958, from 1896 to 1933 in Madagascar. Among his works are *La végétation malgache.* Marseille, Paris 1921, *Les plantes introduites à Madagascar.* Toulouse 1933, *Catalogue des plantes de Madagascar.* Tananarive 1930-1940, *Le Tsaratanana, l'Ankaratra et l'Andringitra.* Tananarive 1927 and *Les Mélastomacées de Madagascar.* Toulouse 1932, contributed to Henri Humbert (1887-1967),

Flore de Madagascar et des Comores. 1936-1967. See Alfred Lacroix, *Notice historique sur quatre botanistes,* etc. [On P.H. Lecomte, H. Perrier de la Bâthie, E.M. Heckel and H. Jumelle.] 1938; R. Zander, F. Encke, G. Buchheim and S. Seybold, *Handwörterbuch der Pflanzennamen.* 14. Aufl. Stuttgart 1993; J.H. Barnhart, *Biographical Notes upon Botanists.* 3: 70. 1965; Ethelyn Maria Tucker, *Catalogue of the Library of the Arnold Arboretum of Harvard University.* Cambridge, Massachusetts 1917-1933.

Perrieranthus Hochr. Malvaceae

Origins:

After the French botanist Joseph Marie Henri Alfred Perrier de la Bâthie, 1873-1958, from 1896 to 1933 in Madagascar; see J.H. Barnhart, *Biographical Notes upon Botanists.* 3: 70. 1965.

Perrierastrum Guillaumin Labiatae

Origins:

After the French botanist Joseph Marie Henri Alfred Perrier de la Bâthie, 1873-1958, from 1896 to 1933 in Madagascar; see J.H. Barnhart, *Biographical Notes upon Botanists.* 3: 70. 1965.

Perrierbambus A. Camus Gramineae

Origins:

After the French botanist Joseph Marie Henri Alfred Perrier de la Bâthie, 1873-1958, from 1896 to 1933 in Madagascar; see J.H. Barnhart, *Biographical Notes upon Botanists.* 3: 70. 1965.

Perrieriella Schltr. Orchidaceae

Origins:

After the French botanist Joseph Marie Henri Alfred Perrier de la Bâthie, 1873-1958, orchid collector, from 1896 to 1933 in Madagascar; see J.H. Barnhart, *Biographical Notes upon Botanists.* 3: 70. 1965.

Perrierodendron Cavaco Sarcolaenaceae

Origins:

After the French botanist Joseph Marie Henri Alfred Perrier de la Bâthie, 1873-1958.

Perrierophytum Hochr. Malvaceae

Origins:

After the French botanist Joseph Marie Henri Alfred Perrier de la Bâthie, 1873-1958; see J.H. Barnhart, *Biographical Notes upon Botanists*. 3: 70. 1965.

Perrierosedum (A. Berger) H. Ohba Crassulaceae

Origins:

For the French botanist Joseph Marie Henri Alfred Perrier de la Bâthie, 1873-1958, from 1896 to 1933 in Madagascar.

Perrottetia Kunth Celastraceae

Origins:

For the French (Swiss-born) gardener George (Georges Guerrard) Samuel Perrottet, 1793-1870 (Pondicherry), sericiculturist, agronomical and government botanist, traveler and plant collector, botanical explorer, in 1824 went to Senegal, West Africa, Director of the Agricultural Station (Senegal), co-author with Jean Baptiste Antoine Guillemin (1796-1842) and Achille Richard (1794-1852) of *Florae Senegambiae tentamen*. [The families arranged after De Candolle, Ranunculaceae to Myrtaceae.] Paris, London 1830-1833. His works include *Catalogue raisonné des plantes introduites dans les colonies françaises de Bourbon et de Cayenne*. Paris 1824 and *Catalogue des plantes du jardin botanique et d'acclimatation du gouvernement à Pondichéry*. Pondichéry 1867; see John H. Barnhart, *Biographical Notes upon Botanists*. 3: 71. 1965; Joseph Vallot, "Études sur la flore du Sénégal." in *Bull. Soc. Bot. de France*. 29: 186. Paris 1882; F.N. Williams, "Collectors of Gambian plants." in *Bull. Herb. Boiss*. sér. 2, 7: 84. Geneva 1907; Ignatz Urban, *Geschichte des Königlichen Botanischen Museums zu Berlin-Dahlem (1815-1913). Nebst Aufzählung seiner Sammlungen*. Dresden 1916; R.W.J. Keay, "Botanical collectors in West Africa prior to 1860." in *Comptes Rendus A.E.T.F.A.T.* 55-68. Lisbon 1962; H.N. Clokie, *Account of the Herbaria of the Department of Botany in the University of Oxford*. 223. Oxford 1964; A. Lasègue, *Musée botanique de Benjamin Delessert*. Paris 1845; E.M. Tucker, *Catalogue of the Library of the Arnold Arboretum of Harvard University*. 1917-1933; Isaac Henry Burkill, *Chapters on the History of Botany in India*. 9-10. Delhi 1965; G. Murray, *History of the Collections Contained in the Natural History Departments of the British Museum*. London 1904; A. White and B.L. Sloane, *The Stapelieae*. Pasadena 1937; Giovanni Borghesi, *Lettera scritta da Pondisceri a' 10 di Febbraio 1704 dal dottore Giovanni Borghesi medico della missione spedita alla*

China dalla Santità di N.S. Papa Clemente XI. Roma, Zenobi 1705; F.N. Hepper and Fiona Neate, *Plant Collectors in West Africa*. 64. 1971; I.C. Hedge and J.M. Lamond, *Index of Collectors in the Edinburgh Herbarium*. Edinburgh 1970.

Species/Vernacular Names:

P. sandwicensis A. Gray

Hawaii: olomea, pua'a olomea, waimea

Persea Miller Lauraceae

Origins:

From the Greek name *persea*, applied by Theophrastus (*HP*. 3.3.5, 4.2.5) and Hippocrates (*De Morbis Mulierum*. 1.90) to an unknown Egyptian tree, possibly *Cordia myxa* L. or a species of *Mimusops*; see Philip Miller, *The Gardeners Dictionary*. Abr. ed. 4. London (28 Jan.) 1754; Helmut Genaust, *Etymologisches Wörterbuch der botanischen Pflanzennamen*. 471. Basel 1996.

Species/Vernacular Names:

P. sp.

Peru: huira palta

Mexico: uxu, yexo, yaxo, yaga yaxo, ya yexru, yrxo, yuccu't'p, cucataj

P. americana Mill. (*Persea gratissima* Gaertn.f.)

English: avocado pear, alligator pear, avocado

French: avocatier

Congo: saboka

Latin America: aguacate, palta

Peru: abacasi, abocate, acapa, aguacate, ahuacate, ahuacatl, apacha, avocado acapa, caí, huira palta, huira palto, parte, palta, palta moena, palto, paltai, palltay, palte, parité, parta, parata

Bolivia: aguacate, palta

Brazil: abacate, abacateiro, louro-abacate, abacate-creme vegetal

Mexico: aguacate, ahuacate, aguacachile, aguacate xinene, ahoacacuáhuitl, ahuacacáhuatl, ahuacat, bashlobó, cucataj, cupandra, cuut'p, cupanda, cuytuim, ohui, on, pagua, palta, shamal, tunuá, uy, ohuacatl, lhpuy, lhpua, tatsan, tzani, tutiti, tzison, tziton, yashusa, yaujca, yaxhu, yéuca-te

Guatemala: ju, oj, un, um, tc'om

Japan: abokado, wani-nashi

The Philippines: abukado

P. borbonia (L.) Spreng.

English: red bay

P. indica (L.) Spreng.

English: Madeira mahogany

Persica Miller Rosaceae

Origins:

Latin *persicus, i* (Persia) *"persica arbor*, the peach tree."

Persicaria Miller Polygonaceae

Origins:

Latin *persicus, i* (Persia) *"persica arbor*, the peach tree," referring to the shape of the leaves; see Philip Miller, *The Gardeners Dictionary.* Abridged edition. 4. London (28 Jan.) 1754.

Species/Vernacular Names:

P. capitata (D. Don) H. Gross (*Polygonum capitatum* Buch.-Ham. ex D. Don)

English: knotweed

P. decipiens (R. Br.) K.L. Wilson (*Polygonum decipiens* R. Br.)

English: knotweed, slender knotweed

P. hydropiper (L.) Spach (*Polygonum hydropiper* L.)

English: water pepper

P. lapathifolia (L.) S.F. Gray (*Polygonum lapathifolium* L.; *Polygonum lapathifolium* L. subsp. *maculatum* (S.F. Gray) T.-Dyer & Trim.; *Polygonum lapathifolium* L. var. *maculatum* T.-Dyer & Trim.)

English: spotted knotweed, pale knotweed

Southern Africa: hanekam, viltige duisendknoop; tolo-la-khongoana (Sotho)

P. limbata (Meisn.) Hara (*Polygonum limbatum* Meisn.; *Polygonum piliferum* Tikovsky; *Polygonum schinzii* C.H. Wr.)

English: knotweed

P. maculosa S.F. Gray (*Polygonum persicaria* L.)

English: red shank, Jesus plant

P. orientalis (L.) Spach (*Polygonum orientale* L.)

English: prince's feathers, prince's plume

P. prostrata (R. Br.) Soják (*Polygonum prostratum* R. Br.)

English: creeping knotweed

P. senegalensis (Meisn.) Soják forma *senegalensis* (*Polygonum sambesiacum* Schuster; *Polygonum senegalense* Meisn. subsp. *senegalense*)

English: snake root

South Africa: duisendknoop, hanekam, slangwortel, snotterbel

P. subsessilis (R. Br.) K.L. Wilson (*Polygonum subsessile* R. Br.)

English: hairy knotweed

Persimon Raf. Ebenaceae

Origins:

English persimmon, see C.S. Rafinesque, *Sylva Telluriana.* 164. 1838; E.D. Merrill, *Index rafinesquianus.* The plant names published by C.S. Rafinesque, etc. 188. 1949; Ernest Weekley, *An Etymological Dictionary of Modern English.* 2: 1073. New York 1967.

Personia Raf. Asteraceae

Origins:

For the South African botanist Christiaan Hendrik Persoon, 1761 (or 1762)-1836 (Paris), mycologist.

Personula Raf. Lentibulariaceae

Origins:

See C.S. Rafinesque, *Flora Telluriana.* 4: 110. 1836 [1838]; E.D. Merrill, *Index rafinesquianus.* The plant names published by C.S. Rafinesque, etc. 222. Jamaica Plain, Massachusetts, USA 1949.

Persoonia J.E. Smith Proteaceae

Origins:

After the South African (b. Cape of Good Hope/Cape Town) botanist Christiaan Hendrik Persoon, 1761/1762-1836 (d. Paris), mycologist, in 1786 studied medicine at Leyden, from 1787 to 1802 studied medicine and natural sciences at Göttingen (never completed his studies), from 1802 to 1836 in Paris. His numerous works include *Observationes mycologicae.* Lipsiae et Lucernae 1796-1799 [-1800], *Species plantarum.* Petropoli [St. Petersburg] 1817-1821, *Synopsis plantarum.* Parisiis lutetiorum [Paris] et Tubingae 1805-1807, *Synopsis methodica fungorum.* Gottingae 1801, *Mycologia europaea.* Erlangae [Erlangen] 1822-1828, *Enumeratio systematica specierum plantarum medicinalium,* e synopsi plantarum C.H. Persoon desumta. [Edited by Gerrit Jan Mulder, 1802-1880] Roterodami [Rotterdam] 1829 and *Traité sur les champignons comestibles.* Paris 1818; see John H. Barnhart, *Biographical Notes upon Botanists.* 3: 72. 1965; Stafleu and Cowan, *Taxonomic Literature.* 4: 178-185. 1983; M.A. Donk, in *Dictionary of Scientific Biography* 10: 530-532. 1981; Gilbert Westacott Reynolds (1895-1967), *The Aloes of South Africa.* 93. Rotterdam 1982; Sir James Edward Smith (1759-1828), in *Transactions of the Linnean Society of London. Botany.* 4: 215. (May) 1798; Mary Gunn and Leslie Edward W. Codd, *Botanical Exploration of Southern Africa.* 279. 1981; T.W. Bossert, *Biographical Dictionary of Botanists Represented in the Hunt*

Institute Portrait Collection. 307. 1972; Jonas C. Dryander, *Catalogus bibliothecae historico-naturalis Josephi Banks*. London 1800; A. Lasègue, *Musée botanique de Benjamin Delessert*. Paris 1845; E.M. Tucker, *Catalogue of the Library of the Arnold Arboretum of Harvard University*. 1917-1933; E. Bretschneider, *History of European Botanical Discoveries in China*. Leipzig 1981; R. Zander, F. Encke, G. Buchheim and S. Seybold, *Handwörterbuch der Pflanzennamen*. 14. Aufl. Stuttgart 1993; William Darlington (1782-1863), *Reliquiae Baldwinianae*. Philadelphia 1843; I.C. Hedge and J.M. Lamond, *Index of Collectors in the Edinburgh Herbarium*. Edinburgh 1970; Frans A. Stafleu, *Linnaeus and the Linnaeans*. The spreading of their ideas in systematic botany, 1735-1789. Utrecht 1971.

Species/Vernacular Names:

P. acerosa Schultes & J.H. Schultes

Australia: mossy geebung (geebung = from the Aboriginal name jibbong)

P. angulata R. Br.

Australia: Blue Mountains geebung

P. arborea F. Muell.

Australia: tree geebung

P. attenuata R. Br.

Australia: narrowed geebung

P. caleyi R. Br.

Australia: Caley's geebung

P. chamaepeuce Meissner

Australia: dwarf geebung

P. chamaepitys Cunn.

Australia: mountain geebung

P. confertiflora Benth.

Australia: cluster-flower geebung

P. cornifolia R. Br.

Australia: horn-leaf geebung, dogwood geebung

P. curvifolia R. Br.

Australia: western geebung

P. fastigiata R. Br.

Australia: rough geebung

P. gunnii Hook.f.

Australia: Gunn's geebung

P. hirsuta Pers.

Australia: hairy geebung

P. juniperina Labill.

Australia: geebung, prickly geebung

P. laevis (Cav.) Domin

Australia: broad-leaved geebung

P. lanceolata Andrews

Australia: lance-leaf geebung

P. laurina Pers.

Australia: laurel geebung

P. linearis Andrews

Australia: narrow-leaved geebung, narrow-leaf geebung

P. marginata R. Br.

Australia: Bathurst geebung

P. media R. Br.

Australia: North Coast geebung

P. mollis R. Br.

Australia: soft geebung

P. myrtilloides Schultes

Australia: myrtle geebung

P. nutans R. Br.

Australia: nodding geebung

P. oblongata R. Br.

Australia: thin-leaf geebung

P. pinifolia R. Br.

Australia: pine-leaved geebung

P. rigida R. Br.

Australia: stiff geebung

P. sericea R. Br.

Australia: silky geebung

P. silvatica L. Johnson

Australia: forest geebung, jungle geebung

P. subvelutina L. Johnson

Australia: velvety geebung

P. tenuifolia R. Br.

Australia: thread-leaf geebung

Persoonia Willd. Meliaceae

Origins:

For the South African botanist Christiaan Hendrik Persoon, 1761 (or 1762)-1836 (Paris), mycologist.

Pertusadina Ridsdale Rubiaceae

Origins:

Fenestrated trunk, Latin *pertusus* "perforated, opening" plus the genus *Adina* Salisb., from *pertundo, tudi, tusum* "to make a hole through, perforate."

Pertya Sch.Bip. Asteraceae

Origins:

Named for the German botanist Joseph Anton Maximilian Perty, 1804-1884, zoologist, entomologist, professor of zoology, naturalist. His works include *Die Anthropologie als die Wissenschaft von dem körperlichen und geistigen Wesen des Menschen*. Leipzig und Heidelberg 1874, *Delectus Animalium articulatorum*, quae in itinere per Brasiliam annis 1817-20 ... collegerunt J.B. de Spix ... et ... C.F.Ph. de Martius, digessit, descripsit, pingenda curavit M. Perty. Monachii 1830-34, *Ueber die Seele*. Bern 1856, *Die mystischen Erscheinungen der menschlichen Natur*. Leipzig und Heidelberg 1861 and *Observationes nonnullae in coleoptera Indiae orientalis*. Monachii 1831; see John H. Barnhart, *Biographical Notes upon Botanists*. 3: 72. 1965.

Perula Raf. Moraceae

Origins:

From the Latin *perula* "a little wallet, pocket," diminutive of *pera*; see C.S. Rafinesque, *Sylva Telluriana*. 58. 1838.

Perula Schreb. Euphorbiaceae

Origins:

Latin *perula* "a little wallet, pocket," diminutive of *pera*, genus *Pera* Mutis.

Perularia Lindl. Orchidaceae

Origins:

Latin *perula* "a little wallet, pocket," indicating the stigmatic surface.

Perulifera A. Camus Gramineae

Origins:

From the Latin *perula* "a little wallet, pocket" and *fero, fers, tuli, latum, ferre* "to bear, carry."

Peruvocereus Akers Cactaceae

Origins:

From Peru and *Cereus*.

Perxo Raf. Labiatae

Origins:

See C.S. Rafinesque, *Autikon botanikon*. Icones plantarum select. nov. vel rariorum, etc. 121. Philadelphia 1840; E.D. Merrill, *Index rafinesquianus*. 209. 1949.

Perymeniopsis H. Robinson Asteraceae

Origins:

Resembling *Perymenium* Schrader.

Perymenium Schrad. Asteraceae

Origins:

From the Greek *peri* "around" and *hymenion* "a small membrane."

Pescatoria Reichb.f. Orchidaceae

Origins:

Dedicated to the Luxemburg-born orchid grower and horticulturist Jean Paul Pescatore, 1793-1855, orchidophile; see G.A. Lueddemann (Lüddemann) (1821-1884), Jules Émile Planchon (1823-1888), Jean Jules Linden (1817-1898) and Heinrich Gustav Reichenbach (1824-1889) co-editors of *Pescatorea*. Iconographie des Orchidées. Bruxelles [1854-] 1860.

Peschiera A. DC. Apocynaceae

Origins:

For Jean Peschier, 1774-1831, physician.

Pesomeria Lindley Orchidaceae

Origins:

Greek *piptein* "to fall off" and *meros* "part," referring to the sepals.

Pessopteris Underw. & Maxon Polypodiaceae

Origins:

Greek *pessos* "any oval body, oval-shaped stone, plug of wool" and *pteris* "fern."

Pestalozzia Zollinger & Moritzi Cucurbitaceae

Origins:
Named for the Swiss pedagogist Johann Heinrich Pestalozzi, 1746-1827.

Petagnaea Caruel Umbelliferae

Origins:
After the Italian botanist Vincenzo Petagna, 1734-1810, physician, entomologist, professor of botany. Among his writings are *Institutiones botanicae*. Neapoli 1785-1787, *Institutiones entomologicae*. Neapoli 1792 and *Specimen insectorum ulterioris Calabriae*. Neapoli 1786 [Editio nova. Lipsiae 1808]. See John H. Barnhart, *Biographical Notes upon Botanists*. 3: 72. 1965; R. Zander, F. Encke, G. Buchheim and S. Seybold, *Handwörterbuch der Pflanzennamen*. 14. Aufl. Stuttgart 1993; Jonas C. Dryander, *Catalogus bibliothecae historico-naturalis Josephi Banks*. London 1800; T.W. Bossert, *Biographical Dictionary of Botanists Represented in the Hunt Institute Portrait Collection*. 307. 1972; E.M. Tucker, *Catalogue of the Library of the Arnold Arboretum of Harvard University*. 1917-1933.

Petagnia Gussone Umbelliferae

Origins:
After the Italian botanist Vincenzo Petagna, 1734-1810.

Petagnia Raf. Solanaceae

Origins:
After the Italian botanist Vincenzo Petagna, 1734-1810.

Petagniana Raf. Fabaceae

Origins:
After the Italian botanist Vincenzo Petagna, 1734-1810.

Petalacte D. Don Asteraceae

Origins:
From the Greek *petalon* "a petal, leaf" and *aktin* "ray," referring to the scales of the receptacle.

Petalactella N.E. Br. Asteraceae

Origins:
Diminutive of the generic name *Petalacte*.

Petaladenium Ducke Fabaceae

Origins:
From the Greek *petalon* "a petal, leaf" and *aden* "gland."

Petalandra F. Muell. ex Boiss. Euphorbiaceae

Origins:
From the Greek *petalon* "a petal" and *aner, andros* "male, stamen."

Petalanthera Nutt. Loasaceae

Origins:
Greek *petalon* "a petal" and *anthera* "anther."

Petalanthera Raf. Acanthaceae

Origins:
From the Greek *petalon* "a petal" and *anthera* "anther"; see C.S. Rafinesque, *Flora Telluriana*. 2: 100. 1836 [1837].

Petalidium Nees Acanthaceae

Origins:
Greek *petalon* "a petal" and *-idium*, a diminutive suffix, the bracts are petal-like.

Species/Vernacular Names:
P. engleranum (Schinz) C.B. Cl. (*Petalidium eurychlamys* Mildbr.; *Petalidium latifolium* (Schinz) C.B. Cl.; *Petalidium ovatum* (Schinz) C.B. Cl.; *Pseudobarleria engleriana* Schinz)
English: petalidium

Petalocentrum Schltr. Orchidaceae

Origins:
From the Greek *petalon* "a petal, leaf" and *kentron* "spur," the spurred petals.

Petalochilus R.S. Rogers Orchidaceae

Origins:

From the Greek *petalon* "a petal" and *cheilos* "lip," the form of the labellus.

Petalodactylis Arènes Rhizophoraceae

Origins:

From the Greek *petalon* "a leaf, a petal" and *daktylos* "a finger."

Petalodiscus Baillon Euphorbiaceae

Origins:

From the Greek *petalon* "a leaf, a petal" and *diskos* "a disc."

Petalogyne F. Muell. Caesalpiniaceae

Origins:

Greek *petalon* "a petal" and *gyne* "a woman, female, female organs," see also *Petalostylis* R. Br.

Petalolepis Cass. Asteraceae

Origins:

Greek *petalon* "a petal" and *lepis* "a scale"; see Alexandre Henri Gabriel Comte de Cassini (1781-1832), in *Bull. Sci. Soc. Philom. Paris*. Année 1817. 138. 1817.

Petalolophus K. Schumann Annonaceae

Origins:

From the Greek *petalon* "a petal" and *lophos* "a crest."

Petalonema Gilg Melastomataceae

Origins:

From the Greek *petalon* "a leaf, a petal" and *nema* "thread."

Petalonema Schltr. Asclepiadaceae

Origins:

Greek *petalon* "a leaf, a petal" and *nema* "thread."

Petalonyx A. Gray Loasaceae

Origins:

Greek *petalon* "a petal" and *onyx, onychos* "a claw, nail."

Petalostelma E. Fourn. Asclepiadaceae

Origins:

From the Greek *petalon* "a petal" and *stelma, stelmatos* "a girdle, belt," see also *Metastelma* R. Br.

Petalostemon Michx. Fabaceae

Origins:

From the Greek *petalon* "a petal" and *stemon* "stamen," petals and stamens are joined, the petals are adnate to the staminal tube.

Petalostigma F. Muell. Euphorbiaceae

Origins:

From the Greek *petalon* "a petal" and *stigma* "a stigma," the stigmas are petal-like.

Species/Vernacular Names:

P. banksii Britten & S. Moore

English: bitter bark

P. pubescens Domin

English: quinine bush, quinine tree, bitter bark, native quince

P. quadriloculare F. Muell.

English: quinine bush

P. triloculare Müll. Arg.

English: long-leaved bitter bark

Petalostylis R. Br. Caesalpiniaceae

Origins:

Greek *petalon* "petal" and *stylos* "style, a pillar," referring to the petaloid styles; see Charles Sturt (1795-1869), *Narrative of an Expedition into Central Australia, Performed ... During the Years 1844, 1845 and 1846*. 2. App. 79. London 1849.

Species/Vernacular Names:

P. labicheoides R. Br.

Australia: butterfly bush, slender petalostylis

Petamenes Salisbury ex J.W. Loudon Iridaceae

Origins:

Derivation doubtful, perhaps from the Greek *petomai, petamai* "to fly, fly abroad" or *petamenos* "tumbler, rope-dancer"; see Jane Wells Loudon (1807-1858), *The Ladies' Flower-Garden.* 42. London 1840.

Petasites Miller Asteraceae

Origins:

Greek *petasitis, petasites* "butter-bur" (Dioscorides and Galenus for a species of *Petasites*), *petasos* "a sun-hat, a hat with a broad brim, broad umbellated leaf," Latin *petasus* "a travelling hat, cap," referring to the large leaves; see Philip Miller, *The Gardeners Dictionary.* Abr. ed. 4. London (28 Jan.) 1754.

Species/Vernacular Names:

P. fragrans (Villars) Presl

English: winter heliotrope

P. hybridus (L.) Gaertner, Meyer & Scherb.

English: bog rhubarb, butter-bur

P. japonicus (Siebold & Zucc.) Maxim. (*Petasites liukiuensis* Kitam.; *Nardosmia japonica* Sieb. & Zucc.)

English: butter-bur, bog rhubarb

Japan: fuki

China: feng dou cai

P. pyrenaicus (Loefl.) G. López

English: winter heliotrope

Petastoma Miers Bignoniaceae

Origins:

Greek *petasos* "broad-brimmed felt hat, broad umbellated leaf" and *stoma* "mouth."

Petchia Livera Apocynaceae

Origins:

For the British (b. Yorks.) botanist Thomas (Tom) Petch, 1870-1948 (d. Norfolk), mycologist, 1905-1925 Peradeniya (Royal Botanic Gardens, Sri Lanka, Ceylon), President of the British Mycological Society, edited and contributed to *Annals Royal Botanic Gardens Peradeniya.* His writings include *Bibliography of Books and Papers Relating to Agriculture and Botany* to the end of the year 1915. Colombo 1925, "The genera *Hypocrella* and *Aschersonia.*" *Ann. Bot. Gard. Peradeniya.* 5: 521-537. 1914, *The Diseases and Pests of the Rubber Tree.* London 1921, *The Diseases of*

the Tea Bush. London 1923, *The Physiology and Diseases of Hevea brasiliensis,* the premier plantation rubber tree. London 1911 and *Plane Geometry Adapted to Heuristic Methods of Teaching.* London [1903], with Guy Richard Bisby (1889-1958) wrote *The Fungi of Ceylon.* Colombo 1950. See J.H. Barnhart, *Biographical Notes upon Botanists.* 3: 72. 1965; S. Lenley et al., *Catalog of the Manuscript and Archival Collections and Index to the Correspondence of John Torrey.* Library of the New York Botanical Garden. 325. 1973; E.M. Tucker, *Catalogue of the Library of the Arnold Arboretum of Harvard University.* 1917-1933; Elmer Drew Merrill, *Contr. U.S. Natl. Herb.* 30(1): 240. 1947; Ray Desmond, *Dictionary of British & Irish Botanists and Horticulturists.* 548. London 1994; E.D. Merrill, in *Bernice P. Bishop Mus. Bull.* 144: 149. 1937; I.C. Hedge and J.M. Lamond, *Index of Collectors in the Edinburgh Herbarium.* Edinburgh 1970.

Petelotiella Gagnepain Urticaceae

Origins:

After the French botanist Paul Alfred Pételot, 1885-(after) 1940, bryologist, in Vietnam, lauréat de l'Académie des Sciences, *chargé de cours* à la Faculté Mixte de Médecine et de Pharmacie de Saigon, Chef de la Division de Botanique du Centre de Recherches Scientifiques et Techniques. His works include "La botanique en Indochine. Bibliographie." *Bull. Econ. Indochine.* 32: 587-632. 1929, *Les plantes médicinales du Cambodge, du Laos et du Viêtnam.* Saigon 1952-1954 and "Bibliographie botanique de l'Indochine." *Arch. Rech. Agron. Past. Viêtnam.* 24: 1-102. 1955, with Charles Cresson wrote *Catalogue des produits de l'Indochine, Produits médicaux.* 1928-1935; see E. Perrot & P. Hurrier, *Matière médicale et pharmacopée sino-annamite.* Paris 1907; Stafleu and Cowan, *Taxonomic Literature.* 4: 189. 1983; L. Menaut, "La matière médicale cambodgienne." *Bull. Econ. Indochine.* 1929; I.C. Hedge and J.M. Lamond, *Index of Collectors in the Edinburgh Herbarium.* Edinburgh 1970.

Petelotoma DC. Rhizophoraceae

Origins:

Greek *petelos* "outspread, stretched, full-grown" or *petelodes* "like a leaf" and *tome* "division, section, cutting."

Peteravenia R.M. King & H. Robinson Asteraceae

Origins:

After Peter Hamilton Raven, botanist and plant collector; see Irving William Knobloch, compil., "A preliminary verified list of plant collectors in Mexico." *Phytologia Memoirs*. VI. 1983; Ida Kaplan Langman, *A Selected Guide to the Literature on the Flowering Plants of Mexico*. 580. University of Pennsylvania Press, Philadelphia 1964.

Peteria A. Gray Fabaceae

Origins:

After the English-born Kentucky geologist, Robert Peter, 1805-1894, wrote *The History of the Medical Department of Transylvania University*. [Prepared for publication by his daughter Miss Johanna Peter.] Louisville 1905. See William Henry Perrin, ed., *History of Bourbon, Scott, Harrison, and Nicholas Counties, Kentucky*. With an outline sketch of the Blue Grass Region, by R. Peter. Chicago 1882; William Henry Perrin, *History of Fayette County, Kentucky*. With an outline sketch of the Blue Grass Region, by R. Peter. Chicago 1882; David Dale Owen, *Second Report of a Geological Reconnoissance of the Middle and Southern Counties of Arkansas*. Made during the years 1859 and 1860. By D.D. Owen ... assisted by Robert Peter, M. Leo Lesquereux, Edward Cox. 1860; Kentucky — Geological Survey, *Report of the Geological Survey in Kentucky*, made during the years 1854 and 1855 by D.D. Owen ... assisted by Robert Peter, Sidney S. Lyon. 1856; Kentucky — Geological Survey, *Second Report of the Geological Survey in Kentucky*, made during the years 1856 and 1857 by D.D. Owen ... assisted by Robert Peter, Sidney S. Lyon. 1857; Kentucky — Geological Survey, *Third Report of the Geological Survey in Kentucky*, made by D.D. Owen ... assisted by Robert Peter, Sidney S. Lyon, Leo Lesquereux, Edward T. Cox. 1857.

Species/Vernacular Names:

P. thompsoniae S. Watson

English: spine-noded milkvetch

Peteria Raf. Rubiaceae

Origins:

Orthographic variant of *Petesia* P. Browne.

Petermannia F. Muell. Philesiaceae (Petermanniaceae, Smilacaceae)

Origins:

After the German cartographer August Heinrich Petermann, 1822-1878, geographer, traveler. His writings include *African Discovery*. London 1854, *The Search for Franklin*. London 1852 and *An Account of the Progress of the Expedition*

to Central Africa ... under Messrs. Richardson, Barth, Overweg and Vogel, in the years 1850, 1851, 1852 and 1853. London 1854, with Bruno Hassenstein wrote *Inner-Afrika nach dem Stande der geographischen Kenntniss in den Jahren 1861 bis 1863*. Gotha 1862. See E. Weller, *Leben und Wirken August Petermanns*. Leipzig 1914; Ferdinand von Mueller, *Fragmenta Phytographiae Australiae*. 2: 92. Melbourne 1860; R. Ericksen, *Ernest Giles, Explorer and Traveler. 1835-1897*. Melbourne 1978; Stafleu and Cowan, *Taxonomic Literature*. 4: 193. Utrecht 1983.

Species/Vernacular Names:

P. cirrosa F. Muell.

Australia: Clarence River vine

Peterodendron Sleumer Flacourtiaceae

Origins:

After the German botanist Gustav Albert Peter, 1853-1937; see John H. Barnhart, *Biographical Notes upon Botanists*. 3: 72. 1965; E.M. Tucker, *Catalogue of the Library of the Arnold Arboretum of Harvard University*. 1917-1933; T.W. Bossert, *Biographical Dictionary of Botanists Represented in the Hunt Institute Portrait Collection*. 307. 1972; R. Zander, F. Encke, G. Buchheim and S. Seybold, *Handwörterbuch der Pflanzennamen*. 14. Aufl. Stuttgart 1993; I.C. Hedge and J.M. Lamond, *Index of Collectors in the Edinburgh Herbarium*. Edinburgh 1970.

Petersia Welwitsch ex Bentham Lecythidaceae

Origins:

For Wilhelm Carl Hartwig Peters, 1815-1883, German professor of medicine (1851) and zoology (1856), naturalist.

Petersianthus Merrill Lecythidaceae

Origins:

After the German entomologist Wilhelm Carl Hartwig Peters, 1815-1883, naturalist, zoologist, physician and traveler, professor of medicine (1851) and zoology (1856), 1842-1848 in South and East Africa and India, joint author of *Naturwissenschaftliche Reise nach Mossambique*. Berlin [1861-] 1862-1864, wrote *Ueber Wohnen und Wandern der Thiere. Vortrag*, etc. Berlin 1867. See F.A. Maximilian Kuhn (1842-1894), *Filices africanae ... Accedunt filices Deckenianae et Petersianae*. Lipsiae [Leipzig] 1868; Otto Kersten (1839-1900), *Geographische Nachrichten für Welthandel und Volkswirtschaft ... unter der ... Redaktion von Dr. O.K.* [Berlin — Central-Verein für Handelsgeographie, etc. Geographische Nachrichten, etc.] Berlin 1879;

Carl Claus von der Decken (1833-1865), *Baron C.C. von der Decken's Reisen in Ost Afrika in 1859-61*. Leipzig & Heidelberg 1869-1879; Friedrich Gerhard Rohlfs (1831-1896), *Kufra ... Reise von Tripolis nach der Oase Kufra*. Leipzig 1881; John H. Barnhart, *Biographical Notes upon Botanists*. 3: 73. 1965; Ida Kaplan Langman, *A Selected Guide to the Literature on the Flowering Plants of Mexico*. 580. University of Pennsylvania Press, Philadelphia 1964; A. Lasègue, *Musée botanique de Benjamin Delessert*. Paris 1845; E.M. Tucker, *Catalogue of the Library of the Arnold Arboretum of Harvard University*. 1917-1933.

Species/Vernacular Names:

P. macrocarpus (P. Beauv.) Liben (*Combretodendron macrocarpum* (P. Beauv.) Keay; *Combretodendron africanum* (Welw.) Exell)

Cameroon: bing, abing, boso

Congo: minzu

Gabon: abing

Ghana: esia

West Africa: essia

Central Africa: minzu, esia, nossoba

Ivory Coast: abale

Nigeria: akasun (Yoruba); owewe (Edo); oze (Ijaw); anwushi (Igbo); okorebeni (Efik); onunun (Boki)

Yoruba: akasun

Petesia P. Browne Rubiaceae

Origins:
Greek *petao, petannymi* "to spread out, unfold," referring to the appearance of the branches.

Petilium Ludw. Liliaceae

Origins:
Latin *petilium* applied by Plinius to an autumnal flower, unknown.

Petimenginia Bonati Scrophulariaceae

Origins:
For the French botanist Marcel Georges Charles Petitmengin, 1881-1908, traveler, plant collector, specialist in Primulaceae, wrote *Contributions à l'étude des Primulacées sino-japonaises*. 1907; see Elmer D. Merrill and Egbert H. Walker, *A Bibliography of Eastern Asiatic Botany*. 384-385. The Arnold Arboretum of Harvard University, Jamaica Plain, Massachusetts, USA 1938; R. Zander, F. Encke, G.

Buchheim and S. Seybold, *Handwörterbuch der Pflanzennamen*. 14. Aufl. Stuttgart 1993.

Petitia Jacq. Labiatae (Verbenaceae)

Origins:
For the French physician François Pourfour du (de) Petit, 1664-1741, botanist, wrote "De la précipitation du sel marin dans la fabrique du salpêtre." Recueil de mémoires, etc. *Académie des Sciences*, Paris 1776, *Lettre de M. Petit ... dans laquelle il démontre que le cristallin est fort près de l'uvée*. Paris 1729, *Lettres d'un Médecin des Hospitaux du Roy à un autre Médecin de ses amis*. 1710 and *Lettres sur les maladies des yeux*. 1729.

Petitmenginia Bonati Scrophulariaceae

Origins:
For the French botanist Marcel Georges Charles Petitmengin, 1881-1908.

Petiveria L. Phytolaccaceae

Origins:
Dedicated to British (b. Warwickshire) botanist Jacob (James) Petiver, 1658-1718 (d. London), apothecary, naturalist, he became a Fellow of the Royal Society in 1695, entomologist; see J.H. Barnhart, *Biographical Notes upon Botanists*. 3: 75. 1965; Ray Desmond, *Dictionary of British & Irish Botanists and Horticulturists*. 549. [b. 1663/1664] 1994; Stafleu and Cowan, *Taxonomic Literature*. 4: 203-204. [b. 1658] 1983; Helmut Genaust, *Etymologisches Wörterbuch der botanischen Pflanzennamen*. 472. [b. 1663/1664] 1996; F. Boerner & G. Kunkel, *Taschenwörterbuch der botanischen Pflanzennamen*. 4. Aufl. 149. 1989; H.N. Clokie, *Account of the Herbaria of the Department of Botany in the University of Oxford*. 223-224. 1964; James Britten, *The Sloane Herbarium*, revised and edited by J.E. Dandy. 1958; S. Lenley et al., *Catalog of the Manuscript and Archival Collections and Index to the Correspondence of John Torrey*. 325. 1973; Ethelyn Maria Tucker, *Catalogue of the Library of the Arnold Arboretum of Harvard University*. 1917-1933; R. Pulteney, *Historical and Biographical Sketches of the Progress of Botany in England*. 2: 31-43. 1790; F.D. Drewitt, *The Romance of the Apothecaries' Garden at Chelsea*. London 1924; Jonas C. Dryander, *Catalogus bibliothecae historico-naturalis Josephi Banks*. London 1800; A. Lasègue, *Musée botanique de Benjamin Delessert*. Paris 1845; William Darlington, *Memorials of John Bartram and Humphry Marshall*. Philadelphia 1849; Blanche Henrey, *British Botanical and Horticultural Literature*

efore 1800. Oxford 1975; Mia C. Karsten, *The Old Company's Garden at the Cape and Its Superintendents*: involving an historical account of early Cape botany. Cape Town 1951; R. Desmond, *The European Discovery of the Indian Flora*. Oxford 1992; Mariella Azzarello Di Misa, a cura di, *Il Fondo Antico della Biblioteca dell'Orto Botanico di Palermo*. 213-214. Regione Siciliana, Palermo 1988; G. Muray, *History of the Collections Contained in the Natural History Departments of the British Museum*. London 1904; Frans A. Stafleu, *Linnaeus and the Linnaeans*. The spreading of their ideas in systematic botany, 1735-1789. 1971; Emil Bretschneider, *History of European Botanical Discoveries in China*. [Reprint of the original edition 1898.] Leipzig 1981; Gilbert Westacott Reynolds, *The Aloes of South Africa*. 29, 30, 80, 82. Rotterdam 1982; Blanche Henrey, *No Ordinary Gardener — Thomas Knowlton, 1691-1781*. Edited by A.O. Chater. British Museum (Natural History). London 1986.

Species/Vernacular Names:

P. sp.

Peru: mucura hembra

P. alliacea L.

Peru: chanviro, micura, mucura, mucará, mucura, mucura-aá, mucura hembra, mucura macho, niwis

Spanish: carricillo silvestre, hierba de las gallinitas, rama de zorrillo, zorrillo, zorrillo silvestre

Mexico: cashni-tlsú, jupachumi, paychée, xpayché, zorro, otzash, pátham

Brazil: mucura-caá, munuca-caá, guiné, raiz-de-guiné, erva de guiné, pipi, raiz de alho, erva de alho, erav pipi, mbaiendo, emboiando, ocoembo, mururacorá, amansa senhor

Central America: anamú, apacin, hierba de gallinitas, pacina, payche, zorrillo

Yoruba: ojuusaju

Petopentia Bullock Asclepiadaceae (Periplocaceae)

Origins:

Referring to the genus *Pentopetia* Decne.

Petracanthus Nees Acanthaceae

Origins:

From the Greek *petros* "rock" and *Acanthus*.

Petradoria E. Greene Asteraceae

Origins:

Greek *petros* "rock" and *dorea, doron, doren, doreia* "a gift."

Petrea L. Verbenaceae

Origins:

For Lord Robert James Petre, 1713-1742, 8th Baron Petre, English patron of botany and horticulture, Fellow of the Royal Society in 1731, introduced *Camellia japonica* into European gardens. See Dawson Turner, *Extracts from the Literary and Scientific Correspondence of R. Richardson, of Bierly, Yorkshire: Illustrative of the State and Progress of Botany* [Edited by D. Turner. Extracted from the memoir of the Richardson family, by Mrs. D. Richardson] Yarmouth 1835; William Darlington, *Memorials of John Bartram and Humphry Marshall*. Philadelphia 1849; F. Boerner & G. Kunkel, *Taschenwörterbuch der botanischen Pflanzennamen*. 4. Aufl. 149. Berlin & Hamburg 1989; Georg Christian Wittstein, *Etymologisch-botanisches Handwörterbuch*. 675. Ansbach 1852.

Species/Vernacular Names:

P. sp.

Brazil: flor de São Miguel, sanango, viuvinha

P. maynensis Huber

Peru: yahuar piri-piri

P. volubilis L.

English: blue petrea, lilac petrea, purple wreath, purple wreath vine, queen's wreath, sandpaper vine, wreath vine, bluebird vine

Portuguese: estrela azul, flor de Jesus, flor de Sta. Maria, viuvinha

Petriella Zotov Gramineae

Origins:

For the Scottish (b. Morayshire) botanist Donald Petrie, 1846-1925, went to Australia in 1868, in New Zealand 1874-1925, in 1894 chief inspector of schools, Auckland, New Zealand, wrote "List of the flowering plants indigenous to Otago." *Trans. Proc. New Zealand Inst.* 1896, "The Gramina of the Subantarctic Islands of New Zealand." *Subantarct. Is N.Z.* 2: 472-481. 1909 and "Some additions to the flora of the Subantarctic Islands of New Zealand." *T.N.Z.I.* 47: 59-60. 1915. See J.H. Barnhart, *Biographical Notes upon Botanists*. 3: 76. 1965; Thomas Frederick Cheeseman, *Manual of the New Zealand Flora*. xxvii. Wellington 1906; I.H. Vegter, *Index Herbariorum*. Part II (5), *Collectors N-R.*

Regnum Vegetabile vol. 109. 1983; I.C. Hedge and J.M. Lamond, *Index of Collectors in the Edinburgh Herbarium.* Edinburgh 1970.

Petrobium R. Br. Asteraceae

Origins:
From the Greek *petros* "rock" and *bios* "life," referring to its habitat.

Petrocallis R. Br. Brassicaceae

Origins:
From the Greek *petros* "rock" and *kalli* "beautiful," *kallos* "beauty," referring to the habitat.

Petrocarya Schreb. Chrysobalanaceae

Origins:
From the Greek *petros* "rock" and *karyon* "a nut," referring to the nature of the fruits.

Petrocodon Hance Gesneriaceae

Origins:
From the Greek *petros* "rock" and *kodon* "a bell."

Petrocoptis A. Braun Caryophyllaceae

Origins:
Greek *petros* "rock" and *kopto* "to cut off, to cut small, to pierce," an allusion to the habitat of these rock-plants.

Petrocosmea Oliver Gesneriaceae

Origins:
From the Greek *petros* "rock" and *kosmos* (*kosmeo* "to rule, adorn, dress") "ornament, decoration," alluding to the habitat.

Petrodoxa Anthony Gesneriaceae

Origins:
Greek *petros* "rock" and *doxa* "glory."

Petrogenia I.M. Johnst. Convolvulaceae

Origins:
From the Greek *petrogenes, petregenes* "rock-born."

Petromecon Greene Papaveraceae

Origins:
From the Greek *petros* "rock" and *mekon* "poppy."

Petronia Barb. Rodr. Orchidaceae

Origins:
For His Majesty Dom Pedro II, 1825-1891 (d. Paris), Emperor of Brazil (1831-1889), a patron of sciences.

Petronymphe H.E. Moore Alliaceae (Liliaceae, Amaryllidaceae)

Origins:
From the Greek *petros* "rock" and *numphai* "nymph."

Petrophila R. Br. Proteaceae

Origins:
See *Petrophile* R. Br. ex Knight.

Petrophile R. Br. ex J. Knight Proteaceae

Origins:
Greek *petros* "rock" and *philos* "lover, loving," referring to the rocky habitats; see Joseph Knight (1777?-1855), *On the Cultivation of the Plants Belonging to the Natural Order of Proteeae.* 92. London 1809.

Species/Vernacular Names:
P. biloba R. Br.
English: granite petrophile
P. canescens A. Cunn. ex R. Br.
English: conesticks
P. heterophylla Lindl.
English: variable-leaved cone bush
P. linearis R. Br.
English: pixie mops
P. longifolia R. Br.
English: long-leaved cone bush

P. multisecta F. Muell.

Australia: cone bush, Kangaroo Island conesticks

P. pedunculata R. Br.

Australia: stalked conesticks

P. pulchella (Schrader & Wendl.) R. Br. (*Petrophile fucifolia* (Salisb.) J. Knight)

Australia: cone bush, conesticks

P. sessilis Schultes

Australia: prickly conesticks, conesticks

P. shirleyae Bailey

Australia: Moreton Bay conesticks

P. teretifolia R. Br.

English: southern pixie mops

Petrophyes Webb & Berthel. Crassulaceae

Origins:
From the Greek *petros* "rock" and *phye* "shape, growth," *petrophyes* "clinging to rock."

Petrophyton Rydb. Rosaceae

Origins:
Greek *petros* "rock" and *phyton* "plant," see *Petrophytum* (Nutt.) Rydb.

Petrophytum (Nutt.) Rydb. Rosaceae

Origins:
From the Greek *petros* "rock" and *phyton* "plant," the plants live in stony places.

Petrorhagia (Ser.) Link Caryophyllaceae

Origins:
Greek *petros* and *rhagas* (*rhegnymi* "to break, break asunder") "a break"; Latin *saxifragus, a, um* (*saxum frango*) "stone-breaking"; see Johann Heinrich Friedrich Link (1767-1851), *Handbuch zur Erkennung der nutzbarsten und am häufigsten vorkommenden Gewächse*. 235. 1829.

Species/Vernacular Names:
P. velutina (Guss.) P. Ball & Heyw. (*Dianthus velutinus* Guss.; *Gypsophila velutina* (Guss.) D. Dietr.; *Tunica velutina* (Guss.) Fischer & C.A. Meyer)

English: velvet pink, hairy pink, childing pink

Petrosavia Beccari Melanthiaceae (Petrosaviaceae)

Origins:
For the Italian botanist Pietro Savi, 1811-1871, professor of botany, 1843-1871 Praefectus of the Pisa Botanic Garden (the Third Orto Botanico), brother of the Italian zoologist and geologist Paolo Savi (1798-1871), son of the Italian botanist Gaetano Savi (1769-1844). See J.H. Barnhart, *Biographical Notes upon Botanists*. 3: 216. 1965; T.W. Bossert, *Biographical Dictionary of Botanists Represented in the Hunt Institute Portrait Collection*. 350. 1972; Ethelyn Maria Tucker, *Catalogue of the Library of the Arnold Arboretum of Harvard University*. Cambridge, Massachusetts 1917-1933; P.E. Tomei and Carlo Del Prete, "The Botanical Garden of the University of Pisa." *Herbarist*. 49: 47-71. Concord, Massachusetts 1983; L. Amadei, "Note sull'*Herbarium Horti Pisani*: l'origine delle collezioni." *Museol. Scient*. 4(1-2): 119-129. 1987; Fabio Garbari, L. Tomasi Tongiorgi and A. Tosi, *Giardino dei Semplici: L'Orto Botanico di Pisa dal XVI al XX secolo*. Pisa 1991; F.A. Stafleu, "Die Geschichte der Herbarien." *Bot. Jahrb. Syst*. 108(2/3): 155-166. 1987; Mariella Azzarello Di Misa, a cura di, *Il Fondo Antico della Biblioteca dell'Orto Botanico di Palermo*. 244-246. Palermo 1988; R. Zander, F. Encke, G. Buchheim and S. Seybold, *Handwörterbuch der Pflanzennamen*. 14. Aufl. Stuttgart 1993.

Petrosciadium Edgew. Umbelliferae

Origins:
Greek *petros* "a rock, stone" and *skiadion, skiadeion* "umbel, parasol."

Petrosedum Grulich Crassulaceae

Origins:
From the Greek *petros* "a rock, stone" plus *Sedum*.

Petroselinum Hill Umbelliferae

Origins:
Greek *petroselinon* "rock-parsley, parsley," *petros* "a rock" and *selinon* "parsley, celery"; Latin *petroselinum* or *petroselinon* "parsley" (Plinius, Palladius Rutilius Taurus Aemilianus et al.); see John Hill (1716-1775), *The British Herbal*. 424. London 1756.

Species/Vernacular Names:

P. crispum (Mill.) Nyman (*Petroselinum sativum* G.F. Hoffm.; *Apium petroselinum* L.; *Apium crispum* Miller; *Carum petroselinum* (L.) Benth.)

English: parsley, garden parsley, common garden parsley

French: persil

South Africa: pieterselie

Japan: paseri

Arabic: maadnous, bagdouness, ma'adnous

P. crispum var. *neapolitanum* Danert

English: Italian parsley

P. crispum var. *tuberosum* (Bernh.) Crov.

English: Hamburg parsley, turnip-rooted parsley

Petteria C. Presl Fabaceae

Origins:

For the Austrian botanist Franz Petter, 1798-1853; see J.H. Barnhart, *Biographical Notes upon Botanists*. 3: 76. Boston 1965; A. Lasègue, *Musée botanique de Benjamin Delessert*. Paris 1845; Stafleu and Cowan, *Taxonomic Literature*. 4: 211-212. Utrecht 1983.

Petunga DC. Rubiaceae

Origins:

A Bengalese name for *Petunga roxburghii*.

Petunia Juss. Solanaceae

Origins:

From the Tupi-Guarani names *petum, petyma, petymbu, petume,* applied to tobacco; see A.L. de Jussieu, in *Annales du Muséum National d'Histoire Naturelle*. 2: 215. Paris 1803; C.T. Onions, *The Oxford Dictionary of English Etymology*. Oxford University Press 1966; Georg Christian Wittstein, *Etymologisch-botanisches Handwörterbuch*. 676. 1852.

Species/Vernacular Names:

P. axillaris (Lam.) Britton, Sterns & Pogg.

English: large white petunia

P. x *hybrida* (Hook.) P.L. Vilmorin

English: petunia, common garden petunia, common petunia

China: bi dong qie

South Africa: oupa-se-hoed

P. integrifolia (Hook.) Schinz & Thell. (*Petunia violacea* Lindl.)

English: violet-flowered petunia, petunia

Japan: tsuku-bane-asa-gao

Peucedanum L. Umbelliferae

Origins:

Latin *peucedanum, i* "hog's-fennel, sulphur-wort" (Plinius, M. Annaeus Lucanus), Greek *peukedanon, peukedanos,* applied by Theophrastus (*HP*. 9.14.1) to a bitter umbelliferous plant, sulphur-wort or hog's-fennel, perhaps from the Greek *peuke* "a pine" and *danos* "parched, burnt, dry," *peukedanos* "bitter"; see Carl Linnaeus, *Species Plantarum*. 245. 1753 and *Genera Plantarum*. Ed. 5. 116. 1754; Salvatore Battaglia, *Grande dizionario della lingua italiana*. XIII: 223. Torino 1986.

Species/Vernacular Names:

P. capense (Thunb.) Sond. var. *capense*

South Africa: lidbossie

P. decursivum (Miq.) Maxim.

English: common hogfennel

China: qian hu, chien hu, tun huo

P. galbanum (L.) Drude (*Bubon galbanum* L.)

English: blister bush, blistering bush, wild celery

South Africa: bergseldery, duinehout, wildeseldery

P. japonicum Thunb.

Japan: botan-bôfû, sakuna, chomiigusa

China: fang kuei

P. officinale L.

English: hog fennel

P. ostruthium (L.) Koch

English: masterwort

P. palustre (L.) Moench

English: milk parsley

P. sandwicense Hillebr. (*Peucedanum kauaiense* Hillebr.; *Peucedanum sandwicense* var. *hiroe* Degener & I. Degener)

Hawaii: makou

P. tenuifolium Thunb.

English: wild parsley

South Africa: wildepieterselie

Peuceluma Baillon Sapotaceae

Origins:

Greek *peuke* "a pine" with *luma*.

Peucephyllum A. Gray Asteraceae

Origins:

From the Greek *peuke* "a pine" and *phyllon* "leaf."

Species/Vernacular Names:

P. schottii A. Gray

English: pygmy-cedar

Peumus Molina Monimiaceae

Origins:

A native plant name in Chile.

Species/Vernacular Names:

P. boldus Molina

Chile: boldo

Central America: boldo, limoncillo

Peyritschia Fournier Gramineae

Origins:

Named for the Austrian botanist Johann (Joannes) Joseph Peyritsch, 1835-1889, physician, M.D. Wien 1864, professor of botany; see J.H. Barnhart, *Biographical Notes upon Botanists*. 3: 77. 1965; T.W. Bossert, *Biographical Dictionary of Botanists Represented in the Hunt Institute Portrait Collection*. 308. 1972; Ida Kaplan Langman, *A Selected Guide to the Literature on the Flowering Plants of Mexico*. 581. Philadelphia 1964; Ethelyn Maria Tucker, *Catalogue of the Library of the Arnold Arboretum of Harvard University*. Cambridge, Massachusetts 1917-1933.

Peyrousea DC. Asteraceae

Origins:

In honor of the French navigator Jean François de Galaup de la Pérouse, 1741-1788, explorer, naturalist, Knight of the Order of St. Louis, 1 August 1785 expedition to the Pacific with two ships, the *Boussole* and the *Astrolabe*, visited Monterey in 1786, June 1788 wreck of the vessels near the island of Vanikoro in the Santa Cruz group. See Edward Weber Allen, *The Vanishing Frenchman: The Mysterious Disappearance of Laperouse*. Rutland, Vermont [1959]; John Dunmore, *Who's Who in Pacific Navigation*. University of Hawaii Press, Honolulu 1991; Ronald Louis Silveira de Braganza, ed., *The Hill Collection of Pacific Voyages*. University of California, San Diego 1974-1983; E.W. Allen, "Jean François Galaup de Lapérouse, a checklist." *California Historical Society Quarterly*. 20: 47-64.

1941; J. Dunmore and M. de Brossard, *Le Voyage de Lapérouse 1785-1788*. Paris 1985; J.J.H. de Labillardière, *Voyage in Search of La Pérouse*, performed by order of the Constituent Assembly, during the years 1791, 1792, 1793 and 1794. Translated from the French. London 1800, the French edition, *Relation du voyage à la recherche de la Pérouse*. Paris [1800]; Jonathan Wantrup, *Australian Rare Books, 1788-1900*. Hordern House, Sydney 1987.

Species/Vernacular Names:

P. umbellata (L.f.) Fourc. (*Cotula umbellata* L.f.; *Peyrousea argentea* Compton; *Peyrousea calycina* DC.; *Peyrousea oxylepis* DC.)

English: peyrousia

Pezisicarpus Vernet Apocynaceae

Origins:

Greek *pezis*, Latin *pezicae* or *pezitae, arum* for mushrooms without a root or without a stalk, and *karpos* "fruit."

Pfeiffera Salm-Dyck Cactaceae

Origins:

For the German botanist Louis (Ludwig, Ludovicus) Karl (Carl) Georg Pfeiffer, 1805-1877, physician, M.D. 1825 Marburg, traveler (Cuba), malacologist, editor of *Zeitschrift für Malakozoologie*. His writings include *Nomenclator botanicus*. Kassel [1871-] 1873-1874, *Universal-Repertorium der deutschen medizinischen, chirurgischen und obstetrizischen* etc. Cassel 1833, *Enumeratio diagnostica Cactearum*. Berlin 1837, *Symbolae ad Historiam Heliceorum*. Cassellis 1841-1846, *Monographia heliceorum viventium*. Lipsiae 1848, *Catalogue of Phaneropneumona ... in the British Museum*. London 1852, *Monographia pneumonopomorum viventium*. Cassellis 1852-1858, *Monographia auriculaceorum viventium*. Cassellis 1856, *Catalogue of Auriculidae, Proserpinidae, and Truncatellidae ... in the British Museum*. London 1857 and *Nomenclator heliceorum viventium*. Cassellis 1881. See J.H. Barnhart, *Biographical Notes upon Botanists*. 3: 79. 1965; T.W. Bossert, *Biographical Dictionary of Botanists Represented in the Hunt Institute Portrait Collection*. 308. 1972; Ida Kaplan Langman, *A Selected Guide to the Literature on the Flowering Plants of Mexico*. 581. Philadelphia 1964; Günther Schmid, *Chamisso als Naturforscher*. Eine Bibliographie. Leipzig 1942; Gordon Douglas Rowley, *A History of Succulent Plants*. Strawberry Press, Mill Valley, California 1997; R. Zander, F. Encke, G. Buchheim and S. Seybold, *Handwörterbuch der Pflanzennamen*. 14. Aufl. Stuttgart 1993; Stafleu and Cowan, *Taxonomic Literature*. 4: 221-225. Utrecht 1983.

Pfeifferago Kuntze Cunoniaceae

Origins:

For the German botanist Louis (Ludwig, Ludovicus) Karl (Carl) Georg Pfeiffer, 1805-1877.

Phaca L. Fabaceae

Origins:

From the Greek *phakos, phakon* "lentil, *Ervum lens* and its fruit," *phake* "a lentil-soup, dish of lentils."

Phacelia Juss. Hydrophyllaceae

Origins:

Greek *phakelos* "a cluster, bundle," in reference to the flowers or to the inflorescence.

Species/Vernacular Names:

P. amabilis Constance

English: Saline Valley phacelia

P. argentea A. Nelson & J.F. Macbr.

English: sand dune phacelia

P. campanularia A. Gray

English: California bluebell

P. congesta Hook.

English: blue curls

P. cookei Constance & Heckard

English: Cooke's phacelia

P. minor (Harv.) Thell.

English: whitlavia

P. mohavensis A. Gray

English: Mojave phacelia

P. orogenes Brand

English: mountain phacelia

P. parishii A. Gray

English: Parish's phacelia

P. tanacetifolia Benth.

English: fiddleneck, tansy phacelia

Phacellanthus Siebold & Zucc. Cyperaceae

Origins:

Greek *phakelos* "a faggott, bundle" and *anthos* "flower."

Phacellanthus Siebold & Zucc. Orobanchaceae (Scrophulariaceae)

Origins:

Greek *phakelos* "a faggott, bundle" and *anthos* "flower."

Phacellaria Benth. Santalaceae

Origins:

From the Greek *phakelos* "a faggott, bundle."

Phacellaria Steudel Gramineae

Origins:

Greek *phakelos* "a faggott, bundle."

Phacellothrix F. Muell. Asteraceae

Origins:

Greek *phakelos* "a cluster, bundle" and *thrix, trichos* "hair," the pappus bristles are in bundles; see Ferdinand von Mueller, *Fragmenta Phytographiae Australiae*. 11: 49. (Nov.) 1878.

Phacelophrynium K. Schumann Marantaceae

Origins:

From the Greek *phakelos* "a cluster, bundle" plus the genus *Phrynium* Willd.

Phacelurus Griseb. Gramineae

Origins:

From the Greek *phakelos* "a cluster, bundle" and *oura* "tail."

Phacocapnos Bernh. Fumariaceae (Papaveraceae)

Origins:

From the Greek *phakos* "a lentil" and *kapnos* "smoke," referring to the black and lenticular seeds.

Phacomene Rydb. Fabaceae

Origins:

From the Greek *phakos* "a lentil" and *mene* "the moon."

Phacopsis Rydb. Fabaceae

Origins:

From the Greek *phakos* "a lentil" and *opsis* "like, resembling."

Phadrosanthus Necker ex Raf. Orchidaceae

Origins:

From the Greek *phaidros* "gay, bright, beaming" and *anthos* "flower," the floral segments.

Phaeanthus Hook.f. & Thomson Annonaceae

Origins:

Greek *phaios* "dark, grey" and *anthos* "flower."

Species/Vernacular Names:

P. ebracteolatus (Presl) Merrill

The Philippines: kalimatas, lanotan, lanotang-itim, banitan, oyoi, kalumatas, takulau, langlangas

Phaedra Klotzsch ex Endl. Euphorbiaceae

Origins:

Greek *phaidros* "gay, bright"; Phaidra (Phaedra) was the daughter of King Minos of Crete.

Phaedranassa Herbert Amaryllidaceae (Liliaceae)

Origins:

From the Greek *phaidros* "gay, bright" and *anassa* "queen, lady, royal," referring to the beauty of the flowers.

Phaedranthus Miers Bignoniaceae

Origins:

Greek *phaidros* "gay, bright" and *anthos* "flower," referring to the beauty of the flowers.

Phaenanthoecium C.E. Hubb. Gramineae

Origins:

From the Greek *phaino*, *phaeino* "bring to light, reveal" and *anthos* "flower," *anthoecium*.

Phaenocoma D. Don Asteraceae

Origins:

Greek *phaino* "to shine, to appear" and *kome* "hair," referring to the appearance of the involucre. See D. Don, "Memoir on the classification and division of *Gnaphalium* and *Xeranthemum* of Linnaeus." *Mem. Wern. Nat. Hist. Soc.* 5(2): 533-563. 1826; Carl Linnaeus, *Species Plantarum.* 858. Holmiae 1753; J. Sims, "*Elichrysum proliferum.* Proliferous Everlasting." *Curtis's Botanical Magazine.* 50: Tab. 2365. 1823.

Species/Vernacular Names:

P. prolifera (L.) D. Don (*Helichrysum proliferum* (L.) Willd.)

English: everlasting, proliferous everlasting

South Africa: rooisewejaartjie

Phaenohoffmannia O. Kuntze Fabaceae

Origins:

For the German botanist Heinrich Karl Hermann Hoffmann, 1819-1891, M.D. Giessen 1841, mycologist, plant geographer, professor of botany, systematist, from 1851 Director of the Botanic Garden at Giessen. His writings include *Grundlinien der physiologischen und pathologischen Chemie*, etc. Heidelberg 1845, *Mykologische Berichte.* Giessen 1870-1872, *Phaenologische Untersuchungen.* Giessen 1887, *Witterung und Wachsthum, oder Grundzüge der Pflanzenklimatologie.* Leipzig 1857 and *Index fungorum*, sistens icones et specimina sicca nuperis temporibus edita. Lipsiae 1863; see J.H. Barnhart, *Biographical Notes upon Botanists.* 2: 189. 1965; Egon Ihne, *Beiträge zur Phänologie.* [II. H. Hoffmann, *Phänologische Beobachtungen aus den Jahren 1879-1882.*] Giessen 1884; T.W. Bossert, *Biographical Dictionary of Botanists Represented in the Hunt Institute Portrait Collection.* 178. 1972; E.M. Tucker, *Catalogue of the Library of the Arnold Arboretum of Harvard University.* 1917-1933.

Phaenopoda Cass. Asteraceae

Origins:

From the Greek *phaino* "to shine, to appear" and *pous, podos* "a foot"; see Alexandre H.G. Comte de Cassini, *Dictionnaire des Sciences Naturelles.* 42: 84. 1826.

Phaenosperma Munro ex Benth. Gramineae

Origins:

From the Greek *phaino* "to shine, to appear" and *sperma* "seed."

Phaeocephalus S. Moore Asteraceae

Origins:
From the Greek *phaios* "dark, dusky" and *kephale* "head."

Phaeomeria Lindl. ex K. Schumann Zingiberaceae

Origins:
From the Greek *phaios* "dark, purple" and *meris* "part."

Phaeoneuron Gilg Melastomataceae

Origins:
From the Greek *phaios* "dark" and *neuron* "nerve."

Phaeonychium O.E. Schulz Brassicaceae

Origins:
From the Greek *phaios* "dark" and *onyx, onychos* "a claw, nail."

Phaeopappus (DC.) Boiss. Asteraceae

Origins:
From the Greek *phaios* "dark" and *pappos* "fluff, pappus."

Phaeoptilum Radlk. Nyctaginaceae

Origins:
Greek *phaios* "dark, grey" and *ptilon* "feather."

Species/Vernacular Names:
P. spinosum Radlk. (*Amphoranthus spinosus* S. Moore; *Nachtigalia protectoratus* Schinz ex Engl.; *Phaeoptilum heimerli* Engl.)

South Africa: brosdoring

Phaeosphaerion Hassk. Commelinaceae

Origins:
From the Greek *phaios* "dark" and *sphaira* "a globe, ball."

Phaeostemma Fourn. Asclepiadaceae

Origins:
From the Greek *phaios* "dark" and *stemma* "a crown."

Phaeostigma Muldashev Asteraceae

Origins:
From the Greek *phaios* "dark" and *stigma* "a stigma."

Phaeostoma Spach Onagraceae

Origins:
From the Greek *phaios* "dark" and *stoma* "mouth."

Phagnalon Cass. Asteraceae

Origins:
Referring to the related genus *Gnaphalium*, from *gnaphalon* "wool scratched or torn off in fulling cloth, flock of wool."

Phaianthes Raf. Iridaceae

Origins:
Greek *phaios* "dark" and *anthos* "flower"; see C.S. Rafinesque, *Flora Telluriana*. 4: 30. 1836 [1838].

Phainantha Gleason Melastomataceae

Origins:
From the Greek *phaios* "dark" and *anthos* "flower."

Phaiophleps Raf. Iridaceae

Origins:
From the Greek *phaios* "dark" and *phleps* "vein," referring to the veins of the flowers; see C.S. Rafinesque, *Flora Telluriana*. 4: 29. 1836 [1838].

Phaius Lour. Orchidaceae

Origins:
Greek *phaios* "dark, grey, swarthy, shining, dusky," referring to the flowers; see João (Joannes) de Loureiro (1717-1791), *Flora cochinchinensis: sistens plantas in regno Cochinchina nascentes*. 2: 517, 529. Ulyssipone [Lisboa] 1790.

Species/Vernacular Names:
P. australis F. Muell.

English: Australian swamp lily, southern swamp orchid

P. bernaysii Rowland ex Reichb.f. (for Dr. Lewis Bernays, collector)

English: yellow swamp orchid

P. flavus (Blume) Lindl. (*Limodorum flavum* Blume; *Phaius maculatus* Lindley)

Japan: hoshi-kei-ran

P. mishmensis (Lindl.) Reichb.f. (*Limatodes mishmensis* Lindl. & Paxt.; *Phaius gracilis* Hayata)

Japan: hime-kaku-ran

P. tankervilliae (L'Hérit.) Blume (*Limodorum tancarvilleae* L'Hérit.; *Limodorum incarvilliae* Pers.; *Phaius bicolor* Lindley; *Phaius grandifolius* Lour.; *Phaius wallichii* Hook.f.) (for the English plant collector Lady Tankerville (Tankarville), d. 1836, wife of Charles, Earl of Tankerville; see Arthur D. Chapman, ed., *Australian Plant Name Index*. 2229. Canberra 1991)

English: Chinese ground orchid, nun's orchid, nun's hood, swamp lily orchid, northern swamp orchid

Japan: kaku-ran

Okinawa: chiru-ran, sarunkwa-bana

Thailand: ueang phrao, chat pra inn

Phalacrachena Iljin Asteraceae

Origins:
Greek *phalakros* "bald, blunt, knobbed" (*phalos* "shining, bright, white" and *akros* "highest, at the top"), *phalakra* "bald bare hill" and *achene*.

Phalacraea DC. Asteraceae

Origins:
Greek *phalakros* "bald, blunt, knobbed," *phalakra* "bald bare hill."

Phalacrocarpum (DC.) Willk. Asteraceae

Origins:
From the Greek *phalakros* "bald, blunt, knobbed" and *karpos* "fruit."

Phalacroderis DC. Asteraceae

Origins:
From the Greek *phalakros* "bald, blunt, knobbed" and *deris* "neck, throat, collar."

Phalacroloma Cass. Asteraceae

Origins:
Greek *phalakros* "bald, blunt, knobbed" and *loma* "border, margin, fringe, edge."

Phalacroseris A. Gray Asteraceae

Origins:
Greek *phalakros* "bald" and *seris* "a species of chicory or endive."

Phalaenopsis Blume Orchidaceae

Origins:
Greek *phalaina* "a moth" and *opsis* "like, appearance, resemblance," indicating the flowers; see Karl Ludwig von Blume, *Bijdragen tot de flora van Nederlandsch Indië*. 294. Batavia 1825.

Species/Vernacular Names:
P. cornu-cervi (Breda) Blume & Reichb.f.

Thailand: kao kwang on

P. rosenstromii F.M. Bailey.

English: Daintree white orchid, moth orchid

Phalangium Miller Liliaceae

Origins:
Latin *phalangium* or *phalangion* for a kind of venomous spider, for spider-root, a species of *anthericum*, for *phalangites*; see Philip Miller (1691-1771), *The Gardeners Dictionary*. Abr. ed. 4. London (28 Jan.) 1754.

Phalarella Boiss. Gramineae

Origins:
The diminutive of *Phalaris*.

Phalaridantha St.-Lag. Gramineae

Origins:
From the genus *Phalaris* and *anthos* "flower."

Phalaridium Nees & Meyen Gramineae

Origins:
Referring to the genus *Phalaris* L.

Phalaris L. Gramineae

Origins:

Greek *phalaris*, *phaleris*, used by Dioscorides for a kind of grass, ribbon grass, canary grass; *phalaros* "having a patch of white, crested," *phalos* "shining, bright, white, a part of the helmet"; Latin *phalaris* or *phaleris*, *idis* for the plant canary-grass; see Carl Linnaeus, *Species Plantarum*. 54. 1753 and *Genera Plantarum*. Ed. 5. 29. 1754.

Species/Vernacular Names:

P. aquatica L. (*Phalaris stenoptera* Hackel; *Phalaris tuberosa* L. var. *stenoptera* (Hackel) A.S. Hitchc.)

English: Toowomba canary grass, phalaris, Harding grass, perennial canary grass, Peruvian winter grass

P. arundinacea L.

English: reed canary grass, gardener's garters, ribbon grass

Southern Africa: langbeenkanariegras, rietgras, rietkanariegras; lekolojane (Sotho)

P. arundinacea L. var. *picta* L.

English: reed canary grass

Japan: shima-yoshi (= striped reed), shima-gaya (= striped grass)

P. canariensis L.

English: canary grass, canary seed grass, common canary grass, birdseed grass

South Africa: gewone kanariegras, kanariesaadgras, kwarrelsaadgras

Japan: kanari-kura-yoshi

P. coerulescens Desf.

English: blue canary grass

P. minor Retz.

English: little-seeded canary grass, small canary grass, lesser canary grass, annual canary grass

Arabic: sha'ir el-far

South Africa: kanariegras, kleinsaadkanariegras

P. paradoxa L.

English: canary grass

P. spinosum Radlk. (*Amphoranthus spinosus* S. Moore; *Nachtigalia protectoratus* Schinz ex Engl.; *Phaeoptilum heimerli* Engl.)

South Africa: brosdoring

Phalaroides Wolf Gramineae

Origins:

Resembling the genus *Phalaris* L.

Phaleria Jack Thymelaeaceae

Origins:

Greek *phalaros* "having a patch of white," *phalos* "shining, bright, white," referring to the flowers; see William Jack (1795-1822), in *Malayan Miscellanies*. 2(7): 59. Bencoolen 1822.

Species/Vernacular Names:

P. chermsideana (F.M. Bailey) C. White (after Lieut.-General Sir Herbert Charles Chermside, 1850-1935, former Governor of Queensland, Australia)

English: scrub daphne

Phallaria Schumach. & Thonn. Rubiaceae

Origins:

From the Greek *phallos* "a penis, wooden club."

Phalocallis Herbert Iridaceae

Origins:

From the Greek *phalos* "shining, bright, white" and *kallos* "beauty."

Phalona Dumort. Gramineae

Origins:

From the Greek *phalos*, *phalon* "shining, bright, white."

Phanera Lour. Caesalpiniaceae

Origins:

Greek *phaneros* "evident, visible, distinct"; see João (Joannes) de Loureiro, *Flora cochinchinensis*: 1: 37. Ulyssipone [Lisboa] 1790.

Phanerandra Stschegl. Epacridaceae

Origins:

From the Greek *phaneros* "evident, visible, distinct" and *aner*, *andros* "male, stamen"; see Serge S. Stscheglejew (fl. 1851), in *Bulletin de la Société Impériale des Naturalistes de Moscou*. 32(1): 20. 1859.

Phanerocalyx S. Moore Olacaceae

Origins:

From the Greek *phaneros* "evident, visible, distinct" and *kalyx* "calyx."

Phanerodiscus Cavaco Olacaceae

Origins:
From the Greek *phaneros* "evident, visible" and *diskos* "disc."

Phaneroglossa Nordenstam Asteraceae

Origins:
From the Greek *phaneros* "evident, visible, distinct" and *glossa* "tongue."

Phanerogonocarpus Cavaco Monimiaceae

Origins:
From the Greek *phaneros* "evident, visible, shining," *gonia* "angle, corner" and *karpos* "fruit."

Phanerophlebia C. Presl Dryopteridaceae (Aspleniaceae)

Origins:
The generic name is based on Greek *phaneros* "evident, shining, manifest" and *phleps, phlebos* "vein."

Species/Vernacular Names:
P. auriculata L. Underwood (*Cyrtomium auriculatum* (L. Underwood) C.V. Morton)

English: Mexican holly fern

Phanerophlebiopsis Ching Dryopteridaceae (Aspleniaceae)

Origins:
Resembling *Phanerophlebia* C. Presl.

Phanerosorus Copel. Matoniaceae

Origins:
From the Greek *phaneros* "evident, shining, remarkable" and *soros* "a vessel, mound, heap."

Phanerostylis (A. Gray) R. King & H. Robinson Asteraceae

Origins:
Greek *phaneros* "evident, shining, remarkable" and *stylos* "style, column, pillar."

Phania DC. Asteraceae

Origins:
From the Greek *phaeinos, phanos* "bright, light," *phane, phanos* "torch."

Phaniasia Blume ex Miq. Orchidaceae

Origins:
Greek *phaeinos, phanos* "bright, light."

Phanopyrum (Raf.) Nash Gramineae

Origins:
Greek *phaeinos, phanos* "bright, light" and *pyros* "grain, wheat."

Pharbitis Choisy Convolvulaceae

Origins:
See Helmut Genaust, *Etymologisches Wörterbuch der botanischen Pflanzennamen.* 475. Basel 1996; F. Boerner & G. Kunkel, *Taschenwörterbuch der botanischen Pflanzennamen.* 4. Aufl. 150. Berlin & Hamburg 1989; Georg Christian Wittstein, *Etymologisch-botanisches Handwörterbuch.* 678. Ansbach 1852.

Pharetrella Salisb. Tecophilaeaceae (Liliaceae)

Origins:
The diminutive of the Greek *pharetra, pharetre* "quiver for arrows," Latin *pharetra, ae* "a quiver for holding arrows," see also *Cyanella* L.

Pharmaceum Kuntze Melastomataceae

Origins:
Greek *pharmakon* "a drug, healing remedy, medicine," *pharmakos* "poisoner, sorcerer, magician."

Pharmacosycea Miq. Moraceae

Origins:
Greek *pharmakon* "a drug, healing remedy" and *sykon* "fig."

Pharnaceum L. Molluginaceae

Origins:
Latin *pharnacion* or *pharnaceon*, *ii* was applied by Plinius to a plant, a species of *panax*, so named after Pharnaces, the name of two kings of Pontus, Greek *pharnakeion*.

Species/Vernacular Names:
P. thunbergii Adamson (*Pharnaceum distichum* Thunb. non L.)
South Africa: droedas kruiden

Pharus P. Browne Gramineae

Origins:
From the Greek *pharos* "mantle, web, a piece of cloth, a wide cloak."

Phaseoloides Duhamel Fabaceae

Origins:
Resembling *Phaseolus*.

Phaseolus L. Fabaceae

Origins:
Greek *phaselos* "a little boat, a light vessel, a light boat," referring to its likeness to a bean-pod; Latin *phaselus* (*phasellus* and *faselus*) or *phaseolus* (*faseolus*) for a kind of bean with an edible pod, French beans, kidney-beans, phasel (Plinius); see Giovanni Semerano, *Le origini della cultura europea. Dizionario della lingua Latina e di voci moderne.* 2(2): 516. Firenze 1994; R. Zander, F. Encke, G. Buchheim and S. Seybold, *Handwörterbuch der Pflanzennamen.* 14. Aufl. Stuttgart 1993; M. Cortelazzo & P. Zolli, *Dizionario etimologico della lingua italiana.* 2: 413. 1980; H. Genaust, *Etymologisches Wörterbuch der botanischen Pflanzennamen.* 475-477. 1996; National Research Council, *Lost Crops of the Incas: Little-Known Plants of the Andes with Promise for Worldwide Cultivation.* National Academy Press, Washington, D.C. 1989; Pierre Fatumbi Verger, *Ewé: The Use of Plants in Yoruba Society.* São Paulo 1995; William W. Megenney, *A Bahian Heritage.* University of North Carolina at Chapel Hill 1978; Celia Blanco, *Santeria Yoruba.* Caracas 1995; Maria Helena Farelli, *Plantas que curam e cortam feitiços.* Rio de Janeiro 1988.

Species/Vernacular Names:
P. sp.
South Laos: (Nya Hön people) plää bbay, plää bbay ddek, plää bbay grong, plää bbay laat, plää bbay riau

Mexico: bizaa guela, bizaa ela, bizaa yela, bizaa hui, bzaa guixe
Yoruba: eree

P. acutifolius A. Gray var. *acutifolius* (*Phaseolus acutifolius* A. Gray var. *latifolius* G. Freeman)
English: tepary bean

P. coccineus L. (*Phaseolus multiflorus* Lam.; *Phaseolus striatus* Brandegee)
English: scarlet runner bean, Dutch case-knife bean, multiflora bean
Japan: hana-sasage

P. filiformis Benth.
English: slender-stemmed bean, Wright's phaseolus

P. lunatus L.
English: Lima bean, civet bean, butter bean, Burma bean
Paraguay: cugué
Japan: aoi-mame
The Philippines: patani, buringi, butingi, buni, gulipatan, kilkilang, kopani, kutakut, puida
Malaya: kachang china, kachang serendeng
Yoruba: eree, awuje, ewuje, ewa, ewe, popondo, kokondo, sese, ewe

P. lunatus L. var. *lunatus* (*Phaseolus inamoenus* L.; *Phaseolus limensis* Macfad.; *Phaseolus lunatus* L. var. *macrocarpus* Benth.; *Phaseolus tunkinensis* Lour.)
English: Lima bean, butter bean

P. maculatus Scheele (*Phaseolus metcalfei* Wooton & Stanley; *Phaseolus retusus* Benth., nom. illeg.; *Phaseolus venosus* Piper)
English: metcalf bean

P. polystachios (L.) Britton et al.
English: thicket bean, beanvine

P. radiatus L.
The Philippines: mungo, balatong
Malaya: kachang chendai, kachang hijau, kachang kedelai

P. vulgaris L.
English: bean, kidney bean, common French bean, Baguio bean, French bean, lima bean
Japan: ingen-mame
China: bai fan dou
Mexico: bizaa, bzaa, bizoono bizaa, pizoono pizaa, nizaa
The Philippines: bitsuwelas, biringi, butingi, inula
Malaya: kachang bunchis, kachang pendek
Tanzania: maharagi

P. vulgaris L. var. *vulgaris* (*Phaseolus compressus* DC. var. *carneus* Martens; *Phaseolus compressus* DC. var. *cervinus*

Martens; *Phaseolus compressus* DC. var. *ferrugineus* Martens; *Phaseolus ellipticus* Martens var. *albus* Martens; *Phaseolus ellipticus* Martens var. *aureolus* Martens; *Phaseolus ellipticus* Martens var. *helvolus* Savi; *Phaseolus ellipticus* Martens var. *mesomelos* Haberle; *Phaseolus ellipticus* Martens var. *pictus* Caval.; *Phaseolus ellipticus* Martens var. *spadiceus* Martens; *Phaseolus gonospermus* Savi var. *oryzoides* Martens; *Phaseolus gonospermus* Savi var. *variegatus* Savi; *Phaseolus oblongus* Savi var. *albus* Martens; *Phaseolus oblongus* Savi var. *spadiceus* Savi; *Phaseolus oblongus* Savi var. *zebrinus* Martens; *Phaseolus sphaericus* Savi var. *atropurpureus* Martens; *Phaseolus sphaericus* Savi var. *minor* Martens; *Phaseolus vulgaris* L. var. *albus* Haberle; *Phaseolus vulgaris* L. var. *nanus* Martens; *Phaseolus vulgaris* L. var. *niger* Martens; *Phaseolus vulgaris* L. var. *ochraceus* Savi; *Phaseolus vulgaris* L. var. *variegatus* DC.; *Phaseolus zebra* Fingerh. var. *purpurascens* Martens; *Phaseolus zebra* Fingerh. var. *carneus* Martens)

English: kidney bean, garden bean, green bean, snap bean, haricot, haricot bean, common bean, French bean, runner bean, string bean, salad bean, wax bean, navy bean

Mexico: frijol, bu'ul, etl

Spanish: frijol

Phaulanthus Ridl. Melastomataceae

Origins:
From the Greek *phaulos* "easy, simple" and *anthos* "flower."

Phaulopsis Willd. Acanthaceae

Origins:
From the Greek *phaulos* "slight, easy, cheap, simple, ordinary" and *opsis* "appearance."

Phaulothamnus A. Gray Achatocarpaceae (Phytolaccaceae)

Origins:
Greek *phaulos* "slight, easy, simple" and *thamnos* "shrub."

Phaylopsis Willd. Acanthaceae

Origins:
Genus *Phaulopsis*, Greek *phaulos* "cheap, simple" and *opsis* "appearance."

Phebalium Ventenat Rutaceae

Origins:
Possibly from a Greek poetic name for the myrtle (*Myrtus communis* L.), or from *phibaleos* (Phibalis was a district of Attica or Megaris), used for a kind of early fig, or from *phoibos* "radiant, Phoebus, bright" and *eileo* "to sun," *eile* "warmth, the sun's warmth"; see Étienne Pierre Ventenat (1757-1808), *Jardin de la Malmaison*. 2: 102, t. 102. Paris 1805; P.G. Wilson, "A taxonomic revision of the genera *Crowea, Eriostemon* and *Phebalium* (Rutaceae)." in *Nuytsia*. 1(1): 6-155. 1970.

Species/Vernacular Names:
P. ambiens (F. Muell.) Maiden & E. Betche

English: forest phebalium

P. brachyphyllum Benth. (*Eriostemon microphyllus* F. Muell., non *Phebalium microphyllum* Turcz.)

English: spreading phebalium

P. bullatum J. Black (*Phebalium sedifolium* F. Muell., partly)

English: silver phebalium, desert phebalium, silvery phebalium

P. carruthersii (F. Muell.) Maiden & E. Betche (after Sir Joseph Carruthers, from 1904 to 1907 Premier of New South Wales)

Australia: Moruya phebalium

P. coxii (F. Muell.) Maiden & E. Betche

Australia: Cox's phebalium

P. dentatum Smith

English: toothed phebalium

P. diosmeum A. Juss.

English: diosma phebalium

P. elatius (F. Muell.) Benth.

English: tall phebalium

P. glandulosum Hook.

English: desert phebalium

P. gracile C. White

English: Scortechini's phebalium

P. hillebrandii (F. Muell.) J.H. Willis

Australia: Hillebrand's phebalium

P. lamprophyllum (F. Muell.) Benth.

English: shining phebalium

P. nottii (F. Muell.) Maiden & E. Betche

English: pink phebalium

P. nudum Hook.

Maori name: mairehau

P. obcordatum Benth.

English: club-leaved phebalium, club-leaf phebalium

P. oldfieldii (F. Muell.) Benth.

Australia: Oldfield's phebalium

P. ovatifolium F. Muell.

English: ovate phebalium

P. phylicifolium F. Muell.

English: mountain phebalium

P. ralstonii (F. Muell.) Benth.

Australia: Ralston phebalium

P. rotundifolium (Endl.) Benth.

English: round-leaved phebalium, round-leaf phebalium

P. squameum (Labill.) Engl. subsp. *squameum*

English: satinwood

P. squamulosum Vent.

English: scaly phebalium

P. stenophyllum (Benth.) Maiden & E. Betche

English: narrow-leaved phebalium, narrow-leaf phebalium

P. sympetalum Paul G. Wilson

Australia: Rylstone bell

P. viridiflorum Paul G. Wilson

English: green phebalium

P. whitei Paul G. Wilson

English: granite phebalium

P. woombye (Bailey) Domin (named for Woombye, near Nambour)

Australia: wallum phebalium

Phedimus Raf. Crassulaceae

Origins:
See Constantine Samuel Rafinesque (1783-1840), *Anal. Nat. Tabl. Univ.* 174. 1815, *Florula ludoviciana.* 168. New York 1817 and *Am. Monthly Mag. Crit. Rev.* 1: 438. 1817; E.D. Merrill, *Index rafinesquianus.* 133. Massachusetts, USA 1949.

Phegopteris (C. Presl) Fée Thelypteridaceae

Origins:
Greek *phegos* "oak" (Theophrastus, *HP.* 3.3.1) and *pteris* "fern"; see Antoine Laurent Apollinaire Fée (1789-1874), *Mémoires sur la famille des Fougères.* Genera Filicum. 5: 242. 1852.

Species/Vernacular Names:
P. connectilis (Michaux) Watt

English: beech fern

Pheidochloa S.T. Blake Gramineae

Origins:
From the Greek *pheidos* "thrifty, sparing" and *chloe, chloa* "grass"; see Stanley Thatcher Blake (1910-1973), in *Proceedings of the Royal Society of Queensland.* 56: 20. 1944.

Pheidonocarpa L.E. Skog Gesneriaceae

Origins:
From the Greek *pheidos* "thrifty, sparing," *pheidon* "oil-can with a narrow neck, thrifty" and *karpos* "fruit."

Phelipaea Desf. Orobanchaceae (Scrophulariaceae)

Origins:
See *Phelypaea* L.

Phelline Labill. Aquifoliaceae (Phellinaceae)

Origins:
Greek *phellos* "cork," *phellinos* "made of cork," referring to the fruits and seeds.

Phellocalyx Bridson Rubiaceae

Origins:
From the Greek *phellos* "cork" and *kalyx* "calyx."

Phellodendron Rupr. Rutaceae

Origins:
Cork trees, from the Greek *phellos* "cork" and *dendron* "tree," an allusion to the corky bark.

Species/Vernacular Names:
P. amurense Rupr.

English: Amur cork-tree, Siberian cork-tree

China: huang bai, huang bo, huang po, po mu

Vietnam: hoang ba

Phellolophium Baker Umbelliferae

Origins:
From the Greek *phellos* "cork" and *lophos* "a crest."

Phellopterus Benth. Umbelliferae

Origins:

From the Greek *phellos* "cork" and *pteron* "wing, feather."

Phellosperma Britton & Rose Cactaceae

Origins:

From the Greek *phellos* "cork" and *sperma* "seed," alluding to the base of the seeds.

Phelpsiella Maguire Rapateaceae

Origins:

Dedicated to William H. Phelps (1875-1965), William H. Phelps, Jr., and Kathleen Deery Phelps, naturalists and ornithologists in Venezuela; see Stafleu and Cowan, *Taxonomic Literature.* 4: 230. Utrecht 1983.

Phelypaea L. Orobanchaceae (Scrophulariaceae)

Origins:

For the French political man Louis Phélypeaux (Count) de Pontchartrain, 1643-1727, see *Mémoires touchant l'établissement des Jésuites dans les Indes d'Espagne*, envoyés à Monseigneur de Pontchartrain Ministre d'État. A Paris le 18 Octobre 1710. [Paris] 1758; F. Hébert, *Les veritables et fausses Lettres des Messieurs l'Évêque d'Agen et Comte de Pontchartrain.* 1712; B. Gigault, *Correspondance du maréchal de Bellefonds*, commandant l'armée royale en Basse-Normandie, *avec M. de Pontchartrain*, ministre de la marine, etc. 1907; Georg Christian Wittstein, *Etymologisch-botanisches Handwörterbuch.* 679. Ansbach 1852.

Phenakospermum Endl. Strelitziaceae (Musaceae, Strelitzioideae)

Origins:

Greek *phenake* "false hair," *phenax, phenakos* "cheat, impostor" and *sperma* "seed."

Phenax Wedd. Urticaceae

Origins:

From the Greek *phenax, phenakos* "cheat, impostor."

Pherolobus N.E. Br. Aizoaceae

Origins:

From the Greek *phero, phoreo* "to bear" and *lobos* "a lobe," the ovary carries five fleshy processes.

Pherosphaera W. Archer Podocarpaceae

Origins:

From the Greek *phero, phoreo* "to bear" and *sphaira* "a globe, ball," the cones are ball-shaped.

Pherotrichis Decne. Asclepiadaceae

Origins:

Greek *phero, phoreo* "to bear" and *thrix, trichos* "hair."

Phialacanthus Benth. Acanthaceae

Origins:

Greek *phiale* and Latin *phiala* for a broad, shallow, drinking-vessel, a saucer, a vial or a bowl and *akantha* "thorn."

Phialanthus Griseb. Rubiaceae

Origins:

From the Greek *phiale* "a vial" and *anthos* "flower."

Phialodiscus Radlk. Sapindaceae

Origins:

Greek *phiale* "a vial" and *diskos* "disc."

Phidiasia Urban Acanthaceae

Origins:

Phidias (Pheidias) was a famous sculptor, contemporary with Pericles.

Philacra Dwyer Ochnaceae

Origins:

Greek *philos* "lover, loving" and *akros* "the summit, highest, at the top."

Philactis Schrader Asteraceae

Origins:

From the Greek *philos* "lover, loving" and *aktis, aktin* "a ray."

Philadelphus L. Hydrangeaceae (Philadelphaceae)

Origins:

Greek *philos* "lover, loving" and *adelphos* "brother," possibly referring to the appearance of the branches; see Carl Linnaeus, *Species Plantarum*. 470. 1753 and *Genera Plantarum*. Ed. 5. 211. 1754.

Species/Vernacular Names:

P. coronarius L.

Mexico: mosqueta, jazmín mosqueta

P. lewisii Pursh

English: wild mock orange

P. mexicanus Schltdl.

English: wild jasmin

Mexico: mosqueta, acuilote, jazmín del monte, jeringuilla, jazmín, jazmín de Hueyapan; acuilotl (Oaxaca)

P. microphyllus A. Gray

English: littleleaf mock orange

Philbornea Hallier f. Linaceae

Origins:

From Philippines and Borneo.

Philesia Comm. ex Juss. Philesiaceae (Smilacaceae, Liliaceae)

Origins:

Greek *philesis* "loving, affection," *philein* "to love," *philesios* "an epithet of Apollo," referring to the flowers; see F. Boerner & G. Kunkel, *Taschenwörterbuch der botanischen Pflanzennamen*. 4. Aufl. 150. Berlin & Hamburg 1989; Georg Christian Wittstein, *Etymologisch-botanisches Handwörterbuch*. 680. Ansbach 1852.

Species/Vernacular Names:

P. magellanica J.F. Gmelin

South America: coicopihue

Philgamia Baillon Malpighiaceae

Origins:

An anagram of *Malpighia* L., or perhaps from the Greek *phileo* "to love" and *gamos* "marriage," *gameo, gamein* "to marry."

Philibertella Vail Asclepiadaceae

Origins:

For the French botanist J.C. Philibert, presumably a pseudonym; see Stafleu and Cowan, *Taxonomic Literature*. 4: 232-233. Utrecht 1983.

Philibertia Kunth Asclepiadaceae

Origins:

Named for the French botanist J.C. Philibert (presumably a pseudonym). His works include *Dictionnaire universel de botanique*. Paris 1804 and *Réflexions philosophiques et critiques sur les couronnes ... Par Frid. ... W.... Traduites de l'allemand* [or, rather, written by J.C. Philibert], etc. 1804.

Philippia Klotzsch Ericaceae

Origins:

After the German botanist Rudolph Amandus (Rodolfo, Rudolf Amando) Philippi, 1808-1904, traveler, botanical explorer and plant collector, studied in Berlin, in 1851 emigrated to Chile, Director of the Museo Nacional de Chile (in Santiago), from 1853 to 1874 professor of botany and zoology at Santiago, 1853-1854 expedition Atacama desert, 1892-1904 founder and editor of the *Anales del Museo nacional* (Chile). His writings include "Observaciones sobre la flora de Juan Fernández." *Anales Univ. Chile*. 13: 157-169. 1856, *Florula atacamensis*. Halis Saxonum [Halle] 1860, "*Sertum mendocinum*. Catálogo de las plantas recojidas cerca de Mendoza ... por don Wenceslao Diaz en los años de 1860 i 1861." *Anales Univ. Chile*. 21: 389-407. 1862, "Comentario sobra las plantas chilenas descritas por el abate D. Juan Ignacio Molina." *Anales Univ. Chile*. 22: 699-741. 1863 and *Plantas nuevas Chilenas*. Santiago de Chile 1892-1896. See C. Muñoz Pizarro, *Las especies de plantas descritas por R.A. Philippi en el siglo xix. Estudio crítico ...* Ediciones de la Universidad de Chile 1960; Charles J. Édouard Morren, *Correspondance botanique*. Liège 1874 and 1884; D.B. Arana, *El doctor don Rodolfo Amando Philippi*. Santiago de Chile 1904; P. Fürstenberg, *Dr. Rudolph Amandus Philippi. Sein Leben und seine Werke*. Santiago de Chile 1906; B. Gotschlich, *Biografía del Dr. Rodulfo Amando Philippi (1808-1904)*. Santiago de

Chile 1906; R. Zander, F. Encke, G. Buchheim and S. Seybold, *Handwörterbuch der Pflanzennamen*. 14. Aufl. Stuttgart 1993; I.C. Hedge and J.M. Lamond, *Index of Collectors in the Edinburgh Herbarium*. Edinburgh 1970; E.D. Merrill, in *Bernice P. Bishop Mus. Bull.* 144: 150. 1937; Elmer Drew Merrill, *Contr. U.S. Natl. Herb.* 30(1): 241. 1947; T.W. Bossert, *Biographical Dictionary of Botanists Represented in the Hunt Institute Portrait Collection*. 309. 1972; A. Lasègue, *Musée botanique de Benjamin Delessert*. Paris 1845; S. Lenley et al., *Catalog of the Manuscript and Archival Collections and Index to the Correspondence of John Torrey*. Library of the New York Botanical Garden. 326. 1973; E.M. Tucker, *Catalogue of the Library of the Arnold Arboretum of Harvard University*. 1917-1933; G. Murray, *History of the Collections Contained in the Natural History Departments of the British Museum*. 1: 173. London 1904.

Species/Vernacular Names:
P. evansii N.E. Br.

English: early-flowering philippia

P. hexandra S. Moore

English: petrolbush

Southern Africa: nyatsiri, dzenga (Shona)

P. mannii (Hook.f.) Alm & Fries (*Philippia pallidiflora* Engl.)

English: hook-bristle philippia

Philippiamra O. Kuntze Portulacaceae

Origins:

For the German botanist Rudolph Amandus Philippi, 1808-1904, traveler and plant collector.

Philippicereus Backeberg Cactaceae

Origins:

For the German botanist Rudolph Amandus Philippi, 1808-1904, traveler and plant collector; see Gordon Douglas Rowley, *A History of Succulent Plants*. Strawberry Press, Mill Valley, California 1997.

Philippiella Spegazzini Caryophyllaceae (Illecebraceae)

Origins:

For the German botanist Rudolph Amandus Philippi, 1808-1904, traveler and plant collector.

Philippinaea Schlechter & Ames Orchidaceae

Origins:

For the Philippines Islands.

Phillipsia Rolfe Acanthaceae

Origins:

For Mrs. E. Lort-Phillips, plant collector in Africa, Somaliland, wife of the explorer Lort-Phillips; see A. White and B.L. Sloane, *The Stapelieae*. Pasadena 1937; M.N. Chaudhri, I.H. Vegter and C.M. De Wal, *Index Herbariorum*, Part II (3), *Collectors I-L*. Regnum Vegetabile vol. 86. 1972.

Phillyraea Moench Oleaceae

Origins:
See *Phyllyrea* L.

Phillyrea L. Oleaceae

Origins:

Greek *philyra* (perhaps from *philos* and *hyron*) used by Theophrastus (*HP.* 1.12.4 and 3.10.4) and Dioscorides for the lime tree, *Tilia platyphyllos* Scop., and silver lime, *Tilia tomentosa* Moench; *philyrea*, Theophrastus (*HP.* 1.9.3) for the mock privet, a species of *Phillyrea*; Latin *philyra* and *philura* for the linden-tree, for the inner bark of the linden-tree and for the skin of the papyrus (Plinius).

Philodendron Schott Araceae

Origins:

Greek *philos* "loving" and *dendron* "tree," referring to the climbing habit; see D.H. Nicolson, "Derivation of aroid generic names." *Aroideana*. 10: 15-25. 1988.

Species/Vernacular Names:
P. sp.

Peru: lluillui-sacha, cipo tracuá, folha de urubu, itininga, mano abierta, patquiña, patquina, savco sacha, tu ёnёti ro

Mexico: juasuchi, lúcuati, cuath

P. bipinnatifidum Endl.

Paraguay: Guembepi

P. camposportoanum G. Barroso

English: Aaron's rod

P. guttiferum Kunth (*Philodendron tessmannii* K. Krause)

Mexico: conté, chupa pito

P. imbe Schott

Peru: imbé, suambé

Brazil: cipóimbe, cipó de imbê, folha de fonte, tajaz de cobra, tracos, cucúba, quimbé, imbé, ombé, bananeira de macaco, cipó imbê, imbê, guiambê

P. lechlerianum Schott

Peru: biston, papa sacha, tuñu

Philodice Martius Eriocaulaceae

Origins:
Greek *philos* "loving" and *dike* "usage, custom," *philodikia* "litigiousness," *philodikos* "litigious."

Philoglossa DC. Asteraceae

Origins:
Greek *philos* "loving" and *glossa* "tongue."

Philonotion Schott Araceae

Origins:
From the Greek *philos* "loving" and *notios, notion* "moist, damp, rainy."

Philotheca Rudge Rutaceae

Origins:
Greek *philos* "loving" and *theke* "a box, case," referring to the knobbed capsules; see Edward Rudge (1763-1846), in *Transactions of the Linnean Society of London*. 11: 298, t. 21. 1816.

Philoxerus R. Br. Amaranthaceae

Origins:
Greek *philos* "loving" and *xeros* "dry," referring to the habitat; see R. Brown, *Prodromus florae Novae Hollandiae et Insulae van-Diemen*. 416. London 1810.

Species/Vernacular Names:
P. wrightii Hook.f. (for the American botanist Charles (Carlos) Wright, 1811-1885, explorer and traveler, botanical collector for Asa Gray in the Mexican Boundary Region, 1853-1855 under Capt. Cadwalader Ringgold (1802-1867) and Comm. John Rodgers (1812-1882) on the five-ship

North Pacific Exploring Expedition; see Asa Gray (1810-1888), *Plantae Wrightianae Texano-Neo-Mexicanae*. 1852-1853; A.H. Dupree, *Asa Gray — American Botanist, Friend of Darwin*. J. Hopkins 1988; August Heinrich Rudolf Grisebach (1814-1879), *Plantae wrightianae, e Cuba orientali*. Cantabrigiae, Nov. Angl. 1860-1862 and *Catalogus Plantarum Cubensium*. Lipsiae 1866; Francisco Adolfo Sauvalle (1807-1879), *Flora Cubana*. Enumeratio nova plantarum cubensium vel revisio catalogi grisebachiani, exhibens descriptiones generum specierumque novarum Caroli Wright et Francisci Sauvalle, etc. Havanae [Habana] [1868-] 1873; J.H. Barnhart, *Biographical Notes upon Botanists*. 3: 523. 1965; W.P. Cummings, S.E. Hillier, D.B. Quinn and G. Williams, *The Exploration of North America 1630-1776*. London 1974; Gordon Douglas Rowley, *A History of Succulent Plants*. Mill Valley, California 1997; Irving William Knobloch, compil., "A preliminary verified list of plant collectors in Mexico." *Phytologia Memoirs*. VI. 1983; Stafleu and Cowan, *Taxonomic Literature*. 7: 464-466. 1988; Elmer Drew Merrill, in *Contr. U.S. Natl. Herb.* 30(1): 318. 1947 and in *Bernice P. Bishop Mus. Bull.* 144: 192. 1937; E. Bretschneider, *History of European Botanical Discoveries in China*. Leipzig 1981; S. Lenley et al., *Catalog of the Manuscript and Archival Collections and Index to the Correspondence of John Torrey*. Library of the New York Botanical Garden. 1973; Joseph Ewan, *Rocky Mountain Naturalists*. 1950; J. Ewan, ed., *A Short History of Botany in the United States*. 1969; R. Zander, F. Encke, G. Buchheim and S. Seybold, *Handwörterbuch der Pflanzennamen*. 1993; T.W. Bossert, *Biographical Dictionary of Botanists Represented in the Hunt Institute Portrait Collection*. 443. 1972; E.M. Tucker, *Catalogue of the Library of the Arnold Arboretum of Harvard University*. Cambridge, Massachusetts 1917-1933)

Japan: iso-fusagi

Philydrella Caruel Philydraceae (Liliidae, Haemodorales)

Origins:
The generic name *Philydrum* and the diminutive suffix *ella*, little water-lover; see Théodore (Teodoro) Caruel (1830-1898), in *Nuovo Giornale Botanico Italiano*. 10: 91. 1878.

Species/Vernacular Names:
P. drummondii L. Adams (for James Drummond, Government botanist, Western Australia)

English: greater butterfly flowers

P. pygmaea (R. Br.) Caruel (*Philydrum pygmaeum* R. Br.; *Hetaeria pygmaea* (R. Br.) Endl. ex Kunth)

English: lesser butterfly flowers

Philydrum Banks ex Gaertner Philydraceae

Origins:
Greek *phileo* "friend" and *hydor* "water," referring to the habitat; see Joseph Gaertner, *De fructibus et seminibus plantarum.* 1: 62. Stuttgart, Tübingen 1788.

Species/Vernacular Names:
P. lanuginosum Gaertner
English: woolly grass, woolly waterlily, frogmouth, frog's mouth, waterwort
Japan: tanuki-ayame
Malaya: kipas
China: tian cong

Philyra Klotzsch Euphorbiaceae

Origins:
From the ancient Greek plant name used by Theophrastus and Dioscorides, *philyra* "lime tree, silver lime."

Philyrea Blume Oleaceae

Origins:
From the ancient Greek plant name used by Theophrastus, *philyrea* "mock privet."

Philyrophyllum O. Hoffm. Asteraceae

Origins:
The classical Greek name *philyra* "lime tree, *Tilia platyphyllos* Scop., silver lime, *Tilia tomentosa* Moench" and *phyllon* "leaf, foliage."

Phinaea Benth. Gesneriaceae

Origins:
An anagram of the generic name *Niphaea* Lindl.

Phippsia (Trinius) R. Br. Gramineae

Origins:
Named after the British Captain Constantine John Phipps, 2nd Baron Mulgrave, 1744-1792, explorer (Arctic), a friend of J. Banks, 1773 Mulgrave's Arctic Expedition (with Israel Lyons, 1739-1775), wrote *A Voyage to the North Pole.* London 1774; see J.H. Barnhart, *Biographical Notes upon Botanists.* 3: 82. 1965; Blanche Henrey, *British Botanical and Horticultural Literature before 1800.* Oxford 1975; A. Lasègue, *Musée botanique de Benjamin Delessert.* Paris 1845; Ray Desmond, *Dictionary of British & Irish Botanists and Horticulturists.* 552. London 1994.

Phlebiogonium Fée Dryopteridaceae (Aspleniaceae)

Origins:
Greek *phlebion* "the smaller vessels, veins" and *gonia* "an angle."

Phlebiophragmus O.E. Schulz Brassicaceae

Origins:
From the Greek *phleps, phlebos* "vein," *phlebion* "the smaller vessels, veins" and *phragma* "a hedge, a fence."

Phlebiophyllum Bosch Hymenophyllaceae

Origins:
From the Greek *phlebion* "the smaller vessels, veins" and *phyllon* "leaf."

Phlebocalymna Griff. ex Miers Icacinaceae

Origins:
From the Greek *phleps, phlebos* "vein" and *kalymna* "a covering."

Phlebocarya R. Br. Haemodoraceae

Origins:
Greek *phleps, phlebos* and *karyon* "a nut," the nut-like fruits are veined; see R. Brown, *Prodromus florae Novae Hollandiae et Insulae van-Diemen.* 301. London 1810.

Phlebochiton Wall. Anacardiaceae

Origins:
Greek *phleps, phlebos* "vein" and *chiton* "a tunic, covering"; see Henri Lecomte, "Sur le genre *Phlebochiton* (Anacardiacées)." *Bull. Soc. Bot. France.* 54: 525-529. 1907.

Phlebodium (R. Br.) J. Smith Polypodiaceae

Origins:

Greek *phlebodes* "full of veins, with large veins, like a vein," referring to the venation; see J. Smith, in *Hooker's Journal of Botany.* 4: 58. 1841.

Species/Vernacular Names:

P. aureum (L.) J. Smith (*Polypodium aureum* L.)

English: golden polypody, goldfoot fern

Phlebolithis Gaertner Sapotaceae

Origins:

Greek *phleps, phlebos* "vein" and *lithos* "stone," *lithis, lithidos, lithiasis* "a callosity."

Phlebolobium O.E. Schulz Brassicaceae

Origins:

From the Greek *phleps, phlebos* "vein" and *lobion* "a little pod."

Phlebophyllum Nees Acanthaceae

Origins:

From the Greek *phleps, phlebos* "vein" and *phyllon* "leaf."

Phlebotaenia Griseb. Polygalaceae

Origins:

From the Greek *phleps, phlebos* "vein" and *tainia* "fillet."

Phlegmariurus (Herter) Holub Lycopodiaceae

Origins:

From the Greek *phlegma* "inflammation, heat, fire" and *oura* "tail."

Phlegmatospermum O. Schulz Brassicaceae

Origins:

Greek *phlegma* (*phlego, phlegein* "to burn") "inflammation, heat, fire" and *sperma* "seed," referring to the mucose when moistened seeds; see Otto Eugen Schulz (1874-1936), in *Botanische Jahrbücher.* 66: 93. 1933.

Species/Vernacular Names:

P. cochlearinum (F. Muell.) O. Schulz (*Eunomia cochlearina* F. Muell.; *Capsella cochlearina* (F. Muell.) F. Muell.; *Thlaspi cochlearinum* (F. Muell.) F. Muell.)

English: oval-podded cress, downy cress

Phleum L. Gramineae

Origins:

Greek *phleos, phlous, phloun, phleon*, ancient name for a kind of grass growing in swamps like reeds, *Arundo ampelodesmon*, or wool-tufted reed, applied by Theophrastus (*HP.* 4.8.1, 4.10.1, 4.10.4) to a species of *Erianthus*; Phleos was an epithet of Dionysus; Latin *pheos, phleos* applied by Plinius to a prickly plant, also called *stoebe* or *stoibe*; see Carl Linnaeus, *Species Plantarum.* 59. 1753 and *Genera Plantarum.* Ed. 5. 29. 1754; Giovanni Semerano, *Le origini della cultura europea.* Dizionario della lingua Greca. 2(1): 304-305, 307. Leo S. Olschki Editore, Firenze 1994; Georg Christian Wittstein, *Etymologisch-botanisches Handwörterbuch.* 682. Ansbach 1852; H. Genaust, *Etymologisches Wörterbuch der botanischen Pflanzennamen.* 478-479. 1996.

Species/Vernacular Names:

P. alpinum L.

English: mountain Timothy

P. pratense L.

English: Timothy grass, cat's tail, meadow cat's tail, Timothy, cultivated Timothy

Phloeophila Hoehne & Schltr. Orchidaceae

Origins:

Greek *phloios* "bark of trees" and *philos* "lover, loving," indicating the epiphytic habit.

Phloga Noronha ex Hook.f. Palmae

Origins:

Greek *phlox, phlogos* "flame," referring to the bright red fruits.

Phlogacanthus Nees Acanthaceae

Origins:

Greek *phlox, phlogos* "flame" and *akantha* "thorn, prickle," referring to the color of the flowers, corolla orange.

Phlogella Baillon Palmae

Origins:
The diminutive of the genus *Phloga*.

Phlojodicarpus Turcz. ex Ledeb. Umbelliferae

Origins:
Greek *phloios* "bark of trees" and *dikarpos* "bearing two fruits, or two crops."

Phlomidoschema (Benth.) Vved. Labiatae

Origins:
From the genus *Phlomis* and *schema, schematos* "appearance, form, shape, figure" (*echo* "to hold, to sustain" and *schein*).

Phlomis L. Labiatae

Origins:
Greek *phlomis, phlomos*, ancient names for some plant, phlome, probably a species of *Phlomis*, or mullein, a species of *Verbascum*, mentioned by Theophrastus (*HP*. 9.12.3), Plinius and Dioscorides; Plinius used Latin *phlomis, phlomidis* and *phlomos, i* for mullein, verbascum; see Carl Linnaeus, *Species Plantarum*. 2: 584. 1753.

Species/Vernacular Names:
P. alpina Pallas
English: alpine Jerusalem sage
China: gao shan cao su
P. atropurpurea Dunn
English: dark-purple Jerusalem sage
China: shen lie cao su
P. dentosa Franchet
English: dentate Jerusalem sage
China: jian chi cao su
P. forrestii Diels
English: Forrest Jerusalem sage
China: cang shan cao su
P. franchetiana Diels
English: Franchet Jerusalem sage
China: da li cao su
P. fruticosa L.
English: Jerusalem sage
China: cheng hua cao su
P. kansuensis C.Y. Wu

English: Gansu Jerusalem sage
China: gan su cao su
P. likiangensis C.Y. Wu
English: Lijiang Jerusalem sage, Likiang Jerusalem sage
China: li jiang cao su
P. maximowiczii Regel
English: Maximowicz Jerusalem sage
China: da ye cao su
P. mongolica Turcz.
English: Mongolian Jerusalem sage
China: chuan ling cao
P. paohsingensis C.Y. Wu
English: Baoxing Jerusalem sage
China: bao xing cao su
P. pratensis Karelin & Kirilov
English: meadow Jerusalem sage
China: cao yuan cao su
P. pygmaea C.Y. Wu
English: dwarf Jerusalem sage
China: ai cao su
P. tibetica Marquand & Airy Shaw
English: Tibet Jerusalem sage
China: xi zang cao su
P. tuberosa L.
English: tuberous-root Jerusalem sage
China: kua gen cao su
P. umbrosa Turcz.
English: shady Jerusalem sage
China: cao su
P. younghusbandii Mukerjee
English: Younghusband Jerusalem sage
China: pang xie jia

Phlomoides Moench Labiatae

Origins:
Resembling the genus *Phlomis* L.

Phlox L. Polemoniaceae

Origins:
Phlox, the Greek name for a flame or plants with flame-colored flowers; Latin *phlox, phlogis* for a flower, otherwise unknown (Plinius).

Species/Vernacular Names:

P. bifida (L.) Beck

English: sand phlox

P. buckleyi Werry

English: sword-leaf phlox

P. caespitosa Nutt.

English: cushion phlox

P. carolina L.

English: thick-leaf phlox

P. dispersa Sharsm.

English: High Sierra phlox

P. divaricata L. (*Phlox canadensis* Sweet)

English: wild sweet William, blue phlox

P. dolichantha A. Gray

English: Bear Valley phlox

P. drummondii Hook.

English: annual phlox, Drummond phlox

Japan: kikyô-nadeshiko

P. glaberrima L.

English: smooth phlox

P. maculata L.

English: wild sweet William, meadow phlox

P. muscoides Nutt.

English: moss phlox

P. nana Nutt.

English: Santa Fe phlox

P. nivalis Lodd. ex Sweet

English: trailing phlox

P. ovata L.

English: mountain phlox

P. paniculata L. (*Phlox decussata* Lyon ex Pursh)

English: perennial phlox, summer phlox, autumn phlox, fall phlox, summer perennial phlox

Japan: kusa-kyo-chiku-tô

P. pilosa L.

English: prairie phlox

P. sibirica L.

English: Siberian phlox

P. speciosa Pursh

English: bush phlox

P. stolonifera Sims

English: creeping phlox

P. subulata L. (*Phlox setacea* L.)

English: moss phlox, mountain phlox, moss pink

Phlyarodoxa S. Moore Oleaceae

Origins:

Probably from the Greek *phlyaros* "nonsense," *phlyarodes* "foolish" and *doxa* "glory."

Phlyctidocarpa Cannon & W.L. Theobald Umbelliferae

Origins:

Greek *phlyktaina* "blister, pustule," *phlyktis, phlyktidos* "boil" and *karpos* "fruit," referring to the mericarps.

Phoberos Lour. Flacourtiaceae

Origins:

Greek *phoberos* "fearful," *phobos* "fear, panic," possibly referring to spiny plants; see João (Joannes) de Loureiro (1717-1791), *Flora cochinchinensis*. 1: 317. Ulyssipone [Lisboa] 1790.

Phocea Seem. Euphorbiaceae

Origins:

Phokaia, Phocaea, a maritime town of Ionia, *phoke* "a seal, sea-dog," Latin *phoce, phoces* and *phoca* for a sea-calf, a seal.

Phoebanthus S.F. Blake Asteraceae

Origins:

From the Greek *phoibos* "pure, bright" and *anthos* "flower."

Phoebe Nees Lauraceae

Origins:

Phoebe (Phoibe), a female Titan, wife of Coeus, mother of Leto and Asteria, grandmother of Apollo and Artemis, daughter of Uranus and Gaea (Gaia), goddess of Moon in Greek mythology, her epithet was Gold-Crowned.

Species/Vernacular Names:

P. sp.

Malaya: medang tanah, medang ketanah, selinchar, medang

Mexico: ya xomul, ya yexru dau

Phoenicanthemum (Blume) Blume Loranthaceae

Origins:

From the Greek *phoinikeos* "red, crimson, purple-red" and *anthemon* "a flower."

Phoenicanthus Alston Annonaceae

Origins:

Greek *phoinix, phoinikos* "purple, red, crimson" and *anthos* "flower."

Phoenicaulis Nutt. Brassicaceae

Origins:

Greek *phoinix, phoinikos* "purple, red, crimson" and *kaulos* "stalk," Latin *caulis* "the stalk, stem of a plant"; some suggest (see James C. Hickman, ed., *The Jepson Manual: Higher Plants of California*. 432. University of California Press, Berkeley 1993) from Greek *phaneros* "evident, conspicuous, visible," *phao* "to shine," *phaino* "come to light, bring to light," visible stem.

Phoenicocissus Martius ex Meissner Bignoniaceae

Origins:

From the Greek *phoinix, phoinikos* "purple, red, crimson" and *kissos* "ivy," referring to the nature of the plant.

Phoenicophorium H.A. Wendland Palmae

Origins:

From *Phoenix* and *phorios* "stolen," stolen palm, thief palm; see H.E. Moore, Jr., in *Principes*. vol. 19, no. 3: 113. 1975.

Species/Vernacular Names:

P. borsigianum (Koch) Stuntz

Brazil: palmeira dourada

Phoenicoseris (Skottsb.) Skottsb. Asteraceae

Origins:

Greek *phoinix, phoinikos* "purple, red" and *seris, seridos* "a chicory or a kind of endive."

Phoenicosperma Miq. Elaeocarpaceae

Origins:

Greek *phoinix, phoinikos* "purple, red" and *sperma* "seed."

Phoenix L. Palmae

Origins:

Phoinix, phoinikos "date palm, date, palm frond, purple, crimson," ancient Greek name used by Theophrastus and Plinius; the Phoenicians, from *Phoenix, nicis, Phoenices*, were the inhabitants of *Phoenice, Phoenicia*, the coastal territory of Syria; *Phoenix* (*Phoi*), in Greek legend, was son of Amyntor and Cleobule (see Homer, *Iliad* ix. 447-480); the *Phoenix* (*Phoi*) was the fabulous sacred bird of Egypt (see Herodotus and Plinius); see Carl Linnaeus, *Species Plantarum*. 1188. 1753 and *Genera Plantarum*. Ed. 5. 496. 1754.

Species/Vernacular Names:

P. canariensis hortorum ex Chabaud

English: Canary date palm, Canary Island date, Canary Island palm, Canary Island date palm, Canary palm

Brazil: tamareira das Canárias, palmeira das Canárias

Japan: Kanari-sote-tsu-shuro

P. dactylifera L.

English: date, date palm

Arabic: nakhl, nekhla, el-nakheil, nakhla

Brazil: tamareira

Mexico: nocuana ticaa yaga ciña

Japan: natsume-yashi

China: wu lou zi

French: palmier dattier

Italian: palma da dattero

P. hanceana Naud. var. *philippinensis* Becc.

The Philippines: voiavoi

P. pusilla Gaertn.

Japan: ô-kami-yashi

P. reclinata Jacq.

English: wild date palm, coffee palm, feather palm, Senegal date palm, dwarf date palm

Brazil: tamareira do Senegal

Japan: kabu-dachi-sotetsu-juro

Southern Africa: wildedadelboom, datelboom, kafferkoffie; iSundu (Xhosa); omuvare (Herero); iDama, iSundu (Zulu); aNkindu (Thonga); liLala (Swazi); mutzhema, mutshema, mutshevho (Venda); kanjedza, isiPuppu, iSundu (Shona); dikindu (Mbukushu)

Nigeria: kijnjiri (Hausa); efu (Nupe); wure (Tiv); ookun (Yoruba); ukukon (Edo); ngala (Igbo); eyup inuen (Efik)

Yoruba: okunkun, elekikobi

P. roebelinii O'Brien

English: miniature date palm, pygmy date palm, Roebelin palm

Brazil: tamareira de jardim, tanareira anã

Japan: Shina-yashi (= Chinese palm)

P. rupicola Anderson

English: wild date palm, cliff date, India date palm, East Indian wine palm

Brazil: tamareira do rochedo

P. sylvestris (L.) Roxb.

English: wild date, India date, wild date palm

Brazil: tamareira silvestre, tamareira selvagem, tamareira da India, tamareira de açúcar

Pholidia R. Br. Myoporaceae

Origins:
Greek *pholis, pholidos* "scale, horny scale"; see Robert Brown (1773-1858), *Prodromus florae Novae Hollandiae et Insulae van-Diemen*. 517. London 1810.

Pholidiopsis F. Muell. Myoporaceae

Origins:
Resembling the genus *Pholidia* R. Br.; see F. von Mueller, in *Linnaea*. 25: 429. (Apr.) 1853.

Pholidocarpus Blume Palmae

Origins:
Corky fruits, Greek *pholis, pholidos* "scale, horny scale" and *karpos* "fruit," referring to the warty covering of the fruits.

Pholidostachys H.A. Wendland ex Hook.f. Palmae

Origins:
From the Greek *pholis, pholidos* "horny scale, bandage" and *stachys* "spike, an ear of wheat," referring to the inflorescence, to the pit bracts on the rachillae.

Pholidota Lindley ex Hooker Orchidaceae

Origins:
From the Greek *pholis, pholidos* "scale, horny scale," *pholidotos* "scaly, clad in scales," probably referring to the bracts of the inflorescence or to the sheaths surrounding the pseudobulbs; see Sir William Jackson Hooker (1785-1865), *Exotic Flora*. 2: 138. Edinburgh 1825.

Species/Vernacular Names:
P. chinensis Lindl.

English: rattlesnake orchid

China: shi xian tao

P. imbricata Hook.f.

English: rattlesnake orchid

Pholisma Nutt. ex Hook. Lennoaceae

Origins:
Possibly from the Greek *pholeia, pholia* "life in a hole," parasitic, living in sand on roots of *Croton, Eriogonum, Ambrosia, Pluchea* or *Hymenoclea* spp.; or from Greek *pholis* "horny scale," referring to the stem, see James C. Hickman, ed., *The Jepson Manual: Higher Plants of California*. 734. 1993.

Species/Vernacular Names:
P. sonorae (A. Gray) Yatskievych

English: sand foot

Pholistoma Lilja Hydrophyllaceae

Origins:
Greek *pholis* "horny scale" and *stoma* "mouth."

Pholiurus Trin. Gramineae

Origins:
From the Greek *pholis, pholidos* "scale, horny scale" and *oura* "tail," referring to the glumes; see Carl Bernhard von Trinius, *Fundamenta Agrostographiae*. 131. Viennae 1820.

Phonus Hill Asteraceae

Origins:
Latin *phonos, phonus* "another name of the plant *atractylis* (Plinius)," Greek *phonos* "murder."

Phoradendron Nutt. Viscaceae

Origins:
Greek *phoros* "bearing, carrying," *phero, phoreo* "to bear" and *dendron* "tree"; see T. Nuttall, in *Journal of the Academy of Natural Sciences of Philadelphia.* 1: 185. (Aug.) 1849.

Species/Vernacular Names:
P. sp.
Peru: cushma, pupa de huarango

P. californicum Nutt.
English: desert mistletoe

P. densum Trel.
English: dense mistletoe

P. juniperinum A. Gray
English: juniper mistletoe

P. laxiflorum Ule (*Phoradendron huallagense* Ule)
Peru: beguefide

P. leucarpon (Raf.) Reveal & M. Johnston
English: mistletoe

P. libocedri (Engelm.) Howell
English: incense-cedar mistletoe

P. macrophyllum (Engelm.) Cockerell
English: big leaf mistletoe

P. mathewsii Trel.
Peru: pishco-ismán

P. pauciflorum Torrey
English: fir mistletoe

P. piperoides (Kunth) Trel. (*Loranthus piperoides* Kunth; *Viscum piperoides* (Kunth) DC.)
Peru: pajar, suelda con suelda
Central America: muerdago, liga, ligamatapalo, matapalo, nigüita

P. quadrangulare (Kunth) Krug & Urban (*Loranthus quadrangularis* Kunth)
Peru: pishco-ismán, suelda con suelda
Central America: muerdago, liga, ligamatapalo, matapalo, nigüita

P. villosum (Nutt.) Nutt.
English: oak mistletoe

Phormangis Schltr. Orchidaceae

Origins:
From the diminutive of *phormos*, the Greek *phormion* "mat, a straw covering" and *angeion, aggeion* "a vessel, cup," the spur of the lip.

Phormium Forst. & Forst.f. Phormiaceae (Agavaceae)

Origins:
Greek *phormion* "mat," referring to the very strong fibers used for textiles, cordage and nets, Latin *phormio, phormionis* and *formio, formionis* for the wicker-work of reeds or rushes, a mat, a straw covering; see Johann Reinhold Forster (1729-1798) and Johann Georg Adam (1754-1794), *Characteres generum plantarum.* 47, t. 34. (Nov.) 1775; H.E. Connor and E. Edgar, "Name changes in the indigenous New Zealand flora, 1960-1986 and Nomina Nova IV, 1983-1986." *New Zealand Journal of Botany.* Vol. 25: 115-170. 1987.

Species/Vernacular Names:
P. cookianum Le Jolis (*Phormium colensoi* Hook.f.; *Phormium hookeri* Hook.f.)
English: mountain flax, New Zealand hemp
Maori name: wharariki

P. tenax Forst. & Forst.f.
English: New Zealand flax, New Zealand hemp, bush flax
Maori names: harakeke, korari

Phorolobus Desv. Pteridaceae (Adiantaceae)

Origins:
From the Greek *phoros* "bearing" and *lobos* "pod, lobe."

Phosanthus Raf. Rubiaceae

Origins:
See Constantine Samuel Rafinesque, *Ann. Gén. Sci. Phys.* 6: 82. 1820; E.D. Merrill, *Index rafinesquianus.* 226. Jamaica Plain, Massachusetts, USA 1949.

Photinia Lindley Rosaceae

Origins:
Greek *photeinos* "shining, bright," *phos* "light," the leaves are shining.

Species/Vernacular Names:
P. glabra (Thunb.) Maxim. (*Crataegus glabra* Thunb.)
English: Japanese photinia
Japan: kaname-mochi, akame-mochi
China: tsu lin tzu

P. pyrifolia (Lam.) K. Robertson & J. Phipps

English: red chokeberry

P. serratifolia (Desf.) Kalkman (*Photinia serrulata* Lindl.)

English: Chinese photinia, photinia

Japan: okaname-mochi

China: shi nan ye

Photinopteris J. Sm. Polypodiaceae

Origins:
From the Greek *photeinos* "shining, bright" and *pteris* "a fern."

Phragmanthera Tieghem Loranthaceae

Origins:
From the Greek *phragma* "a hedge, a fence, screen" and *anthera* "anther."

Species/Vernacular Names:
P. sp.
Congo: nkunkunda ya mulolo

Phragmipedium Rolfe Orchidaceae

Origins:
Greek *phragma* "partition, division" and *pedilon* "a slipper," referring to the divisions of the ovary and to the shape of the lip of these slipper orchids.

Phragmites Adanson Gramineae

Origins:
Greek *phragma* "a hedge, a fence, screen," *phragmites* "of fences," *kalamos phragmites* "reed of hedges," Latin *phragmites, is* for a kind of reed growing in hedges (Plinius); see M. Adanson, *Familles des Plantes.* 2: 34. 1763.

Species/Vernacular Names:
P. australis (Cav.) Steudel (*Phragmites communis* (L.) Trin.; *Arundo australis* Cav.; *Arundo phragmites* L.; *Arundo vulgaris* Lam.)

English: reeds, common reed, reed, bamboo reed, Danube grass, reed grass

Southern Africa: riete, riet, fleikiesriet, fluitjiesriet, gewone fluitjiesriet, sonquasriet, vaderlandsriet, vinkriet, vlakkiesriet; otuu (Herero)

Yoruba: ifu

German: ried, schilfrohr, rohrschilf

Spanish: carrizo

Mexico: caña, holo, acatl, carrizo, taa gui, taa quij, bixilla gui, pixilla qui gui, guii, qui, quij

Japan: yoshi, ashi

China: lu gen, lu, wei, chia

India: nal

Malaya: rumput gedabong, mata burong puding

P. mauritianus Kunth

English: lowveld reed

South Africa: dekriet, fluitjiesriet, laeveldfluitjiesriet; rihlanga (Tsonga)

Rodrigues Island: roseau de pays, jonc

Phragmocarpidium Krapov. Malvaceae

Origins:
Greek *phragma* "a hedge, a division, screen" and *karpos* "fruit," *karpodes* "fruitful," Latin *carpidium* "carpel."

Phragmocassia Britton & Rose Caesalpiniaceae

Origins:
From the Greek *phragma* "a hedge, a division, screen" plus the genus *Cassia*.

Phragmorchis L.O. Williams Orchidaceae

Origins:
The genus *Phragmites* plus *orchis* "orchid," referring to the reed-like stems.

Phragmotheca Cuatrec. Bombacaceae

Origins:
From the Greek *phragma* "a hedge, a division, partition" and *theke* "a box, case, cell."

Phreatia Lindley Orchidaceae

Origins:
Greek *phreatia* "a cistern, a well," alluding to the lateral sepals and labellum or lip; see John Lindley (1799-1865), *The Genera and Species of Orchidaceous Plants.* 63. London 1830;

B.A. Lewis and P.J. Cribb, *Orchids of the Solomon Islands and Bougainville.* Royal Botanic Gardens, Kew 1991.

Species/Vernacular Names:
P. micrantha (A. Rich.) Schltr.
The Solomon Islands: kikilapa ngirisi

Phrissocarpus Miers Apocynaceae

Origins:
From the Greek *phrisso, phritto* "to be rough, bristle" and *karpos* "fruit."

Phrodus Miers Solanaceae

Origins:
Presumably from the Greek *phroudos* "fled, vanished, undone."

Phryganocydia C. Martius ex Bur. Bignoniaceae

Origins:
Greek *phryganon* "dry stick, undershrub" and *kydos* "glory," *kydeis* "glorious," lianes with showy and fragrant flowers.

Phrygilanthus Eichler Loranthaceae

Origins:
From the Greek *phrygilos* "a finch, chaffinch" and *anthos* "a flower"; see Carl (Karl) Friedrich Philipp von Martius (1794-1868), *Flora Brasiliensis.* 5(2): 45. (Jul.) 1868.

Phrygiobureaua Kuntze Bignoniaceae

Origins:
For the French botanist Louis Édouard Bureau, 1830-1918, specialist in Bignoniaceae; see John H. Barnhart, *Biographical Notes upon Botanists.* 1: 283. Boston 1965; T.W. Bossert, *Biographical Dictionary of Botanists Represented in the Hunt Institute Portrait Collection.* 59. 1972; Ida Kaplan Langman, *A Selected Guide to the Literature on the Flowering Plants of Mexico.* 1964; E.M. Tucker, *Catalogue of the Library of the Arnold Arboretum of Harvard University.* Cambridge, Massachusetts 1917-1933; Elmer Drew Merrill, *Bernice P. Bishop Mus. Bull.* 141: 57. 1937; Frans A. Stafleu and Erik A. Mennega, *Taxonomic Literature. Sup-*

plement III. 231-234. 1995; R. Zander, F. Encke, G. Buchheim and S. Seybold, *Handwörterbuch der Pflanzennamen.* 14. Aufl. Stuttgart 1993.

Phryna (Boiss.) Pax & K. Hoffmann Caryophyllaceae

Origins:
From the Greek *phrynos, phryne* "a toad."

Phryne Bubani Brassicaceae

Origins:
From the Greek *phrynos, phryne* "a toad."

Phrynella Pax & K. Hoffmann Caryophyllaceae

Origins:
The diminutive of the genus *Phryna.*

Phrynium Willd. Marantaceae

Origins:
Greek and Latin *phrynion* for a plant, called also *poterion,* Greek *phrynos* "a toad," referring to the marshy habitat.

Phtheirospermum Bunge ex Fischer & C.A. Meyer Scrophulariaceae

Origins:
Greek *phtheir, phtheiros* "a louse" and *sperma* "seed," referring to the appearance of the seeds.

Phthirusa Mart. Loranthaceae

Origins:
From the Greek *phtheir, phtheiros* "a louse," a pest on woody crops.

Species/Vernacular Names:
P. caribaea (Krug & Urban) Britton & P. Wilson
English: West Indies mistletoe

Phucagrostis Cavolini Potamogetonaceae

Origins:
Greek *phykes* "living in seaweed," *phykos* "wrack, seaweed, orchella-weed" and *agrostis, agrostidos* "grass, weed,

couch grass"; see Arthur D. Chapman, ed., *Australian Plant Name Index*. 2249. Canberra 1991.

Phuodendron (Graebner) Dalla Torre & Harms Valerianaceae

Origins:
Latin *phu* or *phun* (Plinius) and Greek *phou* for "a kind of valerian" plus Greek *dendron* "a tree."

Phuopsis (Griseb.) Hook.f. Rubiaceae

Origins:
Greek *phou* "a kind of valerian" and *opsis* "resemblance," resembling to *Valeriana phu* L.

Phycella Lindley Amaryllidaceae (Liliaceae)

Origins:
Greek *phykes* "living in seaweed," *phykos* "wrack, seaweed, orchella-weed."

Phycoschoenus (Asch.) Nakai Cymodoceaceae

Origins:
Greek *phykos* "wrack, seaweed" and *schoinos* "rush, reed, cord."

Phyganthus Poepp. & Endl. Tecophilaeaceae

Origins:
From the Greek *phyge* "flight, escape" and *anthos* "flower."

Phygelius E. Meyer ex Benth. Scrophulariaceae

Origins:
Probably from the Greek *phyge* "flight, escape," *pheugo* "to run away, to shun," referring to its long escaping from the plant hunters, or *pheugo* and *helios* "the sun," a shade-lover, or referring to the distance between this genus and the closest relatives *Digitalis* and *Penstemon*.

Species/Vernacular Names:
P. capensis E. Meyer ex Benth.

English: Cape figwort

Phyla Lour. Verbenaceae

Origins:
Greek *phyle* "a tribe, clan, union," probably referring to the flowers clustered in a tight head or to the spreading mat-like growth, ground-cover; see João (Joannes) de Loureiro, *Flora cochinchinensis*. 1: 63. Ulyssipone [Lisboa] 1790.

Species/Vernacular Names:
P. canescens (Kunth) Greene (*Phyla nodiflora* var. *canescens* (Kunth) Mold.)

English: carpet grass

P. nodiflora (L.) E.L. Greene (*Lippia nodiflora* (L.) Michaux; *Verbena nodiflora* L.; *Zapania nodiflora* (L.) Lam.)

English: Cape weed, daisylawn, frog fruit, fog fruit, matgrass, turkey-tangle, lippia, knotted-flower phyla

Japan: iwa-dare-sô

China: guo jiang teng

Phylacium Bennett Fabaceae

Origins:
From the Greek *phylax, phylakos* "a guardian, protector," *phylake* "prison, guarding," probably referring to the bracts enclosing the inflorescence; see John Joseph Bennett (1801-1876) and Robert Brown, *Plantae Javanicae rariores*. 159, t. 33. London 1840.

Phylactis Schrader Asteraceae

Origins:
See *Philactis*.

Phylanthera Noronha Rubiaceae

Origins:
Perhaps from the Greek *phyle* "a tribe, clan, union" and *anthera* "anther."

Phylica L. Rhamnaceae

Origins:
Greek *philyke* for a shrub, evergreen, privet, a species of *Rhamnus*, see Carl Linnaeus, *Species Plantarum*. 195. 1753 and *Genera Plantarum*. Ed. 5. 90. 1754; referring to the abundant foliage; some suggest from the Greek *phyllikos*

"leafy," see Georg Christian Wittstein, *Etymologisch-botanisches Handwörterbuch*. 685. Ansbach 1852.

Species/Vernacular Names:
P. buxifolia L.
English: phylica with box-like leaves
South Africa: box phylica
P. oleaefolia Vent. (*Phylica oleoides* DC.)
English: phylica with *Olea*-like leaves
South Africa: hardebos
P. paniculata Willd.
English: common hard-leaf, tick-tree
Southern Africa: gewone hardeblaar, umdidi tree, luisboom (= tick-tree); umHlalamithi, umDidi (Zulu)

Phyllacantha Hook.f. Rubiaceae

Origins:
Greek *phyllon* "a leaf" and *akantha* "thorn, prickle," with prickly leaves.

Phyllacanthus Hook.f. Rubiaceae

Origins:
From the Greek *phyllakanthos* "with prickly leaves."

Phyllachne Forst. & Forst.f. Stylidiaceae

Origins:
From the Greek *phyllon* "a leaf" and *achne* "chaff, glume."

Species/Vernacular Names:
P. colensoi (Hook.f.) S. Berggren after the British (Cornish) missionary Rev. (John) William Colenso, 1811-1899 (d. Napier, New Zealand), ethnologist and explorer, botanist, printer, plant collector, North Island, New Zealand, friend and companion of Allan Cunningham, friend and correspondent of Sir William Hooker, 1861 elected to the Parliament, 1865 Fellow of the Linnean Society, 1886 a Fellow of the Royal Society. His works include *Fifty Years Ago in New Zealand*. Napier 1888 and *An Account of Visits to, and Crossing Over, the Ruahine Mountain Range*. Napier 1884; see J.H. Barnhart, *Biographical Notes upon Botanists*. 1: 366. 1965; B. Graham, "Of William Colenso, 1811-1899." *The Cornish Garden*. 1982; Alice Margaret Coats, *The Quest for Plants. A History of the Horticultural Explorers*. 236-239. London 1969; Austin G. Bagnall and George Conrad Petersen, *William Colenso: Printer, Missionary, Botanist, Explorer, Politician*. Wellington 1948; Leonard Huxley, *Life and Letters of Sir J.D. Hooker*. London 1918; G.A. Doumani, ed., *Antarctic Bibliography*. Washington,

Library of Congress 1965-1979; H.N. Clokie, *Account of the Herbaria of the Department of Botany in the University of Oxford*. 147. Oxford 1964; R. Glenn, *The Botanical Explorers of New Zealand*. Wellington 1950; A. Lasègue, *Musée botanique de Benjamin Delessert*. Paris 1845; Audrey le Lièvre, "William Colenso, New Zealand botanist: something of his life and work." *The Kew Magazine*. 7(4): 186-200. November 1990; R. Zander, F. Encke, G. Buchheim and S. Seybold, *Handwörterbuch der Pflanzennamen*. 14. Aufl. Stuttgart 1993; I.C. Hedge and J.M. Lamond, *Index of Collectors in the Edinburgh Herbarium*. Edinburgh 1970; Frans A. Stafleu and Erik A. Mennega, *Taxonomic Literature. Supplement IV: Ce-Cz*. 264-266. Königstein 1997)

English: rock cushion, cushion plant

Phyllactis Pers. Valeraniaceae

Origins:
From the Greek *phyllon* "a leaf" and *aktis, aktin* "a ray," referring to the redical leaves.

Phyllagathis Blume Melastomataceae

Origins:
From the Greek *phyllon* "a leaf" and *agathis* "a ball of thread," an allusion to the large bracts below the flowerheads or referring to the acaulous habit and leaves arising from the base.

Phyllanthera Blume Asclepiadaceae (Periplocaceae)

Origins:
Greek *phyllon* "a leaf" and *anthera* "anther," referring to the nature of the anthers.

Phyllanthodendron Hemsley Euphorbiaceae

Origins:
Greek *phyllon* "a leaf," *anthos* "flower" and *dendron* "tree," the flowers appear on leaf-like cladodes.

Phyllanthus L. Euphorbiaceae

Origins:
Greek *phyllon* "a leaf" and *anthos* "flower," in some species the flowers are produced on leaf-like branches and branchlets, the flowers appear on leaf-like cladodes; see Carl Linnaeus (1707-1778), *Species Plantarum*. 981. 1753 and *Genera Plantarum*. Ed. 5. 422. 1754.

Species/Vernacular Names:

P. sp.

Brazil: quebra-pedras

Mexico: añilillo; panetela (Tabasco); xpbixtdon (Maya l., Yuacatan); palo sonzo (southeast San Luis Potosí); hierba de leche (Fortuño, Veracruz)

Peru: asnac panga, quinilla del tahuampa, sekemo, víbora huasca, arranca pedras, charcapeedia, pachori navi

Colombia: gabellon

Eastern Africa: mtejo-ya-hasi

Yoruba: eyinolobe funfun, aawe, lenkosun

P. acidus (L.) Skeels (*Averrhoa acida* L.; *Cicca disticha* L.; *Cicca acida* (L.) Merr.; *Phyllanthus distichus* (L.) Müll.Arg.)

English: Otaheite gooseberry, gooseberry tree, Indian gooseberry, Malay gooseberry

Malaya: chermai, chermela

India: harfarauri

Colombia: arbolito

Belize: wild plum

Peru: grosella

Mexico: ciruela costeña, ciruelo costeño, cuatelolote; totolole (Oaxaca); manzana estrella (Tamaulipas); pimientillo (Sinaloa)

P. acuminatus M. Vahl (*Conami brasiliensis* Aublet; *Phyllanthus conami* Swartz.)

Belize: ciruelillo

Panama: jobitillo

Peru: barbasco cuartillito, borrachero, barbasco, conabi, conamí, timbó conabi

P. adenodiscus Müll. Arg. (*Diasperus adenodiscus* O. Kuntze)

Mexico: cascabel (Hidalgo); chayacachte (Huejutla, Chililico, Hidalgo)

P. amarus Schum. & Thonn.

Congo: ndiango, mundziri

P. angustifolius (Sw.) Sw.

English: foliage flower

P. arbuscula (Sw.) J.F. Gmel.

English: foliage flower

P. brasiliensis (Aublet) Poiret

Mexico: kahyuk, x-pahul, xpibul (Maya l., Yucatan)

Peru: cantibña, canabi

P. calycinus Labill.

English: false boronia, snowdrop spurge

P. caroliniensis Walter

Mexico: kabalbesikté (Maya l., Yucatan)

P. casticum Willemet f.

Rodrigues Island: castique, bois castique

La Réunion Island: bois de demoiselle, bois de mazelle

Mauritius: bois castique, castique rouge

P. cedrelifolius Verdoorn

English: forest potato bush, feather-leaved phyllanthus

South Africa: bosaartappelbos

P. cuscutiflorus S. Moore

English: pink phyllanthus

P. discoideus (Baill.) Müll.Arg. (*Margaritaria discoidea* (Baill.) Webster)

Central Africa: budela

Ghana: adzadze, apapaya, benkyi, opepea

Sierra Leone: tijoe

Nigeria: asiyin

Togo: konkona

Ivory Coast: bakonko, bon, koe, mousan koe, monsan koe, moussan hoe, musa houe, moussan houe, pepe sia, pepesia, tenouba, tenoura

Cameroon: boudela, kango

P. distichus (L.) Müll.Arg. (*Cicca disticha* L.)

Hawaii: pamakani mahu, pamakani

Indochina: duoc duang

P. emblica L. (*Emblica officinalis* Gaertn.)

English: emblic, myrobalan, Malacca tree

India: amla, amlaki, avula, chyahkya, anula, htaky, lalli, miral, nalli, nelli, nellimara, nilika, suam

Nepal: amala

Tibetan: skyurura

Malaya: laka, laka laka, melaka, asam melaka

Indochina: kam lam, kham, me rung

P. gasstroemi Müll.Arg.

English: blunt spurge

P. glaucescens Kunth (*Diasperus glaucescens* O. Ktze)

Mexico: xpbixtdon (Maya l., Yucatan); palo sonzo (southeast San Luis Potosí)

Belize: monkey rattle, pisch tong

P. gunnii Hook.f.

English: shrubby spurge, scrubby spurge

P. hirtellus F. Muell. ex Müll.Arg.

English: thyme spurge

P. maderaspatensis L.

Southern Africa: skilpadbossie; leêtsane (Tswana)

P. niruri L. (*Phyllanthus filiformis* Pav. ex Baillon; *Phyllanthus lathyroides* Kunth)

English: seed-under-leaf, egg woman

Mexico: dormilona (Tehuantepec, Oaxaca)

Peru: piedra con piedra, chanca piedra, niruri, sacha Foster

Brazil: quebra-pedra, malva-pedra, erva-pombinha

India: bhuniamla

Malaya: dukong anak

The Philippines: sampasampalukan, talikod, kurukalung-gai, malakirum-kirum

P. parvifolius Buch.-Ham. ex D. Don

Nepal: sunpate

P. phillyreifolius Poiret

La Réunion Island: faux bois de demoiselle, bois de négresse, bois de ravine, bois de chien, girimbelle marron

Mauritius: bois balié de la rivière, bois dilo, bois petites feuilles

P. pusillifolius S. Moore

English: stick bush

P. reticulatus Poiret

English: roast potato plant, potato bush

Southern Africa: aartappelbos; mkasiri (Swahili); inTaba yengwe, iNtabayengwe, umChumelo, munyuswane, umTswathiba, uButswamtimi (Zulu); thethenya (Tsonga); makhulu-wamutangauma (Venda)

Eastern Africa: mgogondi (Tabora)

W. Africa: balanbalan

Japan: Taiwan-koban-no-ki

The Philippines: malatinta, tintatintahan, matang-bulud

P. urinaria L.

French: herbe du chagrin, petit tamarin rouge

Congo: passa ndzo

Vietnam: cam kiem, khao ham

China: zhen zhu cao, chen chu tsao

Phyllapophysis Mansf. Melastomataceae

Origins:

From the Greek *phyllon* "leaf" and *apophysis* "side-shoot, branch."

Phyllarthron DC. Bignoniaceae

Origins:

From the Greek *phyllon* "leaf" and *arthron* "a joint," leaves resembling a series of articulated segments.

Phyllaurea Lour. Euphorbiaceae

Origins:

From the Greek *phyllon* "leaf" and Latin *aureus, a, um* "golden," referring to the leaves.

Phyllera Endl. Euphorbiaceae

Origins:

Referring to the Latin *philyra* and *philura* for the linden-tree; see also *Philyra* Klotzsch.

Phyllirea Duhamel Oleaceae

Origins:

See *Phillyrea* L.

Phyllis L. Rubiaceae

Origins:

Latin *phyllis* for an almond-tree (Plinius), Phyllis was the daughter of King Sithon of Thrace, she was changed into an almond-tree; some suggest from the Greek *phyllon* "leaf," alluding to the foliage.

Phyllitis Hill Aspleniaceae

Origins:

Dioscorides applied *phyllitis* to the hart's-tongue, *Scolopendrium*.

Phyllitis Moench Aspleniaceae

Origins:

Greek *phyllon* "leaf," alluding to the fronds; Dioscorides applied *phyllitis* to the hart's-tongue, *Scolopendrium*.

Phyllitis Raf. Pteridaceae

Origins:

For *Pteris* L., Dioscorides applied *phyllitis* to the hart's-tongue, *Scolopendrium*; see Constantine S. Rafinesque, *Am. Monthly Mag. Crit. Rev.* 4: 195. 1819.

Phyllitopsis Reichst. Aspleniaceae

Origins:

Referring to the genus *Phyllitis*.

Phyllobaea Benth. Gesneriaceae

Origins:
See *Phylloboea* Benth.

Phylloboea Benth. Gesneriaceae

Origins:
Greek *phyllon* "leaf" and the genus *Boea* Comm. ex Lam.

Phyllobolus N.E. Br. Aizoaceae

Origins:
From the Greek *phyllon* "leaf" and *bolos* "casting," *phylloboleo* "shed the leaves."

Phyllobotryon Müll. Arg. Flacourtiaceae

Origins:
From the Greek *phyllon* "leaf" and *botrys* "cluster, a bunch of grapes," referring to the epiphyllous inflorescences.

Phyllobotryum Müll. Arg. Flacourtiaceae

Origins:
See *Phyllobotryon*.

Phyllocactus Link Cactaceae

Origins:
Leaf cactus, from the Greek *phyllon* "leaf" plus *cactus*, referring to the stem, see also *Epiphyllum* Haw.

Phyllocalymma Benth. Asteraceae

Origins:
From the Greek *phyllon* "leaf" and *kalymma* "a covering, hood."

Phyllocalyx O. Berg Myrtaceae

Origins:
From the Greek *phyllon* "leaf" and *kalyx* "calyx."

Phyllocarpus Riedel ex Tul. Caesalpiniaceae

Origins:
Greek *phyllon* "leaf" and *karpos* "fruit."

Phyllocephalum Blume Asteraceae

Origins:
Greek *phyllon* and *kephale* "head," referring to the leafy involucre.

Phyllocereus Miq. Cactaceae

Origins:
From the Greek *phyllon* "leaf" plus *cereus*, see also *Epiphyllum* Haw.

Phyllocharis Diels Campanulaceae

Origins:
From the Greek *phyllon* "leaf" and *charis* "delight, grace, beauty."

Phyllochlamys Bureau Moraceae

Origins:
From the Greek *phyllon* "leaf" and *chlamys* "cloak."

Phyllocladus Mirbel Phyllocladaceae

Origins:
Greek *phyllon* "leaf" and *klados* "branch," referring to the flattened branches and leaf-like branchlets; see Charles F.B. de Mirbel (1776-1854), in *Mémoires du Muséum d'Histoire Naturelle*. 13: 48. Paris 1825.

Species/Vernacular Names:
P. aspleniifolius (Labill.) Hook.f.

English: celery-top pine

P. hypophyllus Hook.f.

The Philippines: aransisingit, dalung, galingkinga, salumayag

P. trichomanoides D. Don

English: celery pine

New Zealand: tanekaha (Maori name)

Phylloclinium Baillon Flacourtiaceae

Origins:
Greek *phyllon* "leaf" and *klinion* the diminutive of *kline* "a bed, couch," epiphyllous inflorescences.

Phyllocomos Masters Restionaceae

Origins:
From the Greek *phyllocomos* "thick-leaved."

Phyllocosmus Klotzsch Ixonanthaceae

Origins:
From the Greek *phyllon* "leaf" and *kosmos* (*kosmeo* "to rule, adorn, dress") "ornament, decoration."

Species/Vernacular Names:
P. africanus (Hook.f.) Klotzsch (*Ochthocosmus africanus* Hook.f.)
Nigeria: araba-uji (Igbo)

Phyllocrater Wernham Rubiaceae

Origins:
From the Greek *phyllon* "leaf" and *krater* "a bowl, crater."

Phylloctenium Baillon Bignoniaceae

Origins:
From the Greek *phyllon* "leaf" and *ktenion* "a little comb."

Phyllodesmis Tieghem Loranthaceae

Origins:
Greek *phyllon* "leaf" and *desmis, desmos* "a bond, band, bundle."

Phyllodium Desvaux Fabaceae

Origins:
Greek *phyllon* "a leaf" and *-odes* "resembling, of the nature of, like," referring to the dilated petiole; see Nicaise Auguste Desvaux (1784-1856), in *Journal de Botanique*. 1: 123. Paris 1813.

Species/Vernacular Names:
P. pulchellum (L.) Desv. (*Desmodium pulchellum* (L.) Benth.; *Hedysarum pulchellum* L.; *Meibomia pulchella* (L.) Kuntze)
Japan: uchiwa-tsunagi

Phyllodoce Link Mimosaceae

Origins:
Phyllodoce, daughter of Nereus and Doris, a sea-nymph attending Cyrene.

Phyllodoce Salisb. Ericaceae

Origins:
Phyllodoce, daughter of Nereus and Doris, a sea-nymph attending Cyrene; Greek *phyllon* "a leaf" and *dokeo* "seem," referring to the fruits.

Species/Vernacular Names:
P. breweri (A. Gray) Maxim.
English: mountain heather
P. empetriformis (Smith) D. Don
English: mountain heather

Phyllogeiton (Weberb.) Herzog Rhamnaceae

Origins:
From the Greek *phyllon* "leaf" and *geiton* "a neighbor."

Phylloglossum Kunze Lycopodiaceae

Origins:
Greek *phyllon* "leaf" and *glossa* "a tongue"; see G. Kunze, in *Botanische Zeitung*. 1: 721. 1843.

Species/Vernacular Names:
P. drummondii Kunze
English: pigmy clubmoss

Phyllogonum Coville Polygonaceae

Origins:
From the Greek *phyllon* "leaf" and *gonia* "an angle."

Phyllomelia Griseb. Rubiaceae

Origins:

Greek *phyllon* "leaf" and *meli* "honey" or *melia*, the ancient Greek name for the manna ash or flowering ash tree.

Phyllomphax Schltr. Orchidaceae

Origins:

Greek *phyllon* "leaf" and *omphax, omphakos* "unripe fruit, sour, bitter," indicating the appearance of the foliaceous bracts.

Phyllonoma Willd. ex Schultes Grossulariaceae (Phyllonomaceae, Escalloniaceae)

Origins:

Probably from the Greek *phyllon* "leaf, foliage" and *nomos* "usage, pasture, custom, meadow," inflorescences epiphyllous.

Phyllopappus Walpers Asteraceae

Origins:

From the Greek *phyllon* "leaf" and *pappos* "fluff, pappus"; see Wilhelm G. Walpers (1816-1853), in *Linnaea*. 14: 507. 1841.

Phyllophyton Kudô Labiatae

Origins:

From the Greek *phyllon* "leaf" and *phyton* "plant."

Phyllopodium Benth. Scrophulariaceae

Origins:

From the Greek *phyllon* "leaf" and *podion* "little foot," referring to the stem.

Phyllorachis Trimen Gramineae

Origins:

From the Greek *phyllon* "leaf" and *rhachis* "rachis, axis, midrib of a leaf."

Phyllorchis Thouars Orchidaceae

Origins:

Orthographic variant of *Phyllorkis*.

Phyllorkis Thouars Orchidaceae

Origins:

Greek *phyllon* "leaf" plus *orchis* "orchid," referring to the foliaceous bracts.

Phylloscirpus C.B. Clarke Cyperaceae

Origins:

From the Greek *phyllon* "leaf" plus *Scirpus* L.

Phyllosma L. Bolus ex Schltr. Rutaceae

Origins:

Greek *phyllon* "leaf" and *osme* "scent."

Phyllospadix Hook. Zosteraceae

Origins:

Greek *phyllon* "leaf" and *spadix* "a palm frond, palm branch, spadix," Latin *spadix, spadicis* "a palm branch broken off, together with its fruit," referring to the inflorescence.

Phyllostachys Siebold & Zucc. Gramineae

Origins:

Greek *phyllon* and *stachys* "a spike," referring to the pseudospikelet. See Philipp Franz (Balthasar) von Siebold (1796-1866) and Joseph Gerhard Zuccarini (1797-1848), in *Abhandlungen der Mathematisch-Physikalischen der Königlich Bayerischen Akademie der Wissenschaften*. 3: 745. 1843; F.A. McClure, *Bamboos of the Genus Phyllostachys*. U.S. Department of Agriculture Handbook. no. 114. U.S.D.A. Washington D.C. 1957.

Species/Vernacular Names:

P. spp.

China: zhu ye

P. aurea M.A. & C. Rivière (*Phyllostachys bambusoides* Sieb. & Zucc.)

English: fishpole bamboo

Japan: hotei-chiku, gosan-chiku

P. bambusoides Siebold & Zucc. (*Phyllostachys reticulata* K. Koch)

English: giant timber bamboo, timber bamboo

Japan: ma-dake (= true bamboo)

Okinawa: kara-taki

China: ban zhu gen

P. edulis (Carrière) J. Houz. (*Phyllostachys heterocycla* (Carr.) Matsum.; *Bambusa heterocycla* Carr.; *Phyllostachys pubescens* Mazel ex J. Houz.)

English: moso bamboo

Japan: moso-chiku, mousou-chiku

Okinawa: moso

China: mao sun

P. makinoi Hayata

Japan: kei-chiku

P. nigra (G. Lodd.) Munro

English: black bamboo, partridge cane

Japan: kuro-chiku (= black bamboo)

Okinawa: kuru-chiku

China: zi zhu gen

Phyllostegia Benth. Labiatae

Origins:
From the Greek *phyllon* "leaf" and *stege, stegos* "roof, cover."

Species/Vernacular Names:
P. grandiflora (Gaud.) Benth. (*Prasium grandiflorum* Gaud.)

Hawaii: kapana

P. racemosa Benth.

Hawaii: kiponapona

Phyllostelidium Beauverd Asteraceae

Origins:
Greek *phyllon* "leaf" and *stele* "a pillar, column, trunk, central part of stem," photosynthetic stems.

Phyllostemonodaphne Kosterm. Lauraceae

Origins:
From the Greek *phyllon* "leaf," *stemon* "stamen" and *daphne* "the bay laurel," genus *Daphne*.

Phyllostephanus Tieghem Loranthaceae

Origins:
Greek *phyllon* "leaf" and *stephein* "to crown," *stephanos* "a crown."

Phyllostylon Capanema ex Benth. Ulmaceae

Origins:
From the Greek *phyllon* "leaf" and *stylos* "style, column, pillar."

Species/Vernacular Names:
P. brasiliensis Capanema ex Bentham

English: San Domingo boxwood

Brazil: baitoa

Phyllota (DC.) Benth. Fabaceae

Origins:
Greek *phyllon* "leaf" and *ous, otos* "ear," referring to the leafy bracteoles or to the shape of the leaf; see George Bentham et al., *Enumeratio Plantarum quas in Novae Hollandiae ... collegit C. de Hügel.* 33. Vindobonae 1837.

Species/Vernacular Names:
P. barbata Benth.

English: bearded phyllota

P. humifusa Benth.

English: dwarf phyllota

P. luehmannii F. Muell.

English: mop pea

P. phylicoides (DC.) Benth.

English: heath phyllota

P. pleurandroides F. Muell.

English: healthy phyllota, common phyllota

P. remota J.H. Willis

English: slender phyllota

Phyllotrichum Thorel ex Lecomte Sapindaceae

Origins:
From the Greek *phyllon* "leaf" and *thrix, trichos* "hair."

Phylloxylon Baillon Fabaceae

Origins:
Greek *phyllon* "leaf" and *xylon* "wood."

Phylohydrax Puff Rubiaceae

Origins:

A kind of anagram of the generic name *Hydrophylax*.

Phymaspermum Less. Asteraceae

Origins:

From the Greek *phyma* "a tubercle, swelling, growth" and *sperma* "seed," referring to the papillated achenes.

Phymatarum M. Hotta Araceae

Origins:

From the Greek *phyma, phymatos* "a tubercle, swelling, growth, tumor" plus *Arum*; see D.H. Nicolson, "Derivation of aroid generic names." *Aroideana.* 10: 15-25. 1988.

Phymatidium Lindley Orchidaceae

Origins:

Greek *phyma, phymatos* "swelling, growth," *phymatodes* "full of tumors," the habit of the plants.

Phymatocarpus F. Muell. Myrtaceae

Origins:

Greek *phyma, phymatos* "a tubercle, swelling" and *karpos* "fruit," referring to the warty fruits; see Ferdinand von Mueller, *Fragmenta Phytographiae Australiae.* 3: 120. Melbourne 1862.

Phymatodes Presl Dipteridaceae

Origins:

Greek *phyma, phymatos* "a tubercle, swelling," *phymatodes* "full of tumors"; see Karl (or Carl) B. Presl (1794-1852), *Tentamen Pteridographiae, seu genera Filicacearum.* 195, t. 8. Prague 1836; Arthur D. Chapman, ed., *Australian Plant Name Index.* 2261-2262. Canberra 1991.

Species/Vernacular Names:

P. longissima (Blume) J. Smith (*Polypodium longissimum* Blume; *Microsorium rubidum* (Kunze) Copel.)

Japan: mizu-kazari-shida

Phymatopsis J. Smith Polypodiaceae

Origins:

From the Greek *phyma, phymatos* "a tubercle, swelling" and *opsis* "like, resembling"; see J. Smith, *Hist. Filicum.* 104. 1875 (non Tulasne ex Trevisan 1857).

Phymatopteris Pichi Sermolli Polypodiaceae

Origins:

From the Greek *phyma, phymatos* "a tubercle, swelling" and *pteris* "fern"; see R.E.G. Pichi Sermolli, "Fragmenta Pteridologiae — IV." in *Webbia.* 28(2): 445-477. Dec. 1973.

Phymatosorus Pichi Sermolli Polypodiaceae (Microsoreae)

Origins:

Greek *phyma* "a tubercle, swelling" and *soros* "a vessel for holding anything, a cinerary urn, a coffin, a spore case" but also "a heap" from Akkadian *sarru, zarru* "heap of grain," *zaru* "to winnow," *za'ru, zeru* "seed of cereals." See R.E.G. Pichi Sermolli, "Fragmenta Pteridologiae — IV." in *Webbia.* 28(2): 445-477. Dec. 1973; Arthur D. Chapman, ed., *Australian Plant Name Index.* 2262. ["Dipteridaceae"] Canberra 1991; D.J. Mabberley, *The Plant-Book.* Second edition. 553. ["Polypodiaceae"] Cambridge University Press 1997.

Species/Vernacular Names:

P. diversifolius (Willd.) Pic.Serm. (*Microsorum diversifolium* (Willd.) Copel.)

English: kangaroo fern, hound's tongue fern

P. membranifolius (R. Br.) Pic.Serm.

English: pimple fern

P. pustulatus (Forst.f.) Large, Braggins & P. Green

English: kangaroo fern

P. scandens (Forst.f.) Pic.Serm. (*Microsorum scandens* (Forst.f.) Tind.)

English: fragrant fern

Phymosia Desv. Malvaceae

Origins:

Perhaps from the Greek *phyma* "a tubercle, swelling, a tumor."

Phyodina Raf. Commelinaceae

Origins:
See C.S. Rafinesque, *Flora Telluriana*. 2: 16. 1836 [1837].

Physacanthus Benth. Acanthaceae

Origins:
Perhaps from the Greek *physa* "a bladder, bubble" and *akantha* "thorn."

Physalastrum Monteiro Malvaceae

Origins:
Resembling *physa* "a bladder," *physallis*, *physallidos* "a bladder, bubble, pipe."

Physaliastrum Makino Solanaceae

Origins:
Referring to the genus *Physalis* L.

Species/Vernacular Names:
P. heterophyllum (Hemsley) Migo
English: diversifolious physaliastrum
China: jiang nan san xue dan
P. japonicum (Franch. & Sav.) Honda
English: Japanese physaliastrum
P. kweichouense Kuang & A.M. Lu
English: Kweichow physaliastrum
China: san xue dan
P. sinense (Hemsley) D'Arcy & Z.Y. Zhang
English: Chinese physaliastrum
China: di hai jiao
P. sinicum Kuang & A.M. Lu
English: Chinese physaliastrum
China: hua bei san xue dan
P. yunnanense Kuang & A.M. Lu
English: Yunnan physaliastrum
China: yun nan san xue dan

Physalidium Fenzl Brassicaceae

Origins:
Greek *physa* "a bladder," *physallis*, *physallidos* "a bladder, bubble, pipe."

Physalis L. Solanaceae

Origins:
Greek *physa* "a bladder," *physallis*, *physallidos* "a bladder, bubble, pipe," the calyx is inflated; see Carl Linnaeus, *Species Plantarum*. 182. 1753 and *Genera Plantarum*. Ed. 5. 85. 1754; National Research Council, *Lost Crops of the Incas: Little-Known Plants of the Andes with Promise for Worldwide Cultivation*. National Academy Press, Washington, D.C. 1989.

Species/Vernacular Names:
P. spp.
English: mullaca berry, tomato
Peru: shimon
Mexico: tomate, tumat, sha-mpululu, p'ak-kan, bichooxhe, demshi, dza-ma-lu-can-une, ndemoxu
P. alkekengi L.
English: Chinese lantern, Chinese lantern plant, winter cherry, bladder cherry, Japanese lantern, alkekengi, strawberry tomato, strawberry ground cherry, Jewish cherry
Arabic: kakenedj, kakang
German: Judenkirsche
Japan: hô-zuki, tôfunabii, sôtô, fijichi
China: suan jiang, suan chiang, teng leng tsao (= lantern plant)
P. alkekengi L. var. *franchetii* (Masters) Makino (*Physalis franchetii* Masters)
English: Franchet ground cherry
China: gua jin deng
P. angulata L.
English: gooseberry, wild gooseberry, wild physalis, cut-leaf ground cherry, lance-leaf ground cherry
Peru: bolsa mullaca, camapú, capulí cimarrón, mullaca
Mexico: p'ak-kanil, p'akmuul, tlemoli, tomate, tómatl, tomate de cáscara
Brazil: cumapu, bucho-de-rã, mata-fome, juá-de-capote, campu, juá-poca
South Africa: kalkoengif, klapbessie, wildeappelliefie
Yoruba: amunibimo, koropoo rakuragba, koropon, papo
Japan: sen-nari-hôzuki, kâtôgwa
China: ku zhi, ku chih
P. heterophylla Nees
English: clammy ground cherry
P. lobata Torrey (*Quincula lobata* (Torrey) Raf.)
English: purple ground cherry, lobed ground cherry
P. longifolia Nutt.
English: long-leaf ground cherry

P. minima L.

English: gooseberry, wild gooseberry, little ground cherry, lesser ground cherry

South Africa: kalkoengif, wilde-appelliefie

China: xiao suan jiang, tian zi

Malaya: chepulan

P. obscura Michaux

Brazil: camapu, camaru, joa-poca, jua-poca, joá, juá, juá-de-capote

P. peruviana L.

English: Cape gooseberry, Barbados gooseberry, gooseberry tomato, wild gooseberry, purple ground cherry, ground cherry, winter cherry, cherry tomato, love apple poha, strawberry tomato, husk tomato, Peruvian cherry, Peruvian ground cherry

Peru: aguallu mantu, aguay manto, aguayllumantu, aguaymanto, ahuaimanto(u), pasa capulí, tomate silvestre, capulí

Mexico: miltomate, tomate, tomate de cáscara

Brazil: camapú, bate testa

Madagascar: paokapaoka, voanantsindrana, voanaka, voanakandrivotra, voantsindra, voantsipaoka, groseille du Cap

Mauritius: poquepoque

Southern Africa: appelliefie, wilde-appelliefie, appelderliefde, geel appelliefie, gewone appelliefie, Kaapse nooientjie, Kapseklapbes, makappelliefie, pampelmoertjie, pompelmoertjie, pompelmoesie; kusebere (Sotho); murugudani (Venda); quzumbele (Zulu)

Japan: ke-hôzuki, budô-hôzuki

China: deng long guo

Hawaii: poha, pa'ina

P. philadelphica Lamarck (*Physalis cavaleriei* H. Léveillé)

English: purple ground cherry, jamberry, Mexican husk tomato, ground cherry, purple gooseberry

East Africa: ensobosobo, enyegarori

Central America: miltomate, huevito, tomatillo

Mexico: tomatillo, miltomate, tomate, tomate de cáscara, tomate verde, taxiu-hixi, tulumisi

China: mao suan jiang

P. pruinosa L.

English: dwarf Cape gooseberry, strawberry tomato

Japan: shoku-yô-hôzuki, budô-hôzuki

P. pubescens L.

English: ground cherry, downy ground cherry, strawberry tomato, husk tomato, wild tomato

Central America: miltomate, huevito, tomatillo

Peru: camapú, capulí, juapoca, mullaca, muyaca, toloji, bolsa mullaca, tojolí

Mexico: tomate de cáscara, guajtomate, miltomate, tomate silvestre, tomate verde, tomatillo, tomatillo de campo, tomate de coyote, p'ak-kanil, p'aknul

Brazil: camapu, camaru, joá-poca, juá-poca

China: ku zhi

P. virginiana Miller

English: perennial ground cherry, Virginia ground cherry

P. viscosa L.

English: sticky gooseberry, sticky Cape gooseberry, sticky ground cherry, wild gooseberry, sticky physalis, Cape ground cherry

Mexico: paknul, pahabkan

Southern Africa: klewerige appelliefie, wilde appelliefie

Physalodes Boehm. Solanaceae

Origins:
The genus *Physalis* L. and *odes* "resembling, of the nature of."

Physaloides Moench Solanaceae

Origins:
Resembling the genus *Physalis* L.

Physandra Botsch. Chenopodiaceae

Origins:
From the Greek *physa* "a bladder" and *aner, andros* "man, stamen."

Physanthillis Boiss. Fabaceae

Origins:
Greek *physa* "a bladder" plus the genus *Anthyllis* L., *anthelion, anthyllion* the diminutive of *anthos* "flower, blossom," *anthyllis* is an ancient Greek plant name.

Physaria (Nutt.) A. Gray Brassicaceae

Origins:
From the Greek *physarion* the diminutive of *physa* "bubble, bladder," referring to the inflated fruits.

Physedra Hook.f. Cucurbitaceae

Origins:

From the Greek *physa* "bubble, bladder" and *hedra* "seat, chair," presumably referring to the shape of the edible tubers.

Physematium Kaulf. Dryopteridaceae (Aspleniaceae, Woodsiaceae)

Origins:

Greek *physemation*, the diminutive of *physema, physematos* "blowing, produced by blowing," tufted ferns.

Physena Noronha ex Thouars Physenaceae (Capparidaceae)

Origins:

From the Greek *physao* "inflated, swollen," referring to the fruits.

Physetobasis Hassk. Apocynaceae

Origins:

From the Greek *physao* "inflated, swollen," *physetos* "blown" and *basis* "base, pedestal."

Physinga Lindley Orchidaceae

Origins:

Greek *physinx* "blister, stalk of garlic, a kind of garlic," the shape of the pouch at the base of the lip.

Physocalymma Pohl Lythraceae

Origins:

From the Greek *physa* "a bladder" and *kalymma* "a covering, hood."

Species/Vernacular Names:

P. scaberrima Pohl (Latin *scaber, bra, brum* "rough, scurfy, scabrous")

English: tulipwood

Tropical America: chaquillo, lillo, coloradillo

Physocalymna DC. Lythraceae

Origins:

See *Physocalymma*.

Physocalyx Pohl Scrophulariaceae

Origins:

From the Greek *physa* "a bladder" and *kalyx* "calyx."

Physocardamum Hedge Brassicaceae

Origins:

Greek *physa* "a bladder" and *kardamon* "nasturtium, a kind of cress."

Physocarpus (Cambess.) Maxim. Rosaceae

Origins:

From the Greek *physa* "a bladder" and *karpos* "fruit," an allusion to the inflated follicles or follicular fruits.

Physocaulis (DC.) Tausch Umbelliferae

Origins:

From the Greek *physa* "a bladder" and *kaulos* "stalk."

Physoceras Schltr. Orchidaceae

Origins:

From the Greek *physa* "bladder" and *keras* "horn," the shape of the spur of the lip.

Physochlaina G. Don f. Solanaceae

Origins:

Greek *physa* "a bladder" and *chlaena, chlaenion* "a cloak, blanket," from the inflated calyx.

Species/Vernacular Names:

P. infundibularis Kuang

English: funnelform physochlaina

China: lou dou pao nang cao

P. macrocalyx Pascher

English: large-calyx physochlaina

China: chang e pao nang cao

P. macrophylla Bonati

English: large-leaf physochlaina

China: da ye pao nang cao

P. physaloides (L.) G. Don f.

English: common physochlaina

China: pao nang cao

Physodeira Hanst. Gesneriaceae

Origins:

Greek *physa* "bladder, bubble" and *deire* "neck, throat, collar."

Physogyne Garay Orchidaceae

Origins:

From the Greek *physa* and *gyne* "a woman, female, female organs."

Physokentia Beccari Palmae

Origins:

Greek *physa* "bladder" plus *Kentia*, probably referring to the globose or subglobose fruit.

Physoleucas Jaub. & Spach Labiatae

Origins:

From the Greek *physa* "a bladder" and the genus *Leucas* R. Br., *leukos* "white."

Physolobium Benth. Fabaceae

Origins:

Greek *physa* "a bladder" and *lobos* "pod"; see Karl A.A. von Hügel (1794-1870), *Botanisches Archiv der Gartenbaugesellschaft des Österreichischen Kaiserstaates.* 1. Vienna (Feb.) 1837.

Physolophium Turcz. Umbelliferae

Origins:

From the Greek *physa* "a bladder" and *lophos* "a crest."

Physoplexis (Endl.) Schur Campanulaceae

Origins:

Greek *physa* "a bladder, pipe, bubble" and *plexis* "plaiting, weaving," referring to the divisions of the corolla, basally swollen.

Physopodium Desv. Combretaceae

Origins:

From the Greek *physa* "a bladder" and *podion* "a small foot," *pous, podos* "foot."

Physopsis Turcz. Labiatae (Verbenaceae)

Origins:

Greek *physa* and *opsis* "like, resemblance, aspect"; see Porphir Kiril N.S. Turczaninow (1796-1863), *Bulletin de la Société Impériale des Naturalistes de Moscou.* 22(2): 15. 1849.

Species/Vernacular Names:

P. lachnostachya C.A. Gardner

English: lamb's tail

P. spicata Turcz.

English: Hill River lamb's tail

Physoptychis Boiss. Brassicaceae

Origins:

From the Greek *physa* "a bladder" and *ptyche* "a fold."

Physopyrum Popov Polygonaceae

Origins:

From the Greek *physa* "a bladder" and *pyros* "grain, wheat."

Physorhynchus Hook. Brassicaceae

Origins:

From the Greek *physa* "a bladder" and *rhynchos* "horn, beak."

Species/Vernacular Names:

P. chamaerapistum (Boiss.) Boiss.

Arabic: khophaje

Physosiphon Lindl. Orchidaceae

Origins:

From the Greek *physa* "a bladder, bellows" and *siphon* "tube," referring to the tube of the flower.

Physospermopsis H. Wolff Umbelliferae

Origins:

Resembling *Physospermum*.

Physospermum Cusson Umbelliferae

Origins:

Bladderseed, from the Greek *physa* "a bladder" and *sperma* "seed."

Physostegia Benth. Labiatae

Origins:

From the Greek *physa* "a bladder" and *stege*, *stegos* "roof, shelter," alluding to the calyx.

Species/Vernacular Names:
P. virginiana (L.) Benth.
English: obedient plant

Physostelma Wight Asclepiadaceae

Origins:

From the Greek *physa* "a bladder" and *stelma*, *stelmatos* "a girdle, belt," *stello*, *stellein* "to bring together, to bind, to set," referring to the scales of the corona.

Physostemon C. Martius Capparidaceae (Capparaceae)

Origins:

From the Greek *physa* "a bladder" and *stemon* "stamen."

Physostigma Balf. Fabaceae

Origins:

From the Greek *physa* "a bladder" and *stigma* "a stigma," an allusion to the large hood covering the stigma.

Species/Vernacular Names:
P. venenosum Balf.
English: Calabar bean, ordeal bean
French: fève de calabar
Congo: nouan, nuan

Physothallis Garay Orchidaceae

Origins:

Greek *physa* "a bladder" and *thallos* "blossom, branch," the genera *Physosiphon* and *Pleurothallis*.

Physotrichia Hiern Umbelliferae

Origins:

From the Greek *physa* "a bladder" and *thrix*, *trichos* "hair."

Physurus L. Orchidaceae

Origins:

Greek *physa* "a bladder" and *oura* "tail," alluding to the shape and size of the spur.

Phytelephas Ruíz & Pavón Palmae

Origins:

Greek *phyton* "a plant" and *elephas*, *elephantos* "elephant, ivory," the nut of this palm resembles the true ivory, endosperm very hard, vegetable ivory.

Species/Vernacular Names:
P. aequatorialis Spruce (*Palandra aequatorialis* (Spruce) O.F. Cook)
English: vegetable ivory
Latin America: tagua
P. macrocarpa Ruíz & Pavón
English: ivory nuts
Brazil: jarina, marfim-vegetal
Latin America: tagua

Phyteuma L. Campanulaceae

Origins:

Latin *phyteuma*, *atis* (Plinius) and Greek *phyteuma* for groundsel, Montpellier rocket, *Reseda phyteuma*, perhaps from *phyo* "to grow, to bring forth," see Helmut Genaust,

Etymologisches Wörterbuch der botanischen Pflanzennamen. 482-483. Basel 1996; F. Boerner & G. Kunkel, *Taschenwörterbuch der botanischen Pflanzennamen.* 4. Aufl. 151. Berlin & Hamburg 1989.

Phyteumoides Smeathman ex DC. Rubiaceae

Origins:
Greek *phyteuma, phyteumatos* "that which is planted, a plant, groundsel," resembling *Phyteuma.*

Phytocrene Wallich Icacinaceae

Origins:
Greek *phyton* "a plant" and *krene, krana* "well, spring, fountain, a source," the lianes are a source of potable water.

Phytolacca L. Phytolaccaceae

Origins:
Greek *phyton* "a plant" and the Latin *lacca, ae* (derived from Hindi *lakh,* and referred to a crimson dye); Pseudo Apuleius Barbarus in his *Herbarium* used the word *lacca* for an unknown plant; see Carl Linnaeus, *Species Plantarum.* 441. 1753 and *Genera Plantarum.* Ed. 5. 200. 1754.

Species/Vernacular Names:
P. acinosa Roxb.

English: Indian poke, Indian pokeweed, sweet belladonna

China: shang lu

Tibetan: dpa-bo ser-bo

P. americana L.

English: poke, pokeweed, scoke, garget, pigeonberry, pokeberry, Virginian poke

Portuguese: vinagreira, tintureira, uva dos passarinhos

P. dioica L.

English: pokeberry tree, umbra tree, beautiful shade, tree poke

Southern Africa: belhambra, belambraboom, belombraboom, belhamelboom, ombu, omboe, umbo, koliedruif, Bobbejaandruifboom; uMzimuka (Zulu)

Portuguese: bela sombra

P. dodecandra L'Hérit. (*Phytolacca abyssinica* O. Hoffm.; *Pircunia abyssinica* Moq.)

English: poke

Yoruba: ogbodosun

East Africa: muogo, rutiri

Congo: ite, tidi, tili

P. heptandra Retz. (*Phytolacca stricta* O. Hoffm.; *Pircunia stricta* Hoffm.)

English: inkberry, wild sweet potato

Southern Africa: boesman druiwe, inkbossie; monatja (Sotho); umNyanja (Xhosa)

P. octandra L. (*Phytolacca americana* L. var. *mexicana* L.)

English: inkweed, forest inkberry, inkberry, phytolacca, red inkweed, pokeweed, southern pokeberry

Southern Africa: bobbejaandruif, bobbejaandruiwe, inkbessie, koeliedruif; umNyanja (Xhosa)

Mexico: biaa, piaa

P. sandwicensis Endl. (*Phytolacca abyssinica* sensu Hook. & Arnott; *Phytolacca bogotensis* sensu H. Mann; *Phytolacca brachystachys* Moq.)

Hawaii: popolo ku mai, popolo

Piaranthus R. Br. Asclepiadaceae

Origins:
From the Greek *piaros* "fat" and *anthos* "flower."

Picardaea Urban Rubiaceae

Origins:
For the French plant collector (in Haiti) Louis Picarda.

Picconia A. DC. Oleaceae

Origins:
For J.B. Picconi.

Picea A. Dietrich Pinaceae

Origins:
Latin *picea, ae* "the pitch-pine," *pix, picis* "pitch," Greek *pissa, pitta* and *peuke,* see Giovanni Semerano, *Le origini della cultura europea.* Dizionario della lingua Latina e di voci moderne. 2(2): 519. Firenze 1994; G. Semerano, *Le origini della cultura europea.* Dizionario della lingua Greca. 2(1): 230, 233. 1994.

Species/Vernacular Names:
P. abies (L.) Karsten

English: Norway spruce

P. engelmannii Parry ex Engelmann

English: Engelmann spruce

French: épinette d'Engelmann

Mexico: pino real

P. glauca (Moench) Voss

English: white spruce

P. orientalis (L.) Petermann

English: oriental spruce

P. rubens Sarg.

English: red spruce

P. spinulosa (Griff.) Henry

English: Himalayan spruce

Bhutan: seh shing

Pichi-Sermollia H. Monteiro-Neto Palmae

Origins:

For the Italian botanist Rodolfo Emilio Giuseppe Pichi Ser-molli, b. 1912, pteridologist, botanical explorer, traveler, 1957-1959 Praefectus of the Botanical Garden of Sassari, from 1959 to 1972 Praefectus of the Botanical Garden of Genova. His works include "La vita e le opere di Odoardo Beccari." in O. Beccari, *Nelle foreste di Borneo*. Appendice, 555-582. Longanesi, Milano 1982, "The publication-dates of Colla's 'Plantae rariores in regionibus chilensibus a clar-issimo M.D. Bertero nuper detectae' and 'Herbarium Pedemontanum.'" *Webbia*. 8(1): 123-140. 1951, "Addi-tional notes on the publication-date of Colla's "Plantae rar-iores in regionibus chilensibus a clarissimo M.D. Bertero nuper detectae." *Webbia*. 8(2): 407-411. 1952, "L'Orto Botanico di Genova." *Agricoltura*. 12(4): 87-90. Roma 1963 and "A provisional catalogue of the family names of living Pteridophytes." *Webbia*. 25(1): 219-297. 1970, com-piled *Authors of Scientific Names in Pteridophyta*. [Collab-orators: Maria Paola Bizzarri, Kung-Hsia Shing and Xian-Chun Zhang] Royal Botanic Gardens, Kew 1996, with Maria Paola Bizzarri wrote "The botanical collections (Pteridophyta and Spermatophyta) of the AMF Mares — G.R.S.T.S. Expedition to Patagonia, Tierra del Fuego and Antarctica." *Webbia*. 32(2): 455-534. 1978. See Maria Paola Bizzarri, "L'attività scientifica del prof. Rodolfo E.G. Pichi Sermolli." *Webbia*. 48: 701-733. 1993; H. Monteiro-Neto, in *Rodriguesia*. 41: 198. 1976; T.W. Bossert, *Biographical Dictionary of Botanists Represented in the Hunt Institute Portrait Collection*. 310. 1972; Ida Kaplan Langman, *A Selected Guide to the Literature on the Flowering Plants of Mexico*. 582. University of Pennsylvania Press, Philadel-phia 1964; S. Lenley et al., *Catalog of the Manuscript and Archival Collections and Index to the Correspondence of John Torrey*. Library of the New York Botanical Garden. 327. 1973.

Pichonia Pierre Sapotaceae

Origins:

Possibly named to honor the French botanist Thomas Pichon, fl. 1811, author of *Catalogue raisonné ou tableau analytique et descriptif des plantes cultivées à l'école de botanique du Muséum imperial maritime du Port de Brest*, etc. Brest 1811; see Jean Baptiste Louis Pierre (1833-1905), *Notes Botaniques*. Sapotacées. 22. Paris [1890-1891].

Pickeringia Nuttall ex Torrey & A. Gray Fabaceae

Origins:

For the American naturalist Charles Pickering, 1805-1878, botanist, zoologist, anthropologist, traveler, physician, M.D. Harvard 1826, explorer, plant geographer, historian, ethnologist, 1838-1842 with Charles Wilkes (1798-1877) and William Dunlop Brackenridge (1810-1893) on the U.S. expedition to Antarctic islands and northwest coast of North America, wrote *Chronological History of Plants*. Boston 1879. See D.C. Haskell, *The United States Exploring Expe-dition 1838-1842 and Its Publications 1844-1874*. New York 1942; D.B. Tyler, *The Wilkes Expedition: The First United States Exploring Expedition (1838-1842)*. Philadel-phia 1968; Sydney A. Spence, *Antarctic Miscellany. Books, Periodicals and Maps Relating to the Discovery and Explo-ration of Antarctica*. London 1980; Howard Atwood Kelly and Walter Lincoln Burrage, *Dictionary of American Med-ical Biography*. New York 1928; J.W. Harshberger, *The Botanists of Philadelphia and Their Work*. 190-193. 1899; Joseph Ewan, *Rocky Mountain Naturalists*. [b. 1806] The University of Denver Press 1950; H.N. Clokie, *Account of the Herbaria of the Department of Botany in the University of Oxford*. 224. Oxford 1964; John H. Barnhart, *Biograph-ical Notes upon Botanists*. 3: 84. 1965; T.W. Bossert, *Bio-graphical Dictionary of Botanists Represented in the Hunt Institute Portrait Collection*. 310. 1972; S. Lenley et al., *Catalog of the Manuscript and Archival Collections and Index to the Correspondence of John Torrey*. Library of the New York Botanical Garden. 466. 1973; E.M. Tucker, *Cat-alogue of the Library of the Arnold Arboretum of Harvard University*. 1917-1933; J. Ewan, ed., *A Short History of Botany in the United States*. 1969; Jeannette Elizabeth Graustein, *Thomas Nuttall, Naturalist. Explorations in America, 1808-1841*. 475. Cambridge, Harvard University Press 1967; Elmer Drew Merrill, *Contr. U.S. Natl. Herb.* 30(1): 242. 1947; John Dunmore, *Who's Who in Pacific Navigation*. 265-267. Honolulu 1991; G.A. Doumani, ed., *Antarctic Bibliography*. Washington, Library of Congress 1965-1979; Ida Kaplan Langman, *A Selected Guide to the Literature on the Flowering Plants of Mexico*. Philadelphia 1964.

Species/Vernacular Names:
P. montana Nutt. ex Torrey & A. Gray
English: Chaparral pea, stingaree-bush

Picnomon Adans. Asteraceae

Origins:
Greek *pikros* "pungent, bitter" and *nomos* "meadow," swamp thistle; see M. Adanson, *Familles des Plantes.* 2: 116. (Jul.-Aug.) 1763.

Species/Vernacular Names:
P. acarna (L.) Cass. (*Cnicus acarna* L.; *Cirsium acarna* (L.) Moench)
English: soldier thistle

Picradeniopsis Rydb. ex Britton Asteraceae

Origins:
Greek *pikros* "pungent, bitter, sharp," *aden* "gland" and *opsis* "like."

Picraena Lindley Simaroubaceae

Origins:
Greek *pikros* "pungent, bitter," *pikraino* "irritate, make sharp, make it bitter," perhaps referring to the bark.

Picralima Pierre Apocynaceae

Origins:
Presumably from the Greek *pikros* "bitter" and *lyma* "filth, water used in washing, purgations," the bark used as a febrifuge, seeds as a quinine substitute.

Species/Vernacular Names:
P. nitida (Stapf) Th. & H. Dur.
Nigeria: erin (Yoruba); osu igwe (Igbo)
Yoruba: agege
Cameroon: eban, mototoko
Congo: limeme, ndudi, opati
Ivory Coast: denouain
Gabon: eban, obero

Picramnia Sw. Picramniaceae (Simaroubaceae)

Origins:
Greek *pikros* "bitter" and *amnion* "the amnion, a bowl, the membrane around the fetus," referring to the bark.

Species/Vernacular Names:
P. antidesma Sw.
English: macary bitter
P. spruceana Engl.
Brazil (Amazonas): koeaxi hi, koa akaxi hi

Picrasma Blume Simaroubaceae

Origins:
Greek *pikrasmos* "bitterness," *pikros* "bitter, pungent, sharp," possibly referring to the bark, source of quassia.

Species/Vernacular Names:
P. ailanthoides (Bunge) Planch. (*Picrasma quassioides* (Hamilt.) Benn.; *Simaruba quassioides* D. Don)
English: quassia wood
Japan: niga-ki, njagi
India: tutai, tithai, tithu, hala, hulashi, arkhar

Picria Lour. Scrophulariaceae

Origins:
From the Greek *pikria* "bitterness."

Picricarya Dennst. Oleaceae

Origins:
From the Greek *pikros* "bitter, pungent, sharp" and *karyon* "nut."

Picridium Desf. Asteraceae

Origins:
Latin *picridiae, arum* for a bitter salad, Greek *pikridion* for endive, a kind of *Cichorium*.

Picris L. Asteraceae

Origins:
Greek and Latin *pikris* and *picris* for a bitter lettuce, a kind of salad, a plant that blooms all year round, sour soil (Plinius); see Carl Linnaeus (1707-1778), *Species Plantarum.* 792. 1753 and *Genera Plantarum.* Ed. 5. 347. 1754.

Species/Vernacular Names:
P. echioides L. (*Helmintia echioides* (L.) Gaertner; *Helminthotheca echioides* (L.) Holub)

English: bristly ox-tongue, ox-tongue

South Africa: ostong, stekelpicris, stekelrige beestong

P. hieracioides L.

English: hawkweed picris, hawkweed

Picrocardia Radlk. Simaroubaceae

Origins:

From the Greek *pikros* "bitter, pungent, sharp" and *kardia* "heart," a febrifuge.

Picrodendron Griseb. Euphorbiaceae

Origins:

From the Greek *pikros* "bitter, pungent, sharp" and *dendron* "tree."

Picroderma Thorel ex Gagnepain Meliaceae

Origins:

From the Greek *pikros* "bitter, pungent, sharp" and *derma, dermatos* "skin," alluding to the nature of the bark.

Picrolemma Hook.f. Simaroubaceae

Origins:

Greek *pikros* "bitter, pungent, sharp" and *lemma* "rind, sheath," domatia and fish-poison.

Picrophyta F. Muell. Goodeniaceae

Origins:

Greek *pikros* "bitter, pungent, sharp" and *phyton* "a plant"; see F. von Mueller, in *Linnaea*. 25: 421. (Apr.) 1853.

Picrorhiza Royle ex Benth. Scrophulariaceae

Origins:

From the Greek *pikros* "bitter, pungent, sharp" and *rhiza* "a root," used as a febrifuge.

Species/Vernacular Names:

P. kurrooa Royle ex Benth.

China: hu huang lian

Picrosia D. Don Asteraceae

Origins:

Greek *pikros* "bitter, pungent, sharp."

Picrothamnus Nutt. Asteraceae

Origins:

From the Greek *pikros* "bitter, pungent, sharp" and *thamnos* "shrub."

Piddingtonia A. DC. Campanulaceae

Origins:

For the British (b. Sussex) botanist Henry Piddington, 1797-1858 (d. Calcutta), meteorologist, in India (Calcutta). His writings include "A chemical examination of cotton soils from North America, India, the Mauritius, and Singapore." from the *Transactions of the Agricultural and Horticultural Society of India*. [1839], *Conversations about Hurricanes*: for the use of Plain Sailors. [An *Appendix* published in 1855.] London and Calcutta 1852, *An English Index to the Plants of India*. Calcutta 1832, *The Horn-Book of Storms for the Indian and China Seas*. Calcutta 1844, *A Letter to the European Soldiers in India* on the substitution of coffee for spirituous liquors. [Calcutta 1839] and *The Sailor's Horn-Book for the Law of Storms*. London 1848; see John H. Barnhart, *Biographical Notes upon Botanists*. 3: 84. 1965; E.M. Tucker, *Catalogue of the Library of the Arnold Arboretum of Harvard University*. 1917-1933.

Pierardia Raf. Orchidaceae

Origins:

For the Indian civil servant Francis Pierard; see C.S. Rafinesque (1783-1840), *Flora Telluriana*. 4: 41. Philadelphia 1836 [1838].

Pierardia Roxb. Euphorbiaceae

Origins:

For the Indian civil servant Francis Pierard, flourished 1830s, plant collector; see Ray Desmond, *Dictionary of British & Irish Botanists and Horticulturists*. 552. London 1994; Nathaniel Wallich (1786-1854), *Plantae Asiaticae rariores*. 2: 37. London 1831.

Pieris D. Don Ericaceae

Origins:

Named for the Greek Muses, the Pierides; the cult was introduced into Pieria, a country of Macedonia, from Thrace; Pieris was a daughter of Pierus, a Muse.

Species/Vernacular Names:

P. floribunda (Sims) Benth. (*Andromeda floribunda* Pursh ex Sims)

English: fetter bush

P. japonica (Thunb.) D. Don ex G. Don (*Andromeda japonica* Thunb.)

English: lily of the valley bush, Japanese pieris

Japan: asebi

Pierranthus Bonati Scrophulariaceae

Origins:

After the French botanist Jean Baptiste Louis Pierre, 1833-1905; see John H. Barnhart, *Biographical Notes upon Botanists*. 3: 85. 1965.

Pierrea Hance Flacourtiaceae

Origins:

After the French botanist Jean Baptiste Louis Pierre, 1833-1905.

Pierrebraunia Esteves Cactaceae

Origins:

In honor of Dr. Pierre Braun, specialist in Brazilian cactaceae, author of "On the taxonomy of Brazilian Cereeae (Cactaceae)." in *Bradleya*. 6: 75-89. 1988; see Eddie Esteves Pereira, in *Cactus and Succulent Journal*. 69(6): 296-302. 1997.

Pierreodendron A. Chev. Sapotaceae

Origins:

After the French botanist Jean Baptiste Louis Pierre, 1833-1905.

Species/Vernacular Names:

P. africanum (Hook.f.) Little

Zaire: mukessu, eboke, venda

Pierreodendron Engl. Simaroubaceae

Origins:

After the French (born on Réunion) botanist Jean Baptiste Louis Pierre, 1833-1905, plant collector, traveler and explorer, studied at Paris, 1861 left for India from Réunion, between 1861-1865 at the Calcutta Botanic Garden, from 1865 to 1877 Director of the Saigon Botanical Garden, botanical explorer, he is best remembered for *Flore forestière de la Cochinchine*. Paris [1880-1907], "Sur les plantes à caoutchouc de l'Indochine." *Rev. Cult. Col.* 11: 225-229. 1903 and *Notes Botaniques*. Sapotacées. Paris [1890-1891]. See M.J. Steenis-Kruseman, in *Blumea*. 25: 43. 1979; John H. Barnhart, *Biographical Notes upon Botanists*. 3: 85. 1965; Charles Jacques Édouard Morren (1833-1886), *Correspondance botanique*. Liège 1874 and 1884; A. Chevalier, *Jean Baptiste Louis Pierre 1833-1905*. Paris 1906; T.W. Bossert, *Biographical Dictionary of Botanists Represented in the Hunt Institute Portrait Collection*. 310. 1972; E.M. Tucker, *Catalogue of the Library of the Arnold Arboretum of Harvard University*. 1917-1933; Ida Kaplan Langman, *A Selected Guide to the Literature on the Flowering Plants of Mexico*. Philadelphia 1964; R. Zander, F. Encke, G. Buchheim and S. Seybold, *Handwörterbuch der Pflanzennamen*. 14. Aufl. Stuttgart 1993; I.C. Hedge and J.M. Lamond, *Index of Collectors in the Edinburgh Herbarium*. Edinburgh 1970.

Species/Vernacular Names:

P. africanum (Hook.f.) Little

Nigeria: orokosoro (Yoruba); ekpekukpeku (Edo)

Pierrina Engl. Scytopetalaceae

Origins:

After the French botanist Jean Baptiste Louis Pierre, 1833-1905.

Pigafetta (Blume) Beccari Palmae

Origins:

The generic name may honor Antonio Pigafetta, traveler, who sailed with the Portuguese navigator and explorer Ferdinand Magellan (*circa* 1480-1521, d. Mactan, Philippines) (Fernão de Magalhães, Fernando de Magallanes, Hernando de Magallanes) and returned with Elcano; the earliest European documents on languages of the Austronesian family are two short vocabularies collected by Antonio Pigafetta, the Italian chronicler of the Magellan expedition of 1519-22; see John Dransfield, "*Pigafetta*." *Principes*. 42(1): 34-40. 1998.

Species/Vernacular Names:

P. filaris (Giseke) Beccari

Brazil: wanya

Pigea DC. Violaceae

Origins:

Latin *pigeo, gui, pigitum* "to repent of a thing," species used medically.

Pilea Lindley Urticaceae

Origins:

Latin *pileus* or *pilleus, i* "a cap, felt cap, hat," Greek *pilos* "a cap," referring to the female flowers or to the calyx covering the achene or to the shape of one perianth segment; see John Lindley (1799-1865), in *Collectanea Botanica*; or, figures and botanical illustrations of ... Exotic Plants. 1. (Apr.) London 1821.

Species/Vernacular Names:

P. cadierei Gagnepain & Guillaumin

English: aluminium plant

P. involucrata (Sims) Wright & Dewar

English: friendship plant

Central America: panamica

P. microphylla (L.) Liebm. (*Pilea callitrichoides* (Knuth) Knuth; *Parietaria microphylla* L.)

English: artillery plant, gunpowder plant, pistol plant, rockweed

Japan: kogome-mizu

The Philippines: isang dakot na bigas, alabong

P. repens (Sw.) Wedd.

English: black-leaf panamica

Pileanthus Labill. Myrtaceae

Origins:

Latin *pileus* or *pilleus, i* "a cap" and Greek *anthos* "flower," referring to the structure enclosing the flower; see Jacques J. Houtton de Labillardière (1755-1834), *Novae Hollandiae plantarum specimen*. 2: 11. Parisiis (Feb.) 1806.

Species/Vernacular Names:

P. filifolius Meissner

English: summer copper cups

P. limacis Labill.

English: coastal copper cups

P. peduncularis Endl.

English: copper cups

Pileocalyx Gasp. Cucurbitaceae

Origins:

Greek *pilos* "hat, cap, felt cap" and *kalyx* "calyx."

Pileostegia Hook.f. & Thoms. Hydrangeaceae

Origins:

From the Greek *pilos, pileos* "hat, cap, felt cap" and *stege, stegos* "roof, shelter," referring to the form of the corolla.

Species/Vernacular Names:

P. viburnoides Hook.f. & Thoms.

Japan: shima-yuki-kazura

Pileus Ramírez Caricaceae

Origins:

Latin *pileus* or *pilleus, i* "a cap," referring to the fruits.

Pilgerochloa Eig Gramineae

Origins:

Dedicated to the German botanist Robert Knuds Friedrich Pilger, 1876-1953, traveler, botanical explorer, plant collector in Brazil (Matto Grosso), Director at Botanical Garden Berlin-Dahlem. His works include "Gramineae novae, a cl. K. Skottsberg in Patagonia australi et in Fuegia collectae." *Repert. Spec. Nov. Regni Veg.* 12: 304-308. 1913 and "Sobra algunas gramíneas de América del Sur." *Revista Argent. Agron.* 11(4): 257-264. 1944. See L.R. Parodi, "Robert Pilger." *Revista Argent. Agron.* 20(2): 107-114. 1953; H. Melchior, "Zum Gedächtnis von Robert Pilger." *Bot. Jahrb. Syst.* 76(3): 385-409. 1954; Ida Kaplan Langman, *A Selected Guide to the Literature on the Flowering Plants of Mexico*. University of Pennsylvania Press, Philadelphia 1964; E.D. Merrill, in *Bernice P. Bishop Mus. Bull.* 144: 151. 1937; R. Zander, F. Encke, G. Buchheim and S. Seybold, *Handwörterbuch der Pflanzennamen*. 14. Aufl. Stuttgart 1993; T.W. Bossert, *Biographical Dictionary of Botanists Represented in the Hunt Institute Portrait Collection*. 310. 1972; S. Lenley et al., *Catalog of the Manuscript and Archival Collections and Index to the Correspondence of John Torrey*. Library of the New York Botanical Garden. 327. 1973; E.M. Tucker, *Catalogue of the Library of the*

Arnold Arboretum of Harvard University. 1917-1933; August Weberbauer, *Die Pflanzenwelt der peruanischen Andes in ihren Grundzügen dargestellt.* 39. Leipzig 1911.

Pilgerodendron Florin Cupressaceae

Origins:

For the German botanist Robert Knuds Friedrich Pilger, 1876-1953.

Species/Vernacular Names:

P. uviferum (D. Don) Florin

Chile: ciprés de las Guaytecas

Pilidiostigma Burret Myrtaceae

Origins:

From the Greek *pilos* "hat, cap, felt, felt cloth," *pilidion* "a small cap" and *stigma* "stigma," referring to the shape of the stigma; see (Maximilian) Karl Ewald Burret (1883-1964), in *Notizblatt des Botanischen Gartens und Museums zu Berlin-Dahlem.* 15: 547. 1941.

Species/Vernacular Names:

P. glabrum Burrett

English: plum myrtle

P. tropicum L.S. Sm.

English: apricot myrtle

Pilinophyton Klotzsch Euphorbiaceae

Origins:

From the Greek *pilinos* "made of felt" and *phyton* "plant, tree."

Piliocalyx Brongn. & Gris Myrtaceae

Origins:

Greek *pilos* "hat, cap, felt cap, hair, felt" and *kalyx* "cup, calyx."

Piliostigma Hochst. Caesalpiniaceae

Origins:

Greek *pilos* "hat, cap, felt cap" (or *pilos* "hair") and *stigma*; *pilleus, pileus* was worn by the Romans at festivals, and was given as a sign of freedom to a slave; see Christian

Ferdinand Hochstetter (1787-1860), *Flora oder allgemeine Botanische Zeitung.* 29: 598. 1846.

Pillansia L. Bolus Iridaceae

Origins:

For the South African (b. Rosebank, Cape Town) botanist Neville Stuart Pillans, 1884-1964 (d. Plumstead, Cape Town), from 1918 at the Bolus Herbarium. His writings include "The genus *Phylica* Linn." *J. S. Afr. Bot.* 8: 1-164. 1942, "The African genera and species of Restionaceae." *Trans. R. Soc. S. Afr.* 16: 207-440. 1928 and "Destruction of indigenous vegetation by burning on the Cape Peninsula." *S. Afr. J. Sci.* 21: 348-350. 1924; see John H. Barnhart, *Biographical Notes upon Botanists.* 3: 86. 1965; T.W. Bossert, *Biographical Dictionary of Botanists Represented in the Hunt Institute Portrait Collection.* 311. 1972; Mary Gunn and Leslie Edward W. Codd, *Botanical Exploration of Southern Africa.* 281-282. A.A. Balkema, Cape Town 1981; Gordon Douglas Rowley, *A History of Succulent Plants.* Strawberry Press, Mill Valley, California 1997; John Hutchinson (1884-1972), *A Botanist in Southern Africa.* 53-54. London 1946; A. White and B.L. Sloane, *The Stapelieae.* Pasadena 1937; Gilbert Westacott Reynolds (1895-1967), *The Aloes of South Africa.* Balkema, Rotterdam 1982.

Pillera Endl. Fabaceae

Origins:

After the Austrian botanist Mathias Piller, 1733-1788, clergyman, professor of natural history at the Theresianum in Vienna and later in Buda, author of *Collectio naturalium quae e triplici regno minerali, animali et vegetabili* undequaque completa post obitum ... Mathiae Piller ... reperta est. Graecij 1792, with Ludwig Mitterpacher von Mittenburg (1734-1814) wrote *Iter per Poseganam Sclavoniae provinciam,* etc. Budae [Budapest] 1783. See John H. Barnhart, *Biographical Notes upon Botanists.* 3: 87. 1965; E.M. Tucker, *Catalogue of the Library of the Arnold Arboretum of Harvard University.* 1917-1933.

Pillularia Raf. Marsileaceae

Origins:

Latin *pilula, ae* "a little ball, globule," see also *Pilularia* L.; see C.S. Rafinesque, *Med. Repos.* II. 5: 358. 1808.

Piloblephis Raf. Labiatae

Origins:

See C.S. Rafinesque, *New Fl. N. Am.* 3: 52. 1836 [1837] and *The Good Book.* 45. 1840.

Pilocanthus B.W. Benson & Backeb. Cactaceae

Origins:

Greek *pilos* "cap, hat, felt, hair" and *akantha* "thorn."

Pilocarpus Vahl Rutaceae

Origins:

From the Greek *pilos* "cap, hat, felt cap" and *karpos* "fruit," the shape of the fruit.

Pilocereus K. Schumann Cactaceae

Origins:

Greek *pilos* "cap, hat, felt cap, hair" and *cereus*, see also *Pilosocereus* Byles & Rowley.

Pilocereus Lemaire Cactaceae

Origins:

Greek *pilos* "cap, hat, felt cap, hair" and *cereus*, see also *Cephalocereus* Pfeiff.

Pilocopiapoa F. Ritter Cactaceae

Origins:

Greek *pilos* "cap, hat, felt cap, hair" plus *Copiapoa* Britton & Rose.

Pilogyne Eckl. ex Schrad. Cucurbitaceae

Origins:

Greek *pilos* and *gyne* "a woman, female."

Pilogyne Gagnepain Myrsinaceae

Origins:

From the Greek *pilos* "felt cap, hair" and *gyne* "a woman, female."

Pilophora Jacq. Palmae

Origins:

Greek *pilos* and *phoros* "bearing, carrying."

Pilophyllum Schlechter Orchidaceae

Origins:

From the Greek *pilos* "cap, hat, felt cap, hair" and *phyllon* "leaf."

Pilopleura Schischkin Umbelliferae

Origins:

Greek *pilos* "cap, hat, felt cap, hair" and *pleura, pleuron* "side, rib, lateral."

Pilopsis Y. Ito Cactaceae

Origins:

Greek *pilos* "cap, hat, felt cap, hair" plus *Echinopsis* Zucc.

Pilosella Hill Asteraceae

Origins:

Greek *pilos* "hair, wool," Latin *pilosus* "hairy, shaggy."

Species/Vernacular Names:

P. aurantiaca (L.) F.W. Schultz & Schultz-Bip.

English: devil's paintbrush, fox and cubs

Piloselloides (Less.) C. Jeffrey ex Cufod. Asteraceae

Origins:

Resembling the genus *Pilosella* Hill.

Pilosocereus Byles & G. Rowley Cactaceae

Origins:

Latin *pilosus* "hairy, shaggy" and *Cereus*.

Pilosperma Planchon & Triana Guttiferae

Origins:

Greek *pilos* "hat, felt, cap, hair" and *sperma* "seed."

Pilostaxis Raf. Polygalaceae

Origins:

From the Greek *pilos* and *stachys*, see Constantine Samuel Rafinesque, *New Fl. N. Am.* 4: 87, 89, 111. 1836 [1838]; Elmer D. Merrill, *Index rafinesquianus*. The plant names published by C.S. Rafinesque, etc. 151. Jamaica Plain, Massachusetts, USA 1949.

Pilostemon Iljin Asteraceae

Origins:

Greek *pilos* "hat, felt, cap, hair" and *stemon* "a stamen."

Pilostigma Costantin Asclepiadaceae

Origins:

Greek *pilos* "hat, wool, hair" and *stigma* "stigma."

Pilostigma Tieghem Loranthaceae

Origins:

From the Greek *pilos* "hat" and *stigma* "stigma," or from *pilos* "wool, hair"; see Philippe Édouard Léon van Tieghem (1839-1914), in *Bulletin de la Société Botanique de France*. 41: 488. 1894.

Pilostyles Guillemin Rafflesiaceae

Origins:

Greek *pilos* "hat, cap, felt cap" and *stylos* "style, column, pillar," referring to the central column and disc; or from *pilos* "wool, hair" in reference to the hairy style; see Jean Baptiste Antoine Guillemin (1796-1842), in *Annales des Sciences Naturelles*. 2: 21. (Jul.) Paris 1834.

Species/Vernacular Names:

P. thurberi A. Gray

English: Thurber's pilostyles

Pilothecium (Kiaerskov) Kausel Myrtaceae

Origins:

From the Greek *pilos* "hat, cap, felt cap" and *theke* "a box, case, cell."

Pilularia L. Marsileaceae

Origins:

Latin *pilula, ae* "a little ball, globule," referring to the spherical sporocarps; see Carl Linnaeus, *Species Plantarum*. 1100. 1753 and *Genera Plantarum*. Ed. 5. 486. 1754.

Species/Vernacular Names:

P. americana A. Braun

English: pillwort

P. globulifera L.

English: pillwort

P. novae-hollandiae A. Braun

English: austral pillwort

Pilumna Lindley Orchidaceae

Origins:

Pilumnus and Picumnus were two brother deities of the Romans, the second was a personification of the pestle, Latin *pilum, i* "a pounder, pestle of a mortar," referring to the shape of the column; see Salvatore Battaglia, *Grande dizionario della lingua italiana*. XIII: 485. Torino 1986; Richard Evans Schultes and Arthur Stanley Pease, *Generic Names of Orchids. Their Origin and Meaning*. 244. Academic Press, New York and London 1963.

Pimecaria Raf. Olacaceae

Origins:

From the Greek *pimeles* "fat" and *karyon* "nut"; see C.S. Rafinesque, *Alsographia americana*. 64. Philadelphia 1838.

Pimelandra A. DC. Myrsinaceae

Origins:

From the Greek *pimeles* "fat" and *aner, andros* "man, stamen."

Pimelea Banks & Sol. Thymelaeaceae

Origins:

Greek *pimeles* "fat, soft fat, lard," referring to its richness in oil or to the fleshy cotyledons; see Joseph Gaertner, *De fructibus et seminibus plantarum*. 1: 186. Stuttgart, Tübingen 1788; B.L. Rye, "A revision of Western Australian Thymelaeaceae." *Nuytsia*. 6(2): 129-278. 1988.

Species/Vernacular Names:

P. alpina Meissner

English: alpine riceflower

P. axiflora Meissner

English: tough riceflower, bootlace bush

P. biflora Wakef.

English: matted riceflower

P. curviflora R. Br.

English: curved riceflower

P. flava R. Br.

English: yellow riceflower

P. glauca R. Br.

English: smooth riceflower

P. haematostachya F. Muell. (*Banksia haematostachya* (F. Muell.) Kuntze)

English: pimelea poppy

P. neo-anglica Threlfall

English: poison pimelea

P. octophylla R. Br.

English: woolly riceflower, downy riceflower

P. physodes Hook. (*Banksia physodes* (Hook.) Kuntze; *Macrostegia erubescens* Turcz.)

Australia: Qualup bell

P. rosea R. Br. (*Heterolaena rosea* (R. Br.) C. Meyer; *Banksia rosea* (R. Br.) Kuntze)

Australia: rose banjine

P. spectabilis Lindley (*Banksia spectabilis* (Lindley) Kuntze; *Heterolaena spectabilis* (Lindley) C. Meyer)

Australia: bunjong

P. sulphurea Meissner

Australia: yellow banjine, yellow-flowered pimelea

P. treyvaudii F. Muell. ex Ewart & B. Rees (after the Australian school-master Hector H. Treyvaud)

English: grey riceflower

P. williamsonii J.M. Black (for Herbert B. Williamson, Victorian botanist, Australia)

English: silky riceflower, Williamson's riceflower

Pimelodendron Hassk. Euphorbiaceae

Origins:
Greek *pimele* "fat" and *dendron* "tree"; see Herbert Kenneth Airy Shaw (1902-1985), in *Kew Bull.* 35: 577-700. 1980.

Pimenta Lindley Myrtaceae

Origins:
Spanish *pimento, pimienta, pimiento*, French *piment*, Latin *pigmentum* "a color, pigment, the juice of plants"; see H. Genaust, *Etymologisches Wörterbuch der botanischen Pflanzennamen.* 485. 1996; Ernest Weekley, *An Etymological Dictionary of Modern English.* 2: 1095. 1967; Manlio

Cortelazzo & Paolo Zolli, *Dizionario etimologico della lingua italiana.* 4: 929. Bologna 1985; C.T. Onions, *The Oxford Dictionary of English Etymology.* Oxford University Press 1966; E. Zaccaria, *L'elemento iberico nella lingua italiana.* Bologna 1927; Salvatore Battaglia, *Grande dizionario della lingua italiana.* XIII: 459-460, 486. UTET, Torino 1986; G.L. Beccaria, *Spagnolo e spagnoli in Italia. Riflessi ispanici sulla lingua italiana del Cinque e del Seicento.* Torino 1968; F. D'Alberti di Villanuova, *Dizionario universale, critico, enciclopedico della lingua italiana.* Lucca 1797-1805; Joseph de Acosta, *Historia naturale e morale delle Indie.* Venetia 1596; F. Boerner & G. Kunkel, *Taschenwörterbuch der botanischen Pflanzennamen.* 4. Aufl. 152. 1989; Georg Christian Wittstein, *Etymologisch-botanisches Handwörterbuch.* 694. Ansbach 1852.

Species/Vernacular Names:
P. dioica (L.) Merr.

English: allspice, Jamaica pepper

Spanish: pimento

P. racemosa (Miller) J. Moore

English: bay rum tree

Pimentus Raf. Myrtaceae

Origins:
See *Pimenta* Lindley; see C.S. Rafinesque, *Sylva Telluriana.* 105. 1838.

Pimpinella L. Umbelliferae

Origins:
From the MedLatin *pipinella*, perhaps from *pepo, peponis* "a pumpkin," or derived from Latin *pampinus, i* "a tendril or young shoot of a vine, a vine-leaf"; see Carl Linnaeus, *Species Plantarum.* 263. 1753 and *Genera Plantarum.* Ed. 5. 128. 1754; Helmut Genaust, *Etymologisches Wörterbuch der botanischen Pflanzennamen.* 485-486. Basel 1996; Manlio Cortelazzo & Paolo Zolli, *Dizionario etimologico della lingua italiana.* 4: 929. Bologna 1985; Salvatore Battaglia, *Grande dizionario della lingua italiana.* XIII: 486. Torino 1986; Giovanni Semerano, *Le origini della cultura europea.* Dizionario della lingua Latina e di voci moderne. 2(2): 502, 517. Firenze 1994; Ernest Weekley, *An Etymological Dictionary of Modern English.* 2: 1095-1096. New York 1967.

Species/Vernacular Names:
P. anisum L.

English: anise, sweet cumin

French: anis vert

Italian: anice

China: huai hsiang, huei hsiang, pa yueh chu

Bolivia: anis, pampa anisa

Malaya: jintan manis, anise

Arabic: habbet hléoua, habba helwa

P. major (L.) Hudson

English: burnet saxifrage

P. saxifraga L.

English: burnet saxifrage

Pinacantha Gilli Umbelliferae

Origins:
From the Greek *pinax, pinakos* "board, trencher" and *anthos* "flower."

Pinacopodium Exell & Mendonça Erythroxylaceae

Origins:
Greek *pinax, pinakos* "board, trencher" and *podion* "little foot."

Pinanga Blume Palmae

Origins:
From the Malayan name *pinang*.

Species/Vernacular Names:
P. barnesii Becc.
The Philippines: abisi
P. basilanensis Becc.
The Philippines: buburis
P. copelandii Becc.
The Philippines: abiki
P. coronata (Blume ex Martius) Blume
Brazil: pinanga de coroa, pinanga
P. curranii Becc.
The Philippines: Curran's abiki
P. elmeri Becc.
The Philippines: Elmer abiki
P. heterophylla Becc.
The Philippines: gasigan
P. insignis Becc.
The Philippines: sarawag

P. isabelensis Becc.
The Philippines: takon
P. modesta Becc.
The Philippines: lampigi
P. patula Blume
Brazil: pinanga rabo de peixe
P. sclerophylla Becc.
The Philippines: abiking-tigas
P. sibuyanensis Becc.
The Philippines: tibañgan
P. speciosa Becc.
The Philippines: banisan
P. urdanetensis Becc.
The Philippines: sakolon
P. woodiana Becc.
The Philippines: irar

Pinaropappus Less. Asteraceae

Origins:
From the Greek *pinaros* "dirty" and *pappos* "fluff, pappus."

Pinarophyllon Brandegee Rubiaceae

Origins:
From the Greek *pinaros* "dirty" and *phyllon* "leaf."

Pinckneya Michx. Rubiaceae

Origins:
Presumably dedicated to the US statesman Charles Pinckney (1746-1825) and to the US diplomat and soldier Thomas Pinckney (1750-1828, Governor of South Carolina and Ambassador to Great Britain), born in Charleston.

Species/Vernacular Names:
P. bracteata (Bartram) Raf.
English: fever tree, Georgia bark tree

Pineda Ruíz & Pavón Flacourtiaceae

Origins:
For the Spanish naturalist Antonio de Pineda y Ramírez, 1753-1792, soldier, see *Elogio historico del Señor Don A. de Pineda y Ramirez.* [Madrid? 1795?].

Pinelia Lindley Orchidaceae

Origins:

For M. Pinel, a French botanist.

Pinelianthe Rauschert Orchidaceae

Origins:

The genus *Pinelia* and *anthos* "flower."

Pinellia Tenore Araceae

Origins:

Named for the Italian Giovanni Vincenzo Pinelli, 1535-1601, owner of a botanic garden in Naples; see D.H. Nicolson, "Derivation of aroid generic names." *Aroideana.* 10: 15-25. 1988.

Species/Vernacular Names:

P. ternata (Thunb.) Breitenb. (*Arum ternatum* Thunb.; *Pinellia tuberifera* Ten.)

Japan: karasu-bishaku (karasu = crow)

China: ban xia, pan hsia

P. tripartita (Blume) Schott (*Atherurus tripartitus* Blume)

Japan: ô-hange (= large *Pinellia*)

Pinguicula L. Lentibulariaceae

Origins:

Latin *pinguiculus, a, um* "fattish, somewhat fat," *pinguis, e* "fat," referring to the appearance of the viscid leaves.

Species/Vernacular Names:

P. vulgaris L.

English: butterwort

Pinknea Pers. Rubiaceae

Origins:

See *Pinckneya* Michx.

Pinkneya Raf. Rubiaceae

Origins:

For *Pinckneya* Michx.; see C.S. Rafinesque, *Ann. Gén. Sci. Phys.* 6: 81. 1820.

Pinosia Urban Caryophyllaceae

Origins:

Possibly from the Greek *pinos* "dirty, filth" or a vernacular name.

Pintoa Gay Zygophyllaceae

Origins:

For the General Antonio Pinto, President of Chile; see Claude Gay (1800-1873), *Historia física y política de Chile. Botánica* [*Flora chilena*]. Paris 1845-1852[-1854].

Pinus L. Pinaceae

Origins:

Ancient Latin name *pinus, i*, probably from *pix, picis* "pitch" (Akkadian *pehum* "to caulk," *pihu, pehum* "caulker"); Anglo-Saxon *pin, pinhnutu*, Sanskrit *pitu-daruh* "a kind of pine"; see Carl Linnaeus, *Species Plantarum.* 1000. 1753 and *Genera Plantarum.* Ed. 5. 434. 1754; Manlio Cortelazzo & Paolo Zolli, *Dizionario etimologico della lingua italiana.* 4: 930. 1985; Salvatore Battaglia, *Grande dizionario della lingua italiana.* XIII: 499. 1986; Giovanni Semerano, *Le origini della cultura europea.* Dizionario della lingua Latina e di voci moderne. 2(2): 518. 1994; Ernest Weekley, *An Etymological Dictionary of Modern English.* 2: 1097. 1967.

Species/Vernacular Names:

P. spp. (see Aljos Farjon, *A Field Guide to the Pines of Mexico and Central America.* Royal Botanic Gardens, Kew 1997)

Mexico: pino, tahté, ocotl, ocotes, guere, guiri, queti, yaga guere, quiri bichi, quiri bixhi

P. albicaulis Engelm. (*Apinus albicaulis* (Engelmann) Rydberg)

English: whitebark pine

P. aristata Engelm. (*Pinus balfouriana* var. *aristata* (Engelm.) Engelm.)

English: Rocky Mountains bristlecone pine, Colorado bristlecone pine

P. arizonica Engelm.

English: Arizona pine

P. armandii Franch.

English: Chinese white pine

P. attenuata Lemmon (*Pinus tuberculata* Gordon)

English: knobcone pine

P. ayacahuite Ehrenb. ex Schltdl.

English: Mexican white pine

P. balfouriana Greville & Balfour (*Pinus balfouriana* var. *austrina* (R. Mastrogiuseppe & J. Mastrogiuseppe) Silba; *Pinus balfouriana* subsp. *austrina* R. Mastrogiuseppe & J. Mastrogiuseppe)

English: foxtail pine

P. banksiana Lamb. (*Pinus divaricata* (Ait.) Sudworth)

English: jack pine

French: pin gris

P. bhutanica Grierson, Long & Page

English: East Bhutan pine

P. brutia Ten.

English: Turkish pine, Calabrian pine

P. bungeana Zucc. ex Endl.

English: lacebark pine

China: pai sung

P. californiarum D. Bail.

English: California single-leaf pine

P. canariensis C. Sm.

English: Canary Islands pine

P. caribaea Morelet.

English: Caribbean pine

P. cembra L.

English: Swiss pine, Arolla pine

P. cembroides Zucc. (*Pinus cembroides* var. *bicolor* Little; *Pinus cembroides* var. *remota* Little; *Pinus discolor* D.K. Bailey & Hawksworth; *Pinus remota* (Little) D.K. Bailey & Hawksworth)

English: pinyon pine, Mexican nut pine

Mexico: piñon, pino piñonero

P. chiapensis (Martinez) Andresen

English: Chiapas white pine

P. densiflora Sieb. & Zucc.

English: Japanese red pine

Japan: aka-matsu

China: chih sung

P. edulis Engelm. (*Caryopitys edulis* (Engelm.) Small; *Pinus cembroides* Zucc. var. *edulis* (Engelm.) Voss)

English: pinyon

Mexico: piñon

P. elliottii Engelm. (*Pinus heterophylla* (Elliott) Sudworth, 1893, not K. Koch, 1849; *Pinus taeda* L. var. *heterophylla* Elliott)

English: pine tree, slash pine

South Africa: den

P. engelmannii Carrière (*Pinus macrophylla* Engelm.; *Pinus apacheca* Lemmon; *Pinus latifolia* Sargent)

English: Apache pine

P. halepensis Mill.

English: Aleppo pine, halepensis pine, Jerusalem pine

South Africa: denneboom

P. kesiya Gordon (*Pinus insularis* Endl.)

English: Khasia pine

The Philippines: saleng, parua, balibo, booboo, ol-ol saung, Benguet pine (from Benguet Province)

P. koraiensis Sieb. & Zucc.

English: Korean pine

China: hai song zi, hai sung

P. lambertiana Douglas

English: sugar pine

P. luchuensis Mayr

English: sugar pine, Okinawa pine, Luchu Island pine, Ryukyu pine

Japan: Ryûkyû-matsu, Okinawa-matsu

Okinawa: machi, maachi

P. massoniana Lamb.

English: Masson pine

China: song jie, sung

P. muricata D. Don (*Pinus muricata* var. *borealis* Axelrod; *Pinus muricata* var. *cedrosensis* J.T. Howell; *Pinus muricata* var. *stantonii* Axelrod; *Pinus radiata* var. *binata* (Engelm.) Brewer & S. Watson; *Pinus remorata* H. Mason)

English: Monterey pine, insignis, radiate pine

P. palustris Mill.

English: southern yellow pine, pitch pine, longleaf pine

P. parviflora Sieb. & Zucc.

English: Japanese white pine

Japan: goyo-matsu

P. patula Schltdl. & Cham.

English: Mexican yellow pine, Mexican weeping pine, patula pine, spreading-leaved pine

South Africa: patula den, treurden

P. pinaster Aiton

English: cluster pine, maritime pine, sea pine, star pine

South Africa: denneboom, mannetjiesden, pinasterden, sparden, trosden

Chile: pino marítimo

Portuguese: pinheiro bravo

P. pinea Ait.

English: Italian stone pine, Ponderosa pine, stone pine, umbrella pine, western yellow pine

P. radiata D. Don (*Pinus insignis* Douglas ex Loudon)

English: Monterey pine, insignis, radiate pine

Chile: pino insigne

P. roxburghii Sargent (*Pinus longifolia* Roxb.)

English: chir pine, emodi pine, long-leaved Indian pine

Nepal: khote salla, dhup

P. strobiformis Engelm. (*Pinus ayacahuite* Ehrenberg var. *brachyptera* G.R. Shaw; *Pinus ayacahuite* Ehrenberg var. *reflexa* (Engelm.) Voss; *Pinus ayacahuite* Ehrenberg var. *strobiformis* (Engelm.) Lemmon; *Pinus reflexa* (Engelm.) Engelm.; *Pinus flexilis* E. James var. γ *reflexa* (Engelm.) Engelm.)

English: Southwestern white pine, Mexican white pine

Mexico: pino enano

P. strobus L.

English: Weymouth pine, eastern white pine

P. sylvestris L.

English: Scots pine, Scots fir, Norway fir

P. tabulaeformis Carr.

English: Chinese red pine, Chinese pine

China: song jie

P. taeda L.

English: frankincense pine, loblolly pine, old field pine

P. taiwanensis Hayata

English: Taiwan pine

P. teocote Schiede ex Schltdl. & Cham.

English: twisted leaf pine

P. thunbergii Parl.

English: Japanese black pine

Japan: kuro-matsu

China: hei sung

P. torreyana Parry ex Carr.

English: soledad pine

P. virginiana Mill.

English: scrub pine, Virginia pine

P. wallichiana A.B. Jackson

English: blue pine, Himalayan pine, Bhutan pine

Nepal: gobre salla, dhupi, sa-la

Bhutan: tongphu

Pionandra Miers Solanaceae

Origins:

Greek *pion* "fat" and *aner, andros* "man, stamen, male organs," see also *Cyphomandra* Martius ex Sendtner.

Pionocarpus S.F. Blake Asteraceae

Origins:

From the Greek *pion* "fat" and *karpos* "fruit."

Piper L. Piperaceae

Origins:

Greek *peperi* "pepper," Latin *piper, eris*, Sanskrit *pippali, pipuli* (see *Charaka Samhita*, the *materia medica* of ancient India); see Carl Linnaeus, *Species Plantarum.* 28. 1753 and *Genera Plantarum.* Ed. 5. 18. 1754; P. Fanfani, *Vocabolario dell'uso toscano.* Firenze 1863; Giovanni Semerano, *Le origini della cultura europea.* Dizionario della lingua Latina e di voci moderne. 2(2): 518. Firenze 1994.

Species/Vernacular Names:

P. sp.

Costa Rica: tsögLikö

Mexico: mecaxochitl, chilpatli, tzinacanytlacuatl, jinani, tecualal-itzamal, xalcuáhutl

P. aduncum L.

English: pepper

P. betle L.

English: betle pepper, betel, betel vine, betel pepper, betel leaf pepper

French: poivrier bétel

The Philippines: ikmo, itmo, buyo-anis, buyo, buyo-buyo, buyok, buyu, buyog, gaued, gauod, kanisi, mamon, mamin, samat, gok

Vietnam: trau khong, trau luong, lau, mjau

China: ju jiang

Malaya: sireh china, sireh hudang, sireh melayu

P. caninum Blume

English: common pepper vine

P. capense L.f.

Southern Africa: matimati (Shona)

Yoruba: iyere, iyere gidi, ata iyere

P. clusii DC.

English: West African black pepper

P. cubeba L.f. (from Arabic kubaba, see in *Studi veneziani.* XVII-XVIII: 465. [*piper in chubebe.*] 1975-1976)

English: cubeb pepper, Java pepper, cubebs, West African black pepper, tailed pepper, cubeb

China: bi cheng qie, pi cheng ch'ieh

Arabic: kababah, kebbaba, kababa hindiya, kababa tchini

Malaya: kekumus, kemukus, lada berekor

P. guineense Schum. & Thonn.

English: Guinea cubeb, Ashanti pepper, Benin pepper

French: poivrier d'Afrique

Congo: nkefo

P. kadzura (Choisy) Ohwi (*Ipomoea kadsura* Choisy; *Piper futokadsura* Sieb.)

English: Japanese pepper

Japan: fu-tô-kazura

P. longum L.

English: Indian long pepper, long pepper

India: pippali

China: bi ba, pi po

Tibetan: pi-pi-ling

Nepal: pipla

Malaya: bakek, chabai, kadok, sireh kadok

P. magnificum Trel.

English: laquered pepper

P. mestonii F.M. Bailey

English: long pepper

P. methysticum Forst.f.

Hawaii: 'awa, pu'awa, kava

P. mullesus D. Don

Nepal: pipala, chabo

P. nigrum L.

English: common pepper, pepper plant, black pepper, white pepper, Madagascar pepper, pepper, pepper vine, round pepper

French: poivrier noir, poivrier commun

Arabic: filfil, felfel aswad, felfel akhal

India: maricha, marica-valli, mulaku-koti, milaku, gol mirc, mire, molago-codi, mulaku koti

The Philippines: paminta, pamienta, malisa

Tibetan: na-le-sham, pho-ba-ri

China: hei hu jiao, hu jiao, hu chiao

Vietnam: ho tieu, hat tieu, may loi

Malaya: lada hitam, lada

P. novae-hollandiae Miq.

English: New Holland pepper, giant pepper vine

P. ornatum N.E. Br.

English: Celebes pepper

P. sylvaticum Roxb.

English: mountain long pepper

P. umbellatum L.

English: shrubby pepper

Congo: leleme, elembe, ilelembe

Piperanthera C. DC. Piperaceae

Origins:

From *Piper* and *anthera* "anther," see also *Peperomia* Ruíz & Pav.

Piperia Rydb. Orchidaceae

Origins:

For the American (Canadian born in Victoria, British Columbia) botanist Charles Vancouver Piper, 1867-1926, agronomist, studied at the University of Washington, in 1890 founded the Herbarium at State College (Pullman, Washington), 1893-1903 professor of botany at the State College, 1903-1926 Director of the office of forage crops for US Department of Agriculture. His writings include *The Flora of the Palouse Region*. Pullman, Washington 1901 and *Flora of the State of Washington*. Washington 1906. See John H. Barnhart, *Biographical Notes upon Botanists*. 3: 88. 1965; J. Ewan, ed., *A Short History of Botany in the United States*. 11. 1969; T.W. Bossert, *Biographical Dictionary of Botanists Represented in the Hunt Institute Portrait Collection*. 311. 1972; S. Lenley et al., *Catalog of the Manuscript and Archival Collections and Index to the Correspondence of John Torrey*. Library of the New York Botanical Garden. 327-328. 1973; E.M. Tucker, *Catalogue of the Library of the Arnold Arboretum of Harvard University*. 1917-1933; Joseph Ewan, *Rocky Mountain Naturalists*. The University of Denver Press 1950; Ida Kaplan Langman, *A Selected Guide to the Literature on the Flowering Plants of Mexico*. 584. Philadelphia 1964; R. Zander, F. Encke, G. Buchheim and S. Seybold, *Handwörterbuch der Pflanzennamen*. 14. Aufl. Stuttgart 1993; I.C. Hedge and J.M. Lamond, *Index of Collectors in the Edinburgh Herbarium*. Edinburgh 1970.

Species/Vernacular Names:

P. candida R. Morgan & J. Ackerman

English: white flowered piperia

P. michaelii (E. Greene) Rydb.

English: Michael's piperia

P. yadonii R. Morgan & J. Ackerman

English: Yadon's piperia

Piptadenia Benth. Mimosaceae

Origins:

From the Greek *piptein*, *pipto* "to fall" and *aden* "gland," an allusion to the falling glands of the stamens.

Species/Vernacular Names:

P. sp.

Gabon: ekondjo, esang, essang, mossinga, moussinga, ntoum, ossimiala, ossimiale, tchoumbou, tissalala, tolabitoum, toum, tsoumbou

Cameroon: mbouambo mvoul

Ivory Coast: nete

P. gabunensis (Harms) Robyns

Central Africa: edoum, modouma, ndouma, odouma, adoum, boluma, okan, denya

Ivory Coast: bouémon

Liberia: mbeli-deli

Piptadeniastrum Brenan Mimosaceae

Origins:

A genus allied to *Piptadenia*, *astrum* is a Latin substantival suffix indicating inferiority or incomplete resemblance.

Species/Vernacular Names:

P. africanum (Hook.f.) Brenan (*Piptadenia africana* Hook.f.)

English: African greenheart, African light greenheart

Cameroon: atui, atuij, bohambo, bokombolo, ekombile, jondo, touambo, tombo, toul, wunga, kungu, dabéma, edoundou, ndondon

Central Africa: koungou, mokoungou, agboin, dabéma, daboma, dahoma, ekhimi, itare, iteruku, ohe, ohia, onitoto, osaga, owanghan, sanga, ufi, shaghan, ubam

Congo: n'singa, moussinga, dabema

Gabon: ensale, nchioumbou, nehoumbou, itoumbe, toum

Ghana: dani, dahoma, odahoma, odahuma, odani, ofrafraha

Ivory Coast: abe, ehe, g'bon, kuanga-iniama, kuanguariniama, nainvi, nete, akassanoumou, dabéma, galo

Liberia: gaw, mkeli, mbeli

Yoruba: agbonyin, aga-igi

Nigeria: kiriyar kurmi (Hausa); sanchi kuso (Nupe); agboin (Yoruba); ekhimi (Edo); ufi (Igbo); shagan (Ijaw); owangan (Urhobo); onitoto (Itsekiri); ubam (Efik); ebomme (Ekoi); kachi kabiam (Boki)

W. Africa: bolondo, dabema, edundu

Piptadeniopsis Burkart Mimosaceae

Origins:

Resembling the genus *Piptadenia*.

Piptandra Turcz. Myrtaceae

Origins:

Greek *piptein*, *pipto* "to fall" and *aner*, *andros* "male, stamen"; see Porphir K.N.S. Turczaninow (1796-1863), *Bulletin de la Société Impériale des Naturalistes de Moscou*. 35(2): 323. 1862.

Piptanthocereus (A. Berger) Riccobono Cactaceae

Origins:

Greek *pipto*, *pisso* "to fall, fall down," *anthos* "flower" plus *Cereus*.

Piptanthus Sweet Fabaceae

Origins:

Greek *pipto* "to fall, fall down" and *anthos* "flower."

Species/Vernacular Names:

P. nepalensis (Hook.) Sweet

Nepal: suga phul

Piptatherum P. Beauv. Gramineae

Origins:

Greek *pipto* "to fall" and *ather* "stalk, barb, awn," falling awn; see A. Palisot de Beauvois, *Essai d'une nouvelle Agrostographie*. Paris (Dec.) 1812.

Species/Vernacular Names:

P. micranthum (Trin. & Rupr.) Barkworth

English: small-flowered rice grass

P. miliaceum (L.) Cosson (*Agrostis miliacea* L.; *Oryzopsis miliacea* (L.) Benth. & Hook.f. ex Asch. & Schweinf.)

English: rice millet, many-flowered millet, Smilo grass

Piptocalyx Oliver ex Bentham Trimeniaceae

Origins:

Greek *pipto*, *piptein* "to fall" and *kalyx* "cup, calyx," the calyx is deciduous; see George Bentham, *Flora Australiensis*. 5: 292. (Aug.-Oct.) 1870.

Species/Vernacular Names:

P. moorei Benth.

English: bitter vine

Piptocarpha R. Br. Asteraceae

Origins:

From the Greek *pipto* "to fall" and *karphos* "chip of straw, chip of wood."

Piptochaetium J. Presl Gramineae

Origins:

Greek *pipto* "to fall" and *chaite* "bristle, long hair, foliage."

Piptochlamys C.A. Meyer Thymelaeaceae

Origins:

From the Greek *pipto* "to fall" and *chlamys* "cloak."

Piptocoma Cass. Asteraceae

Origins:

Greek *pipto* and *kome* "hair, hair of the head," referring to the achenes.

Piptolepis Sch.Bip. Asteraceae

Origins:

From the Greek *pipto* "to fall" and *lepis* "a scale."

Piptomeris Turcz. Fabaceae

Origins:

Greek *piptein*, *pipto* and *meris* "a part, portion"; see Porphir K.N.S. Turczaninow, *Bulletin de la Société Impériale des Naturalistes de Moscou.* 26(1): 257. 1853.

Piptophyllum C.E. Hubb. Gramineae

Origins:

From the Greek *pipto* "to fall" and *phyllon* "leaf."

Piptoptera Bunge Chenopodiaceae

Origins:

From the Greek *pipto* "to fall, fall down" and *pteron* "a wing."

Piptospatha N.E. Br. Araceae

Origins:

From the Greek *pipto* "to fall" and *spathe* "spathe," referring to the deciduous top of the spathe, after the fertilization.

Piptostachya (C.E. Hubb.) J.B. Phipps Gramineae

Origins:

Greek *pipto* "to fall" and *stachys* "a spike."

Piptostigma Oliver Annonaceae

Origins:

Greek *pipto* "to fall" and *stigma.*

Species/Vernacular Names:
P. fasciculata (De Wild.) Boutique (*Brieya fasciculata* De Wild.)
Zaire: kingamu, moanda (Mayumbe); usasa (Kinande)
Ghana: dankwakyere (Akan-Twi, Akan-Wasa)
Angola: kididila

Piptothrix A. Gray Asteraceae

Origins:

From the Greek *pipto* "to fall" and *thrix, trichos* "hair."

Pipturus Wedd. Urticaceae

Origins:

Greek *pipto* and *oura* "tail," referring to the deciduous stigma or to the long petiolated leaves or to the long inflorescence; see Hugh Algernon Weddell (1819-1877), in *Annales des Sciences Naturelles.* 1: 196. (Jan.-Jun.) 1854.

Species/Vernacular Names:
P. sp.
Hawaii: mamaki, mamake
P. albidus (Hook. & Arnott) A. Gray (*Boehmeria albida* Hook. & Arnott; *Perlarius albidus* (Hook. & Arnott) Kuntze; *Pipturus taitensis* Wedd.)
Hawaii: waimea
P. arborescens (Link) C.B. Rob. (*Urtica arborescens* Link)
Japan: ô-iwa-ga-ne
Hawaii: mamaki, mamake

P. argenteus (Forst.f.) Wedd. (*Urtica argentea* Forst.f.)

English: false stinger, white nettle, native mulberry

Australia: koomeroo-koomeroo

P. ruber A. Heller (*Pipturus albidus* (Hook. & Arnott) A. Gray var. *kauaiensis* Hochr.)

Hawaii: waimea

Piqueria Cav. Asteraceae

Origins:
For the Spanish physician Andrés Piquér (Andreas Piquerius), 1711-1772. His writings include *Discurso sobre la applicacion de la philosophia a los assuntos de religion para la juventud Española*. Madrid 1757, *Philosophia moral para la juventud Española*. Madrid 1775, *Discurso sobre la enfermedad del Rey Fernando VI*, etc. 1851, A. Piquerii *medicina vetus et nova* postremis curis retractata et aucta. Matritii 1776, *Oratio quam de medicinae experimentalis praestantia et utilitate*, dixit ... Dr. A. Piquer. Matriti [1752], A. Piquerii *Praxis medica*. Matriti 1764-1766, *Tratado de Calenturas*. Madrid 1768 and *Las obras de Hippocrates* mas selectas traducidas en Castellano e ilustradas por A. Piquér. 1788.

Species/Vernacular Names:
P. trinervia Cav.

Spanish: tabardillo

Piqueriella R. King & H. Robinson Asteraceae

Origins:
The diminutive of *Piqueria*, for the Spanish physician Andrés Piquér (Andreas Piquerius), 1711-1772.

Piqueriopsis R. King Asteraceae

Origins:
Resembling *Piqueria* Cav.

Piquetia (Pierre) Hallier f. Theaceae

Origins:
For the British botanist John (Jean) Piquet, 1825-1912 (St. Helier, Jersey, Channel Islands), collected plants, chemist, collector of seaweeds; see John H. Barnhart, *Biographical Notes upon Botanists*. 3: 89. 1965; H.N. Clokie, *Account of the Herbaria of the Department of Botany in the University of Oxford*. 221. Oxford 1964.

Piquetia N.E. Br. Aizoaceae

Origins:
For the British botanist John (Jean) Piquet, 1825-1912.

Piranhea Baillon Euphorbiaceae

Origins:
Piranha or *pirãia*, a caribe, a toothed fish according the Tupi language; see Luíz Caldas Tibiriçá, *Dicionário Tupi-Português*. 159. Traço Editora, Liberdade 1984.

Piratinera Aublet Moraceae

Origins:
A vernacular name.

Piriadacus Pichon Bignoniaceae

Origins:
Anagram of *Cuspidaria* DC.

Pirigara Aublet Lecythidaceae

Origins:
Piriguá, the Guarani name of a medicinal plant, see Luíz Caldas Tibiriçá, *Dicionário Guarani-Português*. 145. Traço Editora, Liberdade 1989.

Piringa Juss. Rubiaceae

Origins:
A vernacular name; see Luíz Caldas Tibiriçá, *Dicionário Tupi-Português*. 159. Traço Editora, Liberdade 1984.

Piriqueta Aubl. Turneraceae

Origins:
A vernacular name; see Luíz Caldas Tibiriçá, *Dicionário Tupi-Português*. 159. Traço Editora, Liberdade 1984.

Pirottantha Spegazzini Mimosaceae

Origins:
Named for the Italian botanist Pietro Romualdo Pirotta, 1853-1936, professor of botany, 1880-1883 Praefectus of the Botanical Garden of Modena, 1883-1925 Praefectus of the 4th Orto Botanico of Roma (Giardini di Villa Corsini, Gianicolo), 1886-1902 editor of vols. 1-16 of *Malpighia*,

editor of *Annuario del Regio Istituto Botanico di Roma*, edited Federico Cesi (1585-1630), *Phytosophicarum tabularum* pars prima ... Ad fidem exemplaris castigatioris iterum edita per R. Pirotta. 1904, wrote "Sul genere Keteleria di Carrière (*Abies Fortunei* Murr.)." *Bull. Soc. Tosc. Ort.* 12: 269-274. 1887. See John H. Barnhart, *Biographical Notes upon Botanists.* 3: 89. 1965; R. Zander, F. Encke, G. Buchheim and S. Seybold, *Handwörterbuch der Pflanzennamen.* 14. Aufl. Stuttgart 1993; T.W. Bossert, *Biographical Dictionary of Botanists Represented in the Hunt Institute Portrait Collection.* 311. 1972; Ethelyn Maria Tucker, *Catalogue of the Library of the Arnold Arboretum of Harvard University.* Cambridge, Massachusetts 1917-1933; Ida Kaplan Langman, *A Selected Guide to the Literature on the Flowering Plants of Mexico.* 585. Philadelphia 1964; L. Pirotta, "Il Regio Orto Botanico di Roma." *Annali di Botanica.* no. 23. 1941.

Piscaria Piper Euphorbiaceae

Origins:

Latin *piscarius, a, um* "fish-," *piscaria, ae* "fish-market."

Piscidia L. Fabaceae

Origins:

Latin *piscis, is* "fish" and *caedo* "to kill, destroy, cut down, slaughter," the roots yield a fish poison.

Species/Vernacular Names:

P. piscipula (L.) Sarg.

English: Jamaica dogwood

Central America: barbasco, chijol, habín, llorasangre, pacaché, palo de zope, zopilote

Piscipula Loefl. Fabaceae

Origins:

From the Latin *piscis, is* "fish," referring to a use as a fish poison, see also *Piscidia* L., see R. Zander, F. Encke, G. Buchheim and S. Seybold, *Handwörterbuch der Pflanzennamen.* 14. Aufl. 653. Stuttgart 1993.

Pisonia L. Nyctaginaceae

Origins:

For the Dutch (b. Leiden) physician Willem (Wilhelm) Piso (Guillaume Le Pois), *circa* 1611-1678 (d. Amsterdam), botanist and traveler, pharmacist, pioneer of tropical medicine,

received M.D. at Caen (1633), from 1636 to 1644 physician of the Dutch settlement in Brazil (with Johan Maurits van Nassau), with the German naturalist and traveler [Georgius Marcgravius, Markgraf, Marcgraf, Georg Marggraff or Margraff etc.] Georg Marcgrave (1610-1644) wrote *Historia naturalis Brasiliae*: *De Medicina Brasiliensi* libri IV (Piso); *Historiae Rerum Naturalium Brasiliae* libri VIII (Margraff). Lugdun. Batavorum (F. Hackius), Amstelodami (L. Elzevir) 1648 and (a second edition, much enlarged, with Margraff and the Dutch physician in the East Jacobus Bontius, 1592 or 1599-1631) *De Indiae utriusque Re Naturali et Medica* libri XIV: libri VI (Piso); libri II (Margraff); libri VI (Bontius), to which is appended *Mantissa Aromatica* (Piso). Amstelaedami (L. & D. Elzevir) 1658. See Peter W. van der Pas, in *Dictionary of Scientific Biography* 10: 621-622. 1981; John H. Barnhart, *Biographical Notes upon Botanists.* 2: 89. Boston 1965; Carl F.P. von Martius, *Versuch eines Commentars über die Pflanzen in den Werken von Marcgrav und Piso über Brasilien. Kryptogamen.* München 1853; Garrison and Morton, *Medical Bibliography.* 5303, 1825. New York 1961; Jonas C. Dryander, *Catalogus bibliothecae historico-naturalis Josephi Banks.* London 1800; A. Lasègue, *Musée botanique de Benjamin Delessert.* Paris 1845; S. Lenley et al., *Catalog of the Manuscript and Archival Collections and Index to the Correspondence of John Torrey.* Library of the New York Botanical Garden. 328. 1973; E.M. Tucker, *Catalogue of the Library of the Arnold Arboretum of Harvard University.* 1917-1933; Carl Linnaeus, *Species Plantarum.* 1026. 1753 and *Genera Plantarum.* Ed. 5. 451. 1754; Mariella Azzarello Di Misa, a cura di, *Il Fondo Antico della Biblioteca dell'Orto Botanico di Palermo.* 215. Regione Siciliana, Palermo 1988.

Species/Vernacular Names:

P. aculeata L.

English: devil's claw pisonia, pisonia vine, thorny pisonia

Japan: toge-kazura (= spiny vine)

Southern Africa: umQopho, uSondesa, Nsuwu (Zulu)

P. grandis R. Br.

English: Malayan lettuce tree, bird catching tree, lettuce tree, cabbage tree, giant pisonia, bird killer tree

Rodrigues Island: bois mapou

Malaya: kemudu, kemudu Siam, kemudu selat, mengkudu, mengkudu Siam, mengkudu selat

P. sandwicensis Hillebr. (*Rockia sandwicensis* (Hillebr.) Heimerl)

Hawaii: aulu, kaulu

P. umbellifera (Forst. & Forst.f.) Seem.

English: bird catcher tree, pisonia tree, bird-lime tree

Japan: ô-kusa-boku

Pisoniella (Heimerl) Standley Nyctaginaceae

Origins:

The diminutive of *Pisonia.*

Pisophaca Rydb. Fabaceae

Origins:

From the Greek *pison* "pea" and *phakos* "a lentil."

Pisosperma Sond. Cucurbitaceae

Origins:

From the Greek *pison* "pea" and *sperma* "seed."

Pistacia L. Anacardiaceae

Origins:

Latin *pistacia* for a pistachio-tree, Latin *pistacium* and *pistaceum* and Greek *pistakion, pistake,* for the fruit of the pistachio-tree; see Ernest Weekley, *An Etymological Dictionary of Modern English.* 2: 1101. New York 1967; P. Sella, *Glossario latino italiano.* Stato della Chiesa–Veneto–Abruzzi. Città del Vaticano 1944.

Species/Vernacular Names:

P. atlantica Desf.

English: mount Atlas mastic

Arabic: botoum

P. lentiscus L.

English: mastic, lentisk, mastic tree

Italian: lentisco

Arabic: dharou, derw, dirw, darw, shagar el-mastika

French: lentisque

P. mexicana Kunth

Mexico: yaga guela, yaga guie gueza, yaga quie queza, yaga laci, yala

P. terebinthus L.

English: terebinth, Cyprus turpentine

Italian: terebinto

Arabic: battoum

French: térébinthe

P. texana Swingle

English: American pistachio, lentisco

P. vera L.

English: green almond, pistachio-nut, pistacio-nut

China: wu ming zi, wu ming tzu, hu chen tzu

Pistaciovitex Kuntze Meliaceae

Origins:

Greek *pistakion, pistake,* Latin *pistacium, ii* and *pistaceum, i* for the fruit of the pistachio tree (Anacardiaceae), and Latin *vitex* for the chaste tree, *Vitex agnus-castus* L. (Labiatae, Verbenaceae); see Tomas Erik von Post (1858-1912) and C.E.O. Kuntze (1843-1907), *Lexicon generum Phanerogamarum.* 442. Stuttgart 1903; Arthur D. Chapman, ed., *Australian Plant Name Index.* 2295. ["Meliaceae"] Canberra 1991.

Pistia L. Araceae

Origins:

Greek *pistos* "drinkable, water" (*pino* "drink"), referring to the aquatic habitat or to the floating habit; see Carl Linnaeus, *Species Plantarum.* 963. 1753 and *Genera Plantarum.* Ed. 5. 411. 1754; Georg Christian Wittstein, *Etymologisch-botanisches Handwörterbuch.* 697. Ansbach 1852; D.H. Nicolson, "Derivation of aroid generic names." *Aroideana.* 10: 15-25. 1988.

Species/Vernacular Names:

P. stratiotes L. (*Pistia stratiotes* var. *cuneata* Engl.)

English: Nile cabbage, Nile lettuce, shell flower, water fern, water lettuce

Japan: botan-uki-kusa (= tree-peony floating herb)

Okinawa: uchikusa

Malaya: kambiang, kiambang

The Philippines: kiapo, loloan, darahuo, daraido, darauo, alaluan, dagaylo, kiupu

South Africa: waterslaai

Madagascar: azafo, hazafo, raizafy, ranomanfaka, rasanjaka, savamanipaka, tsikafonkafona

Congo: okula

Yoruba: oju oro

N. Rhodesia: lungwe

Pistorinia DC. Crassulaceae

Origins:

Etymology uncertain, see Georg Christian Wittstein, *Etymologisch-botanisches Handwörterbuch.* 697. Ansbach 1852; F. Boerner, *Taschenwörterbuch der botanischen Pflanzennamen.* 2. Aufl. 157. Berlin & Hamburg 1966.

Pisum L. Fabaceae

Origins:

Latin *pisum, i* and *pisa, ae* "the pea, a species of leguminous plant" (Plinius), Greek *pison*; Akkadian *pesum, pa'asum*, Hebrew *pasa* "to open wide," *pasam* "to split"; see Carl Linnaeus, *Species Plantarum.* 727. 1753 and *Genera Plantarum.* Ed. 5. 324. 1754; P. Sella, *Glossario latino italiano.* Stato della Chiesa–Veneto–Abruzzi. Città del Vaticano 1944.

Species/Vernacular Names:

P. sativum L.

English: pea, garden pea, green pea, field pea, sugar pea

Japan: saya-endô, shru-indô

The Philippines: sitsaro

Malaya: kachang puteh

Tibetan: sran-ma

China: wan dou, wan tou, jung shu, ching hsiao tou

P. sativum L. subsp. *elatius* (Steven ex M. Bieb.) Asch. & Graebner (*Pisum elatius* Steven ex M. Bieb.)

English: wild pea

P. sativum L. subsp. *sativum* (*Pisum sativum* L. subsp. *hortense* (Neilr.) Asch. & Graebner; *Pisum sativum* L. var. *hortense* Neilr.)

English: canning pea, garden pea, snap pea, snow pea, sugar pea

P. sativum L. subsp. *sativum* var. *arvense* (L.) Poiret (*Pisum arvense* L.)

English: field pea, dun pea, grey pea, mutter pea, partridge pea

Japan: aka-endô, akaindo

Mexico: laa arbeja

Pitardia Battandier ex Pitard Labiatae

Origins:

For the French botanist Charles Joseph Pitard (Pitard-Briau), 1873-1927, traveler, plant collector, wrote *Exploration scientifique du Maroc. Botanique.* Paris 1913-1914; see J.H. Barnhart, *Biographical Notes upon Botanists.* 3: 89. 1965; Ida Kaplan Langman, *A Selected Guide to the Literature on the Flowering Plants of Mexico.* 585. Philadelphia 1964; Stafleu and Cowan, *Taxonomic Literature.* 4: 277-278 1983; E.M. Tucker, *Catalogue of the Library of the Arnold Arboretum of Harvard University.* 1917-1933; I.C. Hedge and J.M. Lamond, *Index of Collectors in the Edinburgh Herbarium.* Edinburgh 1970; R. Zander, F. Encke, G. Buchheim and S. Seybold, *Handwörterbuch der Pflanzennamen.* 14. Aufl. Stuttgart 1993.

Pitcairnia L'Hérit. Bromeliaceae

Origins:

For the English (b. Fife) physician William Pitcairn, 1711(1712?)-1791 (d. London), M.D. Rheims and Oxon (1749), Fellow of the Royal Society in 1750, from 1755 to 1785 President of the Royal College of Physicians, Archibald Menzies (1754-1842) and William Brass (d. at sea 1783) collected plants for him, he was a friend of John Fothergill (1712-1780); see William Munk, *The Roll of the Royal College of Physicians of London.* 2: 172-174. London 1878F. Boerner & G. Kunkel, *Taschenwörterbuch der botanischen Pflanzennamen.* 4. Aufl. 152. Berlin & Hamburg 1989; Ray Desmond, *Dictionary of British & Irish Botanists and Horticulturists.* 554. 1994. Or named for the Scottish (b. Edinburgh) physician Archibald Pitcairne (Archibaldus Pitcarnius), 1652-1713 (d. Edinburgh), M.D. Rheims, professor of medicine at Edinburgh and Leyden. His works include *Roberto Graio Scoto Londini medicinam profitenti, Archibaldus Pitcarnius Scotus S.* [A poem.] [1690?], *Dissertatio de Legibus Historiae Naturalis.* Edinburgi 1696, *Dissertationes medicae.* Roterodami 1701, *Oratio, qua ostenditur medicinam ab omni philosophorum secta esse liberam.* Lugduni Batavorum 1692 and *Elementa Medicinae physico-mathematica.* Londini 1717; see Georg Christian Wittstein, *Etymologisch-botanisches Handwörterbuch.* 697. 1852; Theodore M. Brown, in *Dictionary of Scientific Biography* 11: 1-3. 1981.

Pitcheria Nuttall Fabaceae

Origins:

After the American botanist Zina Pitcher, 1797-1872, physician; see J.H. Barnhart, *Biographical Notes upon Botanists.* 3: 89. 1965; T.W. Bossert, *Biographical Dictionary of Botanists Represented in the Hunt Institute Portrait Collection.* 312. 1972; S. Lenley et al., *Catalog of the Manuscript and Archival Collections and Index to the Correspondence of John Torrey.* Library of the New York Botanical Garden. 466. 1973; Howard Atwood Kelly and Walter Lincoln Burrage, *Dictionary of American Medical Biography.* 917-918. New York 1928; R. Zander, F. Encke, G. Buchheim and S. Seybold, *Handwörterbuch der Pflanzennamen.* 14. Aufl. Stuttgart 1993.

Pithecellobium C. Martius Mimosaceae

Origins:

From the Greek *pithekos* "an ape, monkey" and *ellobion* "ear-ring," (*lobos* "a pod"), referring to the coiled fruits; see Heinrich von Martius (1781-1831), in *Flora oder allgemeine Botanische Zeitung.* 20(2): 114. (Beibl. 8) (Oct.) 1837.

Species/Vernacular Names:

P. sp.

Peru: caballo-usa, chonta-quiro, tamaicaspio, tamia caspi, aguano pashaco, amohua nocco, angalo, angelim rajado, ararandeua, bordão de velho, bushilla, cedro pashaco, chico, esponjeira, faveira do mato, ingarana, llambo pashaco, paricá grande, raya caspi, tamai caspi, tento azul

Mexico: cashitucum, caslstucum, chamacuero, palo de humo, tecui, tzurumbeni, tzuzumban ejéa

Malaya: bengku, jering papan, tuba antu

P. angustifolium (Rusby) Rusby

Bolivia: siraricillo

P. corymbosum (Rich.) Benth.

Bolivia: maní

P. dulce (Roxburgh) Benth. (*Mimosa dulcis* Roxb.; *Inga dulcis* (Roxb.) Willdenow; *Zygia dulcis* (Roxb.) Lyons)

English: Madras thorn, Manila tamarind, Manilla tamarind, soap brak tree, ape's ear ring

Central America: guaymochil, huamuchil, opiuma

Peru: guamuchil

Mexico: becii guii, becigui, pecigui, biciiguii, picijgui, nocuana beguiche, pequiche, piquiche, biguiche, nocuanaguiche, yaga be guiche, yaga-piquiche, pe-qui-che, bebguiche, beguiche, cuamochtl, cuamuchil, guamoche, guamúche, guamúchil, humo, umi, umuh, huamuchil, guaumochtli, chucum blanco, cuamuche, guamuti, macachuni, muchite, múchitl, matúrite, nempa, nipe

Japan: kinki-ju

The Philippines: kamachili, kamatsili, camatsilis, kamantsile, kamatsele, kamansile, kamachilis, kamanchilis, kamachile, kamantilis, kamantiris, komonsili, kamonsili, kamunsil, kamunsili, damortis, komontos, komontres, damulkis, chamultis

Hawaii: 'opiuma

P. flexicaule (Benth.) Coult.

English: Texas ebony

Mexico: ebano, guaypinole, ya'ax-k'iik, acte, ajcte

P. keyense Britt. ex Britt. & Rose

English: black bead

Mexico: xiax-k'aax

P. mathewsii Bentham

Spanish: algarobo

P. pedicellare (DC.) Benth.

Bolivia: arraigán

P. scalare Griseb.

Bolivia: juno, kellu-taku, paraisillo

P. sophorocarpum Benth.

Bolivia: cuchilla

P. unguis-cati (L.) Bentham (*Mimosa unguis-cati* L.; *Zygia unguis-cati* (L.) Sudw.)

English: black bead, cat's claw, black Jessie

Mexico: coralillo, guamuchilillo, otsuiché, t'siu-che, tsimché, tzimché, tzinché

Pithecoctenium Martius ex Meissner Bignoniaceae

Origins:

Greek *pithekos* "an ape, monkey" and *kteis, ktenos* "a comb," referring to the prickly capsules.

Species/Vernacular Names:

P. crucigerum (L.) A. Gentry

English: monkey comb

Pithecollobium C. Martius Mimosaceae

Origins:

Orthographic variant of *Pithecellobium* C. Martius; see Arthur D. Chapman, ed., *Australian Plant Name Index.* 2297. Canberra 1991.

Pithecolobium Bentham Mimosaceae

Origins:

Orthographic variant of *Pithecellobium* C. Martius; see Arthur D. Chapman, ed., *Australian Plant Name Index.* 2297. 1991.

Pithecoseris Martius ex DC. Asteraceae

Origins:

Greek *pithekos* "an ape, monkey" and *seris, seridos* "endive, chicory."

Pithecurus Kunth Gramineae

Origins:

Greek *pithekos* "an ape, monkey" and *oura* "tail."

Pithocarpa Lindley Asteraceae

Origins:

Greek *pithos* "a large jar" and *karpos* "fruit," referring to the shape of the fruits; see John Lindley (1799-1865), *Edwards's Botanical Register*. Appendix to Vols. 1-23: *A Sketch of the Vegetation of the Swan River Colony*. xxiii. London (1 Dec.) 1839.

Pithodes O.F. Cook Palmae

Origins:

From the Greek *pithos* "a large jar," *pithodes* "like a jar."

Pitraea Turcz. Verbenaceae

Origins:

For the Russian botanist Adol'f Samoilovich Pitra, 1830-1889, professor of botany at Kharkov, author of [*On the Knowledge of*] *Spaerolobus stellatus*. Kharkov 1870.

Pittiera Cogniaux Cucurbitaceae

Origins:

After the Swiss-born American botanist Henri (Henry) François Pittier (Pitter de Fábrega), 1857-1950, bryologist, plant collector, traveler, botanical explorer, an authority on the flora of tropical America, 1882-1887 Switzerland (Lausanne), Costa Rica 1887, sent plants to Th. Durand, from 1905 with USDA (Colombia, Venezuela and Central America), 1913 Venezuela (first botanical exploration). His writings include *Manual de las Plantas Usuales de Venezuela y su Suplemento*. Caracas 1971, "Flora venezolana: plantas medicinales." *Memor. 4° Congreso Ven. de Med.* 2: 167-172. 1925, *Leguminosas de Venezuela. I. Papilionáceas*. Venezuela 1944, *Clave analítica de las familias de plantas superiores de la América Tropical*. Caracas 1937, "¿Existe la tagua o marfil vegetal en Venezuela?" *Bol. Com. e Industr.* Año I, no. 4, 103-104. 1920, "La caoba venezolana." *Bol. Com. e Industr.* 18: 582-593. 1921 and "Exploraciones, botánicas y otras, en la cuenca de Maracaibo." *Bol. Com. e Industr.* Año IV, no. 39-40. Caracas 1923, with Tobías Lasser, Ludwig Schnee et al. wrote *Catálogo de la flora venezolana*. Caracas 1945-1947; see John H. Barnhart, *Biographical Notes upon Botanists*. 3: 90. 1965; T.W. Bossert, *Biographical Dictionary of Botanists Represented in the Hunt Institute Portrait Collection*. 312. 1972; E.M. Tucker, *Catalogue of the Library of the Arnold Arboretum of Harvard University*. 1917-1933; S. Lenley et al., *Catalog of the Manuscript and Archival Collections and Index to the Correspondence of John Torrey*. Library of the New York Botanical Garden. 328. 1973; G. Murray, *History of the Collections Contained in the Natural History Departments of the British Museum*. 1: 175. London 1904; Ida Kaplan Langman, *A Selected Guide to the Literature on the Flowering Plants of Mexico*. 585-586. Philadelphia 1964; Irving William Knobloch, compil., "A preliminary verified list of plant collectors in Mexico." *Phytologia Memoirs*. VI. 1983.

Pittierella Schltr. Orchidaceae

Origins:

For the Swiss botanist Henri François Pittier, 1857-1950. His writings include *Manual de las Plantas Usuales de Venezuela y su Suplemento*. Caracas 1971.

Pittierothamnus Steyerm. Rubiaceae

Origins:

For the Swiss botanist Henri François Pittier, 1857-1950.

Pittocaulon H. Robinson & Brettell Asteraceae

Origins:

Greek *pitta, pissa* "pitch, resin" and *kaulos* "stalk."

Pittoniotis Griseb. Rubiaceae

Origins:

After the eminent French botanist Joseph Pitton de Tournefort, 1656-1708 (Paris), physician, naturalist, professor of medicine and botany, studied under Pierre Magnol (1638-1715) at the University of Montpellier, friend of Charles Plumier (1646-1704) and Pierre Joseph Garidel (1658-1737), traveler with Claude Aubriet (1665?-1742) and Andreas v. Gundelsheimer (1668-1715) in the Levant (by order of the King he spent some years in Greece and Asia Minor, 1700-1702), plant collector, he is the father of the generic concept (first to define genus and genera). Among his many works are *Elémens de botanique*. Paris 1694, *Institutiones rei herbariae*. Parisiis 1700, *Materia medica*. London 1708 and *Relation d'un voyage du Levant*. Paris 1717. See René Louiche dit Desfontaines (1750-1833), *Choix de Plantes du Corollaire des Instituts de Tournefort*, publiées d'après son herbier, et gravées sur les dessins originaux d'Aubriet. Paris 1808; Stafleu and Cowan, *Taxonomic Literature*. 6: 412-415. 1986; Jean F. Leroy, *Dictionary of Scientific Biography* 13: 442-444. 1981; Blanche Henrey, *British Botanical and Horticultural Literature before 1800*.

1975; Johann Christian Daniel von Schreber (1739-1810), *Icones et descriptiones plantarum minus cognitarum*. Halae [Halle a.S.] 1765; H.N. Clokie, *Account of the Herbaria of the Department of Botany in the University of Oxford*. 255. Oxford 1964; James Britten, *The Sloane Herbarium*, revised and edited by J.E. Dandy. 1958; John H. Barnhart, *Biographical Notes upon Botanists*. 3: 394. 1965; D.O. Wijnands, *The Botany of the Commelins*. Rotterdam 1983; H. Daudin, *De Linné à Jussieu*. Paris 1926; Alexander B. Adams, *Eternal Quest. The Story of the Great Naturalists*. New York 1969; T.W. Bossert, *Biographical Dictionary of Botanists Represented in the Hunt Institute Portrait Collection*. 404. 1972; M. Colmeiro y Penido, *La Botánica y los Botánicos de la Peninsula Hispano-Lusitana*. Madrid 1858; Jonas C. Dryander, *Catalogus bibliothecae historico-naturalis Josephi Banks*. London 1800; A. Lasègue, *Musée botanique de Benjamin Delessert*. 586. Paris 1845; E.M. Tucker, *Catalogue of the Library of the Arnold Arboretum of Harvard University*. 1917-1933; Emil Bretschneider (1833-1901), *History of European Botanical Discoveries in China*. [Reprint of the original edition 1898.] Leipzig 1981; Frans A. Stafleu, *Linnaeus and the Linnaeans*. The spreading of their ideas in systematic botany, 1735-1789. Utrecht 1971; Mariella Azzarello Di Misa, a cura di, *Il Fondo Antico della Biblioteca dell'Orto Botanico di Palermo*. Soprintendenza per i Beni Culturali e Ambientali. Sezione per i Beni Bibliografici. Regione Siciliana, Palermo 1988.

Pittosporopsis Craib Icacinaceae

Origins:

Greek *pitta* "pitch, resin," *sporos* "a seed" and *opsis* "like," seeds with oily endosperm resembling those of Pittosporaceae.

Pittosporum Banks ex Gaertner
Pittosporaceae

Origins:

Greek *pitta* "pitch, resin" and *sporos* "a seed," the seeds are covered with a resinous, viscid and sticky pulp; see Joseph Gaertner, *De fructibus et seminibus plantarum*. 1: 286, t. 59. Stuttgart, Tübingen 1788; H.E. Connor and E. Edgar, "Name changes in the indigenous New Zealand flora, 1960-1986 and Nomina Nova IV, 1983-1986." *New Zealand Journal of Botany*. Vol. 25: 115-170. 1987; Arthur D. Chapman, ed., *Australian Plant Name Index*. 2297-2300. Canberra 1991; Georg Christian Wittstein, *Etymologisch-botanisches Handwörterbuch*. 698. Ansbach 1852.

Species/Vernacular Names:

P. argentifolium Sherff (*Pittosporum forbesii* Sherff)

Hawaii: ho'awa, ha'awa

P. balfourii Cuf.

Rodrigues Island: bois bécasse, bois bégasse

P. bicolor Hook.

Australia: banyalla

P. confertiflorum A. Gray (*Pittosporum cauliflorum* H. Mann; *Pittosporum halophilum* Rock; *Pittosporum halophiloides* Sherff; *Pittosporum lanaiense* St. John)

Hawaii: ho'awa, ha'awa

P. crassifolium Banks & Sol. ex A. Cunn.

New Zealand: caro, karo

English: evergreen pittosporum

P. eugenioides A. Cunn.

English: commonwood, lemonwood

Maori names: tarata

P. ferrugineum Dryander

Australia: rusty pittosporum

The Solomon Islands: aiofa

Malaya: belalai puak, pulai puak, chabek hantu, chemperai ikan, geramong, kelat tupai, kepialu pajan, kepialu pachau, lusai, medang kelelawak, medang pasir, sapong, sereras, seruras, teranok

P. flocculosum (Hillebr.) Sherff

Hawaii: ho'awa, ha'awa

P. gayanum Rock

Hawaii: ho'awa, ha'awa

P. glabrum Hook. & Arnott (*Pittosporum acuminatum* H. Mann; *Pittosporum dolosum* Sherff; *Pittosporum acutisepalum* (Hillebr.) Sherff)

Hawaii: ho'awa, ha'awa, papahekili

P. hawaiiense Hillebr.

Hawaii: ho'awa, ha'awa

P. hosmeri Rock (*Pittosporum amplectens* Sherff)

Hawaii: ho'awa, ha'awa, 'a'awa, 'a'awa hua kukui

P. kauaiense Hillebr.

Hawaii: ho'awa, ha'awa, ho'awa lau nui

P. melanospermum F. Muell.

Australia: black-seed pittosporum

P. napaliense Sherff

Hawaii: ho'awa, ha'awa

P. oreillyanum C.T. White

English: thorny pittosporum

P. pentandrum (Blanco) Merr.

The Philippines: antoan, pangantoan, pangatoan, basuit, lasuit, oplat, uplai, darayan, dili, pasgik, pasik, mamali, mamalis, taiu, balinkauayan, bolonkoyan, saboagon, marabinga

P. phylliraeoides DC.

Australia: narrow-leaved pittosporum, weeping pittosporum, desert willow, butterbush, berrigan, wild apricot, locket bush, native apricot

P. phylliraeoides DC. var. *microcarpa* S. Moore (*Pittosporum angustifolium* Lodd.; *Pittosporum ligustrifolium* Putterl.; *Pittosporum longifolium* Putterl.; *Pittosporum oleifolium* Putterl.; *Pittosporum roanum* Putterl.; *Pittosporum acacioides* Cunn.; *Pittosporum salicinum* Lindley; *Pittosporum lanceolatum* Cunn.)

Australia: native apricot, weeping pittosporum, native willow, poisonberry tree, apricot tree, berigan, butterbush, meemeei

P. revolutum Dryander

Australia: yellow pittosporum, mock orange, rough-fruit pittosporum

P. rhombifolium Hook.

English: Queensland pittosporum, diamond leaf pittosporum, diamond-leaved pittosporum, diamond laurel, hollywood, white holly, small-fruited pittosporum

P. rubiginosum A. Cunn.

English: red pittosporum

P. senacia Putterlick

Madagascar: ambovisika, mainbovisika, bois-cerf odorant, bois de joli-coeur

P. tenuifolium Gaertner

New Zealand: tawhiwhi, kohuhu, kohukohu

P. terminalioides Planch. ex A. Gray (*Pittosporum kilaueae* St. John)

Hawaii: ho'awa, ha'awa

P. tobira Aiton (*Pittosporum lutchuense* Koidz.; *Pittosporum denudatum* Nakai)

English: mock orange, Japanese pittosporum

Italian: pittosporo, pitosforo

Japan: tobera (= door tree), tubiragi

P. undulatum Vent. (*Pittosporum undulatum* Vent. subsp. *emmettii* W.M. Curtis)

Australia: pittosporum, Victorian box, Victorian laurel, orange berry pittosporum, orange pittosporum, cheesewood, mock orange, sweet pittosporum

South Africa: soet pittosporum

Portuguese: incenseiro, árvore do incenso

P. venulosum F. Muell.

English: rusty pittosporum

P. viridiflorum Sims (*Pittosporum antunesii* Engl.; *Pittosporum commutatum* Putterl.; *Pittosporum floribundum* Wight & Arn.; *Pittosporum malosanum* Bak.; *Pittosporum kruegeri* Engl.; *Pittosporum vosseleri* Engl.; *Pittosporum abyssinicum* Delile var. *angolense* Oliv.; *Pittosporum quartinianum* Cufod.; *Pittosporum sinense* Desf.; *Pittosporum viridiflorum* Sims subsp. *malosanum* (Bak.) Cufod.; *Pittosporum viridiflorum* Sims subsp. *quartinianum* (Cufod.) Cufod.; *Pittosporum viridiflorum* Sims subsp. *angolense* (Oliv.) Cufod.; *Pittosporum viridiflorum* Sims subsp. *commutatum* (Putterl.) Möser; *Pittosporum viridiflorum* Sims subsp. *kruegeri* (Engl.) Möser)

English: cheesewood, white Cape beech, Cape pittosporum

India: yekaddi

Southern Africa: kasuur, bosboekenhout, rooiboekenhout; Mutanzwakhamelo (Venda); muKwenkwe (Shona); umFusamvu, umVusamvu, umKhwenkwe (Zulu); umKhwenkwe (Xhosa); umVusamvu (Swazi); mosetlela, motsosanku, moohluwa, mohatollo, phukgu (South Sotho: Lesotho, Orange Free State, southeast Transvaal)

Pituranthos Viv. Umbelliferae

Origins:
From the Greek *pityron* "chaff, dandruff, bran" and *anthos* "flower," referring to the scaly flowers and fruits.

Pitygentias Gilg Gentianaceae

Origins:
Greek *pitys* "pine," Latin *pitys, idos* "a pine-cone" and *Gentiana*, see also *Gentianella* Moench.

Pityopsis Nutt. Asteraceae

Origins:
From the Greek *pitys* "pine" and *opsis* "resemblance."

Pityopus J.K. Small Ericaceae (Monotropaceae, Pyrolaceae)

Origins:
Greek *pitys* "pine" and *pous, podos* "foot," referring to the habitat.

Species/Vernacular Names:
P. californicus (Eastw.) H. Copel.

English: California pinefoot

Pityothamnus Small Annonaceae

Origins:

From the Greek *pitys* "pine" and *thamnos* "shrub."

Pityphyllum Schltr. Orchidaceae

Origins:

From the Greek *pitys* "pine" and *phyllon* "leaf," indicating the leaves.

Pityranthe Thwaites Tiliaceae

Origins:

From the Greek *pityron* "chaff, dandruff, bran" and *anthos* "flower."

Pityrocarpa (Benth.) Britton & Rose Mimosaceae

Origins:

From the Greek *pityron* "chaff, dandruff, husks of corn" and *karpos* "fruit."

Pityrodia R. Br. Labiatae (Dicrastylidaceae)

Origins:

Greek *pityron* "chaff, scurf, bran" and *eidos, oides* "resemblance," *pityrodes* "chaff-like, scurfy," referring to the scaly leaves; see R. Brown, *Prodromus florae Novae Hollandiae et Insulae van-Diemen.* 513. London (Mar.) 1810.

Species/Vernacular Names:

P. atriplicina (F. Muell.) Benth.

English: saltbush foxglove

P. augustensis Munir

English: Mt. Augustus foxglove

P. axillaris (Endl.) Druce

English: native foxglove, woolly foxglove

P. oldfieldii (F. Muell.) Benth.

English: Oldfield's foxglove

P. terminalis (Endl.) A.S. George

English: native foxglove

Pityrogramma Link Pteridaceae (Adiantaceae)

Origins:

Greek *pityron* "chaff, dandruff, bran" and *gramma* "line, writing, letter," the undersurface of the fronds is powdery

and scaly; see Johann Heinrich Friedrich Link, *Handbuch zur Erkennung der nutzbarsten und am häufigsten vorkommenden Gewächse.* 3: 19. 1833.

Species/Vernacular Names:

P. austroamericana Domin

English: gold fern

P. calomelanos (L.) Link (*Acrostichum calomelanos* L.)

English: silver fern, silverback fern

P. calomelanos (L.) Link var. *aureoflava* (Hook.) Weath. ex Bailey

English: gold fern, golden fern

P. chrysophylla (Swartz) Link

English: gold fern

P. sulphurea (Swartz) Maxon

English: Jamaica gold fern

P. triangularis (Kaulf.) Maxon

English: Californian gold fern

P. viscosa (DC. Eaton) Maxon

English: silverback fern

Placea Miers Amaryllidaceae (Liliaceae)

Origins:

Possibly from the native Chilean name.

Placocarpa Hook.f. Rubiaceae

Origins:

From the Greek *plax, plakos* "a plain, anything flat and broad" and *karpos* "fruit."

Placodiscus Radlk. Sapindaceae

Origins:

Greek *plax, plakos* "a plain, anything flat and broad" and *diskos* "a disc."

Placodium Hook.f. Rubiaceae

Origins:

From the Greek *plax, plakos* "a plain, tablet, flat-top, a flat body," *plakodes* "laminated," see also *Plocama* Aiton.

Placolobium Miq. Fabaceae

Origins:

Greek *plax, plakos* and *lobion* "a small pod."

Placoma J.F. Gmelin Rubiaceae

Origins:

Greek *plax, plakos* "a plain, tablet, flat-top" or derived from the generic name *Plocama* Aiton.

Placopoda Balf.f. Rubiaceae

Origins:

From the Greek *plax, plakos* "a plain, tablet" and *pous* "foot."

Placospermum C.T. White & W.D. Francis Proteaceae

Origins:

Greek *plax, plakos* "a plain, anything flat and broad, a flat round plate" and *sperma* "seed," referring to the disc-like seeds; see Cyril Tenison White (1890-1950) and William Douglas Francis (1889-1959), in *Proceedings of the Royal Society of Queensland*. 35: 79. (Feb.) 1924.

Species/Vernacular Names:

P. coriaceum C.T. White & W.D. Francis

English: rose silky oak

Placostigma Blume Orchidaceae

Origins:

From the Greek *plax, plakos* and *stigma* "stigma."

Placseptalia Espinosa Bromeliaceae

Origins:

Greek *plax, plakos* "a plain, a flat round plate" and Latin *saepio, psi, ptum, ire* "to surround with a hedge, surround," *saeptum* or *septum* "a fence."

Placus Lour. Asteraceae

Origins:

Greek *plakus, plakous* "a flat cake, the seed of the mallow"; see João de Loureiro, *Flora cochinchinensis: sistens plantas in regno Cochinchina nascentes.* 2: 475, 496. Ulyssipone [Lisboa] (Sept.) 1790.

Pladaroxylon (Endl.) Hook.f. Asteraceae

Origins:

Greek *pladaros* "wet, damp, moist" and *xylon* "wood," growing in St. Helena.

Plaesiantha Hook.f. Rhizophoraceae

Origins:

Perhaps from the Greek *plesios* "near, close to" or *plaision* "oblong" and *anthos* "flower."

Plaesianthera (C.B. Clarke) Livera Acanthaceae

Origins:

Probably from the Greek *plesios* or *plaision* plus *anthera* "anther."

Plagiacanthus Nees Acanthaceae

Origins:

Greek *plagios* "oblique" and *akantha* "thorn."

Plagiantha Renvoize Gramineae

Origins:

From the Greek *plagios* "oblique" and *anthos* "flower."

Plagianthera Reichb.f. & Zollinger Euphorbiaceae

Origins:

Greek *plagios* "oblique" and *anthera* "anther"; see Heinrich Zollinger (1818-1859), in *Verhandelingen der Natuurkundige Vereen. in Nederlandsche Indië.* 1: 19. 1856 [*Acta Soc. Regiae Sci. Indo-Neerl.* 1(4): 19. post 1 Sep 1856].

Plagianthus Forst. & Forst.f. Malvaceae

Origins:

Greek *plagios* and *anthos* "a flower," referring to the asymmetrical petals; see J.R. Forster & J.G. Forster, *Characteres generum plantarum.* 85, t. 43. (Nov.) 1775.

Plagiarthron P.A. Duvign. Gramineae

Origins:
From the Greek *plagios* "oblique, placed sideways" and *arthron* "a joint."

Plagiobasis Schrenk Asteraceae

Origins:
From the Greek *plagios* "oblique, placed sideways" and *basis* "base, pedestal."

Plagiobothrys Fischer & C. Meyer Boraginaceae

Origins:
Greek *plagios* and *bothros* "a pit," referring to the scar on the mericarps, to the hollows on the nutlets; see Friedrich Ernst Ludwig von Fischer (1782-1854) and Carl Anton von Meyer (1795-1855), *Index seminum, quae Hortus botanicus imperialis petropolitanus pro mutua commutatione offert*. 2: 46. [St. Petersburg (Jan.) 1836].

Species/Vernacular Names:
P. acanthocarpus (Piper) I.M. Johnston
English: adobe allocarya
P. canescens Benth.
California: Valley popcorn flower
P. elachanthus (F. Muell.) I.M. Johnston (*Heliotropium elachanthum* F. Muell.)
English: hairy forget-me-not
P. glaber (A. Gray) I.M. Johnston
English: hairless popcorn flower
P. hystriculus (Piper) I.M. Johnston
English: bearded popcorn flower
P. kingii (S. Watson) A. Gray
English: Great Basin popcorn flower
P. leptocladus (E. Greene) I.M. Johnston
English: alkali plagiobothrys
P. lithocaryus (E. Greene) I.M. Johnston
California: Mayacamas popcorn flower (Mayacamas Range)
P. myosotoides (Lehm.) Brand
English: forget-me-not popcorn flower
P. nothofulvus (A. Gray) A. Gray
English: popcorn flower
P. orthostatus J. Black
English: brown forget-me-not

P. parishii I.M. Johnston
English: Parish's popcorn flower
P. plurisepalus (F. Muell.) I.M. Johnston (*Maccoya plurisepalea* F. Muell.; *Rochelia maccoya* F. Muell. ex Benth.; *Rochelia plurisepalea* (F. Muell.) Druce; *Allocarya plurisepalea* (F. Muell.) Brand)
English: white forget-me-not, white rochelia
P. strictus (E. Greene) I.M. Johnston
California: Calistoga popcorn flower
P. uncinatus J. Howell
English: hooked popcorn flower

Plagiocarpus Benth. Fabaceae

Origins:
From the Greek *plagios* "oblique" and *karpos* "a fruit."

Plagioceltis Mildbr. ex Baehni Ulmaceae

Origins:
From the Greek *plagios* "oblique" plus *Celtis*.

Plagiocheilus Arn. ex DC. Asteraceae

Origins:
From the Greek *plagios* "oblique" and *cheilos* "lip."

Plagiochloa Adamson & Sprague Gramineae

Origins:
Greek *plagios* and *chloe, chloa* "grass," in reference to the spikelets; see Robert Stephen Adamson (1855-1965) and Thomas Archibald Sprague (1877-1958), in *Journal of South African Botany*. 7: 89. (Apr.) 1941.

Species/Vernacular Names:
P. acutiflora (Nees) Adamson & Sprague (*Brizopyrum acutiflorum* Nees; *Desmazeria acutiflora* (Nees) Hemsl.)
English: desmazeria

Plagiogyria (Kunze) Mett. Plagiogyriaceae

Origins:
Greek *plagios* "oblique" and *gyros* "a ring, round," referring to the annulus of the sporangium; see Georg Heinrich Mettenius (1823-1866), in *Abhandlungen herausgegeben von der Senckenbergischen Naturforschenden Gesellschaft*. 2: 265. Frankfurt a.M. 1858; R.E.G. Pichi Sermolli, "Names and types of fern genera. 3 — Ophioglossaceae,

Osmundaceae, Stromatopteridaceae, Gleicheniaceae, Dipteridaceae, Plagiogyriaceae." *Webbia.* 26(2): 491-536. Apr. 1972.

Species/Vernacular Names:
P. adnata (Bl.) Bedd. (*Lomaria adnata* Blume) (Latin *adnatus, a, um* "joined together, joined to, adnate")

English: Formosan plagiogyria

Japan: takasago-kijo-no-o

Plagiolirion Baker Amaryllidaceae (Liliaceae)

Origins:
From the Greek *plagios* "oblique" and *lirion, leirion* "lily," referring to the form of the flowers.

Plagiolobium Sweet Fabaceae

Origins:
Greek *plagios* and *lobos* "pod," *lobion* "a small pod"; see Robert Sweet (1783-1835), *Flora Australasica.* t. 2. London (Jun.) 1827.

Plagiolophus Greenm. Asteraceae

Origins:
From the Greek *plagios* "oblique" and *lophos* "a crest."

Plagiolytrum Nees Gramineae

Origins:
From the Greek *plagios* "oblique, placed sideways" and *elytron* "sheath, cover, scale, husk."

Plagiopetalum Rehder Melastomataceae

Origins:
Greek *plagios* "oblique, placed sideways" and *petalon* "leaf, petal."

Plagiopteron Griff. Plagiopteraceae (Flacourtiaceae)

Origins:
From the Greek *plagios* "oblique, placed sideways" and *pteron* "wing, feather," referring to the seeds.

Plagiorhegma Maxim. Berberidaceae (Podophyllaceae)

Origins:
Greek *plagios* "oblique, placed sideways" and *rhegma* "a fracture, breakage, cleft, abscess."

Plagioscyphus Radlk. Sapindaceae

Origins:
From the Greek *plagios* "oblique" and *skyphos* "a cup, goblet, a jug."

Plagiosetum Benth. Gramineae

Origins:
From the Greek *plagios* "oblique, placed sideways" and Latin *seta, ae* "bristle."

Species/Vernacular Names:
P. refractum (F. Muell.) Benth. (*Setaria refracta* F. Muell.; *Pennisetum refractum* (F. Muell.) F. Muell.)

English: bristle brush grass

Plagiosiphon Harms Caesalpiniaceae

Origins:
Greek *plagios* "oblique, placed sideways" and *siphon* "tube."

Plagiospermum Oliver Rosaceae

Origins:
From the Greek *plagios* "oblique" and *sperma* "seed."

Plagiostachys Ridley Zingiberaceae

Origins:
From the Greek *plagios* "oblique" and *stachys* "a spike."

Plagiostigma Presl Fabaceae

Origins:
Greek *plagios* "oblique, placed sideways" and *stigma* "stigma."

Plagiostyles Pierre Euphorbiaceae

Origins:
Greek *plagios* "oblique" and *stylos* "style."

Species/Vernacular Names:
P. africana (Müll. Arg.) Prain
Congo: kagoue, mvula, evula, ibula
Gabon: essoula, ngue-ngue, ogourra
Cameroon: essoula, alomba, molubele

Plagiotaxis Wall. ex Kuntze Meliaceae

Origins:
From the Greek *plagios* "oblique" and *taxis* "a series, order, arrangement."

Plagiotheca Chiov. Acanthaceae

Origins:
From the Greek *plagios* "oblique" and *theke* "a box, case, cell."

Plagiotropis F. Muell. Fabaceae

Origins:
Greek *plagios* "oblique" and *trope* "turning, defeat," *trepein* "to turn"; see W.J. Hooker, in *Hooker's Journal of Botany & Kew Garden Miscellany*. 9: 230. London 1857.

Planaltoa Taub. Asteraceae

Origins:
The Brazilian word for plateau.

Plananthus P. Beauv. ex Mirbel Lycopodiaceae

Origins:
Greek *planos* "roaming, deceiving" and *anthos* "flower," presumably referring to the sex of the flowers, or from the Latin *planus* "flat."

Planchonella Pierre Sapotaceae

Origins:
For the French botanist Jules Émile Planchon, 1823-1888, professor of botany and Director of Botanical Garden at Montpellier, co-editor of *Flore des Serres*. 1849-1881, from 1844 to 1848 assistant to William Jackson Hooker at Kew. Among his works are *Les vignes américaines*. Montpellier, Paris 1875 and *Hortus donatensis*. Catalogue des plantes cultivées dans les serres de S.Ex. le prince A. de Démidoff à San Donato, près Florence. Paris 1854-1858, with Louis Benoît Van Houtte (1810-1876) wrote *La Victoria regia*. Gand [Gent] 1850-1851, with J. Lichtenstein wrote *Notes entomologiques sur le Phylloxera Vastatrix*. 1869 and *Le Phylloxera de 1854 à 1873: résumé pratique et scientifique*. Montpellier 1873, with Mr. Saintpierre wrote *Premières Expériences sur la destruction du puceron de la vigne*. Note lue devant la Société d'Agriculture de l'Hérault, au nom d'une Commission spéciale, etc. Montpellier 1868, the French botanist and pharmacist François Gustave Planchon (1833-1900) was his brother, the French botanist Louis David Planchon (1858-1915) was his son. See Charles Henri Marie Flahault (1852-1935), *L'oeuvre de Jules Émile Planchon*. Montpellier 1889; Stafleu and Cowan, *Taxonomic Literature*. 4: 283-289. 1983; John H. Barnhart, *Biographical Notes upon Botanists*. 3: 90. 1965; A. de Démidoff, *Voyage dans la Russie Méridionale et la Crimée par la Hongrie, la Valachie et la Moldavie executé en 1837*. Paris 1840; Jean Baptiste Louis Pierre (1833-1905), *Notes Botaniques. Sapotacées*. 34. Paris [Dec. 1890]; Giulio Giorello & Agnese Grieco, a cura di, *Goethe scienziato*. Einaudi Editore, Torino 1998; G. Schmid, *Goethe und die Naturwissenschaften*. Halle 1940; T.W. Bossert, *Biographical Dictionary of Botanists Represented in the Hunt Institute Portrait Collection*. 312. 1972; E.M. Tucker, *Catalogue of the Library of the Arnold Arboretum of Harvard University*. 1917-1933; Ida Kaplan Langman, *A Selected Guide to the Literature on the Flowering Plants of Mexico*. 586-587. Philadelphia 1964; Leonard Huxley, *Life and Letters of Sir Joseph Dalton Hooker*. London 1918; Elmer Drew Merrill, in *Contr. U.S. Natl. Herb*. 30: 242-243. 1947 and in *Bernice P. Bishop Mus. Bull*. 144: 151. 1937; H.E. Connor and E. Edgar, "Name changes in the indigenous New Zealand flora, 1960-1986 and Nomina Nova IV, 1983-1986." *New Zealand Journal of Botany*. Vol. 25: 115-170. 1987; P.S. Green, *J. Arnold Arbor*. 67: 109-122. 1986; H.H. Allan, *Fl. New Z*. 1: 539. 1961; Emil Bretschneider (1833-1901), *History of European Botanical Discoveries in China*. [Reprint of the original edition 1898.] Leipzig 1981; R. Zander, F. Encke, G. Buchheim and S. Seybold, *Handwörterbuch der Pflanzennamen*. 14. Aufl. Stuttgart 1993.

Species/Vernacular Names:
P. arnhemica (F. Muell.) Royen
English: nutwood
P. australis (R. Br.) Pierre
English: black apple, wild plum, yellow buttonwood
P. clemensii (Lecomte) P. Royen
English: Clemens planchonella

China: xia ye shan lan

P. costata (Endl.) Pierre ex H.J. Lam

Maori name: tawapou

P. obovata (R. Brown) Pierre

English: obovate planchonella, northern yellow boxwood, black ash, yellow teak

China: shan lan

Planchonia Blume Lecythidaceae (Barringtoniaceae)

Origins:

After the French botanist Jules Émile Planchon, 1823-1888.

Species/Vernacular Names:

P. careya (F. Muell.) Knuth

English: cockey apple, cocky apple, billygoat plum

Planera J.F. Gmelin Ulmaceae

Origins:

For the German botanist Johann Jakob (Jacob) Planer, 1743-1789, physician, professor of botany and chemistry at Erfurt, wrote *Index plantarum*, etc. Gothae 1788 and *Observatio oscillationis mercurii in tubo Torricelliano* Erfordiae instituta. [Erfurth?] 1783. See J.H. Barnhart, *Biographical Notes upon Botanists*. 3: 90. 1965; Jonas C. Dryander, *Catalogus bibliothecae historico-naturalis Josephi Banks*. London 1800.

Species/Vernacular Names:

P. aquatica (Walt.) J. Gmelin

English: water elm, planer tree

Planichloa B.K. Simon Gramineae

Origins:

Latin *planus* "flat" and Greek *chloe, chloa* "grass," or from the Greek *planos* "roaming, deceiving."

Planodes E. Greene Brassicaceae

Origins:

From the Greek *planodes* "wandering, rambling."

Plantago L. Plantaginaceae

Origins:

Latin *plantago, inis* for a plantain (Plinius), *planta, ae* "sole of foot," referring to the leaves; see Carl Linnaeus, *Species Plantarum*. 112. 1753 and *Genera Plantarum*. Ed. 5. 52. 1754; Giovanni Semerano, *Le origini della cultura europea. Dizionario della lingua Latina e di voci moderne*. 2(2): 520. Firenze 1994; Manlio Cortelazzo & Paolo Zolli, *Dizionario etimologico della lingua italiana*. 4: 920-921. Zanichelli, Bologna 1985; Georg Christian Wittstein, *Etymologisch-botanisches Handwörterbuch*. 700. Ansbach 1852; Ernest Weekley, *An Etymological Dictionary of Modern English*. 2: 1106. New York 1967; Salvatore Battaglia, *Grande dizionario della lingua italiana*. XIII: 293. Torino 1986.

Species/Vernacular Names:

P. sp.

English: plantain

Maori names: kopakopa, parerarera

P. afra L. (*Plantago psyllium* L.)

English: psyllium, fleawort, Spanish psyllium

Arabic: merwash, bezer

P. alpestris B. Briggs, Carolin & Pulley

English: alpine plantain

P. antarctica Decne.

English: tufted plantain

P. arborescens Poir.

English: Madeira plantain

P. aristata Michx.

English: bracted plantain

P. asiatica L.

English: common plantain, Asiatic plantain, Asian plantain

China: che qian zi, che qian

P. australis Lam.

English: dwarf plantain

Central America: llanten, cola de ardilla, lantén, ractzi

P. bellidioides Decne.

English: coast plantain

P. coronopus L.

English: cut-leaved plantain, buck's-horn plantain, star of the earth

Arabic: wideina, bou djenah

P. cynops L.

English: shrubby plantain

P. debilis R. Br.

English: slender plantain

P. depressa Willd.

English: depressed plantain

China: cheqiancao

P. drummondii Decne.

English: sago weed, dark sago weed

P. glacialis B. Briggs, Carolin & Pulley

English: glacial plantain

P. gunnii Hook.f.

English: cushion plantain

P. hawaiiensis (A. Gray) Pilg. (*Plantago gaudichaudiana* H. Lév.)

Hawaii: laukahi kuahiwi

P. lanceolata L.

English: plantain, rib plantain, ribbed plantain, buckhorn, buckhorn plantain, English plantain, German psyllium, lamb's tongue, lamb's tongues, narrow-leaved plantain, narrow-leaved ribwort, ribgrass, ribwort, ribwort plantain, ripplegrass, small plantain, wild sago

Italian: piantaggine lanceolata, piantaggine lunga, piantaggine minore, arnoglossa, lanciuola, lanciola, piantagine minore

Southern Africa: klein tongblaar, oorpynhoutjie, oorpynwortels, ribbetjiesgras, smalblaarplantago, smalweëblaar, smalweegbree, weeblaar; bolilanyana (Sotho)

Japan: hera-ô-ba-ko

P. major L. (*Plantago dregeana* Decne.)

English: broadleaf, broad-leaved plantain, broad-leaved ribwort, cart-track plant, cart-tract plant, common plantain, giant plantain, greater plantain, large plantain, larger ribwort plantain, ribgrass, ribwort, ripplegrass, rippleseed plantain, way bread, white-man's-foot, wild sago

French: grand plantain

Italian: piantaggine

Central America: llanten, cola de ardilla, lantén, ractzi

Arabic: lisan el-hamal, lisan hamad

Southern Africa: breëblaar, breëblaarplantago, groottongblaar, grootweëblaar, grootweegbree, platvoet, tongblaar, weegbree, weegbreedieblaar; indlebe-ka-tekwane (Zulu)

Japan: Taiwan-ô-ba-ko

Malaya: ekor anjing, ekor angin

Vietnam: xa tien, ma de

China: che chien

The Philippines: lanting, lantin, lanting haba, ilantin, wild saso plantain

Tibetan: tha-ram, be-khur

Hawaii: laukahi, kuhekili

P. maritima L.

English: sea plantain

P. media L.

English: hoary plantain

Italian: piantaggine mezzana, petacciuola pelosa, petacciuola piccola

P. muelleri Pilger

English: star plantain

P. pachyphylla A. Gray (*Plantago glabrifolia* (Rock) Pilg.; *Plantago hillebrandii* Pilg.)

Hawaii: laukahi kuahiwi, manene

P. paradoxa Hook.f.

English: feather plantain

P. princeps Cham. & Schltdl. (*Plantago fauriei* H. Lév.; *Plantago queleniana* Gaud.)

Hawaii: laukahi kuahiwi, ale

P. scabra Moench

English: sand plantain

P. tasmanica Hook.f.

English: Tasman plantain

P. triantha Sprengel

English: rock plantain

P. varia R. Br.

English: variable plantain

P. virginica L.

English: dwarf plantain, pale-seed plantain, sand plantain

Plastolaena Pierre ex Chev. Rubiaceae

Origins:
Greek *plasso, platto* "to form, to mold," *plastos* "formed, molded" and *laina* "cloak, blanket."

Platanocarpum Korth. Rubiaceae

Origins:
Greek *platys* "broad," *platanos, platanistos* "plane tree" and *karpos* "fruit."

Platanocephalus Crantz Rubiaceae

Origins:
Greek *platanos, platanistos* "plane tree," *platys* "broad" and *kephale* "head."

Platanthera Rich. Orchidaceae

Origins:
Greek *platys* "broad" and *anthera* "anther," broad anther.

Species/Vernacular Names:

P. sp.

English: fringed orchid, butterfly orchid

P. hyperborea (L.) Lindley

English: green-flowered bog-orchid

P. japonica (Thunb.) Lindley (*Habenaria japonica* (Thunb.) A. Gray; *Orchis japonica* Thunb.)

Japan: tsure-sagi-sô

P. leucostachys Lindley

English: white-flowered bog-orchid

P. sparsiflora (S. Watson) Schltr.

English: sparse-flowered bog-orchid

P. stricta Lindley

English: slender bog-orchid

Platanus L. Platanaceae

Origins:

Latin *platanus* for the platane or Oriental plane-tree (Plinius), Greek *platanos* used by Theophrastus (*HP.* 4.5.6) and Dioscorides, etymology uncertain, presumably from the Greek *platys* "broad, ample," possibly referring to the shape of the leaves and to the nature of the branches; see G. Semerano, *Le origini della cultura europea*. Dizionario della lingua Latina e di voci moderne. 2(2): 520. 1994; Manlio Cortelazzo & Paolo Zolli, *Dizionario etimologico della lingua italiana*. 4: 941. 1985; Georg Christian Wittstein, *Etymologisch-botanisches Handwörterbuch*. 700. 1852; Ernest Weekley, *An Etymological Dictionary of Modern English*. 2: 1105, 1107. 1967; Salvatore Battaglia, *Grande dizionario della lingua italiana*. XIII: 645-646. Torino 1986.

Species/Vernacular Names:

P. x *acerifolia* (Aiton) Willd.

English: London plane

Japan: momijiba suzukake-no-ki

P. occidentalis L.

English: American sycamore, buttonwood, American plane, sycamore tree, buttonball

Italian: platano occidentale

Japan: Amerika-suzu-kake-ki, Amerika-suzukake-ki

P. orientalis L.

English: Oriental plane

Italian: platano orientale

Japan: suzukake-no-ki

P. racemosa Nutt. (*Platanus californica* Benth.)

English: California sycamore

P. wrightii S. Watson

English: Arizona sycamore

Platea Blume Icacinaceae

Origins:

Greek *platos* "width, plane surface," Latin *platea*, *platalea* "the spoonbill," Greek *plateia* and Latin *platea* for an area, a court-yard, an open space in a house.

Plateilema (A. Gray) Cockerell Asteraceae

Origins:

From *platys* "broad, ample" and *eilema* "a veil, covering, involucre."

Plathymenia Benth. Mimosaceae

Origins:

Greek *platys* "broad, ample" and *hymen* "a membrane."

Platolaria Raf. Bignoniaceae

Origins:

See C.S. Rafinesque, *Sylva Telluriana*. 78. 1838; E.D. Merrill, *Index rafinesquianus*. 220. Jamaica Plain, Massachusetts, USA 1949.

Platonia Martius Guttiferae

Origins:

Dedicated to Plato or Platon, a Greek philosopher, disciple of Socrates; see D.J. Allan, in *Dictionary of Scientific Biography* 11: 22-31. 1981; R. Zander, F. Encke, G. Buchheim and S. Seybold, *Handwörterbuch der Pflanzennamen*. 14. Aufl. 701. Stuttgart 1993.

Platostoma P. Beauv. Labiatae

Origins:

From the Greek *platos* "width, plane surface" and *stoma* "mouth," referring to the mouth of the corolla.

Species/Vernacular Names:

P. africanum P. Beauv.

Congo: erussa, odendemba

Platyadenia B.L. Burtt Gesneriaceae

Origins:
Greek *platys* "broad, ample" and *aden* "gland."

Platyaechmea (Baker) L.B. Sm. & Kress Bromeliaceae

Origins:
Greek *platys* "broad, ample" plus *Aechmea*.

Species/Vernacular Names:
P. fasciata (Lindley) L.B. Sm. & Kress
English: urn plant

Platycalyx N.E. Br. Ericaceae

Origins:
From *platys* "broad, ample" and *kalyx* "calyx," referring to the four-lobed calyx.

Platycapnos (DC.) Bernhardi Fumariaceae (Papaveraceae)

Origins:
Greek *platys* and *kapnos* "smoke," referring to the fruits of *Fumaria*; see Johann Jakob Bernhardi (1774-1850), in *Linnaea*. 8: 471. (Jul.) 1833.

Platycarpha Less. Asteraceae

Origins:
Greek *platys* "broad" and *karphos* "chip of straw, chip of wood, scale, dry stalk," referring to the pappus or to the scales of the involucre.

Platycarpidium F. Muell. Umbelliferae

Origins:
From the Greek *platys* "broad" and the diminutive of *karpos* "fruit."

Platycarpum Bonpl. Rubiaceae (Henriqueziaceae)

Origins:
From the Greek *platys* "broad" and *karpos* "fruit," alluding to the shape of the capsule.

Platycarya Siebold & Zucc. Juglandaceae

Origins:
From the Greek *platys* "broad" and *karyon* "nut," referring to the winged nutlets.

Species/Vernacular Names:
P. strobilacea Siebold & Zucc.
China: huai hsiang, tou lo po hsiang

Platycaulos Linder Restionaceae

Origins:
From the Greek *platys* "broad, flat" and *kaulos* "stalk."

Platycelyphium Harms Fabaceae

Origins:
Greek *platys* "broad, flat" and *kelyphos* "pod, shell, sheath, case, hollow."

Platycentrum Naudin Melastomataceae

Origins:
From the Greek *platys* "broad" and *kentron* "a spur, point," *kenteo* "to prick, torture, torment, sting, spur," having a wide spur.

Platycerium Desv. Polypodiaceae

Origins:
Greek *platykeros* and Latin *platyceros, otis* "broad-horn, having spreading horns, broad-horned, flat-horned," Greek *platys* "broad" and *keras* "horn," referring to the appearance of the fertile fronds; see Nicaise Auguste Desvaux (1784-1856), *Prodrome de la Famille des Fougères*. 213. [*Mém. Soc. Linn. Paris*. 6(2): 213. Mai 1827.] Paris (Jul.-Sept.) 1827; Salvatore Battaglia, *Grande dizionario della lingua italiana*. XIII: 647. [from the Greek *platys* and *kerion* "honeycomb"] 1986.

Species/Vernacular Names:
P. bifurcatum (Cav.) C. Chr. (*Acrostichum bifurcatum* Cav.)
English: common staghorn fern, staghorn fern, elkhorn fern, stag's horn fern, elk's horn fern
Japan: bi-gaku-shida
P. hillii T. Moore
English: northern elkhorn fern

Platychaete Boiss. Asteraceae

Origins:

From the Greek *platys* "broad" and *chaite* "bristle, long hair, foliage."

Platychilum De Launay Fabaceae

Origins:

From the Greek *platys* "broad" and *cheilos* "lip"; see Jean Claude Mien Mordant de Launay (c.1750-1816), *Herbier général de l'amateur.* 3, t. 187. Paris 1819.

Platycladus Spach Cupressaceae

Origins:

Greek *platys* "broad, flat" and *klados* "branch," flattened branches or stems.

Species/Vernacular Names:

P. orientalis (L.) Franco

English: Chinese arbor-vitae, oriental arbor-vitae

French: arbre de vie, thuya d'Orient

China: cebai

Vietnam: co tong pec, trac ba

Platyclinis Benth. Orchidaceae

Origins:

Greek *platys* "broad, flat" and *kline* "a bed," the clinandrium.

Platycodon A. DC. Campanulaceae

Origins:

Greek *platys* "broad" and *kodon* "a bell," referring to the flowers.

Species/Vernacular Names:

P. grandiflorus (Jacq.) A. DC. (*Platycodon glaucus* (Thunb.) Nak.)

English: balloon flower, Chinese bellflower, Japanese bell-flower, bellflower

Japan: kikyô, chichô

China: chieh keng, jie geng

Platycoryne Reichb.f. Orchidaceae

Origins:

Greek *platys* and *koryne* "a club," probably referring to the pendent spur, slightly swollen towards tip, indicating the size of the rostellum.

Platycraspedum O.E. Schulz Brassicaceae

Origins:

From the Greek *platys* "broad" and *kraspedon* "a fringe, border, hem."

Platycrater Siebold & Zucc. Hydrangeaceae

Origins:

From the Greek *platys* "broad" and *krater* "a bowl, crater, large bowl," the calyces of the sterile flowers are saucer-like.

Platycyamus Benth. Fabaceae

Origins:

Greek *platys* "broad" and *kyamos* "a bean."

Platydesma H. Mann Rutaceae

Origins:

From the Greek *platys* "broad" and *desmos* "a bond, band," referring to the filaments forming the staminal tube.

Species/Vernacular Names:

P. rostrata Hillebr.

Hawaii: pilo kea lau li'i

P. spathulata (A. Gray) B. Stone (*Platydesma campanulata* H. Mann)

Hawaii: pilo kea

Platyelasma Kitag. Labiatae

Origins:

From the Greek *platys* "broad" and *elasma, elasmatos, elasmos* "metal plate, flat end."

Platyglottis L.O. Williams Orchidaceae

Origins:

From the Greek *platys* "broad" and *glossa* "tongue," the size of the ligulate lip.

Platygonia Naudin Cucurbitaceae

Origins:

From the Greek *platys* "broad" and *gonia* "an angle."

Platygyna Mercier Euphorbiaceae

Origins:

From the Greek *platys* "broad" and *gyne* "a woman, female."

Platygyria Ching & S.K. Wu Polypodiaceae

Origins:

From the Greek *platys* "broad" and *gyros* "a ring, circle," *gyrios* "circular, round."

Platykeleba N.E. Br. Asclepiadaceae

Origins:

From the Greek *platys* "broad" and *kelebe* "a cup, jar."

Platylepis A. Rich. Orchidaceae

Origins:

From the Greek *platys* "broad" and *lepis* "a scale," perhaps referring to the floral bracts.

Platylepis Kunth Cyperaceae

Origins:

Greek *platys* "broad" and *lepis* "a scale"; see also *Ascolepis* Nees ex Steudel, Greek *askos* "a bag, sac, wine skin" and *lepis*.

Platylobium J.E. Smith Fabaceae

Origins:

Greek *platys* "broad" and *lobos* "pod," referring to the shape of the fruits; see Sir James Edward Smith, *A Specimen of the Botany of New Holland*. 1: 17, t. 6. London 1793.

Species/Vernacular Names:

P. alternifolium F. Muell.

Australia: Grampians flat-pea, Victorian flat-pea

P. formosum Smith

Australia: handsome flat-pea

P. obtusangulum Hook. (*Platylobium obtusangulum* Hook. var. *spinulosum* J.H. Willis)

English: common flat-pea, native holly, eggs and bacon

P. triangulare R. Br.

Australia: ivy flat-pea

Platyloma J. Smith Pteridaceae (Adiantaceae, Sinopteridaceae)

Origins:

Greek *platys* "broad" and *loma* "border, margin, fringe, edge"; see J. Smith, in *Hooker's Journal of Botany*. 4: 160. 1841.

Platylophus D. Don Cunoniaceae

Origins:

From the Greek *platys* "broad" and *lophos* "a crest," referring to the crested fruits.

Species/Vernacular Names:

P. trifoliatus (L.f.) D. Don (*Weinmannia trifoliata* L.f.)

South Africa: witels, white alder

Platyluma Baillon Sapotaceae

Origins:

Greek *platy* and *luma* a vernacular Araucani name, Greek *loma* "border, margin," Latin *luma* "thorn."

Platymerium Bartl. ex DC. Rubiaceae

Origins:

From the Greek *platys* "broad" and *meris, meros* "part, portion, share."

Platymiscium Vogel Fabaceae

Origins:

Probably from the Greek *platys* "broad" and *mischos* "a stalk."

Species/Vernacular Names:
P. pinnatum (Jacq.) Dugand
Colombia: roble

Platymitra Boerl. Annonaceae

Origins:
From the Greek *platys* "broad" and *mitra* "a turban."

Platynema Wight & Arn. Malpighiaceae

Origins:
Greek *platys* and *nema* "thread," with broad threads.

Platyosprion (Maxim.) Maxim. Fabaceae

Origins:
From the Greek *platys* "broad" and *osprion* "pea, bean, vegetables," referring to the pod.

Platypholis Maxim. Orobanchaceae (Scrophulariaceae)

Origins:
From the Greek *platys* "broad" and *pholis, pholidos* "scale, horny scale."

Platypodanthera R.M. King & H. Robinson Asteraceae

Origins:
Greek *platypous* "flat-footed" and *anthera* "anther."

Platypodium Vogel Fabaceae

Origins:
From the Greek *platypous, platypodos* "flat-footed."

Platyptelea J. Drumm. ex Harvey Cunoniaceae

Origins:
From the Greek *platys* "broad" and *ptelea* "an elm tree"; see W.H. Harvey, in *Hooker's Journal of Botany & Kew Garden Miscellany.* 7: 55. (Feb.) 1855.

Platypterocarpus Dunkley & Brenan Celastraceae

Origins:
Greek *platys* "broad," *pteron* "wing" and *karpos* "fruit," winged fruits.

Platypus Small & Nash Orchidaceae

Origins:
Greek *platypous* "flat-footed," the shape of the column.

Platyraphe Miq. Umbelliferae

Origins:
From the Greek *platys* "broad" and *rhaphis, rhaphidos* "a needle, raphe," referring to the bracts.

Platyrhiza Barb. Rodr. Orchidaceae

Origins:
From the Greek *platys* "broad" and *rhiza* "a root," referring to the shape of the roots.

Platyrhodon Hurst Rosaceae

Origins:
From the Greek *platys* "broad" and *rhodon* "rose."

Platysace Bunge Umbelliferae

Origins:
Greek *platys* and *sakos* "a shield," referring to the fruit; see Johann G.C. Lehmann, *Plantae Preissianae.* 1: 285. Hamburgi (Feb.) 1845.

Species/Vernacular Names:
P. clelandii (Maiden & E. Betche) L. Johnson
Australia: fan platysace
P. ericoides (Sprengel) Norman
Australia: heath platysace
P. heterophylla (Benth.) Norman (*Siebera heterophylla* Benth.; *Trachymene heterophylla* (Benth.) F. Muell. ex Tate)
English: parsley, slender platysace
P. lanceolata (Labill.) Druce
Australia: shrubby platysace
P. linearifolia (Cav.) Norman

Australia: narrow-leaf platysace

P. stephensonii (Turcz.) Norman

Australia: Stephenson's platysace

Platyschkuhria (A. Gray) Rydberg Asteraceae

Origins:

Greek *platys* "broad" plus the genus *Schkuhria* A.W. Roth, after the German botanist Christian Schkuhr, 1741-1811; see Gustav Kunze (1793-1851), *Die Farrnkräuter in kolorirten Abbildungen naturgetreu erläutert und beschrieben ... Schkuhr's Farrnkräuter*, Supplement. Leipzig 1840-47; J.H. Barnhart, *Biographical Notes upon Botanists*. 3: 228. 1965; Albrecht Wilhelm Roth (1757-1834), *Catalecta botanica*. 1: 116. Lipsiae 1797; T.W. Bossert, *Biographical Dictionary of Botanists Represented in the Hunt Institute Portrait Collection*. 353. 1972; A. Lasègue, *Musée botanique de Benjamin Delessert*. Paris 1845; R. Zander, F. Encke, G. Buchheim and S. Seybold, *Handwörterbuch der Pflanzennamen*. 14. Aufl. Stuttgart 1993; Mariella Azzarello Di Misa, a cura di, *Il Fondo Antico della Biblioteca dell'Orto Botanico di Palermo*. 249. 1988; Jonas C. Dryander, *Catalogus bibliothecae historico-naturalis Josephi Banks*. London 1800; Ida Kaplan Langman, *A Selected Guide to the Literature on the Flowering Plants of Mexico*. 1964; E.M. Tucker, *Catalogue of the Library of the Arnold Arboretum of Harvard University*. Cambridge, Massachusetts 1917-1933.

Platysepalum Welw. ex Baker Fabaceae

Origins:

Greek *platys* "broad" and Latin *sepalum*, referring to the two upper sepals.

Species/Vernacular Names:

P. violaceum Bak. var. *vanhouttei* (De Wild.) Hauman

Nigeria: katep oshie (Boki)

Platysma Blume Orchidaceae

Origins:

Greek *platysma* "a flat object, tile, plate," indicating the flat lip.

Platyspermation Guillaumin Grossulariaceae (Escalloniaceae)

Origins:

From the Greek *platys* "broad, flat" and *sperma* "a seed," *spermation* "a little seed."

Platyspermum Hook. Brassicaceae

Origins:

From the Greek *platys* "broad, flat" and *sperma* "a seed."

Platystele Schltr. Orchidaceae

Origins:

Greek *platys* "broad, flat" and *stele* "a pillar, column, trunk, central part of stem," a short column dilated above.

Platystemma Wallich Gesneriaceae

Origins:

From the Greek *platys* "broad" and *stemma, stemmatos* "a garland, crown."

Platystemon Benth. Papaveraceae

Origins:

From the Greek *platys* "broad" and *stemon* "a stamen," referring to the expanded filaments.

Species/Vernacular Names:

P. californicus Benth.

English: cream cups

Platystigma Benth. Papaveraceae

Origins:

From the Greek *platys* "broad" and *stigma* "a stigma."

Platystylis Lindl. Orchidaceae

Origins:

From the Greek *platys* "broad" and *stylos* "a style," the shape of the gymnostemium.

Platytaenia Kuhn Pteridaceae (Adiantaceae, Taenitidaceae)

Origins:

Greek *platys* "broad" and *tainia* "fillet."

Platytaenia Nevski & Vved. Umbelliferae

Origins:
Greek *platys* "broad" and *tainia* "fillet."

Platytheca Steetz Tremandraceae

Origins:
Greek *platys* and *theke* "a box, case, cell," referring to the anthers; see Johann G.C. Lehmann (1792-1860), *Plantae Preissianae*. 1: 220. Hamburgi 1845.

Species/Vernacular Names:
P. galioides Steetz
English: platytheca

Platythelys Garay Orchidaceae

Origins:
Greek *platys* "broad" and *thelys* "female, feminine, pertaining to woman," referring to the broad flat rostellum.

Platythyra N.E. Br. Aizoaceae

Origins:
From the Greek *platys* "broad" and *thyra* "a door, entrance, access."

Platytinospora (Engl.) Diels Menispermaceae

Origins:
From the Greek *platys* "broad" plus the genus *Tinospora*.

Platyzoma R. Br. Pteridaceae (Platyzomataceae)

Origins:
Greek *platys* "broad" and *zoma* "a belt," referring to the indusium; see R. Brown, *Prodromus florae Novae Hollandiae et Insulae van-Diemen*. 160. London (Mar.) 1810.

Species/Vernacular Names:
P. microphyllum R. Br.
English: braid fern

Plazia Ruíz & Pav. Asteraceae

Origins:
Named for the Spanish physician Juan Plaza, botanist, a correspondent of Carolus Clusius (1526-1609).

Plecosorus Fée Dryopteridaceae

Origins:
Greek *pleko* "to twist, enfold" and *soros* "a heap."

Plecospermum Trécul Moraceae

Origins:
From the Greek *pleko* "to twist, enfold" and *sperma* "seed."

Plecostachys Hilliard & B.L. Burtt Asteraceae

Origins:
Greek *pleko* "to twist, enfold" and *stachys* "a spike."

Plectaneia Thouars Apocynaceae

Origins:
Greek *plektane* "coil, wreath," referring to the corolla.

Plectocephalus D. Don Asteraceae

Origins:
Greek *plektos* "twisted, plaited" and *kephale* "head"; see J.M. Greenman, "Notes on Southwestern and Mexican Plants I. The Indigenous Centaureas of North America." *Botanical Gazette*. 37: 219-222. 1904.

Species/Vernacular Names:
P. rothrockii (Greenm.) D.J.N. Hind (*Centaurea rothrockii* Greenm.; *Centaurea grandiflora* Sessi & Moc. ex DC.)
English: American basket-flower, Rothrock star-thistle

Plectocomia Martius ex Blume Palmae

Origins:
From the Greek *plektos* "twisted, plaited" and *kome* "hair, hair of the head," referring to the youngest leaves.

Species/Vernacular Names:
P. elmerii Becc.

The Philippines: uñgang

Plectocomiopsis Becc. Palmae

Origins:
Resembling the genus *Plectocomia*.

Plectoma Raf. Lentibulariaceae

Origins:
See Constantine Samuel Rafinesque (1783-1840), *Flora Telluriana*. 4: 110. 1836 [1838]; E.D. Merrill (1876-1956), *Index rafinesquianus*. The plant names published by C.S. Rafinesque, etc. 222. Jamaica Plain, Massachusetts, USA 1949.

Plectopoma Hanst. Gesneriaceae

Origins:
From the Greek *plektos* "twisted, plaited" and *poma* "a lid."

Plectopteris Fée Grammitidaceae

Origins:
Greek *plektos* "twisted, plaited" and *pteris* "a fern."

Plectorrhiza Dockrill Orchidaceae

Origins:
Greek *plektos* "twisted, plaited" (*pleko* "to twist, enfold") and *rhiza* "a root," referring to the tangled roots; see Alick William Dockrill (1915-), *Australasian Sarcanthinae*. 27, t. 14. (Sep.) 1967 and *Australian Indigenous Orchids*. 1: 1-826. Sydney 1969.

Species/Vernacular Names:
P. tridentata (Lindley) Dockr.
English: tangle orchid, tangle root

Plectrachne Henrard Gramineae

Origins:
Greek *plektron* "a spur, cock's spur" and *achne* "chaff, glume," referring to the lemma; see Jan Theodor Henrard (1881-1974), in *Vierteljahrsschrift der Naturforschenden Gesellschaft in Zürich*. 74: 132. (Jun.) 1929; M. Lazarides,

"New taxa of tropical Australian grasses (Poaceae)" in *Nuytsia*. 5(2): 273-303. 1984.

Species/Vernacular Names:
P. helmsii C.E. Hubb.
English: porcupine grass

Plectranthastrum T.C.E. Fr. Labiatae

Origins:
Resembling *Plectranthus*.

Plectranthus L'Hérit. Labiatae

Origins:
Greek *plektron* "a spur, cock's spur" and *anthos* "flower," referring to the shape of the flowers, to the base of the corolla tube; see Charles Louis L'Héritier de Brutelle (1746-1800), *Stirpes novae aut minus cognitae*. 84, t. 41. Parisiis 1788.

Species/Vernacular Names:
P. ambiguus (H. Bol.) Codd (*Orthosiphon ambiguus* H. Bol.; *Plectranthus coloratus* E. Mey. ex Benth. p.p.; *Plectranthus dregei* Codd)

English: plectranthus

P. amboinicus (Lour.) Spreng. (*Coleus amboinicus* Lour.)

English: soup mint, Mexican mint, Indian mint, country borage, French thyme, Spanish thyme, allspice, Indian borage

The Philippines: oregano, suganda, sildu

Vietnam: hung chanh, rau tan la day

P. parviflorus Willd. (*Plectranthus australis* R. Br.)

English: spur flower, cockspur flower

Hawaii: 'ala'ala wai nui, 'ala'ala wai nui pua ki, 'ala'ala wai nui wahine

P. scutellarioides (L.) R. Br. (*Coleus scutellarioides* (L.) Benth.; *Solenostemon scutellarioides* (L.) Codd; *Ocimum scutellarioides* L.; *Coleus blumei* Benth.; *Stenogyne fauriei* H. Lév.)

English: coleus, painted nettle, common coleus

Japan: saya-bana, niwajiku

The Philippines: mayana, maliana, dapoyana, lapunaya, taponaya, patak dugo, saimayu

Hawaii: weleweka

Plectrelminthus Raf. Orchidaceae

Origins:

Greek *plektron* "a spur, cock's spur" and *helmins, helminthos* "worm," referring to the very long spur; see C.S. Rafinesque, *Flora Telluriana*. 4: 42. 1836 [1838].

Plectritis DC. Valerianaceae

Origins:

Greek *plektron* "a spur, cock's spur," referring to the flowers, the tube generally spurred at the base.

Species/Vernacular Names:

P. congesta (Lindley) A. DC.

English: sea blush

Plectrocarpa Gillies ex Hooker & Arnott Zygophyllaceae

Origins:

Greek *plektron, plaktron* "a spur, cock's spur, spear-point" and *karpos* "fruit."

Plectronia L. Rubiaceae

Origins:

Greek *plektron* "a spur, cock's spur," referring to the spiny branches; see Carl Linnaeus, *Systema Naturae*. Ed. 12. 2: 138, 183. 1767 and *Mantissa Plantarum*. 6, 52. 1767.

Plectroniella F. Robyns Rubiaceae

Origins:

The generic name means "the little *Plectronia*"; *Plectronia* is a genus which has now disappeared; the species formerly included in it are now placed in *Canthium*, *Psydrax*, *Pygmaeothamnus*.

Species/Vernacular Names:

P. armata (K. Schumann) Robyns (*Vangueria armata* K. Schumann; *Plectronia ovata* Burtt Davy; *Canthium ovatum* (Burtt Davy) Burtt Davy)

English: false Turkey-berry, bastard Turkey-berry, plectroniella

Southern Africa: basterbokdrol, mufhaladza-tshitangu; mulivhadza-tshitangu (Venda); umPhembedu, umVuthwamini, isiKhwakwane-inkomazi (Zulu)

Plectrophora H. Focke Orchidaceae

Origins:

From the Greek *plektron* "a spur, cock's spur, spear-point" and *phoros* "bearing, carrying," spur formed by the lateral sepals.

Plectrornis Raf. Orchidaceae

Origins:

See Constantine Samuel Rafinesque, *Med. Fl.* 2: 216. 1830; E.D. Merrill, *Index rafinesquianus*. The plant names published by C.S. Rafinesque, etc. 125. 1949.

Plectrurus Raf. Orchidaceae

Origins:

Greek *plektron* "a spur, cock's spur" and *oura* "tail"; see Constantine Samuel Rafinesque (1783-1840), *Neogenyton, or Indication of Sixty-Six New Genera of Plants of North America*. 4. 1825; E.D. Merrill (1876-1956), *Index rafinesquianus*. The plant names published by C.S. Rafinesque, etc. 104. Jamaica Plain, Massachusetts, USA 1949; Richard Evans Schultes and Arthur Stanley Pease (1881-1964), *Generic Names of Orchids. Their Origin and Meaning*. 248. Academic Press, New York and London 1963.

Plecturus Raf. Orchidaceae

Origins:

Greek *plektos* "twisted, plaited" and *oura* "tail," or based on *Plectrurus* Raf., indicating the shape of the spur of the lip; see C.S. Rafinesque, *Herb. Raf.* 73. 1833; E.D. Merrill, *Index rafinesquianus*. 104. Jamaica Plain, Massachusetts, USA 1949.

Plegmatolemma Bremek. Acanthaceae

Origins:

Greek *plegma, plegmatos* "twisted, twined, plaited work, wicker-work" and *lemma* "rind, sheath."

Pleiacanthus (Nutt.) Rydberg Asteraceae

Origins:

From the Greek *pleios* "many, more than one" and *akantha* "thorn."

Pleiadelphia Stapf Gramineae

Origins:
Greek *pleios* "many, more than one" and *adelphos* "brother."

Pleianthemum K. Schumann ex A. Chev. Tiliaceae

Origins:
From the Greek *pleios* "many" and *anthemon* "a flower."

Pleimeris Raf. Rubiaceae

Origins:
From the Greek *pleios* "many" and *meris* "a part, portion"; see C.S. Rafinesque, *Sylva Telluriana*. 21. 1838.

Pleioblastus Nakai Gramineae

Origins:
From the Greek *pleios* "many, more than one" and *blastos* "bud, sprout, germ, ovary, sucker."

Pleiocardia Greene Brassicaceae

Origins:
Greek *pleios* "many, more than one, full of" and *kardia* "heart."

Pleiocarpa Benth. Apocynaceae

Origins:
From the Greek *pleios* "many, full of" and *karpos* "fruit."

Species/Vernacular Names:
P. mutica Benth.
Yoruba: enu marugbo

Pleiocarpidia K. Schumann Rubiaceae

Origins:
From the Greek *pleios* "many, full of" and *karpos* "fruit," Latin *carpidium* "carpel."

Pleioceras Baillon Apocynaceae

Origins:
Greek *pleios* "many, full of" and *keras* "a horn."

Species/Vernacular Names:
P. barteri Baillon
Yoruba: eru ire, efo, dagba, irena kekere, ologbo iyan, afeni

Pleiochasia (Kamienski) Barnhart Lentibulariaceae

Origins:
From the Greek *pleios* "many, full of" and *chasis* "chasm, separation," presumably referring to the photosynthetic appendages.

Pleiochiton Naudin ex A. Gray Melastomataceae

Origins:
From the Greek *pleios* "many" and *chiton* "a tunic, covering."

Pleiococca F. Muell. Rutaceae

Origins:
Greek *pleios* and *kokkos* "a berry"; see Ferdinand Jacob Heinrich von Mueller (1825-1896), *Fragmenta Phytographiae Australiae*. 9: 117. 1875.

Pleiocoryne Rauschert Rubiaceae

Origins:
From the Greek *pleios* "more than one" and *koryne* "a club."

Pleiocraterium Bremek. Rubiaceae

Origins:
Greek *pleios* "more than one" and *krater* "crater, large bowl," Latin *crater, crateris* "a mixing-vessel, a bowl."

Pleiodon Reichb. Gramineae

Origins:
From the Greek *pleios* "more than one" and *odous, odontos* "a tooth."

Pleiogyne K. Koch Menispermaceae

Origins:

Greek *pleios* "many, more than one" and *gyne* "a woman, female"; see Karl (Carl) Heinrich Emil (Ludwig) Koch (1809-1879), in *Botanische Zeitung*. 1: 40. (Jan.) 1843.

Pleiogynium Engl. Anacardiaceae

Origins:

Greek *pleios* and *gyne* "a woman, female," referring to the many female parts or to the large number of carpels; see A. de Candolle and A.C.P. de Candolle, *Monographiae Phanerogamarum*. 4: 255. (Mar.) 1883.

Species/Vernacular Names:

P. timorense (DC.) Leenh. (*Icica timorensis* DC.; *Owenia cerasifera* F. Muell.; *Pleiogynium cerasiferum* (F. Muell.) R. Parker; *Spondias solandri* Benth.; *Pleiogynium solandri* (Benth.) Engl.; *Spondias pleiogyna* F. Muell.)

Australia: Burdekin plum (Burdekin River in Queensland, Australia), tulip plum

Pleiokirkia Capuron Kirkiaceae (Simaroubaceae)

Origins:

Greek *pleios* "many, more than one, more, full of" plus the genus *Kirkia* Oliver; to remember the Scottish explorer Sir John Kirk, 1832-1922, colonial administrator, naturalist, surgeon, philantropist, plant collector, doctor and naturalist on David Livingstone's second Zambesi Expedition, 1864 Fellow of the Linnean Society, 1887 Fellow of the Royal Society. His works include *This Way and That Way. A backward look and a forward look at the problems and progress of child-welfare among the very poor*. London [1917] and *The Zambesi Journal and Letters of Dr. John Kirk, 1858-1863*. Edited by Reginald Foskett. Edinburgh & London 1965; see Sir Reginald Coupland, *Kirk on the Zambesi*. Oxford 1928 and *The Exploitation of East Africa*, etc. (A study of Sir J. Kirk's career at Zanzibar.) London 1939; R. Foskett, ed., *The Zambesi Doctors: David Livingstone's Letters to John Kirk 1859-1872*. Edinburgh 1964; T.W. Bossert, *Biographical Dictionary of Botanists Represented in the Hunt Institute Portrait Collection*. 212. 1972; J.H. Barnhart, *Biographical Notes upon Botanists*. 2: 294. 1965; Ethelyn Maria Tucker, *Catalogue of the Library of the Arnold Arboretum of Harvard University*. Cambridge, Massachusetts 1917-1933; David Williamson, ed., *Sir John Kirk*. [1922]; Ernest Nelmes (1895-1959) and William Cuthbertson (c. 1859-1934), *Curtis's Botanical Magazine Dedications, 1827-1927*. 238. [1931]; R. Zander, F. Encke,

G. Buchheim and S. Seybold, *Handwörterbuch der Pflanzennamen*. 14. Aufl. Stuttgart 1993.

Pleioluma Baillon Sapotaceae

Origins:

Greek *pleios* and *loma* "border, margin, fringe, edge," Latin *luma* "thorn, a kind of cloak," or *luma* a vernacular Araucani name.

Pleiomeris A. DC. Myrsinaceae

Origins:

From the Greek *pleios* "full of" and *meris* "a part, portion."

Pleione D. Don Orchidaceae

Origins:

Named after Pleione, the mother of the seven Pleiades, a daughter of Oceanus and Tethys, wife of Atlas, in Greek mythology.

Species/Vernacular Names:
P. bulbocodioides (Franch.) Rolfe
China: shan ci gu

Pleioneura (C. Hubb.) J. Phipps Gramineae

Origins:
Greek *pleios* "full of" and *neuron* "nerve."

Pleioneura Rech.f. Caryophyllaceae

Origins:

From the Greek *pleios* "full of" and *neuron* "nerve," referring to the leaves.

Pleiosepalum Hand.-Mazz. Rosaceae

Origins:

Greek *pleios* "many, more than one, full, full of" and *sepalon* "sepal."

Pleiospermium (Engl.) Swingle Rutaceae

Origins:

From the Greek *pleios* "many, more than one, full" and *sperma* "seed."

Pleiospilos N.E. Br. Aizoaceae

Origins:

Greek *pleios* "many, full, full of" and *spilos* "a spot, stain," referring to the spotted leaves.

Pleiospora Harv. Fabaceae

Origins:

From the Greek *pleios* "many, more than one, full, full of" and *spora* "seed."

Pleiostachya K. Schum. Marantaceae

Origins:

Greek *pleios* and *stachys* "spike," Latin *pleiostachyus* "having many shoots."

Pleiostachyopiper Trel. Piperaceae

Origins:

Greek *pleios* "many, full of," *stachys* "spike" plus *Piper*.

Pleiostemon Sonder Euphorbiaceae

Origins:

Greek *pleios* "many" and *stemon* "stamen."

Pleiotaenia J.M. Coulter & Rose Umbelliferae

Origins:

From the Greek *pleios* "many" and *tainia* "fillet."

Pleiotaxis Steetz Asteraceae

Origins:

Greek *pleios* "many, more than one, full, full of" and *taxis* "a series, order, arrangement."

Pleistachyopiper Trel. Piperaceae

Origins:

Possibly from the Greek *pleios* "many, more than one," *stachys* "spike" plus *Piper*.

Plenckia Reisseck Celastraceae

Origins:

For the Austrian botanist Joseph Jakob (Jacob) von Plenck (Plenk), 1738-1807, physician, an authority on dermatology. His writings include *Doctrina de morbis cutaneis.* Viennae 1776, *Doctrina de cognoscendis et curandis morbis infantum.* Viennae et Tergesti 1807, *Elementa Artis Obstetriciae.* Viennae 1781, *Elementa Chymiae.* Viennae 1800 and *Physiologia et Pathologia Plantarum.* Viennae 1794. See John H. Barnhart, *Biographical Notes upon Botanists.* 3: 92. 1965; Ida Kaplan Langman, *A Selected Guide to the Literature on the Flowering Plants of Mexico.* 587. University of Pennsylvania Press, Philadelphia 1964; T.W. Bossert, *Biographical Dictionary of Botanists Represented in the Hunt Institute Portrait Collection.* 312. 1972; E.M. Tucker, *Catalogue of the Library of the Arnold Arboretum of Harvard University.* 1917-1933; A. Lasègue, *Musée botanique de Benjamin Delessert.* 540. Paris 1845; Garrison and Morton, *Medical Bibliography.* 3982. 1961; R. Zander, F. Encke, G. Buchheim and S. Seybold, *Handwörterbuch der Pflanzennamen.* 14. Aufl. 1993.

Pleocarphus D. Don Asteraceae

Origins:

From the Greek *pleos* "full, filled" and *karphos* "chip of straw, chip of wood."

Pleocaulus Bremek. Acanthaceae

Origins:

Greek *pleos* "full, filled," *pleon* "many, more" and *kaulos* "stalk."

Pleocnemia C. Presl Dryopteridaceae (Aspidiaceae, Aspleniaceae)

Origins:

Greek *pleon* "many, more" and *kneme* "limb, leg"; see Karl B. Presl (1794-1852), *Tentamen Pteridographiae.* 183. Prague 1836.

Pleodendron Tieghem Canellaceae

Origins:

Greek *pleon, pleos* "filled, full" and *dendron* "tree."

Pleogyne Miers Menispermaceae

Origins:
Greek *pleon* "many, more" and *gyne* "a woman, female, female organs"; see J. Miers, in *Annals and Magazine of Natural History*. Ser. 2, 7: 37. London (Jan.) 1851.

Pleomele Salisb. Dracaenaceae (Agavaceae)

Origins:
Greek *pleon* "many, more," *pleos* "full" and *melon* "an apple" or *meli* "honey," referring to the nature of the fruits; see Richard Anthony Salisbury (1761-1829), *Prodromus stirpium in horto ad Chapel Allerton vigentium*. 245. Londini [London] (Nov.-Dec.) 1796.

Species/Vernacular Names:
P. sp.
Hawaii: hala pepe, le'ie
P. angustifolia (Medik.) N.E. Br.
English: native dracaena

Pleonotoma Miers Bignoniaceae

Origins:
Greek *pleon* "many, more," *pleos* "full" and *tome, tomos, temno* "division, section, to slice," referring to the divided leaves.

Pleopadium Raf. Euphorbiaceae

Origins:
See C.S. Rafinesque, *Autikon botanikon*. Icones plantarum select. nov. vel rariorum, etc. 50. Philadelphia 1840; E.D. Merrill, *Index rafinesquianus*. 156. 1949.

Pleopeltis Humb. & Bonpland ex Willdenow Polypodiaceae

Origins:
Greek *pleos* and *pelte* "a shield," referring to the paraphyses; see Carl L. von Willdenow, *Species Plantarum*. Ed. 4, 5: 211. 1810.

Species/Vernacular Names:
P. astrolepis (Liebmann) E. Fournier (*Polypodium astrolepis* Liebmann; *Grammitis elongata* Swartz (not *Polypodium elongatum* Aiton); *Grammitis lanceolata* Schkuhr; *Grammitis revoluta* Spreng. ex Willdenow; *Pleopeltis revoluta* (Spreng. ex Willdenow) A.R. Smith)

English: star-scaled fern

P. polypodioides (L.) E.G. Andrews & Windham (*Polypodium polypodioides* (L.) Watt; *Acrostichum polypodioides* L.; *Marginaria polypodioides* (L.) Tidestrom)

English: resurrection fern

Pleopogon Nutt. Gramineae

Origins:
Greek *pleos* "full, complete, filled" and *pogon* "beard."

Pleotheca Wall. Rubiaceae

Origins:
From the Greek *pleos* "full, complete, filled" and *theke* "a box, case, capsule."

Plerandra A. Gray Araliaceae

Origins:
Greek *pleroo* "to fill, make full" and *aner, andros* "male, man, stamen."

Plerandropsis R. Vig. Araliaceae

Origins:
Resembling *Plerandra*.

Pleroma D. Don Melastomataceae

Origins:
Greek *pleroo* "to fill, make full," *pleres* "full, full of, filled with, sufficient," *pleos* "satisfied, full," *pleroma, pleromatos* "that which fills, fullness," Latin *pleroma* "fullness," referring to the juicy fruits.

Plesiatropha Pierre ex Hutch. Euphorbiaceae

Origins:
From the Greek *plesios* "near, close to," *plesiastos* "approachable" and *trophe* "food."

Plesioneuron (Holttum) Holttum Thelypteridaceae

Origins:

Greek *plesios* "near, close to" and *neuron* "nerve"; see Richard Eric Holttum (1895-1990), in *Blumea*. 22: 232. (Mar.) 1975.

Plesisa Raf. Lentibulariaceae

Origins:

See C.S. Rafinesque, *Flora Telluriana*. 4: 110. 1836 [1838].

Plesmonium Schott Araceae

Origins:

Greek *plesmone* "being filled, satiety, repletion, abundance."

Plethadenia Urban Rutaceae

Origins:

From the Greek *plethos* "multitude" and *aden* "gland."

Plethiandra Hook.f. Melastomataceae

Origins:

From the Greek *plethos* "multitude" and *aner, andros* "male, stamen."

Plethyrsis Raf. Rubiaceae

Origins:

Greek *plethos* "multitude" and *thyrsos* "a panicle"; see C.S. Rafinesque, *Autikon botanikon*. Icones plantarum select. nov. vel rariorum, etc. 13. Philadelphia 1840; E.D. Merrill, *Index rafinesquianus*. 227. 1949.

Pleuradena Raf. Euphorbiaceae

Origins:

From the Greek *pleura, pleuron* "side, rib, lateral" and *aden* "gland"; see C.S. Rafinesque, *Atl. Jour.* 1: 182. 1833.

Pleuradenia Raf. Labiatae

Origins:

See C.S. Rafinesque, *Neogenyton, or Indication of Sixty-Six New Genera of Plants of North America*. 2. 1825; E.D. Merrill, *Index rafinesquianus*. 209. 1949.

Pleurandra Labill. Dilleniaceae

Origins:

From the Greek *pleura, pleuron* "side, rib, lateral" and *aner, andros* "male, stamen," referring to the one-sided stamens; see J.J.H. de Labillardière (1755-1834), *Novae Hollandiae plantarum specimen*. 2: 5. Parisiis (Feb.) 1806.

Pleurandra Raf. Onagraceae

Origins:

See C.S. Rafinesque, *Florula ludoviciana*. 95. New York 1817; E.D. Merrill, *Index rafinesquianus*. 177. 1949.

Pleurandropsis Baillon Rutaceae

Origins:

Greek *pleura, pleuron* "side, rib, lateral," *aner, andros* "male, stamen" and *opsis* "aspect, appearance, resemblance"; see H.E. Baillon, in *Adansonia*. 10: 305. (Mar.) 1871- (Feb.) 1872.

Pleuranthemum (Pichon) Pichon Apocynaceae

Origins:

From the Greek *pleura, pleuron* "side, lateral" and *anthemon* "a flower," referring to the appearance of the flowers.

Pleuranthium Benth. Orchidaceae

Origins:

Greek *pleura, pleuron* "side, lateral, rib" and *anthos* "a flower," referring to the lateral inflorescences.

Pleuranthodendron L.O. Williams Flacourtiaceae (Tiliaceae)

Origins:

Greek *pleura* "side, lateral," *anthos* "a flower" and *dendron* "tree."

Pleuranthodes Weberb. Rhamnaceae

Origins:

From the Greek *pleura* "side, lateral," *anthos* "flower" and *-odes* "resembling, of the nature of."

Pleuraphis Torrey Gramineae

Origins:

Greek *pleura* "side, lateral" and *raphis* "a needle," referring to the lateral spikelets and the position of awn on lower glume.

Species/Vernacular Names:

P. jamesii Torrey

Spanish: galleta

P. rigida Thurber

English: big galleta

Pleurastis Raf. Amaryllidaceae (Liliaceae)

Origins:

See C.S. Rafinesque, *Flora Telluriana*. 4: 12. 1836 [1837].

Pleureia Raf. Rubiaceae

Origins:

See C.S. Rafinesque, *Sylva Telluriana*. 147. 1838; E.D. Merrill (1876-1956), *Index rafinesquianus*. The plant names published by C.S. Rafinesque, etc. 227. Jamaica Plain, Massachusetts, USA 1949.

Pleuriarum Nakai Araceae

Origins:

Greek *pleura* "side, lateral" plus *Arum*.

Pleuricospora A. Gray Ericaceae (Monotropaceae)

Origins:

Greek *pleurikos* "the sides, of the ribs" and *spora*, *sporos* "a seed, spore," parietal placentas.

Pleuridium (C. Presl) Fée Polypodiaceae

Origins:

Greek *pleuron* "lateral, side," referring to the position of the fruits.

Pleuripetalum T. Durand Annonaceae

Origins:

From the Greek *pleuron*, *pleura* "lateral, side" and *petalon* "leaf, petal."

Pleurisanthes Baillon Icacinaceae

Origins:

Greek *pleuron* and *anthos* "flower," *anthesis* "flowering."

Pleuroblepharis Baill. Acanthaceae

Origins:

Greek *pleuron* "lateral, side" and *Blepharis* Juss.

Pleurobotryum Barb. Rodr. Orchidaceae

Origins:

Greek *pleuron* "lateral, side" and *botrys* "cluster, a bunch of grapes," an allusion to the flowers.

Pleurocalyptus Brongn. & Gris Myrtaceae

Origins:

From the Greek *pleuron* "lateral, side, rib" and *Eucalyptus*.

Pleurocarpaea Benth. Asteraceae

Origins:

Greek *pleura*, *pleuron* "a rib" and *karpos* "fruit," referring to the ribbed achenes; see George Bentham, *Flora Australiensis*. 3: 460. (Jan.) 1867.

Pleurocarpus Klotzsch Rubiaceae

Origins:

From the Greek *pleura*, *pleuron* "a rib" and *karpos* "fruit."

Pleurocitrus Tanaka Rutaceae

Origins:

From the Greek *pleura*, *pleuron* "a rib" plus *Citrus* L.

Pleurocoffea Baill. Rubiaceae

Origins:

Greek *pleura*, *pleuron* "a rib, side" with *Coffea* L.

Pleurocoronis R.M. King & H. Robinson Asteraceae

Origins:

Greek *pleura*, *pleuron* "a rib" and *koronis*, *koronidos* "garland, curved, wreath, bent," side crown, referring to the pappus.

Species/Vernacular Names:

P. pluriseta (A. Gray) R. King & H. Robinson

English: arrow-leaf

Pleuroderris Maxon Dryopteridaceae (Aspleniaceae)

Origins:

Greek *pleura*, *pleuron* "a rib, lateral" and *derris* "a skin."

Pleurofossa Nakai ex H. Ito Vittariaceae

Origins:

Greek *pleura*, *pleuron* "a rib, lateral" and Latin *fossa*, *ae* "a ditch, trench."

Pleurogonium (C. Presl) Lindley Polypodiaceae

Origins:

From the Greek *pleura*, *pleuron* "a rib, lateral" and *gonia* "an angle."

Pleurogramme (Blume) C. Presl Grammitidaceae

Origins:

From the Greek *pleuron* "a rib, lateral" and *gramme* "a line."

Pleurogyna Eschsch. ex Chamisso & Schltdl. Gentianaceae

Origins:

Greek *pleura*, *pleuron* "a rib, lateral" and *gyne* "woman, female, female organs," referring to the ovary.

Pleurogyne Eschsch. ex Griseb. Gentianaceae

Origins:

Greek *pleura*, *pleuron* "a rib, lateral" and *gyne* "woman, female."

Pleurogynella Ikonn. Gentianaceae

Origins:

The diminutive of the genus *Pleurogyne*.

Pleuromanes (C. Presl) C. Presl Hymenophyllaceae

Origins:

Greek *pleura*, *pleuro*, *pleuron* "a rib" and *manes* "a cup, a kind of cup" (or *maino*, *mainomai* "to be furious, rave"); see Karl B. Presl, *Epimeliae botanicae*. 258. Pragae 1849 [reprinted from *Abhandlungen der Königlichen Böhmischen Gesellschaft der Wissenschaften*. 1851].

Species/Vernacular Names:

P. pallidum (Blume) C. Presl

English: floury filmy fern

Pleuropappus F. Muell. Asteraceae

Origins:

Greek *pleura*, *pleuro*, *pleuron* "side, rib, lateral" and *pappos* "fluff, pappus," referring to the oblique pappus; see F. von Mueller, in *Transactions and Proceedings of the Victorian Institute for the Advancement of Science*. 37. (Sep.) 1855.

Pleuropetalum Hook.f. Amaranthaceae

Origins:

Greek *pleura*, *pleuro*, *pleuron* "side, rib, lateral" and *petalon* "petal," referring to the corolla.

Pleurophora D. Don Lythraceae

Origins:

From the Greek *pleuron* "side, rib, lateral" and *phoros* "bearing, carrying."

Pleurophragma Rydb. Brassicaceae

Origins:

From the Greek *pleuron* "side, rib, lateral" and *phragma* "a partition, compartment, wall, screen."

Pleurophyllum Hook.f. Asteraceae

Origins:

From the Greek *pleuron* "side, rib, lateral" and *phyllon* "a leaf."

Pleuropogon R. Br. Gramineae

Origins:

Greek *pleuron* "side, rib, lateral" and *pogon* "beard," referring to the nature of the glumes.

Species/Vernacular Names:

P. hooverianus (L. Benson) J. Hovell

English: north coast semaphore grass

P. refractus (A. Gray) Benth.

English: nodding semaphore grass

Pleuropterantha Franchet Amaranthaceae

Origins:

Greek *pleuron* "side, rib, lateral," *pteron* "wing" and *anthos* "flower."

Pleuropteropyrum Gross Polygonaceae

Origins:

Greek *pleuron* "rib, lateral," *pteron* "wing" and *pyros* "grain, wheat."

Pleuroridgea Tieghem Ochnaceae

Origins:

Greek *pleuron* "rib, lateral" plus *Brackenridgea* A. Gray, genus dedicated to the American (b. Ayr, Scotland) gardener William Dunlop Brackenridge, 1810-1893 (d. Baltimore, Maryland), 1838-1842 assistant botanist and naturalist on Charles Wilkes US expedition to Antarctic islands and northwest coast of North America; see Frans A. Stafleu and Erik A. Mennega, *Taxonomic Literature. Supplement III: Br-Ca.* 2-3. 1995; J.H. Barnhart, *Biographical Notes upon Botanists.* 1: 237. 1965; D.C. Haskell, *The United States Exploring Expedition 1838-1842 and Its Publications 1844-1874.* New York 1942; D.B. Tyler, *The Wilkes Expedition: The First United States Exploring Expedition (1838-1842).* Philadelphia 1968; Sydney A. Spence, *Antarctic Miscellany. Books, Periodicals and Maps Relating to the Discovery and Exploration of Antarctica.* London 1980; G.A. Doumani, ed., *Antarctic Bibliography.* Washington, Library of Congress 1965-1979; T.W. Bossert, *Biographical Dictionary of Botanists Represented in the Hunt Institute Portrait Collection.* 49. 1972; S. Lenley et al., *Catalog of the Manuscript and Archival Collections and Index to the Correspondence of John Torrey.* 448. 1973; E.M. Tucker, *Catalogue of the Library of the Arnold Arboretum of Harvard University.* 1917-1933; Joseph Ewan, *Rocky Mountain Naturalists.* The University of Denver Press 1950; J. Ewan, ed., *A Short History of Botany in the United States.* New York and London 1969.

Pleurosoriopsis Fomin Pteridaceae (Grammitidaceae)

Origins:

Resembling *Pleurosorus*.

Pleurosorus Fée Aspleniaceae

Origins:

From the Greek *pleura, pleuro, pleuron* "side, rib, lateral" and *soros* "a heap, a spore case," referring to the position of the sori; see A.L.A. Fée (1789-1874), *Mémoires sur la famille des Fougères.* Genera Filicum. 5: 179, t. 16.C. Strasbourg & Paris 1850-1852.

Species/Vernacular Names:

P. rutifolius (R. Br.) Fée (*Grammitis rutaefolia* R. Br.)

English: blanket fern

Pleurospa Raf. Araceae

Origins:

See C.S. Rafinesque, *Flora Telluriana.* 4: 8. 1836 [1838]; E.D. Merrill, *Index rafinesquianus.* The plant names published by C.S. Rafinesque, etc. 81. 1949.

Pleurospermopsis C. Norman Umbelliferae

Origins:

Resembling the genus *Pleurospermum*.

Pleurospermum Hoffm. Umbelliferae

Origins:

From the Greek *pleura, pleuron* "side, rib, lateral" and *sperma* "seed," referring to the ridges on the fruits.

Pleurostachys Brongn. Cyperaceae

Origins:

From the Greek *pleura*, *pleuron* "side, lateral" and *stachys* "spike," with spikes at the side.

Pleurostelma Baillon Asclepiadaceae

Origins:

Greek *pleura*, *pleuron* "side, lateral" and *stelma, stelmatos* "a girdle, belt."

Pleurostelma Schltr. Asclepiadaceae

Origins:

Greek *pleura*, *pleuron* and *stelma, stelmatos* "a girdle, belt," see also *Schlechterella* K. Schum.

Pleurostima Raf. Velloziaceae

Origins:

Possibly from the Greek *pleura*, *pleuron* "side, lateral" and *stigma* "stigma," see C.S. Rafinesque, *Flora Telluriana*. 2: 97. 1836 [1837].

Pleurostylia Wight & Arn. Celastraceae

Origins:

Greek *pleura*, *pleuro*, *pleuron* "side, rib, lateral" and *stylos* "style," in the mature ovary the style is borne laterally; see Robert Wight (1796-1872) and G. Arnott Walker Arnott (1799-1868), *Prodromus florae Peninsulae Indiae Orientalis*. 157. London 1834.

Species/Vernacular Names:

P. capensis (Turcz.) Loes. (*Cathastrum capense* Turcz.)

English: coffee pear, bastard saffron, mountain hard pear

Southern Africa: koffiepeer, umkwankwa, bastersaffraanhout, berg hadepeer; umThumelela, umThunyelelwa (Zulu); umNgqangqa (Zulu, Xhosa); umThunyelo, umBovane-ontsaka (Xhosa)

P. putamen Marais

Rodrigues Island: bois d'olive blanc, bois d'olive petites feuilles

Pleurothallis R. Br. Orchidaceae

Origins:

Greek *pleura*, *pleuro* and *thallos* "blossom, branch," perhaps referring to the inflorescence or to the rib-like stems.

Pleurothallopsis Porto & Brade Orchidaceae

Origins:

Resembling the genus of orchids *Pleurothallis* R. Br.

Pleurothyrium Nees Lauraceae

Origins:

Greek *pleura*, *pleuro* "side, rib, lateral" and *athyros* "without door, open," probably referring to the fruit or to the domatia.

Plexaure Endlicher Orchidaceae

Origins:

From the Latin *plexus* "plaiting, twining" and *auris, is* "the ear," possibly referring to the inflorescence.

Plexipus Raf. Verbenaceae

Origins:

From the Latin *plexus* "plaiting, twining," Greek *plektos* "twisted, plaited," *pleko* "to twist, enfold" and *pous* "foot"; see C.S. Rafinesque, *Flora Telluriana*. 2: 104. 1836 [1837].

Plicosepalus Tieghem Loranthaceae

Origins:

Latin *plico* "to fold, to wind together," Greek *pleko* "to twist, enfold," *plektos* "twisted, plaited," Hebrew *pelek* "spindle," Akkadian *pelakku, pilakku* "spindle."

Plicula Raf. Solanaceae

Origins:

Latin *plico* "to fold, to wind together"; see C.S. Rafinesque, *Sylva Telluriana*. 55. 1838.

Plinia L. Myrtaceae

Origins:

For Plinius, Pliny, see David E. Eichholz, in *Dictionary of Scientific Biography* 11: 38-40. 1981.

Plinthanthesis Steudel Gramineae

Origins:

Greek *plinthos, plinth* "pedestal of a column" and *anthesis* "flowering"; some suggest a derivation from *plinthos* and *antithesis* "the opposite," referring to the awn; see E.G. von Steudel (1783-1856), *Synopsis plantarum glumacearum*. 1: 14. Stuttgartiae (Dec.) 1853.

Species/Vernacular Names:

P. paradoxa (R. Br.) S.T. Blake (*Danthonia paradoxa* R. Br.)

English: wiry wallaby-grass

P. rodwayi (C.E. Hubb.) S.T. Blake (*Danthonia rodwayi* C.E. Hubb.) (possibly after the British (b. Devon) botanist Leonard Rodway, 1853-1936 (d. Tasmania), dental surgeon, 1880 to Tasmania, 1896-1932 Hon. Government Botanist, 1928-1932 Director of the Herbarium and Botanic Garden, Hobart. His writings include *The Tasmanian Flora*. Tasmania, Hobart 1903 and *Some Wild Flowers of Tasmania*. Hobart 1910; see John H. Barnhart, *Biographical Notes upon Botanists*. 3: 168. 1965; T.W. Bossert, *Biographical Dictionary of Botanists Represented in the Hunt Institute Portrait Collection*. 335. 1972; E.M. Tucker, *Catalogue of the Library of the Arnold Arboretum of Harvard University*. 1917-1933; S. Lenley et al., *Catalog of the Manuscript and Archival Collections and Index to the Correspondence of John Torrey*. Library of the New York Botanical Garden. 351. 1973; Stafleu and Cowan, *Taxonomic Literature*. 4: 834-836. 1983)

English: wallaby-grass

Plinthus Fenzl Aizoaceae

Origins:

Greek *plinthos, plinth* "pedestal of a column, a tile, brick," Latin *plinthus*, referring to the leaves.

Pliocarpida Post & Kuntze Rubiaceae

Origins:

Greek *pleios* "many, full of" and *karpos* "fruit," Latin *carpidium* "carpel," see also *Pleiocarpidia* K. Schum.

Ploca Lour. ex Gomes Fabaceae

Origins:

From the Greek *plokamos* "lock of hair," *plokos* "folded, a lock of hair, wreath, a braid."

Plocama Aiton Rubiaceae

Origins:

Greek *plokamos* "lock of hair."

Plocaniophyllon Brandegee Rubiaceae

Origins:

Greek *plokanon* "plaited work, strainer" and *phyllon* "leaf."

Plocoglottis Blume Orchidaceae

Origins:

Greek *pleko* "to twist, enfold," *plokos* "folded, a lock of hair, wreath, a braid" and *glottis, glottidos, glotta* "tongue, small tongue," referring to the lip.

Plocosperma Benth. Plocospermataceae (Loganiaceae)

Origins:

From the Greek *plokos* "folded, a lock of hair, wreath" and *sperma* "seed."

Plocostemma Blume Asclepiadaceae

Origins:

From the Greek *plokos* "folded, a lock of hair" and *stemma* "garland, crown."

Ploiarium Korth. Guttiferae (Bonnetiaceae)

Origins:

Greek *ploiarion* "a little skiff, boat," *ploion* "a floating vessel, a ship," referring to the capsule.

Plokiostigma Schuch. Stackhousiaceae

Origins:

Possibly from the Greek *plokion* "a necklace, a chain, a curl, braid," *plokios* "twined" (*pleko* "to twist, enfold") and *stigma*; see (Conrad Gideon) Theodor Schuchardt (1829-1892), in *Linnaea*. 26: 39. (Feb.) 1854.

Plotia Steudel Gramineae

Origins:

For the English (b. Borden, Kent) naturalist Robert Plot, 1640-1696 (d. Kent), chemist, 1677 a Fellow of the Royal Society, 1683 first Keeper of the Ashmolean Museum at Oxford. His writings include *The Natural History of Oxfordshire*. [Folio, first edition, the classic Baconian natural history.] The Theatre: Oxford 1677 and *De origine fontium*. Oxonii 1685. See R. Pulteney, *Historical and Biographical Sketches of the Progress of Botany in England*. 1: 350-352. London 1790; James Britten, *The Sloane Herbarium*, revised and edited by J.E. Dandy. 1958; A.G. Keller, in *Dictionary of Scientific Biography* 11: 40-41. 1981; R.T. Gunther, *Early Science in Cambridge*. Oxford 1937; Elisabeth Leedham-Green, *A Concise History of the University of Cambridge*. Cambridge, University Press 1996.

Pluchea Cass. Asteraceae

Origins:

After the French abbot Noël-Antoine Pluche, 1688-1761 (d. near Paris, France), seminary teacher, naturalist, rejected most of Enlightenment thought. Among his works are *Le spectacle de la nature*. [8 vols.] Paris 1732-1750, *De Linguarum artificio et doctrina*. Paris 1751 and *Histoire du Ciel*, considéré selon les idées des Poëtes, des Philosophes, et de Moïse. Paris 1739-1741 [*The History of the Heavens*. London 1743, second English edition, Engl. transl. by J.B. de Freval.]; see A.H.G. de Cassini, *Bull. Sci. Soc. Philom. Paris*. Année 1817. 31. 1817; Camille Limoges, in *Dictionary of Scientific Biography* 11: 42-44. 1981.

Species/Vernacular Names:

P. sp.

Peru: toñuz

P. baccharioides (F. Muell.) Benth. (*Spiropodium baccharoides* F. Muell.)

English: narrow-leaved Plains-bush

P. camphorata (L.) DC.

English: camphor weed

Mexico: batancual

P. dentex Benth.

English: bowl daisy

P. indica (L.) Less. (*Baccharis indica* L.)

English: Indian fleabane, Indian pluchea

Japan: hiiragi-giku

China: luan xi

The Philippines: kalapini, lagunding late, tulo-lalaki, banig-banig, bauing-bauing

Malaya: beluntas

Vietnam: phat pha, cuc tan, tu bi

P. odorata (L.) Cass.

English: salt marsh fleabane

Mexico: ahuapatli, alinanché, alaa-patli, canela, canelo, cihuapatle, cipatle, comalpatli, chalcay, chalché, flor de ángel, hierba de Santa María, hoja de playa, Santa María, siguapate, teposa, clina, flor de Guadalupe

Latin America: chalche, santa maria

P. sericea (Nutt.) Cov.

English: arrow weed

Spanish: cochinilla

P. symphytifolia (Mill.) Gillis (*Conyza symphytifolia* Mill.)

English: sourbush

P. tetranthera F. Muell.

English: pink Plains-bush

Plukenetia L. Euphorbiaceae

Origins:

For the British physician Leonard Plukenet, 1642-1706 (London), botanist, author of *Almagesti botanici mantissa*. Londini 1700; see John H. Barnhart, *Biographical Notes upon Botanists*. 3: 93. 1965; Stafleu and Cowan, *Taxonomic Literature*. 4: 298-301. Utrecht 1983; T.W. Bossert, *Biographical Dictionary of Botanists Represented in the Hunt Institute Portrait Collection*. 313. 1972; H.N. Clokie, *Account of the Herbaria of the Department of Botany in the University of Oxford*. 225. Oxford 1964; Mary Gunn and Leslie Edward W. Codd, *Botanical Exploration of Southern Africa*. Cape Town 1981; E.M. Tucker, *Catalogue of the Library of the Arnold Arboretum of Harvard University*. 1917-1933; A. Lasègue, *Musée botanique de Benjamin Delessert*. Paris 1845; M. Hadfield et al., *British Gardeners: A Biographical Dictionary*. London 1980; R. Pulteney, *Historical and Biographical Sketches of the Progress of Botany in England*. 2: 18-29. London 1790; J.D. Milner, *Catalogue of Portraits of Botanists Exhibited in the Museums of the Royal Botanic Gardens*. Royal Botanic Gardens, Kew, London 1906; Blanche Henrey, *No Ordinary Gardener — Thomas Knowlton, 1691-1781*. Edited by A.O. Chater. British Museum (Natural History). London 1986; Isaac Henry Burkill (1870-1965), *Chapters on the History of Botany in India*. 9-10. Delhi 1965; Blanche Henrey, *British Botanical and Horticultural Literature before 1800*. 1: 140-145. Oxford 1975; A. White and B.L. Sloane, *The Stapelieae*. Pasadena 1937; Emil Bretschneider, *History of European Botanical Discoveries in China*. [Reprint of the original edition 1898.] Leipzig 1981; James Britten, *The Sloane Herbarium*, revised and edited by J.E. Dandy. 1958; William Darlington, *Memorials of John Bartram and Humphry Marshall*. 1849; J. Ewan, ed., *A Short History of Botany in the United States*. 1969; Georg Christian Wittstein, *Etymologisch-botanisches Handwörterbuch*. 708. Ansbach 1852; Gilbert Westacott Reynolds, *The Aloes of South Africa*. Balkema, Rotterdam 1982.

Species/Vernacular Names:

P. conophora Müll. Arg.

Yoruba: awusa

P. volubilis L.

Peru: amui-o, yuchi, maní del monte, sacha inchic, sacha inchik, sacha yuchi, sacha yachi, sacha yuchiqui, sacha yuchiquio, amuebe, compadre de azeite

Plumbagella Spach Plumbaginaceae

Origins:
The diminutive of the genus *Plumbago* L.

Species/Vernacular Names:
P. micrantha (Ledebour) Spach

English: littleflower plumbagella

China: ji wa cao

Plumbago L. Plumbaginaceae

Origins:
Latin *plumbago, inis* (*plumbum, i* "lead" and the termination *-ago*), Plinius used for a plant called *molybdaena*; see Carl Linnaeus, *Species Plantarum*. 151. 1753 and *Genera Plantarum*. Ed. 5. 75. 1754; Georg Christian Wittstein, *Etymologisch-botanisches Handwörterbuch*. 708. Ansbach 1852; Salvatore Battaglia, *Grande dizionario della lingua italiana*. XIII: 675. Torino 1986.

Species/Vernacular Names:
P. auriculata Lam. (*Plumbago capensis* Thunb.)

English: Cape leadwort, leadwort, plumbago, blueflowered leadwort

Italian: plumbago, piombaggine, piombina

Southern Africa: syselbos; umaBophe, umThi wamadola (Xhosa); utshilitshili (Zulu)

Japan: ruri-matsuri

The Philippines: forget-me-not

P. europaea L.

English: leadwort, toothwort

Arabic: shitaradj

P. indica L. (*Plumbago rosea* L.; *Thela coccinea* Loureiro)

English: Indian leadwort

China: zi hua dan

Malaya: binasa, cheraka, setaka

P. scandens L. (*Plumbago floridana* Nutt.; *Plumbago mexicana* Kunth; *Plumbago occidentalis* Sweet)

English: devil's herb, toothwort

Peru: bela emilia, caapomonga, guapito

Brazil: caaponga, caponga, caajandiva, caataia, louco, folhas de louco, erva de diabo, caajandivas, erva divina, visqueira, erva de amor

Spanish: canutillo, denetalaria, hierba del alacrán, hierba del negro, hierba del pescado, jazmín azul, lagaña de perro, pitillo

P. zeylanica L. (*Plumbago auriculata* Blume; *Plumbago flaccida* Moench; *Plumbago lactea* Salisb.; *Plumbago virginica* Hook.f.; *Plumbago viscosa* Blanco)

English: whiteflower leadwort, Ceylon leadwort

India: chitrak, chitaway, bir kitamuli, chitur, cita, chitwar, sitaparu

Japan: seiron-matsuri

China: bai hua dan

The Philippines: sandikit

Hawaii: 'ilie'e, hilie'e, 'ilihe'e, lauhihi

Yoruba: inabiri, inabii

Plumea Lunan Meliaceae

Origins:
Latin *pluma, plumae* "soft feather," *plumeus, a, um* "downy, filled with down."

Plumeria L. Apocynaceae

Origins:
For the French (b. Marseilles) monk (the Order of Minims) Charles Plumier, 1646-1704 (d. near Cádiz, Spain), Franciscan missionary, naturalist, botanical artist, explorer and botanist, studied mathematics, traveler in the West Indies, (with the physician Joseph Surian) sent to the Caribbean (to explore the French settlements in the West Indies) by Louis XIV, in 1689 and 1690 visited Martinique, Guadeloupe and Haiti. His works include *Description des plantes de l'Amerique*. Paris 1693, *L'art de tourner*. Lyon 1701, *Nova plantarum americanarum genera*. Parisiis 1703 and *Traité des fougères de l'Amerique*. Paris 1705, botanized with Tournefort. See Paul Jovet & J.C. Mallet, in *Dictionary*

of Scientific Biography 11: 47-48. 1981; John H. Barnhart, *Biographical Notes upon Botanists*. 3: 93. 1965; Michael Paiewonsky, *Conquest of Eden 1493-1515*: other voyages of Columbus: Guadeloupe, Puerto Rico, Hispaniola, Virgin Islands. Rome 1991; William B. Griffen, *Indian Assimilation in the Franciscan Area of Nueva Vizcaya*. Tucson 1979; T.W. Bossert, *Biographical Dictionary of Botanists Represented in the Hunt Institute Portrait Collection*. 313. 1972; H.N. Clokie, *Account of the Herbaria of the Department of Botany in the University of Oxford*. 225-226. Oxford 1964; Ida Kaplan Langman, *A Selected Guide to the Literature on the Flowering Plants of Mexico*. 587. Philadelphia 1964; E.M. Tucker, *Catalogue of the Library of the Arnold Arboretum of Harvard University*. 1917-1933; A. Lasègue, *Musée botanique de Benjamin Delessert*. Paris 1845; Jonas C. Dryander, *Catalogus bibliothecae historico-naturalis Josephi Banks*. London 1800; Mariella Azzarello Di Misa, a cura di, *Il Fondo Antico della Biblioteca dell'Orto Botanico di Palermo*. 217. Regione Siciliana, Palermo 1988; Gordon Douglas Rowley, *A History of Succulent Plants*. 1997; Georg Christian Wittstein, *Etymologisch-botanisches Handwörterbuch*. 708. 1852; Blanche Henrey, *No Ordinary Gardener — Thomas Knowlton, 1691-1781*. Edited by A.O. Chater. British Museum (Natural History). London 1986.

Species/Vernacular Names:

P. sp.

Peru: caracocha, cedro pashaco, suche

Mexico: quie-pixi-guij, xuni, guia bigoche, guie bixiguie, quie pixiquij, guia chacha, quia chacha, yo guetzi, xunii

P. alba L.

English: West Indian jasmine

Spanish: flor de Mayo

Mexico: popojoyó, sacnicté, flor de Mayo, nikté ch'om, cacaloxochitl, saknikte (= flor blanca), tlauhquecholxochitl, tizalxochitl

Peru: amancayo, azucena, hamancay, lirio

P. obtusa L.

English: frangipani, great frangipani

Malaya: chempaka

China: du ye ji dan hua

P. rubra L. (*Plumeria acutifolia* Poiret; *Plumeria acuminata* Aiton; *Plumeria bicolor* Ruíz & Pav.; *Plumeria carinata* Ruíz & Pav.; *Plumeria lutea* Ruíz & Pav.; *Plumeria purpurea* Ruíz & Pav.; *Plumeria tricolor* Ruíz & Pav.; *Plumeria incarnata* Miller; *Plumeria rubra* var. *acutifolia* (Poiret) L.H. Bailey)

English: frangipani, red frangipani, common frangipani, Mexican frangipani, red plumeria, temple tree, temple flower, pagoda tree

Spanish: alejandría, campechana, campotonera, flor blanca, flor de mayo, flor de cuervo, tlapalitos, lengua de toro

Peru: aleli, amancayo, amapola, atapaimo, azucena, bellaco caspi, caracucha, caracucho, hamancay, lirio, lirio de la costa, plumeria, suche, suche amarillo, suche blanco, suche rojo, suche turumbaco, suchi, tamaiba

Mexico: acalztatsim, popojoyó, cacajoyó, cacalosúchil, cacalosúchil rojo, jacalosúchil rojo, jacalosúchil blanco, cacaloxóchitl, suchicahue, cundá, nikté, litie, saugran, flor de Mayo, nikté ch'om, chaknicte (= flor roja), tlapalticca-caloxochitl (= flor del cuervo roja), huiloicxitl (= pata de paloma), ayotectli (= vaso de calabazo)

Bolivia: suchi

Japan: Indo-sokei

China: ji dan hua

Vietnam: mien chi tu, hoa cham pa

Malaya: chempaka, chempaka biru, pokok kubur, bunga kubur, bunga kubor, kamboja, kemboja

India: lal golainchi, xenso golainchi, ara golainchi

The Philippines: kalatsutsi, kalanutsi, kachuchi, calacalacutsi

Plumeriopsis Rusby & Woodson Apocynaceae

Origins:
Resembling the genus *Plumeria* L.

Plummera A. Gray Asteraceae

Origins:
For the American plant collector Sarah Allen Lemmon (*née* Plummer), 1836-1923, botanist, 1880 wife of the American botanist and plant collector John Gill Lemmon (1832-1908); see John H. Barnhart, *Biographical Notes upon Botanists*. 3: 93 and 2: 367. 1965; Ida Kaplan Langman, *A Selected Guide to the Literature on the Flowering Plants of Mexico*. University of Pennsylvania Press, Philadelphia 1964; S. Lenley et al., *Catalog of the Manuscript and Archival Collections and Index to the Correspondence of John Torrey*. Library of the New York Botanical Garden. 262. 1973; J.W. Harshberger, *The Botanists of Philadelphia and Their Work*. 1899; E.M. Tucker, *Catalogue of the Library of the Arnold Arboretum of Harvard University*. 1917-1933; R. Zander, F. Encke, G. Buchheim and S. Seybold, *Handwörterbuch der Pflanzennamen*. 14. Aufl. 1993; Joseph Ewan, *Rocky Mountain Naturalists*. The University of Denver Press 1950; Irving William Knobloch, compil., "A preliminary verified list of plant collectors in Mexico." *Phytologia Memoirs*. VI. 1983; Ira L. Wiggins, *Flora of Baja California*. Stanford, California 1980.

Plumosipappus Czerep. Asteraceae

Origins:

Latin *plumosus* "feathered" and *pappus*.

Plutarchia A.C. Sm. Ericaceae

Origins:

Plutarchus, Greek historian, biographer and philosopher.

Pneumatopteris Nakai Thelypteridaceae

Origins:

From the Greek *pneuma, pneumatos* "breath, wind, breathing," *pneo* "to blow, send forth an odor, breathe" and *pteris* "fern"; see Takenoshin Nakai (1882-1952), in *Botanical Magazine*. 47: 179. Tokyo 1933.

Species/Vernacular Names:

P. pennigera (G. Forst.) Holttum

English: lime fern

Pneumonanthe Gled. Gentianaceae

Origins:

From the Greek *pneuma* "breath, wind, breathing," *pneumon, pleumon* "the lungs" and *anthos* "flower."

Poa L. Gramineae

Origins:

From the ancient Greek name *poa, poie, poia* "grass, pasture grass"; see Carl Linnaeus (1707-1778), *Species Plantarum*. 67. 1753 and *Genera Plantarum*. Ed. 5. 31. 1754; Giovanni Semerano, *Le origini della cultura europea. Dizionari Etimologici. Basi semitiche delle lingue indeuropee. Dizionario della lingua Greca*. 2(1): 236. Leo S. Olschki Editore, Firenze 1994.

Species/Vernacular Names:

P. alpina L.

English: bluegrass, alpine meadow grass

P. annua L.

English: annual bluegrass, annual meadow grass, annual poa, dwarf meadow grass, low spear grass, six weeks grass, winter grass, goose grass, sparrow's coat

Southern Africa: eenjarige blougras, straatgras, wintergras; joang-ba-lintja (Sotho)

Japan: suzume-no-katabira

P. atropurpurea Scribner

English: San Bernardino bluegrass

P. bulbosa L.

English: bulbous meadow grass, bulbous bluegrass, bulbous poa

P. chaixii Vill.

English: forest bluegrass, broad-leaved meadow grass

P. compressa L.

English: Canada bluegrass, Canadian bluegrass

P. confinis Vasey

English: beach bluegrass

P. douglasii Nees

English: sand-dune bluegrass

P. drummondiana Nees

English: knotted meadow grass, shaking grass

P. fawcettiae Vick.

English: smooth blue snowgrass

P. fax J.H. Willis & Court (Latin *fax, facis* "a torch")

English: scaly meadow grass, scaly poa

P. fordeana F. Muell.

English: Forde poa

P. glauca Vahl

English: glaucous meadow grass

P. halmaturina J. Black

English: Kangaroo Island poa

P. helmsii Vick.

English: broad-leaved snowgrass

P. hiemata Vick.

English: soft snowgrass

P. infirma Kunth

English: weak bluegrass

P. macrantha Vasey

English: large-flowered sand-dune bluegrass

P. napensis Beetle

English: Napa bluegrass

P. nemoralis L.

English: wood meadow grass, wood bluegrass

P. palustris L.

English: swamp meadow grass, fowl bluegrass

P. pattersonii Vasey

English: Patterson's bluegrass

P. piperi A. Hitchc.

English: Piper's bluegrass

P. poiformis (Labill.) Druce

English: blue tussock grass

P. pratensis L. (*Poa bidentata* Stapf)

English: june grass, Kentucky bluegrass, meadow grass, meadow poa, spear grass, winter grass

P. rhizomata A. Hitchc.

English: timber bluegrass

P. saxicola R. Br.

English: rock poa

P. tenerrima Scribner

English: delicate bluegrass

P. trivialis L.

English: rough meadow grass, rough bluegrass

P. unilateralis Vasey

English: ocean-bluff bluegrass

Poacynum Baillon Apocynaceae

Origins:

An anagram of the generic name *Apocynum* L.

Poaephyllum Ridley Orchidaceae

Origins:

Greek *poa* "grass, pasture grass" and *phyllon* "leaf," narrow leaves.

Poagris Raf. Gramineae

Origins:

Greek *poa* "grass, pasture grass" and *agrios* "wild," Latin *agrius* "wild," for *Poa* L.; see Constantine Samuel Rafinesque (1783-1840), *Flora Telluriana*. 1: 18. 1836 [1837].

Poagrostis Stapf Gramineae

Origins:

The genera *Poa* L. and *Agrostis* L.

Poarion Reichb. Gramineae

Origins:

From the Greek *poa* "grass, pasture grass," the diminutive *poarion*.

Poarium Desv. Scrophulariaceae

Origins:

The diminutive of the Greek *poa* "grass, pasture grass," *poarion*.

Pobeguinea Jacques-Félix Gramineae

Origins:

After the French botanist Charles Henri Oliver Pobéguin, 1856-1951, colonial administrator in French Africa, plant collector in West Africa (French Guinea and Ivory Coast), author of *Essai sur la flore de la Guinée française* produits forestiers, agricoles et industriels. Paris 1906 and *Les plantes médicinales de la Guinée*. Paris 1912; see Auguste Jean Baptiste Chevalier (1873-1956), *Flore vivante de l'Afrique Occidentale Française*. 1938; J.H. Barnhart, *Biographical Notes upon Botanists*. 3: 93. 1965; E.M. Tucker, *Catalogue of the Library of the Arnold Arboretum of Harvard University*. 1917-1933; F.N. Hepper and Fiona Neate, *Plant Collectors in West Africa*. 65. 1971.

Pocillaria Ridl. Icacinaceae

Origins:

From the Latin *pocillum, i* "a little cup," diminutive of *poculum* "a cup, goblet, bowl, beaker."

Poculodiscus Danguy & Choux Sapindaceae

Origins:

From the Latin *poculum* "a cup, goblet, bowl, beaker" and *discus* "a quoit, a disc."

Podachaenium Benth. ex Oersted Asteraceae

Origins:

Greek *pous, podos* "a foot," *a* "without, negative" and *chaino* "to gape, open," referring to the contracted or stalked achenes, Latin *achaenium* or *achenium* "an achene, a dry one-seeded fruit."

Podadenia Thwaites Euphorbiaceae

Origins:

From the Greek *pous, podos* "a foot" and *aden* "gland."

Podaechmea (Mez) L.B. Sm. & Kress Bromeliaceae

Origins:

From the Greek *pous*, *podos* "a foot" plus the genus *Aechmea* Ruíz & Pavón.

Podagrostis (Griseb.) Scribner & Merr. Gramineae

Origins:

From the Greek *pous*, *podos* "a foot" plus *agrostis*, *agrostidos* "grass, weed, couch grass."

Podalyria Willd. Fabaceae

Origins:

After Podalirius or Podaleirios, son of Asklepios or Aesculapius, the god of healing in Greek mythology.

Species/Vernacular Names:

P. calyptrata (Retz.) Willd. var. *calyptrata* (*Podalyria myrtillifolia* Eckl. & Zeyh.; *Podalyria styracifolia* Bot. Mag.) (the specific name from the Latin *calyptratus, a, um* "bearing a calyptra or a caplike"; from the Greek *kalypto* "to hide," *kalyptra* "veil")

English: water blossom pea

South Africa: keur, keurtjie, ertjiebos

P. glauca DC. (*Podalyria buxifolia* Lam.; *Podalyria mundiana* Ecklon & Zeyh.; *Podalyria sparsiflora* Eckl. & Zeyh.)

South Africa: keurtjie

P. sericea (Andr.) R. Br. ex Ait.f. (*Podalyria anomala* Lehm.; *Podalyria canescens* Eckl. & Zeyh.)

English: satin bush

Podandra Baill. Asclepiadaceae

Origins:

From the Greek *pous*, *podos* "a foot" and *aner*, *andros* "man, stamen."

Podandria Rolfe Orchidaceae

Origins:

Greek *pous*, *podos* "a foot" and *aner*, *andros* "man, stamen," referring to the size and shape of the anther.

Podandrogyne Ducke Capparidaceae (Capparaceae)

Origins:

Greek *pous*, *podos* "a foot," *androgynos* "man-woman, hermaphrodite," Latin *androgynus* "having male and female flowers separate but on the same inflorescence."

Podangis Schltr. Orchidaceae

Origins:

Greek *pous*, *podos* and *angeion*, *aggeion* "a vessel, cup," referring to the shape of the spur of the lip.

Podanthera Wight Orchidaceae

Origins:

From the Greek *pous*, *podos* "a foot" and *anthera* "anther."

Podanthes Haw. Asclepiadaceae

Origins:

Greek *pous*, *podos* and *anthos* "flower."

Podanthum Boiss. Campanulaceae

Origins:

From the Greek *pous*, *podos* "a foot" and *anthos, anthemon* "flower."

Podanthus Lagasca Asteraceae

Origins:

From the Greek *pous*, *podos* "a foot" and *anthos* "flower," referring to the stalked flower-heads.

Podistera S. Watson Umbelliferae

Origins:

Greek *podister, podisteros* "foot-entangling," or from *pous* "foot" and *stereos* "solid, firm, tight," solid foot, from the compact habit.

Species/Vernacular Names:

P. nevadensis (A. Gray) S. Watson

English: Sierra podistera

Podocaelia (Benth.) A. Fernandes & R. Fernandes Melastomataceae

Origins:
Probably from the Greek *pous, podos* "a foot" and *koilos* "hollow."

Podocalyx Klotzsch Euphorbiaceae

Origins:
Greek *pous, podos* "a foot" and *kalyx* "calyx."

Podocarpium (Benth.) Y.C. Yang & S.H. Huang Fabaceae

Origins:
From the Greek *pous, podos* "a foot" and *karpos* "fruit."

Podocarpus L'Hérit. ex Pers. Podocarpaceae

Origins:
Greek *pous, podos* and *karpos* "fruit," referring to the length of the fleshy stalks; see Charles Louis L'Héritier de Brutelle, *Synopsis Plantarum.* 2(2): 580. (Sep.) 1807; H.E. Connor and E. Edgar, "Name changes in the indigenous New Zealand flora, 1960-1986 and Nomina Nova IV, 1983-1986." *New Zealand Journal of Botany.* Vol. 25: 115-170. 1987.

Species/Vernacular Names:

P. acutifolius Kirk

English: sharp leaved totara

P. alpinus R. Br. ex Hook.f.

English: Tasmanian podocarp

P. coriaceus Rich.

English: Yacca podocarp

P. elatus Endl.

English: Rockingham podocarp, brown pine, plum pine, yellow pine

P. elongatus (Ait.) L'Hérit. ex Pers. (*Taxus elongata* Ait.; *Podocarpus thunbergii* var. *angustifolia* Sim)

English: Breede river yellowwood, Cape yellowwood, African yellowwood

South Africa: Breëriviergeelhout, westelike geelhout

P. falcatus (Thunb.) Mirb. (*Podocarpus gracilior* sensu Burtt Davy, non Pilg.; *Podocarpus gracillimus* Stapf; *Taxus falcata* Thunb.; *Decussocarpus falcatus* (Thunb.) Laubenf.; *Nageia falcata* (Thunb.) Kuntze)

English: common yellowwood, Outeniqua yellowwood, smooth-barked yellowwood, bastard yellowwood

Southern Africa: Outeniekwageelhout, geelhout, nikolander, kolander, nietlander; umSonti (Swazi); mogobagoba (North Sotho); umSonti, uNomphumelo, umHlenhlane, umGeya, umKhandangoma, umPume (Zulu); umKhoba, umKoleya, umGeya (Xhosa)

P. hallii Kirk (*Podocarpus cunninghamii* Colenso)

English: thin barked totara

P. henkelii Stapf ex Dallim. & Jacks. (*Podocarpus thunbergii* var. *falcata* Sim) (the species was named after Dr J.S. Henkel, 1930s Conservator for Forests in Natal, author of *The Woody Plants of Natal and Zululand*, being a key ... based on leaf characters. Durban and Pietermaritzburg 1934)

English: East Griqualand yellowwood, Natal yellowwood, Henkel's yellowwood, falcate yellowwood

Southern Africa: Henkel-se-geelhout, bastergeelhout; umSonti (Zulu); umSonti (Xhosa)

P. latifolius (Thunb.) R. Br. ex Mirb. (*Taxus latifolia* Thunb.; *Podocarpus thunbergii* Hook.; *Podocarpus milanjianus* Rendle)

English: yellowwood, upright yellowwood, rough-barked yellowwood, broad-leaved yellowwood, true yellowwood, real yellowwood

Southern Africa: opregte geelhout, regte geelhout, Kaapse geelhout, wittegeelhout; monyaunyau, mogobagoba (North Sotho); umKhoba, umSonti, umGeya (Zulu); umCheya, umGeya (this same word used also for giraffe), umKhoba, umSonti (Xhosa); muhovho-hovho (Venda); ruHombge, muNyenza (Shona)

P. lawrencei Hooker

English: mountain plum pine

P. macrocarpus Laubenf.

The Philippines: malakawayan

P. macrophyllus (Thunb.) D. Don (*Taxus macrophylla* Thunb.)

English: bigleaf podocarp, yew podocarp, Japanese yew, southern yew

Japan: inu-maki

Okinawa: chagi, kyangi

China: luo han song shi, lo han sung, lo han mu

P. neriifolius D. Don

English: oleander podocarp, mountain teak

Malaya: jati bukit, setada, sentada, kayu China

The Philippines: dilang butiki, pasuik, huag

P. nivalis Hooker

English: alpine totara, snow totara

P. nubigenus Lindley

English: cloud podocarp, Chilean podocarp

Chile: mañío, mañío de hojas punzantes, pino amarillo

P. pilgeri Foxw.

The Philippines: lubang-lubang, tambiayang

P. polystachyus R. Br. ex Endl.

The Philippines: bantigi, inamagyo

Malaya: setada, sentada, jati laut

P. rumphii Bl.

The Philippines: kasiray

P. salignus D. Don

English: willowleaf podocarp, willow podocarp

Latin America: manio, mañío

Chile: mañío, mañío de hojas largas

P. spinulosus (Smith) Mirb.

English: spinyleaf podocarp

P. totara G. Benn. ex D. Don

New Zealand: totara

Podochilopsis Guillaumin Orchidaceae

Origins:
Resembling *Podochilus* Blume.

Podochilus Blume Orchidaceae

Origins:
Greek *pous, podos* and *cheilos* "lip," referring to the base of the labellum, to the two appendices; see Karl Ludwig von Blume, *Bijdragen tot de flora van Nederlandsch Indië.* 295. Batavia (Sep.-Dec.) 1825; B.A. Lewis and P.J. Cribb, *Orchids of the Solomon Islands and Bougainville.* Royal Botanic Gardens, Kew 1991.

Species/Vernacular Names:
P. bimaculatus Schltr.

Bougainville Island: tukuritukuri

Podochrosia Baillon Apocynaceae

Origins:
From the Greek *pous, podos* "a foot" plus *Ochrosia.*

Podococcus G. Mann & H.A. Wendland Palmae

Origins:
Greek *pous, podos* "a foot" and *kokkos* "berry, seed, grain," this genus has a stalked fruit, bears a spicate and solitary inflorescence, pendulous in fruit, peduncle very slender.

Podocoma Cass. Asteraceae

Origins:
From the Greek *pous, podos* "a foot" and *kome* "hair, hair of the head," referring to the stalked pappus and to the flat achenes; see Alexandre Henri Gabriel Comte de Cassini (1781-1832), in *Bull. Sci. Soc. Philom. Paris.* 137. Paris (Sep.) 1817.

Podocoma R. Br. Asteraceae

Origins:
Greek *pous, podos* and *kome* "hair, hair of the head"; see Charles Sturt (1795-1869), *Narrative of an Expedition into Central Australia, Performed ... During the Years 1844, 1845 and 1846.* London 1849; Arthur D. Chapman, ed., *Australian Plant Name Index.* 2338. Canberra 1991.

Podocytisus Boiss. & Heldr. Fabaceae

Origins:
From the Greek *pous, podos* "a foot" plus *Cytisus* Desf.

Podogynium Taubert Caesalpiniaceae

Origins:
From the Greek *pous, podos* "a foot" and *gyne* "woman, female."

Podolasia N.E. Br. Araceae

Origins:
From the Greek *pous, podos* and the genus *Lasia* Lour., referring to some difference with *Lasia*; see D.H. Nicolson, "Derivation of aroid generic names." *Aroideana.* 10: 15-25. 1988.

Podolepis Labill. Asteraceae

Origins:
Greek *pous, podos* "a foot" and *lepis* "a scale," referring to the involucral bracts, the pedicels are scaly; see Jacques Julien Houtton de Labillardière, *Novae Hollandiae plantarum specimen.* 2: 56, t. 208. Parisiis (Jun.) 1806.

Species/Vernacular Names:
P. arachnoidea (Hook.) Druce (*Rutidosis arachnoidea* Hook.; *Rutidochlamys mitchellii* Sonder; *Podolepis*

rhytidochlamys F. Muell.; *Podolepis rutidochlamys* F. Muell. ex Benth.)

English: clustered copper-wire daisy, cottony podolepis

P. hieracioides F. Muell.

English: long podolepis

P. jaceoides (Sims) Voss

English: showy podolepis

P. lessonii (Cass.) Benth. (after the French botanist René Primevère Lesson, 1794-1849, physician, pharmacist, ornithologist, professor of botany, explorer, from 1822 to 1825 took part in the voyage of the *Coquille* commanded by Louis-Isidor Duperrey (1786-1865), from 1826 to 1829 with Dumont d'Urville on *Astrolabe* expedition. Among his writings are *Flore rochefortine.* Rochefort 1835, *Voyage Médical autour du monde.* Paris 1829, *Notice historique sur l'Admiral Dumont d'Urville.* Rochefort 1846, *Journal d'un Voyage Pittoresque autour du Monde exécuté sur la Corvette La Coquille commandée par M. L.I. Duperrey.* Paris 1830 and *Voyage autour du Monde entrepris par ordre du Gouvernement sur la Corvette La Coquille.* Paris 1838-1839; see J.H. Barnhart, *Biographical Notes upon Botanists.* 2: 372. 1965; T.W. Bossert, *Biographical Dictionary of Botanists Represented in the Hunt Institute Portrait Collection.* 235. 1972; Günther Schmid, *Chamisso als Naturforscher.* Eine Bibliographie. Leipzig 1942; Stafleu and Cowan, *Taxonomic Literature.* 2: 853. 1979; John Dunmore, *Who's Who in Pacific Navigation.* Honolulu 1991)

English: button podolepis

P. monticola R. Henderson

English: rock podolepis

P. muelleri (Sonder) G.L. Davis (*Panaetia muelleri* Sonder; *Podolepis lessonii* sensu J. Black)

English: small copper-wire daisy

P. robusta (Maiden & E. Betche) J.H. Willis

English: alpine podolepis

P. rugata Labill.

English: pleated podolepis

P. tepperi (F. Muell.) D. Cooke (*Helichrysum tepperi* F. Muell.)

English: delicate everlasting

Podolobium R. Br. Fabaceae

Origins:
Greek *pous, podos* and *lobos* "a pod," referring to the pod; see W.T. Aiton, *Hortus Kewensis.* Ed. 2. 3: 9. (Oct.-Nov.) 1811.

Species/Vernacular Names:
P. aciculiferum F. Muell.

English: needle shaggy pea

P. alpestre (F. Muell.) Crisp & P.H. Weston

English: alpine shaggy pea, mountain shaggy pea, alpine oxylobium

Podolotus Royle Fabaceae

Origins:
From the Greek *pous, podos* "a foot" plus *Lotus* L.

Podoluma Baill. Sapotaceae

Origins:
From the Greek *pous, podos* "a foot" with *luma.*

Podonephelium Baill. Sapindaceae

Origins:
Greek *pous, podos* "a foot" and the genus *Nephelium* L., Greek *nephele* "a cloud."

Podonosma Boiss. Boraginaceae

Origins:
From the Greek *pous, podos* "a foot" plus the genus *Onosma* L.

Podoon Baill. Anacardiaceae (Podoaceae)

Origins:
From the Greek *pous, podos* "a foot" and *oon* "egg."

Podopeltis Fée Dryopteridaceae (Aspleniaceae)

Origins:
From the Greek *pous, podos* "a foot" and *peltis* "a shield."

Podopetalum F. Muell. Fabaceae

Origins:
Greek *pous, podos* and *petalon* "a petal," referring to the claws of the petals; see F. von Mueller, *The Chemist and Druggist.* Australasian Supplement 12. (Jun.) 1882.

Podophania Baill. Asteraceae

Origins:
From the Greek *pous, podos* "a foot" and *phane, phanos* "a torch, light, bright."

Podophorus Philippi Gramineae

Origins:
From the Greek *pous, podos* "a foot" and *phoros* "bearing, carrying."

Podophyllum L. Berberidaceae (Podophyllaceae)

Origins:
Derived from *Anapodophyllum* Catesby, Latin *anas, anatis* "the duck," Greek *pous, podos* "a foot" and *phyllon* "a leaf," referring to the shape of the leaves; see Georg Christian Wittstein, *Etymologisch-botanisches Handwörterbuch.* 558. Ansbach 1852; Helmut Genaust, *Etymologisches Wörterbuch der botanischen Pflanzennamen.* 495. Basel 1996; Salvatore Battaglia, *Grande dizionario della lingua italiana.* XIII: 705. Torino 1986.

Species/Vernacular Names:
P. hexandrum Royle
English: may apple
P. peltatum L.
English: may apple, American mandrake
Italian: podofillo
P. versipelle Hance
China: tu chio lien

Podopogon Raf. Gramineae

Origins:
Greek *pous, podos* "a foot" and *pogon* "beard"; see C.S. Rafinesque, *Neogenyton, or Indication of Sixty-Six New Genera of Plants of North America.* 4. 1825; E.D. Merrill, *Index rafinesquianus.* 76. 1949.

Podopterus Bonpl. Polygonaceae

Origins:
From the Greek *pous, podos* "a foot" and *pteris* "wing," referring to the winged outer perianth segments.

Podorungia Baill. Acanthaceae

Origins:
Greek *pous, podos* "a foot" plus *Rungia* Nees.

Podosciadium A. Gray Umbelliferae

Origins:
From the Greek *pous, podos* "a foot" and *skiadion, skiadeion* "umbel, parasol."

Podosemum Desv. Gramineae

Origins:
Probably from the Greek *pous, podos* "a foot" and Latin *semen* "seed."

Podosorus Holttum Polypodiaceae

Origins:
From the Greek *pous, podos* "a foot" and *soros* "a heap, a spore case."

Podosperma Labill. Asteraceae

Origins:
From the Greek *pous, podos* "a foot" and *sperma* "seed"; see Jacques J. Houtton de Labillardière, *Novae Hollandiae plantarum specimen.* 2: 35, t. 177. Parisiis (Apr.) 1806.

Podospermum DC. Asteraceae

Origins:
Greek *pous, podos* "a foot" and *sperma* "seed."

Species/Vernacular Names:
P. laciniatum (L.) DC. (*Scorzonera laciniata* L.)
English: scorzonera

Podostachys Klotzsch Euphorbiaceae

Origins:
From the Greek *pous, podos* "a foot" and *stachys* "spike."

Podostelma K. Schumann Asclepiadaceae

Origins:
Greek *pous, podos* and *stelma, stelmatos* "a girdle, belt."

Podostemma Greene Asclepiadaceae

Origins:
From the Greek *pous*, *podos* "a foot" and *stemma* "a garland, crown."

Podostemon Michaux Podostemaceae

Origins:
Orthographic variant of *Podostemum* Michaux.

Podostemum Michaux Podostemaceae

Origins:
Greek *pous*, *podos* "a foot" and *stemon* "stamen," referring to the nature of the stamens; see André Michaux (1746-1803), *Flora Boreali-Americana*. 2: 164, t. 44. Paris (Mar.) 1803.

Podostigma Elliott Asclepiadaceae

Origins:
From the Greek *pous*, *podos* "a foot" and *stigma* "stigma," referring to the style.

Podotheca Cass. Asteraceae

Origins:
Greek *pous*, *podos* "a foot" and *theke* "a box, case, capsule," referring to the achenes, the fruits are stalked; see Alexandre H.G. Comte de Cassini, *Dictionnaire des Sciences Naturelles*. 23: 561. (Nov.) 1822.

Species/Vernacular Names:
P. angustifolia (Labill.) Less. (*Podosperma angustifolia* Labill.)
English: sticky longheads, sticky heads
P. chrysantha (Steetz) Benth. (*Ixiolaena chrysantha* Steetz)
English: golden longheads

Podranea Sprague Bignoniaceae

Origins:
Anagram of *Pandorea*, an Australian genus with which *Podranea* was once united.

Species/Vernacular Names:
P. brycei (N.E. Br.) Sprague

English: queen of Sheba, Zimbabwe climber, Zimbabwe creeper
Southern Africa: gwebga (Shona)
P. ricasoliana (Tanf.) Sprague (*Tecoma ricasoliana* Tanf.)
English: Port St. John's creeper, pink trumpet vine, podranea, Ricasol podranea

Poecilandra Tul. Ochnaceae

Origins:
From the Greek *poikilos* "spotted, many-colored" and *andros* "man, stamen, male."

Poecilanthe Benth. Fabaceae

Origins:
Greek *poikilos* "spotted, many-colored" and *anthos* "flower."

Poecilocalyx Bremek. Rubiaceae

Origins:
From the Greek *poikilos* "spotted, many-colored" and *kalyx* "calyx."

Poecilocarpus Nevski Fabaceae

Origins:
From the Greek *poikilos* "spotted, many-colored" and *karpos* "fruit."

Poecilochroma Miers Solanaceae

Origins:
From the Greek *poikilos* "spotted, many-colored" and *chroma* "color."

Poecilodermis Schott & Endl. Sterculiaceae

Origins:
Greek *poikilos* "spotted, mottled, many-colored" and *derma*, *dermatos* "skin," with spotted skin; see H.W. Schott and S.F.L. Endlicher, *Meletemata botanica*. 33. Vindobonae [Wien] 1832.

Poecilolepis Grau Asteraceae

Origins:

From the Greek *poikilos* "spotted, mottled, many-colored" and *lepis* "scale."

Poeciloneuron Beddome Guttiferae

Origins:

From the Greek *poikilos* and *neuron* "nerve."

Poecilopteris C. Presl Lomariopsidaceae (Aspleniaceae)

Origins:

From the Greek *poikilos* "spotted, mottled, many-colored" and *pteris* "fern"; see Karl B. Presl, *Tentamen Pteridographiae*. 241. Prague (before Dec.) 1836.

Poecilostachys Hackel Gramineae

Origins:

From the Greek *poikilos* "spotted" and *stachys* "spike."

Poellnitzia Uitewaal Asphodelaceae (Aloaceae)

Origins:

The genus name honors the German botanist Joseph Karl (Carl) Leopoldt Arndt von Poellnitz, 1896-1945 (Germany), agriculturist, specialist in succulent plants. Among his works are *Zur Kenntnis der gattung Echeveria DC ...* Dahlem bei Berlin 1936, "Neue *Anthericum*-Arten aus Südamerika." *Revista Sudamer. Bot.* 7(2-4): 99-104. 1942 and "Versuch einer Monographie der Gattung *Portulaca* L." *Repert. Sp. nov., Fedde.* 37: 240-320. (Dec.) 1934. See John H. Barnhart, *Biographical Notes upon Botanists.* 3: 94. 1965; F. Boerner & G. Kunkel, *Taschenwörterbuch der botanischen Pflanzennamen.* 4. Aufl. 154. Berlin & Hamburg 1989; R. Zander, F. Encke, G. Buchheim and S. Seybold, *Handwörterbuch der Pflanzennamen.* 14. Aufl. 1993; Elmer Drew Merrill, in *Contr. U.S. Natl. Herb.* 30: 243. 1947 and in *Bernice P. Bishop Mus. Bull.* 144: 151. 1937; Ida Kaplan Langman, *A Selected Guide to the Literature on the Flowering Plants of Mexico.* University of Pennsylvania Press, Philadelphia 1964; Gordon Douglas Rowley, *A History of Succulent Plants.* Strawberry Press, Mill Valley, California 1997.

Poeppigia C. Presl Caesalpiniaceae

Origins:

For the German botanist Eduard Friedrich Poeppig, 1798-1868, explorer, naturalist, traveler, zoologist, in Latin America (Cuba, Chile 1826-1829, Peru and Brazil), plant collector. His writings include *Reise in Chile, Peru*. Leipzig [1834-]1835-1836 and "Schreiben des jetzt in Chile reisenden Hrn. Dr. Pöppig. Hütte am Rio Colorado in den Anden Chile's. Decbr. 24. 1827." *Not. Natur-Heilk.* 23(18): 273-282, 23(19): 289-293. 1829. See G. Looser, "El naturalista Poeppig." *Revista Univ.* (Santiago) 15(3): 180-188. 1930; I. Urban, "Biographische Skizzen. IV. 5. Eduard Poeppig (1798-1868)." *Bot. Jahrb. Syst.* 21(4), Beibl. 53: 1-27. 1896; Stafleu and Cowan, *Taxonomic Literature.* 4: 310-312. 1983; John H. Barnhart, *Biographical Notes upon Botanists.* 3: 94. 1965; T.W. Bossert, *Biographical Dictionary of Botanists Represented in the Hunt Institute Portrait Collection.* 313. 1972; H.N. Clokie, *Account of the Herbaria of the Department of Botany in the University of Oxford.* 226. Oxford 1964; E.M. Tucker, *Catalogue of the Library of the Arnold Arboretum of Harvard University.* 1917-1933; A. Lasègue, *Musée botanique de Benjamin Delessert.* Paris 1845; Mary Gunn and Leslie Edward W. Codd, *Botanical Exploration of Southern Africa.* 283. Cape Town 1981; Frederico Carlos Hoehne, M. Kuhlmann and Oswaldo Handro, *O jardim botânico de São Paulo.* 1941; G. Murray, *History of the Collections Contained in the Natural History Departments of the British Museum.* 1: 174. London 1904; August Weberbauer, *Die Pflanzenwelt der peruanischen Andes in ihren Grundzügen dargestellt.* 7-8, 35. Leipzig 1911; Gordon Douglas Rowley, *A History of Succulent Plants.* 1997; R. Zander, F. Encke, G. Buchheim and S. Seybold, *Handwörterbuch der Pflanzennamen.* 14. Aufl. 1993; Ida Kaplan Langman, *A Selected Guide to the Literature on the Flowering Plants of Mexico.* 1964.

Poeppigia Kunze ex Reichb. Tecophilaeaceae

Origins:

For the German botanist Eduard Friedrich Poeppig, 1798-1868.

Poga Pierre Anisophylleaceae (Rosidae, Rosales)

Origins:

A vernacular name from Gabon.

Species/Vernacular Names:

P. oleosa Pierre

Nigeria: imono (Ibo); inoi (Efik); onyo (Boki)

Yoruba: ikujebu

Western Tropical Africa: afo, inoy, inoi nut, poga

Gabon: m'poga, afo, ovoga

Cameroon: mpoi, ngale, pobo, angale, fo

Pogenda Raf. Oleaceae

Origins:
See C.S. Rafinesque, *Sylva Telluriana*. 10. 1838; E.D. Merrill (1876-1956), *Index rafinesquianus*. The plant names published by C.S. Rafinesque, etc. 190. Jamaica Plain, Massachusetts, USA 1949.

Pogochilus Falconer Orchidaceae

Origins:
Greek *pogon* "beard" and *cheilos* "a lip."

Pogochloa S. Moore Gramineae

Origins:
From the Greek *pogon* "beard" and *chloe, chloa* "grass."

Pogogyne Benth. Labiatae

Origins:
Greek *pogon* "beard" and *gyne* "female," referring to the bearded style.

Species/Vernacular Names:
P. abramsii J. Howell

English: San Diego Mesa mint

P. clareana J. Howell

English: Santa Lucia mint

P. floribunda Jokerst

English: profuse flowered pogogyne

P. nudiuscula A. Gray

English: Otay Mesa mint

Pogonachne Bor Gramineae

Origins:
From the Greek *pogon* "beard" and *achne* "chaff, glume."

Pogonanthera (G. Don) Spach Goodeniaceae

Origins:
Greek *pogon* and *anthera* "anther"; see Édouard Spach (1801-1879), *Histoire naturelle des Végétaux*. Phanérogames. 9: 583. Paris (Aug.) 1838.

Pogonanthera Blume Melastomataceae

Origins:
Greek *pogon* "beard" and *anthera* "anther," referring to the nature of the anthers.

Pogonanthus Montrouz. Rubiaceae

Origins:
From the Greek *pogon* "beard" and *anthos* "flower."

Pogonarthria Stapf Gramineae

Origins:
From the Greek *pogon* "beard" and *arthron* "a joint."

Species/Vernacular Names:
P. squarrosa (Roemer & Schultes) Pilg.

English: cross grass, herringbone grass, sickle grass

Southern Africa: pluimsekelgras, sekelgras; lefieloane (Sotho); lefheto (Tswana)

Pogonatherum Beauv. Gramineae

Origins:
From the Greek *pogon* "a beard" and *ather* "awn," referring to the glumes; see Ambroise Marie François Joseph Palisot de Beauvois (1752-1820), *Essai d'une nouvelle Agrostographie*. 56. Paris (Dec.) 1812.

Species/Vernacular Names:
P. crinitum (Thunb.) Kunth (*Andropogon crinitum* Thunb.)

English: bamboo grass

Japan: itachi-gaya

Okinawa: hâmei-kûgii

China: bi zi cao

Pogonetes Lindley Goodeniaceae

Origins:
Greek *pogoniates, pogonites* "bearded."

Pogonia A.L. Juss. Orchidaceae (Pogoniinae)

Origins:
Greek *pogonias* "bearded," Latin *pogonias* "bearded, a kind of comet," an allusion to the fringed lip; see A.L. de Jussieu, *Genera Plantarum*. 65. (Aug.) 1789.

Species/Vernacular Names:
P. ophioglossoides (L.) Ker Gawler
English: beard flower
China: chu lan

Pogonia Andrews Myoporaceae

Origins:
Greek and Latin *pogonias* "bearded."

Pogoniopsis Reichb.f. Orchidaceae (Pogoniinae)

Origins:
Resembling *Pogonia* Juss.

Pogonochloa C.E. Hubb. Gramineae

Origins:
Greek *pogon* "beard" and *chloe, chloa* "grass."

Pogonolepis Steetz Asteraceae

Origins:
Greek *pogon* and *lepis* "a scale," referring to the apices of the bracts; see Johann G.C. Lehmann (1792-1860), *Plantae Preissianae*. 1: 440. Hamburgi (Aug.) 1845.

Species/Vernacular Names:
P. muelleriana (Sonder) P. Short (*Skirrhophorus muellerianus* Sonder; *Angianthus strictus* sensu J. Black)
English: stiff cup-flower, stiff angianthus

Pogonolobus F. Muell. Rubiaceae

Origins:
Greek *pogon* "beard" and *lobos* "a pod"; see Ferdinand von Mueller, *Fragmenta Phytographiae Australiae*. 1: 55. Melbourne (Jul.) 1858.

Pogononeura Napper Gramineae

Origins:
From the Greek *pogon* "beard" and *neuron* "nerve."

Pogonophora Miers ex Benth. Euphorbiaceae

Origins:
From the Greek *pogon* and *phoros* "bearing, carrying."

Pogonophyllum Didr. Euphorbiaceae

Origins:
From the Greek *pogon* "beard" and *phyllon* "leaf."

Pogonopsis C. Presl Gramineae

Origins:
Greek *pogon* "beard" and *opsis* "resembling."

Pogonopus Klotzsch Rubiaceae

Origins:
Greek *pogon* "beard" and *pous* "foot," referring to the shape of the flowers.

Species/Vernacular Names:
P. tubulosus (DC.) Schumann
Bolivia: quina morada, quina, tumparopea

Pogonorrhinum Betsche Scrophulariaceae (Antirr.)

Origins:
Greek *pogon* "beard" and *rhis, rhinos* "snout, nose," the genus *Antirrhinum* L.

Pogonospermum Hochst. Acanthaceae

Origins:
Greek *pogon* "beard" and *sperma* "seed."

Pogonotium J. Dransf. Palmae

Origins:
Greek *pogonias* "bearded" and *ous, otos* "an ear," leaves pinnate, sheath tubular, terminating in two auricles, narrow and erect.

Pogonotrophe Miq. Moraceae

Origins:
Greek *pogon* "beard" and *trophe* "food."

Pogostemon Desf. Labiatae

Origins:
Greek *pogon* "beard" and *stemon* "stamen, thread," alluding to the filaments; see Ferdinand von Mueller, *Fragmenta Phytographiae Australiae*. 1: 55. Melbourne (Jul.) 1858.

Species/Vernacular Names:
P. auricularius (L.) Hasskarl
English: auriculate pogostemon
China: shui zhen zhu cai
P. cablin (Blanco) Bentham
English: patchouly, patchouli, Cablin patchouli
China: huo xiang, guang huo xiang
Vietnam: hoac huong
The Philippines: cablin, cadling, cadlom, kabling, kadling, pacholi
P. brevicorollus Sun ex C.H. Hu
English: short-corolla pogostemon
China: duan guan ci rui cao
P. championii Prain
English: Champion pogostemon
China: duan sui ci rui cao
P. chinensis C.Y. Wu & Y.C. Huang
English: Chinese pogostemon
China: chang bao ci rui cao
P. dielsianus Dunn
English: Diels pogostemon
China: xia ye ci rui cao
P. esquirolii (H. Léveillé) C.Y. Wu & Y.C. Huang
English: Esquirol pogostemon
China: mo ye ci rui cao
P. formosanus Oliver
English: Taiwan pogostemon
China: tai wan ci rui cao
P. glaber Bentham
English: glabrous pogostemon
China: ci rui cao
P. heyneanus Bentham
English: Indian patchouly, Indian patchouli, pachouli
Malaya: nilam bukit, rumput kuku, nilam

P. menthoides Blume
English: mint-like pogostemon
China: xiao ci rui cao
P. septentrionalis C.Y. Wu & Y.C. Huang
English: northern pogostemon
China: bei ci rui cao

Pohliella Engler Podostemaceae

Origins:
For the botanical artist Joseph Pohl, 1864-1939, illustrated many works by Adolf Engler.

Poicilla Griseb. Asclepiadaceae

Origins:
From the Greek *poikillo* "embroider, diversify."

Poicillopsis Schltr. ex Rendle Asclepiadaceae

Origins:
Resembling *Poicilla* Griseb.

Poidium Nees Gramineae

Origins:
Referring to the genus *Poa* L.

Poikilacanthus Lindau Acanthaceae

Origins:
Greek *poikilos* "spotted, many-colored" plus *Acanthus* or *akantha* "thorn."

Poikilogyne Baker f. Melastomataceae

Origins:
From the Greek *poikilos* and *gyne* "a woman, female organs."

Poikilospermum Zipp. ex Miq. Cecropiaceae

Origins:
Greek *poikilos* "spotted, many-colored, varied, variegated" and *sperma* "seed."

Poilanedora Gagnep. Capparidaceae (Capparaceae)

Origins:

For Eugène Poilane, 1887/1888-1964, a plant collector; see I.H. Vegter, *Index Herbariorum*. Part II (5), *Collectors N-R*. Regnum Vegetabile vol. 109. 1983; Clyde F. Reed, *Bibliography to Floras of Southeast Asia*. 146. [Dedicated to E. Poilane and M. Poilane.] Baltimore, Maryland 1969.

Poilaniella Gagnepain Euphorbiaceae

Origins:

For Eugène Poilane, 1887/88-1964, a plant collector. See I.H. Vegter, *Index Herbariorum*. Part II (5), *Collectors N-R*. Regnum Vegetabile vol. 109. 1983; Clyde F. Reed, *Bibliography to Floras of Southeast Asia*. 146. [Dedicated to E. Poilane and M. Poilane.] Baltimore, Maryland 1969.

Poilannammia C. Hansen Melastomataceae

Origins:

For Eugène Poilane, 1887/88-1964, a plant collector (or M. Poilane?). See I.H. Vegter, *Index Herbariorum*. Part II (5), *Collectors N-R*. Regnum Vegetabile vol. 109. 1983; Clyde F. Reed, *Bibliography to Floras of Southeast Asia*. 146. [Dedicated to E. Poilane and M. Poilane.] Baltimore, Maryland 1969.

Poinciana L. Caesalpiniaceae

Origins:

Possibly named for Louis de Poinci (Poincy), French Governor in the West Indies, a patron of botany, author of *Histoire naturelle et morale des Iles Antilles de l'Amérique* ... avec un vocabulaire Caraïbe. Roterdam 1658.

Poincianella Britton & Rose Caesalpiniaceae

Origins:

The diminutive of *Poinciana*.

Poinsettia Graham Euphorbiaceae

Origins:

After the diplomat of South Carolina Joel Roberts Poinsett, 1775-1851, botanist, gardener, the first American Ambassador to Mexico. His works include *Notes on Mexico, Made in the Autumn of 1822* by a citizen of the United States. Philadelphia 1824, *Discursos pronunciados en la Camera de Representantes de los Estados-Unidos de America*. Mexico 1829 and *Notes on Mexico, Made in the Autumn of 1822*, accompanied by an historical sketch of the Revolution, and translations of official reports on the present state of that country. London 1825.

Poiretia Cavanilles Epacridaceae

Origins:

For the French botanist Jean Louis Marie Poiret, 1755-1834.

Poiretia J.F. Gmelin Rubiaceae

Origins:

After the French botanist Jean Louis Marie Poiret, 1755-1834, clergyman.

Poiretia Vent. Fabaceae

Origins:

After the French botanist Jean Louis Marie Poiret, 1755-1834, clergyman, from 1785 to 1786 traveled in North Africa. Among his works are *Voyage en Barbarie*; ou lettres écrites de l'ancienne Numidie pendant les années 1785 et 1786, etc. Paris 1789, *Leçons de flore*. Cours complet de Botanique, etc. Paris 1819-1820 and *Histoire philosophique, littérarire, économiques des plantes de l'Europe*. Paris 1825-1829. See J.H. Barnhart, *Biographical Notes upon Botanists*. 3: 95. 1965; Stafleu and Cowan, *Taxonomic Literature*. 4: 319-321. 1983; François Pierre Chaumeton (1775-1819), *Flore médicale*. [Vols. 3-6: Jean Baptiste Jos. César Tyrbas de Chamberet and Jean L.M. Poiret] Paris 1814-1818; Henri Louis Duhamel du Monceau (1700-1782), *Traité des arbres et arbustes*. Seconde édition ... augmentée [by J.L.M. Poiret, etc.] [1800, etc.]; J.E. Smith, in *Transactions of the Linnean Society of London. Botany*. 9: 304. (Nov.) 1808; R. Zander, F. Encke, G. Buchheim and S. Seybold, *Handwörterbuch der Pflanzennamen*. 14. Aufl. 1993; T.W. Bossert, *Biographical Dictionary of Botanists Represented in the Hunt Institute Portrait Collection*. 313. 1972; S. Lenley et al., *Catalog of the Manuscript and Archival Collections and Index to the Correspondence of John Torrey*. Library of the New York Botanical Garden. 466. 1973; E.M. Tucker, *Catalogue of the Library of the Arnold Arboretum of Harvard University*. 1917-1933; A. Lasègue, *Musée botanique de Benjamin Delessert*. Paris 1845; Ida Kaplan Langman, *A Selected Guide to the Literature on the Flowering Plants of Mexico*. Philadelphia

1964; Emil Bretschneider, *History of European Botanical Discoveries in China.* [Reprint of the original edition 1898.] Leipzig 1981.

Poissonella Pierre Sapotaceae

Origins:

For the French botanist Jules Poisson, 1833-1919.

Poissonia Baillon Fabaceae

Origins:

For the French botanist Jules Poisson, 1833-1919, from 1843 at the Muséum d'Histoire Naturelle de Paris, wrote *Recherches sur les Casuarina.* Paris 1876; see John H. Barnhart, *Biographical Notes upon Botanists.* 3: 95. 1965; R. Zander, F. Encke, G. Buchheim and S. Seybold, *Handwörterbuch der Pflanzennamen.* 14. Aufl. 1993; Ethelyn Maria Tucker, *Catalogue of the Library of the Arnold Arboretum of Harvard University.* Cambridge, Massachusetts 1917-1933; Elmer Drew Merrill, in *Contr. U.S. Natl. Herb.* 30(1): 243. 1947 and *Bernice P. Bishop Mus. Bull.* 144: 151-152. 1937; Ida Kaplan Langman, *A Selected Guide to the Literature on the Flowering Plants of Mexico.* 589. Philadelphia 1964.

Poivrea Comm. ex DC. Combretaceae

Origins:

For the French (b. Lyons) botanist Pierre Poivre, 1719-1786 (d. near Lyons), naturalist, plant collector, traveler (China, Indochina, Moluccas, Philippines, Madagascar, Réunion, Île de France or Mauritius). After two visits to the Île de France he was appointed Intendant (Administrative Officer, in charge of finance) of the island in 1767, in 1770 he took the opportunity to purchase the estate of *Mon Plaisir* for himself, he was the creator of the present gardens of Pamplemousses, he gathered numerous plants from other lands (Africa, India, Malaysia and Polynesia) together with as many indigenous plant species as he could, he introduced the famous *Piper nigrum* and cultivated nutmeg and clove, he was the author of *Voyage d'un Philosophe,* ou observations sur les moeurs ... des peuples de l'Afrique, de l'Asie, etc. 1768 and *De l'Amérique et des Américains.* [Attributed to P. Poivre.] 1771. See Paul Jovet & J.C. Mallet, in *Dictionary of Scientific Biography* 11: 64-65. 1981; T.W. Bossert, *Biographical Dictionary of Botanists Represented in the Hunt Institute Portrait Collection.* 313. 1972; A. Lasègue, *Musée botanique de Benjamin Delessert.* Paris 1845; Madeleine Ly-Tio-Fane, *Mauritius and the Spice Trade. The Odyssey of Pierre Poivre.* Port Louis, Mauritius 1954; Madeleine Ly-Tio-Fane, "Pierre Poivre et l'Expansion

française dans l'Indo-Pacifique." *Bulletin de l'École Française d'Extrême Orient.* 53: 453-511. 1967; Louis Malleret, "Pierre Poivre, l'abbé Galloys et l'introduction d'espèces botaniques et d'oiseaux de Chine à l'Ile Maurice." *Société Royale des Arts et des Sciences.* vol. III, part I. 1968; L. Malleret, *Pierre Poivre.* Paris 1974; J.H. Barnhart, *Biographical Notes upon Botanists.* 1965; Auguste Toussaint and H. Adolphe, *Bibliography of Mauritius, 1502-1954.* Port Louis 1956; Auguste Toussaint, ed., *Dictionnaire de biographie Mauricienne.* [Société de l'Histoire de l'Ile Maurice. Publication no. 2] [Port Louis] 1941-; A.W. Owadally, *A Guide to the Royal Botanic Gardens, Pamplemousses.* 5, 10. Port Louis 1976; Guy Rouillard [with the collaboration of Joseph Guého], *Le Jardin des Pamplemousses 1729-1979. Histoire et Botanique.* Les Pailles 1983; Emil Bretschneider, *History of European Botanical Discoveries in China.* Leipzig 1981; Frans A. Stafleu, *Linnaeus and the Linnaeans.* The spreading of their ideas in systematic botany, 1735-1789. Utrecht 1971.

Pokornya Montrouzier Combretaceae

Origins:

For the Austrian botanist Alois Pokorny, 1826-1886, phytogeographer, naturalist; see J.H. Barnhart, *Biographical Notes upon Botanists.* 3: 95. 1965; T.W. Bossert, *Biographical Dictionary of Botanists Represented in the Hunt Institute Portrait Collection.* 313. 1972; E.M. Tucker, *Catalogue of the Library of the Arnold Arboretum of Harvard University.* 1917-1933; Stafleu and Cowan, *Taxonomic Literature.* 4: 325-327. 1983.

Polakowskia Pittier Cucurbitaceae

Origins:

For the German botanist Helmuth Polakowsky, 1847-1917, traveler, botanical collector, bryologist, author of *La flora de Costa Rica.* San José de Costa Rica 1890 and *Der Chilisalpeter und die Zukunft der Salpeterindustrie.* Berlin 1893. See J.H. Barnhart, *Biographical Notes upon Botanists.* 3: 96. 1965; Ethelyn Maria Tucker, *Catalogue of the Library of the Arnold Arboretum of Harvard University.* Cambridge, Massachusetts 1917-1933; G. Murray, *History of the Collections Contained in the Natural History Departments of the British Museum.* 1: 174. London 1904.

Polanina Raf. Capparidaceae (Capparaceae)

Origins:

See Constantine Samuel Rafinesque, *Western Rev. Miscel. Mag.* 2: 93. 1819.

Polanisia Raf. Capparidaceae (Capparaceae)

Origins:

Greek *polys* "many" and *anisos* (*a* "negative" and *isos* "equal") "unequal," referring to the numbers of the stamens; see Constantine Samuel Rafinesque (1783-1840), *Florula ludoviciana*. 167. New York 1817, *Am. Monthly Mag. Crit. Rev.* 2: 267. 1818 and 4: 188, 207. 1819, *Am. Jour. Sci.* 1: 378. 1819, and in *Journal de Physique, de Chimie, d'Histoire Naturelle et des Arts*. 89: 98. (Jul.) 1819; E.D. Merrill, *Index rafinesquianus*. The plant names published by C.S. Rafinesque, etc. 130-131. 1949.

Polanysia Raf. Capparidaceae (Capparaceae)

Origins:

See Constantine Samuel Rafinesque, *Am. Monthly Mag. Crit. Rev.* 4: 194. 1819.

Polaskia Backeb. Cactaceae

Origins:

Dedicated to an American cacti lover, Mr. Polaski, see F. Boerner & G. Kunkel, *Taschenwörterbuch der botanischen Pflanzennamen*. 4. Aufl. 154. Berlin & Hamburg 1989.

Polemannia Ecklon & Zeyher Umbelliferae

Origins:

After the German chemist Peter Heinrich Poleman (Polemann, Pohlmann), *circa* 1780-1839 (d. Cape Town), apothecary, naturalist, friend of the German physician and naturalist Martin H. Carl (Karl) Lichtenstein (1780-1857); see Mary Gunn and Leslie E. Codd, *Botanical Exploration of Southern Africa*. 285. 1981.

Polemanniopsis B.L. Burtt Umbelliferae

Origins:

Resembling *Polemannia*.

Polemoniella A. Heller Polemoniaceae

Origins:

Diminutive of *Polemonium* L.

Polemonium L. Polemoniaceae

Origins:

Greek *polemonion*, ancient name of a plant, perhaps a species of *Hypericum*, Latin *polemonia, ae* for the Greek valerian, otherwise called *philetaeria* and *philetaeris* (Plinius),

Greek *philetairion*; see Salvatore Battaglia, *Grande dizionario della lingua italiana*. XIII: 730-731. UTET, Torino 1986; Helmut Genaust, *Etymologisches Wörterbuch der botanischen Pflanzennamen*. 496. 1996; Georg Christian Wittstein, *Etymologisch-botanisches Handwörterbuch*. 713. 1852; G. Semerano, *Le origini della cultura europea*. Dizionario della lingua Greca. 2(1): 237. 1994.

Species/Vernacular Names:

P. caeruleum L.

English: Jacob's ladder, Greek valerien, common polemonium

China: hua ren

Italian: polemonio

P. chartaceum H. Mason

English: Mason's sky pilot

P. chinense (Brand) Brand var. *chinense* (*Polemonium liniflorum* Vassil.)

English: Chinese polemonium

China: zhong hua hua ren

P. eximium E. Greene

English: sky pilot

Polevansia de Winter Gramineae

Origins:

After the Welsh (b. near Cardiff, Wales) botanist Illtyd (Iltyd) Buller Pole-Evans, 1879-1968 (d. Umtali, Rhodesia), mycologist, plant collector, traveler, 1907 Fellow of the Linnean Society, 1918-1939 Director Botanical Survey S. Africa, 1921-1939 founder and editor of *Bothalia* (A record of contributions from the National Herbarium, Union of South Africa, Pretoria), editor of volumes 1-19 of *The Flowering Plants of South Africa* (A magazine containing hand-colored figures with descriptions of the flowering plants indigenous to South Africa, London, Johannesburg, and Cape Town. Superseded by *Flowering Plants of Africa*, Pretoria and Ashford). His writings include "A vegetation map of South Africa." *Mem. Bot. Surv. S. Afr.* 15. 1936 and "A reconnaissance trip through the eastern portion of the Bechuanaland Protectorate, April 1931." *Mem. Bot. Surv. S. Afr.* 21: 5-73. 1948; see Mary Gunn and Leslie E. Codd, *Botanical Exploration of Southern Africa*. 283-285. 1981; T.W. Bossert, *Biographical Dictionary of Botanists Represented in the Hunt Institute Portrait Collection*. 314. 1972; A. White and B.L. Sloane, *The Stapelieae*. Pasadena 1937; Gordon Douglas Rowley, *A History of Succulent Plants*. 1997; Ray Desmond, *Dictionary of British & Irish Botanists and Horticulturists*. 557. 1994; I.C. Hedge and J.M. Lamond, *Index of Collectors in the Edinburgh Herbarium*. Edinburgh

1970; Gilbert Westacott Reynolds (1895-1967), *The Aloes of South Africa*. Balkema, Rotterdam 1982.

Polhillia C.H. Stirton Fabaceae

Origins:
For Roger Marcus Polhill, botanist at Kew Gardens, expert in legumes.

Polianthes L. Agavaceae

Origins:
From the Greek *polios* "white, whitish, grey" and *anthos* "a flower."

Species/Vernacular Names:
P. tuberosa L.
English: tuberose
Mexico: omixochitl (= flor doble o de hueso), guia chilla, quije chilla
Japan: chuberosa, gekka-kô
India: sandhyaraga, gulcheri, gulshabba, rajanigandha, undi-mandare, sukandaraji, nelasampenga, virusampenga, nilasampangi, andi-mallery, gulsabo
Malaya: sandarmalam, sundal malam
Italian: tuberosa

Poliomintha A. Gray Labiatae

Origins:
Greek *polios* "white, whitish, grey" and *minthe* "mint."

Poliophyton O.E. Schulz Brassicaceae

Origins:
Greek *polios* "white, whitish, grey" and *phyton* "a plant."

Poliothyrsis Oliver Flacourtiaceae

Origins:
Greek *polios* and *thyrsos* "a panicle," referring to the whitish panicles.

Pollia Thunb. Commelinaceae

Origins:
The genus was named for Jan van der (de) Poll, Dutch Consul; see C.P. Thunberg, *Nova genera Plantarum.* (Genera nova Plantarum) 11. Upsaliae [Uppsala] (Nov.) 1781.

Species/Vernacular Names:
P. crispata (R. Br.) Benth.
English: pollia
P. japonica Thunb.
Japan: yabu-myôga
China: zhu ye lian, tu jo, tu heng

Pollichia Aiton Caryophyllaceae (Illecebraceae)

Origins:
No etymology is given, genus possibly named after the German physician Johann Adam Pollich, 1740-1780, botanist, author of *Historia plantarum in Palatinatu electorali sponte nascentium*. Mannhemii [Mannheim] 1776-1777 and *Dissertatio ... de nutrimento incremento statu ac decremento corporis humani*, etc. Argentorati [Strasbourg, 1763]; see John H. Barnhart, *Biographical Notes upon Botanists*. 3: 96. 1965; T.H. Arnold & B.C. de Wet, eds., *Plants of Southern Africa: Names and Distribution*. National Botanical Institute. Memoirs of the Botanical Survey of South Africa, no. 62. 305. Pretoria 1993; Jonas C. Dryander, *Catalogus bibliothecae historico-naturalis Josephi Banks*. London 1800; R. Zander, F. Encke, G. Buchheim and S. Seybold, *Handwörterbuch der Pflanzennamen*. 14. Aufl. 1993.

Species/Vernacular Names:
P. campestris Ait.
South Africa: kafferdruiwe

Pollichia Medikus Boraginaceae

Origins:
For the German physician Johann Adam Pollich, 1740-1780, botanist, naturalist, author of *Historia plantarum in Palatinatu electorali sponte nascentium*. Mannhemii [Mannheim] 1776-1777 and *Dissertatio ... de nutrimento incremento statu ac decremento corporis humani*, etc. Argentorati [Strasbourg, 1763]. See John H. Barnhart, *Biographical Notes upon Botanists*. 3: 96. 1965; Jonas C. Dryander, *Catalogus bibliothecae historico-naturalis Josephi Banks*. London 1800; R. Zander, F. Encke, G. Buchheim and S. Seybold, *Handwörterbuch der Pflanzennamen*. 14. Aufl. 1993.

Pollinia Sprengel Gramineae

Origins:
After the Italian botanist Ciro Pollini, 1782-1833, physician; see Mariella Azzarello Di Misa, a cura di, *Il Fondo*

Antico della Biblioteca dell'Orto Botanico di Palermo. 218. Regione Siciliana, Palermo 1988.

Pollinia Trinius Gramineae

Origins:

For the Italian botanist Ciro (Cyrus, Siro) Pollini, 1782-1833, physician, lichenologist, bryologist. Among his publications are *Sulle alghe viventi nelle terme Euganee.* Milano 1817, *Flora veronensis.* Veronae 1822-1824, *Elementi di botanica.* Verona 1810-1811, *Discorso istorico sulla Botanica* recitato ... 6 Aprile 1811. Verona 1812, *Catechismo agrario,* coronato dell'Accademia d'Agricoltura ... di Verona. Verona 1807, *Sinonimia botanica moderna,* ossia elenco dei diversi nomi attribuiti dagli autori moderni alle piante. Milano 1805 and *Catalogo delle piante dell'orto botanico veronese per l'anno 1814.* Verona 1814. See G. Moretti, *Intorno alla Flora veronensis.* Milano 1822; John H. Barnhart, *Biographical Notes upon Botanists.* 3: 96. 1965; T.W. Bossert, *Biographical Dictionary of Botanists Represented in the Hunt Institute Portrait Collection.* 314. 1972; E.M. Tucker, *Catalogue of the Library of the Arnold Arboretum of Harvard University.* 1917-1933; Mariella Azzarello Di Misa, a cura di, *Il Fondo Antico della Biblioteca dell'Orto Botanico di Palermo.* 218. Regione Siciliana, Palermo 1988.

Pollinidium Haines Gramineae

Origins:

Resembling *Pollinia.*

Polliniopsis Hayata Gramineae

Origins:

Resembling *Pollinia.*

Pollinirhiza Dulac Orchidaceae

Origins:

From the genus *Pollinia* (a genus of grasses) and Greek *rhiza* "root," or to honor the Italian botanist Ciro Pollini, 1782-1833; see Richard Evans Schultes and Arthur Stanley Pease, *Generic Names of Orchids. Their Origin and Meaning.* 250. 1963.

Polpoda C. Presl Molluginaceae

Origins:

Perhaps from the Greek *polys* "many" and *pous, podos* "a foot," referring to the fringed stipules and perianth segments.

Polyachyrus Lagasca Asteraceae

Origins:

Greek *polys* "many" and *achyron* "chaff, husk," perhaps referring to the ovary.

Polyalthia Blume Annonaceae

Origins:

Greek *polys* and *althos* "a cure, something that heals, a healing," *althaimo* "to heal," referring to its use in native and popular medicine; see Karl Ludwig von Blume (1796-1862), *Flora Javae.* Anonaceae. 68. Bruxelles 1830.

Species/Vernacular Names:

P. longifolia (Sonn.) Thw.

English: Indian willow

India: devadaru, debdari, devidari, asoke, asok, asokan, assothi, asogu, netlingi, asokamu

P. michaelii C.T. White

English: China pine

P. nitidissima (Dunal) Benth.

English: Canary beech, China pine

Polyandra Leal Euphorbiaceae

Origins:

Greek *polys* "many" and *aner, andros* "male, stamen," with many stamens.

Polyandrococos Barb. Rodr. Palmae

Origins:

Greek *polys* "many," *aner, andros* "male, stamen" plus *Cocos,* stamen numerous, approximately 60 to 100, the inflorescence at anthesis is entirely covered by stamens.

Species/Vernacular Names:

P. caudescens (Mart.) Barb. Rodr.

Brazil: buri, palmito amargoso, palmito da folha prateada, palha branca

Polyantherix Nees Gramineae

Origins:

From the Greek *polys* "many" and *antherix, antherikos* "beard of an ear of corn, the ear itself."

Polyanthina R.M. King & H. Robinson Asteraceae

Origins:
Greek *polys* "many" and *anthos* "flower."

Polyanthus C.H. Hu & Y.C. Hu Gramineae

Origins:
From the Greek *polys* "many" and *anthos* "flower."

Polyarrhena Cass. Asteraceae

Origins:
Greek *polys* "many" and *arrhen* "male."

Polyaulax Backer Annonaceae

Origins:
Greek *polys* "many" and *aulax, aulakos* "a furrow"; see Cornelis Andries Backer (1874-1963), in *Blumea*. 5: 492. (Dec.) 1945.

Polybactrum Salisb. Orchidaceae

Origins:
Greek *polys* "many" and *baktron* "a cane, a walking staff."

Polyboea Klotzsch Ericaceae

Origins:
Presumably from the Greek *polyboeios* "covered with many ox hides," some suggest from the Greek *polys* and the genus *Boea* Comm. ex Lam., or from the Greek *baios* "little, insignificant."

Polyboea Klotzsch ex Endl. Euphorbiaceae

Origins:
Presumably from the Greek *polyboeios* "covered with many ox hides," some suggest from the Greek *polys* and the genus *Boea* Comm. ex Lam., or from the Greek *baios* "little, insignificant," referring to the flowers.

Polybotrya Humb. & Bonpland ex Willd. Dryopteridaceae (Aspleniaceae)

Origins:
Greek *polybotrys* "abounding in grapes," referring to the clusters of sporangia.

Polycalymma F. Muell. & Sonder Asteraceae

Origins:
Greek *polys* and *kalymma* "a covering, hood," see also *Myriocephalus*; see O.W. Sonder, in *Linnaea*. 25: 494. (Apr.) 1853.

Species/Vernacular Names:
P. stuartii F. Muell. & Sond.
English: poached-egg daisy, ham and eggs daisy

Polycampium C. Presl Polypodiaceae

Origins:
Perhaps from the Greek *polykampes* "with many curves."

Polycardia Juss. Celastraceae

Origins:
From the Greek *polys* "many" and *kardia* "heart," referring to the leaves.

Polycarena Benth. Scrophulariaceae

Origins:
Greek *polykarenos* "many-headed," *polys* and *karena, karenon, kare* "the head, a mountain-top, the highest part," referring to the terminal clusters of flowers; see W.J. Hooker, *Companion to the Botanical Magazine*. 1: 371. (Jul.) 1836.

Polycarpa L. Caryophyllaceae

Origins:
Orthographic variant of *Polycarpon* L.; see C. Linnaeus, *Systema Naturae*. Ed. 10. 859. (May-June) 1759.

Polycarpaea Lam. Caryophyllaceae

Origins:
Greek *polys* "many" and *karpos* "fruit," *polykarpos* "fruitful, rich in fruit"; see Jean Baptiste Antoine Pierre de Monnet de Lamarck (1744-1829), in *Journal d'Histoire Naturelle*. 2: 3, t. 25. Paris (Jul.) 1792.

Species/Vernacular Names:
P. linearifolia (DC.) DC.
Yoruba: amoritanna, eyin ire, onigba efun

P. spirostylis F. Muell.

English: copper plant

Polycarpon L. Caryophyllaceae

Origins:
Greek *polys* and *karpos* "fruit," Latin *polycarpos* for a plant, also called *polygonous*; see C. Linnaeus, *Systema Naturae*. Ed. 10. 881, 1360. (May-June) 1759.

Species/Vernacular Names:
P. tetraphyllum (L.) L. (*Mollugo tetraphylla* L.; *Polycarpon diphyllum* Cav.)

English: four-leaved allseed, allseed, four-leaf allseed

South Africa: naaldvrug

Polycenia Choisy Selaginaceae

Origins:
Greek *polykenos* "containing much void, porous," *polys* and *kenos* "empty, hollow," referring to the fruit.

Polycephalium Engl. Icacinaceae

Origins:
From the Greek *polys* "many, much" and *kephale* "head."

Polyceratocarpus Engl. & Diels Annonaceae

Origins:
Greek *polys* "many," *keras* "a horn" and *karpos* "a fruit," "many-horned fruit."

Species/Vernacular Names:
P. parviflorus (Bak.f.) Ghesq.

Nigeria: akorsor (Edo)

Polychilos Breda Orchidaceae

Origins:
From the Greek *polys* "many" and *cheilos* "a lip," referring to the lobed lip.

Polychrysum (Tzvelev) Kovalesk. Asteraceae

Origins:
Greek *polychrysos* "rich in gold."

Polyclathra Bertol. Cucurbitaceae

Origins:
Greek *polys* "many" and *klathron, klethron, kleithron* "gate, a lock," *kleio* "to shut, enclose."

Polyclita A.C. Sm. Ericaceae

Origins:
Polyclitus, Polykleitos, was a Greek sculptor contemporary with Pericles; Greek *polykleitos* "far-famed."

Polycnemum L. Chenopodiaceae

Origins:
Greek *polys* and *kneme* "limb, leg," referring to the numerous branches, Greek *polyknemos* means "mountainous, with many mountain-spurs" but is also applied to the field basil, a species of *Ziziphora* L. or *Zizyphora* (Labiatae), Latin *polycnemon* used for a plant otherwise unknown (Plinius); see Carl Linnaeus (1707-1778), *Species Plantarum*. 35. 1753 and *Genera Plantarum*. Ed. 5. 22. 1754.

Polycocca Hill Selaginellaceae

Origins:
From the Greek *polys* "many" and *kokkos* "berry."

Polycoryne Keay Rubiaceae

Origins:
From the Greek *polys* "many" and *koryne* "a club," see also *Pleiocoryne* Rauschert.

Polyctenium Greene Brassicaceae

Origins:
Greek *polys* "many" and *kteis, ktenos* "a comb," alluding to the leaves.

Polycycliska Ridl. Rubiaceae

Origins:
Greek *polys* "many" and *kykliskos* "small circle," *kyklos* "globe, circle, round about, ring."

Polycycnis Reichb.f. Orchidaceae

Origins:
Greek *polys* "many" and *kyknos* "a swan," referring to the lip and column, indicating the resemblance of the flowers to swans.

Polydiclis (G. Don) Miers Solanaceae

Origins:
From the Greek *polys* "many" and *diklis* "double-folding, two-valved."

Polydictyum C. Presl Dryopteridaceae (Aspleniaceae)

Origins:
Greek *polys* "many" and *diktyon* "a net."

Polydragma Hook.f. Euphorbiaceae

Origins:
From the Greek *polys* "many" and *dragma, dragmatos* "handful, uncut corn."

Polygala L. Polygalaceae

Origins:
Greek *polygalon*, ancient name used by Dioscorides, *polys* "much" and *gala* "milk," Latin *polygala, ae* for the herb milkwort (Plinius); see Carl Linnaeus (Carl von Linnaeus, Carl von Linné), *Species Plantarum*. 701. 1753 and *Genera Plantarum*. Ed. 5. 315. 1754.

Species/Vernacular Names:
P. spp.
English: milkwort herb

Mexico: polígala, flor de María; caluante (Totonaca l., El Tajín, Veracruz); chupac, ishpun (Tzotzil l., Simojobel, see Dennis E. Breedlove and Robert M. Laughlin, *The Flowering of Man. A Tzoltzil Botany of Zinacantán*. Smithsonian Contributions to Anthropology. Number 35. Washington 1993); flor de rosario (Jericó, south Tuxtla Gutierrez)

P. arvensis Willd.
English: field milkwort

P. chinensis L.
English: leafy milkwort, Chinese milkwort, Indian senega
Japan: Shinchiku-hime-hagi

China: da jin niu cao

P. cowellii (Britt.) S.F. Blake
English: violet tree
Puerto Rico: violeta, tortuguero

P. heterorhyncha (Barneby) T. Wendt
English: notch-beaked milkwort

P. hottentotta Presl (*Polygala abyssinica* sensu Gibbs)
Southern Africa: lehlokoa-la-tsela (Sotho)

P. japonica Houtt. (*Polygala sibirica* var. *japonica* (Houtt.) T. Itô)
English: milkwort, dwarf milkwort, Japanese milkwort
Japan: hime-hagi
China: gua zi jin

P. longifolia Poir. (*Polygala riukiuensis* Ohwi)
Japan: Riûkyû-hime-hagi

P. lutea L.
English: yellow milkwort, candyweed, yellow bachelor's-button

P. myrtifolia L. (*Polygala pinifolia* Poir.)
English: september bush, parrot bush, myrtle-leaved milkwort, myrtle-leaf milkwort
South Africa: septemberbossie, augusbossie, blouertjie-boom, langelier, langelede (= lange lede= long joints)
Mexico: polígala de jardines

P. paucifolia Willd.
English: flowering wintergreen, bird-on-the-wing, fringed polygala, gay-wings

P. senega L.
English: senega root
Brazil: senega colubrina, poligala-da-virgínia, poligala, senega

P. sibirica L.
English: Siberian milkwort, Siberian senega
China: tian yuan zhi, yuan chih, yao jao, hsiao tsao

P. subspinosa S. Watson
English: spiny milkwort

P. tenuifolia Willd.
English: thinleaf milkwort, Chinese senega
China: yuan zhi

P. virgata Thunb.
English: pride of Manicaland
Southern Africa: furambuku (Shona)

P. virgata Thunb. var. *virgata*
English: purple broom
South Africa: bloukappie, blaauwkappie

P. vulgaris L.

English: gang flower, gand flower, milkwort

Polygaloides Haller Polygalaceae

Origins:

Resembling *Polygala.*

Polygonanthus Ducke Anisophylleaceae (Rosidae, Rosales)

Origins:

Greek *polygonatos* "having many joints" and *anthos* "flower."

Polygonatum Miller Convallariaceae (Liliaceae)

Origins:

Greek *polys* and *gony* "the knee-joint," referring to the rhizomes, Latin *polygonaton, i* for the plant called Solomon's seal, *Convallaria polygonatum* L., another name for *leucacantha* (Plinius).

Species/Vernacular Names:

P. biflorum (Walter) Elliott (*Polygonatum canaliculatum* (Muhlenb.) Pursh; *Polygonatum giganteum* Dietr.)

China: huang ching

P. multiflorum (L.) All.

China: huang ching

P. odoratum (Mill.) Druce (*Polygonatum officinale* All.)

English: aromatic Solomon's seal

China: yu zhu, yu chu, wei jui

P. sibiricum Delaroche

English: Siberian Solomon's seal

China: huang jing

P. verticillatum (L.) All.

Nepal: setak chini

Polygonella Michx. Polygonaceae

Origins:

Referring to the genus *Polygonum.*

Polygonum L. Polygonaceae

Origins:

Greek *polygonon, polygonos*, referring to the many-jointed or swollen stems, Latin *polygonos, polygonus, polygonium* or *polygonon*, for a plant, *herba sanguinalis* or *sanguinaria* (Plinius), knotgrass; see Dioscorides (ed. M. Wellmann, Berlin 1907-14); Carl Linnaeus, *Species Plantarum.* 359. 1753 and *Genera Plantarum.* Ed. 5. 170. 1754.

Species/Vernacular Names:

P. alpinum All.

English: alpine knotweed

P. amphibium L.

English: willow grass, water smartweed

China: tien liao

P. arenastrum Boreau

English: wire weed, common knotweed, doorweed

P. aubertii L. Henry

English: Russian vine, Chine fleece vine, silver lace vine, climbing knotweed

Portuguese: manto de noiva

P. aviculare L.

English: knotweed, knotgrass, bird knotgrass, bird knotweed, door yard knotweed, hogweed, mat grass, prostrate knotweed, wire weed, centinode, goose grass

Arabic: gordhab, gerda, qordab, qoddab

Japan: michi-yanagi

China: bian xu, pien hsu, fen chieh tsao

East Africa: chonge

Southern Africa: duisendknoop, kamferfoelie, knopgras, koperdraadgras, lidjiesgras, varkgras, voëlduisendknoop; lira-ha-li-bonoe (Sotho)

P. baldschuanicum Regel

English: mile-a-minute vine, Russian vine

P. barbatum L.

English: bearded knotweed

The Philippines: subsuban, sigan-lupa, bukakau, kanubsuban, kaykayu, saimbangan tubig

P. bistorta L.

English: bistort, snakeweed, Easter ledges, adderwort, dragonwort

China: quan shen, chuan shen, tzu shen, mou meng

P. bistortoides Pursh

English: western bistort

P. campanulatum Hook.f.

English: lesser knotweed

Tibetan: snyalo

P. capitatum Buch.-Ham. ex D. Don

English: garden knotweed

P. coccineum Muhlenb. ex Willd.

English: water smartweed

P. convolvulus L. (*Bilderdykia convolvulus* (L.) Dumort.)

English: black bindweed, climbing buckwheat

Japan: soba-kazura

P. cuspidatum Siebold & Zucc.

English: Japanese knotweed

China: hu zhang, hu chang, pan chang

P. dibotrys D. Don (*Polygonum cymosum* Trev.)

English: Chinese smartweed

China: huo tan mu cao

P. fagopyrum L.

English: buckwheat

P. glabrum Willd. (*Persicaria densiflora* sensu Degener)

Hawaii: kamole

P. hydropiper L.

English: marsh pepper, water pepper

China: shui liao, yu liao, tse liao

P. hydropiperoides Michaux

English: water pepper

P. japonicum Meissn. (*Persicaria japonica* (Meissn.) Nakai; *Polygonum cuspidatum* Sieb. & Zucc.)

English: Japanese knotweed, Mexican bamboo

Japan: shiro-bana-sakura-tade, ita-dori

China: tsan chien tsao

P. lapathifolium L.

English: willow weed

P. marinense T. Mert. & Raven

English: marin knotweed

P. multiflorum Thunb. (*Pleuropterus multiflorus* (Thunb.) Turcz.; *Reynoutria multiflora* (Thunb.) Moldenke)

English: fleece flower, Chinese cornbind, climbing knotweed

Japan: tsuru-dokudami

Okinawa: kashu

China: he shou wu, ho shou wu

Vietnam: da giao dang

P. orientale L.

English: prince's feather, princess feather, kiss-me-over-the-garden-gate

China: hong cao, hung tsao

P. persicaria L.

English: lady's thumb

China: ma liao, ta liao

P. plebeium R. Br.

English: small knotweed

Arabic: qoutteiah

P. polystachyum Wallich

English: Himalayan knotweed

Tibetan: chu-ma-tsi, tsi stag-mo

P. punctatum Schmidt (*Persicaria hydropiper* sensu Degener & I. Degener)

English: water smartweed

P. sachalinense Maxim.

English: giant knotweed, Sacaline

P. tinctorium Ait. (*Persicaria tinctoria* (Aiton) Spach)

English: Chinese indigo

Japan: ai-tade, ai

China: lan shi, liao lan

P. virginianum L. (*Polygonum filiforme* Thunb.)

China: chin ssu tsao, chin hsien tsao

P. viviparum L.

English: alpine bistort, serpent grass

Tibetan: na-ram

Polylepis Ruíz & Pavón Rosaceae

Origins:
Greek *polys* "many" and *lepis* "scale."

Polylophium Boiss. Umbelliferae

Origins:
Greek *polys* "many" and *lophos* "a crest."

Polylychnis Bremek. Acanthaceae

Origins:
From the Greek *polys* "many" and *lychnis, lychnidos* diminutive of *lychnos* "a lamp."

Polymeria R. Br. Convolvulaceae

Origins:
From the Greek *polys* "many" and *meris* "a part, portion," referring to the branched style or to the branched plants; see Robert Brown, *Prodromus florae Novae Hollandiae et Insulae van-Diemen.* 488. London 1810.

Species/Vernacular Names:

P. calycina R. Br.

English: swamp bindweed, bindweed

P. longifolia Lindley

English: erect bindweed, polymeria

Polymita N.E. Br. Aizoaceae

Origins:

Greek *polymitos, polymiton* "consisting of many threads, damask stuffs, damask robes," Latin *polymita, orum* "damask."

Polymnia L. Asteraceae

Origins:

Polymnia or Polyhymnia, one of the Muses, she of the many hymns; see National Research Council, *Lost Crops of the Incas: Little-Known Plants of the Andes with Promise for Worldwide Cultivation.* National Academy Press, Washington, D.C. 1989.

Polyneura Peter Gramineae

Origins:

Greek *polys* "many" and *neuron* "nerve," ancient Greek *polyneuron* or *arnoglosson* for a species of *Plantago*, Latin *polyneuron* for a plant, a species of *Plantago*, great plantain.

Polyodon Kunth Gramineae

Origins:

From the Greek *polys* "many" and *odous, odontos* "a tooth."

Polyosma Blume Grossulariaceae (Escalloniaceae, Polyosmaceae)

Origins:

Greek *polys* and *osme* "smell, odor, perfume," referring to the strong fragrance; see Karl Ludwig von Blume, *Bijdragen tot de flora van Nederlandsch Indië.* 658. Batavia (Jan.) 1826.

Species/Vernacular Names:

P. alangiaceae F. Muell.

English: white alder

P. cunninghamii Bennett

English: featherwood

Polyotidium Garay Orchidaceae

Origins:

Greek *polyotos* "many-eared," referring to the auricles on the column.

Polyozus Lour. Rubiaceae

Origins:

Greek *polys* "many" and *ozos* "branch, knot."

Polyphlebium Copeland Hymenophyllaceae

Origins:

Greek *polys* "many" and *phleps, phlebos* "vein," having many veins; see Edwin Bingham Copeland (1873-1964), in *Philippine Journal of Science.* 67: 55. 1938.

Species/Vernacular Names:

P. venosum (R. Br.) Copel.

English: veined bristle fern

Polyphragmon Desf. Rubiaceae

Origins:

Greek *polys* "many" and *phragmon* "a thorn-hedge," *phragma* "a partition, compartment, wall, screen," *phrasso* "to enclose, fence in"; see René (Renatus) Louiche Desfontaines (1750-1833), in *Mémoires du Muséum d'Histoire Naturelle.* 6: 5, t. 2. 1820.

Polypleurella Engl. Podostemaceae

Origins:

The diminutive of the genus *Polypleurum*.

Polypleurum (Tul.) Warm. Podostemaceae

Origins:

Greek *polys* "many" and *pleura, pleuro, pleuron* "side, rib."

Polypodiastrum Ching Polypodiaceae

Origins:

Referring to the genus *Polypodium*.

Polypodioides Ching Polypodiaceae

Origins:
Resembling the genus *Polypodium*; see *Acta Phytotax. Sin.* 16(4): 26. 1978.

Polypodiopsis Polypodiaceae

Origins:
Resembling the genus *Polypodium.*

Polypodiopteris C.F. Reed Polypodiaceae

Origins:
Greek *polypodion* plus *pteris* "a fern."

Polypodium L. Polypodiaceae

Origins:
Greek *polypodion, polys* "many" and *podion* "little foot," *pous, podos* "a foot," referring to the scars on the rhizomes; Plinius applied Latin *polypodium (-ion)* to a kind of fern, polypody; see Carl Linnaeus, *Species Plantarum.* 1082. 1753 and *Genera Plantarum.* Ed. 5. 485. 1754.

Species/Vernacular Names:
P. scouleri Hooker & Greville
English: coast polypody, leather polypody
P. vulgare L.
English: common polypody, adder's fern, wall fern, golden maidenhair, European polypody
China: shui long gu

Polypogon Desf. Gramineae

Origins:
Greek *polys* and *pogon* "a beard," much bearded, the panicles are hairy, bristly; see René Louiche Desfontaines, *Flora atlantica,* sive Historia plantarum, quae in Atlante, agro Tunetano et Algeriensi crescunt. 1: 66. Paris (Apr.) 1798.

Species/Vernacular Names:
P. australis Brongn.
English: Chilean beard grass
P. fugax Steudel
Japan: hie-gaeri
P. interruptus Kunth
English: ditch beard grass
P. littoralis Smith

English: perennial beard grass
P. maritimus Willd.
English: Mediterranean beard grass
P. monspeliensis (L.) Desf. (*Alopecurus monspeliensis* L.)
English: annual beard grass, beard grass, rabbit's foot grass, rabbit's foot, Montpellier beard grass
Arabic: deil el-qott
South Africa: brakbaardgras, brakgras
Japan: hama-hie-gaeri
P. viridis (Gouan) Breistr. (*Agrostis semiverticillata* (Forssk.) C. Christ.; *Agrostis viridis* Gouan; *Agrostis verticillata* Villars; *Polypogon semiverticillatus* (Forssk.) Hyl.; *Phalaris semiverticillata* Forssk.)
English: bent grass, water bent grass, water bent

Polypompholyx Lehm. Lentibulariaceae

Origins:
Greek *polys* "many" and *pompholyx, pompholygos* "a bubble," alluding to the small bladders; see Johann G.C. Lehmann (1792-1860), in *Botanische Zeitung.* 2: 109. (Feb.) 1844.

Polyporandra Becc. Icacinaceae

Origins:
Greek *polys* "many," *poros* "opening, pore" and *aner, andros* "male, stamen."

Polypremum Adanson Valerianaceae

Origins:
Greek *polys* "many" and *premnon* "trunk, stem."

Polypremum L. Loganiaceae (Scrophulariales, Buddlejaceae?)

Origins:
Greek *polys* "many" and *premnon* "the stump of a tree, trunk, stem"; see Carl Linnaeus, *Species Plantarum.* 111. 1753.

Polypsecadium O.E. Schulz Brassicaceae

Origins:
Probably from the Greek *polys* "many" and *psekadion, psakadion,* diminutive of *psekas, psakas* "drizzle, drop of rain," *psakazo, psekazo* "rain in small drops, drizzle, drip,"

Latin *psecas, adis* for a female slave who perfumed her mistress's hair.

Polypteris Nutt. Asteraceae

Origins:

Greek *polys* and *pteron* "wing, feather," *polypteros* "many winged," some suggest from *pteris* "fern."

Polyradicion Garay Orchidaceae

Origins:

Latin *polys* "many" and *radix, radicis* "root," gen. plur. *radicium*, referring to the numerous elongate chlorophyllous roots emerging from the stem of these leafless epiphytes orchids; Greek *rhadix, rhadikos* "branch, frond."

Polyraphis (Trin.) Lindl. Gramineae

Origins:

From the Greek *polys* "many, much" and *raphis* "a needle."

Polyrhabda C.C. Towns. Amaranthaceae

Origins:

From the Greek *polys* "many, much" and *rhabdos* "a rod, stick, a magic wand."

Polyrrhiza Pfitzer Orchidaceae

Origins:

Greek *polys* and *rhiza* "a root," Latin adj. *polyrrhizos* "having many roots" (an appellation of several plants, Plinius), subst. *polyrrhizon* for a plant otherwise unknown (Plinius), see also *Polyradicion* Garay.

Polyschistis C. Presl Gramineae

Origins:

From the Greek *polys* "many, much" and *schistos* "cut, divided."

Polyscias Forster & Forster f. Araliaceae

Origins:

Greek *polys* "many" and *skias* "a canopy, pavilion," *skia* "shade, shadow," *polyskios* "very shady," in reference to the foliage or to the umbel of flowers; see J.R. Forster & J.G.

Forster, *Characteres generum plantarum*. 63, t. 32. (Nov.) 1775.

Species/Vernacular Names:

P. sp.

Malaya: puding

P. elegans (C. Moore & F. Muell.) Harms

English: celery wood, Mowbulan whitewood, silver basswood, black pencil cedar

P. farinosa (Del.) Harms (*Aralia farinosa* Del.; *Panax pinnatum* A. Rich.)

English: parasol tree

P. filicifolia (C. Moore ex Fourn.) L.H. Bail., "Chotito"

English: angelica, fern-leaf aralia, fern-leaf polyscias

P. fruticosa (L.) Harms (*Panax fruticosum* L.; *Nothopanax fruticosum* (L.) Miq.)

English: Ming aralia, India polyscias, aralia

Japan: Taiwan momiji

The Philippines: papua, bani, makan

Vietnam: cay goi ca

P. fulva (Hiern) Harms (*Polyscias ferrugineum* Harms; *Polyscias ferruginea* (Hiern) Harms; *Polyscias preussii* Harms; *Panax fulvum* Hiern; *Panax nigericum* A. Chev.)

English: parasol tree

Cameroon: nkogoe, nkoguele, ndongue

P. guilfoylei (Cogn. & March.) L.H. Bail. (*Aralia guilfoylei* Cogn. & March.) (after the Australian botanist William Robert Guilfoyle, 1840-1912, former Director Melbourne Botanic and Domain Gardens, author of *Australian Plants Suitable for Gardens, Parks, Timber Reserves*, etc. Melbourne [1911] and *First Book. Australian Botany: Specially Designed for the Use of Schools*. Melbourne 1878)

English: wild coffee, coffee tree, geranium-leaf aralia, geranium aralia, Guilfoyle polyscias

Sierra Leone: anjelika (Krio)

Japan: arariya

P. malosana Harms

Southern Africa: mupahambgwa, mutengambia, mutengembeya (Shona)

P. murrayi (F. Muell.) Harms

English: pencil cedar, umbrella tree, white basswood, chinky pine

P. nodosa (Bl.) Seem.

English: wild papaya

Malaysia: bingliu, rangit

P. sambucifolius (DC.) Harms

English: elderberry panax, elderberry ash

Polysolen Rauschert Rubiaceae

Origins:

From the Greek *polys* and *solen* "a tube," see also *Indopolysolenia* Bennet.

Polysolenia Hook.f. Rubiaceae

Origins:

Greek *polys* "many" and *solen* "a tube," see also *Indopolysolenia* Bennet.

Polyspatha Benth. Commelinaceae

Origins:

From the Greek *polys* "many" and *spathe* "spathe."

Polysphaeria Hook.f. Rubiaceae

Origins:

Greek *polys* "many" and *sphaira* "a globe, ball."

Polystachya Hooker Orchidaceae

Origins:

Greek *polys* "many" and *stachys* "a spike," the inflorescence is single and terminal usually with short, flowered branches.

Polystemma Decne. Asclepiadaceae

Origins:

Greek *polys* and *stemma* "a crown, garland."

Polystemon D. Don Cunoniaceae

Origins:

From the Greek *polys* "many" and *stemon* "stamen."

Polystemonanthus Harms Caesalpiniaceae

Origins:

Greek *polys* "many," *stemon* "stamen" and *anthos* "flower."

Polystichopsis (J. Sm.) Holttum Dryopteridaceae (Aspleniaceae)

Origins:

Resembling the genus *Polystichum* Roth.

Polystichum Roth Dryopteridaceae (Aspleniaceae)

Origins:

Greek *polys* "many" and *stichos* "a row, series," the sori in many rows; see Albrecht Wilhelm Roth (1757-1834), *Tentamen Florae Germanicae.* 3: 31, 69. Lipsiae (Jun.-Sep.) 1799; Charles Gaudichaud-Beaupré (1789-1854), [Botany of the Voyage.] *Voyage autour du Monde ... sur ... l'Uranie et la Physicienne, pendant ... 1817-1820.* 330. Paris 1826.

Species/Vernacular Names:
P. chilense (Christ) Diels var. *chilense*
Chile: pelomén-lahuén
P. falcinellum Moore
English: shield fern
P. fallax Tindale
English: rock shield fern
P. formosanum Rosenst. (*Polystichum iriomotense* Tagawa)
Japan: ô-mimi-gata-shida (= large ear-shaped fern)
P. formosum Tindale
English: broad shield fern
P. microchlamys (H. Christ) Matsumura (*Aspidium microchlamys* H. Christ)
English: sword fern
P. munitum (Kaulfuss) C. Presl (*Aspidium munitum* Kaulfuss)
English: common sword fern
P. proliferum (R. Br.) C. Presl (*Aspidium proliferum* R. Br.; *Polystichum aculeatum* sensu J. Black)
English: common sword fern, mother shield fern

Polystylus Hasselt ex Hasskarl Orchidaceae

Origins:

Greek *polys* and *stylos* "style, column," lip with appendage.

Polytaenia DC. Umbelliferae

Origins:

From the Greek *polys* "many" and *tainia* "fillet."

Polytaenium Desv. Vittariaceae (Adiantaceae)

Origins:

Greek *polys* and *tainia* "fillet," Latin *taenia, ae* "a band, ribbon, fillet, a head-band."

Polytaxis Bunge Asteraceae

Origins:

Greek *polys* "many" and *taxis* "a series, order, arrangement."

Polytepalum Suess. & Beyerle Caryophyllaceae

Origins:

From the Latin *polys* "many" and *tepalum, tepali* "tepal."

Polythrix Nees Acanthaceae

Origins:

From the Greek *polys* "many" and *thrix, trichos* "hair."

Polythysania Hanst. Gesneriaceae

Origins:

Greek *polys* "many" and *thysanos* "a fringe, tassel."

Polytoca R. Br. Gramineae

Origins:

Greek *polys* "many" and *tokos* "a birth," *polytokia* "fecundity," referring to the numerous offsprings; see John Joseph Bennett (1801-1876) and Robert Brown, *Plantae Javanicae rariores.* 20. London (Jul.) 1838.

Polytoma Loureiro ex Gomes Orchidaceae

Origins:

Greek *polys* "many" and *tome, tomos, temno* "division, section, to slice," indicating the crested lip.

Polytrema C.B. Clarke Acanthaceae

Origins:

Greek *polys* "many" and *trema* "hole, aperture."

Polytrias Hackel Gramineae

Origins:

Greek *polys* "many" and *treis, tria* "three"; see H.G.A. Engler & K.A.E. Prantl (1849-1893), *Die Natürlichen Pflanzenfamilien.* 2(2): 24. (Jul.) 1887.

Species/Vernacular Names:
P. indica (Houtt.) Veldk.
English: Java grass

Polyura Hook.f. Rubiaceae

Origins:

From the Greek *polys* "many" and *oura* "tail."

Polyxena Kunth Hyacinthaceae (Liliaceae)

Origins:

Dedicated to the daughter of Priam and Hecuba.

Polyzone Endlicher Myrtaceae

Origins:

Greek *polys* and *zone* "a belt, armor, girdle," Latin *polyzonos* for a kind of black precious stone with many stripes; see Stephan F. Ladislaus Endlicher, *Stirpium Australasicarum herbarii Hügeliani decades tres.* Vindobonae 1838.

Polyzygus Dalzell Umbelliferae

Origins:

Greek *polys* "many" and *zygon, zygos* "yoke," *polyzygos* "many-benched."

Pomaderris Labill. Rhamnaceae

Origins:

Greek *poma* "a lid" and *derris* "a skin," referring to the membranous valve or to the membranous covering of the capsule; see Jacques Julien Houtton de Labillardière (1755-1834), *Novae Hollandiae plantarum specimen.* 1: 61. Parisiis (Jul.) 1805.

Species/Vernacular Names:
P. andromedifolia Cunn.

English: Andromeda pomaderris

P. angustifolia Wakef.

English: narrow-leaf pomaderris

P. argyrophylla Wakef.

English: silvery pomaderris

P. aspera DC.

English: hazel pomaderris, rough hazel pomaderris

P. aurea Wakef.

English: golden pomaderris

P. betulina Hook.

English: birch pomaderris

P. brunnea Wakef.

English: brown pomaderris

P. cinerea Benth.

English: white pomaderris

P. costata Wakef.

English: veined pomaderris

P. cotoneaster Wakef.

English: cotoneaster pomaderris

P. discolor (Vent.) Poiret

English: eastern pomaderris

P. elachophylla F. Muell.

English: lacy pomaderris, small leaf pomaderris

P. elliptica Labill.

English: oval-leaf pomaderris

New Zealand: poverty weed, gumdigger's soap

Maori names: papapa, kumarahou

P. eriocephala Wakef.

English: woolly-head pomaderris

P. ferruginea Fenzl

English: rusty pomaderris

P. flabellaris (Reissek) J. Black (*Trymalium flabellare* Reissek)

Australia: fan pomaderris

P. kumeraho A. Cunn.

English: poverty weed, gumdigger's soap

Maori names: papapa, kumarahou

P. halmaturina J. Black

Australia: Kangaroo Island pomaderris

P. helianthemifolia (Reissek) Wakef.

English: blunt-leaf pomaderris

P. lanigera (Andrews) Sims

English: woolly pomaderris

P. ledifolia Cunn.

Australia: New South Wales pomaderris

P. ligustrina DC.

English: privet pomaderris

P. nitidula (Benth.) Wakef.

English: shining pomaderris

P. notata S.T. Blake

English: cream pomaderris

P. obcordata Fenzl

English: wedge-leaf pomaderris, pimelea pomaderris

P. oraria F. Muell. ex Reissek (*Pomaderris paniculosa* F. Muell. ex Reissek)

Australia: coast pomaderris, forest pomaderris

P. pallida Wakef.

English: pale pomaderris

P. pauciflora Wakef.

English: mountain pomaderris

P. phylicifolia Link

English: slender pomaderris

P. pilifera Wakef.

English: striped pomaderris

P. prunifolia Fenzl

English: prunus pomaderris

P. queenslandica C. White

English: Queensland pomaderris

P. racemosa Wakef.

English: cluster pomaderris

P. sericea Wakef.

English: silky pomaderris, bent pomaderris

P. subcapitata Wakef.

English: convex pomaderris

P. vacciniifolia Reissek

English: round-leaf pomaderris

P. vellea Wakef.

English: glaucous pomaderris

P. velutina J.H. Willis

English: velvet pomaderris

Pomangium Reinw. Rubiaceae

Origins:

Greek *poma* "a lid, a cover" and *angeion, aggeion* "a vessel, cup."

Pomasterion Miq. Cucurbitaceae

Origins:

Probably from the Greek *poma* "lid, cover, operculum," *pomasteon* "one must cover up," *stereos* "solid, firm, tight."

Pomatium C.F. Gaertner Rubiaceae

Origins:

Greek *poma, pomatos* "lid, operculum," *pomation* "little lid."

Pomatocalpa Breda Orchidaceae

Origins:

Greek *poma, pomatos* "a lid, a cover" and *kalpis* "an urn, jug, a drinking cup," referring to the labellum, to the urceolate lip; see Jacques Gijsbert Samuel van Breda (1788-1867), Heinrich Kuhl (1796-1821), and Johan Coenraad van Hasselt (1797-1823), *Genera et species Orchidearum et Asclepiadearum.* Gent (Aug.) 1829; Alick William Dockrill (1915-), *Australasian Sarcanthinae.* 1-41. 1967 and *Australian Indigenous Orchids.* 1: 1-826. Sydney 1969.

Pomatoderris Schultes Rhamnaceae

Origins:

Greek *poma, pomatos* "a lid, a cover" and *derris* "a skin, a leather covering, leather coat"; see Johann Jakob Roemer (1763-1819) and Josef August Schultes (1773-1831), *Systema Vegetabilium.* Stuttgardtiae (Dec.) 1819.

Pomatophytum M.E. Jones Pteridaceae (Adiantaceae)

Origins:

Greek *poma, pomatos* "a lid, a cover" and *phyton* "plant."

Pomatosace Maxim. Primulaceae

Origins:

Greek *poma, pomatos* "a lid, a cover" and *sakos* "a shield."

Species/Vernacular Names:
P. filicula Maxim.

English: common pomatosace

China: yu ye dian di mei

Pomatostoma Stapf Melastomataceae

Origins:

Greek *poma, pomatos* "a lid, a cover" and *stoma* "mouth."

Pomatotheca F. Muell. Aizoaceae

Origins:

Greek *poma, pomatos* and *theke* "a box, case, capsule"; see F. von Mueller, *Systematic Census of Australian Plants.* 31. 1888.

Pomax Solander ex DC. Rubiaceae

Origins:

Greek *poma* "a lid" and *axon* "axis," referring to the lid by which the fruits dehisce.

Species/Vernacular Names:
P. umbellata (Gaertn.) Benth. (*Opercularia umbellata* Gaertner)

English: pomax

Pomazota Ridley Rubiaceae

Origins:

Greek *pomazo* "furnish with a lid, cover up, seal, cover" and *ous, otos* "an ear."

Pomelia Durando ex Pomel Umbelliferae

Origins:

For the French botanist Auguste Nicolas Pomel, 1821-1898, geologist, in Algeria. His works include *Le Sahara.* Algiers 1872 and *Nouveau Guide de Géologie.* Paris 1869; see John H. Barnhart, *Biographical Notes upon Botanists.* 3: 96. 1965; Ethelyn Maria Tucker, *Catalogue of the Library of the Arnold Arboretum of Harvard University.* Cambridge, Massachusetts 1917-1933; R. Zander, F. Encke, G. Buchheim and S. Seybold, *Handwörterbuch der Pflanzennamen.* 14. Aufl. 1993; in Palermo (Sicily, Italy) *pomelia* is the vernacular name for *Plumeria.*

Pometia Forst. & Forst.f. Sapindaceae

Origins:

After the French botanist Pierre Pomet, 1658-1699, merchant, apothecary to the French Court, druggist, author of *Droguier curieux,* ou catalogue des drogues simples et composées, mis par alphabet. Seconde édition ... corrigée. Paris 1709 and *Histoire générale des drogues,* traitant des plantes, des animaux et des mineraux. [The most comprehensive medical and botanical account of drugs of its time.] Paris 1694.

Species/Vernacular Names:
P. pinnata Forst. & Forst.f.

English: pinnate pometia

Malaya: kasai, langsir, kelisar, asam kuang

Ponapea Beccari Palmae

Origins:
From Ponape, an island in the Carolines.

Ponceletia Thouars Gramineae

Origins:
After Abbé Poncelet or Père Polycarpe or Polycarpe Poncelet, born in Verdun, author of *Histoire naturelle du Froment.* Paris 1779, *La Nature dans la Formation du Tonnerre.* Paris 1766, *Chimie du Goût et de l'Odorat.* Paris 1755 and *Nouvelle Chymie du goût et de l'odorat.* 1800.

Poncirus Raf. Rutaceae

Origins:
From *poncire*, the French name for a species of citron; see C.S. Rafinesque, *Sylva Telluriana.* 143. 1838.

Species/Vernacular Names:
P. trifoliata (L.) Raf.

English: trifoliate orange, bitter orange, Japanese bitter orange

Japan: karatachi

China: zhi shi, gou ju

Ponera Lindley Orchidaceae

Origins:
Greek *poneros* "miserable, worthless, good-for-nothing," referring to the small flowers.

Ponerorchis Reichb.f. Orchidaceae

Origins:
Greek *poneros* "miserable, worthless, good-for-nothing" and *orchis* "orchid," referring to the nature and appearance of the plants.

Pongamia Vent. Fabaceae

Origins:
Pongam is the vernacular Malabar name for *Pongamia pinnata.* See van Rheede in *Hortus Indicus Malabaricus.* 6: t. 3. 1686; Étienne Pierre Ventenat (1757-1808), *Jardin de la Malmaison.* 28, t. 28. Paris (Dec.) 1803.

Species/Vernacular Names:
P. pinnata (L.) Pierre (*Cytisus pinnatus* L.; *Derris indica* (Lam.) Bennet; *Galedupa indica* L.; *Galedupa pinnata* (L.) Taubert; *Pongamia glabra* Vent. nom. illeg.; *Pongamia mitis* Kurz, nom. illeg.)

English: karum tree, pongam, poonga oil tree, Indian beech, thinwin, pongamia

Japan: kuro-yona, okaha, ohbaki

Tibetan: ma-nu shu-zur

China: shui liu dou

Malaya: seashore mempari, mempari, malapari, jador, kachang kayu laut, pari pari, biangsu

India: karanj, karanja, gaanuga, punga, ponga, pongam, honge, pungu, punnu, koranjo, sukhchein, paphri, karchaw

Pongamiopsis R. Viguier Fabaceae

Origins:
Resembling *Pongamia* Vent.

Pongelia Raf. Bignoniaceae

Origins:
See Constantine S. Rafinesque, *Sylva Telluriana.* 78. 1838; E.D. Merrill, *Index rafinesquianus.* The plant names published by C.S. Rafinesque, etc. 220. Jamaica Plain, Massachusetts, USA 1949.

Pontederia L. Pontederiaceae

Origins:
Named for the Italian (b. Vicenza) botanist Giulio Pontedera, 1688-1757 (d. Lonigo, near Vicenza), physician, professor of botany at Padua, plant collector, from 1719 to 1757 Praefectus of the Botanical Garden of Padua. His works include J. Pontederae ... *Compendium tabularum botanicarum*, in quo plantae 272 ab eo in Italia nuper detectae recensentur. Patavii 1718, *J. Pontederae Anthologia,* sive de floris natura libri tres. Patavii 1720 and *Antiquitatum Latinarum Graecarumque* enarrationes atque emendationes, etc. Patavii 1740; see Gunnar Eriksson, in *Dictionary of*

Scientific Biography 11: 83. 1981; A. Béguinot, "Giulio Pontedera." in A. Mieli, ed. *Gli scienziati italiani.* 1: 93-94. 1921; F. Coletti, *Ricordi storici della cattedra e del gabinetto di Materia Medica nella Università di Padova.* Padova 1871; C. Cappelletti, *L'Orto Botanico di Padova.* Roma 1963; C. Gola, *L'Orto Botanico. Quattro secoli di attività (1545-1945).* Padova 1947; I.H. Vegter, *Index Herbariorum.* Part II (5), *Collectors N-R.* Regnum Vegetabile vol. 109. 1983; Carl Linnaeus, *Species Plantarum.* 288. 1753 and *Genera Plantarum.* Ed. 5. 140. 1754; Georg Christian Wittstein, *Etymologisch-botanisches Handwörterbuch.* 720f. Ansbach 1852.

Species/Vernacular Names:

P. cordata L. (*Pontederia lanceolata* Nutt.)

English: pickerel weed

Ponthieva R. Br. Orchidaceae

Origins:

After a French West Indian merchant, Henri de Ponthieu, in 1778 he sent plants to Sir Joseph Banks.

Poponax Raf. Mimosaceae

Origins:

See C.S. Rafinesque, *Sylva Telluriana.* 118. 1838; E.D. Merrill, *Index rafinesquianus.* 147. Jamaica Plain, Massachusetts, USA 1949.

Popoviocodonia Fedorov Campanulaceae

Origins:

For the Russian botanist Mikhail Grigor'evic Popov, 1893-1955, botanical explorer; see S. Lenley et al., *Catalog of the Manuscript and Archival Collections and Index to the Correspondence of John Torrey.* Library of the New York Botanical Garden. 333. 1973; R. Zander, F. Encke, G. Buchheim and S. Seybold, *Handwörterbuch der Pflanzennamen.* 14. Aufl. 1993; I.C. Hedge and J.M. Lamond, *Index of Collectors in the Edinburgh Herbarium.* Edinburgh 1970.

Popowia Endl. Annonaceae

Origins:

For Johannes Siegmund Valentin Popowitsch, 1705-1774, professor of botany in Vienna, author of *Versuch einer Vereinigung der Mundarten in Deutschland.* Wien 1780, see also Christian Gottlieb Schwarz, *Untersuchungen vom Meere.*

Frankfurt und Leipzig 1750, and S.L. Endlicher, *Genera Plantarum.* 831. (Jun.) 1839.

Populina Baillon Acanthaceae

Origins:

Diminutive of the Latin *populus, i* "poplar."

Populus L. Salicaceae

Origins:

Latin *populus* or *popplum* "poplar," Greek *pelea, ptelea, apellon* "black poplar"; Akkadian *papallu* "shoot, bud, sprout," *apellum, alpu, ablu* "son," Greek *apellon*; see Carl Linnaeus, *Species Plantarum.* 1034. 1753 and *Genera Plantarum.* Ed. 5. 447. 1754; E. Weekley, *An Etymological Dictionary of Modern English.* 2: 1125. New York 1967; G. Semerano, *Le origini della cultura europea.* Dizionario della lingua Latina e di voci moderne. 2(2): 524. Firenze 1994.

Species/Vernacular Names:

P. spp.

Mexico: yaga gueza castilla

P. alba L.

English: abele, silver-leaved poplar, white poplar

Southern Africa: popoliri (Sotho)

India: safeda

Tibetan: magal

China: bai bei yang, pai yang

P. angustifolia James

English: narrowleaf cottonwood, willow-leaved poplar, narrow-leaved cottonwood

P. balsamifera L.

English: balsam poplar, hackmatack, tacamahac

China: hai tung, tzu tung

P. x *canadensis* Moench

English: Canadian poplar, hybrid poplar

Nepal: Lahare pipal

P. canescens Smith

English: grey poplar

Southern Africa: gryspopulier, papelierboom, poplier, populier, vaalpopulier; popoliri (Sotho)

P. deltoides Bartr. ex Marshall

English: broad-leaved poplar, Carolina poplar, cottonwood, eastern cottonwood, match poplar, necklace poplar

Southern Africa: vuurhoutjiepopulier; popoliri (Sotho)

P. deltoides Bartr. ex Marshall subsp. *wislizenii* (S. Wats.) Eckenw.

English: poplar, valley cottonwood

P. fremontii S. Watson

English: cottonwood

P. grandidentata Michx.

English: big-toothed aspen, Canadian aspen

P. lasiocarpa Oliv.

English: Chinese necklace poplar

P. laurifolia Ledeb.

English: laurel poplar

P. maximowiczii Henry

English: Japanese poplar

P. nigra L.

English: black poplar, Italian poplar

Arabic: safsaf, asafsaf

P. nigra L. var. *italica* Duroi

English: black poplar, Lombardy poplar, pyramidal poplar, Italian poplar

Japan: Amerika-yama-narashi, popura

Southern Africa: Italiaanse populier, Lombardy populier, regop populier; popoliri (Sotho)

P. sieboldii Miq.

English: Japanese aspen

P. x *tomentosa* Carr.

English: Chinese white poplar

P. tremula L.

English: aspen, quaking aspen

China: i yang, fu i, tang ti, chang ti

P. tremuloides Michaux

English: American aspen, quaking aspen

P. trichocarpa Torrey & A. Gray

English: black cottonwood, western balsam poplar

Porana Burm.f. Convolvulaceae

Origins:
Derivation obscure, probably from the Marathi (Bhauri) name for *Porana paniculata* Roxb., *poranana*; some suggest from the Greek *poreno* "to travel, carry over," referring to the stems; see Nicolaas Laurens (Nicolaus Laurent) Burman (1734-1793), *Flora Indica*. 51. Lugduni Batavorum (Mar.-Apr.) 1768; G.W. Staples, "The genus *Porana* (Convolvulaceae) in Australia." *Nuytsia*. 6(1): 51-59. 1987.

Porandra D.Y. Hong Commelinaceae

Origins:
Greek *poros* "opening, pore" and *aner, andros* "man, stamen."

Poranopsis Roberty Convolvulaceae

Origins:
Resembling *Porana*.

Species/Vernacular Names:
P. paniculata (Roxburgh) Roberty (*Porana paniculata* Roxburgh)

English: bridal bouquet, Christmas-vine porana

China: yuan zhui bai hua ye

P. sinensis (Hand.-Mazz.) Staples (*Porana henryi* Verdcourt)

China: bai hua ye

Poranthera Raf. Gramineae

Origins:
Greek *poros* "opening, pore" and *anthera* "anther"; see C.S. Rafinesque, *Seringe Bull. Bot.* 1: 221. 1830; E.D. Merrill, *Index rafinesquianus*. 76. 1949.

Poranthera Rudge Euphorbiaceae

Origins:
Greek *poros* "opening, pore" and *anthera* "anther," Latin *porus, i* "a pore, a passage in the body," referring to the terminal pores of the anther cells; see Edward Rudge, in *Transactions of the Linnean Society of London. Botany.* 10: 302. 1811.

Species/Vernacular Names:
P. corymbosa Rudge

English: clustered poranthera

P. ericifolia Rudge

English: heath poranthera, heath-leaved poranthera

P. microphylla Brongn.

English: small-leaved poranthera, small poranthera

Porcelia Ruíz & Pav. Annonaceae

Origins:
Possibly for a friend of Ruíz and Pavón, Antonio Porcel.

Porlieria Ruíz & Pav. Zygophyllaceae

Origins:
After the Spanish Ambassador Don Antonio (Antón) Porlier de Baxamar, a promoter of botany.

Porocarpus Gaertn. Rubiaceae

Origins:
From the Greek *poros* "opening" and *karpos* "fruit."

Porocystis Radlk. Sapindaceae

Origins:
From the Greek *poros* "opening" and *kystis* "a bladder, pouch."

Porodittia G. Don f. ex Kraenzlin Scrophulariaceae

Origins:
Greek *poros* "opening" and *dittos* "double."

Porolabium T. Tang & F.T. Wang Orchidaceae

Origins:
Latin *porus, i* "a passage" and *labium* "lip," Greek *poros* "pore," pores near the base of the lip.

Porophyllum Adans. Asteraceae

Origins:
Greek *poros* "opening, pore" and *phyllon* "leaf," referring to the appearance of the gland-dotted leaves.

Species/Vernacular Names:
P. gracile Benth.

Mexico: odora; jestej, hestej (Chihuahua); hierba del venado (Baja California)

Porospermum F. Muell. Araliaceae

Origins:
Greek *poros* "opening, pore" and *sperma* "seed"; see F. von Mueller, *Fragmenta Phytographiae Australiae.* 7: 94. (Apr.) 1870.

Porpax Lindley Orchidaceae

Origins:
Greek *porpax, porpakos* "the handle of a shield," referring to the form of the flowers, or to the oblong leaves or to the pseudobulbs.

Porphyranthus Engl. Pandaceae

Origins:
Greek *porphyra* "purple" and *anthos* "flower."

Porphyrocodon Hook.f. Brassicaceae

Origins:
From the Greek *porphyra* and *kodon* "a bell."

Porphyrocoma Scheidw. ex Hook. Acanthaceae

Origins:
From the Greek *porphyra* "purple" and *kome* "hair," referring to the inflorescence.

Porphyrodesme Schltr. Orchidaceae

Origins:
Greek *porphyra* and *desmis, desmos* "a bond, band, bundle," the inflorescence.

Porphyroglottis Ridley Orchidaceae

Origins:
Greek *porphyra* "purple" and *glotta* "tongue."

Porphyroscias Miq. Umbelliferae

Origins:
Greek *porphyra* and *skias* "a canopy, umbel."

Porphyrospatha Engl. Araceae

Origins:
From the Greek *porphyra* "purple" and *spatha* "spathe."

Porphyrostachys Reichb.f. Orchidaceae

Origins:

Greek *porphyra* "purple" and *stachys* "a spike," indicating the inflorescence.

Porphyrostemma Bentham ex Oliver Asteraceae

Origins:

Greek *porphyra* "purple" and *stemma* "garland."

Porroglossum Schltr. Orchidaceae

Origins:

Greek *porro, proso* "forward, onward, further in" and *glossa* "tongue," referring to the importance of the sensitive lip in the pollination mechanism of this genus; or from Latin *porrum* and *porrus* "leek," always indicating the lip, with rounded callus, cushion-like; see R.E. Schultes and A.S. Pease, *Generic Names of Orchids. Their Origin and Meaning.* 256. 1963; Hubert Mayr, *Orchid Names and Their Meanings.* Vaduz 1998; H. Genaust, *Etymologisches Wörterbuch der botanischen Pflanzennamen.* 502. Basel 1996.

Porrorhachis Garay Orchidaceae

Origins:

Greek *porro, proso* "forward, onward, further in," Latin *porrum* and *porrus* "leek," and *rhachis* "rachis, axis, midrib of a leaf"; see Hubert Mayr, *Orchid Names and Their Meanings.* Vaduz 1998.

Porsildia Á. Löve & D. Löve Caryophyllaceae

Origins:

After the Danish botanist Alf Erling Porsild, 1901-1977, in Canada and Greenland, author of *The Alpine Flora of the East Slope of the Mackenzie Mountains, Northwest Territories.* Ottawa 1945, collaborator of the Danish botanist Morton Pedersen Porsild (1872-1956); see J.H. Barnhart, *Biographical Notes upon Botanists.* 3: 100. 1965; T.W. Bossert, *Biographical Dictionary of Botanists Represented in the Hunt Institute Portrait Collection.* 315. 1972; J. Ewan, ed., *A Short History of Botany in the United States.* 24. 1969.

Portea Brongn. ex K. Koch Bromeliaceae

Origins:

For the French plant explorer Dr. M. Porte, collector, d. 1866; see F. Boerner & G. Kunkel, *Taschenwörterbuch der botanischen Pflanzennamen.* 4. Aufl. 155. Berlin & Hamburg 1989.

Portenschlagia Trattinick Celastraceae

Origins:

For the Austrian botanist Franz [Edler] von Portenschlag-Ledermayer, 1772-1822, botanical explorer and plant collector, author of *J. Edler v. Portenschlag-Ledermayer über den Wasserkopf. Ein Beytrag zu einer Monographie dieser Krankheit. Nebst einem Anhang, verschiedene Anmerkungen, einige Leichenöffnungen, und einen Aufsatz über die Kuhpocke enthaltend.* Wien 1812, with the mycologist Leopold Trattinnick wrote *Enumeratio plantarum in Dalmatia.* Wien 1824; see J.H. Barnhart, *Biographical Notes upon Botanists.* 3: 100. 1965; Leopold Trattinnick (1764-1849), in *Archiv der Gewächskunde.* 5: 16, t. 250. 1818; T.W. Bossert, *Biographical Dictionary of Botanists Represented in the Hunt Institute Portrait Collection.* 315. 1972; E.M. Tucker, *Catalogue of the Library of the Arnold Arboretum of Harvard University.* 1917-1933; A. Lasègue, *Musée botanique de Benjamin Delessert.* Paris 1845; R. Zander, F. Encke, G. Buchheim and S. Seybold, *Handwörterbuch der Pflanzennamen.* 14. Aufl. 1993.

Portenschlagia Vis. Umbelliferae

Origins:

After the Austrian botanist Franz [Edler] von Portenschlag-Ledermayer, 1772-1822.

Portenschlagiella Tutin Umbelliferae

Origins:

The diminutive of *Portenschlagia* R. de Visiani.

Porterandia Ridley Rubiaceae

Origins:

After George Porter (Potter), flourished 1800-1833 (d. Calcutta), plant collector in India and Malaysia, from 1819 to 1822 in charge at the Calcutta Botanic Garden, collector for Nathaniel Wallich. (c. 1822-1823), in 1822 accompanied Wallich to Singapore; see R. Desmond, *The European*

Discovery of the Indian Flora. Oxford 1992; I.H. Vegter, *Index Herbariorum.* Part II (5), *Collectors N-R.* Regnum Vegetabile vol. 109. 1983.

Species/Vernacular Names:

P. cladantha (K. Schum.) Keay

Nigeria: okarakara (Yoruba); ukpakodsa (Edo); ekumalin (Boki)

Yoruba: okarakara, ohaha

Porteranthus Britton Rosaceae

Origins:

For Thomas Conrad Porter, 1822-1901, American botanist.

Porterella Torrey Campanulaceae

Origins:

For the American botanist Thomas Conrad Porter, 1822-1901, clergyman, professor of botany and zoology. His works include *Die Verfasser des Heidelberger Katechismus.* 1863 and *Flora of Pennsylvania.* Edited with the addition of analytical keys by John Kunkel Small (1869-1938). Boston 1903, (with John Merle Coulter, 1851-1928) wrote *Synopsis of the Flora of Colorado.* Washington 1874; see J.H. Barnhart, *Biographical Notes upon Botanists.* 3: 101. 1965; T.W. Bossert, *Biographical Dictionary of Botanists Represented in the Hunt Institute Portrait Collection.* 315. 1972; S. Lenley et al., *Catalog of the Manuscript and Archival Collections and Index to the Correspondence of John Torrey.* Library of the New York Botanical Garden. 334. 1973; E.M. Tucker, *Catalogue of the Library of the Arnold Arboretum of Harvard University.* 1917-1933; J. Ewan, ed., *A Short History of Botany in the United States.* 92. 1969; J.W. Harshberger, *The Botanists of Philadelphia and Their Work.* 236-243. 1899; J.W. von Goethe, *Hermann and Dorothea.* Translated by Thomas Conrad Porter. 1854; Johann Jacob Hottinger, *The Life and Times of U. Zwingli.* Translated from the German by Thomas Conrad Porter. Harrisburg 1856; Joseph William Blankinship, "A century of botanical exploration in Montana, 1805-1905: collectors, herbaria and bibliography." in *Montana Agric. Coll. Sci. Studies Bot.* 1: 1-31. 1904; R. Zander, F. Encke, G. Buchheim and S. Seybold, *Handwörterbuch der Pflanzennamen.* 14. Aufl. 1993; Ida Kaplan Langman, *A Selected Guide to the Literature on the Flowering Plants of Mexico.* 592. University of Pennsylvania Press, Philadelphia 1964; Joseph Ewan, *Rocky Mountain Naturalists.* The University of Denver Press 1950.

Porteresia Tateoka Gramineae

Origins:

For the French botanist Roland Portères, 1906-1974, agronomist, ethnobotanist, traveler, plant collector in West Africa (Senegal, Mali, Ivory Coast, West Cameroon), author of "*Eleusine caracana* Gaertner, céréale des humanités pauvres des pays tropicaux." *Bulletin de l'Institut Français d'Afrique noire.* tom. 13: 1-78. 1951 and *Observations sur les possibilités de culture du soja en Guinée forestière.* Nogent-sur-Marne 1946; see Gilbert Bouriquet et al., "Le vanillier et la vanille." *Encycl. Biol.* vol. 46. [Taxonomy by Roland Portères.] Paris 1954; Réné Letouzey, "Les botanistes au Cameroun." in *Flore du Cameroun.* 7: 1-110. Paris 1968; F.N. Hepper and Fiona Neate, *Plant Collectors in West Africa.* 66. Utrecht 1971; Auguste Jean Baptiste Chevalier, *Flore vivante de l'Afrique Occidentale Française.* 1: xxvii-xxx. Paris 1938.

Portlandia P. Browne Rubiaceae

Origins:

Named after Margaret Cavendish Bentinck, Duchess of Portland (*née* Harley), 1715-1785, in 1734 married William (2nd Duke of Portland); see *Memoir and Correspondence of ... Sir J.E. Smith ...* Edited by Lady Pleasance Smith. London 1832; [Margaret Cavendish Bentinck, Duchess of Portland], *Catalogue of the Portland Museum,* lately the property of the Duchess Dowager of Portland ... which will be sold by auction ... on Monday the 24th of April, 1786, etc. (By John Lightfoot) [1786].

Portulaca L. Portulacaceae

Origins:

Latin *portulaca, porcilaca, porcillaca,* name used by Plinius and Marcus Terentius Varro et al. for *Portulaca oleracea* L., etymology uncertain, possibly from Latin *portula, ae,* the diminutive of *porta, ae* "a door, gate," referring to the capsules, or from *porcus,* referring to the female organs; see Carl Linnaeus, *Species Plantarum.* 445. 1753 and *Genera Plantarum.* Ed. 5. 204. 1754; Salvatore Battaglia, *Grande dizionario della lingua italiana.* XIII: 893, 996. Torino 1986; C.T. Onions, *The Oxford Dictionary of English Etymology.* Oxford University Press 1966; Georg Christian Wittstein, *Etymologisch-botanisches Handwörterbuch.* 723. 1852; Ernest Weekley, *An Etymological Dictionary of Modern English.* 2: 1174. New York 1967.

Species/Vernacular Names:

P. sp.

Peru: berdoloque, chupa tierra, kotspu

P. bicolor F. Muell.

English: yellow purslane

P. grandiflora Hook.

English: rose moss, sun plant, eleven-o'clock, common portulaca, largeflower purslane

Japan: matsu-ba-botan

P. intraterranea J. Black

English: South Australia pigweed

P. lutea Sol. ex Forst.f.

Hawaii: 'ihi

P. molokiniensis Hobdy

Hawaii: 'ihi

P. oleracea L.

English: purslane, pussley, pigweed, common purslane, portulaca, garden purslane

Italian: procacchia, porcacchia, porcachia, porcellana, purciddana

Arabic: rigla, farfena, farfah, blibcha, bleibsha, rashad

Rodrigues Island: pourpier, pourpier rouge

East Africa: akalitete, danga danga, eleketete, gatumia, obwanda, ssezzira

Congo: poli

Yoruba: papasan, segunsete, semolese, akorelowo

Peru: kotspu, llutu-llutu, llutuyuyu, verdolaga

Mexico: nocuana ceeche, zeeche

Bhutan: phagpa jakpo

China: ma chi xian, ma chih hsien

Vietnam: rau sam, ma xi hien

Japan: suberi-hiyu

Okinawa: ninbutukâ

The Philippines: golasiman, olasiman, ulisiman, kolasiman, alusiman, ausiman, gulasiman, sahihan, sahikan, ngalug, bakbakad, lungum, dupdupil

Malaya: gelang pasir, segan jantan

Australia: munyeroo (Aboriginal name in Central Australia), common purslane, common pigweed

Hawaii: 'ihi, 'akulikuli kula, 'akulikuli lau li'i

P. oleracea L. var. *sativa* (Haw.) DC.

English: kitchen garden purslane

P. pilosa L. (*Portulaca cyanosperma* Egler)

Peru: amor crescido, bustirao chama, bustirão chama, flor de las once, flor de medio día, flor de seda, flor de las doce

Japan: ke-tsume-kusa

Hawaii: 'akulikuli

P. sclerocarpa A. Gray

Hawaii: 'ihi, po'e, 'ihi makole

P. villosa Cham. (*Portulaca caumii* F. Brown; *Portulaca hawaiiensis* Degener)

Hawaii: 'ihi

Portulacaria Jacq. Portulacaceae

Origins:

The genus *Portulaca* and the Latin suffix *aria* (it indicates connection), useful as fodder.

Species/Vernacular Names:

P. afra Jacq. (*Claytonia portulacaria* L.; *Portulaca fruticosa* Thunb.)

English: porkbush, elephant's foot, elephant bush, elephant's food

Southern Africa: spekboom (= porkwood); iNtelezi, isiDondwane, isAmbilane, iNdibili, inDibili-enkulu, isiCococo (Zulu); iGqwanitsha (Xhosa)

Posidonia K. König Posidoniaceae (Potamogetonaceae)

Origins:

Referring to the marine habitat of the plants; in classical mythology, the Greek Poseidon, a son of Cronus and Rhea, brother of Zeus and consort of Earth, was the god of the sea and also a god of horses; see Karl Koenig (or Charles Konig) (1774-1851) and John Sims (1749-1831), *Annals of Botany*. 2: 95, t. 6. London (Jun.) 1805.

Species/Vernacular Names:

P. angustifolia Cambridge & Kuo (*Ruppia spiralis* L. ex Dumort.)

English: narrow-leaved tape-weed

P. australis Hook.f.

English: fireball weed, marine-fiber plant

Posoqueria Aublet Rubiaceae

Origins:

From the native name in Guiana.

Species/Vernacular Names:

P. latifolia (Rudge) Roemer & Schultes

English: Brazilian oak, needle flower, stone seed

Peru: raya-caspi, ucullucui, açucena do mato, cedro branco, estrella ucullucui, papa terra

Posoria Raf. Rubiaceae

Origins:
See *Posoqueria* Aublet; see C.S. Rafinesque, *Ann. Gén. Sci. Phys.* 6: 80. 1820.

Postia Boissier & Blanche Asteraceae

Origins:
For the American botanist Rev. George Edward Post, 1838-1909, physician, surgeon, missionary, M.D. New York 1860, professor of surgery in Syria, traveler, plant collector. His writings include *Flora of Syria, Palestine, and Sinai*, from the Taurus to Ras Muhammad, and from the Mediterranean Sea to the Syrian Desert. Syrian Protestant College, Beirut, Syria [1896], *The Botanical Geography of Syria and Palestine*. London 1889 and *Pictures of Medical Work*. London [1894]; see John H. Barnhart, *Biographical Notes upon Botanists*. 3: 102. 1965; H.N. Clokie, *Account of the Herbaria of the Department of Botany in the University of Oxford*. 227. Oxford 1964; Howard Atwood Kelly and Walter Lincoln Burrage, *Dictionary of American Medical Biography*. New York 1928; G. Murray, *History of the Collections Contained in the Natural History Departments of the British Museum*. 1: 174. London 1904; R. Zander, F. Encke, G. Buchheim and S. Seybold, *Handwörterbuch der Pflanzennamen*. 14. Aufl. 1993; I.C. Hedge and J.M. Lamond, *Index of Collectors in the Edinburgh Herbarium*. Edinburgh 1970.

Postuera Raf. Oleaceae

Origins:
See E.D. Merrill (1876-1956), *Index rafinesquianus. The plant names published by C.S. Rafinesque, etc.* 190. Jamaica Plain, Massachusetts, USA 1949; C.S. Rafinesque, *Sylva Telluriana*. 10. 1838.

Potalia Aublet Gentianaceae (Potaliaceae, Loganiaceae)

Origins:
Greek *poton* "drink, drinking water," leaves and green stems used in a very bitter and regurgitive herbal tea.

Potameia Thouars Lauraceae

Origins:
Greek *potamos* "a river, stream" and *meion* "less, smaller."

Potamochloa Griff. Gramineae

Origins:
Greek *potamos* and *chloe, chloa* "grass," see also *Hygroryza* Nees.

Potamoganos Sandwith Bignoniaceae

Origins:
Greek *potamos* and *ganos* "beauty, brightness, ornament."

Potamogeton L. Potamogetonaceae

Origins:
Potamogeiton, from the Greek *potamos* and *geiton* "a neighbor," a neighbor of the river, referring to the aquatic natural habitat; Plinius applied Latin *potamogeton* and *potamogiton* to a water-plant, pondweed, water-milfoil; see Carl Linnaeus (1707-1778), *Species Plantarum*. 126. 1753 and *Genera Plantarum*. Ed. 5. 61. 1754.

Species/Vernacular Names:
P. acutifolius Roemer & Schultes
English: sharp-leaved pondweed, sharp pondweed

P. amplifolius Tuckerman
English: broad-leaved pondweed

P. australiensis A. Bennett
English: thin pondweed

P. coloratus Hornem.
English: fen pondweed

P. crispus L.
English: curled pondweed, curly pondweed, pondweed, crispate-leaved pondweed
South Africa: fonteingraskruid, krulblaarfonteinkruid
Japan: ebi-mo

P. cristatus Regel & Maack
Japan: koba-no-hiru-mushiro (= small-leaved *Potamogeton*)

P. distinctus A. Benn.
Japan: hiru-mushiro
Okinawa: firu-mushiru

P. diversifolius Raf.
English: diverse-leaved pondweed

P. drummondii Benth.
English: Drummond's pondweed

P. filiformis Pers.
English: slender-leaved pondweed

Potamophila R. Br. Gramineae

P. foliosus Raf.

English: leafy pondweed

P. gramineus L.

English: various-leaved pondweed, grass-leaved pondweed

P. illinoensis Morong

English: shining pondweed

P. latifolius (Robb.) Morong

English: Nevada pondweed

P. lucens L.

English: shining pondweed

P. malaianus Miquel (*Potamogeton gaudichaudii* Cham. & Schltdl.; *Potamogeton wrightii* Morong)

Japan: sasa-ba-mo

P. natans L.

English: broad-leaved pondweed, floating-leaved pondweed, pondweed

China: shui an ban

P. nodosus Poir.

English: loddon pondweed, pondweed, long-leaved pondweed

Southern Africa: iKhubalo lomkhondo wempahla

P. ochreatus Raoul

English: blunt pondweed

P. octandrus Poir.

English: pondweed

South Africa: fonteingraskruid

P. pectinatus L.

English: fennel-leaved pondweed, fennel pondweed, potamogeton, sago pondweed, fennel-leaf pondweed

South Africa: fonteingraskruid, fonteinkruid, skedefonteinkruid

Japan: ryu-no-hige-mo

Hawaii: limu alolo

P. perfoliatus L.

English: perfoliate pondweed, clasped pondweed

P. praelongus Wulfen

English: white-stemmed pondweed

P. pusillus L. (*Potamogeton panormitanus* Biv.)

English: small pondweed

Southern Africa: fonteingraskruid, fonteinkruid; joang-ba-metsi-bo-boholo (Sotho)

Japan: ito-mo

P. richardsonii (A. Bennett) Rydb.

English: Richardson's pondweed

P. robbinsii Oakes

English: Robbin's pondweed

P. rutilus Wolfg.

English: Shetland pondweed

P. schweinfurthii A.W. Benn.

English: pondweed

P. thunbergii Cham. & Schltdl. (*Potamogeton natans* sensu Thunb. non L.)

English: floating pondweed, pondweed

Southern Africa: drywende fonteinkruid, fonteinkruid, fonteingraskruid; ntlo-ea-hlapi (Sotho)

P. tricarinatus A. Bennett

English: floating pondweed

P. trichoides Cham. & Schltdl.

English: pondweed

P. zosteriformis Fern.

English: eel-grass pondweed

Potamophila R. Br. Gramineae

Origins:

Greek *potamos* "a river" and *philein* "to love," *philos* "lover, loving," alluding to the habitats; see R. Brown, *Prodromus florae Novae Hollandiae et Insulae van-Diemen*. 211. London (Mar.) 1810.

Potamoxylon Raf. Bignoniaceae

Origins:

Greek *potamos* "a river" and *xylon* "wood"; see C.S. Rafinesque, *Sylva Telluriana*. 78. 1838.

Potaninia Maxim. Rosaceae

Origins:

For the Russian botanist Grigorii Nikolajevic (Grigory Nikolaevich) Potanin, 1835-1920 (Tomsk), explorer and traveler (in Siberia, Mongolia, China and Tibet), plant collector, geographer. See John H. Barnhart, *Biographical Notes upon Botanists*. 3: 102. 1965; Alice Margaret Coats, *The Quest for Plants. A History of the Horticultural Explorers*. London 1969; Kenneth Lemmon, *The Golden Age of Plant Hunters*. London 1968; Elmer D. Merrill and Egbert H. Walker, *A Bibliography of Eastern Asiatic Botany*. 392. The Arnold Arboretum of Harvard University, Jamaica Plain, Massachusetts, USA 1938; Y.A. Demidovich, in *Dictionary of Scientific Biography* 11: 108. 1981; I.C. Hedge and J.M. Lamond, *Index of Collectors in the Edinburgh Herbarium*. Edinburgh 1970.

Potarophytum Sandwith Rapateaceae

Origins:

Latin *potator* "a drinker," *potara* "to drink" and Greek *phyton* "plant."

Potentilla L. Rosaceae

Origins:

Latin *potens, potentis* "powerful, able, mighty" (*possum, potes, potui, posse* "to be able"), referring to the medicinal or astringent properties of some species; see Carl Linnaeus, *Species Plantarum*. 495. 1753 and *Genera Plantarum*. Ed. 5. 219. 1754; Helmut Genaust, *Etymologisches Wörterbuch der botanischen Pflanzennamen*. 503-504. Basel 1996; G.C. Wittstein, *Etymologisch-botanisches Handwörterbuch*. 723. Ansbach 1852; Salvatore Battaglia, *Grande dizionario della lingua italiana*. XIII: 1103. UTET, Torino 1986.

Species/Vernacular Names:

P. anserina L. (pertaining or relating to geese)

English: silverweed, goose grass, goose tansy, silverweed cinquefoil

Tibetan: gro-lo sa-dzin

P. argentea L.

English: silvery cinquefoil, hoary cinquefoil

P. arguta Pursh

English: tall cinquefoil, tall potentilla, white cinquefoil

P. biennis E. Greene

English: biennial cinquefoil

P. canadensis Pursh

English: cinquefoil, dwarf cinquefoil, running fivefingers

P. concinna Richardson

English: alpine cinquefoil

P. crantzii (Crantz) G. Beck.

English: alpine cinquefoil

P. cristae W. Ferlatte & Strother

English: crested potentilla

P. flabellifolia Hook.

English: fan-foil

P. fruticosa L.

English: shrubby cinquefoil, golden hardhack, widdy, bush cinquefoil

Tibetan: spen-ma

P. fulgens Wall. ex Hook.

Nepal: kanthamun

P. hickmanii Eastw.

English: Hickman's cinquefoil

P. intermedia L.

English: downy cinquefoil

P. morefieldii B. Ertter

English: Morefield's cinquefoil

P. newberryi A. Gray

English: Newberry's cinquefoil

P. norvegica L.

English: rough cinquefoil, tall fivefinger

P. pacifica Howell

English: Pacific silverweed, coastal silverweed

P. palustris (L.) Scop.

English: marsh cinquefoil, marsh fivefinger

P. paradoxa Nutt.

English: bushy cinquefoil, diffuse potentilla

P. pensylvanica L.

English: prairie cinquefoil

P. recta L.

English: sulphur cinquefoil, common tormentil, upright cinquefoil, rough-fruited cinquefoil

P. reptans L.

English: creeping cinquefoil, cinquefoil, five finger grass, five leaf

Italian: fragolaria

Arabic: ben tabis, bentalis

P. rimicola (Munz & I.M. Johnston) B. Ertter

English: cliff cinquefoil

P. simplex Michx.

English: old field cinquefoil

P. sterilis (L.) Garcke

English: barren strawberry, strawberry potentilla, strawberry-leaf cinquefoil

P. tridentata Sol.

English: three-tooth cinquefoil, mountain white potentilla

Poteranthera Bong. Melastomataceae

Origins:

Greek *poterion* "a drinking cup, a drinking vessel, a goblet, cup" and *anthera* "anther."

Poteridium Spach Rosaceae

Origins:

Greek *poteridion*, the diminutive of *poterion* "a drinking cup, a goblet, cup."

Poterium L. Rosaceae

Origins:

Old Greek name for a plant also called *phrynion* perhaps *tragacanth*, Greek *poterion* "a drinking cup, a drinking vessel, a goblet, cup," Latin *poterium* and *poterion*; see Carl Linnaeus, *Species Plantarum*. 994. 1753 and *Genera Plantarum*. Ed. 5. 430. 1754.

Pothoidium Schott Araceae

Origins:

Referring to the genus *Pothos*.

Pothomorphe Miq. Piperaceae

Origins:

From the genus *Pothos* and the Greek *morphe* "a form, shape," superficially resembling that genus; see F.A.W. Miquel, in *Bulletin des Sciences Physiques et Naturelles en Néerlande*. 2: 447. Rotterdam 1839.

Species/Vernacular Names:

P. peltata (L.) Miq.

Brazil (Amazonas): mahekoma hanaki

Pothos L. Araceae

Origins:

From the Sinhalese name *potha* for *Pothos scandens* L.; see Carl Linnaeus, *Species Plantarum*. 968. 1753 and *Genera Plantarum*. Ed. 5. 415. 1754; D.H. Nicolson, "Derivation of aroid generic names." *Aroideana*. 10: 15-25. 1988. Latin *pothos*, from Greek *pothos* "desire," used by Plinius for a summer flower, otherwise unknown.

Species/Vernacular Names:

P. brassii B.L. Burtt

English: northern pothos

P. longipes Schott

English: long-footed climbing aroid, native pothos

Potima R.A. Hedw. Rubiaceae

Origins:

From the Greek *potimos* "drinkable," referring to the seeds.

Potoxylon Kosterm. Lauraceae

Origins:

Greek *potos* "for drinking" and *xylon* "wood."

Pottsia Hooker & Arnott Apocynaceae

Origins:

Named for John Potts, d. 1822 (Middx.), plant collector, traveler (China and Bengal), gardener for Horticultural Society of London; see Ray Desmond, *Dictionary of British & Irish Botanists and Horticulturists*. London 1994; Gordon Douglas Rowley, *A History of Succulent Plants*. Strawberry Press, Mill Valley, California 1997; Emil Bretschneider, *History of European Botanical Discoveries in China*. [Reprint of the original edition 1898.] Leipzig 1981.

Species/Vernacular Names:

P. grandiflora Markgraf (*Pottsia pubescens* Tsiang)

English: largeflower pottsia

China: da hua lian zi teng

P. laxiflora (Blume) Kuntze (*Pottsia pubescens* Tsiang)

English: laxflower pottsia, hairy pottsia

China: lian zi teng, hua guai teng gen

Pouchetia A. Rich. ex DC. Rubiaceae

Origins:

For the French (b. Rouen) physician Félix-Archimède Pouchet, 1800-1872 (d. Rouen), naturalist, M.D. Paris 1827, professor of botany. His writings include *Traité élémentaire de zoologie*. Paris 1832, *Hétérogenie, ou traité de la génération spontanée*. Paris 1859, *Histoire des sciences naturelles au moyen Age, ou Albert le Grand et son époque considérés comme point de départ de l'école expérimentale*. Paris 1853, *Moeurs et instincts des animaux*. Paris 1887, *Théorie positive de la fécondation des mammifères*. Paris 1842 and *L'Univers*. Paris 1865; see John H. Barnhart, *Biographical Notes upon Botanists*. 3: 104. 1965; Ida Kaplan Langman, *A Selected Guide to the Literature on the Flowering Plants of Mexico*. 593. University of Pennsylvania Press, Philadelphia 1964; J.K. Crellin, in *Dictionary of Scientific Biography* 11: 109-110. 1981.

Poulsenia Eggers Moraceae

Origins:

Named after the Danish botanist Viggo Albert Poulsen, 1855-1919, professor of botany, 1894-1895 at Bogor (Java). His works include *Botanisk Microkemi*. Copenhagen 1891, "Eriocaulaceae." in J. Schmidt, ed., *Flora of Koh Chang*. Contributions to the knowledge of the vegetation in the Gulf of Siam, VIII. *Bot. Tidsskr*. 26(fasc. 1): 167. 1904 and *Lille Plantelaere*. Copenhagen 1892. See John H. Barnhart, *Biographical Notes upon Botanists*. 3: 104. 1965; T.W. Bossert, *Biographical Dictionary of Botanists Represented in the Hunt Institute Portrait Collection*. 316. 1972; C.F.A. Christensen, *Den danske Botaniks Historie med tilhørende Bibliografi*. Copenhagen 1924-1926.

Species/Vernacular Names:

P. armata (Miquel) Standley

Peru: llanchama, yachapa caspi, yanchama

Pourouma Aubl. Cecropiaceae

Origins:

From the native name in Guiana.

Species/Vernacular Names:

P. sp.

Peru: chimico blanco, guití, imbáuba de cheiro, mapatí, xëxun

P. cecropiifolia C. Martius

Peru: cocura, cucura, uba-uba, uvilla, shuvia, suiya

Brazil: uvilla

Pourretia Ruíz & Pav. Bromeliaceae

Origins:

For the French botanist Pierre André Pourret (Pourret-Figeac), 1754-1818, clergyman.

Pourretia Willd. Bombacaceae

Origins:

For the French botanist Pierre André Pourret (Pourret-Figeac), 1754-1818, clergyman, botanical explorer, professor of botany in Spain, plant collector, author of *Mémoire sur divers volcans éteints de la Catalogne*. 1823; see J.H. Barnhart, *Biographical Notes upon Botanists*. 3: 104. 1965; M. Colmeiro y Penido, *La Botánica y los Botánicos de la Peninsula Hispano-Lusitana*. Madrid 1858; Jonas C. Dryander, *Catalogus bibliothecae historico-naturalis Josephi Banks*. London 1800; A. Lasègue, *Musée botanique de Benjamin Delessert*. Paris 1845; S. Lenley et al., *Catalog*

of the Manuscript and Archival Collections and Index to the Correspondence of John Torrey. Library of the New York Botanical Garden. 466. 1973.

Pouteria Aublet Sapotaceae

Origins:

Pourama pouteri, native Indian name from the Guiana; see J.B.C.F. Aublet, *Histoire des Plantes de la Guiane Françoise*. 1: 85. Paris (Jun.) 1775.

Species/Vernacular Names:

P. sp.

Peru: sacha caimito, caimitillo, cramary, cutiriba, jácana, locma, lucmo, lúcuma, lúcumo, macarandiba, massaranduba, pajurá, pariri, pashiqui, pucuna caspi, quina-quina blanca, quina-quina colorada, quinilla, rucma, rucmo, sacha caimito, tsirincavitiqui, tushmo amarillo, uchpa quinilla, urcu cumala, urcu mala, xumacuti

Mexico: atzapotl, zapote borracho, zapote amarillo

P. annamensis (Pierre) Baehni

English: Annam pouteria

China: tao lan

P. aubrevillei Bernardi (*Eremoluma wurdackii* Aubréville)

Peru: quinilla

P. bangii (Rusby) Penn. (*Sideroxylon bangii* Rusby)

Bolivia: coquino, lucuma

P. bilocularis (H. Winkler) Baehni

Bolivia: coquino

P. caimito (Ruíz & Pav.) Radlkofer (*Achras caimito* Ruíz & Pav.; *Lucuma ternata* Kunth)

Peru: abiú, abui, huangana caspi, caimito, caimo, cauje

Bolivia: aguaycillo

P. campechiana (Kunth) Baehni (*Lucuma campechiana* Kunth; *Lucuma nervosa* A. DC.)

English: eggfruit

Peru: canistel, huevo vegetal

Central America: canistel, sapote borracho, sapote amarillo

Japan: tamago-no-ki

P. castanosperma (C.T. White) Baehni

English: poison plum, milky plum, saffron boxwood

P. cinnamomea Baehni (*Labatia discolor* Diels)

Peru: jaba-jaba, varilla de agua

P. ephedrantha (A.C. Smith) Penn.

Bolivia: rupino colorado

P. grandifolia (Wallich) Baehni

China: long guo

P. guianensis Aublet

Bolivia: balata

P. hispida Eyma

Bolivia: lucuma

P. lucuma (Ruíz & Pav.) Kuntze (*Achras lucuma* Ruíz & Pav.; *Lucuma bifera* Molina) (see National Research Council, *Lost Crops of the Incas: Little-Known Plants of the Andes with Promise for Worldwide Cultivation.* National Academy Press, Washington, D.C. 1989)

Peru: locma, lucma, lucuma, pucuna caspi, oroco, orocu, cumala, rucma, r'ucma

P. lucumifolia (Reissek ex Maxim.) Penn.

Bolivia: aguai

P. macrophylla (Lam.) Eyma

Bolivia: cucuma, coquino, lucuma

P. nemorosa Baehni

Bolivia: coquino

P. obovata (R. Br.) Baehni (*Sersalisia obovata* R. Br.; *Planchonella obovata* (R. Br.) Pierre; *Sideroxylon ferrugineum* Hook. & Arn.; *Sideroxylon liukiuense* Nakai)

English: sea gutta

Japan: aka-tetsu, tumuki, jiiki

Malaya: menasi, misi

P. sandwicensis (A. Gray) Baehni & Degener (*Sapota sandwicensis* A. Gray; *Myrsine molokaiensis* H. Lév.; *Suttonia molokaiensis* (H. Lév.) H. Lév.)

Hawaii: 'ala'a, aulu, 'ela'a, kaulu

P. sapota (Jacq.) H.E. Moore & Stearn (*Pouteria mammosa* Jacq.)

English: mammee sapote, marmalade plum, naseberry

Peru: hison suma, sapote, zapote, mamey colorado, mamey zapote, níspero, zapotillo

Central America: sapote

Mexico: cochitzapotl (= zapote del sueño), iztac tzapotl (= zapote blanco), mamey, tezontzapotl (= zapote como piedra de lava), guela gue, quela que, guela xron, ya xron

P. sericea (Aiton) Baehni

Australia: mongo

P. trilocularis Cronq.

Bolivia: sapito

Pouzolsia Benth. Urticaceae

Origins:
Orthographic variant of *Pouzolzia* Gaudich.

Pouzolzia Gaudich. Urticaceae

Origins:
After the French botanist Pierre Marie Casimir de Pouzolz (Pouzols), 1785-1858, author of *Flore du département du Gard*, etc. Montpellier, Paris 1862 and *Catalogue des plantes qui croissent naturellement dans le Gard*. Nismes [Nîmes] 1842. See John H. Barnhart, *Biographical Notes upon Botanists.* 3: 105. 1965; Charles Gaudichaud-Beaupré (1789-1854), *Voyage autour du Monde ... sur ... l'Uranie et la Physicienne, pendant ... 1817-1820.* [Botany of the Voyage.] Paris 1826 [-1830]; Ethelyn Maria Tucker, *Catalogue of the Library of the Arnold Arboretum of Harvard University.* Cambridge, Massachusetts 1917-1933; I.C. Hedge and J.M. Lamond, *Index of Collectors in the Edinburgh Herbarium.* Edinburgh 1970.

Species/Vernacular Names:
P. guineensis Benth.

Yoruba: aboloko pinran, eemowere

P. mixta Solms (*Pouzolzia hypoleuca* Wedd.)

English: soap nettle

Southern Africa: seepnetel, wildebraam, isikukuku, uDekane (Zulu); nthadzwa (Tsonga); muthanzwa, murovhadembe (Venda); muNanzwa (Shona)

Tropical Africa: tingo

P. sanguinea (Bl.) Merr.

Nepal: lipe

Pozoopsis Benth. Umbelliferae

Origins:
Orthographic variant of *Pozopsis* Hook.

Pozopsis Hook. Umbelliferae

Origins:
Resembling *Pozoa* Lagasca.

Praecereus Buxb. Cactaceae

Origins:
Latin *prae* "before" plus *Cereus*.

Praecitrullus Pang. Cucurbitaceae

Origins:
Latin *prae* "before" plus *Citrullus*.

Pragmotropa Pierre Celastraceae

Origins:
Probably from the Greek *pragma*, *pragmatos* "act, occurrence" and *trope* "turning, defeat."

Prainea King ex Hook.f. Moraceae

Origins:
Named for the British botanist Sir David Prain, 1857-1944 (Surrey), professor of botany at Calcutta, 1884-1887 Indian Medical Service, Royal Botanic Garden at Calcutta 1887-1905, Fellow of the Linnean Society 1888 (President 1916-1919), from 1905 to 1922 Director of Kew, 1905 Fellow of the Royal Society, 1907-1920 editor of *Curtis's Botanical Magazine*. See John H. Barnhart, *Biographical Notes upon Botanists*. 3: 106. 1965; T.W. Bossert, *Biographical Dictionary of Botanists Represented in the Hunt Institute Portrait Collection*. 316. 1972; E.M. Tucker, *Catalogue of the Library of the Arnold Arboretum of Harvard University*. 1917-1933; Ida Kaplan Langman, *A Selected Guide to the Literature on the Flowering Plants of Mexico*. 594. Philadelphia 1964; M. Hadfield et al., *British Gardeners: A Biographical Dictionary*. London 1980; S. Lenley et al., *Catalog of the Manuscript and Archival Collections and Index to the Correspondence of John Torrey*. Library of the New York Botanical Garden. 334-335. 1973; Mea Allan, *The Hookers of Kew*. London 1967; Isaac Henry Burkill, *Chapters on the History of Botany in India*. Delhi 1965; F.D. Drewitt, *The Romance of the Apothecaries' Garden at Chelsea*. London 1924; H.R. Fletcher, *Story of the Royal Horticultural Society, 1804-1968*. Oxford 1969; H.R. Fletcher and W.H. Brown, *Royal Botanic Garden Edinburgh, 1670-1970*. 5-10. Edinburgh 1970; G. Murray, *History of the Collections Contained in the Natural History Departments of the British Museum*. 1: 174. London 1904; Ernest Nelmes and William Cuthbertson, *Curtis's Botanical Magazine Dedications, 1827-1927*. 291-292. [1931]; R. Zander, F. Encke, G. Buchheim and S. Seybold, *Handwörterbuch der Pflanzennamen*. 14. Aufl. 1993; Leonard Huxley, *Life and Letters of Sir Joseph Dalton Hooker*. London 1918.

Pranceacanthus Wasshausen Acanthaceae

Origins:
After Ghillean (Iain) Tolmie Prance, F.L.S., Director of the Royal Botanic Gardens, Kew.

Prasium L. Labiatae

Origins:
Latin *prasion* and *prasium* used by Plinius for an herb, white horehound; Theophrastus (*HP.* 6.2.5) applied Greek *prasion* to horehound, a species of *Marrubium*.

Prasopepon Naudin Cucurbitaceae

Origins:
Greek *prason* "a leek" and *pepon, peponos* "ripe, mild," Latin *pepo, peponis* "a species of large melon," Latin *prason, i* "a marine shrub resembling a leek" (Plinius).

Prasophyllum R. Br. Orchidaceae

Origins:
Greek *prason* "a leek" and *phyllon* "a leaf," referring to the shape of the leek-like sheathing leaf; see R. Brown, *Prodromus florae Novae Hollandiae et Insulae van-Diemen*. 317. London 1810.

Species/Vernacular Names:
P. alpinum R. Br. (*Prasophyllum tadgellianum* (R. Rogers) R. Rogers)

English: alpine leek orchid

P. archeri Hook.f. (*Prasophyllum intricatum* C. Stuart ex Benth.; *Prasophyllum ciliatum* Ewart & Rees)

English: variable midge-orchid

P. australe R. Br.

English: Austral leek orchid

P. brevilabre (Lindley) Hook.f.

English: short-lipped leek orchid

P. elatum R. Br.

English: tall leek orchid, piano orchid

P. flavum R. Br.

English: yellow leek orchid

P. fuscum R. Br.

English: slaty leek orchid

P. macrostachyum R. Br.

English: laughing leek orchid

P. morganii Nicholls

Australia: Cobungra leek orchid

P. odoratum R. Rogers

English: scented leek orchid, sweet leek orchid

P. patens R. Br.

Australia: broad-lipped leek orchid

P. rogersii Rupp

English: march leek orchid

P. striatum R. Br.

English: streaked leek orchid

P. suttonii R. Rogers & B. Rees

English: mauve leek orchid

Prasoxylon M. Roem. Meliaceae

Origins:

From the Greek *prason* "a leek" and *xylon* "wood."

Pratia Gaudich. Campanulaceae (Lobelioideae)

Origins:

Named for the French naval officer Charles Louis Prat-Bernon (died c. 1817, at sea), in September 1817 accompanied Louis de Freycinet on his scientific voyage around the world. See Charles Gaudichaud-Beaupré in L.C.D. de Freycinet, *Voyage autour du Monde entrepris par ordre du Roi ... sur les corvettes de S.M. "L'Uranie" et "La Physicienne".* [Botany of the Voyage.] Paris 1826[-1830]; C. Duplomb, *Campagne de L'Uranie: Journal de Madame Rose de Saulces de Freycinet.* Paris 1937; F. Grille, *Louis de Freycinet: sa Vie de Savant et de Marin.* Paris 1853; Charles Gaudichaud-Beaupré (1789-1854), in *Annales des Sciences Naturelles.* 5: 103. 1825; John Dunmore, *Who's Who in Pacific Navigation.* Honolulu 1991.

Species/Vernacular Names:

P. concolor (R. Br.) Druce (*Lobelia concolor* R. Br.)

English: poison pratia, milky lobelia

P. darlingensis F. Wimmer (named from Darling River)

Australia: Darling pratia

P. gelida (F. Muell.) Benth.

English: snow pratia

P. irrigua (R. Br.) Benth.

Australia: Bass Strait pratia

P. pedunculata (R. Br.) Benth. (*Lobelia pedunculata* R. Br.)

English: matted pratia, trailing pratia

P. platycalyx (F. Muell.) Benth.

English: fleshy pratia

P. puberula Benth.

English: trailing pratia

P. purpurascens (R. Br.) F. Wimmer

English: whiteroot, lobelia pratia

P. surrepens (Hook.f.) F. Wimmer

English: mud pratia

Preissia Opiz Gramineae

Origins:

For the German botanist Balthazar Preiss (Preis), 1765-1850, physician, author of *Rhizographie.* Prag 1823; see John H. Barnhart, *Biographical Notes upon Botanists.* 3: 107. 1965; R. Zander, F. Encke, G. Buchheim and S. Seybold, *Handwörterbuch der Pflanzennamen.* 14. Aufl. 765. 1993.

Premna L. Labiatae (Verbenaceae)

Origins:

Greek *premnon* "the stump of a tree, a tree-trunk," referring to the stature of the species; see C. Linnaeus, *Mantissa Plantarum.* 1: 154, 252. (Oct.) 1767.

Species/Vernacular Names:

P. acutata W. Smith

English: sharp-leaf premna

China: jian chi dou fu chai

P. barbata Wall. ex Schauer

Nepal: ginari

P. bracteata Wallich ex C.B. Clarke

English: bracteate premna

China: bao xu dou fu chai

P. cavaleriei H. Léveillé

English: Cavalerie premna

China: huang yao

P. chevalierei Dop

English: Chevalier premna

China: jian ye dou fu chai

P. confinis P'ei & S.L. Chen ex C.Y. Wu

English: adjoining premna

China: dian gui dou fu chai

P. crassa Handel-Mazzetti

English: thick premna

China: shi shan dou fu chai

P. fordii Dunn

English: Ford premna

China: chang xu xiu huang jing

P. fulva Craib

English: yellow-hairy premna

China: huang mao dou fu chai

P. interrupta Wallich ex Schauer

English: interrupted premna

China: jian xu dou fu chai

P. latifolia Roxburgh

English: broad-leaf premna

China: da ye dou fu chai

P. lignumvitae (Schauer) Pieper

Australia: Queensland lignum-vitae

P. ligustroides Hemsley

English: privet-like premna

China: xiu huang jing, chou huang jing zi

P. microphylla Turcz.

English: Japanese premna

China: dou fu chai, fu bi

P. mooiensis (H. Pearson) Pieper (*Vitex mooiensis* H. Pearson)

English: skunk bush, the stinker

Southern Africa: muishondbos, stinkboom; umTshetshembane, umSuzane, isiDadada, Mqathathongo (Zulu); umTyatyambane, umTyetyembane, umCacabane, umChachambane (Xhosa)

P. obtusifolia R. Br.

English: obtuse-leafed premna, bastard guelder

Rodrigues Island: bois sureau

Malaya: buas buas, bebuas, bebuat

P. oligantha C.Y. Wu

English: few-flowers premna

China: shao hua dou fu chai

P. puberula Pampanini

English: puberulent premna

China: hu xiu chai

P. pyramidata Wallich ex Schauer

English: pyramidal premna

China: ta xu dou fu chai

Malaya: piat, sarang burong

P. racemosa Wallich ex Schauer

English: racemose premna

China: zong xu dou fu chai

P. scandens Roxb.

English: scandent premna

China: teng dou fu chai

P. serratifolia L. (*Premna integrifolia* L.)

English: headache tree

French: arbre de la migraine

Japan: Taiwan-no-kusagi (= Taiwan *Clerodendrum*)

China: san xu xiu huang jing

Vietnam: vong cach, cach nui

P. steppicola Hand.-Mazz.

English: steppe-living premna

China: cao po dou fu chai

P. subcapitata Rehder

English: subcapitate premna

China: jin tou zhuang dou fu chai

P. tomentosa Willd.

English: bastard teak

Malaya: sarang burong, piat, kapiat

P. yunnanensis W. Smith

English: Yunnan premna

China: yun nan dou fu chai

Prenanthella Rydb. Asteraceae

Origins:
The diminutive of the genus *Prenanthes* L.

Prenanthes L. Asteraceae

Origins:
Greek *prenes* "prone, prostrate, with face downward" and *anthos* "flower," referring to the flower-heads.

Prenia N.E. Br. Aizoaceae

Origins:
Greek *prenes* "prone, prostrate, with face downward, bent forward," alluding to the habit.

Preposdesma N.E. Br. Aizoaceae

Origins:
Greek *prepo* "to be clearly seen" and *desmis, desmos* "a bond, band, bundle."

Preptanthe Reichb.f. Orchidaceae

Origins:
Greek *preptos* "distinguished, eminent" and *anthos* "flower," *prepo* "to strike the senses, to catch the eye, to be manifest," alluding to the flowers.

Prepusa Martius Gentianaceae

Origins:
Greek *prepo* "to strike the senses, to catch the eye, to be manifest," alluding to the flowers.

Prescotia Lindley Orchidaceae

Origins:
For the English botanist John D. Prescott, died 1837 (in Russia).

Prescottia Lindley Orchidaceae

Origins:
The generic name honors the English botanist John D. Prescott, died 1837 (in Russia), merchant, traveled widely in the northern regions of Asia, correspondent of W. Hooker and friend of J. Lindley, plant collector in Russia and North Asia; see A. Lasègue, *Musée botanique de Benjamin Delessert*. Paris 1845; *Companion to the Botanical Magazine*. 2: 342-343. 1836; H.N. Clokie, *Account of the Herbaria of the Department of Botany in the University of Oxford*. 228. Oxford 1964; Ray Desmond, *Dictionary of British & Irish Botanists and Horticulturists*. 563. London 1994.

Preslia Opiz Labiatae

Origins:
Dedicated to the brothers Jan Svatopluk (Swatopluk) Presl (1791-1849) and Karel (Karl, Carl, Carel, Carolus) B. Presl (1794-1852), Czech botanists; see A.C. Jermy, in *Dictionary of Scientific Biography* 11: 130. 1981; Stafleu and Cowan, *Taxonomic Literature*. 4: 389-395. Utrecht 1983; Mariella Azzarello Di Misa, ed., *Il Fondo Antico della Biblioteca dell'Orto Botanico di Palermo*. 220-222. Regione Siciliana, Palermo 1988.

Prestoea Hook.f. Palmae

Origins:
After the British botanist Henry Prestoe, 1842-1923 (Sussex), gardener, traveler, plant collector, 1864-1886 Trinidad; see John H. Barnhart, *Biographical Notes upon Botanists*. 3: 108. 1965; E.M. Tucker, *Catalogue of the Library of the Arnold Arboretum of Harvard University*. 1917-1933.

Species/Vernacular Names:
P. allenii H.E. Moore

Panama: maquenque

Prestonia R. Br. Apocynaceae

Origins:
For Charles Preston, 1660-1711, British physician, M.D. Edinburgh 1694, professor of botany, sent plants to H. Sloane; see L. Plukenet, *Almagesti botanici mantissa*. 12. Londini 1700; [John Ray], *The Correspondence of J. Ray*. Edited by E. Lankester. 380-388. London 1848; James Britten, *The Sloane Herbarium*, revised and edited by J.E. Dandy. 1958; Ray Desmond, *Dictionary of British & Irish Botanists and Horticulturists*. 563. London 1994; H.R. Fletcher and W.H. Brown, *Royal Botanic Garden Edinburgh, 1670-1970*. 26-30. Edinburgh 1970; Stafleu and Cowan, *Taxonomic Literature*. 4: 395-396. 1983; F. Boerner & G. Kunkel, *Taschenwörterbuch der botanischen Pflanzennamen*. 4. Aufl. 156. Berlin & Hamburg 1989.

Preussiella Gilg Melastomataceae

Origins:
Dedicated to the German botanist Paul Rudolf Preuss, b. 1861, traveler and explorer (Sierra Leone, Togo, West Cameroon, Southeast Asia and tropical America), plant collector, 1888-1891 Zintgraff Expedition to Cameroon; see John H. Barnhart, *Biographical Notes upon Botanists*. 3: 109. 1965; Ida Kaplan Langman, *A Selected Guide to the Literature on the Flowering Plants of Mexico*. 595. Philadelphia 1964; I.C. Hedge and J.M. Lamond, *Index of Collectors in the Edinburgh Herbarium*. Edinburgh 1970; Ethelyn Maria Tucker, *Catalogue of the Library of the Arnold Arboretum of Harvard University*. Cambridge, Massachusetts 1917-1933; Anthonius Josephus Maria Leeuwenberg, "Isotypes of which holotypes were destroyed in Berlin." *Webbia*. 19(2): 862. 1965; Réné Letouzey, "Les botanistes au Cameroun." in *Flore du Cameroun*. 7: 1-110. Paris 1968; F.N. Hepper and Fiona Neate, *Plant Collectors in West Africa*. 66. 1971; Frank Nigel Hepper (1929-), "Botanical collectors in West Africa, except French territories, since 1860." in *Comptes Rendus de l'Association pour l'étude taxonomique de la flore d'Afrique*, (A.E.T.F.A.T.). 69-75. Lisbon 1962.

Preussiodora Keay Rubiaceae

Origins:
For the German botanist Paul Rudolf Preuss, b. 1861, plant collector, traveler and explorer (Sierra Leone, Southeast Asia and tropical America).

Prevostea Choisy Convolvulaceae

Origins:

For the French painter Jean Louis Prévost (flourished 1760-1810), the Swiss (b. Geneva) plant physiologist Isaac-Bénédict Prévost (1755-1819, d. Montauban, France) and the Swiss professor of physics Pierre Prévost (1751-1839, d. Geneva); see S. Lenley et al., *Catalog of the Manuscript and Archival Collections and Index to the Correspondence of John Torrey.* Library of the New York Botanical Garden. 335. 1973; Stafleu and Cowan, *Taxonomic Literature.* 4: 398. Utrecht 1983; Gloria Robinson, in *Dictionary of Scientific Biography* 11: 131-132. 1981; John G. Burke, in *Dictionary of Scientific Biography* 11: 134-135. 1981.

Priestleya DC. Fabaceae

Origins:

After the English (b. Yorkshire) chemist Joseph Priestley, 1733-1804 (d. Northumberland, Pennsylvania, USA), clergyman, scientist, philosopher, 1766 Fellow of the Royal Society, he was a supporter of the French Revolution, to North America in 1794. He is best known for his research into the chemistry of gases and for his discovery of oxygen. His writings include *The History and Present State of Electricity,* with original experiments. [First edition, proofs corrected by Franklin.] London 1767 and *The History and Present State of Discoveries Relating to Vision, Light, and Colours.* London 1772. See *Memoirs of Dr. Joseph Priestley, to the Year 1795, Written by Himself, With a Continuation to the Time of His Decease, by His Son, Joseph Priestley: and Observations on His Writings by Thomas Cooper.* Northumberland, Pennsylvania 1805 and London 1806; Leonard W. Labaree, ed., *The Papers of Benjamin Franklin.* New Haven, Conn. 1959-; I. Bernard Cohen, *Benjamin Franklin's Science.* 1990; Ronald E. Crook, *A Bibliography of Joseph Priestley, 1733-1804.* London 1904; Robert E. Schofield, in *Dictionary of Scientific Biography* 11: 139-147. 1981; Robert E. Schofield, ed., *A Scientific Autobiography of Joseph Priestley, 1733-1804.* Cambridge, Massachusetts 1963; Alexander B. Adams, *Eternal Quest. The Story of the Great Naturalists.* New York 1969.

Primula L. Primulaceae

Origins:

Primula veris, a medieval name for the daisy; Latin *primus, primulus* "first," Akkadian *pir'u, per'u* "offspring"; see B. Migliorini, *Parole d'autore (onomaturgia).* Firenze 1975; Helmut Genaust, *Etymologisches Wörterbuch der botanischen Pflanzennamen.* 505-506. 1996.

Species/Vernacular Names:

P. egaliksensis Wormsk.

English: Greenland primrose

P. elatior (L.) Hill

English: oxlip

P. sieboldii E. Morren (after the German doctor Philipp Franz (Balthasar) von Siebold, 1796-1866, botanist and naturalist, from 1823 to 1830 and from 1859 to 1862/1863 plant collector in Japan, author of *De historiae naturalis in Japonia statu.* Bataviae 1824, *Flora japonica.* Lugduni Batavorum [Leyden] 1835 [-1841], with Joseph Gerhard Zuccarini (1797-1848) wrote *Florae japonicae familiae naturales.* [München 1845-1846], he was brother of the German physiologist and zoologist Carl Theodor Ernst von Siebold (1804-1885); see John H. Barnhart, *Biographical Notes upon Botanists.* 3: 275. 1965; Frans A. Stafleu and Cowan, *Taxonomic Literature.* 5: 585-592. 1985; T.W. Bossert, *Biographical Dictionary of Botanists Represented in the Hunt Institute Portrait Collection.* 366. 1972; A. Lasègue, *Musée botanique de Benjamin Delessert.* Paris 1845; Ethelyn Maria Tucker, *Catalogue of the Library of the Arnold Arboretum of Harvard University.* Cambridge, Massachusetts 1917-1933; S. Lenley et al., *Catalog of the Manuscript and Archival Collections and Index to the Correspondence of John Torrey.* Library of the New York Botanical Garden. 369. 1973; R. Zander, F. Encke, G. Buchheim and S. Seybold, *Handwörterbuch der Pflanzennamen.* 14. Aufl. 1993)

English: Siebold primrose

Japan: sakura-sô

China: ying cao

P. sinensis Sabine ex Lindley

English: Chinese primrose

Japan: kan-zakura

China: zang bao chun

P. suffrutescens A. Gray

English: Sierra primrose

P. veris L.

English: cowslip

P. vulgaris Huds. (*Primula acaulis* (L.) Hill.)

English: primrose, English primrose

Japan: purimurôzu

Primularia Brenan Melastomataceae

Origins:

Latin *primus, primulus* "the first."

Primulina Hance Gesneriaceae

Origins:
The diminutive of the generic name *Primula*, referring to the likeness.

Princea Dubard & Dop Rubiaceae

Origins:
For a plant collector in Madagascar, Prince.

Principina Uittien Cyperaceae

Origins:
Príncipe island, Sao Tome and Príncipe, Atlantic Ocean.

Pringlea T. Anderson ex Hook.f. Brassicaceae

Origins:
After the Scottish (b. Roxburgh) physician Sir John Pringle, 1707-1782 (London), at Leyden attended Boerhaave, from 1744 to 1752 was Physician-General of the British Army, 1745 elected to the Royal Society. His writings include *Observations on the Diseases of the Army*. London 1752 and *Observations on the Nature and Cure of Jayl-Fevers*. London 1750; see Garrison and Morton, *Medical Bibliography*. 2150, 2156, 5374. 1961; Samuel X. Radbill, in *Dictionary of Scientific Biography* 11: 147-148. 1981; William Munk, *The Roll of the Royal College of Physicians of London*. 2: 252. London 1878; F. Boerner & G. Kunkel, *Taschenwörterbuch der botanischen Pflanzennamen*. 4. Aufl. 156, 339. Berlin & Hamburg 1989.

Pringleochloa Scribner Gramineae

Origins:
After the American botanist Cyrus Guernsey Pringle, 1838-1911, Quaker, plant collector (Pacific States and Mexico), wrote *The Record of a Quaker Conscience. C. Pringle's Diary*. New York 1918. See John H. Barnhart, *Biographical Notes upon Botanists*. 3: 111. 1965; T.W. Bossert, *Biographical Dictionary of Botanists Represented in the Hunt Institute Portrait Collection*. 318. 1972; Ethelyn Maria Tucker, *Catalogue of the Library of the Arnold Arboretum of Harvard University*. Cambridge, Massachusetts 1917-1933; Ezra Brainerd (1844-1924), "Cyrus Guernsey Pringle." *Rhodora*. 13: 225-232. 1911; Irving William Knobloch, compil., "A preliminary verified list of plant collectors in Mexico." *Phytologia Memoirs*. VI. 1983; Helen Burns

Davis, *Life and Work of Cyrus Guernsey Pringle*. Burlington, Vt. 1936; S. Lenley et al., *Catalog of the Manuscript and Archival Collections and Index to the Correspondence of John Torrey*. Library of the New York Botanical Garden. 335-336. 1973; Ira L. Wiggins, *Flora of Baja California*. 42. Stanford, California 1980; Ida Kaplan Langman, *A Selected Guide to the Literature on the Flowering Plants of Mexico*. 596. University of Pennsylvania Press, Philadelphia 1964; Gordon Douglas Rowley, *A History of Succulent Plants*. California 1997; I.C. Hedge and J.M. Lamond, *Index of Collectors in the Edinburgh Herbarium*. Edinburgh 1970.

Pringleophytum A. Gray Acanthaceae

Origins:
For the American botanist Cyrus Guernsey Pringle, 1838-1911.

Prinsepia Royle Rosaceae

Origins:
For the English archeologist James Prinsep, 1799-1840, traveler, Secretary of the Asiatic Society of Bengal, editor of the *Journal of the Asiatic Society*. His works include *An Epitome of Ancient History*. [Calcutta School Book Society; the English compiled by Prinsep and others.] Calcutta 1830, *Benares Illustrated*, in a series of drawings by J. Prinsep. Baptist Mission Press, Calcutta 1831, *Essays on Indian Antiquities*, etc. Edited by E. Thomas. [With a memoir of J. Prinsep by H.T. Prinsep.] London 1858 and *Modification of the Sanskrit Alphabet* from 543 B.C. to 1200 A.D. London 1850; see R. Zander, F. Encke, G. Buchheim and S. Seybold, *Handwörterbuch der Pflanzennamen*. 14. Aufl. 772. 1993.

Printzia Cass. Asteraceae

Origins:
To the memory of Jacob Printz, 1740-1779, a pupil of Linnaeus, author of *Plantae rariores africanae, quas ... praeside ... C. Linnaeo*, etc. Holmiae [Stockholm 1760].

Prionachne Nees Gramineae

Origins:
Greek *prion* "a saw" and *achne* "chaff, glume," see also *Prionanthium* Desv.

Prionanthium Desv. Gramineae

Origins:
Greek *prion* "a saw" and *anthos* "flower."

Prionitis Oerst. Acanthaceae

Origins:
Greek *prionitis* "kestron," the word *kestron* "point, sting" was used by Dioscorides to designate a species of *Stachys*.

Prionium E. Meyer Juncaceae

Origins:
Greek *prionion* "a small saw," *prion* "a saw," referring to the leaves.

Prionophyllum K. Koch Bromeliaceae

Origins:
From the Greek *prion* and *phyllon* "leaf."

Prionoplectus Oerst. Gesneriaceae

Origins:
Greek *prion* and *plektos* "twisted, plaited."

Prionopsis Nutt. Asteraceae

Origins:
Greek *prion* "a saw" and *opsis* "like, resembling," from leaf margin.

Prionosciadium S. Watson Umbelliferae

Origins:
Greek *prion* "a saw" and *skiadion, skiadeion* "umbel, parasol."

Prionosepalum Steudel Restionaceae

Origins:
Greek *prion* "a saw" and Latin *sepalum*, referring to the jagged sepals; see Ernst Gottlieb von Steudel, *Synopsis plantarum glumacearum*. 2: 266. Stuttgartiae (Sep.) 1855.

Prionostemma Miers Celastraceae

Origins:
Greek *prion* "a saw" and *stemma* "garland."

Prionotes R. Br. Epacridaceae

Origins:
Greek *prionotos* "jagged like a saw," *prionion* "a little saw," *prion* "a saw," toothed; see Robert Brown, *Prodromus florae Novae Hollandiae et Insulae van-Diemen*. 552. London 1810.

Prionotrichon Botsch. & Vved. Brassicaceae

Origins:
Greek *prion* "a saw" and *thrix, trichos* "hair."

Prioria Grisebach Caesalpiniaceae

Origins:
For the British (b. Wilts.) botanist Richard Chandler Alexander Prior (*olim* Alexander, in 1859 took name Prior), 1809-1902 (d. London), physician, traveler (South Africa, Canada, USA, Jamaica, West Indies), plant collector, Fellow of the Linnean Society 1851. Among his works are *Notes on Croquet*; and some ancient bat and ball games related to it. London 1872, *On the Popular Names of British Plants*. London and Edinburgh 1863 and *Ancient Danish Ballads*, translated from the originals by R.C.A. Prior. London 1860; see J.H. Barnhart, *Biographical Notes upon Botanists*. 3: 112. 1965; T.W. Bossert, *Biographical Dictionary of Botanists Represented in the Hunt Institute Portrait Collection*. 318. 1972; H.N. Clokie, *Account of the Herbaria of the Department of Botany in the University of Oxford*. 228-229. Oxford 1964; S. Lenley et al., *Catalog of the Manuscript and Archival Collections and Index to the Correspondence of John Torrey*. Library of the New York Botanical Garden. 467. 1973; Mary Gunn and Leslie E. Codd, *Botanical Exploration of Southern Africa*. 287-288. A.A. Balkema Cape Town 1981; G. Murray, *History of the Collections Contained in the Natural History Departments of the British Museum*. 1: 175. London 1904; Ray Desmond, *Dictionary of British & Irish Botanists and Horticulturists*. 565. 1994; James Britten and George E. Simonds Boulger, *A Biographical Index of Deceased British and Irish Botanists*. London 1931.

Priotropis Wight & Arn. Fabaceae

Origins:

Greek *prion* "a saw" and *tropis, tropidos* "a keel."

Prismatocarpus L'Hérit. Campanulaceae

Origins:

From the Greek *prisma, prismatos* "a prism" and *karpos* "a fruit," referring to the ovary and to the fruit.

Prismatomeris Thwaites Rubiaceae

Origins:

Greek *prisma, prismatos* "a prism" and *meris* "a part, portion," referring to the sharply cut petals.

Species/Vernacular Names:

P. labordei (Lév.) Merr.

English: willowleaf prismatomeris

P. tetrandra (Roxb.) K. Schumann

English: fourstamen prismatomeris, Robin's coffee

Malaya: kahwa utan, kopi utan, setulang

Pristidia Thwaites Rubiaceae

Origins:

Greek *pristes* "a sawyer," *prister, pristeros* "saw."

Pristiglottis Cretz. & J.J. Smith Orchidaceae

Origins:

Greek *pristes* "a saw" and *glotta* "tongue," referring to the margin of the lip; see Richard Evans Schultes and Arthur Stanley Pease, *Generic Names of Orchids. Their Origin and Meaning.* 260. 1963; Hubert Mayr, *Orchid Names and Their Meanings.* 171. Vaduz 1998; some suggest from *pristis* "a cup."

Pristimera Miers Celastraceae

Origins:

Greek *prizo, prio* "saw, serrated" and *meris, meros* "part, portion," referring to the parts of the flower.

Pritchardia Seemann & H.A. Wendland Palmae

Origins:

Named to honor William Thomas Pritchard, once British Consul in the Fiji Islands and author of *Polynesian Reminiscences; or, Life in the South Pacific Islands ... Preface* by Dr. Seemann. London 1866; see Luis Pombo, *Mexico: 1876-1892.* [With an English translation by W.T. Pritchard, and maps.] Mexico 1893; Hermann Wendland (1825-1903), *Bonplandia.* 10: 197. 1862; F. Boerner & G. Kunkel, *Taschenwörterbuch der botanischen Pflanzennamen.* 4. Aufl. 156. Berlin & Hamburg 1989.

Species/Vernacular Names:

P. sp.

Hawaii: loulu

P. hillebrandii (Kuntze) Becc. (*Washingtonia hillebrandi* Kuntze)

Hawaii: loulu lelo

P. martii (Gaud.) H.A. Wendl. (*Washingtonia gaudichaudii* (Mart.) Kuntze)

Hawaii: loulu hiwa

P. pacifica Seemann & H.A. Wendl. ex H.A. Wendland

English: Fiji fan palm

Brazil: palmeira leque de Fiji

Pritchardiopsis Beccari Palmae

Origins:

Resembling *Pritchardia* Seemann & Wendland.

Pritzelago O. Kuntze Brassicaceae

Origins:

To remember the German botanist Georg August Pritzel, 1815-1874, eminent botanical bibliographer, archivist and librarian, author of *Thesaurus literaturae botanicae.* Lipsiae [1847-]1851[-1852]; see John H. Barnhart, *Biographical Notes upon Botanists.* 3: 112. 1965; T.W. Bossert, *Biographical Dictionary of Botanists Represented in the Hunt Institute Portrait Collection.* 318. 1972; Ida Kaplan Langman, *A Selected Guide to the Literature on the Flowering Plants of Mexico.* 596. Philadelphia 1964; Frans A. Stafleu and Cowan, *Taxonomic Literature.* 4: 409-414. 1983; R. Zander, F. Encke, G. Buchheim and S. Seybold, *Handwörterbuch der Pflanzennamen.* 14. Aufl. 765. 1993; Mariella Azzarello Di Misa, ed., *Il Fondo Antico della Biblioteca dell'Orto Botanico di Palermo.* 222. Regione Siciliana, Palermo 1988.

Pritzelia F. Mueller Philydraceae

Origins:
For the German botanist Georg August Pritzel, 1815-1874, botanical bibliographer, archivist and librarian.

Pritzelia Schauer Myrtaceae

Origins:
For the German botanist Georg August Pritzel, 1815-1874, botanical bibliographer, archivist and librarian.

Pritzelia Walpers Umbelliferae

Origins:
For the German botanist Georg August Pritzel, 1815-1874.

Priva Adans. Verbenaceae

Origins:
Meaning not clear, perhaps derived from the Latin *privus, a, um* "without, individual, single, one's own, deprived of," or from an Indian vernacular name.

Species/Vernacular Names:
P. cordifolia (L.f.) Druce var. *abyssinica* (Jaub. & Spach) Moldenke (*Priva abyssinica* Jaub. & Spach)

Southern Africa: isinama (Zulu)

P. meyeri Jaub. & Spach var. *meyeri*

South Africa: blaasklits, blasiesklitsbossie

Proatriplex Stutz & G.L. Chu Chenopodiaceae

Origins:
From the Greek *pro* "before, forth" and the genus *Atriplex* L.

Problastes Reinw. Combretaceae

Origins:
Greek *pro* "before, forth" and *blastos* "bud, sprout, germ, ovary, sucker."

Probletostemon K. Schumann Rubiaceae

Origins:
Greek *probletos* "thrown forth, spread" and *stemon* "stamen."

Proboscidea Schmidel Pedaliaceae (Martyniaceae)

Origins:
Proboskis, proboskidos "elephant's trunk," from the Greek *pro* and *bosko* "to graze, feed," referring to the fruit; see Casimir Christoph Schmidel (1718-1792), *Icones Plantarum.* 49, t. 12, 13. (Oct.) 1763.

Species/Vernacular Names:
P. althaeifolia (Benth.) Decne.

English: desert unicorn plant

P. louisianica (Miller) Thell. (*Martynia louisianica* Miller)

English: purple-flowered devil's claw, common unicorn plant

Prochnyanthes S. Watson Agavaceae

Origins:
Probably from the Greek *prochoos, prochon* "a jug, ewer, vase" and *anthos* "flower," referring to the shape of the flower.

Prockiopsis Baillon Flacourtiaceae

Origins:
Resembling *Prockia* P. Browne ex L.

Proclesia Klotzsch Ericaceae

Origins:
Greek *proklesis* "challenge, stimulation"; Procles was the twin brother of Eurysthenes.

Procopiana Gusuleac Boraginaceae

Origins:
For Aurel Procopianu-Procopovici, 1862-1918, botanist, author of *Botanica ilustrata.* Suceava 1897; see John H. Barnhart, *Biographical Notes upon Botanists.* 3: 112. 1965; E.M. Tucker, *Catalogue of the Library of the Arnold Arboretum of Harvard University.* 1917-1933; Al. Borza, *Dictionar etnobotanic.* Bucuresti 1968.

Procopiania Gusuleac Boraginaceae

Origins:
For Aurel Procopianu-Procopovici, 1862-1918.

Procrassula Griseb. Crassulaceae

Origins:

Greek *pro* "before, forth" plus *Crassula*.

Procris Comm. ex Juss. Urticaceae

Origins:

After Procris (-kris), daughter of Erechteus, King of Athens and wife of Cephalus; see A.L. de Jussieu, *Genera Plantarum*. 403. (Aug.) 1789.

Proferea C. Presl Thelypteridaceae

Origins:

Profera, the goddess of Arithmetic, "she that presents, makes known."

Proiphys Herbert Amaryllidaceae (Liliaceae)

Origins:

Greek *pro* "early" and *phyo* "to grow, to bring forth," referring to the premature germination of the seed.

Species/Vernacular Names:

P. cunninghamii (Lindley) Mabb.

English: Brisbane lily

Prolobus R.M. King & H. Robinson Asteraceae

Origins:

Greek *pro* "before, forth" and *lobos* "a pod."

Promenaea Lindley Orchidaceae

Origins:

Greek *pro* and *mene* "moon, the crescent moon," referring to the shape of the viscidium; or dedicated to Promenaea, a Greek priestess of the temple of Dodona, Epirus, see F. Boerner & G. Kunkel, *Taschenwörterbuch der botanischen Pflanzennamen*. 4. Aufl. 156. Berlin & Hamburg 1989; H. Mayr, *Orchid Names and Their Meanings*. 172. Vaduz 1998; H. Genaust, *Etymologisches Wörterbuch der botanischen Pflanzennamen*. 507. ["a gland of an ovate form bearing two double pollen-masses sessile"] 1996; F. Boerner, *Taschenwörterbuch der botanischen Pflanzennamen*. 2. Aufl. 161. Berlin & Hamburg 1966; Richard Evans Schultes and Arthur Stanley Pease, *Generic Names of Orchids. Their Origin and Meaning*. 260. New York and London 1963.

Prometheum (A. Berger) H. Ohba Crassulaceae

Origins:

Prometheus, a son of Iapetus by Climene, he formed men of clay, he ridiculed the gods and deceived Jupiter himself.

Pronaya Endl. Pittosporaceae

Origins:

After the Hungarian naturalist Ladislaus von Pronay.

Pronaya Huegel Pittosporaceae

Origins:

After the Hungarian naturalist Ladislaus von Pronay, died in 1868, Imperial privy councillor, friend of the German-born Austrian plant collector and traveler Baron Karl Alexander Anselm Freiherr von Hügel (Huegel) (1794-1870); see Karl A.A. von Hügel, in *Botanisches Archiv der Gartenbaugesellschaft des Österreichischen Kaiserstaates*. 2, t. 6. Vienna (Oct.) 1837.

Pronephrium C. Presl Thelypteridaceae

Origins:

Greek *pro* "before" and *nephros* "kidney"; see Karl B. Presl, *Epimeliae botanicae*. 258. Pragae 1849 [reprinted from *Abhandlungen der Königlichen Böhmischen Gesellschaft der Wissenschaften*. 1851].

Prosanerpis S.F. Blake Melastomataceae

Origins:

From the Greek *prosanerpo* "creep up to."

Prosaptia C. Presl Grammitidaceae

Origins:

Greek *prosapto* "to fasten, attach."

Species/Vernacular Names:

P. kanashiroi (Hay.) Nakai (*Polypodium kanashiroi* Hayata)

Japan: shima-mukade-shida, Kanagu-suku-shida

Prosartema Gagnep. Euphorbiaceae

Origins:
Greek *prosartao* "fasten, attach," *prosartema, prosartematos* "appendage."

Prosartes D. Don Convallariaceae (Liliaceae)

Origins:
Greek *prosartes* "attached."

Proscephaleium Korth. Rubiaceae

Origins:
Greek *proskephalaion* "cushion, pillow."

Proserpinaca L. Haloragidaceae (Haloragaceae)

Origins:
Latin *proserpinaca, ae* used by Plinius for a plant, also called *polygonon*; Latin *pro-serpo, ere* "to creep forward, to creep along," Greek *proserpo*; see Helmut Genaust, *Etymologisches Wörterbuch der botanischen Pflanzennamen.* 508. Basel 1996.

Prosopanche Bary Hydnoraceae

Origins:
Possibly from the Latin *prosopon* used for a plant, a kind of herb, a wild poppy, and Greek *ancho* "to bind, to strangle," *anchein* "to strangle," *anche* "poison," referring to the parasitic nature.

Prosopidastrum Burkart Mimosaceae

Origins:
Referring to the genus *Prosopis*.

Prosopis L. Mimosaceae

Origins:
From the late Latin *prosopis, idis* and *prosopites, ae*, Greek *prosopis* and *prosopites* (perhaps from *prosopon* "face"), ancient names for the burdock; see C. Linnaeus, *Mantissa Plantarum.* 10, 68. 1767 and *Systema Naturae.* Ed. 12. 2: 282, 293. 1767.

Species/Vernacular Names:
P. sp.

Peru: algaroba, jacarandá

P. africana (Guill. & Perr.) Taub. (*Prosopis oblonga* Benth.; *Coulteria africana* Guill. & Perr.)

Nigeria: kiriya (Hausa); kohi (Fula); sanchi lati (Nupe); kpaye (Tiv); ayan (Yoruba); ubwa (Igbo)

Mali: gele, ngwele, niebere

P. alba Griseb. var. *alba*

Spanish: algarrobo blanco

P. alba Griseb. var. *panta* Griseb. (*Prosopis panta* (Griseb.) Hieron.)

Spanish: algarrobo panta

P. chilensis (Molina) Stuntz (*Prosopis siliquastrum* (Lagasca) DC.; *Acacia siliquastrum* Lagasca; *Ceratonia chilensis* Molina;

Spanish: algarrobo, algarrobo de Chile

Peru: algarroba, algarrobo, garroba, guarango, huaranca, huarancu, huarango, tacco, thacco, thaco

Mexico: bee, pee, be, pe, bihi, yaga bihi, yaga bii, yaga be

P. glandulosa Torrey var. *glandulosa* (*Prosopis chilensis* (Molina) Stuntz var. *glandulosa* (Torrey) Standley; *Prosopis juliflora* (Sw.) DC. var. *glandulosa* (Torrey) Cockerell)

English: mesquite, honey mesquite, prosopis

South Africa: duitwesdoring, muskietboom, suidwesdoring

P. glandulosa Torrey var. *torreyana* (L. Benson) M. Johnston (*Prosopis juliflora* (Sw.) DC. var. *torreyana* L. Benson)

English: mesquite, western honey mesquite, honey mesquite

P. juliflora (Sw.) DC. var. *juliflora* (*Mimosa juliflora* Sw.; *Prosopis vidaliana* Naves)

Spanish: algarrobo, mesquite

Peru: algarroba, garroba, guarango, huarango

Mexico: mezquite, mizquitl

The Philippines: aromang dagat

P. laevigata (Humb. & Bonpl. ex Willd.) M. Johnston (*Prosopis dulcis* Kunth; *Acacia laevigata* Humb. & Bonpl. ex Willd.)

Spanish: mesquite

P. nigra (Griseb.) Hieron. (*Prosopis algarobilla* Griseb. var. *nigra* Griseb.)

Spanish: algarrobo negro

P. pallida (Humb. & Bonpl. ex Willd.) Kunth (*Acacia pallida* Humb. & Bonpl. ex Willd.)

Spanish: algarrobo, algarroba, algaroba, mesquite

Peru: algarrobo, algarrobo paiva, garroba, huarango

Hawaii: kiawe

P. pubescens Benth.

English: screwbean

Spanish: tornillo

P. tamarugo F. Philippi

Spanish: tamarugo

P. velutina Wooton (*Prosopis chilensis* (Molina) Stuntz var. *velutina* (Wooton) Standley; *Prosopis juliflora* (Sw.) DC. var. *velutina* (Wooton) Sarg.)

English: velvet mesquite, prosopis

South Africa: fluweelboontjie

Prosopostelma Baillon Asclepiadaceae

Origins:

Greek *prosopon* "face" and *stelma, stelmatos* "a girdle, belt."

Prosphysis Dulac Gramineae

Origins:

Greek *prosphysis* "growing to, ongrowth, growth of new wood."

Prosphytochloa Schweick. Gramineae

Origins:

Greek *prosphysis* "growing to, growth of new wood" and *chloe, chloa* "grass," indicating the nature of the plant.

Prostanthera Labill. Labiatae

Origins:

Greek *prostheke, prosthema* "appendage, addition, supplement," *prostithemi* "to put besides, add to, join," referring to the appendages borne by the anthers; see J.J.H. de Labillardière, *Novae Hollandiae plantarum specimen.* 2: 18, t. 157. Parisiis (Mar.) 1806; Georg Christian Wittstein, *Etymologisch-botanisches Handwörterbuch.* 727. Ansbach 1852; Salvatore Battaglia, *Grande dizionario della lingua italiana.* XIV: 722. Torino 1988; Barry J. Conn, "A taxonomic revision of *Prostanthera* Labill. Section *Prostanthera* (Labiatae). 1. The species of the Northern Territory, South Australia and Western Australia." *Nuytsia.* 6(3): 351-411. 1988; some suggest from the Greek *prosthen, prosthe* "before" and *anthera* "anther."

Species/Vernacular Names:

P. aspalathoides Benth.

English: scarlet mintbush, pixie caps

P. behriana Schltdl.

English: downy mintbush

P. cuneata Benth.

English: alpine mintbush

P. denticulata R. Br.

English: rough mintbush

P. incana Benth.

English: velvet mintbush

P. incisa R. Br.

English: cut-leaved mintbush

P. lasianthos Labill.

Australia: Victorian Christmas mintbush

P. linearis R. Br.

English: narrow-leaved mintbush

P. nivea Benth.

English: snowy mintbush

P. ringens Benth.

English: gaping mintbush

P. rotundifolia R. Br.

English: round-leaved mintbush

P. serpyllifolia (R. Br.) Briq. subsp. *microphylla* (R. Br.) Conn

English: small-leaved mintbush

P. spinosa F. Muell.

English: spiny mintbush

P. striatiflora F. Muell.

English: striated mintbush, jockey's cap, striped mintbush, streaked mintbush

P. walteri F. Muell.

English: blotchy mintbush

Prosthechea Knowles & Westcott Orchidaceae

Origins:

Greek *prostheke, prosthema* "appendage, addition, supplement," referring to the appendage on the back of the column.

Prosthecidiscus F.D. Sm. Asclepiadaceae

Origins:

Greek *prostheke* "appendage, addition" and *diskos* "a disc."

Protamomum Ridley Lowiaceae

Origins:
Greek *protos* "first, foremost" and *amomon*, an aromatic shrub, an Indian spice plant, the genus *Amomum* Roxb. (Zingiberaceae), see also *Orchidantha* N.E. Br.

Protangiopteris Hayata Marattiaceae (Angiopteridaceae)

Origins:
Greek *protos* "first, foremost" plus *Angiopteris* Hoffm.

Protarum Engl. Araceae

Origins:
Greek *protos* "first, foremost" plus the genus *Arum*; see D.H. Nicolson, "Derivation of aroid generic names." *Aroideana*. 10: 15-25. 1988.

Protasparagus Obermeyer Asparagaceae (Liliaceae)

Origins:
Greek *protos* and the genus *Asparagus* L.; see Anna Amelia Obermeyer (1907-), in *South African Journal of Botany*. 2: 243. 1983.

Species/Vernacular Names:
P. aethiopicus (L.) Oberm.

English: Sprengeri fern, an asparagus fern

P. densiflorus (Kunth) Oberm. (*Asparagus sprengeri* Regel; *Asparagus densiflorus* (Kunth) Jessop; *Asparagopsis densiflora* Kunth)

English: Sprengeri fern, asparagus fern

P. plumosus (Baker) Oberm. (*Asparagus plumosus* Bak.; *Asparagus setaceus* (Kunth) Jessop)

English: climbing asparagus fern, asparagus fern

P. racemosus (Willd.) Oberm. (*Asparagus racemosus* Willd.; *Asparagus saundersiae* Bak.; *Asparagus fasciculatus* R. Br.; *Asparagus acerosus* Roxb.)

English: asparagus fern, native asparagus

India: shadavari

P. suaveolens (Burch.) Oberm. (*Asparagus suaveolens* Burch.)

Southern Africa: wag'n-bietjie; imVane (Xhosa)

P. virgatus (Baker) Oberm. (*Asparagus virgatus* Baker)

English: broom-switch, an asparagus fern

Protea L. Proteaceae

Origins:
After the marine demi-god Proteus, son of Poseidon or Neptune, capable of appearing in many different forms. See C. Linnaeus, *Mantissa Plantarum*. 1: 187. (Oct.) 1767 and *Genera Plantarum*. Ed. 5. 41. 1754.

Species/Vernacular Names:
P. angolensis Welw.

English: northern protea

Southern Africa: muBanda, muBonda (Shona)

P. caffra Meisn. subsp. *caffra* (*Protea bolusii* Phill.; *Protea multibracteata* Phill.; *Protea rhodantha* Hook.f.)

English: Natal sugarbush, common sugarbush, highveld protea

Southern Africa: gewone suikerbos, suikerbos; isiQalaba-sentaba, uHlinkihlane, isiQalaba (Zulu); isiQwane, isiQalaba, iNdlunge (Xhosa); sekila (South Sotho: Lesotho, Orange Free State, southeast Transvaal); sengenge, mohlanko, mogalagala (North Sotho: north and northeast Transvaal)

P. comptonii Beard (after the British (b. Glos.) Prof. Robert Harold Compton, 1886-1979 (d. Cape Town), second Director of National Botanic Gardens at Kirstenbosch, botanical explorer, plant collector (New Caledonia), 1935-1953 founder and editor of *Journal of South African Botany*, author of *Our South African Flora*. Cape Town 1940 and *An Annotated Check List of the Flora of Swaziland*. Cape Town 1966; see Mary Gunn and Leslie E. Codd, *Botanical Exploration of Southern Africa*. Cape Town 1981; J.H. Barnhart, *Biographical Notes upon Botanists*. 1: 371. 1965; J. Hutchinson, *A Botanist in Southern Africa*. London 1946; T.W. Bossert, *Biographical Dictionary of Botanists Represented in the Hunt Institute Portrait Collection*. 81. 1972; Mia C. Karsten, *The Old Company's Garden at the Cape and Its Superintendents: Involving an Historical Account of Early Cape Botany*. Cape Town 1951; Gordon Douglas Rowley, *A History of Succulent Plants*. 1997)

English: Barberton mountain sugarbush

South Africa: Barbertonse bergsuikerbos (the popular name after the town of Barberton, eastern Transvaal, South Africa)

P. coronata Lam. (*Protea incompta* R. Br.; *Protea incompta* R. Br. var. *susannae* Phill.; *Protea macrocephala* Thunb.) (to commemorate Susanna Steyn, the wife of the Scottish naturalist John Muir (1874-1947), physician, plant collector at the Cape; see Gordon Douglas Rowley, *A History of Succulent Plants*. Strawberry Press, Mill Valley, California 1997; Mary Gunn and Leslie E. Codd, *Botanical Exploration of Southern Africa*. 256. Cape Town 1981; Stafleu and Cowan, *Taxonomic Literature*. 3: 658. 1981; Ray Desmond, *Dictionary of British & Irish Botanists and Horticulturists*. 506. London 1994

South Africa: large-headed protea

P. curvata N.E. Br.

South Africa: curved protea

P. cynaroides (L.) L.

English: king protea, giant protea

P. eximia (Salisb. ex Knight) Fourc. (*Protea latifolia* R. Br.; *Protea latifolia* R. Br. var. *auriculata* (Tausch.) O. Kuntze)

English: ray-flowered protea

South Africa: waboom

P. gaguedi Gmelin (*Protea abyssinica* Willd.; *Protea chrysolepis* Engl. & Gilg.; *Protea trigona* Phill.) (the specific name after one of the common names for the tree in Ethiopia)

English: African white sugarbush, sugarbush

Southern Africa: groot suikerbos, Afrikaanse witsuikerbos; sitsuru, sundhla, tshundha, muBanda, muBonda, chiBonja, chiDendere, muMonda, mondo, muNdendere, muOnda (Shona); isiqalaba (Ndebele: central and southern Transvaal); tshizungu (Venda: Southpansberg, northern Transvaal)

P. glabra Thunb.

South Africa: Kaiingbos

P. laurifolia Thunb. (*Protea comigera* Stapf; *Protea marginata* Thunb.)

South Africa: laurel-leaved protea, fringed protea

P. lepidocarpodendron (L.) L. (*Protea fulva* Tausch.; *Protea lepidocarpodendron* (L.) L. var. *villosa* Phill.; *Protea lepidocarpon* Sims) (the specific name means "scaly-fruited tree", from Greek *lepis, lepidos* "scale," *karpos* "fruit" and *dendron* "tree")

English: blackbeard protea, black-flowered protea

South Africa: swartbaardsuikerbos

P. lorifolia (Salisb. ex Knight) Fourc. (*Protea macrophylla* R. Br.)

South Africa: thong-shaped protea

P. madiensis Oliv. (*Protea elliottii* C.H. Wright; *Protea elliottii* var. *angustifolia* Keay; *Protea argyrophaea* Hutch.)

Southern Africa: chiRapadzungu (Shona)

Nigeria: halshensa, halshen-tunkiya (Hausa); dehinbolorun (Yoruba); okwo (Igbo)

Yoruba: dehinkorun, dehinkolorun, dehinbolorun

P. magnifica Link (*Protea barbigera* Meisn.)

English: woolly beard

South Africa: baardsuikerkan, baardsuikerbos

P. neriifolia R. Br. (the specific name is based on *Nerium*, the oleander)

English: blue sugarbush, oleander-leaved protea

South Africa: blousuikerbos, baardsuikerbos, roosboom, waboom

P. nitida Mill. (*Protea arborea* Houtt.; *Protea grandiflora* Thunb.)

Southern Africa: blousuikerbos, brandhout, waboom, wagonboom; isAdlunge (Xhosa)

P. obtusifolia Buek ex Meisn. (*Protea calocephala* Meisn.)

South Africa: Bredasdorp sugarush (the species grows in the Bredasdorp district)

P. repens (L.) L. (*Protea mellifera* Thunb.; *Protea mellifera* Thunb. var. *albiflora* Andr.; *Leucadendron repens* L.)

English: sugarbush, honey-bearing protea

South Africa: suikerbos, suikerbossie, stroopbos

P. roupelliae Meisn. subsp. *roupelliae* (*P. incana* Hort. ex Meisn.) (for the English botanical artist Mrs. Arabella Elizabeth Roupell (*née* Piggott), 1817-1914, painter of *Specimens of the Flora of South Africa* by a Lady. [London 1850] (the botanical text is by William Henry Harvey, 1811-1866); see Allan Bird, *The Lady of the Cape Flowers*. Paintings by Arabella Roupell. Johannesburg. The South African Natural History Publ. [1964]; Allan Bird, *More Cape Flowers by a Lady*. Paintings by Arabella Roupell. Johannesburg. The South African Natural History Publ. 1964; John H. Barnhart, *Biographical Notes upon Botanists*. 3: 184. 1965; Mary Gunn and Leslie E. Codd, *Botanical Exploration of Southern Africa*. 301-302. Cape Town 1981)

English: Drakensberg protea, silver sugarbush

Southern Africa: silwersuikerbos; uQhambathi, isiQalaba (Zulu); isiQalaba (Xhosa); seqalaba (South Sotho: Lesotho, Orange Free State, southeast Transvaal)

P. rubropilosa Beard

South Africa: velvet protea

P. subvestita N.E. Br. (*Protea lacticolor* Salisb. var. *angustata* Phill.; *Protea lacticolor* Salisb. var. *orientalis* (Sim) Phill.) (the specific name from Latin, means "slightly clothed")

English: lip-flower sugarbush, white sugarbush

Southern Africa: witsuikerbos, lippeblomsuikerbos; seqalaba se sesweu (South Sotho); isiQalaba (Zulu), isiQane, isiQalaba (Xhosa)

P. susannae Phill. (the specific name after Mrs. Susan Muir, the wife of the Scottish-born botanist and collector Dr. John Muir, 1874-1947, who wrote "The vegetation of the Riversdale area." *Mem. Bot. Surv. Afr.* 13. 1929)

South Africa: suikerbos

Proteopsis Martius & Zucc. ex Sch. Bip. Asteraceae

Origins:

Resembling the genus *Protea*.

Protium Burm.f. Burseraceae

Origins:
From a Javanese name or from Proteus, referring to the rate of growth; see Nicolaas Laurens Burman, *Flora Indica*. 88. Lugduni Batavorum (Mar.-Apr.) 1768.

Species/Vernacular Names:
P. fimbriatum Sw.
Brazil (Amazonas): weyeri hi, mani hi
P. heptaphyllum (Aublet) Marchand
English: Brazilian elemi, incense tree

Protoceras Joseph & Vajravelu Orchidaceae

Origins:
Greek *protos* "first" and *keras* "a horn."

Protocyrtandra Hosok. Gesneriaceae

Origins:
Greek *protos* "first" plus *Cyrtandra* Forst. & Forst.f.

Protogabunia Boiteau Apocynaceae

Origins:
Greek *protos* "first" plus the genus *Gabunia* Schumann ex Stapf.

Protolindsaya Copel. Dennstaedtiaceae (Lindsaeoideae)

Origins:
Greek *protos* "first" plus the genus *Lindsaea* Dryander ex Sm., after the Jamaican botanist John Lindsay, 1785-1803.

Protolirion Ridley Melanthiaceae (Liliaceae)

Origins:
Greek *protos* "first" and *leirion* "a lily."

Protomarattia Hayata Marattiaceae

Origins:
Greek *protos* plus the genus *Marattia* Sw., dedicated to the Italian botanist (Francesco Giovanni) (Gaetano) Giovanni Francesco Maratti (Joannes Franciscus Marattius), 1723-1777, clergyman, professor at Rome University. His works

include *Descriptio de vera florum existentia vegetatione et forma in plantis dorsiferis*. Romae 1760, *Flora romana ... Opus postumum*. [Edited by M.B. Oliverius.] Romae 1822 and *De plantis zoophytis et lithophytis in Mari mediterraneo viventibus*. Romae 1776. See Pier Andrea Saccardo (1845-1920), *Di un'operetta sulla flora della Corsica di autore pseudonimo e plagiario*. Venezia 1908; Angelo Calogierà, *Nuova raccolta d'opuscoli scientifici e filologici*. [Botanophili Romani ad ... C. Amadutium epistola, qua J.F. Marattium ab Adansonii ... censuris vindicat. 1770] Venezia 1755-1784; R. Zander, F. Encke, G. Buchheim and S. Seybold, *Handwörterbuch der Pflanzennamen*. 14. Aufl. 747. 1993; F. Boerner & G. Kunkel, *Taschenwörterbuch der botanischen Pflanzennamen*. 4. Aufl. 131. Berlin & Hamburg 1989; Jonas C. Dryander, *Catalogus bibliothecae historico-naturalis Josephi Banks*. London 1796-1800; Mariella Azzarello Di Misa, a cura di, *Il Fondo Antico della Biblioteca dell'Orto Botanico di Palermo*. 181. Regione Siciliana, Palermo 1988; E.M. Tucker, *Catalogue of the Library of the Arnold Arboretum of Harvard University*. 1917-1933; John H. Barnhart, *Biographical Notes upon Botanists*. 2: 445. 1965.

Protomegabaria Hutch. Euphorbiaceae

Origins:
Greek *protos* "first" and the genus *Megabaria* Pierre ex Hutch.

Protorhus Engl. Anacardiaceae

Origins:
Greek *protos* "first" and *Rhus*, maybe because *Protorhus* approaches the genus *Rhus*.

Species/Vernacular Names:
P. longifolia (Bernh.) Engl. (*Rhus longifolia* (Bernh.) Sond.; *Anaphrenium longifolium* Bernh.)

English: red beech

Southern Africa: rooiboekenhout, rooibeukehout, harpuisboom; umHlangothi, imFuce (Swazi); isiFice, isiFico, isiFico sehlathi, uNhlangothi, umHlangothi, umHluthi, umHluthi wehlathi (Zulu); uZintlwa, umKhumiso, umKomiso, umKupati, iKhubalo, umHluthi (Xhosa)

Protoschwenckia Solereder Solanaceae

Origins:
After the Dutch physician Martin Wilhelm (Martinus Wilhelmus) Schwencke, 1707-1785, botanist. His works include Dissertatio ... *de operatione inguinali*. Lugduni

Batavorum 1731, *Novae plantae Schwenckia* dictae a cele-berrimo Linnaeo in *Gen. plant.* ed. VI. p. 567 ex celeb. Davidiis van Rooijen Charact. mss. 1761 communicata brevis descriptio et delineatio cum notis characteristicis. Hagae Comitum [The Hague] [typ. van Karnebeek] 1766 and *Officinalium plantarum catalogus.* Hagae-Comitum 1752; see John H. Barnhart, *Biographical Notes upon Botanists.* 3: 250. 1965; Jonas C. Dryander, *Catalogus biblio-thecae historico-naturalis Josephi Banks.* London 1800; H. Heine, in *Kew Bulletin.* 16(3): 465-469. 1963; Frans A. Stafleu and Cowan, *Taxonomic Literature.* 5: 442-443. 1985.

Protoschwenkia Soler. Solanaceae

Origins:
From the Greek *protos* "first" and the genus *Schwenkia* L.

Protowoodsia Ching Dryopteridaceae (Aspleniaceae, Woodsiaceae)

Origins:
Greek *protos* with *Woodsia* R. Br., after the English archi-tect Joseph Woods, 1776-1864, botanist, Fellow of the Lin-nean Society 1807, author of *The Tourist's Flora.* London 1850 and *A Synopsis of the British Species of Rosa.* London 1818; see Dawson Turner and Lewis Weston Dillwyn, *The Botanist's Guide Through England and Wales.* London 1805; H.N. Clokie, *Account of the Herbaria of the Depart-ment of Botany in the University of Oxford.* 267. Oxford 1964; Mariella Azzarello Di Misa, a cura di, *Il Fondo Antico della Biblioteca dell'Orto Botanico di Palermo.* 294. Pal-ermo 1988; R. Zander, F. Encke, G. Buchheim and S. Sey-bold, *Handwörterbuch der Pflanzennamen.* 14. Aufl. 802. 1993; J.H. Barnhart, *Biographical Notes upon Botanists.* 3: 518. 1965; E.M. Tucker, *Catalogue of the Library of the Arnold Arboretum of Harvard University.* Cambridge, Mas-sachusetts 1917-1933.

Provancheria B. Boivin Caryophyllaceae

Origins:
For the Canadian botanist Léon Provancher, 1820-1892, clergyman, naturalist, entomologist, educator, bryologist, traveler, (*rédacteur-propriétaire*) founded *Le Naturaliste Canadien* and edited vols. 1-20 (1869-1891). His works include *Flore canadienne.* Québec 1862, *Une Excursion aux Climats Tropicaux. Voyage aux Iles-du-vent.* Québec 1890, *Les Oiseaux insectivores et les arbres d'ornement et forestiers.* Québec 1874 and *Traité élémentaire de Bota-nique.* Québec 1858; see John H. Barnhart, *Biographical*

Notes upon Botanists. 2: 113. 1965; E.M. Tucker, *Catalogue of the Library of the Arnold Arboretum of Harvard Univer-sity.* 1917-1933.

Provencheria B. Boivin Caryophyllaceae

Origins:
For the Canadian botanist Léon Provancher, 1820-1892.

Prumnopitys Philippi Podocarpaceae

Origins:
Greek *prumna, prumne* "the stern, poop" and *pitys* "the pine, fir-tree"; see Rudolph Amandus Philippi (1808-1904), in *Linnaea.* 30: 731. (Mar.) 1861.

Species/Vernacular Names:
P. amara (Bl.) Laubenf.
The Philippines: pasuig, tumpis
P. andina (Poepp. ex Endl.) Laubenf.
Chile: lleuque, uva de cordillera
P. ferruginea (D. Don) Laubenf.
English: rusty podocarpus
New Zealand: miro
P. taxifolia (D. Don) Laubenf.
English: black pine
New Zealand: matai

Prunella L. Labiatae

Origins:
Corruption of *brunella, brunelle,* used in the 15th and 16th centuries by German herbalists, it probably derives from the German *braun* (Latin *prunum*) "purple" or *Bräune* "quinsy"; see Carl Linnaeus, *Species Plantarum.* 600. 1753 and *Genera Plantarum.* Ed. 5. 261. 1754; Helmut Genaust, *Etymologisches Wörterbuch der botanischen Pflanzenna-men.* 509. Basel 1996.

Species/Vernacular Names:
P. asiatica Nakai
English: Asian self-heal
China: shan bo cai
P. grandiflora (L.) Jacquin (*Prunella grandiflora* (L.) Moench)
English: bigflower self-heal
China: da hua xia ku cao

P. hispida Bentham

English: hispid self-heal

China: ying mao xia ku cao

P. laciniata L.

English: cut-leaf self-heal

P. vulgaris L.

English: heal-all, self-heal, sicklewort

China: xia ku cao, hsia ku tsao

Vietnam: ha kho thao

Prunellopsis Kudô Labiatae

Origins:
Resembling the genus *Prunella*.

Prunus L. Rosaceae

Origins:
From *prunus*, the ancient Latin name for the plum tree, Greek *proumne* "plum tree," *proumnon* plum"; see Carl Linnaeus, *Species Plantarum*. 473. 1753 and *Genera Plantarum*. Ed. 5. 213. 1754; Salvatore Battaglia, *Grande dizionario della lingua italiana*. XIV: 829-832. Torino 1988; Ernest Weekley, *An Etymological Dictionary of Modern English*. 2: 1113, 1162. New York 1967; [Crusca], *Vocabolario degli Accademici della Crusca*. Firenze 1729-1738; G. Volpi, "Le falsificazioni di Francesco Redi nel Vocabolario della Crusca." in *Atti della R. Accademia della Crusca per la lingua d'Italia*. 33-136. 1915-1916; Manlio Cortelazzo & Paolo Zolli, *Dizionario etimologico della lingua italiana*. 4: 995, 996. 1985; N. Tommaseo & B. Bellini, *Dizionario della lingua italiana*. Torino 1865-1879; Giovanni Semerano, *Le origini della cultura europea*. Dizionario della lingua Latina e di voci moderne. 2(2): 533. 1994.

Species/Vernacular Names:
P. spp.

English: cherry blossoms

Japan: sakura

P. africana (Hook.f.) Kalkm. (*Pygeum africanum* Hook.f.)

English: red stinkwood, bitter almond

Cameroon: alumty, vla

Southern Africa: rooistinkhout, bitteramandel, nuwehout; mogotlhori (north and northeast Transvaal); mulala-maanga (Venda); umDumezulu, umDumizula, iNkokhokho, umLalume, nGubozinyeweni, umKhakhazi (Zulu); muChambati, chati (Shona); iNyazangoma, umKakase, umKhakhazi (Xhosa)

Kenya: mueri

P. alleghaniensis Porter

English: northern sloe, Alleghany plum, sloe

P. americana Marsh.

English: wild plum, American red plum, August plum, goose plum, hog plum, sloe

P. andersonii A. Gray

English: desert peach

P. angustifolia Marsh.

English: Chicasa plum, Chickasaw plum, sand plum

P. apetala Franch. & Savat.

English: clove cherry

P. armeniaca L.

English: apricot, common apricot

China: xing ren, hsing, tien mei

Vietnam: mo, mai

Tibetan: kham-bu

Japan: anzu

Arabic: michmèche

P. avium (L.) L.

English: bird cherry, sweet cherry, gean, wild cherry, mazzard cherry, mazzard

P. besseyi Bail.

English: western sand cherry, Rocky Mountains cherry

P. brigantina Vill.

French: Briançon apricot

P. campanulata Maxim.

English: Taiwan cherry, Formosan cherry, bell-flowered cherry

Japan: hi-kan-zakura, sakura

P. canescens Bois.

English: hoary cherry, greyleaf cherry

P. caroliniana (Mill.) Ait.

English: cherry laurel, laurel cherry, wild orange, mock orange, Carolina cherry laurel

P. cerasifera Ehrh.

English: cherry plum, Myrobalan, Myrobalan plum

P. cerasus L.

English: sour cherry, pie cherry

Mexico: belohui naxiñaa, yaga belohui naxiñaa, biziaa nayi xtilla, piziaa nayi castilla

P. cerasus L. var. *austera* L.

English: morello cherry

P. cerasus L. var. *caproniana* L.

English: amarelle cherry, Kentish red cherry

P. cerasus L. var. *frutescens* Neilr.

English: bush sour cherry

P. changyangensis (F.B. Ingram) F.B. Ingram

English: Chinese spring cherry

P. cornuta (Royle) Steud.

English: Himalayan bird cherry, bird cherry

P. davidiana (Carr.) Franch.

English: David's peach, Chinese wild peach

P. domestica L.

English: plum, European plum, garden plum, prune plum

Arabic: aouina

China: li, chia ching tzu

P. dulcis (Miller) D.A. Webb (*Amygdalus dulcis* Miller; *Prunus amygdalus* Batsch)

English: almond, almond tree

Mexico: bizoya xtilla, pizoya castilla

China: ba dan xing ren

P. emarginata (Hook.) Walp.

English: bitter cherry, Oregon cherry

P. fasciculata (Torrey) A. Gray

English: desert almond

P. fremontii S. Watson

English: desert apricot

P. fruticosa Pall.

English: European dwarf cherry, European ground cherry

P. ilicifolia (Nutt.) Walp.

English: evergreen cherry, holly-leaf cherry, islay, mountain holly, wild cherry

P. japonica Thunb.

English: bush cherry, dwarf flowering cherry, Chinese plum tree, Japanese plum

Japan: niwa-ume, kunmi

China: yu li ren, yu li, tang ti, chiao mei, chang ti

P. laurocerasus L.

English: cherry laurel, bay laurel, cherry bay, laurel cherry, English cherry laurel

P. lusitanica L.

English: Portugal laurel, Portuguese cherry bay, Portuguese cherry laurel

Portuguese: gingeira brava

P. mahaleb L.

English: St. Lucie cherry, Mahaleb cherry, perfumed cherry

P. maritima Marsh.

English: beach plum, shore plum

P. mexicana Watson

English: big-tree plum, Mexican plum

P. mume Sieb. & Zucc.

English: Japanese apricot, mume

Japan: ume, umi

China: wu mei, mei

P. padus L.

English: bird cherry, common bird cherry

P. pensylvanica L.f.

English: bird cherry, pin cherry, red cherry

P. persica (L.) Batsch (*Prunus amygdalus* (L.) Batsch; *Prunus vulgaris* Miller)

English: peach, nectarine

Italian: pesca

Arabic: khoukh

Mexico: yaga nocuana naxi castilla

Southern Africa: perskeboom; umumpetshisi (Zulu)

Japan: momo, kii-mumu

China: tao ye, tao ren, tao

Vietnam: dao, may phang, co tao, phieu kiao

P. pseudocerasus Lindl.

English: ying tao cherry

China: ying tao

P. pumila L.

English: sand cherry

P. rivularis Scheele

English: creek plum

P. rufa Hook.f.

English: Himalayan cherry

P. salicifolia Kunth

English: Mexican bird cherry

P. salicina Lindl. (*Prunus triflora* Roxb.)

English: Japanese plum

Japan: su-momo, sumumu, sumomo

China: li, chia ching tzu

P. scopulorum Koehne

English: cliff cherry

P. serotina Ehrh.

English: black cherry, rum cherry, wild black cherry

P. serrula Franch.

English: oriental cherry

P. serrulata Lindley

English: the mountain cherry

Japan: yama-zakura

P. sibirica L.

English: Siberian apricot

P. subhirtella Miq.

English: winter flowering cherry, spring cherry, Higan cherry, rosebud cherry

P. tatsienensis Batal.

English: Kangting cherry

P. tianshanica (Pojark.) Yu & Li

English: Tianshan Mountain cherry

P. tomentosa Thunb.

English: downy cherry, Manchu cherry, nanking Cherry

China: shan ying tao, chu tao, li tao, mei tao

P. trichostoma Koehne

English: West Szechwan cherry

P. undulata Hamil.

English: undulate bird cherry

P. wallichii Steud.

English: Wallich laurel cherry

P. yedoensis Matsum.

English: Tokyo cherry, Yoshino cherry

P. yunnanensis Franch.

English: Yunnan cherry

Przewalskia Maxim. Solanaceae

Origins:
For the great Russian naturalist Nikolay Mikhaylovich (Nikolai Michailowicz or Mikhailovich) Przhevalsky (Prejevalsky, Przewalski, Przheval'sky, Przseválszkij, Prschevalskij), 1839-1888, traveler, geographer, explorer of Central Asia. His writings include *Mongolia, the Tangut Country and the Solitudes of Northern Tibet*. [Translated from the Russian by E.D. Morgan] London 1876 and *From Kulja, across the Tian Shan to Lob-Nor*. [Translated by E.D. Morgan] London 1879 [1878]; see John H. Barnhart, *Biographical Notes upon Botanists*. 3: 114. 1965; Vasiliy A. Esakov, in *Dictionary of Scientific Biography* 11: 180-182. 1981; Donald Rayfield, *The Dream of Lhasa. The Life of Nikolay Przhevalsky (1839-1888) Explorer of Central Asia*. Paul Elek 1976; E.M. Tucker, *Catalogue of the Library of the Arnold Arboretum of Harvard University*. 1917-1933; T.W. Bossert, *Biographical Dictionary of Botanists Represented in the Hunt Institute Portrait Collection*. 319. 1972; Emil Bretschneider (1833-1901), *History of European Botanical Discoveries in China*. [Reprint of the original edition 1898.] Leipzig 1981; I.C. Hedge and J.M. Lamond, *Index of Collectors in the Edinburgh Herbarium*. 1970.

Species/Vernacular Names:
P. tangutica Maxim.

English: Tangut przewalskia

China: ma niao pao

Psacadocalymma Bremek. Acanthaceae

Origins:
Greek *psekadion, psakadion*, diminutive of *psekas, psakas* "drizzle, drop of rain" and *kalymma* "a covering."

Psacadopaepale Bremek. Acanthaceae

Origins:
Greek *psakadion*, diminutive of *psakas* "drizzle, drop of rain" and *paipale* "the finest flour or meal."

Psacaliopsis H. Robinson & Brettell Asteraceae

Origins:
Resembling *Psacalium* Cass.

Psacalium Cass. Asteraceae

Origins:
Greek *psakalon* "new-born, new-born animal," referring to the flowers.

Species/Vernacular Names:
P. decompositum (A. Gray) H. Robinson & Brettell

Mexico: pichichagua, matarique, mataril, chicura, maturi, maturí

Psamma P. Beauv. Gramineae

Origins:
Greek *psammos* "sand," referring to the habitat, see also the genus *Ammophila* Host; see A. Palisot de Beauvois, *Essai d'une nouvelle Agrostographie*. 143, 176. Paris (Dec.) 1812.

Psammagrostis C.A. Gardner & C.E. Hubb. Gramineae

Origins:
Greek *psammos* "sand" and *agrostis, agrostidos* "grass, weed, couch grass"; see Charles Austin Gardner (1896-

1970) and Charles Edward Hubbard (1900-1980), *Hooker's Icones Plantarum*. Ser. 5. (Sep.) 1938.

Psammetes Hepper Scrophulariaceae

Origins:
Greek *psammos, psamme* "sand," *psammites* "from sand."

Psammiosorus C. Chr. Oleandraceae

Origins:
Greek *psammion* "grain of sand" and *soros* "a spore case, a heap."

Psammisia Klotzsch Ericaceae

Origins:
From the Greek *psammiaios* "of the size of a grain of sand."

Psammochloa A. Hitchc. Gramineae

Origins:
From the Greek *psammos* "sand" and *chloe, chloa* "grass."

Psammogeton Edgew. Umbelliferae

Origins:
Greek *psammos* "sand" and *geiton* "a neighbor."

Psammomoya Diels & Loes. Celastraceae

Origins:
Greek *psammos* "sand" and the South American genus *Moya* Griseb.; see F.L.E. Diels and E.G. Pritzel, in *Botanische Jahrbücher*. 35: 339. (Feb.) 1904.

Psammophila Ikonn. Caryophyllaceae

Origins:
Greek *psammos* "sand" and *philos* "lover, loving," see also *Gypsophila* L.

Psammophila Schult. Gramineae

Origins:
Greek *psammos* "sand" and *philos* "lover, loving," see also *Spartina* Schreb.

Psammophiliella Ikonn. Caryophyllaceae

Origins:
The diminutive of the genus *Psammophila* Ikonn., see also *Gypsophila* L.

Psammophora Dinter & Schwantes Aizoaceae

Origins:
From the Greek *psammos* and *phoros* "carrying, bearing," referring to the appearance of the leaves.

Psammopyrum Á. Löve Gramineae

Origins:
Greek *psammos* "sand" and *pyros* "grain, wheat."

Psammotropha Ecklon & Zeyher Molluginaceae

Origins:
Greek *psammos* "sand" and *trophe* "food."

Psathura Comm. ex Juss. Rubiaceae

Origins:
Greek *psathyros* "fragile, friable."

Species/Vernacular Names:
P. borbonica J.F. Gmelin
La Réunion Island: bois cassant

Psathurochaeta DC. Asteraceae

Origins:
Greek *psathyros* and *chaite* "bristle, long hair, mane, foliage."

Psathyranthus Ule Loranthaceae

Origins:
Greek *psathyros* "fragile, friable, crumbling" and *anthos* "flower."

Psathyrostachys (Boiss.) Nevski Gramineae

Origins:
From the Greek *psathyros* "fragile, friable" and *stachys* "a spike."

Species/Vernacular Names:
P. juncea (Fischer) Nevski
English: Russian wild rye

Psathyrotes A. Gray Asteraceae

Origins:
Greek *psathyrotes* "looseness of consistency," brittleness of the stem.

Species/Vernacular Names:
P. ramosissima (Torrey) A. Gray
English: turtleback

Psathyrotopsis Rydberg Asteraceae

Origins:
Resembling the genus *Psathyrotes*.

Psednotrichia Hiern Asteraceae

Origins:
Greek *psednos* "rubbed off, thin, scanty, spare" and *trichion* "small hair."

Pseudabutilon R. Fries Malvaceae

Origins:
From the Greek *pseudes* "false" plus *Abutilon*.

Pseudacanthopale Benoist Acanthaceae

Origins:
Greek *pseudes* plus the genus *Acanthopale* C.B. Clarke.

Pseudacoridium Ames Orchidaceae

Origins:
From the Greek *pseudes* plus the genus of orchids *Acoridium*.

Pseudactis S. Moore Asteraceae

Origins:
From the Greek *pseudes* "false" and *aktis, aktin* "a ray."

Pseudaechmanthera Bremek. Acanthaceae

Origins:
Greek *pseudes* "false" plus the genus *Aechmanthera* Nees.

Pseudaechmea L.B. Sm. & Read Bromeliaceae

Origins:
Greek *pseudes* "false" plus the genus *Aechmea* Ruíz & Pavón.

Pseudaegiphila Rusby Labiatae

Origins:
Greek *pseudes* plus the genus *Aegiphila* Jacq.

Pseudagrostistachys Pax & K. Hoffmann Euphorbiaceae

Origins:
Greek *pseudes* "false" plus the genus *Agrostistachys* Dalz.

Pseudaidia Tirveng. Rubiaceae

Origins:
Greek *pseudes* plus the genus *Aidia* Lour.

Pseudais Decne. Thymelaeaceae

Origins:
From the Greek *pseudes* "false" plus the genus *Dais* L.

Pseudalangium F. Muell. Alangiaceae

Origins:
From the Greek *pseudes* "false" and the genus *Alangium* Lam.; see F. von Mueller, *Fragmenta Phytographiae Australiae.* 2: 84. Melbourne (Aug.) 1860.

Pseudalbizzia Britton & Rose Mimosaceae

Origins:
From the Greek *pseudes* "false" and the genus *Albizia* Durazz.

Pseudalepyrum Dandy Centrolepidaceae

Origins:

Greek *pseudes* "false" and the genus *Alepyrum* R. Br., *lepis* "a scale," *lepyron* "a rind, husk, shell," *lepyrion* "a small husk, thin peel"; see James Edgar Dandy (1903-1976), in *The Journal of Botany*. 70: 334. (Dec.) 1932.

Pseudalthenia Nakai Zannichelliaceae

Origins:

Greek *pseudes* "false" and the genus *Althenia* Petit.

Pseudammi H. Wolff Umbelliferae

Origins:

From the Greek *pseudes* and the genus *Ammi* L.

Pseudanamomis Kausel Myrtaceae

Origins:

Greek *pseudes* "false, untrue" and the genus *Anamomis* Griseb.

Pseudananas (Hassler) Harms Bromeliaceae

Origins:

Like *Ananas*, from the Greek *pseudes* "false, untrue" and the genus *Ananas* Mill.

Pseudannona (Baillon) Saff. Annonaceae

Origins:

Greek *pseudes* "false, untrue" and the genus *Annona* L.

Pseudanthistiria (Hackel) Hook.f. Gramineae

Origins:

From the Greek *pseudes* "false, untrue" plus the genus *Anthistiria* L.f.

Pseudanthus Sieber ex A. Sprengel Euphorbiaceae

Origins:

Greek *pseudes* "false" and *anthos* "flower," referring to the small flowers of *Pseudanthus pimeleoides* Sprengel; see

Kurt Polycarp Joachim Sprengel (1766-1833), *Systema Vegetabilium*. 4(2): 22, 25. (Jan.-Jun.) 1827.

Species/Vernacular Names:

P. divaricatissimus (Müll. Arg.) Benth.

English: tangled pseudanthus

P. micranthus Benth. (*Phyllanthus tatei* F. Muell.; *Micranthemum tatei* (F. Muell.) J. Black)

English: fringed pseudanthus

P. orientalis F. Muell.

English: sandhill pseudanthus

P. pimeleoides Sprengel

English: rice-flower pseudanthus

Pseudarabidella O.E. Schulz Brassicaceae

Origins:

From the Greek *pseudes* "false" and the genus *Arabidella* (F. Muell.) O. Schulz; see Otto Eugen Schulz (1874-1936), *Das Pflanzenreich*. Heft 86. 257. (July) 1924.

Pseudarrhenatherum Rouy Gramineae

Origins:

From the Greek *pseudes* and the genus *Arrhenatherum* P. Beauv.

Pseudartabotrys Pellegrin Annonaceae

Origins:

Greek *pseudes* "false" and the genus *Artabotrys* R. Br.

Pseudarthria Wight & Arnott Fabaceae

Origins:

From the Greek *pseudes* "false" and *arthron* "a joint," the pod is imperfectly articulated.

Pseudatalaya Baillon Sapindaceae

Origins:

Greek *pseudes* and *Atalaya* Blume; see Henri Ernest Baillon, *Histoire des Plantes*. 5: 419. 1873 or 1874.

Pseudatalaza Baillon Sapindaceae

Origins:

Orthographic variant of *Pseudatalaya* Baillon.

Pseudathyrium Newman Dryopteridaceae (Aspleniaceae, Woodsiaceae)

Origins:
From the Greek *pseudes* "false" plus the genus *Athyrium* Roth.

Pseudechinolaena Stapf Gramineae

Origins:
Greek *pseudes* "false" and *Echinolaena* Desv.

Pseudelephantopus Rohr Asteraceae

Origins:
Greek *pseudes* "false" with *Elephantopus* L.; see Julius Philip Benjamin von Rohr (1737-1793), in *Skrifter af Naturhistorie-Selskabet*. 2: 214. Copenhagen 1792.

Species/Vernacular Names:
P. spicatus (Jussieu ex Aublet) Rohr (*Matamoria spicata* (Jussieu ex Aublet) Llave & Lexarza)
Peru: pichi quihua, gopshi kshanate

Pseudelleanthus Brieger Orchidaceae

Origins:
From the Greek *pseudes* "false" and the genus *Elleanthus* Presl.

Pseudellipanthus Schellenb. Connaraceae

Origins:
Greek *pseudes* "false" and the genus *Ellipanthus* Hook.f.

Pseudeminia Verdc. Fabaceae

Origins:
Greek *pseudes* "false" and the genus *Eminia* Taubert.

Pseudephedranthus Aristeg. Annonaceae

Origins:
Greek *pseudes* and the genus *Ephedranthus* S. Moore.

Pseudepidendrum Reichb.f. Orchidaceae

Origins:
From the Greek *pseudes* "false" and the orchid genus *Epidendrum* L.

Pseuderanthemum Radlk. Acanthaceae

Origins:
Greek *pseudes* and *Eranthemum* L.; see Ludwig A.T. Radlkofer (1829-1927), in *Sitzungsberichte der mathematisch-physikalischen Classe der k.b. Akademie der Wissenschaften zu München*. 13: 282. 1884.

Species/Vernacular Names:
P. alatum (Nees) Radlk. (*Chamaeranthemum alatum* hort.)
English: chocolate plant
P. atropurpureum L.H. Bailey (*Eranthemum atropurpureum* hort. Bull., not Hook.f.)
English: purple false eranthemum
Japan: eransemu-modoki
P. leptorhachis Lindau
Peru: chichiac panga
P. reticulatum (Hook.f.) Radlk.
English: speckled eranthemum
Malaya: puding
P. variabile (R. Br.) Radlk.
English: pastel flower

Pseuderemostachys Popov Labiatae

Origins:
From the Greek *pseudes* "false" and the genus *Eremostachys* Bunge.

Pseuderia Schlechter Orchidaceae

Origins:
Greek *pseudes* "false" plus *Eria*; see B.A. Lewis and P.J. Cribb, *Orchids of the Solomon Islands and Bougainville*. Royal Botanic Gardens, Kew 1991.

Species/Vernacular Names:
P. similis (Schltr.) Schltr.
The Solomon Islands: totoa

Pseuderiopsis Reichb.f. Orchidaceae

Origins:

From the Greek *pseudes* "false" and the genus *Eriopsis* Lindl.

Pseuderucaria (Boiss.) O. Schulz Brassicaceae

Origins:

Greek *pseudes* "false" and the genus *Erucaria* Gaertner.

Pseudeugenia Legrand & Mattos Myrtaceae

Origins:

Greek *pseudes* "false" and the genus *Eugenia* L.

Pseudibatia Malme Asclepiadaceae

Origins:

Greek *pseudes* "false" and *batia, batos* "a bush, thicket."

Pseudima Radlk. Sapindaceae

Origins:

Possibly referring to the genus *Dimocarpus* Lour.

Pseudiosma DC. Rutaceae

Origins:

From the Greek *pseudes* "false" and *Diosma* L.

Pseudixora Miq. Rubiaceae

Origins:

From the Greek *pseudes* "false" and *Ixora* L.

Pseudo-Barleria Oerst. Acanthaceae

Origins:

Greek *pseudes* plus the genus *Barleria* L.

Pseudo-Elephantopus Rohr Asteraceae

Origins:

Greek *pseudes* plus *Elephantopus* L.

Pseudo-Lysimachium (Koch) Opiz Scrophulariaceae

Origins:

Greek *pseudes* "false" and *lysimachion* for a plant, otherwise unknown.

Pseudoacanthocereus F. Ritter Cactaceae

Origins:

Greek *pseudes* "false" with the genus *Acanthocereus* (A. Berger) Britton & Rose.

Pseudoanastatica (Boiss.) Grossh. Brassicaceae

Origins:

Greek *pseudes* with the genus *Anastatica* L.

Pseudobaccharis Cabrera Asteraceae

Origins:

From the Greek *pseudes* "false" and *Baccharis* L.

Pseudobaeckea Niedenzu Bruniaceae

Origins:

False *Baeckea*, Greek *pseudes* "false" and *Baeckea* L., Myrtaceae, after the Swedish naturalist Abraham Bäck, 1713-1795.

Pseudobahia (A. Gray) Rydb. Asteraceae

Origins:

Greek *pseudes* "false" plus the genus *Bahia* Lagasca.

Species/Vernacular Names:
P. bahiifolia (Benth.) Rydb.
English: Hartweg's pseudobahia
P. peirsonii Munz
English: Tulare pseudobahia

Pseudobarleria T. Anderson Acanthaceae

Origins:

Greek *pseudes* "false" plus the genus *Barleria* L.

Pseudobartlettia Rydb. Asteraceae

Origins:

Greek *pseudes* "false" plus the genus *Bartlettia* A. Gray.

Pseudobartsia Hong Scrophulariaceae

Origins:

Greek *pseudes* "false" plus *Bartsia* L.

Pseudobastardia Hassl. Malvaceae

Origins:

Greek *pseudes* "false" plus the genus *Bastardia* Kunth.

Pseudoberlinia Duvign. Fabaceae

Origins:

Greek *pseudes* "false" plus the genus *Berlinia* Sol. ex Hook.f. & Benth.

Pseudobersama Verdc. Meliaceae

Origins:

From the Greek *pseudes* "false" and the genus *Bersama*, the plant was first described as a *Bersama*.

Species/Vernacular Names:

P. mossambicensis (Sim) Verdc. (*Bersama mossambicensis* Sim)

English: false white ash

Southern Africa: opho tree, valswitessenhout; umOpho (Zulu)

Pseudobesleria Oerst. Gesneriaceae

Origins:

Greek *pseudes* "false" and the genus *Besleria* L., for the German botanist Basilius Besler, 1561-1629.

Pseudobetckea (Hoeck) Lincz. Valerianaceae

Origins:

For the German botanist Ernst Friedrich Betcke, 1815-1865, physician, wrote *Animadversiones botanicae in Valerianellas*. Rostock 1826; see John H. Barnhart, *Biographical Notes upon Botanists*. 1: 179. 1965; E.M. Tucker, *Catalogue of the Library of the Arnold Arboretum of Harvard University*. 1917-1933.

Pseudoblepharispermum J.-P. Lebrun & Stork Asteraceae

Origins:

From the Greek *pseudes* "false" and the genus *Blepharispermum* Wight ex DC.

Pseudoboivinella Aubrév. & Pellegrin Sapotaceae

Origins:

Greek *pseudes* "false" and the genus *Boivinella* Pierre ex Aubrév. & Pellegrin, for the French botanist Louis Hyacinthe Boivin, 1808-1852 (Brest); see F.N. Hepper and Fiona Neate, *Plant Collectors in West Africa*. 11. 1971; J.H. Barnhart, *Biographical Notes upon Botanists*. 1: 212. 1965; Mary Gunn and Leslie Edward W. Codd, *Botanical Exploration of Southern Africa*. Cape Town 1981; J. Lanjouw and F.A. Stafleu, *Index Herbariorum*. Part II, *Collectors A-D*. Regnum Vegetabile vol. 2. 1954.

Pseudobombax Dugand Bombacaceae

Origins:

Greek *pseudes* "false" and *Bombax* L.

Species/Vernacular Names:

P. munguba (C. Martius & Zuccarini) Dugand

Peru: huina caspi, monguba, punga, punga blanca

Pseudobotrys Moeser Icacinaceae

Origins:

From the Greek *pseudes* "false" and *botrys* "cluster."

Pseudobrachiaria Launert Gramineae

Origins:

Greek *pseudes* "false" and *Brachiaria* (Trin.) Griseb.

Pseudobrassaiopsis R.N. Banerjee Araliaceae

Origins:

From the Greek *pseudes* "false" plus *Brassaiopsis* Decne. & Planchon.

Pseudobravoa Rose Agavaceae

Origins:

Greek *pseudes* "false" plus *Bravoa* Llave & Lex.

Pseudobraya Korsh. Brassicaceae

Origins:

Greek *pseudes* "false" plus *Braya* Sternb. & Hoppe, after the German diplomat Franz Gabriel (François Gabriel) Graf von Bray, 1765-1832.

Pseudobrazzeia Engl. Scytopetalaceae

Origins:

Greek *pseudes* "false" and the genus *Brazzeia* Baillon.

Pseudobrickellia R.M. King & H. Robinson Asteraceae

Origins:

Greek *pseudes* "false" plus the genus *Brickellia* Elliott, after the Irish-born physician John Brickell, 1748-1809.

Pseudobromus K. Schumann Gramineae

Origins:

From the Greek *pseudes* "false" plus *Bromus* L.

Pseudobrownanthus Ihlenf. & Bittrich Aizoaceae

Origins:

Greek *pseudes* "false" plus the genus *Brownanthus* Schwantes, for the British botanist Nicholas Edward Brown, 1849-1934.

Pseudocadia Harms Fabaceae

Origins:

From the Greek *pseudes* "false" and the genus *Cadia* Forssk.

Pseudocadiscus Lisowski Asteraceae

Origins:

From the Greek *pseudes* "false" plus the genus *Cadiscus* E. Meyer ex DC.

Pseudocalymma A. Samp. & Kuhlm. Bignoniaceae

Origins:

Greek *pseudes* "false" and *kalymma* "a covering."

Pseudocalyx Radlk. Acanthaceae

Origins:

Greek *pseudes* "false" and *kalyx* "a calyx."

Pseudocamelina (Boiss.) N. Busch Brassicaceae

Origins:

From the Greek *pseudes* plus the genus *Camelina* Crantz.

Pseudocampanula Kolak. Campanulaceae

Origins:

From the Greek *pseudes* "false" plus *Campanula* L.

Pseudocapsicum Medik. Solanaceae

Origins:

From the Greek *pseudes* "false" plus *Capsicum* L., see also *Solanum* L.

Pseudocarapa Hemsley Meliaceae

Origins:

Greek *pseudes* "false" and the genus *Carapa* Aubl. in the same family; see J.D. Hooker, *Hooker's Icones Plantarum.* 1458. (Sep.) 1884.

Pseudocarpidium Millsp. Labiatae (Verbenaceae)

Origins:

From the Greek *pseudes* "false" and *karpos* "fruit," Latin *carpidium* "carpel."

Pseudocarum C. Norman Umbelliferae

Origins:

From the Greek *pseudes* "false" and the genus *Carum* L.

Pseudocaryophyllus O. Berg Myrtaceae

Origins:

Greek *pseudes* "false" and the genus *Caryophyllus* L., see also *Pimenta* Lindl.

Pseudocassia Britton & Rose Caesalpiniaceae

Origins:

From the Greek *pseudes* "false" plus *Cassia* L., see also *Senna* Mill.

Pseudocassine Bredell Celastraceae

Origins:

Greek *pseudes* "false" and the genus *Cassine* L.

Pseudocatalpa A. Gentry Bignoniaceae

Origins:

Greek *pseudes* "false" and the genus *Catalpa* Scop.

Pseudocedrela Harms Meliaceae

Origins:

Greek *pseudes* "false" and the genus *Cedrela.*

Species/Vernacular Names:

P. kotschyi (Schweinf.) Harms

Nigeria: tuna (Hausa); emigbegi, emigbegeri (Yoruba)

Yoruba: emi gbegi, emi gberi, emi gbegbari, emi gbegberi

Mali: zaza, zega, sinzan

Pseudocentrum Lindley Orchidaceae

Origins:

Greek *pseudes* and *kentron* "a spur, prickle," referring to the false galea, the flowers have a spur-like structure formed from the lateral sepals.

Pseudochaetochloa A. Hitchc. Gramineae

Origins:

From the Greek *pseudes* and the genus *Chaetochloa* Scribn., *chaite* "bristle, long hair" and *chloe, chloa* "grass"; see Albert Spear Hitchcock (1865-1935), in *Journal of the Washington Academy of Sciences.* 14: 492. 1924.

Pseudochamaesphacos Parsa Labiatae

Origins:

From the Greek *pseudes* "false" and the genus *Chamaesphacos* Schrenk ex Fischer & C.A. Meyer.

Pseudochimarrhis Ducke Rubiaceae

Origins:

Greek *pseudes* "false" and the genus *Chimarrhis* Jacq.

Pseudochirita W.T. Wang Gesneriaceae

Origins:

Greek *pseudes* "false" and the genus *Chirita* Buch.-Ham. ex D. Don.

Pseudochrosia Blume Apocynaceae

Origins:

From the Greek *pseudes* "false" plus *Ochrosia* Juss.

Pseudocimum Bremek. Labiatae

Origins:

From the Greek *pseudes* "false" and *Ocimum.*

Pseudocinchona A. Chev. Rubiaceae

Origins:

From the Greek *pseudes* "false" plus *Cinchona* L.

Pseudocladia Pierre Sapotaceae

Origins:

Greek *pseudes* "false" plus *klados* "branch."

Pseudoclappia Rydb. Asteraceae

Origins:

Greek *pseudes* "false" plus the genus *Clappia* A. Gray.

Pseudoclausia Popov Brassicaceae

Origins:

Greek *pseudes* "false" plus *Clausia* Trotzky ex Hayek.

Pseudocoix A. Camus Gramineae

Origins:
Greek *pseudes* "false" plus *Coix* L.

Pseudocolysis Gómez Polypodiaceae

Origins:
From the Greek *pseudes* "false" plus the genus *Colysis* C. Presl.

Pseudoconnarus Radlk. Connaraceae

Origins:
Greek *pseudes* "false" plus *Connarus* L.

Pseudoconyza Cuatrec. Asteraceae

Origins:
Greek *pseudes* "false" plus *Conyza* Less.

Pseudocopaiva Britton & P. Wilson Caesalpiniaceae

Origins:
From *copayba*, the Brazilian (Tupi-Guarani) name; the tree *Copaifera officinalis* (Jacq.) L. is *coppaiba*, *copaive*.

Pseudocorchorus Capuron Tiliaceae

Origins:
From the Greek *pseudes* "false" and the genus *Corchorus* L.

Pseudocranichis Garay Orchidaceae

Origins:
Greek *pseudes* "false" and the genus *Cranichis* Sw.

Pseudocroton Müll. Arg. Euphorbiaceae

Origins:
Greek *pseudes* and the genus *Croton* L.

Pseudocrupina Velen. Asteraceae

Origins:
From the Greek *pseudes* "false" and the genus *Crupina* (Pers.) DC.

Pseudocryptocarya Teschner Lauraceae

Origins:
From the Greek *pseudes* and the genus *Cryptocarya* R. Br.

Pseudoctomeria Kraenzlin Orchidaceae

Origins:
Greek *pseudes* "false" plus the genus of orchids *Octomeria* R. Br.

Pseudocunila Brade Labiatae

Origins:
Greek *pseudes* "false" plus *Cunila* Royen ex L.

Pseudocyclanthera Mart. Crov. Cucurbitaceae

Origins:
Greek *pseudes* "false" with *Cyclanthera* Schrader.

Pseudocyclosorus Ching Thelypteridaceae

Origins:
Greek *pseudes* "false" and *Cyclosorus* Link.

Pseudocydonia (C. Schneider) C. Schneider Rosaceae

Origins:
From the Greek *pseudes* "false" with *Cydonia* Miller.

Species/Vernacular Names:
P. sinensis (Dum.-Cours.) Schneid. (*Cydonia sinensis* (Dum.-Cours.) Thouin)
China: mu kua

Pseudocymopterus Coulter & Rose Umbelliferae

Origins:
From the Greek *pseudes* "false" and the genus *Cymopterus* Raf.

Pseudocynometra Kuntze Caesalpiniaceae

Origins:
Greek *pseudes* "false" and *Cynometra* L.

Pseudocystopteris Ching Dryopteridaceae (Aspleniaceae, Woodsiaceae)

Origins:
Greek *pseudes* and *Cystopteris* Bernh.

Pseudocytisus Kuntze Brassicaceae

Origins:
Greek *pseudes* plus *Cytisus*; see Helmut Genaust, *Etymologisches Wörterbuch der botanischen Pflanzennamen*. 511. Basel 1996.

Pseudodanthonia Bor & C.E. Hubb. Gramineae

Origins:
Greek *pseudes* "false" and *Danthonia* DC., after the French botanist D. (Étienne) Danthoine, fl. 1788, agrostologist.

Pseudodatura Zijp Solanaceae

Origins:
Greek *pseudes* "false" plus *Datura* L.

Pseudodichanthium Bor Gramineae

Origins:
Greek *pseudes* "false" with *Dichanthium* Willemet.

Pseudodicliptera Benoist Acanthaceae

Origins:
Greek *pseudes* "false" and *Dicliptera* Juss.

Pseudodigera Chiov. Amaranthaceae

Origins:
Greek *pseudes* "false" plus *Digera* Forssk.

Pseudodiphasium Holub Lycopodiaceae

Origins:
Greek *pseudes* "false" plus *Diphasium* C. Presl ex Rothm.

Pseudodiphryllum Nevski Orchidaceae

Origins:
Greek *pseudes* plus the genus of orchids *Diphryllum* Raf.

Pseudodissochaeta M.P. Nayar Melastomataceae

Origins:
Greek *pseudes* "false" plus the genus *Dissochaeta* Blume.

Pseudodracontium N.E. Br. Araceae

Origins:
Greek *pseudes* "false" plus *Dracontium* L.; see D.H. Nicolson, "Derivation of aroid generic names." *Aroideana*. 10: 15-25. 1988.

Pseudodrynaria Christensen ex Ching Polypodiaceae

Origins:
From the Greek *pseudes* "false" and the genus *Drynaria* (Bory) J. Sm.; see in *Sunyatsenia*. 6: 10. Jan 1941.

Species/Vernacular Names:
P. coronans (Wall.) Ching (*Polypodium coronans* Wall. ex Mett.)
Japan: kazari-shida

Pseudoechinocereus Buining Cactaceae

Origins:
From the Greek *pseudes* "false" and the genus *Echinocereus* Engelm.

Pseudoechinopepon (Cogn.) Cockerell Cucurbitaceae

Origins:
Greek *pseudes* "false" and the genus *Echinopepon* Naudin.

Pseudoentada Britton & Rose Mimosaceae

Origins:
Greek *pseudes* "false" plus *Entada* Adans.

Pseudoeria (Schltr.) Schltr. Orchidaceae

Origins:
See *Pseuderia* Schltr.

Pseudoeriosema Hauman Fabaceae

Origins:
Greek *pseudes* "false" plus *Eriosema* (DC.) Reichb.

Pseudoernestia (Cogn.) Krasser Melastomataceae

Origins:
Greek *pseudes* "false" plus the genus *Ernestia* DC.

Pseudoespostoa Backeb. Cactaceae

Origins:
Greek *pseudes* "false" and the genus *Espostoa* Britton & Rose.

Pseudoeugenia Scort. Myrtaceae

Origins:
Greek *pseudes* "false" and the genus *Eugenia* L., see also *Syzygium* Gaertner.

Pseudoeurya Yamam. Theaceae

Origins:
Greek *pseudes* "false" and the genus *Eurya* Thunb.

Pseudoeurystyles Hoehne Orchidaceae

Origins:
From the Greek *pseudes* "false" and *Eurystyles* Wawra.

Pseudoeverardia Gilly Cyperaceae

Origins:
From the Greek *pseudes* "false" plus *Everardia* Ridley.

Pseudofortuynia Hedge Brassicaceae

Origins:
From the Greek *pseudes* "false" and the genus *Fortuynia* Shuttlew. ex Boiss.

Pseudofumaria Medik. Fumariaceae (Papaveraceae)

Origins:
From the Greek *pseudes* "false" and the genus *Fumaria* L.

Pseudogaillonia Lincz. Rubiaceae

Origins:
Greek *pseudes* "false" and the genus *Gaillonia* A. Rich. ex DC., named for the French algologist François Benjamin Gaillon, 1782-1839.

Pseudogaltonia (Kuntze) Engl. Hyacinthaceae (Liliaceae)

Origins:
Greek *pseudes* "false" and *Galtonia* Decne., dedicated to the British scientist Sir Francis Galton, 1822-1911 (Haslemere, Surrey), anthropologist.

Pseudogardenia Keay Rubiaceae

Origins:
From the Greek *pseudes* "false" and *Gardenia* Ellis.

Pseudogardneria Racib. Strychnaceae (Loganiaceae)

Origins:
From the Greek *pseudes* "false" and *Gardneria* Wallich, named for the English colonial governmental officer Hon. Edward Gardner (b. 1784).

Pseudoglochidion Gamble Euphorbiaceae

Origins:
Greek *pseudes* "false" and *Glochidion* Forst. & Forst.f.

Pseudoglycine F.J. Herm. Fabaceae

Origins:
Greek *pseudes* "false" and *Glycine* Willd.

Pseudognaphalium Kirpiczn. Asteraceae

Origins:
Greek *pseudes* "false" and the genus *Gnaphalium* L., referring to a superficial resemblance; see Moisey Elevich Kirpicznikov (1913-), in *Trudy Botaniceskogo instituta Akademii nauk SSSR*. Ser. 1. Flora i sistematika vyssih rastenij. 9: 33. Moscow & Leningrad 1950.

Species/Vernacular Names:
P. luteo-album (L.) Hilliard & B.L. Burtt (*Gnaphalium luteo-album* L.)

English: cudweed, Jersey cudweed, Japanese cudweed

Southern Africa: roerkruid, vaalbossie; manku, musuwane (Sotho); mgilane (Zulu)

Japan: Taiwan-haka-ko-gusa

P. oligandrum (DC.) Hilliard & Burtt (*Gnaphalium oligandrum* (DC.) Hilliard & Burtt)

English: undulate cudweed

South Africa: groenbossie

P. undulatum (L.) Hilliard & Burtt (*Gnaphalium undulatum* L.; *Helichrysum montosicolum* Gand.)

English: undulate cudweed

Southern Africa: groenbossie; mothepetelle (Sotho)

Pseudognidia E. Phillips Thymelaeaceae

Origins:
Greek *pseudes* "false" and *Gnidia* L.

Pseudogomphrena R.E. Fries Amaranthaceae

Origins:
Greek *pseudes* "false" plus the genus *Gomphrena* L.

Pseudogonocalyx Bisse & Berazaín Ericaceae

Origins:
Greek *pseudes* plus *Gonocalyx* Planchon & Linden ex Lindley.

Pseudogoodyera Schlechter Orchidaceae

Origins:
Greek *pseudes* "false" and the genus of orchids *Goodyera* R. Br., for the British botanist John Goodyer of Mapledurham, Hants., 1592-1664.

Pseudogynoxys (Greenman) Cabrera Asteraceae

Origins:
Greek *pseudes* "false" plus *Gynoxys* Cass. See A.L. Cabrera, "Notes on the Brazilian Senecioneae." *Brittonia*. 7: 53-74. 1950; H. Robinson & J. Cuatrecasas, "Notes on the genus and species limits of *Pseudogynoxys* (Greenm.) Cabrera (Senecioneae, Asteraceae)." *Phytologia*. 36: 177-192. 1977.

Pseudohamelia Wernham Rubiaceae

Origins:
Greek *pseudes* "false" plus *Hamelia* Jacq., after the French botanist Henri Louis Duhamel du Monceau, 1700-1782.

Pseudohandelia Tzvelev Asteraceae

Origins:
For the Austrian botanist Heinrich von Handel-Mazzetti, 1882-1940, explorer, moss collector, from 1914 to 1919 in China, pupil of Richard Wettstein (1863-1931). His writings include *Monographie der Gattung Taraxacum*. Leipzig und Wien 1907 and *Symbolae sinicae*. Botanische Ergebnisse der Expedition der Akademie der Wissenschaften in Wien nach Südwest-China. 1914/1918. Wien 1929-1937; see John H. Barnhart, *Biographical Notes upon Botanists*. 2: 121. 1965.

Pseudohexadesmia Brieger Orchidaceae

Origins:
Greek *pseudes* and *Hexadesmia* Brongn.

Pseudohomalomena A.D. Hawkes Araceae

Origins:
Greek *pseudes* and *Homalomena* Schott; see D.H. Nicolson, "Derivation of aroid generic names." *Aroideana*. 10: 15-25. 1988.

Pseudohydrosme Engl. Araceae

Origins:
Greek *pseudes* plus *Hydrosme* Schott; see D.H. Nicolson, "Derivation of aroid generic names." *Aroideana*. 10: 15-25. 1988.

Pseudojacobaea (Hook.f.) Mathur Asteraceae

Origins:
Jacobaea is a vernacular name for *Senecio elegans* L. (*Senecio elegans* Willd.; *Senecio pseudo-elegans* Less.).

Pseudokyrsteniopsis R. King & H. Robinson Asteraceae

Origins:
Greek *pseudes* "false" plus the genus *Kyrsteniopsis* R. King & H. Robinson.

Pseudolabatia Aubrév. & Pellegr. Sapotaceae

Origins:
Greek *pseudes* with *Labatia* Sw.

Pseudolachnostylis Pax Euphorbiaceae

Origins:
Greek *pseudes* "false" and the genus *Lachnostylis* Turcz.

Species/Vernacular Names:
P. maprouneifolia Pax var. *maprouneifolia* (*Pseudolachnostylis dekindtii* Pax; *Pseudolachnostylis glauca* (Hiern) Hutch.) (the specific name means "with *Maprounea*-like leaves," *Maprounea* Aubl. is another genus of the Euphorbiaceae)

English: kudu-berry

Southern Africa: duiker food, duiker tree, koedoebessie; nshojowa (Thonga); muTsondzera, muTzondzowa, muPambari, muDyamembgwe, muToto, muTsontsowa, gwaaziyo, muHonjowa, muKuhu, muKuvazwio, muShozhowa, muSonzoa, muShongwa, muSontsoa (Shona); mokonu (western Transvaal, northern Cape, Botswana); mutondowa (Venda); mukunyambambi (Mbukushu: Okavango Swamps and western Caprivi)

Pseudolaelia Porto & Breda Orchidaceae

Origins:
From the Greek *pseudes* "false" and the genus *Laelia* Lindley.

Pseudolarix Gordon Pinaceae

Origins:
Like *Larix*, Greek *pseudes* "false" plus *Larix* Miller.

Species/Vernacular Names:
P. amabilis (Nelson) Rehder (*Pseudolarix kaempferi* (Lamb.) Gordon)
English: golden larch, golden pine
China: tu jing pi

Pseudolasiacis (A. Camus) A. Camus Gramineae

Origins:
Greek *pseudes* "false" plus *Lasiacis* (Griseb.) Hitchc.

Pseudoligandra Dillon & Sagást. Asteraceae

Origins:
Greek *pseudes* "false" with *Oligandra* Less.

Pseudolinosyris Novopokr. Asteraceae

Origins:
Greek *pseudes* "false" and *Linosyris* Cass.

Pseudoliparis Finet Orchidaceae

Origins:

From the Greek *pseudes* "false" plus the genus of orchids *Liparis* Rich.

Pseudolitchi Danguy & Choux Sapindaceae

Origins:

Greek *pseudes* "false" plus the genus *Litchi* Sonn.

Pseudolmedia Trécul Moraceae

Origins:

Greek *pseudes* "false" and the genus *Olmedia* Ruíz & Pav.

Species/Vernacular Names:

P. sp.

Peru: loro micuan

P. laevigata Trécul (*Pseudolmedia mildbraedii* J.F. Macbride)

Peru: chimicua

P. laevis (Ruíz & Pav.) J.F. Macbride (*Pseudolmedia multinervis* Mildbraed; *Olmedia laevis* Ruíz & Pav.)

Peru: congona, chimicua, chimicua colorada, chimicua menuda, chimigua, itahuba, itauba, itauba amarilla

Pseudolobivia (Backeb.) Backeb. Cactaceae

Origins:

Greek *pseudes* "false" plus *Lobivia* Britton & Rose.

Pseudolopezia Rose Onagraceae

Origins:

Greek *pseudes* "false" plus *Lopezia* Cav.

Pseudolophanthus Levin Labiatae

Origins:

Greek *pseudes* "false" plus *Lophanthus* Adans.

Pseudolotus Rech.f. Fabaceae

Origins:

Greek *pseudes* "false" plus *Lotus* L.

Pseudoloxocarya Linder Restionaceae

Origins:

Greek *pseudes* "false" plus the genus *Loxocarya* R. Br.

Pseudoludovia Harling Cyclanthaceae

Origins:

Greek *pseudes* "false" plus *Ludovia* Brongn.

Pseudolycopodiella Holub Lycopodiaceae

Origins:

Greek *pseudes* "false" and *Lycopodiella* Holub.

Pseudolycopodium Holub Lycopodiaceae

Origins:

From the Greek *pseudes* "false" and the genus *Lycopodium* L.

Pseudolysimachion (Koch) Opiz Scrophulariaceae

Origins:

Greek *pseudes* "false" and *lysimachion* for a plant, otherwise unknown.

Pseudomachaerium Hassl. Fabaceae

Origins:

From the Greek *pseudes* "false" and the genus *Machaerium* Pers.

Pseudomacodes Rolfe Orchidaceae

Origins:

Greek *pseudes* "false" and *Macodes* (Blume) Lindley.

Pseudomacrolobium Hauman Caesalpiniaceae

Origins:

Greek *pseudes* "false" plus the genus *Macrolobium* Schreber.

Pseudomammillaria Buxb. Cactaceae

Origins:
Greek *pseudes* "false" plus the genus *Mammillaria* Haw.

Pseudomantalania J. Leroy Rubiaceae

Origins:
Greek *pseudes* plus *Mantalania* Capuron ex J. Leroy.

Pseudomariscus Rauschert Cyperaceae

Origins:
Greek *pseudes* "false" plus *Mariscus* Vahl.

Pseudomarrubium Popov Labiatae

Origins:
Greek *pseudes* "false" plus the genus *Marrubium* L.

Pseudomarsdenia Baillon Asclepiadaceae

Origins:
Greek *pseudes* "false" plus the genus *Marsdenia* R. Br., named for the traveler and plant collector William Marsden, 1754-1836; see Peter Marsden, *The Wreck of the Amsterdam*. New York 1975; Ray Desmond, *Dictionary of British & Irish Botanists and Horticulturists*. 468-469. London 1994; F. Boerner & G. Kunkel, *Taschenwörterbuch der botanischen Pflanzennamen*. 4. Aufl. 132. 1989; Georg Christian Wittstein, *Etymologisch-botanisches Handwörterbuch*. 558. Ansbach 1852; John H. Barnhart, *Biographical Notes upon Botanists*. 2: 450. Boston 1965; R. Brown, "On the Asclepiadeae." *Memoirs of the Wernerian Natural History Society*. 1: 28-29. Edinburgh 1811; M. Archer, *Natural History Drawings in the India Office Library*. London 1962; Warren R. Dawson, *The Banks Letters*, a Calendar of the Manuscript Correspondence of Sir Joseph Banks. London 1958; Jonas C. Dryander, *Catalogus bibliothecae historico-naturalis Josephi Banks*. London 1796-1800; A. Lasègue, *Musée botanique de Benjamin Delessert*. Paris 1845; Ethelyn Maria Tucker, *Catalogue of the Library of the Arnold Arboretum of Harvard University*. Cambridge, Massachusetts 1917-1933; G. Murray, *History of the Collections Contained in the Natural History Departments of the British Museum*. 1: 166. London 1904.

Pseudomaxillaria Hoehne Orchidaceae

Origins:
Greek *pseudes* "false" plus the genus *Maxillaria*.

Pseudomelissitus Ovcz. Fabaceae

Origins:
Greek *pseudes* "false" plus *Melissitus* Medikus.

Pseudomertensia Riedl Boraginaceae

Origins:
Greek *pseudes* "false" plus the genus *Mertensia* Roth, for the German botanist Franz Karl (Carl) Mertens, 1764-1831, professor of botany at Bremen.

Pseudomitrocereus H. Bravo & Buxb. Cactaceae

Origins:
Greek *pseudes* "false" plus *Mitrocereus* (Backeb.) Backeb.

Pseudomorus Bureau Moraceae

Origins:
False mulberry, from the Greek *pseudes* "false" and the genus *Morus* L.; see Louis Édouard Bureau (1830-1918), in *Annales des Sciences Naturelles*. Ser. 5, 11: 371. 1869.

Pseudomuscari Garbari & Greuter Hyacinthaceae (Liliaceae)

Origins:
Greek *pseudes* "false" plus the genus *Muscari* Miller.

Pseudomussaenda Wernham Rubiaceae

Origins:
Greek *pseudes* "false" plus *Mussaenda* L.

Pseudomyrcianthes Kausel Myrtaceae

Origins:
Greek *pseudes* "false" and *Myrcianthes* O. Berg.

Pseudonemacladus McVaugh Campanulaceae

Origins:
Greek *pseudes* "false" with *Nemacladus* Nutt.

Pseudonephelium Radlk. Sapindaceae

Origins:

Greek *pseudes* "false" plus the genus *Nephelium* L.

Pseudonesohedyotis Tennant Rubiaceae

Origins:

Greek *pseudes* plus the genus *Nesohedyotis* (Hook.f.) Bremek.; see J.R. Tennant, in *Kew Bulletin*. 19(2): 277-278. 1965.

Pseudonopalxochia Backeb. Cactaceae

Origins:

Greek *pseudes* "false" plus *Nopalxochia* Britton & Rose.

Pseudonoseris H. Robinson & Brettell Asteraceae

Origins:

Greek *pseudes* "false" and the genus *Onoseris* Willd.

Pseudoorleanesia Rauschert Orchidaceae

Origins:

Greek *pseudes* "false" and the genus *Orleanesia* Barb. Rodr.

Pseudopachystela Aubrév. & Pellegr. Sapotaceae

Origins:

Greek *pseudes* "false" and the genus *Pachystela* Pierre ex Radlk.

Pseudopaegma Urban Bignoniaceae

Origins:

Greek *pseudes* "false" plus *Anemopaegma* C. Martius ex Meissner.

Pseudopanax K. Koch Araliaceae

Origins:

From the Greek *pseudes* and the genus *Panax*; see Karl Heinrich Emil Ludwig Koch (1809-1879), in *Wochenschrift für Gärtnerei und Pflanzenkunde*. 2: 366. Berlin 1859.

Species/Vernacular Names:

P. crassifolius (Cunn.) K. Koch

English: lancewood

Pseudopancovia Pellegr. Sapindaceae

Origins:

From the Greek *pseudes* "false" and the genus *Pancovia* Willd.

Pseudopavonia Hassler Malvaceae

Origins:

For the Spanish botanist José Antonio Pavón, 1754-1844, traveler, explorer, between 1777-1788 he traveled with Hipolito Ruíz Lopez (1754-1815) and Joseph Dombey in Chile and Peru, with H. Ruíz wrote *Flora peruvianae, et chilensis prodromus*. Madrid 1794 and *Flora peruviana, et chilensis*. Madrid 1798-1802; see John H. Barnhart, *Biographical Notes upon Botanists*. 3: 57. 1965.

Pseudopectinaria Lavranos Asclepiadaceae

Origins:

From the Greek *pseudes* "false" plus *Pectinaria* Haw.

Pseudopentameris Conert Gramineae

Origins:

From the Greek *pseudes* "false" with *Pentameris* P. Beauv.

Pseudopentatropis Costantin Asclepiadaceae

Origins:

From the Greek *pseudes* "false" with *Pentatropis* R. Br. ex Wight & Arn.

Pseudopeponidium Homolle ex Arènes Rubiaceae

Origins:

From the Greek *pseudes* "false" plus the genus *Peponidium* Baillon ex Arènes.

Pseudophacelurus A. Camus Gramineae

Origins:

Greek *pseudes* "false" plus the genus *Phacelurus* Griseb.

Pseudophegopteris Ching Thelypteridaceae

Origins:
From the Greek *pseudes* "false" plus *Phegopteris* (C. Presl) Fée.

Pseudophleum Dogan Gramineae

Origins:
Greek *pseudes* "false" plus *Phleum* L.

Pseudophoenix H.A. Wendland ex Sargent Palmae

Origins:
False date palm, Greek *pseudes* "false" plus *Phoenix* L.; see *Botanical Gazette*. 11: 314. 1886.

Species/Vernacular Names:
P. sargentii H.A. Wendland ex Sargent
Brazil: palmeira fuso

Pseudopholidia A. DC. Myoporaceae

Origins:
Greek *pseudes* and *Pholidia* R. Br.; see Alphonse de Candolle, *Prodromus*. 11: 704. (Nov.) 1847.

Pseudopilocereus Buxb. Cactaceae

Origins:
From the Greek *pseudes* "false" plus *Pilocereus* Lemaire, see also *Pilosocereus* Byles & Rowley.

Pseudopinanga Burret Palmae

Origins:
Greek *pseudes* "false" plus *Pinanga* Blume.

Pseudopiptadenia Rauschert Mimosaceae

Origins:
From the Greek *pseudes* "false" plus *Piptadenia* Benth.

Pseudopipturus Skottsb. Urticaceae

Origins:
From the Greek *pseudes* "false" plus *Pipturus* Wedd.

Pseudoplantago Süsseng. Amaranthaceae

Origins:
Greek *pseudes* "false" plus *Plantago* L., Plantaginaceae, referring to the shape of the pollen.

Pseudopogonatherum A. Camus Gramineae

Origins:
From the Greek *pseudes* "false" and the genus *Pogonatherum* Beauv.; see Aimée Antoinette Camus (1879-1965), in *Annales de la société linnéenne de Lyon*. 68: 204. 1922.

Pseudoponera Brieger Orchidaceae

Origins:
From the Greek *pseudes* plus the genus of orchids *Ponera* Lindley.

Pseudoprimula (Pax) O. Schwarz Primulaceae

Origins:
From the Greek *pseudes* "false" with *Primula* L.

Pseudoprosopis Harms Mimosaceae

Origins:
From the Greek *pseudes* "false" plus *Prosopis* L.

Pseudoprotorhus H. Perrier Sapindaceae

Origins:
Greek *pseudes* "false" plus the genus *Protorhus* Engl., Anacardiaceae.

Pseudopteris Baillon Sapindaceae

Origins:
Greek *pseudes* "false" plus *pteris* "fern."

Pseudopyxis Miq. Rubiaceae

Origins:
Greek *pseudes* "false" and *pyxis* "a small box, a small container with lid."

Pseudorachicallis Post & Kuntze Rubiaceae

Origins:
Greek *pseudes* "false" and *Rachicallis* DC.

Pseudoraphis Griffith Gramineae

Origins:
Greek *pseudes* "false" and *rhaphis, rhaphidos* "a needle," referring to a solitary bristle; see William Griffith (1810-1845), *Notulae ad plantas asiaticas*. 3: 29. Calcutta 1851.

Species/Vernacular Names:
P. paradoxa (R. Br.) Pilger (*Panicum paradoxum* R. Br.)
English: slender mud-grass, thorny mud-grass
P. spinescens (R. Br.) Vick. (*Panicum spinescens* R. Br.; *Panicum asperum* C. König; *Pseudoraphis aspera* Pilger)
English: spiny mud-grass

Pseudorchis Séguier Orchidaceae

Origins:
Greek *pseudes* "false" and *Orchis*.

Pseudorhachicallis Hook.f Rubiaceae

Origins:
See *Pseudorachicallis*.

Pseudorhipsalis Britton & Rose Cactaceae

Origins:
Greek *pseudes* "false" plus *Rhipsalis* Gaertner.

Pseudorlaya (Murb.) Murb. Umbelliferae

Origins:
Greek *pseudes* "false" plus *Orlaya* Hoffm., for Johann Orlay, physician in Moscow, *circa* 1770-1829.

Pseudorleanesia Rauschert Orchidaceae

Origins:
Greek *pseudes* "false" plus the genus of orchids *Orleanesia* Barb. Rodr., named after Gaston d'Orléans, 1608-1660, a patron of botany and floriculture; see Abel Brunyer, *Hortus Regius Blesensis*. Paris, Antoine Vitré 1653.

Pseudorobanche Rouy Scrophulariaceae

Origins:
Greek *pseudes* "false" plus *Orobanche* L., Orobanchaceae (Scrophulariaceae).

Pseudoroegneria (Nevski) Á. Löve Gramineae

Origins:
From the Greek *pseudes* "false" plus *Roegneria* K. Koch.

Pseudorontium (A. Gray) Rothm. Scrophulariaceae

Origins:
Greek *pseudes* "false" plus the specific epithet for *Misopates orontium* (L.) Raf. (*Antirrhinum orontium* L.).

Pseudorosularia Gurgen. Crassulaceae

Origins:
Greek *pseudes* with *Rosularia* (DC.) Stapf.

Pseudoruellia Benoist Acanthaceae

Origins:
Greek *pseudes* "false" and *Ruellia* L.

Pseudoryza Griff. Gramineae

Origins:
Greek *pseudes* "false" plus the genus *Oryza* L.

Pseudosabicea N. Hallé Rubiaceae

Origins:
Greek *pseudes* "false" plus *Sabicea* Aublet.

Pseudosagotia Secco Euphorbiaceae

Origins:
Greek *pseudes* plus the genus *Sagotia* Baillon.

Pseudosalacia Codd Celastraceae

Origins:
From the Greek *pseudes* "false" and *Salacia* L.; see L.E. Codd, in *Bothalia*. 10: 565. 30 Aug 1972.

Species/Vernacular Names:
P. streyi Codd (the species honors the South African (b. in Germany) botanist Rudolf Georg Strey, 1907-1988, Curator of the Natal Herbarium, farmer, author of *Conservation Regarding the Preservation of Forest and Open Veld Along the Coast, Especially the St. Lucia Area*. St. Lucia Documentation Centre, Natal Parks Board, St. Lucia Estuary. 1964; in *Bothalia* 10(1): 29-37. 1969, he and Anna Amelia Obermeyer-Mauve described the new *Raphia australis* Obermeyer & Strey from Northern Zululand and Southern Mozambique; see Mary Gunn and Leslie E. Codd, *Botanical Exploration of Southern Africa*. 264, 337. Cape Town 1981)

South Africa: kiplemoen

English: rock lemon

Pseudosamanea Harms Mimosaceae

Origins:
Greek *pseudes* "false" and the genus *Samanea* (Benth.) Merr.

Pseudosantalum Kuntze Araliaceae

Origins:
Greek *pseudes* "false" and *Santalum* L., Santalaceae; see also *Osmoxylon* Miq.

Pseudosaponaria (F.N. Williams) Ikonn. Caryophyllaceae

Origins:
Greek *pseudes* "false" and the genus *Saponaria* L., see also *Gypsophila* L.

Pseudosarcolobus Costantin Asclepiadaceae

Origins:
Greek *pseudes* "false" and *Sarcolobus* R. Br.

Pseudosasa Makino ex Nakai Gramineae

Origins:
Greek *pseudes* "false" with *Sasa* Makino & Shib., referring to the number of the stamens.

Pseudosassafras Lecomte Lauraceae

Origins:
Greek *pseudes* "false" and *Sassafras* Nees & Eberm

Pseudosbeckia A. Fernandes & R. Fernandes Melastomataceae

Origins:
Greek *pseudes* "false" plus *Osbeckia* L., named after the Swedish clergyman Pehr Osbeck, 1723-1805, naturalist and botanist.

Pseudoscabiosa Devesa Dipsacaceae

Origins:
Greek *pseudes* "false" plus *Scabiosa* L.

Pseudoschoenus (C.B. Clarke) Oteng-Yeboah Cyperaceae

Origins:
Greek *pseudes* plus *Schoenus* L.

Pseudosciadium Baillon Araliaceae

Origins:
Greek *pseudes* "false" plus *skias, skiados* "a canopy, umbel," *skiadion, skiadeion* "umbel, parasol."

Pseudoscolopia Gilg Flacourtiaceae

Origins:
Greek *pseudes* "false" and *Scolopia* Schreber.

Species/Vernacular Names:
P. polyantha Gilg (*Pseudoscolopia fraseri* Phill.)
English: false red pear, false-scolopia-with-many-flowers
South Africa: valsrooipeer

Pseudosecale (Godr.) Degen Gramineae

Origins:
From the Greek *pseudes* "false" and *Secale* L.

Pseudosedum (Boiss.) A. Berger Crassulaceae

Origins:
From the Greek *pseudes* "false" and *Sedum* L.

Pseudoselinum C. Norman Umbelliferae

Origins:
Greek *pseudes* "false" and *Selinum* L.

Pseudosempervivum (Boiss.) Grossh. Brassicaceae

Origins:
From the Greek *pseudes* "false" plus the genus *Sempervivum* L., Crassulaceae.

Pseudosericocoma Cavaco Amaranthaceae

Origins:
From the Greek *pseudes* "false" plus *Sericocoma* Fenzl.

Pseudosicydium Harms Cucurbitaceae

Origins:
Greek *pseudes* "false" plus *Sicydium* Schltdl.

Pseudosindora Symington Caesalpiniaceae

Origins:
Greek *pseudes* plus the genus *Sindora* Miq.

Pseudosmelia Sleumer Flacourtiaceae

Origins:
Greek *pseudes* "false" plus the genus *Osmelia* Thwaites.

Pseudosmilax Hayata Smilacaceae

Origins:
Greek *pseudes* "false" plus the genus *Smilax* L.

Pseudosmodingium Engl. Anacardiaceae

Origins:
Greek *pseudes* "false" plus the genus *Smodingium* E. Meyer ex Sonder.

Pseudosolisia Y. Ito Cactaceae

Origins:
Greek *pseudes* "false" plus the genus *Solisia* Britton & Rose.

Pseudosopubia Engl. Scrophulariaceae

Origins:
Greek *pseudes* "false" plus the genus *Sopubia* Buch.-Ham. ex D. Don.

Pseudosorghum A. Camus Gramineae

Origins:
Greek *pseudes* "false" plus *Sorghum* Moench.

Pseudosorocea Baillon Moraceae

Origins:
Greek *pseudes* plus *Sorocea* A. St.-Hil.

Pseudospigelia Klett Strychnaceae (Loganiaceae)

Origins:
Greek *pseudes* "false" and *Spigelia* L., in honor of the Dutch physician Adriaan (Adrian, Adrianus) van der Spiegel (Spigelius, Spieghel), 1578-1625, botanist, professor of anatomy at Padua.

Pseudospondias Engl. Anacardiaceae

Origins:
From the Greek *pseudes* "false" and *Spondias* L.

Species/Vernacular Names:
P. longifolia Engl.
Gabon: ofoss, ozozo
Cameroon: ofod

P. microcarpa (A. Rich.) Engl.

Cameroon: konle, nkanele

Ivory Coast: doveke, tide

Zaire: angombe, butolo, mubulu, bosow, esasau, esosanga, nzuza, suza

Gabon: esongosongo, konghele, konhle

Uganda: muziru

Yoruba: okika aja, ekika aja

Nigeria: kekerakuchi (Nupe); okikan-aja (Yoruba)

Pseudostachyum Munro Gramineae

Origins:

Greek *pseudes* "false" and *stachys* "a spike," see also *Schizostachyum* Nees.

Pseudostelis Schltr. Orchidaceae

Origins:

Greek *pseudes* "false" and the genus *Stelis* Sw.

Pseudostellaria Pax Caryophyllaceae

Origins:

Greek *pseudes* and the genus *Stellaria* L., referring to the placement of the species.

Pseudostenomesson Velarde Amaryllidaceae (Liliaceae)

Origins:

Greek *pseudes* "false" plus *Stenomesson* Herbert.

Pseudostenosiphonium Lindau Acanthaceae

Origins:

Greek *pseudes* and the genus *Stenosiphonium* Nees.

Pseudostifftia H. Robinson Asteraceae

Origins:

Greek *pseudes* "false" and the genus *Stifftia* Mikan.

Pseudostreblus Bureau Moraceae

Origins:

Greek *pseudes* "false" and *Streblus* Lour.

Pseudostreptogyne A. Camus Gramineae

Origins:

Greek *pseudes* plus *Streptogyna* P. Beauv., Greek *streptos* "twisted" and *gyne* "female, ovary," referring to the nature of the ovary.

Pseudostriga Bonati Scrophulariaceae

Origins:

Greek *pseudes* "false" plus *Striga* Lour.

Pseudostrophis T. Durand & B.D. Jacks. Moraceae

Origins:

Greek *pseudes* "false" with *Trophis* P. Browne, some suggest from Greek *strophe* "twist, turning"; see also *Streblus* Lour.

Pseudotaenidia Mackenzie Umbelliferae

Origins:

Greek *pseudes* "false" plus *Taenidia* (Torrey & Gray) Drude.

Pseudotaxus W.C. Cheng Taxaceae

Origins:

Greek *pseudes* "false" plus the genus *Taxus* L.

Pseudotectaria Tardieu Aspleniaceae (Tectarioideae, Dryopteridaceae)

Origins:

Greek *pseudes* "false" plus the genus *Tectaria* Cav., Dryopteridaceae.

Pseudotephrocactus Fric Cactaceae

Origins:

Greek *pseudes* plus the genus *Tephrocactus* Lem., see also *Opuntia* Mill.

Pseudotragia Pax Euphorbiaceae

Origins:

Greek *pseudes* "false" plus the genus *Tragia* L.

Pseudotrewia Miq. Euphorbiaceae

Origins:

Greek *pseudes* "false" with the genus *Trewia* L., see also *Wetria* Baillon; dedicated to the German botanist Christoph Jakob Trew, 1695-1769, physician, traveler, correspondent of the botanical artist Georg Dionysius Ehret (1708-1770), author of *Plantae selectae*, quarum imagines ad exemplaria naturalia Londini in hortis Curiosorum nutrita manu artificiosa doctaque pinxit Georgius Dionysius Ehret ... [Norimbergae] 1750-1773. See John H. Barnhart, *Biographical Notes upon Botanists*. 3: 400. 1965; T.W. Bossert, *Biographical Dictionary of Botanists Represented in the Hunt Institute Portrait Collection*. 406. 1972; Jonas C. Dryander, *Catalogus bibliothecae historico-naturalis Josephi Banks*. London 1800; Blanche Elizabeth Edith Henrey (1906-1983), *British Botanical and Horticultural Literature before 1800*. Oxford 1975; Gordon Douglas Rowley, *A History of Succulent Plants*. Strawberry Press, Mill Valley, California 1997.

Pseudotrimezia R.C. Foster Iridaceae

Origins:

Greek *pseudes* "false" plus *Trimezia* Salisb. ex Herbert.

Pseudotrophis Warb. Moraceae

Origins:

Greek *pseudes* "false" plus the genus *Trophis* P. Browne, see also *Streblus* Lour.

Pseudotsuga Carrière Pinaceae

Origins:

Greek *pseudes* and the genus *Tsuga* (Antoine) Carrière.

Species/Vernacular Names:

P. menziesii (Mirbel) Franco

English: Douglas fir, Oregon pine

French: sapin de Douglas

Pseudourceolina Vargas Amaryllidaceae (Liliaceae)

Origins:

Greek *pseudes* "false" and the genus *Urceolina* Reichb.

Pseudovanilla Garay Orchidaceae

Origins:

From the Greek *pseudes* "false" and the genus *Vanilla* Mill. in the same family; see Leslie Andrews Garay (1924-), in *Botanical Museum Leaflets of Harvard University*. 30: 214. 1986.

Species/Vernacular Names:

P. foliata (F. Muell.) Garay

English: great climbing orchid

P. gracilis (Schltr.) Garay (see B.A. Lewis and P.J. Cribb, *Orchids of the Solomon Islands and Bougainville*. Royal Botanic Gardens, Kew 1991)

The Solomon Islands: bwagoe

Pseudovesicaria (Boiss.) Rupr. Brassicaceae

Origins:

From the Greek *pseudes* and the genus *Vesicaria* Adans.

Pseudovigna (Harms) Verdc. Fabaceae

Origins:

From the Greek *pseudes* "false" and the genus *Vigna* Savi.

Pseudovossia A. Camus Gramineae

Origins:

Greek *pseudes* "false" and the genus *Vossia* Wallich & Griffith, after the German poet Johann Heinrich Voss, 1751-1826.

Pseudovouapa Britton & Rose Caesalpiniaceae

Origins:

Possibly referring to the genus *Vouacapoua* Aublet.

Pseudoweinmannia Engl. Cunoniaceae

Origins:

Greek *pseudes* and the genus *Weinmannia* L., for the German apothecary Johann Wilhelm Weinmann, 1683-1741; see H.G.A. Engler, *Die Natürlichen Pflanzenfamilien*. Ed. 2. 18a: 249. 1928.

Species/Vernacular Names:

P. lachnocarpa (F. Muell.) Engl.

Australia: scrub rosewood, marara, rose marara, mararie, red Carrabeen

Pseudowillughbeia Markgr. Apocynaceae

Origins:

Greek *pseudes* "false" and the genus *Willughbeia* Roxb., for Francis Willughby (Willoughby), 1635-1672, botanist and naturalist.

Pseudowintera Dandy Winteraceae

Origins:

To honor John Winter, traveled with Sir Francis Drake on his first voyage to Virginia in 1577, vice-admiral of Sir Francis Drake's voyage to Tierra del Fuego in 1578, genus *Wintera* Forst.f.; see E.F. Benson, *Sir Francis Drake*. London 1927; H. Suzanne Maxwell and Martin F. Gardner, "The quest for Chilean green treasure: some notable British collectors before 1940." *The New Plantsman*. 4(4): 195-214. December 1997; W. Vink, in *Blumea*. 18: 225-354. 1970; H.E. Connor and E. Edgar, "Name changes in the indigenous New Zealand flora, 1960-1986 and Nomina Nova IV, 1983-1986." *New Zealand Journal of Botany*. Vol. 25: 115-170. 1987.

Species/Vernacular Names:

P. axillaris (Forst. & Forst.f.) Dandy

English: pepper tree

Maori name: horopito

Pseudowolffia Hartog & van der Plas Lemnaceae

Origins:

Greek *pseudes* "false" and the genus *Wolffia* Horkel ex Schleiden, for the German physician Johann Friedrich Wolff, 1778-1806, Dr. med. Altorf 1801, botanist, author of Dissertatio inauguralis de *Lemna*. Altorfii et Norimbergiae 1801 and *Icones Cimicum* descriptionibus illustratae.

Erlangae 1800-1811; see John H. Barnhart, *Biographical Notes upon Botanists*. 3: 514. 1965; Helmut Genaust, *Etymologisches Wörterbuch der botanischen Pflanzennamen*. 693. 1996; F. Boerner & G. Kunkel, *Taschenwörterbuch der botanischen Pflanzennamen*. 4. Aufl. 187. Berlin & Hamburg 1989.

Pseudoxalis Rose Oxalidaceae

Origins:

Greek *pseudes* "false" and the genus *Oxalis* L.

Pseudoxandra R. Fries Annonaceae

Origins:

Greek *pseudes* "false" and the genus *Oxandra* A. Rich.

Pseudoxytenanthera Soderstr. & R.P. Ellis Gramineae

Origins:

Greek *pseudes* "false" and the genus *Oxytenanthera* Munro.

Pseudoxythece Aubrév. Sapotaceae

Origins:

Greek *pseudes* "false" plus the genus *Oxythece* Miq.

Pseudozoysia Chiov. Gramineae

Origins:

Greek *pseudes* "false" plus the genus *Zoysia* Willd., after the Austrian botanist Karl von Zoys, 1756-1800, plant collector.

Pseudozygocactus Backeb. Cactaceae

Origins:

Greek *pseudes* "false" plus the genus *Zygocactus* Schumann.

Pseuduvaria Miq. Annonaceae

Origins:

Greek *pseudes* "false" and the genus *Uvaria* L.; see F.A.W. Miquel, *Flora Indiae Batavae*. 1(2): 32. Amsterdam 1860.

Species/Vernacular Names:

P. froggattii (F. Muell.) Jessup (*Mitrephora froggattii* F. Muell.) (the species collected by Sayer and Froggatt on Mossman's River, Australia; named after the Victorian entomologist and plant collector Walter Wilson Froggatt, 1858-1937, zoologist assistant on the 1885 NG Expedition of the Royal Geographical Society, author of *Australian Insects.* Sydney 1907, *Notes on the Natives of West Kimberly*, N.W. Australia. [Sydney 1888] and *Forest Insects and Timber Borers.* Sydney 1927, collected insects widely)

English: false uvaria

Psiadia Jacq. Asteraceae

Origins:

Greek *psias, psiados* "a drop," referring to the exudation from the leaves.

Species/Vernacular Names:

P. rodriguesiana Balf.f.

Rodrigues Island: bois de ronde

Psiadiella Humbert Asteraceae

Origins:

The diminutive of the genus *Psiadia* Jacq.

Psidiastrum Bello Myrtaceae

Origins:

Resembling *Psidium* L., see also *Eugenia* L.

Psidiomyrtus Guillaumin Myrtaceae

Origins:

The genera *Psidium* L. and *Myrtus*, see also *Rhodomyrtus* (DC.) Reichb.

Psidiopsis O. Berg Myrtaceae

Origins:

Resembling the genus *Psidium* L.

Psidium L. Myrtaceae

Origins:

Latin *psidium*, from the Greek *sidion*, from *side, sida, sibde* "pomegranate," Akkadian *sedum* "red"; see Carl Linnaeus,

Species Plantarum. 470. 1753 and *Genera Plantarum.* Ed. 5. 211. 1754; G.C. Wittstein, *Etymologisch-botanisches Handwörterbuch.* 729. 1852; F. Boerner & G. Kunkel, *Taschenwörterbuch der botanischen Pflanzennamen.* 4. Aufl. 157. 1989; S. Battaglia, *Grande dizionario della lingua italiana.* XIV: 866. Torino 1988; Giovanni Semerano, *Le origini della cultura europea.* Dizionari Etimologici. Basi semitiche delle lingue indeuropee. Dizionario della lingua Greca. 2(1): 261. Leo S. Olschki Editore, Firenze 1994.

Species/Vernacular Names:

P. spp.

Peru: llumich lumi, coca-coca, acca, aka, llo'm

Mexico: behui

P. acutangulum DC. (*Psidium grandiflorum* Ruíz & Pav.)

Peru: ampi yacu, puca yacu, guayaba de agua, guayabo de agua

P. cattleianum Sabine (*Psidium littorale* Raddi)

English: strawberry guava, Brazilian cherry, Cattley guava, purple guava, purple strawberry guava, Calcutta guava, calcuta guava

Central America: goyavier de St. martin, goyavier prune, purple guava

Argentina: araçá saiyu, guayabo amarillo

Brazil: áraçá, araçá do campo, araçá amarelo, araçá vermelho, araçá doce, araçámanteiga, araçá da praia, araçá de coroa, araçá de comer, araçá pera, araçá rosa, araçazeiro

Uruguay: araza

South Africa: aarbeikoejawel

Japan: kiban-zakuro (ki = yellow)

Hawaii: waiawi 'ula'ula

India: pahari payara, konda jamipandu, seemai koyya, bella seebai, malam perakka, pahadi pijuli

P. densicomum DC.

Peru: ampi yacu, guayaba silvestre

P. guajava L. (*Psidium cujavillus* Burm.f.; *Psidium pyriferum* L.; *Psidium pomiferum* L.)

English: common guava, yellow guava, apple guava, guava, wild guava

French: goyavier

Spanish: guayaba, guayaba dulce, guayaba manzana, guayaba perulera, guayabo, guayabo de venado, guayaba blanca

Central America: guayaba, goeajaba, guava, cak, ch'amxuy, coloc, eanandí, ikiec, patá, pataj, pichi, posh

Paraguay: guayaba

Mexico: bui, enandi, jalocote, xalácotl, pata, behui, yaga behui, yaga pehui, pehui, yaga huii, pichi, pojosh, pox, bec

Peru: bimpish, goiaba, guava, guayaba, guayabo, guaya-billo, hoja de guayaba, huallaba, huayaba, huayabo, kima, kumaski, llómy, matos, matus, matus sacha, sacha guayaba, sahuintu, sailla, shahuinto, tehua, tspata, yocaan

Bolivia: guayaba, guayabo, chuará-catoco

Colombia: guaiaba dulce

Argentina: arazapuitá

Brazil: goiaba, goiabeira, guaiaba, araçá-uaçu

Suriname: guave, goejaba

South Africa: koejawel, wilde koejawel

W. Africa: goyaki, byaghe, biaki

Tanzania: mabera

Yoruba: guaba, guafa, guroba, gurofa

India: mansala, amrud, safed safari, amrut, amratafalam, amruta-phalam, peru, pera, paera, peyara, piyara, lal peyara, pyara, piyra, perala, perala-hannu, koyya, koyapalam, etta-jama, jama, jama-phala, jam-pandu, tellajama, goyya, goyya-pandu, goyya-pazham, goaachhi, lal sufrium, goa-chi-phal, jamphal, jamba, tupkel, jamrukh, iamrud, shebe-hannu, sebe hannu, zetton, madhuria

South Laos: mak sidaa (Nya Hön)

Malaya: jambu batu, jambu biji, jambu burong

The Philippines: bayabas, bayauas, bayaya, gayabas, bagabas, bayabo, biabas, gaiyabat, gaiyabit, geyabas, guay-abas, guyabas, kalimbahin, tayabas

China: fan shi liu gan

Japan: banjirô, banshirû, benshirû

Hawaii: kuawa, kuawa ke'oke'o, kwava lemi, kuawa momona, puawa

P. guineense Swartz (*Psidium molle* Bertol.; *Psidium araca* Raddi)

English: guava, Guinea guava

Mexico: guayabo agrio, cashpadan

Peru: guayabillo, guayaba brava, sacha guayaba, huayaba

Bolivia: guayabilla, guayaba

Brazil: araçá, araçá-azedo, araçazeiro

P. littorale Raddi var. *littorale* (*Psidium lucidum* hort.; *Psidium littorale* var. *lucidum* (Deg.) Fosb.), "Waiawi"

English: yellow strawberry guava, yellow Cattley guava, strawberry guava

P. littorale Raddi var. *longipes* (O. Berg) McVaugh (*Psidium cattleianum* Salisb.)

English: purple strawberry guava, purple guava, Cattley guava, strawberry guava

P. montanum Sw.

English: mountain guava, spice guava

P. rostratum McVaugh

Peru: guayabo

P. rutidocarpum Ruíz & Pav. (*Psidium pratense* Poeppig ex O. Berg; *Psidium ruizianum* O. Berg)

Peru: huayabo del monte, monte sahuintu

Psila Philippi Asteraceae

Origins:
Greek *psilos* "bare, naked," Greek *psile* and Latin *psila* "a covering shaggy on one side, a shaggy covering."

Psilactis A. Gray Asteraceae

Origins:
Greek *psilos* "bare, naked" and *aktin* "ray."

Psilantha (K. Koch) Tzvelev Gramineae

Origins:
Greek *psilos* "bare, naked" and *anthos* "flower," see also *Eragrostis* Wolf.

Psilanthele Lindau Acanthaceae

Origins:
Greek *psilos* and *anthele* "a type of inflorescence, a little flower."

Psilanthopsis A. Chev. Rubiaceae

Origins:
Resembling *Psilanthus* Hook.f.

Psilanthus Hook.f. Rubiaceae

Origins:
Greek *psilos* and *anthos* "flower."

Psilathera Link Gramineae

Origins:
From the Greek *psilos* "bare, naked" and *ather* "an awn."

Psilobium Jack Rubiaceae

Origins:
Greek *psilos* "bare, naked, delicate" and *bios* "life," referring to the nature of the plants.

Psilocarphus Nutt. Asteraceae

Origins:
Greek *psilos* "bare, naked" and *karphos* "chip of straw, chip of wood," disk flowers not subtended by chaff scales.

Species/Vernacular Names:
P. elatior (A. Gray) A. Gray
English: tall woolly-heads
P. oregonus Nutt.
English: Oregon woolly-heads

Psilocarya Torrey Cyperaceae

Origins:
From the Greek *psilos* and *karyon* "walnut, nut"; see John Torrey (1796-1873), in *Annals of the Lyceum of Natural History of New York*. 3: 359. 1836.

Psilocaulon N.E. Br. Aizoaceae

Origins:
Greek *psilos* "bare, naked" and *kaulos* "stalk"; see Nicholas Edward Brown (1849-1934), in *The Gardeners' Chronicle*. Ser. 3, 78: 433. (Nov.) 1925.

Species/Vernacular Names:
P. absimile N.E. Br. (Latin *absimilis, absimile* "unlike")
South Africa: asbos
P. rogersiae L. Bol.
South Africa: prenia vygie, asbosvygie
P. tenue (Haw.) Schwantes (*Mesembryanthemum tenue* Haw.)
English: wiry noon-flower, match-head plant

Psilochilus Barb. Rodr. Orchidaceae

Origins:
From the Greek *psilos* and *cheilos* "a lip," the glabrous lip.

Psilochloa Launert Gramineae

Origins:
Greek *psilos* and *chloe, chloa* "grass," see also *Panicum* L.

Psilodigera Suess. Amaranthaceae

Origins:
Greek *psilos* "bare, naked" plus the genus *Digera* Forssk.

Psilodochea Presl Marattiaceae

Origins:
Greek *psilos* and *dochos* "a receptacle."

Psiloesthes Benoist Acanthaceae

Origins:
Greek *psilos* and *esthes, esthos* "a garment, dress."

Psilogramme Kuhn Pteridaceae (Adiantaceae)

Origins:
Greek *psilos* and *gramma* "a letter, line, character."

Psilolemma S.M. Phillips Gramineae

Origins:
Greek *psilos* "bare, naked" and *lemma* "rind, sheath."

Psilolepus Presl Fabaceae

Origins:
Greek *psilos* "bare, naked" and *lepis* "scale."

Psilonema C.A. Meyer Brassicaceae

Origins:
Greek *psilos* and *nema* "thread, filament."

Psilopeganum Hemsley Rutaceae

Origins:
Greek *psilos* and *peganon*, the ancient Greek name for rue.

Psilopogon Hochst. Gramineae

Origins:
From the Greek *psilos* and *pogon* "a beard."

Psilostachys Hochst. Amaranthaceae

Origins:
Greek *psilos* and *stachys* "a spike," see also *Psilotrichum* Blume.

Psilostachys Steud. Gramineae

Origins:
From the Greek *psilos* "bare, naked" and *stachys* "a spike."

Psilostachys Turcz. Euphorbiaceae

Origins:
Greek *psilos* and *stachys* "a spike."

Psilostoma Klotzsch Rubiaceae

Origins:
Greek *psilos* "bare, naked" and *stoma* "mouth."

Psilostrophe DC. Asteraceae

Origins:
Greek *psilos* and *strophe* "twist, turning" or *trophe* "food," receptacle flat and naked.

Species/Vernacular Names:
P. cooperi (A. Gray) E. Greene
English: paper-daisy

Psilothonna E. Meyer ex DC. Asteraceae

Origins:
Greek *psilos* "bare, naked" and the genus *Othonna* L.

Psilotrichopsis C.C. Towns. Amaranthaceae

Origins:
Resembling *Psilotrichum* Blume.

Psilotrichum Blume Amaranthaceae

Origins:
Greek *psilos* and *thrix, trichos* "hair," referring to the leaflets enclosing the fruits; see Karl Ludwig von Blume (1796-1862), *Bijdragen tot de flora van Nederlandsch Indië.* 544. (Jan.) 1826.

Psilotum Swartz Psilotaceae

Origins:
From the Greek *psilos* "naked," referring to the aerial shoots or to the leafless stems or branches; see O.P. Swartz, *Journal für die Botanik.* [Edited by H.A. Schrader] Göttingen 1802.

Species/Vernacular Names:
P. nudum (L.) P. Beauv. (*Lycopodium nudum* L.)
English: whisk-fern
Okinawa: awa-ran
Japan: matsuba-ran (= pine-leaved orchid)
China: shi shua ba

Psiloxylon Thouars ex Tul. Myrtaceae (Psiloxylaceae)

Origins:
Greek *psilos* "naked" and *xylon* "wood."

Species/Vernacular Names:
P. mauritianum Bouton ex Hook.f.
La Réunion Island: bois de gouyave marron, bois à gratter, bois de pêche marron, bois sans écorce
Mauritius: bois bigaignon

Psilurus Trin. Gramineae

Origins:
Greek *psilos* "naked, slender" and *oura* "tail," referring to the slender spikes; see Carl Bernhard von Trinius, *Fundamenta Agrostographiae.* 93. Viennae (Jan.) 1820; H. Schinz and Albert Thellung (1881-1928), in *Vierteljahrsschrift der Naturforschenden Gesellschaft in Zürich.* 58: 40, adnot. (Sep.) 1913.

Species/Vernacular Names:
P. incurvus (Gouan) Schinz & Thell. (*Nardus incurvus* Gouan; *Nardus aristata* L.; *Psilurus aristatus* (L.) Duval-Jouve; *Psilurus nardoides* Trin.)
English: bristle-tail grass

Psittacanthus C. Martius Loranthaceae

Origins:

Greek *psittakos* "a parrot" and *anthos* "flower."

Psittacoglossum La Llave & Lexarza Orchidaceae

Origins:

Greek *psittakos* "a parrot" and *glossa* "tongue," the lip resembles the tongue of a parrot, see also *Maxillaria* Ruíz & Pav.

Psittacoschoenus Nees Cyperaceae

Origins:

From the Greek *psittakos* and *schoinos* "rush, reed, cord," Latin *schoenus, i* for a rush; see J.G.C. Lehmann, *Plantae Preissianae*. 2: 87. Hamburgi (Nov.) 1846.

Psomiocarpa C. Presl Dryopteridaceae (Aspleniaceae)

Origins:

From the Greek *psomos* "morsel, bit" and *karpos* "food."

Psophocarpus Necker ex DC. Fabaceae

Origins:

Greek *psophos* "a noise, sound" and *karpos* "fruit," referring to the opening of the capsules or to the seeds and the rattling noise in the pods.

Species/Vernacular Names:

P. tetragonolobus (L.) DC. (*Dolichos tetragonolobus* L.)

English: asparagus pea, Goa bean, winged bean, winged pea, prince's pea

Japan: hane-mi-sasage

The Philippines: sigadilyas, cigarillas, segadilla, kalamismis, kamaluson, pallang, palag, palam, parupagulong, amale, buligan, beyed, serenella, batung-baimbing

Psoralea L. Fabaceae

Origins:

Greek *psoraleos* "scabby, warty, warted, scurfy," *psora* "mange," referring to glandular and resinous dots on plants

of the genus; see Carl Linnaeus, *Species Plantarum*. 762. 1753 and *Genera Plantarum*. Ed. 5. 336. 1754.

Species/Vernacular Names:

P. adscendens F. Muell.

English: dusky scurf-pea

P. cinerea Lindley

English: grey scurf-pea

P. esculenta Pursh

English: Indian breadroot

P. pinnata L.

English: blue-pea, fountain tree, fountain bush, African scurf-pea, North American prairie turnip

Southern Africa: fonteinbos, fonteinhout, bloukeur, pinwortel, penwortel (= taproot); umHlonishwa (Zulu)

P. tenax Lindley

English: tough scurf-pea

Psoralidium Rydb. Fabaceae

Origins:

Referring to the genus *Psoralea*.

Species/Vernacular Names:

P. lanceolatum (Pursh) Rydb.

English: scurf-pea

Psorobatus Rydb. Fabaceae

Origins:

Greek *psora* and *batos* "a bush, thicket, a kind of cup, a thorn, bramble."

Psorodendron Rydb. Fabaceae

Origins:

From the Greek *psora* "mange" and *dendron* "tree," see also *Psorothamnus* Rydb.

Psorospermum Spach Guttiferae

Origins:

From the Greek *psora* and *sperma* "a seed," referring to the nature of the seeds.

Species/Vernacular Names:

P. febrifugum Spach

English: Rhodesian holly, Christmas berry

French: millepertuis vert

Southern Africa: muLapamasi, muMenu, muMimu, muSwaswa, muTskwatsgwa (Shona)

Yoruba: legun oko, iyun orisa, legun kuro

N. Rhodesia: muhota, kavandula

W. Africa: karijakuma

Psorothamnus Rydb. Fabaceae

Origins:

From the Greek *psoros* "mangy, scabby" and *thamnos* "bush."

Species/Vernacular Names:

P. spinosus (A. Gray) Barneby

English: smoke tree

Pstathura Raf. Rubiaceae

Origins:

Perhaps from the Greek *psathyros* "friable," see also *Psathura* Comm. ex Juss.; no evidence of the genus *Pstathura* in E.D. Merrill, *Index rafinesquianus*. The plant names published by C.S. Rafinesque, etc. Jamaica Plain, Massachusetts, USA 1949.

Psychanthus (K. Schumann) Ridley Zingiberaceae

Origins:

From the Greek *psyche* "butterfly, moth" and *anthos* "flower," referring to the shape of the calyx; Psyche was a maiden beloved by Cupid, made immortal by Jupiter.

Psychanthus Raf. Polygalaceae

Origins:

See Constantine S. Rafinesque, *Specchio delle Scienze*. 1: 116. Palermo 1814, *New Fl. N. Am*. 4: 87. 1836 [1838] and *Autikon botanikon*. Icones plantarum select. nov. vel rariorum, etc. 6. Philadelphia 1840; E.D. Merrill, *Index rafinesquianus*. 152. 1949.

Psychechilos Breda Orchidaceae

Origins:

From the Greek *psyche* "butterfly" and *cheilos* "a lip," referring to the shape of the labellum, see also *Zeuxine* Lindley.

Psychilis Raf. Orchidaceae

Origins:

Greek *psyche* and *cheilos* "a lip," referring to the colored lip; see Constantine Samuel Rafinesque, in *Flora Telluriana*. 4: 40. 1836 [1838]; E.D. Merrill, *Index rafinesquianus*. 104. 1949.

Psychine Desf. Brassicaceae

Origins:

Greek *psyche* "butterfly," referring to the shape of the fruit.

Psychopsiella Lückel & Braem Orchidaceae

Origins:

The diminutive of the genus *Psychopsis*.

Psychopsis Raf. Orchidaceae

Origins:

Greek *psyche* "butterfly" and *opsis* "aspect, appearance, resemblance," alluding to the large flowers, similar to some butterfly species; see Constantine Samuel Rafinesque, in *Flora Telluriana*. 4: 40. 1836 [1838]; E.D. Merrill, *Index rafinesquianus*. 104. 1949.

Psychotria L. Rubiaceae

Origins:

Probably from the Greek *psychotria* "vivifying, exhilarating" or *psyche* "soul, life" and *iatria* "therapy, medicine," referring to the healing properties of some species; or modified and coined by Linnaeus from the Greek word *psychotrophon, psychros* "cold" and *trophe* "food," a name already applied by Patrick Browne (1720-1790) to describe a Jamaican taxon; Latin *psychotrophon, i* used by Plinius for a plant, betony. See C. Linnaeus, *Systema Naturae*. Ed. 10. 929, 1364. (May-Jun.) 1759.

Species/Vernacular Names:

P. sp.

Hawaii: kopiko

Peru: batsicawa, dahippiri, gorowa, ija, quina-quina

Nigeria: medzim-kouro, bodikido

Yoruba: akowo, gwawobo

P. balfouriana Verdc.

Rodrigues Island: bois lubine

P. brachybotrya Müll. Arg. (*Psychotria iquitosensis* Standley)

Peru: mitir-ey

P. capensis (Eckl.) Vatke subsp. *capensis* var. *capensis* (*Grumilea capensis* (Eckl.) Sond.; *Grumilea capensis* (Eckl.) Sond. var. *angustifolia* Sond.)

English: black bird-berry, izele tree

Southern Africa: swartvoëlbessie; uDzilidzili omhlophe (Swazi); iZele, isiThitibala, isThithibala, umGongono, uSinga lwamadoda, uManyanya (Zulu); umGono-gono, umGonogono (Xhosa); tshidiri (Venda)

P. carthagenensis Jacquin (*Psychotria alba* Ruíz & Pav.; *Mapouria alba* (Ruíz & Pav.) Muell.; *Uragoga alba* (Ruíz & Pav.) Kuntze; *Psychotria foveolata* (Ruíz & Pav.) Gómez Maza)

Peru: cawa, chacruna, rami eppe, mito micunan, rumi caspi, ucumi micuna

P. emetica L.f.

English: false ipecac

P. hawaiiensis (A. Gray) Fosb. (*Straussia hawaiiensis* A. Gray; *Psychotria hawaiiensis* var. *glomerata* (Rock) Fosb.)

Hawaii: kopiko 'ula, 'opiko

P. kaduana (Cham. & Schltdl.) Fosb. (*Coffea kaduana* Cham. & Schltdl.; *Coffea chamissonis* Hook. & Arnott; *Psychotria leptocarpa* (Hillebr.) Fosb.; *Psychotria longissima* (Rock) St. John)

Hawaii: kopiko kea

P. loniceroides DC.

English: rusty psychotria

P. mahonii C.H. Wright (*Grumilea punicea* S. Moore; *Psychotria megistosticta* (S. Moore) var. *punicea* (S. Moore) Petit)

English: large psychotria

Southern Africa: Bwanyfupa (Shona)

P. mauiensis Fosb. (*Straussia oncocarpa* Hillebr.; *Psychotria hawaiiensis* var. *molokaiensis* (Rock) Fosb.)

Hawaii: 'opiko

P. nematopoda F. Muell.

English: smooth psychotria

P. nervosa Sw.

English: wild coffee

P. peduncularis (Salisb.) Steyerm.

Yoruba: efun kojiya

P. zombamontana (Kuntze) Petit (*Grumilea kirkii* Hiern; *Uragoga zombamontana* Kuntze) (the type species was collected on Mount Zomba in Malawi)

Southern Africa: Zomba psychotria; bwanishupa (Shona)

Psychotrophum P. Browne Rubiaceae

Origins:

Greek *psyche* "butterfly" and *trophe* "food," or *psychros* "cold" and *trophe* "food," Latin *psychotrophon, i* used by Plinius for a plant, betony, see *Psychotria* L.

Psychrogeton Boiss. Asteraceae

Origins:

Greek *psychros* "cold" and *geiton* "a neighbor," referring to the habitat.

Psychrophila (DC.) Bercht. & J.S. Presl Ranunculaceae

Origins:

Greek *psychros* "cold" and *philos* "lover, loving"; see Friedrich Berchtold (1781-1876) and Jan Swatopluk Presl (1791-1849), *O prirozenosti rostlin.* 1: 79. Praha 1823.

Psychrophyton Beauverd Asteraceae

Origins:

From the Greek *psychros* and *phyton* "plant," referring to the habitat; see H.H. Allan, *Fl. New Z.* 1: 702, 712, 713. 1961.

Psydrax Gaertner Rubiaceae

Origins:

Greek *psydrax, psydrakos* "blister, bump," in allusion to the warted and wrinkled fruits of some species or to the pimply seeds.

Species/Vernacular Names:

P. arnoldiana (De Wild. & Th. Dur.) Bridson (*Canthium arnoldianum* (De Wild. & Th. Dur.) Hepper)

Nigeria: ebrenze (Edo)

P. livida (Hiern) Bridson (*Canthium huillense* Hiern; *Canthium lividum* Hiern; *Canthium wildii* (Suess.) Codd; *Plectronia junodii* Burtt Davy)

Southern Africa: groenboom (= green tree); monyonyana (Western Transvaal, northern Cape, Botswana); modumelantswe (Ngwaketse dialect, Botswana); muvhibvela-shadani (= ripens on shoulder), mukavha-mahunguvhu (= crow's perch) (Venda); muDandashoko (Shona)

P. locuples (K. Schumann) Bridson (*Canthium locuples* (K. Schumann) Codd; *Plectronia locuples* K. Schumann) (from the Latin *locuples, locupletis* "rich, opulent, wealthy")

English: whipstick canthium

Southern Africa: kranskwar, Krantz quar; isiKhondlwane (Zulu)

P. obovata (Ecklon & Zeyher) Bridson subsp. *obovata* (*Canthium obovatum* Eckl. & Zeyh.; *Canthium obovatum* Eckl. & Zeyh. var. *pyrifolium* (Eckl. & Zeyh.) Sond.)

English: black alder

Southern Africa: kwar, quar, elandshoorn, basterolienhout; umZilambuzi, umHlehlela, umBonemfama, uKukisaketelo (= make the kettle boil) (Zulu); umBombomfene, umBomomfene (= baboon's nose), umGupe (Xhosa)

P. subcordata (DC.) Bridson (*Canthium subcordatum* DC.)

Nigeria: ajelara (Yoruba); atan (Edo)

Yoruba: ajeleera, ajileera, igi eleera

Psygmium C. Presl Polypodiaceae

Origins:
Greek *psygma, psygmatos* "a fan, anything that cools," *psygmos* "dampness."

Psygmorchis Dodson & Dressler Orchidaceae

Origins:
Greek *psygmos* "dampness," *psygma, psygmatos* "a fan" plus *orchis* "orchid," referring to the habitat or habit of these epiphytes with large flowers and loving moderate shade and humidity.

Psylliostachys (Jaub. & Spach) Nevski Plumbaginaceae

Origins:
Greek *psyllion* "fleawort" and *stachys* "a spike," Dioscorides used *psyllion* for a species of *Plantago*; Latin *psyllion* applied by Plinius to a plant, fleabane, fleawort.

Psyllocarpus Martius & Zucc. Rubiaceae

Origins:
From the Greek *psylla* "flea" and *karpos* "fruit," referring to the seeds.

Ptaeroxylon Ecklon & Zeyher Ptaeroxylaceae

Origins:
Greek *ptairo, ptairein* "to sneeze" and *xylon* "wood," the red heartwood contains peppery irritating oil.

Species/Vernacular Names:
P. obliquum (Thunberg) Radlk. (*Ptaeroxylon utile* Ecklon & Zeyher)

English: sneezewood

Southern Africa: nieshout, muandara mahogany, umtata, umTati; umThathe, uBhaqa (= torch) (Zulu); umThathi (Xhosa); munukha-vhaloi (= tree smelling evilly of witches) (Venda); umThathi (Swazi); tati (Western Transvaal, northern Cape, Botswana); Ambhandadzwidzwi (Thonga)

East Africa: mwandara

Ptelea L. Rutaceae

Origins:
Greek *ptelea, ptelee* "an elm tree," Akkadian *petelu, patalu* "to wind, entwine," referring to the fruits; see G. Semerano, *Le origini della cultura europea*. Dizionario della lingua Greca. 2(1): 242. 1994; Giovanni Semerano. *Dizionario della lingua Latina e di voci moderne*. 2(2): 598. [Latin *ulmus* "an elm, elm-tree," Akkadian *u'ulum* "to bind," *u'iltu* "binding."] Firenze 1994.

Species/Vernacular Names:
P. trifoliata L.
English: hop tree

Pteleocarpa Oliver Boraginaceae (Ehretiaceae)

Origins:
Greek *ptelea* "an elm tree" and *karpos* "fruit."

Pteleopsis Engl. Combretaceae

Origins:
Greek *ptelea* "an elm tree" and *opsis* "aspect, appearance, resemblance."

Species/Vernacular Names:
P. anisoptera (Welw. ex Laws.) Engl. & Diels

English: four-winged pteleopsis

Southern Africa: muFundabulu (Shona)

N. Rhodesia: mufunyi

P. diptera (Welw.) Engl. & Diels

Zaire: lupala

P. hylodendron Mildbr.

Zaire: mukala, osanga

Central Africa: miong, nka, sikon

Cameroon: rissiehe, sikon, mobito

Congo: miong

Ivory Coast: koframire

Gabon: nka

P. myrtifolia (Laws.) Engl. & Diels (*Combretum myrtifolium* Laws.)

English: stink bushwillow, pteleopsis, myrtle-leaved pteleopsis, two-winged pteleopsis

Southern Africa: stinkboswilg, mirteboswilg; muSanganyemba (Shona); uMwandla, uMandla (Zulu); mwanda (Venda); mwanzabelo, mafungi (Kololo, Barotseland)

P. suberosa Engl. & Diels

Nigeria: wuyan giwa (Hausa)

W. Africa: nyanyanga

Mali: tereni, nanyinge

Ptelidium Thouars Celastraceae

Origins:
Resembling *ptelea* "the elm tree."

Pteracanthus (Nees) Bremek. Acanthaceae

Origins:
Greek *pteron* "wing, feather" and *akantha* "thorn."

Pterachenia (Benth.) Lipschitz Asteraceae

Origins:
From the Greek *pteron* "wing, feather" and *achaenium* or *achenium* "achene."

Pteralyxia K. Schumann Apocynaceae

Origins:
From the Greek *pteron* plus the genus *Alyxia* Banks ex R. Br.

Pterandra A. Juss. Malpighiaceae

Origins:
Greek *pteron* and *aner, andros* "male, stamen."

Pteranthus Forssk. Caryophyllaceae (Illecebraceae)

Origins:
Greek *pteron* "wing, feather" and *anthos* "flower."

Pteretis Raf. Dryopteridaceae (Woodsiaceae)

Origins:
Based on *Struthiopteris* Willd.; see C.S. Rafinesque, *Am. Monthly Mag. Crit. Rev.* 2: 268. 1818; E.D. Merrill (1876-1956), *Index rafinesquianus*. The plant names published by C.S. Rafinesque, etc. 71, 72. Jamaica Plain, Massachusetts, USA 1949.

Pterichis Lindley Orchidaceae

Origins:
From the Greek *pteron* "wing, feather" or *pteris* "fern" and *orchis* "orchid," referring to the shape of the labellum; see Richard Evans Schultes and Arthur Stanley Pease, *Generic Names of Orchids. Their Origin and Meaning.* 264. Academic Press, New York and London 1963; Hubert Mayr, *Orchid Names and Their Meanings.* 173. Vaduz 1998.

Pteridanetium Copel. Vittariaceae (Adiantaceae)

Origins:
Greek *pteridios* "feathered, winged" plus the genus *Anetium* (Kunze) Splitg.

Pteridella Mett. ex Kuhn Pteridaceae (Adiantaceae)

Origins:
From the Greek *pteris, pteridos* "a fern, male-fern" plus the diminutive.

Pteridium Gled. ex Scopoli Dennstaedtiaceae (Pteridaceae)

Origins:
Pteridion, diminutive of the Greek *pteris* "fern," from *pteron* "a wing, a feather," from Sanskrit *pat* "to fly," *patara* "flying"; see Giovanni Antonio Scopoli (1723-1788), *Flora Carniolica.* 169. Viennae 1760.

Species/Vernacular Names:

P. aquilinum (L.) Kuhn (*Pteris aquilinum* L.)

English: bracken, eagle fern, bracken fern, brake, hog-pasture brake, pasture brake, hog brake

French: fougère des aigles, fougère aigle

Congo: kungu, koungou, esiela

Southern Africa: adelaarsvaring; ukozani (Zulu); hombewe, hombge, mondgio (Shona); muvanguluvha (Venda)

German: adlerfarn, Christus-wurzel

Japan: warabi (= bracken)

China: jue

P. esculentum (Forst.f.) Nak. (*Pteris esculenta* Forst.f)

English: bracken, Austral bracken, common bracken

Maori names: rahurahu, rarahu, rarauhe, marohi, takata

Pteridium Raf. Pteridaceae

Origins:
Pteridion, diminutive of the Greek *pteris* "fern," for *Pteris* L.; see C.S. Rafinesque, *Princ. Somiol.* 26. 1814.

Pteridoblechnum Hennipman Blechnaceae (Blechnoideae)

Origins:
From *pteridion* a diminutive of the Greek *pteris, pteridos* "fern" and *blechnon* a word used for a species of fern; see Elbert Hennipman (1937-), in *Blumea*. 13: 397. (Feb.) 1966.

Pteridocalyx Wernham Rubiaceae

Origins:
Greek *pteris, pteridos* "fern" and *kalyx* "calyx," some suggest from the Greek *pteridios* "feathered, winged."

Pteridophyllum Siebold & Zucc. Pteridophyllaceae (Fumariaceae, Papaveraceae)

Origins:
Greek *pteris, pteridos* "fern" and *phyllon* "a leaf."

Pteridrys C. Chr. & Ching Dryopteridaceae (Aspleniaceae)

Origins:
Probably from the Greek *pteris, pteridos* "fern" and *drys* "the oak."

Pterigeron (DC.) Benth. Asteraceae

Origins:
Greek *pteron* "wing, feather" and the genus *Erigeron* L. (*geron* "an old man"), referring to the pappus; see George Bentham (1800-1884), *Flora Australiensis*. 3: 531. London 1867.

Pteriglyphis Fée Dryopteridaceae (Aspleniaceae, Woodsiaceae)

Origins:
Greek *pteris* "fern" and *glypho* "to carve, engrave," *glyphe* "a carving."

Pterilis Raf. Dryopteridaceae (Woodsiaceae)

Origins:
Greek *pteris* "fern," for *Struthiopteris* Willd. = *Matteuccia* Tod.; see C.S. Rafinesque, *Am. Monthly Crit. Rev.* 4: 195. 1819.

Pterilis Raf. Pteridaceae

Origins:
Greek *pteris* "fern," for *Pteris* L.; see C.S. Rafinesque, *Med. Fl.* 2: 254. 1830 and *Flora Telluriana*. 1: 83. 1836 [1837].

Pterinodes Siegesb. ex Kuntze Dryopteridaceae (Aspleniaceae, Woodsiaceae)

Origins:
Greek *pterinos* "made of feathers," *pteris, pteridos* "fern."

Pteris Gled. ex Scop. Dryopteridaceae

Origins:
Greek *pteris* "fern" (Theophrastus), Latin *pteris, pteridis* for a species of fern (Plinius), see also *Dryopteris* Adans.

Pteris L. Pteridaceae (Adiantaceae)

Origins:
Referring to the feathery fronds, from the Greek *pteris* "fern" (Theophrastus), from *pteron* "wing, feather," from Sanskrit *pat* "to fly," *patara* "flying"; Latin *pteris, idis*, a kind of fern (Plinius); see Carl Linnaeus, *Species Plantarum*. 1073. 1753.

Species/Vernacular Names:
P. sp.

Yoruba: omu osun, omu

P. bahamensis (J. Agardh) Fée (*Pteris diversifolia* Swartz var. *bahamensis* J. Agardh; *Pteris longifolia* L. var. *bahamensis* (J. Agardh) Hieronymus; *Pycnodoria bahamensis* (J. Agardh) Small; *Pycnodoria pinetorum* Small)

English: Bahama ladder brake, plumy ladder brake, Bahama brake

P. biaurita L.

Nepal: hade unyu

P. comans Forst.f.

English: netted brake, coastal brake

P. ensiformis Burm.f.

English: sword brake

Japan: hoko-shida

The Philippines: pakong parang

P. macilenta A. Rich.

English: sweet fern

P. multifida Poir. (*Pycnodoria multifida* (Poir.) Small)

English: spider brake, spider fern, Chinese brake, Huguenot fern, saw-leaved bracken

Japan: i-no-moto-sô (= herb-near-the-well)

China: feng wei cao

P. mutilata L.

English: sword brake

The Philippines: pakong parang

P. tremula R. Br.

English: tender brake, shaking brake, Australian bracken, tender bracken, poor-man's cibotium

P. tripartita Swartz

English: giant bracken, giant brake (*Litobrochia tripartita* (Swartz) C. Presl)

P. umbrosa R. Br.

English: jungle brake

P. vittata L. (*Pycnodoria vittata* (L.) Small; *Pteris longifolia* L.)

English: ladder brake, Chinese brake, Chinese ladder brake, rusty brake

Japan: moe-jima-shida

China: wu gong cao

Pterisanthes Blume Vitaceae

Origins:

Greek *pteron* "wing, feather" and *anthos* "flower," an allusion to the winged axis of the inflorescence.

Pterium Desv. Gramineae

Origins:

From *pterion*, the diminutive of the Greek *pteron* "feather."

Pternandra Jack Melastomataceae

Origins:

Greek *pterna, pteren* "heel" and *aner, andros* "male, stamen," the anthers are heeled; see William Jack, in *Malayan Miscellanies*. Bencoolen [Benkulen, Indonesia] 2(7): 60. 1822.

Species/Vernacular Names:

P. sp.

Malaya: sial menahun

P. coerulescens Jack

English: cursed shade

Malaya: sial menahun, lidah katak, benyut paya, bunut paya, delek puteh, kelat biru, kulit nipis, menaun, nipis kulit, sial menaun, ubah merkatak

Pternopetalum Franchet Umbelliferae

Origins:

Greek *pterna* "heel" and *petalon* "leaf, petal."

Pterobesleria C. Morton Gesneriaceae

Origins:

From the Greek *pteron* "wing" plus the genus *Besleria* L.

Pterocactus K. Schumann Cactaceae

Origins:

Greek *pteron* "wing, feather" plus *cactus*, referring to the dispersal of the seeds.

Pterocarpus Jacq. Fabaceae

Origins:

Greek *pteron* "wing" and *karpos* "fruit," referring to the broadly winged pod.

Species/Vernacular Names:

P. sp.

English: bloodwood, swamp bloodwood, white ebony, corkwood, African padauk

French: bois l'étang, bois de corail tendre, corail, santal rouge

Brazil: angu, mututy, mututy da terra firma, mututy da varzea, tachyzeiro, tachy de flor amarella, sapupira amarella; pau sangue (Bahia)

Venezuela: haro, lagunero

Argentina: ybira ra, iba ra

Colombia: sangre drago, sangregao

Guatemala: kaway

Honduras: sangre

Mexico: guayabillo, llora sangre

Puerto Rico: palo pallo

Guyana: hiburu, itchiki-boura, itik-boura, warau

French Guiana: liège de pays, moutouchi, mutushi, palétuvier, coumate

Suriname: bebe, bigbe, djoekabebe, itiki boeroe, itiki boro, itiki boro hororadikoro, itiki boura, matosirian, waata gwegwe, bebe hoedoe, itjoeroe tanomoetoesi, mattoe gwegwe, moetoesi, moetoesiran, moutouchi

Philippines Islands: naga, nala, nara, narra blancos, odiau, tagga, taggat, tagka, vitali, antagan, apalit, asana, bitali, dungon, hagad, kamarag, vitali; nara (Luzon); sagat, taggat (Cayagan)

Sumatra: tarpandi

Indochina: dang huong, giang huong, knong, santal rouge, hoang ba, douk, duc, maidou, hue moe

Cambodia: thnong

India: mai padu

Gabon: gula, ngula, hula, lucula, tacula

Zambia: mwangula, mwangura

Liberia: zahn

Nigeria: agbolosun, uhiye, ukpa

Yoruba: ogun malarere

Ghana: kerewowo, apehiripe, asante, esukatia, gedar kurumi, gyamantoa, nsaki

Ivory Coast: totohote, ouokisse

Zaire: katondo tondo, takula, nkula; mukula (Shaba)

East Africa: mgando-mkalate, mkurungu

West Africa: batwi

P. angolensis DC. (*Pterocarpus dekindtianus* Harms; *Pterocarpus bussei* Harms)

English: bloodwood, Transvaal teak, wild teak, African teak, Rhodesian teak, sealing-wax tree, Matabeleland deal

Southern Africa: mukwa, kiaat (= from old Dutch name for teak, kajaten), kehatenhout, kajatenhout, Transvaal kajatenhout, greinhout, lakboom, kajat, moroto, munaabenaabe, ngillasondo; mukwa (Zimbabwe); muBvamakova, muBvamaropa, muBvangazi, muKwirambira, muKurambira, iMvangasi, muVamaropa, muVunzamaropa (Shona); umVangazi, umBilo, inGozina, inDlandlovu (Zulu); umVangatzi, umVangati (Swazi); mokwa, morotomadi (= exudes blood) (western Transvaal, northern Cape, Botswana); moroto (North Sotho: north and northeast Transvaal); mutondo (Venda); mulombe (Subya: Botswana, eastern Caprivi); uguva, muguva (Deiriku); moowa (Mbukushu: Okavango Swamps and western Caprivi); ugruva (Sambui: Okavango Native Territory)

Zambia: mukwa, mulambi, mulombe, mulombwa, muzwamwloa, ndombe

Zimbabwe: mukwa, muBuaropa, muKurambira, umVagazi, muBvinza-maropa, muBvamaropa, mvangasi, umvagaz, umvangazi

Portuguese East Africa: ambila, umbila, gulomnila

East Africa: mininga, mninga

Mozambique: imbilo, moquombire-bire, shuiaan, thondo

Zaire: mulombo, mulombua, mulombwa, mulumba, musekeh, mutondo

Tanzania: titwego (Zigua)

Namibia (Ovamboland): omuva, omuuva (Kwanyama)

Malawi: mlombwa (Chichewa, Nyanja); mtumbati (Nkhonde); mtumbali (Yao)

Angola: njila-sonde, mutete, mudilahonde, mun haneca; kaionga (Kimbundu); mukula (Ganguela); omuliahond, omupaku (Lunyaneka); muva (Xikuanhama)

P. dalbergioides Roxb. ex DC.

English: Andaman padauk, Indian mahogany, Andaman redwood, padauk

India: yerravegisa, vengai, chalangada

P. erinaceus Poiret

English: barwood, West African kino, African rosewood, Gambian rosewood, African kino, Senegal rosewood, West African rosewood

French: palissandre du Senegal, santal d'Afrique

Ghana: doti, krayie, segbe, senya, tandasi, tim

East Africa: mininga, mninga, mtumball

South Africa: kajatenhout

West Africa: ven, vene

Dahomey: ven

Mali: genu, gweni, bani, nangerenge

North Nigeria: madobia

Nigeria: ara, apepe, madobia, arun, buma, vene, venni, uviara, tulum, mokoli-koli, akume, bekaka, eyiyi, oshun, osun, yabmatchal, uffe, uhie, ukpa, ume, panatan, muengi; madobiya (Hausa); banuhi (Fula); zanchi (Nupe); apepe, osun dudu (Yoruba); upeka (Edo); ufilarha (Etsako); aze egu (Igbo)

Yoruba: arira, ariraju, apepe, aara, osun dudu

Sudan: goni, guenou

Zaire: moolumbwa, mulombwa, mutondo

Zambia: mulombwa

French Guiana: M'gouin, vene

P. indicus Willd.

English: amboyna-wood, narra, Burmese rosewood, Papua New Guinea rosewood, red narra, Malay padauk

Malaya: angsana, sena

India: yerravegisa, vengai

Java: sono kembang

Thailand: mai pradoo

Japan: yaeyama-shi-tan

China: zi tan

Philippine Islands: naga, nala, nara, narra, red narra, bital, hagad, vitali, daitanag, odiau, Philippine padauk, Manila padouk, agana, asana, sagat, kamarag, tagga, tagka, balauning

Burma: angsanah, padouk

Fiji Islands: cibi cibi

P. indicus Willd.f. *echinatus* (Pers.) Rojo (*Pterocarpus echinatus* Pers.; *Pterocarpus vidalianus* Rolfe)

Philippines Islands: naga, nala, narra, prickly narra, odiau

P. lucens Guill. & Perr. subsp. *antunesii* (Taub.) Rojo (*Pterocarpus antunesii* (Taub.) Harms; *Calpurnia antunesii* Taub.; *Pterocarpus stevensonii* Burtt Davy) (after the Portuguese priest José Maria Antunes, 1856-1928, a botanical collector in Angola from 1889-1903)

South Africa: Antune's Kiaat

W. Africa: dakala, bala, dabakala

P. macrocarpus Kurz

English: Burma padauk, Tenasserim mahogany, Burmese rosewood

Burma: mai chi tawk, mai pi tawk, padauk, Burma padauk, inland padauk, padauk-ni, padauk-nyo, padauk-pyo, red padauk, padauk-sal, padauk-sat, padauk-wa, padoo, padouk, Burma padouk, padu

East India: brown padauk, inland padauk, mixed-colored padauk, white padauk, yellow padauk, padoo, padu, figured padouk, brown padouk

Indochina: dang, giang huong

P. marsupium Roxb.

English: Malabar kino, Indian kino tree, Malabar kino tree

India: bija sal, bijasal, bija, bibla, biyo, byasa, gammala, pitasara, pitshal, dhorbenla, asan, hiradakhan, yegi, peddagi,

vengai, honne, bange, venga; bibla (Central & South); dhorbioza, hid, hond, honne, ragatbera, venga, vengur (East); murga, piasal (Bengal); netra (Deccan); pedegi, vengai, yegi (Madras)

Sri Lanka: gammala

P. mildbraedii Harms

Yoruba: gbodogbodo

Nigeria: gennigar, uruhe, urube, uru-kho, urugho, panatan, ire, uluhe, yabmatchal; madobiyar rafi (Hausa); urube (Edo); geneghar (Ijaw); urhuko (Urhobo); kakupupu (Boki)

Cameroon: mbel afum

Ivory Coast: aguaya

P. officinalis Jacq.

Central America: kaway, sangre de drago

Puerto Rico: palo de pollo

Venezuela: sangre de drago

Nicaragua: sangregado

P. osun Craib

Cameroon: mohingué mossoumbé, mbel oswé

Nigeria: osun, uke, akume, bekaka, mokoli-koli, ume; osun (Yoruba); ume (Edo); urheri isele (Urhobo); eyiyi (Itsekiri); isele (Ijaw); ubie (Igbo); ukpa (Efik); boko anya (Boki)

Yoruba: irosun, osun, osun pupa

P. rohrii Vahl

Bolivia: verdolago blanco

P. rotundifolius (Sonder) Druce (*Pterocarpus martinii* Dunkley)

English: round-leaved bloodwood

Southern Africa: white mukwa, round-leaved kiaat, dopperkiaat; muChinane, muChirara, muHungu, muKuhutu, muMbhungu, muMungu, muWayawaya (Shona)

P. rotundifolius (Sonder) Druce subsp. *polyanthus* (Harms) Mendonça & Sousa (*Pterocarpus polyanthus* Harms)

South Africa: blinkblaar

P. rotundifolius (Sonder) Druce subsp. *rotundifolius* (*Dalbergia rotundifolia* Sonder; *Pterocarpus sericeus* Benth.)

English: round-leaved kiaat, round-leaved bloodwood, round-leaved teak

Southern Africa: dopperkiaat, blinkblaarboom, kiaathout, wildekweper; iNdlandlovu (Zulu); inDlebezindlovu (Swazi); miyataha, nshelela (Thonga or Tsonga or Shangaan); ndleve-ya-ndlopfu (Tsonga); mushusha-phombwe (= startles the adulter), muataha (Venda); mpanda (Kalanga: Northern Botswana); modianzovu (northern Botswana)

Southern Zambia: mulianzoha

P. santalinoides L'Hérit. ex DC.

W. Africa: jako, jagu, jegu

Mali: jawu, jagu

Nigeria: totohoti, gbingbin, gbingbindo, aku-emzi, uruhe, nja, gedar-kurumi, gunduru, pori-pori, ositua; gunduru, gyadar kurmi (Hausa); maganchi (Nupe); kereke (Tiv); gbengbe (Yoruba); akumeze (Edo); nturukpa (Igbo); nja (Efik)

Yoruba: gbengben, gbengbendo, gbingbin

P. santalinus L.f.

English: red sandalwood, red sanderswood, rubywood, Indian sandalwood, red sanders, red saunderswood

Java: almug

India: raktachandana, rakta tchandana, gosircha chandana, raktachandan, rakta gandhamu, lalchandan, tambada chandana, yerra chandanamu, chandanam, ratanjali, agarugandhamu, agaru, honne, kempugandha, chekke, patrangam, tilaparnni, tailaparni, atti, sivappu

East India: chandanam, panaka; lal chandan, rakta chandan, yerra chandanam (Madras)

Tibetan: tsan-dan dmar-po

China: tzu tan

P. soyauxii Taubert

English: barwood, African padauk, African padouk, West African padauk, West African padouk, camwood

French: bois corail, bois rouge, padouk d'Afrique

Gabon: corail, ebel, ebeul, ezigo, issipou, igoungou, mbel, ohinego, tiseze, tisseze, mogonda

Cameroon: ebeul, epion, mbe, mbe miki, m'bea, mbel, mbele, mbie, mbil, mbili, bo, ngele, ngola, mohingue, mohingue mossoumbe, mouengue moussoumbe, muenge, padouk

Central Africa: koula

Congo: kisese, tizeze, kisesi, m'bio, ngo, onguele, ongouele

Zaire: boi sulu, bosulu, boisolu, bosolo, ngele, ngula, nkula, nzali, wele

Nigeria: atu, boku, boko, auchi, akume, mbea, ebel, eba, igbuli, ihie, mbondi, mbonde, ebe, mohingui, muenge, muengi, nkohen, sako, uhiye, osunpupa, osun pupa, ukpa, ume, wosoka, padouk, mbe, mbe-miki, nkui-yang, ekuiyong; osun pupa (Yoruba); akume (Edo); awo (Igbo); ukpa (Efik); boku (Boki)

West Africa: ndimbo

Spanish Guinea: enve, palo rojo

Pterocarya Kunth Juglandaceae

Origins:

From the Greek *pteron* "wing, feather" and *karyon* "a nut."

Species/Vernacular Names:
P. stenoptera C. DC.

China: chu, chu liu

Pterocassia Britton & Rose Caesalpiniaceae

Origins:

From the Greek *pteron* "wing, feather" plus the genus *Cassia*, see also *Senna* Mill.

Pterocaulon Elliott Asteraceae

Origins:

Greek *pteron* and *kaulos* "a stem, a branch or stalk," referring to the decurrent leaves; see Stephen Elliott (1771-1830), *A Sketch of the Botany of South Carolina and Georgia*. 2: 323. Charleston 1823.

Species/Vernacular Names:
P. sphacelatum (Labill.) F. Muell. (*Monenteles sphacelatus* Labill.)

English: apple-bush

Pterocelastrus Meissner Celastraceae

Origins:

Greek *pteron* "wing" and *Celastrus* L., *kelastron*, *kelastros*, an ancient name used by Theophrastus for an evergreen tree whose fruits were on the tree throughout winter; the seeds are winged.

Species/Vernacular Names:
P. echinatus N.E. Br. (*Pterocelastrus galpinii* Loes.; *Pterocelastrus rehmannii* Davison; *Pterocelastrus variabilis* sensu Sim; *Gymnosporia nyasica* Burtt Davy & Hutch.) (Latin *echinatus, a, um* "echinate, armed with hairs or spines or prickles"; Latin *echinus* and Greek *echinos* "sea-urchin," Indoeuropean *echis* "snake")

English: white candlewood, hedgehog pterocelastrus

Southern Africa: witkershout; iNqayi-elibomvu, uSahlulamanye, isiHlulamanye, uGobandlovu (Zulu); iBholo, umGobandlovu (Xhosa); mutongola (Venda)

P. rostratus Walp. (*Celastrus rostratus* Thunb.; *Asterocarpus rostratus* (Thunb.) Eckl. & Zeyh.)

English: red candlewood, beaked pterocelastrus

Southern Africa: rooikershout; uSahlulamanye, uGobandlovu (Zulu)

P. tricuspidatus (Lam.) Sond. (*Pterocelastrus litoralis* Walp.; *Pterocelastrus stenopterus* Walp.; *Pterocelastrus*

tetrapterus Walp.; *Pterocelastrus variabilis* Sond.; *Celastrus tricuspidatus* Lam.)

English: candlewood, cherrywood

Southern Africa: kershout; uSahlulamanye (Zulu); uTywina, uGobandlovu, iBholo (Xhosa)

Pteroceltis Maxim. Ulmaceae

Origins:

From the Greek *pteron* "wing" and the genus *Celtis* L., referring to the winged fruits.

Pterocephalidium G. López Gonz. Dipsacaceae

Origins:

Referring to the genus *Pterocephalus*.

Pterocephalus Adans. Dipsacaceae

Origins:

Greek *pteron* "wing" and *kephale* "head," alluding to the fruiting head.

Pteroceras Hasselt ex Hassk. Orchidaceae

Origins:

From the Greek *pteron* "wing" and *keras* "a horn," referring to the appendages at the base of the labellum; see Justus Carl Hasskarl (1811-1894), *Flora oder allgemeine Botanische Zeitung*. 25. (July-Dec.) 1842.

Pterocereus MacDougall & Miranda Cactaceae

Origins:

From the Greek *pteron* "wing" plus *Cereus*.

Pterochaeta Steetz Asteraceae

Origins:

Greek *pteron* and *chaite* "bristle, long hair"; see Johann G.C. Lehmann (1792-1860), *Plantae Preissianae*. 1: 455. Hamburgi (Aug.) 1845.

Pterochilus Hooker & Arnott Orchidaceae

Origins:

From the Greek *pteron* "wing" and *cheilos* "a lip," an allusion to the lip, see also *Maxillaria*.

Pterochlaena Chiov. Gramineae

Origins:

From the Greek *pteron* "wing" and *chlaena* "a cloak, blanket."

Pterochlamys Roberty Convolvulaceae

Origins:

Greek *pteron* "wing" and *chlamys, chlamydos* "cloak."

Pterochloris (A. Camus) A. Camus Gramineae

Origins:

From the Greek *pteron* plus *Chloris* Swartz.

Pterochrosia Baillon Apocynaceae

Origins:

From the Greek *pteron* "wing" plus the genus *Ochrosia* Juss.

Pterocissus Urban & Ekman Vitaceae

Origins:

Greek *pteron* "wing" and *kissos* "ivy."

Pterocladon Hook.f. Melastomataceae

Origins:

From the Greek *pteron* "wing" and *klados* "branch."

Pterococcus Hassk. Euphorbiaceae

Origins:

From the Greek *pteron* "wing" and *kokkos* "berry," referring to the winged cocci.

Pterococcus Pall. Polygonaceae

Origins:

Greek *pteron* "wing" and *kokkos* "berry."

Pterocyclus Klotzsch Umbelliferae

Origins:

From the Greek *pteron* "wing" and *kyklos* "a circle, ring."

Pterocymbium R. Br. Sterculiaceae

Origins:

Greek *pteron* "wing" and *kymbe* "boat," *kymbos* "cavity," referring to the ovaries.

Pterocypsela C. Shih Asteraceae

Origins:

From the Greek *pteron* "wing" and *kypsele* "a hollow vessel, beehive, basket."

Pterodiscus Hook. Pedaliaceae

Origins:

From the Greek *pteron* "wing" and *diskos* "a disc," from the broadly winged disk of the fruit.

Pterodon Vogel Fabaceae

Origins:

From the Greek *pteron* and *odous, odontos* "a tooth."

Pterogaillonia Lincz. Rubiaceae

Origins:

Greek *pteron* "wing" plus the genus *Gaillonia* A. Rich. ex DC., after the French algologist François Benjamin Gaillon, 1782-1839; see J.H. Barnhart, *Biographical Notes upon Botanists.* 2: 23. 1965; J.D. Milner, *Catalogue of Portraits of Botanists Exhibited in the Museums of the Royal Botanic Gardens.* Royal Botanic Gardens, Kew, London 1906.

Pterogastra Naudin Melastomataceae

Origins:

From the Greek *pteron* "wing" and *gaster* "belly, paunch."

Pteroglossa Schltr. Orchidaceae

Origins:

From the Greek *pteron* "wing" and *glossa* "tongue," the wing-like dilatation of the lip.

Pteroglossaspis Reichb.f. Orchidaceae

Origins:

From the Greek *pteron* "wing," *glossa* "tongue" and *aspis* "a shield," referring to the wings of the column.

Pterogonum Gross Polygonaceae

Origins:

From the Greek *pteron* "wing" plus *Eriogonum* Michaux.

Pterogyne Tul. Caesalpiniaceae

Origins:

From the Greek *pteron* "wing" and *gyne* "a woman, female organs."

Pterolepis (DC.) Miq. Melastomataceae

Origins:

Greek *pteron* "wing, feather" and *lepis* "a scale."

Pterolepis Endl. Melastomataceae

Origins:

Greek *pteron* and *lepis* "a scale."

Pterolobium R. Br. ex Wight & Arnott Caesalpiniaceae

Origins:

Greek *pteron* "wing" and *lobos* "a pod," the winged pods; see Robert Wight and G. Arnott Walker Arnott, *Prodromus florae Peninsulae Indiae Orientalis.* 283. London 1834.

Species/Vernacular Names:

P. stellatum (Forssk.) Brenan (*Pterolobium exosum* (J.F. Gmel.) Bak.f.; *Pterolobium lacerans* R. Br.; *Acacia stellata* (Forssk.) Willd.; *Cantuffa exosa* J.F. Gmelin; *Mimosa stellata* Forssk.)

Southern Africa: katdoring, rank-wag'n-bietjie, vlam-wag-'n-bietjie; luhakangue (Venda); igado (Shona)

Pteroloma Desv. ex Benth. Fabaceae

Origins:

From the Greek *pteron* and *loma* "border, margin, fringe, edge."

Pteroloma Hochst. & Steud. Capparidaceae (Capparaceae)

Origins:

Greek *pteron* and *loma* "border, margin, fringe, edge," fruit a samara, see also *Dipterygium* Decne.

Pteromanes Pic. Serm. Hymenophyllaceae

Origins:

Greek *pteron* and *manes* "a cup, a kind of cup," or *manos* "loose, sparse, flaccid," or *maino, mainomai* "to be furious, rave," etc., see also *Crepidomanes*; see R.E.G. Pichi Sermolli, "Fragmenta Pteridologiae–VI." in *Webbia*. 31(1): 237-259. Apr. 1977.

Pteromimosa Britton Mimosaceae

Origins:

From the Greek *pteron* "wing" plus *Mimosa* L.

Pteromischus Pichon Bignoniaceae

Origins:

From the Greek *pteron* "wing" and *mischos* "a stalk."

Pteroneura Fée Davalliaceae

Origins:

Greek *pteron* "wing" and *neuron* "nerve."

Pteronia L. Asteraceae

Origins:

Greek *pteron* "a wing," referring to the seeds; see C. Linnaeus, *Species Plantarum*. Ed. 2. 2: 1176. 1763.

Species/Vernacular Names:

P. incana (Burm.) DC. (*Pteronia xantholepis* DC.; *Chrysocoma incana* Burm.)

English: blue bush

South Africa: asbossie, bloubos, bitterbos, laventelbossie, perdebossie, ribbokbos, vaalbossie

P. pallens L.f.

English: Scholtz bush

South Africa: aasvoëlbossie, armoedsbossie, bloubossie, gombossie, mierbossie, Scholtz-bossie, swaelbos, witbas, witbossie, witgatbossie, joggemscholtzbos, skitterybossie, stolsbos, witstambos

P. sordida N.E. Br. (*Pteronia chlorolepis* Dinter; *Pteronia glomerata* sensu Range non L.f.)

South Africa: swartkaroo

Pteropavonia Mattei Malvaceae

Origins:

From the Greek *pteron* "a wing" plus *Pavonia* Cav.

Pteropentacoilanthus Rappa & Camarrone Aizoaceae

Origins:

Greek *pteron* "a wing," *pente* "five," *koilos* "hollow" and *anthos* "flower"; for the Italian botanist and pharmacist Francesco Rappa (1880-1963) see Gordon Douglas Rowley, *A History of Succulent Plants*. Strawberry Press, Mill Valley, California 1997.

Pteropepon (Cogn.) Cogn. Cucurbitaceae

Origins:

Greek *pteron* and *pepon, peponos* "ripe, mild," Latin *pepo, peponis* "a species of large melon."

Pteropetalum Pax Capparidaceae (Capparaceae)

Origins:

From the Greek *pteron* "a wing" and *petalon* "a petal, leaf," see also *Euadenia* Oliv.

Pterophacos Rydb. Fabaceae

Origins:

From the Greek *pteron* "a wing" and *phakos* "a lentil."

Pterophora Harv. Asclepiadaceae

Origins:
Greek *pteron* and *phoros* "carrying, bearing."

Pterophylla D. Don Cunoniaceae

Origins:
From the Greek *pteron* "wing" and *phyllon* "leaf."

Pteropodium DC. ex Meissner Bignoniaceae

Origins:
From the Greek *pteron* and *podion* "a little foot, stalk."

Pteropodium Steud. Gramineae

Origins:
Greek *pteron* "wing" and *podion* "a little foot, stalk."

Pteropogon DC. Asteraceae

Origins:
Greek *pteron* "a wing" and *pogon* "a beard"; see A.P. de Candolle, *Prodromus.* 6: 245. 1838.

Pteropsis Desv. Polypodiaceae

Origins:
Greek *pteron* "a wing" and *opsis* "aspect, resemblance, appearance, with the form of."

Pteroptychia Bremek. Acanthaceae

Origins:
From the Greek *pteron* "a wing" and *ptyche* "a fold."

Pteropyrum Jaub. & Spach Polygonaceae

Origins:
From the Greek *pteron* and *pyros* "grain, wheat."

Pterorhachis Harms Meliaceae

Origins:
Greek *pteron* "a wing" and *rhachis* "rachis, axis, midrib of a leaf."

Pteroscleria Nees Cyperaceae

Origins:
From the Greek *pteron* "a wing" and *skleros* "hard, dry."

Pteroselinum Reichb. Umbelliferae

Origins:
Greek *pteron* "a wing" and *selinon* "parsley, celery."

Pterosicyos Brandegee Cucurbitaceae

Origins:
Greek *pteron* "a wing" with the genus *Sicyos* L., Greek *sikyos* "wild cucumber, gourd."

Pterosiphon Turcz. Meliaceae

Origins:
From the Greek *pteron* "a wing" and *siphon* "tube."

Pterospermum Schreber Sterculiaceae

Origins:
Greek *pteron* and *sperma* "seed," with reference to the winged seeds; see J.C.D. von Schreber, *Genera Plantarum.* 2: 461. 1791.

Species/Vernacular Names:
P. acerifolium (L.) Willd.

English: maple-leaved bayur, mapleleaf wingseedtree

Tropical Asia: bayur

India: karnikara, kanak-champa, kanako champa, kaniar, katha-champa, muchkund, muskunda, taun-poewun, matsa kanda, mayeng, hatipeala, hathipaila, morra, moragos, dieng-khong-swet, dieng-tharo-masi, waisip-thing, num-bong

Nepal: hattipaila

P. suberifolium Willd. (*Pentapetes suberifolia* L.; *Pterospermum canescens* Roxb.)

English: cork-leaved bayur, fishing rod tree

Sri Lanka: welan, valangu, velung, taddaemarum

India: much kand, vinangu

Pterospora Nutt. Ericaceae (Monotropaceae)

Origins:
Greek *pteron* "a wing" and *spora* "seed," with reference to the winged seeds.

Species/Vernacular Names:
P. andromedea Nutt.
English: pinedrops

Pterostegia Fischer & C.A. Meyer Polygonaceae

Origins:
Winged cover, Greek *pteron* "a wing" and *stege, stegos* "roof, shelter," referring to the involucre.

Pterostelma Wight Asclepiadaceae

Origins:
Greek *pteron* and *stelma, stelmatos* "a girdle, belt."

Pterostemma Kraenzlin Orchidaceae

Origins:
From the Greek *pteron* "a wing" and *stemma* "garland, wreath," indicating the leaves.

Pterostemon Schauer Grossulariaceae (Pterostemonaceae)

Origins:
Greek *pteron* "a wing" and and *stemon* "stamen."

Pterostephus C. Presl Rubiaceae

Origins:
From the Greek *pteron* "a wing" and *stephein* "to crown," *stephanos* "a crown," see also *Spermacoce* L.

Pterostylis R. Br. Orchidaceae

Origins:
Greek *pteron* and *stylos* "style, column," referring to the laterally winged column; see Robert Brown (1773-1858), *Prodromus florae Novae Hollandiae et Insulae van-Diemen*. 326. London 1810; Arthur D. Chapman, ed., *Australian Plant Name Index*. 2432-2443. Canberra 1991.

Species/Vernacular Names:
P. acuminata R. Br.
English: pointed greenhood, sharp greenhood

P. alata (Labill.) Reichb.f. (*Disperis alata* Labill.; *Pterostylis praecox* Lindley)
English: striped green-orchid, striped greenhood, straited greenhood, purplish greenhood

P. alpina R. Rogers
English: alpine greenhood

P. alveata Garnet
English: jug greenhood

P. baptistii Fitzg.
English: king greenhood, giant greenhood

P. barbata Lindley
English: bearded greenhood

P. biseta Blackmore & Clemesha
English: rustyhood, veined greenhood, fringed rustyhood

P. boormanii Rupp
English: Boorman's greenhood

P. coccinea Fitzg.
English: redhood, scarlet greenhood

P. concinna R. Br.
English: trim greenhood

P. cucullata R. Br.
English: leafy greenhood

P. curta R. Br.
English: blunt greenhood, brown-nose

P. cycnocephala Fitzg. (= swan-headed, Greek *kyknos* "a swan" and *kephale* "head")
English: swan greenhood

P. daintreana Benth.
English: Daintree's greenhood

P. decurva R. Rogers
English: summer greenhood

P. dubia R. Br.
English: blue-tongued greenhood

P. fischii Nicholls (after the original collector Paul Fisch, at Woodside, in S. Gippsland, Australia)
English: winter greenhood

P. foliata Hook.f. (*Pterostylis vereenae* R. Rogers) (found by Miss Vereena Jacobs, at Cherry Gardens, South Australia)
English: slender greenhood, large-leaf greenhood

P. furcata Lindley
English: sickle greenhood, forked greenhood

P. gibbosa R. Br.
English: northern ruddyhood

P. grandiflora R. Br.

English: cobra greenhood, superb greenhood, long-tongue greenhood

P. hildae Nicholls

English: rainforest greenhood, Curtis' greenhood

P. longifolia R. Br.

English: tall greenhood

P. mutica R. Br.

English: midget greenhood

P. nana R. Br.

English: dwarf greenhood

P. nutans R. Br.

English: nodding greenhood

P. ophioglossa R. Br.

English: snake tongue greenhood, northern trim greenhood

P. parviflora R. Br.

English: tiny greenhood, baby greenhood

P. pedoglossa Fitzg.

English: prawn greenhood, tailed greenhood

P. pedunculata R. Br.

English: maroonhood, red ridinghood

P. plumosa Cady

English: bearded greenhood

P. pusilla R. Rogers

English: ruddyhood

P. rufa R. Br.

English: rusty greenhood, rufous greenhood

P. toveyana Ewart & Sharman (after J.R. Tovey, a plant collector in Victoria, Australia)

English: Mentone greenhood

P. truncata Fitzg.

English: little dumpies, little Dumpy, brittle greenhood

P. woollsii Fitzg.

English: long-tailed greenhood

Pterostyrax Siebold & Zucc. Styracaceae

Origins:
Greek *pteron* "wing" and the genus *Styrax* L., winged fruits.

Species/Vernacular Names:
P. corymbosus Siebold & Zuccarini

English: corymbose epaulette tree

China: xiao ye bai xin shu

P. psilophyllus Diels ex Perkins

English: glabrous-leaf epaulette tree

China: bai xin shu

Pterotaberna Stapf Apocynaceae

Origins:
From the Greek *pteron* "wing" and the genus *Tabernaemontana* L.

Pterothrix DC. Asteraceae

Origins:
From the Greek *pteron* "wing" and *thrix* "hair," referring to the plumose pappus bristles.

Pterotropia W.F. Hillebr. Araliaceae

Origins:
Greek *pteron* and *trepo, trope* "a turning," *tropias* "turned," *tropos* "way, manner, direction," see also *Tetraplasandra* A. Gray.

Pteroxygonum Dammer & Diels Polygonaceae

Origins:
From the Greek *pteron* "wing," *oxys* "sharp" and *gonia* "an angle."

Pterozonium Fée Pteridaceae (Adiantaceae)

Origins:
Greek *pteron* "wing" and *zone* "a belt, armor, girdle," *zonion* is the diminutive of *zone*.

Pterygiella Oliver Scrophulariaceae

Origins:
Pterygion is the diminutive of the Greek *pteryx, pterygos* "a wing."

Pterygiosperma O.E. Schulz Brassicaceae

Origins:
From the Greek *pterygion* "a small wing" and *sperma* "seed."

Pterygocalyx Maxim. Gentianaceae

Origins:

Greek *pteryx, pterygos* "a wing" and *kalyx* "calyx," calyx campanulate, tube 4-winged.

Species/Vernacular Names:

P. volubilis Maxim.

English: twining pterygocalyx

China: yi e man

Pterygodium Swartz Orchidaceae

Origins:

Greek *pteryx, pterygos* "a wing," *pterygodes, pterygoeides* "like a wing," the flower has a wing-like appearance.

Species/Vernacular Names:

P. hallii (Schelpe) Kurzweil & Linder (*Anochilus hallii* Schelpe) (for Harry Hall, 1906-1986 (d. South Africa), employed at the Kirstenbosch National Botanic Gardens, collector of succulent plants, Curator of the Darrah Collection of Succulents, Manchester; see Gordon Douglas Rowley, *A History of Succulent Plants*. Strawberry Press, Mill Valley, California 1997; Mary Gunn and Leslie E. Codd, *Botanical Exploration of Southern Africa*. 176-177. Cape Town 1981)

English: pterygodium

Pterygoloma Hanst. Gesneriaceae

Origins:

Greek *pteryx, pterygos* "a wing" and *loma* "border, margin, fringe."

Pterygopappus Hook.f. Asteraceae

Origins:

Greek *pteryx, pterygos* "wing, small wing, feather" and *pappos* "fluff, downy appendage, hairy tufts on achenes and fruits," referring to the feathery pappus; see J.D. Hooker, *Hooker's London Journal of Botany*. 6: 120. 1847.

Pterygopleurum Kitagawa Umbelliferae

Origins:

Greek *pteryx, pterygos* "wing, feather" and *pleura, pleuron* "side, rib."

Pterygopodium Harms Caesalpiniaceae

Origins:

From the Greek *pteryx, pterygos* "wing, feather" and *podion* "a little foot, stalk."

Pterygostachyum Steud. Gramineae

Origins:

Greek *pteryx, pterygos* and *stachys* "a spike."

Pterygostemon V.V. Botschantzeva Brassicaceae

Origins:

Greek *pteryx, pterygos* "wing, feather" and *stemon* "stamen."

Pterygota Schott & Endlicher Sterculiaceae

Origins:

Greek *pterygotos* "winged," *pteron* "wing," referred to the winged seeds.

Species/Vernacular Names:

P. alata (Roxb.) R. Br.

English: winged pterygota

Malaya: kasah

P. bequaertii De Wild.

W. Africa: awari

Central Africa: ake, mauya, kakende

Cameroon: efok ayus, mauya

Ivory Coast: akodiakede

Gabon: aké

P. macrocarpa K. Schumann

Yoruba: poroporo, obuburu, oporoporo

Nigeria: oporoporo, poroporo-funfun, okoko, ebututu, efog, efungi, poroporo; oporoporo (Yoruba)

Congo: nakomepok

Cameroon: muyali, efok ayus

Ivory Coast: koto, botie

Zaire: mulaba

Pteryxia (Torrey & A. Gray) J. Coulter & Rose Umbelliferae

Origins:

Greek *pteryx, pterygos* "wing, feather, feathery foliage."

Ptilagrostis Griseb. Gramineae

Origins:
Greek *ptilon* "a wing, feather" and *agrostis, agrostidos* "grass, couch grass."

Species/Vernacular Names:
P. kingii (Bolander) Barkworth
English: King's ricegrass

Ptilanthelium Steudel Cyperaceae

Origins:
Greek *ptilon* "a wing, feather" and *anthele* "a type of inflorescence, a little flower," *anthelion, anthyllion* is a diminutive of *anthos* "flower, blossom"; see Ernst Gottlieb von Steudel (1783-1856), *Synopsis plantarum glumacearum.* 2: 166. Stuttgartiae 1855.

Ptilanthus Gleason Melastomataceae

Origins:
From the Greek *ptilon* "a wing, feather" and *anthos* "a flower."

Ptilepida Raf. Asteraceae

Origins:
Greek *ptilon* "a wing, feather" and *lepis, lepidos* "scale"; see C.S. Rafinesque, *Am. Monthly Mag. Crit. Rev.* 2: 268. 1818 and 4: 189. 1819, *Jour. Phys. Chim. Hist. Nat.* 89: 261. 1819, *New Fl. N. Am.* 1: 60. 1836 and *Herb. Raf.* 39. 1833.

Ptileris Raf. Asteraceae

Origins:
See C.S. Rafinesque, *Am. Monthly Mag. Crit. Rev.* 2: 268. 1818 and 4: 195. 1819.

Ptilimnium Raf. Umbelliferae

Origins:
Greek *ptilon* and *limnion*, the diminutive of *limne* "a marsh"; see C.S. Rafinesque, *Am. Monthly Mag. Crit. Rev.* 4: 192. 1819, *Jour. Phys. Chim. Hist. Nat.* 89: 258. 1819, *Neogen.* 2. 1825, *Seringe Bull. Bot.* 1: 217. 1830, *New Fl. N. Am.* 4: 33. 1836 [1838] and *The Good Book.* 55. 1840;

E.D. Merrill, *Index rafinesquianus.* 182. Jamaica Plain, Massachusetts, USA 1949.

Ptilocalyx Torrey & A. Gray Boraginaceae (Ehretiaceae)

Origins:
From the Greek *ptilon* "a wing, feather" and *kalyx* "a calyx."

Ptilochaeta Turcz. Malpighiaceae

Origins:
Greek *ptilon* and *chaite* "bristle, mane."

Ptilocnema D. Don Orchidaceae

Origins:
Greek *ptilon* "feather" and *kneme* "limb, leg," the shape of the peduncle, see also *Pholidota* Lindley ex Hook.

Ptiloneilema Steudel Gramineae

Origins:
From the Greek *ptilon* "feather" and *eilema* "a veil, covering, involucre," see also *Melanocenchris* Nees.

Ptilonema Hook.f. Gramineae

Origins:
Greek *ptilon* and *nema* "thread, filament," see also *Melanocenchris* Nees.

Ptilophyllum Bosch Hymenophyllaceae

Origins:
Greek *ptilon* "feather, wing, down" and *phyllon* "a leaf," see also *Trichomanes* L.

Ptilopteris Hance Monachosoraceae (Adiantaceae, Dennstaedtiaceae)

Origins:
Greek *ptilon* "feather, down" and *pteris* "a fern."

Ptilostemon Cass. Asteraceae

Origins:
From the Greek *ptilon* "feather, down, soft feathers" and *stemon* "stamen."

Ptilothrix K.A. Wilson Cyperaceae

Origins:
Greek *ptilon* "feather, down, soft feathers" and *thrix, trichos* "hair."

Ptilotrichum C.A. Meyer Brassicaceae

Origins:
Greek *ptilon* "feather" and *thrix, trichos* "hair," referring to the starry down.

Ptilotus R. Br. Amaranthaceae

Origins:
Greek *ptilotos* "winged," referring to the bottlebrush inflorescence or to the flowers, the calyx is often hairy; see Robert Brown, *Prodromus florae Novae Hollandiae*. 415. London 1810; Arthur D. Chapman, ed., *Australian Plant Name Index*. 2443-2453. Canberra 1991.

Species/Vernacular Names:
P. exaltatus Nees (*Trichinium exaltatum* (Nees) Benth.)

Australia: pink mulla mulla

P. exaltatus Nees var. *exaltatus*

Australia: tall mulla mulla

P. gaudichaudii (Steudel) J. Black (*Trichinium gaudichaudii* Steudel)

Australia: paper foxtail

P. helipteroides (F. Muell.) F. Muell. (*Trichinium helipteroides* F. Muell.; *Trichinium brachytrichum* F. Muell.)

Australia: hairy mulla mulla

P. macrocephalus (R. Br.) F. Muell. (*Trichinium macrocephalum* R. Br.; *Trichinium angustifolium* Moq.; *Ptilotus pachocephalus* (Moq) F. Muell.)

Australia: large green pussy tail, feather heads, squareheaded foxtail, green pussy tails

P. manglesii (Lindley) F. Muell.

English: pussy tail

P. obovatus (Gaudich.) F. Muell.

Australia: smoke bush, cotton bush

Ptycanthera Decne. Asclepiadaceae

Origins:
From the Greek *ptyche* "a fold" and *anthera* "anther."

Ptychandra Scheffer Palmae

Origins:
Greek *ptyche* "a fold" and *andros* "man, male, stamen," the male flowers with anthers bent over in bud; see also *Heterospathe* Scheff.

Ptychocarpa (R. Br.) Spach Fabaceae

Origins:
Greek *ptyche* and *karpos* "fruit"; see Édouard Spach (1801-1879), *Histoire naturelle des Végétaux*. 10: 402. Paris 1841.

Ptychocarpus Kuhlm. Flacourtiaceae

Origins:
Greek *ptyche* "a fold" and *karpos* "fruit," see also *Neoptychocarpus* Buchheim.

Ptychochilus Schauer Orchidaceae

Origins:
Greek *ptyche* "a fold" and *cheilos* "lip," the lip encloses the column.

Ptychococcus Beccari Palmae

Origins:
Greek *ptyche* "a fold" and *kokkos* "berry," referring to the albumen of the seeds, indicating the grooved seeds.

Ptychodea Willd. ex Chamisso & Schltdl. Rubiaceae

Origins:
From the Greek *ptyche, ptyx* "a fold," *ptychodes* "in folds or layers."

Ptychodon Klotzsch ex Reichb. Lythraceae

Origins:
From the Greek *ptyche, ptyx* "a fold" and *odous, odontos* "a tooth."

Ptychogyne Pfitzer Orchidaceae

Origins:
Greek *ptyche, ptyx* and *gyne* "woman, female, column," referring to a fold at the base of the lip, see also *Coelogyne* Lindl.

Ptycholobium Harms Fabaceae

Origins:
From the Greek *ptyche* "a fold" and *lobos* "a pod," an allusion to the shape of the pod.

Ptychomanes Hedw. Hymenophyllaceae

Origins:
Greek *ptyche* and *manes* "a cup, a kind of cup," see also *Hymenophyllum* Sm.

Ptychomeria Benth. Burmanniaceae

Origins:
Greek *ptyche* "a fold" and *meris* "a portion, part," see also *Gymnosiphon* Blume.

Ptychopetalum Benth. Olacaceae

Origins:
From the Greek *ptyche* "a fold" and *petalon* "petal."

Ptychophyllum C. Presl Hymenophyllaceae

Origins:
From the Greek *ptyche* "a fold" and *phyllon* "a leaf," see also *Hymenophyllum* Sm.

Ptychopyxis Miq. Euphorbiaceae

Origins:
From the Greek *ptyche* "a fold" and *pyxis* "a small box."

Ptychoraphis Beccari Palmae (Arecaceae)

Origins:
Greek *ptyche* "a fold" and *rhaphis, rhaphidos* "a needle, raphe," grooved seeds along the raphe, see also *Rhopaloblaste* R. Scheff.

Species/Vernacular Names:
P. cagayanensis Becc.
The Philippines: Cagayan bilisan (from Luzon, Cagayan)
P. intermedia Becc.
The Philippines: lubi-lubi

Ptychosema Benth. Fabaceae

Origins:
Greek *ptyche* and *sema* "sign, standard"; see John Lindley, *Edwards's Botanical Register.* Appendix to Vols. 1-23: *A Sketch of the Vegetation of the Swan River Colony.* xvi. London (November) 1839.

Ptychosperma Labill. Palmae

Origins:
Greek *ptyche* "a fold" and *sperma* "seed," referring to the albumen of the seeds, *Ptychosperma* is characterized by its longitudinally 5-grooved seeds; see J.J.H. de Labillardière (1755-1834), in *Mémoires de la classe des sciences mathématiques et physiques de l'Institut National de France.* Paris 1808(2): 252. 1809.

Species/Vernacular Names:
P. elegans (R. Br.) Blume
English: Alexander palm, solitaire palm
Brazil: palmeira solitária
Japan: yusura-yashi
P. macarthurii (H. Wendl.) Nicholson (*Actinophloeus macarthurii* (Wendl.) Beccari; *Kentia macarthurii* H. Wendl.)
English: Macarthur cluster palm
Brazil: palmeira de macarthur
Japan: yahazu-yashi
P. sanderianum Ridl. (*Actinophloeus sanderianus* (Ridl.) Burret)
Japan: hoso-ba-kencha
P. salomonense Burret
Brazil: pticosperma de salomão

Ptychostigma Hochst. Rubiaceae

Origins:
From the Greek *ptyche* "a fold" and *stigma* "a stigma."

Ptychostylus Tieghem Loranthaceae

Origins:

From the Greek *ptyche* and *stylos* "pillar, style," see also *Struthanthus* Mart.

Ptychotis W.D.J. Koch Umbelliferae

Origins:

Greek *ptyche* "a fold" and *ous, otos* "an ear."

Ptyssiglottis T. Anderson Acanthaceae

Origins:

From the Greek *ptysso* "fold, double up" and *glossa* "tongue."

Pubistylus Thoth. Rubiaceae

Origins:

Latin *pubes, is* "the hair, fullness, ripeness," *pubens* "in full vigor, juicy, flourishing" and *stilus* "a stake, style," referring to the nature of the style.

Puccinellia Parlatore Gramineae

Origins:

After the Italian botanist Benedetto Luigi Puccinelli, 1808-1850, professor of botany, from 1830 to 1850 Director of the Botanical Garden of Lucca. His works include *Synopsis plantarum in Agro Lucensi sponte nascentium.* Lucae [Lucca] 1841 [-1848] and *Osservazioni sui funghi dell'Agro Lucchese.* Lucca 1841. See Filippo Parlatore (1816-1877), *Flora Italiana.* 1: 366. 1848; John H. Barnhart, *Biographical Notes upon Botanists.* 3: 114. 1965; Paolo Emilio Tomei, "Benedetto Puccinelli botanico." *Riv. Arch. St. Econ. Cost.* 3(4): 28-30. 1975; P.E. Tomei, "L'opera micologica di Benedetto Puccinelli, lucchese." *Micologia Italiana.* 5(2): 33-35. 1976; P.E. Tomei and M.E. Seghieri, "I cultori di botanica in Lucchesia dal XVI al XIX secolo." *Atti Ist. Bot. Lab. Critt. Univ. Pavia.* 14(6): 249-271. 1981; P.E. Tomei and S. Lucchesi, *Le indagini micologiche in Lucchesia.* Lucca 1980; G. Murray, *History of the Collections Contained in the Natural History Departments of the British Museum.* 1: 175. 1904; T.W. Bossert, *Biographical Dictionary of Botanists Represented in the Hunt Institute Portrait Collection.* 319. 1972; E.M. Tucker, *Catalogue of the Library of the Arnold Arboretum of Harvard University.*
1917-1933; F. Boerner & G. Kunkel, *Taschenwörterbuch der botanischen Pflanzennamen.* 4. Aufl. 157. Berlin & Hamburg 1989.

Species/Vernacular Names:

P. angusta (Nees) C.A. Sm. & C.E. Hubb.

English: finch alkali grass

South Africa: brakgras, vinkbrakgras, vinkgras

P. distans (Jacq.) Parl.

English: European alkali grass

P. fasciculata (Torrey) E. Bickn.

English: Borrer's saltmarsh grass

P. howellii J.I. Davis

English: Howell's alkali grass

P. lemmonii (Vasey) Scribner

English: Lemmon's alkali grass

P. nutkaensis (Presl) Fern. & Weath.

English: Alaska alkali grass

P. nuttalliana (Schultes) A. Hitchc.

English: Nuttall's alkali grass

P. parishii A. Hitchc.

English: Parish's alkali grass

P. pumila (Vasey) A. Hitchc.

English: dwarf alkali grass

P. stricta (Hook.f.) C.H. Blom

English: Australian saltmarsh grass, marsh grass

Puelia Franchet Gramineae

Origins:

After the French botanist Timothée Puel, 1812-1890, physician, wrote *Catalogue des plantes vasculaires qui croissent dans le département Lot.* Cahors 1845-1853; see J.H. Barnhart, *Biographical Notes upon Botanists.* 3: 114. 1965; E.M. Tucker, *Catalogue of the Library of the Arnold Arboretum of Harvard University.* 1917-1933; J.T. Timothée Puel, *Essai sur les causes locales de la différence de taille qu'on observe chez les habitants des deux cantons de Latronquière et de Livernon.* in Collection des thèses soutenues à la Faculté de Médecine de Paris. an 1839-1878. Paris 1839-1878.

Pueraria DC. Fabaceae

Origins:

For the Swiss botanist Marc Nicolas Puerari, 1766-1845, teacher, pupil of Martin Vahl (1749-1804) in Copenhagen.

See A.P. de Candolle, *Hist. bot. genev.* 48. 1830 [= *Mém. Soc. Phys. Genève.* 5: 48. 1830] and *Annales des Sciences Naturelles.* 4: 97. 1825; J.H. Barnhart, *Biographical Notes upon Botanists.* 3: 114. 1965; Hans West (1758-1811), *Bidrag til Beskrivelse over Ste Croix med en kort Udsigt over St. Thomas, St. Jean, Tortola, Spanishtown og Crabeneiland.* Kjöbenhaven 1793; A. Lasègue, *Musée botanique de Benjamin Delessert.* Paris 1845; Stafleu and Cowan, *Taxonomic Literature.* 4: 423. Utrecht 1983; Georg Christian Wittstein, *Etymologisch-botanisches Handwörterbuch.* 740. Ansbach 1852.

Species/Vernacular Names:

P. montana (Lour.) Merr. var. *lobata* (Willd.) Maesen & S. Almeida (*Dolichos hirsutus* Thunb.; *Dolichos lobatus* Willd.; *Neustanthus chinensis* Benth.; *Pachyrhizus thunbergianus* Siebold & Zucc.; *Pueraria hirsuta* (Thunb.) Matsum., nom. illeg.; *Pueraria lobata* (Willd.) Ohwi; *Pueraria thunbergiana* (Siebold & Zucc.) Benth.)

English: kudzu vine, kudsu, kudzu, Japanese arrowroot, lobed kudzu vine, kudzu hemp

Japan: kuzu

China: ge gen, ge gan

P. phaseoloides (Roxb.) Benth.

English: tropical kudzu

P. tuberosa (Willd.) DC.

India: siali, vidari kand

Pugionium Gaertner Brassicaceae

Origins:
Latin *pugio, pugionis* "a dagger, dirk."

Pukanthus Raf. Solanaceae

Origins:
See C.S. Rafinesque, *Sylva Telluriana.* 53, 158. 1838; E.D. Merrill (1876-1956), *Index rafinesquianus.* 212. Jamaica Plain, Massachusetts 1949.

Pulchranthus Baum, Reveal & Nowicke Acanthaceae

Origins:
Latin *pulcher, pulchra, pulchrum* "beautiful, handsome" and Greek *anthos* "flower."

Pulicaria Gaertner Asteraceae

Origins:
Latin subst. *pulicaria* for a plant called also *psyllion*, Latin adj. *pulicarius* and *pulicaris* "belonging to fleas," *herba pulicaria, ae* (*pulex, icis* "a flea"), reputed to repel fleas.

Species/Vernacular Names:
P. arabica (L.) Cass.

Arabic: abu 'ain safra

P. scabra (Thunb.) Druce (*Erigeron scabrum* Thunb.; *Inula capensis* Spreng.; *Pulicaria capensis* DC.)

Southern Africa: aambeibos; umkhathula (Zulu)

Puliculum Haines Gramineae

Origins:
Latin *pulex, pulicis* "a flea," *pulico* "to produce fleas."

Pullea Schlechter Cunoniaceae

Origins:
After the Netherlands botanist August Adriaan Pulle, 1878-1955, plant taxonomist, botanical explorer and plant collector in Suriname and New Guinea, 1914-1949 professor of systematic botany at the University of Utrecht, from 1932 to 1955 editor of the *Flora of Suriname* (Netherlands Guyana). Amsterdam 1932 etc. Among his numerous publications are *An Enumeration of the Vascular Plants Known from Surinam.* Leiden 1906, *Naar het sneeuwgebergte van Nieuw-Guinea met de derde Nederlandsche expeditie.* Amsterdam [1914] and *Compendium van de terminologie, nomenclatuur en systematiek der zaadplanten.* Utrecht 1938. See André Joseph Guillaume Henri Kostermans (b. 1907), *Studies in South American Malpighiaceae, Lauraceae and Hernandiaceae, Especially of Surinam.* Amsterdam 1936; Stafleu and Cowan, *Taxonomic Literature.* 4: 425-442. 1983; John H. Barnhart, *Biographical Notes upon Botanists.* 3: 115. 1965; Friedrich Richard Rudolf Schlechter (1872-1925), in *Botanische Jahrbücher.* 52: 164. 1914; T.W. Bossert, *Biographical Dictionary of Botanists Represented in the Hunt Institute Portrait Collection.* 319. 1972; E.M. Tucker, *Catalogue of the Library of the Arnold Arboretum of Harvard University.* 1917-1933; Ida Kaplan Langman, *A Selected Guide to the Literature on the Flowering Plants of Mexico.* 598. University of Pennsylvania Press, Philadelphia 1964.

Pulmonaria L. Boraginaceae

Origins:

Latin *pulmo, pulmonis* "lung," adj. *pulmonarius, a, um* "diseased in the lungs, curative of the lungs."

Species/Vernacular Names:

P. mollissima A. Kerner

English: glandular-hair lungwort

China: xian mao fei cao

P. officinalis L.

English: lungwort

Pulsatilla Miller Ranunculaceae

Origins:

Latin *pulso, avi, atum* "to push, strike."

Species/Vernacular Names:

P. alpina (L.) Delarb.

English: alpine Pasque flower

P. ambigua Turcz. ex Pritz.

English: Mongolian pulsatilla

P. chinensis (Bunge) Regel

English: Chinese flower, Chinese anemone

China: bai tou weng

P. dahurica (Fisch.) Spreng.

English: Dahurian pulsatilla

P. patens (L.) Mill.

English: eastern Pasque flower, spreading Pasque flower

P. vulgaris Mill.

English: Pasque flower

Pultenaea J.E. Smith Fabaceae

Origins:

For the British botanist Richard Pulteney, 1730-1801 (Dorset), physician, M.D. Edinburgh 1764, naturalist, historian of science and biographer of Linnaeus, he became a Fellow of the Royal Society in 1762 and of the Linnean Society in 1789. His works include *A General View of the Writings of Linnaeus*. London 1781, *Catalogues of the Birds, Shells, and Some of the More Rare Plants of Dorsetshire*. London 1799 and *Historical and Biographical Sketches of the Progress of Botany in England*. London 1790 [the German translation: *Geschichte der Botanik* bis auf die neueren Zeiten mit besonderer Rücksicht auf England. Leipzig 1798]. See J.H. Barnhart, *Biographical Notes upon Botanists*. 3: 115. 1965; G.C. Gorham, *Memoirs of John Martyn* ... *and Thomas Martyn*. London 1830; William Munk, *The Roll of the Royal College of Physicians of London*. 1878; G. Murray, *History of the Collections Contained in the Natural History Departments of the British Museum*. 1: 175. 1904; Claus Nissen, *Die Botanische Buchillustration*. 1951; T.W. Bossert, *Biographical Dictionary of Botanists Represented in the Hunt Institute Portrait Collection*. 319. 1972; E.M. Tucker, *Catalogue of the Library of the Arnold Arboretum of Harvard University*. 1917-1933; Jonas C. Dryander, *Catalogus bibliothecae historico-naturalis Josephi Banks*. London 1800; A. Lasègue, *Musée botanique de Benjamin Delessert*. Paris 1845; Blanche Henrey, *British Botanical and Horticultural Literature before 1800*. 1975; Ida Kaplan Langman, *A Selected Guide to the Literature on the Flowering Plants of Mexico*. 598. Philadelphia 1964; J.D. Milner, *Catalogue of Portraits of Botanists Exhibited in the Museums of the Royal Botanic Gardens*. Royal Botanic Gardens, Kew, London 1906; Ray Desmond, *Dictionary of British & Irish Botanists and Horticulturists*. 567. 1994; D.E. Allen, *The Botanists*: A History of the Botanical Society of the British Isles through a Hundred and Fifty Years. Winchester 1986; Blanche Henrey, *No Ordinary Gardener — Thomas Knowlton, 1691-1781*. Edited by A.O. Chater. British Museum (Natural History). London 1986; Georg Christian Wittstein, *Etymologisch-botanisches Handwörterbuch*. 741. Ansbach 1852.

Species/Vernacular Names:

P. acerosa R. Br. ex Benth. (*Pultenaea acerosa* R. Br. ex Benth. var. *acicularis* H.B. Williamson)

English: bristly bush-pea

P. involucrata Benth.

Australia: Mount Lofty bush-pea

P. trifida J. Black

Australia: Kangaroo Island bush-pea

Pulteneya Hoffsgg. Fabaceae

Origins:

Orthographic variant of *Pultenaea* Smith.

Pultnaea Graham Fabaceae

Origins:

Orthographic variant of *Pultenaea* Smith.

Pulvinaria Fourn. Asclepiadaceae

Origins:

Latin *pulvinar, pulvinaris* "a sofa, cushion."

Pumilo Schltdl. Asteraceae

Origins:

From the Latin *pumilus, i* "dwarf"; see Diederich Franz Leonhard von Schlechtendal, in *Linnaea.* 21: 448. (Aug.) 1848.

Punctillaria N.E. Br. Aizoaceae

Origins:

Latin *punctillum* "a little point, dot, spot," see also *Pleiospilos* N.E. Br.

Punica L. Punicaceae (Lythraceae)

Origins:

The Latin name, *malum punicum* "Carthaginian apple," *Punicus, a, um,* from *Poenus, i* "a Carthaginian," *Poenus, a, um* "Punic, Carthaginian," *Poeni, orum* "the Phoenicians, the Carthaginians," Greek *Phoinix* "Phoenician"; see Carl Linnaeus (Carl von Linnaeus, Carl von Linné) (1707-1778), *Species Plantarum.* 472. 1753 and *Genera Plantarum.* Ed. 5. 212. 1754.

Species/Vernacular Names:

P. sp.

Peru: cashahuaro

P. granatum L.

English: pomegranate

Arabic: ruman, rommen, romman

India: dadam, dadima, dadiman, dadimah, dadima-phalam, dalimgachh, dalimba, dalimbay, dalimbu-hannu, darimba, dalimma, danimma, dalim, dalimb, dhalim, darim, daru, dhaun, delumgaha, madalai, madulai, madalam, madalng-kai, matalam, matala-narakam, anar, anara, anar-ke-per, anar-dakum, kuchaphala, shukadana, naspal, jaman, gulnar, darakte-nar

Tibetan: se-'bru

Mexico: yaga tini, zehe castilla, xoba zehe

The Philippines: granada

Malaya: delima

Japan: zakuro, hime-zakuro (for the dwarf variety)

China: shi liu pi, an shih liu, shan shih liu

Vietnam: an thach luu, mac liu

Nigeria: rimani

Punicella Turcz. Myrtaceae

Origins:

The diminutive of *Punica* L., see also *Balaustion* Hooker; see Porphir Kiril N.S. Turczaninow, in *Bulletin de la classe physico-mathématique de l'Académie Impériale des sciences de Saint-Pétersbourg.* 10: 333. St. Petersburg 1852.

Pupalia A.L. Juss. Amaranthaceae

Origins:

From *pupali,* a vernacular name; see A.L. de Jussieu, in *Annales du Muséum National d'Histoire Naturelle.* 2: 132. 1803.

Species/Vernacular Names:

P. lappacea (L.) A. Juss. var. *lappacea* (*Achyranthes atropurpurea* Lam.; *Achyranthes lappacea* L.; *Desmochaeta atropurpurea* DC.; *Pupalia atropurpurea* (Lam.) Moq.)

English: sweethearts

Southern Africa: beesklits, klits; isinama-esibomvuse-hlathi (Zulu)

Yoruba: eemo agbo, eemagbo, eemo agbotomo, agbongbon

Purdiaea Planchon Cyrillaceae

Origins:

Named for the Scottish botanist William Purdie, 1817-1857 (Trinidad), plant collector in tropical South America, gardener, 1846-1857 Botanical Garden Trinidad, wrote "Journal of Botanical Mission to West Indies in 1843-1844." in *London J. Bot.* 3: 501-533. 1844 and 4: 14-27. 1845. See J.H. Barnhart, *Biographical Notes upon Botanists.* 3: 115. 1965; Ray Desmond, *Dictionary of British & Irish Botanists and Horticulturists.* 567. London 1994; Mea Allan, *The Hookers of Kew.* London 1967; Henri François Pittier, *Manual de las Plantas Usuales de Venezuela* y su Suplemento. Caracas 1971; F.N. Hepper and Fiona Neate, *Plant Collectors in West Africa.* 66. 1971; Joseph Vallot, "Études sur la flore du Sénégal." in *Bull. Soc. Bot. de France.* 29: 191-192. Paris 1882; I.C. Hedge and J.M. Lamond, *Index of Collectors in the Edinburgh Herbarium.* Edinburgh 1970.

Purdieanthus Gilg Gentianaceae

Origins:

Dedicated to the Scottish botanist William Purdie, 1817-1857.

Purpurella Naudin Melastomataceae

Origins:

Latin *purpura* "purple color."

Purpureostemon Gugerli Myrtaceae

Origins:

From the Latin *purpureus* "purple colored" and Greek *stemon* "stamen."

Purpusia Brandegee Rosaceae

Origins:

Dedicated to the German botanical explorer Carl (Karl) Albert (Alberto) Purpus, 1851-1941, botanist, botanical and plant collector in Mexico and the southwestern United States, physician, wrote *Die Kakteen der Grand Mesa in West-Colorado*. 1893, he was the brother of the gardener Joseph Anton Purpus (1860-1932); see J.H. Barnhart, *Biographical Notes upon Botanists*. 3: 116. 1965; S. Lenley et al., *Catalog of the Manuscript and Archival Collections and Index to the Correspondence of John Torrey*. Library of the New York Botanical Garden. 337. 1973; Ida Kaplan Langman, *A Selected Guide to the Literature on the Flowering Plants of Mexico*. 598-601. Philadelphia 1964; Stafleu and Cowan, *Taxonomic Literature*. 4: 445-446. Utrecht 1983; Gordon Douglas Rowley, *A History of Succulent Plants*. Strawberry Press, Mill Valley, California 1997; Joseph Ewan, *Rocky Mountain Naturalists*. The University of Denver Press 1950; Irving William Knobloch, compil., "A preliminary verified list of plant collectors in Mexico." *Phytologia Memoirs*. VI. 1983; Mario S. Sousa, "Las colecciónes botánicas de C.A. Purpus en México, periodo 1898-1925." *Univ. Calif. Publ. Bot.* 51: i-ix, 1-36. 1969; R. Zander, F. Encke, G. Buchheim and S. Seybold, *Handwörterbuch der Pflanzennamen*. 14. Aufl. 766. Stuttgart 1993; I.C. Hedge and J.M. Lamond, *Index of Collectors in the Edinburgh Herbarium*. Edinburgh 1970.

Purshia DC. ex Poiret Rosaceae

Origins:

For the German (b. Saxony) botanist Frederick (Friedrich, Fredric, Frederic) Traugott Pursh (Pursch), 1774-1820 (d. Montreal, Canada), traveler, gardener, plant collector. Among his writings is *Flora Americae septentrionalis*. London 1814. See Thomas Potts James (1803-1882), *Journal of a Botanical Excursion in the Northeastern Parts of the States of Pennsylvania and New York During the Year 1807*. By Frederick Pursh. Philadelphia 1869; Joseph Ewan, in

Dictionary of Scientific Biography 11: 217-219. 1981; R. Zander, F. Encke, G. Buchheim and S. Seybold, *Handwörterbuch der Pflanzennamen*. 14. Aufl. 766. Stuttgart 1993; Blanche Henrey, *British Botanical and Horticultural Literature before 1800*. 1975; J.H. Barnhart, *Biographical Notes upon Botanists*. 3: 116. 1965; H.N. Clokie, *Account of the Herbaria of the Department of Botany in the University of Oxford*. 229. Oxford 1964; Ethelyn Maria Tucker, *Catalogue of the Library of the Arnold Arboretum of Harvard University*. Cambridge, Massachusetts 1917-1933; A. Lasègue, *Musée botanique de Benjamin Delessert*. Paris 1845; Joseph Ewan, ed., *A Short History of Botany in the United States*. New York and London 1969; Jeannette Elizabeth Graustein, *Thomas Nuttall, Naturalist. Explorations in America, 1808-1841*. Cambridge, Harvard University Press 1967; J.W. Harshberger, *The Botanists of Philadelphia and Their Work*. 113-117. 1899; Edwin M. Betts, ed., "Thomas Jefferson's Garden Book, 1766-1824." in *Mem. Amer. Phil. Soc.* 22: 1-704. Philadelphia 1944; Joseph Ewan and Nesta Ewan, "John Lyon, nurseryman, and plant hunter, and his journal, 1799-1814." in *Transactions of the American Philosophical Society*. 53(2): 1-69. 1963; Ida Kaplan Langman, *A Selected Guide to the Literature on the Flowering Plants of Mexico*. 1964; William Darlington (1782-1863), *Reliquiae Baldwinianae*. Philadelphia 1843.

Species/Vernacular Names:

P. tridentata (Pursh) DC.

English: antelope bush

Purshia Raf. Haloragidaceae (Haloragaceae)

Origins:

An orthographic variant of *Burshia* Raf.; see C.S. Rafinesque, *Anal. Nat. Tabl. Univ.* 195. 1815, *Am. Monthly Mag. Crit. Rev.* 4: 191, 195. 1819, *Jour. Phys. Chim. Hist. Nat.* 89: 257. 1819, *Med. Repos.* II. 3: 422. 1806 and 5: 357. 1808; E.D. Merrill (1876-1956), *Index rafinesquianus. The plant names published by C.S. Rafinesque, etc.* 177. Jamaica Plain, USA 1949.

Puschkinia M.F. Adams Hyacinthaceae

Origins:

For the Russian plant collector Apollos Apollossowitsch (alt. name: Apollinaire von) Muschkin-Puschkin (Mussin-Pus[c]hkin), 1760-1805, in Siberia and Caucasus; see I.H. Vegter, *Index Herbariorum*. Part II (4), *Collectors M*. Regnum Vegetabile vol. 93. 1976; R. Zander, F. Encke, G. Buchheim and S. Seybold, *Handwörterbuch der Pflanzennamen*. 14. Aufl. 766. Stuttgart 1993.

Putoria Pers. Rubiaceae

Origins:

Stinking shrubs, Latin *putor, putoris* "a strong smell, stench," *puteo, es, ere* "to stink, to be rotten, putrid," referring to the leaves; Akkadian *pasaru* "to loosen," *pasasu* "to destroy, to remove," *pasatu* "to destroy."

Putranjiva Wallich Euphorbiaceae

Origins:

A Sanskrit name, *putra* "son" and *juvi* "prosperity, life," *Ficus benjamina* L. in India is also *putra-juvi*; see M.P. Nayar, *Meaning of Indian Flowering Plant Names*. 290. Dehra Dun 1985; Shri S.P. Ambasta, ed., *The Useful Plants of India*. 505. Council of Scientific & Industrial Research, New Delhi 1986.

Species/Vernacular Names:

P. roxburghii Wall. (see also *Drypetes*)

India: putranjiva, putrajiva, putijia, putajan, jiaputa, juti, jewanputr, kudrajuwi, kuduru, putrajivika, irukolli, karupalai, amani, pongalam, poitundia

Putterlickia Endlicher Celastraceae

Origins:

After the Austrian botanist Aloys (Alois, Aloysio) Putterlick, 1810-1845, bryologist, physician, from 1840 to 1845 in charge at the Natural History Museum of Vienna, author of *Synopsis Pittosporearum* ... Vindobonae [Wien] 1839. See Theodor F.L. Nees von Esenbeck (1787-1837), *Genera plantarum Florae germanicae*. Bonnae [1833-] 1835-1860; J.H. Barnhart, *Biographical Notes upon Botanists*. 3: 117. 1965; R. Zander, F. Encke, G. Buchheim and S. Seybold, *Handwörterbuch der Pflanzennamen*. 14. Aufl. 766. Stuttgart 1993; Johann G.C. Lehmann (1792-1860), *Plantae Preissianae* ... Plantarum quas in Australasia occidentali et meridionali-occidentali annis 1838-41 collegit L. Preiss. Hamburgi 1844-1847[-1848]; Ethelyn Maria Tucker, *Catalogue of the Library of the Arnold Arboretum of Harvard University*. Cambridge, Massachusetts 1917-1933.

Species/Vernacular Names:

P. pyracantha (L.) Szyszyl. (*Celastrus pyracanthus* L.; *Celastrus saxatilis* Burch.; *Gymnosporia integrifolia* (L.f.) Glover; *Gymnosporia saxatilis* (Burch.) Davison)

English: false spike-thorn, fire-thorn putterlickia

South Africa: basterpendoring

P. retrospinosa Van Wyk & Mostert

South Africa: large-leaved bastard spike-thorn

P. verrucosa (E. Mey. ex Sond.) Szyszyl. (*Celastrus verrucosus* Sond.)

English: false forest spike-thorn, warted bastard spike-thorn, warted putterlickia

Southern Africa: basterbospendoring; umHlabankonkani, uHlinzanyaka (Zulu); iNthlangwana (Xhosa)

Puya Molina Bromeliaceae

Origins:

Spanish *puya* "goad, lance head."

Species/Vernacular Names:

P. sp.

Peru: ahuarancu

P. raimondii Harms (see Timothy J. Killeen, Emilia García E. and Stephan G. Beck, eds., *Guía de Arboles de Bolivia*. 166. Herbario Nacional de Bolivia and Missouri Botanical Garden 1993)

Bolivia: cinco, junco, chuqui kayara (Aymara); ilakuash (Quechua)

Peru: ckara, cunco, junco, llacuash, santon, tica-tica, titanca

Pycnandra Benth. Sapotaceae

Origins:

Greek *pyknos* "dense, compact" and *aner, andros* "male, stamen."

Pycnanthemum Michaux Labiatae

Origins:

From the Greek *pyknos* "dense, numerous, compact, crowded" and *anthemon* "a flower," the flowers are densely crowded.

Pycnanthus Warb. Myristicaceae

Origins:

Greek *pyknos* and *anthos* "flower," the flowers are numerous and packed together.

Species/Vernacular Names:

P. angolensis (Welw.) Warb. (*Pycnanthus microcephalus* (Benth.) Warb.)

English: cardboard, false nutmeg, white cedar, African nutmeg

French: faux muscadier, arbre à suif

West Africa: eteng, ilomba, kombo, akomu, oto

Central Africa: calabo, bakondo, bosamba, eteng, nkomo, téngé, gele, akomu, oti, oualélé, gboyei

Zaire: banga, bohondo, bodenga, djadja, gwanga, likoka, lolako, iomba, lukalakala, mudilampwepwe, tshilonbe, tshimbuku

Congo: nlomba, ilomba

Gabon: dilomba, ilomba, ekombo, nkoma, nkombo, n-komo, nlombo, geomba, ikoum, illomba, kombo, lilombo, lomba, lombo, moulomba, mulomba, sombo, eteng, etan, etang, etengui, eting, lating'e, lengye

Ivory Coast: adria, anakue, djilo, edua, effoi, epoi, etama, olele, oualélé, walelé, teke

Ghana: atenli, atta-bini, bini, bove, etsu, oti, otie, tika, walele

Cameroon: bakondo, bokondo, ilimba, kiang, nasamba, ten, teng, tengé, eteng, tian, tombe

Togo: obala

Yoruba: akomu

Nigeria: abae, aba-oro, okujaoti, abora, awka-mille, abakan, akomu, awka-minni, abo, bapulo, bakondo, bokondo, bocham, etan, eteng, itang, ilomba, mili, mile, nasamba, n'gosam, ndababa, ndodabo, ngosa, ngwasama, obabi, ebubi, esamba, obenazi, tengo, teng, tamarkwa, umoghan, ibicho, nkpanti, etana; akomu (Yoruba); umoghan (Edo); abora (Itsekiri); bapulo (Ijaw); abaoror (Urhobo); tamakwa (Kwale Igbo); akwa-mili (Igbo); kpokogi (Nupe); bucham (Boki); abakang (Ibibo)

Pycnarrhena Miers ex Hook.f. & Thomson Menispermaceae

Origins:
Greek *pyknos* and *arrhen* "male," the male flowers are fascicled together; see J.D. Hooker & Thomas Thomson (1817-1878), *Flora Indica*. 1: 206. London 1855.

Pycnobolus Willd. ex O.E. Schulz Brassicaceae

Origins:
Greek *pyknos* "dense, crowded" and *bolos* "casting."

Pycnobotrya Benth. Apocynaceae

Origins:
From the Greek *pyknos* and *botrys* "cluster."

Species/Vernacular Names:
P. nitida Benth.

Yoruba: abesokoro

Pycnobregma Baillon Asclepiadaceae

Origins:
Greek *pyknos* "dense, crowded" and *bregma, bregmatos* "front part of the head."

Pycnocephalum (Less.) DC. Asteraceae

Origins:
Greek *pyknos* "compact, crowded" and *kephale* "head."

Pycnocoma Benth. Euphorbiaceae

Origins:
From the Greek *pyknos* "close-packed, crowded" and *kome* "hair of the head."

Pycnocomon Hoffmannsegg & Link Dipsacaceae

Origins:
Greek *pyknos* "close-packed, close, strong" and *kome* "hair of the head."

Pycnocycla Lindley Umbelliferae

Origins:
Greek *pyknos* "close-packed, close, strong" and *kyklos* "a circle," referring to the seeds.

Pycnodoria C. Presl Pteridaceae (Adiantaceae)

Origins:
Perhaps from the Greek *pyknos* "close-packed, close, crowded" and *dorea, doron* "a gift."

Pycnolachne Turcz. Labiatae (Verbenaceae, Dicrastylidaceae)

Origins:
Greek *pyknos* "dense" and *lachne* "wool, woolly hair, downy"; see Porphir Kiril N.S. Turczaninow, in *Bulletin de la Société Impériale des Naturalistes de Moscou*. 36(2): 214. 1863.

Pycnoloma C. Chr. Polypodiaceae

Origins:

From the Greek *pyknos* and *loma* "border, margin, fringe, edge."

Pycnoneurum Decne. Asclepiadaceae

Origins:

Greek *pyknos* and *neuron* "nerve."

Pycnonia L. Johnson & B. Briggs Proteaceae

Origins:

Greek *pyknos* "dense," alluding to an affinity with the genus *Persoonia* Smith in the same family; see Lawrence Alexander Sidney Johnson (1925-) and Barbara Gillian Briggs (1934-), in *Botanical Journal of the Linnean Society.* 70: 175. (Sep.) 1975.

Pycnophyllopsis Skottsb. Caryophyllaceae

Origins:

Resembling *Pycnophyllum* Remy.

Pycnophyllum J. Remy Caryophyllaceae

Origins:

Greek *pyknos* "dense" and *phyllon* "leaf."

Pycnoplinthopsis Jafri Brassicaceae

Origins:

Resembling the genus *Pycnoplinthus* O. Schulz.

Pycnoplinthus O.E. Schulz Brassicaceae

Origins:

From the Greek *pyknos* "dense" and *plinthos, plinth* "pedestal of a column."

Pycnopteris T. Moore Dryopteridaceae (Aspleniaceae)

Origins:

Greek *pyknos* "compact, crowded" and *pteris* "a fern."

Pycnorhachis Benth. Asclepiadaceae

Origins:

From the Greek *pyknos* "dense" and *rhachis* "rachis, axis, midrib of a leaf."

Pycnosandra Blume Euphorbiaceae

Origins:

From the Greek *pyknos* "dense, compact" and *aner, andros* "stamen, male."

Pycnosorus Benth. Asteraceae

Origins:

Greek *pyknos* and *soros* "a heap, a spore case"; see George Bentham et al., *Enumeratio Plantarum quas in Novae Hollandiae* ... collegit C. de Hügel. 63, in nota. Vindobonae 1837.

Pycnospatha Thorel ex Gagnepain Araceae

Origins:

Greek *pyknos* "dense, crowded" and *spatha* "spathe." See J. Bogner, "Die Gattung *Pycnospatha* Thorel ex Gagnep. (Araceae)." *Oesterreichische Botanische Zeitschrift.* 122: 199-216. 1973; S.Y. Hu, "Studies in the flora of Thailand 41. Araceae." *Dansk Botanisk Arkiv.* 23: 409-457. 1968; D.H. Nicolson, "Derivation of aroid generic names." *Aroideana.* 10: 15-25. 1988.

Pycnosphace Rydb. Labiatae

Origins:

Greek *pyknos* "dense, compact" and *sphakos* "sage," see also *Salvia* L.

Pycnosphaera Gilg Gentianaceae

Origins:

From the Greek *pyknos* "dense" and *sphaira* "a globe, ball, sphere."

Pycnospora R. Br. ex Wight & Arn. Fabaceae

Origins:

Greek *pyknos* and *spora, sporos* "a seed, spore," referring to the packed-together seeds; see Robert Wight and G.

Arnott Walker Arnott, *Prodromus florae Peninsulae Indiae Orientalis*. 197. London 1834.

Species/Vernacular Names:
P. lutescens (Poiret) Schindler (*Pycnospora hedysaroides* R. Br. ex Baker; *Hedysarum lutescens* Poiret) (see Arthur D. Chapman, ed., *Australian Plant Name Index*. 2472. Canberra 1991)

Japan: kin-chaku-mame

Pycnostachys Hooker Labiatae

Origins:
From the Greek *pyknos* "dense" and *stachys* "a spike," referring to the flower-spikes.

Species/Vernacular Names:
P. urticifolia Hook.
English: blue boys

Pycnostelma Bunge ex Decne. Asclepiadaceae

Origins:
Greek *pyknos* "dense" and *stelma, stelmatos* "a girdle, belt."

Pycnostylis Pierre Menispermaceae

Origins:
From the Greek *pyknos* and *stylos* "pillar, column."

Pycnothymus (Benth.) Small Labiatae

Origins:
From the Greek *pyknos* "dense, compact" plus *Thymus*.

Pycreus P. Beauv. Cyperaceae

Origins:
An anagram of the generic name *Cyperus*; see A. Palisot de Beauvois, *Flore d'Oware et de Benin en Afrique*. 2: t. 86. Paris (Aug.) 1816.

Species/Vernacular Names:
P. sp.
Hawaii: mau'u, kaluhaluha
P. filicinus (Vahl) T. Koyama
English: Nuttall's cyperus
P. flavescens (L.) Reichb. (Latin *flaveo* "to become golden yellow")

Southern Africa: motaoataoane (Sotho)
P. intactus (Vahl) J. Raynal (*Pycreus ferrugineus* C.B. Cl.)
Southern Africa: qoqotho (Sotho)
P. nitidus (Lam.) J. Raynal (*Pycreus lanceus* Turrill; *Pycreus umbrosus* Nees)
Southern Africa: waterbiesie; motobane (Sotho)

Pygeum Gaertner Rosaceae

Origins:
Greek *pyge* "the rump, buttock," referring to the shape of the fruits, see also *Prunus* L.; see Joseph Gaertner (1732-1791), *De fructibus et seminibus plantarum*. 1: 218, t. 46, fig. 4. Stuttgart, Tübingen 1788.

Pygmaea B.D. Jackson Scrophulariaceae

Origins:
Orthographic variant of *Pygmea* Hook.

Pygmaeocereus J.H. Johnson & Backeb. Cactaceae

Origins:
Greek *pygmaios* "dwarfish" plus *Cereus*.

Pygmaeopremna Merrill Labiatae (Verbenaceae)

Origins:
Greek *pygmaios* "dwarfish" and the genus *Premna* L., Greek *premnon* "the stump of a tree, a tree-trunk"; see E.D. Merrill, in *Philippine Journal of Science*. 5: 225. Manila (Aug.) 1910.

Pygmaeorchis Brade Orchidaceae

Origins:
From the Greek *pygmaios* "dwarfish" plus *orchis* "orchid," referring to the size of the plant.

Pygmaeothamnus Robyns Rubiaceae

Origins:
Greek *pygmaios* "dwarfish" and *thamnos* "bush," suffrutices.

Species/Vernacular Names:

P. zeyheri (Sond.) Robyns var. *zeyheri* (*Pachystigma zeyheri* Sond.; *Pachystigma zeyheri* (Sond.) Robyns var. *oatesii* Robyns; *Vangueria zeyheri* (Sond.) Sond.)

English: sand apple

Southern Africa: goorappeltjie, gousiektebos, maidrek, sandappel; mosisá, mothlabelo (Tswana); umkukuzela (Ndebele)

Pygmea Hook.f. Scrophulariaceae

Origins:

Greek *pygmaios* "dwarfish," Latin *pygmaeus, a, um* "dwarf," referring to the habit of growth; see J.D. Hooker, *Handbook of the New Zealand Flora*. 217. London 1864.

Pyracantha M. Roemer Rosaceae

Origins:

From the Greek *pyrakantha, pyr* "fire" and *akantha* "thorn," thorny branches and red fruits; see Max Joseph Roemer (or Römer) (1791-1849), *Familiarum naturalium regni vegetabilis synopses monographicae*. 3: 104, 219. Vimariae [Weimar] (Apr.) 1847.

Species/Vernacular Names:

P. angustifolia (Franchet) C. Schneider (*Cotoneaster angustifolius* Franchet)

English: firethorn, orange firethorn, yellow firethorn

South Africa: geelvuurdoring

Japan: tachibana-modoki

P. coccinea M. Roemer (*Cotoneaster pyracantha* (L.) Spach)

English: pyracanth, firethorn, red firethorn

French: buisson ardent

P. crenatoserrata (Hance) Rehd. (*Pyracantha fortuneana* (Maxim.) Li)

China: chi yang zi

P. crenulata (Roxb. & Lindley) M. Roemer (*Crataegus crenulata* Roxb. & Lindley)

English: Nepalese white thorn

Nepal: ghangharu

Pyragra Bremek. Rubiaceae

Origins:

Greek *pyragra* "pair of fire-tongs, forceps."

Pyramidanthe Miq. Annonaceae

Origins:

Greek *pyramis, pyramidos* "a pyramid" and *anthos* "flower."

Pyramidium Boiss. Brassicaceae

Origins:

From the Greek *pyramis, pyramidos* "a pyramid."

Pyramidocarpus Oliv. Flacourtiaceae

Origins:

Greek *pyramis, pyramidos* and *karpos* "fruit."

Pyramidoptera Boiss. Umbelliferae

Origins:

Greek *pyramis, pyramidos* "a pyramid" and *pteron* "a wing."

Pyrenacantha Wight Icacinaceae

Origins:

Greek *pyren* "a kernel, a fruit stone" and *akantha* "thorn," referring to the inner walls of the fruits.

Species/Vernacular Names:

P. sp.

Nigeria: abere (Yoruba)

Pyrenaria Blume Theaceae

Origins:

Greek *pyren* "a kernel," referring to the fruit.

Species/Vernacular Names:

P. acuminata Planchon

English: bat's apple

Malaya: lidah kerbau, lidah lembu, chempahong, gelugor gajah, kelat jambu arang, lidah lidah, medang gelugor, perepat bukit, perupok, samak jantan, tampoi dada, berembang

P. championi H. Keng

Hong Kong: tutcheria

P. yunnanensis Hu

English: Yunnan pyrenaria

Pyrenocarpa H.T. Chang & R.H. Miau Myrtaceae

Origins:

Greek *pyren* "a kernel" and *karpos* "fruit."

Pyrenoglyphis Karsten Palmae

Origins:

From the Greek *pyren* "a kernel" and *glypho* "to carve, engrave," *glyphe* "a carving."

Pyrethrum Zinn Asteraceae

Origins:

Greek *pyrethron* "pellitory," *pyretos* "fever, burning heat," *pyr* "fire," referring to the roots of *Anacyclus pyrethrum*, Latin *pyrethrum* or *pyrethron* for a plant, Spanish chamomile, pellitory, a species of *Anthemis*; see Johann Gottfried Zinn (1727-1759), *Catalogus plantarum horti academici et agri gottingensis*. 414. Gottingae [Göttingen] 1757; Ernest Weekley, *An Etymological Dictionary of Modern English*. 2: 1177. New York 1967; Helmut Genaust, *Etymologisches Wörterbuch der botanischen Pflanzennamen*. 521. Basel 1996.

Pyrgophyllum (Gagnepain) T.L. Wu & Z.Y. Chen Zingiberaceae

Origins:

Greek *pyrgos* "tower" and *phyllon* "a leaf."

Pyriluma Baillon ex A. Aubrév. Sapotaceae

Origins:

Probably from the Greek *pyr* "fire" or Latin *pirum, pyrum, pyrus, pirus* "pear, a pear-tree" plus *luma*.

Pyrogennema Lunell Onagraceae

Origins:

Some suggest from the Greek *pyrogenes* "fire-born, made from wheat," *gennao* "to generate."

Pyrola L. Ericaceae (Pyrolaceae)

Origins:

The diminutive of the Latin *pirum, pyrum, pyrus, pirus* "pear, a pear-tree," referring to the foliage.

Species/Vernacular Names:

P. chlorantha Sw.

English: green-flowered wintergreen

P. minor L.

English: lesser wintergreen

P. picta Smith

English: white-veined wintergreen

P. rotundifolia L.

English: shin leaf, wintergreen

China: lu xiao cao, lu ti tsao

Pyrolirion Herbert Amaryllidaceae (Liliaceae)

Origins:

Fire-lily, from the Greek *pyr* "fire" and *leirion* "a lily," referring to the flowers.

Pyrospermum Miq. Celastraceae

Origins:

Greek *pyr* "fire" and *sperma* "seed."

Pyrostegia C. Presl Bignoniaceae

Origins:

Greek *pyr* "fire" and *stege, stegos* "roof, shelter," alluding to upper lip of the flower; see Karl (Carl) B. Presl (1794-1852), in *Abhandlungen der Königlichen Böhmischen Gesellschaft der Wissenschaften*. Ser. 5, 3: 523. (Jul.-Dec.) 1845.

Species/Vernacular Names:

P. venusta (Ker Gawler) Miers

English: flame vine, golden shower

Paraguay: Yvyratî

Pyrostria Comm. ex Juss. Rubiaceae

Origins:

Greek *pyr* "fire" and *ostreios, ostrios* "purple," referring to the color of flowers of some species; see Helmut Genaust, *Etymologisches Wörterbuch der botanischen Pflanzennamen*. 522. [from *Pyrus*] Basel 1996.

Species/Vernacular Names:

P. hystrix (Brem.) Bridson (*Dinocanthium hystrix* Brem.)

English: porcupine bush

South Africa: ystervarkbos, dinocanthium

Pyrrhanthera Zotov Gramineae

Origins:
From the Greek *pyrros* "flame-colored, red" and *anthera* "anther."

Pyrrhanthus Jack Combretaceae

Origins:
Greek *pyrros* and *anthos* "flower."

Pyrrheima Hassk. Commelinaceae

Origins:
Greek *pyrros* "flame-colored, red" and *heima* "clothing, cloak," see also *Siderasis* Raf.

Pyrrhocactus (A. Berger) Backeb. & F.M. Knuth Cactaceae

Origins:
From the Greek *pyrros* plus *Cactus*.

Pyrrhopappus DC. Asteraceae

Origins:
Greek *pyrros* and *pappos* "fluff, downy appendage."

Pyrrocoma Hook. Asteraceae

Origins:
Greek *pyrros* and *kome* "hair of the head," reddish pappus.

Species/Vernacular Names:
P. *uniflora* (Hook.) E. Greene var. *gossypina* (E. Greene) J. Kartesz & K. Gandhi
English: Bear Valley haplopappus

Pyrrorhiza Maguire & Wurdack Haemodoraceae

Origins:
From the Greek *pyrros* "red" and *rhiza* "a root."

Pyrrosia Mirbel Polypodiaceae

Origins:
Greek *pyrros* "flame-colored, reddish yellow, red, tawny"; see Jean Baptiste Antoine Pierre de Monnet de Lamarck (1744-1829) and Charles François Brisseau de Mirbel (1776-1854), *Histoire naturelle des Végétaux.* 3: 471. and 5: 91. 1802; P. Hovenkamp, *A Monograph of the Fern Genus Pyrrosia.* Leiden Botanical Series, vol. 9. 1986.

Species/Vernacular Names:
P. adnascens (Sw.) Ching (*Polypodium adnascens* Swartz; *Cyclophorus adnascens* (Swartz) Desv.) (Latin *adnascens* "growing to, growing upon")
Japan: hito-tsuba-mame-zuta
P. confluens (R. Br.) Ching (*Polypodium confluens* R. Br.)
English: horseshoe felt fern
P. lingua (Thunb.) Farwell (*Acrostichum lingua* Thunb.)
English: felt fern, tongue fern, Japanese felt fern
Japan: hito-tsu-ba
China: shi wei
P. piloselloides (L.) Price
The Philippines: pagong-pagongan
P. rupestris (R. Br.) Ching (*Polypodium rupestris* R. Br.)
English: rock felt fern

Pyrrothrix Bremek. Acanthaceae

Origins:
Greek *pyrros* "flame-colored, reddish yellow, red, tawny" and *thrix, trichos* "hair."

Pyrularia Michaux Santalaceae

Origins:
A diminutive from the Latin *pyrum, pirum*, referring to the shape of the fruit.

Species/Vernacular Names:
P. pubera Michaux
English: buffalo nut, elk nut

Pyrus L. Rosaceae

Origins:
Latin *pirum, pyrum* "a pear," *pirus, pyrus* "a pear-tree," Akkadian *pir'um, per'um*, Hebrew *peri* "fruit, offspring"; see Carl Linnaeus, *Species Plantarum.* 479. 1753 and

Genera Plantarum. Ed. 5. 214. 1754; Giovanni Semerano, *Le origini della cultura europea.* Dizionario della lingua Latina e di voci moderne. 2(2): 518. Firenze 1994.

Species/Vernacular Names:

P. betulifolia Bunge

English: Chinese pear

China: tang li

P. calleryana Decne.

English: Callery pear

P. communis L.

English: pear, common pear, western pear

Japan: seiyô-nashi

India: bagugosha

Arabic: anjas

P. kawakamii Hayata

English: evergreen pear

P. nivalis Jacq.

English: snow pear

P. pashia Buch.-Ham. ex D. Don

Nepal: mayal

P. pyrifolia (Burm.f.) Nakai (*Ficus pyrifolia* Burm.; *Pyrus montana* Nakai; *Pyrus serotina* Rehd.)

English: sand pear, Japanese pear, Chinese pear, country pear

Japan: nashi

India: nashpati, berikai

Pythius B.D. Jacks. Euphorbiaceae

Origins:

Greek *Pythios* "Delphian," Latin *Pythius* "Apollonian."

Pyxidanthera Michaux Diapensiaceae

Origins:

Greek *pyxis* "a small box" and *anthera* "anther," referring to the anthers.

Pyxidanthus Naudin Melastomataceae

Origins:

Greek *pyxis* "a small box" and *anthos* "flower."

Pyxidaria Gled. Hymenophyllaceae

Origins:

From the Greek *pyxis* "a small box, a small container with lid."

Pyxidaria Kuntze Scrophulariaceae

Origins:

Greek *pyxis* "a small box, a small container with lid"; see Otto Kuntze (1843-1907), *Revisio generum Plantarum Vascularium.* 2: 464. Leipzig (Nov.) 1891.

Q

Qaisera Omer Gentianaceae

Origins:
For Mohammad Qaiser, b. 1946.

Quadrangula Baum.-Bod. Casuarinaceae

Origins:
Latin *quadrangulum, i* (*quattuor angulus*) "a quadrangle," *quadriangulus, a, um* "four-cornered."

Quadricosta Dulac Onagraceae

Origins:
Latin *quadri-* "four" and *costa, ae* "a rib, side, wall."

Quadripterygium Tardieu Celastraceae

Origins:
Latin *quadri-* "four" and *pterygos, pterygion* "a small wing."

Quamoclidion Choisy Nyctaginaceae

Origins:
Probably from the Greek *kuamos* "bean" and *kleidion, klidion*, diminutive of *kleis* "lock, key," or from the Aztec and Mexican names *cuamóchitl, guamúchil, cuamúchil, quamóchitl*.

Quamoclit Miller Convolvulaceae

Origins:
Derivation of the name is obscure, possibly from the Aztec and Mexican (also dedicated to some species of *Pithecolobium, Acacia,* etc.) *cuamóchitl, guamúchil, cuamúchil, quamóchitl*; some suggest from the Greek *kuamos* "bean" or *kuamos* and *klitos* "the lower part, sloping," or Sanskrit *kama-lata* for *Quamoclit pennata* Boj.; see Philip Miller, *The Gardeners Dictionary.* Abr. ed. 4. London (28 Jan.) 1754; Maximino Martínez, *Catálogo de nombres vulgares y científicos de plantas mexicanas.* 238, 384, 1199. México 1987; William T. Stearn, *Stearn's Dictionary of Plant Names for Gardeners.* 253. Cassell 1993; Georg Christian Wittstein, *Etymologisch-botanisches Handwörterbuch.* 746. 1852; Helmut Genaust, *Etymologisches Wörterbuch der botanischen Pflanzennamen.* 523. 1996; F. Boerner & G. Kunkel, *Taschenwörterbuch der botanischen Pflanzennamen.* 4. Aufl. 158. 1989; Salvatore Battaglia, *Grande dizionario della lingua italiana.* XV: 54. 1994.

Quamoclit Moench Convolvulaceae

Origins:
See *Quamoclit* Miller.

Quamoclita Raf. Convolvulaceae

Origins:
Orthographic variant of *Quamoclit*; see Constantine Samuel Rafinesque (1783-1840), *Flora Telluriana.* 4: 74. 1836 [1838].

Quamoctita Raf. Convolvulaceae

Origins:
Based on *Quamoclita* and *Quamoclit*; see Constantine Samuel Rafinesque, *Flora Telluriana.* 4: 74. 1836 [1838].

Quaqua N.E. Br. Asclepiadaceae

Origins:
According to Sir Henry Barkly (1815-1898) *quaqua* was the nearest equivalent to the Hottentot name for the plant; according to some other authors the generic name after the place in Zimbabwe where it was found.

Species/Vernacular Names:
Q. mammillaris (L.) Bruyns (*Caralluma mammillaris* (L.) N.E. Br.; *Caralluma winkleriana* (Dinter) White & Sloane)
South Africa: aroena

Quararibea Aublet Bombacaceae

Origins:

A vernacular name, *guarariba*, for *Quararibea guyanensis* Aubl.

Quartinia Endlicher Lythraceae

Origins:

For the French botanist Léon Richard Quartin-Dillon, d. 1841, physician, explorer, plant collector, wrote *Des différences appréciables entre le sang de la veine porte et celui des autres veins*. in Collection des thèses soutenues à la Faculté de Médecine de Paris. an 1839-1878, tom. 14. Paris 1839-1878. See T. Lefebvre, *Voyage en Abyssinie* exécuté … par une Commission scientifique, composée de MM. T. Lefebvre, A. Petit et Quartin-Dillon, etc. [1845]; J.H. Barnhart, *Biographical Notes upon Botanists*. 3: 118. 1965; A. Lasègue, *Musée botanique de Benjamin Delessert*. Paris 1845; Joseph Vallot, "Études sur la flore du Sénégal." in *Bull. Soc. Bot. de France*. 29: 187. Paris 1882; A. White and B.L. Sloane, *The Stapelieae*. Pasadena 1937.

Quassia L. Simaroubaceae

Origins:

Named to honor Graman Quasi (or Quassi or Kwasi, from *kwasida* or *kwasi* = Sunday or the first day of the week), a Negro slave of Carl Gustav Dahlberg when he explored Suriname (Dutch Guiana), *Quashee* is the nickname for negro (from Ashantee or Fantee *kwasi*, name commonly given to child born on Sunday); see C. Linnaeus, *Species Plantarum*. Ed. 2. 1: 553. 1762; Ernest Weekley, *An Etymological Dictionary of Modern English*. 2: 1183. New York 1967; C.T. Onions, *The Oxford Dictionary of English Etymology*. Oxford University Press 1966; F. D'Alberti di Villanuova, *Dizionario universale, critico, enciclopedico della lingua italiana*. Lucca 1797-1805; Georg Christian Wittstein, *Etymologisch-botanisches Handwörterbuch*. 746. 1852; Salvatore Battaglia, *Grande dizionario della lingua italiana*. XV: 96-97. Torino 1994.

Species/Vernacular Names:

Q. africana Baillon

Congo: simbikali, mupessi, otapaa

Q. amara L. (*Simarouba amara* Aublet)

English: bitter wood, Surinam quassia wood, stave wood, bitter quassia, bitter ash, South American bitter wood

Brazil: quassia, murubá, murupá, quina-de-caiena, simaruba, maruba, marupá

Bolivia: lucumo, amargo negro, chuña-chuña, chiriguaná

Central America: guavito, palo isidoro, cuasia, cuasia de Surinam, hombre grande, limoncillo, palo de hombre

The Philippines: corales, kuasia

Q. cedron (Planchon) D. Dietr.

Brazil: cedron, pau-paratudo, paratudo, paracatá

Central America: quassia

Q. grandiflora (Engl.) Nooteb.

Yoruba: oro kosoro

Q. indica (Gaertner) Nooteb.

India: lokhandi, nibam, niepa, karinjottei, nipa, samdera, karinjotta

Q. simarouba L.f.

English: paradise tree

Brazil: marupá, marubá, arubá, marupá-verdadeiro, paraíba, simaruba

Central America: aceituna, negrito, jocote de mico, simarouba, simaúba, simaruba

Q. undulata D. Dietr.

Gabon: nkourangueuk

Cameroon: ataf, obek, eseng

Central Africa: babolo

Nigeria: igbo

Yoruba: oja, igbo, igigun, orisi

Ivory Coast: apohia, effeu

Mali: jafulate, kolonson

Quaternella Pedersen Amaranthaceae

Origins:

Latin *quater* "four times," *quaterni, ae* "four each."

Quebrachia Griseb. Anacardiaceae

Origins:

Spanish *quebrar* "to break, to crush," *quebracho* "quebracho, breakax wood," Latin *crepo, as, ui, itum, are* "to crack, to break with a crash," the vernacular name *quebracho* for *Schinopsis* spp., *quebracha* for *Aspidosperma tomentosum* Martius & Zucc.; see Giovanni Semerano, *Le origini della cultura europea*. Dizionario della lingua Latina e di voci moderne. 2(2): 376. Firenze 1994

Queenslandiella Domin Cyperaceae

Origins:

From the Australian province of Queensland; see Karel Domin, *Beiträge zur Flora und Pflanzengeographie*

Australiens. [Bibliotheca Botanica Heft 85 (Dec. 1915) 415] 1915).

Quekettia Lindley Orchidaceae

Origins:

For the British (b. Somerset) plant anatomist Edwin John Quekett, 1808-1847, botanist, microscopist, surgeon, in 1836 became a Fellow of the Linnean Society, founder member of the Royal Microscopical Society, with I.J. Goodfellow edited *The London Physiological Journal*, brother of the British surgeon and microscopist John Thomas Quekett (1815-1861); see Ray Desmond, *Dictionary of British & Irish Botanists and Horticulturists.* 569. 1994; Georg Christian Wittstein, *Etymologisch-botanisches Handwörterbuch.* 746. Ansbach 1852.

Quercifilix Copel. Dryopteridaceae (Aspleniaceae)

Origins:

From *quercus* "oak" and *filix* "fern," alluding to the form of the fronds.

Quercus L. Fagaceae

Origins:

The ancient Latin name for this tree, *quercus, us*; Akkadian *daru* "everlasting, enduring, durable: said of materials, constructions," *kassu* "strong"; see Carl Linnaeus, *Species Plantarum.* 994. 1753 and *Genera Plantarum.* Ed. 5. 431. 1754; Salvatore Battaglia, *Grande dizionario della lingua italiana.* XV: 113-116. 1994; Giovanni Semerano, *Le origini della cultura europea.* Dizionario della lingua Latina e di voci moderne. 2(2): 538-539. Firenze 1994.

Species/Vernacular Names:

Q. spp.

Mexico: encino, ch'alol, ahuatl, bigaga, pigaga, yaga bigaga, yaga pigaga, yaga bixohui, yaga zachi, yaga xoo, yaga zoo, haga lluchi, hica vixue, yaga yecho, yaga niza

Q. acatenangensis Trelease

Central America: encino, bans, chicharro, col, huite, masket, malcote, pitán, roble, sical, sunuj, zinuh

Q. acutissima Carruth.

English: sawtooth oak, fibrous oak

Japan: kunu-gi

Tibetan: mon-cha-ra

China: xiang shi

Q. agrifolia Nee

English: coast live oak, encina

Q. chrysolepis Liebm.

English: maul oak, canyon live oak

Q. cornelius-mulleri K. Nixon & K. Steele

English: Muller's oak

Q. douglasii Hooker & Arnott

English: blue oak

Q. dumosa Nutt.

English: Nuttall's scrub oak

Q. durata Jepson

English: leather oak

Q. engelmannii E. Greene

English: Engelmann oak, mesa oak

Q. garryana Hook.

English: Oregon oak

Q. glauca Thunb. (*Cyclobalanopsis glauca* (Thunb.) Oerst.)

English: blue Japanese oak

Nepal: sano falant, phalat

India: siri, pharonj, phaniant, bran, banni, imbri, banku, phalat

Japan: ara-kashi

Okinawa: koka, kashigi

Q. ilex L.

English: holm oak, holly oak, ilex

India: brechur, irri

Arabic: ballout

Q. john-tuckeri K. Nixon & C.H. Muller

English: Tucker's oak

Q. kelloggii Newb.

English: California black oak

Q. lanata Sm.

Nepal: banjh

Q. lobata Née

English: valley oak

Q. oglethorpensis W.H. Duncan (named after General James Edward Oglethorpe, 1696-1785, British Army Officer, in 1732 founded the colony of Georgia in North America; see Allen J. Coombes and W. Nigel Coates, "Oglethorpe and the Oglethorpe Oak." in *The New Plantsman.* 2(4): 226-234. December 1995)

English: Oglethorpe's oak, Oglethorpe oak

Q. palmeri Engelm.

English: Palmer's oak

Q. phillyreoides Gray

Japan: uba-megashi

Okinawa: urufugi

Q. robur L.

English: English oak, oak tree, pedunculate oak, truffle oak, common oak

South Africa: akkerboom

Q. salicina Bl. (*Cyclobalanopsis stenophylla* (Bl.) Schott)

Japan: urajiro-kashi

Q. suber L.

English: cork oak

Mexico: yaga quebe, yaga xoga, yaga zacho

Q. tomentella Engelm.

English: island oak

Q. turbinella E. Greene

English: shrub live oak

Q. vaccinifolia Kellogg

English: huckleberry oak

Q. wislizenii A. DC.

English: interior live oak

Queria L. Caryophyllaceae

Origins:

For the Spanish botanist José Quer y Martinez, 1695-1764, physician. His writings include Dissertacion ... *sobre el uso de la Cicuta.* Madrid 1764, Dissertacion ... *sobre la passion nephritica y su verdadero especifico; la uva-ursi, ó gayubas.* Madrid 1763 and *Flora Española.* Tom. 1-4. Madrid 1762-1784. See J.H. Barnhart, *Biographical Notes upon Botanists.* 3: 119. 1965; T.W. Bossert, *Biographical Dictionary of Botanists Represented in the Hunt Institute Portrait Collection.* 320. 1972; E.M. Tucker, *Catalogue of the Library of the Arnold Arboretum of Harvard University.* 1917-1933; M. Colmeiro y Penido, *La Botánica y los Botánicos de la Peninsula Hispano-Lusitana.* Madrid 1858; Jonas C. Dryander, *Catalogus bibliothecae historico-naturalis Josephi Banks.* London 1800; Ida Kaplan Langman, *A Selected Guide to the Literature on the Flowering Plants of Mexico.* 602. Philadelphia 1964; A. Lasègue, *Musée botanique de Benjamin Delessert.* Paris 1845.

Quesnelia Gaudich. Bromeliaceae

Origins:

For a certain M. E. Quesnel, a French (from Le Havre) horticulturist and botanist, plant collector, in South America (Cayenne, French Guiana); see Mulford Foster (1888-1978), "*Quesnelia quesneliana.*" in *Bromeliad Society Bulletin.* XI: 22-24. 1961; F. Boerner, *Taschenwörterbuch*

der botanischen Pflanzennamen. 2. Aufl. 164. Berlin & Hamburg 1966.

Queteletia Blume Orchidaceae

Origins:

Dedicated to the Belgian (b. Ghent) natural scientist Lambert-Adolphe-Jacques (Adolph Jacob) Quételet, 1796-1874 (d. Bruxelles/Brussels), mathematician, historian of science, statistician. His works include *Éléments d'Astronomie.* Paris 1847, *Du système social et des lois qui le régissent.* Paris 1848, *Histoire des Sciences Mathématiques et Physiques chez les Belges.* Bruxelles 1864, *Anthropométrie, ou mesure des différentes facultés de l'homme.* Bruxelles 1870 and *Physique Sociale, ou Essai sur le développement des facultés de l'homme.* Bruxelles 1869. See J.H. Barnhart, *Biographical Notes upon Botanists.* 3: 119. 1965; Stafleu and Cowan, *Taxonomic Literature.* 4: 457. 1983; Leonard Huxley, *Life and Letters of Sir Joseph Dalton Hooker.* London 1918; Hans Freudenthal, in *Dictionary of Scientific Biography* (Editor in Chief Charles Coulston Gillispie.) 11: 236-238. New York 1981; E. Mailly, *Essai sur la vie et les ouvrage de L.-A.-J. Quételet.* Brussels 1875; Garrison and Morton, *Medical Bibliography.* 171. 1961; Giulio Giorello & Agnese Grieco, a cura di, *Goethe scienziato.* Einaudi Editore, Torino 1998; G. Schmid, *Goethe und die Naturwissenschaften.* Halle 1940.

Quetzalia Lundell Celastraceae

Origins:

A trogon, the national symbol of Guatemala and anciently regarded as a deity by the Mayas; Quetzalcoatl, a traditional god and heroic figure of the Aztecs.

Quezelia Scholz Brassicaceae

Origins:

For Pierre Ambrunaz Quézel, b. 1926.

Quezeliantha Scholz ex Rauschert Brassicaceae

Origins:

For Pierre Ambrunaz Quézel, b. 1926.

Quidproquo Greuter & Burdet Brassicaceae

Origins:

Latin *quid pro quo*, something for something, substitution, prescription.

Quiina Aublet Quiinaceae

Origins:

A vernacular name.

Quillaja Molina Rosaceae

Origins:

Spanish *quillay*, from the vernacular Araucani name *quillai*, *killai*; see H. Genaust, *Etymologisches Wörterbuch der botanischen Pflanzennamen*. 524. 1996; S. Battaglia, *Grande dizionario della lingua italiana*. XV: 152. 1994.

Species/Vernacular Names:

Q. brasiliensis (A. St.-Hil. & Tulasne) C. Martius

Peru: quillai, quillay

Q. saponaria Molina

Chile: quillai

Quinchamalium Molina Santalaceae

Origins:

A vernacular Araucani name.

Species/Vernacular Names:

Q. lomae Pilger

Peru: chinchimato

Q. procumbens Ruíz & Pav. (*Quinchamalium linifolium* Meyen ex Walpers; *Quinchamalium raimondii* Pilger)

Peru: chinchamala, chinchamali, chinchimani, hierba del toro, khenchamali

Quincula Raf. Solanaceae

Origins:

See C.S. Rafinesque, *Atl. Jour.* 1: 145. 1832; E.D. Merrill, *Index rafinesquianus*. The plant names published by C.S. Rafinesque, etc. 212. Jamaica Plain, Massachusetts, USA 1949.

Quinetia Cass. Asteraceae

Origins:

After the French intellectual Edgar Quinet, 1803-1875, politician, historician and poet, traveler. Among his works are *Le Christianisme et la Révolution Française*. Paris 1845, *L'Expédition du Mexique*. London 1862, *La Question romaine devant l'histoire, 1848 à 1867*. Paris 1868 and *Lettres d'exil à Michelet et à divers amis*. [Edited by Madame Quinet, Hermione Asachi, 1821-1900] Paris 1885.

See Edgar Quinet, *Oeuvres complètes*. Paris [1877-1882]; C.L. Chassin, *Edgar Quinet, sa vie et son oeuvre*. Genève 1970; Alexandre Henri Gabriel Comte de Cassini (1781-1832), in *Annales des Sciences Naturelles*. 17: 415. 1829.

Species/Vernacular Names:

Q. urvillei Cass.

English: quinetia

Quinqueremulus Paul G. Wilson Asteraceae

Origins:

Latin *quinque* "five" and *remulus, i* "a small oar," *remus, i* "an oar," referring to the shape of the five pappus scales; see Paul Graham Wilson, "*Quinqueremulus linearis*, a new genus and species in the Australian Asteraceae (tribe Inuleae)." in *Nuytsia*. 6(1): 1-5. 1987.

Quinquina Boehm. Rubiaceae

Origins:

See *Cinchona* L.

Quintinia A. DC. Grossulariaceae (Escalloniaceae)

Origins:

After the French horticulturist Jean (Johannis) de la Quintinie (Quintinye), 1626-1688 (or 1685, see Brian Halliwall in *Curtis's Botanical Magazine*. Volume 14. 3: 173. August 1997), botanist, he was appointed Director-General of the Royal vegetable gardens by King Louis XIV, wrote *L'Art ou la Manière particulière et seure de tailler des arbres fruitiers*. Amsterdam 1699, *Instruction pour les Jardins Fruitiers et Potagers, avec un Traité des Orangers*. Amsterdam 1692 and *Le Parfait Jardinier*. Paris, Genève 1695; see A.P. de Candolle, *Prodromus*. 4: 5. 1830; Mariella Azzarello Di Misa, a cura di, *Il Fondo Antico della Biblioteca dell'Orto Botanico di Palermo*. 223. Palermo 1988.

Species/Vernacular Names:

Q. sieberi A. DC.

Australia: opossum wood, possumwood

Quisqualis L. Combretaceae

Origins:

Which and of what kind, Latin *quis* "which? who?" and *qualis* "what kind? what?"; the name is based on Rumphius (Rumpf) (Georg Eberhard or Everard, 1628-1702) observation and astonishment of the variable growth habit of the plant. See G. Ballintijn, *Rumphius de blinde ziener van*

Ambon. Utrecht 1944; C. Linnaeus, *Species Plantarum.* Ed. 2. 1: 556. 1762; [Rumphius], *Rumphius Gedenkboek* 1702-1902. Haarlem, Koloniaal Museum 1902; Mariella Azzarello Di Misa, a cura di, *Il Fondo Antico della Biblioteca dell'Orto Botanico di Palermo.* 239. Regione Siciliana, Palermo 1988; Blanche Henrey, *No Ordinary Gardener — Thomas Knowlton, 1691-1781.* Edited by A.O. Chater. British Museum (Natural History). London 1986.

Species/Vernacular Names:

Q. indica L. (*Quisqualis indica* var. *villosa* C.B. Clarke)

English: Rangoon creeper, Chinese honeysuckle, liane-vermifuge

The Philippines: niogniogan, niyog-niyogan, niog-niogan, tangolan, balitadham, kasumbal, tartarau, sagasi, sagisi, tagisi

Malaya: akar pontianak, akar suloh

China: shih chun tzu, shi jun zi

Vietnam: qua nac, day giun

Japan: Indo-shikunshi

Yoruba: ogan funfun, ogan igbo

Q. parviflora Gerr. ex Harv.

English: quisqualis

Southern Africa: hoedanig (= how, what; from the Malay name udani); umBondwe-wehlathi, iHlaphu elincane (Zulu); umQuotho (Xhosa)

Quisqueya D. Don Orchidaceae

Origins:

The Caribbean name for the Antillan island Hispaniola (Haiti); see H. Mayr, *Orchid Names and Their Meanings.* 174. Vaduz 1998.

Quisumbingia Merrill Asclepiadaceae

Origins:

After the Philippine botanist Eduardo Quisumbing (y Argüelles), 1895-1986, botanical explorer, plant collector, specialized in herbarium taxonomy, Director of the National Museum of Manila, collaborated with the orchidologist Oakes Ames (1874-1950). Among his most important writings is the volume on *Medicinal Plants of the Philippines.* Katha Publishing, Quezon City 1978 and

"Studies of Philippine bananas." *Philipp. Agric. Rev.* 12: 1-90, t. 1-30. 1919, "Philippine plants used for arrow and fish poisons." *Philip. Jour. Sci.* 77: 127-177. 1947, "Vegetable poisons of the Philippines." *Philip. Jour. For.* 5: 145-171. 1947. See J.H. Barnhart, *Biographical Notes upon Botanists.* 3: 120. 1965; T.W. Bossert, *Biographical Dictionary of Botanists Represented in the Hunt Institute Portrait Collection.* 321. 1972.

Quivisia Cav. Meliaceae

Origins:

Possibly from the Latin *quivis* "what you please."

Quivisianthe Baillon Meliaceae

Origins:

The genus *Quivisia* Cav. and Greek *anthos* "flower."

Quoya Gaudich. Verbenaceae

Origins:

After the French (b. Vandée) naturalist and zoologist Jean René Constant Quoy, 1790-1869 (d. St.-Jean-de-Liversay, France), a naval surgeon, from 1817 to 1820 a zoologist (with J.P. Gaimard, 1790-1858) on Freycinet's *Uranie* voyage, from 1826 to 1829 with Dumont d'Urville on *Astrolabe* expedition to Australia, New Zealand and New Guinea, wrote *Epistola Domine G. ... de nonnullis pavoris effectibus.* Monspelii 1814. See Toby A. Appel, in *Dictionary of Scientific Biography* 11: 242-244. 1981; L.C.D. de Freycinet, *Voyage autour du Monde entrepris par ordre du Roi ... sur les corvettes de S.M. "L'Uranie" et "La Physicienne".* Zoologie, par MM. Quoy et Gaimard. [Zoology of the Voyage, written with the French zoologist Joseph Paul Gaimard; for the botanical results of Freycinet's voyages see Charles Gaudichaud-Beaupré.] Paris 1826[-1830]; Jules S.C. Dumont d'Urville (1790-1842), *Voyage de Découvertes autour du Monde ... sur la corvette L'Astrolabe pendant les Années 1826-1829.* Zoologie, par MM. Quoy et Gaimard. Paris 1832-1848.

BIBLIOGRAPHY

Eleanor B. Adams, *A Bio-Bibliography of Franciscan Authors in Colonial Central America.* Washington, D.C. 1953.

H.M. Adams, *Catalogue of Books Printed on the Continent of Europe, 1501-1600 in Cambridge Libraries.* Compiled by H.M. Adams. Cambridge University Press, Cambridge 1967.

R.F.G. Adams, "Efik vocabulary of living things." *Nigerian Field.* 11, 1943, II(12): 23-24. 1947 and III(13): 61-67. 1948.

Edouard J. Adjanohoun et al., *Contribution aux études ethnobotaniques et floristiques au Togo.* Editions agence de coopération culturelle et technique (A.C.C.T.), Paris 1986.

Z.M. Agha, *Bibliography of Islamic Medicine and Pharmacy.* London 1983.

W. Aiton, *Hortus kewensis.* London 1789.

J.A. Akinniyi and M.U.S. Sultanbawa, "A glossary of Kanuri names of plants, with botanical names, distribution and uses." *Ann. Borno.* 1. 1983.

J. Albany, *P'tit glossaire. Le piment des mots créoles.* Paris 1974 and 1983.

Albinal and Maljac, *Dictionnaire malgache-français.* Fianarantsoa 1987.

Janis B. Alcorn, *Huastec Mayan Ethnobotany.* University of Texas Press, Austin 1984.

Janis B. Alcorn and Cándido Hernández V., "Plants of the Huastecan region with an analysis of their Huastec names." *Journal of Mayan Linguistics.* 4: 11-18. 1983.

John Alden and Dennis Channing Landis, Eds., *European Americana: A Chronological Guide to Works Printed in Europe Relating to the Americas, 1493-1776.* Six volumes. New York 1980-1997.

Francisco Javier Alegre, S.J., *Historia de la provincia de la Compañia de Jesús de Nueva España.* New edition edited by Ernest J. Burrus and Felix Zubillaga. Rome 1956-1960.

G. Alessio, *Lexicon etymologicum.* Napoli 1976.

J.J.G. Alexander and A.C. de la Mare, *The Italian Manuscripts in the Library of Major J.R. Abbey.* London 1969.

Mea Allan, *The Hookers of Kew.* London 1967.

A. Amiaud and L. Méchineau, *Tableau comparé des écritures babylonienne et assyrienne, archaïques et modernes.* Paris 1902.

Fr. Francisco Antolín, O.P., *Notices of the Pagan Igorots in the Interior of the Island of Manila.* Manila 1988.

Archives. *Archives d'Études Orientales,* publiés par J.A. Lundell. Roma, Lund etc. 1911-1934.

Fernando de Armas Medina, *Cristianización del Perú (1532-1600).* Escuela de Estudios Hispano-Americanos de Sevilla. Sevilla 1953.

T.H. Arnold and B.C. de Wet, Eds., *Plants of Southern Africa: Names and Distribution.* Pretoria 1993.

M. Arunachalam, *Festivals of Tamil Nadu.* Tiruchitrambalam 1980.

F. Ascarelli, *La tipografia cinquecentina italiana.* Sansoni, Firenze 1953.

F. Ascarelli, *Le cinquecentine romane.* Milano 1972.

G. Aschieri, *Dizionario compendiato di geologia e mineralogia.* Milano 1855.

P.M. Ashburn, *The Ranks of Death: A Medical History of the Conquest of America.* New York 1947.

William Ashworth and Bruce Bradley, *Jesuit Science in the Age of Galileo.* Linda Hall Library, Kansas City 1986.

The Assyrian Dictionary of the Oriental Institute of the University of Chicago. Chicago 1963.

Jean Baptiste Christophe Fusée Aublet, *Histoire des Plantes de la Guiane Françoise.* Paris 1775.

Gabriel Austin, *The Library of Jean Grolier: A Preliminary Catalogue.* New York 1971.

João Lucio de Azevedo, *Os Jesuítas no Grão Pará, Suas missões e colonização.* Coimbra 1930.

Oreste Badellino and Ferruccio Calonghi, *Dizionario Italiano-Latino e Latino-Italiano.* Third edition. Rosenberg & Sellier, Torino 1990.

R.P. Ch. Bailleul, Ed., *Petit dictionnaire français-bambara, bambara-français.* Avebury Publishing Company 1981.

M.J. Balick and P.A. Cox, *Plants, People, and Culture: The Science of Ethnobotany.* Scientific American Press 1996.

S.C. Banerjee, *Flora and Fauna in Sanskrit Literature.* Calcutta 1980.

Antoine Alexandre Barbier, *Dictionnaire des ouvrages anonymes et pseudonymes.* Paris 1872-1879.

C. Bardesono di Rigras, *Vocabolario marinaresco.* Roma 1932.

John H. Barnhart, *Biographical Notes upon Botanists.* Boston 1965.

J. Barrau, "Notes on the significance of some vernacular names of food plants in the South Pacific Islands." *Proc. 9th Pacific Sci. Cong.* 4: 296. 1957 [1962].

A. Barrera Marín, Alfredo Barrera Vázquez and Rosa María López Franco, *Nomenclatura etnobotánica maya: una interpretación taxonómica.* México, D.F. 1976.

Katharine Bartlett, *Prehistoric Pueblo Foods.* Flagstaff, Arizona, Museum Notes, vol. 4, no. 4, October, 1931.

Fray D. Basalenque, *Historia de la Provincia de San Nicolás Tolentino de Michoacán* del Orden de Nuestro Padre Santísimo San Augustín. México 1963.

Henri Louis Baudrier and Julien Baudrier, *Bibliographie lyonnaise*: Recherches sur les imprimeurs, libraires, relieurs et fondeurs de lettres de Lyon au XVIe siècle. VII. Lyons, Paris 1908.

Cecil Beaton, *Indian Diary & Album.* [Originally published by B.T. Batsford Ltd. London as *Far East,* 1945 and *Chinese Album,* Winter 1945-1946] Oxford University Press, Oxford, New York, and Delhi 1991.

Marcos E. Becerra, *Nombres geográficos indígenas del Estado de Chiapas.* Chiapas, México: Tuxtla Gutiérrez 1932.

James Ford Bell, *Jesuit Relations and Other Americana in the Library of James F. Bell*. A catalogue compiled by Frank K. Walter and Virginia Doneghy. Minneapolis 1950.

Eric Temple Bell, *Men of Mathematics*. Simon & Schuster, New York 1965.

G. Bellini, *La tipografia del Seminario di Padova*. Libreria Gregoriana editrice, Padova 1927.

E.C. Bénézit, *Dictionnaire critique et documentaire des peintres, sculpteurs, dessinateurs et graveurs de tous les pays*. Paris 1976.

E. Benveniste, *Origine de la formation des noms en indo-européen*. Paris 1962.

Brent Berlin, "Speculations on the growth of ethnobotanical nomenclature." *Journal of Language in Society*. 1: 63-98. 1972.

Brent Berlin, Dennis E. Breedlove and Peter H. Raven, *Principles of Tzeltal Plant Classification*. Academic Press, New York and London 1974.

G. Bertho, "Quatre dialects Mandé du Nord-Dahomey et de la Nigeria anglaise." *Bull. Inst. Franç. Afr. Noire*. 13: 126-171. 1951.

Elsdon Best, *Maori Agriculture*. Wellington 1976.

N.N. Bhattacharya, *Glossary of Hindu Religious Terms and Concepts*. South Asia Publications, Columbia 1990.

B. Bhattacharyya, *An Introduction to Buddhist Esoterism*. Oxford 1932.

Bibliothèque Nationale, Paris, *Catalogue général des livres imprimés de la Bibliothèque Nationale. Auteurs*. Paris 1897-1981.

Blodwen Binns, *A First Check List of the Herbaceous Flora of Malawi*. Zomba, Malawi 1968.

Blodwen Binns, *Dictionary of Plant Names in Malawi*. Zomba, Malawi 1972.

H. Birnbaum and J. Puhvel, *Ancient Indo-European Dialects*. 1965.

Dor Bahadur Bista, *The People of Nepal*. Kathmandu 1972.

Paul Black, *Aboriginal Languages of the Northern Territory*. Darwin 1983.

E.H. Blair and J.A. Robertson, *The Philippine Islands 1493-1803 [-1898]*. Cleveland 1903-1909.

Joseph Blumenthal, *Art of the Printed Book, 1455-1955*: Masterpieces of Typography through Five Centuries from the Collections of The Pierpont Morgan Library, New York. New York and Boston 1973.

F. Boerner and G. Kunkel, *Taschenwörterbuch der botanischen Pflanzennamen*. 4. Aufl. Berlin & Hamburg 1989.

J. Bondet de la Bernadie, "Le dialecte des Kha Boloven." *Bulletin de la Societé des Études Indochinoises*. tome XXIV, no. 3, 1949.

Sarita Boodhoo, *Kanya Dan — The Why's of Hindu Marriage Rituals*. Mauritius Bhojpuri Institute, Port Louis 1993.

Franz Bopp, *Glossarium Sanscritum*. Second edition. Berlin 1847.

Woodrow Borah, *Early Trade and Navigation between Mexico and Peru*. Berkeley 1954.

Rubens Borba de Moraes, *Bibliographia Brasiliana*. Los Angeles and Rio de Janeiro 1983.

G. Borsa, *Clavis typographorum librariorumque Italiae 1465-1600*. Aureliae Aquensis [Baden-Baden] 1980.

L. Bossi, *Spiegazione di alcuni vocaboli geologici*. Milano 1817.

L. Bouquiaux, "Les noms des plantes chez les Birom." *Afrika und Übersee*. 55. 1971-1972.

C.R. Boxer, *The Dutch Seaborne Empire 1600-1800*. Harmondsworth 1973.

W.J. Boyd, *Satana and Mara. Christian and Buddhist Symbols of Evil*. Leiden 1975.

Pascal Boyeldieu, *La langue lua, groupe Boua, Moyen-Chari, Tchad*. CUP 1985.

Leonard E. Boyle, O.P., *Medieval Latin Palaeography: A Bibliography*. Toronto Medieval Bibliographies 8. University of Toronto Press, Toronto 1984.

A. Brebion, *Bibliographie des voyages dans l'Indochine française du IXe au XIXe siècle*. Paris 1910.

J.B. Brebner, *The Explorers of North America 1492-1806*. Cleveland 1964.

Dennis E. Breedlove, *History of Botanical Exploration in Chiapas, Mexico*. 1971.

Dennis E. Breedlove and Robert M. Laughlin, *The Flowering of Man — A Tzotzil Botany of Zinacantán*. Smithsonian Institution Press, Washington, D.C. 1993.

J.A. Brendon, *Great Navigators and Discoverers*. Books for Libraries Press, Freeport 1967.

Emil Bretschneider, *Mediaeval Researches from Eastern Asiatic Sources*. Fragments towards the knowledge of the geography and history of Central and Western Asia from the 13th to the 17th century. [Reprint of the 1888 edition.] London [1950?].

Emil Bretschneider, *History of European Botanical Discoveries in China*. [Reprint of the original edition, St. Petersburg 1898.] Leipzig 1981.

British Library, *The British Library General Catalogue of Printed Books to 1975*. London 1979.

British Library, *Catalogue of Books from the Low Countries 1601-1621 in the British Library*. Compiled by Anna E.C. Simoni. London 1990.

British Museum, *Short-title Catalogue of Books Printed in France and of French Books Printed in Other Countries from 1470 to 1600 now in the British Museum*. London 1966.

British Museum, *Short-title Catalogue of Books Printed in German-speaking Countries and German Books Printed in Other Countries from 1455 to 1600 now in the British Museum*. London 1958.

British Museum, *Short-title Catalogue of Books Printed in Italy and of Italian Books Printed in Other Countries from 1465 to 1600 now in the British Museum*. London 1958.

British Museum, *Short-title Catalogue of Books Printed in the Netherlands and Belgium and of Dutch and Flemish Books Printed in Other Countries from 1470 to 1600 now in the British Museum*. London 1965.

British Museum (Natural History), *Catalogue of the Books, Manuscripts, Maps and Drawings in the British Museum*. London 1903-1933.

C. Brockelmann, *Lexicon Syriacum*. Halis Saxon 1928.

J.S. Bromley and E.H. Kossmann, Eds., *Britain and the Netherlands in Europe and Asia*. London 1968.

H.C. Brooks, *Compendiosa bibliografia di edizioni bodoniane*. Firenze 1927.

R. Brough Smith, *Aborigines of Victoria*. Ferres, Melbourne 1878.

Patrick Browne, *The Civil and Natural History of Jamaica*. London 1756.

R.K. Brummitt, comp., *Vascular Plant Families and Genera*. Royal Botanic Gardens, Kew 1992.

R.K. Brummitt and C.E. Powell, *Authors of Plant Names*. Royal Botanic Gardens, Kew 1992.

J.C. Brunet, *Manuel du libraire et de l'amateur de livres*. 6 vols. Paris 1860-1865 (two vols. of supplements Paris 1878-1880).

Otto Brunner, *Land und Herrschaft*. Wien 1965.

A.T. Bryant, "Zulu medicine and medicine-men." *Annals of the Natal Museum*. 2(1): 1-104. 1909.

H.M. Burkill, *The Useful Plants of West Tropical Africa*. Edition 2. Royal Botanic Gardens, Kew 1985-1997.

I.H. Burkill, *A Dictionary of the Economic Products of the Malay Peninsula*. Kuala Lumpur 1966.

J. Burney, *A Chronological History of Discoveries in the South Sea or Pacific Ocean*. Amsterdam 1967.

Ernest J. Burrus, S.J., Ed., *Misiones Norteñas Mexicanas de la Compañia de Jesús, 1751-1757*. México 1963.

Robert Byron, *The Road to Oxiana*. Oxford University Press, Oxford 1982.

Antoine Cabaton, "Dix dialectes indochinois recueillis par Prosper Oden'hal, étude linguistique." *Journal Asiatique*, 10e série, 5: 265-344. 1905.

Lydia Cabrera, *Vocabulario Congo* (el Bantu que se habla en Cuba). Downtown Book Center, Miami 1984.

Armando Cáceres, *Plantas de uso medicinal en Guatemala*. Universidad de San Carlos de Guatemala 1996.

L. Cadière, *Croyances et pratiques religieuses des Annamites dans les environs de Hué*. Hanoi 1944.

María de los Angeles Calatayud Arinero, *Catalogo de las Expediciones y Viajes Científicas a America y Filipinas* (siglos XVIII y XIX). Fondos del Archivo del Museo Nacional de Ciencias Naturales. Madrid 1984.

J. Callander, *Terra Australis Cognita or Voyages to the Terra Australis*. London 1768 [Amsterdam 1967].

J.C. Campbell, *An Illustrated Encyclopaedia of Traditional Symbols*. Thames and Hudson Ltd., London 1982.

John F. Campbell, *History and Bibliography of the New American Practical Navigator and the American Coast Pilot*. Peabody Museum, Salem 1964.

Joseph Campbell, *The Mythic Image*. Princeton University Press, Princeton, New Jersey 1974.

Marinus F.A. Campbell, *Annales de la Typographie Néerlandaise au XVe siècle*. The Hague 1874.

Elias Canetti, *Masse und Macht*. Hamburg 1960.

M.A. Canini, *Dizionario etimologico dei vocaboli italiani d'origine ellenica*. UTET, Torino 1925.

Adriano Capelli, *Cronologia, cronografia e calendario perpetuo, dal principio dell'èra cristiana ai nostri giorni*. Edd. Daniela Stroppa Salina, Diego Squarcialupi, Enrico Quigini Puliga. Milano 1978.

Auguste Carayon, *Bibliographie historique de la Compagnie de Jésus*, ou catalogue des ouvrages relatifs à l'histoire des jésuites depuis leur origine jusqu'à nos jours. Paris 1864.

G.G. Carlson and V.H. Jones, "Some notes on uses of plants by the Comanche Indians." *Pap. Michigan Acad. Sci. Arts, and Letters*, vol. 25: 517-542. 1940.

A. Carnoy, *Dictionnaire étymologyque des noms grecs des plantes*. Louvain 1959.

Jean-Marie Carré, *Voyageurs et écrivains français en Égypte*. Le Caire 1956.

John Carter and Percy Muir, *Printing and the Mind of Man*. Compiled and edited by ..., assisted by Nicholas Barker, H.A. Feisenberger, Howard Nixon and S.H. Feinberg. With an introductory essay by D. Hay. Munich 1983.

Carlo Castellani, *La stampa in Venezia dalla sua origine alla morte di Aldo Manuzio seniore*. Venezia 1889.

A.C. Cavicchioni, *Vocabolario Italiano-Swahili*. Zanichelli Editore, Bologna 1923.

Adriano Ceresoli, *Bibliografia delle opere italiane latine e greche su la caccia, la pesca e la cinologia*. Bologna 1969.

L. Chabuis, *Petite histoire naturelle de la Polynésie français*. Papeete n.d.

John Chadwick, *The Decipherment of Linear B*. Cambridge University Press, New York 1958.

Lokesh Chandra, *Buddhist Iconography*. New Delhi 1987.

Lokesh Chandra, *Tibetan–Sanskrit Dictionary*. New Delhi 1958-1961.

Chandra Moti, "Cosmetics and coiffure in ancient India." *Journal of the Indian Society of Oriental Art*, VIII. 1940.

P. Chantraine, *Dictionnaire étymologique de la langue grecque*. 1968-1980.

P. José Chantre y Herrera, *Historia de las Misiones de la Compañia de Jesús en el Marañon Español*. Madrid 1901.

Arthur D. Chapman, ed., *Australian Plant Name Index*. Canberra 1991.

Louis Charbonneau-Lassay, *Il bestiario di Cristo*. Edizioni Arkeios, Roma 1994.

Louis Charbonneau-Lassay, *Il giardino del Cristo ferito*. Il Vulnerario e il Florario di Cristo. Edizioni Arkeios, Roma 1995.

Carlos E. Chardón, *Los naturalistas en la América Latina*. Ciudad Trujillo 1949.

K.N. Chaudhuri, *The English East India Company ... 1600-1640*. London 1965.

K.N. Chaudhuri, *Trade and Civilisations in the Indian Ocean*. Cambridge University Press, Cambridge 1985.

Nanimadhab Chaudhuri, "A pre-historic tree cult." *India Historical Quarterly*, XIX, 4. 1943.

Huguette and Pierre Chaunu, *Séville et l'Atlantique (1504-1650)*. Paris 1955-1960.

U. Chevalier, *Répertoire des sources historiques du Moyen Age. Bio-bibliographie*. Paris 1905-1907.

U. Chevalier, *Répertoire des sources historiques du Moyen Age. Topo-bibliographie*. Montbéliard 1903.

C. Christensen et al., *Index filicum*, with four supplements. Hagerup, Copenhagen, etc. 1906-1965.

C.M. Churchward, *Tongan Dictionary*. Government Printing Press, Nuku'alofa, Tonga 1959.

Paolo Bartolomeo Clarici, *Istoria e coltura delle piante*. Venezia 1726.

Inga Clendinnen, *Ambivalent Conquests — Maya and Spaniard in Yucatan, 1517-1570*. Cambridge University Press, Cambridge 1987.

Terry Clifford, *Tibetan Buddhist Medicine and Psychiatry: The Diamond Healing*. York Beach, Maine 1986.

Michael D. Coe, *The Maya Scribe and His World*. The Grolier Club, New York 1973.

M.R. Cohen and I.E. Drabkin, *A Source Book in Greek Science*. New York 1948.

George S. Cole, *A Complete Dictionary of Dry Goods and History of Silk, Cotton, Linen, Wool and Other Fibrous Substances*. Chicago 1892.

Don Miguel Colmeiro, *Diccionario de los diversos nombres vulgares de muchas plantas usuales o notables del antiguo y nuevo mundo*. Madrid 1871.

Georges Condominas, "Enquête linguistique parmi les populations montagnardes du Sud-Indochinois." *Bulletin de l'École Française d'Extrême-Orient*, tome XLVI, 2, 1954.

Giovanni Consolino, *Vocabolario del dialetto di Vittoria*. Pisa 1986.

W.A. Copinger, *Supplement to Hain's Repertorium bibliographicum*. London 1895-1902.

Henri Cordier, *Bibliotheca Indosinica*. Dictionnaire bibliographique des ouvrages relatifs à la péninsule indochinoise. Paris 1912-1932.

Henri Cordier, *Bibliotheca Japonica*. Dictionnaire bibliographique des ouvrages relatifs à l'Empire Japonais. Paris 1912.

Henri Cordier, *Bibliotheca Sinica*. Dictionnaire bibliographique des ouvrages relatifs à l'Empire chinois. Paris 1904-1924.

Diego de Córdoba Salinas, O.F.M., *Crónica franciscana de las provincias del Perú*. Ed. by Lino G. Canedo, O.F.M. Washington 1957.

Nuñez Corona, *Mitología tarasca*. México 1957.

Corpus of Maya Hieroglyphic Inscriptions. Vols. 1-6. Cambridge, Massachusetts: Peabody Museum of Archaeology and Ethnology. Harvard University 1977-.

Paul Alan Cox and Sandra Anne Banack, Eds., *Islands, Plants, and Polynesians*. An introduction to Polynesian ethnobotany. Dioscorides Press, Portland, Oregon 1991.

G. Crabb, *Universal Historical Dictionary*, or explanation of the names of persons and places in the departments of biblical, political, and ecclesiastical history, mythology, heraldry, biography, bibliography, geography, and numismatics. Enlarged edition. Baldwin & Cradock, London 1833.

J. Cretinau-Joly, *Histoire religieuse, politique et littéraire de la Compagnie de Jésus*. Paris 1844-1846.

A.J. Cronquist, *An Integrated System of Classification of Flowering Plants*. Columbia University Press, New York 1981.

A.J. Cronquist, *The Evolution and Classification of Flowering Plants*. Second edition. The New York Botanical Garden, New York 1988.

Laureano de la Cruz, *Nuevo descubrimiento del Marañon, 1651: Varones ilustres de la Orden Seráfica en el Ecuador*. Quito 1885.

J.S. Cumpston, *Shipping Arrivals and Departures, Sydney*. Canberra 1977.

Eustachio D'Afflitto, *Memorie degli scrittori del Regno di Napoli*. Napoli 1782.

G. D'Africa, *Il R. Istituto-Orto Botanico ed il R. Giardino Coloniale di Palermo*. Palermo 1945.

E.W. Dahlgren, *Les Relations commerciales et maritimes entre la France et les Côtes de l'Océan Pacifique ...* Paris 1909.

R.M. Dahlgren, H.T. Clifford and P.F. Yeo, *The Families of the Monocotyledons: Structure, Evolution and Taxonomy*. Berlin 1985.

E.G. Dale, "Literature on the history of botany and botanic gardens 1730-1840." *Huntia*. 6: 1-121. 1985.

Dam Bo, *Les populations montagnardes du Sud-Indochinois*. Lyon 1950.

William Darlington, *Memorials of John Bartram and Humphry Marshall*. Philadelphia 1849.

Vaidya Bhagvan Dash, *Alchemy and Metallic Medicines in Ayurveda*. Concept Publishing Company, New Delhi 1986.

Hugh William Davies, *Devices of the Early Printers, 1457-1560: Their History and Development*. London 1935.

Warren R. Dawson, *The Banks Letters*. London 1958.

J. Day, "Agriculture in the life of Pompeii." *Yale Classical Studies*. 3: 166-208. 1932.

A. De Baker and A. Carayon, *Bibliothèque de la Compagnie de Jésus*. Paris 1890-1900.

William T. De Bary, *Sources of the Indian Tradition*. New York 1964.

Gerard Decorme, *La Obra de Los Jesuítas Mexicanos Durante la Epoca Colonial, 1572-1767*. México 1941.

V.M. De Grandis, *Dizionario etimologico-scientifico*. Napoli 1824.

F.C. Deighton, *Vernacular Botanical Vocabulary for Sierra Leone*. London 1957.

F. Delitzsch, *Assyrisches Handwörterbuch*. Reprint of the 1896 edition. Leipzig 1968.

William Denevan, Ed., *The Native Population of the Americas in 1492*. University of Wisconsin Press, Madison 1976.

G.R. Dent and C.L.S. Nyembezi, *Scholar's Zulu Dictionary*. Shuter and Shooter, Pietermaritzburg 1995.

Desmond C. Derbyshire and Geoffrey K. Pullum, Eds., *Handbook of Amazonian Languages*. Berlin 1986.

L. Dermigny, *La Chine et l'Occident: le commerce à Canton au XVIIIe siècle*. Paris 1964.

Giovanni Bernardo De Rossi, *Annales hebraeo-typographici sec XV*. Parmae 1795.

Ray Desmond, *Dictionary of British & Irish Botanists and Horticulturists*. London 1994.

Dictionary of Scientific Biography. Editor in Chief Charles Coulston Gillispie. New York 1981.

G.W. Dimbleby, *Plants and Archaeology*. London 1978.

Wilfred Douglas, *The Aboriginal Languages of the South-West of Australia*. Canberra 1976.

P. Dourisboure, *Les sauvages Ba-Hnars* (Cochinchine orientale). Paris 1929.

John Dowson, *A Classical Dictionary of Hindu Mythology and Religion, Geography, History, and Literature*. London 1968.

John Dreyfus et al., Eds., *Printing and the Mind of Man: An Exhibition of Fine Printing in the King's Library of the British Museum*. The British Museum 1963.

A.H. Driver, *Catalogue of Engraved Portraits in the Royal College of Physicians of London.* London 1952.

Jonas C. Dryander, *Catalogus bibliothecae historico-naturalis Josephi Banks.* London 1796-1800.

Abbé Dubois, *Hindu Manners, Customs and Ceremonies.* Oxford University Press, Oxford 1906.

James Alan Duke and Rodolfo Vasquez, *Amazonian Ethnobotanical Dictionary.* CRC Press, Boca Raton, Florida 1994.

James Alan Duke, *CRC Handbook of Medicinal Herbs.* CRC Press, Boca Raton, Florida 1985.

Peter M. Dunne, *Pioneer Jesuits in Northern Mexico.* Berkeley 1944.

Peter M. Dunne, *Early Jesuit Missions in Tarahumara.* Berkeley 1948.

Denis I. Duveen, *Bibliotheca Alchemica et Chemica.* London 1949.

Ursula Dyckerhoff, "Mexican toponyms as a source in regional ethnohistory," in H.R. Harvey and Hanns J. Prem, Eds., *Explorations in Ethnohistory. Indians of Central Mexico in the Sixteenth Century.* 229-252. University of New Mexico Press, Albuquerque 1984.

R.A. Dyer, *The Genera of Southern Africa Flowering Plants.* Pretoria 1975-1976.

Frederic Adolphus Ebert, *A General Bibliographical Dictionary.* Oxford University Press, Oxford 1837.

James Edge-Partington, *Ethnographical Album of the Pacific Islands.* [Originally published as *Album of the Weapons, Tools, Ornaments, Articles of Dress of Natives of the Pacific Islands*] Second edition expanded and edited by Bruce L. Miller. SDI Publications, Bangkok 1996.

M.B. Emeneau, *The Strangling Figs in Sanskrit Literature.* California 1949.

Iris H.W. Engstrand, *Spanish Scientists in the New World.* University of Washington Press, Seattle 1981.

A. Ernout, *Morphologie historique du latin.* Paris 1974.

A. Ernout and A. Meillet, *Dictionnaire étymologique de la langue latine.* Paris 1985.

P. Lucas Espinosa, *Los tupí del oriente peruano*: estudio lingüístico y etnográfico. Madrid 1935.

E.E. Evans-Pritchard, *Witchcraft, Oracles and Magic among the Azande.* Oxford 1937.

Filippo Evola, *Storia tipografico-letteraria del secolo XVI in Sicilia.* Palermo 1878.

Alfonso Fabila, *Las tribus yaquis de Sonora*, su cultura y anhelada autodeterminación. Instituto Nacional Indigenista, México 1978.

William Falconer, *An Universal Dictionary of the Marine.* London 1781.

D.B. Fanshawe, revised by, *Check List of Vernacular Names of the Woody Plants of Zambia.* Lusaka 1965.

D.B. Fanshawe and C.D. Hough, *Poisonous Plants of Zambia.* Forest Research Bulletin, no. 1, revised. Ministry of Rural Development, Government Printer, Lusaka 1967.

D.B. Fanshawe and J.M. Mutimushi, *A Check List of Plant Names in the Nyanja Languages.* Forest Research Bulletin, no. 21, Ministry of Rural Development, Government Printer, Lusaka 1969.

E. Farr, J.A. Leussink and F.A. Stafleu, Eds., *Index nominum genericorum (plantarum).* 3 vols. Utrecht, etc. 1979.

E. Farr, J.A. Leussink and G. Zijlstra, Eds., *Index nominum genericorum (plantarum).* Supplementum I. Utrecht, etc. 1986.

J.A. Farrell, "A Hlengwe botanical dictionary of some trees and shrubs in Southern Rhodesia." *Kirkia.* 4: 165-172. 1964.

Mariano Fava and Giovanni Bresciano, *La stampa a Napoli nel XV secolo.* Leipzig 1912.

John Ferguson, *Bibliotheca Chemica.* Glasgow 1906.

L. Ferrari, *Onomasticon. Repertorio bibliografico degli scrittori italiani dal 1501 al 1850.* Milano 1947.

Francisco de Figueroa, *Relación de las Misiones de la Compañia de Jesús en el país de los Maynas.* Madrid 1904.

H.R. Fletcher, *Story of the Royal Horticultural Society, 1804-1968.* Oxford University Press, Oxford 1969.

H.R. Fletcher and W.H. Brown, *Royal Botanic Garden Edinburgh, 1670-1970.* Edinburgh 1970.

Lázaro Flury, *Tradiciones, leyendas y vida de los Indios del norte.* Editorial Ciordia y Rodríguez, Buenos Aires 1951.

Egidio Forcellini, *Lexicon totius Latinitatis.* Patavii 1940.

Karen Cowan Ford, *Las yerbas de la gente: A Study of Hispano-American Medicinal Plants.* Ann Arbor, Michigan 1975.

Richard I. Ford, Ed., *The Nature and Status of Ethnobotany.* Ann Arbor, Michigan 1978.

C. Fossey, *Manuel d'assyriologie. Fouilles, écriture, langues, littérature, géographie, histoire, religion, institutions, art.* Paris 1904-26.

F.W. Foxworthy, *I. Timbers of British North Borneo. II. Minor Forest Products and Jungle Produce.* Sandakan 1916.

Dora Franceschi Spinazzola, *Catalogo della biblioteca di Luigi Einaudi.* Opere economiche e politiche dei secoli XVI-XIX. Fondazione Luigi Einaudi, Torino 1981.

Fathers Franciscan, *An Ethnologic Dictionary of the Navaho Language.* St. Michaels, Arizona 1910.

J.D. Freeman, *Iban Agriculture.* London 1955.

H. Friis, Ed., *The Pacific Basin: A History of Its Geographical Exploration.* New York 1967.

G. Fumagalli, *Lexicon typographicum Italiae.* Dictionnaire géographique d'Italie pour servir à l'histoire de l'imprimerie dans ce pays. Leo S. Olschki, Firenze 1966.

V.A. Funk and S.A. Mori, "A bibliography of plant collectors in Bolivia." *Smithsonian Contr. Bot.* 70: 1-20. 1989.

M.M. Gabriel, *Livia's Garden Room at Prima Porta.* New York 1955.

F.N. Gachathi, *Kikuyu Botanical Bictionary of Plant Names and Uses.* Nairobi 1993.

Andrew Thomas Gage, *A History of the Linnean Society of London.* London 1938 (revised by W.T. Stearn 1985).

A. Galletti et al., *The Dutch in Malabar.* Madras 1911.

G. Garbini, *Le lingue semitiche.* Napoli 1984.

H. García-Barriga, *Flora medicinal de Colombia — botánica médica.* Bogotá 1974-1975.

Z.O. Gbile, *Vernacular Names of Nigerian Plants (Hausa).* For. Res. Inst. Nigeria, Ibadan 1980.

Z.O. Gbile and M.O. Soladoye, "Plants in traditional medicine in West Africa." *Monographs in Systematic Botany from Missouri Botanical Garden.* 25: 343-349. 1985.

Deno John Geanakoplos, *Greek Scholars in Venice: Studies in Dissemination of Greek Learning from Byzantium to Western Europe.* Harvard University Press, Cambridge 1962.

Deno John Geanakoplos, *Interaction of the "Sibling" Byzantine and Western Cultures in the Middle Ages and Italian Renaissance (330-1600).* Yale University Press, New Haven 1976.

I.J. Gelb, *Glossary of Old Akkadian.* The University of Chicago Press, Chicago 1957.

I.J. Gelb, *Materials for the Assyrian Dictionary.* Chicago 1952-70.

Helmut Genaust, *Etymologisches Wörterbuch der botanischen Pflanzennamen.* Birkhäuser Verlag, Basel 1996.

J/F.M. Genibrel, *Dictionnaire Annamite-Français.* Imprimerie de la Mission à Tan Dinh, Saigon 1898.

Alwyn H. Gentry, *A Field Guide to the Families and Genera of Woody Plants of Northwest South America (Colombia, Ecuador, Peru).* Washington, D.C. 1993.

P. Gerhard, *A Guide to the Historical Geography of New Spain.* Berkeley 1972.

J. Gildemeister, *Bibliothecae sanskritae.* Bonn 1847.

W. Gilges, *Some African Poison Plants and Medicines of Northern Rhodesia.* The occasional papers of the Rhodes–Livingstone Museum, no. 11. Livingstone, Northern Rhodesia 1955.

M.R. Gilmore, *Uses of Plants by the Indians of the Missouri River Region.* 33rd Ann. Rep. Bur. Amer. Ethnol. 1919.

Monique Girardin, *Bibliographie de l'Ile de la Réunion 1973-1992.* 1994.

Lorenzo Giustiniani, *Saggio storico critico sulla tipografia del Regno di Napoli.* Napoli 1793.

Heidi K. Gloria, *The Bagobos: Their Ethnohistory and Acculturation.* New Day Publishers, Quezon City, Philippines 1987.

Harry E. Godwin, *The History of the British Flora.* Cambridge University Press, Cambridge 1956.

Frederick R. Goff, *Incunabula in American Libraries.* A third census. Bibliographical Society of America, New York 1964.

Peter H. Goldsmith, *A Brief Bibliography of Books in English, Spanish and Portuguese, Relating to the Republics Commonly Called Latin American,* with comments. New York 1915.

V.F. Goldsmith, *A Short Title Catalogue of Spanish and Portuguese Books 1601-1700 in the Library of the British Museum* (The British Library–Reference Division). Folkestone and London 1974.

Mario Gongora, *Los grupos de conquistadores en Tierra Firme.* Santiago, Chile 1962.

M. Gongora, *Studies in the Colonial History of Spanish America.* Cambridge 1975.

E.R. Goodenough, *Jewish Symbols in the Greco-Roman Period.* Bollingen Series XXXVII. Pantheon Books, New York and Princeton University Press, Princeton 1953-1965.

E.J. Goodman, *The Explorers of South America.* London 1972.

Jean S. Gottlieb, *A Checklist of the Newberry Library's Printed Books in Science, Medicine, Technology, and the Pseudosciences circa 1460-1750.* New York and London 1992.

S.W. Gould and D.C. Noyce, *International Plant Index.* New York 1962, 1965.

Johann Georg Theodor Graesse, *Trésor de livres rares et précieux.* Dresde 1859-1869.

G. Grandidier, *Bibliographie de Madagascar.* Paris 1906-1957.

Edward Lee Greene, *Landmarks of Botanical History.* Edited by Frank N. Egerton. Stanford, California 1983.

P.J. Greenway, *A Swahili–Botanical–English Dictionary of Plant Names.* Dar es Salaam 1940.

R. Gregorio, *L'Orto botanico di Palermo.* Palermo 1821.

Pierre Grimal, *Les Jardins Romains.* Fayard 1984.

Barbara F. Grimes, Ed., *Ethnologue. Languages of the World.* Twelfth Edition. Dallas, Texas 1992.

J.W. Grimes and B.S. Parris, *Index Thelypteridaceae.* Royal Botanic Gardens, Kew 1986.

Arthur Eric Gropp, *A Bibliography of Latin American Bibliographies.* Scarecrow Press, Metuchen, New Jersey 1968-1971.

J.A. Gruys and C. De Wolf, *Thesaurus 1473-1800.* Dutch Printers and Booksellers with places and years of activity. Nieuwkoop 1989.

Antonio Guasch and Diego Ortiz, *Diccionario Castellano-Guarani, Guarani-Castellano.* Centro de Estudios Paraguayos Antonio Guasch, Asunción 1996.

T. Guignard, *Dictionnaire laotien-français.* Hong Kong 1912.

Mary Gunn and Leslie E. Codd, *Botanical Exploration of Southern Africa.* Cape Town 1981.

N. Gunson, *Messengers of Grace: Evangelical Missionaries in the South Seas.* Melbourne 1972.

H.B. Guppy, "The Polynesians and their plant-names." *Victoria Inst., Journ. of Trans.,* vol. 29: 135-170. 1895-1896.

Shaki M. Gupta, *Plant, Myths and Traditions in India.* Leiden 1971.

R. Gusmani, "Di alcuni prestiti greci in latino." *Bollettino di studi latini.* III: 76-88. 1973.

Malcolm Guthrie, *The Classification of the Bantu Languages.* Oxford University Press, Oxford 1948.

Malcolm Guthrie, *Comparative Bantu.* Farnborough 1971.

Miles Hadfield et al., *British Gardeners: A Biographical Dictionary.* London 1980.

K. Haebler, *Bibliografía iberica del siglo XV.* La Haya, Leipzig 1903-1917.

Ludwig Hain, *Repertorium bibliographicum* in quo libri omnes ab arte typographica inventa usque ad annum MD typis expressi ordine alphabetico vel simpliciter enumerantur vel adcuratius recensentur. Stuttgardiae 1826-1838.

S. Halkett and J. Laing, *A Dictionary of the Anonymous and Pseudonymous Literature of Great Britain.* Edinburgh 1882-1888.

J.B. Hall, P. Pierce and G. Lawson, *Common Plants of the Volta Lake.* Dept. of Botany, University of Ghana, Legon 1971.

E. Charles B. Hallam, *Oriya Grammar for English Students*. Baptist Mission Press, Calcutta 1874.

A. von Haller, *Bibliotheca Botanica*. Zürich 1771-1772.

Carl A. Hanson, *Dissertations on Iberian and Latin American History*. New York 1975.

C.H. Haring, *Trade and Navigation between Spain and the Indies in the Time of the Habsburgs*. Cambridge, Massachusetts 1918.

G.W. Harley, *Native African Medicine with Special Reference to Its Practice in the Mano Tribe of Liberia*. Cambridge, Massachusetts 1941.

Michael J. Harner, Ed., *Hallucinogens and Shamanism*. Oxford University Press, New York 1973.

Regina Harrison, *Signs, Songs, and Memory in the Andes — Translating Quechua Language and Culture*. University of Texas Press, Austin 1989.

John Harvey, *Early Gardening Catalogues*. Phillimore, London 1972.

J.G. Hawkes, "On the origin and meaning of South American Indian potato names." *Journal of the Linnean Society of London*, Botany. 53: 205-250. 1947.

Nicola Francesco Haym, *Biblioteca Italiana o sia Notizia dei Libri Rari nella Lingua Italiana*. Venezia 1728.

Charles B. Heiser, "Cultivated plants and cultural diffusion in nuclear America." *American Anthropologist*. 67: 930-949. 1965.

Blanche Elizabeth Edith Henrey, *British Botanical and Horticultural Literature before 1800*. Oxford University Press, Oxford 1975.

G. Heyd, *Storia del commercio del Levante nel Medio Evo*. Torino 1913.

L.R. Hiatt, Ed., *Australian Aboriginal Mythology*. Essays in honour of W.E.H. Stanner. Australian Institute of Aboriginal Studies, Canberra 1975.

M.E. Hoare, *The Tactless Philosopher*. Melbourne 1976.

George M. Hocking, "The doctrine of signatures." *Quarterly Journal of Crude Drug Research*. 15: 198-200. 1977.

Harry Hoijer, *Navaho Phonology*. University of New Mexico Publications in Anthropology. 1945.

William R. Holland, *Medicina Maya en los Altos de Chiapas*. Instituto Nacional Indigenista, México n.d.

J. Hollyman and A. Pawley, Eds., *Studies in Pacific Languages and Cultures*. Auckland 1981.

M. Holmes, *Captain James Cook: A Bibliographical Excursion*. London 1952.

J.H. Hospers, *A Basic Bibliography for the Study of the Semitic Languages*. Leiden 1973-74.

A.W. Howitt, *The Native Tribes of South-East Australia*. Macmillan, London 1904.

C. Huelsen, "Piante iconografiche encise in marmo." *Mitteilungen des deutschen archäologischen Instituts, Römische Abteilung*. 5: 46-63. 1890.

The Hunt Botanical Library, *Catalogue of Botanical Books in the Collection of Rachel McMasters Miller Hunt*. Pittsburgh 1958-1961.

Francis Huxley, *Affable Savages*. The Viking Press, New York 1957.

Index Aureliensis, *Index aureliensis*: catalogus librorum sedecimo saeculo impressorum [volumes I-IX, A-Coq]. Baden-Baden 1962-1991.

Benjamin Daydon Jackson, *A glossary of botanic terms* with their derivation and accent. Fourth edition. London 1928.

Benjamin Daydon Jackson, *Guide to the Literature of Botany*. Otto Koeltz Science Publishers, Königstein 1974.

Benjamin Daydon Jackson et al., *Index Kewensis plantarum phanerogamarum*. Oxford University Press, Oxford [1893-] 1895, 15 supplements 1901-1974.

T. Jacobsen, *The Sumerian King Lists*. Chicago University Press, Chicago 1939.

S.K. Jain, "Studies in Indian ethnobotany — origin and utility of some vernacular plant names." *Proc. Natl. Acad. Sci.* 33B: 525-530. 1963.

S.K. Jain, Ed., *Glimpses of Indian Ethnobotany*. Oxford and IBH Publishing Co., New Delhi 1981.

F.M. Jarrett et al., *Index Filicum. Supplementum Quintum pro annis 1961-1975*. Oxford University Press, Oxford 1985.

Wilhelmina F. Jashemski, *The Gardens of Pompeii*. Herculaneum and the Villas Destroyed by Vesuvius. New Rochelle, New York 1979.

Ch. F. Jean, *Sumer et Akkad*. Paris 1923.

Susan Jellicoe, "The development of the Mughal Garden." in *The Islamic Garden*. Dumbarton Oaks Colloquium on the History of Landscape Architecture. IV: 107-129, 134-135. Edited by Elisabeth B. MacDougall and Richard Ettinghausen. Washington 1976.

Hans Jensen, *Sign, Symbol, and Script*. New York 1969.

Volney H. Jones, "The nature and status of ethnobotany." *Chronica Botanica*. 6(10): 219-221. 1941.

Elviro Jorde Pérez, *Catálogo bio-bibliográfico de los Religiosos Agustinos de la Provincia del Santísimo Nombre de Jesús de las Islas Filipinas*. Manila 1901.

José Juanen, *Historia de la Compañia de Jesús en la Antigua Provincia de Quito, 1570-1774*. Quito 1941-1943.

M.W. Jurriaanse, *Catalogue of the Archives of the Dutch Central Government of Coastal Ceylon 1640-1796*. Colombo 1943.

H. and R. Kahane and L. Bremner, *Glossario degli antichi portolani italiani*. Traduzione e note di M. Cortelazzo. Firenze 1968.

Lama Dawasamdup Kazi, *English–Tibetan Dictionary*. Baptist Mission Press, Calcutta 1919.

Howard Kelly, *Some American Medical Botanists Commemorated in our Botanical Nomenclature*. Southworth, Troy 1914.

Hayward Keniston, *List of Works for the Study of Hispanic-American History*. The Hispanic Society of America, New York 1920 [Reprint New York 1967 Kraus Reprint Society].

K. Kerényi, *Asklepios: Archetypal Image of the Phisician's Existence*. Bollingen Series LXV. 3. Pantheon Books, New York 1959.

J. Kerharo and J.G. Adam, *La pharmacopée sénégalaise traditionnelle. Plantes médicinales et toxiques*. Paris 1974.

J. Kerharo and A. Bouquet, *Plantes médicinales et toxiques de la Côte d'Ivoire — Haute Volta*. Paris 1950.

M.C. Kiemen, *The Indian Policy of Portugal in the Amazon Region, 1614-1693*. New York 1973.

Konrad Kingshill, *Ku Daeng — The Red Tomb*. A village study in northern Thailand. Suriyaban Publishers, Bangkok, Thailand 1976.

E.F.K. Koerner and R.E. Asher, Eds., *Concise History of the Language Sciences from the Sumerians to the Cognitivists*. Pergamon, Cambridge 1995.

J.O. Kokwaro, *Luo-English Botanical Dictionary*. E. African Publishing House, Nairobi 1972.

Mamadou Koumaré et al., *Contribution aux études ethno-botaniques et floristiques au Mali*. Agence de coopération culturelle et technique (A.C.C.T.), Paris 1984.

Samuel N. Kramer, *Sumerian Mythology*. The American Philosophical Society, Philadelphia 1944.

A.L. Kroeber, *Cultural and Natural Areas of Native America*. University of California Press, Berkeley 1939.

A.L. Kroeber, *Peoples of the Philippines*. Handbook, American Museum of Natural History, no. 8. New York 1919.

G. Kunkel, *Geography through Botany. A Dictionary of Plant Names with Geographical Meanings*. The Hague 1990.

Peter Kunstadter, E.C. Chapman and Sanga Sabhasri, Eds., *Farmers in the Forest*. The University Press of Hawaii, Honolulu 1978.

W. LaBarre, "Potato taxonomy among the Aymara Indians of Bolivia." *Acta Americana*. 5: 83-102. 1947.

R. Labat, *Manuel d'épigraphie akkadienne (signes, syllabaire, idéogrammes)*. Troisième édition. [Reprint of the 1948 lithographed edition]. Paris 1963.

R. Labat, *Traité akkadien de diagnotics et pronostics médicaux*. Leiden 1951.

Y. Laissus, "Catalogue des manuscrits de Philibert Commerson (1727-1773) conservés à la Bibliothèque centrale du Muséum nationale d'Histoire Naturelle (Paris)." *Rev. Hist. Sci., Paris*. 31: 131-162. 1978.

Imre Lakatos, *Proofs and Refutations*. Cambridge University Press, New York 1976.

John Landwehr, *Studies in Dutch Books with Coloured Plates Published 1662-1875*. The Hague 1976.

J. Lang, *Conquest and Commerce: Spain and England in the Americas*. New York 1975.

H. Lanjouw and H. Uittien, "Un nouvel herbier de Fusée Aublet découvert en France." *Recueil des Travaux botaniques néerlandais*. 37: 133-170. 1940.

Daniel N. Lapedes, Editor in Chief, *McGraw-Hill Dictionary of Scientific and Technical Terms*. McGraw-Hill, New York 1978.

Antoine Lasègue, *Musée botanique de M. Benjamin Delessert*. Paris, Leipzig 1845.

G.H.M. Lawrence et al., *B-P-H Botanico-Periodicum-Huntianum*. Hunt Botanical Library, Pittsburgh 1968.

W.F. Leemans, *Foreign Trade in the Old Babylonian Period as Revealed by Texts from Southern Mesopotamia*. Leiden 1960.

A.J.M. Leeuwenberg, compiler, *Medicinal and Poisonous Plants of the Tropics*. Proceedings of Symposium 5-35 of the 14th International Botanical Congress, Berlin, 24 July – 1 August 1987.

Emile Legrand, *Bibliographie hellénique, ou description raisonnée des ouvrages publiés en grec par des grecs au XVe et XVIe siècles*. Paris 1885-1906.

J. Lemoine, *Un village Hmong vert du Haut-Laos*. Paris 1972.

Rodolfo Lenz, *Diccionario etimolójico de las voces chilenas derivadas de lenguas indijenas americanas*. Santiago de Chile 1904-1910.

Lewis and Short, *A Latin Dictionary*. Oxford University Press, Oxford 1995.

Henry George Liddell and Robert Scott, *A Greek-English Lexicon*. With a Revised Supplement. New edition, revised and augmented by H.S. Jones and R. McKenzie. Oxford University Press, Oxford 1996.

G. Liebert, *Iconographic Dictionary of the Indian Religions — Hinduism, Buddhism, Jainism*. Leiden 1976.

Carl Linnaeus (Carl von Linnaeus, Carl von Linné, Karl af Linné), *Species Plantarum*. Stockholm 1753 and *Genera Plantarum*. Ed. 5. Stockholm 1754.

Linnean Society of London, *Catalogue of the Printed Books and Pamphlets in the Library*. London 1925.

Jorge A. Lira, *Medicina Andina*. Peru, Cusco 1985.

Emilio Lisson Chávez, ed., *La Iglesia de España en el Perú*. Sevilla 1943-1947.

Alfredo López Austin, *Textos de Medicina Náhuatl*. México 1971.

Elias Avery Lowe, *Codices Latini Antiquiores*. III. Clarendon Press, Oxford 1938 and XI. Clarendon Press, Oxford 1966.

Katharine Luomala, *Ethnobotany of the Gilbert Islands*. Bernice P. Bishop Museum, Bulletin 213, Honolulu, Hawaii 1953.

Madeleine Ly-Tio-Fane, *Mauritius and the Spice Trade: The Odyssey of Pierre Poivre*. Port Louis 1958.

Ma Touan-Lin, *Ethnographie des peuples étrangers à la Chine*. Geneva 1876.

D.J. Mabberley, *The Plant-Book*. Second edition. Cambridge 1997.

D. Macdonald, *Moeurs et Coutumes des Tibétains*. Paris 1930.

Arthur Anthony Macdonell, *A Practical Sanskrit Dictionary*. Oxford University Press, Oxford 1976.

Arthur Anthony Macdonell and Arthur Berriedale Keith, *Vedic Index of Names and Subjects*. Motilal Banarsidass, Delhi, Varanasi (U.P.) and Patna (Bihar) 1967.

Elisabeth B. MacDougall, "Ars hortulorum: Sixteenth century garden iconography and literary theory in Italy." in *The Italian Garden*. Dumbarton Oaks Colloquium on the History of Landscape Architecture. I. Edited by David R. Coffin. Washington 1972.

Peter MacPherson, "The doctrine of signatures." *Glasgow Naturalist*. 20(3): 191-210. 1982.

José María Magalli de Pred, *Colección de Cartas sobre las Misiones Dominicanas del Oriente*. Quito 1890.

R.E. Magill, Ed., *Glossarium Polyglottum Bryologiae. A Multilingual Glossary for Bryology*. St. Louis, Missouri 1990.

R. Maire, "Études sur la flore et la végétation du Sahara central." *Mém. Soc. Hist. Nat. Afr. du Nord*. 65. Algiers 1933.

R. Maire and Th. Monod, "Études sur la flore et la végétation du Tibesti." *Mém. Inst. Franç. Afr. Noire*. 8. 1950.

G.P. Majumdar, *Vanaspati. Plants and Plant-life as in Indian Treatises and Traditions*. Calcutta 1927.

Trilok Chandra Majupuria and Rohit Kumar (Majupuria), *Gods and Goddesses — An Illustrated Account of Hindu, Buddhist, Tantric, Hybrid and Tibetan Deities*. Lalitpur Colony, Lashkar (Gwalior) 1994.

M. Maldonado-Koerdell, *Bibliografía Geológica y Paleontológica de América Central*. Mexico 1958.

M.E.L. Mallowan, *Early Mesopotamia and Iran*. McGraw-Hill, New York 1965.

M.A. Mandango and M.B. Bandole, "Contribution à la connaissance des plantes médicinales des Turumbu de la zone de Basoko (Zaire)." *Monographs in Systematic Botany from Missouri Botanical Garden.* 25: 373-384. 1988.

Clements Robert Markham, "A list of the tribes of the valley of the Amazons, ..." *Journal of the Anthropological Institute.* 40: 73-140. 1910.

Sidney David Markman, compiled and annotated by, *Colonial Central America. A Bibliography.* Arizona State University, Center for Latin American Studies, Tempe, Arizona 1977.

A.W. Martin and P. Wardle, *Members of the Legislative Assembly of New South Wales, 1856-1901.* Australian National University, Canberra 1959.

J.E. Martin-Allanic, *Bougainville navigateur et les Découvertes de son Temps.* Paris 1964.

John Alden Mason and George Agogino, "The ceremonialism of the Tepecan." Eastern New Mexico University, Paleo-Indian Institute, *Contributions in Anthropology.* vol. 4(1), October 1972.

C. Massin, *La médecine Tibétaine.* Paris 1982.

R.H. Mathews, *Ethnological Notes on the Aboriginal Tribes of New South Wales and Victoria.* F.W. White, Sydney 1905.

N. Mathur, *Red Fort and Mughal Life.* New Delhi 1964.

S. Matsushita, *A Comparative Vocabulary of Gwandara Dialects.* Tokyo 1974.

Washington Matthews, "Navaho names and uses for plants." *American Naturalist.* 20(9): 767-777. 1886.

M.G. de la Maza, *Diccionario botánico de los nombres vulgares cubanos y puertorriqueños.* 1889.

E.H. McCormick, *Tasman and New Zealand: A Bibliographical Study.* Wellington 1959.

Isaac McCoy, *History of Baptist Indian Missions.* New York 1840.

E. McKay, Ed., *Studies in Indonesian History.* Carlton (Vic.) 1976.

Norman A. McQuown, "The classification of the Mayan languages." *International Journal of American Linguistics.* 22: 191-195. 1956.

Norman A. McQuown, "The indigenous languages of Latin America." *American Anthropologist.* 57: 501-570. 1955.

José Toribio Medina, *Biblioteca hispano-americana (1493-1810).* Santiago de Chile 1897-1907.

José Toribio Medina, *Diccionario biográfico colonial de Chile.* Santiago de Chile 1906.

José Toribio Medina, *Los aboríjenes de Chile.* Santiago 1882.

C.K. Meek, *Tribal Studies in Northern Nigeria.* Oxford University Press, Oxford 1925.

Betty J. Meggers and Clifford Evans, edited by, *Aboriginal Cultural Development in Latin America: An Interpretative Review.* The Smithsonian Institution, Washington, D.C. 1963.

M.A.P. Meilink-Roelofsz, *Asian Trade and European Influence in the Indonesian Archipelago between 1500 and about 1630.* The Hague 1962.

Gaetano Melzi, *Dizionario di opere anonime e pseudonime di scrittori italiani o come che sia aventi relazione all'Italia.* Milano 1848-1859.

Sidney Mendelssohn, *Mendelssohn's South African Bibliography.* London 1910.

Manuel de Mendiburu, *Diccionario histórico-biográfico del Perú.* Lima 1931-1935.

Elmer D. Merrill, *A Dictionary of the Plant Names of the Philippine Islands.* Manila 1903.

Ellen Messer, *Zapotec Plant Knowledge. Memoirs of the Museum of Anthropology,* University of Michigan Press, Ann Arbor 1978.

W. Meyer-Lübke, *Romanisches etymologisches Wörterbuch.* Heidelberg 1935.

S.P. Michel and P.H. Michel, *Répertoire des ouvrages imprimés en langue italienne au XVIIe siècle conservés dans les bibliothèques de France.* Paris 1967-1984.

Frederick William Hugh Migeod, *Mende Natural History Vocabulary.* London 1913.

G.B. Milner, *Samoan Dictionary.* Oxford University Press, London 1966.

J.D. Milner, *Catalogue of Portraits of Botanists Exhibited in the Museums of the Royal Botanic Gardens.* Royal Botanic Gardens, Kew, London 1906.

Giuseppe Maria Mira, *Bibliografia siciliana* ovvero Gran dizionario bibliografico delle opere edite e inedite, antiche e moderne di autori siciliani e di argomento siciliano stampate in Sicilia e fuori. Palermo 1875-1881.

J. Mirsky, *The Westwards Crossings: Balboa, Mackenzie, Lewis and Clark.* Chicago 1970.

Sarat Chandra Mitra, "Worship of the Pipal Tree in North Bihar." *Journal of the Bihar and Orissa Society.* VI. 1920.

A. Mongitore, *Bibliotheca Sicula* sive De scriptoribus Siculis qui tum vetera, tum recentiora saecula illustrarunt, ... notitiae locupletis-simae. Panormi 1708-1714.

M. Monier-Williams, *A Sanskrit–English Dictionary.* Motilal Banarsidass, Delhi 1990.

M. Monier-Williams, *English–Sanskrit Dictionary.* Munshiram Manoharlal, New Delhi 1976.

Franco Montanari, *Vocabolario della lingua Greca.* Loescher Editore, Torino 1995.

François-Auguste de Montequin, Ed., *Aspects of Ancient Maya Civilization.* Hamline University, St. Paul, Minn. 1976.

Vayaskara N.S. Mooss, *Ayurvedic Flora Medica.* Kottayam, S. India 1978.

L. Moranti, *Le cinquecentine della biblioteca universitaria di Urbino.* Firenze 1977.

Magnus Mörner, Ed., *The Expulsion of the Jesuits from Latin America.* New York 1967.

Edward E. Morris, *Austral English.* London 1898.

H.B. Morse, *The Chronicles of the East India Company Trading to China 1635-1834.* Oxford University Press, Oxford 1926-1929.

S. Moscati et al., *An Introduction to Comparative Grammar of the Semitic Language: Phonology and Morphology.* Edited by S. Moscati. Wisbaden 1969.

G. Murray, *History of the Collections Contained in the Natural History Departments of the British Museum*. London 1904.

Gary Paul Nabhan, *Gathering the Desert*. The University of Arizona Press, Tucson, Arizona 1985.

A. Narbone, *Bibliografia sicola sistematica*. Palermo 1850-1855.

National Library of Scotland, *A Short-Title Catalogue of Foreign Books, Printed up to 1600*. Edinburgh 1970.

The National Union Catalogue, pre 1965 imprints. London 1968-1981.

D.H. Nicolson, "Orthography of names and epithets." *Taxon*. 23: 549-561, 843-851. 1974.

D.H. Nicolson and R.A. Brooks, "Orthography of names and epithets." *Taxon*. 23: 163-177. 1974.

[Nigeria], *Vocabulary of Nigerian Names of Trees, Shrubs and Herbs*. Government Printer, Lagos 1936.

C. Nigra, *Saggio lessicale di basso latino curiale compilato su Estratti di Statuti medievali piemontesi*. Torino 1920.

Claus Nissen, *Die Botanische Buchillustration*. Ihre Geschichte und Bibliographie. Stuttgart 1966.

Erland Nordenskiöld, *Analyse ethno-géographique de la culture matérielle de deux tribus Indiennes du Gran Chaco*. Paris 1929.

Haskell F. Norman, *The Haskell F. Norman Library of Science and Medicine*. Compiled by Diana H. Hook and Jeremy M. Norman. San Francisco 1991.

Kalervo Oberg, *Indian Tribes of Northern Mato Grosso, Brazil*. Smithsonian Institution, Institute of Social Anthropology, Publication no. 15. Washington 1953.

[Catalogo del Fondo Fiammetta Olschki], *Viaggi in Europa. Secoli XVI-XIX*. Gabinetto scientifico e letterario G.P. Viesseux. Firenze 1990.

Martha Ornstein, *The Role of Scientific Societies in the Seventeenth Century*. University of Chicago Press, Chicago [1938].

F.L. Orpen, "Botanical–Vernacular and Vernacular–Botanical Names of Some Trees and Shrubs in Matabeleland." *Rhod. Agr. Journ.* XLVIII, 2. 1951.

Bernard R. Ortiz de Montellano, "Empirical Aztec Medicine." *Science*. 188: 215-220. 1975.

G. Ortolani, *Biografia degli uomini illustri della Sicilia*. Napoli 1827-1831.

William Osler, *Bibliotheca Osleriana*. Clarendon Press, Oxford 1929.

José I. Otero, Rafael A. Toro and Lydia Pagán de Otero, *Catálogo de los Nombres Vulgares y Científicos de Algunas Plantas Puertor-riqueñas*. Universidad de Puerto Rico, Rio Piedras, Puerto Rico 1945.

Giuseppe Ottino and Giuseppe Fumagalli, eds., *Bibliotheca bibliographica italica*. Roma and Torino 1889-1902.

Keith Coates Palgrave, *Trees of Southern Africa*. Struik Publishers, Cape Town 1990.

Eve Palmer and Norah Pitman, *Trees of Southern Africa*. Cape Town 1972.

M. Parenti, *Dizionario dei luoghi di stampa falsi, inventati o supposti*. Sansoni, Firenze 1951.

M. Parenti, *Esercitazioni filologiche*. Modena 1844-1857.

A. Parrot, *Sumer, the Dawn of Art*. Golden Press, New York 1961.

J.H. Parry, *Trade and Dominion: The European Overseas Empires in the Eighteenth Century*. London 1971.

Claire D.F. Parsons, ed., *Healing Practices in the South Pacific*. The Institute for Polynesian Studies 1985.

Michele Pasqualino, *Vocabolario etimologico siciliano italiano e latino*. Palermo 1783-1795.

Giambattista Passano, *Dizionario di opere anonime e pseudonime in supplemento a quello di Gaetano Melzi*. Ancona 1887.

Edward R. Pease, *The History of the Fabian Society*. London 1925.

G.B. Pellegrini, *Gli arabismi nelle lingue neolatine con speciale riguardo all'Italia*. Brescia 1972.

Antonio Penafiel, *Nombres geographicos de Mexico*. Mexico 1885.

Campbell W. Pennington, *The Tepehuan of Chihuahua — Their Material Culture*. University of Utah Press, Salt Lake City 1969.

Campbell W. Pennington, Ed., *The Pima Bajo of Central Sonora, Mexico*. University of Utah Press, Salt Lake City 1979-1980.

Rev. Father Fray Angel Perez of the Order of San Agustin, *Igorots, Geographic and Ethnographic Study of Some Districts of Northern Luzon*. Cordillera Studies Center, University of Philippines, Baguio City 1988.

David Perini, *Bibliographia augustiniana. Scriptores itali*. Firenze 1929-1935.

Luciano Petech, *I missionari italiani nel Tibet e nel Nepal*. Roma 1952-1956.

A. Pételot, "Bibliographie botanique de l'Indochine." *Arch. Regn. Agron. Pastor. Vietnam* no. 24, Saigon 1955.

W.J. Peters, *Landscape in Romano–Campanian Painting*. Assen 1963.

J. Petzholdt, *Bibliotheca bibliographica*. Leipzig 1866.

J.L. Phelan, *El Reino Milenario de los Franciscanos en el Nuevo Mundo*. México 1972.

Sandro Piantanida, Lamberto Diotallevi and Giancarlo Livraghi, eds., *Autori Italiani del '600*. Libreria Vinciana, Milano 1948-1951.

Rodolfo E.G. Pichi Sermolli et collab., "Index Filicum, Supplementum quartum pro annis 1934-1960." *Regnum Veget.* 37: I-XIV + 1-370. Utrecht 1965.

Rodolfo E.G. Pichi Sermolli, comp., *Authors of Scientific Names in Pteridophyta*. Royal Botanic Gardens, Kew 1996.

Arpad Plesch, *The Magnificent Botanical Library of the Stiftung für Botanik, Vaduz Liechtenstein*, collected by the late Arpad Plesch. Part I [-III]. 3 vols. London, Sotheby & Co., 16 June 1975 – 15 March 1976.

L. Polgar, *Bibliography of the History of the Society of Jesus*. Rome 1957.

Alfred W. Pollard and G.R. Redgrave, Eds., *A Short-Title Catalogue of Books Printed in England, Scotland and Ireland, and of English Books Printed Abroad 1475-1640*. Second edition, revised and enlarged, begun by W.A. Jackson and F.S. Ferguson, completed by Katharine F. Pantzer. Bibliographical Society, London 1976-1991.

R. Portères, "La sombre aroidée cultivée: *Colocasia antiquorum* Schott ou taro de Polynésie: essai d'etymologie sémantique." *J. Agric. Trop. Botan. Appl.* 7: 169-192. 1960.

R. Portères, "Les appelations des céréales en Afrique." *J. Agric. Trop. Botan. Appl.* 5 and 6. 1958-1959.

F.N.C. Poynter, Ed., *Medicine and Culture*. London 1969.

Efren C. del Pozo, "Estudios farmacológicos de algunas plantas usadas en medicina azteca." *Bol. Indigenista*. 6: 350-365. 1946.

A.H.J. Prins, *The Swahili-speaking Peoples of Zanzibar and the East African Coast (Arabs, Shirazi and Swahili)*. London 1961.

E.H. Pritchard, *Anglo-Chinese Relations during the Seventeenth and Eighteenth Centuries*. Urbana (Ill.) 1929.

E.H. Pritchard, *The Crucial Years of Early Anglo–Chinese Relations 1750-1800*. New York 1936.

G.A. Pritzel, *Thesaurus literaturae botanicae omnium gentium*. Lipsiae 1871-1877.

Robert Proctor, *The Printing of Greek in the Fifteenth Century*. Bibliographical Society Illustrated Monograph 8. Oxford University Press for The Bibliographical Society, London 1900.

Brian Pullan, *Rich and Poor in Renaissance Venice: The Social Institutions of a Catholic State, to 1620*. Oxford, Cambridge 1971.

Quaritch, *Catalogue of a Most Important Collection of Publications of the Aldine Press, 1494-1595*. Quaritch, London 1929.

Joseph-Marie Quérard, *La France littéraire*. Paris 1827-1864.

Jacobus Quétif and Jacobus Echard, *Scriptores Ordinis Praedicatorum*. Lutetiae Parisiorum 1719-1721.

A. Rainaud, *Le Continent Austral*: Hypothèses et Découvertes. Paris 1894.

Sri Ramakrishna Centenary Committee, Ed., *Cultural Heritage of India*. Calcutta 1937.

José Ramírez, *Sinonimia vulgar y científica de las plantas mexicanas*. México 1902.

W.D. Raymond, "Native poisons and native medicines of Tanganyika." *J. Trop. Med. & Hyg.* 42: 295-303. 1939.

C.F. Reed, "Index Thelypteridis." *Phytologia*. 17(4): 249-328. 1968.

Gerardo Reichel-Dolmatoff, *Amazonian Cosmos*. The University of Chicago Press, Chicago 1971.

Gerardo and Alicia Reichel-Dolmatoff, *The People of Aritama*. The University of Chicago Press, Chicago 1970.

Dietrich Reichling, *Appendices ad Hainii–Copingeri Repertorium bibliographicum*. Additiones et emendationes. Monachii 1905-1911.

Blas Pablo Reko, *Mitobotánica Zapoteca*. Tacubaya, D.F. 1945.

Jane Renfrew, *Palaeoethnobotany*. Columbia University Press, New York 1973.

Antoine-Augustin Renouard, *Annales de l'imprimerie des Aldes*. Paris 1834.

Antoine-Augustin Renouard, *Annales de l'imprimerie des Estienne*. Paris 1843.

Philippe Renouard, *Imprimeurs parisiens*: Libraires, fondeurs de caractères et correcteurs d'imprimerie; depuis l'introduction de l'imprimerie à Paris (1470) jusqu'à la fin du XVIe siècle. Paris 1898.

T.W. Rhys-Davids and William Stede, *The Pali Text Society's Pali–English Dictionary*. London 1979.

Darcy Ribeiro and Mary Ruth Wise, *Los grupos étnicos de la Amazonía Peruana*. Yarinacocha, Pucallpa 1978.

Massimo Ricciardi and G.G. Aprile, "Preliminary data on the floristic components of some carbonized plant remains found in the archaeological area of Oplontis near Naples." *Annali della Facoltà di Scienze Agrarie dell'Università di Napoli in Portici*, ser. 4, 12: 204-212. 1978.

Tania Maura Nora Riccieri, *Bibliografia de Plantas Medicinais*. Jardim Botânico do Rio de Janeiro. Rio de Janeiro 1989.

Roberto Ridolfi, *La stampa in Firenze nel secolo XV*. Firenze 1958.

C.L. Riley, J.C. Kelley, C.W. Pennington and R.L. Rands, Eds., *Man Across the Sea: Problems of Pre-Columbian Contacts*. University of Texas Press, Austin 1971.

Rechung Rinpoche, *Tibetan Medicine*. Berkeley 1973.

C.H. Robinson, *Dictionary of the Hausa Language*. 2 vols. Cambridge University Press, Cambridge 1913.

R.J. Rodin, *The Ethnobotany of the Kwanyama Ovambos*. Missouri Botanical Garden 1985.

F.L.O. Roehrig and J.G. Cogswell, compiled by, *Catalogue of Books in the Astor Library Relating to the Languages and Literature of Asia, Africa and the Oceanic Islands*. New York 1854.

G.N. Roerich, *Le parler de l'Amdo*. Roma 1958.

J. Roscoe, *The Northern Bantu: An Account of Some Central Africa Tribes of the Uganda Protectorate*. Cambridge 1915.

Walter E. Roth, *Ethnological Studies Among the North-West-Central Queensland Aborigines*. Brisbane 1897.

Leslie B. Rout, Jr., *The African Experience in Spanish America*. 1502 to the present day. Cambridge University Press, Cambridge 1976.

The Royal College of Physicians of London. *Portraits*. Edited by G.W. Wolstenholme. Described by David Piper. London 1964.

Ralph L. Roys, *The Ethnobotany of the Mayas*. Tulane University, Middle American Research Institute, New Orleans 1931.

J. Rzedowski, "Nombres regionales de algunas plantas de la Huasteca Potosina." *Actas Científicas Potosinas*. 6: 7-58. 1966.

P.A. Saccardo, *Cronologia della flora italiana*. Padova 1909.

Ch. Sacleux, *Dictionnaire Français-Swahili*. Zanzibar, Paris 1891.

R.N. Salaman, *The History and Social Influence of the Potato*. Cambridge 1949.

Contessa Anna di San Giorgio, *née* Harley d'Oxford, *Catalogo poliglotto delle piante*. Stabilimento tipografico di Giuseppe Pellas, Firenze 1870.

G. Sansom, *The Western World and Japan, 1334-1615*. London 1950.

G. Sansom, *A History of Japan*. London 1961.

Francisco J. Santamaría, *Diccionario de mejicanismos*. México 1978.

G. Saporetti, *Onomastica medio-assira*. Roma 1970.

G. Sapori, *Il fondo di medicina antica della Biblioteca Ginecologica di Milano*. Milano 1975.

G. Sapori, *Le cinquecentine dell'Università di Milano*. Milano 1969.

B.K. Sarkar, *The Folk Elements in Hindu Culture*. New Delhi 1981.

Swami Satchidananda, *Dictionary of Sanskrit Names*. Integral Yoha Publications, Buckingham, Virginia 1989.

Linda Schele, *Maya Glyphs: The Verbs*. University of Texas Press, Austin 1982.

G. Schembari, *La scienza orientale*. Torino 1924.

Amelie Schenk and Holger Kalweit, Eds., *Heilung des Wissens*. Munich 1987.

Stephan Schuhmacher and Gert Woerner, Eds., *The Encyclopedia of Eastern Philosophy and Religion. Buddhism, Hinduism, Taoism, Zen*. Shambhala, Boston 1994.

Hans Wolfgang Schumann, *Immagini buddhiste*. Edizioni Mediterranee. Roma 1989.

D. Scinà, *Prospetto della storia letteraria di Sicilia nel secolo XVIII*. Palermo 1969.

William Henry Scott, *The Discovery of the Igorots — Spanish Contacts with the Pagans of Northern Luzon*. New Day Publishers, Quezon City 1987.

Giovanni Semerano, Le origini della cultura europea. Dizionari Etimologici. Basi semitiche delle lingue indeuropee. *Dizionario della lingua Greca*. Leo S. Olschki Editore, Firenze 1994.

Giovanni Semerano, Le origini della cultura europea. *Dizionario della lingua Latina e di voci moderne*. Leo S. Olschki Editore, Firenze 1994.

María Teresa Sepúlveda y H., *La medicina entre los purépecha prehispánicos*. Universidad Nacional Autónoma de México, México 1988.

R. Shafer, *Ethnography of Ancient India*. Wiesbaden 1954.

J. Shapera, ed., *The Bantu-speaking Tribes of South Africa*. London 1937.

P.V. Sharma, *Indian Medicine in the Classical Age*. India 1972.

H. Sharp et al., *Siamese Rice Village: A Preliminary Study of Bang Chan, 1948-1949*. Bangkok 1953.

Moodeen Sheriff, *A Catalogue of Indian Synonymes of the Medicinal Plants, Products, Inorganic Substances, etc., Proposed to be Included in the Pharmacopoeia of India*. [Published in 1869, 1st reprint edition 1978] Printed at Jayyed Press, Billimaran, Delhi 1978.

Dorothy Shineberg, *They Came for Sandalwood*. Melbourne 1967.

Davi Shorto, Judith Jacob and E.H.S. Simmons, *Bibliographies of Mon-Khmer and Thai Linguistics*. Oxford University Press, Oxford 1963.

M.F. da Silva, P.L. Braga Lisbôa and R.C. Lobato Lisbôa, *Nomes vulgares de plantas amazonicas*. Manaus, Brasil 1977.

D.R. Simmons, *The Great New Zealand Myth*. Wellington 1967.

C. Earle Smith, Jr., Ed., *Man and His Foods: Studies in the Ethnobotany of Nutrition — Contemporary, Primitive and Prehistoric Non-European Diets*. University of Alabama Press, Alabama 1973.

Christo Albertyn Smith, *Common Names of South African Plants*. Edited by E. Percy Phillips and Estelle van Hoepen. Botanical Survey Memoir, no. 35. 1966.

William Henry Smyth and Sir E. Belcher, *Dictionary of Nautical Terms*. Glasgow 1867.

W. von Soden and W. Rollig, *Das akkadische Syllabar*. Roma 1967.

C. Sommervogel, S.J., *Bibliothèque de la Compagnie de Jésus*. Bruxelles and Paris 1890-1916.

C. Sommervogel, S.J., *Dictionnaire des ouvrages anonymes et pseudonymes publiés par des religieux de la Compagnie de Jésus*. Paris 1884.

Jacques Soustelle, *The Four Suns*. Grossman Publishers, New York 1971.

O.H.K. Spate, *The Spanish Lake*. University of Minnesota Press, Minnesota 1979 [Italian translation: *Storia del Pacifico — Il lago spagnolo*. A cura di Gianluigi Mainardi. Giulio Einaudi editore. Torino 1987].

O.H.K. Spate, *Monopolists and Freebooters*. 1983 [Italian translation: *Storia del Pacifico — Mercanti e bucanieri*. A cura di Gianluigi Mainardi. Giulio Einaudi editore. Torino 1988].

O.H.K. Spate, *Paradise Found and Lost*. 1988 [Italian translation: *Storia del Pacifico — Un paradiso trovato e perduto*. A cura di Gianluigi Mainardi. Giulio Einaudi editore. Torino 1993].

Jonathan Spence, *The Memory Palace of Matteo Ricci*. New York 1983.

Baldwin Spencer, *Native Tribes of the Northern Territory of Australia*. Macmillan, London 1914.

Baldwin Spencer and F.J. Gillen, *The Native Tribes of Central Australia*. London 1899.

Frans A. Stafleu and Richard S. Cowan, *Taxonomic Literature*. 7 vols. 1976-1988.

Frans A. Stafleu and E.A. Mennega, *Taxonomic Literature*. 5 suppls. 1992-1998.

Paul C. Standley, *Trees and Shrubs of Mexico*. Washington, D.C. 1920-1926.

Frederick Starr, *Notes upon the Ethnography of Southern Mexico*. [Davenport] 1900-1902.

W.T. Stearn, *Botanical Latin*. London 1966.

W.T. Stearn, *Stearn's Dictionary of Plant Names for Gardeners*. London 1993.

M.J. van Steenis-Kruseman, "Malaysian plant collectors and collections." Supplement 2. Page I-CXV. in *Flora Malesiana* 8(1). Nordhoof International Publishing, Leyden, The Netherlands 1974.

George Steiner, *After Babel: Aspects of Language and Translation*. Oxford University Press, New York 1975.

E. Steinkellner and H. Tauscher, Eds., *Contributions on Tibetan Language History and Culture*. Wien 1983 (Reprint Delhi 1995).

A.F. Stenzler, *Primer of the Sanskrit Language*. Translated into English with some revision by Renate Söhnen. School of Oriental and African Studies. University of London 1992.

Julian Haynes Steward, Ed., *Handbook of South American Indians*. U.S. Government Printing Office, Washington, D.C. 1946-.

George Stewart, *Names on the Globe*. Oxford University Press, Oxford 1975.

Margaret Bingham Stillwell, *Incunabula and Americana 1450-1800. A Key to Bibliographical Study*. Columbia University Press, New York 1931.

Margaret Bingham Stillwell, *The Awakening Interest in Science during the First Century of Printing 1450-1550: An Annotated Checklist of First Editions.* New York 1970.

R. Stillwell, Ed., *The Princeton Encyclopedia of Classical Sites.* Princeton University Press, Princeton 1976.

Margaret Stutley and James Stutley, *A Dictionary of Hinduism.* London 1977.

M. Swadesh, *Indian Linguistic Groups of Mexico.* México, D.F. 1961.

M. Swadesh, "Interrelaciones de las lenguas Mayas." *Anales del Instituto Nacional de Antropología e Historia.* Mexico 1961.

E. Taillemite, *Bougainville et ses Compagnons autour du monde.* Paris 1977.

Avenir Tchemerzine, *Bibliographie d'éditions originales et rares d'auteurs français des XV, XVI, XVII et XVIII siècles.* Paris 1927-1933.

Saladin S. Teo, *The Life-style of the Badjaos — A Study of Education and Culture.* Manila 1989.

H. Ternaux-Compans, *Bibliothèque asiatique et africaine.* Paris 1841.

Cyrus Thomas, assisted by John R. Swanton, *Indian Languages of Mexico and Central America and Their Geographical Distribution.* Smithsonian Institution, Bureau of American Ethnology, Bulletin 44. Washington 1911.

P.L. Thomas, "A Chitonga-Botanical dictionary of some species occurring in the vicinity of the Mwenda Estuary, Lake Kariba, Rhodesia." *Kirkia.* 7, 2: 269-284. 1970.

Dorothy Burr Thompson, "Ancient gardens in Greece and Italy." *Archaeology.* 4. Spring 1951.

Dorothy Burr Thompson, "The Garden of Hephaistos." *Hesperia.* 6: 396-425. 1937.

J. Eric S. Thompson, *Ethnology of the Mayas of Southern and Central British Honduras.* Chicago 1930.

R.C. Thompson, *A Dictionary of Assyrian Botany.* First edition. London 1949.

R.C. Thompson, *A Dictionary of Assyrian Chemistry and Geology.* Oxford University Press, Oxford 1936.

K. Thomson and G. Serle, *A Biographical Register of the Victorian Parliament, 1859-1900.* Australian National University Press, Canberra 1972.

F. Thureau-Dangin, *Le Syllabaire Accadien.* Librairie Orientaliste Paul Geuthner, Paris 1926.

N.B. Tindale, *Aboriginal Tribes of Australia.* University of California Press, Berkeley 1974.

Julio Tobar Donoso, *Historiadores y Cronistas de las Misiones.* Puebla, Mexico 1960.

S. Tolkowsky, *Hesperides, A History of the Culture and Use of Citrus Fruits.* London 1938.

G. Toscano, *Alla scoperta del Tibet. Relazioni dei missionari del sec. XVII.* Bologna 1977.

Margaret A. Towle, *The Ethnobotany of Pre-Columbian Peru.* Aldine Publishing Company, Chicago 1961.

L. Trabut, *Répertoire des noms indigènes des plantes spontanées, cultivées et utilisées dans le Nord de l'Afrique.* Alger 1935.

Hugh Trevor-Roper, *Hermit of Peking. The Hidden Life of Sir Edmund Backhouse.* 1978.

Rolla Milton Tryon and Alice Faber Tryon, *Ferns and Allied Plants.* New York, Heidelberg, Berlin 1982.

Ethelyn (Daliaette) Maria Tucker, *Catalogue of the Library of the Arnold Arboretum of Harvard University.* Cambridge, Massachusetts 1917-1933.

Peter J. Ucko and G.W. Dimbleby, Eds., *The Domestication and Exploitation of Plants and Animals.* Aldine Publishing Co., Chicago 1969.

Ruth M. Underhill, "The Papago Indians of Arizona and their relatives the Pima." *Sherman Pamphlet* no. 3 (Indian Life and Customs), U.S. Office of Indian Affairs, Lawrence, Kansas 1940.

Emerenziana Vaccaro, *Le marche dei tipografi ed editori italiani del sec. XVI nella Biblioteca Angelica di Roma.* Leo S. Olschki, Firenze 1983.

Luis E. Valcárcel, *Mirador Indio.* Apuntes para una filosofía de la cultura incaica. Lima 1937-1941.

Paul Valkema Blouw, *Typographia Batava 1541-1600.* A repertorium of books printed in the Northern Netherlands between 1541 and 1600. Nieuwkoop 1998.

Alvaro Valladares, *Cartas sobre las Misiones Dominicanas del Oriente del Ecuador.* Quito 1912.

David Vancil, *Catalog of Dictionaries, Word Books, and Philological Texts, 1440-1900: Index of the Cordell Collection Indiana State University.* Westport [1993].

John Venn and J.A. Venn, *Alumni Cantabrigienses. A Biographical List.* Cambridge 1922-1954.

F. Verdoorn, Ed., *Plants and Plant Science in Latin America.* Waltham, Massachusetts 1945.

Dominique D. Vérut, *Precolombian Dermatology and Cosmetology in Mexico.* Schering Corporation USA 1973.

J.E. Vidal, "Bibliographie botanique indochinoise de 1955 à 1969." *Bull. Soc. Étud. Indoch.* n.s. 47: 657-748. Saigon 1972.

J.E. Vidal and B. Wall, "Contribution à l'ethnobotanique des Nya Hön." *Journal d'agriculture tropicale et de botanique appliquée*, tome XV(7-8): 243-264. 1968.

Jules Vidal, *La végétation du Laos.* Toulouse 1960.

J.E. Vidal, Y. Vidal and Pham Hoang Hô, *Bibliographie botanique indochinoise de 1970 à 1985.* Paris 1988.

Odette Viennot, *Le culte de l'arbre dans l'Inde ancienne.* Annales du Musée Guimet, LIX. Paris 1954.

Evon Z. Vogt, *Zinacantán: A Maya Community in the Highlands of Chiapas.* Cambridge, Massachusetts 1969.

H.R. Wagner, *The Cartography of the Northwest Coast of America to the Year 1800.* Berkeley 1937.

H.R. Wagner, *Spanish Voyages to the Northwest Coast of America in the Sixteenth Century.* Amsterdam 1966.

A. Walker and R. Sillans, *Les plantes utiles de Gabon.* Paris 1961.

Barbara Wall, *Les Nya Hön.* Étude ethnographique d'une population du Plateau des Bolovens (Sud Laos). Vithagna, Vientiane, Laos 1975.

Alan C. Wares, *Bibliography of the Summer Institute of Linguistics 1935-1985.* Summer Institute of Linguistics, Dallas 1979, 1985 and 1986.

David Watkins Waters, *The Art of Navigation in England in Elizabethan and Early Stuart Times.* Yale University Press, New Haven 1958.

D.B. Waterson, *A Biographical Register of the Queensland Parliament, 1860-1929*. Australian National University Press, Canberra 1972.

J. Watt et al., *Starving Sailors*. London 1981.

S.H. Weber, *Voyages and Travels in the Near East during the XIX Century*. Princeton 1952-1953.

Waldo R. Wedel, "Environment and Native Subsistence Economies in the Central Great Plains." *Smithsonian Miscellaneous Collections*, volume 101, number 3, August 20, 1941.

Ernest Weekley, *An Etymological Dictionary of Modern English*. In two volumes. Dover Publications, New York 1967.

John M. Weeks, comp., *Mesoamerican Ethnohistory in United States Libraries*. Reconstruction of the William E. Gates Collection of Historical and Linguistic Manuscripts. Culver City, California, Labyrinthos 1990.

D. Westermann and M.A. Bryan, *Handbook of African Languages*. London 1970.

Keith Whinnom, *Spanish Contact Vernaculars in the Philippine Islands*. Hong Kong 1956.

K.D. White, *Agricultural Implements of the Roman World*. Cambridge 1967.

R.A. White et al., *Bibliography of American Pteridology*. 1976-1987.

J.H. Wiersema, J.H. Kirkbride and C.R. Gunn, *Legume (Fabaceae) Nomenclature in the USDA Germplasm System*. USDA Tech. Bull. 1757. 1990.

H. Wild, *A Rhodesian Botanical Dictionary of African and English Plant Names*. Revised and enlarged by H.M. Biegel and S. Mavi. Salisbury, Rhodesia 1972.

H. Wild, *A Southern Rhodesian Botanical Dictionary of Native and English Plant Names*. Salisbury, Southern Rhodesia 1953.

W. Willcocks, *Plans of Irrigation of Mesopotamia*. Survey Department, Cairo 1911.

Herbert W. Williams, *A Dictionary of the Maori Language*. Wellington 1975.

Jessie Williamson, *Useful Plants of Malawi*. Zomba, Malawi 1972.

Jessie Williamson, *Useful Plants of Nyasaland*. Zomba, Nyasaland 1955.

K. Williamson, "Some food-plant names in the Niger Delta." *Int. J. Amer. Linguistics*. 36: 156-167. 1970.

J.E. Wills, *Pepper, Guns and Parleys: The Dutch East India Company and China 1622-1681*. Harvard University Press, Cambridge, Massachusetts 1974.

Simon Winchester, *The Professor and the Madman*. 1998.

G.P. Winship, "The Coronado expedition, 1540-1542." *Bureau of American Ethnology, Annual Report 14*, 1: 329-613. 1892-1893 [1896].

L. Wittmack, "Die in Pompeji gefundenen pflanzlichen Reste." *Beiblatt zu den Botanischen Jahrbuchern*. 73. 1903.

Johannes Christophorus Wolff, *Bibliotheca hebraea*. Hamburgi, Lipsiae 1715-1733.

G. Marshall Woodroow, *Gardening in the Tropics*. London n.d.

Yale University, School of Medicine, *Collected Papers on Bibliography and the History of Medicine*. New Haven 1940-1945.

S. Yerasimos, *Les voyageurs dans l'Empire Ottoman (XIVe — XVIe siècles). Bibliographie, itinéraires et inventaire des lieux habités*. Ankara 1991.

Francisco Zambrano, *Diccionario bio-bibliográfico de la Compañia de Jesús en México*. Mexico 1961-1969.

R. Zander, F. Encke, G. Buchheim and S. Seybold, *Handwörterbuch der Pflanzennamen*. 14. Aufl. Stuttgart 1993.

G. Zappella, *Le marche dei tipografi e degli editori italiani del Cinquecento*. Repertorio di figure, simboli e soggetti e dei relativi motti. Milano 1986.

H. Zimmer, *Myths and Symbols in Indian Art and Civilization*. Bollingen Series VI. Pantheon Books, New York and Princeton University Press, Princeton, New Jersey 1946.

Printed and bound by CPI Group (UK) Ltd, Croydon, CR0 4YY

23/10/2024

01778248-0020